RADIOACTIVE GEOCHRONOMETRY

FROM THE TREATISE ON GEOCHEMISTRY

Editors

H. D. Holland
University of Pennsylvania, Philadelphia, PA, USA

K. K. Turekian
Yale University, New Haven, CT, USA

AMSTERDAM • BOSTON • HEIDELBERG • LONDON • NEW YORK • OXFORD
PARIS • SAN DIEGO • SAN FRANCISCO • SINGAPORE • SYDNEY • TOKYO
Academic Press is an imprint of Elsevier

ACADEMIC PRESS

Front cover photograph courtesy of Sam Bowring, MIT

Academic Press is an imprint of Elsevier
32 Jamestown Road, London NW1 7BY, UK
Radarweg 29, PO Box 211, 1000 AE Amsterdam, The Netherlands
30 Corporate Drive, Suite 400, Burlington, MA 01803, USA
525 B Street, Suite 1900, San Diego, CA 92101-4495, USA

First edition 2011

British Library Cataloguing in Publication Data
A catalogue record for this book is available from the British Library

Library of Congress Cataloging-in-Publication Data
A catalog record for this book is available from the Library of Congress

ISBN: 978-0-08-096708-0

For information on all Academic Press publications
visit our website at elsevierdirect.com

Printed and bound in Italy

11 12 13 14 10 9 8 7 6 5 4 3 2 1

Working together to grow
libraries in developing countries
www.elsevier.com | www.bookaid.org | www.sabre.org

ELSEVIER   BOOK AID International   Sabre Foundation

CONTENTS

Introduction

Interest in the chronology of the Earth has a long history. By the time of Shakespeare's allusions to the age of the Earth ('The poor world is almost six thousand years old,' *As You Like It*), the Biblical age of the Earth had been established. Between 1650 and 1654, Archbishop James Usher published his Annales Veteris and Novi Testamenti. In these he propounded a detailed scheme of biblical chronology, whose dates were later inserted in the margin of reference editions of the Authorized Version of the Bible (The King James Version). His date for the creation of heaven and earth, based on the chronologies of the Old Testament, was pronounced to be 4004 BC (This is the date subscribed to by many 'Creationists' of today).

In the eighteenth century the concept of an age for the Earth was considered irrelevant by many geologists. James Hutton (1795) concluded that 'The result, therefore, of the present inquiry is that we find no vestige of a beginning, no prospect of an end.' The early nineteenth century, however, was populated by geologists who were concerned with making sense of the geologic record as evidence of biblical events such as the creation and the Noahic Flood (Gillispie, 1951). Gradually, especially with the influence of geologists like Charles Lyell and the biologist (originally trained as a geologist) Charles Darwin, the concept of a longer period of time than biblical constraints allowed was shown to be required.

By 1905 Ernest Rutherford, a New Zealander working at the University of Manchester, had devised a radioactive dating technique based on the accumulation of helium generated by the radioactive decay of uranium and thorium. He presented these ideas at Yale University as the Silliman Lecturer at the invitation of Bertram B. Boltwood. In 1907 Boltwood, a physical chemist at Yale University, interested in rare earth elements among other things, made the first radioactive date determinations using the chemical ratio of lead to uranium in a uraninite from Connecticut and a thorianite from Ceylon (now Sri Lanka). He assumed that all of the lead in these minerals was due to the radioactive decay of uranium. The results gave what was a reasonable age of the order of about 500 My for the Connecticut pegmatite but an excessively large age for the thorianite (because he did not believe that thorium was radioactive and produced lead).

The discovery, in 1914, that neon had two isotopes, by J. J. Thomson at Cambridge, using a primitive mass spectrograph, was almost immediately improved by F. W. Aston (1919) in his laboratory and J. J. Dempster (1918) at the University of Chicago. These improvements made the prospect of the determination of the isotopic composition of elements a reality. It remained for the detailed work of A. O. C. Nier, from the University of Minnesota, working at Harvard to show both the ratio of ^{235}U to ^{238}U as being constant and the large variability of Pb isotopes in lead ores – both observations providing the fundamental platform of radioactive geochronometry. He actually determined the first age of a rock, using Pb isotopes relative to U isotopic abundances, thereby vindicating Boltwood's purely elementally based age determination with an accurate approach.

Using the U–Pb method of dating, the age of the oldest known rocks has increased from 2.7 Ga in the late 1940s to 4.0 Ga. The oldest known zircons, found in a sedimentary setting, have an age of 4.40 Ga.

The age of the Earth as part of the Solar System has been sought using the U–Pb system in a variety of ways (Dalrymple, 1991). The great pioneer in seeking the age of the Earth, Arthur Holmes, struggled with this problem during much of his life. In the opening paragraph of his 1913 book, *'The Age of the Earth,'* he ventured 'It is perhaps a little indelicate to ask of our Mother Earth her age, but science acknowledges no shame and from time to time has boldly attempted to wrest from her a secret which is proverbially well guarded.' Holmes and later, Fritz Houtermans used the Pb isotope data from galenas measured by Nier, to arrive at a first radioactivity based estimate of the age of the Earth at about 3.3 Ga. The result contributed significantly to an understanding of the great antiquity of Earth. Both men lived to see the problem solved using the analysis of meteorites by Clair Patterson of Caltech (and formerly of the University of Chicago) in 1955. Patterson refined his original estimate of 4.5 Ga for the age of meteorites (and by inference, the Earth) to 4.55 Ga and this figure has been changed very little into the twenty-first century. However, the entire field of cosmochemistry to which the early history of Earth is inextricably linked has burgeoned since the 1950s. Much of the progress in defining the chronology of the early solar system, and the origin and earliest history of the Earth is summarized in Chapters 1–4.

Since 1920 dates of varying quality associated with stratigraphic rock systems have become more and more refined with the passage of time. These dates have been attached to the Phanerozoic time scale (see Chapter 5 and Appendix 3), and a very large number of radiometric dates have defined the evolution of the Earth and its biota during the Precambrian (see Chapters 6–8). The number of radioactive decay schemes used in dating ancient rocks and minerals has increased considerably, and the analytical techniques have proliferated magnificently.

The radioactive decay schemes that have proved so useful in defining the long sweep of Earth history are based on the decay of very long-lived radioisotopes. Their use for dating events and processes on short time scales is quite limited, although radioactive disequilibrium in the uranium decay series has proved very helpful in solving geological and geochemical problems on time scales of decades to hundreds of thousand of years, especially those involving the geochronometry of marine deposits (Chapters 9 and 10) and the tracing of atmospheric chemistry (Chapter 11).

The discovery, shortly after the end of World War II, of radionuclides produced by cosmic rays in the Earth's atmosphere launched a major revolution in the dating of short-term events and processes. Measurements of ^{14}C (Chapter 12), the most famous of these radionuclides, has had a major impact in fields as disparate as anthropology, ground water dating (Chapter 13), and marine chemistry (Chapter 9). Other radionuclides such as ^{10}Be have turned out to be useful in studies of ground water dating (Chapter 13), marine geochronometry (Chapter 9), and rates of weathering and erosion (Chapter 14).

Man-made radioisotopes generated during atomic bomb testing in the atmosphere between 1945 and 1966 have clarified many processes that operate on decadal time scales. The excess ^{14}C and ^{3}H produced during these tests have found use in chemical oceanography, and ^{137}Cs has been of importance in tracing the movement of heavy metals in the environment (Chapter 15).

Radioactivity was discovered only a little more than a century ago. The application of radioactive geochronometry and environmental tracing since then has solved many major geological problems on time scales that range from decades to billions of years. Deep time has been quantified and contemporary processes tracked. The successes in this field are surely among the brightest achievements of the Earth Sciences. This book, as a collection of chapters from the Treatise on Geochemistry dealing with radioactive geochronometry, may be used as a supplement to texts such as Claude Allègre's 'Isotope Geology,' Gunter Faure's and T. M. Mensing's 'Isotopes, Principles and Applications,' and Alan Dickin's 'Radiogenic Isotope Geology.'

H.D. Holland and K.K. Turekian

REFERENCES

Boltwood B. B. (1907) On the ultimate disintegration products of the radio-active elements. Part II; The disintegration products of uranium. *American Journal of Science 4th ser* **11**: 77–88.
Dalrymple G. B. (1991) *The Age of the Earth.* Stanford, CA: Stanford University Press.
Gillispie C. C. (1951) *Genesis and Geology.* Cambridge: Harvard University Press.
Hutton J. (1795) *Theory of the Earth with Proofs and Illustrations.* London and Edinburgh.
Patterson C. C., Tilton G. R., and Ingraham M. G. (1955) Age of the earth. *Science* **121**: 69–75.

CONTRIBUTORS

M. P. Bacon
Woods Hole Oceanographic Institution, MA, USA

W. B. N. Berry
University of California, Berkeley, CA, USA

W. S. Broecker
Lamont-Doherty Earth Observatory, Palisades, NY, USA

E. Callender
US Geological Survey, Westerly, RI, USA

M. C. Castro
University of Michigan, Ann Arbor, USA

A. M. Davis
University of Chicago, Chicago, IL, USA

C. M. R. Fowler
Royal Holloway, University of London, Egham, UK

D. E. Granger
Purdue University, West Lafayette, IN, USA

W. C. Graustein
Yale University, New Haven, CT, USA

A. N. Halliday
Eidgenössische Technische Hochschule, Zürich, Switzerland

G. F. Herzog
Rutgers University, Piscataway, NJ, USA

K. V. Hodges
Massachusetts Institute of Technology, Cambridge, MA, USA

C. Koeberl
University of Vienna, Vienna, Austria

K. D. McKeegan
University of California, Los Angeles, CA, USA

E. G. Nisbet
Royal Holloway, University of London, Egham, UK

P. J. Patchett
University of Arizona, Tucson, AZ, USA

F. M. Phillips
New Mexico Tech, Socorro, NM, USA

C. S. Riebe
Stillwater Sciences, Berkeley, CA, USA

S. D. Samson
Syracuse University, NY, USA

K. K. Turekian
Yale University, New Haven, CT, USA

M. Wadhwa
Arizona State University, Tempe, AZ, USA

1

Cosmic-Ray Exposure Ages of Meteorites

G. F. Herzog

Rutgers University, Piscataway, NJ, USA

1.1   INTRODUCTION

The classic idea of a cosmic-ray exposure (CRE) age for a meteorite is based on a simple but useful picture of meteorite evolution, the one-stage irradiation model. The precursor rock starts out on a parent body, buried under a mantle of material many meters thick that screens out cosmic rays. At a time t_i, a collision excavates a precursor rock—a "meteoroid." The newly liberated meteoroid, now fully exposed to cosmic rays, orbits the Sun until a time t_f, when it strikes the Earth, where the overlying blanket of air (and possibly of water or ice) again shuts out almost all cosmic rays (cf., Masarik and Reedy, 1995). The quantity t_f–t_i is called the CRE age, t. To obtain the CRE age of a meteorite, we measure the concentrations in it of one or more cosmogenic nuclides (Table 1), which are nuclides that cosmic rays produce by inducing nuclear reactions. Many shorter lived radionuclides excluded from Table 1 such as ^{22}Na ($t_{1/2} = 2.6$ years) and ^{60}Co ($t_{1/2} = 5.27$ years) can also furnish valuable information, but

Table 1   Cosmogenic nuclides used for calculating exposure ages.

Nuclide	Half-life[a] (Myr)
Radionuclides	
^{14}C	0.005730
^{59}Ni	0.076
^{41}Ca	0.1034
^{81}Kr	0.229
^{36}Cl	0.301
^{26}Al	0.717
^{10}Be	1.51
^{53}Mn	3.74
^{129}I	15.7
Stable nuclides	
^{3}He	
^{21}Ne	
^{38}Ar	
^{83}Kr	
^{126}Xe	

[a] http://atom.kaeri.re.kr

can be measured only in meteorites that fell within the last few half-lives of those nuclides (see, e.g., Leya *et al.*, 2001 and references therein).

CRE ages have implications for several inter-related questions. From how many different parent bodies do meteorites come? How well do meteorites represent the population of the asteroid belt? How many distinct collisions on each parent body have created the known meteorites of each type? How often do asteroids collide? How big and how energetic were the collisions that produced meteoroids? What factors control the CRE age of a meteorite and how do meteoroid orbits evolve through time? We will touch on these questions below as we examine the data.

By 1975, the CRE ages of hundreds of meteorites had been estimated from noble gas measurements. Histograms of the CRE age distributions pointed to several important observations:

1. The CRE ages of meteorites increase in the order stones < stony irons < irons.
2. The CRE ages of stones rarely exceed 100 Myr; the average ages of stony irons are typically between 50 and 200 Myr; the CRE ages of irons vary with group but more often than not exceed 200 Myr.
3. The CRE ages of stones and of irons are neither uniformly distributed nor tightly clustered.

These early conclusions imply first that meteoroid production does not take place uniformly through time, for if it did, then we ought to see a distribution of CRE ages without peaks. Second, they imply that mechanical toughness contributes to the survival ability of meteoroids, a hypothesis that helps explain the greater fraction of irons with high-CRE ages and the much shorter CRE ages of, for example, the relatively fragile carbonaceous chondrites. Third, comparisons of the CRE age distributions of different types of stones point to the importance of orbits. Although aubrites and CI carbonaceous chondrites, for example, are both fairly fragile, aubrites have much larger CRE ages. This difference (along with dynamical calculations) suggested early on that the original orbit of the parent body affects CRE ages.

Since the early 1970s, several developments have brought the landscape of CRE ages into sharper focus. The number of meteorites available for analyses has increased greatly, by a factor of ~ 10, thanks to abundant finds in the Antarctic, northern Africa/Arabia, and Australia. With increased sampling, the statistical properties of CRE age distributions have become more convincing. Further, the world's collection of meteorites has become more diverse. In this respect, the lunar and the martian meteorites take pride of place but leave ample room for R, CH, and CB chondrites, new angrites, and other unusual specimens. At the same time, better experimental methods have lowered detection limits for cosmogenic nuclides and the modeling calculations needed to interpret the measurements have improved.

With greater analytical power has come the ability to recognize and, increasingly, to characterize more complex irradiation histories. As it turns out, many meteorites retain the effects not only of recent irradiation but also of irradiations that took place at earlier times, in different settings:

1. Collisions in space reduced the sizes and changed the shapes of some meteoroids. The cosmogenic nuclide inventories in such meteorites may record the two distinct periods of exposure.
2. Certain components of polymict meteoritic breccias (rocks that consist of unlike grains cemented together) spent time at the surfaces of their parent bodies before they were buried in parent bodies or perhaps in meteoroids. While at the parent-body surface these components must have been exposed directly not only to galactic cosmic rays (GCRs) (the high-energy particles from outside the solar system that are responsible for most of the production of cosmogenic nuclides), but also to lower-energy cosmic rays from the Sun.
3. Selected petrologic phases—the chondrules and the calcium- and aluminum-rich inclusions found in some meteorites—may have been irradiated just after forming in the very early solar system. One proposed mechanism is irradiation by the so-called X-wind, an intense outflow of nuclear-active particles hypothesized for the primitive Sun (Shu *et al.*, 1996).
4. Interstellar grains isolated from certain meteorites retain cosmogenic nuclides made by irradiation in interstellar space or, perhaps, close to other stars, at a time predating the formation of the solar system. We will interweave a few examples of multistage exposures into the discussion, but our main emphasis will be on the most recent one.

Honda and Arnold (1967), Wasson (1974), Reedy *et al.* (1983), Caffee *et al.* (1988), Vogt *et al.* (1990), Tuniz *et al.* (1998), Wieler and Graf (2001), and Eugster (2003) have published general reviews of CRE ages.

1.2 CALCULATION OF EXPOSURE AGES

1.2.1 Basic Equations

During a one-stage exposure, cosmogenic nuclides accumulate in meteoroids much as rain collects in a bucket. We can determine the length of time the bucket is outside (its exposure age) by measuring the height of the water that collects. The calculation requires that the rain falls at a known rate, preferably constant. Further, the bucket must be empty initially, be of known size and shape, and lose water only in known and reproducible ways. In our imperfect analogy, the bucket plays the role of the meteoroid while the rain doubles as both cosmic rays and their products, the cosmogenic nuclides. Translated into the language of CRE ages, the standard calculations for a one-stage irradiation require: a constant flux of cosmic rays; and either (1) known production rates, P (and known loss rates if radioactive decay or diffusion matter) for the cosmogenic nuclides of interest; or (2) known production rate ratios. When losses are negligible, we have for the concentration of a stable cosmogenic nuclide, s,

$$s = P_s t \tag{1}$$

and for the concentration of a radioactive cosmogenic nuclide r, with decay constant λ,

$$r = \frac{P_r\left(1 - e^{-\lambda t}\right)}{\lambda} \tag{2}$$

On combining Equations (1) and (2), we have

$$\frac{s}{r} = \frac{P_s}{P_r}\frac{\lambda t}{\left(1 - e^{-\lambda t}\right)} \tag{3}$$

Given either s and P_s, or r and P_r, or the ratios s/r and P_s/P_r, one may calculate a CRE age. The problem is how to get the production rates or the ratio P_s/P_r.

1.2.2 Factors Influencing Production Rates

The production rate of each nuclide, i, depends on numerous factors unique to each meteorite. At any time t', the full expression for the production rate at a location with coordinates x, y, and z in the meteoroid is given by

$$P_i(x, y, z, t', \text{composition})$$
$$= \sum_k \sum_j N_j \int_E \{\phi_k(E, x, y, z, t')\}\sigma_{j,k}(E)\ dE$$

$$\tag{4a}$$

where ϕ_k is the flux of nuclear-active particles of type k (mainly protons and neutrons, but also some alpha particles and pions) and energy E; $\sigma_{j,k}$ the nuclear cross section for the production of i from chemical element j by particle k; and N_j the concentration of element j in the meteoroid (assumed constant). In the absence of information about x, y, and z, one customarily assumes a spherical meteoroid with radius R. Under spherical symmetry and with uniform composition, the depth of the sample below the meteoroid surface, d, then sets the production rate:

$$P_i(d, R, t', \text{composition})$$

$$= \sum_k \sum_j N_j \int_E \{\phi_k(E, d, R, t')\} \, \sigma_{j,k}(E) \, dE$$

$$(4b)$$

For a constant flux of GCRs (for further discussion of this point, see Lavielle *et al.*, 1999), the time dependence vanishes. The parameters size and depth, remain along with the elemental abundances and nuclear cross sections; in principle, the elemental abundances and nuclear cross sections can be measured directly.

During the 1990s, teams led by Janet Sisterson at Harvard and by Rolf Michel at Hannover measured many nuclear cross sections relevant to the calculation of production rates. Cross sections for nuclear reactions induced by energetic neutrons, although not measured directly in most cases, have been inferred from thick target irradiations. Modern modeling calculations incorporate all these results. Given the preatmospheric size and shape of a meteoroid and the location of a sample within it, these calculations can now reproduce the production rates of many cosmogenic nuclides to within 15% with the introduction of just one universal parameter, a measure of the intensity of GCRs (Leya *et al.*, 2000; Masarik *et al.*, 2001). Even these best estimates, however, ultimately derive from absolute CRE age calibrations that are based on radioactive cosmogenic nuclides.

Cosmogenic nuclides can be divided into three broad and imperfect categories according to the energies of the particles that produce them (e.g., Tuniz *et al.*, 1998). At the highest energies, nuclear spallation reactions dominate. The concentrations of nuclides produced in such reactions—^{21}Ne in iron meteoroids, for example—tend to decrease with increasing depth. At intermediate energies, simpler two-body nuclear reactions and compound nucleus reactions are common. The nuclear-active particles with the energies in this range are for the most part secondary particles, themselves generated by reactions of higher energy cosmic rays. The concentrations of the cosmogenic nuclides produced in reactions induced by these

particles—^{21}Ne in stony meteoroids for example—typically increase and pass through a maximum as the depth in the meteoroid increases. At the lowest energies, we have thermal neutrons. Eventually, through inelastic collisions, secondary neutrons slow and may reach thermal velocities, $\sqrt{2kT/m}$, where k is Boltzmann's constant, T the temperature, and m the neutron mass. Special importance attaches to the fluxes of thermal neutrons, because certain isotopes have extraordinarily large cross sections for their capture, and hence for making detectable concentrations of cosmogenic nuclides. Examples include ^{59}Co(n, γ) ^{60}Co, ^{35}Cl$(n, \gamma)^{36}$Cl, ^{40}Ca$(n, \gamma)^{41}$Ca, ^{58}Ni(n, γ) ^{59}Ni, ^{79}Br$(n, \gamma)^{80}$Br$\rightarrow^{80}$Kr, ^{149}Sm$(n, \gamma)^{150}$Sm, and ^{157}Gd$(n, \gamma)^{158}$Gd. Nuclear modeling calculations and direct measurements of depth profiles show that the production rates of these nuclides peak at greater depths in larger bodies relative to cosmogenic nuclides produced at higher energies (Eberhardt *et al.*, 1963; Spergel *et al.*, 1986).

1.2.3 Measurement Units and Quantities

The concentrations of (stable) ^{3}He, ^{21}Ne, and ^{38}Ar constitute by far the largest and best bank of data for the calculation of CRE ages (Schultz and Franke, 2002). Noble gas concentrations are often reported in 10^{-8} cm^3 STP (g sample)$^{-1}$, where STP refers to a temperature of 273.15 K and a pressure of 1 atm. To convert the concentration of an isotope from 10^{-8} cm^3 STP (g sample)$^{-1}$ to atoms per gram of sample, one multiplies by N_A/RT, where N_A is Avogadro's number and R the gas constant, i.e., by 2.687×10^{11}:

$$s[\text{atom(g sample)}^{-1}]$$

$$= 2.687 \times 10^{11} s[10^{-8}\text{cm}^3 \text{ STP (g sample)}^{-1}]$$

$$(5)$$

To a reasonable first approximation, the production rate of ^{21}Ne in a very common type of meteorite, an H-chondrite, is 8×10^{10} atom g^{-1} Myr^{-1}. Thus, an irradiation lasting 7.5 Myr produces $\sim 6 \times 10^{11}$ atom g^{-1} of ^{21}Ne, or $\sim 2 \times 10^{-8}$ cm^3 STP g^{-1} of ^{21}Ne.

Radionuclide concentrations are normally reported as activities, A, in units of disintegrations per minute per kilogram (dpm kg^{-1}). Nishiizumi (1987) compiled a comprehensive source of data for ^{36}Cl, ^{26}Al, ^{10}Be, and ^{53}Mn in meteorites. Table 2 shows a few activities that are typical for stony meteorites of small to moderate size (preatmospheric radius < 40 cm). To convert from dpm kg^{-1} into atom g^{-1}, one uses the radioactive decay law, $\lambda r = -dr/dt = A$. From the expression for the half-life, $t_{1/2} = \ln(2)/\lambda$, we have after conversion of units (1 year = 365.25 days)

Table 2 Typical activities (dpm kg^{-1}) of selected cosmogenic radionuclides in stony meteorites.

Nuclide	Activity
^{14}C	50
^{41}Ca	24[a]
^{81}Kr	0.003
^{36}Cl	22[a]
^{26}Al	60
^{10}Be	20
^{53}Mn	400[a]
^{129}I	0.0003

[a] Metal phase.

$$r\left(\text{atom g}^{-1}\right) = 7.588 \times 10^8 \times A\left(\text{dpm kg}^{-1}\right)$$
$$\times\, t_{1/2}(\text{Myr}) \qquad (6)$$

As shown in the next section, we often want the ratios of noble gas concentrations (atom g^{-1}) to radionuclide activities (atom Myr^{-1}g^{-1}); such ratios have dimensions of time and are closely related to CRE ages. In converting from measurement units, the half-life does not enter and we have, for all noble gas–radionuclide pairs,

$$\frac{s\left(\text{atom g}^{-1}\right)}{r\left(\text{atom Myr}^{-1}\text{g}^{-1}\right)}$$
$$= 510.9 \times \frac{s\left(10^{-8}\text{cm}^3\ \text{STP g}^{-1}\right)}{r\left(\text{dpm kg}^{-1}\right)} \qquad (7)$$

1.2.4 Calibration of Production Rates

To infer CRE ages from cosmogenic noble gas contents, we must know their absolute production rates. We now discuss two examples that illustrate how to obtain this information.

1.2.4.1 ^{26}Al versus ^{21}Ne calibration

Inspection of Equation (2) shows that in the limit of CRE ages long compared with $1/\lambda$, the concentration of a cosmogenic radionuclide approaches the constant value P/λ. Hence, in this case, the activity at the time of a meteorite fall (or the collection of a lunar sample) is equal to the production rate. Once the production rate is known for one meteorite, one may analyze for the same radionuclide in a second meteorite with, perhaps, a shorter CRE age. If P was the same in the second meteorite (an important assumption), then we can solve Equation (2) for the CRE age. In a kind of bootstrapping process, measurements of other cosmogenic nuclides in the second meteorite then can serve as the basis for calculating their production rates, and so on.

Herzog and Anders (1971) tried to calibrate the ^{21}Ne production rate through measurements of the activity of ^{26}Al, $A(^{26}$Al), and the concentration of ^{21}Ne, [^{21}Ne], for a suite of meteorites. From Equation (1), we have $T = {}^{21}$Ne/P^{21}. We substitute this result in Equation (2), written for ^{26}Al, and rearrange

$$A\left(^{26}\text{Al}\right) = P_{26}\left[1 - \exp\left(-\lambda_{26}\frac{[^{21}\text{Ne}]}{P_{21}}\right)\right] \qquad (8)$$

A nonlinear fitting routine applied to Equation (8) gives the parameters P_{26} and P_{21}. In practice, the important information about P_{21} comes from meteorites with low ^{21}Ne contents. Ironically, this method originally gave too large a value for P_{21}, because several of the meteorites with low ^{21}Ne contents had complex rather than one-stage exposure histories. In other words, the meteorites chosen for study contained ^{21}Ne but not much ^{26}Al left over from an earlier irradiation period, because the ^{26}Al created in the earlier period had largely decayed away.

The principle of the method is sound, however, and with careful sample selection and the addition of shielding adjustments the procedure works well for several pairs of cosmogenic nuclides. Once P_{21} and P_{26} are known, we can calculate the ratio P_{21}/P_{26}. On expanding the exponential in Equation (8), we then have, for any meteorite with $t \gg 1/\lambda_{26}$ the relation

$$\frac{[^{21}\text{Ne}]}{A\left(^{26}\text{Al}\right)} \approx \frac{P_{21}t}{P_{26}}\left(t \gg \frac{1}{\lambda_{26}}\right) \qquad (9)$$

where units are atom per unit mass or atom per unit mass per unit time. For L-chondrites, the value of the production rate ratio P_{21}/P_{26} is ~ 2.5 (atom ^{21}Ne/atom ^{26}Al), where the units of time cancel out, or 0.005 (10^{-8} cm^3 STP g^{-1} Myr^{-1})/(dpm kg^{-1}) and varies slightly with shielding.

1.2.4.2 ^{83}Kr/^{81}Kr calibration

Marti (1967) introduced the ^{81}Kr/Kr method for determining CRE ages. The ^{26}Al/^{21}Ne method sketched above rests only on analyses of meteorites; in contrast, the ^{81}Kr/Kr method also demands knowledge of the relative cross sections for certain nuclear reactions. To see how the method works, we write Equation (1) for ^{83}Kr, a stable isotope produced mainly by spallation, and Equation (2) for ^{81}Kr, which has a half-life of 0.229 Myr, and then divide one by the other, obtaining

$$\frac{^{83}\text{Kr}}{^{81}\text{Kr}} = \frac{\lambda_{81}P_{83}t}{P_{81}(1 - e^{-\lambda_{81}t})} \qquad (10)$$

Note that the appearance of λ_{81} here reflects the fact that it is the concentration ratio of the two nuclides, rather than a concentration/activity ratio that is measured. As CRE ages of less than a few half-lives of ^{81}Kr are rare, for the great majority of meteorites Equation (10) simplifies to

$$\frac{^{83}\text{Kr}}{^{83}\text{Kr}} \approx \frac{\lambda_{81}P_{83}t}{P_{81}} \quad (t \gg \lambda_{81}) \qquad (11)$$

Provided that the ratio P_{83}/P_{81} is known, the calculation of a CRE age reduces to the measurement of the ^{83}Kr/^{81}Kr (atom/atom) ratio. In the earliest calculations, P_{81}/P_{83} was set by the relation

$$\frac{P_{81}}{P_{83}} = 0.95 \left(\frac{^{80}\text{Kr} + {}^{82}\text{Kr}}{2 \times {}^{83}\text{Kr}} \right)_{\text{spallogenic}} \qquad (12)$$

The factor of 0.95 was chosen to take account of the isobaric yield, which, in turn, was based on cross section measurements made for the element silver. Since then laboratory simulations of meteorite irradiation have furnished more relevant cross section data (Gilabert *et al.*, 2002). Although these authors recommend no changes to the calculation of ^{81}Kr/Kr ages, the new data may make it necessary to revise Equation (12).

The use of ^{80}Kr and ^{82}Kr to arrive at P_{81}/P_{83} (Equation (12)) is unsatisfactory for samples that contain appreciable amounts of Br, from which neutron-capture produces ^{80}Kr and ^{82}Kr, but not ^{81}Kr. To avoid this problem, P_{81}/P_{83} has often been calculated from an empirical relation obtained by selecting samples with negligible neutron-capture effects and then regressing calculated values of P_{81}/P_{83} (from Equation (12)) on directly measured values of ^{78}Kr/^{83}Kr (Marti and Lugmair, 1971). The regression gave

$$\frac{P_{81}}{P_{83}} = 1.262 \left(\frac{^{78}\text{Kr}}{^{83}\text{Kr}} \right)_{\text{spallogenic}} + 0.381 \qquad (13)$$

Underlying this equation is the additional assumption that ratios of production rate ratios (e.g., $P_{81}/P_{83} \div P_{78}/P_{83}$) are insensitive to changes in composition and in shielding.

Eugster (1988) and Eugster and Michel (1995) have used ^{81}Kr/Kr CRE ages to derive equations for ^{3}He, ^{21}Ne, and ^{22}Ne production rates in chondrites. These empirical correlations are embodied in production rate equations for ^{3}He, ^{21}Ne, and ^{38}Ar that are now in wide use.

1.2.5 Equations for Calculating One-Stage CRE Ages

Here we describe and comment critically on a few of the many methods used to calculate CRE ages.

1.2.5.1 *^{21}Ne–^{22}Ne/^{21}Ne ages*

The age equation is a specialized form of Equation (1):

$$t = \frac{^{21}\text{Ne}}{P_{21}} \qquad (14)$$

Eugster (1988) and Eugster and Michel (1995) give formulas for the numerical calculation of P_{21} in ordinary chondrites and certain achondrites, respectively. The absolute timescale is set by reference to the ^{81}Kr/Kr ages of selected meteorites. The formula for the production rate of ^{21}Ne has the general form

$$P = F \times S \times A \qquad (15)$$

Here, F (dimensionless) corrects for composition over a limited range; S (dimensionless) corrects for shielding through the cosmogenic ^{22}Ne/^{21}Ne ratio, also over a limited range; A is a normalizing constant with units of 10^{-8} cm^3 STP g^{-1} Myr^{-1}, or more specifically, the production rate for average L-chondrite composition (Avg L) and shielding. By definition, "average" shielding means the shielding associated with a cosmogenic ^{22}Ne/^{21}Ne ratio of 1.11. For the case of ordinary chondrites, we have

$$F = \frac{(1.63[\text{Mg}] + 0.6[\text{Al}] + 0.32[\text{Si}])_{\text{sample}}}{(1.63[\text{Mg}] + 0.6[\text{Al}] + 0.32[\text{Si}])_{\text{Avg L}}} \qquad (16)$$

and

$$S = \left(\frac{1}{4.494(^{22}\text{Ne}/^{21}\text{Ne}) - 3.988} \right) \qquad (17)$$

where we normally express elemental concentrations as mass fractions, although mass percentages will do as well. Equation (17) may be used over a range of ^{22}Ne/^{21}Ne ratios extending from ~ 1.08 ($S = 1.16$) to 1.21 ($S = 0.69$), and perhaps for ^{22}Ne/^{21}Ne ratios outside this range, but with lower reliability (Masarik *et al.*, 2001). The normalization constant (average production rate) is given by

$$A = 0.332 \times 10^{-8} \text{ cm}^3 \text{ STP g}^{-1} \text{ Myr}^{-1} \qquad (18)$$

1.2.5.1.1 Measurements needed. Cosmogenic ^{21}Ne content and ^{22}Ne/^{21}Ne ratio of the sample. Although seldom made, direct measurements of the elemental composition of a sample aliquot are highly desirable.

1.2.5.1.2 Range of applicability. Stones and the stony portions of stony iron meteorites: best for ^{22}Ne/^{21}Ne ratios between 1.09 and 1.18.

1.2.5.1.3 Limitations. (1) The presence of large amounts of trapped neon may prevent an accurate determination of the cosmogenic ^{22}Ne/^{21}Ne ratio, and, less frequently, of the ^{21}Ne content itself; (2) for ratios $\lesssim 1.09$ (i.e., in the interiors of large meteorites), the ^{22}Ne/^{21}Ne ratio does not determine the ^{21}Ne production rate uniquely (cf., Masarik *et al.*, 2001); (3) the behavior of the ^{22}Ne/^{21}Ne ratio for samples containing unusually large concentrations of sodium is not known;

(4) reliability is not well established in samples with high ^{22}Ne/^{21}Ne ratios (> 1.22); (5) while ^{21}Ne does not usually leak out of meteorites, it may do so in cases of exceptionally intense heating; and (6) the ^{22}Ne/^{21}Ne ratio depends on the Mg/Si ratio. Corrections for variations in the Mg/Si ratio are possible, but in practice it is usually necessary to recalibrate the production rates (through ^{81}Kr or some other method).

1.2.5.2 ^{38}Ar–^{22}Ne/^{21}Ne ages

The age equation is again a specialized form of Equation (1):

$$t_{38} = \frac{^{38}\text{Ar}}{P_{38}} \qquad (19)$$

and has been calibrated in the same way as Equation (14) for ^{21}Ne–^{22}Ne/^{21}Ne ages (Eugster, 1988; Eugster and Michel, 1995). The cosmogenic ^{38}Ar production rate has the same form as Equation (15) with

$$F = \frac{(11[\text{K}] + 1.58[\text{Ca}] + 0.33[\text{Ti} + \text{Cr} + \text{Mn}] + 0.086[\text{Fe} + \text{Ni}])_{\text{sample}}}{(11[\text{K}] + 1.58[\text{Ca}] + 0.33[\text{Ti} + \text{Cr} + \text{Mn}] + 0.086[\text{Fe} + \text{Ni}])_{\text{AvgL}}} \qquad (20)$$

and

$$S = 2.706 - 1.537 \left(\frac{^{22}\text{Ne}}{^{21}\text{Ne}}\right)_{\text{cosmogenic}} \qquad (21)$$

Note that for ^{38}Ar the (empirical) shielding correction varies linearly with the cosmogenic ^{22}Ne/^{21}Ne ratio, rather than inversely as for ^{21}Ne. The normalization constant is given by

$$A = 0.0462 \times 10^{-8} \text{ cm}^3 \text{ STP g}^{-1} \text{ Myr}^{-1} \qquad (22)$$

1.2.5.2.1 Measurements needed. Cosmogenic ^{21}Ne, ^{22}Ne, and ^{38}Ar. Direct measurements of elemental composition are also desirable.

1.2.5.2.2 Range of applicability. Best for stony meteorites and the stony phases of stony iron meteorites.

1.2.5.2.3 Limitations. (1) The remarks made above concerning the calculation of cosmogenic ^{22}Ne/^{21}Ne ratios for ^{21}Ne CRE ages apply. While with ^{21}Ne the corrections for trapped gases are usually small, with ^{38}Ar the corrections are often 20% or more. (2) In the normal deconvolution of spallogenic ^{38}Ar, one assumes that the trapped argon component has a ^{36}Ar/^{38}Ar ratio between 5.32 and 5.36 and that the (high-energy) spallogenic component has a ratio of 0.63–0.65. In some meteorites, however, the argon may comprise a

third component, namely, ^{36}Ar produced by (low-energy) neutron capture on ^{35}Cl (Göbel *et al.*, 1982). This effect is restricted mainly to interior portions of large meteorites. Failure to account for any neutron-produced ^{36}Cl leads to an underestimate of the cosmogenic ^{38}Ar content. (3) Compositional variations are a more serious problem. In stony meteorites, the ^{38}Ar production rate depends sensitively on the concentrations of potassium and calcium, which can vary and are rarely well known.

1.2.5.3 ^{3}He ages

Again, the age equation is a specialized form of Equation (1):

$$t_3 = \frac{^3\text{He}}{P_3} \qquad (23)$$

Eugster (1988) and Eugster and Michel (1995) give formulas for the numerical calculation of P_3

---

in different kinds of stony meteorites. The approach is similar to the one taken for ^{21}Ne. For ^{3}He, however, we have

$$F = \frac{(2.66 - 0.0096[\text{Ti} + \text{Cr} + \text{Mn} + \text{Fe} + \text{Ni}])_{\text{sample}}}{(2.66 - 0.0096[\text{Ti} + \text{Cr} + \text{Mn} + \text{Fe} + \text{Ni}])_{\text{Avg L}}} \qquad (24)$$

and

$$S = 2.09 - 0.43 \times \left(\frac{^{22}\text{Ne}}{^{21}\text{Ne}}\right) \qquad (25)$$

Here the ^{3}He production rate is a relatively weak function of shielding: S varies by only 5–10% over the normal range of cosmogenic ^{22}Ne/^{21}Ne ratios (Wright *et al.*, 1973). The normalization constant is given by

$$A = 1.61 \times 10^{-8} \text{ cm}^3 \text{ STP g}^{-1} \text{ Myr}^{-1} \qquad (26)$$

1.2.5.3.1 Range of applicability. All stones and the silicate phases of stony irons.

1.2.5.3.2 Measurements needed. When trapped corrections are small as is usually the case, we need only the total measured ^{3}He content to calculate t_3. Within fairly broad limits, the value of P_3 is not very sensitive to elemental composition.

1.2.5.3.3 Limitations. (1) Meteorites are susceptible to ^{3}He and ^{3}H (a precursor of ^{3}He) losses on heating, which probably explains why so many ^{3}He CRE ages are smaller than the corresponding ^{21}Ne ages. Losses of ^{3}He can often, but not always, be identified by plotting ^{3}He/^{21}Ne ratios against ^{22}Ne/^{21}Ne concentrations and seeing whether the meteorite of interest conforms to a well-established general trend (Eberhardt *et al.*, 1966) or falls well below it. The method does not infallibly identify diffusion losses because both complex exposure histories and very heavy shielding can also cause a data point to lie below the trend line (Masarik *et al.*, 2001). It is fair to say, however, that if a sample appears to have a low ^{3}He/^{21}Ne ratio for any reason, the corresponding ^{3}He age should be regarded with caution. (2) Contrary to the implication of Equation (25), with increasing meteoroid size and sample depth, there must come a point where shielding depresses P_3, appreciably. We know from direct measurement that ^{3}He concentrations decrease with depth in several large iron meteorites. On the other hand, they decrease little or not at all in the L5 chondrite, Knyahinya (Graf *et al.*, 1990a), which had a preatmospheric radius of 40–50 cm. We conclude that shielding corrections for ^{3}He are probably small in most stones but may be appreciable in stony irons and irons.

1.2.5.4 $^{36}Cl/^{36}Ar$ ages

A ^{36}Cl/^{36}Ar age may be calculated from Equation (3) with due allowance for the decay of ^{36}Cl into ^{36}Ar (cf., Begemann *et al.*, 1976; Albrecht *et al.*, 2000):

$$\frac{t}{1 - e^{-\lambda t}} = (430 \pm 16)$$
$$\times \frac{^{36}\mathrm{Ar}\left[10^{-8}\mathrm{cm}^3\ \mathrm{STP\ g}^{-1}\right]}{^{36}\mathrm{Cl}\left[\mathrm{dpm}/(\mathrm{kg\ metal})\right]} \quad (27)$$

Other authors give values of the constant between 425 and 433 (Lavielle *et al.*, 1999; Terribilini *et al.*, 2000a; Leya *et al.*, 2000). For the few meteorites with very short exposure ages, this equation must be solved iteratively. Most meteoroids, however, orbit in space for times long compared with the half-life of ^{36}Cl, ~ 300 kyr. Thus, $e^{-\lambda t}$ is usually negligible and we have

$$t\ (\mathrm{Myr}) = 430\frac{^{36}\mathrm{Ar}}{^{36}\mathrm{Cl}} \quad (28)$$

where ^{36}Ar is in 10^{-8} cm^3 STP (g metal)$^{-1}$, ^{36}Cl in dpm (kg metal)$^{-1}$ and the constant 430 has the units necessary to convert the ratio of measured quantities to million years.

1.2.5.4.1 Measurements needed. ^{36}Ar, ^{38}Ar, and ^{36}Cl in clean metal. A measurement of the ^{36}Ar/^{38}Ar ratio makes it possible to correct for trapped ^{36}Ar.

1.2.5.4.2 Range of applicability. Metal or magnetite under all shielding conditions; all falls and finds for which the terrestrial age is known or known to be short compared with the half-life of ^{36}Cl.

1.2.5.4.3 Limitations. (1) As with cosmogenic ^{38}Ar, the deconvolution of the cosmogenic component of ^{36}Ar requires corrections for the presence of trapped ^{36}Ar. (2) Long terrestrial ages will lower appreciably the measured ^{36}Cl contents. An independent measure of terrestrial age may be necessary to correct for this effect. (3) The measurements are sensitive to the presence of calcium- and of rare potassium-bearing impurities in the metal phase.

1.2.5.5 $^{81}Kr/Kr$ ages

We presented in Section 1.2.4.2 the equations needed for calculating ^{81}Kr/Kr CRE ages.

1.2.5.5.1 Measurements needed. ^{81}Kr/^{83}Kr and either ^{78}Kr/^{83}Kr or ^{80}Kr/^{83}Kr and ^{82}Kr/^{83}Kr when neutron capture on bromine is negligible.

1.2.5.5.2 Range of applicability. Falls and finds for which terrestrial age is known. For exposures long compared with the half-life of ^{81}Kr, ~ 229 kyr, the term $1-e^{-\lambda 81 t}$ reduces to 1 and the age equation becomes easy to solve. For shorter ages the equation must be solved iteratively. Production from strontium dominates and explains why the early measurements of cosmic-ray krypton focused on eucrites and lunar samples, both of which are rich in strontium. In cases where the terrestrial age is not known and the exposure age is long, one may use the measured ^{81}Kr content to obtain a terrestrial age.

1.2.5.5.3 Limitations. (1) As with other radionuclide-based ages, the terrestrial age of the sample must be known. (2) Concentrations of ^{81}Kr are quite low in most meteorites, typically just 5×10^5 atom g^{-1} in chondrites. For this reason, ^{81}Kr measurements are still scarce and their uncertainties can be relatively large, often $\sim 20\%$. (3) Production rates for krypton isotopes may vary with the abundances of rubidium, yttrium, and zirconium relative to strontium. It should be understood that the original basis for the calculation of P_{81}/P_{83} was a set of relative cross section measurements for the production of krypton from silver (Marti, 1967).

1.2.5.6 $^{40}K/K$ ages

Cosmic rays produce stable ^{39}K and ^{41}K along with radioactive ^{40}K ($t_{1/2} = 1.27$ Gyr) in meteoroids. At first glance, it would seem straightforward to write an equation similar to Equation (3) with ^{40}K playing the role of r, and, say ^{41}K playing the role of s. An immediate wrinkle is that virtually all meteoritic materials contain some nonspallogenic (native or primordial) potassium, and perhaps some potassium introduced by terrestrial contamination. Native potassium in the stony portions of meteorites overwhelms any spallogenic component. Even irons and the metal phases of stony irons contain enough native (or contaminant) potassium to make the necessary corrections significant, and especially so for relatively short CRE ages (i.e., ≲100 Myr).

By itself, the need to correct for native potassium would not seem terribly problematic. After all, such corrections are routine in calculating CRE ages with the noble gases, for example, ^{21}Ne and ^{38}Ar. In the case of potassium, however, an experimental constraint introduces a second wrinkle. Specifically, the mass spectrometry done to date for potassium does not yield absolute concentrations of the isotopes as it does for noble gases, but rather a pair of isotope ratios, conventionally ^{39}K/^{40}K and ^{41}K/^{40}K. Without the absolute concentrations of ^{x}K, the separate calculation of the ratio s/r becomes impossible and we must give up on any strict analogue to Equation (3).

Fortunately, as Voshage and coworkers were able to show, the available data do allow the construction algebraically of a mathematical expression that removes the native component and provides a measure of spallogenic potassium production. We begin with the following relations:

$$^{39}K = {}^{39}K_{sp} + {}^{39}K_p \quad (29a)$$

$$^{40}K = {}^{40}K_{sp} \quad (29b)$$

$$^{41}K = {}^{41}K_{sp} + {}^{41}K_p \quad (29c)$$

$$a = {}^{41}K_p/{}^{39}K_p \sim 0.07 \quad (30)$$

The subscripts sp and p denote spallogenic and primordial, respectively. The (primordial) ratio $^{41}K_p/{}^{39}K_p$ is assumed to have the terrestrial value. Note that Equation (29b) builds in two additional assumptions: that irradiation took place in one stage and that $^{40}K_p = 0$. The four equations can be combined to eliminate $^{39}K_p$ and $^{41}K_p$, yielding the result

$$\frac{^{41}K_{sp}}{^{40}K_{sp}} - a\frac{^{39}K_{sp}}{^{40}K_{sp}} = \frac{^{41}K}{^{40}K} - a\frac{^{39}K}{^{40}K} \quad (31)$$

Voshage refers to the *measurable* quantity on the right-hand side as M. To first order, we have $^{41}K_{sp}/{}^{40}K_{sp} \sim {}^{39}K_{sp}/{}^{40}K_{sp} \sim 1$. Because a (~0.07) is small compared with $^{x}K_{sp}/{}^{40}K_{sp}$ (~1),

when the primordial correction is also small, M approaches $^{41}K_{sp}/{}^{40}K_{sp}$, i.e., the ratio s/r in the terminology of Equation (3). By general analogy with Equation (3), it remains to take account of production rates. To do so we write

$$^{39}K_{sp} = P_{39}t \quad (32a)$$

$$^{40}K_{sp} = P_{40}/\lambda_{40} \times (1 - e^{\lambda t}) \quad (32b)$$

$$^{41}K_{sp} = P_{41}t \quad (32c)$$

From these equations we can construct an expression equal to the left-hand side of Equation (31). Skipping the algebraic details, we find

$$\frac{^{41}K_{sp}}{^{40}K_{sp}} - a\frac{^{39}K_{sp}}{^{40}K_{sp}} = \left(\frac{P_{41}}{P_{40}} - a\frac{P_{39}}{P_{40}}\right)\frac{\lambda t}{1 - e^{-\lambda t}} \quad (33)$$

where λ is the decay constant of ^{40}K, taken as $0.546\,\text{Gyr}^{-1}$ by Voshage and co-workers. Voshage refers to the terms in parentheses on the right-hand side of Equation (33) as N. Finally, combining Equations (31) and (33), we obtain the equation for ^{40}K/K ages

$$\frac{^{41}K}{^{40}K} - a\frac{^{39}K}{^{40}K} = \left(\frac{P_{41}}{P_{40}} - a\frac{P_{39}}{P_{40}}\right)\frac{\lambda t}{1 - e^{-\lambda t}} \quad (34)$$

When the primordial correction is small, N approaches the production rate ratio P_{41}/P_{40} to within 10% or so.

The evaluation of N presents a last major hurdle. With the half-life of ^{40}K so long, with exposure ages of meteorites too short for ^{40}K to saturate (see below), and with the absolute concentrations of ^{40}K unknown, we cannot get directly at ^{40}K production rates from the available data as we can do for shorter lived nuclides such as ^{26}Al and ^{81}Kr. At the same time, the lack of measured cross sections for the relevant reactions (e.g., Fe $(p, X)^{40}$K)—a lack that persists to this day—poses a problem for obtaining the production rates from modeling. To make matters still more complicated, N varies with irradiation hardness or shielding, so that it is not enough to evaluate N once and apply that result to numerous meteorites, as is done for ^{36}Cl/^{36}Ar dating.

Space does not permit a detailed discussion of the two basic methods used to address these difficulties. In brief, both of them start with measurements of ^{4}He/^{21}Ne ratios, which provide a measure of the irradiation hardness in metal (but not in stony phases). One method then relies on modeling calculations. Voshage and Hintenberger (1963) used those of Arnold *et al.* (1961), first to find the conditions needed to reproduce the observed spallogenic ^{4}He/^{21}Ne ratios, and then to obtain production rates for the potassium isotopes from "estimated" cross sections. As it turned out, the calculated values of N correlate linearly with the ^{4}He/^{21}Ne ratios, thereby providing a basis for calculating N for any other iron with a known ^{4}He/^{21}Ne ratio. The values of N found in this way range from ~ 1.45 to ~ 1.57.

In discussing the second method, we pass over the work of Voshage and Hintenberger (1963) in favor of a more recent treatment by Lavielle *et al.* (1999). These authors made three assumptions: first, ^{40}K/K CRE ages should closely approximate ^{36}Cl/^{36}Ar CRE ages, at least for the last 500 Myr; second, ^{36}Cl/^{36}Ar CRE ages of irons can be written in the form $t_{36} = k\ ^{36}$Ar/^{36}Cl, i.e., as in Equation (28), but with the constant k allowed temporarily to float; and third, N can be written in the form

$$N = a_1 \left[1 + a_2 \left(\exp\left\{ a_3 \times \frac{^4\text{He}}{^{21}\text{Ne}} \right\} \right) \right] \quad (35)$$

They then assembled measured potassium isotope ratios, ^{36}Cl/^{36}Ar ratios, and ^{4}He/^{21}Ne ratios for selected iron meteorites and calculated the values of the parameters a_n and k that made the ^{40}K/K and ^{36}Cl/^{36}Ar CRE ages agree best. This method ultimately recovers to within a few percent the values of N estimated by Voshage and Hintenberger (1963). It also gives the result $k = 590$ (measurement units) that differs substantially from 430, the value shown in Equation (28). We will return to this point later.

1.2.5.6.1 Measurements needed. ^{39}K/^{40}K, ^{41}K/^{40}K, and ^{4}He/^{21}Ne ratios.

1.2.5.6.2 Range of applicability. Irons and the metal phases of other meteorites with CRE ages greater than 100 Myr and preferably greater than 200 Myr.

1.2.5.6.3 Limitations. The ^{40}K/K method of calculating CRE ages works poorly for ages < 100 Myr because of the need for large corrections for native potassium. The absolute calibration of the method is indirect and rests on educated guesses and assumptions. Nickel content may influence the results but is not normally considered explicitly.

1.2.6 The Importance of Half-Lives

The half-lives (decay constants) that appear in many of the equations used to calculate CRE ages undergo revision from time to time. For example, at this writing (October 2006), the half-lives of ^{10}Be and of ^{53}Mn are under active scrutiny. In all cases, the absolute production rates depend inversely on the half-life of the radionuclide of interest. Thus, an increase of 10% in the half-life of ^{81}Kr would decrease by 10% any value of P_s based on ^{81}Kr measurements. It follows that all CRE ages would then shift upward by 10%. While

such changes may be important for understanding the dynamics of meteoroid delivery to Earth, they do not affect the relative values of the CRE ages, and hence the characteristic shapes of CRE age distributions.

1.3 CARBONACEOUS CHONDRITES

1.3.1 CI, CM, CO, CV, and CK Chondrites

Mazor *et al.* (1970) presented the first extensive set of CRE ages for carbonaceous chondrites, based mainly on ^{21}Ne. Figure 1 shows ^{21}Ne exposure ages for carbonaceous chondrites calculated from the compilation of (Schultz and Franke, 2002), assuming that the neon was in each case a mixture of solar and cosmogenic gas. Some cautions are in order for CI and CM chondrites in particular.

- CI and CM chondrites often have large concentrations of interfering noncosmogenic components. The process of stripping away these contributions to obtain cosmogenic ^{21}Ne adds to the uncertainty of the result. The presence of noncosmogenic components also undermines the one-stage irradiation scenario.
- The measurement uncertainties of the cosmogenic ^{22}Ne/^{21}Ne ratio, which is a durable indicator of shielding in ordinary chondrites, are increased in making corrections for noncosmogenic components.
- The CI and CM chondrites retain some cosmogenic noble gases from early, regolith irradiations.

For these reasons, we expect CRE ages of CI and CM meteorites to have a precision no better than ~ 30–40%. Carbonaceous chondrites belonging to the other classes generally have lower concentrations of noncosmogenic gas and may give

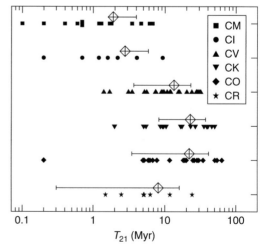

Figure 1 ^{21}Ne exposure ages (Myr) of carbonaceous chondrites. Large symbols show group averages.

better precision, perhaps 10–15%. Inspection of Figure 1 reveals several interesting features.

The CI and CM chondrites have unusually short CRE ages, many of them <1 Myr. The group averages are 1.8 ± 2.1 and 2.8 ± 3.1 Myr, respectively. Except for lunar meteorites, no other group of meteorites has such a high proportion (1/4–1/2) of short-lived objects. Scherer and Schultz (2000) list three possible reasons why meteorites may have short exposure ages: the parent body orbited close to a resonance; the parent body was in an Earth crossing orbit when a collision released the meteoroid; and the meteoroids are so fragile that collisions destroy them if they fail to reach Earth quickly.

For CI and CM chondrites, cosmogenic radionuclides provide an excellent alternative way to calculate CRE ages. Based on ^{26}Al and ^{10}Be measurements, Caffee and Nishiizumi (1997) present a preliminary age distribution for selected C2 carbonaceous chondrites, which are, for the most part, CM chondrites. Remarkably, they find a strong peak at 0.2 Myr, a peak absent from the distribution of ^{21}Ne ages. Perhaps this difference merely reflects the large uncertainties of the noble gas CRE ages. It seems equally likely, however, that many of the CI and CM chondrites retain significant fractions of ^{21}Ne from earlier irradiations. Within the CM2 chondrites (e.g., Murchison, Murray, and Cold Bokkeveld), the concentrations of ^{21}Ne in individual "irradiated" grains (grains with track densities that show previous exposure in a regolith) vary by factors of 10–30 (Hohenberg *et al.*, 1990). Even within the "unirradiated" grains (grains with low track densities, which presumably had minimal previous exposure in a regolith), ^{21}Ne ages vary by a factor of 2, from 1.4 to 2.8 Myr.

The average ^{21}Ne ages for CV, CK, and CO chondrites, not corrected for pairing, are 13 ± 10, 23 ± 14, and 22 ± 18 Myr, respectively. These averages exceed the CRE ages of the CI and CM chondrites by a factor of ~ 10. The data hint but do not clearly establish that the CV chondrites have shorter CRE ages than the CK and CO chondrites. Goswami *et al.* (2001) have shown that at least one of the CK meteorites, Kobe, had a complex exposure history, for which the most recent stage lasted only ~ 1 Myr.

1.3.2 The CR Clan

The CR clan comprises ~ 20 meteorites subgrouped as CR, CH, or CB meteorites. Noble gas analyses have been reported for five of them: Acfer 094 (paired with El Djouf 001), Al Rais, GRA 95229, Loongana, and the type specimen, Renazzo (Scherer and Schultz, 2000; Mazor *et al.*, 1970). Because of the presence of trapped gases, for only one of these meteorites, Loongana, can the cosmogenic ^{22}Ne/^{21}Ne ratio be determined with sufficient precision to make a reliable

shielding correction. Accordingly, we have calculated the ^{21}Ne ages shown in Figure 1 by assuming average shielding, average CR composition (Lodders and Fegley, 1998), and solar neon isotope ratios. The results have large uncertainties— 30% or so—in part because the isotopic composition of the trapped component is not entirely certain. Nishiizumi *et al.* (1996b) present ^{10}Be, ^{26}Al, and ^{36}Cl activities for the CR chondrites (EET 87770, MAC 87320, and PCA 91082). PCA 91082 has an unusually low ^{10}Be content, only 10.9 dpm kg^{-1}, which is consistent with an age of 1.5 Myr. The activities for the other meteorites appear to be saturated and give only lower bounds on the exposure ages. At this early stage, the age distribution of CR chondrites seems to be most like that of the CV chondrites both in terms of average age, ~ 8 Myr, and range, from 1 to 25 Myr.

Noble gas analyses and CRE ages have been reported for all but one (NWA 470) of the six known CH chondrites: the CRE ages are shown in Table 3. Weber *et al.* (2001) note apparent systematic differences between CRE ages based on noble gases, T_{Gas}, or on cosmogenic radionuclides, T_{Radio}, and suggest that multiple stages of exposure may explain the differences. At first glance, it appears that the CRE ages of CH chondrites seem most similar to those of the CR and CV chondrites. This group needs more work.

Five bencubbinites (CB chondrites) are known—Bencubbin, Gujba, Weatherford, Hammada al Hamra 237, and QUE 94411 (paired with QUE 94627)—three of them found since 1994. Begemann *et al.* (1976) obtained a ^{36}Cl/^{36}Ar age of 36 ± 5 Myr for Bencubbin after making a substantial correction for trapped argon. From the noble gas contents of Gujba silicates, Rubin *et al.* (2003) calculate a CRE age of 26 ± 7 Myr with the unusually large uncertainty reflecting a lack of a compositional analysis and an unusually low ^{22}Ne/^{21}Ne ratio of ~ 1.05. An earlier noble gas analysis of Weatherford by Stauffer (1962) is difficult to interpret because of the lack of compositional information. In sum, the state of knowledge about CRE ages for this interesting group of meteorites is inadequate.

Table 3 CRE ages (Myr) of CH chondrites.

Meteorite	T_{Gas}	T_{Radio}[a]
Acfer 182 (+ 207 + 214)	12[b]	
ALH 85085	1.7 ± 0.8[c]	0.9
PAT 91546	> 8[d]	> 3.5
PCA 91328 (+ 452 + 467)	4.3[d]	1.1
RKP 92435	1.5[d]	0.6

[a] Nishiizumi *et al.* (1996a). [b] Bischoff *et al.* (1993). [c] Eugster and Niedermann (1990). [d] Weber *et al.* (2001).

1.4 H-CHONDRITES

Among ordinary chondrites, $\sim 50\%$ (well over 7,000 stones) are classified as H-chondrites and three fourths of those belong to just two petrologic groups, H5 or H6. Graf and Marti (1995) have reviewed noble gas data for more than 400 H-chondrites, extracted CRE ages from those data, and examined their distributions. The CRE ages rest ultimately on the production rates of Eugster (1988), which is to say on an absolute calibration based on the ^{81}Kr/Kr ages of ~ 20 ordinary chondrites. Graf *et al.* (2001) subsequently presented ^{36}Cl/^{36}Ar ages for 16 H-chondrites. Below, we summarize some conclusions from these articles.

The CRE age distribution for all H-chondrites spans an apparent range from < 1 to ~ 80 Myr (Figure 2). For stones at the low end of the distribution, the possibility of complex irradiation looms large, and indeed, that possibility has been confirmed in several cases (Herzog *et al.*, 1997). Graf and Marti (1995) conclude that "it appears there are very few H-chondrites with short exposure ages." As noted by Wieler (2002), this observation has two implications—first, that collisions producing H-chondrites are rare events, occurring with a frequency of no more than a few per million years and second, that the collisions take place so far away that the fragments take several million years to get to Earth.

An enormous peak between about 6 and 10 Myr in the CRE age distribution encompasses nearly half the H-chondrites. The H5- and H4-chondrites populate the peak more heavily, both in an absolute and a relative sense, than do the H3- or H6-chondrites. Interestingly, the position of the maxima may differ slightly among the groups. The effect is not statistically significant at present, but, if verified, would signal two events close in time, one of them leading to the production of H5 meteorites and a second one to other petrologic types of H-chondrites. Graf and Marti (1995) also note other distinguishing features of the H5 chondrites. One subset has lower ^{3}He/^{38}Ar ratios, which may reflect greater heating in orbit, and a different distribution of fall times (see also Wieler and Graf, 2001).

A second peak in the CRE age distribution for all H-chondrites crops up at ~ 33 Myr and seems fairly robust, but the uncertainties are appreciable. The distribution for H6-chondrites has a unique peak at ~ 24 Myr.

What do these ages tell us about the history of the H-chondrites? First, as has been known for some time, the age distribution is not consistent with continuous delivery, but rather is dominated by a very small number of events, consistent with a correspondingly small number of parent bodies. Second, the existence of a peak at 33 Myr common to H-chondrites of all petrologic types suggests that most types of H-chondrites are present in at least some asteroids. As Graf and Marti (1995) note, material that was at the surface of the parent body for some period of time (as indicated by the presence of trapped noble gases) occurs about as often in one petrologic type as another. Thus, at the time of the major collision at 6–10 Myr, the source of H-chondrites was not an ancient layered object with, say, matter of higher petrographic type sheltered in the interior. In this context, it would be interesting to reexamine the meteorites with an age of 33 Myr to see whether trapped gas contents correlate with petrographic type.

1.5 L-CHONDRITES

Numbering well over 6,000, the L-chondrites are dominated by two petrographic types, the L5 (20%) and L6 (65%). Marti and Graf (1992) have reviewed L-chondrite exposure ages calculated from the light noble gases. The collective CRE age distribution looks at first to have an exponential envelope that rises from 0 to a peak at or near 40 Myr with a fairly sharp decrease beyond (Figure 3). On closer examination, the CRE age distributions of the L5- and L6-chondrites appear to be neither completely monotonic nor the same for all petrologic types. The strongest peak occurs at ~ 40 Myr. With a little imagination one can make out a peak at 5 Myr and a broad hump

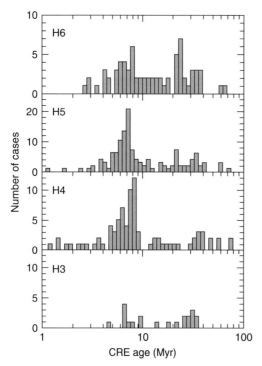

Figure 2 Exposure ages (Myr) of H chondrites. Source: Graf and Marti (1995).

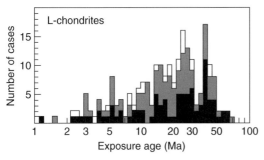

Figure 3 Exposure ages of L-chondrites. Different fills indicate varying degrees of precision. Reproduced by permission of Annual Reviews from Marti and Graf (1992).

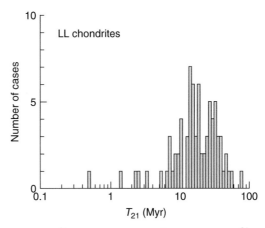

Figure 4 ^{21}Ne exposure ages of LL chondrites. ^{21}Ne ages recalculated with the formulas of Eugster (1988). Multiple analyses for meteorites were averaged. Meteorite finds from Antarctica and northern Africa with similar exposure ages and collected at the same site were treated as paired. Source: Schultz and Franke (2002).

running from 20 to 30 Myr with a maximum somewhere in the vicinity of 28 Myr. At present, we do not seem to have any L3- or L4-chondrites with exposure ages of 50 Myr or more.

A handful of L-chondrites have the low ^{21}Ne contents, $< \sim 0.4 \times 10^{-8}$ cm^3 STP g^{-1}, associated with an exposure age of 1 Myr or less, namely, Farmington (L5), Pampa (L4), Shaw (L6-7), and Ladder Creek. Herzog *et al.* (1997) concluded that Ladder Creek had a complex history, but found no evidence for one in Shaw. Marti and Matthew (2002) conclude that Farmington, too, had a simple exposure history in the sense that the material was deeply shielded until ~ 25 kyr before the present.

The authors of many early studies of L-chondrites searched for trends relating CRE ages to losses of the radiogenic gases ^{40}Ar and ^{4}He. Interestingly, among L-chondrites with significant losses of ^{40}Ar, the peaks at 5 and 28 Myr noted above emerge much more clearly (Marti and Graf, 1992) than they do in Figure 3. That we have samples from (at least) two events in heated portions of a parent body seems clear; it is not obvious whether the events took place on the same or different parent bodies.

1.6 LL CHONDRITES

Graf and Marti (1994) compiled light-noble-gas exposure ages for ~ 60 LL chondrites. We have recalculated ^{21}Ne ages for 77 LL chondrites by using the formula of Eugster (1988) without culling the data and simply averaging where analyses for more than one sample were available (Figure 4). The figure includes one LL chondrite with a short exposure age, Hunter, which was not in the data set of Graf and Marti (1994). In general, Figure 4 is similar to figure 1 of Graf and Marti (1994). The distribution runs from 0.5 Myr for Hunter to just under 80 Myr for Soko Banja. The strongest peak in the distribution, with 20% of the L-chondrites, occurs at ~ 15 Myr. Whether

different petrologic types of LL chondrite populate the peak to different degrees seems to us debatable given the small numbers of meteorites in the LL3 category. In our larger data set, we have two meteorites, Richfield (LL 3.7) and Acfer 160 (LL3.8-6), with ^{21}Ne ages close to 15 Myr. A broad peak spreads from 27 to 33 Myr, overlapping the 28 Myr peak in the L-chondrite CRE age distribution.

1.7 E-CHONDRITES

Patzer and Schultz (2001) calculated ^{21}Ne exposure ages of ~ 60 E-chondrites with shielding corrections based on ^{22}Ne/^{21}Ne ratios (Eugster, 1988) and adjustments for composition based on average compositions of one of the following groups: EH3; EH4,5; EL3; and EL5,6 (Kong *et al.*, 1997). In shape, the distribution of ^{21}Ne ages (Figure 5) resembles most closely that of the L-chondrites, although it has relatively more members with short CRE ages and fewer with ages > 40 Myr. Similarities to the age distribution of the enstatite achondrites or aubrites (Figure 6) are limited.

Patzer and Schultz (2001) identified possible clusters of ages at about 3.5, 8, and 25 Myr with a well-advised caution that confirmation is needed. While the experimental data are not in doubt, the bellwether ^{21}Ne CRE ages do not agree particularly well with either ^{3}He or ^{38}Ar CRE ages, the typical spread being $\sim 25\%$, with many larger excursions. Reasons for the disagreements may include weathering, particularly of calcium-bearing sulfides, target element variations in the case of ^{38}Ar, and diffusion losses for ^{3}He.

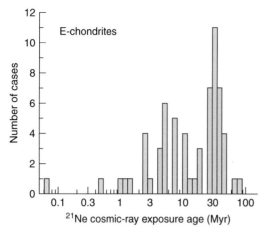

Figure 5 Cosmic-ray exposure ages of E-chondrites. Sources: Patzer and Schultz (2001), Patzer *et al.* (2001), and Okazaki *et al.* (2000).

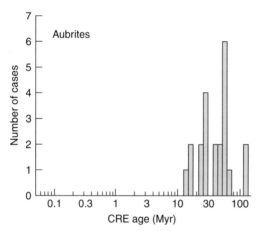

Figure 6 CRE ages of aubrites (for references, see Section 1.14).

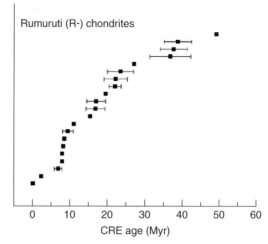

Figure 7 Cosmic-ray exposure ages of R-chondrites, based on noble gas contents. Source: Weber and Schultz (2001).

1.8 R-CHONDRITES

Figure 7 shows CRE ages for a relatively new group of ordinary chondrites, the R-chondrites, named for the type specimen, Rumuruti. In tabulating ages, we have either quoted directly from the literature or, where possible, recalculated ^{21}Ne ages using the formalism of Eugster (1988) and the average compositional data of Lodders and Fegley (1998). The presence of large concentrations of trapped noble gases limits the accuracy of the shielding corrections in several cases. The CRE ages range from 0.2 Myr for Northwest Africa and 053 to ~ 50 Myr for Hughes 030. Five of the 21 R-chondrites in Figure 7 have CRE ages close to 7.5 Myr: Carlisle Lakes, 6.9 ± 1.0; Northwest Africa 755, 8.0; Ouzina, 8.0; Y79357, 8.3 and 8.6 (shown separately in Figure 7); and Dar al Gani 013, 9.5 ± 1.4. Although it would be premature to identify these five meteorites with a cluster,

we cannot resist pointing out the coincidence with the major peaks in the respective CRE age distributions of the H-chondrites and of the acapulcoite/lodranite group (see below).

1.9 LODRANITES AND ACAPULCOITES

Several lines of evidence point to a common parent body for lodranites and acapulcoites: overlapping oxygen-isotope composition, identical mineral constituents, complex thermal histories, and, as we will see, a peaked distribution of CRE ages (Clayton and Mayeda, 1996; McCoy *et al.*, 1997). Terribilini *et al.* (2000a), Weigel *et al.* (1999), and McCoy *et al.* (1996) report analyses of noble gases in lodranites and acapulcoites and give references to earlier work. Xue *et al.* (1994) and Terribilini *et al.* (2000a) report analyses of cosmogenic radionuclides. The CRE ages of lodranites and acapulcoites fall in a fairly narrow range running from ~ 4 to ~ 10 Myr (Figure 8). Terribilini *et al.* (2000a) note that H-chondrites also have a peak in this region. They do not suggest a common parent body for the two groups of meteorites but raise the possibility of unusually high collisional activity in the asteroid belt, due perhaps to the passage through the asteroid belt of a third party of impactors.

The lodranites and acapulcoites present two kinds of problems for the calculation of CRE ages. First, the lodranites are coarse grained and many of them contain a large fraction of metal. Special attention to elemental composition is, therefore, desirable in interpreting the data. Second, as it turns out, many lodranites and acapulcoites have exceptionally high cosmogenic ^{22}Ne/^{21}Ne ratios, which normally indicate near-surface locations, probably in small meteoroids, where solar cosmic rays (SCRs) may have left

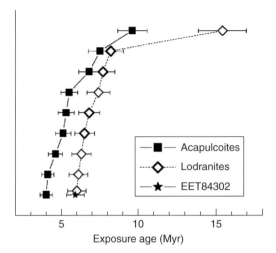

Figure 8 CRE ages of lodranites: Gibson, Y791491, LEW 88280, FRO 90011, GRA 95209, Y74357, Lodran, MAC 88177, and QUE93148; and of acapulcoites: ALH 77081, ALH 81261, FRO 95029, GRA 98028, ALH 81187, ALH 84190, Monument Draw, Acapulco, and Y74063 and of EET 84302.

their mark on the samples. The quantitation of effects due to SCRs introduces another level of difficulty in the calculation of CRE ages. The ages shown in Figure 8 were calculated in various ways: mostly from ^{21}Ne and ^{22}Ne/^{21}Ne ratios, but with a few ^{36}Cl/^{36}Ar ages, one ^{81}Kr/^{83}Kr age, and model-dependent estimates of cosmogenic nuclide contributions due to SCRs.

1.10 LUNAR METEORITES

1.10.1 Overview

We summarize results pertaining to the CRE histories of lunar meteorites in Table 4. The meanings of the column headers are discussed in the table notes. Eugster (1989) and Warren (1994, 2001) have written specialized reviews dealing with questions related to the exposure histories of lunar meteorites. Some general observations and conclusions from this work follow.

- Most lunar meteorites travel to Earth in < 1 Myr and some in < 0.1 Myr; other kinds of meteorites generally take longer. As the Moon is close by, the short transit times are intuitively appealing. They also fall naturally out of theoretical calculations, as does the expectation that a small fraction of the lunar meteorites will have longer exposure ages (cf., Gladman *et al.*, 1995).
- Most lunar meteorites contain detectable concentrations of cosmogenic nuclides attributable to irradiation within a few meters of the lunar surface. This observation sets lunar meteorites apart from the general run of meteorites in

which the signs of parent-body irradiation are rarer and generally harder to detect. The near-surface irradiations prior to launch also hint at the importance of "smaller" impact events in launching lunar meteorites. We will return to the meaning of "smaller" below.

- Many lunar meteorites are regolith breccias, meaning that over time at least some of their constituent grains resided not only below the surface waiting for launch as just described, but also *at* the very surface, where the grains could collect solar wind ions, and presumably at other depths as well. Such grains may have extremely complex exposure histories. For this reason, if considered alone, the concentrations of stable cosmogenic nuclides in lunar meteorites can seldom be interpreted uniquely. For lunar meteorites especially, the combination of radioactive and stable cosmogenic nuclides is important. Even with both kinds of measurements, however, the reconstruction of a grain-by-grain irradiation history is probably beyond reach. In any case, if detailed exposure histories on the Moon are of interest, it is probably simpler to study nonregolithic rocks that astronauts have brought back from the Moon.
- Most lunar meteorites are breccias with low porosity. Warren (2001) notes that low-porosity breccias constitute a larger fraction of lunar meteorites than they do of the lunar surface as a whole. The low porosity of lunar meteorites seems to correlate with high mechanical strength and by implication, with resistance to collisional destruction. By extension, this observation points strongly to the importance of the mechanical properties of the target rock in determining the kinds of material that we sample from asteroids and Mars.

1.10.2 Construction of CRE Histories

Sorting out the complex irradiation histories of lunar meteorite requires some art. As a first step, we usually assume that irradiation on the Moon has left no measurable traces and try to explain the observations with a simple, one-stage or 4π irradiation history in space. Here, the modifier 4π refers to the assumption, almost always true, that the meteoroid was small enough in space for cosmic rays to reach interior material from all solid angles. In the idealized world of spherical meteoroids, a one-stage history has four important parameters: the size of the meteoroid, the depth of the sample within it, the duration of the irradiation in space, and the terrestrial age. In principle, during transit to Earth, SCRs could also affect matter within a few millimeters of the surface. Ablation losses usually remove such material, but in rare instances the surface material survives. When the one-stage irradiation model fails to explain the

Table 4 Exposure histories of lunar meteorites.

Meteorite	Pair	Mass (g)	$D_{2\pi}$ (g cm^{-2})	$T_{2\pi}$ (Myr)	$R_{4\pi}$ (g cm^{-2})	$T_{4\pi}$ (Myr)	T_{Terr} (kyr)	Notes	References
Cumulate olivine norite with regolith breccia									
NWA 773		633		160					a
Feldspathic fragmental breccia									
Dhofar 081		174	190–210	100					a
Dhofar 081		174		> 4		Short	200	(i)	b
Dhofar 280	081	251							
Dhofar 489		34.4							
Feldspathic fragmental/regolith breccia									
Yamato 82192	82193, 86032	712	> 1,000			9	~ 90		c
Feldspathic impact melt breccia									
NWA 482		1,015			15–19	0.9 ± 0.2	60–120	(ii)	d
Dhofar 026		148				< ~ 0.003			d
Dhofar 026		148	1,100–1,300	10		~ 0			d
Dhofar 026		148				< 0.01			e
Dhofar 301	025?	9							
Dhofar 302		3.83							
Dhofar 303		4.15							
Feldspathic regolith breccia									
Dar al Gani 400		1,425		< 3		< 1			f
"1153" alleged									z
ALH 81005		31.4	164			0.0025	9		c
Dar al Gani 262		513	50			0.5	300		g
Dar al Gani 262		513	75–85	Long	> 3	Short	50–60		g
Dar al Gani 262		513	55–85	Long		< 0.15			h
Dar al Gani 262		513	50–80	500–1,000	= 10				i
Dhofar 025	025	751				4–20	500–600		j
MAC 88104	88105	724	370			~ 0.04	210–250		c

Name	Pair	Mass							Ref
MAC 88104	88105	724	85	630 ± 200		< 0.24	100–600	(iii)	k
MAC 88104	88105	724	360–400	> 5		0.04–0.05	210–250		l
MAC 88104	88105	724	390–500	> 5		0.04–0.11	100–190		m
QUE 93069	94269	24.5	~ 90	1,000 ± 400	= 40	0.15 ± 0.02	< 15		n
QUE 93069	94269	24.5	65–90	> 500	Small	~ 0.02–0.05	5–10		o
Y 791197		52.4	5			< 0.019	30–90		c
Y 791197		52.4		450					p
Y 983885		289							
Feldspathic/mare regolith breccia									
Calcalong Creek		19	40–50	> 300		< 0.2	< 70		q
QUE 94281		23	270–320	400 ± 60			150–200		r
QUE 94281		23							s
Y 79374/981031		194.7	165			< 0.02	< 20		c
Y 79374/981031		194.7	40	510 ± 140					t
Y 79374/981031		194.7	150–190	700 ± 200		< 0.12			u
Y 79374/981031		194.7	35 ± 15						v
Mare basalt									
NWA 032/479		~456	> 1100		< 12	0.042	< 80		b
Y 793169		6.1	Deep			1.1 ± 0.2	< 50		w
Y 793169		6.1	500	50 ± 10					n
Dhofar 287		154							
Asuka 881757		442	> 3 m			0.9	< 50		c
Mare polymict breccia									
EET 87521	96008	84	565	26		< 0.1	15–50		x
EET 87521	96008	84	540–600	26		< 0.01	80 ± 30		y
EET 96008	87521	84	200–600					(iv)	i

Notes: The petrographic descriptions are from List of lunar meteorites (2006). Pair refers to paired meteorites from the same locality. Mass is the recovered mass. $D_{2\pi}$ is the depth at which irradiation on the Moon took place, $T_{2\pi}$ is the duration of the lunar irradiation, $R_{4\pi}$ is the radius of the meteoroid while in transit to Earth, $T_{4\pi}$ is the duration of transit to Earth, and T_{Terr} is the terrestrial age. (i) Greshake et al. (2001) note similarities to MAC 88104/5. (ii) Depth calculations assume a density of 2.7 g cm^{-3}. (iii) $T_{2\pi}$ before compaction. (iv) Full model has three stages. References: (a) Eugster and Lorenzetti (2001); (b) Nishiizumi and Caffee (2001a); (c) Warren (1994); (d) Nishiizumi and Caffee (2001b); (e) Shukolyukov et al. (2001); (f) Scherer et al. (2001); (g) Nishiizumi et al. (1998); (h) Bischoff et al. (1998); (i) Eugster et al. (2000); (j) Nishiizumi and Caffee (2000); (k) Eugster et al. (2001b); (l) Nishiizumi et al. (1991a); (m) Vogt et al. (1991); (n) Thalmann et al. (1996); (o) Nishiizumi et al. (1996b); (p) Ostertag et al. (1986); (q) Swindle et al. (1995); (r) Nishiizumi and Caffee (1996b); (s) Polnau and Eugster (1998); (t) Takaoka and Yoshida (1992); (u) Nishiizumi et al. (1991b); (v) Eugster et al. (1992); (w) Nishiizumi et al. (1992a); (x) Vogt et al. (1993); (y) Nishiizumi et al. (1999); (z) Yanai (2000).

results for a lunar meteorite, we precede it with an earlier stage of irradiation on the Moon. In the approximation that the local topography of the lunar surface did not much affect the GCR flux and that erosion of the lunar surface was negligible, the earlier stage introduces two more parameters— the depth of the sample while on the Moon and the duration of the irradiation there. Neither of these approximations probably holds in reality. Thus, the inferred conditions of early exposure will generally have large uncertainties.

As all known lunar meteorites are finds (and therefore have nonzero terrestrial ages), we need at least four measured quantities to determine the four parameters of a simple one-stage history. Similarly, for a simple two-stage history, we need at least six measured quantities. Typically, the data set available comprises ^{3}He, ^{21}Ne, ^{22}Ne, ^{38}Ar, ^{36}Cl, ^{26}Al, and ^{10}Be. Occasionally we may have other information—the concentrations of spallogenic krypton isotopes, spallogenic xenon isotopes, ^{14}C, ^{41}Ca, and ^{53}Mn, the densities of nuclear tracks (tracks/unit area), and the concentrations of certain isotopes produced by thermal neutrons, for example, ^{36}Ar (from ^{36}Cl) and ^{158}Gd.

In practice, the important information about terrestrial age derives mainly from ^{14}C, ^{36}Cl, or ^{81}Kr. Chlorine-36, although suboptimal for terrestrial ages <100 kyr, is the most widely used of the three and the potential of ^{41}Ca has yet to be realized routinely. In the determination of the other parameters of the irradiation histories, the possibility of diffusion loss compromises ^{3}He with the consequence that it is respected when it confirms other results but rejected when it does not. As noted above (Section 1.2.5.2), the calculation of ^{38}Ar CRE ages leaves room for improvement and for this reason results based on ^{38}Ar may generate skepticism. Commonly, then, we end up with five pieces of information on which to build the exposure histories of lunar meteorites: ^{21}Ne, ^{22}Ne, ^{36}Cl, ^{26}Al, and ^{10}Be, and the hope, sometimes realized, that the elemental composition is known for the particular sample analyzed for cosmogenic nuclides. The importance of spallogenic ^{22}Ne arises mainly through the spallogenic ^{22}Ne/^{21}Ne ratio, which as noted above provides some measure of shielding conditions. In lunar samples, unfortunately, the presence of large concentrations of trapped (solar) gases increases the uncertainties of the calculated spallogenic ^{22}Ne contents. Problems aside, armed with this information, we can proceed by an iterative process to find the set of parameters that best matches the data, provided that we also have adequate knowledge of production rate systematics for the cosmogenic nuclides.

Space does not allow us to discuss lunar (or meteoritic) production rate systematics in detail. In brief, several groups have studied this issue theoretically, with modeling calculations, two of

them in ways that have proved accessible and useful for meteoriticists (cf., Leya *et al.*, 2000; Masarik and Reedy, 1994). The calculated production rates of spallogenic nuclides are probably good to within ∼ 15%. Both sets of calculations incorporate a parameter that, in effect, specifies the intensity of the GCR flux. This flux is smaller by perhaps 3–5% at 1 AU than at 2–5 AU (Reedy *et al.*, 1993; Reedy and Masarik, 1994; Michel *et al.*, 1996). Additional important information used to calibrate the production rates comes from another source: the empirical studies of depth profiles in meteorites and lunar samples. In daily practice, researchers are likely to rely on a combination of results from modeling calculations and from measured depth profiles.

1.10.3 Production Rate of Lunar Meteorites

Vogt *et al.* (1991) quote J. Melosh to the effect that an impactor striking the Moon must have a radius >10 m to accelerate rocks to escape velocity. The cumulative rate of influx onto the Moon's surface for impactors of $R \geq 10$ m is 30 Myr^{-1} (Melosh, 1989, p. 189). As Table 4 shows, of 18 ± 6 meteorites thought to have left the Moon independently and for which we have estimates of the transit time to Earth, 16 have transit times $(T_{4\pi}) \lesssim 1$ Myr. If a similar proportion holds for the lunar meteorites not yet analyzed, then the arrival rate of lunar meteorites will have bumped up against the estimated production rate of 30 even though we have sampled effectively only a tiny fraction of the Earth's surface. Vogt *et al.* (1991) inferred as much on the basis of a much smaller sample set. They, therefore, suggested that the impact events on the Moon must typically launch not one or even a few meteoroids, but large numbers of them. The alternative explanation is that smaller impactors, for which fluxes are higher, also produce lunar meteorites. The common occurrence of lunar meteorites with signs of lunar irradiation would seem to point in this direction in the sense that smaller impacts probably excavate material from shallower depths.

1.11 HOWARDITE–EUCRITE– DIOGENITE METEORITES

Just two major impact events, one at ∼ 20 Myr and the other at ∼ 40 Myr, on one body could account for the great majority of the eucrites, howardites, and diogenites reaching Earth today. One or more events at earlier times also seem likely.

1.11.1 Eucrites

Among the various types of exposure ages, ^{81}Kr/Kr CRE ages appear to be the most reliable for eucrites (plagioclase–pigeonite achondrites). Eucrites contain strontium ($\sim 80\,ppm$), yttrium ($\sim 15\,ppm$), and zirconium (20–$90\,ppm$) at levels considerably higher than those that occur in ordinary chondrites. The high concentrations of these three elements enhance the production of spallogenic krypton and so facilitate the determination of $^{81}Kr/Kr$ exposure ages. Argon-38 CRE ages may also be trustworthy when adjusted for shielding and sample composition (Aylmer *et al.*, 1988). Most eucrites have $\sim 7\,wt.\%$ of calcium, a prime target for production of ^{38}Ar. Unfortunately, disappointingly few reliable CRE ages can be calculated from well over 200(!) analyses of cosmogenic ^{3}He and ^{21}Ne in eucrites. Difficulties arise partly because of diffusion losses from feldspar (Megrue, 1966; Heymann *et al.*, 1968). Feldspar typically constitutes 30–40% of eucrites (Kitts and Lodders, 1998), and any small, randomly chosen sample may contain more or less of that mineral. Further, low magnesium contents make the $^{22}Ne/^{21}Ne$ shielding monitor nearly useless.

Figure 9 summarizes the $^{81}Kr/Kr$ ages of 19 of the ~ 200 known eucrites (The 1977 *Appendix to the Catalog of Meteorites* lists only 31 eucrites!). All 19 of them have values between 4 and 60 Myr. Eugster and Michel (1995) assign these $^{81}Kr/Kr$ CRE ages to five clusters at ages of 6 ± 1, 12 ± 2, 21 ± 4, 38 ± 8, and $73 \pm 3\,Myr$. We note also that a suite of Yamato meteorites that may be paired with each other may populate a separate cluster at 73 Myr (Miura *et al.*, 1993). These $^{81}Kr/Kr$ CRE ages are uncertain because the terrestrial ages of the meteorites are not known. In independent work, Shukolyukov and Begemann (1996) reassigned essentially the same eucrites to five clusters at 7 ± 1, 10 ± 1, 14 ± 1, 22 ± 2, and $37 \pm 1\,Myr$.

Shukolyukov and Begemann have provided a muscular criterion for deciding when meteorites belong in a cluster. In our view, the clusters of five eucrites near 20 Myr (30% of the cases) and of four meteorites at 37 Myr (20% of the cases) are plausible; the identification of groupings below 20 Myr seems doubtful at this time. Recent estimates of the ^{38}Ar ages of the eucrites (Piplia Kalan and Vissannapeta) appear to place both meteorites in the cluster between 20 and 25 Myr (Bhandari *et al.*, 1998; Mahajan *et al.*, 2000).

1.11.2 Diogenites

With strontium, yttrium, and zirconium concentrations much lower than in eucrites, the diogenites present a less-attractive target for $^{81}Kr/Kr$ measurements and we are aware of only three such ages (Eugster and Michel, 1995). Fortunately, the higher magnesium contents in diogenites reinstate the value of the $^{22}Ne/^{21}Ne$ ratio as a shielding monitor. Welten *et al.* (1997, 2001a) present a careful assessment of the exposure ages of ~ 20 diogenites, based on results for light noble gases with shielding corrections from $^{22}Ne/^{21}Ne$ and compositional corrections based on same-sample elemental analyses.

The diogenites, as the eucrites, define a range of CRE ages that lie mostly between about 5 and 60 Myr. Figure 10 lends unambiguous support to a cluster of ages at $\sim 22\,Myr$ (1/3–1/2 of the cases, depending on pairing), as suggested by Eugster and Michel (1995). Perhaps one-sixth of the diogenites have CRE ages near 37 Myr. With only five diogenites plotted for CRE ages between 4 and 20 Myr and only one possible coincidence among those five (Aioun el Atrouss and TIL 82410 at 12 Myr), we regard the identification of clusters below 20 Myr as premature for diogenites.

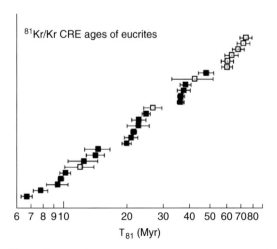

Figure 9 Eucrite exposure ages. Data shown as closed squares from Marti (1967), Hudson (1981), Freundel *et al.* (1986), Shukolyukov and Begemann (1996), and Miura *et al.* (1998). Data shown as open squares from Miura *et al.* (1993). Additional results may be found in Paetsch *et al.* (2001).

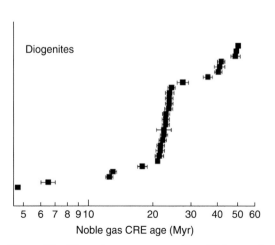

Figure 10 Diogenite exposure ages from Welten *et al.* (1997, 2001a). Other results from Miura and Nagao (2000) and from Paetsch *et al.* (2001) agree.

1.11.3 Howardites

Relatively precise CRE ages are available for only a few of the more than 100 howardites known. Welten *et al.* (1997) tabulate CRE ages from ^{21}Ne and ^{38}Ar for 19 howardites. In 10 cases, the CRE ages agree within the uncertainties; in nine cases T_{21} and T_{38} disagree by more than 50%. With feldspar a major mineral and same-sample elemental analyses uncommon, these discrepancies are not surprising. Eugster and Michel (1995) recommend CRE ages for 17 howardites among which they include ^{81}Kr/Kr ages for two stones, namely, Bholghati (17 Myr) and Petersburg (21.5 Myr). Published ^{26}Al activities (Nishiizumi, 1987) suggest generally normal levels of shielding. Measurements of other cosmogenic radionuclides are scarce.

Figure 11 shows the CRE ages mentioned above along with recent data for Lohawat. All but two or three of the CRE ages may well belong to just two clusters at about 20 and 40 Myr, consistent with the best-defined peaks in the CRE age distributions of the eucrites and of the diogenites. At present, it is hard to tell whether Luotolax (70 Myr) or Lohawat (110 Myr?) has the largest CRE age for a howardite. In either case, the value is larger than the maxima seen among diogenites, although some Yamato eucrites may have CRE ages as high as 70 Myr (see above). While it is easy to identify the howardite with the smallest CRE age (Chaves), it is hard to tell what, exactly, that CRE age may be: the error bars allow a range from 6 to 19 Myr! In contrast, both the diogenites

and the eucrites clearly include younger members; $\sim 15\%$ of the diogenites and up to 50% of the eucrites have CRE ages < 19 Myr.

1.11.4 Kapoeta

An interesting controversy has arisen over the CRE history of Kapoeta. Inclusions in Kapoeta contain variable concentrations of noble gases produced by cosmic rays. Neither (1) shielding effects nor (2) compositional variations can explain the variations in the cosmogenic gases inasmuch as (1) only a few centimeters separate the inclusions and (2) the effects of elemental composition are explicitly taken into account. By elimination, the variations must record an earlier period of irradiation. To characterize this earlier irradiation we first remove the effects of the most recent stage of irradiation. From measurements of cosmogenic radionuclides, Caffee and Nishiizumi (2001) persuasively conclude that Kapoeta was last irradiated as a body with a diameter 20 cm for ~ 3 Myr prior to striking Earth in 1942. With this information and the application of modeling calculations, one can estimate the concentrations of noble gases attributable to the last 3 Myr of cosmic-ray irradiation and subtract them from the observed totals. The balance represents the contributions from earlier irradiation. It is the characterization of the early irradiation that has occasioned disagreement. Wieler *et al.* (2000) account for the early neon with a GCR-only irradiation lasting, perhaps, tens to hundreds of

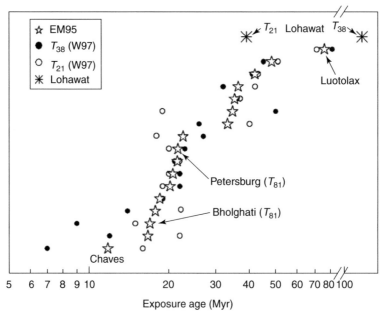

Figure 11 CRE ages of howardites. EM = best estimates from Eugster and Michel (1995). T_{38} = ^{38}Ar CRE age from W97 = Welten *et al.* (1997). T_{21} = ^{21}Ne CRE age from Welten *et al.* (1997). Data for Lohawat from Sisodia *et al.* (2001). Analytical uncertainties of the CRE ages are $\sim 10\%$, although as shown the results of different methods may disagree by larger percentages.

million years at a depth of several hundred grams per square centimeter in a large body (2π irradiation). Rao *et al.* (1997) do not agree, suggesting instead that SCRs produced as much as 80% of the early neon, with GCR producing the rest. In more modern (lunar) samples, Rao *et al.* argue that SCRs produce 40% or less of the cosmogenic neon. Among several possible explanations for the higher apparent proportion of SCR-produced neon in Kapoeta, Rao *et al.* favor an SCR flux enhanced by a factor of 10 or more, probably generated by an early active Sun (see also Hohenberg *et al.*, 1990). Wieler *et al.* (2000) observe that the divergence of views hinges on a technical point, namely, the deconvolution of SCR and GCR neon. The results reflect the (model-dependent) neon isotopic ratios adopted for SCR and GCR neon in the phases analyzed (feldspars and pyroxenes). We regard this argument as unsettled.

1.12 ANGRITES

Oxygen isotope systematics tie the angrites to the howardite–eucrite diogenite (HED) complex (Clayton and Mayeda, 1996), but CRE ages are dissimilar. CRE ages are known for five, and a very uncertain lower limit for one (LEW 87051) of the angrites (Figure 12). The better-defined ages range from 5.4 Myr for Asuka 881371 to 56 Myr for Angra dos Reis. As far as we know, the angrites show no signs of earlier irradiation as a small body or in a regolith.

1.13 UREILITES

Goodrich *et al.* (2002) suggest that more than 110 ureilites known today were excavated by a single collision and then reassembled in a single parent body. Figure 13 shows the distribution of approximate ^{21}Ne ages of 26 ureilites calculated from the data compiled by Schultz and Franke (2002) and a ^{21}Ne production rate of

0.412×10^{-8} cm^3 STP g^{-1} Myr^{-1}. Four of the samples have unusually short ages—< 2 Myr. ALH 78019 is remarkable in having a CRE age of < 100 kyr; such low values are common only among lunar meteorites and, to a lesser degree, CM chondrites. Kenna has the maximum CRE age for a ureilite, ~ 35 Myr. The age distribution may have a cluster in the vicinity of 10 Myr. The peak sharpens for CRE ages based on ^{21}Ne/^{10}Be ratios (Graf *et al.*, 1990b) and data from Aylmer *et al.* (1990). If real, a cluster would lend credibility to the idea that many ureilites come from one parent body (Goodrich *et al.*, 2002).

1.14 AUBRITES (ENSTATITE ACHONDRITES)

Early on, Eberhardt *et al.* (1965) presented the irradiation histories of aubrites and identified the essential features of their CRE age distribution. More recently, Lorenzetti *et al.* (2003) compiled available data, reported new measurements (for Mt. Egerton, Mayo Belwa, and what are probably five distinct Antarctic meteorites), and updated the calculations of CRE ages. We show the results in Figure 14 (and in Figure 6).

Whether considered individually or on average, the aubrites have the largest CRE ages of all stony meteorites. For example, the Norton County fall has a CRE age of over 100 Myr. It is an often-remarked curiosity that this atypically large CRE age was also the first one ever reported for a stone (Begemann *et al.*, 1957). Even after 40 years of analyses, Norton County, along with one other aubrite, Mayo Belwa, ranks among the stones with the longest CRE ages. At the other extreme, the smallest CRE age for aubrites, ~ 12 Myr, belongs to the type specimen for the group, the Aubres fall.

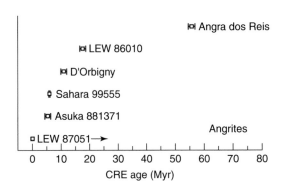

Figure 12 CRE ages of angrites. Sahara 99555, Bischoff *et al.* (2000); D'Orbigny, Kurat *et al.* (2001); Asuka 881371, Weigel *et al.* (1997); Angra dos Reis, Lugmair and Marti (1977); LEW 86010 and LEW 87051, Eugster *et al.* (1991a).

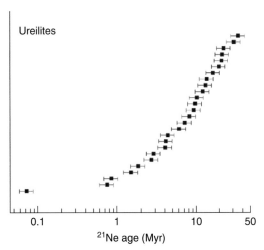

Figure 13 Approximate ^{21}Ne CRE ages of 26 ureilites.

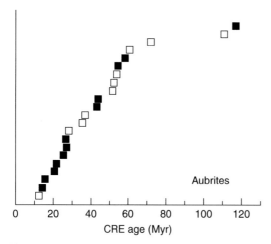

Figure 14 CRE ages of aubrites from Eberhardt *et al.* (1965) (open squares) and Lorenzetti *et al.* (2003) (closed squares). Analytical uncertainties of the ages are ~ 10%; the true uncertainties are larger because of the possibility that some aubrites had multistage exposures.

Eberhardt *et al.* (1965) noted an apparent cluster between 40 and 50 Myr in the CRE age distribution of the aubrites. The original cluster comprised six meteorites—Bishopville, Bustee, Cumberland Falls, Khor Temiki, Peña Blanca Spring, and Pesyanoe. Lorenzetti *et al.* (2003) reevaluated these CRE ages and in doing so increased them by ~ 20%, introduced more spread among them, and added a backdrop of five Antarctic aubrites with lower CRE ages. Thus, taken as a group, the aubrite ages no longer show any strong tendency to cluster, although we suspect that one collision produced both Norton County and Mayo Belwa because of their unusually large CRE ages. The large CRE ages of the aubrites are often attributed to storage in orbits free from objects likely to cause destructive collisions. Gaffey *et al.* (1992) have associated the aubrites with E-type Apollo asteroid 3103 based on their spectral properties.

Eberhardt *et al.* (1965) noted two features important in interpreting the irradiation histories of aubrites, namely the presence of solar noble gases and of isotopic anomalies associated with irradiation by thermal neutrons. Six of the aubrites contain solar noble gases (Lorenzetti *et al.*, 2003), a fact that has significance for their CRE ages (Caffee *et al.*, 1988). First, the presence of solar noble gases in interior samples of meteorites implies a two-stage (at least) irradiation by GCRs. Second, the presence of solar wind noble gases implies irradiation "at a surface" by solar energetic particles or SCRs as well as by GCRs. These observations raise interesting and largely unanswered questions about how to specify the duration and intensity of the surface irradiations.

The long CRE ages of aubrites make them good samples to search for the production of "stable" isotopes by thermal neutrons. At the same time, their compositions (low iron contents) allow the development of higher fluxes of thermal neutrons, which also support higher production rates of "radioactive" nuclides (Spergel *et al.*, 1986). Hidaka *et al.* (1999) have shown that the neutron fluences (10^{16} n cm^{-2}) in aubrites vary from < 0.01 in Happy Canyon to 1.19 in Norton County to 3.99 in Cumberland Falls. Working from the other direction, i.e., from published ^{41}Ca activities (Fink *et al.*, 1992) and a CRE age of 100 Myr, Fink *et al.* (2002) have calculated similar neutron fluences in various Norton County samples.

1.15 BRACHINITES

At this time, seven brachinites and one close relation are known. CRE ages based on noble gases are available for three of these meteorites, but may be problematic for one reason or another: Brachina, 3 Myr (Ott *et al.*, 1985); ALH 84025, 10 Myr (Ott *et al.*, 1987); and Eagles Nest, 57 Myr (Swindle *et al.*, 1998). Mittlefehldt and Berkley (2002) have noted that all the brachinites come from the southern hemisphere, either Australia or the Antarctic. Notwithstanding the geographical closeness of the find locations, if the CRE ages hold up, they suggest at least three independent falls.

1.16 MARTIAN METEORITES

Table 5 shows CRE ages of 24 meteorites identified as martian as of March 2002. The results are based on a critical review of the literature referenced in the notes to Table 5. The distribution of CRE ages for martian meteorites appears in Figure 15. The ages range from a low of ~ 0.6 Myr for EET 79001 to a high no greater than 20 Myr. Thus, the CRE ages of martian meteorites tend to be short relative to those of asteroidal meteorites. Most workers agree that one impact ~ 12 Ma produced five nakhlites and that the CRE age difference between the nakhlites and the orthopyroxenite ALH 84001 is big enough to infer an event. Opinion is divided as to whether Chassigny, with its distinct mineralogy, came out of the same martian crater as the nakhlites (Terribilini *et al.*, 1998; Nyquist *et al.*, 1998).

Much ink has flowed over the number of impacts required to produce the shergottites. The controversy arises partly out of technical issues concerning production rates and partly because of the apparent scarcity on Mars of target sites with rocks that appear likely to have the crystallization ages measured for the martian meteorites, which tend to be young, ~ 200 Myr (Nyquist *et al.*, 1998, 2001). Clearly, if martian terrain with the right crystallization age is rare, then each new discovery of a shergottite with that crystallization age

Table 5 CRE ages of martian meteorites.

	Pairs	Total mass (kg)	Number	CRE age (Myr)
Shergottites/basalts			10	
Dar al Gani 476	489, 670, 735, 876	6.37		1.20 ± 0.15
Dhofar 019		1.06		17 ± 6
EETA 79001		7.90		0.6 ± 0.1
Los Angeles 001	002	0.70		3.0 ± 0.3
Northwest Africa 480		0.028		2.4 ± 0.2
Northwest Africa 856		0.320		
QUE 94201		0.01		2.5 ± 0.6
Sayh Al Uhaymir 005	008, 051, 194	10.51		1.3 ± 0.2
Shergotty		5		2.5 ± 0.6
Zagami		18.1		2.8 ± 0.2
Shergottites/unspecified			3	
Dhofar 378		0.015		
GRV 9927		0.010		
Northwest Africa 1068	1,110	0.772		
Shergottites/lherzolites			4	
ALHA 77005		0.48		3.3 ± 0.6
LEW 88516		0.013		4 ± 1
Y793605		0.016		4.5 ± 0.5
YA 1075		0.055		
Nakhlites/N-clinopyroxenites			5	
Governador Valadares		0.16		10.1 ± 22
Lafayette		0.8		11.4 ± 2.1
Nakhla		10		10.8 ± 0.4
Northwest Africa 817		0.104		11 ± 1
Y000593	Y000749	13.7		12 ± 1
Chassignite/Dunite			1	
Chassigny		4		12 ± 1
Orthopyroxenite			1	
ALH 84001		1.9		14 ± 1

References: Dar al Gani 476: Zipfel *et al.* (2000); Scherer and Schultz (1999); Park *et al.* (2001); Folco *et al.* (2000). Dhofar 019: Shukolyukov *et al.* (2000); Park *et al.* (2001); Nishiizumi *et al.* (2002). EET 79001: Eugster *et al.* (1997); Schnabel *et al.* (2001); Nishiizumi *et al.* (1986); Jull and Donahue (1988). Los Angeles: Garrison and Bogard (2000); Nishiizumi *et al.* (2000a). Northwest Africa 480: Marty *et al.* (2001). QUE 94201: Eugster *et al.* (1997); Garrison and Bogard (1998); Nishiizumi and Caffee (1996a); Schnabel *et al.* (2001). Sayh al Uhaymir 005: Paetsch *et al.* (2000); Park *et al.* (2001); Nishiizumi *et al.* (2001). Shergotty: Eugster *et al.* (1997); Garrison and Bogard (1998); Terribiliini *et al.* (2000b); Nishiizumi (1987); Nishiizumi and Caffee (1996a); Zagami: Eugster *et al.* (1997); Terribiliini *et al.* (2000b); Schnabel *et al.* (2001). Allan Hills 77005: Eugster *et al.* (1997); Garrison *et al.* (1995); Nyquist *et al.* (1998). Nishiizumi (1987); Nishiizumi *et al.* (1986, 1994); Schnabel *et al.* (2001); Schultz and Freundel (1984). Lewis Cliffs 88516: Eugster *et al.* (1997); Jull *et al.* (1994); Nishiizumi *et al.* (1992b); Schnabel *et al.* (2001). Y793605: Terribilini *et al.* (1998); Nagao *et al.* (1997, 1998); Nishiizumi and Caffee (1997). Governador Valadares, Lafayette, and Nakhla: Eugster *et al.* (1997); Bogard and Husain (1977); Terribilini *et al.* (2000b); Jull *et al.* (1999). Northwest Africa 817: Marty *et al.* (2001). Yamato 000593: Imae *et al.* (2002). Chassigny: Eugster *et al.* (1997); Terribilini *et al.* (1998, 2000b); Matthew and Marti (2001). Allan Hills 84001: Eugster *et al.* (1997); Matthew and Marti (2001); Garrison and Bogard (1998); Jull *et al.* (1994); Nishiizumi *et al.* (1994).

makes it harder to imagine how separate impactors could have found their way to that terrain. The hypothesis of one impact dispels this problem. A sufficiently large impact might well sample several different lithologies, for example, nakhlites and Chassigny. EET 79001 provides direct evidence that different lithologies may coexist at close quarters.

In our view the one-crater hypothesis for shergottites has two weaknesses. First, measured crystallization ages must be equated to the absolute ages of martian provinces, which are inferred from crater counts. Nyquist *et al.* (2001) discuss several difficulties associated with this exercise. Second, it requires a complex reinterpretation of the cosmogenic nuclide data, i.e., either the construction

of multistage exposure histories where one-stage histories would seem to suffice or an *ad hoc* treatment of production rates. Based on inspection of Figure 15, we favor five distinct events and times for launching the shergottites: (1) one < 1 Myr; (2) two between 1 and 2 Myr; (3) five between 2 and 3 Myr; (4) three between 3 and 4 Myr; and (5) one (Dhofar 019) > 10 Myr. Even the observation that Dhofar 019 has a CRE age more than twice as large as does any other shergottite, however, will not end this debate until we can be certain of minimal preirradiation on Mars and have better calculations of production rates. Measurements of gadolinium and neodymium isotope systematics such as those of Hidaka *et al.* (2001) should help with respect to the former. We would

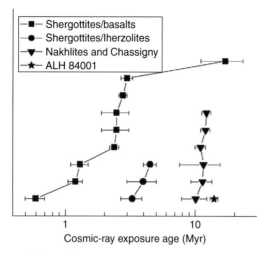

Figure 15 CRE ages of martian meteorites.

note in this context that the number of shergottites with known CRE ages, 12, is still fairly small; no doubt, surprises are waiting.

Numerous authors (e.g., Warren, 1994; Wieler, 2002; Nyquist *et al.*, 2001) have contrasted the exposure histories and other properties of lunar and martian meteorites. On average, we would expect key systematic differences to relate to their respective distances from the Earth (or more precisely how easily their ejecta could attain Earth-crossing orbits), the respective depths of their gravitational wells, the mechanical properties of their regoliths, and the relative fluxes of impacting bodies.

- Martian meteorites have larger average recovered masses than lunar meteorites. According to Mileikowsky *et al.* (2000), ejecta size increases with the size of the impact event. With more energy required (and hence available) for launch from Mars than from the Moon, it seems reasonable to expect larger martian fragments. Once ejected, larger fragments would be more likely to survive collisional destruction en route to Earth.
- More martian than lunar meteorites appear to have come to Earth per crater. According to Mileikowsky *et al.* (2000), the total mass of the ejecta increases with the size of the impact event. With more energy required for launch from Mars, and with launches rare, a greater likelihood of the pairing of source craters for martian meteorites seems reasonable.
- The numbers of lunar and of martian meteorites are nearly equal. The near equality at first seems odd in light of the closeness and smaller gravitational field of the Moon. The absolute number of martian meteorites may be misleading, however. As suggested by the grouping of CRE ages above, that number is almost certainly larger than the number of events on Mars

that produced the meteorites. Further, asteroids may strike Mars more often than they do the Moon, producing more ejecta (Wieler, 2002).

- Many lunar meteorites retain a record of cosmic-ray irradiation on the Moon while martian meteorites show few obvious signs of preirradiation near the surface of Mars. The former result points to a launch depth within a few meters of the lunar surface. The latter observation suggests that martian meteorites came from depths of a few meters or more below the surface although further investigation may alter this view. Warren (1994) argues that the mechanical properties of the lunar regolith favor near-surface lunar objects while weathering on Mars may have destroyed smaller, suitable candidates near the martian surface. In considering closely related issues, Nyquist *et al.* (2001) conclude, "The launch mechanism for the martian meteorites is sufficiently uncertain that a number of possible mechanisms should continue to be evaluated."

1.17 MESOSIDERITES

Begemann *et al.* (1976) presented the first comprehensive survey of CRE ages for mesosiderites. They calculated $^{36}Cl/^{36}Ar$ ages for metal phases and ^3He, ^{21}Ne, and ^{38}Ar ages for the silicates. Nagai *et al.* (1993) investigated production rate systematics of cosmogenic nuclides in mesosiderites. Since then, Terribilini *et al.* (2000c); Albrecht *et al.* (2000); Nishiizumi *et al.* (2000b), and Welten *et al.* (2001b) have reported new cosmogenic nuclide data for various members of the group.

The $^{36}Cl/^{36}Ar$ CRE ages for the metal phases of mesosiderites are probably the most reliable. Figure 16, therefore, shows the distribution of CRE ages for mesosiderites based mainly on metal-phase $^{36}Cl/^{36}Ar$, and bulk krypton isotopic measurements. The average, ~ 90 Myr, is intermediate between CRE ages typical for the generally younger stones and the generally older irons. Five of the mesosiderite ages surpass the CRE ages of the oldest stones. The oldest CRE age, 340 Myr for EET 87500, may be compared with CRE age of the oldest iron, ~ 2,500 Myr for Deep Springs. The age distribution does not have any strong clusters, although Welten *et al.* (2001b) note that three of 19 mesosiderites (Chinguetti, Crab Orchard, and Estherville) have CRE ages close to 70 Myr. Terribilini *et al.* (2000c) call attention to the dissimilarity of the CRE age distributions for mesosiderites and HED meteorites.

Despite their brecciated nature, only two mesosiderites—Veramin (Begemann *et al.*, 1976) and Eltanin (Nishiizumi *et al.*, 2000b)—seem to contain appreciable concentrations of solar wind gases. The meteorite Eltanin deserves special consideration. Fragments of this object were

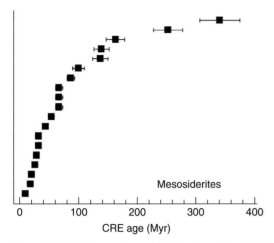

Figure 16 CRE ages of mesosiderites from Terribilini *et al.* (2000c), Albrecht *et al.* (2000), Welten *et al.* (2001b), Lavielle *et al.* (1998), and Nishiizumi *et al.* (2001).

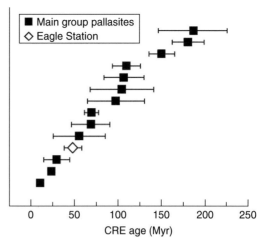

Figure 17 CRE ages of pallasites. Sources: Begemann *et al.* (1976), Honda *et al.* (2002), Megrue (1968), Miura (1995), Nagao *et al.* (1983), Schultz and Hintenberger (1967), and Shukolyukov and Petaev (1992).

recovered from deep-ocean sediments beneath stormy southern seas. The meteoroid was probably enormous (Gersonde *et al.*, 1997). Silicates dominate and metal is rare in the material recovered. The silicates seem to be intermediate in character between those of eucrites and of mesosiderites (Kyte *et al.*, 2000). Eltanin's exposure age of 20 Myr lies close to the CRE ages of several eucrites.

1.18 PALLASITES

Most CRE ages and histories of pallasites rest shakily on noble gas analyses alone. We summarize the results in Figure 17. Pallasite exposure ages are longer than those of most stony meteorites and comparable with those of the other major stony iron group, the mesosiderites.

1.19 IRONS

Figure 18 shows (1) the ^{40}K/K CRE ages of ~ 80 iron meteorites and (2) the ^{38}Ar ages of ~ 160 iron meteorites based on shielding-corrected production rates calculated according to a method not discussed in detail here, that of Lavielle *et al.* (1985). These authors set the parameters of the ^{38}Ar production rate so that the ^{40}K/K and the ^{38}Ar ages would agree for as many irons as possible. Most of the ^{40}K/K CRE ages lie between 200 and 1,000 Myr. The ^{38}Ar fill in nicely the low-age gaps in the ^{40}K/K CRE age distributions, gaps that partly reflect the limitations of the ^{40}K/K dating method. A few irons have CRE ages larger than 1,000 Myr with the championship held by Deep Springs at over 2,000 Myr. For group I irons, the ^{40}K/K ages define a cluster at ~ 900 Myr. For group III irons, both the ^{40}K/K and the ^{38}Ar-based CRE ages establish a cluster at ~ 650 Myr; in addition, the ^{38}Ar-based CRE ages suggest a second peak in the vicinity of 450 Myr

(Lavielle *et al.*, 1985). In considering these plots, it is worth noting that the chemical groupings/classifications of iron meteorites may change. For example, Wasson and Kallemeyn (2002) recommend reclassifying the IIIC and IIID subgroups and making them part of the IAB complex.

A comparison of the ^{40}K/K and ^{36}Cl/^{36}Ar ages of 17 irons (Lavielle *et al.*, 1999) raises two unsettled and related issues. First, as succinctly summarized by Eugster (2003), "there is still some unresolved bias in the age scales for the different methods." In particular, CRE ages based on shorter-lived radionuclides including ^{36}Cl are $\sim 30\%$ lower than ^{40}K/K CRE ages. The standard explanation is that the intensity of the cosmic-ray flux increased at some time within the last 100 Myr or so. The increase raised the activities and hence the production rates for the short-lived radionuclides, but had a smaller effect on the longer lived ^{40}K. Compared with the shorter-lived radionuclides, ^{40}K should give a better measure of the average cosmic-ray flux because its half-life is closer to the typical CRE ages of irons. If the cosmic-ray flux indeed changed, then the production rates (but not necessarily the total concentrations of cosmogenic nuclides!) should have increased by the same fractions and at the same time for all meteoroids.

The related unresolved issue concerns the prevalence of multiple exposures. For example, Lavielle *et al.* (1999) identify Bendego as an iron that may have undergone multiple periods of exposure, presumably initiated by collisions, and there are others such as Canyon Diablo (Michlovich *et al.*, 1994). In general, we would expect collisions to reduce the sizes of meteoroids and thereby increase production rates, i.e., to have the same qualitative effect as would an increase in

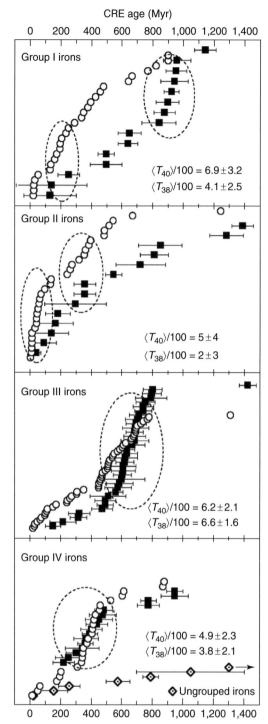

Figure 18 ^{40}K/K CRE ages of iron meteorites as closed squares (Voshage, 1967; Voshage and Feldmann, 1979; Voshage *et al.*, 1983), ^{38}Ar–^{4}He/^{21}Ne ages of iron meteorites as open circles after Lavielle *et al.* (1985) with noble gas data taken from the compilation of Schultz and Franke (2002). Possible clusters of ages are marked with dashed circles (see also Eugster, 2003). The angular brackets denote group averages of the two types of ages for groups I–IV. For the ungrouped irons $\langle T_{40} \rangle = 732 \pm 292$ Myr and $T_{38} = 469 \pm 567$ Myr.

the cosmic-ray flux. Such changes, however, should change production rates at random times and by random increments. The relative importance of multiple exposures in irons is difficult to assess from the available data.

1.20 THE SMALLEST PARTICLES: MICROMETEORITES, INTERPLANETARY DUST PARTICLES, AND INTERSTELLAR GRAINS

1.20.1 Background

We will use the term micrometeorite in the most general sense to include all particles with a maximum dimension < 0.5 mm, i.e., those that melted (the cosmic spherules), those that did not (the "unmelted" micrometeorites), and the interplanetary dust particles (IDPs). According to theory, several kinds of physical forces act on small particles and should limit their expected transit times to Earth (Burns *et al.*, 1979; Gustafson, 1994) and hence their exposure to cosmic rays. In the inner solar system, Poynting–Robinson drag slows the smaller (< 100 μm) particles down to the point that they fall into the Sun. The timescale, < 10^5 years, for this process is much shorter than the one defined by CRE ages of asteroidal meteoroids. For larger particles, those in the 100–500 μm range, collisional destruction may limit the transit times to 50–60 kyr (cf., Kortenkamp and Dermott, 1998).

1.20.2 Micrometeorites and IDPs

At present, quantitative assessments of the CRE ages of small particles are few. The measurements, especially for the tiny IDPs, are difficult, and even heroic. Solar noble gases (solar wind and solar energetic particles) occur widely in micrometeorites in concentrations that are consistent with relatively short periods of exposure in space as small bodies (Nier, 1995). In contrast, what we know from the cosmogenic nuclides is either ambiguous or seems to require much longer periods of irradiation. Olinger *et al.* (1990) analyzed the neon in individual, unmelted micrometeorites in the size range between 0.1 and 0.5 mm. Then, with the aid of various assumptions, they estimated CRE ages, obtaining values from < 0.5 to 20 Myr. The range of CRE ages was well over a factor of 10 regardless of the method of calculation. Among the IDPs, which are smaller, several have ^{3}He/^{4}He ratios that indicate appreciable concentrations of cosmogenic ^{3}He (Nier, 1995). Pepin *et al.* (2001) measured ^{3}He concentrations in samples from a single IDP that correspond to a CRE age ~ 10^9 years! To explain the long period of exposure, these authors consider the possibility that the samples came not from the inner solar system, but from much farther away, in the Edgeworth–Kuiper Belt (30–120 AU), where collision rates were lower and particle survival times longer. Transport of the IDPs as part of larger objects might preserve them from collisional destruction along the way. Pepin *et al.* also discuss several other imaginative explanations for the large concentrations of ^{3}He, including implantation by solar energetic particles, interstellar pickup, and an unusual primordial component rich in ^{3}He. The view that a parent-body irradiation produced ^{3}He and ^{21}Ne seems to us the most appealing.

Turning now to larger particles, Figure 19 shows the ^{10}Be and ^{26}Al activities of micrometeorites harvested from the deep sea and from the Antarctic (Raisbeck *et al.*, 1985a; Nishiizumi *et al.*, 1991c, 1995). The gray region of Figure 19 shows the "allowed" range of ^{26}Al and ^{10}Be activities expected for small particles that have undergone a one-stage irradiation by GCRs and varying degrees of irradiation by SCRs. The lower curve close to the *x*-axis traces the expected growth of the two nuclides in small "metallic" particles based on production rates for iron meteorites (Albrecht *et al.*, 2000). The lower curve terminates at the point where the activities reach the production rates, $P_{10} \sim 6$ and $P_{26} \sim 4.2$. Contrary to expectations, several I-type spherules have ^{10}Be activities larger than the maximum production rate of $\sim 6 \, \text{dpm kg}^{-1}$. These experimental results would have plotted at even higher values had we corrected for the terrestrial addition of oxygen to the samples. SCRs offer no way to bring the straying data points into the "allowed" region for SCR and do not produce appreciable quantities of ^{26}Al or ^{10}Be from iron and nickel (Reedy, 1987b). In sum, the high-measured ^{10}Be activities are not consistent with the presumed preatmospheric composition of the spherules—metallic—or with what we think we know about irradiation conditions in the solar system. Other than experimental error, possible explanations include an origin in a location where production rates were higher (Raisbeck *et al.*, 1985b; Pepin *et al.*, 2001); the presence of siliceous material in the preatmospheric object; or terrestrial addition of ^{10}Be in an aqueous environment.

The upper curve in Figure 19 traces the expected growth of ^{10}Be and ^{26}Al in stony material exposed to both solar and GCRs (Reedy, 1987a, 1990). Numerous data points for S-type spherules lie to the right of the allowed field. Here an explanation does not require much of a stretch, although the construction of unambiguous exposure histories is not yet possible. If the particles had short lifetimes in space, less than say 1 Myr, then they should preserve cosmogenic radionuclides produced if they lay within 50 cm or so of the surface of their precursor bodies. There, ^{10}Be chondritic production rates of 20 dpm kg^{-1} or more could have prevailed. Unclear is what fraction of the observed ^{26}Al derives from the earlier irradiation. The extraordinarily high ^{26}Al activities (relative to chondritic values of $\sim 50 \, \text{dpm kg}^{-1}$) and ^{26}Al/^{10}Be ratios observed in virtually all the S-type particles require significant SCR exposure. In view of the prevalence of ^{26}Al activities well above chondritic values, it seems likely that the S-type spherules in the "forbidden" field acquired most of their SCR-produced ^{26}Al recently, as small bodies. We cannot yet rule out a contribution from earlier irradiation at the surface of a larger body, however. In this connection, it would be interesting to know the depth range in the S-type parent bodies from which collisions excavate the S-type particles. Although a few S-type spherules appear to the left of the "allowed" region, their error bars reach into it. Experimental uncertainties probably account for these observations.

We turn finally to the allowed region itself. The simplest interpretation is that the particles follow intermediate curves of growth (not shown), where ^{26}Al production from SCRs varied because of variations in particle size and geometry. In view of the fairly uniform distribution of data points, however, we think it more likely that most of these particles, like their siblings to the right of the allowed region, retain some record of earlier irradiation. Thus, the sorting out of their exposure histories will require deconvolution of effects due to SCR and GCR in at least two stages of irradiation. It will be an interesting task made worthwhile by the clues of high-flux irradiations at large perihelia.

1.20.3 Interstellar Grains

Several types of grains separated from carbonaceous chondrites have unusual isotopic

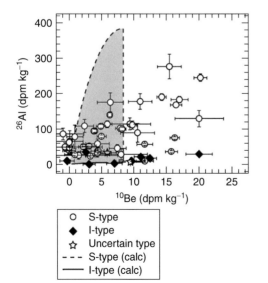

Figure 19 ^{10}Be and ^{26}Al activities of micrometeorites.

compositions thought to reflect an origin outside the solar system. As these interstellar grains traveled through interstellar space, it seems certain that cosmic rays would have produced cosmogenic nuclides in them. Vogt *et al.* (1990) reviewed early treatments of this question. They concluded that the uncertainties of ^{21}Ne CRE ages, $\sim 40\,$Myr, found for SiC grains were too large to permit meaningful comparisons with (much longer) theoretical estimates for the lifetimes of interstellar grains. In particular, the CRE age calculation (Equation (1)) requires (1) a production rate, which depends on the unknown intensity cosmic-ray flux more than 4.5 Ga in far away places and (2) an estimate of the cosmogenic ^{21}Ne content. To arrive at the cosmogenic ^{21}Ne content, one must separate it from trapped components (e.g., solar wind) and from other, local cosmogenic components (e.g., ^{21}Ne produced in an asteroidal regolith or later, during exposure as a meteoroid). Further, because of the small grain sizes—only a few micrometers—of the interstellar SiC grains, one must compensate explicitly for recoil losses.

Viewed in the light of new developments, the prospect of having reliable ^{21}Ne CRE ages for interstellar grains has dimmed further. Ott and Begemann (2000) measured recoil losses of ^{21}Ne from SiC in the laboratory. They found that the recoil losses were much larger (over 90%) for small ($< 1\,\mu$m) grains than originally estimated. Thus, analyses of ^{21}Ne in small grains provide little reliable information about total exposure. As if this result were not discouraging enough, Ott and Begemann reopened the question of how to distinguish the cosmogenic ^{21}Ne content of interstellar origin. They concluded that the ^{21}Ne previously classified as cosmogenic might, in fact, be nucleosynthetic, i.e., a special type of trapped component. On the bright side, Ott and Begemann suggest that cosmogenic ^{126}Xe could offer a more solid base for calculating exposure ages of interstellar grains. With various assumptions, they estimate presolar xenon ages for finer grains to be 100–300 Myr and infer that the larger grains are younger than the smaller ones.

1.21 CONCLUSIONS

By and large, the generalizations about CRE ages made in the early 1970s have stood up well against a tide of new data. The field has also progressed considerably since then. With a few notable exceptions, the exposure age distributions of the well-established meteorite groups are now known and it is clear that exposure ages differ systematically among different groups of meteorites. As Wieler and Graf (2001) conclude, the clumpiness of the CRE age distributions strongly suggests that a small number of collisions

produced a very large fraction of the known meteorites. It follows, these authors continue, that we may have sampled only a tiny portion of the potential parent bodies in the asteroid belt. We have identified a few of those parent bodies and have hints about the identities of others. Figure 20 plots the average ages of lunar meteorites ($0.2 \pm 0.3\,$Myr excluding only Dhofar 025), of martian meteorites ($7 \pm 5\,$Myr), and of HED meteorites ($25 \pm 8\,$Myr), which are tied to the asteroid Vesta and its kin, against aphelion of the respective parent bodies. Also shown are five ordinary chondrite falls with known orbital parameters. The scatter of the data for the H5 chondrites alerts us to the dangers of oversimplifying (see below). Nonetheless, it is hard to resist the conclusion that on average, the further away the source of a meteorite, the longer that meteorite takes to get to Earth.

Recently, theorists seem to have come to grips with all the factors needed to account for the principal features of CRE age distributions. A decade ago, in the reigning view, the Earth competed for meteoroids against collisional destruction and ejection from the solar system or incineration by the Sun (Greenberg and Nolan, 1989). Earth capture depended on injection of the meteoroid into an orbit close to or in a "chaotic resonance in the inner main belt" (Bottke *et al.*, 2000). This picture could accommodate qualitatively the relatively long exposure ages of stony irons and irons (resistance to collisional destruction) and depressed ^{3}He ages of many meteorites (^{3}He losses at small perihelion) and those elements of the model remain. The model failed, however, on several counts having to do with the observed properties of asteroids (Bottke *et al.*, 2002). Further and of more direct relevance here, the calculations of Gladman *et al.* (1997) showed that meteoroid transport through the resonances took much less time than

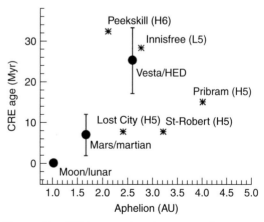

Figure 20 CRE ages of various meteorites versus aphelion of parent body or meteoroid.

actually clocked by the CRE ages. Consideration of the Yarkovsky effect now seems to have resolved the order-of-magnitude discrepancies between theory and measured CRE ages (Farinella *et al.*, 1998; Vokrouhlický *et al.*, 2000; Spitale and Greenberg, 2002). The Yarkovsky effect refers to the asymmetric heating of meteoroids by solar radiation, which leads to small, instantaneous forces of acceleration that change the principal elements of the meteoroid orbit. The changes occur slowly enough that a meteoroid may spend considerable time—tens of millions of years or more—in the main belt before it reaches a resonance that brings it to Earth, always provided that collisions do not destroy the object first. Interestingly, the physical properties of irons give rise to a smaller Yarkovsky effect for irons allowing a longer random-walk passage through the main belt. While much work in modeling the Yarkovsky effect remains to be done (Spitale and Greenberg, 2002), the essential physical elements that control CRE age distributions now seem to be understood. We should not forget, however, that the creation of a meteoroid begins with a collision on a parent body. Crater counting and the general clumpiness of the CRE age distributions suggest that these meteoroid-producing events are rare and that chance is a crucial factor in determining the CRE ages.

Many interesting questions remain for future research. Cosmic rays have irradiated, albeit weakly at times, the stuff of meteorites ever since parent bodies formed 4.5 Ga. In our view, the ultimate goal of CRE studies is to document the exposure histories of meteorites and their components over that entire period. We have tools for examining the earliest irradiations and the most recent. The period in between presents a major challenge. We conclude with a few questions addressable in the nearer term.

1. What are the CRE age distributions of the CHs, the CBs, the pallasites, and the micrometeorites?
2. How tight are the clusters of CRE ages? Can we resolve CRE ages that differ by $< 5\%$?
3. Can we determine separately the CRE histories of chondrules and other small components of meteorites (e.g., Polnau *et al.*, 2001)?
4. What is the overall likelihood that meteoroids experienced multiple, recent periods of irradiation?
5. How has the flux of galactic cosmic rays varied in time?
6. Can we construct CRE histories from analyses of cosmogenic nuclides made aboard spacecraft?
7. Can we reduce the mass needed for analyzing cosmogenic nuclides so that it will be possible to construct CRE histories for materials returned from the surfaces of asteroids and comets?

ACKNOWLEDGMENTS

I thank Ludolf Schultz for making available a compilation of noble gas data and am grateful to Robert Reedy, Rainer Wieler, Donald Bogard, and Andrew Davis for helpful comments and editorial suggestions. This work was supported in part by NASA grant NNG05GF82G.

REFERENCES

Albrecht A., Schnabel C., Vogt S., Xue S., Herzog G. F., Begemann F., Weber H. W., Middleton R., Fink D., and Klein J. (2000) Light noble gases and cosmogenic radionuclides in Estherville, Budulan and other mesosiderites: implications for exposure histories and production rates. *Meteorit. Planet. Sci.* **35**, 975–986.

Arnold J. R., Honda M., and Lal D. (1961) Record of cosmicray intensity in meteorites. *J. Geophys. Res.* **66**, 3519–3531.

Aylmer D., Herzog G. F., Klein J., and Middleton R. (1988) ^{10}Be and ^{26}Al contents of eucrites: implications for production rates and exposure ages. *Geochim. Cosmochim. Acta* **52**, 1691–1698.

Aylmer D., Vogt S., Herzog G. F., Klein J., Fink D., and Middleton R. (1990) Low ^{10}Be and ^{26}Al contents of ureilites: production at meteoroid surfaces. *Geochim. Cosmochim. Acta* **54**, 1775–1784.

Begemann F., Geiss J., and Hess D. C. (1957) Radiation age of a meteorite from cosmic-ray produced He^3 and H^3. *Phys. Rev.* **107**, 540–542.

Begemann F., Weber H. W., Vilcsek E., and Hintenberger H. (1976) Rare gases and ^{36}Cl in stony-iron meteorites: cosmogenic elemental production rates, exposure ages, diffusion losses and thermal histories. *Geochim. Cosmochim. Acta* **40**, 353–368.

Bhandari N., Murty S. V. S., Suthar K. M., Shukla A. D., Ballabh G. M., Sisodia M. S., and Vaya V. K. (1998) The orbit and exposure history of the Piplia Kalan eucrite. *Meteorit. Planet. Sci.* **33**, 455–461.

Bischoff A., Clayton R. N., Markl G., Mayeda T. K., Palme H., Schultz L., Srinivasan G., Weber H. W., Weckwerth G., and Wolf D. (2000) Mineralogy, chemistry, noble gases and oxygen- and magnesium-isotopic compositions of the angrite Sahara 99555. *Meteorit. Planet. Sci.* **35**, A27.

Bischoff A., Palme H., Schultz L., Weber D., Weber H. W., and Spettel B. (1993) Acfer 182 and paired samples, an iron-rich carbonaceous chondrite: similarities with ALH 85085 and relationship to CR chondrites. *Geochim. Cosmochim. Acta* **57**, 2631–2648.

Bischoff A., Weber D., Clayton R. N., Faestermann T., Franchi I. A., Herpers U., Knie K., Korschinek G., Kubik P. W., Mayeda T. K., Merchel S., Michel R., Neumann S., Palme H., Pillinger C. T., Schultz L., Sexton A. S., Spettel B., Verchovsky A. S., Weber H. W., Weckwerth G., and Wolf D. (1998) Petrology, chemistry and isotopic compositions of the lunar highland regolith breccia Dar al Gani 262. *Meteorit. Planet. Sci.* **33**, 1243–1257.

Bogard D. D. and Husain L. (1977) A new 1.3-aeon-young achondrite. *Geophys. Res. Lett.* **4**, 69–71.

Bottke W. F., Rubincam D. P., and Burns J. A. (2000) Dynamical evolution of main belt meteoroids: numerical simulations incorporating planetary perturbations and Yarkovsky thermal forces. *Icarus* **145**, 301–331.

Bottke W. F., Vokrouhlický D., Rubincam D., and Brož M. (2002) The effect of Yarkovsky thermal forces on the dynamical evolution of asteroids and meteoroids. In *Asteroids III* (eds. W. F. Bottke, A. Cellino, P. Paolicchi, R. Binzel). pp. 395–408. University of Arizona Press, Tucson.

Burns J. A., Lamy P. L., and Soter S. (1979) Radiation forces on small particles in the solar system. *Icarus* **40**, 1–48.

Caffee M. W., Goswami J. N., Hohenberg C. M., Marti K., and Reedy R. C. (1988) Irradiation records in meteorites. In *Meteorites and the Early Solar System* (eds. J. F. Kerridge, M. S. Matthews). pp. 205–245. University of Arizona Press, Tucson.

Caffee M. W. and Nishiizumi K. (1997) Exposure ages of carbonaceous chondrites: II. *Meteorit. Planet. Sci.* **32**, A26.

Caffee M. W. and Nishiizumi K. (2001) Exposure history of separated phases from the Kapoeta meteorite. *Meteorit. Planet. Sci.* **36**, 429–437.

Clayton R. N. and Mayeda T. K. (1996) Oxygen isotope studies of achondrites. *Geochim. Cosmochim. Acta* **60**, 1999–2017.

Eberhardt P., Eugster O., and Geiss J. (1965) Radiation ages of aubrites. *J. Geophys. Res.* **70**, 4427–4434.

Eberhardt P., Eugster O., Geiss J., and Marti K. (1966) Rare gas measurements in 30 stone meteorites. *Naturforsch* **21A**, 414–426.

Eberhardt P., Geiss J., and Lutz H. (1963) Neutrons in meteorites. In *Earth Science and Meteoritics* (eds. J. Geiss, E. D. Goldberg). pp. 143–168. North-Holland, Amsterdam.

Eugster O. (1988) Cosmic-ray production rates for ^{3}He, ^{21}Ne, ^{38}Ar, ^{83}Kr, and ^{126}Xe in chondrites based on ^{81}Kr–Kr exposure ages. *Geochim. Cosmochim. Acta* **52**, 1649–1659.

Eugster O. (1989) History of meteorites from the Moon collected in Antarctica. *Science* **245**, 1197–1202.

Eugster O. (2003) Cosmic-ray exposure ages of meteorite and lunar rocks and their significance. *Chem. Erde* **63**, 3–30.

Eugster O., Beer J., Burger M., Finkel R. C., Hofmann H. J., Krähenbühl U., Michel Th., Synal H. A., and Wölfli W. (1991a) History of paired lunar meteorites MAC 88104 and MAC 88105 derived from noble gas isotopes, radionuclides and some chemical abundances. *Geochim. Cosmochim. Acta* **55**, 3139–3148.

Eugster O. and Lorenzetti S. (2001) Exposure history of some differentiated and lunar meteorites. *Meteorit. Planet. Sci.* **36**, A54.

Eugster O. and Michel T. (1995) Common asteroid break-up events of eucrites, diogenites and howardites, and cosmicray production rates for noble gases in achondrites. *Geochim. Cosmochim. Acta* **59**, 177–199.

Eugster O., Michel Th., and Niedermann S. (1991b) ^{244}Pu–Xe formation and gas retention age, exposure history and terrestrial age of angrites LEW 86010 and LEW 87051: comparison with Angra dos Reis. *Geochim. Cosmochim. Acta* **55**, 2957–2964.

Eugster O., Michel Th., and Niedermann S. (1992) Solar wind and cosmic ray exposure history of lunar meteorite Yamato-793274. *Proc. NIPR Symp. Antarct. Meteorit.* **5**, 23–35.

Eugster O. and Niedermann S. (1990) Solar noble gases in the unique chondritic breccia Allan Hills 85085. *Earth Planet. Sci. Lett.* **101**, 139–147.

Eugster O., Polnau E., Salerno E., and Terribilini D. (2000) Lunar surface exposure models for meteorites Elephant Moraine 96008 and Dar al Gani 262 from the Moon. *Meteorit. Planet. Sci.* **35**, 1177–1181.

Eugster O., Weigel A., and Polnau E. (1997) Ejection times of martian meteorites. *Geochim. Cosmochim. Acta* **61**, 2749–2757.

Farinella P., Vokrouhlický D., and Hartmann W. (1998) Meteorite delivery via Yarkovsky orbital drift. *Icarus* **132**, 378–387.

Fink D., Klein J., Middleton R., Dezfouly-Arjomandy B., Herzog G. F., and Albrecht A. (1992) ^{41}Ca in the Norton County aubrite. In *Lunar Planet. Sci.* **XXIII**, Lunar and Planetary Institute, Houston, pp. 355–356.

Fink D., Ma P., Herzog G. F., Albrecht A., Garrison D. H., Bogard D. D., Reedy R. C., and Masarik J. (2002) ^{10}Be, ^{26}Al, ^{36}Cl, and non-spallogenic ^{36}Ar in the Norton County aubrite. *Meteorit. Planet. Sci.* **37**(suppl. A46).

Folco L., Franchi I. A., D'Orazio M., Rocchi S., and Schultz L. (2000) A new martian meteorite from the Sahara: the shergottite Dar al Gani 489. *Meteorit. Planet. Sci.* **35**, 827–839.

Freundel M., Schultz L., and Reedy R. C. (1986) ^{81}Kr–Kr ages of Antarctic meteorites. *Geochim. Cosmochim. Acta* **50**, 2663–2673.

Gaffey M. J., Reed K. L., and Kelley M. S. (1992) Relationship of E-type Apollo asteroid 3103 (1982 BB) to the enstatite achondrite meteorites and the Hungaria asteroids. *Icarus* **100**, 95–109.

Garrison D. H. and Bogard D. D. (1998) Isotopic composition of trapped and cosmogenic noble gases in several martian meteorites. *Meteorit. Planet. Sci.* **33**, 721–726.

Garrison D. H. and Bogard D. D. (2000) Cosmogenic and trapped noble gases in the Los Angeles martian meteorite. *Meteorit. Planet. Sci.* **35**, A58.

Garrison D. H., Rao M. N., and Bogard D. D. (1995) Solarproton-produced neon in shergottite meteorites and implications for their origin. *Meteoritics* **30**, 738–747.

Gersonde R., Kyte F. T., Bleil U., Diekmann B., Flores J. A., Gohl K., Grahl G., Hagen R., Kuhn G., Sierro F. J., Völker D., Abelmann A., and Bostwick J. A. (1997) Geological record and reconstruction of the late Pliocene impact of the Eltanin asteroid in the Southern Ocean. *Nature* **390**, 357–363.

Gilabert E., Lavielle B., Michel R., Leya I., Neumann S., and Herpers U. (2002) Production of krypton and xenon isotopes in thick stony and iron targets isotropically irradiated with 1600 MeV protons. *Meteorit. Planet. Sci.* **37**, 951–976.

Gladman B. J., Burns J. A., Duncan M. J., and Levinson H. F. (1995) The dynamical evolution of lunar impact ejecta. *Icarus* **118**, 302–321.

Gladman B. J., Migliorini F., Morbidelli A., Zappalà V., Michel P., Cellino A., Froeschlé C., Levison H. F., Bailey M., and Duncan M. (1997) Dynamical lifetimes of objects injected into asteroid belt resonances. *Science* **277**, 197–201.

Göbel R., Begemann F., and Ott U. (1982) On neutron-induced and other noble gases in Allende inclusions. *Geochim. Cosmochim. Acta* **46**, 1777–1792.

Goodrich C. A., Krot A. N., Scott E. R. D., Taylor G. J., Fioretti A. M., and Keil K. (2002) Formation and evolution of the ureilite parent body and its offspring. In *Lunar Planet. Sci.* **XXXIII**, #1379. Lunar and Planetary Institute, Houston (CD-ROM).

Goswami J. N., Sinha N., Nishiizumi K., Caffee M. W., Komura K., and Nakamura K. (2001) Cosmogenic records in Kobe (CK4) meteorite: implications of transport of meteorites from the asteroid belt. *Meteorit. Planet. Sci.* **36**, A70–A71.

Graf Th., Baur H., and Signer P. (1990b) A model for the production of cosmogenic nuclides in chondrites. *Geochim. Cosmochim. Acta* **54**, 2521–2534.

Graf Th., Caffee M. W., Marti K., Nishiizumi K., and Ponganis K. V. (2001) Dating collisional events: ^{36}Cl–^{36}Ar exposure ages of chondritic metals. *Icarus* **150**, 181–188.

Graf Th. and Marti K. (1994) Collisional records in LL-chondrites. *Meteoritics* **29**, 643–648.

Graf Th. and Marti K. (1995) Collisional history of H chondrites. *J. Geophys. Res. (Planets)* **100**, 21247–21263.

Graf Th., Signer P., Wieler R., Herpers U., Sarafin R., Vogt S., Fieni Ch., Pellas P., Bonani G., Suter M., and Wölfli W. (1990a) Cosmogenic nuclides and nuclear tracks in the chondrite Knyahinya. *Geochim. Cosmochim. Acta* **54**, 2511–2520.

Greenberg R. and Nolan M. C. (1989) Delivery of asteroids and meteorites to the inner solar system. In *Asteroids II* (eds. R. P. Binzel, T. Gehrels, M. S. Matthews). pp. 778–804. University of Arizona Press, Tucson.

Greshake A., Schmitt R. T., Stöffler D., Pätsch M., and Schultz L. (2001) Dhofar 081: a new lunar highland meteorite. *Meteorit. Planet. Sci.* **36**, 459–470.

Gustafson B. A. S. (1994) Physics of zodiacal dust. *Annu. Rev. Earth Planet. Sci.* **22**, 553–595.

Herzog G. F. and Anders E. (1971) Absolute scale for radiation ages of stony meteorites. *Geochim. Cosmochim. Acta* **35**, 605–611.

Herzog G. F., Vogt S., Albrecht A., Xue S., Fink D., Klein J., Middleton R., Weber H. W., and Schultz L. (1997) Complex exposure histories for meteorites with "short" exposure ages. *Meteorit. Planet. Sci.* **32**, 413–422.

Heymann D., Mazor E., and Anders E. (1968) Ages of calcium-rich achondrites. I: Eucrites. *Geochim. Cosmochim. Acta* **32**, 1241–1268.

Hidaka H., Ebihara M., and Yoneda S. (1999) High fluences of neutrons determined from Sm and Gd isotopic compositions in aubrites. *Earth Planet. Sci. Lett.* **173**, 41–51.

Hidaka H., Yoneda S., and Nishiizumi K. (2001) Neutron capture effects on Sm and Gd isotopes in martian meteorites. *Meteorit. Planet. Sci.* **36**, A80–A81.

Hohenberg C. M., Nichols R. H., Jr., Olinger C. T., and Goswami J. N. (1990) Cosmogenic neon from individual grains of CM meteorites: extremely long pre-compaction exposure histories or an enhanced early particle flux. *Geochim. Cosmochim. Acta* **54**, 2133–2140.

Honda M. and Arnold J. R. (1967) Effects of cosmic rays on meteorites. *Handb. phys.* **46**, 613–632.

Honda M., Caffee M. W., Miura Y. N., Nagai H., Nagao K., and Nishiizumi K. (2002) Cosmogenic products in the Brenham pallasite. *Meteorit. Planet. Sci.* **37**, 1711–1728.

Hudson B. G. (1981) Noble gas retention chronologies for the St. Severin meteorite. PhD Thesis, Washington University, Saint Louis, MO.

Imae N., Okazaki R., Kojima H., Nagao K. (2002) The first nakhlite from Antarctica. In *Lunar Planet. Sci.* **XXXIII**, #1483. Lunar and Planetary Institute, Houston (CD-ROM).

Jull A. J. T., Cielaszyk E., Brown S. T., and Donahue D. J. (1994) ^{14}C terrestrial ages of achondrites from Victoria Land, Antarctica. In *Lunar Planet. Sci.* **XXV**, Lunar and Planetary Institute, Houston, pp. 647–648.

Jull A. J. T., Clandrud S. E., Schnabel C., Herzog G. F., Nishiizumi K., and Caffee M. W. (1999) Cosmogenic radio-nuclide studies of the Nakhlites. In *Lunar Planet. Sci.* **XXX**, #1004. Lunar and Planetary Institute, Houston (CD-ROM)..

Jull A. J. T. and Donahue D. J. (1988) Terrestrial ^{14}C age of the Antarctic shergottite EETA 79001. *Geochim. Cosmochim. Acta* **52**, 1309–1311.

Kitts K. and Lodders K. (1998) Survey and evaluation of eucrite bulk compositions. *Meteorit. Planet. Sci.* **33**, A197–A213.

Kong P., Mori T., and Ebihara M. (1997) Compositional continuity of enstatite chondrites and implications for heterogeneous accretion of the enstatite chondrite parent body. *Geochim. Cosmochim. Acta* **61**, 4895–4914.

Kortenkamp S. J. and Dermott S. F. (1998) Accretion of interplanetary dust particles by the Earth. *Icarus* **135**, 469–495.

Kurat G., Brandstätter F., Clayton R., Nazarov M. A., Palme H., Schultz L., Varela M. E., Wäsch E., Weber H. W., and Weckwerth G. (2001) D'Orbigny: a new and unusual angrite. In *Lunar Planet. Sci.* **XXXII**, #1753. Lunar and Planetary Institute, Houston (CD-ROM).

Kyte F. T., Langenhorst F., and Tepley F. J., III. (2000) The Eltanin meteorite: large messenger from the HED or meso-siderite parent body? In *Lunar Planet. Sci.* **XXXI**, #1811, Lunar and Planetary Institute, Houston (CD-ROM).

Lavielle B., Gilabert E., Soares M. R., Vasconcellos M. A. Z., Poupeau G., Canut de Bon C., Cisternas M. E., and Scorzelli R. B. (1998) Noble-gas and metal studies in the Vaca Muerta mesosiderite. *Meteorit. Planet. Sci.* **33**, A91–A92.

Lavielle B., Marti K., Jeannot J.-P., Nishiizumi K., and Caffee M. (1999) The ^{36}Cl–^{36}Ar–^{40}K–^{41}K records and cosmic ray production rates in iron meteorites. *Earth Planet. Sci. Lett.* **170**, 93–104.

Lavielle B., Marti K., and Regnier S. (1985) Ages d'exposition des meteorites de fer: histoires multiples et variations d'intensité du rayonnement cosmique. In *CNES Isotopic Ratios in the Solar System.* Cepadues-Éditions, Toulouse, France, pp. 15–20 (SEE N86-332164 24-88).

Leya I., Lange H.-J., Neumann S., Wieler R., and Michel R. (2000) The production of cosmogenic nuclides in stony meteoroids by galactic cosmic ray particles. *Meteorit. Planet. Sci.* **35**, 259–286.

Leya I., Wieler R., Aggrey K., Herzog G. F., Schnabel C., Metzler K., Hildebrand A. R., Bouchard M., Jull A. J. T.,

Andrews H. R., Wang M.-S., Ferko T. E., Lipschutz M. E., Wacker J. F., Neumann S., and Michel R. (2001) Exposure history of the St-Robert (H5) fall. *Meteorit. Planet. Sci.* **36**, 1479–1494.

List of lunar meteorites, http://epsc.wustl.edu/admin/resources/meteorites/moon_meteorites_list.html (accessed August 2006).

Lodders K. and Fegley B.Jr., (1998) *The Planetary Scientist's Companion.* New York: Oxford University Press.

Lorenzetti S., Eugster O., Busemann H., Marti K., Burbine T., and McCoy T. (2003) History and origin of enstatite achondrites. *Geochim. Cosmochim. Acta* **67**, 557–571.

Lugmair G. and Marti K. (1977) Sm–Nd–Pu time pieces in the Angra dos Reis meteorite. *Earth Planet. Sci. Lett.* **35**, 273–284.

Mahajan R. R., Murty S. V. S., and Ghosh S. (2000) Exposure ages of Lohawat (howardite) and Vissannapeta (eucrite), the recent falls in India. *Meteorit. Planet. Sci.* **35**, A101.

Marti K. (1967) Mass spectrometric detection of cosmic-ray produced ^{81}Kr in meteorites and the possibility of Kr–Kr dating. *Phys. Rev. Lett.* **18**, 264–266.

Marti K. and Graf T. (1992) Cosmic-ray exposure history of ordinary chondrites. *Annu. Rev. Earth Planet. Sci.* **20**, 221–243.

Marti K. and Lugmair G. (1971) Kr^{81}–Kr and K–Ar^{40} ages, cosmic-ray spallation products and neutron effects in lunar samples from Oceanus Procellarum. *Proc. 2nd Lunar Sci. Conf.* **2**, 1591–1605.

Marti K. and Matthew K. J. (2002) Near-Earth asteroid origin for the Farmington meteorite. In *Lunar Planet. Sci.* **XXXIII**, #1132. Lunar and Planetary Institute, Houston (CD-ROM).

Marty B., Marti K., Barrat J. A., Birck J. L., Blichert-Toft J., Chaussidon M., Deloule E., Gillet P., Göpel C., Jambon A., Manhès G., and Sautter V. (2001) Noble gases in new SNC meteorites NWA 817 and NWA 480. *Meteorit. Planet. Sci.* **36**, A122–A123.

Masarik J., Nishiizumi K., and Reedy R. C. (2001) Production rates of ^{3}He, ^{21}Ne and ^{22}Ne in ordinary chondrites and the lunar surface. *Meteorit. Planet. Sci.* **36**, 643–650.

Masarik J. and Reedy R. C. (1994) Effects of bulk composition on nuclide production processes in meteorites. *Geochim. Cosmochim. Acta* **58**, 5307–5317.

Masarik J. and Reedy R. C. (1995) Terrestrial cosmogenic-nuclide production systematics calculated from numerical simulations. *Earth Planet. Sci. Lett.* **136**, 381–395.

Matthew K. J. and Marti K. (2001) Early evolution of martian volatiles: nitrogen and noble gas components in ALH84001 and Chassigny. *J. Geophys. Res.* **106**, 1401–1422.

Mazor E., Heymann D., and Anders E. (1970) Noble gases in carbonaceous chondrites. *Geochim. Cosmochim. Acta* **34**, 781–824.

McCoy T. J., Keil K., Clayton R. N., Mayeda T. K., Bogard D. D., Garrison D. H., Huss G. R., Hutcheon I. D., and Wieler R. (1996) A petrologic, chemical, and isotopic study of Monument Draw and comparison with other acapulcoites: evidence for formation by incipient partial melting. *Geochim. Cosmochim. Acta* **60**, 2681–2708.

McCoy T. J., Keil K., Clayton R. N., Mayeda T. K., Bogard D. D., Garrison D. H., and Wieler R. (1997) A petrologic and isotopic study of lodranites: evidence for early formation as partial melt residues from heterogeneous precursors. *Geochim. Cosmochim. Acta* **61**, 621–637.

Megrue G. H. (1966) Rare-gas chronology of calcium-rich achondrites. *J. Geophys. Res.* **71**, 4021–4027.

Megrue G. H. (1968) Rare gas chronology of hypersthene achondrites and pallasites. *J. Geophys. Res.* **73**, 2027–2033.

Melosh H. J. (1989) *Impact Cratering.* Oxford University Press, New York.

Michel R., Leya I., and Borges L. (1996) Production of cosmogenic nuclides in meteoroids: accelerator experiments and model calculations to decipher the cosmic ray record in extraterrestrial matter. *Nucl. Instr. Meth.: Phys. Res. B* **113**, 434–444.

Michlovich E. S., Vogt S., Masarik J., Reedy R. C., Elmore D., and Lipschutz M. E. (1994) ^{26}Al, ^{10}Be, and ^{36}Cl depth profiles in the Canyon Diablo iron meteorite. *J. Geophys. Res.: Planets* **99**, 23187–23194.

Mileikowsky C., Cucinotta F. A., Wilson J. W., Gladman B., Horneck G., Lindegren L., Melosh J., Rickman H., Valtonen M., and Zheng J. Q. (2000) Natural transfer of viable microbes in space. *Icarus* **145**, 391–427.

Mittlefehldt D. W. and Berkley J. L. (2002) Petrology and geochemistry of paired brachinites EET 99402 and EET 99407. In *Lunar Planet. Sci.* **XXXIII**, #1008. Lunar and Planetary Institute, Houston (CD-ROM).

Miura Y. and Nagao K. (2000) Noble gases in Y-791192, Y 75032-type diogenites and A-881838. *Antarct. Meteorit.* **XXV**, 85–87.

Miura Y., Nagao K., and Fujitani T. (1993) ^{81}Kr terrestrial ages and grouping of Yamato eucrites based on noble gas and chemical compositions. *Geochim. Cosmochim. Acta* **57**, 1857–1866.

Miura, Y. N. (1995) Studies on differentiated meteorites: evidence from ^{244}Pu-derived fission Xe, ^{81}Kr, other noble gases and nitrogen. PhD Dissertation, University of Tokyo, Tokyo, Japan.

Miura Y. N., Nagao K., Sugiura N., Fujitani T., and Warren P. H. (1998) Noble gases, ^{81}Kr–Kr exposure ages and ^{244}Pu-Xe ages of six eucrites, Béréba, Binda, Camel Donga, Juvinas, Millbillillie, and Stannern. *Geochim. Cosmochim. Acta* **62**, 2369–2387.

Nagai H., Honda M., Imamura M., and Kobayashi K. (1993) Cosmogenic ^{10}Be and ^{26}Al in metal, carbon and silicate of meteorites. *Geochim. Cosmochim. Acta* **57**, 3705–3723.

Nagao K., Nakamura T., Miura Y. N., and Takaoka N. (1997) Noble gases and mineralogy of primary igneous materials of the Yamato-793605 shergottite. *Antarct. Meteorit. Res.* **10**, 125–142.

Nagao K., Nakamura T., Okazaki R., Miura Y. R., and Takaoka N. (1998) Two-stage irradiation of the Yamato 793605 martian meteorite. *Meteorit. Planet. Sci.* **33**, A114.

Nagao K., Takaoka N., and Saito K. (1983) Rare gas studies on the Antarctic Meteorites. *Abstr. 8th Symp. Ant. Meteorit.*, Tokyo, pp. 83–84.

Nier A. O. (1995) Helium and neon in interplanetary dust particles. In *Analysis of Interplanetary Dust, AIP Conference Proceedings* (eds. M. E. Zolensky, T. L. Wilson, F. J. M. Rietmeijer, G. J. Flynn). vol. 310, pp. 115–126. Springer, New York.

Nishiizumi K. (1987) ^{53}Mn, ^{26}Al, ^{10}Be, and ^{36}Cl in meteorites: data compilation. *Nucl. Tracks Radiat. Meas.* **13**, 209–273.

Nishiizumi K., Arnold J. R., Brownlee D. E., Caffee M. W., Finkel R. C., and Harvey R. P. (1995) Beryllium-10 and aluminum-26 in individual cosmic spherules from Antarctica. *Meteoritics* **30**, 728–732.

Nishiizumi K., Arnold J. R., Caffee M. W., Finkel R. C., and Southon J. (1992a) Exposure histories of Calcalong Creek and LEW 88516 meteorites. *Meteoritics* **27**, 270.

Nishiizumi K., Arnold J. R., Caffee M. W., Finkel R. C., Southon J., and Reedy R. C. (1992b) Cosmic ray exposure histories of lunar meteorites Asuka 881757, Yamato 793169 and Calcalong Creek. In *17th Symp. Antarctic Meteorites*. Natl. Inst. Polar Res., Tokyo, Japan, pp. 129–132.

Nishiizumi K., Arnold J. R., Fink D., Klein J., Middleton R., Brownlee D. E., and Maurette M. (1991c) Exposure history of individual cosmic particles. *Earth Planet. Sci. Lett.* **104**, 315–324.

Nishiizumi K., Arnold J. R., Klein J., Fink D., Middleton R., Kubik P. W., Sharma P., Elmore D., and Reedy R. C. (1991a) Exposure histories of lunar meteorites: ALH81005, MAC88104, and Y791197. *Geochim. Cosmochim. Acta* **55**, 3149–3155.

Nishiizumi K., Arnold J. R., Klein J., Fink D., Middleton R., Sharma P., and Kubik P. W. (1991b) Cosmic ray exposure history of lunar meteorite Yamato 793274. In *16th Symp.*

Antarctic Meteorites. Natl. Inst. Polar Res., Tokyo, Japan, pp. 188–191.

Nishiizumi K. and Caffee M. (1996a) Exposure history of shergottite Queen Alexandra Range 94201. In *Lunar Planet. Sci.* **XXVII**, Lunar and Planetary Institute, Houston, pp. 961–962.

Nishiizumi K. and Caffee M. (1996b) Exposure histories of lunar meteorites Queen Alexandra Range 94281 and 94269. In *Lunar Planet. Sci.* **XXVII**, Lunar and Planetary Institute, Houston, pp. 959–960.

Nishiizumi K. and Caffee M. (1997) Exposure history of shergottite Yamato 793605. *Antarct. Meteorit.* **XXII**, 149–151.

Nishiizumi K. and Caffee M. (2001b) Exposure histories of lunar meteorites Dhofar 025, 026 and Northwest Africa 482. *Meteorit. Planet. Sci.* **36**, A148–A149.

Nishiizumi K. and Caffee M. W. (2001a) Exposure histories of lunar meteorites Northwest Africa 032 and Dhofar 081. In *Lunar Planet. Sci.* **XXXII**, #2101. Lunar and Planetary Institute, Houston (CD-ROM).

Nishiizumi K., Caffee M. W., Bogard D. D., Garrison D. H., and Kyte F. T. (2000b) Noble gases and cosmogenic radionuclides in the Eltanin meteorite. In *Lunar Planet. Sci.* **XXXI**, #2070. Lunar and Planetary Institute, Houston (CD-ROM).

Nishiizumi K., Caffee M. W., and Finkel R. C. (1994) Exposure histories of ALH 84001 and ALHA 77005. *Meteoritics* **29**, 511.

Nishiizumi, K., Caffee, M. W., and Jull, A. J. T. (1998) Exposure histories of Dar al Gani 262 lunar meteorites. In *Lunar Planet. Sci.* **XXIX**, #1957. Lunar and Planetary Institute, Houston (CD-ROM).

Nishiizumi, K., Caffee, M. W., Jull, A. J. T., and Klandrud, S. E. (2001) Exposure history of shergottites Dar al Gani 476/489/670/735 and Sayh al Uhaymir 005. In *Lunar Planet. Sci.* **XXXII**, #2117. Lunar and Planetary Institute, Houston (CD-ROM).

Nishiizumi K., Caffee M. W., Jull A. J. T., and Reedy R. C. (1996a) Exposure history of lunar meteorites Queen Alexandra Range 93069 and 94269. *Meteorit. Planet. Sci.* **31**, 893–896.

Nishiizumi K., Caffee M. W., and Masarik J. (2000a) Cosmogenic radionuclides in the Los Angeles martian meteorite. *Meteorit. Planet. Sci.* **35**, A120.

Nishiizumi K., Caffee M. W., Nagai H., and Imamura M. (1996b) Multiple breakup of ALH 85085-like and CR group chondrites. *Meteorit. Planet. Sci.* **31**, A99–A100.

Nishiizumi K., Klein J., Middleton R., Elmore D., Kubik P. W., and Arnold J. R. (1986) Exposure history of shergottites. *Geochim. Cosmochim. Acta* **50**, 1017–1021.

Nishiizumi, K., Masarik, J., Caffee, M. W., and Jull, A. J. T. (1999) Exposure histories of pair lunar meteorites EET 96008 and EET 87521. In *Lunar Planet. Sci.* **XXX**, #1980. Lunar and Planetary Institute, Houston (CD-ROM).

Nishiizumi, K., Okazaki, R., Park, J., Nagao, K., Masarik, J., and Finkel, R. C. (2002) Exposure and terrestrial histories of Dhofar 019 martian meteorite. In *Lunar Planet. Sci.* **XXXIII**, #1366. Lunar and Planetary Institute, Houston (CD-ROM).

Nyquist, L. E., Bogard, D. D., Garrison, D. H., and Reese, Y. (1998) A single crater origin for martian shergottites: resolution of the age paradox? In *Lunar Planet. Sci.* **XXIX**, #1688. Lunar and Planetary Institute, Houston (CD-ROM).

Nyquist L. E., Bogard D. D., Shih C.-Y., Greshake A., Stöffler D., and Eugster O. (2001) Ages and geologic histories of martian meteorites. *Space Sci. Rev.* **96**, 105–164.

Okazaki R., Takaoka N., Nakamura T., and Nagao K. (2000) Cosmic-ray exposure ages of enstatite chondrites. *Antarct. Meteorit. Res.* **13**, 153–169.

Olinger C. T., Maurette M., Walker R. M., and Hohenberg C. M. (1990) Neon measurements of individual Greenland sediment particles: proof of an extraterrestrial origin and comparison with EDX and morphological analyses. *Earth Planet. Sci. Lett.* **100**, 77–93.

Ostertag R., Stöffler D., Bischoff A., Palme H., Schultz L., Spettel B., Weber H., Weckwerth G., and Wänke H. (1986) Lunar Meteorite Yamato-791197: petrography, shock history and chemical composition. *Mem. Natl. Inst. Polar. Res. (Tokyo), Spec. Issue* **41**, 17–44.

Ott U. and Begemann F. (2000) Spallation recoil and age of presolar grains in meteorites. *Meteorit. Planet. Sci.* **35**, 53–63.

Ott U., Löhr H.-P., and Begemann F. (1985) Noble gases and the classification of Brachina. *Meteoritics* **20**, 69–78.

Ott U., Löhr H.-P., and Begemann F. (1987) Noble gases in ALH 84025: like Brachina, unlike Chassigny. *Meteoritics* **22**, 476–477.

Paetsch M., Altmaier M., Herpers U., Kosuch H., Michel R., and Schultz L. (2000) Exposure age of the new SNC meteorite Sayh al Uhaymir 005. *Meteorit. Planet. Sci.* **35**, A124–A125.

Paetsch, M., Weber, H. W., and Schultz, L. (2001) Noble gas investigations of new meteorites from Africa. In *Lunar Planet. Sci.* **XXXII**, #1526. Lunar and Planetary Institute, Houston (CD-ROM).

Park J., Okazaki R., and Nagao K. (2001) Noble gases in the SNC meteorites: Dar al Gani 489, Sayh al Uhaymir 005 and Dhofar 019. *Meteorit. Planet. Sci.* **36**, A157.

Patzer A., Franke L., and Schultz L. (2001) New noble gas data of four enstatite chondrites and Zaklodzie. *Meteorit. Planet. Sci.* **36**, A157–A158.

Patzer A. and Schultz L. (2001) Noble gases in enstatite chondrites. I: Exposure ages, pairing and weathering effects. *Meteorit. Planet. Sci.* **36**, 947–961.

Pepin R. O., Palma R. L., and Schlutter D. J. (2001) Noble gases in interplanetary dust particles. II: Excess helium-3 in cluster particles and modeling constraints on interplanetary dust exposures to cosmic-ray irradiation. *Meteorit. Planet. Sci.* **36**, 1515–1534.

Polnau E. and Eugster O. (1998) Cosmic-ray produced, radiogenic and solar noble gases in lunar meteorites Queen Alexandra Range 94269 and 94281. *Meteorit. Planet. Sci.* **33**, 313–319.

Polnau E., Eugster O., Burger M., Krähenbühl U., and Marti K. (2001) Precompaction exposure of chondrules and implications. *Geochim. Cosmochim. Acta* **65**, 1849–1866.

Raisbeck G. M., Yiou F., and Brownlee D. (1985b) Unusually high concentration of ^{10}Be in a cosmic spherule: possible evidence for irradiation outside the planetary solar system. *Meteoritics* **20**, 734–735.

Raisbeck G. M., Yiou F., Klein J., Middleton R., and Brownlee D. (1985a) ^{26}Al/^{10}Be in deep sea spherules as evidence of cometary origin. In *Properties and Interactions of Interplanetary Dust* (eds. R. H. Giese, P. Lamy). pp. 169–174. D. Reidel, Dordrecht, The Netherlands.

Rao M. N., Garrison D. H., Palma R. L., and Bogard D. D. (1997) Energetic proton irradiation history of the howardite parent body regolith and implications for ancient solar activity. *Meteorit. Planet. Sci.* **32**, 531–543.

Reedy R. C. (1987a) Nuclide production by primary cosmic ray protons. *Proc. 17th Lunar Planet. Sci. Conf. Part 2: J. Geophys Res.* **92**, E697–E702.

Reedy R. C. (1987b) Cosmogenic nuclide production in small metallic spherules. In *Lunar Planet. Sci.* **XVIII**, Lunar and Planetary Institute, Houston, pp. 820–821.

Reedy R. C. (1990) Cosmogenic-radionuclide production rates in mini-spherules. In *Lunar Planet. Sci.* **XXI**, Lunar and Planetary Institute, Houston, pp. 1001–1002.

Reedy R. C., Arnold J. R., and Lal D. (1983) Cosmic-ray record in solar system matter. *Annu. Rev. Nucl. Part. Sci.* **33**, 505–537.

Reedy R. C. and Masarik J. (1994) Cosmogenic-nuclide depth profiles in the lunar surface. In *Lunar Planet. Sci.* **XXV**, Lunar and Planetary Institute, Houston, pp. 1119–1120.

Reedy R. C., Masarik J., Nishiizumi K., Arnold J. R., Finkel R. C., Caffee M. W., Southon J., Jull A. J. T., and Donahue D. J. (1993) Cosmogenic-radionuclide profiles in Knyahinya.

In *Lunar Planet. Sci.* **XXIV**, Lunar and Planetary Institute, Houston, pp. 1195–1196.

Rubin A. E., Kallemeyn G. W., Wasson J. T., Clayton R. N., Mayeda T. K., Grady M., Verchovsky A. B., Eugster O., and Lorenzetti S. (2003) Formation of metal and silicate globules in Gujba: a new Bencubbin-like meteorite fall from Nigeria. *Geochim. Cosmochim. Acta* **67**(17): 3283–3298.

Scherer P., Paetsch A., and Schultz L. (1998) Noble gas study of the new highland meteorite Dar al Gani 400. *Meteorit. Planet. Sci.* **32**, A135.

Scherer, P. and Schultz, L. (1999) Noble gases in the SNC meteorite Dar al Gani 476. In *Lunar Planet. Sci.* **XXX**, #1144. Lunar and Planetary Institute, Houston (CD-ROM).

Scherer P. and Schultz L. (2000) Noble gas record, collisional history and pairing of CV, CO, CK and other carbonaceous chondrites. *Meteorit. Planet. Sci.* **35**, 145–153.

Schnabel C., Ma P., Herzog G. F., Faestermann T., Knie K., and Korschinek K. (2001) ^{10}Be, ^{26}Al, and ^{53}Mn in martian meteorites. In *Lunar Planet. Sci.* **XXXII**, #1353. Lunar and Planetary Institute, Houston (CD-ROM).

Schultz L. and Franke L. (2002) Helium, neon, and argon in meteorites: a data collection. Update 2002. CD-ROM Max-Planck-Institut für Chemie, Mainz.

Schultz L. and Freundel M. (1984) Terrestrial ages of Antarctic meteorites. *Meteoritics* **19**, 310.

Schultz L. and Hintenberger H. (1967) Edelgasmessungen an Eisenmeteoriten. *Z. Naturforsch.* **22a**, 773–779.

Shu F. H., Shang H., and Lee T. (1996) Toward an astrophysical theory of chondrites. *Science* **271**, 1545–1552.

Shukolyukov A. and Begemann F. (1996) Cosmogenic and fissiogenic noble gases and Kr-81-Kr exposure age clusters of eucrites. *Meteorit. Planet. Sci.* **31**, 60–72.

Shukolyukov, Yu. A., Nazarov, M. A., Pätsch, M., and Schultz, L. (2001) Noble gases in three meteorites from Oman. In *Lunar Planet. Sci.* **XXXII**, #1502. Lunar and Planetary Institute, Houston (CD-ROM).

Shukolyukov Yu. A., Nazarov M. A., and Schultz L. (2000) Dhofar 019: a shergottite with an approximately 20-million-year exposure age. *Meteorit. Planet. Sci.* **35**, A147.

Shukolyukov Yu. A. and Petaev M. I. (1992) Noble gases in the Omolon pallasite. In *Lunar Planet. Sci.* **XXIII**, Lunar and Planetary Institute, Houston, pp. 1297–1298.

Sisodia M. S., Shukla A. D., Suthar K. M., Mahajan R. R., Murty S. V. S., Shukla P. N., Bhandari N., and Natarajan R. (2001) Lohawat howardite: mineralogy, chemistry and cosmogenic effects. *Meteorit. Planet. Sci.* **36**, 1457–1466.

Spergel M. S., Reedy R. C., Lazareth O. W., Levy P. W., and Slatest A. (1986) Cosmogenic neutron-capture-produced nuclides in stony meteorites. *Proc. 16th Lunar Planet. Sci. Conf.: J. Geophys. Res. Suppl.* **91**, D484–D494.

Spitale J. and Greenberg R. (2002) Numerical evaluation of the general Yarkovsky effect: effects on eccentricity and longitude of periapse. *Icarus* **156**, 211–222.

Stauffer H. (1962) On the production ratios of rare gas isotopes in stone meteorites. *J. Geophys. Res.* **67**, 2023–2028.

Swindle T. D., Burkland M. K., and Grier J. A. (1995) Noble gases in the lunar meteorites Calcalong Creek and Queen Alexandra Range 93069. *Meteoritics* **30**, 584–585.

Swindle T. D., Kring D. A., Burkland M. K., Hill D. H., and Boynton W. F. (1998) Noble gases, bulk chemistry, and petrography of olivine-rich achondrites Eagles Nest and Lewis Cliff 88763: comparison to brachinites. *Meteorit. Planet. Sci.* **33**, 31–48.

Takaoka N. and Yoshida Y. (1992) Noble gases in Yamato-793274 and 86032 lunar meteorites. *Proc. Natl. Inst. Polar Res. Symp. Antarct. Meteorit.* **5**, 36–48.

Terribilini D., Busemann H., and Eugster O. (2000b) Krypton-81-krypton cosmic-ray exposure ages of the martian meteorites including the new shergottite Los Angeles. *Meteorit. Planet. Sci.* **35**, A155–A156.

Terribilini D., Eugster O., Burger M., Jakob A., and Krähenbühl U. (1998) Noble gases and chemical composition of Shergotty mineral fractions, Chassigny and Yamato 793605:

the trapped argon-40/argon-36 ratio and ejection times of martian meteorites. *Meteorit. Planet. Sci.* **33**, 677–684.

Terribilini D., Eugster O., Herzog G. F., and Schnabel C. (2000a) Evidence for common break-up events of the acapulcoites/lodranites and chondrites. *Meteorit. Planet. Sci.* **35**, 1043–1050.

Terribilini D., Eugster O., Mittlefehldt D. W., Diamond L. W., Vogt S., and Wang D. (2000c) Mineralogical and chemical composition and cosmic-ray exposure history of two mesosiderites and two iron meteorites. *Meteorit. Planet. Sci.* **35**, 617–628.

Thalmann C., Eugster O., Herzog G. F., Klein J., Krähenbühl U., Vogt S., and Xue S. (1996) History of lunar meteorites Queen Alexandra Range 93069, Asuka 881757 and Yamato 793169 based on noble gas isotopic abundances, radionuclide concentrations and chemical composition. *Meteorit. Planet. Sci.* **31**, 857–868.

Tuniz C., Bird J., Fink D., and Herzog G. F. (1998) *Accelerator Mass Spectrometry: Ultrasensitive Analysis for Global Science.* CRC Press, Boca Raton, FL, chap. 9, pp. 155–176.

Vokrouhlický D., Milani A., and Chesley S. R. (2000) Yarkovsky effect on small near-Earth asteroids: mathematical formulation and examples. *Icarus* **148**, 118–138.

Vogt S., Herzog G. F., Eugster O., Michel Th., Niedermann S., Krähenbühl U., Middleton R., Dezfouly-Arjomandy B., Fink D., and Klein J. (1993) Exposure history of the lunar meteorite Elephant Moraine 87521. *Geochim. Cosmochim. Acta* **57**, 3793–3799.

Vogt S., Herzog G. F., Fink D., Klein J., Middleton R., Dockhorn B., Korschinek G., and Nolte E. (1991) Exposure histories of the lunar meteorites MacAlpine Hills 88104, MacAlpine Hills 88105, Yamato 791197 and Yamato 86032. *Geochim. Cosmochim. Acta* **55**, 3157–3165.

Vogt S., Herzog G. F., and Reedy R. C. (1990) Cosmogenic nuclides in extraterrestrial materials. *Rev. Geophys.* **28**, 253–275.

Voshage H. (1967) Bestrahlungsalter und Herkunft der Eisenmeteorite. *Z. Naturforsch* **22a**, 477–506.

Voshage H. and Feldmann H. (1979) Investigations on cosmicray-produced nuclides in iron meteorites. 3: Exposure ages, meteoroid sizes and sample depths determined by mass spectrometric analyses of potassium and rare gases. *Earth Planet. Sci. Lett.* **45**, 293–308.

Voshage H., Feldmann H., and Braun O. (1983) Investigations of cosmic-ray-produced nuclides in iron meteorites. 5: More data on the nuclides of potassium and noble gases, on exposure ages and meteoroid sizes. *Z. Naturforsch.* **38a**, 273–280.

Voshage H. and Hintenberger H. (1963) The cosmic-ray exposure ages of iron meteorites. as derived from the isotopic composition of potassium and the production rates of cosmogenic nuclides in the past. In *Radioactive Dating.* International Atomic Energy Agency, Vienna, pp. 367–379.

Warren P. (1994) Lunar and martian meteorite delivery services. *Icarus* **111**, 338–353.

Warren P. H. (2001) Porosities of lunar meteorites: strength, porosity and petrologic screening during the meteorite delivery process. *J. Geophys. Res.* **106**, 10101–10111.

Wasson J. T. (1974) *Meteorites.* Berlin: Springer.

Wasson J. T. and Kallemeyn G. W. (2002) The IAB ironmeteorite complex: a group, five subgroups, numerous grouplets, closely related, mainly formed by crystal segregation in rapidly cooling melts. *Geochim. Cosmochim. Acta* **66**, 2445–2473.

Weber H. W., Franke L., and Schultz L. (2001) Subsolar noble gases in metal-rich carbonaceous (CH) chondrites. *Meteorit. Planet. Sci.* **36**, A220–A221.

Weber, H. W., and Schultz, L. (2001) Noble gases in five new Rumuruti chondrites. In *Lunar Planet. Sci.* **XXXII**, #1500. Lunar and Planetary Institute, Houston (CD-ROM).

Weigel A., Eugster O., Koeberl C., and Krähenbühl U. (1997) Differentiated achondrites Asuka 881371, an angrite and Divnoe: noble gases, ages, chemical composition and relation to other meteorites. *Geochim. Cosmochim. Acta* **61**, 239–248.

Weigel A., Eugster O., Koeberl C., Michel R., Krähenbühl U., and Neumann S. (1999) Relationships among lodranites and acapulcoites: Noble gas isotopic abundances, chemical composition, cosmic-ray exposure ages and solar cosmic ray effects. *Geochim. Cosmochim. Acta* **63**, 175–192.

Welten K. C., Bland P. A., Russell S. S., Grady M. M., Caffee M. W., Masarik J., Jull A. J. T., Weber H. W., and Schultz L. (2001b) Exposure age, terrestrial age and pre-atmospheric radius of the Chinguetti mesosiderite: not part of a much larger mass. *Meteorit. Planet. Sci.* **36**, 939–946.

Welten K. C., Lindner L., van der Borg K., Loeken T., Scherer P., and Schultz L. (1997) Cosmic-ray exposure ages of diogenites and the recent collisional history of the howardite, eucrite, and diogenite parent body/bodies. *Meteorit. Planet. Sci.* **32**, 891–902.

Welten K. C., Nishiizumi K., Caffee M. W., and Schultz L. (2001a) Update on exposure ages of diogenites: the impact history of the HED parent body and evidence of space erosion and/or collisional disruption of stony meteoroids. *Meteorit. Planet. Sci.* **36**, A223.

Wieler R. (2002) Cosmic-ray-produced noble gases in meteorites. In *Noble Gases in Geochemistry and Cosmo-chemistry* (eds. D. Porcelli, C. J. Ballertine, and R. Wieler). *Rev. Mineral. Geochem.*, Mineralogical Society of America, vol. 47, pp. 125–170.

Wieler R. and Graf T. (2001) Cosmic ray exposure history of meteorites. In *Accretion of Extraterrestrial Material on Earth over Time* (eds. B. Peucker-Ehrenbrink, B. Schmitz). pp. 221–240. Kluwer Academic/Plenum, New York.

Wieler R., Pedroni A., and Leya I. (2000) Cosmogenic neon in mineral separates from Kapoeta: no evidence for an irradiation of its parent body regolith by an early active Sun. *Meteorit. Planet. Sci.* **35**, 251–257.

Wright R. J., Simms L. A., Reynolds M. A., and Bogard D. D. (1973) Depth variation of cosmogenic noble gases in the 120-kg Keyes chondrite. *J. Geophys. Res.* **78**, 1308–1318.

Xue S., Herzog G. F., Klein J., and Middleton R. (1994) ^{26}Al And ^{10}Be activities and exposure ages of lodranites, acapulcoites, Kakangari, and Pontlyfni. In *Lunar Planet. Sci.* **XXV**, Lunar and Planetary Institute, Houston, pp. 1523–1524.

Yanai K. (2000) Achondrite polymict breccia 1153: a new lunar meteorite classified to anorthositic regolith breccia. In *Lunar Planet. Sci.* **XXXI**, #1101. Lunar and Planetary Institute, Houston (CD-ROM).

Zipfel J., Scherer P., Spettel B., Dreibus G., and Schultz L. (2000) Petrology and chemistry of the new shergottite Dar al Gani 476. *Meteorit. Planet. Sci.* **35**, 95–106.

Radioactive Geochronometry
ISBN: 978-0-08-096708-0

2

Early Solar System Chronology

K. D. McKeegan

University of California, Los Angeles, CA, USA

and

A. M. Davis

University of Chicago, Chicago, IL, USA

2.1 INTRODUCTION

2.1.1 Chondritic Meteorites as Probes of Early Solar System Evolution

The evolutionary sequence involved in the formation of relatively low-mass stars, such as the Sun, has been delineated in recent years through impressive advances in astronomical observations at a variety of wavelengths, combined with improved numerical and theoretical models of the physical processes thought to occur during each stage. From the models and the observational statistics, it is possible to infer in a general way how our solar system ought to have evolved through the various stages from gravitational collapse of a fragment of a molecular cloud to the accretion of planetary-sized bodies (e.g., Cameron, 1995; Shu *et al.*, 1987; André *et al.*, 2000; Alexander *et al.*, 2001; see Chapter 3). However, the details of these processes remain obscured, literally, from an astronomical perspective, and the dependence of such models on various parameters requires data to constrain the specific case of our solar system's origin.

Fortunately, the chondritic meteorites sample aspects of this evolution. The term "chondrite" (or chondritic) was originally applied to meteorites-bearing chondrules, which are approximately millimeter-sized solidified melt droplets consisting largely of mafic silicate minerals and glass commonly with included metal or sulfide. However, the meaning of chondritic has been expanded to encompass all extraterrestrial materials that are "primitive," that is, are undifferentiated samples having nearly solar elemental composition. Thus, the chondrites represent a type of cosmic sediment, and to a first approximation can be thought of as "hand samples" of the condensable portion of the solar nebula. The latter is a general term referring to the phase(s) of solar system evolution intermediate between molecular cloud collapse and planet formation. During the nebular phase, the still-forming Sun was an embedded young-stellar object (YSO) enshrouded by gas and dust, which was distributed first in an extended envelope that later evolved into an accretion disk that ultimately defined the ecliptic plane. The chondrites agglomerated within this accretion disk, most likely close to the position of the present asteroid belt from whence meteorites are currently derived. In addition to chondrules, an important component of some chondrites are inclusions containing refractory oxide and silicate minerals, so-called calcium- and aluminum-rich inclusions (CAIs) that also formed as free-floating objects within the solar nebula. These constituents are bound together by a "matrix" of chondrule fragments and fine-grained dust (which includes a tiny fraction of dust grains that predate the solar nebula). It is important to realize that, although these materials accreted together at a specific time in some planetesimal, the individual components of a given chondrite can, and probably do, sample different places and/or times during the nebular phase of solar system formation. Thus, each grain in one of these cosmic sedimentary rocks potentially has a story to tell regarding aspects of the early evolution of the solar system.

Time is a crucial parameter in constructing any story. Understanding of relative ages allows placing events in their proper sequence, and measures of the duration of events are critical to developing an understanding of the process. If disparate observations can be related temporally, then structure (at any one time) and evolution of the solar system can be better modeled; or, if a rapid succession of events can be inferred, it can dictate a cause and effect relationship. This chapter is concerned with understanding the timing of different physical and chemical processes that occurred in the solar nebula and possibly on early accreted planetesimals that existed during the nebula stage. These events are "remembered" by the components of chondrites and recorded in the chemical, and especially, isotopic compositions of the host mineral assemblages; the goal is to decide which events were witnessed by these ancient messengers and to decipher those memories recorded long ago.

2.1.2 Short-Lived Radioactivity at the Origin of the Solar System

The elements of the chondritic meteorites, and hence of the terrestrial planets, were formed in previous generations of stars. Their relative abundances represent the result of the general chemical evolution of the galaxy, possibly enhanced by recent local additions from one or more specific sources just prior to collapse of the solar nebula $\sim$4.56 Ga. A volumetrically minor, but nevertheless highly significant part of this chemical inventory, is comprised of radioactive elements, from which this age estimate is derived. The familiar long-lived radionuclides, such as ^{238}U, ^{235}U, ^{232}Th, ^{87}Rb, ^{40}K, and others, provide the basis for geochronology and the study of large-scale differentiation amongst geochemical reservoirs over time (see Chapter 4). They also provide a major heat source to drive chemical differentiation on a planetary scale (e.g., terrestrial plate tectonics).

A number of short-lived radionuclides also existed at the time that the Sun and the rocky bits of the solar system were forming (Table 1). These nuclides are sufficiently long-lived that they could exist in appreciable quantities in the earliest solar system rocks, but their mean lives are short enough that they are now completely decayed from their primordial abundances. In this sense they are referred to as extinct nuclides. Although less familiar than the still-extant radionuclides, these

Table 1 Short-lived radioactive nuclides once existing in solar system objects.

Fractionation[a]	Parent nuclide	Half-life[b]	Daughter nuclide	Estimated initial solar system abundance	Objects found in	References
Neb	^{7}Be	53.1 days	^{7}Li	10^{-3} ^{9}Be	CAIs	(1)
Neb	^{41}Ca	102 kyr	^{41}K	10^{-8} ^{40}Ca	CAIs	(2)
Plan	^{36}Cl	301 kyr	^{36}S, ^{36}Ar	$(\sim 4 \times 10^{-6})$ ^{35}Cl	CAIs, chondrites	(3)
Neb	^{26}Al	717 kyr	^{26}Mg	(6.33×10^{-5}) ^{27}Al	CAIs, chondrules, achondrite	(4)
Neb, Plan	^{60}Fe	1.5 Myr	^{60}Ni	$(\sim 5-10 \times 10^{-7})$ ^{56}Fe	Achondrites, chondrites	(6)
Neb	^{10}Be	1.51 Myr	^{10}B	$(\sim 1.0 \times 10^{-3})$ ^{9}Be	CAIs	(5)
Neb, Plan	^{53}Mn	3.74 Myr	^{53}Cr	(1.0×10^{-5}) ^{55}Mn	CAIs, chondrules, carbonates, achondrites	(7)
Plan	^{107}Pd	6.5 Myr	^{107}Ag	$(\sim 5 \times 10^{-5})$ ^{108}Pd	Iron meteorites, pallasites	(8)
Plan	^{182}Hf	8.90 Myr	^{182}W	(1.07×10^{-4}) ^{180}Hf	Planetary differentiates	(9)
Plan	^{129}I	15.7 Myr	^{129}Xe	10^{-4} ^{127}I	Chondrules, secondary minerals	(10)
Plan	^{205}Pb	17.3 Myr	^{205}Tl	$(\sim 1-2 \times 10^{-4})$ ^{204}Pb	Iron meteorites	(11)
Plan	^{92}Nb	34.7 Myr	^{92}Zr	10^{-4} ^{93}Nb	Chondrites, mesosiderites	(12)
Plan	^{244}Pu	80.0 Myr	Fission products	(7×10^{-3}) ^{238}U	CAIs, chondrites	(13)
Plan	^{146}Sm	103 Myr	^{142}Nd	(9×10^{-4}) ^{147}Sm	Chondrites	(14)

References: (1) Chaussidon *et al.* (2006); (2) Srinivasan *et al.* (1994, 1996); (3) Lin *et al.* (2005), Hsu *et al.* (2006); (4) Bizzarro *et al.* (2004, 2005a), this work; (5) Chaussidon *et al.* (2006), this work; (6) Mostefaoui *et al.* (2005), Tachibana *et al.* (2006); (7) Dauphas *et al.*, 2005; (8) Chen and Wasserburg (1990); (9) Kleine *et al.* (2005a); (10) Jeffery and Reynolds (1961); (11) Nielsen *et al.* (2006); (12) Schönbachler *et al.* (2002); (13) Hudson *et al.* (1988); (14) Lugmair *et al.* (1983).

[a] Environment in which most significant parent–daughter fractionation processes occur.

[b] Half-lives from National Nuclear Data Center, Brookhaven National Laboratory (National Nuclear Data Center, 2006).

short-lived isotopes potentially play similar roles: their relative abundances can, in principle, form the basis of various chronometers that constrain the timing of early chemical fractionations, and the more abundant radioisotopes can possibly provide sufficient heat to drive differentiation (i.e., melting) of early accreted planetesimals. The very rapid rate of decay of the short-lived isotopes, however, means that inferred isotopic differences translate into relatively short amounts of time, that is, these potential chronometers have inherently high precision (temporal resolution). The realization of these possibilities is predicated upon understanding the origin(s) and distributions of the now-extinct radioactivity. While this is a comparatively easy task for the long-lived, still existing radionuclides, it poses a significant challenge for the studies of the early solar system. However, this represents the best chance at developing a quantitative high-resolution chronology for events in the solar nebula and, moreover, the question of the origins of the short-lived radioactivity has profound implications for the mechanisms of formation of the solar system (as being, possibly, quite different from that for solar-mass stars in general).

2.1.3 A Brief History and the Scope of the Present Review

That short-lived radioactive isotopes existed in the early solar system has been known since the 1960s, when ^{129}Xe excesses were first shown to be correlated with the relative abundance of iodine, implicating the former presence of its parent nuclide, ^{129}I (Jeffery and Reynolds, 1961). Because the half-life of ^{129}I ($\sim$16 Myr) is not so short, its presence in the solar system can be understood as primarily a result of the ambient, quasisteady-state abundance of this nuclide in the parental molecular cloud due to continuous r-process nucleosynthesis in the galaxy (Wasserburg, 1985). The situation changed dramatically in the mid-1970s when it was discovered that CAIs from the Allende meteorite exhibited apparent excesses of ^{26}Mg (Gray and Compston, 1974; Lee and Papanastassiou, 1974) and that the degree of excess ^{26}Mg correlated with Al/Mg in CAI mineral separates (Lee *et al.*, 1976) in a manner indicative of the *in situ* decay of ^{26}Al ($t_{1/2} = 0.73$ Myr).

The high abundance inferred for this short-lived isotope ($\sim 5 \times 10^{-5}$ ^{27}Al) demanded that it had been produced within a few million years of CAI formation, possibly in a single stellar source, which "contaminated" the nascent solar system with freshly synthesized nuclides (Wasserburg and Papanastassiou, 1982). Because of the close time constraints an attractively parsimonious idea arose, whereby the very same dying star that threw out new radioactivity into the interstellar medium may also have served to initiate gravitational collapse of the molecular cloud fragment that would become the solar system, through the shock wave created by its expanding ejecta (Cameron and Truran, 1977). An alternative possibility that the new radioactive elements were produced "locally" through nuclear reactions between energetic solar particles and the surrounding nebular material was also quickly recognized (Heymann and Dziczkaniec, 1976; Clayton *et al.*, 1977; Lee, 1978). However, many of the early models were unable to produce sufficient amounts of ^{26}Al by irradiation within the constraints of locally available energy sources and the lack of correlated isotopic effects in other elements (see discussion in Wadhwa and Russell, 2000). Almost by default, "external seeding" scenarios and the implied supernova trigger became the preferred class of models for explaining the presence of ^{26}Al and its distribution in chondritic materials.

In the intervening 30 years, as indicated in Table 1, many other short-lived isotopes have been found to have existed in early solar system materials. Several of these have been discovered in recent years, and the record of the distribution of ^{26}Al and other nuclides in a variety of primitive and evolved materials has been documented with much greater clarity. Significant progress has been made since the first edition of this chapter was written. It now seems that both stellar and local production are necessary to explain the full range of short-lived radionuclide abundances. In part due to improvements in mass spectrometry, new data are being generated at an increasing pace, and in some cases, interpretations that seemed solid only a short time ago are now being revised. For further details, the reader is directed to several excellent reviews (Wasserburg, 1985; Swindle *et al.*, 1996; Podosek and Nichols, 1997; Gilmour, 2000; Wadhwa and Russell, 2000; Russell *et al.*, 2001; Kita *et al.*, 2005; Wadhwa *et al.*, 2006a, b).

Development of a quantitative understanding of the source, or sources, of the now-extinct radionuclides is important for constraining the distribution of these radioactive species throughout the early solar system and, thus, is critical for chronology. For the major part of this review, we will tacitly adopt the prevailing point of view, namely that external seeding for the most important short-lived isotopes dominates over possible local additions from nuclear reactions with energetic particles associated with the accreting Sun. This approach permits examination of timescales for self-consistency with respect to major chemical or physical "events" in the evolution of the solar system; the issues of the scale of possible isotopic heterogeneity within the nebula and assessment of local irradiation effects will be explicitly addressed following an examination of the preserved record.

2.2 DATING WITH ANCIENT RADIOACTIVITY

In "normal" radioactive dating, the chemical fractionation of a parent isotope from its radiogenic daughter results, after some decay of the parent, in a linear correlation of excesses of the daughter isotope with the relative abundance of the parent. For a cogenetic assemblage, such a correlation is an isochron and its slope permits the calculation of the time since the attainment of isotopic closure, that is, since all relative transport of parent or daughter isotopes effectively ceased. If the fractionation event is magmatic, and the rock quickly cooled, then this time corresponds to an absolute crystallization age.

In a manner similar to dating by long-lived radioisotopes, the former presence of short-lived radioactivity in a sample is demonstrated by excesses of the radiogenic daughter isotope that correlate with the inferred concentration of the parent. However, because the parent isotope is extinct, a stable isotope of the respective parent element must serve as a surrogate with the same geochemical behavior (see Wasserburg, 1985; figure 2). The correlation line yields the initial concentration of radioactive parent relative to its stable counterpart and may represent an isochron; however, its interpretation in terms of "age" for one sample relative to another requires an additional assumption. The initial concentrations of a short-lived radionuclide among a suite of samples can correspond to relative ages only if the samples are all derived from a reservoir that at one time had a uniform concentration of the radionuclide. Under these conditions, differences in concentration correspond to differences in time only. As before, if the fractionation event corresponds to mineral formation and isotopic closure is rapidly achieved and maintained, then relative crystallization ages are obtained.

One further complication potentially arises that is unique to the now-extinct nuclides. In principle, excesses of a radiogenic daughter isotope could be "inherited" from an interstellar (grain) component, in a manner similar to what is known to have occurred for some stable isotope anomalies in CAIs and other refractory phases of chondrites (e.g., Begemann, 1980; Niederer *et al.*, 1980; Niemeyer and Lugmair, 1981; Fahey *et al.*, 1987). In such a case, the correlation of excess daughter isotope with radioactive parent would represent a mixing line rather than *in situ* decay from the time of last chemical fractionation. Such "fossil" anomalies (in magnesium) have, in fact, been documented in *bona fide* presolar grains (Zinner, 1998). These grains of SiC, graphite, and corundum crystallized in the outflows of evolved stars, incorporating very high abundances of newly synthesized radioactivity with $^{26}Al/^{27}Al$ sometimes approaching unity. However, because these grains did not form in the solar nebula from a uniform isotopic reservoir, there is no chronological constraint that can be derived. Probably, the radioactivity in such grains decayed during interstellar transit, and hence arrived in the solar nebula as a "fossil."

Even before the discovery of presolar materials, Clayton championed a fossil origin for the magnesium isotope anomalies in CAIs in a series of papers (e.g., Clayton, 1982, 1986). A significant motivation for proposing a fossil origin was, in fact, to obviate chronological constraints derived from Al–Mg systematics in CAIs that apparently required a late injection and fast collapse timescales along with a long (several million years) duration of small dust grains in the nebula. Although some level of inheritance may be present, and can possibly even be the dominant signal in a few rare samples or for specific isotopes (discussed below), for the vast majority of early solar system materials it appears that most of the inventory of short-lived isotopes did indeed decay following mineral formation in the solar nebula. MacPherson *et al.* (1995) summarized the arguments against a fossil origin for the ^{26}Mg excesses in their comprehensive review of the Al–Mg systematics in early solar system materials. In addition to the evidence regarding chemical partitioning during igneous processing of CAIs, the number of short-lived isotopes known (Table 1) and a general consistency of the isotopic records in a wide variety of samples must now be added. The new observations buttress the previous conclusions of MacPherson *et al.* (1995), such that the overwhelming consensus of current opinion is that correlation lines indicative of the former presence of the now-extinct isotopes are truly isochrons representing *in situ* radioactive decay. This is a necessary, but not sufficient, condition for developing a chronology based on these systems.

2.3 "ABSOLUTE" AND "RELATIVE" TIMESCALES

To tie high-resolution relative ages to an "absolute" chronology, a correlation must be established between the short- and long-lived chronometers, that is, the ratio of the extinct nuclide to its stable partner isotope must be established at some known time (while it was still alive). This time could correspond to the "origin of the solar system," which, more precisely defined, means the crystallization age of the first rocks to have formed in the solar system, or it could refer to some subsequent well-defined fractionation event, for example, large-scale isotopic homogenization and fractionation occurring during planetary melting and differentiation. Both approaches for reconciling relative and absolute chronologies have been investigated in recent years, for example, utilizing the ^{26}Al–^{26}Mg and Pb–Pb systems in CAIs and chondrules for constraining the timing and duration of events in the nebula, and the ^{53}Mn–^{53}Cr and Pb–Pb systems in differentiated meteorites to pin the timing of early planetary melting. The consistency of the deduced chronologies may be evaluated to give confidence (or not) that the assumptions necessary for a temporal interpretation of the record of short-lived radioactivity are, indeed, fulfilled.

2.3.1 An Absolute Timescale for Solar System Formation

The early evolution of the solar system is characterized by significant thermal processing of original presolar materials. This processing typically results in chemical fractionation that may potentially be dated by isotopic means in appropriate samples, for example, nebular events such as condensation or distillation fractionate parent and daughter elements according to differing volatility. Likewise, chemical differentiation during melting and segregation leads to unequal rates of radiogenic ingrowth in different planetary reservoirs (e.g., crust, mantle, and core) that can constrain the nature and timing of early planetary differentiation. Several long-lived and now-extinct radioisotope systems have been utilized to delineate these various nebular and parent-body processes; however, it is only the U–Pb system that can record the absolute ages of the earliest volatility-controlled fractionation events, corresponding to the formation of the first refractory minerals, as well as the timing of melt generation on early planetesimals with sufficiently high precision as to provide a quantitative link to the short-lived isotope systems.

The U–Pb system represents the premier geochronometer because it inherently contains two long-lived isotopic clocks that run at different rates: ^{238}U decays to ^{206}Pb with a half-life of 4,468 Myr, and ^{235}U decays to ^{207}Pb with a much shorter half-life of 704 Myr. This unique circumstance provides a method for checking isotopic disturbance (by either gain or loss of uranium or lead) that is revealed by discordance in the ages derived from the two independent isotopic clocks with the same geochemical behavior (Wetherill, 1956; Tera and Wasserburg, 1972). Such an approach is commonly used in evaluating the ages of magmatic or metamorphic events in terrestrial samples. For obtaining the highest precision ages of volatility-controlled fractionation events in the solar nebula, the U–Pb concordance approach is of limited utility, however, and instead one utilizes $^{207}Pb/^{206}Pb$ and $^{204}Pb/^{206}Pb$ variations in a suite of cogenetic samples to evaluate crystallization ages. The method has a significant analytical advantage since only isotope ratios need to be determined in the mass spectrometer, but equally important is the high probability that the age obtained represents a true crystallization age, because the system is relatively insensitive to recent gain or loss of lead (or, more generally, recent fractionation of U/Pb). Moreover, this age is fundamentally based on the isotopic evolution of uranium, a refractory element whose isotopic composition is thought to be invariant throughout the solar system (Chen and Wasserburg, 1980, 1981), and the radiogenic $^{207}Pb/^{206}Pb$ evolves rapidly at 4.5 Ga because of the relatively short half-life of ^{235}U. In principle, ancient lead loss or redistribution (e.g., owing to early metamorphic or aqueous activity on asteroids, the parent bodies of meteorites) can confound the interpretation of lead isotopic ages as magmatic ages, but such closure effects are usually considered to be insignificant for the most primitive meteorite samples. Whether or not this is a valid assumption is an issue that is open to experimental assessment and interpretation (see discussions in Tilton, 1988 and Tera and Carlson, 1999). The Pb–Pb method can have precisions of 0.5–1.0 Myr in favorable cases using the most advanced current techniques (Amelin, 2006). There are uncertainties in the decay constants of the uranium isotopes that give an uncertainty of 9.3 Myr for early solar system ages. These uncertainties are not included when Pb–Pb ages are quoted. This is not a problem when comparing relative chronologies determined by Pb–Pb with those from short-lived chronometers, but needs to be considered when comparing absolute Pb–Pb ages with those determined from other long-lived chronometers (Amelin, 2006). One further potential complication of the U–Pb system is the possible presence of extinct ^{247}Cm in the early solar system. This isotope decays to ^{235}U by three α- and two β-decays and could perturb Pb–Pb ages. A recent high-precision study by Stirling *et al.* (2005) has established that $^{235}U/^{238}U$ ratios in bulk meteorites are uniform to within 2 ε units (parts in 10^4), so the precision of Pb–Pb dates for the early solar system remains robust.

Absolute crystallization ages have been calculated for refractory samples, CAIs that formed with very high depletions of volatile lead, by modeling the evolution of $^{207}Pb/^{206}Pb$ from primordial common (i.e., unradiogenic) lead found in early formed sulfides from iron meteorites.

Such "model ages" can be determined with good precision (typically a few million years ago), but accuracy depends on the correctness of the assumption of the isotopic composition of initial lead. Sensitivity to this correction is relatively small for fairly radiogenic samples such as CAIs where almost all the lead is due to *in situ* decay, nevertheless, depending on the details of data reduction and sample selection, even the best early estimates of Pb–Pb model ages for CAI formation ranged over ~15 Ma, from 4,553 to 4,568 Ma, with typical uncertainties in the range of 4–5 Ma (see discussions in Tilton, 1988 and Tera and Carlson, 1999). By progressively leaching samples to remove contaminating lead (probably introduced from the meteorite matrix), Allègre *et al.* (1995) were able to produce highly radiogenic ($^{206}Pb/^{204}Pb > 150$) fractions from four CAIs from the Allende CV3 chondrite, which yielded Pb–Pb model ages of $4,566 \pm 2$ Ma. Accuracy problems associated with initial lead corrections can also be addressed by an isochron approach where no particular composition of common lead needs to be assumed, only that a suite of samples are cogenetic and incorporated varying amounts of the same initial lead on crystallization (Tera and Carlson, 1999). Utilizing this approach, Tera and Carlson (1999) reinterpreted previous lead isotopic data obtained on nine Allende coarse-grained CAIs that had indicated a spread of ages (Chen and Wasserburg, 1981) to instead fit a single lead isochron of age equal to $4,566 \pm 8$ Ma which, however, is evolved from an initial lead isotopic composition that is unique to CAIs. More recently, Amelin *et al.* (2002) used the isochron method to determine absolute ages of formation for two CAIs from the Efremovka CV3 carbonaceous chondrite. Both samples are consistent with a mean age of $4,567.2 \pm 0.6$ Ma (Figure 1). Amelin *et al.* (2006) reported further isotopic analyses on one of the two Efremovka CAIs, E60, refining the age to $4,567.11 \pm 0.16$ Ma, which is the most precise absolute age obtained on CAIs. Because the previous best ages on Allende CAIs are consistent, within their relatively larger errors, with this new lead isochron age we adopt this value of $4,567.11 \pm 0.16$ Ma as the best estimate for the absolute formation age for coarse-grained (igneous) CAIs from CV chondrites. There is an alternative view; however, Baker *et al.* (2005) reported high-precision lead and magnesium isotopic data for two angrites, SAH99555 and NWA1296, that showed a Pb–Pb age only 1.0 ± 0.6 Myr younger than the Amelin *et al.* (2006) CAI age. Magnesium and initial strontium isotopic compositions indicate a time difference of 3.3–3.8 Myr between CAI formation and angrite formation. Baker *et al.* (2005) suggested that CAI leachates, which are dominated by lead introduced during terrestrial exposure, should be used to represent the common lead

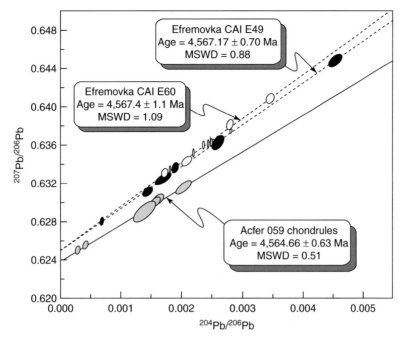

Figure 1 Pb–Pb isochrons for acid-washed fractions of two CAIs from CV3 Efremovka and for the six most radiogenic fractions of acid-washed chondrules from the CR chondrite Acfer 059. The $^{207}Pb/^{206}Pb$ data are not corrected for any assumed common lead composition; 2σ error ellipses are shown. Isochron ages for the two CAIs overlap with a weighted mean age of $4,567.2 \pm 0.6$ Ma, which is ~2.5 Myr older than the chondrules. Data and figure reproduced from Amelin *et al.* (2002).

component in Efremovka E60, which would change the Pb–Pb age to $4,569.5 \pm 0.4$ Ma. Thus, further lead isotope work on the absolute formation age of CAIs remains to be done before this age can be considered to be robust.

To the extent that this high-precision, high-accuracy result represents the absolute age of crystallization of CAIs generally, it provides a measure of the age of formation of the solar system since several lines of evidence, in addition to the absolute Pb–Pb ages, indicate that CAIs are the first solid materials to have formed in the solar nebula (for a review, see Podosek and Swindle, 1988). In fact, it is the relative abundances of the short-lived radionuclides, especially ^{26}Al, which provides the primary indication that CAIs are indeed these first local materials. Other evidence is more circumstantial, for example, the prevalence of large stable isotope anomalies in CAIs compared with other materials of solar system origin. We will return to the issue of antiquity of CAIs when we examine the distribution of short-lived isotopes among different CAI types.

Other volatility-controlled, long-lived parent–daughter isotope systems (e.g., Rb–Sr) yield absolute ages that are compatible with the coupled U–Pb systems, albeit with poorer precision. Because the chondrites are unequilibrated assemblages of components that may not share a common history, whole-rock or even mineral separate "ages" are not very meaningful for providing a very useful

constraint on accretion timescales. High-precision age determinations, approaching 1 Ma resolution, can in principle be obtained from initial $^{87}Sr/^{86}Sr$ in low Rb/Sr phases, such as CAIs (e.g., Podosek *et al.*, 1991). However, such ages depend on deriving an accurate model of the strontium isotopic evolution of the reservoir from which these materials formed. The latter is a very difficult requirement, because it is not likely that a strictly chondritic Rb/Sr ratio was always maintained in the nebular regions from which precursor materials that ultimately formed CAIs, chondrules, and other meteoritic components condensed. Thus, initial strontium "ages," while highly precise, may be of little use in terms of quantitatively constraining absolute ages of formation of individual nebular objects and are best interpreted as only providing a qualitative measure of antiquity (Podosek *et al.*, 1991). It is possible that initial $^{87}Sr/^{86}Sr$ ratios of similar nebular components, for example, type B CAIs, could provide relative formation ages under the assumption that such objects share a common long-term Rb/Sr heritage; however, this has not yet been demonstrated.

2.3.2 An Absolute Timescale for Chondrule Formation

Although chondrule formation is thought to be one of the most significant thermal processes to have occurred in the solar nebula, in the sense of affecting the majority of planetary materials in the inner solar system, the mechanism(s) responsible remains hotly debated after many years of investigation. Similarly, it has long been recognized that obtaining good measurements of chondrule ages would be extremely useful for possibly constraining formation mechanisms and environments, as well as setting important limits on the duration of the solar nebula and, thus, on accretion timescales. However, determination of crystallization ages of chondrules is very difficult because their mineralogy is typically not amenable to large parent–daughter fractionation. Several short-lived isotope systems (discussed below) have been explored in recent years to try to delimit relative formation times for chondrules, for example, compared with CAIs, but high-precision absolute Pb–Pb ages have been measured for only a single meteorite. Amelin *et al.* (2002) used aggressive acid washing of a suite of chondrules from the unequilibrated CR chondrite Acfer 059 to remove unradiogenic lead (from both meteorite matrix and terrestrial contamination). Isochron ages ranged from 4,563 to nearly 4,565 Ma, with a preferred value of 4,564.7 ± 0.6 Ma (Figure 1) for six of the most radiogenic samples ($^{206}Pb/^{204}Pb > 395$). It is argued that this result dates chondrule formation because lead-closure effects are thought to be insignificant for these pristine samples. If these CR chondrules are representative of chondrules generally, then the data of Amelin *et al.* (2002) imply an interval of ~2.5 Ma between the formation of CV CAIs and chondrules in the nebula. Krot *et al.* (2005) reported Pb–Pb ages for chondrules from the CB$_a$ chondrite Gujba (4,562.7 ± 0.5 Ma) and CB$_b$ chondrite Hammadah al Hamra 257 (4,562.8 ± 0.9 Ma). These ages are ~5 Myr younger than CAIs. Krot *et al.* (2005) argued that this time difference was too great for nebular processes and suggested that CB chondrite chondrules were formed in a giant impact between planetary embryos.

2.3.3 An Absolute Timescale for Early Differentiation of Planetesimals

Time markers for tying short-lived chronometers to an absolute timescale can potentially be provided by early planetary differentiates. The basic requirements are that appropriately ancient samples would have to have evolved from a reservoir (magma) that had achieved isotopic equilibrium with respect to daughter elements of both long- and short-lived systems (i.e., lead and chromium or magnesium, respectively), then cooled rapidly following crystallization, and remained isotopically closed until analysis in the laboratory. In practice, the latter requirement means that samples should be undisturbed by shock and free of terrestrial contamination. No sample is perfect in all these respects, but the angrites are considered to

be nearly ideal (the major problem being terrestrial lead contamination). By careful cleaning, Lugmair and Galer (1992) determined high-precision Pb–Pb model ages for the angrites Lewis Cliff 86010 (LEW) and Angra dos Reis (ADOR). The results are concordant in U/Pb and with other isotopic systems as well as with each other, and provide an absolute crystallization age of 4,557.8 ± 0.5 Ma for the angrites (Lugmair and Galer, 1992). This is a significant time marker ("event") because angrite mineralogy also provides large Mn/Cr fractionation that is useful for accurate $^{53}Mn/^{55}Mn$ determination. Baker *et al.* (2005) reported a highly precise and ancient Pb–Pb age for the Sahara 99555 angrite of 4,566.2 ± 0.1 Ma, indicating crystallization of basalt on the surface of the angrite parent body only 1 Myr after the formation of CAIs. Amelin *et al.* (2006) reported a Pb–Pb age for the Asuka 881394 eucrite of 4,566.52 ± 0.33 Ma, only 0.59 ± 0.36 Myr after CAI formation.

The eucrites are highly differentiated (basaltic) achondrites that, along with the related howardites and diogenites, may have originated from the asteroid 4 Vesta (Binzel and Xu, 1993). Unfortunately, the U/Pb systematics of eucrites appear to be disturbed, yielding Pb–Pb ages up to ~220 Myr younger than angrites (Galer and Lugmair, 1996). This compromises the utility of the eucrites as providing independent tie points between long- and short-lived chronometers.

Evidence for an extended thermal history of equilibrated ordinary chondrites is provided by U–Pb analyses of phosphates (Göpel *et al.*, 1994). The phosphates (merrillite and apatite) are metamorphic minerals produced by the oxidation of phosphorus originally present in metal grains. Phosphate mineral separates obtained from chondrites of metamorphic grade 4 and greater have Pb–Pb model ages (Göpel *et al.*, 1994) from 4,563 (for H4, Ste. Marguerite) to 4,502 Ma (for H6, Guareña). The oldest ages are nearly equivalent to Pb–Pb ages from CR chondrules (Amelin *et al.*, 2002) and only a few million years younger than CAIs, indicating that accretion and thermal processing was rapid for the H4 chondrite parent body. The relatively long time interval of ~60 Myr has implications for the nature of the H chondrite parent body and the heat sources responsible for long-lived metamorphism (Göpel *et al.*, 1994).

2.4 THE RECORD OF SHORT-LIVED RADIONUCLIDES IN EARLY SOLAR SYSTEM MATERIALS

Here, we discuss the evidence for the prior existence of the now-extinct isotopes in meteoritic materials and, in the better-studied cases, what is known about the distribution of that isotope in the early solar system. Table 1 summarizes the basic facts regarding those short-lived radioisotopes that are unequivocally known to have existed as live radioactivity in rocks formed in the early solar system and provides an estimate of their initial abundances compared with a reference isotope. The table is organized in terms of increasing half-life and according to the main environment for parent–daughter chemical fractionation. The latter property indicates what types of events can potentially be dated and largely dictates what types of samples record evidence that a certain radioisotope once existed. Note that there is only a small degree of overlap demonstrated thus far for a few of the isotope systems. For example, it is well-documented that the Mn–Cr system is sensitive to fractionation in both nebular and parent-body environments and, as we see below, new high-precision magnesium isotopic measurements are making possible application of the Al–Mg system to parent-body processes, but other systems which might similarly provide linkages from the nebula through accretion to early differentiation have not been fully developed due to either analytical difficulties (e.g., Fe–Ni) and/or difficulties in constraining mineral hosts and closure effects (e.g., I–Xe and ^{244}Pu). The initial abundances refer to the origin of the solar system, which, as discussed previously, means the time of CAI formation, and hence these can only be measured directly in nebular samples. The initial abundances of those isotopes that are found only in differentiated meteorites also refer back to the time of CAI formation, but such a calculation necessarily requires a chronological framework and is underpinned by the absolute time markers provided by the Pb–Pb system.

We now describe, in order of half-life, short-lived radionuclides whose presence has been searched for, and in most cases, confirmed, in the early solar system.

2.4.1 Beryllium-7

Beryllium-7 decays by electron capture to ^{7}Li with a half-life of 53.3 days. Chaussidon *et al.* (2006) found variations of ~25% in ^{7}Li/^{6}Li ratios within an Allende CAI, and suggested that they may be due to *in situ* decay of ^{7}Be. Boron isotopes were also measured in this CAI and it has a well-defined ^{10}Be–^{10}B isochron with a slope corresponding to an initial ^{10}Be/^{9}Be ratio of $(1.04 \pm 0.09) \times 10^{-3}$ (see Section 2.4.5). There are significant difficulties associated with lithium isotopic measurements in CAIs, in that lithium can be introduced by secondary alteration processes long after the decay of short-lived ^{7}Be. Chaussidon *et al.* (2006) rejected analyses that did not plot along trajectories expected for closed system crystal fractionation on a Be versus Li concentration plot. There are further difficulties caused by the fact that lithium diffuses very rapidly and can

mass fractionate during diffusion (Richter *et al.*, 2004). Corrections also had to be made for spallation production of lithium isotopes by galactic cosmic rays. After correction for these effects and rejection of "contaminated" analyses, Chaussidon *et al.* (2006) inferred an initial ^{7}Be/^{9}Be ratio of 0.0061 ± 0.0013 for Allende CAI 3529-41. In their calculation of the slope of the isochron, they did not weight data points by their uncertainties. A weighted fit gives a slightly higher initial ^{7}Be/^{9}Be ratio of 0.0100 ± 0.0007. The identification of ^{7}Be as a short-lived radionuclide in the early solar system remains tentative, given how large the correction effects are and the necessity of rejecting a significant fraction of the measured data. If ^{7}Be is present, it has profound importance for short-lived radionuclide production within the early solar system. The half-life is so short that it cannot have been produced elsewhere.

2.4.2 Calcium-41

Calcium-41 decays by electron capture to ^{41}K with a half-life of only 103 kyr. It has the distinction of being the shortest-lived isotope for which firm evidence exists in early solar system materials, and this fact makes it key for constraining the timescale of last nucleosynthetic addition to solar system matter (in the external seeding scenario). It also makes ^{41}Ca exceedingly difficult to detect experimentally, because it can only be found to have existed in the oldest materials and then in only very small concentrations. Fortunately, its daughter potassium is rather volatile and calcium is concentrated in refractory minerals (the "C" in CAI) leading to large fractionations. Hutcheon *et al.* (1984) found hints for ^{41}Ca in Allende refractory inclusions, but could not clearly resolve ^{41}K excesses above measurement uncertainties.

The first unambiguous evidence of live ^{41}Ca came with the demonstration of correlated excesses of ^{41}K/^{39}K with Ca/K in Efremovka CAIs by Srinivasan *et al.* (1994, 1996). Subsequent measurements by the PRL group have established that ^{41}Ca was also present in refractory oxide phases (hibonite) of CM and CV chondrites (Sahijpal *et al.*, 1998, 2000). The CM hibonite grains are generally too small to permit enough multiple measurements to define an isochron on individual objects, even by ion probe; however, hibonite crystals from Allende CAIs show good correlation lines (Sahijpal *et al.*, 2000) consistent with that found for Efremovka and indicating that ^{41}Ca decayed *in situ*. Most of the isolated CM hibonite grains also show ^{41}K/^{39}K excesses that are consistent with the isochrons obtained on silicate minerals of CAIs, except ~1/3 of the hibonite grains appear to have crystallized with "dead" calcium (i.e., they have normal ^{41}K/^{39}K compositions). The ensemble isochron (Figure 2) yields an initial value of ^{41}Ca/^{40}Ca = 1.4×10^{-8}

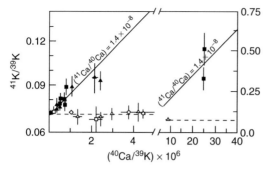

Figure 2 Potassium isotopic compositions measured in individual hibonite grains (Sahijpal *et al.*, 1998) plotted as a function of Ca/K ratio. Hibonite grains from the carbonaceous chondrites Murchison, Allende, and Efremovka, which formed with close to canonical levels of ^{26}Al are indicated as filled symbols, whereas hibonite grains that crystallized with no ^{26}Al are open squares and triangles. Terrestrial standards are plotted as open diamonds; error bars are 1σ. The isochron corresponding to live ^{41}Ca at the level ^{41}Ca/^{40}Ca $= 1.4 \times 10^{-8}$, determined for Efremovka CAIs (Srinivasan *et al.*, 1996), is also shown. Those hibonite grains that contained ^{26}Al are seen to plot on the same ^{41}Ca isochron as the CAIs, but grains lacking ^{26}Al are also lacking ^{41}Ca and plot on the horizontal dashed line corresponding to terrestrial ^{41}K/^{39}K. Data and figure adapted from Sahijpal *et al.* (1998).

with a formal error of ~10% relative and a statistical scatter that is commensurate with the measurement uncertainties. Such a small uncertainty would correspond to a very tight timescale (~15 kyr) for the duration of formation of these objects; however, possible systematic uncertainties in the mass spectrometry may increase this interval somewhat. The hibonite grains that contain no excess ^{41}K/^{39}K are unlikely to have lost that signal and, thus, must either have formed well after the other samples, or else they never incorporated live ^{41}Ca during their crystallization. An important clue is that these same grains also never contained ^{26}Al (Sahijpal and Goswami, 1998; Sahijpal *et al.*, 1998, 2000); we will return to the significance of this correlation in discussing the scale of isotopic heterogeneity in the nebula and the source of ^{41}Ca and ^{26}Al. The recent discovery that bulk CAIs lie along an isochron consistent with a somewhat higher ^{26}Al/^{27}Al ratio than that inferred from internal isochrons of individual CAIs raises the possibility that the initial solar system ^{41}Ca/^{40}Ca ratio was higher than the value inferred by Sahijpal *et al.* (1998). The half-life of ^{41}Ca is so short that the early solar system ^{41}Ca/^{40}Ca increases by a factor 10 for each 26% increase in the early solar system ^{26}Al/^{27}Al. Correcting the inference of Sahijpal *et al.* (1998), ^{41}Ca/^{40}Ca $= 1.4 \times 10^{-8}$ for hibonite with canonical ^{26}Al/^{27}Al (5×10^{-5}), to the current best estimate of early solar system ^{26}Al/^{27}Al (6.33×10^{-5}, Table 1) gives initial ^{41}Ca/^{40}Ca ratios of 7×10^{-8}. Further

potassium isotope measurements with high-precision magnesium isotopic measurement on CAIs are needed to better constrain this value.

2.4.3 Chlorine-36

Chlorine-36 has a half-life of 300 kyr and decays by β-decay (98.1%) to ^{36}Ar and by electron capture and positron emission (1.9%) to ^{36}S. Murty *et al.* (1997) reported ^{36}Ar in the matrix of the Efremovka CV chondrite in excess of the amount expected from trapped and cosmogenic components and attributed it to *in situ* decay of ^{36}Cl. They inferred an initial ^{36}Cl/^{35}Cl ratio of $(1.4 \pm 0.2) \times 10^{-6}$. Lin *et al.* (2005) used an ion microprobe to measure sulfur isotopes in sodalite, a chlorine-bearing mineral commonly found as a secondary alteration product in CAIs and matrix in CV chondrites. They found well-defined isochrons in four sodalite-rich regions in a CAI from the Ningqiang CV chondrite, leading to inferred initial ^{36}Cl/^{35}Cl ratios of $(5-11) \times 10^{-6}$. From the fact that these areas had ^{26}Al/^{27}Al of $<5 \times 10^{-6}$, they inferred an initial solar system ^{36}Cl/^{35}Cl ratio of $\geq 1.6 \times 10^{-4}$. Hsu *et al.* (2006) reported a combined ^{36}Cl–^{36}S and ^{26}Al–^{26}Mg study of an altered Allende CAI named the Pink Angel. From the inferred ^{36}Cl/^{35}Cl ratio and the upper limit on ^{26}Al/^{27}Al, they calculated an early solar system ^{36}Cl/^{35}Cl ratio of $>10^{-3}$. It is not plausible to produce this level of ^{36}Cl in supernovae or AGB stars, so Hsu *et al.* (2006) concluded that ^{36}Cl must have been produced by a late episode of particle bombardment within the solar system.

2.4.4 Aluminum-26

Aluminum-26 decays by positron emission and electron capture to ^{26}Mg with a half-life of ~730 kyr. The discovery circumstances of ^{26}Al have already been discussed (Section 2.1.3) and since those early measurements, a large body of data has grown to include analyses of CAIs from all major meteorite classes (carbonaceous, ordinary, and enstatite) as well as important groups within these classes (e.g., CM, CV, CH, CR, CO); sparse data also exist for aluminum-rich phases from several differentiated meteorites and in chondrules. Data obtained prior to 1995 were the subject of a comprehensive review by MacPherson *et al.* (1995); for the most part, their analysis relied heavily on the extensive record in the large, abundant CAIs from CV chondrites, although significant numbers of refractory phases from other carbonaceous chondrite groups were also considered. Between that time and the first edition of this chapter, work generally concentrated on extending the database to include smaller CAIs from underrepresented meteorite groups and, especially, chondrules (mostly from ordinary chondrites). Most measurements were performed by ion microprobe because of the need to localize analysis on mineral phases with high Al/Mg ratios to resolve the addition of radiogenic ^{26}Mg*; this capability was particularly important for revealing internal Al–Mg isochrons in chondrules by examining small regions of trapped melt or glassy mesostasis in between the larger ferromagnesian minerals that dominate chondrules (Russell *et al.*, 1996; Kita *et al.*, 2000; McKeegan *et al.*, 2000b; Mostefaoui *et al.*, 2002). There have been two significant technical developments in that past 3 years that have profoundly changed understanding of the Al–Mg system, both of which resulted in much higher precision magnesium isotopic analyses: (1) high-precision isotopic analysis by multicollector inductively coupled plasma mass spectrometry (MC-ICPMS), both on dissolved samples and using laser ablation sampling devices for spot analyses; and (2) multiple collector development on large-radius ion microprobes.

With the new level of precision of magnesium isotopic analyses, additional care must be taken in treating the data. Magnesium has three stable isotopes, ^{24}Mg, ^{25}Mg, and ^{26}Mg. Isotopic mass fractionation of magnesium can occur in nature, during chemical separation of magnesium from samples, and in mass spectrometers. During mass fractionation, the ^{26}Mg/^{24}Mg ratio varies by about twice as much as the ^{25}Mg/^{24}Mg ratio. When magnesium isotopic compositions were measured with precision of $\sim$1‰ or worse, the exact relationship between ^{26}Mg/^{24}Mg fractionation and ^{25}Mg/^{24}Mg fractionation did not matter much when determining the amount of radiogenic ^{26}Mg. However in more recent results with precisions of 0.1‰ or better on samples with low Al/Mg, the fractionation law used becomes important. Isotopic mass fractionation during chemical separation is minimized by ensuring that chemical yields are high; fractionation during mass spectrometry can be corrected using standards of known isotopic composition such that the instrumental mass fractionation law is well calibrated.

Magnesium isotopic compositions are usually expressed in δ notation, i.e., δ^{25}Mg $= [(^{25}$Mg/^{24}Mg$)_{sample}/(^{25}$Mg/^{24}Mg$)_{standard}-1]\times 1,000$. On a plot of δ^{25}Mg versus δ^{26}Mg, mass fractionation due to equilibrium fractionation or kinetic effects lie along lines that have a slope of $\sim$0.5. Although these relationships appear to be linear over a narrow range of mass fractionation, in fact they are curves. Since most mass fractionation mechanisms are exponential processes, it is convenient to express isotope ratios as another related quantity, $1000\times\ln[(^{25}$Mg/^{24}Mg$)_{sample}/(^{25}$Mg/^{24}Mg$)_{standard} - 1]$, denoted as ϕ^{25}Mg by Davis *et al.* (2005) and as δ^{25}Mg$'$ by Young *et al.* (2005). On a plot of ϕ^{25}Mg versus ϕ^{26}Mg, the different mass fractionation processes plot as straight lines, but with differing slopes depending on the nature of the fractionation process. Fractionation due to the kinetic isotope effect gives a slope of 0.51101 and that due to equilibrium isotope partitioning gives 0.52100 (Young *et al.*, 2002; Davis *et al.*, 2005). Most CAIs have magnesium that is mass-fractionated by a few ‰ amu^{-1}).

The mass fractionation in CAIs is believed to have been caused by the kinetic isotope effect during high-temperature evaporation in the solar nebula. Davis *et al.* (2005) evaporated melts of CAIs in vacuum and measured magnesium isotopic compositions by MC-ICPMS. On a plot of ϕ^{25}Mg versus ϕ^{26}Mg, their data give a slope of 0.51400 ± 0.00024. Several papers have been published with high-precision magnesium isotopic data, using different slopes to correct for mass fractionation: Bizzarro and coworkers use 0.511, Young and coworkers use 0.521, and McKeegan and coworkers use 0.514. As an example of the effect of these different fractionation laws, consider a spinel grain, with ^{27}Al/^{24}Mg $= 2.53$, a typical degree of mass fractionation for a CAI, δ^{25}Mg $= 5$‰, and an initial ^{26}Al/^{27}Al value of 6.33×10^{-5} (see below). Excesses or deficits in ^{26}Mg due to ^{26}Al decay are usually expressed as Δ^{26}Mg $= \phi^{26}$Mg $- \phi^{26}$Mg $\times$ slope. In this example, Δ^{26}Mg should be 1.15‰ after complete decay of ^{26}Al. If this is the value obtained with the slope we have adopted, 0.514, recalculating with the kinetic value, 0.511, gives 1.07‰, and the equilibrium value, 0.521, gives 1.26‰. These shifts may seem small, but for an isochron passing through the origin and the spinel in our example, the two slopes would imply initial ^{26}Al/^{27}Al values of 5.89×10^{-5} and 6.93×10^{-5}, a range of 15% that corresponds to a time difference of 168 kyr. The effect of ϕ^{25}Mg versus ϕ^{26}Mg slope on Δ^{26}Mg depends only on the degree of mass fractionation, not on Al/Mg ratio. It has only become important recently with the development of high-precision magnesium isotopic methods applied to low-Al/Mg samples.

To first order, the larger data set prior to the high-precision measurements extends and confirms the general assessments of MacPherson *et al.* (1995). The distribution of inferred initial ^{26}Al/^{27}Al in CAIs is bimodal (Figure 3a), with the dominant peak at the so-called canonical value of 4.5×10^{-5}, and a second peak at dead aluminum (i.e., ^{26}Al/^{27}Al $= 0$). MacPherson *et al.* (1995) demonstrated that this pattern applied to all classes of carbonaceous chondrites, although the relative heights of the two peaks varied among different meteorites (mostly reflecting a difference in CAI types). The dispersion of the canonical peak (amounting to $\sim$1 $\times 10^{-5}$, FWHM) was considered to represent a convolution of measurement error and geologic noise; there was no robust data indicating that any CAIs formed with $(^{26}$Al/^{27}Al$)_0$ significantly above the canonical ratio.

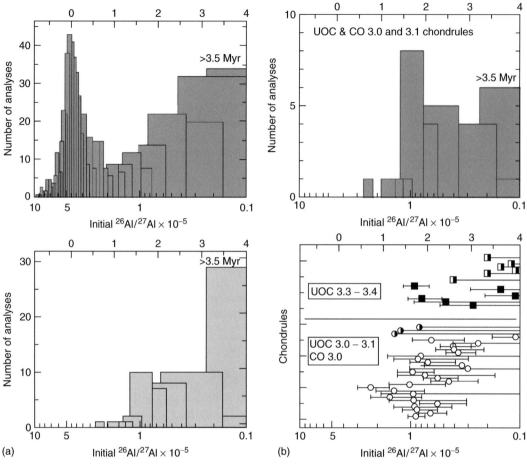

Figure 3 (a) Top panel: Histogram of initial $^{26}Al/^{27}Al$ inferred for CAIs; the number of analyses (taken to be representative of the number of samples) is plotted versus time after CAI formation (top axis), where time 0 is taken as the "canonical" $^{26}Al/^{27}Al = 4.5 \times 10^{-5}$ peak of the distribution for CAIs. In addition to the canonical value, a significant number of CAIs do not preserve any evidence for having formed with live ^{26}Al; samples with only upper limits are summed in the last bin, indicating the achievement of isotopic closure at least 3.5 Myr after time 0, or alternatively, never having incorporated ^{26}Al at all (see text). Data sources are summarized by MacPherson *et al.* (1995). Bottom panel: Similar histogram summarizing data on plagioclase-olivine-inclusions (POIs) and chondrules (both aluminum-rich and ferromagnesian). In contrast to CAIs, there is no peak at $\sim 5 \times 10^{-5}$ and most chondrules show no evidence for having incorporated ^{26}Al. Some chondrules do show evidence for $^{26}Al/^{27}Al$ initial values at the level of $\sim 1 \times 10^{-5}$ or lower, indicating the formation 1.5 to several million years after CAIs. Data sources are those summarized by MacPherson *et al.* (1995), supplemented by more recent data (Russell *et al.*, 1996; Kita *et al.*, 2000; McKeegan *et al.*, 2000b; Huss *et al.*, 2001; Mostefaoui *et al.*, 2002; Hsu *et al.*, 2003; Kunihiro *et al.*, 2004). (b) Top panel: Histogram similar to the bottom panel of (a), except showing the inferred $^{26}Al/^{27}Al$ distribution for only those chondrules from the most unequilibrated meteorites, that is, POIs and chondrules from metamorphic grades >3.1 have been removed from the plot. Also, this plot now shows the number of chondrules with that distribution, as opposed to the number of analyses considering each datum as a model isochron. Chondrules for which ^{26}Mg excesses are not well resolved (i.e., only upper limits are obtained or Al–Mg isochron slopes are within 2σ error of 0) are accumulated in the last histogram bin. A peak in the distribution may be discerned at $^{26}Al/^{27}Al \sim 1 \times 10^{-5}$, which corresponds to 1.5–2 Myr after time 0. Bottom panel: Inferred $^{26}Al/^{27}Al$ ratios for individual ferromagnesian and aluminum-rich chondrules with 2σ errors. Chondrules from the lowest metamorphic grades (3.0, 3.1) of unequilibrated ordinary (LL) and carbonaceous (CO) chondrites are shown in open circles, those from metamorphic grades 3.3 and above are shown in filled squares. Chondrules for which only upper limits are obtained are shown in half-open/half-filled symbols. It is apparent that chondrules from more intensely metamorphosed meteorites display apparently lower $^{26}Al/^{27}Al$ initial values. Among the most unequilibrated samples, an interval of >1 Myr is implied for the duration of chondrule formation. Data sources as in (a).

The first hint that CAIs formed with $(^{26}Al/^{27}Al)_0$ significantly above $\sim 5 \times 10^{-5}$ came from the data for one Allende CAI (Galy *et al.*, 2000): a model isochron yields $(^{26}Al/^{27}Al)_0 = (6.01 \pm 0.22) \times 10^{-5}$ (corrected with a $\phi^{25}Mg$ versus $\phi^{26}Mg$ slope of 0.514), which is marginally higher than any previously determined value. Another strong hint came from MC-ICPMS measurements of a number of bulk CAIs, which gave an isochron with a slope corresponding to $(6.85 \pm 0.85) \times 10^{-5}$, reported by Galy *et al.* (2004). These data were collected in two laboratories, each of which used a different mass fractionation law (I. D. Hutcheon, personal communication). Correction of all data to the 0.514 slope yields $(^{26}Al/^{27}Al)_0 = (6.36 \pm 0.13) \times 10^{-5}$. This work suggested the possibility that nebular fractionation established the bulk Al/Mg ratios of CAIs, but that internal isochrons determined

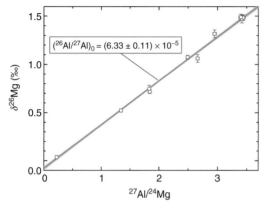

Figure 4 $^{26}Al–^{26}Mg$ isochron diagram for several Allende CAIs. The data are as reported by Bizzarro *et al.* (2004, 2005), but corrected to $\phi^{25}Mg$ versus $\phi^{26}Mg$ slope of 0.514 (see text). This data set provides the best current estimate of the early solar system $^{26}Al/^{27}Al$ ratio, which represents the time of volatility fractionation of aluminum from magnesium in the solar nebula.

by ion microprobe recorded later remelting events. Bizzarro *et al.* (2004) strongly confirmed this with a very precise isochron computed for a suite of bulk Allende CAIs. Their data are shown in Figure 4, which is a corrected version of their published plot: Bizzarro *et al.* (2005) published an erratum correcting Al/Mg ratios and we have recorrected their magnesium isotopic data for mass fractionation using the 0.514 slope (they used 0.511, M. Bizzarro, personal communication). The slope corresponds to $(^{26}Al/^{27}Al)_0 = (6.33 \pm 0.11) \times 10^{-5}$, in remarkable agreement with the Galy *et al.* (2004) value, and we adopt it as the best current estimate of the initial solar system $^{26}Al/^{27}Al$ ratio.

In most ion microprobe measurements of internal isochrons for CAIs, the slope of the isochron is largely determined by analyses of anorthitic plagioclase. This phase is susceptible to mobilization of magnesium during metamorphism (LaTourrette and Wasserburg, 1998) or, possibly, during nebular events (Podosek *et al.*, 1991). However, with the recent availability of high-precision magnesium measurements on other phases, there are now indications of a real spread in initial $^{26}Al/^{27}Al$ ratios within CAIs. Young *et al.* (2005) reported over 200 laser ablation MC-ICPMS analyses of eight CAIs. The data are shown in Figure 5, and are corrected to a $\phi^{25}Mg$ versus $\phi^{26}Mg$ slope of 0.514 (Young *et al.* used 0.521). Three things are clear from this data set: (1) most of the points lie above the old canonical $(^{26}Al/^{27}Al)_0$ value of 4.5×10^{-5}; (2) few data points are above the new early solar

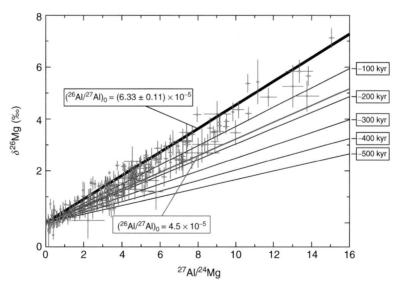

Figure 5 $^{26}Al–^{26}Mg$ isochron diagram for over 200 laser ablation analysis spots on eight CAIs reported by Young *et al.* (2005) and corrected to $\phi^{25}Mg$ versus $\phi^{26}Mg$ slope of 0.514 (see text). Also shown are the best estimate for the early solar system $^{26}Al/^{27}Al$ ratio, 6.33×10^{-5}, an earlier estimate of this ratio, 4.5×10^{-5} (MacPherson *et al.*, 1995), and several isochrons drawn at intervals of 100 kyr younger than the early solar system $^{26}Al/^{27}Al$ ratio. There are no data points significantly above the early solar system $^{26}Al/^{27}Al$ ratio, but there is a distribution below this ratio, implying that these CAIs were recrystallized, perhaps many times, over a few hundred thousand years after initial formation.

system $(^{26}Al/^{27}Al)_0$ value of 6.33×10^{-5}; and (3) the data have significant scatter that corresponds to a time period of several times 10^5 years.

High-precision multicollector ion microprobe techniques have been applied recently to a variety of CAIs. These give internal isochrons corresponding to initial $^{26}Al/^{27}Al$ ratios of 3.9–6.26×10^{-5} (Taylor *et al.*, 2005; Cosarinsky *et al.*, 2006; Liu *et al.*, 2006). No ages significantly in excess of the new canonical early solar system $^{26}Al/^{27}Al$ ratio of 6.33×10^{-5} have been found. Some work has also been done on the Wark–Lovering rims commonly found around CAIs), showing that they formed 0–300 kyr after the interior of the host CAI (Cosarinsky *et al.*, 2005; Simon *et al.*, 2005). A fairly consistent picture appears to be forming: a nebular Al/Mg volatility fractionation occurred at a well-defined time corresponding to the new canonical early solar system $^{26}Al/^{27}Al$ ratio, establishing the bulk CAI isochron of Bizzarro *et al.* (2004) (corrected), followed by several hundred thousand years of reheating events that established internal isochrons of individual CAIs.

The existence of a canonical $(^{26}Al/^{27}Al)_0$ value was previously based on analyses of CAIs only from carbonaceous chondrites; refractory inclusions from ordinary and enstatite chondrites are rare and often very small, and thus few had been discovered and none analyzed. There are now data for four CAIs from unequilibrated ordinary chondrites (Russell *et al.*, 1996; Huss *et al.*, 2001) and for 11 hibonite-bearing inclusions from enstatite chondrites (Guan *et al.*, 2000); all are consistent with $(^{26}Al/^{27}Al)_0$ in the range $\sim(3.5$–$5.5) \times 10^{-5}$, except for four of the (very small) hibonite grains for which $^{26}Mg^*$ could not be resolved. Thus, the same canonical value characterizes CAIs from all major meteorite classes. The possible meaning of this confirmation in terms of nebular chronology based on ^{26}Al is not completely straightforward, however.

The idea that many CAIs, whether they originally formed by melt crystallization or by condensation, have suffered some degree of disturbance to their Al–Mg isotopic system is well documented via correlated petrographic and isotopic evidence (MacPherson *et al.*, 1995 and references therein). For example, *in situ* isotopic measurements have demonstrated that certain anorthite crystals within a CAI can record resetting events ~ 1 Myr or more following CAI formation. In general, it seems to be the large type B CAIs from CV chondrites that are the most prone to have suffered multiple thermal events capable of at least partially resetting the Al–Mg system (Podosek *et al.*, 1991; Caillet *et al.*, 1993; MacPherson and Davis, 1993; MacPherson *et al.*, 1995); the protracted and complex thermal histories of type B CAIs are also evident in other chemical and isotopic systems, particularly the microdistribution of oxygen isotopes

within individual inclusions (Clayton and Mayeda, 1984; Young and Russell, 1998; Yurimoto *et al.*, 1998; McKeegan and Leshin, 2001). MacPherson *et al.* (1995) have argued that the trailing distribution of $^{26}Al/^{27}Al$ values downward from the canonical peak primarily represents a protracted period of thermal processing of CAIs, possibly accompanied by secondary mineral formation, over a few million years residence time in the solar nebula. Recently, Hsu *et al.* (2000) documented multiple isochrons within a single type B Allende CAI that they interpreted as signifying three discrete melting events separated in time by a few hundred thousand years. Such observations set lower bounds on the duration of the lifetime of the nebula and of significant heat sources, capable of producing CAIs, within the regions of the nebula.

The duration of high-temperature processes in the solar nebula is closely related to the age difference between CAIs and chondrules, and it is in this area that some of the most significant new data have been developed in recent years. The first evidence for radiogenic $^{26}Mg^*$ in non-CAI material was found in a plagioclase-bearing chondrule from the highly unequilibrated ordinary chondrite Semarkona (Hutcheon and Hutchison, 1989); the isochron implies an initial abundance of $(^{26}Al/^{27}Al)_0 = (7.7 \pm 2.1) \times 10^{-6}$. In most cases, however, only upper limits on ^{26}Al abundances could be determined in a handful of plagioclase grains from chondrules in ordinary chondrites (Hutcheon *et al.*, 1994; Hutcheon and Jones, 1995). Today, initial $^{26}Al/^{27}Al$ ratios have been determined in ~ 50 chondrules from several unequilibrated ordinary and carbonaceous chondrites. Chondrules with abundant aluminum-rich minerals (plagioclase-rich chondrules) and those with "normal" ferromagnesian mineralogy have been analyzed (Figure 3a, bottom panel). Chondrules have distinctly lower $(^{26}Al/^{27}Al)_0$ than CAIs, mostly by a factor of 5 or more. A significant number of chondrules show no resolvable $^{26}Mg^*$, implying that if they evolved from the same canonical $(^{26}Al/^{27}Al)_0$ that characterized the nebular regions where many CAIs formed, then chondrules achieved isotopic closure of the Al–Mg system at least 3–4 Myr (and possibly significantly more) after CAI formation. A closer inspection of the record, however, indicates that those chondrules from meteorites that are more extensively metamorphosed tend to have lower $(^{26}Al/^{27}Al)_0$ values (Figure 3b). This would indicate that metamorphic redistribution, on an asteroid, could be obscuring the nebular record of $^{26}Mg^*$ in these meteorites.

Chondrules that have been analyzed from some of the most pristine meteorites (e.g., Semarkona, Bishunpur, Yamato 81020) tend to show detectable ^{26}Mg excesses that imply $(^{26}Al/^{27}Al)_0$ values $\sim 1 \times 10^{-5}$, with some significant spread in this peak of the distribution (Russell *et al.*, 1996; Kita *et al.*, 2000; McKeegan *et al.*, 2000b; Huss *et al.*, 2001; Mostefaoui *et al.*, 2002; Hsu *et al.*, 2003; Kunihiro *et al.*, 2004; Hutcheon and Hutchison, 1989). A couple of chondrules have $(^{26}Al/^{27}Al)_0$ values that approach the range seen in some CAIs, and Galy *et al.* (2000) report one chondrule (not plotted in Figure 3b) with $(^{26}Al/^{27}Al)_0 = (3.7 \pm 1.2) \times 10^{-5}$, which overlaps the canonical CAI value within uncertainty. Bizzarro *et al.* (2005) reported that ^{26}Al–^{26}Mg model ages for a number of whole chondrules from the Allende CV chondrites covered a range corresponding to 0–1 Myr after the new initial CAI $^{26}Al/^{27}Al$ value. These data are for ICPMS measurement of whole chondrules, and there are currently no data showing internal Al–Mg isochrons for chondrules that fall within error of the CAI value. The overall data imply that chondrule formation began ~1 Myr after the formation of most CAIs and then continued for another ~2 Myr or more. Some chondrules may have formed later still, or, more likely, only achieved closure temperatures for magnesium diffusion following parent-body cooling at times exceeding ~4 Myr after CAIs. That mild metamorphism in chondrites could delay isotopic closure of the Al–Mg system is further evidenced by analyses of plagioclase grains from the H4 chondrites Ste. Marguerite and Forest Vale (Zinner and Göpel, 2002). The inferred $^{26}Al/^{27}Al$ ratios indicate retention of $^{26}Mg*$ by ~5–6 Myr following CAIs, which is consistent with timescales of parent-body metamorphism implied by absolute Pb–Pb ages of (secondary) phosphates in these meteorites.

A similar temporal interpretation is generally not invoked for those CAIs that exhibit an apparent lack of initial ^{26}Al (Figure 3a). As pointed out by MacPherson *et al.* (1995), many of the inclusions in the low $(^{26}Al/^{27}Al)_0$ peak are not mineralogically altered, which argues against late metamorphism. Moreover, these inclusions are typically hosts for very significant isotopic anomalies in a variety of elements, which argues strongly for their antiquity. Included in this group are the so-called FUN (fractionated and unknown nuclear isotopic effects) inclusions (e.g., Lee *et al.*, 1977, 1980) and the platelet hibonite crystals, which are extremely refractory grains from CM chondrites that are characterized by huge isotopic anomalies in the subiron group elements like titanium and calcium (Fahey *et al.*, 1987; Ireland, 1988). Because of their preservation of extreme stable isotope anomalies, these refractory phases are best understood as having formed at an early time in the nebula, but from an isotopic reservoir (or precursor minerals) that was missing the ^{26}Al inventory sampled by other "normal" refractory materials. The scope of this heterogeneity, both spatially and temporally, is the focus

of much conjecture and research, as this is a key issue for the utility of ^{26}Al as a high-resolution chronometer for nebular events (see discussion in Section 2.6).

Relatively few data exist for the former presence of ^{26}Al in differentiated (i.e., melted) meteorites, even though there is a widespread assumption that ^{26}Al provided a significant, if not the dominant, heat source for melting of early accreted planetesimals (e.g., Grimm and McSween, 1994; Schramm *et al.*, 1970). Plagioclase crystals in the eucrite Piplia Kalan have significant excess ^{26}Mg (Srinivasan *et al.*, 1999); however, the correlation of $^{26}Mg*$ with Al/Mg in the plagioclase is poor, indicating that the system has suffered partial reequilibration of magnesium isotopes following crystallization. A best-fit correlation through plagioclase and pyroxene yields an apparent $(^{26}Al/^{27}Al)_0 = (7.5 \pm 0.9) \times 10^{-7}$, which would correspond to ~4 Myr after the CAI canonical value.

Recently, several abstracts have reported Al–Mg data for achondrites, which can potentially be tied to the ^{53}Mn–^{53}Cr system. The petrographically unique eucrite Asuka 881394 exhibits a good Al–Mg isochron with well-resolved $^{26}Mg*$ in its anorthitic plagioclase that yields $^{26}Al/^{27}Al = (1.19 \pm 0.13) \times 10^{-6}$, corresponding to ~4 Myr after CAIs (Nyquist *et al.*, 2001b). In contrast, the eucrite Juvinas shows only an upper limit of $^{26}Al/^{27}Al \sim 10^{-7}$ (Wadhwa *et al.*, 2003). Basaltic clasts in the ultramafic ureilite DaG-319 all lie on a single Al–Mg isochron with slope $^{26}Al/^{27}Al = (3.95 \pm 0.59) \times 10^{-7}$ indicating that they achieved isotopic closure ~5 Myr after CAI formation (Kita *et al.*, 2003). The data for two angrites (Nyquist *et al.*, 2003) yield a two-point isochron with somewhat lower slope, corresponding to $^{26}Al/^{27}Al = (2.3 \pm 0.8) \times 10^{-7}$. Wadhwa *et al.* (2005) have tied together the Al–Mg, Mn–Cr, and Pb–Pb chronometers for the eucrite Asuka 881394, which has internal isochrons corresponding to $^{26}Al/^{27}Al = (1.34 \pm 0.05) \times 10^{-6}$, $^{53}Mn/^{55}Mn = (4.02 \pm 0.26) \times 10^{-6}$, and an absolute Pb–Pb age of 4.56503 ± 0.00085 Ga. We will return to the interpretation of these data later.

2.4.5 Iron-60

^{60}Fe β-decays to ^{60}Ni with a half-life of 1.5 Myr. Unlike the other short-lived nuclides with half-lives of a few million years or less, and in particular contrast to ^{10}Be, ^{60}Fe is not produced by spallation because there are no suitable target elements, and therefore all of its solar system inventory must reflect recent stellar nucleosynthesis. The first plausible evidence for the existence of ^{60}Fe in the solar system was provided by ^{60}Ni excesses found in bulk samples of the eucrites Chervony Kut and Juvinas (Shukolyukov and Lugmair, 1993a, b). These are basaltic achondrites, the result of planetary-scale melting and differentiation (possibly on the asteroid Vesta) that fractionated nickel into the core. Thus, the excess ^{60}Ni cannot represent nucleogenetic isotope anomalies of the iron-group elements, as is seen in CAIs, and its presence in such a large volume material indicates wide-scale occurrence of ^{60}Fe in the solar system (Shukolyukov and Lugmair, 1993a). However, internal mineral isochrons could not be obtained on the eucrite samples because of element redistribution after the decay of ^{60}Fe (Shukolyukov and Lugmair, 1993b). Moreover, the inferred initial ^{60}Fe/^{56}Fe differs by an order of magnitude between these eucrites for which other isotopic systems (e.g., ^{53}Mn–^{53}Cr) indicate a similar formation age (Lugmair and Shukolyukov, 1998). These inconsistencies point out problems with interpreting eucrite ^{60}Fe/^{56}Fe abundances in chronologic terms and indicate that estimates of a solar system initial ^{60}Fe/^{56}Fe, based on an absolute age of eucrite formation, is likely subject to large systematic uncertainties.

Recent *in situ* measurements on high Fe/Ni phases in chondrites help to constrain this initial value. Tachibana and Huss (2003) found good correlations of excess ^{60}Ni with Fe/Ni ratios in sulfide minerals of the (LL3.1) unequilibrated ordinary chondrites Bishunpur and Krymka (Figure 6), which imply ^{60}Fe/^{56}Fe ratios of between 1.0×10^{-7} and 1.8×10^{-7}. Mostefaoui *et al.* (2005) reported similar good correlations in troilite from Semarkona (LL3.0) with a slope corresponding to ^{60}Fe/^{56}Fe $= (9.2 \pm 2.4) \times 10^{-7}$; a weighted regression through their data excluding data points they discounted gives ^{60}Fe/^{56}Fe $= (9.5 \pm 1.3) \times 10^{-7}$

(Figure 6). They also measured nickel isotopes in magnetite and found a correlation implying ^{60}Fe/^{56}Fe $= (1.1 \pm 0.4) \times 10^{-7}$. If the two correlations found by Mostefaoui *et al.* (2005) are isochrons, magnetite is ~5 Myr younger than troilite. Although it is somewhat ambiguous whether sulfides achieved isotopic closure in the solar nebula or on an asteroidal parent body, it is likely that they have suffered significantly less disturbance of their Fe–Ni isotopic system than have eucrites, making an extrapolation back to the time of CAI formation more robust. With plausible assumptions, Tachibana and Huss (2003) estimate $(^{60}$Fe/^{56}Fe$)_0$ for solar system formation between 1×10^{-7} and 6×10^{-7} with a probable value (depending on the age of the sulfides relative to CAIs) of (~3–4) $\times 10^{-7}$. Mostefaoui *et al.* (2005) preferred to simply consider their measured ^{60}Fe/^{56}Fe value of 9.2×10^{-7} as a lower limit to the solar system initial value. Further progress has come with the measurement by Tachibana *et al.* (2006) of nickel isotopes in ferromagnesian silicates in chondrules in Semarkona and Bishunpur. They found correlations of excess ^{60}Ni with Fe/Ni ratio consistent with ^{60}Fe/^{56}Fe $= (2$–$3.7) \times 10^{-7}$ (Figure 6). Since ^{26}Al–^{26}Mg chronometry indicates that these chondrules are 1.5–2 Myr older than CAIs, they estimated an initial solar system ^{60}Fe/^{56}Fe ratio of (5–10) $\times 10^{-7}$. These data are consistent with an upper limit of $(^{60}$Fe/^{56}Fe$)_0 = \sim 3.5 \times 10^{-7}$ derived from the analyses of nickel isotopes in FeO-rich olivine from a (LL3.0) Semarkona chondrule which exhibited $(^{26}$Al/^{27}Al$)_0 = 0.9 \times 10^{-5}$ (Kita *et al.*, 2000). The early solar system estimate of Tachibana *et al.* (2006) is significantly lower than

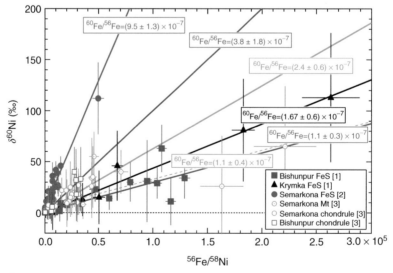

Figure 6 ^{60}Fe–^{60}Ni isochron diagram showing all chondrite data at this writing. Data for all chondrules, all troilite, or all magnetite grains in each meteorite have been grouped and weighted regressions calculated. There is a range of nearly a factor of 10 among the different isochrons, implying a time difference of ~5 Myr among formation ages of these various objects. References: [1] Tachibana and Huss (2003); [2] Mostefaoui *et al.* (2005); [3] Tachibana *et al.* (2006).

a value of $(^{60}Fe/^{56}Fe)_0 = (1.6 \pm 0.5) \times 10^{-6}$ inferred for an Allende CAI (Birck and Lugmair, 1988), indicating that the ^{60}Ni excesses in this CAI are probably of a nucleosynthetic origin and are not due to *in situ* decay of ^{60}Fe. It would be desirable to have a direct measure of a $^{60}Fe/^{56}Fe$ isochron in a CAI; however, as a volatile element, iron is generally depleted in refractory inclusions and samples containing appropriate mineralogy for this determination may not be found.

The $^{60}Fe-^{60}Ni$ system is beginning to be applied to differentiated meteorites again, more than 10 years after the pioneering work of Shukolyukov and Lugmair (1993a, b). Moynier *et al.* (2005) reported a correlation for iron meteorites corresponding to a remarkably high $^{60}Fe/^{56}Fe$ value of $(3.0 \pm 0.2) \times 10^{-6}$, but Cook *et al.* (2005, 2006) found no correlation between Fe/Ni and $^{60}Ni/^{58}Ni$ and ^{60}Ni excesses in some of the same samples. Cook *et al.* (2006) reported $^{60}Fe/^{56}Fe = (5.4 \pm 4.2) \times 10^{-7}$, which they did not take as evidence for live ^{60}Fe at the time iron meteorites formed. Bizzarro *et al.* (2006) also performed high-precision MC-ICPMS measurements of nickel isotopes in iron meteorites, finding small deficits of 0.02–0.03‰ in $\delta^{60}Ni$. Using Fe/Ni ratios calculated for initial liquid core compositions, they derive model initial $^{60}Fe/^{56}Fe$ ratios of $(1.48 \pm 0.87) \times 10^{-6}$ and $(1.09 \pm 0.14) \times 10^{-7}$ for IIAB and IIIAB iron meteorites, respectively. If $^{60}Fe/^{56}Fe$ was $\sim 1 \times 10^{-6}$ at the time of CAI formation, this implies core formation before CAI formation, perhaps by as much as 1 Myr. However, Bizzarro *et al.* (2006) also found that the iron meteorites they studied exhibited deficits in $\delta^{62}Ni$, which raises the question of whether the small $\delta^{60}Ni$ deficits could be nucleosynthetic in origin. All of this work is preliminary, but it is clear that further nickel isotopic studies of differentiated meteorites will yield interesting results. The interpretation, however, will require that our understanding of issues of accuracy and isotopic homogeneity of solar system reservoirs advance to a level commensurate with the precision offered by the new analytical methods.

2.4.6 Beryllium-10

^{10}Be β-decays to ^{10}B with a half-life of 1.51 Myr. Evidence for its former existence in the solar system is provided by excesses of $^{10}B/^{11}B$ correlated with Be/B ratio (Figure 7), first found within coarse-grained (type B) CAIs from Allende (McKeegan *et al.*, 2000a). From the slope of the correlation line, McKeegan *et al.* calculated an initial $^{10}Be/^{9}Be = (9.5 \pm 1.9) \times 10^{-4}$ at the time corresponding to isotopic closure of the Be–B system. This discovery was rapidly confirmed and extended by analyses of a variety of CAIs of types A and B, and a FUN inclusion from

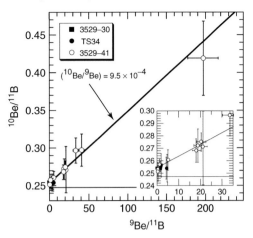

Figure 7 Boron isotopic composition of individual minerals from Allende CAIs as a function of Be/B ratio in the same material; error bars are 2σ. The $^{10}B/^{11}B$ values from various spots of CAI 3529-41 show ^{10}B excesses that are correlated with the Be/B ratio in a manner indicative of the *in situ* decay of ^{10}Be. The slope of the correlation line corresponds to an initial $^{10}Be/^{9}Be = (9.5 \pm 1.9) \times 10^{-4}$ at the time of crystallization. The intercept indicates $^{10}B/^{11}B = 0.254 \pm 0.002$, which is higher than $^{10}B/^{11}B$ for CI chondrites (shown by the horizontal line). Inset figure shows the same data at an expanded scale; data for CAIs 3529-30 and TS-34 are consistent with the Be–B isotope systematics of 3529-41. Data and figure reproduced from McKeegan *et al.* (2000a).

various CV3 chondrites, including Allende, Efremovka, Vigarano, Leoville, and Axtell (MacPherson and Huss, 2001; McKeegan *et al.*, 2001; Sugiura *et al.*, 2001; Chaussidon *et al.*, 2003; MacPherson *et al.*, 2003). The most robust Be–B isochron is that of the Allende CAI 3529-41 (Chaussidon *et al.*, 2006), which is based on 66 ion microprobe spot analyses. Chaussidon *et al.* (2006) reported a slope corresponding to $^{10}Be/^{9}Be = (8.8 \pm 0.6) \times 10^{-4}$, but in their regression, they did not weight data points by their uncertainties. A weighted fit yields $^{10}Be/^{9}Be = (1.038 \pm 0.092) \times 10^{-3}$, which makes this the highest precisely determined initial $^{10}Be/^{9}Be$ ratio found so far. Of the nearly two dozen CAIs that have been examined so far, in every case for which high Be/B ratios could be found in a sample (i.e., except where boron contamination is prevalent), excesses of $^{10}B/^{11}B$ are measured, implying that the existence of live ^{10}Be was rather widespread in the solar nebula, at least at the locale of CAI formation. Some spread in initial $^{10}Be/^{9}Be$ ratios is apparent, but overall it is remarkably uniform, especially considering the difficulties of the measurements and the susceptibility of samples to contamination by trace amounts of boron (cf., Chaussidon *et al.*, 1997). Calculated initial $^{10}Be/^{9}Be$ ratios for "normal" CV CAIs range only over a factor of 2 from $(\sim 4.5-10.0) \times 10^{-4}$, with no difference seen between type B CAIs (mean of 12 samples: ^{10}Be/^{9}Be $= (6.3 \pm 0.4) \times 10^{-4}$) and type A CAIs (mean of five samples: ^{10}Be/^{9}Be $= (6.7 \pm 0.6) \times 10^{-4}$). The one FUN inclusion measured, a type A from Axtell (MacPherson *et al.*, 2003), has the lowest initial ^{10}Be/^{9}Be $= (3.6 \pm 0.9) \times 10^{-4}$, but even this value is within error of the lower values measured on "normal" (i.e., non-FUN) CAIs. One CAI, Efremovka E44, has been measured independently in two laboratories with excellent agreement (McKeegan *et al.*, 2001; Sugiura *et al.*, 2001), indicating that potential systematic uncertainties are not significant compared with statistical errors. The initial boron isotopic composition (prior to any ^{10}Be decay) is the same among these various CAIs, with a small degree of relative scatter. However, the mean value, ^{10}B/^{11}B $= 0.250 \pm 0.001$, is distinct from a chondritic value ($= 0.248$) measured for CI chondrites (Zhai *et al.*, 1996).

The former presence of ^{10}Be was extended to another important class of refractory objects, hibonite from the CM2 Murchison meteorite (Marhas *et al.*, 2002). Hibonite [CaAl$_{12-2x}$ (Mg$_x$Ti$_x$)O$_{19}$] is one of the most refractory minerals calculated to condense from a gas of solar composition, and is known to host numerous isotopic anomalies, especially in the heavy isotopes of calcium and titanium (Ireland *et al.*, 1985; Zinner *et al.*, 1986; Fahey *et al.*, 1987). Curiously, when these anomalies are of an exceptionally large magnitude (in the ~several to 10% range), the hibonite grains show a distinct lack of evidence for having formed with ^{26}Al (e.g., Ireland, 1988, 1990) or ^{41}Ca (Sahijpal *et al.*, 1998, 2000). Marhas *et al.* (2002) found excesses of ^{10}B/^{11}B in three such hibonite grains that are each devoid of either ^{26}Mg* or ^{41}K* from the decay of ^{26}Al and ^{41}Ca, respectively. Collectively, the Be–B data imply ^{10}Be/^{9}Be $= (5.2 \pm 2.8) \times 10^{-4}$ when these hibonites formed. This initial ^{10}Be/^{9}Be is in the same range as for other refractory inclusions and indicates that existence of ^{10}Be is decoupled from the other two short-lived nuclides that partition into refractory objects, namely ^{26}Al and ^{41}Ca. Even more striking evidence for decoupling of the ^{26}Al–^{26}Mg and ^{10}Be–^{10}B systems came with the report of Marhas and Goswami (2003) that hibonite in the well-known FUN CAI HAL had an initial ^{10}Be/^{9}Be ratio in the same range as other CAIs, yet had an initial ^{26}Al/^{27}Al ratio three orders of magnitude lower than the canonical early solar system ratio. The significance of this lack of correlation, for both chronology and source of radionuclides, is discussed further below.

Convincing evidence of live ^{10}Be has so far only been found in refractory inclusions because these samples exhibit large volatility-controlled Be–B fractionation. A tantalizing hint for ^{10}Be was found in one anorthite-rich chondrule from a highly unequilibrated (CO3) chondrite: the Be–B

correlation diagram displays a large amount of scatter, but an initial ^{10}Be/^{9}Be ratio of $7.2 \pm 2.9 \times 10^{-4}$ may be calculated (Sugiura, 2001). This value is similar to that seen in CAIs, but needs to be confirmed by further measurements.

2.4.7 Manganese-53

^{53}Mn decays by electron capture to ^{53}Cr with a half-life of 3.7 Myr. This relatively long half-life, and the fact that manganese and chromium are reasonably abundant elements that undergo relative fractionation in evaporation/condensation processes as well as magmatic processes, make the ^{53}Mn–^{53}Cr system particularly interesting for bridging the time period from nebular events to accretion and differentiation of early-formed planetesimals. Accordingly, this system has been intensively investigated and evidence of live ^{53}Mn has now been found in nebular components such as (1) CAIs (Birck and Allègre, 1985, 1988; Papanastassiou *et al.*, 2002) and (2) chondrules (Nyquist *et al.*, 2001a), as well as (3) bulk ordinary chondrites (Nyquist *et al.*, 2001a; Lugmair and Shukolyukov, 1998), (4) bulk carbonaceous chondrites (Birck *et al.*, 1999), (5) CI carbonates (Endress *et al.*, 1996; Hutcheon and Phinney, 1996; Hutcheon *et al.*, 1999b), (6) enstatite chondrite sulfides (Wadhwa *et al.*, 1997), and (7) various achondrites including angrites, eucrites, diogenites, pallasites, and SNC meteorites (Lugmair and Shukolyukov, 1998; Nyquist *et al.*, 2001b, 2003). Owing to the wealth of high-quality data, an impressively detailed high-resolution relative chronometry can be developed (e.g., Lugmair and Shukolyukov, 2001), however interpretation of the ^{53}Mn–^{53}Cr system with respect to other chronometers is complex, particularly with respect to nebular events. The primary reasons for these complexities are difficulty in evaluating the initial ^{53}Mn/^{55}Mn of the solar system and in establishing its homogeneity in the nebula (see discussions in Birck *et al.*, 1999; Lugmair and Shukolyukov, 2001; and Nyquist *et al.*, 2001a).

As with ^{26}Al, ^{41}Ca, and ^{10}Be, the obvious samples in which to try to establish the solar system initial value for ^{53}Mn/^{55}Mn are CAIs. However, in this case there are three factors which work against this goal: (1) volatility-controlled fractionation is not favorable when the parent (^{53}Mn) is more volatile than the daughter (^{53}Cr); (2) both manganese and chromium are moderately volatile elements and significantly depleted in CAIs; and (3) the daughter element is known to exhibit nucleogenetic anomalies in most CAIs (e.g., Papanastassiou, 1986). Together, these properties mean that there are no mineral phases with large Mn/Cr in CAIs, and it is not feasible to find large ^{53}Cr excesses that are uniquely and fully attributable to ^{53}Mn decay. Birck and Allègre (1988) first demonstrated the *in situ* decay of ^{53}Mn by correlating ^{53}Cr excesses with Mn/Cr in mineral separates of an Allende inclusion, deriving an initial ^{53}Mn/^{55}Mn = $(3.7 \pm 1.2) \times 10^{-5}$. Comparison with other Allende CAIs led these authors to estimate $\sim 4.4 \times 10^{-5}$ as the best initial ^{53}Mn/^{55}Mn for CAIs; however, Nyquist *et al.* (2001a) prefer a somewhat lower value $(2.8 \pm 0.3) \times 10^{-5}$ based on the same mineral separate analyses plus consideration of nonradiogenic chromium in a spinel separate from an Efremovka CAI. In recent work, Birck *et al.* (1999) have emphasized that refractory inclusions are inconsistent with the solar system evolution of the ^{53}Mn–^{53}Cr system, noting that the inferred chronology is necessarily model-dependent. Lugmair and Shukolyukov (1998) reach a similar assessment, describing the "chronological meaning of ^{53}Mn/^{55}Mn ratios in CAIs" as "tentative." Papanastassiou *et al.* (2002) also studied Mn–Cr systematics of CAIs and concluded that although spinel preserved the initial ^{53}Cr/^{52}Cr ratio, manganese with live ^{53}Mn was introduced during secondary alteration, so it was not clear what event was being dated in CAIs.

Whole chondrule Mn–Cr isochrons (Figure 8) have been reported for the ordinary chondrites Chainpur (LL3.4) and Bishunpur (LL3.1) by Nyquist *et al.* (2001a). The chondrules from both meteorites are consistent with a single isochron with $(^{53}$Mn/^{55}Mn$)_0 = (8.8 \pm 1.9) \times 10^{-6}$ and an intercept $\varepsilon(^{53}$Cr$) = -0.03 \pm 0.06$ (Figure 5). If the chondrule data are considered with Mn–Cr data for whole chondrites (Nyquist *et al.*, 2001a), then the slope increases slightly to $(^{53}$Mn/^{55}Mn$)_0 = (9.5 \pm 1.7) \times 10^{-6}$, which Nyquist and colleagues interpret as reflecting the time of Mn/Cr fractionation during the condensation of chondrule precursors. If this occurred in the same nebular environments as CAI mineral condensation characterized by the preferred $(^{53}$Mn/^{55}Mn$)_0 = 2.8 \times 10^{-5}$, this implies a time difference of 5.8 ± 2.7 Myr. This is significantly longer than the CAI-chondrule timescale inferred from ^{26}Al/^{27}Al (also for Bishunpur chondrules); however, it is not clear that the two chronometers are dating the same events (see discussion in Nyquist *et al.*, 2001a).

A more straightforward interpretation of Mn–Cr ages can, in principle, be achieved for planetary differentiates since these certainly homogenized chromium isotopes during melting and also likely underwent Mn/Cr fractionation at a well-defined nebular locale (the asteroid belt). Although Lugmair and Shukolyukov (1998) have argued for heterogeneity of ^{53}Mn/^{55}Mn as a function of heliocentric distance, such effects would be negligible considered over the probable distances of

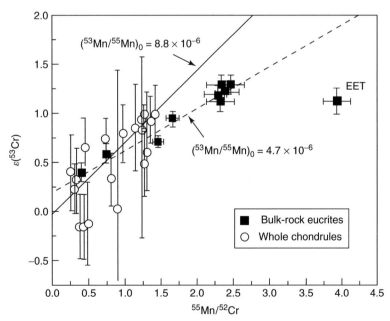

Figure 8 ^{53}Mn–^{53}Cr evolution diagram for nebular components (whole chondrules from ordinary chondrites Bishunpur and Chainpur; Nyquist *et al.*, 2001) and for planetary differentiates (whole-rock eucrites; Lugmair and Shukolyukov, 1998). Plotted are measured values of $\varepsilon(^{53}$Cr$)$, the deviation of ^{53}Cr/^{52}Cr in a sample from the terrestrial standard value in parts per 10^4, as a function of ^{55}Mn/^{52}Cr. The correlation is interpreted as an isochron indicating the *in situ* decay of ^{53}Mn; the slope for the eucrites (dashed line) corresponds to an initial ^{53}Mn/^{55}Mn = $(4.7 \pm 0.5) \times 10^{-6}$ and that for chondrules (solid line) indicates $(^{53}$Mn/^{55}Mn$)_0 = (8.8 \pm 1.9) \times 10^{-6}$, implying that Mn/Cr fractionation in chondrule precursors preceded global fractionation of the eucrite parent body by approximately one half-life, or ~ 3.5 Myr. All data are replotted from Lugmair and Shukolyukov (1998) and Nyquist *et al.* (2001a); 2σ error bars are indicated and the datum for EET87520 is excluded from the fit for the eucrite whole-rock isochron.

formation for the asteroids (meteorite parent bodies). The rapidly cooled angrites provide the anchor point between ^{53}Mn–^{53}Cr and the absolute age determined by Pb–Pb since both isotopic systems should have closed contemporaneously (Lugmair and Shukolyukov, 1998). The olivine fraction of LEW has a high Mn/Cr and thus provides a good precision for the isochron, with $^{53}Mn/^{55}Mn = (1.25 \pm 0.07) \times 10^{-6}$ and $\varepsilon(^{53}Cr) = +0.40 \pm 0.16$ (Lugmair and Shukolyukov, 1998), which is tied to the Pb–Pb age of $4{,}557.8 \pm 0.5$ Ma (Lugmair and Galer, 1992).

As alluded to above in the discussion of absolute ages of differentiated objects, the eucrites have suffered a more prolonged and complex thermal and shock history, which is reflected in their internal ^{53}Mn–^{53}Cr systematics. Despite this, excesses of ^{53}Cr in bulk samples of eucrites are well correlated with Mn/Cr (Figure 8) indicating large-scale differentiation on the eucrite parent body prior to the decay of ^{53}Mn (Lugmair and Shukolyukov, 1998). The slope of the correlation line yields $^{53}Mn/^{55}Mn = (4.7 \pm 0.5) \times 10^{-6}$, which is nearly two half-lives of ^{53}Mn steeper (older) than the 1.25×10^{-6} value obtained for angrites. Thus, these data indicate that the parent asteroid of the eucrites (Vesta?) was totally molten, probably during mantle–core differentiation, at 7.1 ± 0.8 Ma prior to the crystallization of angrite LEW. By calibration with the absolute Pb–Pb chronology of angrites, this indicates igneous differentiation of the eucrite parent body at $4{,}564.8 \pm 0.9$ Ma (Lugmair and Shukolyukov, 1998). It should be clear that this time does not necessarily represent the crystallization age of individual eucrite meteorites, but the last time of global chromium isotope equilibration and Mn/Cr fractionation. In fact, internal ^{53}Mn–^{53}Cr isochrons for individual cumulate and noncumulate eucrites show a range of apparent $^{53}Mn/^{55}Mn$ values, from close to the global fractionation event (e.g., 3.7×10^{-6} for Chervony Kut) to essentially "dead" ^{53}Mn (e.g., Caldera; Wadhwa and Lugmair, 1996). It is not certain whether these ages, especially the young ones, reflect prolonged igneous activity over a period of tens of millions of years, or cooling ages, or disturbance of the Mn–Cr system by impacts, or some combination of the above (Lugmair and Shukolyukov, 1998). The ^{53}Mn–^{53}Cr ages for individual eucrites do not correlate particularly well with Pb–Pb ages, for example, Chervony Kut with an $^{53}Mn/^{55}Mn$ initial ratio indicating isotopic closure at $\sim 4{,}564$ Ma (almost contemporaneous with mantle differentiation) has a Pb–Pb age of $4{,}312.6 \pm 1.6$ Ma (Galer and Lugmair, 1996). This discrepancy can be attributed to the U–Pb system being more easily disturbed than Mn–Cr (Lugmair and Shukolyukov, 1998); however, as discussed in more detail by Tera and Carlson (1999), it also means that the eucrites cannot serve as an independent check on the validity of coupling ^{53}Mn–^{53}Cr model ages to an absolute timescale based on the Pb–Pb ages of angrites.

Chromium has four stable isotopes, ^{50}Cr, ^{52}Cr, ^{53}Cr, and ^{54}Cr. Mass fractionation in isotopic measurements can alter $^{53}Cr/^{52}Cr$ ratios, but these are corrected by normalizing measured data for all isotopes to the terrestrial $^{50}Cr/^{52}Cr$ ratio. Under the assumption that there are no nucleosynthetic anomalies in ^{54}Cr, Lugmair and colleagues routinely use the small deviations in the mass fractionation-corrected $^{54}Cr/^{52}Cr$ ratio to make a "second-order" correction on $^{53}Cr/^{52}Cr$ (e.g., Lugmair and Shukolyukov, 1998). It has been known for some time that CI and CM chondrites have nucleosynthetic ^{54}Cr anomalies (Rotaru et al., 1992; Podosek et al., 1997). This was recognized for carbonaceous chondrites and ^{54}Cr anomalies were even recognized at the Cretaceous-Tertiary boundary (Shukolyukov and Lugmair, 1998). Trinquier et al. (2005a, b) have recently analyzed a number of meteorites, correcting chromium isotopic data only for mass fractionation using $^{50}Cr/^{52}Cr$ and found that eucrites, diogenites, mesosiderites, and pallasites have a uniform deficit in ^{54}Cr of 0.73 ± 0.02 ε units (parts in 10^4) and that carbonaceous chondrites are enriched in ^{54}Cr by 0.6–1.5ε. Trinquier et al. (2005b) reported a new ^{53}Mn–^{53}Cr isochron for basaltic achondrites without making a second-order correction. Their slope corresponded to $^{53}Mn/^{55}Mn = (4.53 \pm 0.17) \times 10^{-5}$, in excellent agreement with Lugmair and Shukolyukov (1998), but the intercept was at $\varepsilon^{53}Cr = -0.15 \pm 0.06$, rather than $+0.25$, a shift of 0.4ε. Lugmair and Shukolyukov (1998) had asserted that a range in intercepts implied a radial gradient in $^{53}Mn/^{55}Mn$ in the early solar system, but the new data of Trinquier et al. (2005a, b) are consistent with $^{53}Mn/^{55}Mn$ being homogeneous in the early solar system.

The ^{53}Mn–^{53}Cr system has also proved useful in constraining the timescales of earliest aqueous activity on the parent bodies of some carbonaceous chondrites by dating Mn/Cr fractionation associated with the formation of aqueously precipitated minerals. Carbonates from the CI chondrites Orgueil and Ivuna show very large ^{53}Cr excesses correlated with Mn/Cr; inferred initial $^{53}Mn/^{55}Mn$ ratios range from 1.42×10^{-6} to 1.99×10^{-6} (Endress et al., 1996). Carbonates from other carbonaceous chondrites show a wider range extending to significantly higher initial $^{53}Mn/^{55}Mn$ ratios: $(6.4 \pm 1.2) \times 10^{-6}$ in CM chondrites Nogoya and Y791198, and $(9.4 \pm 1.6) \times 10^{-6}$ in the unusual carbonaceous chondrite Kaidun (Hutcheon et al., 1999a, b). The latter values are similar to $^{53}Mn/^{55}Mn$ found in ordinary chondrite chondrules (Nyquist et al., 2001a). Fayalite (FeO-rich olivine) from the Mokoia oxidized and aqueously altered CV3 chondrite formed with very high $^{55}Mn/^{52}Cr$ ratios ($>10^4$) and exhibits

^{53}Mn/^{55}Mn = (2.32 ± 0.18) × 10^{-6} (Hutcheon *et al.*, 1998), similar to CI carbonates and eucrites. Mn–Cr data for fayalite from the Kaba chondrite yields the same ^{53}Mn/^{55}Mn within uncertainty (Hua *et al.*, 2002).

2.4.8 Palladium-107

^{107}Pd β-decays to ^{107}Ag with a half-life of 6.5 Myr. Evidence for this now-extinct nuclide is found in metallic phases of iron meteorites, since large Pd/Ag fractionations occur during magmatic partitioning of metal (Kelly and Wasserburg, 1978; see also review by Wasserburg, 1985). Kaiser and Wasserburg (1983) demonstrated that a linear correlation exists between excess ^{107}Ag/^{109}Ag and Pd/Ag in different fractions of metal and sulfide from the group IIIB iron meteorite Grant and from the isochron inferred an initial ^{107}Pd/^{108}Pd = ~1.7 × 10^{-5} at the time of crystallization of this meteorite. Extrapolation back to the time of CAI formation would yield an initial ^{107}Pd/^{108}Pd of approximately twice this value for the solar system, though with considerable uncertainty. Further isochrons were determining in other many iron and stony-iron meteorites, showing that there is a wide range of initial ^{107}Pd/^{108}Pd ratios, but that many samples have ratios in the range (1.5–2.5) × 10^{-5} (Chen and Wasserburg, 1996; Chen *et al.*, 2002). Recently, Carlson and Hauri (2001) have developed ICPMS methods for determining silver isotope ratios with high precision, thus permitting the investigation of phases with more moderate Pd/Ag fractionation. They found good isochrons for the pallasite (stony-iron) Brenham and the IIIB iron Grant, both with inferred initial ^{107}Pd/^{108}Pd = 1.6 × 10^{-5}. A two-point correlation between metal and sulfide was also determined for Canyon Diablo (group IA iron), yielding an apparent initial ^{107}Pd/^{108}Pd essentially identical to that previously found for Gibeon (Chen and Wasserburg, 1990). Interpreted chronologically, the data imply that Brenham and Grant formed some 3.5 Myr following Canyon Diablo and Gibeon. Small (5ε) ^{107}Ag/^{109}Ag anomalies were also documented for the carbonaceous chondrite Allende (Carlson and Hauri, 2001), which, given its relatively low Pd/Ag content, would imply an enormous initial ^{107}Pd/^{108}Pd (~39 × 10^{-5}) if this anomaly had evolved from the most unradiogenic sample (Canyon Diablo sulfide) due to ^{107}Pd decay only. However, no internal isochron is obtained for Allende and considering its unequilibrated nature (i.e., it hosts many isotopic anomalies) there is no compelling reason to assume that this value represents a solar nebular abundance of live ^{107}Pd.

2.4.9 Hafnium-182

^{182}Hf β-decays to ^{182}W with a half-life of 9 Myr. This has been recognized as an extremely important isotopic system in recent years (e.g., Lee and Halliday, 1996; Halliday and Lee, 1999) because it is almost uniquely sensitive to metal–silicate fractionation and its rather long half-life makes it a useful probe for both nebular and planetary processes. Specifically, tungsten is highly siderophile whereas hafnium is retained in silicates during melting and metal segregation. Thus, tungsten isotope compositions could be very different in silicates and metal from distinct planetary objects depending on whether or not metal/silicate fractionation in those objects predated significant decay of ^{182}Hf. Internal isochrons, demonstrating good correlations of ^{182}W/^{180}W with Hf/W, are found for several separates of ordinary chondrites (Kleine *et al.*, 2002a, b; Yin *et al.*, 2002); samples of whole-rock carbonaceous chondrites and a CAI from Allende also fall within error of these isochrons (Yin *et al.*, 2002). The Pb–Pb ages of phosphates in the ordinary chondrites (Kleine *et al.*, 2002a) and the coincidence of the CAI data (Yin *et al.*, 2002) allow a robust estimate of the initial ^{182}Hf/^{180}Hf of the solar system of 1.0–1.1 × 10^{-4} with an initial ^{182}W/^{180}W significantly (~−3ε) lower than terrestrial mantle samples. A regression through data for two bulk CAIs, several fragments of a single CAI, and bulk carbonaceous chondrites yields the most robust currently available early solar system ^{182}Hf/^{180}Hf value, (1.07 ± 0.10) × 10^{-4}, and ε^{182}W value, −3.47 (Kleine *et al.*, 2005).

Iron meteorites have ε^{182}W values similar or below the early solar system estimated value of ε^{182}W = −3.47 (Kleine *et al.*, 2005a; Markowski *et al.*, 2006; Qin *et al.*, 2006). The values below the early solar system value apparently result from cosmic ray exposure effects, but it does appear that a number of iron meteorites have the same tungsten isotopic composition as the early solar system. This implies that metal–silicate segregation occurred no later than 1 Myr after the formation of CAIs (Kleine *et al.*, 2005a; Markowski *et al.*, 2006; Qin *et al.*, 2006). Tungsten isotopes have also been used to show that most eucrites experienced a thermal event 16 ± 2 Myr after mantle–crust differentiation in the eucrite parent body (Kleine *et al.*, 2005b).

The meaning of tungsten isotopes with regard to timescales of accretion and core formation of the Earth and formation of the Moon is discussed in Chapter 3.

2.4.10 Iodine-129

^{129}I β-decays to ^{129}Xe with a half-life of 15.7 Myr. As mentioned in the historical introduction (Section 2.1.3), ^{129}I was the first extinct isotope whose presence in the early solar system was inferred from excesses of its daughter ^{129}Xe in meteorites (Jeffery and Reynolds, 1961). Both parent and daughter are mobile elements, and coupled with the relatively long half-life, this means that closure effects on the I–Xe system likely limit its utility to parent-body processes (e.g., Swindle *et al.*, 1996), although arguments have been advanced that I–Xe can date nebular events in favorable circumstances (Whitby *et al.*, 2001). New analytical techniques that enable the investigation of single mineral phases (Gilmour, 2000; Gilmour and Saxton, 2001) have helped in the understanding of apparent I–Xe isochrons (as differentiated from mixing lines of multiple phases) and enabled more confident chronological interpretations, particularly of secondary mineral phases formed on asteroidal parent bodies. Brazzle *et al.* (1999) demonstrated concordancy between I–Xe and Pb–Pb chronometers for chondrite phosphates over a timescale of tens of millions of years. At another extreme, Whitby *et al.* (2000) found an initial ratio of $^{129}I/^{127}I = (1.35 \pm 0.05) \times 10^{-4}$ in halite from a relatively unequilibrated ordinary chondrite. This result is close to the estimated initial value for the solar system ($\sim 10^{-4}$), implying that the aqueous activity responsible for precipitating the halite occurred immediately upon accretion, probably within a few million years of CAI formation (Whitby *et al.*, 2000).

2.4.11   Lead-205

^{205}Pb decays by electron capture to ^{205}Tl with a half-life of 17.3 Myr. It is unique among the short-lived radionuclides present in the early solar system in being produced only by s-process nucleosynthesis. Nielsen *et al.* (2006) reported a correlation between $^{205}Tl/^{203}Tl$ and $^{204}Pb/^{203}Tl$ ratios among metal and troilite from the IAB iron meteorites Toluca and Canyon Diablo that was consistent with $^{205}Pb/^{204}Pb = (7.4 \pm 1.0) \times 10^{-5}$. The range in $^{205}Tl/^{203}Tl$ is $\sim 5‰$ and Nielsen *et al.* (2006) conclude that mixing of mass-fractionated components is unlikely to be the cause of this variation. Nielsen *et al.* (2006) used the I–Xe age of IAB silicate inclusions to calculate an early solar system $^{205}Pb/^{204}Pb$ value of $(1.0–2.1) \times 10^{-4}$.

2.4.12   Niobium-92

^{92}Nb decays by electron capture to ^{92}Zr with a half-life of 36 Ma. ^{92}Nb is a p-process nuclide. The first hint that this isotope was present in the early solar system was based on an $8.8 \pm 1.7\varepsilon$ excess in ^{92}Zr in a niobium-rich rutile grain from the Toluca IAB iron meteorite (Harper, 1996). This corresponded to an initial $^{92}Nb/^{93}Nb$ ratio of $(1.6 \pm 0.3) \times 10^{-5}$, but the time of formation of Toluca rutile is not known. Three subsequent studies that used MC-ICPMS to measure zirconium isotopic composition reported that the initial solar system $^{92}Nb/^{93}Nb$ was $\sim 10^{-3}$, higher by two orders of magnitude (Yin *et al.*, 2000; Münker *et al.*, 2000;

Sanloup *et al.*, 2000). This initial $^{92}Nb/^{93}Nb$ was nearly one quarter of the p-process production ratio (Harper, 1996) and was difficult to understand, as most ^{93}Nb is made by the s-process. The situation was resolved with the work of Schönbachler *et al.* (2002), who reported internal Nb–Zr isochrons for the Estacado H6 chondrite and for a clast from the Vaca Muerta mesosiderite, both of which give an initial solar system $^{92}Nb/^{93}Nb$ of $\sim 10^{-5}$, a much more plausible value in terms of nucleosynthetic considerations. This lower initial ratio limits the utility of the ^{92}Nb–^{92}Zr for chronometry (see Chapter 3 for further discussion).

2.4.13   Plutonium-244 and Samarium-146

These relatively long-lived isotopes are mentioned here for completeness since both have been shown to have existed in the early solar system. However, neither ^{244}Pu nor ^{146}Sm has been developed for chronological applications, for very practical reasons. ^{244}Pu suffers from the fact that there are no long-lived isotopes of plutonium against which to normalize its abundance, and its primary application in meteorite studies is for obtaining cooling rates from the annealing of fission tracks in appropriate minerals. The half-life of ^{146}Sm (103 Myr) is too long and its abundance and relative fractionation from daughter ^{142}Nd are insufficient for it to constitute a useful chronometer for early solar system processes. Its primary interest is for nuclear astrophysics (e.g., Prinzhofer *et al.*, 1989), because this isotope is on the neutron-deficient side of the valley of β-stability. Interested readers are referred to Stewart *et al.* (1994) and review by Podosek and Swindle (1988) and Wasserburg (1985) for more information.

2.5   ORIGINS OF THE SHORT-LIVED NUCLIDES IN THE EARLY SOLAR SYSTEM

The ability of short-lived radioisotopes to function as chronometers for the early solar system is critically dependent on there having been an initially uniform distribution of the radioactivity throughout the nebula, or at least in those regions from which meteoritic components are derived. Only in this circumstance can differences in initial abundances of a radionuclide compared with a stable counterpart, as inferred by the excesses of the respective daughter isotope, be interpreted as due to radioactive decay from the initial inventory. The homogeneity of the distribution of radionuclides in the solar nebula depends, in turn, on the processes that created those isotopes some time before the formation of early solar system materials. For the longer-lived isotopes listed in Table 1 (e.g., ^{182}Hf, ^{129}I, ^{205}Pb, ^{92}Nb, ^{146}Sm, and ^{244}Pu), continuous nucleosynthesis may have been sufficient to produce a quasiequilibrium abundance of these species that was inherited by the solar nebula. However, the shorter half-life isotopes require a more immediate source (e.g., Meyer and Clayton, 2000; Wasserburg et al., 1996).

In principle, new (radioactive) isotopes could have been created by nuclear processes within the solar nebula itself, or they could have originated from sources external to the nebula. In the latter case, the most likely source is stellar nucleosynthesis in the interiors of nearby mass-losing stars (e.g., Cameron, 2001a, b; Cameron et al., 1995; Wasserburg et al., 1994, 1996, 1998), although spallation reactions in the molecular cloud parental to the solar nebula are also a possibility. If short-lived radioactivity is produced locally, for example, by spallation reactions with nuclear particles (protons and alphas) accelerated by interaction with an active young Sun (e.g., Gounelle et al., 2001; Lee et al., 1998), then it is unlikely that the products of those reactions will be distributed uniformly throughout the accretion disk. Homogeneity over nebular scale-lengths is much more likely for an "external seeding" scenario, although even in this case strong isotopic heterogeneity is possible at the very early stages following injection, before local mixing can act to smooth out the memory of the particular mechanism for "contamination" of the nebula by the new isotopes. The injection of radioactive stellar debris in a "triggered" collapse scenario for solar system formation is reviewed by Boss and Vanhala (2001); later, we consider the possible implications of this model for understanding isotopic heterogeneities in certain refractory inclusions.

The possible stellar sources of the short-lived isotopes, as well as constraints on nuclear spallation processes that could have produced them, are reviewed in detail by Goswami and Vanhala (2000). Since that work, three new developments have occurred: the discovery of evidence for live ^{10}Be in CAIs (McKeegan et al., 2000a); the observation of in situ ^{60}Fe decay in chondrites (Tachibana and Huss, 2003; Mostefaoui et al., 2005; Tachibana et al., 2006) that leads to a factor of ~20 increase in the estimated $(^{60}$Fe/^{56}Fe$)_0$ for the solar system initial; and the observation of in situ ^{36}Cl decay in secondary alteration phases in CAIs (Lin et al., 2005; Hsu et al., 2006). These isotopes are particularly significant because their respective modes of origin are much more tightly constrained than those of the other extinct nuclides. ^{10}Be is not produced by stellar nucleosynthesis, thus its existence in the early solar system is strong evidence for a spallogenic source of some short-lived nuclides. The amount of ^{36}Cl inferred for the early solar system is higher than is plausible for stellar sources, implying a late episode of irradiation in the early solar system

(Hsu et al., 2006). However, ^{60}Fe is not produced by spallation reactions, but it is produced in core-collapse supernovae and in intermediate mass asymptotic giant branch (AGB) stars (Wasserburg et al., 1994, 2006; Busso et al., 1999). The existence of ^{60}Fe in the relatively high abundance of $(5–10) \times 10^{-7}$ is therefore compelling evidence that stellar debris seeded the early solar system with new radioactivity. A recently proposed hypothesis considers that the source of spallogenic ^{10}Be is actually magnetically trapped cosmic rays in the interstellar medium prior to the collapse of a molecular cloud to form the solar system (Desch et al., 2004). An alternative model considers ^{10}Be to be produced during supernova explosions (Cameron, 2001a, b), but there are problems in coproducing ^{10}Be with other short-lived isotopes (see below). Both modes of origin are doubtful and would be firmly ruled out if the existence of live ^{7}Be in early solar system objects, as suggested by the analyses of an Allende CAI by Chaussidon et al. (2006), can be corroborated by Li–Be–B studies of other samples. The abundance of ^{10}Be in CAIs is consistent with expectations based on observations of X-ray luminosity in young, solar-like stars (Feigelson et al., 2002a, b; Preibisch et al., 2005) and models of particle acceleration due to magnetic flare activity near the protosun (Lee et al., 1998; Leya et al., 2003). In summary, the most likely scenario implied by the new meteoritic data is that the overall inventory of extinct nuclides contained both a spallogenic component, probably produced locally, and a nucleogenetic component, probably produced in a supernova, although contributions from AGB and other rapidly evolving mass-losing stars are also possible. Supernovae are often associated with star-forming regions (e.g., Hester et al., 2004; Ouellette et al., 2005), whereas low- and intermediate-mass AGB stars take much longer to evolve and are more or less randomly distributed relative to star-forming regions. Busso et al. (1999) and Wasserburg et al. (2006) have considered these issues in detail.

Although ^{10}Be and ^{60}Fe are interesting isotopes for delimiting possible origins of short-lived radioactivity, it is ^{41}Ca, ^{26}Al, and ^{53}Mn that are potentially most useful for chronology. Thus, a key task is to sort out, quantitatively, what sources are responsible for these isotopes in the early solar system. This can be addressed theoretically for both stellar and spallogenic sources; however, a clear consensus is lacking (e.g., Goswami et al., 2001; Gounelle et al., 2001; Leya et al., 2003) since production models can be tweaked by adjustable parameters (e.g., energy spectrum and target compositions) that are poorly constrained by observation. Another approach is to examine the isotopic record in meteoritic components for correlations that may indicate common sources (and distributions) for these nuclides.

The refractory inclusions provide the best samples since they incorporated all three of these radioisotopes as well as ^{10}Be. It has already been mentioned that ^{41}Ca and ^{26}Al are highly correlated in CAIs and hibonite grains (Figure 2). At face value, this would imply the same source for both these refractory elements. A problem with ^{41}Ca, however, is that its abundance is only marginally above detection limits and it decays very quickly, so that there is essentially no chance to test for concordant decay between the ^{41}Ca and ^{26}Al systems. This is not the case for ^{26}Al and ^{10}Be, which exist in much higher abundances and which have half-lives that differ by only a factor of 2.

The initial ^{26}Al/^{27}Al and ^{10}Be/^{9}Be values have been measured in a variety of refractory phases from both CV and CM carbonaceous chondrites (Figure 9). "Normal" CAIs of both petrologic types A and B have inferred ^{26}Al/^{27}Al values that plot within error of $\sim 5 \times 10^{-5}$ that is typical for CAI crystallization; even for cases where the Al–Mg system is disturbed in anorthite, other phases in the inclusion plot near this value (e.g., Sugiura *et al.*, 2001). As noted above, initial ^{10}Be/^{9}Be ratios for "normal" CV CAIs also show no discrimination based on petrology and the total range covered is approximately a factor of 2, which is only marginally outside of experimental uncertainty. Thus, for normal CAIs it is difficult to claim that the two isotopic systems are definitively discordant since the resolution of the data is not quite good enough.

However, the situation is different when one considers hibonites-rich and FUN inclusions (Figure 9). For most of these objects only an upper limit on initial ^{26}Al/^{27}Al ($< \sim 10^{-5}$) is obtained, yet they have initial ^{10}Be/^{9}Be similar to most of the other refractory inclusions (MacPherson *et al.*, 2003; Marhas *et al.*, 2002; Marhas and Goswami, 2003). The data are still not completely convincing until one includes the famous FUN inclusion "HAL" (Lee *et al.*, 1978, 1980). Recent analyses by Marhas and Goswami (2003) demonstrate that this hibonite-rich Allende CAI has ^{10}B/^{11}B

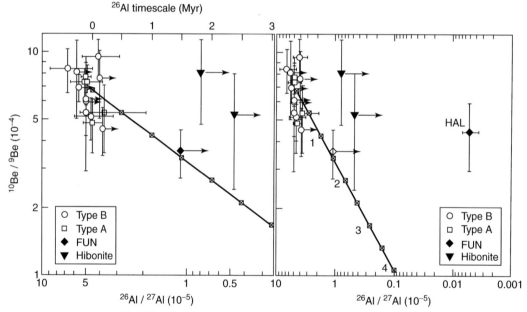

Figure 9 Inferred initial ^{26}Al/^{27}Al versus initial ^{10}Be/^{9}Be for refractory inclusions in CM and CV carbonaceous chondrites. All data are plotted with 2σ errors; upper limits are indicated by arrows. The locus of concordant ages by free decay from assumed solar system initial values of ^{26}Al/^{27}Al $= 4.5 \times 10^{-5}$ and ^{10}Be/^{9}Be $= 6.7 \times 10^{-4}$ is shown by the heavy line with 0.5 Myr tick marks. Left panel: the ^{26}Al/^{27}Al timescale is also shown on the top axis. It may be seen that "normal" CAIs of both petrologic types A and B have maximal ^{26}Al/^{27}Al values that plot within error of the "canonical" solar system initial, but the FUN inclusion (Axtell 2771; MacPherson *et al.*, 2003) and the CM hibonite grains (Marhas *et al.*, 2002) are depleted in ^{26}Al, with upper limits $< \sim 10^{-5}$. For nearly half of the type B CAIs, the Al–Mg system shows evidence of secondary disturbance; in these cases the maximum inferred ^{26}Al/^{27}Al is plotted as an upper limit (i.e., the inclusions are assumed to have formed with close to these values). With this approximation, the normal CAIs are relatively tightly clustered in ^{26}Al/^{27}Al, but show a range of approximately a factor 2 in ^{10}Be/^{9}Be, which is resolved at the 2σ level for several cases. Right panel: expanded scale showing new data from Marhas and Goswami (2003). In contrast to other FUN inclusions, HAL shows resolved ^{26}Mg excesses (Fahey *et al.*, 1987) implying a very low initial ^{26}Al/^{27}Al $= (5.2 \pm 1.7) \times 10^{-8}$, but it also has ^{10}Be/^{9}Be similar to other refractory inclusions (Marhas and Goswami, 2003), demonstrating that ^{10}Be and ^{26}Al are decoupled. Data sources: Fahey *et al.* (1987), Podosek *et al.* (1991), McKeegan *et al.* (2000a, 2001), Srinivasan (2001), Sugiura *et al.* (2001), MacPherson *et al.* (2003), Marhas *et al.* (2002), and Marhas and Goswami (2003).

excesses that imply $^{10}Be/^9Be = {\sim}4\times10^{-4}$, close to that of other CAIs, yet HAL has a well-resolved, but exceedingly low, initial $^{26}Al/^{27}Al = 5\times10^{-8}$ (Fahey *et al.*, 1987). These data clearly demonstrate that HAL formed from a reservoir with a characteristic $^{10}Be/^9Be$ similar to that of other refractory materials, but that it was almost completely lacking in $^{26}Al/^{27}Al$. The low value of $^{26}Al/^{27}Al$ that it does have may, in fact, be commensurate with ambient background in the molecular cloud, that is, independent of any specific additional source of ^{26}Al that spiked the CAI-forming regions of the solar nebula (Marhas and Goswami, 2003). Because the ^{10}Be is clearly spallogenic, this provides strong evidence that the vast majority of the ^{26}Al cannot have been produced that way and therefore that essentially all ^{26}Al is derived from external seeding of the nebula. The correlation of ^{26}Al with ^{41}Ca, even though it is not temporally quantitative, is then further evidence for the coproduction and injection of these nuclides into the solar nebula as freshly synthesized stellar debris.

Unfortunately, similar arguments cannot be advanced for ^{53}Mn, primarily because of the poor constraints on initial $^{53}Mn/^{55}Mn$ in CAIs. As discussed further below, the Mn–Cr systematics of nebular components are difficult to interpret in terms of a reasonable chronology, and one possible reason for this could be a significant contribution to the ^{53}Mn inventory by local production processes.

2.6 IMPLICATIONS FOR CHRONOLOGY

In principle, the record of each of the now-extinct isotopes can be interpreted to infer a chronology for various events that caused chemical fractionations in early solar system materials. Here we evaluate the consistency of these records, both internally and with each other, as well as with the Pb–Pb chronometer, to determine what quantitative constraints can be confidently inferred for the sequence and duration of processes in the solar nebula and on earliest planetesimals (planetary-scale differentiation, for example, relative to the Earth, is considered in Chapter 3). To obtain reference points for cross-calibrating relative and absolute chronologies, we require samples which achieved rapid isotopic closure following a well-defined fractionation event and for which a robust and high-precision data set exists. By these criteria, only two anchor points are possible for the cross-calibration: (1) the Pb–Pb and Al–Mg records in CAIs and (2) the Pb–Pb and Mn–Cr records in angrites. As demonstrated in Figure 10, the former provides a reasonably self-consistent, high-resolution record for nebular events, and the latter yields unique temporal information regarding early planetary differentiation processes, but that global consistency between the Al–Mg and Mn–Cr systems is problematic. The existing

record for the other short-lived radionuclides is either not well-preserved across different types of samples (e.g., ^{41}Ca, ^{10}Be, and ^{182}Hf), or is insufficiently precise or uncertain as to the nature of isotopic closure (e.g., ^{60}Fe and ^{129}I) so that cross-calibrations spanning the nebular and planetary accretion timescales are not yet possible.

2.6.1 Formation Timescales of Nebular Materials

A consistent timescale for fractionation events that occurred during high-temperature processing of nebular materials is obtained (Figure 10) by fixing the canonical $^{26}Al/^{27}Al$ value (4.5×10^{-5}) measured in CAIs to the absolute timescale provided by the recent high-precision Pb–Pb isochron age of $4{,}567.11\pm0.16$ Ma (Amelin *et al.*, 2006). There is an uncertainty in tying these timescales together here, in that it is not clear whether the Pb–Pb age of CAIs is the crystallization age, where $^{26}Al/^{27}Al = 4.5\times10^{-5}$ would be appropriate, or the time of nebular Al/Mg fractionation, in which case $^{26}Al/^{27}Al = 6.3\times10^{-5}$ would be a better choice. These two $^{26}Al/^{27}Al$ choices differ in time by 350 kyr. By this calibration, the initial $^{26}Al/^{27}Al$ values inferred for chondrules from the most unequilibrated chondrites ($\sim1\times10^{-5}$; Figure 3) indicate that chondrule formation began by at least $\sim4{,}565$ Ma and continued probably for another ~1–2 Myr. This time frame fits with a high-precision Pb–Pb isochron for chondrules (from CR chondrites) which yields $4{,}564.7\pm0.6$ Ma (Figure 1). Chondrule ages which appear younger than $\sim4{,}563$ by Al–Mg probably reflect metamorphic cooling rather than nebular formation, still younger Pb–Pb ages may reflect alternative, late formation scenarios, such as protoplanetary collisions (Krot *et al.*, 2005).

The same is not true for the majority of "anomalous CAIs," those that apparently formed lacking any significant live ^{26}Al. These refractory inclusions, which are often hibonite-rich, typically exhibit very large anomalies in "stable" isotopes (e.g., calcium or titanium) that are most readily interpreted as indicating a lack of mixing with average solar nebula materials. Because isotopic homogenization is expected to be an ongoing process during nebular evolution, the preservation of these anomalies argues strongly for a very "primitive" nature of these materials, that is, they probably formed early (not late) and also they escaped any significant isotopic reequilibration from later heating (MacPherson *et al.*, 1995; Sahijpal and Goswami, 1998). Sahijpal and Goswami (1998) suggested that the highly anomalous CM hibonite grains might have formed in a triggered collapse scenario just prior to injection of the radionuclides (^{41}Ca and ^{26}Al), which could theoretically trail the shock front (Foster and Boss, 1997). It would be

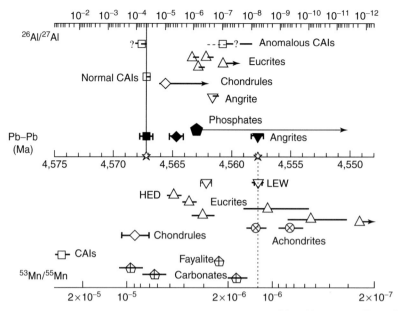

Figure 10 Timeline for early solar system events integrating the ^{26}Al–^{26}Mg and ^{53}Mn–^{53}Cr short-lived chronometers with the absolute timescale provided by the Pb–Pb chronometer. The anchor points (vertical dashed lines) are (1) the Pb–Pb age of CAIs (Amelin *et al.*, 2002) with "canonical" ^{26}Al/^{27}Al and (2) the Pb–Pb age of angrites (Lugmair and Galer, 1992) with the ^{53}Mn/^{55}Mn ratio in LEW (Lugmair and Shukolyukov, 1998). Pb–Pb ages are indicated for the filled symbols read against the absolute timescale (central axis); the top axis shows the initial ^{26}Al/^{27}Al values measured in various phases (open symbols) and the bottom axis refers to initial ^{53}Mn/^{55}Mn for the open symbols in the bottom panel. Squares—CAIs, diamonds—chondrules, upright triangles—eucrites (basaltic achondrites), inverted triangles—angrites, crossed circles—pallasites and Acapulco, and pentagons—secondary minerals in chondrites (phosphates, carbonates, and fayalite). The datum labeled "HED" represents the Mn–Cr correlation line for bulk eucrites. "LEW" refers to the anchor point for ^{53}Mn/^{55}Mn and Pb–Pb; the remaining angrite datum represents Mn–Cr and Al–Mg analyses of D'Orbigny (Nyquist *et al.*, 2003). "Anomalous CAIs" refers to those that apparently formed with no live ^{26}Al—see text for discussion.

useful to demonstrate the plausibility of this scenario by measuring an absolute Pb–Pb age on a suite of these objects; even if such a measurement might lack the precision to resolve the prearrival interval, it could at least demonstrate that the samples were not anomalously young.

There are other refractory inclusions, for example, grossite-bearing CAIs from CH chondrites (Weber *et al.*, 1995), which do not fit this model since they lack calcium and titanium isotopic anomalies as well as ^{26}Al. One interpretation of such objects could be that CAI formation lasted several million years, but this is not supported by any independent evidence and there could well be other reasons for the lack of both short-lived radioactivity and large isotopic anomalies (aside from ^{16}O excesses; Sahijpal *et al.*, 1999) in these inclusions. Circumstantial arguments against a long time period for CAI formation are that it leads to problems with understanding the distribution of the oxygen isotope anomalies in nebular components (see also McKeegan and Leshin, 2001) and with calculations of dynamical lifetimes of CAIs as independent objects in the nebula (Weidenschilling, 1977). Alternative explanations must invoke spatial heterogeneity within the

nebula, either with respect to radionuclide distribution or CAI distribution, or both. It is beyond the scope of this review to critically assess models of turbulence and mixing in the solar nebula or evidence regarding the provenance of various CAI types; see Shu *et al.* (2001), Cuzzi *et al.* (2003), McKeegan *et al.* (2000a), Krot *et al.* (2002) and Alexander *et al.* (2001) for discussions.

Difficulties in interpreting an absence of ^{26}Al in some samples notwithstanding, on the basis of the good concordance of the Al–Mg and Pb–Pb systems the first-order conclusion is that ^{26}Al/^{27}Al records do have chronological significance for most CAIs and chondrules. Taking the conventional (and reasonable) point of view that chondrules are nebular products, their formation ages relative to normal CAIs imply a duration of at least ~2–3 Myr for the solar nebula. Such a duration is plausible from an astrophysical viewpoint (Podosek and Cassen, 1994; Cameron, 1995), and it has interesting implications for timescales of accretion and radioactive heating of early-formed planetary bodies.

The nebular chronology inferred from initial ^{53}Mn/^{55}Mn (Figure 10) is not consistent with the Al–Mg and Pb–Pb systems either in terms of intervals or absolute ages (when Mn–Cr is anchored by the absolute Pb–Pb age of the LEW 86020 angrite, Lugmair and Galer, 1992). Because $(^{53}Mn/^{55}Mn)_0$ is poorly defined for CAIs (see above and discussion in Nyquist *et al.*, 2001a), the inferred interval between CAI and chondrule formation is rather uncertain, but is at least 4 Myr, with a more likely minimum value of ~6 Myr (Nyquist *et al.*, 2001a). The angrite-calibrated $^{53}Mn/^{55}Mn$ age of CAIs is too old by a minimum of 7 Myr compared with the measured Pb–Pb age, and chondrules are calculated to be ~1–2 Myr older than their measured Pb–Pb absolute age.

The discrepancies due to aberrantly old Mn–Cr ages of CAIs and chondrules were recognized by Lugmair and Shukolyukov (2001), who argued that a 4,571 Ma absolute age of the solar system, with $(^{53}Mn/^{55}Mn)_0 = 1.4 \times 10^{-5}$, would resolve the difficulties. In this case, $^{53}Mn/^{55}Mn$ could not be used to date CAI formation. More significantly, this would imply that Pb–Pb ages of CAIs could not be crystallization ages but must (based on the time interval) represent metamorphic cooling times. A problem with such an interpretation is the apparently unique composition of initial lead in CAIs (Tera and Carlson, 1999), which could not be maintained in a parent-body setting above the closure temperature for lead diffusion. Additionally, this interpretation (based on a model of chromium isotopic evolution in the solar nebula) runs counter to the good concordance of the Al–Mg system with Pb–Pb. At this time, it seems more reasonable to conclude that Mn–Cr does not provide a consistent high-resolution chronology for nebular events because one or more of the assumptions (initial homogeneity, isotopic closure, etc.) regarding the behavior of this short-lived chronometer is not satisfied within nebular components of chondrites.

A relatively long interval (>4 Myr) between CAIs and chondrules can be inferred on the basis of I–Xe dating (see Swindle *et al.*, 1996 for a review). At face value, this might be seen as support for an Mn–Cr age for chondrule formation; however, in detail it does not work. The siting of ^{129}I is uncertain in both CAIs and chondrules and isotopic closure effects are evidenced by I–Xe apparent ages of chondrules that span an interval of up to several tens of millions of years, implicating asteroidal rather than nebular processes (e.g., Swindle *et al.*, 1991).

2.6.2 Timescales of Planetesimal Accretion and Early Chemical Differentiation

Although the interpretation of apparent initial $^{53}Mn/^{55}Mn$ values in terms of a chronology for nebular fractionation events is problematic, the Mn–Cr system seems amenable to timing chemical fractionations associated with "geologic" activity on early-formed planetary bodies. A timescale is presented in the bottom panel of Figure 10, following the suggestion of Lugmair and Shukolyukov (1998) to utilize the LEW86010 angrite as a reference point to cross-calibrate the Mn–Cr and Pb–Pb systems. Thus, $^{53}Mn/^{55}Mn = 1.25 \times 10^{-6}$ is tied to an absolute Pb–Pb age of 4,557.8 Ma. By this reckoning, the "global" differentiation of the Howardite–Eucrite–Diogenite (HED) parent body is pinned by the ensemble eucrite Mn–Cr isochron to 4,565 Ma. As mentioned previously, individual eucrites show internal Mn–Cr isochrons that indicate attainment of isotopic closure from just slightly after this time to significantly later, implying an extended (>10^7 years) history of thermal activity on the HED asteroid. This is qualitatively in agreement with the young U–Pb ages of eucrites; however, a quantitative correlation between Mn–Cr and U–Pb ages is lacking (Tera and Carlson, 1999).

The $^{53}Mn–^{53}Cr$ isochron for the HED parent body is generally consistent with the timing of other indicators of early planetary processes. The Pb–Pb age for the oldest phosphates, from the least metamorphosed (H4) chondrites studied, postdates HED differentiation by ~2 Myr. This is approximately equivalent to the Mn–Cr closure age for Chervony Kut, the noncumulate eucrite with the highest individual $^{53}Mn/^{55}Mn$ initial ratio. Other achondrites, including a pallasite and the unusual basaltic achondrite Acapulco, have Mn–Cr ages ~8–10 Myr after the HED differentiation event. These timescales are consistent with the notion that a variety of differentiated meteorites sample various depths in asteroids of various sizes during this early epoch following accretion.

A problem arises with the apparent chronology of aqueous activity on carbonaceous chondrite parent bodies. The formation time of fayalite is reasonable from the Mn–Cr point of view; however, carbonates from CM chondrites and from the unique chondrite Kaidun have $^{53}Mn/^{55}Mn$ initial values commensurate with those of chondrules. Although we have argued above that there are problems in understanding the temporal meaning of Mn–Cr systematics in CAIs and chondrules, we note that if $^{53}Mn–^{53}Cr$ can serve as an accurate chronometer at least for chondrules, then it implies that aqueous activity on some chondrite parent bodies was contemporaneous with chondrule formation elsewhere. The data that exist thus far are for carbonates from carbonaceous chondrites and for chondrules from ordinary chondrites. There are no data to suggest that chondrules from carbonaceous chondrites might be older than those from ordinary chondrites; in fact limited Al–Mg data could be interpreted to suggest the opposite (Kunihiro *et al.*, 2004). Clearly, carbonates formed very early, but whether chondrule formation was still ongoing during this aqueous activity will have to be decided by further study, preferably of chondrules and secondary minerals from the same meteorites.

The experimental record documenting the prior existence of ^{26}Al in differentiated meteorites is mostly based on very recent data, the majority of which have only been published in abstract form. Certainly, the finding of ^{26}Mg* and of good Al–Mg isochrons in eucrites is consistent with the early age of igneous activity inferred from Mn–Cr systematics of bulk eucrites. This is strong confirmation that planetary-scale melting began not more than a few million years following CAI crystallization, and quite possibly, while chondrule formation was still ongoing.

Amelin *et al.* (2006) have compared Pb–Pb, Al–Mg, and Mn–Cr ages for a number of differentiated meteorites, tying Pb–Pb with Al–Mg for CAIs and Pb–Pb with Mn–Cr for LEW86010 as we discuss above. They find internal consistency with two exceptions: the Pb–Pb ages of the Asuka 881394 eucrite and the Sahara 99555 angrite are ~3 Myr older than both Al–Mg and Mn–Cr ages. Amelin *et al.* (2006) suggest that this discordance may be caused by differences in closure temperatures of the three chronometers. Asuka 881394 has a slow cooling rate allowing this explanation, but mineralogical and textural indicators suggest that Sahara 99555 cooled too fast to allow a 3 Myr discordance in ages. Progress is being made in tying together long- and short-lived chronometers, but a completely self-consistent picture has not yet emerged.

2.7 CONCLUSIONS

Both chondrites and differentiated meteorites preserve records of short-lived radionuclides which are now-extinct, but which were present when the solar system formed (Table 1). These isotopic records yield information on the amount of radioactivity contained by ancient solar system minerals, from which the relative timing of chemical fractionations between parent and daughter elements can be inferred (assuming that the short-lived radionuclides were originally distributed homogenously). The fractionation events can often be related to thermal processes occurring in the solar nebula or on early-accreted planetesimals, thus allowing a high-resolution relative chronology to be delineated (Figure 10).

The existence of ^{10}Be, ^{36}Cl, and ^{60}Fe in various early solar system materials provides strong evidence for a multiplicity of sources for short-lived isotopes. The first two isotopes most likely result from local production by energetic particle irradiation, perhaps near the forming Sun, whereas the latter is evidence for seeding of the solar nebula by freshly synthesized stellar ejecta. In principle, the inventory of other radioisotopes may contain contributions from both these sources in addition to other nondiscrete ("background") sources such as galactic stellar nucleosynthesis or spallogenic nuclear reactions in the protosolar molecular cloud. However, correlations of radiogenic isotope signatures in CAIs and hibonite grains indicate that spallogenic contributions to the abundances of the shortest-lived isotopes, ^{41}Ca and ^{26}Al, are minor and that these refractory isotopes arrived together in the solar nebula.

2.7.1 Implications for Solar Nebula Origin and Evolution

The short lifetimes of ^{26}Al and, especially, ^{41}Ca, coupled with the evidence for an external origin of these nuclides, have important implications for the origin of the solar system. On the basis of estimated production rates and isotope mixing during interstellar transit and injection into the solar system, a duration of at most ~1 Myr can be accommodated for the total time between nucleosynthetic production and incorporation of these isotopes into crystalline solids in the early solar system. Such a rapid timescale implies a triggering mechanism for fragmentation and collapse of a portion of the presolar molecular cloud to form the early Sun and its accretion disk. Although it is known that many AGB stars contributed dust to the early solar nebula and that a wind from such a star could theoretically provide a sufficient shock to initiate collapse, astrophysical considerations of stellar lifetimes suggest a nearby type II supernova as a more likely trigger.

Supernovae can be the source of most of the short-lived radionuclides (except ^{10}Be); however there are difficulties in reconciling relative abundances of all species with a single event (see review by Goswami and Vanhala, 2000). While this may be aesthetically desirable, it is not required, especially for the longer-lived isotopes of Table 1. Other evidence indicates that it is probably not correct and that the truth is more complex than a single supernova triggering and injection. The "last" supernova is not the source of large stable isotope anomalies in oxygen, calcium, or titanium, demonstrating that isotopic memories of other presolar components survived to be incorporated into early solar system minerals. Additionally, the evidence for pervasive ^{10}Be signatures in CAIs, the strong hint for ^{7}Be, and the abundant astronomical evidence for copious X-ray activity of YSO indicate that early-formed solar system materials were most likely strongly irradiated if they were not shielded. Further work is required to quantitatively assess the proportion of those radionuclides (besides ^{26}Al) that were produced locally by solar energetic particles.

Cross-calibration of the initial ^{26}Al/^{27}Al records inferred for nebular components of chondrites with the absolute Pb–Pb ages of CAIs results in a self-consistent high-resolution chronology for the high-temperature phases of solar nebula evolution. A plausible scenario and timeline can be constructed:

1. at nearly 4,568 Ma, a shock wave, probably initiated by a "nearby" supernova, triggers fragmentation and gravitational collapse of a portion of a molecular cloud;
2. near the central, hot regions of the nebula the first refractory minerals form by evaporation and/or recondensation and melting of mixtures of presolar dust grains from various interstellar heritages; these hibonite grains and FUN inclusions incorporate ^{10}Be produced by irradiation of the dust grains by solar energetic particles, but they do not sample the radioactivity accompanying the supernova shock wave;
3. shortly afterward, at ~4,567 Ma, the fresh radioactivity arrives in the inner nebula and most CAIs form over a short interval incorporating ^{26}Al and ^{41}Ca; if the new high-precision Al–Mg data on large CV CAIs is representative of most refractory inclusions, then this interval may be as short as 15 kyr;
4. high-temperature processing of some CAIs continues for a few hundred thousand years, but most of those that do not accrete to the Sun are removed from high-temperature regions of the nebula, perhaps by entrainment in bipolar outflows, and survive for a long period of time in undetermined nebular locations;
5. at ~4,566 Ma, chondrule formation begins and continues for ~1–2 Myr; CAIs are largely absent from the nebular regions where chondrule melting occurs; and
6. at ~4,565–4,564 Ma, CAIs have joined chondrules and nebular dust in accreting to planetesimals in the asteroid belt. If the latter process is considered as the termination of the nebular phase of solar system evolution, then its lifetime is ~4 Ma as recorded by radionuclides in nebular materials.

The timescales for accretion and early evolution of these planetesimals are also constrained by short-lived radioactivity. This record is best elucidated with the $^{53}Mn–^{53}Cr$ isotopic system, even though as discussed previously the record of $^{53}Mn/^{55}Mn$ in solar nebula objects does not yield a consistently interpretable chronology. The $^{182}Hf–^{182}W$ isotopic system indicates that core separation on differentiated meteorites occurred within 1 Myr of CAI formation. Accretion of some planetesimals started very early, perhaps even before the bulk of chondrule formation began. By ~4,565–4,564 Ma, large-scale melting and differentiation occurred on the HED parent body, most likely the asteroid 4 Vesta. Some eucrites crystallized soon after mantle differentiation, quickly cooling

through isotopic closure for magnesium and chromium by ~4,564–4,563 Ma. Energy from ^{26}Al and ^{60}Fe decay probably contributed substantially to the heat required for melting, but the asteroid was large enough that igneous activity continued for several tens of million years. Some angrites appear to have erupted early, cooling by ~4,561 Ma, but ADOR and LEW did not crystallize until 4,558 Ma. Other asteroidal bodies, from which chondrites are derived, either accreted somewhat later than Vesta or remained as relatively small bodies for several million years. Absolute Pb–Pb ages of phosphates indicate that metamorphic temperatures were reached on some ordinary chondrite asteroids by ~4,563 Ma; this timescale is consistent with the $^{26}Al/^{27}Al$ records of chondrules. Metamorphism on chondrite parent bodies continued for up to tens of millions of years as indicated by Pb–Pb and I–Xe dating. Aqueous activity (formation of carbonate) happened very early, perhaps "too" early, on the parent asteroids of some carbonaceous chondrites. Calibration of the $^{53}Mn–^{53}Cr$ chronometer by the Pb–Pb age of angrites implies formation of the earliest of these carbonates by ~4,567 Ma, which is not compatible with the nebular chronology discussed above. Accretion and differentiation of planetary embryos continued from this early epoch for a period of several tens of millions of years (see Chapter 3).

2.7.2 Future Directions

The quantitative comparison of various short-lived radionuclide systems with each other and with Pb–Pb chronology has only been made possible by new data obtained during the last decade, or in many cases, the last few years. Over this same time period, evidence for the decay of several important new short-lived isotopes in the early solar system has been discovered. The record of now-extinct isotopes in early solar system materials is becoming sufficiently well defined to allow construction of a plausible timeline and scenario for solar system origin. However, even though broad areas of consistency have been revealed, there are significant problems that will require further investigation. One of the most important is trying to understand the role of energetic particle irradiation in the early solar system. Energetic processes associated with magnetic flare activity of the young Sun almost certainly occurred; the question is what effect these had on isotopic and mineralogical records of early-formed solar system rocks. Could solar system irradiation be responsible for some of the confusion of the nebular record of $^{53}Mn/^{55}Mn$? There appears to be large-scale inhomogeneity in the $^{53}Mn–^{53}Cr$ systematics: could some of this be explicable in terms of solar system production and/or large-scale radial transport of nebular components? Or, is ^{53}Mn/^{55}Mn homogeneous after all (Trinquier *et al.*, 2005b)?

It has been recently hypothesized (Desch *et al.*, 2004) that ^{10}Be may result from magnetic trapping of cosmic radiation in molecular cloud material, such that all short-lived nuclides predate solar system formation. However, little attention has so far been paid to the role of magnetic fields in triggered collapse mechanisms. It is clear that magnetic pressure cannot substantially inhibit collapse, otherwise the delay would cause extinction of the signal of ^{41}Ca in CAIs. The correlation of ^{41}Ca with ^{26}Al needs to be better quantified, and even the canonical ^{26}Al record more closely examined to sort out the intrinsic dispersion in the distribution from the effects of secondary heating and alteration of CAI minerals. As it stands, the duration of CAI production seems implausibly short compared with CAI longevity in the nebula, but this is largely a model-dependent result. A better understanding of the locales and formation mechanisms of CAIs and chondrules, and their relationships to each other, will help in constraining such models. Finally, it can be anticipated that in the near future much more data will be gathered by *in situ* methods and high-precision bulk methods that will greatly improve our knowledge of the distributions of ^{10}Be and ^{60}Fe in a wide range of early materials. So far, these isotopes have been primarily exploited as semiquantitative indicators of process; perhaps with a more robust data set, it will be possible to employ them as further chronological tools for understanding solar nebula origin and evolution.

ACKNOWLEDGMENTS

We thank Y. Amelin, G. Huss, G. J. MacPherson, and S. Tachibana for providing figures and data used herein.

REFERENCES

Alexander C. M. O. D., Boss A. P., and Carlson R. W. (2001) The early evolution of the inner solar system: a meteoritic perspective. *Science* **293**, 64–68.

Allègre C. J., Manhès G., and Göpel C. (1995) The age of the Earth. *Geochim. Cosmochim. Acta* **59**, 1445–1456.

Amelin Y. (2006) The prospect of high-precision Pb isotopic dating of meteorites. *Meteorit. Planet. Sci.* **41**, 7–17.

Amelin Y., Krot A. N., Hutcheon I. D., and Ulyanov A. A. (2002) Lead isotopic ages of chondrules and calcium–aluminum-rich inclusions. *Science* **297**, 1678–1683.

Amelin Y., Wadhwa M., and Lugmair G. W. (2006) Pb-isotopic dating of meteorites using ^{202}Pb–^{205}Pb double spike: comparison with other high resolution chronometers. In *Lunar Planet. Sci.* XXXVII, #1970. The Lunar and Planetary Institute, Houston (CD-ROM).

André P., Ward-Thompson W., and Barsony M. (2000) From prestellar cores to protostars: the initial conditions of star formation. In *Protostars and Planets IV* (eds. V. Mannings, A. P. Boss, and S. S. Russell). University of Arizona Press, Tucson, AZ, pp. 59–96.

Baker J., Bizzarro M., Wittig N., Connelly J., and Haack H. (2005) Early planetesimal melting from an age of 4.5662 Gyr for differentiated meteorites. *Nature* **436**, 1127–1130.

Begemann F. (1980) Isotopic anomalies in meteorites. *Rep. Prog. Phys.* **43**, 1309–1356.

Binzel R. P. and Xu S. (1993) Chips off of asteroid 4 Vesta: evidence for the parent body of basaltic achondrite meteorites. *Science* **260**, 186–191.

Birck J. L. and Allègre C. J. (1985) Evidence for the presence of Mn-53 in the early solar-system. *Geophys. Res. Lett.* **12**, 745–748.

Birck J. L. and Allègre C. J. (1988) Manganese–chromium isotope systematics and the development of the early solar-system. *Nature* **331**, 579–584.

Birck J. L. and Lugmair G. W. (1988) Nickel and chromium isotopes in Allende inclusions. *Earth Planet. Sci. Lett.* **90**, 131–143.

Birck J. L., Rotaru M., and Allègre C. J. (1999) ^{53}Mn–^{53}Cr evolution of the early solar system. *Geochim. Cosmochim. Acta* **63**, 4111–4117.

Bizzarro M., Baker J. A., and Haack H. (2004) Mg isotope evidence for contemporaneous formation of chondrules and refractory inclusions. *Nature* **431**, 275–278.

Bizzarro M., Baker J. A., and Haack H. (2005) Corrigendum to Bizzarro *et al.* (2004). *Nature* **435**, 1280.

Bizzarro M., Ulfbeck D., and Thrane K. (2006) Nickel isotopes in meteorites: evidence for live ^{60}Fe and distinct ^{62}Ni isotope reservoirs in the early solar system. In *Lunar Planet. Sci.* XXXVII, #2020. The Lunar and Planetary Institute, Houston (CD-ROM).

Boss A. P. and Vanhala H. A. T. (2001) Injection of newly synthesized elements into the protosolar cloud. *Phil. Trans. Roy Soc. Lond. A* **359**, 2005–2016.

Brazzle R. H., Pravdivtseva O. V., Meshik A. P., and Hohenberg C. M. (1999) Verification and interpretation of the I–Xe chronometer. *Geochim. Cosmochim. Acta* **63**, 739–760.

Busso M., Gallino R., and Wasserburg G. J. (1999) Nucleosynthesis in asymptotic giant branch stars: relevance for galactic enrichment and solar system formation. *Annu. Rev. Astron. Astrophys.* **37**, 239–309.

Caillet C., MacPherson G. J., and Zinner E. K. (1993) Petrologic and Al–Mg isotopic clues to the accretion of two refractory inclusions onto the Leoville parent body: one was hot, the other wasn't. *Geochim. Cosmochim. Acta* **57**, 4725–4743.

Cameron A. G. W. (1995) The first ten million years in the solar nebula. *Meteorit. Planet. Sci.* **30**, 133–161.

Cameron A. G. W. (2001a) Extinct radioactivities, core-collapse supernovas, jets, and the r-process. *Nucl. Phys. A* **688**, 289C–296C.

Cameron A. G. W. (2001b) Some properties of r-process accretion disks and jets. *Astrophys. J.* **562**, 456–469.

Cameron A. G. W., Höflich P., Myers P. C., and Clayton D. D. (1995) Massive supernovae, Orion gamma rays, and the formation of the solar system. *Astrophys. J.* **447**, L53–L57.

Cameron A. G. W. and Truran J. W. (1977) The supernova trigger for formation of the solar system. *Icarus* **30**, 447–461.

Carlson R. W. and Hauri E. (2001) Extending the ^{107}Pd–^{107}Ag chronometer to low Pd/Ag meteorites with MC-ICP-MS. *Geochim. Cosmochim. Acta* **65**, 1839–1848.

Chaussidon M., Robert F., Mangin D., Hanon P., and Rose E. F. (1997) Analytical procedures for the measurement of boron isotope compositions by ion microprobe in meteorites and mantle rocks. *Geostand. Newslett.* **21**, 7–17.

Chaussidon M., Robert F., and McKeegan K. D. (2006) Li and B isotopic variations in an Allende CAI: evidence for the *in situ* decay of short-lived ^{10}Be and for the possible presence of the short-lived nuclide ^{7}Be in the early solar system. *Geochim. Cosmochim. Acta* **70**, 224–245.

Chaussidon M., Robert F., Russell S. S., Gounelle M., and Ash R. D. (2003) Variations of apparent ^{10}Be/^{9}Be ratios in

Leoville MRS-06 type B1 CAI: constraints on the origin of ^{10}Be and ^{26}Al. In *Lunar Planet. Sci.* **XXXIV**, #1347. The Lunar and Planetary Institute, Houston (CD-ROM).

Chen J. H., Papanastassiou D. A., and Wasserburg G. J. (2002) Re–Os and Pd–Ag systematics in group IIIAB irons and in pallasites. *Geochem. Cosmochim. Acta* **66**, 3793–3810.

Chen J. H. and Wasserburg G. J. (1980) A search for isotopic anomalies in uranium. *Geophys. Res. Lett.* **7**, 275–278.

Chen J. H. and Wasserburg G. J. (1981) The isotopic composition of uranium and lead in Allende inclusions and meteorite phosphates. *Earth Planet. Sci. Lett.* **52**, 1–15.

Chen J. H. and Wasserburg G. J. (1987) A search for evidence of extinct lead 205 in iron meteorites. *Lunar Planet. Sci.* **XVIII**, 165–166.

Chen J. H. and Wasserburg G. J. (1990) The isotopic composition of Ag in meteorites and the presence of ^{107}Pd in protoplanets. *Geochim. Cosmochim. Acta* **54**, 1729–1743.

Chen J. H. and Wasserburg G. J. (1996) Live ^{107}Pd in the early solar system and implications for planetary evolution. In *Earth Processes: Reading the Isotopic Code* (eds. A. Basu and S. Hart). Geophys. Monogr. 95. American Geophysical Union, Washington, DC, pp. 1–20.

Clayton D. D. (1982) Cosmic chemical memory—a new astronomy. *Quart. J. Roy. Astron. Soc.* **23**, 174–212.

Clayton D. D. (1986) Interstellar fossil ^{26}Mg and its possible relation to excess meteoritic ^{26}Mg. *Astrophys. J.* **310**, 490–498.

Clayton D. D., Dwek E., and Woosley S. E. (1977) Isotopic anomalies and proton irradiation in the early solar system. *Astrophys. J.* **214**, 300–315.

Clayton R. N. and Mayeda T. K. (1984) The oxygen isotope record in Murchison and other carbonaceous chondrites. *Earth Planet. Sci. Lett.* **67**, 151–161.

Cook D. L., Wadhwa M., Clayton R. N., Janney P. E., Dauphas N., and Davis A. M. (2005) Nickel isotopic composition of meteoritic metal: implications for the initial ^{60}Fe/^{56}Fe ratio in the early solar system. *Meteorit. Planet. Sci.* **40**, A33.

Cook D. L., Wadhwa M., Janney P. E., Dauphas N., Clayton R. N., and Davis A. M. (2006) High precision measurements of nickel isotopes in metallic samples via multi-collector ICPMS. *Anal. Chem.* (submitted).

Cosarinsky M., Taylor D. J., and McKeegan K. D. (2006) Aluminum-26 model ages of hibonite and spinel from type A inclusions in CV chondrites. In *Lunar Planet. Sci.* **XXXVII**, #2357. The Lunar and Planetary Institute, Houston (CD-ROM).

Cosarinsky M., Taylor D. J., McKeegan K. D., and Hutcheon I. D. (2005) Mg isotopic study of Wark–Lovering rims in type A inclusions from CV chondrites: formation mechanisms and timing. *Meteorit. Planet. Sci.* **40**, A34.

Cuzzi J. N., Davis S. S., and Dobrovolskis A. R. (2003) Creation and distribution of CAIs in the protoplanetary nebula. In *Lunar Planet. Sci.* **XXXIV**, #1749. The Lunar and Planetary Institute, Houston (CD-ROM).

Dauphas N., Foley C. N., Wadhwa M., Davis A. M., Janney P. E., Qin L., Göpel C., and Birck J.-L. (2005) Protracted core differentiation in asteroids from ^{182}Hf–^{182}W systematics in the Eagle Station pallasite. In *Lunar Planet. Sci.* **XXXVI**, #1100. The Lunar and Planetary Institute, Houston (CD-ROM).

Davis A. M., Richter F. M., Mendybaev R. A., Janney P. E., Wadhwa M., and McKeegan K. D. (2005) Isotopic mass fractionation laws and the initial solar system ^{26}Al/^{27}Al ratio. In *Lunar Planet. Sci.* **XXXVI**, #2334. The Lunar and Planetary Institute, Houston (CD-ROM).

Desch S. J., Connolly H. C., Jr., and Srinivasan G. (2004) An interstellar origin for the beryllium 10 in calcium-rich, aluminum-rich inclusions. *Astrophys. J.* **602**, 528–542.

Endress M., Zinner E., and Bischoff A. (1996) Early aqueous activity on primitive meteorite parent bodies. *Nature* **379**, 701–703.

Fahey A. J., Goswami J. N., McKeegan K. D., and Zinner E. (1987) ^{26}Al, ^{244}Pu, ^{50}Ti, REE, and trace element abundances

in hibonite grains from CM and CV meteorites. *Geochim. Cosmochim. Acta* **51**, 329–350.

Feigelson E. D., Broos P., Gaffney J. A., Garmire G., Hillenbrand L. A., Pravdo S. H., Townsley L., and Tsuboi Y. (2002a) X-ray emitting young stars in the Orion Nebula. *Astrophys. J.* **574**, 258–292.

Feigelson E. D., Garmire G. P., and Pravdo S. H. (2002b) Magnetic flaring in the pre main-sequence sun and implications for the early solar system. *Astrophys. J.* **572**, 335–349.

Foster P. N. and Boss A. P. (1997) Injection of radioactive nuclides from the stellar source that triggered the collapse of the presolar nebula. *Astrophys. J.* **489**, 346–357.

Galer S. J. G. and Lugmair G. W. (1996) Lead isotope systematics of non-cumulate eucrites. *Meteoritics* **31**, A47–A48.

Galy A., Hutcheon I. D., and Grossman L. (2004) $(^{26}$Al/^{27}Al$)_0$ of the solar nebula inferred from Al–Mg systematics in bulk CAIs from CV3 chondrites. In *Lunar Planet. Sci.* **XXXV**, #2334. The Lunar and Planetary Institute, Houston (CD-ROM).

Galy A., Young E. D., Ash R. D., and O'Nions R. K. (2000) The formation of chondrules at high gas pressures in the solar nebula. *Science* **290**, 1751–1753.

Gilmour J. D. (2000) The extinct radionuclide timescale of the solar system. *Space Sci. Rev.* **192**, 123–132.

Gilmour J. D. and Saxton J. M. (2001) A timescale of formation of the first solids. *Phil. Trans. Roy Soc. Lond. A* **359**, 2037–2048.

Göpel C., Manhès G., and Allègre C. J. (1994) U–Pb systematics of phosphates from equilibrated ordinary chondrites. *Earth Planet. Sci. Lett.* **121**, 153–171.

Goswami J. N., Marhas K. K., and Sahijpal S. (2001) Did solar energetic particles produce the short-lived nuclides present in the early solar system? *Astrophys. J.* **549**, 1151–1159.

Goswami J. N. and Vanhala H. A. T. (2000) Extinct radionuclides and the origin of the solar system. In *Protostars and Planets IV* (eds. V. Mannings, A. P. Boss, and S. S. Russell). University of Arizona Press, Tucson, AZ, pp. 963–994.

Gounelle M., Shu F. H., Shang H., Glassgold A. E., Rehm K. E., and Lee T. (2001) Extinct radioactivities and protosolar cosmic rays: self-shielding and light elements. *Astrophys. J.* **548**, 1051–1070.

Gray C. M. and Compston W. (1974) Excess ^{26}Mg in the Allende meteorite. *Nature* **251**, 495–497.

Grimm R. E. and McSween H. Y. (1994) Heliocentric zoning of the asteroid belt by aluminum-26 heating. *Science* **259**, 653–655.

Guan Y., Huss G. R., MacPherson G. J., and Wasserburg G. J. (2000) Calcium–aluminum-rich inclusions from enstatite chondrites: indigenous or foreign? *Science* **289**, 1330–1333.

Halliday A. and Lee D. C. (1999) Tungsten isotopes and the early development of the Earth and Moon. *Geochim. Cosmochim. Acta* **63**, 4157–4179.

Harper C. L., Jr. (1996) Evidence for ^{92g}Nb in the early solar system and evaluation of a new p-process cosmochronometer from ^{92g}Nb/^{92}Mo. *Astrophys. J.* **466**, 437–456.

Hester J. J., Desch S. J., Healy K. R., and Leshin L. A. (2004) The cradle of the solar system. *Science* **304**, 1116–1117.

Heymann D. and Dziczkaniec M. (1976) Early irradiation of matter in the solar system: magnesium (proton, neutron) scheme. *Science* **191**, 79–81.

Hsu W., Guan Y., Leshin L. A., Ushikubo T., and Wasserburg G. J. (2006) A late episode of irradiation in the early solar system: evidence from extinct ^{36}Cl and ^{26}Al in meteorites. *Astrophys. J.* **640**, 525–529.

Hsu W., Huss G. R., and Wasserburg G. J. (2003) Al–Mg systematics of CAIs, POI, and ferromagnesian chondrules from Ningqiang. *Meteorit. Planet. Sci.* **38**, 35–48.

Hsu W. B., Wasserburg G. J., and Huss G. R. (2000) High time resolution by use of the Al-26 chronometer in the multistage formation of a CAI. *Earth Planet. Sci. Lett.* **182**, 15–29.

Hua J., Huss G. R., and Sharp T. G. (2002) ^{53}Mn–^{53}Cr dating of fayalite formation in the Kaba CV3 carbonaceous chondrite. *Lunar Planet. Sci.* **XXXIII**, 1660.

Hudson G. B., Kennedy B. M., Podosek F. A., and Hohenberg C. M. (1988) The early solar system abundance of ^{244}Pu as inferred from the St. Severin chondrite. *Proc. 19th Lunar Planet. Sci. Conf.* pp. 547–557.

Huss G. R., MacPherson G. J., Wasserburg G. J., Russell S. S., and Srinivasan G. (2001) Aluminum-26 in calcium–aluminum-rich inclusions and chondrules from unequilibrated ordinary chondrites. *Meteorit. Planet. Sci.* **36**, 975–997.

Hutcheon I. D., Armstrong J. T., and Wasserburg G. J. (1984) Excess ^{41}K in Allende CAI: confirmation of a hint. In *Lunar Planet. Sci.* The Lunar and Planetary Institute, Houston, pp. 387–388.

Hutcheon I. D., Browning L., Keil K., Krot A. N., Phinney D. L., Prinz M., and Weisberg M. K. (1999a) Timescale of aqueous activity in the early solar system. In *Ninth Goldschmidt Conf. Abstr.*, #971. Lunar and Planetary Institute, Houston (CD-ROM).

Hutcheon I. D., Huss G. R., and Wasserburg G. J. (1994) A search for ^{26}Al in chondrites: chondrule formation times. In *Lunar Planet. Sci.* The Lunar and Planetary Institute, Houston, pp. 587–588.

Hutcheon I. D. and Hutchison R. (1989) Evidence from the Semarkona ordinary chondrite for ^{26}Al heating of small planets. *Nature* **337**, 238–241.

Hutcheon I. D. and Jones R. H. (1995) The ^{26}Al–^{26}Mg record of chondrules: clues to nebular chronology. In *Lunar Planet. Sci.* The Lunar and Planetary Institute, Houston, pp. 647–648.

Hutcheon I. D., Krot A. N., Keil K., Phinney D. L., and Scott E. R. D. (1998) ^{53}Mn–^{53}Cr dating of fayalite formation in the CV3 chondrite Mokoia: evidence for asteroidal alteration. *Science* **282**, 1865–1867.

Hutcheon I. D. and Phinney D. L. (1996) Radiogenic ^{53}Cr* in Orgueil carbonates: chronology of aqueous activity on the CI parent body. In *Lunar Planet. Sci.* The Lunar and Planetary Institute, Houston, pp. 577–578.

Hutcheon I. D., Weisberg M. K., Phinney D. L., Zolensky M. E., Prinz M., and Ivanov A. V. (1999b) Radiogenic ^{53}Cr in Kaidun carbonates: evidence for very early aqueous activity. In *Lunar Planet. Sci.* **XXX**, #1722. The Lunar and Planetary Institute, Houston (CD-ROM).

Ireland T. R. (1988) Correlated morphological, chemical, and isotopic characteristics of hibonites from the Murchison carbonaceous chondrite. *Geochim. Cosmochim. Acta* **52**, 2827–2839.

Ireland T. R. (1990) Presolar isotopic and chemical signatures in hibonite-bearing refractory inclusions from the Murchison carbonaceous chondrite. *Geochim. Cosmochim. Acta* **54**, 3219–3237.

Ireland T. R., Compston W., and Heydegger H. R. (1985) Titanium isotopic anomalies in hibonites from the Murchison carbonaceous chondrite. *Geochim. Cosmochim. Acta* **49**, 1989–1993.

Jeffery P. M. and Reynolds J. H. (1961) Origin of excess Xe^{129} in stone meteorites. *J. Geophys. Res.* **66**, 3582–3583.

Kaiser T. and Wasserburg G. J. (1983) The isotopic composition and concentration of Ag in iron meteorites. *Geochim. Cosmochim. Acta* **47**, 43–58.

Kelly W. R. and Wasserburg G. J. (1978) Evidence for the existence of ^{107}Pd in the early solar system. *Geophys. Res. Lett.* **5**, 1079–1082.

Kita N. T., Huss G. R., Tachibana S., Amelin Y., Nyquist L. E., and Hutcheon I. D. (2005) Constraints on the origin of chondrules and CAIs from short-lived and long-lived radionuclides. In *Chondrites and the Protoplanetary Disk* (eds. A. N. Krot, E. R. D. Scott, and B. Reipurth), Astron. Soc. Pacific Conf. Ser. 341. Astronomical Society of the Pacific, San Francisco, pp. 558–587.

Kita N. T., Ikeda Y., Shimoda H., Morshita Y., and Togashi S. (2003) Timing of basaltic volcanism in ureilite parent body inferred from the ^{26}Al ages of plagioclase-bearing clasts in DAG-319 polymict ureilite. In *Lunar Planet. Sci.* **XXXIV**, #1557. The Lunar and Planetary Institute, Houston (CD-ROM).

Kita N. T., Nagahara H., Togashi S., and Morshita Y. (2000) A short duration of chondrule formation in the solar nebula: evidence from ^{26}Al in Semarkona ferromagnesian chondrules. *Geochim. Cosmochim. Acta* **64**, 3913–3922.

Kleine T., Mezger K., Palme H., Scherer E., and Münker C. (2005a) Early core formation in asteroids and late accretion of chondrite parent bodies: evidence from ^{182}Hf–^{182}W in CAIs, metal-rich chondrites, and iron meteorites. *Geochim. Cosmochim. Acta* **69**, 5805–5818.

Kleine T., Mezger K., Palme H., Scherer E., and Münker C. (2005b) The W isotope composition of eucrite metals: constraints on the timing and cause of thermal metamorphism of basaltic eucrites. *Earth Planet. Sci. Lett.* **231**, 41–52.

Kleine T., Munker C., Mezger K., and Palme H. (2002a) Rapid accretion and early core formation on asteroids and the terrestrial planets from Hf–W chronometry. *Nature* **418**, 952–955.

Kleine T., Munker C., Mezger K., Palme H., and Bischoff A. (2002b) Revised Hf–W ages for core formation in planetary bodies. *Geochim. Cosmochim. Acta* **66**, A404–A404.

Krot A. N., Amelin Y., Cassen P., and Meibom A. (2005) Young chondrules in CB chondrites from a giant impact in the early solar system. *Nature* **436**, 989–992.

Krot A. N., McKeegan K. D., Leshin L. A., MacPherson G. J., and Scott E. R. D. (2002) Existence of an ^{16}O-rich gaseous reservoir in the solar nebula. *Science* **295**, 1051–1054.

Kunihiro T., Rubin A. E., McKeegan K. D., and Wasson J. T. (2004) Initial ^{26}Al/^{27}Al in carbonaceous-chondrite chondrules: too little ^{26}Al to melt asteroids. *Geochim. Cosmochim. Acta* **68**, 2947–2957.

LaTourrette T. and Wasserburg G. J. (1998) Mg diffusion in anorthite: implications for the formation of early solar system planetesimals. *Earth Planet. Sci. Lett.* **158**, 91–108.

Lee D. C. and Halliday A. N. (1996) Hf–W isotopic evidence for rapid accretion and differentiation in the early solar system. *Science* **274**, 1876–1879.

Lee T. (1978) A local proton irradiation model for isotopic anomalies in the solar system. *Astrophys. J.* **224**, 217–226.

Lee T., Mayeda T. K., and Clayton R. N. (1980) Oxygen isotopic anomalies in Allende inclusion HAL. *Geophys. Res. Lett.* **7**, 493–496.

Lee T. and Papanastassiou D. A. (1974) Mg isotopic anomalies in the Allende meteorite and correlation with O and Sr effects. *Geophys. Res. Lett.* **1**, 225–228.

Lee T., Papanastassiou D. A., and Wasserburg G. J. (1976) Demonstration of ^{26}Mg excess in Allende and evidence for ^{26}Al. *Geophys. Res. Lett.* **3**, 109–112.

Lee T., Papanastassiou D. A., and Wasserburg G. J. (1977) Aluminum-26 in the early solar system: fossil or fuel? *Astrophys. J* **211**, L107–L110.

Lee T., Russell W. A., and Wasserburg G. J. (1978) Calcium isotopic anomalies and the lack of aluminum-26 in an unusual Allende inclusion. *Astrophys. J.* **228**, L93–L98.

Lee T., Shu F. H., Shang H., Glassgold A. E., and Rehm K. E. (1998) Protostellar cosmic rays and extinct radioactivities in meteorites. *Astrophys. J.* **506**, 898–912.

Leya I., Halliday A. N., and Wieler R. (2003) The predictable collateral consequences of nucleosynthesis by spallation reactions in the early solar system. *Astrophys. J.* **594**, 605–616.

Lin Y., Guan Y., Leshin L. A., Ouyang Z., and Daode W. (2005) Short-lived chlorine-36 in a Ca- and Al-rich inclusion from the Ningqiang carbonaceous chondrite. *Proc. Natl. Acad. Sci.* **102**, 1306–1311.

Liu M.-C., McKeegan K. D., and Davis A. M. (2006) Magnesium isotopic compositions of CM hibonite grains. In *Lunar Planet. Sci.* **XXXVII**, #2428. The Lunar and Planetary Institute, Houston (CD-ROM).

Lugmair G. W. and Galer S. J. G. (1992) Age and isotopic relationships among the angrites Lewis Cliff 86010 and Angra dos Reis. *Geochim. Cosmochim. Acta* **56**, 1673–1694.

Lugmair G. W., Shimamura T., Lewis R. S., and Anders E. (1983) Samarium-146 in the early solar system: evidence from neodymium in the Allende meteorite. *Science* **222**, 1015–1018.

Lugmair G. W. and Shukolyukov A. (1998) Early solar system timescales according to ^{53}Mn–^{53}Cr systematics. *Geochim. Cosmochim. Acta* **62**, 2863–2886.

Lugmair G. W. and Shukolyukov A. (2001) Early solar system events and timescales. *Meteorit. Planet. Sci.* **36**, 1017–1026.

MacPherson G. J. and Davis A. M. (1993) A petrologic and ion microprobe study of a Vigarano type B refractory inclusion: evolution by multiple stages of alteration and melting. *Geochim. Cosmochim. Acta* **57**, 231–243.

MacPherson G. J., Davis A. M., and Zinner E. K. (1995) The distribution of aluminum-26 in the early solar system—a reappraisal. *Meteoritics* **30**, 365–386.

MacPherson G. J. and Huss G. R. (2001) Extinct ^{10}Be in CAIs from Vigarano, Leoville, and Axtell. In *Lunar Planet. Sci.* **XXXII**, #1882. The Lunar and Planetary Institute, Houston (CD-ROM).

MacPherson G. J., Huss G. R., and Davis A. M. (2003) Extinct ^{10}Be in type A calcium–aluminum-rich inclusions from CV chondrites. *Geochim. Cosmochim. Acta* **67**, 3165–3179.

Marhas K. K. and Goswami J. N. (2003) Be–B systematics in CM and CV hibonites: implications for solar energetic particle production of short-lived nuclides in the early solar system. In *Lunar Planet. Sci.*, #1303. The Lunar and Planetary Institute, Houston (CD-ROM).

Marhas K. K., Goswami J. N., and Davis A. M. (2002) Short-lived nuclides in hibonite grains from Murchison: evidence for solar system evolution. *Science* **298**, 2182–2185.

Markowski A., Quitté G., Halliday A. N., and Kleine T. (2006) Tungsten isotopic compositions of iron meteorites: chronological constraints vs. cosmogenic effects. *Earth Planet. Sci. Lett.* **242**, 1–15.

McKeegan K. D., Chaussidon M., Krot A. N., Robert F., Goswami J. N., and Hutcheon I. D. (2001) Extinct radionuclide abundances in Ca, Al-rich inclusions from the CV chondrites Allende and Efremovka: a search for synchronicity. In *Lunar Planet. Sci.*, #2175. The Lunar and Planetary Institute, Houston (CD-ROM).

McKeegan K. D., Chaussidon M., and Robert F. (2000a) Incorporation of short-lived Be-10 in a calcium–aluminum-rich inclusion from the Allende meteorite. *Science* **289**, 1334–1337.

McKeegan K. D., Greenwood J. P., Leshin L. A., and Cosarinsky M. (2000b) Abundance of ^{26}Al in ferromagnesian chondrules of unequilibrated ordinary chondrites. In *Lunar Planet. Sci.* **XXXI**, #2009. The Lunar and Planetary Institute, Houston (CD-ROM).

McKeegan K. D. and Leshin L. A. (2001) Stable isotope variations in extraterrestrial materials. *Rev. Mineral. Geochem.* **43**, 279–318.

Meyer B. and Clayton D. D. (2000) Short-lived radioactivities and the birth of the Sun. *Space Sci. Rev.* **92**, 133–152.

Mostefaoui S., Kita N. T., Togashi S., Tachibana S., Nagahara H., and Morishita Y. (2002) The relative formation ages of ferromagnesian chondrules inferred from their initial aluminum-26/aluminum-27 ratios. *Meteorit. Planet. Sci.* **37**, 421–438.

Mostefaoui S., Lugmair G. W., and Hoppe P. (2005) ^{60}Fe: a heat source for planetary differentiation from a nearby supernova explosion. *Astrophys. J.* **625**, 271–277.

Moynier F., Telouk P., and Albarede F. (2005) Excesses of ^{60}Ni in chondrites and iron meteorites. In *Lunar Planet. Sci.* **XXXVI**, #1593. The Lunar and Planetary Institute, Houston (CD-ROM).

Münker C., Weyer S., Mezger K., Rehkämper M., Wombacher F., and Bischoff A. (2000) ^{92}Nb–^{92}Zr and the early differentiation history of planetary bodies. *Science* **289**, 1538–1542.

Murty S. V. S., Goswami J. N., and Shukolyukov Y. A. (1997) Excess Ar-36 in the Efremovka meteorite: a strong hint for the presence of Cl-36 in the early solar system. *Astrophys. J.* **475**, L65–L68.

National Nuclear Data Center, http://www.nndc.bnl.gov (accessed on August 2006).

Niederer F. R., Papanastassiou D. A., and Wasserburg G. J. (1980) Endemic isotopic anomalies in titanium. *Astrophys. J.* **240**, L73–L77.

Nielsen S. G., Rehkämper M., and Halliday A. N. (2006) Large thallium isotopic variations in iron meteorites and evidence for lead-205 in the early solar system. *Geochim. Cosmochim. Acta* **70**, 2643–2657.

Niemeyer S. and Lugmair G. W. (1981) Ubiquitous isotopic anomalies in Ti from normal Allende inclusions. *Earth Planet. Sci. Lett.* **53**, 211–225.

Nyquist L., Lindstrom D., Mittlefehldt D., Shih C. Y., Wiesmann H., Wentworth S., and Martinez R. (2001a) Manganese–chromium formation intervals for chondrules from the Bishunpur and Chainpur meteorites. *Meteorit. Planet. Sci.* **36**, 911–938.

Nyquist L. E., Reese Y., Wiesmann H., Shih C.-Y., and Takeda H. (2001b) Live ^{53}Mn and ^{26}Al in a unique cumulate eucrite with very calcic feldspar (An-98). *Meteorit. Planet. Sci.* **36**, A151–A152.

Nyquist L. E., Shih C.-Y., Wiesmann H., and Mikouchi T. (2003) Fossil ^{26}Mg and ^{53}Mn in D'Orbigny and Sahara 99555 and the timescale for angrite magmatism. In *Lunar Planet. Sci.* **XXXIV**, #1338. The Lunar and Planetary Institute, Houston (CD-ROM).

Ouellette N., Desch S. J., Hester J. J., and Leshin L. A. (2005) A nearby supernova injected short-lived radionuclides into our protoplanetary disk. In *Chondrites and the Protoplanetary Disk* (eds. A. N. Krot, E. R. D. Scott, and B. Reipurth), Astron. Soc. Pacific Conf. Ser. 341. Astronomical Society of the Pacific, San Francisco, pp. 527–538.

Papanastassiou D. A. (1986) Chromium isotopic anomalies in the Allende meteorite. *Astrophys. J.* **308**, L27–L30.

Papanastassiou D. A., Bogdanovski O., and Wasserburg G. J. (2002) ^{53}Mn–^{53}Cr systematics in Allende refractory inclusions. *Meteorit. Planet. Sci.* **37**, A114.

Podosek F. A. and Cassen P. (1994) Theoretical, observational, and isotopic estimates of the lifetime of the solar nebula. *Meteoritics* **29**, 6–25.

Podosek F. A. and Nichols R. H., Jr. (1997) Short-lived radionuclides in the solar nebula. In *Astrophysical Implications of the Laboratory Study of Presolar Materials* (eds. T. J. Bernatowicz and E. Zinner). American Institute of Physics, Woodbury, pp. 617–647.

Podosek F. A., Ott U., Brannon J. C., Neal C. R., Bernatowicz T. J., Swan P., and Mahan S. (1997) Thoroughly anomalous chromium in Orgueil. *Meteorit. Planet. Sci.* **32**, 617–627.

Podosek F. A. and Swindle T. D. (1988) Extinct radionuclides. In *Meteorites and the Early Solar System* (eds. J. F. Kerridge and M. S. Matthews). University of Arizona Press, Tucson, AZ, pp. 1093–1113.

Podosek F. A., Zinner E., MacPherson G. J., Lundberg L. L., Brannon J. C., and Fahey A. J. (1991) Correlated study of initial ^{87}Sr/^{86}Sr and Al–Mg isotopic systematics and petrologic properties in a suite of refractory inclusions from the Allende meteorite. *Geochim. Cosmochim. Acta* **55**, 1083–1110.

Preibisch T., Kim Y. C., Favata F., Feigelson E. D., Flaccomio E., Getman K., Micela G., Sciortino S., Stassun K., Stelzer B., and Zinnecker H. (2005) The origin of T tauri X-ray emission: new insights from the Chandra Orion Ultradeep Project. *Astrophys. J. Suppl.* **160**, 401–422.

Prinzhofer A., Papanastassiou D. A., and Wasserburg G. J. (1989) The presence of ^{146}Sm in the early solar system and implications for its nucleosynthesis. *Astrophys. J.* **344**, 81–84.

Qin L., Dauphas N., Wadhwa M., Janney P. E., Davis A. M., and Mazarik J. (2006) Evidence of correlated cosmogenic effects in iron meteorites: implications for the timing of metal–silicate differentiation in asteroids. In *Lunar Planet. Sci.* **XXXVII**, #1771. The Lunar and Planetary Institute, Houston (CD-ROM).

Richter F. M., Davis A. M., DePaolo D. J., and Watson E. B. (2003) Isotope fractionation by chemical diffusion between molten basalt and rhyolite. *Geochim. Cosmochim. Acta* **67**, 3905–3923.

Rotaru M., Birck J.-L., and Allègre C. J. (1992) Clues to early solar system history from chromium isotopes in carbonaceous chondrites. *Nature* **358**, 465–470.

Russell S. S., Gounelle M., and Hutchison R. (2001) Origin of short-lived radionuclides. *Phil. Trans. Roy Soc. Lond.* **359**, 1991–2004.

Russell S. S., Srinivasan G., Huss G. R., Wasserburg G. J., and MacPherson G. J. (1996) Evidence for widespread ^{26}Al in the solar nebula and constraints for nebula timescales. *Science* **273**, 757–762.

Sahijpal S. and Goswami J. N. (1998) Refractory phases in primitive meteorites devoid of ^{26}Al and ^{41}Ca: representative samples of first solar system solids? *Astrophys. J.* **509**, L137–L140.

Sahijpal S., Goswami J. N., and Davis A. M. (2000) K, Mg, Ti, and Ca isotopic compositions and refractory trace element abundances in hibonites from CM and CV meteorites: implications for early solar system processes. *Geochim. Cosmochim. Acta* **64**, 1989–2005.

Sahijpal S., Goswami J. N., Davis A. M., Grossman L., and Lewis R. S. (1998) A stellar origin for the short-lived nuclides in the early solar system. *Nature* **391**, 559.

Sahijpal S., McKeegan K. D., Krot A. N., Weber D., and Ulyanov A. A. (1999) Oxygen isotopic compositions of Ca–Al-rich inclusions from the CH chondrites, Acfer 182 and Pat91546. *Meteorit. Planet. Sci.* **34**, A101.

Sanloup C., Blicher-Toft J., Télouk P., Gillet P., and Albarède F. (2000) Zr isotope anomalies in chondrites and the presence of ^{92}Nb in the early solar system. *Earth Planet. Sci. Lett.* **184**, 75–81.

Schönbachler M., Rehkämper M., Halliday A. N., Lee D.-C., Bourot-Denise M., Zanda B., Hattendorf B., and Günther D. (2002) Niobium–zirconium chronometry and early solar system development. *Science* **295**, 1705–1708.

Schramm D. N., Tera F., and Wasserburg G. J. (1970) The isotopic abundance of ^{26}Mg and limits on ^{26}Al in the early solar system. *Earth Planet. Sci. Lett.* **10**, 44–59.

Shu F. H., Adams F. C., and Lizano S. (1987) Star formation in molecular clouds: observation and theory. *Annu. Rev. Astron. Astrophys.* **25**, 23–81.

Shu F. H., Shang H., Gounelle M., Glassgold A. E., and Lee T. (2001) The origin of chondrules and refractory inclusions in chondritic meteorites. *Astrophys. J.* **548**, 1029–1050.

Shukolyukov A. and Lugmair G. W. (1993a) Fe-60 in eucrites. *Earth Planet. Sci. Lett.* **119**, 159–166.

Shukolyukov A. and Lugmair G. W. (1993b) Live iron-60 in the early solar system. *Science* **259**, 1138–1142.

Shukolyukov A. and Lugmair G. W. (1998) Isotopic evidence for the Cretaceous–Tertiary impactor and its type. *Science* **282**, 927–929.

Simon J. I., Young E. D., Russell S. S., Tonui E. K., Dyl K. A., and Manning C. E. (2005) A short timescale for changing oxygen fugacity in the solar nebula revealed by high-resolution ^{26}Al–^{26}Mg dating of CAI rims. *Earth Planet. Sci. Lett.* **238**, 272–283.

Srinivasan G. (2001) Be–B isotope systematics in CAI E65 from Efremovka CV3 chondrite. *Meteorit. Planet. Sci.* **36**, A195–A196.

Srinivasan G., Goswami J. N., and Bhandari N. (1999) Al-26 in eucrite Piplia Kalan: plausible heat source and formation chronology. *Science* **284**, 1348–1350.

Srinivasan G., Sahijpal S., Ulyanov A. A., and Goswami J. N. (1996) Ion microprobe studies of Efremovka CAIs: potassium isotope composition and ^{41}Ca in the early solar system. *Geochim. Cosmochim. Acta* **60**, 1823–1835.

Srinivasan G., Ulyanov A. A., and Goswami J. N. (1994) ^{41}Ca in the early solar system. *Astrophys. J.* **431**, L67–L70.

Stewart B. W., Papanastassiou D. A., and Wasserburg G. J. (1994) Sm–Nd chronology and petrogenesis of mesosiderites. *Geochim. Cosmochim. Acta* **58**, 3487–3509.

Stirling C. H., Halliday A. N., and Porcelli D. (2005) In search of live ^{247}Cm in the early solar system. *Geochim. Cosmochim. Acta* **69**, 1059–1071.

Sugiura N. (2001) Boron isotopic compositions in chondrules: anorthite-rich chondrules in the Yamato 82094 (CO3) chondrite. In *Lunar Planet. Sci.* **XXXII**, #1277. The Lunar and Planetary Institute, Houston (CD-ROM).

Sugiura N., Shuzou Y., and Ulyanov A. (2001) Beryllium–boron and aluminum–magnesium chronology of calcium–aluminum-rich inclusions in CV chondrites. *Meteorit. Planet. Sci.* **36**, 1397–1408.

Swindle T. D., Caffee M. W., Hohenberg C. M., Lindstrom M. M., and Taylor G. J. (1991) Iodine–xenon studies of petrographically and chemically characterized Chainpur chondrules. *Geochim. Cosmochim. Acta* **55**, 861–880.

Swindle T. D., Davis A. M., Hohenberg C. M., MacPherson G. J., and Nyquist L. E. (1996) Formation times of chondrules and Ca–Al-rich inclusions: constraints from short-lived radionuclides. In *Chondrules and the Protoplanetary Disk* (eds. R. H. Hewins, R. H. Jones, and E. R. D. Scott). Cambridge University Press, New York, pp. 77–86.

Tachibana S. and Huss G. R. (2003) The initial abundance of ^{60}Fe in the solar system. *Astrophys. J.* **588**, L41–L44.

Tachibana S., Huss G. R., Kita N. T., Shimoda G., and Morishita Y. (2006) ^{60}Fe in chondrites: debris from a nearby supernova in the early solar system? *Astrophys. J.* **639**, L87–L90.

Taylor D. J., Cosarinsky M., Liu M.-C., McKeegan K. D., Krot A. N., and Hutcheon I. D. (2005) Survey of initial ^{26}Al in type A and type B CAIs: evidence for an extended formation period for refractory inclusions. *Meteorit. Planet. Sci.* **40**, A151.

Tera F. and Carlson R. W. (1999) Assessment of the Pb–Pb and U–Pb chronometry of the early solar system. *Geochim. Cosmochim. Acta* **63**, 1877–1889.

Tera F. and Wasserburg G. J. (1972) U–Th–Pb systematics in three Apollo 14 basalts and the problem of initial lead in lunar rocks. *Earth Planet. Sci. Lett.* **14**, 281–304.

Tilton G. R. (1988) Age of the solar system. In *Meteorites and the Early Solar System* (eds. J. F. Kerridge and M. S. Matthews). University of Arizona Press, Tucson, AZ, pp. 259–275.

Trinquier A., Birck J.-L., and Allègre C. J. (2005a) ^{54}Cr anomalies in the solar system: their extent and origin. In *Lunar Planet. Sci.* **XXXVI**, #1259. The Lunar and Planetary Institute, Houston (CD-ROM).

Trinquier A., Birck J.-L., and Allègre C. J. (2005b) Reevaluation of the ^{53}Mn–^{53}Cr systematic in the basaltic achondrites. In *Lunar Planet. Sci.* **XXXVI**, #1946. The Lunar and Planetary Institute, Houston (CD-ROM).

Wadhwa M., Amelin Y., Bogdanovski O., Shukolyukov A., Lugmair G. W., and Janney P. (2005) High precision relative and absolute ages for Asuka 881394, a unique and ancient basalt. In *Lunar Planet. Sci.* **XXXVI**, #2126. The Lunar and Planetary Institute, Houston (CD-ROM).

Wadhwa M., Amelin Y., Davis A. M., Lugmair G. W., Meyer B., Gounelle M., and Desch S. J. (2006a) From dust to planetesimals: implications for the solar protoplanetary disk from short-lived radionuclides. In *Protostars and Planets V* (eds. B. Reipurth, D. Jewitt, and K. Keil). University of Arizona Press, Tucson, AZ.

Wadhwa M., Foley C. N., Janney P., and Beecher N. A. (2003) Magnesium isotopic composition of the Juvinas eucrite: implications for concordance of the Al–Mg and Mn–Cr chronometers and timing of basaltic volcanism on asteroids. In *Lunar Planet. Sci.* **XXXIV**, #2055. The Lunar and Planetary Institute, Houston (CD-ROM).

Wadhwa M. and Lugmair G. W. (1996) Age of the eucrite "Caldera" from convergence of long-lived and short-lived chronometers. *Geochim. Cosmochim. Acta* **60**, 4889–4893.

Wadhwa M. and Russell S. S. (2000) Timescales of accretion and differentiation in the early solar system: the meteoritic evidence. In *Protostars and Planets IV* (eds. V. Mannings, A. P. Boss, and S. S. Russell). University of Arizona Press, Tucson, AZ, pp. 995–1018.

Wadhwa M., Srinivasan G., and Carlson R. W. (2006b) Timescales of planetary differentiation in the early solar system. In *Meteorites and the Early Solar System II* (eds. D. S. Lauretta and H. Y. Jr. McSween). University of Arizona Press, Tucson, AZ, pp. 715–731.

Wadhwa M., Zinner E. K., and Crozaz G. (1997) Manganese–chromium systematics in sulfides of unequilibrated enstatite chondrites. *Meteorit. Planet. Sci.* **32**, 281–292.

Wasserburg G. J. (1985) Short-lived nuclei in the early solar system. In *Protostars and Planets II* (eds. D. C. Black and M. S. Matthews). University of Arizona Press, Tucson, AZ, pp. 703–737.

Wasserburg G. J., Busso M., and Gallino R. (1996) Abundances of actinides and short-lived nonactinides in the interstellar medium: diverse supernova sources for the r-processes. *Astrophys. J.* **466**, L109–L113.

Wasserburg G. J., Busso M., Gallino R., and Nollett K. M. (2006) Short-lived nuclei in the early solar system: possible AGB sources. *Nucl. Phys. A* (in press).

Wasserburg G. J., Busso M., Gallino R., and Raiteri C. M. (1994) Asymptotic giant branch stars as a source of short-lived radioactive nuclei in the solar nebula. *Astrophys. J.* **424**, 412–428.

Wasserburg G. J., Gallino R., and Busso M. (1998) A test of the supernova trigger hypothesis with Fe-60 and Al-26. *Astrophys. J.* **500**, L189–L193.

Wasserburg G. J. and Papanastassiou D. A. (1982) Some short-lived nuclides in the early solar system—a connection with the placental ISM. In *Essays in Nuclear Astrophysics* (ed. D. N. Schramm). Cambridge University Press, Cambridge, MA, pp. 77–140.

Weber D., Zinner E., and Bischoff A. (1995) Trace element abundances and magnesium, calcium, and titanium isotopic compositions of grossite-containing inclusions from the carbonaceous chondrite Acfer 182. *Geochim. Cosmochim. Acta* **59**, 803–823.

Weidenschilling S. J. (1977) Aerodynamics of solid bodies in the solar nebula. *Mon. Not. Roy. Astron. Soc.* **180**, 57–70.

Wetherill G. W. (1956) Discordant uranium–lead ages. *Trans. Am. Geophys. Union* **37**, 320–326.

Whitby J., Burgess R., Turner G., Gilmour J., and Bridges J. (2000) Extinct ^{129}I in halite from a primitive meteorite: evidence for evaporite formation in the early solar system. *Science* **288**, 1819–1821.

Whitby J., Gilmour J., Turner G., Prinz M., and Ash R. D. (2001) Iodine–xenon dating of chondrules from the Quinzhen and Kota enstatite chondrites. *Geochim. Cosmochim. Acta* **66**, 347–359.

Yin Q. Z., Jacobsen S. B., McDonough W. F., and Horn I. (2000) Supernova sources and the ^{92}Nb–^{92}Zr p-process chronometer. *Astrophys. J.* **535**, L49–L53.

Yin Q. Z., Jacobsen S. B., Yamashita K., Blichert-Toft J., Telouk P., and Albarede F. (2002) A short timescale for terrestrial planet formation from Hf–W chronometry of meteorites. *Nature* **418**, 949–952.

Young E. D. and Russell S. S. (1998) Oxygen reservoirs in the early solar nebula inferred from an Allende CAI. *Science* **282**, 452–455.

Young E. D., Galy A., and Nagahara H. (2002) Kinetic and equilibrium mass-dependent isotope fractionation laws in nature and their geochemical and cosmochemical significance. *Geochim. Cosmochim. Acta* **66**, 1095–1104.

Young E. D., Simon J. I., Galy A., Russell S. S., Tonui E., and Lovera O. (2005) Supra-canonical ^{26}Al/^{27}Al and the residence time of CAIs in the solar protoplanetary disk. *Science* **308**, 223–227.

Yurimoto H., Ito M., and Nagasawa H. (1998) Oxygen isotope exchange between refractory inclusion in Allende and solar nebula gas. *Science* **282**, 1874–1877.

Zhai M., Nakamura E., Shaw D. M., and Nakano T. (1996) Boron isotope ratios in meteorites and lunar rocks. *Geochim. Cosmochim. Acta* **60**, 4877–4881.

Zinner E. (1998) Stellar nucleosynthesis and the isotopic composition of presolar grains from primitive meteorites. *Annu. Rev. Earth Planet. Sci.* **26**, 147–188.

Zinner E., Fahey A. J., Goswami J. N., Ireland T. R., and McKeegan K. D. (1986) Large ^{48}Ca anomalies are associated with ^{50}Ti anomalies in Murchison and Murray hibonites. *Astrophys. J.* **311**, L103–L107.

Zinner E. and Göpel C. (2002) Aluminum-26 in H4 chondrites: implications for its production and its usefulness as a fine-scale chronometer for early solar system events. *Meteorit. Planet. Sci.* **37**, 1001–1013.

Radioactive Geochronometry
ISBN: 978-0-08-096708-0

pp. 35–70

3

The Origin and Earliest History of the Earth

A. N. Halliday

Eidgenössische Technische Hochschule, Zürich, Switzerland

3.1   INTRODUCTION

The purpose of this chapter is to explain the various lines of geochemical evidence relating to the origin and earliest development of the Earth, while at the same time clarifying current limitations on these constraints. The Earth's origins are to some extent shrouded in greater uncertainty than those of Mars or the Moon because, while vastly more accessible and extensively studied, the geological record of the first 500 Myr is almost entirely missing. This means that we have to rely heavily on theoretical modeling and geochemistry to determine the mechanisms and timescales involved. Both of these approaches have yielded a series of, sometimes strikingly different, views about Earth's origin and early evolution that have seen significant change every few years. There has been a great deal of discussion and debate in the past few years in particular, fueled by new kinds of data and more powerful computational codes.

The major issues to address in discussing the origin and early development of the Earth are as follows:

(i) What is the theoretical basis for our understanding of the mechanisms by which the Earth accreted?

(ii) What do the isotopic and bulk chemical compositions of the Earth tell us about the Earth's accretion?

(iii) How are the chemical compositions of the early Earth and the Moon linked? Did the formation of the Moon affect the Earth's composition?

(iv) Did magma oceans exist on Earth and how can we constrain this from geochemistry?

(v) How did the Earth's core form?

(vi) How did the Earth acquire its atmosphere and hydrosphere and how have these changed?

(vii) What kind of crust might have formed in the earliest stages of the Earth's development?

(viii) How do we think life first developed and how might geochemical signatures be used in the future to identify early biological processes?

Although these issues could, in principle, all be covered in this chapter, some are dealt with in more detail in other chapters and, therefore, are given only cursory treatment here. Furthermore, there are major gaps in our knowledge that render a comprehensive overview unworkable. The nature of the early crust (item (vii)) is poorly constrained, although some lines of evidence will be mentioned. The nature of the earliest life forms (item (viii)) is so loaded with projections into underconstrained hypothetical environments that not a great deal can be described as providing a factual basis suitable for inclusion in a reference volume at this time. Even in those areas in which geochemical constraints are more plentiful, it is essential to integrate them with astronomical observations and dynamic (physical) models of planetary growth and primary differentiation. In some cases, the various theoretical dynamic models can be tested with isotopic and geochemical methods. In other cases, it is the Earth's composition itself that has been used to erect specific accretion paradigms. Therefore, much of this background is provided in this chapter.

All these models and interpretations of geochemical data involve some level of assumption in scaling the results to the big picture of the Earth. Without this, one cannot erect useful concepts that address the above issues. It is one of the main goals of this chapter to explain what these underlying assumptions are. As a consequence, this chapter focuses on the range of interpretations and uncertainties, leaving many issues "open." The chapter finishes by indicating where the main sources of uncertainty remain and what might be done about these in the future.

3.2   OBSERVATIONAL EVIDENCE AND THEORETICAL CONSTRAINTS PERTAINING TO THE NEBULAR ENVIRONMENT FROM WHICH EARTH ORIGINATED

3.2.1   Introduction

The starting place for all accretion modeling is the circumstellar disk of gas and dust that formed during the collapse of the solar nebula. It has been theorized for a long time that a disk of rotating circumstellar material will form as a normal consequence of transferring angular momentum during cloud collapse and star formation. Such disks now are plainly visible around young stars in the Orion nebula, thanks to the Hubble Space Telescope (McCaughrean and O'Dell, 1996). However, circumstellar disks became clearly detectable before this by using ground-based interferometry. If the light of the star is canceled out, excess infrared can be seen being emitted from the dust around the disk. This probably is caused by radiation from the star itself heating the disk.

Most astronomers consider nebular timescales to be of the order of a few million years (Podosek and Cassen, 1994). However, this is poorly constrained because unlike dust, gas is very difficult to detect around other stars. It may be acceptable to assume that gas and dust stay together for a portion of nebular history. However, the dust in some of these disks is assumed to be the secondary product of planetary accretion. Colliding planetesimals and planets are predicted to form at an early stage, embedded in the midplane of such optically thick disks (Wetherill and Stewart, 1993; Weidenschilling, 2000). The age of Beta Pictoris (Artymowicz, 1997; Vidal-Madjar *et al.*, 1998) is rather unclear but it is probably more than ~20 Myr old (Hartmann, 2000) and the dust in this case probably is secondary, produced as a consequence of collisions. Some disks around younger (<10 Myr) stars like HR 4796A appear to show evidence of large inner regions entirely swept clear of dust. It has been proposed that in these regions the dust already may be incorporated into planetary objects (Schneider *et al.*, 1999). The Earth probably formed by aggregating planetesimals and small planets that had formed in the midplane within such a dusty disk.

3.2.2 Nebular Gases and Earth-like versus Jupiter-like Planets

What features of Earth's composition provide information on this early circumstellar disk of dust that formed after the collapse of the solar nebula? The first and foremost feature of the Earth that relates to its composition and accretion is its size and density. Without any other information, this immediately raises questions about how Earth could have formed from the same disk as Jupiter and Saturn. The uncompressed density of the terrestrial planets is far higher than that of the outer gas and ice giant planets. The four most abundant elements making up ~90% of the Earth are oxygen, magnesium, silicon, and iron. Any model of the Earth's accretion has to account for this. The general explanation is that most of the growth of terrestrial planets postdated the loss of nebular gases from the disk. However, this is far from certain.

Some solar-like noble gases were trapped in the Earth and although other explanations are considered (Trieloff *et al.*, 2000; Podosek *et al.*, 2003), the one that is most widely accepted is that the nebula was still present at the time of Earth's accretion (Harper and Jacobsen, 1996a). How much nebular gas was originally present is unclear. There is xenon-isotopic evidence that the vast majority (>99%) of Earth's noble gases were lost subsequently (Ozima and Podosekm, 1999; Porcelli and Pepin, 2000). The dynamics and timescales for accretion will be very different in the presence or absence of nebular gas. In fact, one needs to consider the possibility that even Jupiter-sized gas giant planets may have formed in the terrestrial planet-forming region and were subsequently lost by being ejected from the solar system or by migrating into the Sun (Lin *et al.*, 1996). More than half the extrasolar planets detected are within the terrestrial planet-forming region of their stars, and these all are, broadly speaking, Jupiter-sized objects (Mayor and Queloz, 1995; Boss, 1998; Lissauer, 1999). There is, of course, a strong observational bias: we are unable to detect Earth-sized planets, which are not massive enough to induce a periodicity in the observed Doppler movement of the associated star or large enough to significantly occult the associated star (Boss, 1998; Seager, 2003).

For many years, it had been assumed that gas dissipation is a predictable response to radiative effects from an energetic young Sun. For example, it was theorized that the solar wind would have been ~100 times stronger than today and this, together with powerful ultraviolet radiation and magnetic fields, would have driven gases away from the disk (e.g., Hayashi *et al.*, 1985). However, we now view disks more as dynamic "conveyor belts" that transport mass *into* the star. The radiative effects on the materials that form the terrestrial planets may in fact be smaller than previously considered. Far from being "blown off" or "dissipated," the gas may well have been lost largely by being swept into the Sun or incorporated into planetary objects—some of which were themselves consumed by the Sun or ejected (Murray *et al.*, 1998; Murray and Chaboyer, 2001). Regardless of how the solar nebula was lost, its former presence, its mass, and the timing of Earth's accretion relative to that of gas loss from the disk will have a profound effect on the rate of accretion, as well as the composition and physical environment of the early Earth.

3.2.3 Depletion in Moderately Volatile Elements

Not only is there a shortage of nebular gas in the Earth and terrestrial planets today but the moderately volatile elements also are depleted (Figure 1) (Gast, 1960; Wasserburg *et al.*, 1964; Cassen, 1996). As can be seen from Figure 2, the depletion in the moderately volatile alkali elements, potassium and rubidium in particular, is far greater than that found in any class of chondritic meteorites (Taylor and Norman, 1990; Humayun and Clayton, 1995; Halliday and Porcelli, 2001; Drake and Righter, 2002). The traditional explanation is that the inner "terrestrial" planets accreted where it was hotter, within the so-called "ice line" (Cassen, 1996; Humayun and Cassen, 2000). For several reasons, it has long been assumed that the solar nebula in the terrestrial planet-forming region started as a very hot, well-mixed gas from which all of the solid and liquid Earth materials condensed. The geochemistry

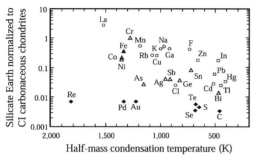

Figure 1 The estimated composition of the silicate portion of the Earth as a function of condensation temperature normalized to CI values in Anders and Grevesse (1989). Open circles: lithophile elements; shaded squares: chalcophile elements; shaded triangles: moderately siderophile elements; solid diamonds: highly siderophile elements. The spread in concentration for a given temperature is thought to be due to core formation. The highly siderophile element abundances may reflect a volatile depleted late veneer. Condensation temperatures are from Newsom (1995).

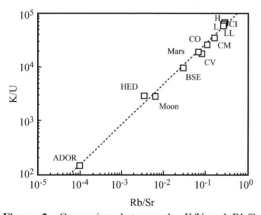

Figure 2 Comparison between the K/U and Rb/Sr ratios of the silicate Earth compared with other solar system objects. ADOR: Angra dos Reis; HED: howardite–eucrite–diogenite parent body; BSE: bulk silicate Earth; CI, CM, CV, CO, H, L, and LL are all classes of chondrites (source Halliday and Porcelli, 2001).

literature contains many references to this hot nebula, as well to major *T-Tauri* heating events that may have further depleted the inner solar system in moderately volatile elements (e.g., Lugmair and Galer, 1992). Some nebula models predict early temperatures that were sufficiently high to prevent condensation of moderately volatile elements (Humayun and Cassen, 2000), which somehow were lost subsequently. To what extent these volatile elements condensed on grains that are now in the outer solar system and may be represented by interplanetary dust particles (Jessberger *et al.*, 1992) is unclear.

Nowadays, inner solar system models are undergoing major rethinking because of new observations of stars, theoretical models, and data from meteorites. It is important to keep track of the models and observational evidence on stars and disks as this continuously changes with greater resolution and detectability. The new data provide important insights into how our solar system may have developed. As of early 2000s, the linkage between temperature in the disk and accretion dynamics is anything but clear. There is no question that transient heating was important on some scale. But a large-scale hot nebula now is more difficult to accommodate. The depletion in volatile elements in the Earth is probably the result of several different processes and the latest astronomical evidence for these is summarized below. To understand these processes one has to have some idea of how solar mass stars and their disks are thought to "work."

3.2.4 Solar Mass Stars and Heating of the Inner Disk

Solar mass stars are thought to accrete rapidly. The pre-main-sequence solar mass *protostar* probably forms from collapse of a portion of a molecular cloud onto a "cloud core" in something like 10^5 yr (Hartmann, 2000). Strong outflows and jets are sometimes observable. Within a few hundred thousand years such protostars have developed into class I young stellar objects, as can be seen in the Orion nebula. These objects already have disks and are called *proplyds*. Remaining material from the cloud will accrete onto both the disk and onto the star itself. The disk also accretes onto the star and, as it does so, astronomers can track the accretion rate from the radiation produced at the innermost margin of the disk. In general terms, the accretion rate shows a very rough decrease with age of the star. From this, it can be shown that the mass of material being accreted from the disk onto the star is about the same as the minimum mass solar nebula estimated for our solar system (Hartmann, 2000).

This "minimum mass solar nebula" is defined to be the minimum amount of hydrogen–helium gas with dust, in bulk solar system proportions, that is needed in order to form our solar system's planets (Hoyle, 1960; Weidenschilling, 1977a). It is calculated by summing the assumed amount of metal (in the astronomical sense, i.e., elements heavier than hydrogen and helium) in all planets and adding enough hydrogen and helium to bring it up to solar composition. Usually, a value of 0.01 solar masses is taken to be the "minimum mass" (Boss, 1990). The strongest constraint on the value is the abundance of heavy elements in Jupiter and Saturn. This is at least partially independent of the uncertainty of whether these elements are hosted in planetary cores. Such estimates for the minimum mass solar nebula indicate that the disk was at least a factor of 10 more massive than the total mass of the current planets. However, the mass may have been much higher and because of this the loss of metals during the planet-forming process sometimes is factored in. There certainly is no doubt that some solids were consumed by the protosun or ejected into interstellar space. This may well have included entire planets. Therefore, a range of estimates for the minimum mass of 0.01–0.1 solar masses can be found. The range reflects uncertainties that can include the bulk compositions of the gas and ice giant planets (Boss, 2002), and the amount of mass loss from, for example, the asteroid belt (Chambers and Wetherill, 2001).

Some very young stars show enormous rapid changes in luminosity with time. These are called *FU Orionis* objects. They are young Sun-like stars that probably are temporarily accreting material at rapid rates from their surrounding disks of gas and dust; they might be consuming planets, for example (Murray and Chaboyer, 2001). Over a year to a decade, they brighten by a hundred times, then stay bright for a century or so before fading again (Hartmann and Kenyon, 1996). A protostar may go through this sequence many times before the accretion disk and surrounding cloud are dispersed. Radiation from the star on to the disk during this intense stage of activity could be partially responsible for volatile depletions in the inner solar system (Bell *et al.*, 2000), but the relative importance of this versus other heating processes has not been evaluated. Nor is it known if our Sun experienced such dramatic behavior.

T-Tauri stars also are pre-main-sequence stars. They are a few times 10^5 yr to a few million years in age and the *T-Tauri* effect appears to develop after the stages described above. They have many of the characteristics of our Sun but are much brighter. Some have outflows and produce strong stellar winds. Many have disks. The *T-Tauri* effect itself is poorly understood. It has long been argued that this is an early phase of heating of the inner portions of the disk. However, such disks are generally thought to have inclined surfaces that dip in toward the star (Chiang and Goldreich, 1997, 1999). It is these surfaces that receive direct radiation from the star and produce the infrared excess observed from the dust. The *T-Tauri* stage may last a few million years. Because it heats the disk surface it may not have any great effect on the composition of the gas and dust in the accretionary midplane of the disk, where planetesimal accretion is dominant.

Heating of inner solar system material in the midplane of the disk *will* be produced from compressional effects. The thermal effects can be calculated for material in the disk being swept into an increasingly dense region during migration toward the Sun during the early stages of disk development. Boss (1990) included compressional heating and grain opacity in his modeling and showed that temperatures in excess of 1,500 K could be expected in the terrestrial planet-forming region. The main heating takes place at the midplane, because that is where most of the mass is concentrated. The surface of the disk is much cooler. More recent modeling includes the detailed studies by Nelson *et al.* (1998, 2000), which provide a very similar overall picture. Of course, if the material is being swept into the Sun, one has to ask how much of the gas and dust would be retained from this portion of the disk. This process would certainly be very early. The timescales for subsequent cooling at 1 AU would have been very short (10^5 yr). Boss (1990), Cassen (2001), and Chiang *et al.* (2001) have independently modeled the thermal evolution of such a disk and conclude that in the midplane, where planetesimals are likely to accrete, temperatures will drop rapidly. Even at 1 AU, temperatures will be ~300 K after only 10^5 yr (Chiang *et al.*, 2001). Most of the dust settles to the midplane and accretes to form planetesimals over these same short timescales (Hayashi *et al.*, 1985; Lissauer, 1987; Weidenschilling, 2000); the major portion of the solid material may not be heated externally strongly after 10^5 yr.

Pre-main-sequence solar mass stars can be vastly (10^4 times) more energetic in terms of X-ray emissions from solar flare activity in their earliest stages compared with the most energetic flare activity of the present Sun (Feigelson *et al.*, 2002a). With careful sampling of large populations of young solar mass stars in the Orion nebula it appears that this is the normal behavior of stars like our Sun. This energetic solar flare activity is very important in the first million years or so, then decreases (Feigelson *et al.*, 2002a). From this it has been concluded that the early Sun had a 10^5-fold enhancement in energetic protons which may have contributed to short-lived nuclides (Lee *et al.*, 1998; McKeegan *et al.*, 2000; Gounelle *et al.*, 2001; Feigelson *et al.*, 2002b; Leya *et al.*, 2003).

Outflows, jets, and X-winds may produce a flux of material that is scattered across the disk from the star itself or the inner regions of the disk (Shu *et al.*, 1997). The region between the outflows and jets and the disk may be subject to strong magnetic fields that focus the flow of incoming material from the disk as it is being accreted onto the star and then project it back across the disk. These "X-winds" then produce a conveyor belt that cycle material through a zone where it is vaporized before being condensed and dispersed as grains of high-temperature condensates across the disk. If material from areas close to the Sun is scattered across the disk as proposed by Shu *et al.* (1997) it could provide a source for early heated and volatile depleted objects such as calcium-, aluminum-rich refractory inclusions (CAIs) and chondrules, as well as short-lived nuclides, regardless of any direct heating of the disk at 1 AU.

Therefore, from all of the recent examples of modeling and observations of circumstellar disks a number of mechanisms can be considered that might contribute to very early heating and depletion of moderately volatile elements at 1 AU. However, some of these are localized processes and the timescales for heating are expected to be short in the midplane.

It is unclear to what extent one can relate the geochemical evidence of extreme volatile depletion in the inner solar system (Figure 2) to these observations of processes active in other disks. It has been argued that the condensation of iron grains would act as a thermostat, controlling temperatures and evening out gradients within the inner regions of the solar nebula (Wood and Morfill, 1988; Boss, 1990; Wood, 2000). Yet the depletion in moderately volatile elements between different planetary objects is highly variable and does not even vary systematically with heliocentric distance (Palme, 2000). The most striking example of this is the Earth and Moon, which have very different budgets of moderately volatile elements. Yet they are at the same heliocentric distance and appear to have originated from an identical mix of solar system materials as judged from their oxygen isotopic composition (Figure 3) (Clayton and Mayeda, 1975; Wiechert *et al.*, 2001). Oxygen isotopic compositions are highly heterogeneous among inner solar system objects (Clayton *et al.*, 1973; Clayton, 1986, 1993). Therefore, the close agreement in oxygen isotopic composition between the Earth and Moon (Clayton and Mayeda, 1975), recently demonstrated to persist to extremely high precision (Figure 3), is a striking finding that provides good evidence that the Earth and Moon were formed from material of similar origin and presumably similar composition (Wiechert *et al.*, 2001). The very fact that chondritic materials are

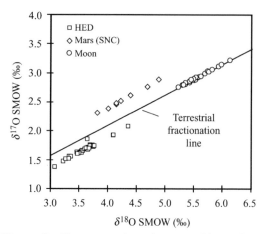

Figure 3 The oxygen isotopic compositions of the Earth and Moon are identical to extremely high precision and well resolved from the compositions of meteorites thought to come from Mars and Asteroid 4 Vesta (sources Wiechert *et al.*, 2001, 2003).

not as heavily depleted in moderately volatile elements as the Earth and Moon provides evidence that other mechanisms of volatile loss must exist. Even the enstatite chondrites, with exactly the same oxygen isotopic composition as the Earth and Moon, are not as depleted in moderately volatile alkali elements (Newsom, 1995). The geochemical constraints on the origins of the components that formed the Earth are discussed below. But first it is necessary to review some of the history of the theories about how the Earth's chemical constituents were first incorporated into planetary building material.

3.2.5 The "Hot Nebula" Model

The current picture of the early solar system outlined above, with a dynamic dusty disk, enormous gradients in temperature, and a rapidly cooling midplane, is different from that prevalent in geochemistry literature 30 yr ago. The chemical condensation sequences modeled thermodynamically for a nebular gas cooling slowly and perhaps statically from 2,000 K were long considered a starting point for understanding the basic chemistry of the material accreting in the inner solar system (Grossman and Larimer, 1974). These traditional standard hot solar nebula models assumed that practically speaking *all* of the material in the terrestrial planet-forming region resulted from gradual condensation of such a nebular gas. Because so many of the concepts in the cosmochemistry literature relate to this hot nebula model, it is important to go through the implications of the newer ways of thinking about accretion of chemical components in order to better understand how the Earth was built.

Here are some of the lines of evidence previously used to support the theory of a large-scale hot nebula that now are being reconsidered.

(i) The isotopic compositions of a wide range of elements have long been known to be broadly similar in meteorites thought to come from Mars and the asteroid belt on the one hand and the Earth and Moon on the other. Given that stars produce huge degrees of isotopic heterogeneity it was assumed that the best way to achieve this homogenization was via a well-mixed gas from which all solids and liquids condensed (Suess, 1965; Reynolds, 1967). However, we know now that chondrites contain presolar grains that cannot have undergone the heating experienced by some of the other components in these meteorites, namely CAIs and chondrules. Presolar grains are unstable in a silicate matrix above a few hundred degrees Celsius (e.g., Mendybaev et al., 2002). The ubiquitous former presence of presolar grains (Huss, 1997; Huss and Lewis, 1995; Nittler, 2003; Nittler et al., 1994) provides unequivocal evidence of dust that has been physically admixed after the formation of the other components (CAIs and chondrules). It is this well-mixed cold dust that forms the starting point for the accretion of chondrite parent bodies, and probably the planets.

(ii) The models of Cameron (1978) using a 1 solar mass disk produced extremely high temperatures ($T > 2,000$ K) throughout most of the nebular disk. Such models fueled the hot nebula model but have been abandoned in favor of minimum mass nebula models. Some such viscous accretion disk models produced very low temperatures at 1 AU because these did not include compressional heating. However, Boss (1990) provided the first comprehensive thermal model including compressional heating and grain opacity, and this model does produce temperatures in excess of 1,500 K in the terrestrial planet region.

(iii) CAIs were found to have the composition of objects that condensed at high temperatures from a gas of solar composition (Grossman, 1972; Grossman and Larimer, 1974). Their old age confirmed that they were the earliest objects to form in the solar system (Göpel et al., 1991, 1994; Amelin et al., 2002). Although most CAIs have bulk compositions broadly consistent with high-temperature condensation (Wänke et al., 1974), nearly all of them have been melted and recrystallized, destroying any textural record of condensation. It now appears that they condensed and then were reheated and possibly partially evaporated, all within a short time. It is suspected by some that these objects condensed at very high temperatures close to the Sun and that they were scattered across the disk to be admixed with other components (Gounelle et al., 2001; Shu et al., 1997). This is far from certain and some "FUN"

CAIs have isotopic compositions that cannot be easily reconciled with such a model (MacPherson et al., 1988). However, the important point is that their old age and refractory nature can be explained in ways other than just with a large-scale hot solar nebula.

(iv) The overall composition of the Earth is volatile-element depleted and this depletion is broadly consistent with that predicted from condensation theory (Cassen, 1996; Humayun and Cassen, 2000; Allègre et al., 2001). However, this agreement has rather little genetic significance. Why should chondrites be less depleted in volatiles like potassium and rubidium than the terrestrial planets and asteroids (Figure 2) if this is a nebular phenomenon? One explanation is that the chondrites accreted at 2–3 AU, where Boss (1990) shows that the nebula was cooler ($<1,000$ K). However, this provides no explanation for the extreme depletion in alkalis in eucrites and the Moon. The latter could be related to impact-induced losses (Halliday and Porcelli, 2001) but then the question arises as to whether the Earth's depletion in alkalis also relates to this in part. There is as yet no basis for distinguishing the volatile depletion that might be produced in planetary collisions (O'Neill, 1991a,b; Halliday and Porcelli, 2001) from that predicted to occur as a result of incomplete condensation of nebular gas.

(v) Strontium isotope differences between early very rubidium-depleted objects and planetesimals such as CAIs, eucrites and angrites have long been thought to provide evidence that they must have been created within a high-Rb/Sr environment such as the solar nebula but at a temperature above the condensation of rubidium (Gray et al., 1973; Wasserburg et al., 1977b; Lugmair and Galer, 1992; Podosek et al., 1991). The timescales over which the solar nebula has to be maintained above the condensation temperature of rubidium for this to work are a few million years. However, there is growing evidence that both cooling of the inner nebula and planetesimal growth may be very fast. Excluding the thermal effects from dense planetary atmospheres and the effects of planetary collisions, the timescale for major direct heating of the inner disk itself may be rather short (10^5 yr), but this view could change again with new observational data.

3.2.6 The "Hot Nebula" Model and Heterogeneous Accretion

It was at one time thought that even the terrestrial planets themselves formed directly by condensation from a hot solar nebula. This led to a class of models called heterogeneous accretion models, in which the composition of the material accreting to form the Earth changed with time as the nebula cooled. Eucken (1944) proposed such a heterogeneous accretion model in which early condensed metal formed a core to the Earth around which silicate accreted after condensation at lower temperatures. In this context the silicate-depleted, iron-enriched nature of Mercury makes sense as a body that accreted in an area of the solar nebula that was kept too hot to condense the same proportion of silicate as is found in the Earth (Lewis, 1972; Grossman and Larimer, 1974). Conversely, the lower density of Mars could partly reflect collection of an excess of silicate in cooler reaches of the inner solar nebula. So the concept of heliocentric "feeding zones" for accretion fitted this nicely. The discovery that iron metal condenses at a lower temperature than some refractory silicates made these models harder to sustain (Levin 1972). Nevertheless, a series of models involving progressive heterogeneous accretion at successively lower condensation temperatures were developed for the Earth (e.g., Turekian and Clark, 1969; Smith, 1977, 1980).

These models "produced" a zoned Earth with an early metallic core surrounded by silicate, without the need for a separate later stage of core formation. The application of condensation theory to the striking variations in the densities and compositions of the terrestrial planets, and how metal and silicate form in distinct reservoirs has been seen as problematic for some time. Heterogeneous accretion models require fast accretion and core formation if these processes reflect condensation in the nebula and such timescales can be tested with isotopic systems. The timescales for planetary accretion now are known to be far too long for an origin by partial condensation from a hot nebular gas. Nevertheless, heterogeneous accretion models have become embedded in the textbooks in Earth sciences (e.g., Brown and Mussett, 1981) and astronomy (e.g., Seeds, 1996).

An important development stemming from heterogeneous accretion models is that they introduced the concept that the Earth was built from more than one component and that these may have been accreted in separate stages. This provided an apparent answer to the problem of how to build a planet with a reduced metallic core and an oxidized silicate mantle. However, heterogeneous accretion is hard to reconcile with modern models for the protracted dynamics of terrestrial planet accretion compared with the shortness of nebular timescales. Therefore, they have been abandoned by most scientists and are barely mentioned in modern geochemistry literature any more.

3.3 THE DYNAMICS OF ACCRETION OF THE EARTH

3.3.1 Introduction

Qualitatively speaking, all accretion involves several stages, although the relative importance must differ between planets and some mechanisms are only likely to work under certain conditions that currently are underconstrained. Although the exact mechanisms of accretion of the gas and ice giant planets are poorly understood (Boss, 2002), all such objects need to accrete very rapidly in order to trap large volumes of gas before dissipation of the solar nebula. Probably this requires timescales of $<10^7$ yr (Podosek and Cassen, 1994). In contrast, the most widely accepted dynamic models advocated for the formation of the terrestrial planets (Wetherill, 1986), involve protracted timescales $\sim 10^7$–10^8 yr. Application of these same models to the outer planets would mean even longer timescales. In fact, some of the outermost planets would not have yet formed. Therefore, the bimodal distribution of planetary density and its striking spatial distribution appear to require different accretion mechanisms in these two portions of the solar system. However, one simply cannot divide the accretion dynamics into two zones. A range of rate-limiting processes probably controlled accretion of both the terrestrial and Jovian planets and the debates about which of these processes may have been common to both is far from resolved. There almost certainly was some level of commonality.

3.3.2 Starting Accretion: Settling and Sticking of Dust at 1 AU

In most models of accretion at 1 AU, the primary process being studied is the advanced stage of gravitationally driven accretion. However, one first has to consider how accretion got started and in many respects this is far more problematic. Having established that the disk was originally dominated by gas and dust, it must be possible to get these materials to combine and form larger objects on a scale where gravity can play a major role. The starting point is gravitational settling toward the midplane. The dust and grains literally will "rain" into the midplane. The timescales proposed for achieving an elevated concentration of dust in the midplane of the disk are rapid, $\sim 10^3$ yr (Hayashi *et al.*, 1985; Weidenschilling, 2000). Therefore, within a very short time the disk will form a concentrated midplane from which the growth of the planets ultimately must be fed.

Laboratory experiments on sticking of dust have been reviewed by Blum (2000), who concluded that sticking microscopic grains together with static and Van der Waals forces to build millimeter-sized compact objects was entirely feasible. However, building larger objects (fist- to football-pitch-sized) is vastly more problematic. Yet it is only when the objects are roughly kilometer-sized that gravity plays a major role. Benz (2000) has reviewed the dynamics of accretion of the larger of such intermediate-sized objects. The accretion of smaller objects is unresolved.

One possibility is that there was a "glue" that made objects stick together. Beyond the ice line, this may indeed have been relatively easy. But in the terrestrial planet-forming region in which early nebular temperatures were >1,000 K such a cement would have been lacking in the earliest stages. Of course, it already has been pointed out that cooling probably was fast at 1 AU. However, even this may not help. The baseline temperature in the solar system was then, and is now, above 160 K (the condensation temperature of water ice), so that no matter how rapid the cooling rate, the temperature would not have fallen sufficiently. The "stickiness" required rather may have been provided by carbonaceous coatings on silicate grains which might be stable at temperatures of >500 K (Weidenschilling, 2000). Waiting for the inner solar nebula to cool before accretion proceeds may not provide an explanation, anyway, because dynamic simulations provide evidence that these processes must be completed extremely quickly. The early Sun was fed with material from the disk and Weidenschilling (1977b, 2000) has argued that unless the dust and small debris are incorporated into much larger objects very quickly (in periods of less than $\sim 10^5$ yr), they will be swept into the Sun. Using a relatively large disk, Cuzzi *et al.* (2003) propose a mechanism for keeping a small fraction of smaller CAIs and fine debris in the terrestrial planet-forming region for a few million years. Most of the dust is lost. Another way of keeping the solids dust from migrating would have been the formation of gaps in the disk, preventing transfer to the Sun. The most obvious way of making gaps in the disk is by planet formation. So there is a "chicken and egg problem." Planets cannot form without gaps. Gaps cannot form without planets. This is a fundamental unsolved problem of terrestrial planetary accretion dynamics that probably deserves far more attention than has been given so far. Some, as yet uncertain, mechanism must exist for sticking small bodies together at 1 AU.

3.3.3 Starting Accretion: Migration

One mechanism to consider might be planetary migration (Lin *et al.*, 1996; Murray *et al.*, 1998). Observations of extrasolar planets provide strong evidence that planets migrate after their formation (Lin *et al.*, 1996). Resonances are observed in extrasolar planetary systems possessing multiple Jupiter-like planets. These resonances can only be explained if the planets migrated after their formation (Murray *et al.*, 1998). Two kinds of models can be considered.

(i) If accretion could not have started in the inner solar system, it might be that early icy and gas rich planets formed in the outer solar system and then migrated in toward the Sun where they

opened up gaps in the disk prior to being lost into the Sun. They then left isolated zones of material that had time to accrete into planetesimals and planets.

(ii) Another model to consider is that the terrestrial planets themselves first started forming early in the icy outer solar system and migrated in toward the Sun, where gaps opened in the disk and prevented further migration. There certainly is evidence from noble gases that Earth acquired volatile components from the solar nebula and this might be a good way to accomplish this.

Both of these models have difficulties, because of the evidence against migration in the inner solar system. First, it is hard to see why the migrating planets in model (i) would not accrete most of the material in the terrestrial planet-forming region, leaving nothing for subsequent formation of the terrestrial planets themselves. Therefore, the very existence of the terrestrial planets would imply that such migration did not happen. Furthermore, there is evidence against migration in general in the inner solar system, as follows. We know that Jupiter had to form fast (<10 Myr) in order to accrete sufficient nebular gas. Formation of Jupiter is thought to have had a big effect (Wetherill, 1992) causing the loss of >99% of the material from the asteroid belt (Chambers and Wetherill, 2001). Therefore, there are good reasons for believing that the relative positions of Jupiter and the asteroid belt have been maintained in some approximate sense at least since the earliest history of the solar system. Strong supporting evidence against inner solar system migration comes from the fact that the asteroid belt is zoned today (Gaffey, 1990; Taylor, 1992). [26]Al heating is a likely cause of this (Grimm and McSween, 1993; Ghosh and McSween, 1999). However, whatever the reason it must be an early feature, which cannot have been preserved if migration were important.

Therefore, large-scale migration from the outer solar system is not a good mechanism for initiating accretion in the terrestrial planet-forming region unless it predates formation of asteroid belt objects or the entire solar system has migrated relative to the Sun. The outer solar system provides some evidence of ejection of material and migration but the inner solar system appears to retain much of its original "structure."

3.3.4 Starting Accretion: Gravitational Instabilities

Sticking together of dust and small grains might be aided by differences between gas and dust velocities in the circumstellar disk (Weidenschilling, 2000). However, the differential velocities of the grains are calculated to be huge and nobody has been able to simulate this adequately. An early solution that was proposed by Goldreich and Ward (1973) is that gravitational instabilities built up in the disk. This means that sections of the swirling disk built up sufficient mass to establish an overall gravitational field that prevented the dust and gas in that region from moving away. With less internal differential movement there would have been more chance for clumping together and sticking. A similar kind of model has been advocated on a much larger scale for the rapid growth of Jupiter (Boss, 1997). Perhaps these earlier models need to be looked at again because they might provide the most likely explanation for the onset of terrestrial planet accretion. This mechanism has recently been reviewed by Ward (2000).

3.3.5 Runaway Growth

Whichever way the first stage of planetary accretion is accomplished, it should have been followed by *runaway gravitational growth* of these kilometer-scale planetesimals, leading to the formation of numerous Mercury- to Mars-sized planetary embryos. The end of this stage also should be reached very quickly according to dynamic simulations. Several important papers study this phase of planetary growth in detail (e.g., Lin and Papaloizou, 1985; Lissauer, 1987; Wetherill and Stewart, 1993; Weidenschilling, 2000; Kortenkamp *et al.*, 2000). With runaway growth, it is thought that Moon-sized "planetary-embryos" are built over timescales $\sim 10^5$ yr (Wetherill, 1986; Lissauer, 1993; Wetherill and Stewart, 1993). Exhausting the supply of material in the immediate vicinity prevents further runaway growth. However, there are trade-offs between the catastrophic and constructive effects of planetesimal collisions. Benz and Asphaug (1999) calculate a range of "weakness" of objects with the weakest in the solar system being ~ 300 km in size. Runaway growth predicts that accretion will be completed faster, closer to the Sun where the "feeding zone" of material will be more confined. On this basis material in the vicinity of the Earth would accrete into Moon-sized objects more quickly than material in the neighborhood of Mars, for example.

3.3.6 Larger Collisions

Additional growth to form Earth-sized planets is thought to require collisions between these "planetary embryos." This is a stochastic process such that one cannot predict in any exact way the detailed growth histories for the terrestrial planets. However, with Monte Carlo simulations and more powerful computational codes the models have become quite sophisticated and yield similar and apparently robust results in terms of the kinds of

timescales that must be involved. The mechanisms and timescales are strongly dependent on the amount of nebular gas. The presence of nebular gas has two important effects on accretion mechanisms. First, it provides added friction and pressure that speeds up accretion dramatically. Second, it can have the effect of reducing eccentricities in the orbits of the planets. Therefore, to a first approximation one can divide the models for the overall process of accretion into three possible types that have been proposed, each with vastly differing amounts of nebular gas and therefore accretion rates:

(i) *Very rapid accretion in the presence of a huge nebula.* Cameron (1978) argued that the Earth formed with a solar mass of nebular gas in the disk. This results in very short timescales of $< 10^6$ yr for Earth's accretion.

(ii) *Protracted accretion in the presence of a minimum mass solar nebula.* This is known as the Kyoto model and is summarized nicely in the paper by Hayashi *et al.* (1985). The timescales are 10^6–10^7 yr for accretion of all the terrestrial planets. The timescales increase with heliocentric distance. The Earth was calculated to form in ~ 5 Myr.

(iii) *Protracted accretion in the absence of a gaseous disk.* This model simulates the effects of accretion via planetesimal collisions assuming that all of the nebular gas has been lost. Safronov (1954) first proposed this model. He argued that the timescales for accretion of all of the terrestrial planets then would be very long, in the range of 10^7–10^8 yr.

Safronov's model was confirmed with the Monte Carlo simulations of Wetherill (1980), who showed that the provenance of material would be very broad and only slightly different for each of the terrestrial planets (Wetherill, 1994). The timescales for accretion of each planet also would be very similar. By focusing on the solutions that result in terrestrial planets with the correct (broadly speaking) size and distribution and tracking the growth of these objects, Wetherill (1986) noted that the terrestrial planets would accrete at something approaching exponentially decreasing rates. The half-mass accretion time (time for half of the present mass to accumulate) was comparable (~ 5–7 Myr), and in reality indistinguishable, for Mercury, Venus, Earth, and Mars using such simulations. Of course these objects, being of different size, would have had different absolute growth rates.

These models did not consider the effects of the growth of gas giant planets on the terrestrial planet-forming region. However, the growth of Jupiter is unlikely to have slowed down accretion at 1 AU (Kortenkamp and Wetherill, 2000).

Furthermore, if there were former gas giant planets in the terrestrial planet-forming region they probably would have caused the terrestrial planets to be ejected from their orbits and lost.

In order to distinguish between these models one has to know the amount of nebular gas that was present at the time of accretion. For the terrestrial planets this is relatively difficult to estimate. Although the Safronov–Wetherill model, which specifically assumes no nebular gas, has become the main textbook paradigm for Earth accretion, the discovery that gas giant planets are found in the terrestrial planet-forming regions of other stars (Mayor and Queloz, 1995; Boss, 1998; Lissauer, 1999; Seager, 2003) has fueled re-examination of this issue. Furthermore, recent attempts of accretion modeling have revealed that terrestrial planets can indeed be formed in the manner predicted by Wetherill but that they have high eccentricities (Canup and Agnor, 1998). Thus, they depart strongly from circular orbits. The presence of even a small amount of nebular gas during accretion has the effect of reducing this eccentricity (Agnor and Ward, 2002). This, in turn, would have sped up accretion. As explained below, geochemical data provide strong support for a component of nebular-like gases during earth accretion.

The above models differ with respect to timing and therefore can be tested with isotopic techniques. However, not only are the models very different in terms of timescales, they also differ with respect to the environment that would be created on Earth. In the first two cases the Earth would form with a hot dense atmosphere of nebular gas that would provide a ready source of solar noble gases in the Earth. This atmosphere would have blanketed the Earth and could have caused a dramatic buildup of heat leading to magma oceans (Sasaki, 1990). Therefore, the evidence from dynamic models can also be tested with compositional data for the Earth, which provide information on the nature of early atmospheres and melting.

3.4 CONSTRAINTS FROM LEAD AND TUNGSTEN ISOTOPES ON THE OVERALL TIMING, RATES, AND MECHANISMS OF TERRESTRIAL ACCRETION

3.4.1 Introduction: Uses and Abuses of Isotopic Models

Radiogenic isotope geochemistry can help with the evaluation of the above models for accretion by determining the rates of growth of the silicate reservoirs that are residual from core formation. By far the most useful systems in this regard have been the $^{235}U/^{238}U$–$^{207}Pb/^{206}Pb$ and ^{182}Hf–^{182}W systems. These are discussed in detail below.

Other long-lived systems, such as ^{87}Rb–^{87}Sr, ^{147}Sm–^{143}Nd, ^{176}Lu–^{176}Hf, and ^{187}Re–^{187}Os, have provided more limited constraints (Tilton, 1988; Carlson and Lugmair, 2000), although in a fascinating piece of work, McCulloch (1994) did attempt to place model age constraints on the age of the earth using strontium isotope data for Archean rocks (Jahn and Shih, 1974; McCulloch, 1994). The short-lived systems ^{129}I–^{129}Xe and ^{244}Pu–^{136}Xe have provided additional constraints (Wetherill, 1975a; Allègre *et al.*, 1995a; Ozima and Podosek, 1999; Pepin and Porcelli, 2002). Other short-lived systems that have been used to address the timescales of terrestrial accretion and differentiation are ^{53}Mn–^{53}Cr (Birck *et al.*, 1999), ^{92}Nb–^{92}Zr (Münker *et al.*, 2000; Jacobsen and Yin, 2001), ^{97}Tc–^{97}Mo (Yin and Jacobsen, 1998), and ^{107}Pd–^{107}Ag (Carlson and Hauri, 2001). None of these now appear to provide useful constraints. Either the model deployed currently is underconstrained (as with Mn–Cr) or the isotopic effects subsequently have been shown to be incorrect or better explained in other ways.

Hf–W and U–Pb methods both work well because the mechanisms and rates of accretion are intimately associated with the timing of core formation and this fractionates the parent/daughter ratio strongly. For a long while, however, it was assumed that accretion and core formation were completely distinct events. It was thought that the Earth formed as a cold object in less than a million years (e.g., Hanks and Anderson, 1969) but that it then heated up as a result of radioactive decay and later energetic impacts. On this basis, it was calculated that the Earth's core formed rather gradually after tens or even hundreds of millions of years following this buildup of heat and the onset of melting (Hsui and Toksöz, 1977; Solomon, 1979).

In a similar manner isotope geochemists have at various times treated core formation as a process that was distinctly later than accretion and erected relatively simple lead, tungsten and, most recently, zirconium isotopic model ages that "date" this event (e.g., Oversby and Ringwood, 1971; Allègre *et al.*, 1995a; Lee and Halliday, 1995; Galer and Goldstein, 1996; Harper and Jacobsen, 1996b; Jacobsen and Yin, 2001; Dauphas *et al.*, 2002; Kleine *et al.*, 2002; Schöenberg *et al.*, 2002). A more complex model was presented by Kramers (1998). Detailed discussions of U–Pb, Hf–W, and Nb–Zr systems are presented later in this chapter. However, some generalities should be mentioned first.

In looking at these models the following "rules" apply:

(i) Both U–Pb and Hf–W chronometry are unable to distinguish between early accretion with late core formation, and late accretion with concurrent late core formation because it is dominantly core formation that fractionates the parent/daughter ratio.

(ii) If accretion or core formation or both are protracted, the isotopic model age does not define any particular event. In the case of U–Pb it could define a kind of weighted average. In the case of short-lived nuclides, such as the ^{182}Hf–^{182}W system with a half-life of ~ 9 Myr, it cannot even provide this. Clearly, if a portion of the core formation were delayed until after ^{182}Hf had become effectively extinct, the tungsten isotopic composition of the residual silicate Earth would not be changed. Even if >50% of the mass of the core formed yesterday it would not change the tungsten isotopic composition of the silicate portion of the Earth! Therefore, the issue of how long core formation persisted is completely underconstrained by Hf–W but is constrained by U–Pb data. It also is constrained by trace element data (Newsom *et al.*, 1986).

(iii) Isotopic approaches such as those using Hf–W can only provide an indication of how quickly core formation may have started if accretion was early and very rapid. Clearly this is not a safe assumption for the Earth. If accretion were protracted, tungsten isotopes would provide only minimal constraints on when core formation started.

Tungsten and lead isotopic data can, however, be used to define the timescales for accretion, simply by assuming that core formation, the primary process that fractionates the parent/daughter ratio, started very early and that the core grew in constant proportion to the Earth (Halliday *et al.*, 1996, 2000; Harper and Jacobsen, 1996b; Jacobsen and Harper, 1996; Halliday and Lee, 1999; Halliday, 2000). There is a sound basis for the validity of this assumption, as follows.

(i) The rapid conversion of kinetic energy to heat in a planet growing by accretion of planetesimals and other planets means that it is inescapable that silicate and metal melting temperatures are achieved (Sasaki and Nakazawa, 1986; Benz and Cameron, 1990; Melosh, 1990). This energy of accretion would be sufficient to melt the entire Earth such that in all likelihood one would have magma oceans permitting rapid core formation.

(ii) There is strong observational support for this view that core formation was quasicontinuous during accretion. Iron meteorites and basaltic achondrites represent samples of small planetesimals that underwent core formation early. A strong theoretical basis for this was recently provided by Yoshino *et al.* (2003). Similarly, Mars only reached one-eighth of the mass of the Earth but clearly its size did not limit the opportunity for core formation. Also, most of the Moon is

thought to come from the silicate-rich portion of a Mars-sized impacting planet, known as "Theia" (Cameron and Benz, 1991; Canup and Asphaug, 2001; Halliday, 2000), which also was already differentiated into core and silicate. The amount of depletion in iron in eucrites, martian meteorites, and lunar samples provides support for the view that the cores of all the planetesimals and planets represented were broadly similar in their proportions to Earth's, regardless of absolute size. The slightly more extensive depletion of iron in the silicate Earth provides evidence that core formation was more efficient or protracted, but not that it was delayed.

There is no evidence that planetary objects have to achieve an Earth-sized mass or evolve to a particular state (other than melting), before core formation will commence. It is more reasonable to assume that the core grew with the accretion of the Earth in roughly the same proportion as today. If accretion were protracted, the rate-limiting parameter affecting the isotopic composition of lead and tungsten in the silicate Earth would be the timescale for accretion. As such, the "age of the core" is an average time of formation of the Earth itself. Therefore, simple tungsten and lead isotopic model ages do not define an event as such. The isotopic data instead need to be integrated with models for the growth of the planet itself to place modeled limits for the rate of growth.

The first papers exploring this approach were by Harper and Jacobsen (1996b) and Jacobsen and Harper (1996). They pointed out that the Monte Carlo simulations produced by Wetherill (1986) showed a trend of exponentially decreasing planetary growth with time. They emulated this with a simple expression for the accretionary mean life of the Earth, where the mean life is used in the same way as in nuclear literature as the inverse of a time constant. This model is an extension of the earlier model of Jacobsen and Wasserburg (1979) evaluating the mean age of the continents using Sm–Nd. Jacobsen and Harper applied the model to the determination of the age of the Earth based on (then very limited) tungsten isotope data. Subsequent studies (Halliday *et al.*, 1996, 2000; Halliday and Lee, 1999; Halliday, 2000; Yin *et al.*, 2002), including more exhaustive tungsten as well as lead isotope modeling, are all based on this same concept. However, the data and our understanding of the critical parameters have undergone major development.

Nearly all of these models assume that:

(i) accretion proceeded at an exponentially decreasing rate from the start of the solar system;

(ii) core formation and its associated fractionation of radioactive parent/radiogenic daughter ratios was coeval with accretion;

(iii) the core has always existed in its present proportion relative to the total Earth;

(iv) the composition of the accreting material did not change with time;

(v) the accreting material equilibrated fully with the silicate portion of the Earth just prior to fractionation during core formation; and

(vi) the partitioning of the parent and daughter elements between mantle and core remained constant.

The relative importance of these assumptions and the effects of introducing changes during accretion have been partially explored in several studies (Halliday *et al.*, 1996, 2000; Halliday and Lee, 1999; Halliday, 2000). The issue of metal–silicate equilibration has been investigated recently by Yoshino *et al.* (2003). However, the data upon which many of the fundamental isotopic and chemical parameters are based are in a state of considerable uncertainty.

3.4.2 Lead Isotopes

Until recently, the most widely utilized approach for determining the rate of formation of the Earth was U–Pb geochronology. The beauty of using this system is that one can deploy the combined constraints from both ^{238}U–^{206}Pb ($T_{1/2} = 4,468$ Myr) and ^{235}U–^{207}Pb ($T_{1/2} = 704$ Myr) decay. Although the atomic abundance of both of the daughter isotopes is a function of the U/Pb ratio and age, combining the age equations allows one to cancel out the U/Pb ratio. The relative abundance of ^{207}Pb and ^{206}Pb indicates when the fractionation took place. Patterson (1956) adopted this approach in his classic experiment to determine the age of the Earth. Prior to his work, there were a number of estimates of the age of the Earth based on lead isotopic data for terrestrial galenas. However, Patterson was the first to obtain lead isotopic data for early low-U/Pb objects (iron meteorites) and this defined the initial lead isotopic composition of the solar system. From this, it was clear that the silicate Earth's lead isotopic composition required between 4.5 Gyr and 4.6 Gyr of evolution as a high-U/Pb reservoir. Measurements of the lead isotopic compositions of other high-U/Pb objects such as basaltic achondrites and lunar samples confirmed this age for the solar system.

In detail it is now clear that most U–Pb model ages of the Earth (Allègre *et al.*, 1995a) are significantly younger than the age of early solar system materials such as chondrites (Göpel *et al.*, 1991) and angrites (Wasserburg *et al.*, 1977b; Lugmair and Galer, 1992). Such a conclusion has been reached repeatedly from consideration of the lead isotope compositions of early Archean rocks (Gancarz and Wasserburg, 1977; Vervoort *et al.*, 1994),

conformable ore deposits (Doe and Stacey, 1974; Manhès *et al.*, 1979; Tera, 1980; Albarède and Juteau, 1984), average bulk silicate Earth (BSE) (Galer and Goldstein, 1996) and mid-ocean ridge basalts (MORBs) (Allègre *et al.*, 1995a), all of which usually yield model ages of <4.5 Ga. Tera (1980) obtained an age of 4.53 Ga using Pb–Pb data for old rocks but even this postdates the canonical start of the solar system by over 30 Myr.

The reason why nearly all such approaches yield similar apparent ages that postdate the start of the solar system by a few tens of millions of years is that there was a very strong U/Pb fractionation that took place during the protracted history of accretion. The U–Pb model age of the Earth can only be young if U/Pb is fractionated at a late stage. This fractionation was of far greater magnitude than that associated with any later processes. Thus it has left a clear and irreversible imprint on the ^{207}Pb/^{206}Pb and ^{207}Pb/^{204}Pb isotope ratios of the silicate portion of the Earth. Uranium, being lithophile, is largely confined to the silicate portion of the Earth. Lead is partly siderophile and chalcophile such that >90% of it is thought to be in the core (Allègre *et al.*, 1995a). Therefore, it was long considered that the lead isotopic "age of the Earth" dates core formation (Oversby and Ringwood, 1971; Allègre *et al.*, 1982).

The exact value of this fractionation is poorly constrained, because lead also is moderately volatile, so that the U/Pb ratio of the total Earth (mantle, crust, and core combined) also is higher than chondritic. In fact, some authors even have argued that the dominant fractionation in U/Pb in the BSE was caused by volatile loss (Jacobsen and Harper, 1996b; Harper and Jacobsen, 1996b; Azbel *et al.*, 1993). This is consistent with some compilations of data for the Earth, which show that lead is barely more depleted in the bulk silicate portion of the Earth than lithophile elements of similar volatility (McDonough and Sun, 1995). It is, therefore, important to know how much of the lead depletion is caused by accretion of material that was depleted in volatile elements at an early (nebular) stage. Galer and Goldstein (1996) and Allègre *et al.* (1995a) have compellingly argued that the major late-stage U/Pb fractionation was the result of core formation. However, the uncertainty over the U/Pb of the total Earth and whether it changed with accretion time remains a primary issue limiting precise application of lead isotopes.

Using exponentially decreasing growth rates and continuous core formation one can deduce an accretionary mean life assuming a ^{238}U/^{204}Pb for the total Earth of 0.7 (Halliday, 2000). This value is based on the degree of depletion of moderately volatile lithophile elements, as judged from the K/U ratio of the BSE (Allègre *et al.*, 1995a). Application of this approach to the lead isotopic compositions of the Earth (Halliday, 2000) provides evidence that the Earth accreted with an accretionary mean life of between 15 Myr and 50 Myr, depending on which composition of the BSE is deployed (Figure 4). A similar figure to this in Halliday (2000) is slightly different (i.e., incorrect) because of a scaling error. The shaded region covers the field defined by eight estimates for the composition of the BSE as summarized by Galer and Goldstein (1996). The star shows the estimate of Kramers and Tolstikhin (1997). The thick bar shows the recent estimate provided by Murphy *et al.* (2003). Regardless of which of these 10 estimates of the lead isotopic composition is used, the mean life for accretion is at least 15 Myr. Therefore, there is no question that the lead isotopic data for the Earth provide evidence of a protracted history of accretion and concomitant core formation as envisaged by Wetherill (1986).

3.4.3 Tungsten Isotopes

While lead isotopes have been useful, the ^{182}Hf–^{182}W chronometer ($T_{1/2} = 9$ Myr) has been at least as effective for defining rates of accretion (Halliday, 2000; Halliday and Lee, 1999; Harper and Jacobsen, 1996b; Jacobsen and Harper, 1996; Lee and Halliday, 1996, 1997; Yin *et al.*, 2002). Like U–Pb, the Hf–W system has been used more for defining a model age of core formation

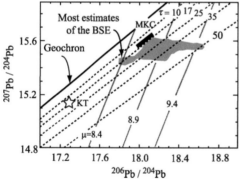

Figure 4 Lead isotopic modeling of the composition of the silicate Earth using continuous core formation. The principles behind the modeling are as in Halliday (2000). See text for explanation. The field for the BSE encompasses all of the estimates in Galer and Goldstein (1996). The values suggested by Kramers and Tolstikhin (1997) and Murphy *et al.* (2003) also are shown. The mean life (τ) is the time required to achieve 63% of the growth of the Earth with exponentially decreasing rates of accretion. The μ values are the ^{238}U/^{204}Pb of the BSE. It is assumed that the μ of the total Earth is 0.7 (Allègre *et al.*, 1995a). It can be seen that the lead isotopic composition of the BSE is consistent with protracted accretion over periods of 10^7–10^8 yr.

(Kramers, 1998; Horan *et al.*, 1998; Kleine *et al.*, 2002; Lee and Halliday, 1995, 1996, 1997; Quitté *et al.*, 2000; Dauphas *et al.*, 2002; Schöenberg *et al.*, 2002). As explained above this is not useful for an object like the Earth.

The half-life renders ^{182}Hf as ideal among the various short-lived chronometers for studying accretionary timescales. Moreover, there are two other major advantages of this method.

(i) Both parent and daughter elements (hafmium and tungsten) are refractory and, therefore, are in chondritic proportions in most accreting objects. Therefore, unlike U–Pb, we think we know the isotopic composition and parent/daughter ratio of the entire Earth relatively well.

(ii) Core formation, which fractionates hafmium from tungsten, is thought to be a very early process as discussed above. Therefore, the rate-limiting process is simply the accretion of the Earth.

There are several recent reviews of Hf–W (e.g., Halliday and Lee, 1999; Halliday *et al.*, 2000), to which the reader can refer for a comprehensive overview of the data and systematics. However, since these were written it has been shown that chondrites, and by inference the average solar system, have tungsten isotopic compositions that are resolvable from that of the silicate Earth (Kleine *et al.*, 2002; Lee and Halliday, 2000a; Schoenberg *et al.*, 2002; Yin *et al.*, 2002). Although the systematics, equations, and arguments have not changed greatly, this has led to considerable uncertainty over the exact initial ^{182}Hf abundance in the early solar system. Because this is of such central importance to our understanding of the timescales of accretion that follow from the data, it is discussed in detail below. Similarly, some of the tungsten isotopic effects that were once considered to reflect radioactive decay within the Moon (Lee *et al.*, 1997; Halliday and Lee, 1999) are now thought to *partly* be caused by production of cosmogenic ^{182}Ta (Leya *et al.*, 2000; Lee *et al.*, 2002).

The differences in tungsten isotopic composition are most conveniently expressed as deviations in parts per 10,000, as follows:

$$\varepsilon_W = \left[\frac{(^{182}W/^{184}W)_{sample}}{(^{182}W/^{184}W)_{BSE}} - 1 \right] \times 10^4$$

where the BSE value $(^{182}W/^{184}W)_{BSE}$ is the measured value for an NIST tungsten standard. This should be representative of the BSE as found by comparison with the values for terrestrial standard rocks (Lee and Halliday, 1996; Kleine *et al.*, 2002; Schoenberg *et al.*, 2002). If ^{182}Hf was sufficiently abundant at the time of formation (i.e., at an early age), then minerals, rocks, and reservoirs with higher Hf/W ratios will produce tungsten that is significantly more radiogenic (higher $^{182}W/^{184}W$ or ε_W) compared with the initial tungsten isotopic composition of the solar system. Conversely, metals with low Hf/W that segregate at an early stage from bodies with chondritic Hf/W (as expected for most early planets and planetesimals) will sample unradiogenic tungsten.

Harper *et al.* (1991a) were the first to provide a hint of a tungsten isotopic difference between the iron meteorite Toluca and the silicate Earth. It is now clear that there exists a ubiquitous clearly resolvable deficit in ^{182}W in iron meteorites and the metals of ordinary chondrites, relative to the atomic abundance found in the silicate Earth (Lee and Halliday, 1995, 1996; Harper and Jacobsen, 1996b; Jacobsen and Harper, 1996; Horan *et al.*, 1998). A summary of most of the published data for iron meteorites is given in Figure 5. It can be seen that most early segregated metals are deficient by $\sim(3-4)\varepsilon_W$ units (300–400 ppm) relative to the silicate Earth. Some appear to be even more negative, but the results are not well resolved. The simplest explanation for this difference is that the metals, or the silicate Earth, or both, sampled early solar system tungsten before live ^{182}Hf had decayed.

The tungsten isotopic difference between early metals and the silicate Earth reflects the time integrated Hf/W of the material that formed the Earth and its reservoirs, during the lifetime of ^{182}Hf. The Hf/W ratio of the silicate Earth is considered to be in the range of 10–40 as a result of an intensive study by Newsom *et al.* (1996). This is an order of magnitude higher than in carbonaceous and ordinary chondrites and a consequence of terrestrial core formation. More recent estimates are provided in Walter *et al.* (2000).

If accretion and core formation were early, an excess of ^{182}W would be found in the silicate Earth, relative to average solar system (chondrites). However, the tungsten isotopic difference between early metals and the silicate Earth on its own does not provide constraints on timing. One needs to know the atomic abundance of ^{182}Hf at the start of the solar system (or the $(^{182}Hf/^{180}Hf)_{BSSI}$, the "bulk solar system initial") and the composition of the chondritic reservoirs from which most metal and silicate reservoirs were segregated. In other words, it is essential to know to what extent the "extra" ^{182}W in the silicate Earth relative to iron meteorites accumulated in the accreted chondritic precursor materials or proto-Earth with an Hf/W ~1 prior to core formation, and to what extent it reflects an accelerated change in isotopic composition because of the high Hf/W (~15) in the silicate Earth.

For this reason some of the first attempts to use Hf–W (Harper and Jacobsen, 1996b; Jacobsen and Harper, 1996) gave interpretations that are now known to be incorrect because the $(^{182}Hf/^{180}Hf)_{BSSI}$ was underconstrained. This is a central concern in Hf–W chronometry that does not apply to U–Pb; for the latter system, parent abundances can still be measured today. In order to determine the $(^{182}Hf/^{180}Hf)_{BSSI}$ correctly one can use several approaches with varying degrees of reliability:

(i) The first approach is to model the expected $(^{182}Hf/^{180}Hf)_{BSSI}$ in terms of nucleosynthetic processes. Wasserburg *et al.* (1994) successfully predicted the initial abundances of many of the short-lived nuclides using a model of nucleosynthesis in AGB stars. Extrapolation of their model predicted a low $(^{182}Hf/^{180}Hf)_{BSSI}$ of $<10^{-5}$, assuming that ^{182}Hf was indeed produced in this manner. Subsequent to the discovery that the $(^{182}Hf/^{180}Hf)_{BSSI}$ was $>10^{-4}$ (Lee and Halliday, 1995, 1996), a number of new models were developed based on the assumption that ^{182}Hf is produced in the same kind of r-process site as the actinides (Wasserburg *et al.*, 1996; Qian *et al.*, 1998; Qian and Wasserburg, 2000).

(ii) The second approach is to measure the tungsten isotopic composition of an early high-Hf/W phase. Ireland (1991) attempted to measure the amount of ^{182}W in zircons (with very high hafnium content) from the mesosiderite Vaca Muerta, using an ion probe, and from this deduced

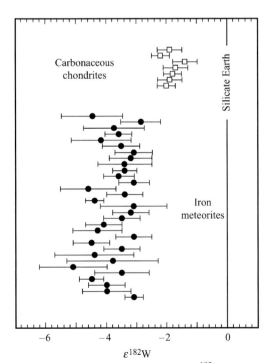

Figure 5 Well-defined deficiency in ^{182}W in early metals and carbonaceous chondrites relative to the silicate Earth (source Lee and Halliday, 1996; Horan *et al.*, 1998; Kleine *et al.*, 2002).

that the $(^{182}Hf/^{180}Hf)_{BSSI}$ was $<10^{-4}$. Unfortunately, these zircons are not dated with sufficient precision (Ireland and Wlotzka, 1992) to be very certain about the time extrapolation of the exact hafnium abundances. Nevertheless, on the basis of this work and the model of Wasserburg *et al.* (1994), Jacobsen and Harper (1996) assumed that the $(^{182}Hf/^{180}Hf)_{BSSI}$ was indeed low ($\sim 10^{-5}$). It was concluded that the difference in tungsten isotopic composition between the iron meteorite Toluca and the terrestrial value could only have been produced by radioactive decay within the silicate Earth with high Hf/W. Therefore, the fractionation of Hf/W produced by terrestrial core formation had to be early. They predicted that the Earth accreted very rapidly with a two-stage model age of core formation of $<15\,Myr$ after the start of the solar system.

(iii) The third approach is to simply assume that the consistent unradiogenic tungsten found in iron meteorites and metals from ordinary chondrites represents the initial tungsten isotopic composition of the solar system. This is analogous to the way in which the lead isotopic composition of iron meteorites has been used for decades. The difference between this and the present-day value of carbonaceous chondrites represents the effects of radiogenic ^{182}W growth *in situ* with chondritic Hf/W ratios. This in turn will indicate the $(^{182}Hf/^{180}Hf)_{BSSI}$. The difficulty with this approach has been to correctly determine the tungsten isotopic composition of chondrites. Using multiple collector ICPMS, Lee and Halliday (1995) were the first to publish such results and they reported that the tungsten isotopic compositions of the carbonaceous chondrites Allende and Murchison could not be resolved from that of the silicate Earth. The $(^{182}Hf/^{180}Hf)_{BSSI}$, far from being low, appeared to be surprisingly high at $\sim 2 \times 10^{-4}$ (Lee and Halliday, 1995, 1996). Subsequently, it has been shown that these data must be incorrect (Figure 5). Three groups have independently shown (Kleine *et al.*, 2002; Schoenberg *et al.*, 2002; Yin *et al.*, 2002) that there is a small but clear deficiency in ^{182}W ($\varepsilon_W = -1.5$ to -2.0) in carbonaceous chondrites similar to that found in enstatite chondrites (Lee and Halliday, 2000a) relative to the BSE. Kleine *et al.* (2002), Schoenberg *et al.* (2002), and Yin *et al.* (2002) all proposed a somewhat lower $(^{182}Hf/^{180}Hf)_{BSSI}$ of $\sim 1.0 \times 10^{-4}$. However, the same authors based this result on a solar system initial tungsten isotopic composition of $\varepsilon_W = -3.5$, which was derived from their own, rather limited, measurements. Schoenberg *et al.* (2002) pointed out that if instead one uses the full range of tungsten isotopic composition previously reported for iron meteorites (Horan *et al.*, 1998; Jacobsen and Harper, 1996; Lee and Halliday, 1995, 1996) one obtains a $(^{182}Hf/^{180}Hf)_{BSSI}$ of $>1.3 \times 10^{-4}$ (Table 1).

Table 1 Selected W-isotope data for iron meteorites.

Iron meteorite	$\varepsilon^{182}W$	$^{182}Hf/^{180}Hf$
Bennett Co.	-4.6 ± 0.9	$(1.74 \pm 0.57) \times 10^{-4}$
Lombard	-4.3 ± 0.3	$(1.55 \pm 0.37) \times 10^{-4}$
Mt. Edith	-4.5 ± 0.6	$(1.68 \pm 0.46) \times 10^{-4}$
Duel Hill-1854	-5.1 ± 1.1	$(2.07 \pm 0.68) \times 10^{-4}$
Tlacotepec	-4.4 ± 0.4	$(1.68 \pm 0.41) \times 10^{-4}$

Source: Horan *et al.* (1998). The calculated difference between the initial and present-day W-isotopic composition of the solar system is equal to the $^{180}Hf/^{184}W$ of the solar system multiplied by the $(^{182}Hf/^{180}Hf)_{BSSI}$. The W-isotopic compositions of iron meteorites are maxima for the $(\varepsilon^{182}W)_{BSSI}$, and therefore provide a limit on the minimum $(^{182}Hf/^{180}Hf)_{BSSI}$. The $^{182}Hf/^{180}Hf$ at the time of formation of these early metals shown here is calculated from the W-isotopic composition of the metal, the W-isotopic composition of carbonaceous chondrites (Kleine *et al.*, 2002), and the average $^{180}Hf/^{184}W$ for carbonaceous chondrites of 1.34 (Newsom *et al.*, 1996).

(iv) The fourth approach is to determine an internal isochron for an early solar system object with a well-defined absolute age (Swindle, 1993). The first such isochron was for the H4 ordinary chondrite Forest Vale (Lee and Halliday, 2000a). The best-fit line regressed through these data corresponds to a slope $(=^{182}Hf/^{180}Hf)$ of $(1.87 \pm 0.16) \times 10^{-4}$. The absolute age of tungsten equilibration in Forest Vale is unknown but may be 5 Myr younger than the CAI inclusions of Allende (Göpel *et al.*, 1994). Kleine *et al.* (2002) and Yin *et al.* (2002) both obtained lower initial $^{182}Hf/^{180}Hf$ values from internal isochrons, some of which are relatively precise. The two meteorites studied by Yin *et al.* (2002) are poorly characterized, thoroughly equilibrated meteorites of unknown equilibration age. The data for Ste. Marguerite obtained by Kleine *et al.* (2002) were obtained by separating a range of unknown phases with very high Hf/W. They obtained a value closer to 1.0×10^{-4}, but it is not clear whether the phases studied are the same as those analyzed from Forest Vale by Lee and Halliday (2000b) with lower Hf/W. Nevertheless, they estimated that their isochron value was closer to the true $(^{182}Hf/^{180}Hf)_{BSSI}$.

Both, the uncertainty over $(^{182}Hf/^{180}Hf)_{BSSI}$ and the fact that the tungsten isotopic composition of the silicate Earth is now unequivocally resolvable from a now well-defined chondritic composition (Kleine *et al.*, 2002; Lee and Halliday, 2000a; Schoenberg *et al.*, 2002; Yin *et al.*, 2002), affect the calculated timescales for terrestrial accretion. It had been argued that accretion and core formation were fairly protracted and characterized by equilibration between accreting materials and the silicate Earth (Halliday, 2000; Halliday *et al.*, 1996, 2000; Halliday and Lee, 1999). In other words, the tungsten isotope data provide very strong confirmation of the models of Safronov (1954) and Wetherill (1986). This general scenario remains the same with the new data but in detail there are changes to the exact timescales.

Previously, Halliday (2000) estimated that the mean life, the time required to accumulate 63% of the Earth's mass with exponentially decreasing accretion rates, must lie in the range of 25–40 Myr based on the combined constraints imposed by the tungsten and lead isotope data for the Earth. Yin *et al.* (2002) have argued that the mean life for Earth accretion is more like ~11 Myr based on their new data for chondrites. The lead isotope data for the Earth are hard to reconcile with such rapid accretion rates as already discussed (Figure 4). Therefore, at present there is an unresolved apparent discrepancy between the models based on tungsten and those based on lead isotope data. Resolving this discrepancy highlights the limitations in both the tungsten and the lead isotope modeling. Here are some of the most important weaknesses to be aware of:

(i) The U/Pb ratio of the total Earth is poorly known.

(ii) In all of these models it is assumed that the Earth accretes at exponentially decreasing rates. Although the exponentially decreasing rate of growth of the Earth is based on Monte Carlo simulations and makes intuitive sense given the ever decreasing probability of collisions, the reality cannot be this simple. As planets get bigger, the average size of the objects with which they collide also must increase. As such, the later stages of planetary accretion are thought to involve major collisions. This is a stochastic process that is hard to predict and model. It means that the current modeling can only provide, at best, a rough description of the accretion history.

(iii) The Moon is thought to be the product of such a collision. The Earth's U/Pb ratio conceivably might have increased during accretion if a fraction of the moderately volatile elements were lost during very energetic events like the Moon-forming giant impact (Figure 6).

(iv) Similarly, as the objects get larger, the chances for equilibration of metal and silicate would seem to be less likely. This being the case, the tungsten and lead isotopic composition of the silicate Earth could reflect only partial equilibration with incoming material such that the tungsten and lead isotopic composition is partly inherited. This has been modeled in detail by Halliday (2000) in the context of the giant impact and more recently has been studied by Vityazev *et al.* (2003) and Yoshino *et al.* (2003) in the context of equilibration of asteroidal-sized objects. If correct, it would mean accretion was even slower than can be deduced from tungsten or lead isotopes. If lead equilibrated more readily than tungsten did, for whatever reason, it might help explain some of the discrepancy. One possible way to decouple lead from tungsten would be by their relative volatility. Lead could have been

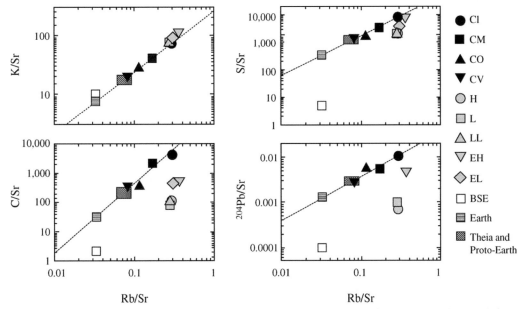

Figure 6 Volatile/refractory element ratio–ratio plots for chondrites and the silicate Earth. The correlations for carbonaceous chondrites can be used to define the composition of the Earth, the Rb/Sr ratio of which is well known, because the strontium isotopic composition of the BSE represents the time-integrated Rb/Sr. The BSE inventories of volatile siderophile elements carbon, sulfur, and lead are depleted by more than one order of magnitude because of core formation. The values for Theia are time-integrated compositions, assuming time-integrated Rb/Sr deduced from the strontium isotopic composition of the Moon (Figure 8) can be used to calculate other chemical compositions from the correlations in carbonaceous chondrites (Halliday and Porcelli, 2001). Other data are from Newsom (1995).

equilibrated by vapor-phase exchange, while tungsten would not have been able to do this and would require intimate physical mixing and reduction to achieve equilibrium.

(v) $(^{182}Hf/^{180}Hf)_{BSSI}$ is, at present, very poorly defined and could be significantly higher (Table 1). This would result in more protracted accretion timescales deduced from Hf–W (Table 2), which would be more consistent with the results obtained from U–Pb. An improved and more reliable Hf–W chronometry will depend on the degree to which the initial hafnium and tungsten isotopic compositions at the start of the solar system can be accurately defined (Tables 1 and 2). Techniques must be developed for studying very early objects like CAIs.

(vi) There is also a huge range of uncertainty in the Hf/W ratio of the silicate Earth (Newsom *et al.*, 1996), with values ranging between 10 and 40. The tungsten isotope age calculations presented in the literature tend to assume a value at the lower end of this range. Adoption of higher values also would result in more protracted accretion timescales based on Hf–W. The most recent independent estimates (Walter *et al.*, 2000) are significantly higher than those used by Yin *et al.* (2002) and Kleine *et al.* (2002) in support of their proposed timescales.

(vii) The lead isotopic composition of the BSE is not well defined (Figure 4). If the correct value lies closer to the Geochron than previously recognized, then the apparent accretion timescales for the Earth would be shortened.

(viii) Lastly, the decay constant for ^{182}Hf has a reported uncertainty of $\pm 22\%$ and really accurate determinations of accretion timescales require a significant reduction in this uncertainty.

Having made all of these cautionary statements, one still can state something useful about the overall accretion timescales. All recent combined accretion/continuous core formation models (Halliday, 2000; Halliday *et al.*, 2000; Yin *et al.*, 2002) are in agreement that the timescales are in the range 10^7–10^8 yr, as predicted by Wetherill (1986). Therefore, we can specifically evaluate the models of planetary accretion proposed earlier as follows.

If the Earth accreted very fast, in $<10^6$ yr, as proposed by Cameron (1978), the silicate Earth would have a tungsten isotopic composition that is vastly more radiogenic than that observed today (Figure 5). Such objects would have $\varepsilon_W > +10$, rather than 0 (just two ε units above average solar system). Therefore, we can say with some confidence that this model does not describe the accretion of the Earth. Protracted accretion in the absence of nebular gas, as proposed by Safronov and Wetherill, is very consistent with the close agreement between chondrites and the silicate Earth (Figure 4). To what extent the Kyoto model, which involves a significant amount of nebular gas (Hayashi *et al.*, 1985), can be confirmed or discounted is unclear at present. However, even the timescales presented by Yin *et al.* (2002) are long compared with the 5 Myr for accretion of the Earth predicted by the Kyoto model. This could change somewhat with further tungsten and lead isotope data and the definition of the critical parameters and modeling. However, the first-order conclusion is that nebular gas was at most somewhat limited during accretion and that there must have been much less gas than that implied by the minimum mass solar nebula scenario proposed in the Kyoto model.

Table 2 The apparent mean life of formation of the Earth, as well as the predicted W-isotopic compositions of the lunar mantle and silicate Earth.

$\varepsilon^{182}W_{BSSI}$	-5.0	-4.5	-4.0	-3.5
$(^{182}Hf/^{180}Hf)_{BSSI}$	2.0×10^{-4}	1.7×10^{-4}	1.4×10^{-4}	1.0×10^{-4}
Accretionary mean life of the earth (yr)	15×10^6	14×10^6	13×10^6	12×10^6
$\varepsilon^{182}W$ for: lunar initial/present-day lunar mantle/present-day silicate Earth				
Giant impact at 30×10^6 yr	$+0.3/+7.1/+0.4$	$-0.1/+5.6/0.0$	$-0.4/+4.2/-0.4$	$-0.8/+2.8/-0.7$
Giant impact at 40×10^6 yr	$-0.3/+2.9/+0.8$	$\mathbf{-0.6/+2.0/+0.3}$	$\mathbf{-0.8/+1.3/-0.1}$	$-1.1/+0.6/-0.5$
Giant impact at 45×10^6 yr	$\mathbf{-0.6/+1.6/0.0}$	$\mathbf{-0.8/+1.0/-0.3}$	$-1.0/+0.5/-0.6$	$-1.2/-0.1/-0.9$
Giant impact at 50×10^6 yr	$-0.8/+0.7/-0.6$	$-1.0/+2.0/-0.8$	$-1.2/-0.2/-1.0$	$-1.3/-0.6/-1.2$
Giant impact at 60×10^6 yr	$-1.2/-0.9/-1.2$	$-1.3/-1.0/-1.3$	$-1.4/-1.0/-1.4$	$-1.5/-1.2/-1.6$
Giant impact at 70×10^6 yr	$-1.3/-1.2/-1.5$	$-1.4/-1.3/-1.6$	$-1.5/-1.4/-1.7$	$-1.6/-1.5/-1.7$

All parameters are critically dependent on the initial W- and Hf- isotopic composition of the solar system, which at present are poorly known (see text). All calculated values assume that the W depletion in the silicate Earth and the lunar mantle are as given in Walter *et al.* (2000). The $(^{182}Hf/^{180}Hf)_{BSSI}$, model ages, and Earth and Moon compositions are all calculated using the W-isotopic composition of carbonaceous chondrites (Kleine *et al.*, 2002) and the average $^{180}Hf/^{184}W$ for carbonaceous chondrites of 1.34 (Newsom, 1990). The principles behind the modeling are as in Halliday (2000). The giant impact is assumed to have occurred when the Earth had reached 90% of its current mass, the impactor adding a further 9%. Exponentially decreasing accretion rates are assumed before and after (Halliday, 2000). The solutions in bold type provide the best match with the current data for the W-isotopic compositions of the initial Moon (~ 0), and the present-day lunar mantle (<2) and silicate Earth (0).

3.5 CHEMICAL AND ISOTOPIC CONSTRAINTS ON THE NATURE OF THE COMPONENTS THAT ACCRETED TO FORM THE EARTH

3.5.1 Chondrites and the Composition of the Disk from Which Earth Accreted

A widespread current view is that chondrites represent primitive undifferentiated material from which the Earth accreted. In reality, chondrites are anything but simple and, although they contain early and presolar objects, how primitive and how representative of the range of very early planetesimals they are is completely unclear. In all probability the Earth was built largely from more extensively differentiated materials (Taylor and Norman, 1990). Nevertheless, chondrites do represent a useful set of reference reservoirs in chemical and isotopic terms from which one can draw some conclusions about the "average stuff" from which the Earth may have been built (e.g., Ganapathy and Anders, 1974; Anders, 1977; Wolf et al., 1980). Indeed, if geochemists did not have chondrite samples to provide a reference, many geochemical arguments about Earth's origin, not to say present-day interior structure, would be far weaker.

The degree to which different kinds of chondrites reflect bulk Earth composition has been extensively debated, since none of them provide a good match. CI carbonaceous chondrites define a reference reservoir for *undepleted* solar system compositions because they are strikingly similar in composition to the Sun when normalized to an element such as silicon (Grevesse and Sauval, 1998). However, they are not at all similar to the Earth's volatile-depleted composition. Nor are they much like the vast majority of other chondrites! Confirmation of the primitive nature of CIs is most readily demonstrated by comparing the relative concentrations of a well-determined moderately volatile major element to a well-determined refractory major element with the values in the Sun's photosphere. For example, the concentrations of sodium and calcium are both known to better than a percent in the photosphere (Grevesse and Sauval, 1998) and are identical to within a percent with the values found in CIs despite a big difference in volatility between the two elements (Anders and Grevesse, 1989; Newsom, 1995). It is tempting to ascribe CIs to a complete condensation sequence. However, they also contain dispersed isotopic anomalies in Cr, for example, that indicate a lack of homogenization on a fine scale (Podosek et al., 1997). This cannot be reconciled with nebular condensation from high temperatures.

Volatile-element depletion patterns in other (e.g., CV, CM, or CO) carbonaceous chondrites (Larimer and Anders, 1970; Palme et al., 1988;

Takahashi et al., 1978; Wolf et al., 1980) and ordinary chondrites (Wasson and Chou, 1974) have long been considered to partly reflect incomplete condensation (Wasson, 1985), with the more volatile elements removed from the meteorite formation regions before cooling and the termination of condensation. However, this depletion must at least partially reflect the incorporation of volatile-depleted CAIs (Guan et al., 2000) and chondrules (Grossman, 1996; Meibom et al., 2000). The origins of CAIs have long been enigmatic, but recently it has been proposed that CAIs may be the product of rather localized heating close to the young Sun (Shu et al., 1997) following which they were scattered across the disk. Regardless of whether this model is exactly correct, the incorporation of volatile-depleted CAIs must result in some degree of volatile-depleted composition for chondritic meteorites. Chondrules are the product of rapid melting of chondritic materials (Connolly et al., 1998; Connolly and Love, 1998; Desch and Cuzzi, 2000; Jones et al., 2000) and such events also may have been responsible for some degree of volatile-element depletion (Vogel et al., 2002).

The discovery that pristine presolar grains are preserved in chondrites indicates that they were admixed at a (relatively) late stage. Many of the presolar grain types (Mendybaev et al., 2002; Nittler, 2003) could not have survived the high temperatures associated with CAI and chondrule formation. Therefore, chondrites must include a widely dispersed presolar component (Huss, 1988, 1997; Huss and Lewis, 1995) that either settled or was swept into the chondrule-forming and/or chondrule-accumulating region and would have brought with it nebular material that was not depleted in volatiles. Therefore, volatile depletion in chondrites must reflect, to some degree at least, the mixing of more refractory material (Guan et al., 2000; Kornacki and Fegley, 1986) with CI-like matrix of finer grained volatile-rich material (Grossman, 1996). CIs almost certainly represent accumulations of mixed dust inherited from the portion of the protostellar molecular cloud that collapsed to form the Sun within a region of the nebula that was never greatly heated and did not accumulate chondrules and CAIs (for whatever reason). Though undifferentiated, chondrites, with the possible exception of CIs, are not primitive and certainly do not represent the first stages of accretion of the Earth.

Perhaps the best current way to view the disk of debris from which the Earth accreted was as an environment of vigorous mixing. Volatile-depleted material that had witnessed very high temperatures at an early stage (CAIs) mixed with material that had been flash melted (chondrules) a few million years later. Then presolar grains that had escaped these processes rained into the midplane or more likely were swept in from the outer regions of the solar system. It is within these environments that dust, chondrules and CAIs mixed and accumulated into planetesimals. There also would have been earlier formed primary planetesimals (planetary embryos) that had accreted rapidly by runaway growth but were subsequently consumed. The earliest most primitive objects probably differentiated extremely quickly aided by the heat from runaway growth and live ^{26}Al.

It is not clear whether we have *any* meteorites that are samples of these very earliest planetesimals. Some iron meteorites with very unradiogenic tungsten (Horan *et al.*, 1998) might be candidates (Table 1). Nearly all silicate-rich basaltic achondrites that are precisely dated appear to have formed later (more than a million years after the start of the solar system). Some of the disk debris will have been the secondary product of collisions between these planetesimals, rather like the dust in Beta Pictoris. Some have proposed that chondrules formed in this way (Sanders, 1996; Lugmair and Shukolyukov, 2001), but although the timescales appear right this is not generally accepted (Jones *et al.*, 2000). A working model of the disk is of a conveyor belt of material rapidly accreting into early planetesimals, spiraling in toward the Sun and being fed by heated volatile-depleted material (Krot *et al.*, 2001) scattered outwards and pristine volatile-rich material being dragged in from the far reaches of the disk (Shu *et al.*, 1996). This mixture of materials provided the raw ingredients for early planetesimals and planets that have largely been destroyed or ejected. One of the paths of destruction was mutual collisions, and it was these events that ultimately led to the growth of terrestrial planets like the Earth.

3.5.2 Chondritic Component Models

Although it is now understood quite well that chondrites are complicated objects with a significant formation history, many have proposed that they should be used to define Earth's bulk composition. More detailed discussion of the Earth's composition is provided elsewhere in this treatise and only the essentials for this chapter are covered here. A recent overview of some of the issues is provided by Drake and Righter (2002). The carbonaceous chondrites often are considered to provide the best estimates of the basic building blocks (Agee and Walker, 1988; Allègre *et al.*, 1995a,b, 2001; Anders, 1977; Ganapathy and Anders, 1974; Herzberg, 1984; Herzberg *et al.*, 1988; Jagoutz *et al.*, 1979; Jones and Palme, 2000; Kato *et al.*, 1988a,b; Newsom, 1990). Because the Earth is formed from collisions between differentiated material, it is clear that one is simply

using these reference chondrites to provide little more than model estimates of the total Earth's composition. Allègre *et al.* (1995a,b, 2001), for example, have plotted the various compositions of chondrites using major and trace element ratios and have shown that the refractory lithophile elements are most closely approximated by certain specific kinds of chondrites. Relative abundances of the platinum group elements and osmium isotopic composition of the silicate Earth have been used extensively to evaluate which class of chondrites contributed the late veneer (Newsom, 1990; Rehkämper *et al.*, 1997; Meisel *et al.*, 2001; Drake and Righter, 2002), a late addition of volatile rich material. Javoy (1998, 1999) has developed another class of models altogether based on enstatite chondrites.

If it is assumed that one can use chondrites as a reference, one can calculate the composition of the total Earth and predict the concentrations of elements that are poorly known. For example, in theory one can predict the amount of silicon in the total Earth and determine from this how much must have gone into the core. The Mainz group produced many of the classic papers pursuing this approach (Jagoutz *et al.*, 1979). However, it was quickly realized that the Earth does not fit any class of chondrite. In particular, Jagoutz presented the idea of using element ratios to show that the Earth's upper mantle had Mg/Si that was non-chondritic. Jagoutz *et al.* (1979) used CI, ordinary and enstatite chondrites to define the bulk Earth. Anders always emphasized that all chondrites were fractionated relative to CIs in his series of papers on "Chemical fractionations in meteorites" (e.g., Wolf *et al.*, 1980). That is, CMs and CVs, etc., are also fractionated, as are planets. Jagoutz *et al.* (1979) recognized that the fractionations of magnesium, silicon, and aluminum were found in both the Earth and in CI, O, and E chondrites, but not in C2–C3 chondrites, which exhibit a fixed Mg/Si but variable Al/Si ratio, probably because of addition of CAIs.

One also can use chondrites to determine the abundances of moderately volatile elements such as potassium. Allègre *et al.* (2001) and Halliday and Porcelli (2001) pursued this approach to show that the Earth's K/U is $\sim 10^4$ (Figure 2). One can calculate how much of the volatile chalcophile and siderophile elements such as sulfur, cadmium, tellurium, and lead may have gone in to the Earth's core (Figure 6; Yi *et al.*, 2000; Allègre *et al.*, 1995b; Halliday and Porcelli, 2001). Allègre *et al.* (1995a) also have used this approach to estimate the total Earth's ^{238}U/^{204}Pb.

From the budgets of potassium, silicon, carbon, and sulfur extrapolated from carbonaceous chondrite compositions, one can evaluate the amounts of various light elements possibly incorporated in the core (Allègre *et al.*, 1995b; Halliday and

Porcelli, 2001) and see if this explains the deficiency in density relative to pure iron (Ahrens and Jeanloz, 1987). Such approaches complement the results of experimental solubility measurements (e.g., Wood, 1993; Gessmann *et al.*, 2001) and more detailed comparisons with the geochemistry of meteorites (Dreibus and Palme, 1996).

It should be noted that the Jagoutz *et al.* (1979) approach and the later studies by Allègre and co-workers (Allègre *et al.*, 1995a,b, 2001) are mutually contradictory, a fact that is sometimes not recognized. Whereas Jagoutz *et al.* (1979) used CI, ordinary, and enstatite chondrites, Allègre and co-workers used CI, CM, CV, and CO chondrites (the carbonaceous chondrite mixing line) to define the bulk earth, ignoring the others.

These extrapolations and predictions are based on two assumptions that some would regard as rendering the approach flawed:

(i) The carbonaceous chondrites define very nice trends for many elements that can be used to define the Earth's composition, but the ordinary and enstatite chondrites often lie off these trends (Figure 6). There is no obvious basis for simply ignoring the chemical compositions of the ordinary and enstatite chondrites unless they have undergone some parent body process of loss or redistribution in their compositions. This is indeed likely, particularly for volatile chalcophile elements but is not well understood at present.

(ii) The Earth as well as the Moon, Mars, and basaltic achondrite parent bodies are depleted in moderately volatile elements, in particular the alkali elements potassium and rubidium, relative to *most* classes of chondrite (Figure 2). This needs to be explained. Humayun and Clayton (1995) performed a similar exercise, showing that since chondrites are all more volatile rich than terrestrial planets, the only way to build planets from chondrite precursors was to volatilize alkalis and other volatile elements. This was not considered feasible in view of the identical potassium isotope compositions of chondrites, the Earth and the Moon. Taylor and Norman (1990) made a similar observation that planets formed from volatile-depleted and differentiated precursors and not from chondrites. Clearly, there must have been other loss mechanisms (Halliday and Porcelli, 2001) beyond those responsible for the volatile depletion in chondrites unless the Earth, Moon, angrite parent body and Vesta simply were accreted from parts of the disk that were especially enriched in some highly refractory component, like CAIs (Longhi, 1999).

This latter point and other similar concerns represent such a major problem that some scientists abandoned the idea of just using chondrites as a starting point and instead invoked the former presence of completely hypothetical components in the inner solar system. This hypothesis has been explored in detail as described below.

3.5.3 Simple Theoretical Components

In addition to making comparisons with chondrites, the bulk composition of the Earth also has been defined in terms of a "model" mixture of highly reduced, refractory material combined with a much smaller proportion of a more oxidized volatile-rich component (Wänke, 1981). These models follow on from the ideas behind earlier heterogeneous accretion models. According to these models, the Earth was formed from two components. Component A was highly reduced and free of all elements with equal or higher volatility than sodium. All other elements were in CI relative abundance. The iron and siderophile elements were in metallic form, as was part of the silicon. Component B was oxidized and contained all elements, including those more volatile than sodium in CI relative abundance. Iron and all siderophile and lithophile elements were mainly in the form of oxides.

Ringwood (1979) first proposed these models but the concept was more fully developed by Wänke (1981). In Wänke's model, the Earth accretes by heterogeneous accretion with a mixing ratio A:B $\sim$ 85:15. Most of component B would be added after the Earth had reached about two thirds of its present mass. The oxidized volatile-rich component would be equivalent to CI carbonaceous chondrites. However, the reduced refractory rich component is hypothetical and never has been identified in terms of meteorite components.

Eventually, models that involved successive changes in accretion and core formation replaced these. How volatiles played into this was not explained except that changes in oxidation state were incorporated. An advanced example of such a model is that presented by Newsom (1990). He envisaged the history of accretion as involving stages that included concomitant core formation stages (discussed under core formation).

3.5.4 The Nonchondritic Mg/Si of the Earth's Primitive Upper Mantle

It has long been unclear why the primitive upper mantle of the Earth has a nonchondritic proportion of silicon to magnesium. Anderson (1979) proposed that the mantle was layered with the lower mantle having higher Si/Mg. Although a number of coeval papers presented a similar view (Herzberg, 1984; Jackson, 1983) it is thought by many geochemists these days that the major element composition of the mantle is, broadly speaking, well mixed and homogeneous as a result of 4.5 Gyr of mantle convection (e.g., Hofmann, 1988; Ringwood, 1990). A notable exception is the enstatite chondrite model of Javoy (1999).

Wänke (1981) and Allègre *et al.* (1995b) have proposed that a significant fraction of the Earth's silicon is in the core. However, this is not well supported by experimental data. To explain the silicon deficiency this way conditions have to be so reducing that niobium would be siderophile and very little would be left in the Earth's mantle (Wade and Wood, 2001).

Ringwood (1989a) proposed that the nonchondritic Mg/Si reflected a radial zonation in the solar system caused by the *addition* of more volatile silicon to the outer portions of the solar system. That is, he viewed the Earth as possibly more representative of the solar system than chondrites.

Hewins and Herzberg (1996) proposed that the midplane of the disk from which the Earth accreted was dominated by chondrules. Chondrules have higher Mg/Si than chondrites and if the Earth accreted in a particular chondrule rich area (as it may well have done) the sorting effect could dominate planetary compositions.

3.5.5 Oxygen Isotopic Models and Volatile Losses

Some have proposed that one can use the oxygen isotopic composition of the Earth to identify the proportions of different kinds of chondritic components (e.g., Lodders, 2000). The isotopic composition of oxygen is variable, chiefly in a mass-dependent way, in terrestrial materials. However, meteorites show mass-independent variations as well (Clayton, 1986, 1993). Oxygen has three isotopes and a three-isotope system allows discrimination between mass-dependent planetary processes and mass-independent primordial nebular heterogeneities inherited during planet formation (Clayton and Mayeda, 1996; Franchi *et al.*, 1999; Wiechert *et al.*, 2001, 2003). Clayton pursued this approach and showed that the mix of meteorite types based on oxygen isotopes would not provide a chemical composition that was similar to that of the Earth. In particular, the alkalis would be too abundant. In fact, the implied relative proportions of volatile-rich and volatile-poor constituents are in the opposite sense of those derived in the Wänke–Ringwood mixing models mentioned above (Clayton and Mayeda, 1996). Even the differences in composition between the planets do not make sense. Earth would need to be less volatile-depleted than Mars, whereas the opposite is true (Clayton, 1993; Halliday *et al.*, 2001). The oxygen isotope compositions of Mars (based on analyses of SNC meteorites) show a *smaller* proportion of carbonaceous-chondrite-like material in Mars than in Earth (Clayton, 1993; Halliday *et al.*, 2001). It thus appears that,

unless the Earth lost a significant proportion of its moderately volatile elements after it formed, the principal carrier of moderately volatile elements involved in the formation of the terrestrial planets was not of carbonaceous chondrite composition. This, of course, leads to the viewpoint that the Earth was built from volatile-depleted material, such as differentiated planetesimals (Taylor and Norman, 1990), but it still begs the question of how volatile depletion occurred (Halliday and Porcelli, 2001).

Whether it is plausible, given the dynamics of planetary accretion discussed above, that the Earth lost a major fraction of its moderately volatile elements during its accretion history is unclear. This has been advanced as an explanation by Lodders (2000) and is supported by strontium isotope data for early solar system objects discussed below (Halliday and Porcelli, 2001). The difficulty is to come up with a mechanism that does not fractionate K isotopes (Humayun and Clayton, 1995) and permits loss of heavy volatile elements. The degree to which lack of fractionation of potassium isotopes offers a real constraint depends on the mechanisms involved (Esat, 1996; Young, 2000). There is no question that as the Earth became larger, the accretion dynamics would have become more energetic and the temperatures associated with accretion would become greater (Melosh and Sonett, 1986; Melosh and Vickery, 1989; Ahrens, 1990; Benz and Cameron, 1990; Melosh, 1990; Melosh *et al.*, 1993). However, the gravitational pull of the Earth would have become so large that it would be difficult for the Earth to lose these elements even if they were degassed into a hot protoatmosphere. This is discussed further below.

3.6 EARTH'S EARLIEST ATMOSPHERES AND HYDROSPHERES

3.6.1 Introduction

The range of possibilities to be considered for the nature of the earliest atmosphere provides such a broad spectrum of consequences for thermal and magmatic evolution that it is better to consider the atmospheres first before discussing other aspects of Earth's evolution. Therefore, in this section a brief explanation of the different kinds of early atmospheres and their likely effects on the Earth are given in cursory terms. More comprehensive information on atmospheric components is found elsewhere in Volume 4 of this treatise.

3.6.2 Did the Earth Have a Nebular Protoatmosphere?

Large nebular atmospheres have at various times been considered a fundamental feature of the early Earth by geochemists (e.g., Sasaki and Nakazawa, 1988; Pepin, 2000; Porcelli *et al.*, 1998). Large amounts of nebular gases readily explain why the Earth has primordial ^{3}He (Clarke *et al.*, 1969; Mamyrin *et al.*, 1969; Lupton and Craig, 1975; Craig and Lupton, 1976) and why a component of solar-like neon with a solar He/Ne ratio can be found in some plume basalts (Honda *et al.*, 1991; Dixon *et al.*, 2000; Moreira *et al.*, 2001). Evidence for a solar component among the heavier noble gases has been more scant (Moreira and Allègre, 1998; Moreira *et al.*, 1998). There is a hint of a component with different ^{38}Ar/^{36}Ar in some basalts (Niedermann *et al.*, 1997) and this may reflect a solar argon component (Pepin and Porcelli, 2002). Also, Caffee *et al.* (1988, 1999) have made the case that a solar component of xenon can be found in well gases.

The consequences for the early behavior of the Earth are anticipated to be considerable if there was a large nebular atmosphere. Huge reducing protoatmospheres, be they nebular or impact-induced can facilitate thermal blanketing, magma oceans and core formation. For example, based on the evidence from helium and neon, Harper and Jacobsen (1996b) suggested that iron was reduced to form the core during a stage with a massive early H_2–He atmosphere. A variety of authors had already proposed that the Earth accreted with a large solar nebular atmosphere. Harper and Jacobsen's model builds upon many earlier such ideas primarily put forward by the Kyoto school (Hayashi *et al.*, 1979, 1985; Mizuno *et al.*, 1980; Sasaki and Nakazawa, 1988; Sasaki, 1990). The thermal effects of a very large protoatmosphere have been modeled by Hayashi *et al.* (1979), who showed that surface temperatures might reach >4,000 K. Sasaki (1990) also showed that incredibly high temperatures might build up in the outer portions of the mantle, leading to widespread magma oceans. Dissolving volatiles like noble gases into early silicate and metal liquids may have been quite easy under these circumstances (Mizuno *et al.*, 1980; Porcelli *et al.*, 2001; Porcelli and Halliday, 2001).

Porcelli and Pepin (2000) and Porcelli *et al.* (2001) recently summarized the noble gas arguments pointing out that first-order calculations indicate that significant amounts of noble gases with a solar composition are left within the Earth's interior but orders of magnitude more have been lost, based on xenon isotopic evidence. Therefore, one requires a relatively large amount of nebular gas during Earth's accretion. To make the Earth this way, the timescales for accretion need to be characteristically short ($\sim$$10^6$–$10^7$ yr) in order to trap such amounts of gas before the remains of the solar nebula are accreted into the Sun or other planets. Therefore, a problem may exist reconciling the apparent need to acquire a large nebular atmosphere with the longer time-scales ($\sim$$10^7$–$10^8$ yr) for accretion implied by Safronov–Wetherill models and by tungsten and lead isotopic data. It is hard to get around this problem because the nebular model is predicated on the assumption that the Earth grows extremely fast such that it can retain a large atmosphere (Pepin and Porcelli, 2002).

Therefore, other models for explaining the incorporation of solar-like noble gases should be considered. The most widely voiced alternative is that of accreting material that formed earlier elsewhere that already had acquired solar-like noble gases. For example, it has been argued that the neon is acquired as "Ne–B" from accretion of chondritic material (Trieloff *et al.*, 2000). With such a component in meteorites, one apparently could readily explain the Earth's noble gas composition, which may have a ^{20}Ne/^{22}Ne ratio that is lower than solar (Farley and Poreda, 1992). However, this component is no longer well defined in meteorites and the argument is less than certain because of this. In fact, Ne–B is probably just a fractionated version of solar (Ballentine *et al.*, 2001). This problem apart, the idea that the Earth acquired its solar-like noble gases from accreting earlier-formed objects is an alternative that is worth considering. Chondrites, however, mainly contain very different noble gases that are dominated by the so-called "Planetary" component (Ozima and Podosek, 2002). This component is nowadays more precisely identified as "Phase Q" (Wieler *et al.*, 1992; Busemann *et al.*, 2000). In detail, the components within chondrites show little sign of having incorporated large amounts of solar-like noble gases. Early melted objects like CAIs and chondrules are, generally speaking, strongly degassed (Vogel *et al.*, 2002, 2003).

Podosek *et al.* (2003) recently have proposed that the noble gases are incorporated into early formed planetesimals that are irradiated by intense solar-wind activity from the vigorous early Sun. With the new evidence from young solar mass stars of vastly greater flare activity (Feigelson *et al.*, 2002a,b), there is strong support for the notion that inner solar system objects would have incorporated a lot of solar-wind implanted noble gases. The beauty of this model is that small objects with larger surface-to-volume ratio trap more noble gases. Unlike the nebular model it works best if large objects take a long time to form. As such, the model is easier to reconcile with the kinds of long timescales for accretion implied by tungsten and lead isotopes. Podosek *et al.* (2003) present detailed calculations to illustrate the feasibility of such a scenario.

To summarize, although large nebular atmospheres have long been considered the most likely explanation for primordial solar-like noble gases in the Earth, the implications of such models appear hard to reconcile with the accretionary timescales determined from tungsten and lead isotopes. Irradiating planetesimals with solar wind currently appears to be the most promising alternative. If this is correct, the Earth may still at one time have had a relatively large atmosphere, but it would have formed by degassing of the Earth's interior.

3.6.3 Earth's Degassed Protoatmosphere

The discovery that primordial ^{3}He still is being released from the Earth's interior (Clarke *et al.*, 1969; Mamyrin *et al.*, 1969; Lupton and Craig, 1975; Craig and Lupton, 1976) is one of the greatest scientific contributions made by noble gas geochemistry. Far from being totally degassed, the Earth has deep reservoirs that must supply ^{3}He to the upper mantle and thence to the atmosphere. However, the idea that the majority of the components in the present-day atmosphere formed by degassing of the Earth's interior is much older than this. Brown (1949) and Rubey (1951) proposed this on the basis of the similarity in chemical composition between the atmosphere and hydrosphere on the one hand and the compositions of volcanic gases on the other. In more recent years, a variety of models for the history of degassing of the Earth have been developed based on the idea that a relatively undegassed lower mantle supplied the upper mantle with volatiles, which then supplied the atmosphere (Allègre *et al.*, 1983, 1996; O'Nions and Tolstikhin, 1994; Porcelli and Wasserburg, 1995).

Some of the elements in the atmosphere provide specific "time information" on the degassing of the Earth. Of course, oxygen was added to the atmosphere gradually by photosynthesis with a major increase in the early Proterozoic (Kasting, 2001). However, the isotopic composition and concentration of argon provides powerful evidence that other gases have been added to the atmosphere from the mantle over geological time (O'Nions and Tolstikhin, 1994; Porcelli and Wasserburg, 1995). In the case of argon, a mass balance can be determined because it is dominantly (>99%) composed of ^{40}Ar, formed by radioactive decay of ^{40}K. From the Earth's potassium concentration and the atmosphere's argon concentration it can be shown that roughly half the ^{40}Ar is in the atmosphere, the remainder presumably being still stored in the Earth's mantle (Allègre *et al.*, 1996). Some have argued that this cannot be correct on geophysical grounds, leading to the proposal that the Earth's K/U ratio has been overestimated (Davies, 1999). However, the relationship with Rb/Sr (Figure 2) provides strong evidence that the Earth's K/U ratio is $\sim 10^4$ (Allègre *et al.*, 2001; Halliday and Porcelli, 2001). Furthermore, support for mantle reservoirs that are relatively

undegassed can be found in both helium and neon isotopes (Allègre *et al.*, 1983, 1987; Moreira *et al.*, 1998, 2001; Moreira and Allègre, 1998; O'Nions and Tolstikhin, 1994; Niedermann *et al.*, 1997; Porcelli and Wasserburg, 1995). The average timing of this loss is poorly constrained from argon data. In principle, this amount of argon could have been supplied catastrophically in the recent past. However, it is far more reasonable to assume that because the Earth's radioactive heat production is decaying exponentially, the amount of degassing has been decreasing with time. The argon in the atmosphere is the time-integrated effect of this degassing.

Xenon isotopes provide strong evidence that the Earth's interior may have undergone an early and catastrophic degassing (Allègre *et al.*, 1983, 1987). Allègre *et al.* discovered that the MORB-source mantle had elevated ^{129}Xe relative to atmospheric xenon, indicating that the Earth and atmosphere separated from each other at an early stage. These models are hampered by a lack of constraints on the xenon budgets of the mantle and by atmospheric contamination that pervades many mantle-derived samples. Furthermore, the model assumes a closed system. If a portion of the xenon in the Earth's atmosphere was added after degassing (Javoy, 1998, 1999) by cometary or asteroidal impacts (Owen and Bar-Nun, 1995, 2000; Morbidelli *et al.*, 2000), the model becomes underconstrained. The differences that have been reported between the elemental and atomic abundances of the noble gases in the mantle relative to the atmosphere may indeed be explained by heterogeneous accretion of the atmosphere (Marty, 1989; Caffee *et al.*, 1988, 1999).

Isotopic evidence aside, many theoretical and experimental papers have focused on the production of a steam atmosphere by impact-induced degassing of the Earth's interior (Abe and Matsui, 1985, 1986, 1988; O'Keefe and Ahrens, 1977; Lange and Ahrens, 1982, 1984; Matsui and Abe, 1986a, b; Sasaki, 1990; Tyburczy *et al.*, 1986; Zahnle *et al.*, 1988). Water is highly soluble in silicate melts at high pressures (Righter and Drake, 1999; Abe *et al.*, 2000). As such, a large amount could have been stored in the Earth's mantle and then released during volcanic degassing.

Ahrens (1990) has modeled the effects of impact-induced degassing on the Earth. He considers that the Earth probably alternated between two extreme states as accretion proceeded. When the Earth was degassed it would accumulate a large reducing atmosphere. This would provide a blanket that also allowed enormous surface temperatures to be reached: Ahrens estimates $\sim 1,500$ K. However, when an impact occurred this atmosphere would be blown off. The surface of the Earth would become cool and oxidizing again (Ahrens, 1990). There is strong isotopic evidence for such early losses of the early atmospheres as explained in the next section.

3.6.4 Loss of Earth's Earliest Atmosphere(s)

The xenon isotope data provide evidence that much (>99%) of the Earth's early atmosphere was lost within the first 100 Myr. Several papers on this can be found in the literature and the most recent ones by Ozima and Podosek (1999) and Porcelli *et al.* (2001) are particularly useful. The basic argument for the loss is fairly simple and is not so different from the original idea of using xenon isotopes to date the Earth (Wetherill, 1975a). We have a rough idea of how much iodine exists in the Earth's mantle. The Earth's current inventory is not well constrained but we know enough (Déruelle *et al.*, 1992) to estimate the approximate level of depletion of this volatile element. It is clear from the degassing of noble gases that the present ratio of I/Xe in the Earth is orders of magnitude higher than chondritic values. We know that at the start of the solar system ^{129}I was present with an atomic abundance of $\sim 10^{-4}$ relative to stable ^{127}I (Swindle and Podosek, 1988). All of this ^{129}I formed ^{129}Xe, and should have produced xenon that was highly enriched in ^{129}Xe given the Earth's I/Xe ratio. Yet instead, the Earth has xenon that is only slightly more radiogenic than is found in meteorites rich in primordial noble gases; the ^{129}Xe excesses in the atmosphere and the mantle are both minute by comparison with that expected from the Earth's I/Xe. This provides evidence that the Earth had a low I/Xe ratio that kept its xenon isotopic compositions close to chondritic. At some point xenon was lost and by this time ^{129}I was close to being extinct such that the xenon did not become very radiogenic despite a very high I/Xe.

The xenon isotopic arguments can be extended to fissionogenic xenon. The use of the combined $^{129}I–^{129}Xe$ and $^{244}Pu–^{132,134,136}Xe$ (spontaneous fission half-life ~ 80 Myr) systems provides estimates of ~ 100 Myr for loss of xenon from the Earth (Ozima and Podosek, 1999). The fission-based models are hampered by the difficulties with resolving the heavy xenon that is formed from fission of ^{244}Pu as opposed to longer-lived uranium. The amount of ^{136}Xe that is expected to have formed from ^{244}Pu within the Earth should exceed that produced from uranium as found in well gases (Phinney *et al.*, 1978) and this has been confirmed with measurements of MORBs (Kunz *et al.*, 1998). However, the relative amount of plutonogenic Xe/uranogenic Xe will be a function of the history of degassing of the mantle. Estimating these amounts accurately is very hard.

This problem is exacerbated by the lack of constraint that exists on the initial xenon isotopic composition of the Earth. The xenon isotopic composition of the atmosphere is strikingly different from that found in meteorites (Wieler *et al.*, 1992; Busemann *et al.*, 2000) or the solar wind (Wieler *et al.*, 1996). It is fractionated relative to solar, the light isotopes being strongly depleted (Figure 7). One can estimate the initial xenon isotopic composition of the Earth by assuming it was strongly fractionated from something more like the composition found in meteorites and the solar wind. By using meteorite data to determine a best fit to atmospheric Xe one obtains a composition called "U–Xe" (Pepin, 1997, 2000). However, this is based on finding a composition that is consistent with the present-day atmosphere. Therefore caution is needed when using the fissionogenic xenon components to estimate an accretion age for the Earth because the arguments become circular (Zhang, 1998).

The strong mass-dependent fractionation of xenon has long been thought to be caused by hydrodynamic escape (Hunten *et al.*, 1987; Walker, 1986) of the atmosphere. Xenon probably was entrained in a massive atmosphere of light gases presumably dominated by hydrogen and helium that was lost (Sasaki and Nakazawa, 1988). This is consistent with the view based on radiogenic and fissionogenic xenon that a large fraction of the Earth's atmosphere was lost during the lifetime of ^{129}I.

Support for loss of light gases from the atmosphere via hydrodynamic escape can be found in other "atmophile" isotopic systems. The $^{20}Ne/^{22}Ne$ ratio in the mantle is elevated relative to the atmosphere (Trieloff *et al.*, 2000). The mantle value is close to solar (Farley and Poreda, 1992), but the atmosphere plots almost exactly

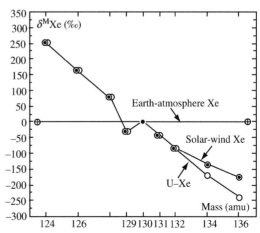

Figure 7 Mass fractionation of xenon in the atmosphere relative to the solar value (Pepin and Porcelli, 2002) (reproduced by permission of Mineralogical Society of America from *Rev. Mineral. Geochem.* **2002**, *47*, 191–246).

according to a heavy mass-dependent fractionated composition. Similarly, the hydrosphere has a D/H ratio that is heavy relative to the mantle. However, argon and krypton isotopes provide no support for the theory (Pepin and Porcelli, 2002). Therefore, this model cannot be applied in any simple way to all of the atmophile elements.

It is possible that the atmosphere was blown off by a major impact like the Moon-forming giant impact, but this is far from clear at this stage. Another mechanism that often is considered is the effect of strong ultraviolet wavelength radiation from the early Sun (Zahnle and Walker, 1982). This might affect Xe preferentially because of the lower ionization potential. It is of course possible that the Earth simply acquired an atmosphere, with xenon, like today's (Marty, 1989; Caffee *et al.*, 1999). However, then it is not clear how to explain the strong isotopic fractionation relative to solar and meteorite compositions.

Taken together, the noble gas data provide evidence that Earth once had an atmosphere that was far more massive than today's. If true, this would have had two important geochemical consequences. First, there would have been a blanketing effect from such an atmosphere. This being the case, the temperature at the surface of the Earth would have been very high. There may well have been magma oceans, rock vapor in the atmosphere, and extreme degassing of moderately volatile as well as volatile elements. Blow-off of this atmosphere may have been related to the apparent loss of moderately volatile elements (Halliday and Porcelli, 2001). With magma oceans there would be, at best, a weak crust, mantle mixing would have been very efficient, and core formation would have proceeded quickly.

The second consequence is that it would have been relatively easy to dissolve a small amount of this "solar" gas at high pressure into the basaltic melts at the Earth's surface (Mizuno *et al.*, 1980). This "ingassing" provides a mechanism for transporting nebular gases into the Earth's interior. The ultimate source of solar helium and neon in the mantle is unknown. At one time, it was thought to be the mantle (Allègre *et al.*, 1983) and this still seems most likely (Porcelli and Ballentine, 2002) but the core also has been explored (Porcelli and Halliday, 2001) as a possible alternative.

3.7 MAGMA OCEANS AND CORE FORMATION

Core formation is the biggest differentiation process that has affected the planet, resulting in a large-scale change of the distribution of density and heat production. One would think that such a basic feature would be well understood. However, the very existence of large amounts of iron metal at the center of an Earth with an oxidized mantle is problematic (Ringwood, 1977). Large reducing atmospheres and magma oceans together provide a nice explanation. For example, Ringwood (1966) considered that the iron metal in the Earth's core formed by reduction of iron in silicates and oxides and thereby suggested a huge CO atmosphere. Clearly, if the Earth's core formed by reduction of iron in a large atmosphere, the process of core formation would occur early and easily.

However, there is very little independent evidence from mantle or crustal geochemistry to substantiate the former presence of magma oceans. There has been no shortage of proposals and arguments for and against on the basis of petrological data for the Earth's upper mantle. The key problem is the uncertainty that exists regarding the relationship between the present day upper mantle and that that may have existed in the early Earth. Some have argued that the present day lower mantle is compositionally distinct in terms of major elements (e.g., Herzberg, 1984; Jackson, 1983), whereas others (e.g., Hofmann, 1988; Hofmann *et al.*, 1986; Davies, 1999) have presented strong evidence in favor of large-scale overall convective interchange. Although several papers have used the present-day major, trace, and isotopic compositions of the upper mantle to provide constraints on the earliest history of the Earth (Allègre *et al.*, 1983; Agee and Walker, 1988; Kato *et al.*, 1988a,b; McFarlane and Drake, 1990; Jones *et al.*, 1992; Drake and McFarlane, 1993; Porcelli and Wasserburg, 1995; O'Nions and Tolstikhin, 1994; Righter and Drake, 1996), it is unclear whether or not this is valid, given 4.5 Gyr of mantle convection.

For example, the trace element and hafnium isotopic compositions of the upper mantle provide no sign of perovskite fractionation in a magma ocean (Ringwood, 1990; Halliday *et al.*, 1995) and even the heavy REE pattern of the upper mantle appears to be essentially flat (e.g., Lee *et al.*, 1996). This is not to be expected if majorite garnet was a liquidus phase (Herzberg *et al.*, 1988). One explanation for the lack of such evidence is that the magma ocean was itself an efficiently mixed system with little if any crystal settling (Tonks and Melosh, 1990). Another possibility is that the entire mantle has been rehomogenized since this time. Ringwood (1990) suggested this possibility, which leaves some arguments regarding the relevance of the composition of the upper mantle in doubt. Of course, the subsequent introduction of heterogeneities by entrainment and radioactive decay and the development of an isotopically stratified mantle (Hofmann *et al.*, 1986, Moreira and Allègre, 1998) are not inconsistent with this.

Whether or not magma oceans are a necessary prerequisite for core formation is unclear. It is necessary to understand how it is possible for metallic iron to migrate through the silicate mantle (Shaw, 1978). Many have assumed that a part of the mantle at least was solid during core formation. A variety of mechanisms have been studied including grain boundary percolation (Minarik *et al.*, 1996; Rushmer *et al.*, 2000; Yoshino *et al.*, 2003) and the formation of large-scale metal structures in the upper mantle that sink like diapirs into the center of the Earth (Stevenson, 1981, 1990). Under some circumstances these may break up into small droplets of metal (Rubie *et al.*, 2003). To evaluate these models, it is essential to have some idea of the physical state of the early Earth at the time of core formation. All of these issues are addressed by modern geochemistry but are not yet well constrained. Most effort has been focused on using the composition of the silicate Earth itself to provide constraints on models of core formation.

The major problem presented by the Earth's chemical composition and core formation models is providing mechanisms that predict correctly the siderophile element abundances in the Earth's upper mantle. It long has been recognized that siderophile elements are more abundant in the mantle than expected if the silicate Earth and the core were segregated under low-pressure and moderate-temperature equilibrium conditions (Chou, 1978; Jagoutz *et al.*, 1979). Several explanations for this siderophile "excess" have been proposed, including:

(i) partitioning into liquid metal alloy at high pressure (Ringwood, 1979);

(ii) equilibrium partitioning between sulfur-rich liquid metal and silicate (Brett, 1984);

(iii) inefficient core formation (Arculus and Delano, 1981; Jones and Drake, 1986);

(iv) heterogeneous accretion and late veneers (Eucken, 1944; Turekian and Clark, 1969; Clark *et al.*, 1972; Smith, 1977, 1980; Wänke *et al.*, 1984; Newsom, 1990);

(v) addition of material to the silicate Earth from the core of a Moon-forming impactor (Newsom and Taylor, 1986);

(vi) very high temperature equilibration (Murthy, 1991); and

(vii) high-temperature equilibrium partitioning in a magma ocean at the upper/lower mantle boundary (Li and Agee, 1996; Righter *et al.*, 1997).

Although the abundances and partition coefficients of some of the elements used to test these models are not well established, sufficient knowledge exists to render all of them problematic. Model (vii) appears to work well for some moderately siderophile elements. Righter and Drake (1999) make the case that the fit of the siderophile element metal/silicate partion coefficient data is best achieved with a high water content (per cent level) in the mantle. This would have assisted the formation of a magma ocean and provided a ready source of volatiles in the Earth. Walter *et al.* (2000) reviewed the state of the art in this area. However, the number of elements with well-established high-pressure partition coefficients for testing this model is still extremely small.

To complicate chemical models further, there is some osmium isotopic evidence that a small flux of highly siderophile elements from the core could be affecting the abundances in the mantle (Walker *et al.*, 1995; Brandon *et al.*, 1998). This model has been extended to the interpretation of PGE abundances in abyssal peridotites (Snow and Schmidt, 1998). The inventories of many of these highly siderophile elements are not that well established and may be extremely variable (Rehkämper *et al.*, 1997, 1999b). In particular, the use of abyssal peridotites to assess siderophile element abundances in the upper mantle appears to be problematic (Rehkämper *et al.*, 1999a). Puchtel and Humayun (2000) argued that if PGEs are being fluxed from the core to the mantle then this is not via the Walker *et al.* (1995) mechanism of physical admixture by entrainment of outer core, but must proceed via an osmium isotopic exchange, since the excess siderophiles were not found in komatiite source regions with radiogenic ^{187}Os. Taken together, the status of core to mantle fluxes is very vague at the present time.

3.8 THE FORMATION OF THE MOON

The origin of the Moon has been the subject of intense scientific interest for over a century but particularly since the Apollo missions provided samples to study. The most widely accepted current theory is the giant impact theory but this idea has evolved from others and alternative hypotheses have been variously considered. Wood (1984) provides a very useful review. The main theories that have been considered are as follows:

Co-accretion. This theory proposes that the Earth and Moon simply accreted side by side. The difficulty with this model is that it does not explain the angular momentum of the Earth–Moon system, nor the difference in density, nor the difference in volatile depletion (Taylor, 1992).

Capture. This theory (Urey, 1966) proposes that the Moon was a body captured into Earth's orbit. It is dynamically difficult to do this without the Moon spiraling into the Earth and colliding. Also the Earth and Moon have indistinguishable oxygen isotope compositions (Wiechert *et al.*, 2001) in a solar system that appears to be highly heterogeneous in this respect (Clayton, 1986).

Fission. This theory proposes that the Moon split off as a blob during rapid rotation of a molten Earth. George Howard Darwin, the son of Charles

Darwin originally championed this idea (Darwin, 1878, 1879). At one time (before the young age of the oceanfloor was known) it was thought by some that the Pacific Ocean might have been the residual space vacated by the loss of material. This theory is also dynamically difficult. Detailed discussions of the mechanisms can be found in Binder (1986). However, this model does have certain features that are attractive. It explains why Earth and Moon have identical oxygen isotope compositions. It explains why the Moon has a lower density because the outer part of the Earth would be deficient in iron due to core formation. It explains why so much of the angular momentum of the Earth–Moon system is in the Moon's motion. These are key features of any successful explanation for the origin of the Moon.

Impact models. Mainly because of the difficulties with the above models, alternatives were considered following the Apollo missions. Hartmann and Davis (1975) made the proposal that the Moon formed as a result of major impacts that propelled sufficient debris into orbit to produce the Moon. However, an important new facet that came from sample return was the discovery that the Moon had an anorthositic crust implying a very hot magma ocean. Also it was necessary to link the dynamics of the Moon with that of the Earth's spin. If an impact produced the Moon it would be easier to explain these features if it was highly energetic. This led to a series of single giant impact models in which the Moon was the product of a glancing blow collision with another differentiated planet (Cameron and Benz, 1991). A ring of debris would have been produced from the outer silicate portions of the Earth and the impactor planet (named "Theia," the mother of "Selene," the goddess of the Moon). Wetherill (1986) calculated that there was a realistic chance of such a collision. This model explains the angular momentum, the "fiery start," the isotopic similarities and the density difference.

The giant-impact theory has been confirmed by a number of important observations. Perhaps most importantly, we know now that the Moon must have formed tens of millions of years after the start of the solar system (Lee *et al.*, 1997; Halliday, 2000). This is consistent with a collision between already formed planets. The masses of the Earth and the impactor at the time of the giant impact have been the subject of major uncertainty. Two main classes of models are usually considered. In the first, the Earth was largely (90%) formed at the time of the impact and the impacting planet Theia was roughly Mars-sized (Cameron and Benz, 1991). A recent class of models considers the Earth to be only half-formed at the time of the impact, and the mass ratio Theia/proto-Earth to be 3:7 (Cameron, 2000). The latter model is no longer considered likely; the most recent simulations have reverted to a Mars-sized impactor at the end of Earth accretion (Canup and Asphaug, 2001). The tungsten isotope data for the Earth and Moon do not provide a unique test (Halliday *et al.*, 2000).

The giant-impact model, though widely accepted, has not been without its critics. Geochemical arguments have been particularly important in this regard. The biggest concern has been the similarities between chemical and isotopic features of the Earth and Moon. Most of the dynamic simulations (Cameron and Benz, 1991; Cameron, 2000; Canup and Asphaug, 2001) predict that the material that forms the Moon is derived from Theia, rather than the Earth. Yet it became very clear at an early stage of study that samples from the Moon and Earth shared many common features that would be most readily explained if the Moon was formed from material derived from the Earth (Wänke *et al.*, 1983; Wänke and Dreibus, 1986; Ringwood, 1989b, 1992). These include the striking similarity in tungsten depletion despite a strong sensitivity to the oxidation state of the mantle (Rammensee and Wänke, 1977; Schmitt *et al.*, 1989). Other basaltic objects such as eucrites and martian meteorites exhibit very different siderophile element depletion (e.g., Treiman *et al.*, 1986, 1987; Wänke and Dreibus, 1988, 1994). Therefore, why should the Earth and Moon be identical if the Moon came from Theia (Ringwood, 1989b, 1992)? In a similar manner, the striking similarity in oxygen isotopic composition (Clayton and Mayeda, 1975), still unresolvable to extremely high precision (Wiechert *et al.*, 2001), despite enormous heterogeneity in the solar system (Clayton *et al.*, 1973; Clayton, 1986, 1993; Clayton and Mayeda, 1996), provides support for the view that the Moon was derived from the Earth (Figure 3).

One can turn these arguments around, however, and use the compositions of lunar samples to define the composition of Theia, assuming the impactor produced most of the material in the Moon (MacFarlane, 1989). Accordingly, the similarity in oxygen isotopes and trace siderophile abundances between the Earth and Moon provides evidence that Earth and Theia were neighboring planets made of an identical mix of materials with similar differentiation histories (Halliday and Porcelli, 2001). Their similarities could relate to proximity in the early solar system, increasing the probability of collision.

Certain features of the Moon may be a consequence of the giant impact itself. The volatile-depleted composition of the Moon, in particular, has been explained as a consequence of the giant impact (O'Neill, 1991a; Jones and Palme, 2000). It has been argued (Kreutzberger *et al.*, 1986; Jones and Drake, 1993) that the Moon could not have formed as a volatile depleted residue of material from the Earth because it has Rb/Cs that is lower than that of the Earth and caesium supposedly is more volatile. However, the assumptions regarding the Earth's Rb/Cs upon which this is based are rather weak (McDonough *et al.*, 1992). Furthermore, the exact relative volatilities of the alkalis are poorly known. Using the canonical numbers, the Earth, Moon, and Mars are all more depleted in less volatile rubidium than more volatile potassium (Figure 2). From the slope of the correlation, it can be seen that the terrestrial depletion in rubidium (50% condensation temperature $\sim$1,080 K) is $\sim$80% greater than that in potassium (50% condensation temperature $\sim$1,000 K) (Wasson, 1985). Similar problems are found if one compares sodium depletion, or alkali concentrations more generally for other early objects, including chondrites.

Attempts to date the Moon were initially focused on determining the ages of the oldest rocks and therefore providing a lower limit. These studies emphasized precise strontium, neodymium, and lead isotopic constraints (Tera *et al.*, 1973; Wasserburg *et al.*, 1977a; Hanan and Tilton, 1987; Carlson and Lugmair, 1988; Shih *et al.*, 1993; Alibert *et al.*, 1994). At the end of the Apollo era, Wasserburg *et al.* (1977a) wrote "The actual time of aggregation of the Moon is not precisely known, but the Moon existed as a planetary body at 4.45 Ga, based on mutually consistent Rb–Sr and U–Pb data. This is remarkably close to the ^{207}Pb–^{206}Pb age of the Earth and suggests that the Moon and the Earth were formed or differentiated at the same time." Although these collective efforts made a monumental contribution, such constraints on the age of the Moon still leave considerable scope (>100 Myr) for an exact age.

Some of the most precise and reliable early ages for lunar rocks are given in Table 3. They provide considerable support for an age of >4.42 Ga. Probably the most compelling evidence comes from the early ferroan anorthosite 60025, which defines a relatively low first-stage μ (or ^{238}U/^{204}Pb) and an age of $\sim$4.5 Ga. Of course, the ages of the oldest lunar rocks only date igneous events. Carlson and Lugmair (1988) reviewed all of the most precise and concordant data and concluded that the Moon had to have formed in the time interval 4.44–4.51 Ga. This is consistent with the estimate of 4.47 $\pm$ 0.02 Ga of Tera *et al.* (1973).

Model ages can provide upper and lower limits on the age of the Moon. Halliday and Porcelli (2001) reviewed the strontium isotope data for early solar system objects and showed that the initial strontium isotopic compositions of early lunar highlands samples (Papanastassiou and Wasserburg, 1976; Carlson and Lugmair, 1988) are all slightly higher than the best estimates of the solar system initial ratio (Figure 8). The conservative estimates of the strontium isotope data indicate that the difference between the ^{87}Sr/^{86}Sr of the bulk solar system initial at 4.566 Ga is 0.69891 $\pm$ 2 and the Moon at $\sim$4.515 Ga is 0.69906 $\pm$ 2 is fully resolvable. An Rb–Sr model age for the Moon can be calculated by assuming that objects formed from material that separated from a solar nebula reservoir with the Moon's current Rb/Sr ratio. Because the Rb/Sr ratios of the lunar samples are extremely low, the uncertainty in formation age does not affect the calculated initial strontium isotopic composition, hence the model age, significantly. The CI chondritic Rb/Sr ratio (^{87}Rb/^{86}Sr = 0.92) is assumed to represent the solar nebula. This model provides an upper limit on the formation age of the object, because the solar nebula is thought to represent the most extreme Rb/Sr reservoir in which the increase in strontium isotopic composition could have been accomplished. In reality, the strontium isotopic composition probably evolved in a more complex manner over a longer time. The calculated time required to generate the difference in strontium isotopic composition in a primitive solar nebula environment is 11 $\pm$ 3 Myr. This is, therefore, the earliest point in time at which the Moon could have formed (Halliday and Porcelli, 2001).

A similar model-age approach can be used with the Hf–W system. In fact, Hf–W data provide the most powerful current constraints on the exact age of the Moon. The tungsten isotopic compositions of bulk rock lunar samples range from $\varepsilon_W \sim 0$ like the silicate Earth to $\varepsilon_W > 10$ (Lee *et al.*, 1997, 2002). This was originally interpreted as the result of radioactive decay of formerly live ^{182}Hf within the Moon, which has a variable but generally high Hf/W ratio in its mantle (Lee *et al.*, 1997). Now we know that a major portion of the ^{182}W excess in lunar samples is cosmogenic and the result of the reaction ^{181}Ta(n,γ)^{182}Ta(β^-)^{182}W while these rocks were exposed on the surface of the Moon (Leya *et al.*, 2000; Lee *et al.*, 2002). This can be corrected using (i) estimates of the cosmic ray flux from samarium and gadolinium compositions, (ii) the exposure age and Ta/W ratio, or (iii) internal isochrons of tungsten isotopic composition against Ta/W (Lee *et al.*, 2002). The best current estimates for the corrected compositions are shown in Figure 9. The spread in the data is reduced and the stated uncertainties are greater relative to the raw tungsten isotopic compositions (Lee *et al.*, 1997). Most data are within error of the Earth. A small excess ^{182}W is still resolvable for some samples, but these should be treated with caution.

The most obvious and clear implication of these data is that the Moon, a high-Hf/W object, must have formed late (Lee *et al.*, 1997). Halliday (2000) argued that the tungsten isotopic composition was hard to explain if the Moon formed

Table 3 Recent estimates of the ages of early solar system objects and the age of the Moon.

Object	Sample(s)	Method	References	Age (Ga)
Earliest solar system	Allende CAIs	U–Pb	Göpel et al. (1991)	4.566 ± 0.002
Earliest solar system	Efremovka CAIs	U–Pb	Amelin et al. (2002)	4.5672 ± 0.0006
Chondrule formation	Acfer chondrules	U–Pb	Amelin et al. (2002)	4.5647 ± 0.0006
Angrites	Angra dos Reis and LEW 86010	U–Pb	Lugmair and Galer (1992)	4.5578 ± 0.0005
Early eucrites	Chervony Kut	Mn–Cr	Lugmair and Shukolyukov (1998)	4.563 ± 0.001
Earth accretion	Mean age	U–Pb	Halliday (2000)	≤4.55
Earth accretion	Mean age	U–Pb	Halliday (2000)	≥4.49
Earth accretion	Mean age	Hf–W	Yin et al. (2002)	≥4.55
Lunar highlands	Ferroan anorthosite 60025	U–Pb	Hanan and Tilton (1987)	4.50 ± 0.01
Lunar highlands	Ferroan anorthosite 60025	Sm–Nd	Carlson and Lugmair (1988)	4.44 ± 0.02
Lunar highlands	Norite from breccia 15445	Sm–Nd	Shih et al. (1993)	4.46 ± 0.07
Lunar highlands	Ferroan noritic anorthosite in breccia 67016	Sm–Nd	Alibert et al. (1994)	4.56 ± 0.07
Moon	Best estimate of age	U–Pb	Tera et al. (1973)	4.47 ± 0.02
Moon	Best estimate of age	U–Pb, Sm–Nd	Carlson and Lugmair (1988)	4.44–4.51
Moon	Best estimate of age	Hf–W	Halliday et al. (1996)	4.47 ± 0.04
Moon	Best estimate of age	Hf–W	Lee et al. (1997)	4.51 ± 0.01
Moon	Maximum age	Hf–W	Halliday (2000)	≤4.52
Moon	Maximum age	Rb–Sr	Halliday and Porcelli (2001)	≤4.55
Moon	Best estimate of age	Hf–W	Lee et al. (2002)	4.51 ± 0.01
Moon	Best estimate of age	Hf–W	Kleine et al. (2002)	4.54 ± 0.01

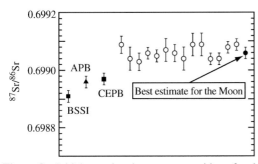

Figure 8 Initial strontium isotope composition of early lunar highland rocks relative to other early solar system objects. APB: Angrite Parent Body; CEPB: Cumulate Eucrite Parent Body; BSSI: Bulk Solar System Initial (source Halliday and Porcelli, 2001).

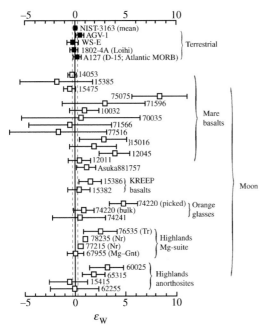

Figure 9 The tungsten isotopic compositions of lunar samples after calculated corrections for cosmogenic contributions (source Leya *et al.*, 2000).

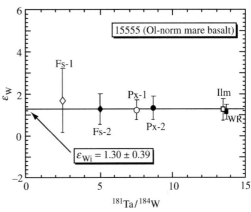

Figure 10 The tungsten isotopic composition of Apollo 15 basalt 15555 shows no internal variation as a function of Ta/W, consistent with its low exposure age (source Lee *et al.*, 2002).

system (Halliday, 2000; Lee *et al.*, 2002). The Moon must also have formed by the time defined by the earliest lunar rocks. The earliest most precisely determined crystallization age of a lunar rock is that of 60025 which has a Pb–Pb age of close to 4.50 Ga (Hanan and Tilton, 1987). Therefore, the Moon appears to have formed before ~4.50 Ga.

Defining the age more precisely is proving difficult at this stage. First, more precise estimates of the Hf–W systematics of lunar rocks are needed. The amount of data for which the cosmogenically produced ^{182}W effects are well resolved is very limited (Lee *et al.*, 2002) and analysis is time consuming and difficult. Second, the $(^{182}Hf/^{180}Hf)_{BSSI}$ is poorly defined, as described above. If one uses a value of 1.0×10^{-4} (Kleine *et al.*, 2002; Schöenberg *et al.*, 2002; Yin *et al.*, 2002), the small tungsten isotopic effects of the Moon probably were produced ≥ 30 Myr after the start of the solar system (Kleine *et al.*, 2002). If, however, the $(^{182}Hf/^{180}Hf)_{BSSI}$ is slightly higher, as discussed above, the model age would be closer to 40–45 Myr (Table 3). The uncertainty in the ^{182}Hf decay constant ($\sim\pm22\%$) also limits more precise constraints.

Either way, these Hf–W data provide very strong support for the giant-impact theory of lunar origin (Cameron and Benz, 1991). It is hard to explain how the Moon could have formed at such a late stage unless it was the result of a planetary collision. The giant-impact theory predicts that the age of the Moon should postdate the origin of the solar system by some considerable amount of time, probably tens of millions of years if Wetherill's predictions are correct. This is consistent with the evidence from tungsten isotopes.

The giant impact can also be integrated into modeling of the lead isotopic composition of the Earth (Halliday, 2000). Doing so, one can

before ~50 Myr after the start of the solar system. The revised parameters for the average solar system (Kleine *et al.*, 2002) mean that the Moon, like the Earth, has a well-defined excess of ^{182}W. This may have been inherited from the protolith silicate reservoirs from which the Moon formed. Alternatively, a portion might reflect ^{182}Hf decay within the Moon itself. Assuming that the Moon started as an isotopically homogeneous reservoir the most likely explanation for the small but well-defined excess ^{182}W found in Apollo basalts such as 15555 (Figure 10) relative to some of the other lunar rocks is that the Moon formed at a time when there was still a small amount of live ^{182}Hf in the lunar interior. This means that the Moon had to have formed within the first 60 Myr of the solar

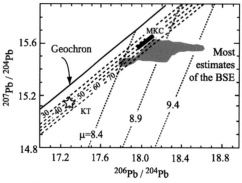

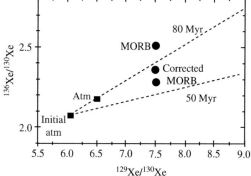

Figure 11 Lead isotopic modeling of the composition of the silicate Earth using continuous core formation and a sudden giant impact when the Earth is 90% formed. The impactor adds a further 9% to the mass of the Earth. The principles behind the modeling are as in Halliday (2000). See text for explanation. The field for the BSE encompasses all of the estimates in Galer and Goldstein (1996). The values suggested by Kramers and Tolstikhin (1997) and Murphy *et al.* (2002) are also shown. The figure is calibrated with the time of the giant impact (Myr). The μ values are the $^{238}U/^{204}Pb$ of the BSE. It is assumed that the μ of the total Earth is 0.7 (Allègre *et al.*, 1995a). It can be seen that the lead isotopic composition of the BSE is hard to reconcile with formation of the Moon before ~45 Myr after the start of the solar system.

Figure 12 The relationship between the amounts of radiogenic ^{129}Xe inferred to come from ^{129}I decay within the Earth and fissionogenic ^{136}Xe thought to be dominated by decay of ^{244}Pu within the Earth. The differences in composition between the atmosphere and upper mantle relate to the timing of atmosphere formation. The compositions of both reservoirs are not very different from solar or initial U–Xe. This provides evidence that the strong depletion of xenon, leading to very high I/Xe, for example, was late. The data are shown modeled as a major loss event 50–80 Myr after the start of the solar system (Porcelli and Ballentine, 2002). The exact correction for uranium-derived fission ^{136}Xe in the MORB-mantle is unclear. Two values are shown from Phinney *et al.* (1978) and Kunz *et al.* (1998).

constrain the timing. Assuming a Mars-sized impactor was added to the Earth in the final stages of accretion (Canup and Asphaug, 2001) and that, prior to this, Earth's accretion could be approximated by exponentially decreasing rates (Halliday, 2000), one can calibrate the predicted lead isotopic composition of the Earth in terms of the time of the giant impact (Figure 11). It can be seen that all of the many estimates for the lead isotopic composition of the BSE compiled by Galer and Goldstein (1996) plus the more recent estimates of Kramers and Tolstikhin (1997) and Murphy *et al.* (2003) appear inconsistent with a giant impact that is earlier than ~45 Myr after the start of the solar system.

Xenon isotope data have also been used to argue specifically that the Earth lost its inventory of noble gases as a consequence of the giant impact (Pepin, 1997; Porcelli and Pepin, 2000). The timing of the "xenon loss event" looks more like 50–80 Myr on the basis of the most recent estimates of fissionogenic components (Porcelli and Pepin, 2000) (Figure 12). This agrees nicely with some estimates for the timing of the giant impact (Tables 2 and 3, and Figure 11) based on tungsten and lead isotopes. If the value of ~30 Myr is correct (Kleine *et al.*, 2002), there appears to be a problem linking the xenon loss event with the Moon forming giant impact, yet it is hard to decouple these. If the tungsten chronometry recently proposed by Yin *et al.* (2002) and

Kleine *et al.* (2002) is correct it would seem that the giant impact cannot have been the last big event that blew off a substantial fraction of the Earth's atmosphere. On the basis of dynamic simulations, however, it is thought that subsequent events cannot have been anything like as severe as the giant impact (Canup and Agnor, 2000; Canup and Asphaug, 2001). It is, of course, conceivable that the protoatmosphere was lost by a different mechanism such as strong UV radiation (Zahnle and Walker, 1982). However, this begs the question of how the (earlier) giant impact could have still resulted in retention of noble gases. It certainly is hard to imagine the giant impact without loss of the Earth's primordial atmosphere (Melosh and Vickery, 1989; Ahrens, 1990; Benz and Cameron, 1990; Zahnle, 1993).

3.9 MASS LOSS AND COMPOSITIONAL CHANGES DURING ACCRETION

Collisions like the giant impact may well represent an important growth mechanism for terrestrial planets in general (Cameron and Benz, 1991; Canup and Agnor, 2000). These collisions are extraordinarily energetic and the question arises as to whether it is to be expected that accretion itself will lead to losses ("erosion") of material from the combined planetary masses. If this is the case it is only to be expected that the composition of the Earth does not add up to what one might expect from "chondrite building blocks" and, for example, the oxygen isotope composition (Clayton and Mayeda, 1996). There are several indications that the Earth may have lost a significant fraction of some elements during accretion and earliest development; all are circumstantial lines of evidence:

(i) As discussed above it has been argued that impact processes in particular are responsible for eroding protoatmospheres (Melosh and Vickery, 1989; Ahrens, 1990; Benz and Cameron, 1990; Zahnle, 1993). If these atmospheres were very dense and hot they may have contained a significant fraction of Earth's moderately volatile elements.

(ii) "Glancing blow" collisions between already differentiated planets, as during the giant impact, might be expected to preferentially remove major portions of the outer silicate portions of the planet as it grows. The Fe/Mg ratio of a planet is the simplest chemical parameter relating to planetary density and indicates the approximate size of the core relative to the silicate mantle. Mercury, with its high density, is a prime candidate for a body that lost a great deal of its outer silicate material by giant impacts (Benz *et al.*, 1987). Therefore, by analogy the proportional size of the Earth's core may have increased as a consequence of such impact erosion. Conversely, Mars (Halliday *et al.*, 2001) with a density lower than that of the Earth, may actually be a closer approximation of the material from which Earth accreted than Earth itself is (Halliday *et al.*, 2001; Halliday and Porcelli, 2001).

(iii) Such late collisional loss of Earth's silicate is clearly evident from the low density of the Moon. The disk of material from which the Moon accreted during the giant impact was silicate-rich. Most simulations predict that Theia provided the major source of lunar material. The density difference between silicate and metal leads to a loss of silicate from the combined Theia–protoEarth system, even when the loss primarily is from the impactor. Note, however, that the giant-impact simulations retain most of the mass overall (Benz and Cameron, 1990). Very little is lost to space from a body as large as the Earth.

(iv) If there was early basaltic crust on the Earth (Chase and Patchett, 1988; Galer and Goldstein, 1991) or Theia or other impacting planets, repeated impact erosion could have had an effect on the Si/Mg ratio of the primitive mantle. However, the maximum effect will be very small because silicon is a major element in the mantle. Earth's Si/Mg ratio is indeed low, as discussed above, but this may instead represent other loss processes. Nevertheless, the erosion effects could have been accentuated by the fact that silicon is relatively volatile and there probably was a

magma ocean. With extremely high temperatures and a heavy protoatmosphere (Ahrens, 1990; Sasaki, 1990; Abe *et al.*, 2000), it seems possible that one could form a "rock atmosphere" by boiling the surface of the magma ocean. This atmosphere would in turn be very vulnerable to impact-induced blow-off.

(v) The budgets of other elements that are more heavily concentrated in the outer portions of the Earth, such as the highly incompatible lithophile elements caesium, barium, rubidium, thorium, uranium, niobium, and potassium and the light rare earths, were possibly also depleted by impact erosion. Indeed some people have argued that the primitive mantle is slightly higher than chondritic in Sm/Nd (Nägler and Kramers, 1998). However, it is worth noting that there is no evidence for europium anomalies in the BSE as might be expected from impact loss of feldspathic flotation cumulates. These issues are further complicated by the fact that the early Earth (or impacting planets) may have had magma oceans with liquidus phases such as majorite or calcium- or magnesium-perovskite (Kato *et al.*, 1988a,b). This in turn could have led to a very variable element distribution in a stratified magma ocean. There is no compelling evidence in hafnium isotopes, yet, for loss of a portion of the partially molten mantle fractionated in the presence of perovskite (Ringwood, 1990).

(vi) The oxygen isotopic compositions of lunar samples have been measured repeatedly to extremely high precision (Wiechert *et al.*, 2001) using new laser fluorination techniques. The $\Delta^{17}O$ of the Moon relative to the terrestrial fractionation line can be shown to be zero (Figure 3). Based on a 99.7% confidence interval (triple standard error of the mean) the lunar fractionation line is, within $\pm 0.005‰$, identical to that of the Earth. There now is no doubt that the mix of material that accreted to form the Earth and the Moon was effectively identical in its provenance. Yet there is a big difference in the budgets of moderately volatile elements (e.g., K/U or Rb/Sr). One explanation is that there were major losses of moderately volatile elements during the giant impact (O'Neill, 1991a,b; Halliday and Porcelli, 2001).

(vii) The strontium isotopic compositions of early lunar highland rocks provide powerful support for late losses of alkalis (Figure 8). Theia had a time integrated Rb/Sr that was more than an order of magnitude higher than the actual Rb/Sr of the Moon, providing evidence that the processes of accretion resulted in substantial loss of alkalis (Halliday and Porcelli, 2001). Some of the calculated time-integrated compositions of the precursors to the present Earth and Moon are shown in Figure 6.

(viii) The abundances of volatile highly siderophile elements in the Earth are slightly depleted relative to refractory highly siderophile elements (Yi *et al.*, 2000). This appears to reflect the composition of a late veneer. If this is representative of the composition of material that accreted to form the Earth as a whole it implies that there were substantial losses of volatile elements from the protoplanets that built the Earth (Yi *et al.*, 2000).

Therefore, there exist several of lines of evidence to support the view that impact erosion may have had a significant effect on Earth's composition. However, in most cases the evidence is suggestive rather than strongly compelling. Furthermore, we have a very poor idea of how this is possible without fractionating potassium isotopes (Humayun and Clayton, 1995), unless the entire inventory of potassium is vaporized (O'Neill, 1991a,b; Halliday *et al.*, 1996; Halliday and Porcelli, 2001). We also do not understand how to lose heavy elements except via hydrodynamic escape of a large protoatmosphere (Hunten *et al.*, 1987; Walker, 1986). Some of the loss may have been from the proto-planets that built the Earth.

3.10 EVIDENCE FOR LATE ACCRETION, CORE FORMATION, AND CHANGES IN VOLATILES AFTER THE GIANT IMPACT

There are a number of lines of evidence that the Earth may have been affected by additions of further material subsequent to the giant impact. Similarly, there is limited evidence that there was additional core formation. Alternatively, there also are geochemical and dynamic constraints that strongly limit the amount of core formation and accretion since the giant impact. This is a very interesting area of research that is ripe for further development. Here are some of the key observations:

(i) It has long been recognized that there is an apparent excess of highly siderophile elements in the silicate Earth (Chou, 1978; Jagoutz *et al.*, 1979). These excess siderophiles already have been discussed above in the context of core formation. However, until Murthy's (1991) paper, the most widely accepted explanation was that there was a "late veneer" of material accreted after core formation and corresponding to the final one percent or less of the Earth's mass. Nowadays the effects of high temperatures and pressures on partitioning can be investigated and it seems clear that some of the excess siderophile signature reflects silicate-metal equilibration at depth. The volatile highly siderophile elements carbon, sulfur, selenium, and tellurium are more depleted in the silicate Earth than the refractory siderophiles (Figure 1; Yi *et al.*, 2000). Therefore,

if there was a late veneer it probably was material that was on average depleted in volatiles, but not as depleted as would be deduced from the lithophile volatile elements (Yi *et al.*, 2000).

(ii) The light xenon isotopic compositions of well gases may be slightly different from those of the atmosphere in a manner that cannot be easily related to mass-dependent fractionation (Caffee *et al.*, 1988). These isotopes are not affected by radiogenic or fissionogenic additions. The effect is small and currently is one of the most important measurements that need to be made at higher precision. Any resolvable differences need to then be found in other reservoirs (e.g., MORBs) in order to establish that this is a fundamental difference indicating that a fraction of atmospheric xenon is not acquired by outgassing from the interior of the Earth. It could be that the Earth's atmospheric xenon was simply added later and that the isotopic compositions have no genetic link with those found in the Earth's interior.

(iii) Support for this possibility has come from the commonly held view that the Earth's water was added after the giant impact. Having lost so much of its volatiles by early degassing, hydrodynamic escape and impacts the Earth still has a substantial amount of water. Some have proposed that comets may have added a component of water to the Earth, but the D/H ratio would appear to be incorrect for this unless the component represented a minor fraction (Owen and Bar-Nun, 1995, 2000). An alternative set of proposals has been built around volatile-rich chondritic planetary embryos (Morbidelli *et al.*, 2000).

(iv) Further support for this latter "asteroidal" solution comes from the conclusion that the asteroid belt was at one time relatively massive. More than 99% of the mass of the asteroid belt has been ejected or added to other objects (Chambers and Wetherill, 2001). The Earth, being *en route* as some of the material preferentially travels toward the Sun may well have picked up a fraction of its volatiles in this way.

(v) The Moon itself provides a useful monitor of the amount of late material that could have been added to the Earth (Ryder *et al.*, 2000). The Moon provides an impact history (Hartmann *et al.*, 2000) that can be scaled to the Earth (Ryder *et al.*, 2000). In particular, there is evidence of widespread and intense bombardment of the Moon during the Hadean and this can be scaled up, largely in terms of relative cross-sectional area, to yield an impact curve for the Earth (Sleep *et al.*, 1989).

(vi) However, the Moon also is highly depleted in volatiles and its surface is very depleted in highly siderophile elements. Therefore, the Moon also provides a limit on how much can be added to the Earth. This is one reason why the more recent models of Cameron (2000) involving a giant impact that left an additional third are hard to accommodate. However, the database for this currently is poor (Righter *et al.*, 2000).

(vii) It has been argued that the greater depletion in iron and in tellurium in the silicate Earth relative to the Moon reflects an additional small amount of terrestrial core formation following the giant impact (Halliday *et al.*, 1996; Yi *et al.*, 2000). It could also simply reflect differences between Theia and the Earth. If there was further post-giant-impact core formation on Earth, it must have occurred prior to the addition of the late veneer.

3.11 THE HADEAN

3.11.1 Early Mantle Depletion

3.11.1.1 Introduction

Just as the I–Pu–Xe system is useful for studying the rate of formation of the atmosphere and U–Pb and Hf–W are ideal for studying the rates of accretion and core formation, lithophile element isotopic systems are useful for studying the history of melting of the silicate Earth. Two in particular, ^{92}Nb ($T_{1/2} = 36$ Myr) and ^{146}Sm ($T_{1/2} = 106$ Myr), have sufficiently long half-lives to be viable but have been explored with only limited success. Another chronometer of use is the long-lived chronometer ^{176}Lu ($T_{1/2} = 34$–38 Gyr).

3.11.1.2 ^{92}Nb–^{92}Zr

^{92}Nb decays by electron capture to ^{92}Zr with a half-life of 36 ± 3 Myr. At one time it was thought to offer the potential to obtain an age for the Moon by dating early lunar ilmenites and the formation of ilmenite-rich layers in the lunar mantle. Others proposed that it provided constraints on the timescales for the earliest formation of continents on Earth (Münker *et al.*, 2000). In addition, it was argued that it would date terrestrial core formation (Jacobsen and Yin, 2001). There have been many attempts to utilize this isotopic system over the past few years. To do so, it is necessary to first determine the initial ^{92}Nb abundance in early solar system objects accurately and various authors have made claims that differ by two orders of magnitude.

Harper *et al.* (1991b) analyzed a single niobium-rutile found in the Toluca iron meteorite and presented the first evidence for the former existence of ^{92}Nb from which an initial ^{92}Nb/^{93}Nb of $(1.6 \pm 0.3) \times 10^{-5}$ was inferred. However, the blank correction was very large. Subsequently, three studies using multiple collector inductively coupled plasma mass spectrometry proposed that the initial ^{92}Nb/^{93}Nb ratio of the solar system was more than two orders of magnitude higher

(Münker *et al.*, 2000; Sanloup *et al.*, 2000; Yin *et al.*, 2000). Early processes that should fractionate Nb/Zr include silicate partial melting because niobium is more incompatible than zirconium (Hofmann *et al.*, 1986). Other processes relate to formation of titanium-rich (hence niobium-rich) and zirconium-rich minerals, the production of continental crust, terrestrial core formation (Wade and Wood, 2001) and the differentiation of the Moon. Therefore, on the basis of the very high ^{92}Nb abundance proposed, it was argued that, because there was no difference between the zirconium isotopic compositions of early terrestrial zircons and chondrites, the Earth's crust must have formed relatively late (Münker *et al.*, 2000). Similarly, because it is likely that a considerable amount of the Earth's niobium went into the core it was argued that core formation must have been protracted or delayed (Jacobsen and Yin, 2001). We now know that these arguments are incorrect. Precise internal isochrons have provided evidence that the initial abundance of ^{92}Nb in the early solar system is indeed low and close to 10^{-5} (Schönbächler *et al.*, 2002).

Therefore, rather than proving useful, the ^{92}Nb–^{92}Zr has no prospect of being able to provide constraints on these issues because the initial ^{92}Nb abundance is too low.

3.11.1.3 ^{146}Sm–^{142}Nd

High-quality terrestrial data now have been generated for the ^{146}Sm–^{142}Nd (half-life = 106 Myr) chronometer (Goldstein and Galer, 1992; Harper and Jacobsen, 1992; McCulloch and Bennett, 1993; Sharma *et al.*, 1996). Differences in ^{142}Nd/^{144}Nd in early Archean rocks would indicate that the development of a crust on Earth was an early process and that subsequent recycling had failed to eradicate these effects. For many years, only one sample provided a hint of such an effect (Harper and Jacobsen, 1992) although these data have been questioned (Sharma *et al.*, 1996). Recently very high precision measurements of Isua sediments have resolved a 15 ± 4 ppm effect (Caro *et al.*, 2003).

Any such anomalies are clearly small and far less than might be expected from extensive, repeated depletion of the mantle by partial melting in the Hadean. It seems inescapable that there was melting on the early Earth. Therefore, the interesting and important result of these studies is that such isotopic effects must largely have been eliminated. The most likely mechanism is very efficient mantle convection. In the earliest Earth convection may have been much more vigorous (Chase and Patchett, 1988; Galer and Goldstein, 1991) because of the large amount of heat left from the secular cooling and the greater radioactive heat production.

3.11.1.4 ^{176}Lu–^{176}Hf

A similar view is obtained from hafnium isotopic analyses of very early zircons. The ^{176}Lu–^{176}Hf isotopic system ($T_{1/2} = 34$–38 Gyr) is ideally suited for studying early crustal evolution, because hafnium behaves in an almost identical fashion to zirconium. As a result, the highly resistant and easily dated mineral zircon typically contains ~1 wt% Hf, sufficient to render hafnium isotopic analyses of single zircons feasible using modern methods (Amelin et al., 1999). The concentration of lutetium in zircon is almost negligible by comparison. As a result, the initial hafnium isotopic composition is relatively insensitive to the exact age of the grain and there is no error magnification involved in extrapolating back to the early Archean. Furthermore, one can determine the age of the single zircon grain very precisely using modern U–Pb methods. One can obtain an extremely precise initial hafnium isotopic composition for a particular point in time on a single grain, thereby avoiding the problems of mixed populations. The U–Pb age and hafnium isotopic compositions of zircons also are extremely resistant to resetting and define a reliable composition at a well-defined time in the early Earth. The hafnium isotopic composition that zircon had when it grew depends on whether the magma formed from a reservoir with a time-integrated history of melt depletion or enrichment. Therefore, one can use these early zircons to search for traces of early mantle depletion.

Note that this is similar to the approach adopted earlier with ^{147}Sm–^{143}Nd upon which many ideas of Hadean mantle depletion, melting processes and early crust were based (Chase and Patchett, 1988; Galer and Goldstein, 1991). However, the difficulty with insuring closed-system behavior with bulk rock Sm–Nd in metamorphic rocks and achieving a robust age correction of long-lived ^{147}Sm over four billion years has meant that this approach is now viewed as suspect (Nägler and Kramers, 1998). The ^{176}Lu–^{176}Hf isotopic system and use of low-Lu/Hf zircons is far more reliable in this respect (Amelin et al., 1999, 2000). In practice, however, the interpretation is not that simple, for two reasons:

(i) The hafnium isotopic composition and Lu/Hf ratio of the Earth's primitive mantle is poorly known. It is assumed that it is broadly chondritic (Blichert-Toft and Albarède, 1997), but which exact kind of chondrite class best defines the isotopic composition of the primitive mantle is unclear. Without this information, one cannot extrapolate back in time to the early Earth and state with certainty what the composition of the primitive mantle reservoir was. Therefore, one cannot be sure what a certain isotopic composition means in terms of the level of time-integrated depletion.

(ii) The half-life of ^{176}Lu is *not* well established and is the subject of current debate and research (Scherer et al., 2001). Although the determination of the initial hafnium isotopic composition of zircon is not greatly affected by this, because the Lu/Hf ratio is so low that the age correction is tiny, the correction to the value for the primitive mantle is very sensitive to this uncertainty.

With these caveats, one can deduce the following. Early single grains appear to have recorded hafnium isotopic compositions that provide evidence for chondritic or enriched reservoirs. There is no evidence of depleted reservoirs in the earliest (Hadean) zircons dated thus far (Amelin et al., 1999). Use of alternative values for the decay constants or values for the primitive mantle parameters increases the proportion of hafnium with an enriched signature (Amelin et al., 2000), but does not provide evidence for early mantle depletion events. Therefore, there is little doubt that the Hadean mantle was extremely well mixed. Why this should be is unclear, but it probably relates in some way to the lack of preserved continental material from prior to 4.0 Ga.

3.11.2   Hadean Continents

Except for the small amount of evidence for early mantle melting we are in the dark about how and when Earth's continents first formed (Figure 13). We already have pointed out that in its early stages Earth may have had a magma ocean, sustained from heat from accretion and the blanketing effects of a dense early atmosphere. With the loss of the early atmosphere during planetary collisions, the Earth would have cooled quickly, the outer portions would have solidified and it would thereby have developed its first primitive crust.

We have little evidence of what such a crust might have looked like. Unlike on the Moon and Mars, Earth appears to have no rock preserved that is more than 4.0 Gyr old. There was intense bombardment of the Moon until ~3.9 Gyr ago (Wetherill, 1975b; Hartmann et al., 2000; Ryder et al., 2000). Earth's earlier crust may therefore have been decimated by concomitant impacts. It may also be that a hotter Earth had a surface that was inherently unstable. Some argued that the earliest crust was like the lunar highlands—made from a welded mush of crystals that had previously floated on the magma ocean. Others have suggested that it was made of denser rocks more like those of the Earth's present oceanfloor (Galer and Goldstein, 1991). But firm evidence has so far been sparse.

Froude et al. (1983) reported the exciting discovery of pre-4.0 Ga zircon grains formed on Earth. The host rock from which these grains were recovered is not so old. The pre-4.0 Gyr

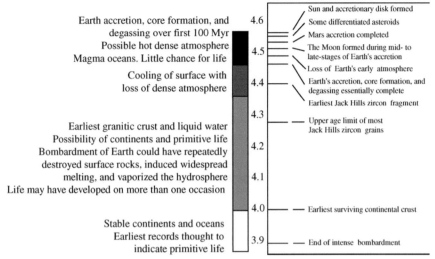

Figure 13 Schematic showing the timescales for various events through the "Dark Ages" of the Hadean.

rocks were largely destroyed but the zircons survived by becoming incorporated in sands that formed a sedimentary rock that now is exposed in Australia as the Jack Hills Metaconglomerate. By measuring uranium-lead ages a sizeable population of grains between 4.1 Gyr and 4.2 Gyr was discovered (Froude *et al.*, 1983). Subsequently, Wilde *et al.* (2001) and Mojzsis *et al.* (2001) reported uranium–lead ages and oxygen isotopic compositions of further old zircon grains. A portion of one grain appears to have formed 4.40 Gyr ago and this is the oldest terrestrial solid yet identified. More recent work has been published by Peck *et al.* (2001) and Valley *et al.* (2002).

These zircons provide powerful evidence for the former existence of some unknown amount of continental crust in the Hadean. Nearly all zircons grew from granite magmas, not similar at all to those forming the oceanfloor or the lunar highlands. Granite magmas usually form at >700 °C and >20 km depth, mainly by melting preexisting crust above subduction zones. Being buoyant they are typical of continental mountain regions such as the Andes—sites of very active erosion. The existence of such zircons would be consistent with continental crust as far back as 4.40 Gyr–~100 Myr after the formation of the Moon and the earliest atmosphere. It would be nice to have more data and one is extrapolating through many orders of magnitude in mass when inferring extensive continents from single zircons. Large ocean islands like Iceland have small volumes of granitic magma and so one can conceive of protocontinents that started from the accumulation of such basaltic nuclei, the overall mass of silicic crustal material increasing gradually.

Oxygen isotopic measurements can be used to infer the presence of liquid water on the Earth's early surface. The oxygen isotopic composition reflects that of the magma from which the zircon crystallized, which in turn reflects that of the rocks that were melted to form the magma. Heavy oxygen (with a high proportion of ^{18}O) is produced by low-temperature interactions between a rock and liquid water such as those that form clay by weathering. The somewhat heavy oxygen of these zircons provides evidence that the rocks that were melted to form the magma included components that had earlier been at the surface in the presence of liquid water. Early on, when Earth was hotter, they might also have formed by melting of wet oceanfloor basalt that was taken back into the mantle by a process potentially comparable to modern subduction. Either way, the data indicate that surface rocks affected by low-temperature fluids were probably being transported to significant depths and melted, as occurs today.

These grains represent a unique archive for information on the early Earth. The potential is considerable. For example, Wilde *et al.* (2001) and Peck *et al.* (2001) also used trace elements and tiny inclusions to reconstruct the composition of the parent magma. In all of this work there is a need to be very aware that the grains may have been disturbed after they formed. Wet diffusion of oxygen could lead to an ^{18}O-rich composition that was acquired during subsequent metamorphic or magmatic histories. Thus, ancillary information on diffusivities and the degree to which compositions might have been perturbed by later metamorphism must be acquired. Most importantly, however, there exists an urgent need of many more grains.

3.11.3 The Hadean Atmosphere, Hydrosphere, and Biosphere

The nature of the Earth's atmosphere and hydrosphere after the Earth had cooled, following the cessation of the main stages of accretion, has been the subject of a fair amount of research (Wiechert, 2002), particularly given the *lack* of data upon which to base firm conclusions. An interesting and important question is to what extent it may have been possible for life to develop during this period (Mojzsis *et al.*, 1996; Sleep *et al.*, 2001; Zahnle and Sleep, 1996). A great deal has been written on this and because it is covered elsewhere in this treatise only cursory background is provided here.

The first-order constraints that exist on the nature of Earth's early exosphere (Sleep *et al.*, 1989; Sleep and Zahnle, 2001) are as follows:

(i) The Sun was fainter and cooler than today because of the natural start-up of fusion reactions that set it on the Main Sequence (Kasting and Grinspoon, 1991; Sagan and Chyba, 1997; Pavlov *et al.*, 2000). Therefore, the level of radiation will have been less.

(ii) Earth's interior was a few 100 K hotter because of secular cooling from accretion and far greater radiogenic heat production. The Earth's heat flow was 2–3 times higher. Therefore, one can assume that more heat was escaping via mantle melting and production of oceanic crust.

(iii) This, in turn, means more mantle-derived volatiles such as CO_2 were being released.

(iv) It also means that there was more hydrothermal alteration of the oceanfloor. Therefore, CO_2 was converted to carbonate in altered basalt and returned to the mantle at subduction zones (if those really existed already).

(v) There may have been far less marine carbonate. We can infer this from the geological record for the Archean. Therefore, it appears that atmospheric CO_2 levels were low—most of the CO_2 was being recycled to the mantle.

(vi) Because atmospheric CO_2 exerts a profound effect on temperature as a greenhouse gas, low concentrations of CO_2 imply that atmospheric temperatures were cold, unless another greenhouse gas such as methane (CH_4) was very abundant (Pavlov *et al.*, 2000). However, clear geochemical evidence for a strong role of methane in the Archean currently is lacking.

(vii) Impacts, depending on their number and magnitude (Hartmann *et al.*, 2000; Ryder *et al.*, 2000), may have had a devastating effect on the early biosphere. Impact ejecta will react with atmospheric and oceanic CO_2 and thereby lower atmospheric CO_2 levels reducing atmospheric temperatures still further.

The Hadean is fast becoming one of the most interesting areas of geochemical research. With so little hard evidence (Figure 13), much of the progress probably will come from theoretical modeling and comparative planetology.

3.12 CONCLUDING REMARKS—THE PROGNOSIS

Since the early 1990s, there has been great progress in understanding the origins and early development of the Earth. In some cases, this has been a function of improved modeling. This is particularly true for noble gases. However, in most cases it has been the acquisition of new kinds of data that has proven invaluable. The most obvious examples are in isotope geochemistry and cosmochemistry.

It is perhaps worth finishing by pointing out the kinds of developments that can be expected to have an impact on our understanding of the early Earth. One can ask a question like "what if we could measure...?" Here are some things that would be very interesting and useful to explore:

(i) If planets such as the Earth formed by very energetic collisions that were sufficient to cause vaporization of elements and compounds that normally are solid, it may be possible to find evidence for the kinetic effects of boiling in isotopic fractionations. Humayun and Clayton (1995) explored potassium at per mil levels of precision and found no significant evidence of fractionation. Poitrasson *et al.* (2003) have found evidence that the iron on the Moon may be very slightly heavy relative to other planetary objects. Note that this has nothing to do with the fractionations produced in the lunar surface during irradiation and implantation (Wiesli *et al.*, 2003). Perhaps boiling during the giant impact caused this. This is just a preliminary inference at this stage. But if so, small fractionations also should be found in other elements of similar volatility such as lithium, magnesium, silicon, and nickel. There is much to be done to explore these effects at very high precision.

(ii) A vast amount of work still is needed on core formation, understanding the depletion of siderophile elements in the Earth's mantle (Walter *et al.*, 2000), and determining the abundances of light elements in the core (Gessmann *et al.*, 2001). Much of this depends on using proxy elements that are very sensitive to pressure, temperature, water content or f_{O_2} (e.g., Wade and Wood, 2001). A problem at present is that too many of the elements of interest are sensitive to more than one parameter and, therefore, the solutions are under constrained. Also, new experiments are needed at very high pressures—close to those of the core–mantle boundary.

(iii) The origins of Earth's water and volatiles more generally are the subject of considerable debate (e.g., Anders and Owen, 1977; Carr and Wänke, 1992; Owen and Bar-Nun, 1995, 2000; Javoy, 1998; Caffee *et al.*, 1999; Righter and Drake, 1999; Abe *et al.*, 2000; Morbidelli *et al.*, 2000). A great deal probably will be learned from further modeling. For example, it is very important to understand what kinds of processes in the early Earth could have caused loss of early atmospheres. We do not understand how volatiles are retained during a Moon-forming giant impact (Melosh and Sonett, 1986; Melosh, 1990; Melosh et al., 1993) or what the early Sun might have done to the atmosphere. We need to acquire more reliable data on the isotopic compositions of volatiles in the deep Earth. We might be able to learn much more about Earth's volatile history from more precise measurements of the volatile components in other planets. In this respect more detailed studies of Mars and how closely it resembles the Earth (Carr and Wänke, 1992) could prove critical.

(iv) Mojzsis et al. (2001) and Wilde et al. (2001) have made major advances in studying the Hadean using single zircons. Apart from needing many more such samples, one has to ask what other kinds of information might be extractable from such zircons. The oxygen isotopic composition provides evidence for early low-temperature water. Exploring the melt inclusions and the trace-element concentrations also has been shown to have potential (Peck et al., 2001). Zircons are also iron-rich and conceivably could eventually be used to provide evidence for biological processes in the Hadean. However, the sensitivities of the techniques need to be improved vastly for this to be achieved with single grains. Furthermore, the current status of the rapidly expanding field of iron isotope geochemistry provides no clear basis for assuming a distinctive signal of biotic effects will be realizable. Also, the really interesting zircons are so precious that one should use minimally destructive techniques like SIMS on single grains. However, the required precision for measuring isotopic ratios in a useful manner is not available for trace elements in minerals using this method at present. Similarly, hafnium isotopes on single zircon grains provide the most reliable and powerful constraints on the extent of mantle depletion (Amelin et al., 1999) but require destruction of a part of the grain. Developing improved methods that achieve far higher overall sensitivity is critical.

(v) Determining the rates of accretion of the Earth and Moon more reliably will be critically dependent on the correct and precise determination of the initial tungsten and hafnium isotopic compositions of the solar system, and the ^{182}Hf decay constant. The initial isotopic compositions really require the more widespread application of negative ion thermal ionization mass spectrometry (N-TIMS) (Quitté et al., 2000). The decay constant work is going to require the acquisition of ^{182}Hf, probably from neutron-irradiated ^{180}Hf.

(vi) Similarly, progress in using hafnium isotopes to study the degree of early mantle depletion is being thwarted by the uncertainties associated

with the ^{176}Lu decay constant (Scherer et al., 2001), and some new experimental work is needed in this area.

(vii) A major need is for a closer integration of the modeling of different isotopic systems. As of early 2000s, this has really only been attempted for tungsten and lead (Halliday, 2000). In the future it will be essential to integrate xenon isotopes in with these and other accretion models.

(viii) Finally, another major area of modeling has to occur in the area of "early Earth system science." There needs to be integrated modeling of the evolution of the atmosphere, oceans, surface temperature, mantle convection, and magma oceans. This is now being attempted. For example, the studies by Sleep and Zahnle (2001) and Sleep et al. (1989, 2001) are paving the way for more comprehensive models that might involve the fluid dynamics of mantle convection.

ACKNOWLEDGMENTS

This chapter benefited enormously from discussion with, and comments and criticism received from Tom Ahrens, Alan Boss, Pat Cassen, Andy Davis, Martin Frank, Tim Grove, Munir Humayun, Don Porcelli, Norm Sleep, Mike Walter, Uwe Wiechert, Rainer Wieler, Kevin Zahnle, and two anonymous reviewers.

REFERENCES

Abe Y. and Matsui T. (1985) The formation of an impact-generated H_2O atmosphere and its implications for the early thermal history of the Earth. *Proc. 15th Lunar Planet. Sci. Conf.: J. Geophys. Res.* **90**(suppl.), C545–C559.

Abe Y. and Matsui T. (1986) Early evolution of the earth: accretion, atmosphere formation, and thermal history. *Proc. 17th Lunar Planet. Sci. Conf.: J. Geophys. Res.* **91** (no. B13), E291–E302.

Abe Y. and Matsui T. (1988) Evolution of an impact-generated H_2O–CO_2 atmosphere and formation of a hot proto-ocean on Earth. *Am. Meteorol. Soc.* **45**, 3081–3101.

Abe Y., Ohtani E., and Okuchi T. (2000) Water in the early earth. In *Origin of the Earth and Moon* (eds. R. M. Canup and K. Righter). University of Arizona Press, Tucson, pp. 413–433.

Agee C. B. and Walker D. (1988) Mass balance and phase density constraints on early differentiation of chondritic mantle. *Earth Planet. Sci. Lett.* **90**, 144–156.

Agnor C. B. and Ward W. R. (2002) Damping of terrestrial-planet eccentricities by density-wave interactions with a remnant gas disk. *Astrophys. J.* **567**, 579–586.

Ahrens T. J. (1990) Earth accretion. In *Origin of the Earth* (eds. H. E. Newsom and J. H. Jones). Oxford University Press, Oxford, pp. 211–217.

Ahrens T. J. and Jeanloz R. (1987) Pyrite shock compression, isentropic release and composition of the Earth's core. *J. Geophys. Res.* **92**, 10363–10375.

Albarède F. and Juteau M. (1984) Unscrambling the lead model ages. *Geochim. Cosmochim. Acta* **48**, 207–212.

Alibert C., Norman M. D., and McCulloch M. T. (1994) An ancient Sm–Nd age for a ferroan noritic anorthosite clast from lunar breccia 67016. *Geochim. Cosmochim. Acta* **58**, 2921–2926.

Allègre C. J., Dupré B., and Brévart O. (1982) Chemical aspects of formation of the core. *Phil. Trans. Roy. Soc. London* **A306**, 49–59.

Allègre C. J., Staudacher T., Sarda P., and Kurz M. (1983) Constraints on evolution of Earth's mantle from rare gas systematics. *Nature* **303**, 762–766.

Allègre C. J., Staudacher T., and Sarda P. (1987) Rare gas systematics: formation of the atmosphere, evolution, and structure of the Earth's mantle. *Earth Planet. Sci. Lett.* **81**, 127–150.

Allègre C. J., Manhès G., and Göpel C. (1995a) The age of the Earth. *Geochim. Cosmochim. Acta* **59**, 1445–1456.

Allègre C. J., Poirier J.-P., Humler E., and Hofmann A. W. (1995b) The chemical composition of the Earth. *Earth Planet. Sci. Lett.* **134**, 515–526.

Allègre C. J., Hofmann A. W., and O'Nions R. K. (1996) The argon constraints on mantle structure. *Geophys. Res. Lett.* **23**, 3555–3557.

Allègre C. J., Manhès G., and Lewin E. (2001) Chemical composition of the Earth and the volatility control on planetary genetics. *Earth Planet. Sci. Lett.* **185**, 49–69.

Amelin Y., Lee D.-C., Halliday A. N., and Pidgeon R. T. (1999) Nature of the Earth's earliest crust from hafnium isotopes in single detrital zircons. *Nature* **399**, 252–255.

Amelin Y., Lee D.-C., and Halliday A. N. (2000) Early-middle Archean crustal evolution deduced from Lu–Hf and U–Pb isotopic studies of single zircon grains. *Geochim. Cosmochim. Acta* **64**, 4205–4225.

Amelin Y., Krot A. N., Hutcheon I. D., and Ulyanov A. A. (2002) Lead isotopic ages of chondrules and calcium–aluminum-rich inclusions. *Science* **297**, 1678–1683.

Anders E. (1977) Chemical compositions of the Moon, Earth, and eucrite parent body. *Phil. Trans. Roy. Soc. London* **A295**, 23–40.

Anders E. and Grevesse N. (1989) Abundances of the elements: meteoritic and solar. *Geochim. Cosmochim. Acta* **53**, 197–214.

Anders E. and Owen T. (1977) Mars and Earth: origin and abundance of volatiles. *Science* **198**, 453–465.

Anderson D. L. (1979) Chemical stratification of the mantle. *J. Geophys. Res.* **84**, 6297–6298.

Arculus R. J. and Delano J. W. (1981) Siderophile element abundances in the upper mantle: evidence for a sulfide signature and equilibrium with the core. *Geochim. Cosmochim. Acta* **45**, 1331–1344.

Artymowicz P. (1997) Beta Pictoris: an early solar system? *Ann. Rev. Earth Planet. Sci.* **25**, 175–219.

Azbel I. Y., Tolstikhin I. N., Kramers J. D., Pechernikova G. V., and Vityazev A. V. (1993) Core growth and siderophile element depletion of the mantle during homogeneous Earth accretion. *Geochim. Cosmochim. Acta* **57**, 2889–2898.

Ballentine C. J., Porcelli D., and Wieler R. (2001) A critical comment on Trieloff *et al.. Science* **291**, 2269 (online).

Bell K. R., Cassen P. M., Wasson J. T., and Woolum D. S. (2000) The FU Orionis phenomenon and solar nebular material. In *Protostars and Planets IV* (eds. V. Mannings, A. P. Boss, and S. S. Russell). University of Arizona Press, Tucson, pp. 897–926.

Benz W. (2000) Low velocity collisions and the growth of planetesimals. *Space Sci. Rev.* **92**, 279–294.

Benz W. and Asphaug E. (1999) Catastrophic disruptions revisited. *Icarus* **142**, 5–20.

Benz W. and Cameron A. G. W. (1990) Terrestrial effects of the giant impact. In *Origin of the Earth* (eds. H. E. Newsom and J. H. Jones). Oxford University Press, Oxford, pp. 61–67.

Benz W., Cameron A. G. W., and Slattery W. L. (1987) Collisional stripping of Mercury's mantle. *Icarus* **74**, 516–528.

Binder A. B. (1986) The binary fission origin of the Moon. In *Origin of the Moon* (eds. W. K. Hartmann, R. J. Phillips, and G. J. Taylor). Lunar and Planetary Institute, Houston, pp. 499–516.

Birck J.-L., Rotaru M., and Allègre C. J. (1999) ^{53}Mn–^{53}Cr evolution of the early solar system. *Geochim. Cosmochim. Acta* **63**, 4111–4117.

Blichert-Toft J. and Albarède F. (1997) The Lu–Hf isotope geochemistry of chondrites and the evolution of the mantle–crust system. *Earth Planet. Sci. Lett.* **148**, 243–258.

Blum J. (2000) Laboratory experiments on preplanetary dust aggregation. *Space Sci. Rev.* **92**, 265–278.

Boss A. P. (1990) 3D Solar nebula models: implications for Earth origin. In *Origin of the Earth* (eds. H. E. Newsom and J. H. Jones). Oxford University Press, Oxford, pp. 3–15.

Boss A. P. (1997) Giant planet formation by gravitational instability. *Science* **276**, 1836–1839.

Boss A. P. (1998) *Looking for Earths.* Wiley, New York, 240pp.

Boss A. P. (2002) The formation of gas and ice giant planets. *Earth Planet. Sci. Lett.* **202**, 513–523.

Brandon A. D., Walker R. J., Morgan J. W., Norman M. D., and Prichard H. M. (1998) Coupled ^{186}Os and ^{187}Os evidence for core–mantle interaction. *Science* **280**, 1570–1573.

Brett R. (1984) Chemical equilibrium between Earth's core and upper mantle. *Geochim. Cosmochim. Acta* **48**, 1183–1188.

Brown G. C. and Mussett A. E. (1981) *The Inaccessible Earth.* Allen and Unwin, London.

Brown H. (1949) Rare gases and the formation of the Earth's atmosphere. In *The Atmospheres of the Earth and Planets* (ed. G. P. Kuiper). University of Chicago Press, Chicago, pp. 258–266.

Busemann H., Baur H., and Wieler R. (2000) Primordial noble gases in "phase Q" in carbonaceous and ordinary chondrites studied by closed-system stepped etching. *Meteorit. Planet. Sci.* **35**, 949–973.

Caffee M. W., Hudson G. B., Velsko C., Alexander E. C., Jr., Huss G. R., and Chivas A. R. (1988) Non-atmospheric noble gases from CO_2 well gases. In *Lunar Planet. Sci.* **XIX.** The Lunar and Planetary Institute, Houston, pp. 154–155.

Caffee M. W., Hudson G. B., Velsko C., Huss G. R., Alexander E. C., Jr., and Chivas A. R. (1999) Primordial noble gases from Earth's mantle: identification of a primitive volatile component. *Science* **285**, 2115–2118.

Cameron A. G. W. (1978) Physics of the primitive solar accretion disk. *Moons and Planets* **18**, 5–40.

Cameron A. G. W. (2000) Higher-resolution simulations of the Giant Impact. In *Origin of the Earth and Moon* (eds. K. Righter and R. Canup). University of Arizona Press, Tucson, pp. 133–144.

Cameron A. G. W. and Benz W. (1991) Origin of the Moon and the single impact hypothesis: IV. *Icarus* **92**, 204–216.

Canup R. M. and Agnor C. (1998) Accretion of terrestrial planets and the earth–moon system. In *Origin of the Earth and Moon,* LPI Contribution No. **597** Lunar and Planetary Institute, Houston, pp. 4–7.

Canup R. M. and Asphaug E. (2001) Origin of the Moon in a giant impact near the end of the Earth's formation. *Nature* **412**, 708–712.

Carlson R. W. and Hauri E. H. (2001) Extending the ^{107}Pd–^{107}Ag chronometer to Low Pd/Ag meteorites with the MC-ICPMS. *Geochim. Cosmochim. Acta* **65**, 1839–1848.

Carlson R. W. and Lugmair G. W. (1988) The age of ferroan anorthosite 60025: oldest crust on a young Moon? *Earth Planet. Sci. Lett.* **90**, 119–130.

Carlson R. W. and Lugmair G. W. (2000) Timescales of planetesimal formation and differentiation based on extinct and extant radioisotopes. In *Origin of the Earth and Moon* (eds. K. Righter and R. Canup). University of Arizona Press, Tucson, pp. 25–44.

Caro G., Bourdon B., Birck J. L., and Moorbath S. (2003) ^{146}Sm–^{142}Nd evidence from Isua metamorphosed sediments for early differentiation of the Earth's mantle. *Nature* **423**, 428–431.

Carr M. H. and Wänke H. (1992) Earth and Mars: water inventories as clues to accretional histories. *Icarus* **98**, 61–71.

Cassen P. (1996) Models for the fractionation of moderately volatile elements in the solar nebula. *Meteorit. Planet. Sci.* **31**, 793–806.

Cassen P. (2001) Nebular thermal evolution and the properties of primitive planetary materials. *Meteorit. Planet. Sci.* **36**, 671–700.

Chambers J. E. and Wetherill G. W. (2001) Planets in the asteroid belt. *Meteorit. Planet. Sci.* **36**, 381–399.

Chase C. G. and Patchett P. J. (1988) Stored mafic/ultramafic crust and Early Archean mantle depletion. *Earth Planet. Sci. Lett.* **91**, 66–72.

Chiang E. I. and Goldreich P. (1997) Spectral energy distributions of T Tauri stars with passive circumstellar disks. *Astrophys. J.* **490**, 368.

Chiang E. I. and Goldreich P. (1999) Spectral energy distributions of passive T Tauri disks: inclination. *Astrophys. J.* **519**, 279.

Chiang E. I., Joung M. K., Creech-Eakman M. J., Qi C., Kessler J. E., Blake G. A., and van Dishoeck E. F. (2001) Spectral energy distributions of passive T Tauri and Herbig Ae disks: grain mineralogy, parameter dependences, and comparison with ISO LWS observations. *Astrophys. J.* **547**, 1077.

Chou C. L. (1978) Fractionation of siderophile elements in the Earth's upper mantle. *Proc. 9th Lunar Sci. Conf.* 219–230.

Clark S. P., Jr., Turekian K. K., and Grossman L. (1972) Model for the early history of the Earth. In *The Nature of the Solid Earth* (ed. E. C. Robertson). McGraw-Hill, New York, pp. 3–18.

Clarke W. B., Beg M. A., and Craig H. (1969) Excess ^{3}He in the sea: evidence for terrestrial primordial helium. *Earth Planet. Sci. Lett.* **6**, 213–220.

Clayton R. N. (1986) High temperature isotope effects in the early solar system. In *Stable Isotopes in High Temperature Geological Processes* (eds. J. W. Valley, H. P. Taylor, and J. R. O'Neil). Mineralogical Society of America, Washington, DC, pp. 129–140.

Clayton R. N. (1993) Oxygen isotopes in meteorites. *Ann. Rev. Earth Planet. Sci.* **21**, 115–149.

Clayton R. N. and Mayeda T. K. (1975) Genetic relations between the moon and meteorites. *Proc. 11th Lunar Sci. Conf.* 1761–1769.

Clayton R. N. and Mayeda T. K. (1996) Oxygen isotope studies of achondrites. *Geochim. Cosmochim. Acta* **60**, 1999–2017.

Clayton R. N., Grossman L., and Mayeda T. K. (1973) A component of primitive nuclear composition in carbonaceous meteorites. *Science* **182**, 485–487.

Connolly H. C., Jr. and Love S. G. (1998) The formation of chondrules: petrologic tests of the shock wave model. *Science* **280**, 62–67.

Connolly H. C., Jr., Jones B. D., and Hewins R. G. (1998) The flash melting of chondrules: an experimental investigation into the melting history and physical nature of chondrules. *Geochim. Cosmochim. Acta* **62**, 2725–2735.

Craig H. and Lupton J. E. (1976) Primordial neon, helium, and hydrogen in oceanic basalts. *Earth Planet. Sci. Lett.* **31**, 369–385.

Cuzzi J. N., Davis S. S., and Dobrovolskis A. R. (2003) Creation and distribution of CAIs in the protoplanetry nebula. In *Lunar Planet. Sci.* **XXXIV**, #1749. The Lunar and Planetary Institute, Houston (CD-ROM).

Darwin G. H. (1878) On the precession of a viscous spheroid. *Nature* **18**, 580–582.

Darwin G. H. (1879) On the precession of a viscous spheroid and on the remote history of the Earth. *Phil Trans. Roy. Soc. London* **170**, 447–538.

Dauphas N., Marty B., and Reisberg L. (2002) Inference on terrestrial genesis from molybdenum isotope systematics. *Geophys. Res. Lett.* **29**, 1084 doi:10.1029/2001GL014237.

Davies G. F. (1999) Geophysically constrained mass flows and the ^{40}Ar budget: a degassed lower mantle? *Earth Planet. Sci. Lett.* **166**, 149–162.

Déruelle B., Dreibus G., and Jambon A. (1992) Iodine abundances in oceanic basalts: implications for Earth dynamics. *Earth Planet. Sci. Lett.* **108**, 217–227.

Desch S. J. and Cuzzi J. N. (2000) The generation of lightning in the solar nebula. *Icarus* **143**, 87–105.

Dixon E. T., Honda M., McDougall I., Campbell I. H., and Sigurdsson I. (2000) Preservation of near-solar neon isotopic ratios in Icelandic basalts. *Earth Planet. Sci. Lett.* **180**, 309–324.

Doe B. R. and Stacey J. S. (1974) The application of lead isotopes to the problems of ore genesis and ore prospect evaluation: a review. *Econ. Geol.* **69**, 755–776.

Drake M. J. and McFarlane E. A. (1993) Mg-perovskite/silicate melt and majorite garnet/silicate melt partition coefficients in the system CaO–MgO–SiO$_2$ at high temperatures and pressures. *J. Geophys. Res.* **98**, 5427–5431.

Drake M. J. and Righter K. (2002) Determining the composition of the Earth. *Nature* **416**, 39–44.

Dreibus G. and Palme H. (1996) Cosmochemical constraints on the sulfur content in the Earth's core. *Geochim. Cosmochim. Acta* **60**, 1125–1130.

Esat T. M. (1996) Comment on Humayun and Clayton (1995). *Geochim. Cosmochim. Acta* **60**, 2755–2758.

Eucken A. (1944) Physikalisch-Chemische Betrachtungen über die früheste Entwicklungsgeschichte der Erde. *Nachr. Akad. Wiss. Göttingen, Math-Phys. Kl.* **Heft 1**, 1–25.

Farley K. A. and Poreda R. (1992) Mantle neon and atmospheric contamination. *Earth Planet. Sci. Lett.* **114**, 325–339.

Feigelson E. D., Broos P., Gaffney J. A., III, Garmire G., Hillenbrand L. A., Pravdo S. H., Townsley L., and Tsuboi Y. (2002a) X-ray-emitting young stars in the Orion nebula. *Astrophys. J.* **574**, 258–292.

Feigelson E. D., Garmire G. P., and Pravdo S. H. (2002b) Magnetic flaring in the pre-main-sequence Sun and implications for the early solar system. *Astrophys. J.* **572**, 335–349.

Franchi I. A., Wright I. P., Sexton A. S., and Pillinger C. T. (1999) The oxygen isotopic composition of Earth and Mars. *Meteorit. Planet. Sci.* **34**, 657–661.

Froude D. O., Ireland T. R., Kinny P. D., Williams I. S., Compston W., Williams I. R., and Myers J. S. (1983) Ion microprobe identification of 4,100–4,200 Myr-old terrestrial zircons. *Nature* **304**, 616–618.

Gaffey M. J. (1990) Thermal history of the asteroid belt: implications for accretion of the terrestrial planets. In *Origin of the Earth* (eds. H. E. Newsom and J. H. Jones). Oxford University Press, Oxford, pp. 17–28.

Galer S. J. G. and Goldstein S. L. (1991) Early mantle differentiation and its thermal consequences. *Geochim. Cosmochim. Acta* **55**, 227–239.

Galer S. J. G. and Goldstein S. L. (1996) Influence of accretion on lead in the Earth. In *Isotopic Studies of Crust–Mantle Evolution* (eds. A. R. Basu and S. R. Hart). American Geophysical Union, Washington, DC, pp. 75–98.

Ganapathy R. and Anders E. (1974) Bulk compositions of the Moon and Earth, estimated from meteorites. *Proc. 5th Lunar Conf.* 1181–1206.

Gancarz A. J. and Wasserburg G. J. (1977) Initial Pb of the Amîtsoq Gneiss, West Greenland, and implications for the age of the Earth. *Geochim. Cosmochim. Acta* **41**, 1283–1301.

Gast P. W. (1960) Limitations on the composition of the upper mantle. *J. Geophys. Res.* **65**, 1287–1297.

Gessmann C. K., Wood B. J., Rubie D. C., and Kilburn M. R. (2001) Solubility of silicon in liquid metal at high pressure: implications for the composition of the Earth's core. *Earth Planet. Sci. Lett.* **184**, 367–376.

Ghosh A. and McSween H. Y., Jr. (1999) Temperature dependence of specific heat capacity and its effect on asteroid thermal models. *Meteorit. Planet. Sci.* **34**, 121–127.

Goldreich P. and Ward W. R. (1973) The formation of planetesimals. *Astrophys. J.* **183**, 1051–1060.

Goldstein S. L. and Galer S. J. G. (1992) On the trail of early mantle differentiation: ^{142}Nd/^{144}Nd ratios of early Archean rocks. *Trans. Am. Geophys. Union*, 323.

Göpel C., Manhès G., and Allègre C. J. (1991) Constraints on the time of accretion and thermal evolution of chondrite parent bodies by precise U–Pb dating of phosphates. *Meteoritics* **26**, 73.

Göpel C., Manhès G., and Allègre C. J. (1994) U–Pb systematics of phosphates from equilibrated ordinary chondrites. *Earth Planet. Sci. Lett.* **121**, 153–171.

Gounelle M., Shu F. H., Shang H., Glassgold A. E., Rehm K. E., and Lee T. (2001) Extinct radioactivities and protosolar cosmic-rays: self-shielding and light elements. *Astrophys. J.* **548**, 1051–1070.

Gray C. M., Papanastassiou D. A., and Wasserburg G. J. (1973) The identification of early condensates from the solar nebula. *Icarus* **20**, 213–239.

Grevesse N. and Sauval A. J. (1998) Standard solar composition. *Space Sci. Rev.* **85**, 161–174.

Grimm R. E. and McSween H. Y., Jr. (1993) Heliocentric zoning of the asteroid belt by aluminum-26 heating. *Science* **259**, 653–655.

Grossman J. N. (1996) Chemical fractionations of chondrites: signatures of events before chondrule formation. In *Chondrules and the Protoplanetary Disk* (eds. R. H. Hewins, R. H. Jones, and E. R. D. Scott). Cambridge University of Press, Cambridge, pp. 243–253.

Grossman L. (1972) Condensation in the primitive solar nebula. *Geochim. Cosmochim. Acta* **36**, 597–619.

Grossman L. and Larimer J. W. (1974) Early chemical history of the solar system. *Rev. Geophys. Space Phys.* **12**, 71–101.

Guan Y., McKeegan K. D., and MacPherson G. J. (2000) Oxygen isotopes in calcium–aluminum-rich inclusions from enstatite chondrites: new evidence for a single CAI source in the solar nebula. *Earth Planet. Sci. Lett.* **183**, 557–558.

Halliday A. N. (2000) Terrestrial accretion rates and the origin of the Moon. *Earth Planet. Sci. Lett.* **176**, 17–30.

Halliday A. N. and Lee D.-C. (1999) Tungsten isotopes and the early development of the Earth and Moon. *Geochim. Cosmochim. Acta* **63**, 4157–4179.

Halliday A. N. and Porcelli D. (2001) In search of lost planets—the paleocosmochemistry of the inner solar system. *Earth Planet. Sci. Lett.* **192**, 545–559.

Halliday A. N., Lee D.-C., Tomassini S., Davies G. R., Paslick C. R., Fitton J. G., and James D. (1995) Incompatible trace elements in OIB and MORB and source enrichment in the sub-oceanic mantle. *Earth Planet. Sci. Lett.* **133**, 379–395.

Halliday A. N., Rehkämper M., Lee D.-C., and Yi W. (1996) Early evolution of the Earth and Moon: new constraints from Hf–W isotope geochemistry. *Earth Planet. Sci. Lett.* **142**, 75–89.

Halliday A. N., Lee D.-C., and Jacobsen S. B. (2000) Tungsten isotopes, the timing of metal-silicate fractionation and the origin of the Earth and Moon. In *Origin of the Earth and Moon* (eds. R. M. Canup and K. Righter). University of Arizona Press, Tucson, pp. 45–62.

Halliday A. N., Wänke H., Birck J.-L., and Clayton R. N. (2001) The accretion, bulk composition and early differentiation of Mars. *Space Sci. Rev.* **96**, 197–230.

Hanan B. B. and Tilton G. R. (1987) 60025: Relict of primitive lunar crust? *Earth Planet. Sci. Lett.* **84**, 15–21.

Hanks T. C. and Anderson D. L. (1969) The early thermal history of the earth. *Phys. Earth Planet. Int.* **2**, 19–29.

Harper C. L. and Jacobsen S. B. (1992) Evidence from coupled [147]Sm–[143]Nd and [146]Sm–[142]Nd systematics for very early (4.5-Gyr) differentiation of the Earth's mantle. *Nature* **360**, 728–732.

Harper C. L. and Jacobsen S. B. (1996a) Noble gases and Earth's accretion. *Science* **273**, 1814–1818.

Harper C. L. and Jacobsen S. B. (1996b) Evidence for [182]Hf in the early solar system and constraints on the timescale of terrestrial core formation. *Geochim. Cosmochim. Acta* **60**, 1131–1153.

Harper C. L., Völkening J., Heumann K. G., Shih C.-Y., and Wiesmann H. (1991a) [182]Hf–[182]W: new cosmochronometric constraints on terrestrial accretion, core formation, the astrophysical site of the r-process, and the origin of the solar

system. In *Lunar Planet. Sci.* **XXII**. The Lunar and Planetary Science, Houston, pp. 515–516.

Harper C. L., Wiesmann H., Nyquist L. E., Howard W. M., Meyer B., Yokoyama Y., Rayet M., Arnould M., Palme H., Spettel B., and Jochum K. P. (1991b) [92]Nb/[93]Nb and [92]Nb/[146]Sm ratios of the early solar system: observations and comparison of p-process and spallation models. In *Lunar Planet. Sci.* **XXII**. The Lunar and Planetary Institute, Houston, pp. 519–520.

Hartmann L. (2000) Observational constraints on transport (and mixing) in pre-main sequence disks. *Space Sci. Rev.* **92**, 55–68.

Hartmann L. and Kenyon S. J. (1996) The FU Orionis phenomena. *Ann. Rev. Astron. Astrophys.* **34**, 207–240.

Hartmann W. K. and Davis D. R. (1975) Satellite-sized planetesimals and lunar origin. *Icarus* **24**, 505–515.

Hartmann W. K., Ryder G., Dones L., and Grinspoon D. (2000) The time-dependent intense bombardment of the primordial Earth/Moon system. In *Origin of the Earth and Moon* (eds. R. Canup and K. Righter). University of Arizona Press, Tucson, pp. 493–512.

Hayashi C., Nakazawa K., and Mizuno H. (1979) Earth's melting due to the blanketing effect of the primordial dense atmosphere. *Earth Planet. Sci. Lett.* **43**, 22–28.

Hayashi C., Nakazawa K., and Nakagawa Y. (1985) Formation of the solar system. In *Protostars and Planets II* (eds. D. C. Black and D. S. Matthews). University of Arizona Press, Tucson, pp. 1100–1153.

Herzberg C. (1984) Chemical stratification in the silicate Earth. *Earth Planet. Sci. Lett.* **67**, 249–260.

Herzberg C., Jeigenson M., Skuba C., and Ohtani E. (1988) Majorite fractionation recorded in the geochemistry of peridotites from South Africa. *Nature* **332**, 823–826.

Hewins R. H. and Herzberg C. (1996) Nebular turbulence, chondrule formation, and the composition of the Earth. *Earth Planet. Sci. Lett.* **144**, 1–7.

Hofmann A. W. (1988) Chemical differentiation of the Earth: the relationship between mantle, continental crust and oceanic crust. *Earth Planet. Sci. Lett.* **90**, 297–314.

Hofmann A. W., Jochum K. P., Seufert M., and White W. M. (1986) Nb and Pb in oceanic basalts: new constraints on mantle evolution. *Earth Planet. Sci. Lett.* **79**, 33–45.

Honda M., McDougall I., Patterson D. B., Doulgeris A., and Clague D. A. (1991) Possible solar noble-gas component in Hawaiian basalts. *Nature* **349**, 149–151.

Horan M. F., Smoliar M. I., and Walker R. J. (1998) [182]W and [187]Re–[187]Os systematics of iron meteorites: chronology for melting, differentiation, and crystallization in asteroids. *Geochim. Cosmochim. Acta* **62**, 545–554.

Hoyle F. (1960) On the origin of the solar nebula. *Quart. J. Roy. Astron. Soc.* **1**, 28–55.

Hsui A. T. and Toksöz M. N. (1977) Thermal evolution of planetary size bodies. *Proc. 8th Lunar Sci. Conf.* 447–461.

Humayun M. and Cassen P. (2000) Processes determining the volatile abundances of the meteorites and terrestrial planets. In *Origin of the Earth and Moon* (eds. R. M. Canup and K. Righter). University of Arizona Press, Tucson, pp. 3–23.

Humayun M. and Clayton R. N. (1995) Potassium isotope cosmochemistry: genetic implications of volatile element depletion. *Geochim. Cosmochim. Acta* **59**, 2131–2151.

Hunten D. M., Pepin R. O., and Walker J. C. G. (1987) Mass fractionation in hydrodynamic escape. *Icarus* **69**, 532–549.

Huss G. R. (1988) The role of presolar dust in the formation of the solar system. *Earth Moon Planet.* **40**, 165–211.

Huss G. R. (1997) The survival of presolar grains during the formation of the solar system. In *Astrophysical Implications of the Laboratory Study of Presolar Materials*, American Institute of Physics Conf. Proc. 402, Woodbury, New York (eds. T. J. Bernatowicz and E. Zinner), pp. 721–748.

Huss G. R. and Lewis R. S. (1995) Presolar diamond, SiC, and graphite in primitive chondrites: abundances as a function of meteorite class and petrologic type. *Geochim. Cosmochim. Acta* **59**, 115–160.

Ireland T. R. (1991) The abundance of ^{182}Hf in the early solar system. In *Lunar Planet. Sci.* **XXII.** The Lunar and Planetary Institute, Houston, pp. 609–610.

Ireland T. R. and Wlotzka F. (1992) The oldest zircons in the solar system. *Earth Planet. Sci. Lett.* **109**, 1–10.

Jackson I. (1983) Some geophysical constraints on the chemical composition of the Earth's lower mantle. *Earth Planet. Sci. Lett.* **62**, 91–103.

Jacobsen S. B. and Harper C. L., Jr. (1996) Accretion and early differentiation history of the Earth based on extinct radionuclides. In *Earth Processes: Reading the Isotope Code* (eds. E. Basu and S. Hart). American Geophysical Union, Washington, DC, pp. 47–74.

Jacobsen S. B. and Wasserburg G. J. (1979) The mean age of mantle and crust reservoirs. *J. Geophys. Res.* **84**, 7411–7427.

Jacobsen S. B. and Yin Q. Z. (2001) Core formation models and extinct nuclides. In *Lunar Planet. Sci.* **XXXII**, #1961. The Lunar and Planetary Institute, Houston (CD-ROM).

Jagoutz E., Palme J., Baddenhausen H., Blum K., Cendales M., Drebus G., Spettel B., Lorenz V., and Wänke H. (1979) The abundances of major, minor, and trace elements in the Earth's mantle as derived from primitive ultramafic nodules. *Proc. 10th Lunar Sci. Conf.*, 2031–2050.

Jahn B.-M. and Shih C. (1974) On the age of the Onverwacht group, Swaziland sequence, South Africa. *Geochim. Cosmochim. Acta* **38**, 873–885.

Javoy M. (1998) The birth of the Earth's atmosphere: the behaviour and fate of its major elements. *Chem. Geol.* **147**, 11–25.

Javoy M. (1999) Chemical earth models. *C.R. Acad. Sci., Ed. Sci. Méd. Elseviers SAS* **329**, 537–555.

Jessberger E. K., Bohsung J., Chakaveh S., and Traxel K. (1992) The volatile enrichment of chondritic interplanetary dust particles. *Earth Planet. Sci. Lett.* **112**, 91–99.

Jones J. H. and Drake M. J. (1986) Geochemical constraints on core formation in the Earth. *Nature* **322**, 221–228.

Jones J. H. and Drake M. J. (1993) Rubidium and cesium in the Earth and Moon. *Geochim. Cosmochim. Acta* **57**, 3785–3792.

Jones J. H. and Palme H. (2000) Geochemical constraints on the origin of the Earth and Moon. In *Origin of the Earth and Moon* (eds. R. M. Canup and K. Righter). University of Arizona Press, Tucson, pp. 197–216.

Jones J. H., Capobianco C. J., and Drake M. J. (1992) Siderophile elements and the Earth's formation. *Science* **257**, 1281–1282.

Jones R. H., Lee T., Connolly H. C., Jr., Love S. G., and Shang H. (2000) Formation of chondrules and CAIs: theory vs. observation. In *Protostars and Planets IV* (eds. V. Mannings, A. P. Boss, and S. S. Russell). University of Arizona Press, Tucson, pp. 927–962.

Kasting J. F. (2001) The rise of atmospheric oxygen. *Science* **293**, 819–820.

Kasting J. F. and Grinspoon D. H. (1991) The faint young Sun problem. In *The Sun in Time* (eds. C. P. Sonett, M. S. Gimpapa, and M. S. Matthews). University of Arizona Press, Tucson, pp. 447–462.

Kato T., Ringwood A. E., and Irifune T. (1988a) Experimental determination of element partitioning between silicate perovskites, garnets and liquids: constraints on early differentiation of the mantle. *Earth Planet. Sci. Lett.* **89**, 123–145.

Kato T., Ringwood A. E., and Irifune T. (1988b) Constraints on element partition coefficients between MgSiO$_3$ perovskite and liquid determined by direct measurements. *Earth Planet. Sci. Lett.* **90**, 65–68.

Kleine T., Münker C., Mezger K., and Palme H. (2002) Rapid accretion and early core formation on asteroids and the terrestrial planets from Hf–W chronometry. *Nature* **418**, 952–955.

Kornacki A. S. and Fegley B., Jr. (1986) The abundance and relative volatility of refractory trace elements in Allende Ca, Al-rich inclusions: implications for chemical and physical processes in the solar nebula. *Earth Planet. Sci. Lett.* **79**, 217–234.

Kortenkamp S. J., Kokubo E., and Weidenschilling S. J. (2000) Formation of planetary embryos. In *Origin of the Earth and Moon* (eds. R. M. Canup and K. Righter). University of Arizona Press, Tucson, pp. 85–100.

Kortenkamp S. J. and Wetherill G. W. (2000) Terrestrial planet and asteroid formation in the presence of giant planets. *Icarus* **143**, 60–73.

Kramers J. D. (1998) Reconciling siderophile element data in the Earth and Moon, W isotopes and the upper lunar age limit in a simple model of homogeneous accretion. *Chem. Geol.* **145**, 461–478.

Kramers J. D. and Tolstikhin I. N. (1997) Two terrestrial lead isotope paradoxes, forward transport modeling, core formation and the history of the continental crust. *Chem. Geol.* **139**, 75–110.

Kreutzberger M. E., Drake M. J., and Jones J. H. (1986) Origin of the Earth's Moon: constraints from alkali volatile trace elements. *Geochim. Cosmochim. Acta* **50**, 91–98.

Krot A. N., Meibom A., Russell S. S., Alexander C. M. O'D., Jeffries T. E., and Keil K. (2001) A new astrophysical setting for chondrule formation. *Science* **291**, 1776–1779.

Kunz J., Staudacher T., and Allègre C. J. (1998) Plutonium-fission xenon found in Earth's mantle. *Science* **280**, 877–880.

Lange M. A. and Ahrens T. J. (1982) The evolution of an impact-generated atmosphere. *Icarus* **51**, 96–120.

Lange M. A. and Ahrens T. J. (1984) FeO and H$_2$O and the homogeneous accretion of the earth. *Earth Planet. Sci. Lett.* **71**, 111–119.

Larimer J. W. and Anders E. (1970) Chemical fractionations in meteorites: III. Major element fractionation in chondrites. *Geochim. Cosmochim. Acta* **34**, 367–387.

Lee D.-C. and Halliday A. N. (1995) Hafnium–tungsten chronometry and the timing of terrestrial core formation. *Nature* **378**, 771–774.

Lee D.-C. and Halliday A. N. (1996) Hf–W isotopic evidence for rapid accretion and differentiation in the early solar system. *Science* **274**, 1876–1879.

Lee D.-C. and Halliday A. N. (1997) Core formation on Mars and differentiated asteroids. *Nature* **388**, 854–857.

Lee D.-C. and Halliday A. N. (2000a) Accretion of primitive planetesimals: Hf–W isotopic evidence from enstatite chondrites. *Science* **288**, 1629–1631.

Lee D.-C. and Halliday A. N. (2000b) Hf–W isotopic systematics of ordinary chondrites and the initial ^{182}Hf/^{180}Hf of the solar system. *Chem. Geol.* **169**, 35–43.

Lee D.-C., Halliday A. N., Davies G. R., Essene E. J., Fitton J. G., and Temdjim R. (1996) Melt enrichment of shallow depleted mantle: a detailed petrological, trace element and isotopic study of mantle derived xenoliths and megacrysts from the Cameroon line. *J. Petrol.* **37**, 415–441.

Lee D.-C., Halliday A. N., Snyder G. A., and Taylor L. A. (1997) Age and origin of the Moon. *Science* **278**, 1098–1103.

Lee D.-C., Halliday A. N., Leya I., Wieler R., and Wiechert U. (2002) Cosmogenic tungsten and the origin and earliest differentiation of the Moon. *Earth Planet. Sci. Lett.* **198**, 267–274.

Lee T., Shu F. H., Shang H., Glassgold A. E., and Rehm K. E. (1998) Protostellar cosmic rays and extinct radioactivities in meteorites. *Astrophys. J.* **506**, 898–912.

Levin B. J. (1972) Origin of the earth. *Tectonophysics* **13**, 7–29.

Lewis J. S. (1972) Metal/silicate fractionation in the solar system. *Earth Planet. Sci. Lett.* **15**, 286–290.

Leya I., Wieler R., and Halliday A. N. (2000) Cosmic-ray production of tungsten isotopes in lunar samples and meteorites and its implications for Hf–W cosmochemistry. *Earth Planet. Sci. Lett.* **175**, 1–12.

Leya I., Wieler R., and Halliday A. N. (2003) The influence of cosmic-ray production on extinct nuclide systems. *Geochim. Cosmochim. Acta* **67**, 527–541.

Li J. and Agee C. B. (1996) Geochemistry of mantle–core differentiation at high pressure. *Nature* **381**, 686–689.

Lin D. N. C. and Papaloizou J. (1985) On the dynamical origin of the solar system. In *Protostars and Planets II* (eds. D. C. Black and M. S. Matthews). University of Arizona Press, Tucson, pp. 981–1072.

Lin D. N. C., Bodenheimer P., and Richardson D. C. (1996) Orbital migration of the planetary companion of 51 Pegasi to its present location. *Nature* **380**, 606–607.

Lissauer J. J. (1987) Time-scales for planetary accretion and the structure of the protoplanetry disk. *Icarus* **69**, 249–265.

Lissauer J. J. (1993) Planet formation. *Ann. Rev. Astron. Astrophys.* **31**, 129–174.

Lissauer J. J. (1999) How common are habitable planets? *Nature* **402**(suppl.2), C11–C14.

Lodders K. (2000) An oxygen isotope mixing model for the accretion and composition of rocky planets. *Space Sci. Rev.* **92**, 341–354.

Longhi J. (1999) Phase equilibrium constraints on angrite petrogenesis. *Geochim. Cosmochim. Acta* **63**, 573–585.

Lugmair G. W. and Galer S. J. G. (1992) Age and isotopic relationships between the angrites Lewis Cliff 86010 and Angra dos Reis. *Geochim. Cosmochim. Acta* **56**, 1673–1694.

Lugmair G. W. and Shukolyukov A. (1998) Early solar system timescales according to ^{53}Mn–^{53}Cr systematics. *Geochim. Cosmochim. Acta* **62**, 2863–2886.

Lugmair G. W. and Shukolyukov A. (2001) Early solar system events and timescales. *Meteorit. Planet. Sci.* **36**, C17–C26.

Lupton J. E. and Craig H. (1975) Excess ^3He in oceanic basalts: evidence for terrestrial primordial helium, *Earth Planet. Sci. Lett.* **26**, 133–139.

MacFarlane E. A. (1989) Formation of the Moon in a giant impact: composition of impactor. *Proc. 19th Lunar Planet. Sci. Conf.* 593–605.

MacPherson G. J., Wark D. A., and Armstrong J. T. (1988) Primitive material surviving in chondrites: refractory inclusions. In *Meteorites and the Early Solar System* (eds. J. F. Kerridge and M. S. Matthews). University of Arizona Press, Tucson, pp. 746–807.

Mamyrin B. A., Tolstikhin I. N., Anufriev G. S., and Kamensky I. L. (1969) Anomalous isotopic composition of helium in volcanic gases. *Dokl. Akad. Nauk. SSSR* **184**, 1197–1199 (in Russian).

Manhès G., Allègre C. J., Dupré B., and Hamelin B. (1979) Lead–lead systematics, the "age of the Earth" and the chemical evolution of our planet in a new representation space. *Earth Planet. Sci. Lett.* **44**, 91–104.

Marty B. (1989) Neon and xenon isotopes in MORB: implications for the earth-atmosphere evolution. *Earth Planet. Sci. Lett.* **94**, 45–56.

Matsui T. and Abe Y. (1986a) Evolution of an impact-induced atmosphere and magma ocean on the accreting Earth. *Nature* **319**, 303–305.

Matsui T. and Abe Y. (1986b) Impact-induced atmospheres and oceans on Earth and Venus. *Nature* **322**, 526–528.

Mayor M. and Queloz D. (1995) A Jupiter-mass companion to a solar-type star. *Nature* **378**, 355–359.

McCaughrean M. J. and O'Dell C. R. (1996) Direct imaging of circumstellar disks in the Orion nebula. *Astronom. J.* **111**, 1977–1986.

McCulloch M. T. (1994) Primitive $^{87}Sr/^{86}Sr$ from an Archean barite and conjecture on the Earth's age and origin, *Earth Planet. Sci. Lett.* **126**, 1–13.

McCulloch M. T. and Bennett V. C. (1993) Evolution of the early Earth: constraints from ^{143}Nd–^{142}Nd isotopic systematics. *Lithos* **30**, 237–255.

McDonough W. F. and Sun S.-S. (1995) The composition of the Earth. *Chem. Geol.* **120**, 223–253.

McDonough W. F., Sun S.-S., Ringwood A. E., Jagoutz E., and Hofmann A. W. (1992) Potassium, rubidium, and cesium in the Earth and Moon and the evolution of the mantle of the Earth, *Geochim. Cosmochim. Acta* **53**, 1001–1012.

McFarlane E. A. and Drake M. J. (1990) Element partitioning and the early thermal history of the Earth. In *Origin of the Earth* (eds. H. E. Newsom and J. H. Jones). Lunar and Planetary Institute, Houston, pp. 135–150.

McKeegan K. D., Chaussidon M., and Robert F. (2000) Incorporation of short-lived Be-10 in a calcium–aluminum-rich inclusion from the Allende meteorite. *Science* **289**, 1334–1337.

Meibom A., Desch S. J., Krot A. N., Cuzzi J. N., Petaev M. I., Wilson L., and Keil K. (2000) Large-scale thermal events in the solar nebula: evidence from Fe, Ni metal grains in primitive meteorites. *Science* **288**, 839–841.

Meisel T., Walker R. J., Irving A. J., and Lorand J.-P. (2001) Osmium isotopic compositions of mantle xenoliths: a global perspective. *Geochim. Cosmochim. Acta* **65**, 1311–1323.

Melosh H. J. (1990) Giant impacts and the thermal state of the early Earth. In *Origin of the Earth* (eds. H. E. Newsom and J. H. Jones). Oxford University Press, Oxford, pp. 69–83.

Melosh H. J. and Sonett C. P. (1986) When worlds collide: jetted vapor plumes and the Moon's origin. In *Origin of the Moon* (eds. W. K. Hartmann, R. J. Phillips, and G. J. Taylor). Lunar and Planetary Institute, Houston, pp. 621–642.

Melosh H. J. and Vickery A. M. (1989) Impact erosion of the primordial atmosphere of Mars. *Nature* **338**, 487–489.

Melosh H. J., Vickery A. M., and Tonks W. B. (1993) Impacts and the early environment and evolution of the terrestrial planets. In *Protostars and Planets III* (eds. E. H. Levy and J. I. Lunine). University of Arizona Press, Tucson, pp. 1339–1370.

Mendybaev R. A., Beckett J. R., Grossman L., Stolper E., Cooper R. F., and Bradley J. P. (2002) Volatilization kinetics of silicon carbide in reducing gases: an experimental study with applications to the survival of presolar grains in the solar nebula. *Geochim. Cosmochim. Acta* **66**, 661–682.

Minarik W. G., Ryerson F. J., and Watson E. B. (1996) Textural entrapment of core-forming melts. *Science* **272**, 530–533.

Mizuno H., Nakazawa K., and Hayashi C. (1980) Dissolution of the primordial rare gases into the molten Earth's material. *Earth Planet. Sci. Lett.* **50**, 202–210.

Mojzsis S. J., Arrhenius G., McKeegan K. D., Harrison T. M., Nutman A. P., and Friend C. R. L. (1996) Evidence for life on Earth before 3,800 million years ago. *Nature* **384**, 55–59.

Mojzsis S. J., Harrison T. M., and Pidgeon R. T. (2001) Oxygen isotope evidence from ancient zircons for liquid water at the Earth's surface 4300 Myr ago Jack Hills, evidence for more very old detrital zircons in Western Australia. *Nature* **409**, 178–181.

Morbidelli A., Chambers J., Lunine J. I., Petit J. M., Robert F., Valsecchi G. B., and Cyr K. E. (2000) Source regions and time-scales for the delivery of water to the Earth. *Meteorit. Planet. Sci.* **35**, 1309–1320.

Moreira M. and Allègre C. J. (1998) Helium–neon systematics and the structure of the mantle. *Chem. Geol* **147**, 53–59.

Moreira M., Kunz J., and Allègre C. J. (1998) Rare gas systematics in Popping Rock: isotopic and elemental compositions in the upper mantle. *Science* **279**, 1178–1181.

Moreira M., Breddam K., Curtice J., and Kurz M. D. (2001) Solar neon in the Icelandic mantle: new evidence for an undegassed lower mantle. *Earth Planet. Sci. Lett.* **185**, 15–23.

Münker C., Weyer S., Mezger K., Rehkämper M., Wombacher F., and Bischoff A. (2000) ^{92}Nb–^{92}Zr and the early differentiation history of planetary bodies. *Science* **289**, 1538–1542.

Murphy D. T., Kamber B. S., and Collerson K. D. (2003) A refined solution to the first terrestrial Pb-isotope paradox. *J. Petrol.* **44**, 39–53.

Murray N. and Chaboyer B. (2001) Are stars with planets polluted? *Ap. J.* **566**, 442–451.

Murray N., Hansen B., Holman M., and Tremaine S. (1998) Migrating planets. *Science* **279**, 69–72.

Murthy V. R. (1991) Early differentiation of the Earth and the problem of mantle siderophile elements: a new approach. *Science* **253**, 303–306.

Nägler Th. F. and Kramers J. D. (1998) Nd isotopic evolution of the upper mantle during the Precambrian: models, data and the uncertainty of both. *Precambrian Res.* **91**, 233–252.

Nelson A. F., Benz W., Adams F. C., and Arnett D. (1998) Dynamics of circumstellar disks. *Ap. J.* **502**, 342–371.

Nelson A. F., Benz W., and Ruzmaikina T. V. (2000) Dynamics of circumstellar disks: II. Heating and cooling. *Ap. J.* **529**, 357–390.

Newsom H. E. (1990) Accretion and core formation in the Earth: evidence from siderophile elements. In *Origin of the Earth* (eds. H. E. Newsom and J. H. Jones). Oxford University Press, Oxford, pp. 273–288.

Newsom H. E. (1995) Composition of the solar system, planets, meteorites, and major terrestrial reservoirs. In *Global Earth Physics: A Handbook of Physical Constants*, AGU Reference Shelf 1 (ed. T. J. Ahrens). American Geophysical Union, Washington, DC.

Newsom H. E. and Taylor S. R. (1986) The single impact origin of the Moon. *Nature* **338**, 29–34.

Newsom H. E., White W. M., Jochum K. P., and Hofmann A. W. (1986) Siderophile and chalcophile element abundances in oceanic basalts, Pb isotope evolution and growth of the Earth's core. *Earth Planet. Sci. Lett.* **80**, 299–313.

Newsom H. E., Sims K. W. W., Noll P. D., Jr., Jaeger W. L., Maehr S. A., and Bessera T. B. (1996) The depletion of W in the bulk silicate Earth. *Geochim. Cosmochim. Acta* **60**, 1155–1169.

Niedermann S., Bach W., and Erzinger J. (1997) Noble gas evidence for a lower mantle component in MORBs from the southern East Pacific Rise: decoupling of helium and neon isotope systematics. *Geochim. Cosmochim. Acta* **61**, 2697–2715.

Nittler L. R. (2003) Presolar stardust in meteorites: recent advances and scientific frontiers. *Earth Planet. Sci. Lett.* **209**, 259–273.

Nittler L. R., Alexander C. M. O'D., Gao X., Walker R. M., and Zinner E. K. (1994) Interstellar oxide grains from the Tieschitz ordinary chondrite. *Nature* **370**, 443–446.

O'Keefe J. D. and Ahrens T. J. (1977) Impact-induced energy partitioning, melting, and vaporization on terrestrial planets. *Proc. 8th Lunar Sci. Conf.* 3357–3374.

O'Neill H. St. C. (1991a) The origin of the Moon and the early history of the Earth—a chemical model: Part I. The Moon. *Geochim. Cosmochim. Acta* **55**, 1135–1158.

O'Neill H. St. C. (1991b) The origin of the Moon and the early history of the Earth—a chemical model: Part II. The Earth. *Geochim. Cosmochim. Acta* **55**, 1159–1172.

O'Nions R. K. and Tolstikhin I. N. (1994) Behaviour and residence times of lithophile and rare gas tracers in the upper mantle. *Earth Planet. Sci. Lett.* **124**, 131–138.

Oversby V. M. and Ringwood A. E. (1971) Time of formation of the Earth's core. *Nature* **234**, 463–465.

Owen T. and Bar-Nun A. (1995) Comets, impacts and atmosphere. *Icarus* **116**, 215–226.

Owen T. and Bar-Nun A. (2000) Volatile contributions from icy planetesimals. In *Origin of the Earth and Moon* (eds. R. M. Canup and K. Righter). University of Arizona Press, Tucson, pp. 459–471.

Ozima M. and Podosek F. A. (1999) Formation age of Earth from $^{129}I/^{127}I$ and $^{244}Pu/^{238}U$ systematics and the missing Xe. *J. Geophys. Res.* **104**, 25493–25499.

Ozima M. and Podosek F. A. (2002) *Noble Gas Geochemistry.* 2nd edn. Cambridge University Press, Cambridge, 286p.

Palme H. (2000) Are there chemical gradients in the inner solar system? *Space Sci. Rev.* **92**, 237–262.

Palme H., Larimer J. W., and Lipschultz M. E. (1988) Moderately volatile elements. In *Meteorites and the Early Solar System* (eds. J. F. Kerridge and M. S. Matthews). University of Arizona Press, Tucson, pp. 436–461.

Papanastassiou D. A. and Wasserburg G. J. (1976) Early lunar differentiates and lunar initial $^{87}Sr/^{86}Sr$. In *Lunar Sci.* **VII**. The Lunar Science Institute, Houston, pp. 665–667.

Patterson C. C. (1956) Age of meteorites and the Earth. *Geochim. Cosmochim. Acta* **10**, 230–237.

Pavlov A. A., Kasting J. F., Brown L. L., Rages K. A., and Freedman R. (2000) Greenhouse warming by CH_4 in the atmosphere of early Earth. *J. Geophys. Res.* **105**, 11981–11990.

Peck W. H., Valley J. W., Wilde S. A., and Graham C. M. (2001) Oxygen isotope ratios and rare earth elements in 3.3. to 4.4 Ga zircons: ion microprobe evidence for high $\delta^{18}O$ continental crust and oceans in the Early Archean. *Geochim Cosmochim. Acta* **65**, 4215–4229.

Pepin R. O. (1997) Evolution of Earth's noble gases: consequences of assuming hydrodynamic loss driven by giant impact. *Icarus* **126**, 148–156.

Pepin R. O. (2000) On the isotopic composition of primordial xenon in terrestrial planet atmospheres. *Space Sci. Rev.* **92**, 371–395.

Pepin R. O. and Porcelli D. (2002) Origin of noble gases in the terrestrial planets. In *Noble Gases in Geochemistry and Cosmochemistry,* Rev. Mineral. Geochem. 47 (eds. D. Porcelli, C. J. Ballentine, and R. Wieler). Mineralogical Society of America, Washington, DC, pp. 191–246.

Phinney D., Tennyson J., and Frick U. (1978) Xenon in CO_2 well gas revisited. *J. Geophys. Res.* **83**, 2313–2319.

Podosek F. A. and Cassen P. (1994) Theoretical, observational, and isotopic estimates of the lifetime of the solar nebula. *Meteoritics* **29**, 6–25.

Podosek F. A., Zinner E. K., MacPherson G. J., Lundberg L. L., Brannon J. C., and Fahey A. J. (1991) Correlated study of initial Sr-87/Sr-86 and Al–Mg isotopic systematics and petrologic properties in a suite of refractory inclusions from the Allende meteorite. *Geochim. Cosmochim. Acta* **55**, 1083–1110.

Podosek F. A., Ott U., Brannon J. C., Neal C. R., Bernatowicz T. J., Swan P., and Mahan S. E. (1997) Thoroughly anomalous chromium in Orgueil. *Meteorit. Planet. Sci.* **32**, 617–627.

Podosek F. A., Woolum D. S., Cassen P., Nicholls R. H., Jr., and Weidenschilling S. J. (2003) Solar wind as a source of terrestrial light noble gases. *Geochim. Cosmochim. Acta* (in press).

Poitrasson F., Halliday A. N., Lee D.-C., Levasseur S., and Teutsch N. (2003) Iron isotope evidence for formation of the Moon through partial vaporisation. In *Lunar Planet. Sci.* **XXXIV**, #1433. The Lunar and Planetary Institute, Houston (CD-ROM).

Porcelli D. and Ballentine C. J. (2002) Models for the distribution of terrestrial noble gases and evolution of the atmosphere. In *Noble Gases in Geochemistry and Cosmochemistry,* Rev. Mineral. Geochem. 47 (eds. D. Porcelli, C. J. Ballentine, and R. Wieler). Mineralogical Society of America, Washington, DC, pp. 411–480.

Porcelli D. and Halliday A. N. (2001) The possibility of the core as a source of mantle helium. *Earth Planet. Sci. Lett.* **192**, 45–56.

Porcelli D. and Pepin R. O. (2000) Rare gas constraints on early earth history. In *Origin of the Earth and Moon* (eds. R. M. Canup and K. Righter). University of Arizona Press, Tucson, pp. 435–458.

Porcelli D. and Wasserburg G. J. (1995) Mass transfer of helium, neon, argon and xenon through a steady-state upper mantle. *Geochim. Cosmochim. Acta* **59**, 4921–4937.

Porcelli D., Cassen P., Woolum D., and Wasserburg G. J. (1998) Acquisition and early losses of rare gases from the deep Earth. In *Origin of the Earth and Moon,* LPI Contribution No. 597. Lunar and Planetary Institute, Houston, pp. 35–36.

Porcelli D., Cassen P., and Woolum D. (2001) Deep Earth rare gases: initial inventories, capture from the solar nebula and losses during Moon formation. *Earth Planet. Sci. Lett* **193**, 237–251.

Puchtel I. and Humayun M. (2000) Platinum group elements in Kostomuksha komatiites and basalts: implications for oceanic crust recycling and core–mantle interaction. *Geochim. Cosmochim. Acta* **64**, 4227–4242.

Qian Y. Z. and Wasserburg G. J. (2000) Stellar abundances in the early galaxy and two r-process components. *Phys. Rep.* **333–334**, 77–108.

Qian Y. Z., Vogel P., and Wasserburg G. J. (1998) Diverse supernova sources for the r-process. *Astrophys. J.* **494**, 285–296.

Quitté G., Birck J.-L., and Allègre C. J. (2000) ^{182}Hf–^{182}W systematics in eucrites: the puzzle of iron segregation in the early solar system. *Earth Planet. Sci. Lett.* **184**, 83–94.

Rammensee W. and Wänke H. (1977) On the partition coefficient of tungsten between metal and silicate and its bearing on the origin of the Moon. *Proc. 8th Lunar Sci. Conf.* 399–409.

Rehkämper M., Halliday A. N., Barfod D., Fitton J. G., and Dawson J. B. (1997) Platinum group element abundance patterns in different mantle environments. *Science* **278**, 1595–1598.

Rehkämper M., Halliday A. N., Alt J., Fitton J. G., Zipfel J., and Takazawa E. (1999a) Non-chondritic platinum group element ratios in abyssal peridotites: petrogenetic signature of melt percolation? *Earth Planet. Sci. Lett.* **172**, 65–81.

Rehkämper M., Halliday A. N., Fitton J. G., Lee D.-C., and Wieneke M. (1999b) Ir, Ru, Pt, and Pd in basalts and komatiites: new constraints for the geochemical behavior of the platinum-group elements in the mantle. *Geochim. Cosmochim. Acta* **63**, 3915–3934.

Reynolds J. H. (1967) Isotopic abundance anomalies in the solar system. *Ann. Rev. Nuclear Sci.* **17**, 253–316.

Righter K. and Drake M. J. (1996) Core formation in Earth's Moon, Mars, and Vesta. *Icarus* **124**, 513–529.

Righter K. and Drake M. J. (1999) Effect of water on metal-silicate partitioning of siderophile elements: a high pressure and temperature terrestrial magma ocean and core formation. *Earth Planet. Sci. Lett.* **171**, 383–399.

Righter K., Drake M. J., and Yaxley G. (1997) Prediction of siderophile element metal/silicate partition coefficients to 20 GPa and 2800 °C: the effects of pressure, temperature, oxygen fugacity, and silicate and metallic melt compositions. *Phys. Earth Planet. Int.* **100**, 115–142.

Righter K., Walker R. J., and Warren P. W. (2000) The origin and significance of highly siderophile elements in the lunar and terrestrial mantles. In *Origin of the Earth and Moon* (eds. R. M. Canup and K. Righter). University of Arizona Press, Tucson, pp. 291–322.

Ringwood A. E. (1966) The chemical composition and origin of the Earth. In *Advances in Earth Sciences* (ed. P. M. Hurley). MIT Press, Cambridge, MA, pp. 287–356.

Ringwood A. E. (1977) Composition of the core and implications for origin of the Earth. *Geochem. J.* **11**, 111–135.

Ringwood A. E. (1979) *Origin of the Earth and Moon.* Springer, New York, 295p.

Ringwood A. E. (1989a) Significance of the terrestrial Mg/Si ratio. *Earth Planet. Sci. Lett.* **95**, 1–7.

Ringwood A. E. (1989b) Flaws in the giant impact hypothesis of lunar origin. *Earth Planet. Sci. Lett.* **95**, 208–214.

Ringwood A. E. (1990) Earliest history of the Earth–Moon system. In *Origin of the Earth* (eds. A. E. Newsom and J. H. Jones). Oxford University Press, Oxford, pp. 101–134.

Ringwood A. E. (1992) Volatile and siderophile element geochemistry of the Moon: a reappraisal. *Earth Planet. Sci. Lett.* **111**, 537–555.

Rubey W. W. (1951) Geological history of seawater. *Bull. Geol. Soc. Am.* **62**, 1111–1148.

Rubie D. C., Melosh H. J., Reid J. E., Liebske C., and Righter K. (2003) Mechanisms of metal-silicate equilibration in the terrestrial magma ocean. *Earth Planet. Sci. Lett.* **205**, 239–255.

Rushmer T., Minarik W. G., and Taylor G. J. (2000) Physical processes of core formation. In *Origin of the Earth and Moon* (eds. R. M. Canup and K. Righter). University of Arizona Press, Tucson, pp. 227–243.

Ryder G., Koeberl C., and Mojzsis S. J. (2000) Heavy bombardment of the Earth at ~3.85 Ga: the search for petrographic and geochemical evidence. In *Origin of the Earth and Moon* (eds. R. M. Canup and K. Righter). University of Arizona Press, Tucson, pp. 475–492.

Safronov V. S. (1954) On the growth of planets in the protoplanetary cloud. *Astron. Zh.* **31**, 499–510.

Sagan C. and Chyba C. (1997) The early faint Sun paradox: organic shielding of ultraviolet-labile greenhouse gases. *Science* **276**, 1217–1221.

Sanders I. S. (1996) A chondrule-forming scenario involving molten planetesimals. In *Chondrules and the Protoplanetary Disk* (eds. R. H. Hewins, R. H. Jones, and E. R. D. Scott). Cambridge University Press, Cambridge, UK, pp. 327–334.

Sanloup C., Blichert-Toft J., Télouk P., Gillet P., and Albarède F. (2000) Zr isotope anomalies in chondrites and the presence of live ^{92}Nb in the early solar system. *Earth Planet. Sci. Lett.* **184**, 75–81.

Sasaki S. (1990) The primary solar-type atmosphere surrounding the accreting Earth: H_2O-induced high surface temperature. In *Origin of the Earth* (eds. H. E. Newsom and J. H. Jones). Oxford University Press, Oxford, pp. 195–209.

Sasaki S. and Nakazawa K. (1986) Metal-silicate fractionation in the growing Earth: energy source for the terrestrial magma ocean. *J. Geophys. Res.* **91**, B9231–B9238.

Sasaki S. and Nakazawa K. (1988) Origin of isotopic fractionation of terrestrial Xe: hydrodynamic fractionation during escape of the primordial H_2–He atmosphere. *Earth Planet. Sci. Lett.* **89**, 323–334.

Scherer E. E., Münker C., and Mezger K. (2001) Calibration of the lutetium–hafnium clock. *Science* **293**, 683–687.

Schmitt W., Palme H., and Wänke H. (1989) Experimental determination of metal/silicate partition coefficients for P, Co, Ni, Cu, Ga, Ge, Mo, and W and some implications for the early evolution of the Earth. *Geochim. Cosmochim. Acta* **53**, 173–185.

Schneider G., Smith B. A., Becklin E. E., Koerner D. W., Meier R., Hines D. C., Lowrance P. J., Terrile R. I., and Rieke M. (1999) Nicmos imaging of the HR 4796A circumstellar disk. *Astrophys. J.* **513**, L127–L130.

Schönberg R., Kamber B. S., Collerson K. D., and Eugster O. (2002) New W isotope evidence for rapid terrestrial accretion and very early core formation. *Geochim. Cosmochim. Acta* **66**, 3151–3160.

Schönbächler M., Rehkämper M., Halliday A. N., Lee D. C., Bourot-Denise M., Zanda B., Hattendorf B., and Günther D. (2002) Niobium–zirconium chronometry and early solar system development. *Science* **295**, 1705–1708.

Seager S. (2003) The search for Earth-like extrasolar planets. *Earth Planet. Sci. Lett.* **208**, 113–124.

Seeds M. A. (1996) *Foundations of Astronomy.* Wadsworth Publishing Company, Belmont, California, USA.

Sharma M., Papanastassiou D. A., and Wasserburg G. J. (1996) The issue of the terrestrial record of ^{146}Sm. *Geochim. Cosmochim. Acta* **60**, 2037–2047.

Shaw G. H. (1978) Effects of core formation. *Phys. Earth Planet. Int.* **16**, 361–369.

Shih C.-Y., Nyquist L. E., Dasch E. J., Bogard D. D., Bansal B. M., and Wiesmann H. (1993) Age of pristine noritic clasts from lunar breccias 15445 and 15455. *Geochim. Cosmochim. Acta* **57**, 915–931.

Shu F. H., Shang H., and Lee T. (1996) Toward an astrophysical theory of chondrites. *Science* **271**, 1545–1552.

Shu F. H., Shang H., Glassgold A. E., and Lee T. (1997) X-rays and fluctuating x-winds from protostars. *Science* **277**, 1475–1479.

Sleep N. H. and Zahnle K. (2001) Carbon dioxide cycling and implications for climate on ancient Earth. *J. Geophys. Res.* **106**, 1373–1399.

Sleep N. H., Zahnle K. J., Kasting J. F., and Morowitz H. J. (1989) Annihilation of ecosystems by large asteroid impacts on the early Earth. *Nature* **342**, 139–142.

Sleep N. H., Zahnle K., and Neuhoff P. S. (2001) Initiation of clement surface conditions on the earliest Earth. *Proc. Natl. Acad. Sci.* **98**, 3666–3672.

Smith J. V. (1977) Possible controls on the bulk composition of the Earth: implications for the origin of the earth and moon. *Proc. 8th Lunar Sci. Conf.* 333–369.

Smith J. V. (1980) The relation of mantle heterogeneity to the bulk composition and origin of the Earth. *Phil. Trans. Roy. Soc. London A* **297**, 139–146.

Snow J. E. and Schmidt G. (1998) Constraints on Earth accretion deduced from noble metals in the oceanic mantle. *Nature* **391**, 166–169.

Solomon S. C. (1979) Formation, history and energetics of cores in the terrestrial planets. *Earth Planet. Sci. Lett.* **19**, 168–182.

Stevenson D. J. (1981) Models of the Earth's core. *Science* **214**, 611–619.

Stevenson D. J. (1990) Fluid dynamics of core formation. In *Origin of the Earth* (eds. H. E. Newsom and J. H. Jones). Oxford University Press, Oxford, pp. 231–249.

Suess (1965) Chemical evidence bearing on the origin of the solar system. *Rev. Astron. Astrophys.* **3**, 217–234.

Swindle T. D. (1993) Extinct radionuclides and evolutionary timescales. In *Protostars and Planets III* (eds. E. H. Levy and J. I. Lunine). University of Arirona Press, Tucson, pp. 867–881.

Swindle T. D. and Podosek (1988) Iodine–xenon dating. In *Meteorites and the Early Solar System* (eds. J. F. Kerridge and M. S. Matthews). University of Arizona Press, Tucson, pp. 1127–1146.

Takahashi H., Janssens M.-J., Morgan J. W., and Anders E. (1978) Further studies of trace elements in C3 chondrites. *Geochim. Cosmochim. Acta* **42**, 97–106.

Taylor S. R. (1992) *Solar System Evolution: A New Perspective*. Cambridge University Press, New York.

Taylor S. R. and Norman M. D. (1990) Accretion of differentiated planetesimals to the Earth. In *Origin of the Earth* (eds. H. E. Newsom and J. H. Jones). Oxford University Press, Oxford, pp. 29–43.

Tera F. (1980) Reassessment of the "age of the Earth." *Carnegie Inst. Wash. Yearbook* **79**, 524–531.

Tera F., Papanastassiou D. A., and Wasserburg G. J. (1973) A lunar cataclysm at ~3.95 AE and the structure of the lunar crust. In *Lunar Sci.* **IV**. The Lunar Science Institute, Houston, pp. 723–725.

Tilton G. R. (1988) Age of the solar system. In *Meteorites and the Early Solar System* (eds. J. F. Kerridge and M. S. Matthews). University of Arizona Press, Tucson, pp. 259–275.

Tonks W. B. and Melosh H. J. (1990) The physics of crystal settling and suspension in a turbulent magma ocean. In *Origin of the Earth* (eds. H. E. Newsom and J. H. Jones). Oxford University Press, Oxford, pp. 17–174.

Treiman A. H., Drake M. J., Janssens M.-J., Wolf R., and Ebihara M. (1986) Core formation in the Earth and shergottite parent body (SPB): chemical evidence from basalts. *Geochim. Cosmochim. Acta* **50**, 1071–1091.

Treiman A. H., Jones J. H., and Drake M. J. (1987) Core formation in the shergottite parent body and comparison with the Earth. *J. Geophys. Res.* **92**, E627–E632.

Trieloff M., Kunz J., Clague D. A., Harrison D., and Allègre C. J. (2000) The nature of pristine noble gases in mantle plumes. *Science* **288**, 1036–1038.

Turekian K. K. and Clark S. P., Jr. (1969) Inhomogeneous accumulation of the earth from the primitive solar nebula. *Earth Planet. Sci. Lett.* **6**, 346–348.

Tyburczy J. A., Frisch B., and Ahrens T. J. (1986) Shock-induced volatile loss from a carbonaceous chondrite: implications for planetary accretion. *Earth Planet. Sci. Lett.* **80**, 201–207.

Urey H. C. (1966) The capture hypothesis of the origin of the Moon. In *The Earth–Moon System* (eds. B. G. Marsden and A. G. W. Cameron). Plenum, New York, pp. 210–212.

Valley J. W., Peck W. H., King E. M., and Wilde S. A. (2002) A cool early Earth. *Geology* **30**, 351–354.

Vervoort J. D., White W. M., and Thorpe R. I. (1994) Nd and Pb isotope ratios of the Abitibi greenstone belt: new evidence for very early differentiation of the Earth. *Earth Planet. Sci. Lett.* **128**, 215–229.

Vidal-Madjar A., Lecavelier des Etangs A., and Ferlet R. (1998) β Pictoris, a young planetary system? A review. *Planet. Space Sci.* **46**, 629–648.

Vityazev A. V., Pechernikova A. G., Bashkirov A. G. (2003) Accretion and differentiation of terrestrial protoplanetary bodies and Hf–W chronometry. In *Lunar Planet. Sci.* **XXXIV**, #1656. The Lunar and Planetary Institute, Houston (CD-ROM).

Vogel N., Baur H., Bischoff A., and Wieler R. (2002) Noble gases in chondrules and metal-sulfide rims of primitive chondrites—clues on chondrule formation. *Geochim. Cosmochim. Acta* **66**, A809.

Vogel N., Baur H., Leya I., and Wieler R. (2003) No evidence for primordial noble gases in CAIs. *Meteorit. Planet. Sci.* **38**(suppl.), A75.

Wade J. and Wood B. J. (2001) The Earth's "missing" niobium may be in the core. *Nature* **409**, 75–78.

Walker J. C. G. (1986) Impact erosion of planetary atmospheres. *Icarus* **68**, 87–98.

Walker R. J., Morgan J. W., and Horan M. F. (1995) Osmium-187 enrichment in some plumes: evidence for core–mantle interaction? *Science* **269**, 819–822.

Walter M. J., Newsom H. E., Ertel W., and Holzheid A. (2000) Siderophile elements in the Earth and Moon: Metal/silicate partitioning and implications for core formation. In *Origin of the Earth and Moon* (eds. R. M. Canup and K. Righter). University of Arizona Press, Tucson, pp. 265–289.

Wänke H. (1981) Constitution of terrestrial planets. *Phil. Trans. Roy. Soc. London* **A303**, 287–302.

Wänke H. and Dreibus G. (1986) Geochemical evidence for formation of the Moon by impact induced fission of the proto-Earth. In *Origin of the Moon* (eds. W. K. Hartmann, R. J. Phillips, and G. J. Taylor). Lunar and Planetary Institute, Houston, pp. 649–672.

Wänke H. and Dreibus G. (1988) Chemical composition and accretion history of terrestrial planets. *Phil. Trans. Roy. Soc. London A* **325**, 545–557.

Wänke H. and Dreibus G. (1994) Chemistry and accretion of Mars. *Phil. Trans. Roy. Soc. London A* **349**, 285–293.

Wänke H., Baddenhausen H., Palme H., and Spettel B. (1974) On the chemistry of the Allende inclusions and their origin as high temperature condensates. *Earth Planet. Sci. Lett.* **23**, 1–7.

Wänke H., Dreibus G., Palme H., Rammensee W., and Weckwerth G. (1983) Geochemical evidence for the formation of the Moon from material of the Earth's mantle. In *Lunar Planet. Sci.* **XIV**. The Lunar and Planetary Institute, Houston, pp. 818–819.

Wänke H., Dreibus G., and Jagoutz E. (1984) Mantle chemistry and accretion history of the Earth. In *Archaean Geochemistry* (eds. A. Kroner, G. N. Hanson, and A. M. Goodwin). Springer, New York, pp. 1–24.

Ward W. R. (2000) On planetesimal formation: the role of collective particle behavior. In *Origin of the Earth and Moon* (eds. R. M. Canup and K. Righter). University of Arizona Press, Tucson, pp. 75–84.

Wasserburg G. J., MacDonald F., Hoyle F., and Fowler W. A. (1964) Relative contributions of uranium, thorium, and potassium to heat production in the Earth. *Science* **143**, 465–467.

Wasserburg G. J., Papanastassiou D. A., Tera F., and Huneke J. C. (1977a) Outline of a lunar chronology. *Phil. Trans. Roy. Soc. London A* **285**, 7–22.

Wasserburg G. J., Tera F., Papanastassiou D. A., and Huneke J. C. (1977b) Isotopic and chemical investigations on Angra dos Reis. *Earth Planet. Sci. Lett.* **35**, 294–316.

Wasserburg G. J., Busso M., Gallino R., and Raiteri C. M. (1994) Asymptotic giant branch stars as a source of short-lived radioactive nuclei in the solar nebula. *Astrophys. J.* **424**, 412–428.

Wasserburg G. J., Busso M., and Gallino R. (1996) Abundances of actinides and short-lived nonactinides in the interstellar medium: diverse supernova sources for the r-processes. *Astrophys. J.* **466**, L109–L113.

Wasson J. T. (1985) *Meteorites: Their Record of Early Solar-system History*. W. H. Freeman, New York, 251p.

Wasson J. T. and Chou C.-L. (1974) Fractionation of moderately volatile elements in ordinary chondrites. *Meteoritics* **9**, 69–84.

Weidenschilling S. J. (1977a) The distribution of mass in the planetary system and solar nebula. *Astrophys. Space Sci.* **51**, 153–158.

Weidenschilling S. J. (1977b) Aerodynamics of solid bodies in the solar nebula. *Mon. Not. Roy. Astron. Soc.* **180**, 57–70.

Weidenschilling S. J. (2000) Formation of planetesimals and accretion of the terrestrial planets. *Space Sci. Rev.* **92**, 295–310.

Wetherill G. W. (1975a) Radiometric chronology of the early solar system. *Ann. Rev. Nuclear Sci.* **25**, 283–328.

Wetherill G. W. (1975b) Late heavy bombardment of the moon and terrestrial planets. *Proc. 6th Lunar Sci. Conf.* 1539–1561.

Wetherill G. W. (1980) Formation of the terrestrial planets. *Ann. Rev. Astron. Astrophys.* **18**, 77–113.

Wetherill G. W. (1986) Accumulation of the terrestrial planets and implications concerning lunar origin. In *Origin of the Moon* (eds. W. K. Hartmann, R. J. Phillips, and G. J. Taylor). Lunar and Planetary Institute, Houston, pp. 519–551.

Wetherill G. W. (1992) An alternative model for the formation of the Asteroids. *Icarus* **100**, 307–325.

Wetherill G. W. (1994) Provenance of the terrestrial planets. *Geochim. Cosmochim. Acta* **58**, 4513–4520.

Wetherill G. W. and Stewart G. R. (1993) Formation of planetary embryos: effects of fragmentation, low relative velocity, and independent variation of eccentricity and inclination. *Icarus* **106**, 190–209.

Wiechert U. (2002) Earth's early atmosphere. *Science* **298**, 2341–2342.

Wiechert U., Halliday A. N., Lee D.-C., Snyder G. A., Taylor L. A., and Rumble D. A. (2001) Oxygen isotopes and the Moon-forming giant impact. *Science* **294**, 345–348.

Wiechert U., Halliday A. N., Palme H., and Rumble D. (2003) Oxygen isotopes in HED meteorites and evidence for rapid mixing in planetary embryos. *Earth Planet. Sci. Lett.* (in press).

Wieler R., Anders E., Baur H., Lewis R. S., and Signer P. (1992) Characterisation of Q-gases and other noble gas components in the Murchison meteorite. *Geochim. Cosmochim. Acta* **56**, 2907–2921.

Wieler R., Kehm K., Meshik A. P., and Hohenberg C. M. (1996) Secular changes in the xenon and krypton abundances in the solar wind recorded in single lunar grains. *Nature* **384**, 46–49.

Wiesli R. A., Beard B. L., Taylor L. A., Welch S. A., and Johnson C. M. (2003) Iron isotope composition of the lunar mare regolith: implications for isotopic fractionation during production of single domain iron metal. In *Lunar Planet Sci.* **XXXIV**, #1500. The Lunar and Planetary Institute, Houston (CD-ROM).

Wilde S. A., Valley J. W., Peck W. H., and Graham C. M. (2001) Evidence from detrital zircons for the existence of continental crust and oceans on the Earth 4.4 Gyr ago. *Nature* **409**, 175–178.

Wolf R., Richter G. R., Woodrow A. B., and Anders E. (1980) Chemical fractionations in meteorites: XI. C2 chondrites. *Geochim. Cosmochim. Acta* **44**, 711–717.

Wood B. J. (1993) Carbon in the core. *Earth Planet. Sci. Lett.* **117**, 593–607.

Wood J. A. (1984) Moon over Mauna Loa: a review of hypotheses of formation of Earth's moon. In *Origin of the Moon* (eds. W. K. Hartmann, R. J. Phillips, and G. J. Taylor). Lunar and Planetary Institute, Houston, pp. 17–55.

Wood J. A. (2000) Pressure ands temperature profiles in the solar nebula. *Space Sci. Rev.* **92**, 87–93.

Wood J. A. and Morfill G. E. (1988) A review of solar nebula models. In *Meteorites and the Early Solar System* (eds. J. F. Kerridge and M. S. Matthews). University of Arizona Press, Tucson, pp. 329–347.

Yi W., Halliday A. N., Alt J. C., Lee D.-C., Rehkämper M., Garcia M., Langmuir C., and Su Y. (2000) Cadmium, indium, tin, tellurium and sulfur in oceanic basalts: implications for chalcophile element fractionation in the earth. *J. Geophys. Res.* **105**, 18927–18948.

Yin Q. Z. and Jacobsen S. B. (1998) The ^{97}Tc–^{97}Mo chronometer and its implications for timing of terrestrial accretion and core formation. In *Lunar Planet Sci.* **XXIX**, #1802. The Lunar and Planetary Institute, Houston (CD-ROM).

Yin Q. Z., Jacobsen S. B., McDonough W. F., Horn I., Petaev M. I., and Zipfel J. (2000) Supernova sources and the ^{92}Nb–^{92}Zr p-process chronometer: *Astrophys. J.* **535**, L49–L53.

Yin Q. Z., Jacobsen S. B., Yamashita K., Blicher-Toft J., Télouk P., and Albarède F. (2002) A short timescale for terrestrial planet formation from Hf–W chronometry of meteorites. *Nature* **418**, 949–952.

Yoshino T., Walter M. J., and Katsura T. (2003) Core formation in planetesimals triggered by permeable flow. *Nature* **422**, 154–157.

Young E. D. (2000) Assessing the implications of K isotope cosmochemistry for evaporation in the preplanetary solar nebula. *Earth Planet. Sci. Lett.* **183**, 321–333.

Zahnle K. J. (1993) Xenological constraints on the impact erosion of the early Martian atmosphere. *J. Geophys. Res.* **98**, 10899–10913.

Zahnle K. J. and Sleep N. H. (1996) Impacts and the early evolution of life. In *Comets and the Origin and Evolution of Life* (eds. P. J. Thomas, C. F. Chyba, and C. P. McKay). Springer, Heidelberg, , pp. 175–208.

Zahnle K. J. and Walker J. C. G. (1982) The evolution of solar ultraviolet luminosity. *Rev. Geophys.* **20**, 280.

Zahnle K. J., Kasting J. F., and Pollack J. B. (1988) Evolution of a steam atmosphere during Earth's accretion. *Icarus* **74**, 62–97.

Zhang Y. (1998) The young age of the Earth. *Geochim. Cosmochim. Acta* **62**, 3185–3189.

4

Long-Lived Chronometers

M. Wadhwa

Arizona State University, Tempe, AZ, USA

4.1 INTRODUCTION

4.1.1 Basic Principles

Long-lived radioactive isotopes, defined here as those that have half-lives in excess of a few hundred million years, have been utilized for chronology since the early part of the twentieth century. The decay of a radioactive ("parent") isotope involves its spontaneous transformation, sometimes through other intermediate radioisotopes, into a stable ("daughter") isotope at a rate proportional to the number of atoms of the radioisotope at any given time, such that

$$P = P_0 e^{-\lambda t} \tag{1}$$

where P is the number of atoms of the parent isotope remaining at present, P_0 the initial abundance of the parent isotope at the time of isotopic closure, t the time elapsed since isotopic closure (e.g., crystallization age for a rock) and λ the decay constant. Equation (1) may be rewritten in terms of the abundance of the radiogenic daughter isotope ($D*$) as follows:

$$D^* = P(e^{\lambda t} - 1) \qquad (2)$$

However, since the total number of atoms of the daughter isotope (D) is the sum of the radiogenic ($D*$) and the nonradiogenic (D_0) components,

$$D = D_0 + P(e^{\lambda t} - 1) \qquad (3)$$

Normalizing to a stable isotope of the daughter element (D_s),

$$D/D_s = D_0/D_s + P/D_s(e^{\lambda t} - 1) \qquad (4)$$

As such, the slope in an isochron plot for a long-lived chronometer (i.e., where D/D_s is plotted versus P/D_s) is given by $(e^{\lambda t}-1)$, from which the age (t) may be determined.

The past several decades have seen significant improvements in the precision and accuracy of chronological information based on the decay of long-lived radioisotopes. These have resulted particularly from advances in the mass spectrometric techniques for measurement of isotope ratios and better constraints on the relevant decay constants. Chronometers based on the decay of radioisotopes essentially date the time of isotopic closure following a chemical event that fractionated the parent element from the daughter element. Assuming that parent/daughter isotope ratios can be determined accurately and precisely and that the decay constant is known, meaningful age information based on such chronometers may only be obtained if: (1) there was complete equilibration of the isotopic composition of the daughter element prior to fractionation of the parent element from the daughter element; and (2) there has been no disturbance of isotope systematics following the isotopic closure event that is to be dated.

4.1.2 Application to Meteorites and Planetary Materials: A Historical Perspective

Clair Patterson's analyses of terrestrial and meteoritic lead isotopic compositions (Patterson, 1955, 1956) heralded the modern age of isotope chronology. He obtained a $^{207}Pb/^{206}Pb$ age from three stony meteorites of 4.55 ± 0.07 Ga and suggested that this represented the time of formation of the solar system and the Earth. Since that time, (1) advances in analytical instrumentation (allowing more precise isotopic ratio measurements), (2) more accurately determined decay constants, and (3) more appropriate sample selection have led to increasingly refined and precise estimates of this age. By chance, changes in these three factors have compensated one another in such a way that half a century later, Patterson's initial estimate of the age of the solar system still agrees with the current best estimate of this age. The $^{207}Pb/^{206}Pb$ systematics in the refractory calcium-, aluminum-rich inclusions (CAIs), believed to be among the first solids formed in the early history of the solar system, have been utilized to provide an estimate of the (minimum) age of the solar system. As will be discussed in more detail in the section below, the most recent analyses of lead-isotope systematics in CAIs from the Efremovka carbonaceous (CV3) chondrite yield a highly precise age of $4,567.1 \pm 0.2$ Ma (Amelin *et al.*, 2002, 2006).

The various long-lived radioisotopes that have thus far been used for chronological investigations of meteorites and their components are given in Table 1. Among these, the ones that have been most commonly applied are the $^{40}K–^{40}Ar$, $^{87}Rb–^{87}Sr$, $^{147}Sm–^{143}Nd$, and $^{235,238}U$, $^{232}Th–^{207,206,208}Pb$ chronometers. These have mostly been used for determining the crystallization and secondary alteration (e.g., by shock metamorphism) ages of various classes of meteorites. For the same meteorites, different chronometers may date different events in their histories, depending on the geochemical behaviors of the parent and daughter elements and their ease of equilibration. For example, while the $^{40}K–^{40}Ar$ system in most basaltic eucrites is partially or totally reset as a result of shock metamorphism at 3.4–4.1 Ga (Bogard, 1995), the $^{147}Sm–^{143}Nd$ ages of several samples belonging to this class of meteorites still reflect their crystallization at ~4.5 Ga.

Table 1 Long-lived radioisotopes used for chronological studies of meteorites.

Radioisotope	*Daughter isotope*	*Reference stable isotope*	*Half-life* (10^9 years)
^{40}K	^{40}Ar, ^{40}Ca	^{36}Ar	1.27
^{87}Rb	^{87}Sr	^{86}Sr	48.8
^{147}Sm	^{143}Nd	^{144}Nd	106
^{176}Lu	^{176}Hf	^{177}Hf	35.7
^{187}Re	^{187}Os	^{188}Os	41.6
^{190}Pt	^{186}Os	^{188}Os	489
^{232}Th	^{208}Pb	^{204}Pb	14.01
^{235}U	^{207}Pb	^{204}Pb	0.704
^{238}U	^{206}Pb	^{204}Pb	4.469

Of all the long-lived chronometers applied to meteorites so far, the combined $^{235,238}U-^{207,206}$ Pb systems provide the highest time resolution. This is so because the combination of two chronometers that involve the same parent and daughter elements effectively allows the determination of a time "*t*" (or a $^{207}Pb-^{206}Pb$ age) without having to measure the parent/daughter elemental ratio and based only on the isotopic composition ($^{207}Pb/^{206}Pb$ ratio) of the daughter element, which can be very precisely measured. Moreover, the relatively short half-life of ^{235}U compared to the other radioisotopes in Table 1 implies that, following a parent/daughter fractionation event, the $^{207}Pb/^{206}Pb$ ratio evolves rapidly over geologic timescales, thereby allowing sub-Myr time resolution. The $^{207}Pb-^{206}Pb$ age for a sample can either be a single-stage model age, which is determined by subtracting an assumed isotopic composition for "common Pb" (which includes the initial Pb and any extraneous Pb of terrestrial or extraterrestrial origin) from the measured composition, or an isochron age. The latter is obtained from a regression of the data for multiple samples, or components of a sample, on a Pb–Pb isochron plot (i.e., $^{207}Pb/^{206}Pb$ versus $^{204}Pb/^{206}Pb$) to obtain the purely radiogenic $^{207}Pb/^{206}Pb$ ratio (i.e., the intercept of this isochron plot) from which an age is calculated. As long as it is reasonable to assume that all samples plotted on a Pb–Pb isochron plot shared the same common lead component, the isochron method of calculating the age is the preferable one since no assumption of a common lead composition need be made.

Although much valuable chronological information is now being obtained from chronometers based on the decay of short-lived radionuclides that were present in the early solar system (see Chapter 2), long-lived chronometers (particularly those based on the $^{235,238}U-^{207,206}Pb$ systems) provide the only means of anchoring the relative ages provided by the extinct chronometers to an absolute timescale. In this review, an overview is presented of the chronological constraints that have been obtained so far for events occurring in the early history of the solar system based on long-lived radionuclides. Although results from earlier studies are briefly summarized, the focus of this review will be on more recent reports (i.e., those published within the last decade or so) and their implications. For additional details on previous studies of early solar system chronology based on both long- and short-lived radionuclides, the reader is referred to several excellent reviews (e.g., Wasserburg, 1985; Tilton, 1988; Podosek and Nichols, 1997; Carlson and Lugmair, 2000; Kita *et al.*, 2005; Chapter 2).

4.2 CHONDRITES AND THEIR COMPONENTS

4.2.1 Formation Ages of Chondritic Components

4.2.1.1 Calcium-, aluminum-rich inclusions

CAIs are refractory millimeter- to centimeter-sized objects found in primitive chondrite meteorite groups. They are thought to represent some of the first solids that formed in the solar protoplanetary disk. The earliest lead-isotope studies of CAIs (Chen and Tilton, 1976; Tatsumoto *et al.*, 1976) indicated that these were indeed ancient objects that formed in the earliest history of the solar system, close to 4.56 Ga. Subsequently, Chen and Wasserburg (1981) reported the lead-isotope compositions of several CAIs from the Allende carbonaceous (oxidized CV3) chondrite. Considering the most radiogenic of these samples and regressing these data through the Canyon Diablo lead-isotope composition (assumed here as the initial lead composition for the solar system), these authors reported an age of 4.559 Ga for Allende CAIs. However, if all of the data for CAIs from Chen and Wasserburg (1981) are taken together, they fall along a single linear array in a Pb–Pb isochron plot that (although it does not pass through the Canyon Diablo lead-isotope composition, implying that these CAIs contain a common lead component with a composition distinct from this) yields a $^{207}Pb-^{206}Pb$ age of 4,566 ± 8 Ma (Tera and Carlson, 1999) (Figure 1). Following this work, U–Pb analyses of several other Allende CAIs gave a consistent, but more precise, $^{207}Pb-^{206}Pb$ age of 4,566 ± 2 Ma (Göpel *et al.*, 1991; Allègre *et al.*, 1995). In more recent years, several studies have demonstrated the importance of the removal of common lead for obtaining high-precision $^{207}Pb-^{206}Pb$ ages for meteorites and their components (e.g., Lugmair and Galer, 1992; Amelin *et al.*, 2002, 2005). In particular, using extensive acid leaching to remove the common lead component, Amelin *et al.* (2002) obtained a precise $^{207}Pb-^{206}Pb$ age of 4,567.2 ± 0.6 Ma for two CAIs (E49 and E60) from the Efremovka carbonaceous (reduced CV3) chondrite (Figure 2). Additional analyses of the E60 CAI using step-leaching and $^{202}Pb-^{205}Pb$ double-spike in combination with the results reported by Amelin *et al.* (2002) have yielded the most precise $^{207}Pb-^{206}Pb$ age of 4,567.1 ± 0.2 Ma (Amelin *et al.*, 2006). However, E60 is a relatively rare type of CAI (forsterite-bearing Type B; Amelin *et al.*, 2002), and it is unclear whether its age (the most precisely defined though it is) is indeed representative of that of the more common CAI types. Nevertheless, at present, this represents the best estimate for time of formation of the earliest solids

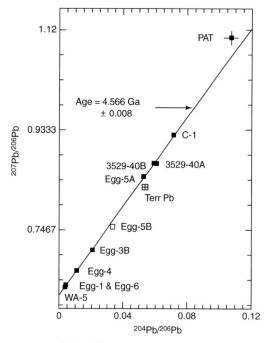

Figure 1 ^{207}Pb–^{206}Pb isochron diagram of the data of Chen and Wasserburg (1981) for Allende CAIs. The data define a single array that corresponds to a ^{207}Pb–^{206}Pb age of 4.566 Ga. PAT = Pb isotope composition of Canyon Diablo troilite. Reproduced by permission of Elsevier from Tera and Carlson (1999).

in the solar nebula and, therefore, the best estimate of the minimum age of the solar system.

The ^{87}Sr/^{86}Sr ratio has also been used as a tracer for the formation time of CAIs. In this approach, a formation time interval is estimated based on the measured initial ^{87}Sr/^{86}Sr ratio of a particular sample with a low Rb/Sr ratio (such as a CAI) and the time that is taken to evolve to this omposition from a less radiogenic strontium-isotope composition (such as the starting composition inferred for the solar nebula) in an environment with a given Rb/Sr ratio. The high Rb/Sr ratio in the solar nebula (Anders and Grevesse, 1989) implies that the ^{87}Sr/^{86}Sr ratio would increase rapidly in material evolving in such an environment until a major Rb/Sr fractionation event (such as CAI formation) defines the initial ^{87}Sr/^{86}Sr ratio of the object formed during this event. Comparison of the initial ^{87}Sr/^{86}Sr ratios for solar system materials can potentially resolve time differences of the order of a million years or so. The antiquity of CAIs is indicated by their extremely unradiogenic strontium isotopic compositions. A CAI (D7) from Allende has the lowest reported ^{87}Sr/^{86}Sr ratio of any solar system material (Gray *et al.*, 1973). However, there is some complexity in the strontium isotopic composition of CAIs since other Allende inclusions analyzed by Gray *et al.* (1973) and

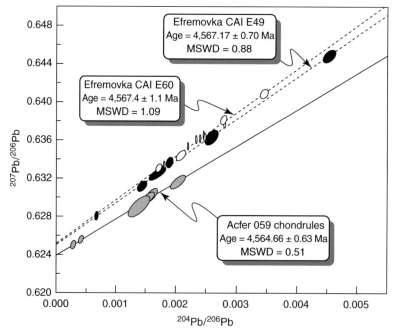

Figure 2 ^{207}Pb–^{206}Pb isochron diagram for acid-washed fractions from two Efremovka CAIs (E49 and E60); the weighted average of the ^{207}Pb–^{206}Pb ages obtained for these two CAIs is 4,567.2 ± 0.6 Ma. Also shown are the Pb-isotope data for the six most radiogenic analyses of acid-washed chondrules From the CR chondrite Acfer 059. Reproduced by permission of American Association for the Advancement of Science from Amelin *et al.* (2002).

Podosek *et al.* (1991) have $^{87}Sr/^{86}Sr$ ratios that are slightly higher than in D7 (translating to a time span of up to ~3 Myr between these and D7). Furthermore, more recent analyses of the strontium-isotope composition of the D7 inclusion are also slightly higher than the initial $^{87}Sr/^{86}Sr$ ratio that was reported for this CAI by Gray *et al.* (1973). Potential factors affecting this issue may be the presence of nucleosynthetic anomalies in the strontium-isotope composition or the disturbance of the strontium isotopic composition by secondary events or by contamination. In this regard, nucleosynthetic anomalies in strontium isotopes have been reported in CAIs that have been shown to record other fractionation and unknown nuclear (FUN) effects (Papanastassiou and Wasserburg, 1978; Loss *et al.*, 1994).

4.2.1.2 Chondrules

Chondrules are sub-millimeter to centimeter-sized ferromagnesian silicate spherules found in chondrites. Although, in detail, there are several hypotheses for the exact mechanism involved in chondrule formation (currently, the two leading ones being the X-wind and the shock-wave models; see Ciesla, 2005 and references therein): they are generally considered to have resulted from transient heating events in the solar nebula. The earliest lead-isotope study of chondrules was performed on those separated from the Allende chondrite and gave an age of $4,560 \pm 67$ Ma (Chen and Tilton, 1976). The low precision of this date was due to the relatively large unradiogenic Pb component in these chondrules and small spread in the $^{207}Pb/^{206}Pb$ ratios. More recently, using extensive leaching procedures to remove the common lead component, the chondrules from a carbonaceous chondrite belonging to the CR group were precisely dated at $4,564.7 \pm 0.6$ (Amelin *et al.*, 2002) (Figure 2), which indicates that these were formed 2.4 ± 0.6 Myr after CAIs. Lead-isotope compositions of chondrules from Allende have also been recently analyzed and yield an older age of $4,566.7 \pm 1.0$ Ma (Amelin *et al.*, 2004), which overlaps within the uncertainties with the $^{207}Pb-^{206}Pb$ age of CAIs. As such, chondrules from these primitive chondrite groups define ages that suggest that they began forming almost contemporaneously with CAIs and continued to form for at least another 2–3 Myr afterwards. Recently, Amelin *et al.* (2005) reported a $^{207}Pb-^{206}Pb$ age of $4,562.7 \pm 1.7$ Ma from pyroxene-rich chondrules and chondrule fragments from the Richardton H5 equilibrated ordinary chondrite. Given the equilibrated nature of this chondrite, these authors argued that this age was the minimum age for the formation of chondrules in this sample (possibly corresponding to the time of cessation of lead loss).

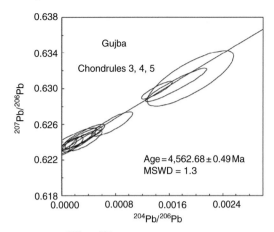

Figure 3 $^{207}Pb-^{206}Pb$ isochron diagram for three Gujba chondrules. Reproduced by permission of Nature Publishing Group from Krot *et al.* (2005).

Chondrules from the metal-rich CB chondrites Gujba and Hammadah al Hamarah 237 gave $^{207}Pb-^{206}Pb$ ages of $4,562.7 \pm 0.5$ Ma (Figure 3) and $4,562.8 \pm 0.9$ Ma, respectively (Krot *et al.*, 2005). It had previously been suggested that chondrules in the CB chondrites originated as a result of an energetic impact between large planetesimals (Rubin *et al.*, 2003) and these relatively young ages have been used in support of this hypothesis (Krot *et al.*, 2005).

4.2.2 Ages of Secondary Events Recorded in Chondrites

Chondritic meteorites record a variety of secondary alteration processes, including aqueous alteration, thermal and shock metamorphism, and brecciation. These events spanned a long time interval, beginning almost close to the formation of the solar system at ~4.56 Ga and extending over a period of several millions of years thereafter. The following is a discussion of the timescales for these different secondary alteration processes inferred from various long-lived chronometers, although the boundaries between these processes are not always well defined (e.g., an energetic impact event on a planetesimal may result in shock metamorphism as well as thermal processing).

4.2.2.1 Aqueous alteration

Aqueous alteration of chondrites and their components is thought to have occurred in a variety of settings, including the solar nebula and on accreted parent bodies (Zolensky and McSween, 1988; Brearley, 2006). The CI carbonaceous chondrites contain several secondary mineral phases (particularly carbonates and sulfates) that provide a record of aqueous alteration on their parent body.

The carbonates are a good candidate for dating using the initial $^{87}Sr/^{86}Sr$ approach since this mineral is typically characterized by low Rb/Sr ratios. Macdougall *et al.* (1984) reported the strontium isotopic compositions of several carbonate separates (dolomite and breunnerite) from the Orgueil CI chondrite. These authors showed that Orgueil carbonates have a range of $^{87}Sr/^{86}Sr$ ratios, with the lowest one being similar to the most primitive Sr-isotope composition measured in Allende CAIs (Gray *et al.*, 1973; Podosek *et al.*, 1991). This implies that the onset of aqueous alteration on the CI chondrite parent body occurred essentially contemporaneously with its formation.

4.2.2.2 Thermal metamorphism

Most chondrites have experienced some degree of thermal metamorphism, defined here as alteration resulting from heating at temperatures in the range of 400–1,000 °C at low lithostatic pressure for extended time periods (McSween *et al.*, 1988; Huss *et al.*, 2006). Calcium phosphates in chondritic meteorites are minor but uranium-rich secondary minerals that were formed during this thermal processing of the parent bodies of these meteorites, most likely by oxidation of phosphorus-rich metal (Perron *et al.*, 1988). As such, U–Pb systematics in secondary phosphates from chondrite groups that have experienced different degrees of metamorphic equilibration can provide constraints on the timescales involved in thermal processing of planetesimals following their accretion. The first studies of U–Pb systematics in phosphates from a chondrite were performed on the LL6 equilibrated ordinary chondrite Saint-Sèverin (Manhès *et al.*, 1978; Chen and Wasserburg, 1981). The results from these studies are consistent with a later investigation of phosphates from this same meteorite (Göpel *et al.*, 1994) and together yield a ^{207}Pb–^{206}Pb age of 4,558 ± 6 Ma. In addition to Saint-Sèverin, Göpel *et al.* (1994) reported U–Pb systematics in phosphates from 14 other equilibrated ordinary chondrites belonging to the H4, H5, H6, L5, L6, LL5, and LL6 groups. For the H chondrites, these authors noted a correlation between the ^{207}Pb–^{206}Pb ages of the phosphates and their degree of metamorphism. Specifically, ^{207}Pb–^{206}Pb model ages (with a typical precision of ± 1 Ma) ranged from 4,563 Ma for the Ste. Marguerite H4 chondrite to 4,504 Ma for the Guareña H6 chondrite, thereby indicating that thermal processing of the H chondrite parent body(ies) extended over a period of ~60 Myr. The ^{207}Pb–^{206}Pb age of phosphate fractions from the Richardton H5 chondrite (4,550.7 ± 2.6 Ma; Amelin *et al.*, 2005) falls within this time span. In the case of the L chondrites, ^{207}Pb–^{206}Pb model

ages for their phosphates ranged from 4,543 to 4,511 Ma, while for the LL chondrites these ages ranged from 4,557 to 4,536 Ma. These ages also suggest that thermal metamorphism of the L and LL parent bodies had extended for tens of millions of years in the early history of the solar system.

Although less precise than the U–Pb chronometer, the ^{39}Ar–^{40}Ar technique has also been applied toward constraining the duration of thermal processing of chondrite parent bodies. Some equilibrated (but unshocked) ordinary chondrites show a range of ^{39}Ar–^{40}Ar ages from ~4.5 to ~4.4 Ga (with a typical precision of ± 30 Ma) (Turner *et al.*, 1978; Hohenberg *et al.*, 1981) which, to first order at least, is comparable to the duration indicated by the U–Pb systematics in phosphates from the ordinary chondrites. The initial $^{87}Sr/^{86}Sr$ method (described in the previous section) may additionally be used to assess the duration of thermal processing of chondrite parent bodies. Two examples of the application of this approach (i.e., to phosphates from Beaver Creek and Guareña chondrites) toward obtaining chronological information regarding formation of secondary phosphates are illustrated in Figure 4. Time intervals of tens of millions of years are obtained based on the initial $^{87}Sr/^{86}Sr$ isotopic compositions reported for ordinary chondrite phosphates (Wasserburg *et al.*, 1969; Manhès *et al.*, 1978; Brannon *et al.*, 1988; Podosek and Brannon, 1991) and assuming evolution from a primitive strontium isotopic value similar to the average value for Allende inclusions (Gray *et al.*, 1973; Podosek *et al.*, 1991) and equilibration of strontium isotopes on the whole-rock scale. This is again broadly consistent with the duration of metamorphic events as indicated by the ^{207}Pb–^{206}Pb and ^{39}Ar–^{40}Ar ages discussed above. However, even though all three of these chronometers seem to be indicating generally similar timescales of tens of millions of years for the thermal processing on the chondrite parent bodies, when specific ages from these three chronometers are considered in detail (as was done by Göpel *et al.*, 1994), there does not appear to be any correlation between them. This could be indicative of complex histories of ordinary chondrite parent bodies following their initial accretion, which might affect these three chronometers differently.

Finally, ^{87}Rb–^{87}Sr investigations of chondrules additionally indicate that thermal processing of chondrite parent bodies extented to ~4.4 Ga. Chondrules from the Richardton ordinary equilibrated (H5) chondrite yield a ^{87}Rb–^{87}Sr age of 4.45 ± 0.03 Ga (Evensen *et al.*, 1979). Studies of chondrules from the Allende carbonaceous chondrite also indicate that the ^{87}Rb–^{87}Sr system was significantly affected by late thermal processing (Gray *et al.*, 1973; Tatsumoto *et al.*, 1976;

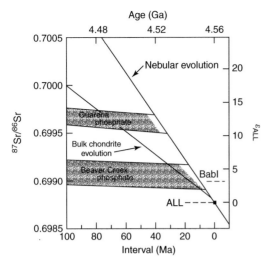

Figure 4 Two examples (phosphates from the Beaver Creek and Guareña chondrites) illustrating the application of the initial $^{87}Sr/^{86}Sr$ approach toward obtaining chronological constraints. ALL = initial $^{87}Sr/^{86}Sr$ ratio for the D7 Allende CAI with the most primitive composition (Gray *et al.*, 1973). Steeper solid line shows evolution of $^{87}Sr/^{86}Sr$ in a nebular environment ($^{87}Rb/^{86}Sr \sim 1.5$); shallower solid line shows evolution with bulk chondritic $^{87}Rb/^{86}Sr$ (~ 0.75). Shaded gray bands are extrapolations based on the measured Rb–Sr systematics in phosphates from the Beaver Creek and Guareña chondrites (the width of the bands indicating the uncertainties). As shown here, the time interval between formation of Allende CAIs and formation of secondary phosphates in these two chondrites (i.e., indicated on the *x*-axis by the points of the intersection of the gray bands with the bulk chondrite evolution line) is tens of millions of years. Reproduced by permission of Meteoritics and Planetary Science from Podosek and Brannon (1991).

Shimoda *et al.*, 2005). In particular, Shimoda *et al.* (2005) suggest that the mesostasis-rich chondrules (considered to be most susceptible to disturbance, particularly in terms of alkali elements such as rubidium) would best record this late processing. As such, consideration of only the mesostasis-rich chondrules from their study and from previous investigations (Gray *et al.*, 1973; Tatsumoto *et al.*, 1976) yields a $^{87}Rb–^{87}Sr$ age of $4.36 \pm 0.08\,\text{Ga}$. Within errors, this age is similar to that obtained for the Richardton chondrules.

4.2.2.3 Shock metamorphism

Impacts between solar system bodies have played an important role in their evolutionary histories and many chondritic meteorites preserve a record of these events (Stöffler *et al.*, 1988). The timing of impact events affecting chondrite parent bodies have been determined predominantly with the $^{39}Ar–^{40}Ar$ dating method (e.g., Turner,

1969; Bogard *et al.*, 1976; Bogard and Hirsch, 1980; Kaneoka, 1981; Stephan and Jessberger, 1988; Kring *et al.*, 1996; Grier *et al.*, 2004), although in some cases other isotope systems such as Rb–Sr, Pb–Pb, and Sm–Nd (roughly in that order of susceptibility to disturbance) are also affected by such events. Most experimental work on the effects of shock alone on various isotopic chronometers indicates that shock pressures up to $\sim 60\,\text{GPa}$ are usually insufficient to reset these chronometers (Deutsch and Schärer, 1994). However, if the shock event is associated with thermal annealing, it can easily affect some isotopic systems (particularly the K–Ar system, but also Rb–Sr) (e.g., Nyquist *et al.*, 1991). In the particular case of the $^{39}Ar–^{40}Ar$ dating method, if the shocked sample is insufficiently heated or if it cooled rapidly, there may be only a small amount of the radiogenic ^{40}Ar lost from the sample and thus the K–Ar system would only be partially reset. On the other hand, if the shock is accompanied by extended thermal annealing (and also if the sample had small grain sizes facilitating diffusional loss of argon), then the K–Ar system may be almost totally reset by this event (at least in some minerals of a shocked sample that may be more susceptible to being reset). In any case, the $^{39}Ar–^{40}Ar$ age of the shock event is determined by using a stepwise temperature release of argon, which helps to separate the K–Ar chronologies of different minerals of the shocked sample.

Bogard (1995) summarized the impact ages of various chondrite classes, that are mostly based on the $^{39}Ar–^{40}Ar$ method. Most chondritic meteorites have $^{39}Ar–^{40}Ar$ ages that are younger than $\sim 1.3\,\text{Ga}$ (Figure 5). These young ages are considered to be reflecting relatively few impact events and, in fact, the peak at $\sim 0.5\,\text{Ga}$ for the L class of ordinary chondrites in the histogram shown in Figure 5 is thought to be the result of a single large impact that catastrophically disrupted the parent body of these meteorites (Haack *et al.*, 1996a). At least ~ 4 distinct impact events are required to account for the Ar–Ar ages of most chondrites. The ~ 0.3 and $\sim 0.5\,\text{Ga}$ events are the best defined and affected the L and H ordinary chondrite parent bodies. Additional events at $\sim 0.9\,\text{Ga}$ (affecting the L and H parent bodies) and at $\sim 1.2\,\text{Ga}$ (affecting the LL parent body) are also indicated.

The Ar–Ar and Rb–Sr ages of some chondrites, however, indicate significantly older impact ages of ~ 3.5–$4.0\,\text{Ga}$ (e.g., Keil *et al.*, 1980; Stephan and Jessberger, 1988; Nakamura *et al.*, 1994). These ages most likely reflect the time of heavy bombardment experienced by bodies in the inner solar system. This event has also been recorded in the impact-reset ages of many achondrites

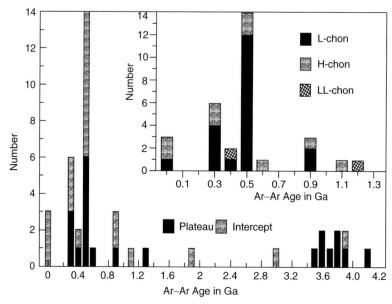

Figure 5 Histograms of impact-reset ^{39}Ar–^{40}Ar ages of ordinary chondrites, with plotted age interval of 0.1 Ga. Inset shows an expanded scale figure with data for chondrites having impact ages of <1.3 Ga. In main figure, ages defined by a significant age plateau ("plateau") are in black while those defined primarily from the age intercept of a diffusion profile ("intercept") are in gray. Reproduced by permission of Meteoritics and Planetary Science from Bogard (1995).

(see Section 4.3.2.3) and lunar samples (see Section 4.4.1.2).

4.3 DIFFERENTIATED METEORITES

4.3.1 Primitive Achondrites: Timing of Incipient Differentiation on Planetesimals

Primitive achondrites, such as acapulcoites, lodranites, winonaites, and brachinites, are considered to be the products of the earliest stages of melting and igneous processing on planetesimals (see Mittlefehldt et al., 1998 and references therein). The acapulcoites and lodranites are thought to be the residual products of partial melting of chondritic precursors (Mittlefehldt et al., 1996; McCoy et al., 1997a, b). ^{147}Sm–^{143}Nd systematics determined by Prinzhofer et al. (1992) for Acapulco gave a very old age (4.60 ± 0.03 Ga), which the authors interpreted as the time of recrystallization immediately following its formation event. Uranium–lead systematics in phosphates from Acapulco give a ^{207}Pb–^{206}Pb model age of 4,557 ± 2 Ma (Göpel et al., 1992, 1994), indicating that this achondrite formed approximately ~10 Myr after the formation of CAIs (at 4,567.1 ± 0.2 Ma; Amelin et al., 2002, 2006). More recently, a ^{207}Pb–^{206}Pb isochron age for Acapulco phosphates and mixed grain fractions of 4,556.52 ± 0.78 Ma (or 10.6 ± 0.8 Myr after CAI formation) has been reported (Amelin and Pravdivtseva, 2005; Amelin et al., 2006). This ^{207}Pb–^{206}Pb age for Acapulco is marginally

younger, but much more precise, than the ^{147}Sm–^{143}Nd age. McCoy et al. (1996) have argued that the older ^{147}Sm–^{143}Nd age could be due to disturbance during extensive later metamorphism experienced by this meteorite. The Divnoe meteorite is an ultramafic primitive achondrite whose relationship with other primitive achondrite groups is as yet unclear (Petaev et al., 1994; Weigel et al., 1996). This meteorite also has an old ^{147}Sm–^{143}Nd age of 4.62 ± 0.07 Ga. Although uncertainties are large (Bogdanovski and Jagoutz, 1996), the youngest formation time indicated by this ^{147}Sm–^{143}Nd age is ~17 Myr after CAI formation. The above discussion shows that the onset of melting on some planetesimals, as evidenced by primitive achondrites such as Acapulco and Divnoe, occurred within ~10–20 Myr of the beginning of the solar system.

4.3.2 Basaltic and Other Achondrites: Timing of Asteroidal Differentiation and Cataclysm

4.3.2.1 Crust-formation timescales from chronology of achondrites and their components

Primary crystallization ages of individual members of achondrite groups such as the angrites and noncumulate eucrites, which represent basaltic rocks that formed in asteroidal near-surface environments, provide constraints on the timing of silicate differentiation and crust formation on planetesimals during the early history of the solar system.

Angrites are a small group of mineralogically unique basalts composed mostly of Ca–Al–Ti-rich pyroxenes (fassaite), olivine and anorthitic plagioclase (see Mittlefehldt *et al.*, 1998 and references therein). ^{147}Sm–^{143}Nd systematics in Angra dos Reis (ADOR) and LEW 86010 (LEW) are well-behaved and give old crystallization ages between 4.53 ± 0.04 and 4.56 ± 0.04 Ga (Lugmair and Marti, 1977; Wasserburg *et al.*, 1977; Jacobsen and Wasserburg, 1984; Lugmair and Galer, 1992; Nyquist *et al.*, 1994). ^{147}Sm–^{143}Nd systematics have also been determined in the more recently discovered angrite D'Orbigny, and, despite some disturbance evident in the plagioclase, possibly due to late metamorphism and/or terrestrial weathering, are generally consistent with earlier results for ADOR and LEW (Nyquist *et al.*, 2003a; Tonui *et al.*, 2003). The most precise estimate of the crystallization age of the angrites is offered by their ^{207}Pb–^{206}Pb systematics. A ^{207}Pb–^{206}Pb isochron defined by LEW minerals gives an age of $4,558.2 \pm 3.4$ Ma, concordant with the highly precise model age of $4,557.8 \pm 0.5$ Ma obtained from the extremely radiogenic lead compositions in the pyroxenes of ADOR and LEW (Lugmair and Galer, 1992). Preliminary model ages derived from the lead-isotope compositions of the D'Orbigny pyroxenes appeared to be in agreement with those derived from ADOR and LEW pyroxenes (Jagoutz *et al.*, 2003). However, a recent reevaluation of these data by Jagoutz and colleagues has resulted in a somewhat older age of $4,563 \pm 1$ Ma for D'Orbigny (G. W. Lugmair, personal communication). This revised age is in agreement with the results from a more recent study that reported a ^{207}Pb ^{206}Pb model age of $4,563.8 \pm 0.6$ Ma for D'Orbigny (Zartman *et al.*, 2006). Finally, Baker *et al.* (2005) reported a highly precise and extremely ancient ^{207}Pb–^{206}Pb isochron age of $4,566.18 \pm 0.14$ Ma for the Sahara 99555 angrite (Figure 6). Given the ^{207}Pb–^{206}Pb age for CAIs of $4,567.1 \pm 0.2$ Ma (Amelin *et al.*, 2002, 2006), this suggests that basalts were forming on the surface of the angrite parent body within ~ 1 Myr of CAI formation. The above discussion shows that the ^{207}Pb–^{206}Pb ages of the angrites span a time interval of almost ~ 8 Myr, the youngest (LEW and ADOR) having an age of 4,558 Ma and the oldest (Sahara 99555) being 4,566 Ma.

Like the angrites, the noncumulate eucrites are pyroxene–plagioclase rocks. However, there are significantly greater numbers of known noncumulate eucrites than there are angrites. Recent high-precision oxygen-isotope data of Wiechert *et al.* (2004) demonstrate that most noncumulate eucrites along with the cumulate eucrites, diogenites, and howardites have uniform Δ^{17}O (within $\pm 0.02\permil$) and thus lie on a single mass fractionation line,

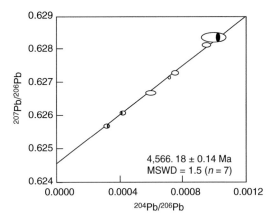

Figure 6 ^{207}Pb–^{206}Pb isochron diagram for fragments of Sahara 99555 and NWA 1296 angrites and acid-washed pyroxene from Sahara 99555. Reproduced by permission of Nature Publishing Group from Baker *et al.* (2005).

consistent with their origin on a single parent body. Therefore, these basalts are the most numerous crustal rocks available from any single solar system body other than the Earth and the Moon. A handful of the noncumulate eucrites (in particular Ibitira, but possibly also Caldera, Pasamonte, and ALHA 78132) have oxygen-isotope compositions distinct from the others, implying either that these samples originated on different parent bodies or that isotopic heterogeneity was preserved on the Howardite–Eucrite–Diogenite (HED) parent body (Wiechert *et al.*, 2004). Unlike angrites (which did not undergo any significant degree of recrystallization or metamorphism), the noncumulate eucrites appear to record a protracted history of extensive thermal processing on their parent body subsequent to their original crystallization. As a result, many of the chronometers investigated in these samples appear to record varying degrees of disturbance from secondary thermal events. Nevertheless, there are several lines of evidence that suggest that these basalts crystallized very early in the history of the solar system. Although typically characterized by large uncertainties, the ^{147}Sm–^{143}Nd ages of several of these samples such as Chervony Kut ($4,580 \pm 30$ Ma; Wadhwa and Lugmair, 1995), Juvinas ($4,560 \pm 80$ Ma; Lugmair, 1974), Pasamonte ($4,580 \pm 120$ Ma; Unruh *et al.*, 1977), Piplia Kalan ($4,570 \pm 23$ Ma; Kumar *et al.*, 1999), and Yamato 792510 ($4,570 \pm 90$ Ma; Nyquist *et al.*, 1997) are close to ~ 4.56 Ga. In some cases where ^{147}Sm–^{143}Nd systematics appear to record ages younger than ~ 4.56 Ga, the ^{146}Sm–^{142}Nd systematics hint at early crystallization of basaltic eucrites such as Caldera (^{146}Sm/^{144}Sm $= 0.0073 \pm 0.0011$; Wadhwa and Lugmair, 1996) and Ibitira (^{146}Sm/^{144}Sm $\sim 0.009 \pm 0.001$; Prinzhofer *et al.*, 1992). This may be explained by a model proposed by Prinzhofer *et al.* (1992), according to which the apparent discrepancy between the long-lived ^{147}Sm–^{143}Nd and the short-lived ^{146}Sm–^{142}Nd systems can be interpreted in terms of a short episodic disturbance resulting in partial reequilibration of the rare earth elements (REEs), predominantly between plagioclase (which has very low REE abundances) and phosphates (which are the primary REE carriers). As shown by the modeling results of these authors, such a disturbance could partially reset the ^{147}Sm–^{144}Nd isochron, without having a resolvable effect on the ^{146}Sm–^{142}Nd system.

The Rb–Sr chronometer also generally indicates ancient formation ages for the noncumulate eucrites (e.g., Allègre *et al.*, 1975; Nyquist *et al.*, 1986). The most precise of the absolute chronometer, that is, the U–Pb system, appears to have been affected by postcrystallization events and terrestrial Pb contamination in most noncumulate eucrites, recording mineral isochron ages in the range of 4,128–4,530 Ma (Tatsumoto *et al.*, 1973; Unruh *et al.*, 1977; Galer and Lugmair, 1996; Tera *et al.*, 1997). However, Ibitira whole-rock samples with radiogenic Pb isotopic compositions gave old ^{207}Pb–^{206}Pb model ages of $4,556 \pm 6$ Ma (Chen and Wasserburg, 1985) and $4,560 \pm 3$ Ma (Manhès *et al.*, 1987). Furthermore, a recent determination of the ^{207}Pb–^{206}Pb mineral isochron for the Asuka 881394 eucrite yielded a precise and ancient age of $4,566.5 \pm 0.3$ Ma (Amelin *et al.*, 2006). This is only 0.6 ± 0.4 Myr younger than the time of CAI formation ($4,567.1 \pm 0.2$ Ma; Amelin *et al.*, 2002, 2006) and, as in the case of the Sahara 99555 angrite discussed earlier, indicates that crust formation on differentiated planetesimals occurred within ~1 Myr of the formation of CAIs.

Basaltic noncumulate eucrites thus show clear evidence of having formed close to ~4.56 Ga in the crust of their parent planetesimal. In contrast, cumulate eucrites, which formed in the crust of the same parent planetesimal as the noncumulate eucrites (Clayton and Mayeda, 1996; Wiechert *et al.*, 2004), have significantly younger concordant Sm–Nd and Pb–Pb ages, ranging from the oldest of $4,456 \pm 25$ Ma (Sm–Nd) and $4,484 \pm 19$ Ma (Pb–Pb) for Moore County (Tera *et al.*, 1997) to the youngest of $4,410 \pm 20$ Ma (Sm–Nd; Lugmair *et al.*, 1977) and $4,399 \pm 35$ Ma (Pb–Pb; Tera *et al.*, 1997) for Serra de Magé. Thus, Sm–Nd and Pb–Pb systematics in the cumulate eucrites indicate that isotopic closure occurred up to ~150 Myr after the noncumulate eucrites (Lugmair *et al.*, 1977; Jacobsen and Wasserburg, 1984; Lugmair *et al.*, 1991; Tera *et al.*, 1997), possibly implying that active magmatism persisted on the eucrite parent body (EPB) for this extended period (Tera *et al.*, 1997).

Since, as discussed above, the oldest basaltic meteorites formed within ~10 Myr of CAI formation (and some, specifically the angrite Sahara

99555 and the eucrite Asuka 881394, within only ~1 Myr), the decay of short-lived radionuclides such as ^{26}Al and ^{60}Fe is the likely heat source for the early and extensive differentiation experienced by their parent planetesimal. Energy sources that can account for later igneous activity (i.e., tens of millions of years after CAI formation) on small planetesimals are not obvious unless the cumulate eucrites are the crystallization products of impact melting on the EPB. Alternatively (and perhaps more likely), since the cumulate eucrites are slowly cooled rocks that possibly formed deeper within the crust of their parent body than noncumulate eucrites, the long-lived chronometers could be recording the long cooling times required to achieve subsolidus temperatures. This is supported by the modeling results of Ghosh and McSween (1998), which show that, assuming reasonable parameters, it is possible to maintain temperatures in excess of the solidus temperature for basalt at a depth of ~100 km for over ~100 Myr in a Vesta-sized planetesimal.

Besides the angrites and the eucrites, there are other types of achondrites, such as aubrites and ureilites, whose origins are somewhat enigmatic but which were nevertheless formed in the crusts of extensively differentiated asteroidal bodies. There are very few chronological investigations of the aubrites, which are essentially monomineralic rocks composed of coarse-grained enstatite. U–Th–Pb and Sm–Nd isotope systematics in the ureilites generally indicate early formation, although there are apparent complications resulting from later disturbance during a metasomatic event on the ureilite parent body and/or by recent terrestrial contamination (Goodrich and Lugmair, 1995; Torigoye-Kita *et al.*, 1995a, b).

4.3.2.2 *Global differentiation timescales based on whole-rock isochrons and initial* $^{87}Sr/^{86}Sr$

While an internal mineral isochron can provide constraints on the timing of formation of an individual achondrite in the crust of a planetesimal, a whole-rock isochron of a particular achondrite group can provide limits on the timing of the major fractionation event that established the parent/daughter element ratio in the whole rocks (which could have pre-dated the formation of an individual sample). Depending on the geochemical behavior of the parent and daughter elements, this major parent/daughter fractionation recorded in the whole rocks may reflect a global silicate fractionation event (possibly associated with crystallization of a magma ocean) that established the source characteristics for these rocks, or it could simply reflect crystal fractionation from a parental melt, which resulted in the formation of these rocks. Whole-rock Rb–Sr isochrons for the basaltic eucrites established early on that Rb–Sr fractionation in the mantle source reservoir of these achondrites occurred close to ~4.6 Ga (Papanastassiou and Wasserburg, 1969; Birck and Allègre, 1978). Smoliar (1993) evaluated all available Rb–Sr data for the eucrites and obtained a whole-rock isochron age of 4.55 ± 0.06 Ga for the noncumulate eucrites (Figure 7). A whole-rock

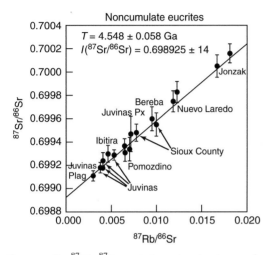

Figure 7 ^{87}Rb–^{87}Sr whole-rock isochron for noncumulate eucrites. All data points indicate whole-rocks samples or clasts, with the exception of two mineral fractions from Juvinas (Px = pyroxene; Plag = plagioclase) that are also plotted here. Data are from various literature sources given in Smoliar (1993). Adapted by permission of Meteoritics and Planetary Science from Smoliar (1993).

^{147}Sm–^{143}Nd isochron for 18 noncumulate and cumulate eucrites was reported by Blichert-Toft *et al.* (2002) and gave a relatively young age of $4,464 \pm 75$ Ma (Figure 8). Since the slope of this isochron is controlled by the cumulate eucrites (which have the most fractionated whole-rock Sm/Nd ratios), the authors interpreted this to imply that cumulate eucrites were formed ~100 Myr after solar system formation. This is supported by the relatively young ^{207}Pb–^{206}Pb ages and ^{147}Sm–^{143}Nd internal isochron ages of the cumulate eucrites (see earlier discussion in this section). Patchett and Tatsumoto (1980) reported the first whole-rock ^{176}Lu–^{176}Hf isochron for the eucrites. At the time, the half-life of ^{176}Lu was not well-constrained and these authors assigned an age of 4.55 Ga for this whole-rock isochron and thus estimated a half-life of 35.3 Gyr (corresponding to a decay constant for ^{176}Lu or $\lambda_{176_{Lu}}$ of 1.96×10^{-11} yr^{-1}). A more recent study of ^{176}Lu–^{176}Hf systematics in whole rocks of eucrites by Blichert-Toft *et al.* (2002) reported an errorchron corresponding to an age of $4,604 \pm 39$ Ma for the noncumulate eucrites (Figure 9). The cumulate eucrites, on the other hand, defined a ^{176}Lu–^{176}Hf isochron with an age of $4,470 \pm 22$ Ma (Figure 9), indistinguishable from their ^{147}Sm–^{143}Nd whole-rock age. Based on these whole-rock ^{147}Sm–^{143}Nd and ^{176}Lu–^{176}Hf systematics in the eucrites, these authors suggested that cumulate eucrites were formed ~100 Myr after the noncumulate eucrites, while the latter were formed close to the beginning of the solar system. In their study, Blichert-Toft *et al.* (2002) assumed a

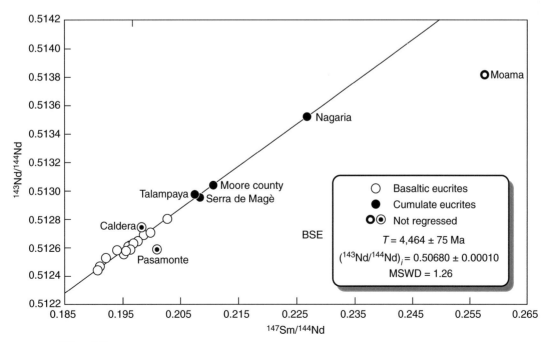

Figure 8 ^{147}Sm–^{143}Nd whole-rock isochron for eucrites. Reproduced by permission of Elsevier from Blichert-Toft *et al.* (2002).

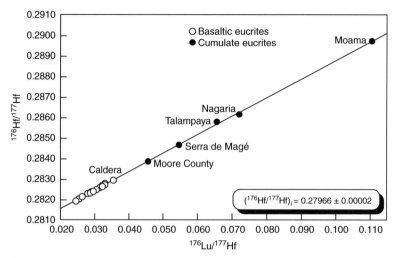

Figure 9 ^{176}Lu–^{176}Hf whole-rock isochron for the eucrites. A statistically significant isochron cannot be obtained if all data are considered together. If only noncumulate eucrites are considered (with the exception of one sample, Palo Blanco Creek) an errorchron corresponding to an age of $4,604 \pm 39$ Ma is obtained. If only the three cumulate eucrites Moama, Moore County, and Serra de Magé are regressed together, they yield an isochron corresponding to an age of $4,470 \pm 22$ Ma. Reproduced by permission of Elsevier from Blichert-Toft *et al.* (2002).

$\lambda_{176_{Lu}}$ value of $\sim 1.93 \times 10^{-11} \, \mathrm{yr}^{-1}$ (similar to that suggested by Patchett and Tatsumoto, 1980). However, studies of terrestrial samples on the one hand and meteoritic samples on the other indicate that the half-life of ^{176}Lu may be either 37.2 Gyr ($\lambda_{176_{Lu}} \sim 1.86 \times 10^{-11} \, \mathrm{yr}^{-1}$; Scherer *et al.*, 2001) or 34.9 Gyr ($\lambda_{176_{Lu}} \sim 1.98 \times 10^{-11} \, \mathrm{yr}^{-1}$; Bizzarro *et al.*, 2003), respectively. Amelin and Davis (2005) have shown that this apparent discrepancy cannot be accounted for by a possible branched decay of ^{176}Lu. Although other possible explanations were evaluated by these authors, none were considered to be plausible. As such, the half-life of ^{176}Lu still remains uncertain, thereby limiting the usefulness of the ^{176}Lu–^{176}Hf chronometer at the present time.

Finally, the ancient ages for the establishment of the highly volatile-depleted mantle source reservoirs of the angrites and the eucrites on their respective parent bodies can be inferred from their initial ^{87}Sr/^{86}Sr ratios. Papanastassiou and Wasserburg (1969) first estimated the initial ^{87}Sr/^{86}Sr of the EPB (basaltic achondrite best initial or BABI). At the time, this was the most primitive known strontium isotopic composition. Subsequently, however, an even more primitive strontium isotopic composition was reported for an Allende CAI (ALL) (Gray *et al.*, 1973). Assuming that the initial ^{87}Sr/^{86}Sr ratio at the beginning of the solar system is represented by the average initial ^{87}Sr/^{86}Sr ratio measured in Allende CAIs (Gray *et al.*, 1973; Podosek *et al.*, 1991), and that subsequent evolution of radiogenic strontium occurred in an environment with solar Rb/Sr ratios, the initial ^{87}Sr/^{86}Sr ratios of the angrites (Lugmair and Galer, 1992; Nyquist *et al.*, 1994, 2003a; Tonui *et al.*, 2003) translate

to an age difference of ~ 4 Myr between CAI formation and the timing of Rb/Sr depletion event that established the angrite source characteristics. The very low initial ^{87}Sr/^{86}Sr ratios of the eucrites similarly indicate that the volatile depletion characterizing the EPB may have occurred early in the history of the solar system (Carlson and Lugmair, 2000, and references therein). A re-evaluation of the strontium-isotope data for the eucrites by Smoliar (1993) shows that whole-rock Rb–Sr isochrons for the noncumulate and the cumulate eucrites define slightly, but resolvably, different initial ^{87}Sr/^{86}Sr ratios (that are both distinctly lower than the eucrite initial, BABI, previously defined by Papanastassiou and Wasserburg, 1969). In fact, the initial ^{87}Sr/^{86}Sr ratio for the cumulate eucrites proposed by Smoliar (1993) is, within errors, similar to that for the angrites (e.g., Lugmair and Galer, 1992), suggesting that the volatile depletion in their sources was established at similar times (possibly ~ 4 Myr after CAI formation; see above). However, the slighter higher initial ^{87}Sr/^{86}Sr ratio defined by the noncumulate eucrites is potentially problematic since the simplest interpretation would be that their source evolved with a chondritic Rb/Sr ratio for ~ 4 Myr longer (and is thus younger) than that of the cumulate eucrites, further implying that these two types of eucrites originated on distinct parent bodies (Smoliar, 1993). This is inconsistent with recent high-precision oxygen-isotope data (Wiechert *et al.*, 2004) that suggest that the cumulate and noncumulate eucrites (with the possible exception of Ibitira) originated on a common parent planetesimal. As discussed by Carlson and Lugmair (2000), a possible explanation could be that the severe volatile depletion on the EPB did not occur in a single-step process, but rather took place over the course of its accretionary and early differentiation history. Subsequently, the process of magma ocean crystallization may have resulted in a slighter higher Rb/Sr ratio in the source of the noncumulate eucrites compared to that of the cumulate eucrites, thereby resulting in the higher initial $^{87}Sr/^{86}Sr$ ratio indicated by the whole-rock isochron for the noncumulate eucrites.

4.3.2.3 Inner solar system bombardment history based on reset ages

As discussed previously for chondritic meteorites, the $^{39}Ar–^{40}Ar$ technique has also been widely applied toward dating the thermal histories of achondritic meteorites, and particularly for determining the ages of impact events on their parent bodies (e.g., Podosek and Huneke, 1973; Bogard et al., 1990; Takeda et al., 1994; Bogard and Garrison, 2003). As is the case for severely shocked chondritic samples, other isotopic chronometers such as Rb–Sr, Pb–Pb, and Sm–Nd in some differentiated achondrites that have experienced postshock thermal annealing record varying degrees of disturbance (e.g., Unruh et al., 1977; Nyquist et al., 1990). The impact ages of most achondrites (particularly the HED meteorites) fall within a relatively narrow time interval of ~3.4–4.1 Ga (Bogard, 1995; Bogard and Garrison, 2003) (Figure 10). As discussed in the earlier sections, these achondrites have crystallization ages that are close to ~4.56 Ga and so the age distribution shown in Figure 10 may be reasonably interpreted to reflect the timing of thermal metamorphism on the HED parent body resulting from large impacts (which are considered to be the most plausible heat source for these late events). Most of the HEDs are brecciated rocks that preserve clear textural, mineralogical, and chemical evidence of shock metamorphism (e.g., Stöffler et al., 1988; Metzler et al., 1995), further supporting the above interpretation.

Based on the age distribution shown in Figure 10, it has been suggested that there was a peak in the impactor flux (i.e., a cataclysm) on the HED parent body at ~3.7 Ga, which decreased sharply and tailed down to ages slightly younger than ~3.4 Ga (Bogard, 1995; Bogard and Garrison, 2003). Therefore, the $^{39}Ar–^{40}Ar$ ages of the HED meteorites suggest that the region where their parent body resided in the inner solar system experienced a period of heavy bombardment during the time interval extending from ~4.1 Ga until at least ~3.4 Ga. The peak at ~4.48 Ga in Figure 10 has been explained by Bogard and Garrison (2003) in terms of a single large impact that excavated these eucritic meteorites from their parent body.

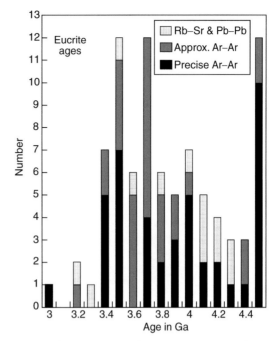

Figure 10 Histogram of impact-reset ages of the eucrites, with plotted age interval of 0.1 Ga. In the case of $^{39}Ar–^{40}Ar$ ages, those reported with uncertainties ("precise Ar–Ar") are shown in black while less precise ones reported without uncertainties ("approximate Ar–Ar") are shown in gray. Rb–Sr and Pb–Pb ages <4.3 Ga are shown as the stipled areas. Reproduced by permission of Elsevier from Bogard and Garrison (2003).

4.3.3 Iron Meteorites and Pallasites: Timescales of Core Crystallization on Planetesimals

Limits on the timescales involved in metallic core formation and crystallization on planetesimals may be obtained from chronological investigations of iron-rich meteorites, such as magmatic irons (that represent the cores of differentiated asteroidal bodies) and pallasites (considered to have formed near the core–mantle boundary). Once the metal has segregated into the core of a planetesimal, the timescales involved in the crystallization of this metal may be constrained by isochrons based on bulk samples and mineral phases of magmatic iron meteorites and pallasites. In recent years, precise Re–Os isochrons have been obtained for various groups of the iron meteorites (Figure 11). Nevertheless, one of the main problems affecting the ability to obtain accurate and precise ages based on such isochrons has been the large (~±3%) uncertainty in the measured decay constant of ^{187}Re (Lindner et al., 1989). Recent Re–Os studies (e.g., Smoliar et al., 1996; Horan et al., 1998; Chen et al., 2002) have assumed a more precise value for the ^{187}Re decay constant under the assumption that the Re–Os system attained closure in the IIIAB irons

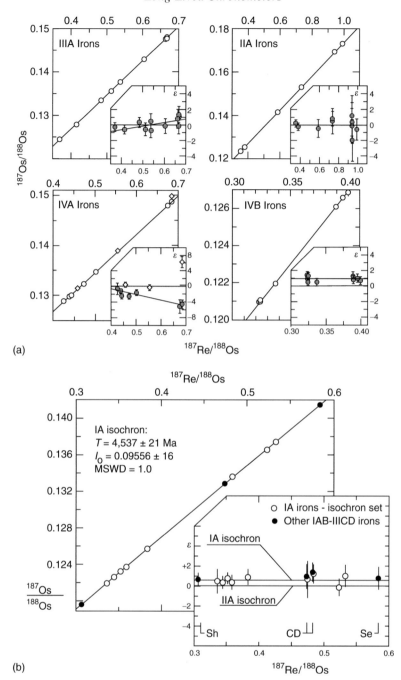

Figure 11 ^{187}Re–^{187}Os isochrons for meteorites from (a) the magmatic IIIA (4,557 ± 12 Ma), IIA (4,537 ± 8 Ma), IVA (4,456 ± 26 Ma; open diamonds indicate three samples that were omitted from the regression), and IVB (4,527 ± 29 Ma) groups, and (b) the nonmagmatic IAB–IIICD (4,537 ± 21 Ma) groups (labeled data points are: CD—Canyon Diablo, Se—Seeläsgen, and Sh—Shrewsbury). Ages are calculated assuming a decay constant of 1.666 × 10^{-11} yr^{-1}. Insets show the deviation in parts per 10^4 (i.e., in ϵ units) of the data points from the best-fit line; these deviations are calculated relative to the IIA isochron (shown as the horizontal line in each of the insets). (a) Reproduced by permission of American Association for the Advancement of Science from Smoliar *et al.* (1996) (b) Reproduced by permission of Elsevier from Horan *et al.* 1998).

almost contemporaneously with the formation of the angrites ADOR and LEW at 4,557.8 ± 0.5 Ma (Lugmair and Galer, 1992). Indeed, the Mn–Cr model ages for the IIIAB irons (Hutcheon and Olsen, 1991; Hutcheon *et al.*, 1992; Sugiura and

Hoshino, 2003) and these angrites (Nyquist *et al.*, 1994; Lugmair and Shukolyukov, 1998) are approximately coincident within ~±5 Myr and, therefore, this assumption may be valid at this level of uncertainty.

Assuming then that the IIIAB magmatic irons attained closure of the Re–Os system at 4,558 Ma, most iron meteorite groups give relatively old Re–Os ages (i.e., older than ~4.5 Ga) (Shen *et al.*, 1996; Smoliar *et al.*, 1996; Horan *et al.*, 1998; Cook *et al.*, 2004). There is an apparent discrepancy, however, in the Re–Os systematics reported in the IVA magmatic irons. While Shen *et al.* (1996) report a Re–Os age for the IVA irons that is 60 ± 45 Myr older than for IIAB irons, the data of Smoliar *et al.* (1996) give an age of $4,456 \pm 25$ Ma, significantly younger than other iron meteorite groups, which these authors attributed to later disturbance of the Re–Os system. As Horan *et al.* (1998) have pointed out, while the Re–Os isotopic compositions of most of the IVA irons analyzed by Smoliar *et al.* (1996) do lie on the $4,456 \pm 25$ Ma isochron, some (specifically, Duel Hill and Bushman Land) are consistent with the older age reported by Shen *et al.* (1996). It is possible that the Re–Os isotope systematics in different IVA irons are variably disturbed, perhaps by processes such as breakup and reassembly of the parent planetesimal (a process that has been invoked to explain the range of metallographic cooling rates of the IVA irons, e.g., Haack *et al.*, 1996b). This may have resulted in the resetting of the Re–Os system in some IVA irons but not in others.

As noted earlier, Re–Os ages reported so far use a [187]Re decay constant that was determined by assuming that the IIIAB isochron should give the same age as the U–Pb age of the angrites, so the accuracy of these ages is only as good as the validity of this assumption. Nevertheless, the age range indicated by Re–Os isochrons is largely independent of the precise value of half-life, so the

results for various magmatic iron meteorite groups suggest that core crystallization (or more specifically, Re–Os isotopic closure) in planetesimals spanned a period of ~30 Myr (although the relatively large errors certainly allow this time interval to be significantly narrower than this). This time interval for core crystallization (the process that is most likely to have produced Re–Os fractionation among the different members of a particular group of the magmatic irons) is distinct from that for core formation (or metal-silicate segregation) on the iron meteorite parent bodies. The latter has been inferred (predominantly from the short-lived [182]Hf–[182]W chronometer) to have occurred over a relatively short time period of only a few Myr after CAI formation (e.g., Markowski *et al.*, 2006).

Re–Os systematics in pallasites indicate that they may be younger than iron meteorites by ~60 Myr (Figure 12). However, this apparently young age may be due to later reequilibration of the Re–Os system (Chen *et al.*, 2002). Cook *et al.* (2004) recently reported the first high-precision [190]Pt–[186]Os isochrons for the IIAB and IIIAB magmatic irons, and estimated ages for these meteorite groups of $4,323 \pm 80$ Ma and $4,325 \pm 26$ Ma, respectively. These ages are somewhat younger than Re–Os ages for iron meteorites, and these authors suggested that this discrepancy could reflect an error in the decay constant for [190]Pt.

4.4 PLANETARY MATERIALS

Besides the Earth, the Moon and Mars are the only other planetary bodies from which there are samples currently available for chronological

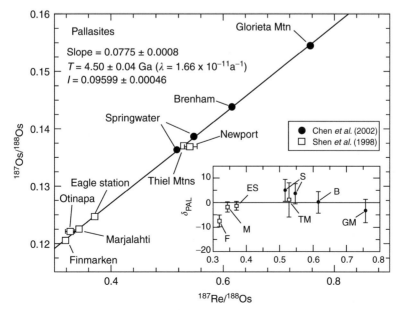

Figure 12 [187]Re–[187]Os isochron diagram for pallasite meteorites. Inset shows deviations from the best-fit line (δ_{PAL}) versus [187]Re/[188]Os ratios. Reproduced by permission of Elsevier from Chen *et al.* (2002).

investigations. In the case of the Moon, there are the samples that were returned by the Apollo and Luna missions, as well as ~40 distinct meteorites of basaltic and feldspathic compositions that are thought to have originated from a variety of terrains on the Moon. In the case of Mars, there are currently about three dozen or so distinct meteorites of mafic and ultramafic compositions thought to have originated from this planet. The following provides a brief summary of the chronology of these samples and the inferred differentiation and evolutionary histories of the Moon and Mars.

4.4.1 Timing of Lunar Differentiation and Cataclysm from Chronology of Lunar Samples

4.4.1.1 *Lunar differentiation history*

Lunar samples returned by the Apollo and Luna missions as well as the lunar meteorites broadly fall within two compositional categories: mare basalts and feldspathic (highlands) rocks. The mare basalts are a geochemically diverse group and are comprised of high-titanium (9–14 wt.% TiO_2), low-titanium (1–5 wt.% TiO_2), and very low-titanium (<1 wt.% TiO_2) basaltic samples. The feldspathic rocks are also geochemically and petrologically diverse and are composed of a variety of pristine igneous rocks as well as polymict breccias. Among the pristine igneous samples are the ferroan anorthosites and magnesium-rich rocks. The latter are composed of subgroups of magnesian-suite, alkali-suite, granites/felsites, and KREEP basalt and quartz monzodiorite rocks.

Among known lunar samples, the ones that have yielded the oldest crystallization ages are the ferroan anorthosites, which are thus inferred to be remnants of the earliest-formed lunar crust (Hanan and Tilton, 1987; Carlson and Lugmair, 1988; Borg et al., 1999a; Alibert et al., 1994; Norman et al., 2003). These ages have been determined mostly using the Sm–Nd chronometer (in one case, with the U–Pb chronometer), and span a time interval from ~4.29 Ga ± 0.06 (Borg et al., 1999a) to ~4.56 Ga ± 0.07 (Alibert et al., 1994). Norman et al. (2003) have argued that plagioclase in these anorthosites may have been subject to later disturbance by impact metamorphism and if only the mafic minerals are considered, these yield an Sm–Nd age for four ferroan anorthosites of 4.46 ± 0.04 Ga. This is then likely to be the best estimate of the crystallization age of the oldest lunar crustal samples.

The crystallization ages of the magnesium-rich highlands rocks (determined predominantly with Sm–Nd, but also with the Rb–Sr and U–Pb chronometers) are also generally ancient. Among these, the oldest (~4.1–4.5 Ga) are the magnesian-suite norites, troctolites, and dunites, some of which may have formed contemporaneously with the

ferroan anorthosites. These were followed by the alkali-suite (~4.0–4.3 Ga), granites/felsites (~3.8–4.3), KREEP basalts (~3.8–4.0 Ga), and quartz monzodiorite rocks (~4.3 Ga) (see Nyquist and Shih, 1992; Papike et al., 1998; Snyder et al., 2000; and references therein).

The formation ages of the mare basalts and lunar pyroclastic glasses based on the Sm–Nd, Rb–Sr, and U–Pb chronometers are summarized in table 4 of Chapter 1.21. Among the oldest dated basaltic lunar material are the mare-like clasts in highland breccias from the Apollo 14 landing site that have ages as old as ~4.2 Ga (Taylor et al., 1983). Most mare volcanism, however, postdated the period of heavy bombardment at ~3.9 Ga (see discussion of impact ages of lunar samples in the following section). Thus far, one of the youngest mare basalts to be dated is the KREEP-rich basaltic lunar meteorite NWA 773 that has $^{39}Ar-^{40}Ar$ and Sm–Nd ages of ~2.9 Ga (Fernandes et al., 2003; Borg et al., 2004). Fernandes et al. (2003) reported a similarly young $^{39}Ar-^{40}Ar$ age of ~2.8 Ga for another lunar meteorite, NWA 032, an unbrecciated mare basalt, and suggested that this also represented the time of crystallization for this basalt. Figure 16 of Chapter 1.21 summarizes the crystallization ages of the variety of lunar samples discussed here.

The Moon is thought to have undergone major global differentiation early in its history that resulted in the formation of the earliest crust (represented by the feldspathic highlands rocks) as well as the mantle source reservoirs of the basaltic lunar samples. The timing of this lunar global differentiation has been inferred from a variety of methods. The crystallization ages of the feldspathic highlands rocks, particularly the ferroan anorthosites, discussed earlier have been used to estimate a minimum age of ~4.3 to ~4.5 Ga for this event (e.g., Carlson and Lugmair, 1988; Borg et al., 1999a; Norman et al., 2003). Estimates for the formation of the KREEP component in the lunar mantle (thought to represent the residuum from Moon-wide differentiation; e.g., Wood, 1972; Warren and Wasson, 1979) also place limits on the timing of this event. This was estimated to be at ~4.6 Ga from the Rb–Sr model age of lunar soils (Papanastassiou et al., 1970), ~4.42 Ga from U–Pb systematics (Tera and Wasserburg, 1974), and ~4.36 Ga from Sm–Nd model ages of KREEP samples (Lugmair and Carlson, 1978). Nyquist and Shih (1992) estimated an average value from various Rb–Sr model ages for KREEP to be 4.42 ± 0.14 Ga. They suggested this as the best value for the timing of lunar global differentiation, with the uncertainty reflecting the possibility that this event did not occur at a sharply defined time and that some Rb–Sr fractionation may have occurred during the petrogenesis of KREEP basalts. Finally, combined $^{147}Sm-^{143}Nd$

and ^{146}Sm–^{142}Nd systematics for lunar basaltic samples indicate that the sources of these basalts were established in the lunar mantle $\sim$200 Myr after the beginning of the solar system (Nyquist *et al.*, 1995b; Rankenburg *et al.*, 2006), a timescale that is consistent with the others discussed above.

4.4.1.2 Lunar bombardment history

The flux of impactors on the Moon as a function of time is a topic that is highly debated and is of great interest since it has implications for the bombardment history of the Earth, which may in turn have played a role in the evolution of the Earth's atmosphere and in the initiation and evolution of life on this planet. A distinct spike in the bombardment history of the Moon or a "lunar cataclysm" at $\sim$3.9 Ga was first explicitly hypothesized on the basis of ^{39}Ar–^{40}Ar, U–Pb, and Rb–Sr systematics in rocks from the Apollo 15, 16, and 17 landing sites that appeared to have been reset or disturbed at this time (Tera *et al.*, 1974). Subsequent studies of ^{39}Ar–^{40}Ar ages of the Apollo and Luna impact melt rocks further demonstrated the lack of ages significantly older than $\sim$4.0 Ga (e.g., Dalrymple and Ryder, 1991, 1993, 1996; Swindle *et al.*, 1991), thus supporting this hypothesis. However, since the Apollo and Luna samples represent only a very small proportion of the lunar crust, they may be dominated by the effects of a few large impacts. Lunar meteorites provide a potential source of a more random, possibly less biased, selection of material from the Moon. Recent ^{39}Ar–^{40}Ar investigations of impact melt clasts from several lunar meteorites have yielded ages that are typically <3.9 Ga (Fernandes *et al.*, 2000; Cohen *et al.*, 2000, 2005; Gnos *et al.*, 2004). Although these studies support the cataclysm hypothesis in that there are no ages older than $\sim$4.0 Ga, they also indicate that, rather than dropping sharply to a nearly constant rate (as had been suggested by some studies of the Apollo samples; e.g., Guinness and Arvidson, 1977; Bogard *et al.*, 1994), the impact flux was declining gradually in the $\sim$3 or so billion years following the enhanced period of bombardment at $\sim$3.9 Ga. A recent ^{39}Ar–^{40}Ar study of glass spherules from the Apollo 14 soils also suggests a gradual decline in the impact flux (by a factor of $\sim$2–3) between $\sim$3.5 and $\sim$0.5 Ga (Culler *et al.*, 2000), followed by a marked enhancement in the impact flux (by a factor of $\sim$3–5) within the last $\sim$400–500 Myr. Another similar study of glass spherules from Apollo 12 soils (Levine *et al.*, 2005) was consistent with, but did not require a recent increase in the bombardment rate. However, this interpretation has been questioned (Horz, 2000), and remains one of the controversial issues related to the lunar bombardment history that has yet to be resolved. A more thorough discussion of the

impact ages of a variety of lunar samples obtained with various chronometers based on long-lived radionuclides (particularly the ^{39}Ar–^{40}Ar method), including the implications for the bombardment history of the Moon, has been provided in Chapter 1.21.

4.4.2 Timescales for the Evolution of Mars from Chronology of Martian Meteorites

The martian meteorites represent a variety of volcanic to subvolcanic as well as plutonic igneous rocks that are broadly categorized, based on their petrologic and geochemical characteristics, into four groups: the basaltic and lherzolitic shergottites, the clinopyroxenitic nakhlites, the dunitic chassignites, and the orthopyroxenite ALH84001. Chronological investigations of these martian meteorites and their implications for the evolution of Mars have been thoroughly reviewed by Nyquist *et al.* (2001) and Borg and Drake (2005). As such, only a brief summary is presented here.

Figure 13 summarizes the absolute ages for various events in martian history based on the radiometric dating of the martian meteorites and their components. The oldest formation age of $\sim$4.5 Ga is yielded by a ^{147}Sm–^{143}Nd internal isochron for the ALH84001 orthopyroxenite (Jagoutz *et al.*, 1994; Nyquist *et al.*, 1995a). All other dated martian meteorites have internal isochron ages (based mostly on the Sm–Nd and Rb–Sr chronometers) that are younger than $\sim$1.3 Ga, with the youngest samples being only $\sim$170 Ma. Based on their geochemical features (particularly the trace- and minor-element zonations in their minerals; e.g., Jones, 1986), these ages are generally interpreted to reflect the crystallization ages of these samples, thus indicating that Mars may still be a geologically active planet. A recent study has suggested that the young ages of the shergottites may in fact reflect the timing of secondary alteration, and that their formation ages are close to $\sim$4.0 Ga (Bouvier *et al.*, 2005). This interpretation requires that the late-stage minerals (i.e., phosphates) in these shergottites be affected by this secondary alteration event. However, studies of trace-element microdistributions have shown that these phosphates were formed by closed system fractional crystallization of the shergottite parent melts and that they preserve their original igneous compositions (Wadhwa *et al.*, 1994). As such, at the present time, the interpretation of the Sm–Nd and Rb–Sr internal isochrons ages of the shergottites as their crystallization ages is the most likely. Therefore, as discussed by Borg and Drake (2005) and summarized in Figure 13, there is evidence from the martian meteorites for igneous events on Mars at $4,300 \pm 130$, $1,327 \pm 29$, 575 ± 7, 474 ± 11, 332 ± 9, and 174 ± 2 Ma.

Besides primary igneous events, the martian meteorites also provide evidence for secondary

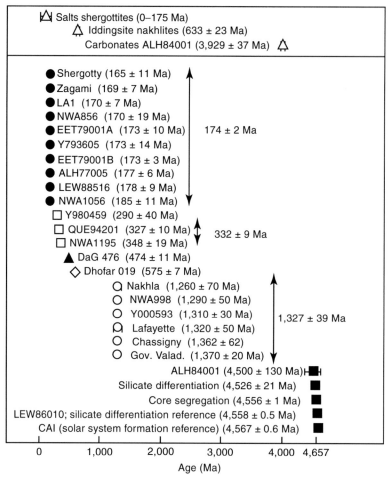

Figure 13 Absolute ages of events in Mars' history based on radiometric dating of martian meteorites and their components. Crystallization ages of the shergottites (~170–575 Ma; filled circles, open squares, filled triangle, and open diamond), the nakhlites (~1.3 Ga; open circles), and the orthopyroxenite ALH84001 (~4.5 Ga) are shown. The ages of aqueous alteration events recorded by secondary products in the martian meteorites are shown as the open triangles. Also shown are the ages of global differentiation events (core formation and silicate differentiation) based on the $^{147,146}Sm–^{143,142}Nd$ and $^{182}Hf–^{182}W$ systematics of the martian meteorites. Reproduced with permission from Borg and Drake (2005).

events on Mars. A $^{39}Ar–^{40}Ar$ age in the range of ~3.8–4.1 Ga for the ALH84001 orthopyroxenite has been interpreted as possibly reflecting the time of heavy bombardment on Mars coincident with a similar event on the Moon (Ash et al., 1996; Turner et al., 1997). Moreover, carbonates in this rock, thought to be precipitated as a result of interaction with an aqueous fluid in a near-surface environment, have also been dated at ~3.9 Ga based on Rb–Sr and U–Pb systematics (Borg et al., 1999b). In addition to the carbonates in ALH84001, there are other products of near-surface aqueous alteration in the other martian meteorites, in particular, a hydrous mineral (iddingsite) in the nakhlites (e.g., Bunch and Reid, 1975; Treiman et al., 1993) and a variety of secondary minerals (including carbonates, sulfates, and clays) in the shergottites (e.g., Gooding et al., 1988, 1990). Based on Rb–Sr and K–Ar systematics

in iddingsite-rich fractions (Shih et al., 1998, 2002; Swindle et al., 2000), it has been argued that this hydrous secondary phase was formed at ~600–700 Ma. The only age constraint on the secondary alteration products in the shergottites may be obtained from consideration of the crystallization ages of the particular samples (~175 Ma) in which these are found; as such, it is estimated that these secondary minerals were formed at some time <175 Ma. Based on the abundances of alteration products in the martian meteorites and their discrete formation time intervals, Borg and Drake (2005) argued that aqueous fluids were present episodically, and not continuously, in the near-surface environment on Mars.

Finally, whole-rock Rb–Sr and U–Pb systematics of the martian meteorites indicate that global silicate differentiation on Mars occurred close to ~4.5 Ga (Shih et al., 1982; Chen and

Wasserburg, 1986). Using an approach similar to that applied by Nyquist *et al.* (1995b) to the lunar basaltic samples (i.e., combined ^{147}Sm–^{143}Nd and ^{146}Sm–^{142}Nd systematics), a more precise estimate of $4,525 \pm 20$ Ma for the timing of major silicate differentiation on Mars that established the source reservoir of the shergottites has been made (Borg *et al.*, 2003; Foley *et al.*, 2005). The short-lived ^{146}Sm–^{142}Nd and ^{182}Hf–^{182}W chronometers further indicate that the nakhlite mantle source may have been established contemporaneously with that of the shergottites, or it may have pre-dated this event by a few tens of millions of years (Foley *et al.*, 2005).

4.5 CONCLUSIONS

4.5.1 A Timeline for Solar System Events

From the application of various chronometers based on the long-lived radionuclides to meteoritic and planetary materials, the following may be inferred as the sequence of events that occurred in the history of the solar system:

1. The earliest solids to form in the solar protoplanetary disk were the refractory CAIs, which formed at $4,567.1 \pm 0.2$ Ma. This represents the minimum age of the solar system.
2. Chondrules from CV and CR chondrites began forming more or less contemporaneously with CAIs ($4,566.7 \pm 1.0$ Ma for Allende chondrules), and continued to form for at least another 2–3 Myr.
3. Accretion and differentiation of planetesimals also began almost contemporaneously with CAIs, with basalts forming on the surfaces of these bodies well within ~ 1 Myr of CAI formation. The accretion process for planetesimals is also likely to have continued for a few million years (with at least some chondrite parent bodies possibly being assembled a few million years after CAIs).
4. Energetic collisions between the accreted planetary embryos resulted in the formation of impact-generated chondrules several millions of years after CAI formation. Specifically, chondrules from the CB chondrites, which are thought to result from such a process, were formed ~ 5 Myr after CAIs.
5. Thermal processing of accreted undifferentiated planetesimals began early (e.g., at $\sim 4,563$ Ma for the parent body of the Ste. Marguerite H4 chondrite), but this process continued for tens of millions of years after the beginning of the solar system. Metamorphism on chondrite parent bodies is likely to have extended for at least ~ 60 Myr after CAI formation. Aqueous alteration of the CI chondrite parent body is likely to have begun almost as soon as it was accreted.
6. Basaltic melts continued to be generated on differentiated planetesimals for a period of ~ 10 Myr after solar system formation. In the case of the EPB, these basalts were then subjected to a complex and protracted (lasting tens of millions of years) history of thermal processing, most likely resulting from large impacts on the surface of the EPB. Based on the ages of the cumulate eucrites, it is inferred that either igneous activity may have lasted for ~ 100–150 Myr on the EPB or isotopic systems in these samples record slow cooling at depth in the EPB.
7. The iron-nickel cores of some differentiated planetesimals began crystallization and solidification within ~ 10 Myr of CAI formation, but this process on other differentiated parent bodies may have extended for another tens of millions of years.
8. Global silicate differentiation on the Moon and Mars and establishment of mantle source reservoirs for lunar and martian basalts occurred ~ 200 Myr and within ~ 50 Myr after the beginning of the solar system, respectively.
9. The earliest crustal (highlands) rocks on the Moon was formed at ~ 4.5 Ga. The oldest basaltic lunar materials have ages of ~ 4.2 Ga, although most mare basalts crystallized after ~ 3.9. Mare volcanism continued at least till ~ 2.9 Ga.
10. Crystallization ages of martian meteorites range from ~ 4.5 Ga for the ALH84001 orthopyroxenite to ~ 170 Ma for some shergottites, suggesting that magmatic activity on Mars began early and may still continue. Aqueous alteration events recorded in the martian meteorites are estimated to be have occurred at ~ 3.9 Ga, 600–700 Ma, and <175 Ma. These discrete formation times suggest that aqueous fluids were present episodically, and not continuously, in the near-surface environment on Mars.
11. Members belonging to several groups of meteorites (eucrites and chondrites) and planetary materials (lunar samples and the martian meteorite ALH84001) record impact ages that are consistent with a period of heavy bombardment in the inner solar system that peaked close to ~ 4 Ga. Peaks in the impact ages at <1.3 Ga for the chondrites and close to ~ 4.5 Ga for the eucrites may point toward a few large impact events that occurred on their parent bodies.

4.5.2 Outlook and Future Prospects

Among the long-lived chronometers, the one based on the ^{235,238}U–^{207,206}Pb systems provides the best precision. With extensive leaching procedures to effectively remove the common lead component and use of a double spike to improve the analytical precision, it is now possible to attain a precision better than $\pm\,0.5$ Myr on ^{207}Pb–^{206}Pb ages for radiogenic samples. As such, more future studies of early solar system chronology are likely to focus on the utilization of this chronometer as the absolute anchor for other chronometers based on short-lived radionuclides, which also have the potential for providing sub-Myr time resolution for events occurring in the earliest history of the solar system. Nevertheless, there are numerous assumptions and complexities involved in the utilization of chronometers based on the short-lived radionuclides, and the U–Pb system is not an appropriate chronometer for many types of meteoritic materials (e.g., those that have relatively unradiogenic lead-isotope compositions). As such, obtaining the absolute ages of a variety of meteorites and their components using various long-lived chronometers will remain a high priority for understanding their formation histories in the early solar system, particularly if future developments allow higher precision and accuracy for these absolute ages. These future developments would include analytical advances in mass spectrometric techniques for isotope ratio measurements and the precise and accurate determination of decay constants of some of the long-lived radionuclides. Furthermore, there are as yet unresolved issues that are likely to be only addressed through chronological studies based on the long-lived radionuclides. In particular, better constraints on the impact flux in the inner solar system will only be possible through additional and more extensive studies of the impact ages (mostly from ^{39}Ar–^{40}Ar, but possibly also using other chronometers such as ^{87}Rb–^{87}Sr and ^{235,238}U–^{207,206}Pb) of a variety of meteoritic and planetary materials.

REFERENCES

Allègre C. J., Birck J.-L., Fourcade S., and Semet M. P. (1975) Rubidium-87/strontium-87 age of Juvinas basaltic achondrite and early igneous activity in the solar system. *Science* **187**, 436–438.

Allègre C. J., Manhès G., and Göpel C. (1995) The age of the Earth. *Geochim. Cosmochim. Acta* **59**, 1445–1456.

Alibert C., Norman M. D., and McCulloch (1994) An ancient Sm–Nd age for a ferroan noritic anorthosite clast from lunar breccia 67016. *Geochim. Cosmochim. Acta* **58**, 2921–2926.

Amelin Y. and Davis W. J. (2005) Geochemical test for branching decay of ^{176}Lu. *Geochim. Cosmochim. Acta* **69**, 465–473.

Amelin Y., Ghosh A., and Rotenberg E. (2005) Unraveling the evolution of chondrite parent asteroids by precise U–Pb dating and thermal modeling. *Geochim. Cosmochim. Acta* **69**, 505–518.

Amelin Y., Krot A. N., Hutcheon I. D., and Ulyanov A. A. (2002) Lead isotopic ages of chondrules and calcium-aluminum-rich inclusions. *Science* **297**, 1678–1683.

Amelin Y., Krot A. N., and Twelker E. (2004) Pb isotopic ages of the CB chondrite Gujba, and the duration of the chondrule formation interval. *Geochim. Cosmochim. Acta* **68**, A759.

Amelin Y. and Pravdivtseva O. (2005) U–Pb age of the Acapulco phosphate: testing the calibration of the I–Xe chronometer. *Meteorit. Planet. Sci* **40**, A16.

Amelin Y., Wadhwa M., and Lugmair G. (2006) Pb-isotopic dating of meteorites using ^{202}Pb–^{205}Pb double-spike: comparison with other high-resolution chronometers. In *Lunar Planet. Sci.* **XXXVII**, #1970. The Lunar and Planetary Institute, Houston (CD-ROM).

Anders E. and Grevesse N. (1989) Abundances of the elements: meteoritic and solar. *Geochim. Cosmochim. Acta* **53**, 197–214.

Ash R. D., Knott S. F., and Turner G. (1996) A 4 Gyr shock age for a martian meteorite and implications for the cratering history of Mars. *Nature* **380**, 57–59.

Baker J., Bizzarro M., Wittig N., Connelly J., and Haack H. (2005) Early planetesimal melting from an age of 4.5662 Gyr for differentiated meteorites. *Nature* **436**, 1127–1130.

Birck J.-L. and Allègre C. J. (1978) Chronology and chemical history of the parent body of the basaltic achondrites studied by the ^{87}Rb–^{87}Sr method. *Earth Planet. Sci. Lett.* **39**, 37–51.

Bizzarro M., Baker J. A., Haack H., Ulfbeck D., and Rosing M. (2003) Early history of Earth's crust-mantle system inferred from hafnium isotopes in chondrites. *Nature* **421**, 931–933.

Blichert-Toft J., Boyet M., Télouk P., and Albarède F. (2002) ^{147}Sm–^{143}Nd and ^{176}Lu–^{176}Hf in eucrites and the differentiation of the HED parent body. *Earth Planet. Sci. Lett.* **204**, 167–181.

Bogard D. D. (1995) Impact ages of meteorites: a synthesis. *Meteoritics* **30**, 244–268.

Bogard D. D. and Garrison D. H. (2003) ^{39}Ar–^{40}Ar ages of eucrites and thermal history of asteroid 4 Vesta. *Meteorit. Planet. Sci.* **38**, 669–710.

Bogard D. D., Garrison D. H., Jordan J. L., and Mittlefehldt D. W. (1990) ^{39}Ar–^{40}Ar dating of mesosiderites: evidence for major parent disruption <4 Ga ago. *Geochim. Cosmochim. Acta* **54**, 2549–2564.

Bogard D. D., Garrison D. H., Shih C.-Y., and Nyquist L. E. (1994) ^{39}Ar–^{40}Ar dating of two lunar granites—the age of Copernicus. *Geochim. Cosmochim. Acta* **58**, 3093–3100.

Bogard D. D. and Hirsch W. C. (1980) ^{39}Ar–^{40}Ar dating, Ar diffusion properties, and cooling rate determinations of severely shocked chondrites. *Geochim. Cosmochim. Acta* **44**, 1667–1682.

Bogard D. D., Husain L., and Wright R. J. (1976) ^{40}Ar–^{39}Ar dating of collisional events in chondrite parent bodies. *J. Geophys. Res.* **81**, 5664–5678.

Bogdanovski O., Jagoutz E., (1996) Sm–Nd system in the Divnoe meteorite. In *Lunar Planet. Sci.* **XXVII**. The Lunar and Planetary Institute, Houston, pp. 129–130.

Borg L. and Drake M. J. (2005) A review of meteorite evidence for the timing of magmatism and of surface or near-surface liquid water on Mars. *J. Geophys. Res.* **110**, E12S03.

Borg L., Norman M., Nyquist L., Bogard D., Snyder G., Taylor L., and Lindstrom M. (1999a) Isotopic studies of ferroan anorthosite 62236: a young lunar crustal rock from a light rare-earth-element-depleted source. *Geochim. Cosmochim. Acta* **63**, 2679–2691.

Borg L. E., Connelly J. N., Nyquist L. E., Shih C.-Y., Wiesmann H., and Reese Y. (1999b) The age of the carbonates in martian meteorite ALH84001. *Science* **268**, 90–94.

Borg L. E., Nyquist L. E., Wiesmann H., Shih C.-Y., and Reese Y. (2003) The age of Dar al Gani 476 and the differentiation history of the martian meteorites inferred from their radiogenic isotope systematics. *Geochim. Cosmochim. Acta* **67**, 3519–3536.

Borg L. E., Shearer C. K., Asmerom Y., and Papike J. J. (2004) Prolonged KREEP magmatism on the Moon indicated by the youngest dated lunar igneous rock. *Nature* **432**, 209–211.

Bouvier A., Blichert-Toft J., Vervoort J. D., and Albarede F. (2005) The age of SNC meteorites and the antiquity of the martian surface. *Earth Planet. Sci. Lett.* **240**, 221–233.

Brannon J. C., Podosek F. A., and Lugmair G. W. (1988) Initial ^{87}Sr/^{86}Sr and Sm–Nd chronology of chondritic meteorites. *Proc. 18th Lunar Planet. Sci. Conf.* 555–564.

Brearley A. J. (2006) The action of water. In *Meteorites and the Early Solar System Vol. II* (eds. D. Lauretta and H. Y. McSween Jr.). University of Arizona Press, Tucson, AZ (in press).

Bunch T. E. and Reid A. M. (1975) The nakhlites. 1: Petrology and mineral chemistry. *Meteoritics* **10**, 303–315.

Carlson R. W. and Lugmair G. W. (1988) The age of ferroan anorthosite 60025: oldest crust on a young Moon? *Earth Planet. Sci. Lett.* **90**, 119–130.

Carlson R. W. and Lugmair G. W. (2000) Timescales of planetesimal formation and differentiation based on extinct and extant radionuclides. In *The Origin of the Earth and Moon* (eds. R. M. Canup and K. Righter). University of Arizona Press, Tucson, AZ, pp. 25–44.

Chen J. H., Papanastassiou D. A., and Wasserburg G. J. (2002) Re–Os and Pd–Ag systematics in Group IIAB irons and in pallasites. *Geochim. Cosmochim. Acta* **66**, 3793–3810.

Chen J. H. and Tilton G. R. (1976) Isotopic lead investigations on the Allende carbonaceous chondrite. *Geochim. Cosmochim. Acta* **40**, 635–643.

Chen J. H. and Wasserburg G. J. (1981) The isotopic composition of uranium and lead in Allende inclusions and meteoritic phosphates. *Earth Planet. Sci. Lett.* **5**, 21–51.

Chen J. H. and Wasserburg G. J. (1985). U–Th–Pb isotopic studies on meteorite ALHA 81005 and Ibitira. In *Lunar Planet. Sci.* **XVI**. The Lunar and Planetary Institute, Houston, pp. 119–120.

Chen J. H. and Wasserburg G. J. (1986) Formation ages and evolution of Shergotty and its parent planet from U–Th–Pb systematics. *Geochim. Cosmochim. Acta* **50**, 955–968.

Ciesla F. J. (2005) Chondrule forming processes—an overview. In *Chondrites and the Protoplanetary Disk* (eds. A. N. Krot, E. R. D. Scott, and B. Reipurth), ASP Conference Series. Astronomical Society of the Pacific, San Francisco, vol. 341 pp. 558–587.

Clayton R. N. and Mayeda T. K. (1996) Oxygen isotope studies of achondrites. *Geochim. Cosmochim. Acta* **60**, 1999–2017.

Cohen B. A., Swindle T. D., and Kring D. A. (2000) Support for the lunar cataclysm hypothesis from lunar meteorite impact melt ages. *Science* **290**, 1754–1756.

Cohen B. A., Swindle T. D., and Kring D. A. (2005) Geochemistry and ^{40}Ar–^{39}Ar geochronology of impact-melt clasts in feldspathic lunar meteorites: implications for lunar bombardment history. *Meteorit. Planet. Sci.* **40**, 755–777.

Cook D. L., Walker R. J., Horan M. F., Wasson J. T., and Morgan J. W. (2004) Pt–Re–Os systematics of group IIAB and IIIAB iron meteorites. *Geochim. Cosmochim. Acta* **68**, 1413–1431.

Culler T. S., Becker T. A., Muller R. A., and Renne P. R. (2000) Lunar impact history from $^{40}Ar/^{39}Ar$ dating of glass spherules. *Science* **287**, 1785–1788.

Dalrymple G. B. and Ryder G. (1991) $^{40}Ar/^{39}Ar$ ages of six Apollo 15 impact melt rocks by laser step heating. *Geophys. Res. Lett.* **18**, 1163–1166.

Dalrymple G. B. and Ryder G. (1993) $^{40}Ar/^{39}Ar$ age spectra of Apollo 15 impact melt rocks by laser step heating and their bearing on the history of lunar basin formation. *J. Geophys. Res.* **98**, 13085–13095.

Dalrymple G. B. and Ryder G. (1996) Argon-40/argon-39 age spectra of Apollo 17 highlands breccia samples by laser step heating and the age of the Serenitatis basin. *J. Geophys. Res.* **101**, 26069–26084.

Deutsch A. and Schärer U. (1994) Dating terrestrial impact events. *Meteoritics* **29**, 301–322.

Evensen N. M., Carter S. R., Hamilton P. J., O'Nions R. K., and Ridley W. I. (1979) A combined chemical-petrological study of separated chondrules from the Richardton meteorite. *Earth Planet. Sci. Lett.* **42**, 223–236.

Fernandes V. A., Burgess R., and Turner G. (2000) Laser argon-40–argon-39 age studies of Dar al Gani 262 lunar meteorite. *Meteorit. Planet. Sci.* **25**, 1355–1364.

Fernandes V. A., Burgess R., and Turner G. (2003) ^{40}Ar–^{39}Ar chronology of lunar meteorites Northwest Africa 032 and 773. *Meteorit. Planet. Sci.* **38**, 555–564.

Foley C. N., Wadhwa M., Borg L. E., Janney P. E., Hines R., and Grove T. L. (2005) The early differentiation history of Mars from ^{182}W–^{142}Nd isotope systematics in the SNC meteorites. *Geochim. Cosmochim. Acta* **69**, 4557–4571.

Galer S. J. G. and Lugmair G. W. (1996) Lead isotope systematics of noncumulate eucrites. *Meteorit. Planet. Sci.* **31**, A47–A48.

Ghosh A. and McSween H. Y. M., Jr. (1998) A thermal model for the differentiation of Asteroid 4 Vesta, based on radiogenic heating. *Icarus* **134**, 187–206.

Gnos E., Hofmann B. A., Al A- K., Lorenzetti S., Eugster O., Whitehouse M. J., Villa I. M., Jull A. J. T., Eikenberg J., Spettel B., Krähenbühl U., Franchi I. A., and Greenwood R. C. (2004) Pinpointing the source of a lunar meteorite: implications for the evolution of the Moon. *Science* **305**, 657–659.

Gooding J. L., Aggrey K. E., and Muenow D. W. (1990) Volatile compounds in shergottites and nakhlite meteorites. *Meteoritics* **25**, 281–289.

Gooding J. L., Wentworth S. J., and Zolensky M. E. (1988) Calcium carbonate and sulfate of possible extraterrestrial origin in the EETA 79001 meteorite. *Geochim. Cosmochim. Acta* **52**, 909–915.

Goodrich C. A. and Lugmair G. W. (1995) Stalking the LREE-enriched component in ureilites. *Geochim. Cosmochim. Acta* **59**, 2609–2620.

Göpel C., Manhès G., and Allègre C. J. (1991) Constraints on the time of accretion and thermal evolution of chondrite parent body. *Meteoritics* **26**, 338.

Göpel C., Manhès G., and Allègre C. J. (1992) U–Pb study of the Acapulco meteorite. *Meteoritics* **27**, 226.

Göpel C., Manhès G., and Allègre C. J. (1994) U–Pb systematics of phosphates from equilibrated ordinary chondrites. *Earth Planet. Sci. Lett.* **121**, 153–171.

Gray C. M., Papanastassiou D. A., and Wasserburg G. J. (1973) The identification of early condensates from the solar nebula. *Icarus* **20**, 213–239.

Grier J. A., Kring D. A., Swindle T. D., Rivkin A. S., Cohen B. A., and Britt D. T. (2004) Analyses of the chondritic meteorite Orvino (H6): insights into the origins and evolution of shocked H chondrite material. *Meteorit. Planet. Sci.* **39**, 1475–1493.

Guinness E. A. and Arvidson R. E. (1977) On the constancy of the lunar cratering flux over the past 3.3×10^9 yr. *Proc. 8th Lunar Sci. Conf.* 3475–3494.

Haack H., Farinella P., Scott E. R. D., and Keil K. (1996a) Meteoritic, asteroidal and theoretical constraints on the 500 Ma disruption of the L chondrite parent body. *Icarus* **119**, 182–191.

Haack H., Scott E. R. D., Love S. G., Brearley A. J., and McCoy T. J. (1996b) Thermal histories of IVA stony-iron meteorites: evidence for asteroid fragmentation and reaccretion. *Geochim. Cosmochim. Acta* **60**, 3103–3113.

Hanan B. B. and Tilton G. R. (1987) 60025: Relict of primitive lunar crust? *Earth Planet. Sci. Lett.* **84**, 15–21.

Hohenberg C. M., Hudson B., Kennedy B. M., and Podosek F. A. (1981) Noble gas retention chronologies for the St Séverin meteorite. *Geochim. Cosmochim. Acta* **45**, 535–546.

Horan M. F., Smoliar M. I., and Walker R. J. (1998) ^{182}W and ^{187}Re–^{187}Os systematics of iron meteorites: chronology for melting, differentiation, and crystallization in asteroids. *Geochim. Cosmochim. Acta* **62**, 545–554.

Horz F. (2000) Time variable cratering rates? *Science* **288**, 2095a.

Huss G. R., Rubin A. E., and Grossman J. N. (2006) Thermal metamorphism in chondrites. In *Meteorites and the Early Solar System Vol. II* (eds. D. Lauretta and H. Y. McSween Jr.). University of Arizona Press, Tucson, AZ (in press).

Hutcheon I. D. and Olsen E. J. (1991) Cr isotopic composition of differentiated meteorites: a search for ^{53}Mn. In *Lunar Planet. Sci.* **XXII**. The Lunar and Planetary Institute, Houston, pp. 605–606.

Hutcheon I. D., Olsen E., Zipfel J., and Wasserburg G. J. (1992) Chromium isotopes in differentiated meteorites: evidence for ^{53}Mn. In *Lunar Planet. Sci.* **XXIII**. The Lunar and Planetary Institute, Houston, pp. 565–566.

Jacobsen S. B. and Wasserburg G. J. (1984) Sm–Nd isotopic evolution of chondrites and achondrites, II. *Earth Planet. Sci. Lett.* **67**, 137–150.

Jagoutz E., Jotter R., Kubny A., Varela M. E., Zartman R., Kurat G., and Lugmair G. W. (2003) Cm?–U–Th–Pb isotopic evolution of the D'Orbigny angrite. *Meteorit. Planet. Sci.* **38**, A81.

Jagoutz E., Sorowka A., Vogel J. D., and Wänke H. (1994) ALH84001: alien or progenitor of the SNC family? *Meteoritics* **29**, 478–479.

Jones J. H. (1986) A discussion of isotopic systematics and mineral zoning in the shergottites: evidence for a 180 m.y. igneous crystallization age. *Geochim. Cosmochim. Acta* **50**, 969–977.

Kaneoka I. (1981) ^{40}Ar–^{39}Ar ages of Antarctic meteorites: Y-74191, Y-75258, Y-7308, Y-74450, and ALH-765. *Proc. 6th NIPR Symp. Antarct. Met.* 250–263.

Keil K., Fodor R. V., Starzyk P. M., Schmidt R. A., Bogard D. D., and Husain L. (1980) A 3.6 b.y.-old impact melt rock fragment in the Plainview chondrite: implications for the age of the H-group chondrite parent body regolith formation. *Earth Planet. Sci. Lett.* **51**, 235–247.

Kita N. T., Huss G. R., Tachibana S., Amelin Y., Nyquist L. E., and Hutcheon I. D. (2005) Constraints on the origin of chondrules and CAIs from short-lived and long-lived radionuclides. In *Chondrites and the Protoplanetary Disk* (eds. A. N. Krot, E. R. D. Scott, and B. Reipurth), ASP Conference Series. Astronomical Society of the Pacific, San Francisco, vol. 341, pp. 558–587.

Kring D. A., Swindle T. D., Britt D. T., and Grier J. A. (1996) Cat Mountain: a meteoritic sample of an impact-melted asteroid regolith. *J. Geophys. Res.* **101**, 29353–29371.

Krot A. N., Amelin Y., Cassen P., and Meibom A. (2005) Young chondrules in CB chondrites from a giant impact in the early solar system. *Nature* **436**, 989–992.

Kumar A., Gopalan K., and Bhandari N. (1999) ^{147}Sm–^{143}Nd and ^{87}Rb–^{87}Sr ages of the eucrite Piplia Kalan. *Geochim. Cosmochim. Acta* **63**, 3997–4001.

Levine J., Becker T. A., Muller R. A., and Renne P. R. (2005) 40Ar/39Ar dating of Apollo 12 impact spherules. *Geophys. Res. Lett.* **32**, L15201.

Lindner M., Leich D. A., Russ G. P., Bazan J. M., and Borg R. J. (1989) Direct determination of the half-life of ^{187}Re. *Geochim. Cosmochim. Acta* **53**, 1597–1606.

Loss R. D., Lugmair G. W., Davis A. M., and MacPherson G. J. (1994) Isotopically distinct reservoirs in the solar nebula: isotopic anomalies in Vigarano meteorite inclusions. *Astrophys. J.* **436**, L193–L196.

Lugmair G. W. (1974) Sm–Nd ages: a new dating method. *Meteoritics* **9**, 369.

Lugmair G. W. and Carlson R. W. (1978) The Sm–Nd history of KREEP. *Proc. 9th Lunar Planet. Sci. Conf.* 689–704.

Lugmair G. W. and Galer S. J. G. (1992) Age and isotopic relationships among angrites Lewis Cliff 86010 and Angra dos Reis. *Geochim. Cosmochim. Acta* **56**, 1673–1694.

Lugmair G. W., Galer S. J. G., and Carlson R. W. (1991) Isotope systematics of cumulate eucrite EET-87520. *Meteoritics* **26**, 368.

Lugmair G. W. and Marti K. (1977) Sm–Nd–Pu timepieces in the Angra dos Reis meteorite. *Earth Planet. Sci. Lett.* **35**, 273–284.

Lugmair G. W., Scheinin N. B., and Carlson R. W. (1977) Sm–Nd systematics of the Serra de Magé eucrite. *Meteoritics* **10**, 300–301.

Lugmair G. W. and Shukolyukov A. (1998) Early solar system timescales according to ^{53}Mn–^{53}Cr systematics. *Geochim. Cosmochim. Acta* **62**, 2863–2886.

Macdougall J. D., Lugmair G. W., and Kerridge J. F. (1984) Early solar system aqueous activity: Sr isotope evidence from the Orgueil CI meteorite. *Nature* **307**, 249–251.

Manhès G., Göpel C., and Allègre C. J. (1987) High resolution chronology of the early solar system based on lead isotopes. *Meteoritics* **22**, 453–454.

Manhès G., Minster J. F., and Allègre C. J. (1978) Comparative U–Th–Pb and Rb–Sr study of the Saint-Sèverin amphoterite: consequences for early solar system chronology. *Earth Planet. Sci. Lett.* **39**, 14–24.

Markowski A., Quitté G., Kleine T., and Halliday A. N. (2006) Tungsten isotopic composition of iron meteorite: chronological constraints vs. cosmogenic effects. *Earth Planet. Sci. Lett.* **214**, 1–15.

McCoy T. J., Keil K., Clayton R. N., Mayeda T. K., Bogard D. D., Garrison D. H., Huss G. R., Hutcheon I. D., and Wieler R. (1996) A petrologic, chemical, and isotopic study of Monument Draw and comparison with other acapulcoites: evidence for formation by incipient partial melting. *Geochim. Cosmochim. Acta* **60**, 2681–2708.

McCoy T. J., Keil K., Clayton R. N., Mayeda T. K., Bogard D. D., Garrison D. H., and Wieler R. (1997a) A petrologic and isotopic study of lodranites: evidence of early formation as partial melt residues from heterogeneous precursors. *Geochim. Cosmochim. Acta* **61**, 623–637.

McCoy T. J., Keil K., Muenow D. W., and Wilson L. (1997b) Partial melting and melt migration in the acapulcoite-lodranite parent body. *Geochim. Cosmochim. Acta* **61**, 639–650.

McSween H. Y., Jr., Sears D. W. G., and Dodd R. T. (1988) Thermal metamorphism. In *Meteorites and the Early Solar System* (eds. J. F. Kerridge and M. S. Matthews). University of Arizona Press, Tucson, AZ, pp. 102–113.

Metzler K., Bobe K. D., Palme H., Spettel B., and Stoffler D. (1995) Thermal and impact metamorphism on the HED parent asteroid. *Planet. Space Sci.* **43**, 499–525.

Mittlefehldt D. W., Lindstrom M. M., Bogard D. D., Garrison D. H., and Field S. W. (1996) Acapulco- and Lodran-like achondrites: petrology, geochemistry, chronology and origin. *Geochim. Cosmochim. Acta* **60**, 867–882.

Mittlefehldt D. W., McCoy T. J., Goodrich C. A., and Kracher A. (1998) Non-chondritic meteorites from asteroidal bodies. In *Planetary Materials Planetary Materials* (ed. J. J. Papike). *Rev. Mineral.*, Mineralogical Society of America, Washington, DC, vol. 36, pp. 4-1–4-195.

Nakamura N., Morikawa N., Hutchison R., Clayton R. N., Mayeda T., Nagao K., Misawa K., Okano O., Yamamoto K., Yanai K., and Matsumoto Y. (1994) Trace element and isotopic characteristrics of inclusions in the Yamato ordinary chondrites Y-75097, Y-793241, and Y-794046. *Proc. 7th NIPR Symp. Antarct. Met.* 125–143.

Norman M. D., Borg L. E., Nyquist L. E., and Bogard D. D. (2003) Chronology, geochemistry, and petrology of a ferroan noritic anorthosites clast from Descartes breccia 67215: clues to the age, origin, structure, and impact history of the lunar crust. *Meteorit. Planet. Sci.* **38**, 645–661.

Nyquist L. E., Bansal B., Wiesmann H., and Shih C.-Y. (1994) Neodymium, strontium and chromium isotopic studies of the LEW 86010 and Angra dos Reis meteorites and the chronology of the angrite parent body. *Meteoritics* **29**, 872–885.

Nyquist L. E., Bansal B. M., Wiesmann H., and Shih C.-Y. (1995a) "Martians" young and old: Zagami and ALH84001. In *Lunar Planet. Sci.* **XXVI**. The Lunar and Planetary Institute, Houston, pp. 1065–1066.

Nyquist L. E., Bogard D. D., Garrison D. H., Bansal B., Wiesmann H., and Shih C.- Y. (1991) Thermal resetting of radiometric ages. In *Lunar Planet. Sci.* **XXII**. The Lunar and Planetary Institute, Houston, pp. 985–988.

Nyquist L. E., Bogard D. D., Shih C.-Y., Greshake A., Stöffler D., and Eugster E. (2001) Ages and geologic histories of martian meteorites. *Space Sci. Rev.* **96**, 105–164.

Nyquist L., Bogard D., Takeda H., Bansal B., Wiesmann H., and Shih C.-Y. (1997) Crystallization, recrystallization, and impact-metamorphic ages of eucrites Y792510 and Y791186. *Geochim. Cosmochim. Acta* **61**, 2119–2138.

Nyquist L. E., Bogard D. D., Wiesmann H., Bansal B., Shih C.-Y., and Morris R. M. (1990) Age of a eucrite clast from the Bholghati howardite. *Geochim. Cosmochim. Acta* **54**, 2195–2206.

Nyquist L. E. and Shih C.-Y. (1992) The isotopic record of lunar volcanism. *Geochim. Cosmochim. Acta* **56**, 2213–2234.

Nyquist L. E., Shih C. Y., Wiesmann H., and Mikouchi T. (2003a) Fossil ^{26}Al and ^{53}Mn in D'Orbigny and Sahara 99555 and the timescale for angrite magmatism. In *Lunar Planet. Sci.* **XXXIV**, #1388. The Lunar and Planetary Institute, Houston (CD-ROM).

Nyquist L. E., Takeda H., Bansal B. M., Shih C.-Y., Wiesmann H., and Wooden J. L. (1986) Rb-Sr and Sm-Nd internal isochron ages of a subophitic basalt clast and a matrix sample from the Y75011 eucrite. *J. Geophys. Res.* **91**, 8137–8150.

Nyquist L. E., Wiesmann H., Bansal B., Shih C.-Y., Keith J. E., and Harper C. L. (1995b) ^{46}Sm–^{142}Nd formation interval for the lunar mantle. *Geochim. Cosmochim. Acta* **59**, 2817–2837.

Papanastassiou D. A. and Wasserburg G. J. (1969) Initial strontium isotopic abundances and the resolution of small time differences in the formation of planetary objects. *Earth Planet. Sci. Lett.* **5**, 361–376.

Papanastassiou D. A. and Wasserburg G. J. (1978) Strontium isotopic anomalies in the Allende meteorite. *Geophys. Res. Lett.* **5**, 595–598.

Papanastassiou D. A., Wasserburg G. J., and Burnett D. S. (1970) Rb-Sr ages of lunar rocks from the Sea of Tranquility. *Earth Planet. Sci. Lett.* **8**, 1–19.

Papike J. J., Ryder G., and Shearer C. K. (1998) Lunar samples. In *Planetary Materials* (ed. J. J. Papike). *Rev. Mineral.*, Mineralogical Society of America, Washington, DC, vol. 36, 5-1–5-234.

Patchett P. J. and Tatsumoto M. (1980) Lu-Hf total-rock isochron for the eucrite meteorites. *Nature* **288**, 571–574.

Patterson C. C. (1955) The $^{207}Pb/^{206}Pb$ ages of some stone meteorites. *Geochim. Cosmochim. Acta* **7**, 151–153.

Patterson C. C. (1956) Age of meteorites and the Earth. *Geochim. Cosmochim. Acta* **10**, 230–237.

Perron C., Bourot-Denise M., Marti K., Kim S., and Crozaz G. (1988) The metal-phosphate connection in chondrites. *Meteoritics* **27**, 275.

Petaev M. I., Baruskova L. D., Lipschutz M. E., Wang M.-S., Arsikan A. A., Clayton R. N., and Mayeda T. K. (1994) The Divnoe meteorite: petrology, chemistry, oxygen isotopes and origin. *Meteoritics* **29**, 182–199.

Podosek F. A. and Brannon J. C. (1991) Chondrite chronology by initial $^{87}Sr/^{86}Sr$ in phosphates. *Meteoritics* **26**, 145–152.

Podosek F. A. and Huneke J. C. (1973) Argon 40-argon 39 chronology of four calcium-rich achondrites. *Geochim. Cosmochim. Acta* **37**, 667–684.

Podosek F. A. and Nichols R. H., Jr. (1997) Short-lived radionuclides in the solar nebula. In *Astrophyical Implications of the Laboratory Study of Presolar Materials* (eds. T. J. Bernatowicz and E. Zinner). American Institute of Physics, Woodbury, pp. 617–647.

Podosek F. A., Zinner E. K., MacPherson G. J., Lundberg L. L., Brannon J. C., and Fahey A. J. (1991) Correlated study of initial $^{87}Sr/^{86}Sr$ and Al/Mg isotopic systematics and petrologic properties in a suite of refractory inclusions from the Allende meteorite. *Geochim. Cosmochim. Acta* **55**, 1083–1110.

Prinzhofer A., Papanastassiou D. A., and Wasserburg G. J. (1992) Samarium-neodymium evolution of meteorites. *Geochim. Cosmochim. Acta* **56**, 797–815.

Rankenburg K., Brandon A. D., and Neal C. R. (2006) Neodymium isotope evidence for a chondritic composition of the Moon. *Science* **312**, 1369–1372.

Rubin A. E., Kallemeyn G. W., Wasson J. T., Clayton R. N., Mayeda T. K., Grady M., Verchovsky A. B., Eugster O., and Lorenzetti S. (2003) Formation of metal and silicate globules in Gujba: a new Bencubbin-like meteorite fall. *Geochim. Cosmochim. Acta* **67**, 3283–3298.

Scherer E., Münker C., and Mezger K. (2001) Calibration of the lutetium-hafnium clock. *Science* **293**, 683–687.

Shen J. J., Papanastassiou D. A., and Wasserburg G. J. (1996) Precise Re–Os determinations and systematics of iron meteorites. *Geochim. Cosmochim. Acta* **60**, 2887–2900.

Shih C.-Y., Nyquist L. E., Bogard D. D., McKay G. A., Wooden J. L., Bansal B., and Wiesmann H. (1982) Chronology and petrology of young achondrites, Shergotty, Zagami and ALHA77005: late magmatism on a geologically active planet. *Geochim. Cosmochim. Acta* **46**, 2323–2344.

Shih C.-Y., Nyquist L. E., Reese Y., and H. Wiesmann (1998) The chronology of the nakhlite Lafayette: Rb-Sr and Sm-Nd isotopic ages. In *Lunar Planet. Sci.* **XXXIV**, #1145. The Lunar and Planetary Institute, Houston (CD-ROM).

Shih C.-Y., Wiesmann H., Nyquist L. E., and Misawa K. (2002) Crystallization age of Antarctic nakhlite Y000593: further evidence for nakhlite launch pairing. *Antarct. Meteorites* **XXVII**, 151–153.

Shimoda G., Nakamura N., Kimura M., Kani T., Nohda S., and Yamamoto K. (2005) Evidence from the Rb–Sr system for 4.4 Ga alteration of chondrules in the Allende (CV3) parent body. *Meteorit. Planet. Sci.* **40**, 1059–1072.

Smoliar M. I. (1993) A survey of Rb–Sr systematics of eucrites. *Meteoritics* **28**, 105–113.

Smoliar M. I., Walker R. J., and Morgan J. W. (1996) Re-Os ages of group IIA, IIIA, IVA, and IVB iron meteorites. *Science* **271**, 1099–1102.

Snyder G. A., Borg L. E., Nyquist L. E., and Taylor L. A. (2000) Chronology and isotopic constraints on lunar evolution. In *Origin of the Earth and Moon* (eds. R. Canup and K. Righter). University of Arizona Press, Tucson, AZ, pp. 361–395.

Stephan T. and Jessberger E. K. (1988) ^{40}Ar–^{39}Ar ages of types 3 and 4, L and H chondrites from Antarctica. *Meteoritics* **23**, 373–377.

Stöffler D., Bischoff A., Buchvald V., and Rubin A. E. (1988) Shock effects in meteorites. In *Meteorites and the Early Solar System* (eds. J. F. Kerridge and M. S. Matthews). University of Arizona Press, Tucson, AZ, pp. 165–202.

Sugiura N. and Hoshino H. (2003) Mn–Cr chronology of five IIIAB iron meteorites. *Meteorit. Planet. Sci.* **38**, 117–143.

Swindle T. D., Spudis P. D., Taylor G. J., Korotev R. L., and Nichols R. H. (1991) Searching for Crisium basin ejecta—chemistry and ages of Luna 20 impact melts. *Proc. 21st Lunar Planet. Sci. Conf.* 167–181.

Swindle T. D., Treiman A. H., Lindstrom D. L., Burkland M. K., Cohen B. A., Grier J. A., Li B., and Olsen E. K. (2000) Noble gasses in iddingsite from the Lafayette meteorite: evidence for liquid water on Mars in the last few hundred million years. *Meteorit. Planet. Sci.* **35**, 107–115.

Takeda H., Morei H., and Bogard D. D. (1994) Mineralogy and ^{39}Ar–^{40}Ar age of an old pristine basalt: thermal history of the HED parent body. *Earth Planet. Sci. Lett.* **122**, 183–194.

Tatsumoto M., Knight R. J., and Allègre C. J. (1973) Time differences in the formation of meteorites as determined from the ratio of lead-207 to lead-206. *Science* **180**, 1279–1283.

Tatsumoto M., Unruh D. M., and Desborough G. A. (1976) U–Th–Pb and Rb–Sr systematics of Allende and U–Th–Pb systematics of Orgueil. *Geochim. Cosmochim. Acta* **40**, 617–634.

Taylor L. A., Shervais J. W., Hunter R. H., Shih C. Y., Nyquist L., Bansal B., Wooden J., and Laul J. C. (1983) Pre-4.2 AE mare basalt volcanism in the lunar highlands. *Earth Planet. Sci. Lett.* **66**, 33–47.

Tera F. and Carlson R. W. (1999) Assessment of the Pb–Pb and U–Pb chronometry of the early Solar System. *Geochim. Cosmochim. Acta* **63**, 1877–1889.

Tera F., Carlson R. W., and Boctor N. Z. (1997) Radiometric ages of basaltic achondrites and their relation to the early history of the solar system. *Geochim. Cosmochim. Acta* **61**, 1713–1731.

Tera F., Papanastassiou D. A., and Wasserburg G. J. (1974) Isotopic evidence for a terminal lunar cataclysm. *Earth Planet. Sci. Lett.* **22**, 1–21.

Tera F. and Wasserburg G. J. (1974) U–Th–Pb systematics on lunar rocks and inferences about lunar evolution and the age of the moon. *Proc. 5th Lunar Sci. Conf.* 1571–1599.

Tilton G. R. (1988) Age of the solar system. In *Meteorites and the Early Solar System* (eds. J. F. Kerridge and M. S. Matthews). University of Arizona Press, Tucson, AZ, pp. 249–258.

Tonui E. K., Ngo H. H., and Papanastassiou D. A. (2003) Rb–Sr and Sm–Nd study of the D'Orbigny angrite. In *Lunar Planet. Sci.* **XXXIV**, #1812. Lunar and Planetary Institute, Houston (CD-ROM).

Torigoye-Kita N., Misawa K., and Tatsumoto M. (1995a) U–Th–Pb and Sm–Nd isotopic systematics of the Goalpara ureilites: resolution of terrestrial contamination. *Geochim. Cosmochim. Acta* **59**, 381–390.

Torigoye-Kita N., Tatsumoto M., Meeker G. P., and Yanai K. (1995b) The 4.56 Ga age of the MET 78008 ureilite. *Geochim. Cosmochim. Acta* **59**, 2319–2329.

Treiman A. H., Barrett R. A., and Gooding J. L. (1993) Preterrestrial aqueous alteration of the Lafayette (SNC) meteorite. *Meteoritics* **28**, 86–97.

Turner G. (1969) Thermal histories of meteorites by the ^{39}Ar–^{40}Ar method. In *Meteorite Research* (ed. P. M. Milliman). Springer-Verlag, New York, pp. 407–417.

Turner G., Enright M. C., and Cadogan P. H. (1978) The early history of chondrite parent bodies inferred from ^{40}Ar–^{39}Ar ages. *Proc. 9th Lunar Planet. Sci. Conf.* 989–1025.

Turner G., Knott S. F., Ash R. D., and Gilmour J. D. (1997) Ar–Ar chronology of the martian meteorite ALH84001: evidence for the timing of early bombardment of Mars. *Geochim. Cosmochim. Acta* **61**, 3835–3850.

Unruh D. M., Nakamura N., and Tatsumoto M. (1977) History of the Pasamonte achondrite: relative susceptibility of the Sm–Nd, Rb–Sr, and U–Pb systems to metamorphic events. *Earth Planet. Sci. Lett.* **37**, 1–12.

Wadhwa M. and Lugmair G. W. (1995) Sm–Nd systematics of the eucrite Chervony Kut. In *Lunar Planet. Sci.* **XXVI**. The Lunar and Planetary Institute, Houston, pp. 1453-1454.

Wadhwa M. and Lugmair G. W. (1996) Age of the eucrite "Caldera" from convergence of long-lived and short-lived chronometers. *Geochim. Cosmochim. Acta* **60**, 4889–4893.

Wadhwa M., McSween H. Y., Jr., and Crozaz G. (1994) Petrogenesis of the shergottite meteorites inferred from their minor and trace element microdistributions. *Geochim. Cosmochim. Acta* **58**, 4213–4229.

Warren P. H. and Wasson J. T. (1979) The origin of KREEP. *Rev. Geophys. Space Phys.* **17**, 73–88.

Wasserburg G. J. (1985) Short-lived nuclei in the early solar system. In *Protostars and Planets II* (eds. D. C. Black and M. S. Matthews). The University of Arizona Press, Tucson, AZ, pp. 703–737.

Wasserburg G. J., Papanastassiou D. A., and Sanz H. G. (1969) Initial strontium for a chondrite and the determination of a metamorphism or formation interval. *Earth Planet. Sci. Lett.* **7**, 33–43.

Wasserburg G. J., Tera F., Papanastassiou D. A., and Huneke J. C. (1977) Isotopic and chemical investigations on Angra dos Reis. *Earth Planet. Sci. Lett.* **35**, 294–316.

Weigel A., Eugster O., Koeberl C., and Krähenbühl U. (1996) Primitive differentiated achondrite Divnoe and its relationship to brachinites. In *Lunar Planet. Sci.* **XXVII**. The Lunar and Planetary Institute, Houston, pp. 1403–1404.

Wiechert U. H., Halliday A. N., Palme H., and Rumble D. (2004) Oxygen isotope evidence for rapid mixing of the HED meteorite parent body. *Earth Planet. Sci. Lett.* **221**, 373–382.

Wood J. (1972) Fragments of terra rock in the Apollo 12 soil samples and a structural model of the Moon. *Icarus* **16**, 461–501.

Zartman R. E., Jagoutz E., and Bouring S. A. (2006) Pb–Pb dating of the D'Orbigny and Asuka 881371 angrites and a second absolute time calibration of the Mn–Cr chronometer. In *Lunar Planet. Sci.* **XXXVII**, #1580. The Lunar and Planetary Institute, Houston (CD-ROM).

Zolensky M. and McSween H. Y., Jr. (1988) Aqueous alteration. In *Meteorites and the Early Solar System* (eds. J. F. Kerridge and M. S. Matthews). University of Arizona Press, Tucson, AZ, pp. 114–143.

5

The Geochemistry and Cosmochemistry of Impacts

C. Koeberl

University of Vienna, Vienna, Austria

5.1 INTRODUCTION: THE USE OF GEOCHEMISTRY IN IMPACT STUDIES

The geochemistry and cosmochemistry of impacts (i.e., of impact craters and impact processes) is a rapidly developing research area that encompasses such wide-ranging topics as the simple chemical characterization of the various rock types involved (target rocks, impact breccias, melt rocks, etc.), the identification of extraterrestrial components in impact ejecta, the determination of the impactor (projectile) composition, and the determination of the causes of environmental changes from chemolithostratigraphic analyses.

This chapter is divided into three general areas. At the beginning, to set the stage and to introduce the relevant background information and terminology, a brief introduction to impact craters and processes is provided. Then, the main geochemical methods employed in the study of impact craters and processes are described. Finally, a number of examples are given in which geochemical methods in the study of impacts were used. An extensive reference list, intended to expand the usefulness of this chapter, is included. Parts of this chapter are updated from sections in Koeberl (1998) and Montanari and Koeberl (2000). It should be noted that this chapter is almost exclusively concerned with the study of terrestrial impacts (with the exception of a few lunar examples), mainly due to accessibility of rocks for study.

5.2 BACKGROUND ON IMPACT CRATERS AND PROCESSES

This section provides a very short introduction to the importance and characteristics of impact structures (for the most part following reviews by Koeberl, 2001, 2002 and Koeberl and Martinez-Ruiz, 2003). These few paragraphs should introduce the reader to the general topic of impact cratering and to various terms used in later sections that can only be understood in the context of impact cratering studies.

All bodies in the solar system (planets, moons, asteroids, etc.) that have solid surfaces are covered by craters. In contrast to many other planets and moons in the solar system, the recognition of impact craters on the Earth is difficult, because active geological and atmospheric processes on our planet tend to obscure or erase the impact record in geologically short time periods. Impact craters must be verified from the study of their rocks—remote sensing and geophysical investigations can only provide initial hints at the possible presence of an impact crater or supporting information. Petrographic studies of rocks at impact craters can lead to the confirmation of impact-characteristic shock metamorphic effects, and

geochemical studies may yield information on the presence of meteoritic components in these rocks. Craters of any type and morphology are not a common landform on Earth. About 170 impact structures are currently (2006) known on Earth (updates are available on the internet, see the Earth Impact Database, 2006). Considering that some impact events demonstrably affected the geological and biological evolution on Earth (e.g., papers in Koeberl and MacLeod, 2002), and that even small impacts can disrupt the biosphere and lead to local and regional devastation (e.g., Chapman and Morrison, 1994), the understanding of impact structures and the processes by which they form should be of interest not only to earth scientists, but also to society in general.

When discussing morphological aspects, we need to mention the distinction between an impact crater, that is, the feature that results from the impact, and an impact structure, which is what we observe today (i.e., long after formation and modification of the crater). Thus, unless a feature is fairly fresh and unaltered by erosion, it should be called an "impact structure" rather than an "impact crater." Impact craters (before postimpact modification by erosion and other processes) occur on Earth in two distinctly different morphological forms. They are known as simple craters (small bowl-shaped craters, e.g., Figure 1) with diameters up to about 2–4 km, and complex craters, which have larger diameters.

Complex craters are characterized by a central uplift in the form of either a central peak or a central ring of hills (Figure 2). Craters of both types have an outer rim and are filled by a mixture of fallback ejecta and material slumped in from the walls and crater rim during the early phases of cratering. Such crater infill may include brecciated and/or fractured rocks, and impact melt rocks. For craters formed in a marine setting, the infill may include a significant component of material reworked and transported into the crater by strong resurge currents or tsunami. Fresh simple craters have an apparent depth (crater rim to present-day crater floor) that is about one-third of the crater diameter. For complex craters, this value is closer to one-fifth or one-sixth. The central structural uplift in complex craters commonly exposes rocks that are usually uplifted from considerable depth and thus contrasts with the stratigraphic sequence of the environs around the impact structure. On average, the actual stratigraphic uplift amounts to about one-tenth of the crater diameter (e.g., Melosh, 1989).

Remote sensing and morphological observations may yield important initial data regarding the recognition of a potential impact structure (such as annular drainage patterns or topographic ring structures), but cannot provide confirming evidence. Geological structures with a circular

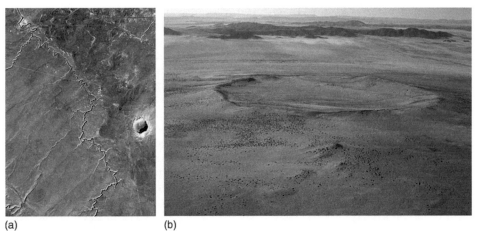

(a)   (b)

Figure 1 Simple meteorite impact craters. (a) Meteor crater (Barringer Crater) in Arizona, ∼1.2 km in diameter, age 50 ka. Landsat image. (b) Roter Kamm crater in Namibia, ca. 2.5 km in diameter, 3.7 Ma old. Aerial photograph (CK).

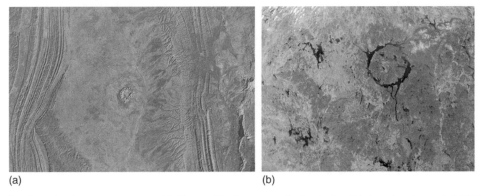

(a)   (b)

Figure 2 Complex impact structures. (a) Gosses Bluff structure in Australia; this is a deeply eroded 143 Ma old structure, of which only the eroded central uplift (in form of a 6 km diameter ring of hills) remains. NASA ISS photo E-5697. (b) Manicouagan structure in Canada, age ∼214 Ma, originally ∼100 km in diameter. Landsat image.

outline that are located in places with no other obvious mechanism for producing near-circular features may be of impact origin and at least deserve further attention. Geophysical methods are also useful in identifying candidate sites for further studies, especially for subsurface features. In complex craters, the central uplift usually consists of dense basement rocks and frequently contains severely shocked material. This uplift is often more resistant to erosion than the rest of the crater, and, thus, in old eroded structures, may be the only remnant of the crater that can still be identified.

The geophysical characteristics of impact structures can differ from those of surrounding, unaffected areas in a number of different ways including gravity, magnetic properties, distinct reflection and/or refraction seismic signatures, electrical resistivity anomalies, and others (see, e.g., Pilkington and Grieve, 1992; Grieve and Pilkington, 1996, for reviews). In general, simple craters have negative gravity anomalies due to the lower density of the brecciated rocks compared to the unbrecciated target rocks outside of the structure, whereas complex craters often have a positive

gravity anomaly associated with the central uplift of dense rocks originally located lower in the Earth's crust. This central positive gravity anomaly may be surrounded by an annular negative anomaly. Magnetic anomalies can be more varied than gravity anomalies (e.g., Henkel and Reimold, 2002). The target rocks may have been magnetically diverse, but the impact event may also cause anomalies related to impact-induced remanence. Seismic investigations of impact structures often show the loss of seismic coherence due to structural disturbance, slumping, and brecciation. Such geophysical surveys are important for the recognition of anomalous subsurface structural features, which may be deeply eroded craters or impact structures entirely covered by postimpact sediments. In the past decades, a large number of impact structures have been identified in the course of geophysical and drilling surveys related to hydrocarbon and other economic exploration (Grieve and Masaitis, 1994; Reimold *et al.*, 2005).

Final confirmation of an impact origin can only come from petrographic and geochemical study of actual rocks from the potential impact structure.

In the case of a structure that is not exposed on the surface, drill core samples are essential. Various types of breccia and melt rocks are good materials for the recognition of impact evidence. These rocks often carry unambiguous evidence for the impact origin of a structure, in the form of shocked mineral and lithic clasts or of contamination from the extraterrestrial projectile.

The fill of both main impact crater types consists of a variety of breccia types, as shown in Figure 3. Fragmental impact breccia is a "monomict or polymict impact breccia with clastic matrix containing shocked and unshocked mineral and lithic clasts, but lacking cogenetic impact melt particles" (Stöffler and Grieve, 1994). These rocks have also been termed lithic breccia (French, 1998). Impact melt breccia has been defined by Stöffler and Grieve (1994) as an "impact melt rock containing lithic and mineral clasts displaying variable degrees of shock metamorphism in a crystalline, semihyaline, or hyaline matrix (crystalline or glassy impact melt breccias)" (with an impact melt rock being a "crystalline, semihyaline, or hyaline rock solidified from impact melt"). Suevite (or suevitic breccia) is defined as a "polymict breccia with clastic matrix containing lithic and mineral clasts in various stages of shock metamorphism including cogenetic impact melt particles, which are in a glassy or crystallized state." The distribution of the rock types is a function of their formation and the order in which they formed. For example, lithic breccias can occur not only inside (fallback breccia or injected into the crater floor), but also outside a crater (fallout breccia). Figure 4 shows an example

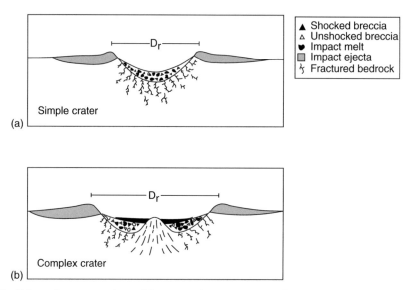

Figure 3 Schematic cross-sections of the two main crater types. After Montanari and Koeberl (2000).

Figure 4 Cliff of suevitic breccia with large clasts overlain by a sheet of fine-grained (brown) impact melt rock, at the ca. 100 km diameter, 35 Ma-old Popigai impact structure in Russia. Photo courtesy P. Claeys, Free Univ. Brussels.

of a suevite deposit (with large clasts) overlain by an impact melt sheet—it is rare that such a well-preserved occurrence is exposed.

Material ejected early in the crater formation will still be in the air after the crater is completely excavated and after the more massive local ejecta have been deposited to form the continuous deposits around the crater. Material that is ejected late during the crater formation (when the shock wave has already decayed somewhat) is ejected at lower velocities and will be deposited close to the crater rim, in contrast to early ejecta that will travel farther. Ejecta deposited at the crater rim often preserve the initial stratigraphic relationship, which leads to the inverted stratigraphy (the overturned flap) at the crater rim (because material from greater depth at the target is deposited last). This relationship also indicates that material from greater stratigraphic depth at a crater location is, in general, deposited close to the crater, whereas distal ejecta more commonly consist of the uppermost stratigraphic layers of the target. Because of the higher energies available early in the crater formation sequence, the earliest ejecta may be molten or at least strongly shocked, whereas later ejecta may only be slightly shocked or not at all. On Earth the interaction with the atmosphere leads to early ejecta being entrained in the rapidly expanding fireball and, if the event is large enough, they will be ejected outside of the atmosphere, leading to a global distribution. Size sorting will result in larger particles settling out first and finer dust being deposited later. Distal ejecta even from moderately sized craters (a few tens of kilometers in diameter) may travel several thousand kilometers from the impact site.

For the identification of meteorite impact structures, suevites and impact melt breccias (or impact melt rocks) are the most commonly studied units. It is easy to distinguish between the two impact formations, as suevites are polymict breccias that contain inclusions of melt rock (or impact glass) in a clastic groundmass (see examples in Figure 5), and impact melt breccias have a melt matrix with a variable amount of (often shocked) rock fragments as clasts (they are also referred to in the literature as "melt-matrix breccias"). Whether or not these various breccia types are present and/or preserved in a crater depends on factors including the size of the crater, the target composition (e.g., crystalline or sedimentary rocks), the degree of porosity of the target, and the level of erosion for an impact structure. In cases of very deeply eroded structures, only remnants of injected impact breccias in the form of veins or dikes may remain. Besides injections of suevite and impact melt rock, and local (*in situ*) formations of monomict or polymict clastic impact breccia, this may involve veins and pods of the so-called pseudotachylitic breccia that are recorded from a number of impact structures. This material may closely resemble what is known as "pseudotachylite," the term for "friction melt," and indeed has been referred to as this extensively in the literature (see, e.g., papers in Koeberl and Henkel, 2005). However, it has become clear in recent years that not all of the formations of such appearance actually represent friction melt, but may also include impact melt rock and even tectonically produced fault breccias (friction melt, mylonite, or cataclasite). Thus, it is prudent to verify the nature of any such material before labeling it with a genetic lithological term (Reimold and Gibson, 2005).

The rocks in the crater rim zone are usually only subjected to relatively low shock pressures (commonly <2 GPa), leading mostly to fracturing and brecciation, and often do not show shock-characteristic deformation. Even at craters of several kilometers in diameter, crater rim rocks that are *in situ* rarely show evidence for shock deformation. However, there may be injections of impact breccias that may contain shock-metamorphosed mineral and rock fragments. In well-preserved impact structures, the area

(a)   (b)

Figure 5 Examples of suevitic impact breccia. (a) Drill core from outside the northern rim of the Bosumtwi impact structure in Ghana (ca. 1.1 Ma old, 11 km in diameter), showing characteristic inclusions of frothy glass within a polymict breccia. (b) Cut slab of a ca. 25-cm-wide piece of suevite from the Bosumtwi structure, Ghana, showing the wide variety of rock types and clasts sizes that make up this breccia.

directly outside the crater rim is covered by a (vertical) sequence of different impactite deposits, which often allow the identification of these structures as being of impact origin.

The presence of shock metamorphic effects constitutes confirming evidence for impact processes. In nature, shock metamorphic effects are uniquely characteristic of shock levels associated with hypervelocity impact. Shock metamorphic effects are best studied in the various breccia types that are found within and around a crater structure, as well as in the formations exhumed in the central uplift area. During impact, shock pressures of $\geq 100\,GPa$ and temperatures $\geq 3,000\,°C$ are produced in large volumes of target rock. These conditions are significantly different from conditions for endogenic metamorphism of crustal rocks, with maximum pressures of usually $<2\,GPa$ and temperatures of $<1,200\,°C$. Shock compression is not a thermodynamically reversible process, and most of the structural and phase changes in minerals and rocks are uniquely characteristic of the high pressures (diagnostic shock effects are known for the range from 8 to $>50\,GPa$) and extreme strain rates (10^6–$10^8\,s^{-1}$) associated with impact. The products of static compression, as well as those of volcanic or tectonic processes, differ from those of shock metamorphism, because of lower peak pressures and strain rates that are different by many orders of magnitude.

A wide variety of microscopic shock metamorphic effects have been identified. The most common ones include planar microdeformation features; optical mosaicism; changes in refractive index, birefringence, and optical axis angle; isotropization (e.g., formation of diaplectic glasses); and phase changes (high pressure phases; melting). Kink bands (mainly in micas) have also been described as a result of shock metamorphism, but can also be the result of normal tectonic deformation (for reviews and images of examples,

refer to, for example, Stöffler and Langenhorst, 1994; Grieve *et al.*, 1996; French, 1998).

Planar microstructures are the most characteristic expressions of shock metamorphism and occur as planar fractures (PFs) and planar-deformation features (PDFs). The presence of PDFs in rock-forming minerals (e.g., quartz, feldspar, or olivine) provides diagnostic evidence for shock deformation, and, thus, for the impact origin of a geological structure or ejecta layer (see, e.g., Stöffler, 1972, 1974; Stöffler and Langenhorst, 1994; Montanari and Koeberl, 2000, and references therein). PFs, in contrast to irregular non-PFs are thin fissures, spaced about $20\,\mu m$ or more apart. While they are not considered shock diagnostic *per se*, should they be observed in significant abundance and in particularly densely spaced sets of multiple orientations, they can provide a strong indication of shock pressures around 5–10 GPa. To an inexperienced observer, it is not always easy to distinguish "true" PDFs from other lamellar features (fractures, fluid-inclusion trails, and tectonic-deformation bands).

The most important characteristics of PDFs are that they are extremely narrow, closely and regularly spaced, completely straight, parallel, extend often (although not always) through a whole crystal, and, at shock pressures above about 15 GPa, occur in more than one set of specific crystallographic orientation per grain (see examples in Figures 6a and 6b). This way, they can be distinguished from features that are produced at lower strain rates, such as the tectonically formed Böhm lamellae, which are not completely straight, occur only in one set, usually consist of bands that are $>10\,\mu m$ wide, and are spaced at distances of $>10\,\mu m$. Transmission electron microscopy (TEM) studies demonstrate that PDFs consist of amorphous silica (i.e., they are planes of amorphous quartz that extend through the quartz crystal). This allows them to be preferentially etched by,

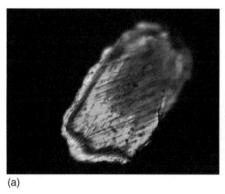 

(a)   (b)

Figure 6 Examples of shocked quartz grains. (a) Shocked quartz grain (ca. $150\,\mu m$ in size) from distal ejecta of the Manson impact structure (USA). Note the multiple sets of PDFs, with straight lamellae that are very closely spaced and penetrate the whole grain. (b) Thin section showing shocked quartz from the Woodleigh impact structure, Australia. Width of image 1 mm, crossed polarizers.

for example, hydrofluoric acid, thus accentuating the PDFs (e.g., Bohor *et al.*, 1993; Montanari and Koeberl, 2000). PDFs occur in planes that correspond to specific rational crystallographic orientations (for details, see, e.g., Stöffler and Langenhorst, 1994). With increasing shock pressure, the distances between the planes decrease, and the PDFs become more closely spaced and more homogeneously distributed through the grain, until at about 30–35 GPa the grains show complete isotropization. Depending on the peak pressure, PDFs are observed in about 2–10 orientations per grain. To confirm the presence of PDFs, it is necessary to measure their crystallographic orientations by using either a universal stage or a spindle stage with an optical microscope, or to characterize them by TEM. Because PDFs are well developed in quartz (Stöffler and Langenhorst, 1994), a very widely observed rock-forming mineral, and because their crystallographic orientations are easy to measure in this mineral, most studies report only shock features in quartz. However, other rock-forming minerals, as well as accessory minerals, also develop PDFs.

Higher shock pressures than those recorded in PDFs in quartz and other rock-forming minerals lead to shock-induced amorphization (without melting) of the minerals (producing "diaplectic" minerals, such as diaplectic quartz or feldspar), thermal decomposition or melting of selected minerals (e.g., the monomineralic melt of quartz is lechatelierite), and whole-rock melting (e.g., Stöffler, 1984). Depending on the thermal history of the rock after melting, recrystallization can occur (e.g., Figure 7a), or, if quenched, impact glasses (often showing compositional flow-banding, also called "schlieren"—see Figure 7b) may be preserved. Impact glasses can form directly at a crater, or be ejected to great distances (tektites, see Section 5.4.3). These melt rocks and glasses are often the objects of geochemical investigations. For example, the detection of small amounts of meteoritic matter in breccias and melt rocks can

also provide confirming evidence of impact, but this is extremely difficult, and is the topic of later sections in this review.

5.3 METHODS

5.3.1 General Geochemistry: Major and Trace Elements

The measurement of the chemical composition of rocks in both impact craters (target rocks/source rocks as well as breccias and their components, impact melt rocks, and impact glasses) and distal impact ejecta are part of a standard characterization of these rocks, which would normally involve a petrographic and a geochemical study. Besides helping to classify the rock types present at an impact structure, such basic geochemical data can also be used to determine if any unusual or extraneous components are present in breccias or melt rocks (as well as in distal ejecta), and to determine the origin of impact glasses (e.g., Taylor, 1967). Also, once all rock types present at a particular impact site are found and analyzed, mixing calculations (see, e.g., Reimold, 1982; Stöckelmann and Reimold, 1989; French *et al.*, 1997) allow the reconstruction of the proportions of the different rock types that comprise breccias or melt rocks. Such mixing calculations are important to define the indigenous contents of siderophile elements in breccias and melt rocks, which is essential for the determination of the possible presence of the admixture of extraterrestrial components (see the following sections).

The study of trace-element compositions of the various rock types at impact structures also allows investigation of alteration processes that are either part of the natural environment or due to impact-induced (e.g., hydrothermal) alterations. Changes in the contents of, for example, volatile elements may constrain physical and chemical processes that are operating during the impact event—usually connected to the high temperatures that

(a)

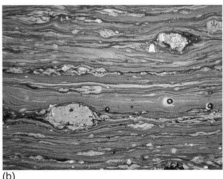
(b)

Figure 7 (a) Thin-section image of impact melt rock, partly recrystallized, in a breccia from the Woodleigh impact structure, Australia (width of image 1 mm, crossed polarizers). (b) Schlieren in impact glass from the El'gygytgyn impact structure, Russia (width of image 1.7 mm, parallel polars).

prevail during such events—and can lead to elemental fractionations. Another aspect of major and trace-element analyses concerns the compositions of individual minerals within rocks or melt rocks, or of compositional changes within impact glasses, as in, for example, tektites.

As described in detail in chapter 6 of the book by Montanari and Koeberl (2000), a crucial part of geochemical analyses of impactites starts with sampling and sample preparation, where it has to be decided which types of analyses will be done (e.g., siderophile trace elements such as the platinum-group elements (PGEs), or isotopic analysis, in addition to basic major and trace-element studies). As a general rule, the less sample preparation is required, the better. Each preparation and treatment step increases the chance of contamination or loss. In geochemical analyses, a compromise must be reached between available sample mass and what constitutes a representative sample. In the study of impactoclastic (distal ejecta) layers, this problem is even more severe, because of the low abundance of the impact-derived debris within a large amount of local matrix. It is necessary to consider if the goal of the analyses is to search for an extraterrestrial component in the first place, or to study in detail a known impactoclastic layer. In the first case, a large number of samples need to be scanned for impact-characteristic signatures, whereas the second case usually requires breaking down a large sample from a known location into its components. If proximal ejecta, or crater rocks, are studied, care has to be taken to obtain a representative set of target rock and impactite lithologies for chemical analyses and to search for shock effects. In some cases, the characteristics of rocks from a particular crater structure are to be compared with the corresponding characteristics of distal ejecta to determine if there is a connection between the two.

Taking all these considerations into account leads to a variety of sample preparation requirements, but they all have some points in common. The most important requirement is to avoid contamination. The search for siderophile element anomalies commonly involves very low elemental abundances, and cross contamination from other samples, from crushers and mills, and during chemical treatment, can introduce severe problems that may not occur during standard geochemical work. In general, it is advantageous to avoid using steel jaw crushers and any mills involving metal parts (e.g., those that use tungsten carbide or similar components and alloys, which are very common in standard swing mills). Coarse crushing can be done with the samples wrapped in thick sheets of plastic foil, followed by crushing in jaw crushers with ceramic jaws (e.g., alumina). Powdering of the samples to the required particle size for mineral separation (sieving) or bulk analysis is best done

in alumina ceramic, agate, or boron carbide mills (of various designs). Boron carbide is brittle and very expensive, but it is useful for very hard materials, and our own experience has shown that this seems to be the mill material with the least amount of abrasion, which, therefore, leads to the least amount of sample contamination. Nevertheless, it is always a good idea to run blanks for each crusher and mill (using, e.g., pure quartzite), and to insert a cleaning step in between each sample (e.g., by grinding clean, commercially available sea sand). No equipment (e.g., in central sample processing facilities) that has been used in the preparation of mineralized samples should be employed for the crushing and grinding of impact-derived material, as it is very easy to introduce contamination at the $\leq 10^{-9}$ g g^{-1} (subparts per billion) levels of interest for such analyses.

The methods that are used to measure the various major- and trace-element compositions are varied and comprise the "usual" range of techniques. Almost all whole-rock analyses require that the rocks are powdered. The most common methods for bulk rock major- and trace-element analysis are X-ray fluorescence spectrometry (XRF, with sample powder pressed as pills, or with glass disks made by fusion of sample powder and a flux material), inductively coupled plasma atomic emission spectrometry (ICP-AES), inductively coupled plasma mass spectrometry (ICP-MS), and instrumental neutron activation analysis (INAA). All of these methods allow the simultaneous (rarely sequential) determination of a large number of major and trace elements. Requirements for sample mass vary somewhat, depending on the method (XRF tends to need several grams of material, whereas a few tens to hundreds of milligrams are sufficient for the latter three methods). For mineral and microscale analyses, the common electron and ion beam methods are used.

5.3.2 Rb–Sr and Sm–Nd Isotopes

The "classic" isotope systems Rb–Sr and Sm–Nd are commonly used in impact studies for dating of impact events—mostly by separating different phases from impact melt rocks and constructing isochron diagrams (e.g., Deutsch and Schärer, 1994). These two systems are based on the β^--decay of ^{87}Rb to ^{87}Sr with a half-life of 48.8 Gyr, and the α-decay of ^{147}Sm to ^{143}Nd with a half-life of 106 Gyr, respectively. As age dating is, however, not the topic of this review, it should be noted that these (and other) isotope systems can be used also as tracers of source rocks. For example, the isotopic systematics can demonstrate that impact melt rocks have been derived from near-surface crustal rocks and not from the deep crust or the mantle (as is the case for most "normal" igneous rocks). Comparatively high ^{87}Sr/^{86}Sr ratios indicate that melt rocks in young impact structures were derived from older continental crust as such high ratios show that the melt rocks were produced from melting of older and more radiogenic crustal rocks. Oceanic crust would show somewhat different compositions but still be distinct. Another example is that the neodymium isotopic composition of the melt rocks at the large Sudbury impact structure (originally about 250 km in diameter; age ~1.86 Ga) in Canada showed that the precursor rocks were crustal rocks without any obvious mantle component, as had been suspected earlier (cf. French, 1998), and as was confirmed by osmium isotopic studies (Walker *et al.*, 1991; Dickin *et al.*, 1992). For the evaluation of the strontium and neodymium isotope systems in source rock studies plots of ε(Sr) versus ε(Nd) are usually utilized. Examples would also be the use of such data in the determination of source craters of tektites (see Section 5.4.3). In the case of the Cretaceous–Tertiary boundary, strontium isotopic studies have been used to determine the presence of excess runoff of continental material into the oceans.

In terms of methodology, Rb–Sr and Sm–Nd isotope analyses are fairly routine procedures: whole-rock powders or mineral separates are spiked with an isotopic standard, dissolved, the various elements separated mostly by ion exchange, and the isolated elemental fractions are subjected to thermal ionization mass spectrometry to determine the isotopic ratios.

5.3.3 Siderophile Element Studies

As mentioned in Section 5.2, the detection of small amounts of meteoritic matter in breccias and melt rocks can provide confirming evidence of the impact origin of a particular crater structure, but such a determination is not easy. Only elements that have high abundances in meteorites, but low contents in terrestrial crustal rocks (both continental and oceanic crust; cf. Taylor and McLennan, 1985), are useful. A complication is the existence of a variety of meteorite groups and types (the three main groups are stony meteorites, iron meteorites, and stony–iron meteorites, in order of decreasing abundance), which have widely varying siderophile element compositions. Distinctly higher siderophile element contents in impact melts, compared to target rock abundances, can be indicative of the presence of either a chondritic or an iron meteoritic component. Achondritic projectiles are much more difficult to discern, because they have significantly lower abundances of the key siderophile elements. It is also necessary to sample all possible target rocks to determine the so-called indigenous component (i.e., the contribution to the siderophile element content of the impact melt rocks from the target),

to ascertain that no possibly siderophile element-rich mantle-derived target rock has remained undetected. So far, meteoritic components have been identified for just over 40 impact structures out of the more than 170 impact structures currently identified on Earth.

Siderophile (and related) elements that have often been used are nickel, cobalt, and chromium. In addition, the interelement ratios of these elements are an effective discriminator. If the meteoritic contribution exceeds 0.1% by weight, it is possible to distinguish between stony meteorites (chondrites) and iron meteorites, because chondrites have high concentrations of Cr (typically about 0.26 wt.%), whereas iron meteorites have Cr concentrations that are much more variable, but typically about 100 times lower than those of chondrites, and low Ni/Cr or Co/Cr ratios. Therefore, the Cr concentration and evaluation of the Ni/Cr or Co/Cr ratios in impact melts can be used to distinguish between iron and chondritic projectiles (e.g., Palme *et al.*, 1978; Evans *et al.*, 1993). Unfortunately, Co, Cr, and Ni do not show particularly low concentrations in terrestrial rocks. Thus, there may be a relatively high (and variable) indigenous component that is derived from the target rocks (cf. Palme, 1980). If any conclusions are to be drawn from the abundances of elements such as chromium, nickel, and cobalt, a detailed study of their abundances in the target rocks needs to be done. From mixing calculations, the relative proportions of the various target rocks types that are involved with producing a breccia or melt rock should be determined (see previous section on mixing calculations); from this mixture an indigenous concentration can be determined and subtracted from the abundances found in the impact melt rocks, to yield the "pure" meteoritic abundance ratios (e.g., Morgan and Wandless, 1983). Thus, it is necessary to analyze all rock types that are known or suspected to be present in the preimpact target area (in some cases, clasts within breccias can be used). This procedure gets complicated if too many target rocks with variable compositions are involved. In reality, it is difficult to ascertain all target rocks that were involved in forming impact melt rocks or breccia (e.g., loss of some rock types to erosion, or lack of exposure), or because of very low or highly variable indigenous PGE concentrations (cf., e.g., Schmidt and Pernicka, 1994; Pernicka *et al.*, 1996).

The first detailed studies aimed at projectile identification were spawned—like so many other techniques—from the Apollo program (i.e., from analyses of lunar rocks in the early 1970s). It was shown that the PGEs (ruthenium, rhodium, palladium, osmium, iridium, and platinum) and gold are better suited for identifying a meteoritic component than chromium, nickel, and cobalt alone (e.g., Morgan *et al.*, 1975; Gros *et al.*,

1976; Palme *et al.*, 1978, 1979). The abundances of the PGEs in chondrites and most iron meteorites are several orders of magnitude higher than those in terrestrial crustal rocks (continental and oceanic), as illustrated for iridium in Figure 8. The range of Ir and Os concentrations in chondrites is about 400–800 ppb (e.g., Wasson and Kallemeyn, 1988; Anders and Grevesse, 1989), while iron meteorites show a much wider range (e.g., Buchwald, 1975; cf. also Koeberl, 1998). In contrast, continental crustal rocks contain on the order of 0.02–0.03 ppb Ir or Os (e.g., Taylor and McLennan, 1985, 1995; Peucker-Ehrenbrink and Jahn, 2001; see Table 1). Thus, the signal-to-background ratio is very high for the PGEs (i.e., low indigenous concentrations; high concentrations in the "contaminating" meteorite).

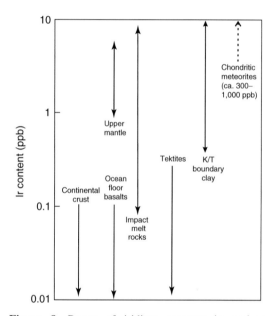

Figure 8 Range of iridium contents in various terrestrial and extraterrestrial materials. After Koeberl (1998).

The determination of the PGE iridium is the most commonly used scanning technique in the search for chemical impact markers (see the section on the K–T boundary). Although this point has been made before, it should be emphasized that the reason that often only the Ir content is measured is because this element can be determined with greater sensitivity and more ease than any of the other PGEs (see also Sawlowicz, 1993; Kramar *et al.*, 2001). Thus, iridium acts as a marker for the other PGEs, which, if an iridium anomaly is found, may then also be determined in a limited subset of samples (given the much greater analytical efforts for complete PGE analyses). Typical ranges of Ir concentrations in terrestrial and extraterrestrial rocks are shown in Figure 8.

There are a variety of techniques for measuring Ir contents, with the most commonly used ones based on neutron activation. The reason is the very high neutron capture cross-section of the isotope [191]Ir, resulting in the radioactive ^{192}Ir isotope, which β-decays with a half-life of 74.2 days, emitting several γ-ray lines in the energy range to which γ-ray detectors are most sensitive. Detection limits of routine INAA procedures vary according to the neutron flux during irradiation, duration of irradiation, counting duration, detector sensitivity, and sample composition (the latter influencing the background of the spectrum), but are commonly on the order of 0.5–2 ppb Ir (see, e.g., Koeberl, 1993a). Radiochemical methods (e.g., Palme *et al.*, 1978, 1979; French *et al.*, 1989) can be used to measure lower abundances, but are tedious. Detection limits can be significantly improved by using γ–γ coincidence spectrometry, as described by, for example, Koeberl and Huber (2000). The technique makes use of the coincident decay of ^{192}Ir by counting a radioactive sample simultaneously with two detectors and plotting the resulting three-dimensional coincidence spectrum. The coincidence method allows the analysis of a fair number of samples with

Table 1 PGEs and Re and Au contents in selected meteorite types, the continental crust, and two meteoritic-component-bearing melt rocks from the Moroweng impact structure.

	Re	Os	Ir	Ru	Pt	Rh	Pd	Au
50% Condensation temperature (K)	1,821	1,812	1,603	1,551	1,408	1,392	1,324	1,060
Orgueil C1 chondrite	37	490	480	690	1,050	130	530	140
H5	99	814	770	820	1,700	210	1,140	260
L6	37	530	490	710	1,390	250	680	220
E4	62	589	500	970	1,300	250	790	440
Continental crust	0.198	0.031	0.022	0.210	0.510	0.38	0.520	1.8
Morokweng-43	1.5	16	18	22	46	4.9	25	6.2
Morokweng-48	1.7	18	20	27	43.5	6.5	23	7.4

Note: 50% Condensation temperatures are given for a gas of solar composition at a total pressure of 10^{-4} bar (Lodders, 2003). Concentration data in ppb, from Taylor and McLennan (1985), Koeberl and Reimold (2003), Koeberl *et al.* (2000), and Peucker-Ehrenbrink and Jahn (2001) and references therein.

detection limits that are on the order of 5–30 ppt (10^{-12} g g^{-1} or pg g^{-1}) Ir, again depending on the irradiation, instrumentation, counting, and sample parameters (e.g., Koeberl and Huber, 2000).

Recently, ICP-MS techniques have been used for iridium analyses, but these require larger samples, chemical sample preparation (dissolution and separation), and reach the detection limits obtained by coincidence counting only with great analytical effort. Isotope-dilution techniques (with the measurements made by thermal ionization mass spectrometry (TIMS) or ICP-MS) are very precise and yield very low detection limits, but the great experimental effort makes them impractical to use for reconnaissance purposes. Also, after completion of the counting for INAA the samples can be reused for other analyses.

Most methods for PGE analysis used in geochemistry have been developed for the study of, or reconnaissance for, commercially important ore deposits; more recently, some methods were developed for the study of mantle rocks. The abundance levels of interest in impactite studies are several orders of magnitude lower than those in ore deposit studies. Chondritic meteorites have average abundances of (for example) about 700 ppb Ir, 1,800 ppb Pt, 700 ppb Pd, and 900 ppb Ru (see Anders and Grevesse, 1989). Typical mantle xenoliths have abundances on the order of 2 ppb Ir, 5 ppb Pt, 2 ppb Pd, and 4 ppb Ru. The upper continental crust, into which impacts of meteoritic projectiles occur, has abundances that are about a factor of 100 below those of mantle rocks. Of course, impacts can occur into oceanic crust as well, but these would have to be much larger than into continental crust because of the screening effect of the ocean water for small impacts, and also rarer because of the required impactor size. In addition, oceanic crust is young, obliterating older impact events. As discussed above, impactites contain very small meteoritic contributions (usually well below 1 wt.%). Assuming, for example, that 0.2% by weight of a chondritic projectile (the most common meteorite type) is mixed in with upper crustal rocks: this yields, for Ir, $700 \times 0.002 = 1.4$ ppb in the resulting impactite (and we note, as mentioned in the previous section, that Ir is the easiest of the PGEs to determine). Abundances in impactoclastic layers are commonly much lower (cf. Montanari and Koeberl, 2000). Quantitative PGE determinations at sub-parts per billion abundance levels are analytically very challenging.

Earlier studies (e.g., Morgan *et al.*, 1975, 1979; Morgan, 1978; Palme *et al.*, 1978, 1979, 1981; Wolf *et al.*, 1980) used radiochemical methods, where the PGEs were separated from the sample matrix, and in some cases from each other, after neutron irradiation (i.e., radiochemical neutron activation analysis (RNAA)). This procedure involved (for example) dissolution of the radioactive sample by alkali fusion or acid digestion, followed by ion exchange, liquid extraction, and/or precipitation steps that may be repeated several times for better separation efficiency. Counting and data reduction were then done by standard INAA techniques. Such procedures are currently not much used due to the reluctance of handling radioactive samples in liquid form, and because of the radioactive decay of short-lived isotopes during the time-consuming separation steps, which lead to a loss of signal. It is also possible (but not easy—Dai *et al.*, 2000) to separate the PGEs before the irradiation.

More recently, the development of stable ICP-MS instruments, both with quadrupole and high-resolution MS, has provided the geochemist with another tool for low-level PGE work. The sample digestion and separation procedures are still similar to those used before in RNAA work. One of the most common methods involves a NiS fire assay with Te coprecipitation, followed by ICP-MS measurements. Detection limits obtained with this method by Koeberl *et al.* (2000) are (in parts per billion) 0.02 Ir, 0.05 Ru, 0.02 Rh, 0.14 Pt, 0.10 Pd, and 0.02 Au. A somewhat similar procedure was recently developed by Hassler *et al.* (2000), which allows not only the determination of the PGEs, but also of the Os isotopic ratio in the same sample (see Section 5.3.4). Their method uses PGE isotope spikes, which are introduced at the dissolution step, providing the advantages of isotope dilution methods (e.g., elimination of yield determination).

Several problems with all these methods result from sample size and detection limits. First, due to their low abundances the PGEs are often inhomogeneously distributed in rocks, which, together with sensitivity limits of the measurements, require a large enough sample for a representative analysis. This causes problems with complete dissolution of such large samples, and, because of the larger amount of chemicals required in the following steps, leads directly to the second problem: potentially interfering blank levels of the PGEs in the reagents used. For example, great care has to be employed to obtain NiS that has low-enough levels of the PGEs so that blank values do not interfere with sample values. However, all such determinations are at the current limits of analytical geochemistry and require a dedicated laboratory and great analytical experience. At the sub-parts per billion abundance level, there are no routine PGE measurements.

5.3.4 Osmium Isotopes

The use of the osmium isotopic system is based on the formation of ^{187}Os by β-decay of ^{187}Re (half-life of 42.3 ± 1.3 Gyr). In principle, it is necessary to determine the ratio of the radiogenic osmium isotope, ^{187}Os, compared to a nonradiogenic osmium isotope. Osmium has seven naturally occurring isotopes, which are all stable.

The amount of ^{187}Os increases with time as a result of the decay of ^{187}Re. This decay can be described by normalizing to an Os isotope not affected by radioactive decay:

$$^{187}\text{Os}/^{188}\text{Os} = (^{187}\text{Os}/^{188}\text{Os})_i$$
$$+ (^{187}\text{Re}/^{188}\text{Os})(e^{\lambda t} - 1)$$

where ^{187}Os/^{188}Os and ^{187}Re/^{188}Os are the measured ratios of these isotopes, $(^{187}\text{Os}/^{188}\text{Os})_i$ is the initial isotopic ratio at the time when the system became closed for Re and Os, λ the decay constant for ^{187}Re, and t the time elapsed since system closure for Re and Os. Meteorites (and the terrestrial mantle) have Os and other PGEs contents that are higher than those in terrestrial crustal rocks by factors of 10^2–10^5. In addition, meteorites have relatively low Re and high Os concentrations, resulting in Re/Os ratios less or equal to 0.1, whereas the Re/Os ratio of terrestrial crustal rocks is usually no less than 10 (Figure 9). Similar to the Rb–Sr and Sm–Nd isotopic methods, the abundance of radiogenic ^{187}Os is normalized to the abundance of nonradiogenic ^{188}Os. The use of the method is based on the fact that the ^{187}Os/^{188}Os isotopic ratios for meteorites and terrestrial crustal rocks are significantly different (Allègre and Luck, 1980). As a result of the high Re and low Os concentrations in old crustal rocks, their ^{187}Os/^{188}Os ratio increases rapidly with time (average upper crustal ^{187}Os/^{188}Os = 1–1.2). In contrast, meteorites have low ^{187}Os/^{188}Os ratios of about 0.11–0.18, and, as osmium is much more abundant in meteorites than rhenium, only small changes in the meteoritic ^{187}Os/^{188}Os ratio occur with time (see Figure 10).

The first application of this isotope system to impact-related materials was suggested by Turekian (1982) and data reported by Luck and Turekian (1983), who measured the osmium isotopic composition of K–T boundary clays, which was feasible because of the relatively high Os abundances (tens of parts per billion) in these rocks. On the other hand, the first application for impact crater studies, by Fehn *et al.* (1986), was hampered by a lack in sensitivity, as target rocks and typical impactites have very low (sub-parts per billion) Os abundances. This changed with the introduction of the negative thermal ionization mass spectrometry (NTIMS) technique (Völkening *et al.*, 1991), which allowed the measurement of osmium isotopic ratios in samples of a few grams mass containing sub-parts per billion amounts of Os. The first successful application of this method dealt with the Bosumtwi impact structure in Ghana (Koeberl and Shirey, 1993), followed by other similar studies (e.g., Koeberl *et al.*, 1994a, b, c).

As a result of the relatively high Os abundances in meteorites, the addition of even a small amount of meteoritic matter to the crustal target rocks leads to an almost complete change of the osmium isotopic signature of the resulting impact melt or breccia. Thus, the Re–Os isotopic diagram is not used in the "usual" sense, to extract age information as in an Rb–Sr diagram, but as a mixing

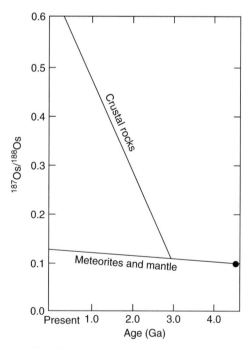

Figure 10 Development of the osmium isotopic ratio with time, in rocks that have low Re and high Os (mantle rocks, meteorites), resulting in little change of the isotopic ratio, and in crustal rocks that have high Re and low Os concentrations, leading to a rapid change in the isotopic ratio with time. After Faure (1986).

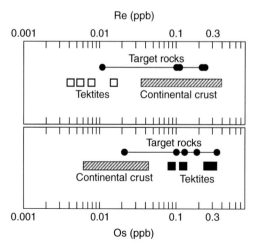

Figure 9 Range of rhenium and osmium contents in various terrestrial and extraterrestrial materials. After Koeberl and Shirey (1997).

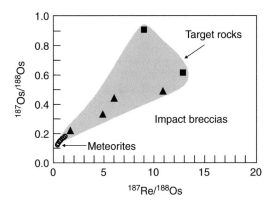

Figure 11 A Re–Os isotopic diagram for target rocks and impact breccias from the Kalkkop impact crater in South Africa. The triangles are different breccia samples, the squares represent the two main target rock types (sandstone and shale), and meteorite data are indicated by open circles. The diagram illustrates the use of this isotopic diagram as a mixing diagram. After Koeberl *et al.* (1994a).

diagram, as shown in Figure 11. It should also be noted that contamination from an achondritic meteorite requires much higher percentages of meteoritic addition due to the much lower PGE abundances in achondrites compared to chondritic and iron meteorites. In addition, the present-day $^{187}Os/^{188}Os$ ratio of mantle rocks is about 0.13, which is similar to meteoritic values. Thus, contribution of a significant mantle component (e.g., ultramafic rocks) to an impact breccia or melt rock has to be ruled out. Fortunately, the PGE abundances in typical rocks from the upper mantle are at least two orders of magnitude lower than those in (chondritic and iron) meteorites. Thus, to achieve the same disturbance of the Os isotope value, at least 100 times higher mantle contributions (i.e., 10 or more wt.%) than meteoritic contributions (e.g., 0.1 wt.%) would be needed. The presence of such a significant mantle component would be easily discernable from petrographic studies of the clast population in breccias, and/or from geochemical data (especially strontium and neodymium isotopes) of melt rocks. Compared to the use of PGE elemental abundances and ratios, the osmium isotope method is superior with respect to detection limit and selectivity, as discussed in detail (with several case histories) by Koeberl and Shirey (1997) and Koeberl *et al.* (2002b). There are, however, a few disadvantages; the laboratory procedures (sample preparation, digestion, and measurement) are complex, and the osmium isotopic method does not allow determination of the projectile type.

In the laboratory, the concentrations of Os (in some cases also of Re) and the $^{187}Os/^{188}Os$ isotopic ratio are measured by sensitive NTIMS (e.g., Creaser *et al.*, 1991; Völkening *et al.*, 1991). Most modern analyses involve a high-temperature acid digestion step for dissolution of the rocks and isotope equilibration between the spike and the sample Re and Os (e.g., the Carius tube method; cf. Shirey and Walker, 1995), followed by separation of the Os in some sort of distillation, and isolation of Re by an ionexchange chromatography (e.g., Birck *et al.*, 1997). The low abundances of Os in terrestrial samples require the use of fairly large samples, about 2–10 g of sample powder. For the NTIMS measurements, rhenium and osmium fractions are loaded (separately) on filaments (commonly Pt) that have to be cleaned to remove any rhenium background, and are measured as negative ReO_4^- and OsO_3^- ions, respectively. Measurements can also be done by ICP-MS. A different method involves a nickel sulfide fire assay preconcentration, dissolution, followed by purging (in an argon stream) volatile OsO_4 directly in an ICP-MS instrument (without a nebulizer), allowing direct, rapid (measurement time is a few minutes), and very sensitive measurement of osmium abundance and isotopic composition. The remaining solution can then be used for "normal" ICP-MS measurement of the abundances of the PGEs (Ravizza and Pyle, 1997; Hassler *et al.*, 2000). This method is faster than the standard NTIMS osmium isotope method, but has lower analytical precision, which is, however, of little concern in impact studies where order of magnitude changes are observed.

Apart from the use of osmium isotopic compositions in mixing relations between meteorites and basement rocks, as described above, osmium isotopes can also reveal the global effect of large-scale impact events. Over the past decade, it has been shown that the marine osmium isotopic record of the past 80 Myr is a mixture of terrestrial and extraterrestrial sources (e.g., Peucker-Ehrenbrink *et al.*, 1995; Peucker-Ehrenbrink, 1996; Klemm *et al.*, 2005), and that there are clear indications of the K–T and the late Eocene impact events in the marine Os isotopic record (Figure 12).

5.3.5 Chromium Isotopes

This new method provides evidence not only for the presence of an extraterrestrial component in impactites, but also information regarding the type of projectiles (Shukolyukov and Lugmair, 1998). It is based on the determination of the $^{53}Cr/^{52}Cr$ ratio, which can be affected by the decay of the extinct radionuclide ^{53}Mn (half-life = 3.7 Myr). The chromium isotopic compositions are expressed as the deviations of the $^{53}Cr/^{52}Cr$ ratio in a sample relative to the standard terrestrial $^{53}Cr/^{52}Cr$ ratio by high-precision TIMS. These deviations are usually expressed in ε units (1ε is 1 part in 10^4, or 0.01%). Terrestrial rocks are not expected to show (and, indeed, do not reveal) any variation in $^{53}Cr/^{52}Cr$ ratio, because the

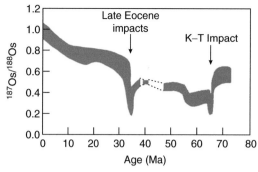

Figure 12 Evolution of the marine Os isotopic record with time. Distinct anomalies in the isotopic record, with low Os isotopic ratios, are clearly evident at about 65 and 35 Ma, coinciding with the K–T boundary Chicxulub impact event and the Late Eocene Chesapeake Bay and Popigai impact events. Range of data from Peucker-Ehrenbrink *et al.* (1995), Peucker-Ehrenbrink (1996), and Klemm *et al.* (2005).

homogenization of the Earth was completed long after all ^{53}Mn (which was injected into, or formed within, the solar nebula; see Chapter 2) had decayed. In contrast, most meteorite groups analyzed so far (Shukolyukov and Lugmair, 1998, 2000), such as carbonaceous, ordinary, and enstatite chondrites, primitive achondrites, and other differentiated meteorites (including the SNC meteorites, which originated from Mars) show a variable excess of ^{53}Cr relative to terrestrial samples. The range for meteorites is about +0.1 to +1.3ε, depending on the meteorite type, except for carbonaceous chondrites, which show an apparent deficit in ^{53}Cr of about −0.4ε. These differences reflect a heterogeneous distribution of ^{53}Mn in the early solar system and early Mn/Cr fractionation in the solar nebula and in meteorite parent bodies (Lugmair and Shukolyukov, 1998). The negative ε value for carbonaceous chondrites is an artifact of using the ^{54}Cr/^{52}Cr ratio for a second-order fractionation correction (Lugmair and Shukolyukov, 1998; Trinquier *et al.*, 2006), because these meteorites carry a presolar ^{54}Cr component (Shukolyukov *et al.*, 2000). The actual, unnormalized ^{53}Cr/^{52}Cr ratio is similar to that of other undifferentiated meteorites, and the apparent ^{53}Cr deficit in the carbonaceous chondrites is actually due to an excess of ^{54}Cr (Trinquier *et al.*, 2006). However, the presence of ^{54}Cr excesses in bulk carbonaceous chondrites allows distinguishing these meteorites from the other meteorite classes.

Shukolyukov and Lugmair (1998) have used the chromium isotopic method on samples from the K–T boundary in Denmark and Spain and found that about 80% of the chromium in these samples originated from an impactor with a carbonaceous chondritic composition. More recently, Shukolyukov *et al.* (2000) found a similar (carbonaceous chondritic) signal for some samples from Archean spherule layers in the Barberton

Mountain Land, South Africa. Shukolyukov and Lugmair (2000) reported on analyses of impact melt rocks from the East Clearwater (Canada), Lappajärvi (Finland), and Rochechouart (France) impact structures. In all three cases, the impactors were found to have had ordinary chondritic composition. A summary of some chromium isotope data is given in Figure 13.

The chromium isotopic method has, thus, the advantage over both the use of osmium isotopes or PGE abundance ratios of being selective not only regarding the chromium source (terrestrial versus extraterrestrial), but also regarding the meteorite type. A disadvantage of this method is the complicated and time-consuming analytical procedure. Also, a significant proportion of the chromium in an impactite, compared to the abundance in the target, has to be of extraterrestrial origin. The detection limit of the chromium isotopic method, assuming a chondritic composition, is a function of the chromium content in the (terrestrial) target rocks that were involved in the formation of the impact breccias or melt rocks (Koeberl *et al.*, 2002b). For example, if the average Cr concentration in the target is ~185 ppm (the average Cr concentration in the bulk continental crust; Taylor and McLennan, 1985), only an extraterrestrial component of more than 1.2% by weight can be detected (Koeberl *et al.*, 2002b).

5.3.6 Tungsten Isotopes

The tungsten isotopic composition, which is based on the extinct ^{182}Hf–^{182}W isotope system, has been used very recently and less in context of impact crater-scale studies, but rather on a large scale, to constrain the timing of Earth's accretion and core formation (e.g., Halliday *et al.*, 1996; Halliday, 2004). This relates to probably the largest collision of all that the Earth endured—the Moon-forming impact event (cf. Lee *et al.*, 1997; Canup and Righter, 2000; Canup and Asphaug, 2001; Canup, 2004). This use is possible because (1) the ^{182}Hf half-life of 9 Myr is on the order of early solar system accretionary processes; (2) both hafnium and tungsten are refractory elements and thus are not fractionated by nebular processes; and (3) hafnium is lithophile, whereas tungsten is siderophile, resulting in strong Hf/W fractionations during metal/silicate separation (i.e., core formation) (cf. Kleine *et al.*, 2004). As discussed by these authors, as a result of core formation, the bulk silicate Earth has an elevated Hf/W ratio (Hf/W = 10–40) relative to chondrites (Hf/W~1), and if the Earth's core formed early (when ^{182}Hf was still present), an excess in the ^{182}W atomic abundance in the bulk silicate Earth relative to chondrites would be generated. Kleine *et al.* (2004) showed that the tungsten isotope composition of the bulk silicate Earth exhibits a small but

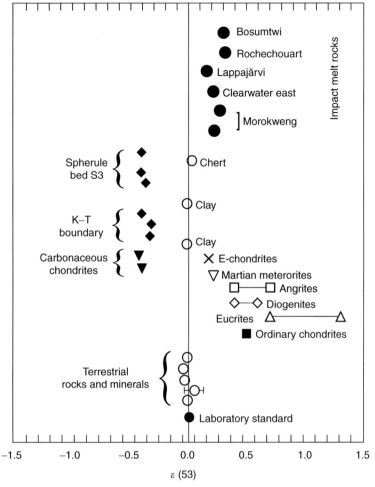

Figure 13 Chromium isotopic composition of various terrestrial standards, meteorites, K–T boundary sediments, Early Archean spherule samples from Barberton (South Africa), and melt rocks/glasses from several impact craters. It is evident that the Early Archean spherules and the K–T boundary data indicate a carbonaceous chondritic projectile, whereas the various impact craters have ordinary chondrite precursors. Data from Shukolyukov and Lugmair (2000), Shukolyukov *et al.* (2000), and Koeberl *et al.* (2004b, in preparation).

resolvable excess in the abundance of ^{182}W relative to that of chondrites, confirming that the Earth's core formed within the lifetime of now extinct ^{182}Hf. Furthermore, tungsten isotope data indicate that the Moon cannot have formed before ~40 Myr, which results in an age of the Earth and the Moon of 40–70 Myr after the start of the solar system.

Because the tungsten contents in meteorites are not particularly high compared to terrestrial rocks, and the differences between terrestrial and extraterrestrial materials are small and difficult to measure: the use of this isotopic system in terrestrial cratering studies is limited (but see the discussion in Section 5.4.8).

5.3.7 Stable Isotopes

The most commonly measured stable isotopes in impact-related materials are those of carbon,

oxygen, and sulfur. The background for these isotopic methods is well known and described in a variety of different references (e.g., Faure, 1986, and references therein). The measurement of such isotopic ratios may have three main goals. First, if isotopic compositions of impactites themselves are measured, source rocks can be determined and alteration processes can be studied (e.g., Chaussidon *et al.*, 1994). For example, oxygen isotopic measurements have shed light on the sources of tektites (e.g., Taylor and Epstein, 1966; Engelhardt *et al.*, 1987, 2005; Blum and Chamberlain, 1992; Koeberl *et al.*, 1997b). Second, the geochemistry of carbon in impactites can be best determined from isotopic studies of carbon components (e.g., Gilmour, 1998; Gilmour *et al.*, 1992, 2003; Parnell *et al.*, 2005). Third, the direct lithological or secondary paleoenvironmental effects of impact events in the geological record can be studied (e.g., in case of the K–T boundary, cf. De Paolo *et al.*, 1983; Schmitz *et al.*, 1988; Gilmour *et al.*, 1990, and many others).

5.3.8   Other Methods

A variety of other methods that involve geochemical data have been used in the study of impact craters and events. A few examples are presented here.

5.3.8.1   Helium-3

This helium isotope is extremely rare in terrestrial crustal rocks, but relatively abundant in interplanetary dust particles (IDPs). However, particles larger than a few tens of micrometers would heat up entering the Earth's atmosphere to the point of releasing their helium, which, being very light, would be lost into the outer space. Farley *et al.* (1998) found much enhanced levels of ^{3}He coinciding with the two late Eocene impactoclastic layers that are correlated to the Popigai (Masaitis, 1994; Bottomley *et al.*, 1997; Vishnevsky and Montanari, 1999) and Chesapeake Bay (Poag *et al.*, 1994, 2004; Koeberl *et al.*, 1996a; Poag, 1997) impact events, which also had a climatic influence (e.g., Bodiselitsch *et al.*, 2004) and shows up in the marine Os record (Dalai *et al.*, 2006). Helium-3 is a proxy for the influx of extraterrestrial dust (e.g., Mukhopadhyay

et al., 2001), and Farley *et al.* (1998) interpreted the broad rise in ^{3}He abundance to indicate that during the late Eocene there was a time of enhanced comet activity in the inner solar system, probably resulting in a higher impact rate than usual (see Figure 14). In contrast, Tagle and Claeys (2004) cited PGE data of Popigai impact melt rocks that indicate an L-chondritic, and thus asteroidal and not cometary, composition of the Popigai projectile. More recently Farley *et al.* (2006) found a ^{3}He signal in deep-sea sediments that seems to be correlated to an asteroidal breakup event in the mid-Miocene; this may resolve the discrepancy by indicating that particles derived from both cometary and asteroidal sources carry a ^{3}He signal.

5.3.8.2   Beryllium-10

Studies of Australasian tektites used measurements of the concentrations of the cosmogenic radionuclide ^{10}Be, which forms by interaction of cosmic rays with nitrogen in the atmosphere and is concentrated in the top of any sediment column, to constrain the location and characteristics of the source material. Beryllium-10 has a half-life of about 1.5 Myr; thus, it can only be studied in reasonably young impact structures and glasses. The average value of ^{10}Be (corrected to time of formation at 0.77 Ma) in Australasian tektites is $143 \pm 50 \times 10^6$ atoms g^{-1}. Such a value is

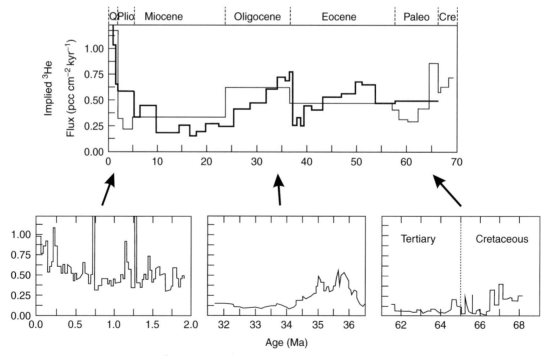

Figure 14   Diagram showing the ^{3}He abundance in deep-sea sediments over the past 70 million years, as a proxy for the influx of extraterrestrial dust. The insets show no large-scale accretion event of extraterrestrial dust at the K–T boundary (indicating a single impact event, not related to any asteroidal breakup event), a broad rise in ^{3}He abundance during the late Eocene, coinciding with multiple impact events, and some more recent (Pliocene) anomalies. Data from Farley *et al.* (1998, 2005, 2006) and Mukhopadhyay *et al.* (2001).

comparable to those measured in near-surface source materials, such as soils (terrestrial) or sediments (marine and terrestrial). It was also found (Ma *et al.*, 2004) that within the Australasian strewn field there is a correlation between the tektite type and [10]Be concentration. Aerodynamically shaped tektites (Australites) found farther from the presumed impact site in southeast Asia have higher contents than layered (Muong Nong-type) tektites that have not been transported far. This has implications on the formation mechanism of tektites (see below). The Ivory Coast tektites are regular (splash-form; see, e.g., Koeberl *et al.*, 1997b) tektites related to the Bosumtwi Crater formed at 1.07 Ma, and if Ivory Coast tektites also formed from surficial materials as did the Australasian tektites, then they are expected to contain similar concentrations of [10]Be. Measurements by Serefiddin *et al.* (2005) show that [10]Be concentrations of Ivory Coast tektites are consistent with formation from mostly near-surface sediments or soils. These authors found that Ivory Coast tektites have [10]Be values on average 77% lower than the Australasian tektites and concluded that several factors may have contributed to the variability in tektite [10]Be concentration: age of the tektites; nature and age of the source materials; depth of the sample in the soil column; and environmental conditions at time of formation. Individually these different factors seem unlikely to account for the full difference in [10]Be concentrations between the Ivory Coast and Australasian tektites. However, a combination of these factors can readily explain up to 80% of this difference. A mixture of sediments and soils with low [10]Be materials such as deposits eroded from bedrock may also explain the lower values in the Ivory Coast tektites.

5.3.8.3 Remote sensing

As mentioned above, only the presence of diagnostic shock metamorphic effects and, in some cases, the discovery of meteorites, or traces thereof, are generally accepted as providing unambiguous evidence for an impact origin. However, many structures exist on Earth that superficially might resemble (eroded) impact craters. Remote sensing studies, at high resolution and with multispectral information, can help to weed out structures that are clearly not of impact origin. But even though remote sensing is an extremely useful tool, it does not allow (so far) the detection of unique and unambiguous criteria for the confirmation of the impact origin of geological structures. Nevertheless, in terms of chemical characterization, it is possible to use remote sensing techniques on Earth (e.g., Ramsey, 2002) or other planets and satellites in the solar system for geochemical mapping (e.g., on the Moon,

see Figure 15). On Earth, the use of γ-ray spectrometry in aerogeophysical studies can determine the distribution of such radioactive elements as potassium, thorium, and uranium, which helps to provide maps of the distribution of certain rock types (Figure 16), and which in turn can help to guide more detailed explorations, such as drilling (e.g., Boamah and Koeberl, 2002; Pesonen *et al.*, 2003).

5.4 EXAMPLES

The following sections give a selection of examples of the use of geochemical methods and studies of impact craters and processes. Space limitations restrict that selection.

5.4.1 Meteorite Craters: Source Rocks and Impactites

It is important to understand how the compositions of various impact lithologies (i.e., melt-matrix breccia, black-matrix breccia, defined below, and suevites) can be explained as mixtures of different proportions of target rocks. A good example from Meteor Crater was given by Hörz *et al.* (2002). This usually involves mixing calculations. In the case of Meteor Crater, Hörz *et al.* (2002) found that preserved impact glasses are generally clear and texturally homogeneous, but unlike typical impact melts, they have unusually heterogeneous compositions, both within individual particles and from sample to sample. The meteoritic content (at Meteor Crater an iron meteorite) is unusually high and it is distributed bimodally, with specific samples containing either 5–10 or 20–30% FeO. According to Hörz *et al.* (2002) these compositional heterogencities may reflect the high carbonate content of the target rocks and the release of CO_2 during melting. The high projectile content and the CO_2-depleted residue of purely sedimentary rocks produced mafic melts that show three compositional groups, reflecting variable proportions of the major target rock formations (the Moenkopi, Kaibab, and Coconino formations). Least-square mixing calculations indicated that one group of impact glasses was made from 55% Moenkopi, 40% quartz-rich upper Kaibab, and 5% meteorite, suggesting a source depth of <30 m from the preimpact surface. The other two impact melt groups have higher contents of meteorite (15–20%) and Kaibab (50–70%), implying melt depths >90 m or <30 m, respectively. The work by Hörz *et al.* (2002) is a good example of the usefulness of mixing calculations.

For such calculations, the average compositions (with standard deviations) of the main target rock types present should be used. To illustrate the procedure in more detail, I describe another example of an application of mixing calculations, for

Figure 15 Multispectral image, taken by the Clementine spacecraft, of the Aristarchus impact crater on the Moon. This is a relatively young impact crater, and the multispectral image shows different colors for rocks of different compositions. The impact event excavated anorthositic (plagioclase-rich) lunar highland rocks that stand out (in blue in this false-color image) from the mature lunar regolith.

the Gardnos impact crater, Norway (French *et al.*, 1997). At Gardnos, three basement (target) lithologies have been distinguished (French *et al.*, 1997): (1) a variable suite of granitic gneisses that form the majority (>75–80 area%) of the outcrop area; (2) amphibolite, generally present as crosscutting dikes in the granitic gneisses; and (3) a coarse-grained metamorphic orthoquartzite, which is white and massive outside the structure and becomes black and highly fractured within it. Impactites at Gardnos are: (1) shocked quartzite, which is black and highly fractured within the structure; (2) lithic breccias: (a) the well-known subcrater "Gardnos Breccia," which consists of fragments of white granitic gneiss in a pulverized black matrix; (b) a "black-matrix breccia," which is similar to the "Gardnos Breccia," but contains fewer fragments and a higher percentage of generally darker (carbon-rich) matrix; and (3) melt-bearing breccias: (a) suevite, in which fragments of melt and basement rock occur in the clastic matrix; (b) melt-matrix breccias (or impact melt breccias) in which crystallized melt forms a matrix to rock and mineral clasts. In the region there are several carbon-rich shales (Alum shale and Biri shale) that could be the source of the high carbon contents of the impactites (Gilmour *et al.*, 2003).

Mixing calculations were performed with the Harmonic Least-Squares (HMX) mixing calculation program (Stöckelmann and Reimold, 1989), which allows the usage of any number of components (in this case, target rock components) and component or mixture parameters (here elemental abundances). An advantage of this particular program is that analytical uncertainties can be included in the model computation. Furthermore, it is possible to select from a number of so-called refinement control parameters for the purpose of setting different constraints on such calculations. It is, for example, possible to constrain any results (component proportions) to total exactly to 100% or, if this option is not chosen, it is possible to determine whether additional, not yet considered, components are needed to obtain a result of 100%. It is also possible to determine *a priori* that a so-called pivot component, one of the offered components, is considered a "fixed" component of the mixture.

Four calculation runs were performed for each of the mean mixture compositions and with all the

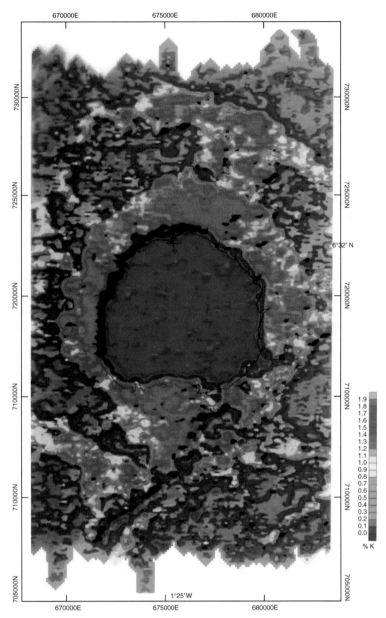

Figure 16 Radiometry image of the Bosumtwi impact structure, Ghana, showing the distribution of potassium, from an aerogeophysical study of Pesonen *et al.* (2003). Such maps are useful to study the distribution of different rock types at impact structures.

average component compositions (French *et al.*, 1997). All of these led to the conclusion that gneiss is the dominating target rock component of all three impactites. Before the first calculation was attempted, the average compositions were scrutinized to determine which parameters show large-enough variations between the target rock groups to be useful for distinguishing the resulting mixtures. Some element abundances were likely to have undergone postimpact changes, making them of little use for constraining target rock proportions. Some calculations were performed with major element parameters only, others with major elements in combination with a number of trace

elements that appeared to be sufficiently different in abundance from component to component. The conditions for the various computation runs were: Run 1, nine major elements without P_2O_5, all component proportions were forced to total to 100%; Run 2, seven major elements without P_2O_5, Na_2O, and K_2O, but including carbon, totals = 100%; Run 3, seven major elements, as in Run 2, but without carbon, and including scandium, yttrium, lanthanum, ytterbium, hafnium, and thorium, totals = 100%; Run 4, as in Run 2, but without carbon and with quartzite as a forced component; Run 5, seven major and eight trace elements, without carbon, total = 100%.

The carbon-bearing shale component was inferred from the high carbon content of the breccias, but is not observed to occur in any of the analyzed target rocks. Results are summarized in Table 2a.

In Table 2b, the observed and calculated mixture compositions for the three main impactite groups are compared, and the discrepancies observed for the various parameters are listed. Discrepancy factors, as given in Table 2b, are a calculated measure for the validity of the results: the better and statistically more valid a result, the closer the corresponding discrepancy value approaches 0. Some results are not satisfactory. For example, due to the (postimpact) mobility of the alkalis, the calculations containing Na_2O and K_2O concentrations (Run 1) may give unreliable results. Addition of trace-element data (scandium, chromium, cobalt, lanthanum, ytterbium, hafnium, throrium, and uranium) to the parameter list led to a slight increase in the discrepancy factor. For the melt-matrix breccias and the black-matrix breccias, runs with good discrepancy factors, and for the suevites, with reasonable discrepancy factors, were obtained. The high carbon content in the impactites can be reproduced by adding a carbon-shale component similar to Biri shale. For the best runs, the discrepancies between the calculated and observed compositions are small for all elements in the melt- and black-matrix breccias, with somewhat higher deviations for CaO and MgO in the suevites. Discrepancy factors do not change significantly when a carbon-rich shale component is included. The results show that gneiss is the dominant target rock component for all three impactites, with significant (10–30%) amphibolite contributions for the melt-bearing breccias, and 3–19% carbon-shale contributions in all three breccia types.

Such calculations allow the determination of the rock types that can explain the compositions of breccias and melt rocks. This is important if the indigenous compositions need to be determined for the study of possible meteoritic components, as described in the next section.

5.4.2 Extraterrestrial Components in Impactites

As discussed above, the admixture of minor quantities of meteoritic material (chondrite and iron meteorite) to crustal target rocks yields significantly elevated abundances of the PGEs in the resulting impact melt rocks or breccias (Table 3). For example, if only 1 wt.% of a chondritic

Table 2a Comparison of measured breccia compositions from the Gardnos Crater with those obtained from mixing calculations.

	Impact melt breccia		Black-matrix breccia		Suevites	
	Run 2	$\Delta_{obs-calc}$	Run 2	$\Delta_{obs-calc}$	Run 3	$\Delta_{obs-calc}$
SiO_2	66.21	−0.11	74.31	−0.33	63.31	−0.70
TiO_2	0.834	0.006	0.425	−0.05	0.92	0
Al_2O_3	12.92	0.14	10.88	0	13.31	0.04
Fe_2O_3	5.07	−0.009	3.43	−0.13	6.52	−0.02
MnO	0.075	<0.001	0.052	0.006	0.135	0.014
MgO	1.32	0.02	0.57	0.14	2.38	0.57
CaO	1.41	−0.06	0.76	−0.02	1.94	−0.25

From French *et al.* (1997).Note: All data in wt.%. The runs giving the best results were used for this comparison. $\Delta_{obs-calc}$ = observed value (Table 1) minus calculated value.

Table 2b Mixing calculations to reproduce Gardnos impact melt breccia and suevite compositions.

		Target rock components				Discrepancy factor
		Quartzite	Amphibolite	Gneiss	C-Shale	
Impact melt breccia	1	0.0	6.5 ± 1.8	93.5 ± 2.9	—	6.5
Impact melt breccia	2	0.0	12.1 ± 1.1	84.9 ± 1.2	2.9 ± 0.08	1.06
Impact melt breccia	3	4.6 ± 1.7	17.1 ± 1.8	78.3 ± 2.7	—	4.6
Black-matrix breccia	2	7.7 ± 1.2	<0.5	80.4 ± 2.2	11.9 ± 1.2	0.38
Black-matrix breccia	3	12.8 ± 1.8	<0.5	87.2 ± 1.8	—	1.3
Suevites	2	<0.5	25.1 ± 2.3	56.1 ± 3.2	18.8 ± 0.5	6.7
Suevites	3	<0.2	25.7 ± 2.6	74.3 ± 2.6	—	8.9
Suevites	4	1.2 ± 1.4	28.0 ± 2.9	70.8 ± 3.8	—	5.2

From French *et al.* (1997).
Run numbers: 1, all major elements, components total 100%; 2, major elements, except Na and K, plus C, components total 100%; 3, major elements, except Na and K, plus Sc, Y, La, Yb, Hf, Th; 4, as 2, but with quartz as a forced component.

Table 3 Terrestrial impact structures with meteoritic components and inferred impactor types.

Name	Country	Diameter (km)	Location (center)	Bolide type	Evidence	References
Wabar	Saudia Arabia	0.10[a]	21°30′N, 50°28′E	IIIAB iron	M, S	Morgan et al. (1975), Mittlefehldt et al. (1992b)
Kaalijärvi	Estonia	0.11[a]	58°24′N, 22°40′E	IAB	M	Buchwald (1975)
Henbury	Australia	0.16[a]	24°35′S, 133°09′E	IIIAB	M, S	Taylor (1967)
Odessa	USA	0.17[a]	31°45′N, 102°29′W	IIIAB	M	Buchwald (1975)
Boxhole	Australia	0.17	22°37′S, 135°12′E	IIIAB	M	Buchwald (1975)
Macha	Russia	0.3[a]	59°59′N, 118°00′E	Iron	MS	Gurov (1996)
Aouelloul	Mauritania	0.4	20°15′N, 12°41′W	Iron (IIIB, IID?)	S, Os	Morgan et al. (1975), Koeberl et al. (1998)
Monturaqui	Chile	0.46	23°56′S, 68°17′W	IAB	M, S	Bunch and Cassidy (1972), Buchwald (1975)
Kalkkop	South Africa	0.64	32°42′S, 24°26′E	Chondrite?	S, Os	Koeberl et al. (1994a), Reimold et al. (1998)
Wolfe Creek	Australia	0.9	19°18′S, 127°46′E	IIIAB	M, S	Attrep et al. (1991)
Meteor (Barringer)	USA	1.2	35°02′N, 111°01′W	IAB	M, S	Morgan et al. (1975), Mittlefehldt et al. (1992a)
Saltpan	South Africa	1.2	25°24′S, 28°05′E	Chondrite	S, Os	Koeberl et al. (1994a)
Roter Kamm	Namibia	2.5	27°46′S, 16°18′E	Chondrite?	S	Reimold et al. (unpublished data)
New Quebec	Canada	3.4	61°17′N, 73°40′W	Chondrite (L?)	S	Grieve et al. (1991), Evans et al. (1993)
Brent	Canada	3.8	46°05′N, 78°29′W	Chondrite	S	Palme et al. (1981), Evans et al. (1993)
Gow Lake	Canada	4	56°27′N, 104°29′W	Iron?	(S)	Wolf et al. (1980)
Rio Cuarto	Argentina	4.5[a]	30°52′S, 64°14′W	Chondrite (H)	M, S, Os	Schultz et al. (1994), Koeberl (unpublished data)
Ilyinets	Ukraine	4.5	49°06′N, 29°12′E	Iron?	S	Grieve and Shoemaker (1994)
Süüksjürvi	Finland	5	61°24′N, 22°24′E	Stony-iron, iron?	S	Palme et al. (1980), Schmidt et al. (1997)
Gardnos	Norway	6	60°40′N, 09°00′E	Chondrite	S, Os	French et al. (1997)
Wanapitei	Canada	7.5	46°45′N, 80°45′W	Chondrite	S	Wolf et al. (1980), Evans et al. (1993)
Mien	Sweden	9	56°25′N, 14°52′E	Stone?	S	Palme et al. (1980)
Bosumtwi	Ghana	11	06°30′N, 01°25′W	L Chondrite	S, Os, Cr	Koeberl and Shirey (1993), Koeberl et al. (2004)
Ternovka	Ukraine	12	48°01′N, 33°05′E	Chondrite?	S	Grieve and Shoemaker (1994)
Nicholson Lake	Canada	12.5	62°40′N, 102°41′W	Achondrite	S	Wolf et al. (1980)
Zhamanshin	Kazakhstan	13.5	48°20′N, 60°58′E	Chondrite (Iron?)	S	Glass et al. (1983), Palme et al. (1978)
Dellen	Sweden	15	61°55′N, 16°39′E	Stone?	S	Palme et al. (1980)
Obolon	Ukraine	15	49°30′N, 32°55′E	Iron?	S	Grieve and Shoemaker (1994)

(Continued)

Table 3 Continued

Name	Country	Diameter (km)	Location (center)	Bolide type	Evidence	References
Lappajärvi	Finland	17	63°12'N, 23°42'E	Chondrite	S	Göbel et al. (1980)
Elgygytgyn	Russia	18	67°30'N, 172°00'E	Achondrite?	S	Grieve and Shoemaker (1994)
Clearwater East	Canada	22	56°05'N, 74°07'W	L Chondrite	S, Cr	McDonald (2002); Shukolyukov and Lugmair (2000)
Rochechouart	France	23	45°50'N, 00°56'E	Chondrite? Iron?	S, Cr	Janssens et al. (1977), Wolf et al. (1980), Lambert (1982), Shukolyukov and Lugmair (2000)
Ries	Germany	24	48°53'N, 10°37'E	Achondrite?	S	Morgan et al. (1979), Schmidt and Pernicka (1994)
Boltysh	Ukraine	25	48°45'N, 32°10'E	Chondrite?	S	Grieve and Shoemaker (1994)
Strangways	Australia	25	15°12'S, 133°35'E	Achondrite	S	Morgan and Wandless (1983)
Mistastin	Canada	28	55°53'N, 63°18'W	Iron?	S	Wolf et al. (1980)
Manson	USA	38	42°35'N, 94°33'W	Chondrite	S, Os	Pernicka et al. (1996), Koeberl and Shirey (1996)
Mjølnir	Norway	40	73°48'N, 29°40'E	Iron	S	Dypvik and Attrep (1999)
Morokweng	USA	70	26°20'S, 23°32'E	L Chondrite	S, Os, Cr	Koeberl et al. (1997a, 2002b), McDonald et al. (2001), Koeberl and Reimold (2003)
Chesapeake Bay	USA	85	37°16'N, 76°01'W	Chondrite?	S, Os	Lee et al. (2006)
Acraman	Australia	90	32°01'S, 135°27'E	Chondrite	Es	Gostin et al. (1989), Wallace et al. (1990)
Kara	Russia	65–100	69°12'N, 65°00'E	Chondrite?	S	Nazarov et al. (1989, 1990)
Popigai	Russia	100	71°30'N, 111°00'E	L Chondrite	S	Masaitis and Raikhlin (1985), Masaitis (1994), Tagle and Claeys (2004)
Chicxulub	Mexico	180	21°20'N, 89°30'W	CM2 Chondrite	S, Os, Es, Cr	Koeberl et al. (1994b), Schuraytz et al. (1996), Shukolyukov and Lugmair (1998), Trinquier et al. (2006)
Vredefort	South Africa	300	27°00'S, 27°30'E	Chondrite	S, Os	Koeberl et al. (1996b)

Updated after Koeberl (1998).

[a] Crater field; largest dimension of largest structure is given

Evidence: S, siderophile element enrichment and/or pattern; Os, osmium isotopic ratio; Cr, chromium isotopic ratio; M, meteorite fragments; MS, metallic spherules; Es, siderophile element enrichment in ejecta. Only craters larger than 0.1 km in diameter are listed.

meteorite is mixed into terrestrial crustal target rocks, the resulting breccia or melt will contain about 4 ppb of Ir. Even 0.1 wt.% of a chondritic component still results in 0.4 ppb of Ir (or Os) in the breccia, which is at least one order of magnitude higher than average background values. Achondrites have a much wider range of Ir and Os contents, and, in general, lower abundances than for chondrites. Ranges of contents of about 0.1–760 ppb Ir and 0.02–5 ppb Os have been reported (Mason, 1979). It is not clear if the apparent differences between the Ir and Os contents are real or are the result of analytical problems, because Gros *et al.* (1976) reported Ir and Os to have comparable concentrations in the Juvinas eucrite (0.028 and 0.018 ppb, respectively).

Confirmed enrichments of iridium and other PGEs in impact-derived rocks provides good evidence for the presence of a meteoritic component (see, e.g., Morgan *et al.*, 1975; Palme *et al.*, 1978; Palme, 1982). In a number of these studies, PGE abundances and interelement ratios were used in an attempt to resolve the type or class of meteorite (e.g., Morgan *et al.*, 1975; Palme *et al.*, 1978, 1979; Evans *et al.*, 1993).

Using siderophile element abundances and interelement ratios, the projectile type was determined for a number of terrestrial craters. A list of such projectile identifications for 45 impact structures is given in Table 3. Some comments regarding this table, which is updated from Koeberl (1998), are required. First, only impact craters/structures with diameters larger than 100 m are listed. Several smaller-impact craters with known projectile types exist, and in all these cases iron meteorites have been found as remnants of the impactor. For such small craters, the change from a simple excavation hole to an actual hypervelocity impact crater is not well defined. As a general rule, the larger the crater, the more likely it is that the projectile has been vaporized. Some of the smaller structures listed are in fact crater fields consisting of several more or less circular structures, such as Wabar, Henbury, Odessa, and Rio Cuarto. The latter one consists of at least 10 rimmed, oblong structures ranging in size between about 0.1×0.3 and 4.5×1.1 km, which formed by a very low-angle (grazing) impact of a chondritic body, with the top of the impactor being decapitated and producing the down-range crater structures (e.g., Schultz *et al.*, 1994).

In Table 3, a few points are worth noticing. Almost all of the smaller-impact craters (less than about 1.2 km in diameter) were formed by iron projectiles of known composition, with the exception of the Saltpan crater (Koeberl *et al.*, 1994a) and possibly even of the 0.65 km crater Kalkkop (cf. Reimold *et al.*, 1998), for which chondritic projectiles seem more consistent with the data. The types of most of these iron meteorites are known, because fragments of the

meteorites were still present at the crater sites. All the craters with meteorite fragments are relatively young (less than about 0.1 Ma), whereas at older craters (e.g., Aouelloul at 3.1 Ma) no remnants of the actual impactor were recovered. While the Saltpan crater is only 0.22 Myr old, no remnants of the impacting chondrite were found, most likely due to erosion of any remaining fragments. In contrast, a small mass of the chondritic impactor (of H4/5 type) that formed the much larger Rio Cuarto crater field is still preserved, as this structure is quite young (probably <4,000 years).

It should also be pointed out that conflicting identifications have been made for a number of impact structures, and that many identifications are highly uncertain. For example, both iron and chondritic projectiles were proposed for the Bosumtwi (Ghana), Zhamanshin (Kazakhstan), and Rochechouart (France) impact structures. For others, the question mark indicates a greater uncertainty in identifications (e.g., those that are based only on the study of Ni, Co, and Cr, without determination of the PGEs). The reasons for some of this greater uncertainty are discussed in the following section.

Apart from the seemingly straightforward cases in which meteorite fragments are found at a crater, inferences about the projectile type have to be made from the abundances and interelement ratios of various siderophile elements in impactites. However, several complications are introduced by complex fractionation processes that seem to take place during the formation of impact glasses and melts.

In recent studies of impact glasses from some small craters for which the meteorite has been partly preserved (e.g., Meteor Crater, Wabar, and some Australian craters), it was found that the siderophile elements show strong (and highly variable!) fractionation in the interelement ratios compared to the initial ratios in the impacting meteorite. The changes do not seem to be correlated in a straightforward way with any physical or chemical properties of the elements (Attrep *et al.*, 1991; Mittlefehldt *et al.*, 1992a, b; Hörz *et al.*, 2002). Mittlefehldt *et al.* (1992b) ruled out simple vapor fractionation of the pure elements, or post-depositional fractionation. Instead, they proposed that the siderophile elements may have been fractionated from each other during the early phases of the impact, while the projectile was undergoing decompression and before mixing with the target materials, although they were unable to explain the fractionation with a specific model.

In general, though, there is a correlation between the amount of melt and vapor generated and increasing impact velocity, which may, thus, be the most important factor controlling the incorporation of a meteoritic component, the amount of which is known to vary widely between craters of similar size, and also within impact breccias and melt rocks from a single crater. However, this energy scaling relationship cannot explain the variations in fractionation within a single crater. This observation does not bode well for any attempts to directly infer projectile types of small craters from siderophile element ratios in impactites. And indeed, the element ratios at, for example, Aouelloul or Brent do not readily conform to those of any known meteorite types, but the discrepancies can be interpreted in various ways.

The Au/Ir ratio has also been used to distinguish between cosmic and terrestrial signatures (Palme, 1982). However, gold is much more mobile than iridium under a number of terrestrial conditions (e.g., weathering, diagenesis, and metamorphism), which may lead to nonchondritic Ir/Au ratios even if a meteoritic component is present in impact-derived rocks. Specifically, gold often shows high indigenous abundances in some terrestrial rocks, yielding low (nonchondritic) Ir/Au ratios. Matters are further complicated by the fact that PGEs are enriched in certain terrestrial rock types (see below in the section on Ivory Coast tektites). The PGE patterns of mantle and some mantle-derived rocks may be similar to those of chondrites. The PGE abundances in mantle rocks are also higher than those in the crust by factors of about 10^2–10^3. This makes it difficult, if not impossible, to distinguish between an exposed mantle section or a component in the impactites derived from an ultramafic precursor rock and a meteoritic component.

A further complication is the possibility of siderophile element fractionation in the impact melt while it is still molten. This effect may be significant in larger craters, because there the melt can stay hot for several thousand years. Different mineral phases, such as sulfides or oxides (e.g., magnetite and chromite), may take up various proportions of the PGEs or other siderophile elements, leading to an irregular distribution of these elements and possibly fractionated interelement ratios and patterns. Such irregular distribution of siderophiles is known from, for example, the East and West Clearwater impact structures (Palme *et al.*, 1979) and the Chicxulub impact structure (Koeberl *et al.*, 1994c; see also below). Hydrothermal processes associated with the hot impact melt may also change PGE abundances. In contrast to a widely held opinion, actual data show that meteoritic components are often inhomogeneously distributed in impact melt rocks and breccias (e.g., data in Palme *et al.*, 1979; Koeberl *et al.*, 1994a b, c; Schuraytz *et al.*, 1994).

A variety of fractionation effects have also been documented for distal ejecta, for example, from the K–T boundary impact at various localities around the world (e.g., Evans *et al.*, 1993). High PGE abundances were discovered in impact ejecta from the Acraman structure in Australia

(Williams, 1986, 1994); however, these show deviations from chondritic patterns due to low-temperature hydrothermal alteration (e.g., Gostin *et al.*, 1986, 1989; Wallace *et al.*, 1990; cf. also Colodner *et al.*, 1992). All these problems make it difficult to properly identify a meteoritic component based only on PGE abundances and ratios.

Thus, in the following paragraphs three examples are given in which a combination of various elemental and isotopic methods are shown to be successful.

5.4.2.1 Vredefort impact structure (South Africa)

The Vredefort structure in South Africa, centered about 140 km southwest of Johannesburg, currently has a diameter of about 100 km, which is believed to represent the central uplift of a much larger, deeply eroded impact structure. The whole impact structure could initially have been as large as 300 km, comprising the entire Witwatersrand Basin (Henkel and Reimold, 2002). The origin of the structure has been debated during most of the twentieth century, but recent data confirmed the existence of impact-characteristic shock metamorphic effects in Vredefort rocks, such as basal Brazil twins in quartz (Leroux *et al.*, 1994) and shock-characteristic PDFs in zircon (Kamo *et al.*, 1996), supporting an impact origin as opposed to an internal origin of the structure (see Reimold and Gibson, 1996, for a review). The Vredefort event is well dated at 2023 ± 4 Ma (Kamo *et al.*, 1996). Dikes of granophyric rock, the so-called Vredefort Granophyre, occur in the basement core of the structure and along the boundary between the core and the supracrustal rocks of the collar. Previous studies indicated that the granophyre could have formed by impact melting of granite, shale, and quartzite. A major mafic contribution is unlikely due to the scarcity of mafic clasts and because the granophyre composition can be perfectly modeled from mixing of felsic crustal and supracrustal rocks (French *et al.*, 1989; French and Nielsen, 1990; Reimold *et al.*, 1990; Koeberl *et al.*, 1996b; Reimold and Gibson, 1996). All these authors agreed that the Vredefort Granophyre represents an impact melt rock, probably material that gravitationally settled into impact-generated fractures in the crater basement after the impact event.

French *et al.* (1989) analyzed the iridium content of seven Granophyre samples and reported a range of 57–130 ppt Ir. These authors also analyzed a suite of country rocks and found that the Granophyre is enriched in iridium by a factor of about 20–50 compared to granitic country rocks (target rocks?). However, they also found that some Witwatersrand shale samples have Ir contents of 160–330 ppt. French *et al.* (1989) suggested that all iridium in the Granophyre could be explained by admixture of iridium from shales and similar rocks during impact melting. However, this would require that at least one-third of the Granophyre composition be derived from shale, which contradicts mixing calculations of French and Nielsen (1990), who only find a 10% shale contribution. All other relevant country rocks have Ir contents of 3–62 ppt. Thus, it seems that there is an iridium excess in the Granophyre; nevertheless, this ambiguity made the Granophyre, and Vredefort in general, an ideal target for an osmium isotopic study.

Such a study was undertaken by Koeberl *et al.* (1996b) to search for a meteoritic component in Vredefort Granophyre, and to confirm that the granophyre represents an impact melt rock. The Os abundances in the granophyre range from 0.11 to 1.11 ppb, which is significantly higher than the average of the source rock values, indicating a distinct enrichment of osmium in most of the granophyre samples compared to the country rocks. In addition, the $^{187}Os/^{188}Os$ ratios of the granophyre samples are significantly lower than those of the supracrustal country rocks. The $^{187}Re/^{188}Os$ and $^{187}Os/^{188}Os$ ratios of the Vredefort Granophyre scatter about a 2 Ga isochron, with the majority of the initial $^{187}Os/^{188}Os$ ratios (at 2 Ga) ranging from 0.13 to 0.22. These values overlap the meteoritic data range and indicate that all measured granophyre samples contain some meteoritic osmium. In addition, the Re–Os isotopic composition of the granophyre is significantly different from that of any of the target rocks. Shale samples were also analyzed and yielded values of 37–162 ppt Os, but the osmium isotopic compositions are significantly different from those of the granophyre samples.

Thus, a meteoritic component in the Vredefort Granophyre is the only explanation that is in agreement with these observations. This conclusion is supported by some enrichments in chromium, cobalt, nickel, and iridium in the granophyre compared to the country rocks, although these enrichments are not as unambiguous as the Re–Os data. Assuming chondritic meteorite Os abundances (about 500 ppb), Koeberl *et al.* (1996b) concluded that the Vredefort Granophyre contains ≤0.2% of a chondritic component. However, the meteoritic component is not homogeneously distributed, probably due to a nugget effect, as is evident from a spread in $^{187}Re/^{188}Os$ ratios, in agreement with observations from other impact melt rocks (cf. Koeberl, 1998). Meteorites contain about 10 times less rhenium than osmium, indicating that the rhenium contribution from the meteoritic material to the granophyre was ≤30%, subordinate to the osmium, which is almost exclusively of meteoritic origin. The almost constant Re concentrations found in the granophyre samples indicate homogenization during the impact. Thus, the granophyre is a mixture of a large amount of low-osmium, high-rhenium material (crustal rocks) with a small contribution of high-osmium, low-rhenium meteoritic material. The osmium isotope results showed that all analyzed Vredefort Granophyre samples contain some meteoritic osmium, confirming that the granophyre is an impact melt rock.

Koeberl *et al.* (2002b) reported on the chromium isotopic composition of Vredefort Granophyre. In contrast to the osmium isotope study, the chromium isotope study did not reveal the presence of extraterrestrial chromium. The two Vredefort Granophyre samples yielded $^{53}Cr/^{52}Cr$ values of $-0.01 \pm 0.06\varepsilon$ and $+0.03 \pm 0.03\varepsilon$ ($2\sigma_{mean}$), which are indistinguishable from the terrestrial mean value ($\equiv 0\varepsilon$). The average Cr content of the Granophyre is about 420 ppm (Koeberl *et al.*, 1996b; Reimold *et al.*, 1999), whereas Cr contents in the various target rocks range from 6 to 750 ppm, with a possible average Cr content on the order of 300 ppm. Based on the chromium data, the upper limit for a chondritic component in these samples is ~2% (Figure 2), which is consistent with the better-constrained limit (ca. 0.2%) provided by the osmium isotopes. An abundance of terrestrial chromium masks the cosmic chromium and, in this case, the sensitivity of the chromium isotopic method is not sufficient to resolve the meteoritic component.

5.4.2.2 Morokweng impact structure (South Africa)

The ca. 70–80 km diameter Morokweng impact structure is centered at $23°32'$ E and $26°20'$ S, close to the border with Botswana, in the Northwest Province of South Africa. The structure was recognized as a circular positive magnetic anomaly of up to 350 nT above regional background. This anomaly forms a central 30-km-diameter near-circular area, which is surrounded by a concentric, magnetically quiet zone that is 20 km wide. Refined processing of the gravity and aeromagnetic data revealed the possible presence of a larger circular structure (Corner *et al.*, 1997; Reimold *et al.*, 1999; Andreoli *et al.*, 1999). The discovery of impact-characteristic shock metamorphic effects in rocks from the Morokweng area (e.g., Corner *et al.*, 1997; Koeberl *et al.*, 1997a; Hart *et al.*, 1997) confirmed the presence of a large meteorite impact structure. Several boreholes in the structure were sampled (see, e.g., Koeberl *et al.*, 1997a; Reimold *et al.*, 1999; Koeberl and Reimold, 2003).

In the boreholes, Tertiary to Holocene Kalahari Group calcrete is directly underlain by a darkbrown melt rock. In a couple of holes the lower contact of the melt rock was intersected and granitic rocks were reached. Most of the melt rock appears fresh and homogeneous, except for a large number of lithic clasts. The clast population includes many gabbro fragments, but microscopic studies reveal that felsic, clearly granitoid-derived, clasts are dominant. The granites drilled below the melt body are locally brecciated and pervasively recrystallized, probably due to the heart of the overlying melt sheet. Some primary minerals are preserved and display shock deformation in the form of PDFs in quartz, plagioclase, alkali feldspar, and potassium feldspar (Koeberl *et al.*, 1997a; Reimold *et al.*, 1999). SHRIMP ion probe dating of zircons from the melt rock yielded an age of 146.2 ± 1.5 Ma, which is indistinguishable from that of the Jurassic–Cretaceous boundary (Koeberl *et al.*, 1997a; Reimold *et al.*, 1999).

With the exception of a few obviously altered melt rock samples, the Morokweng melt body is extremely homogeneous in composition. Variations for major elements do not exceed 2–5 rel.%. The Morokweng melt rock contains relatively high proportions of CaO (on average, 3.41 wt.%), MgO (3.70 wt.%), and Fe_2O_3 (5.87 wt.%), and an average SiO_2 content of 65.75 wt.%. Siderophile elements are consistently enriched (the variation between samples is less than a factor of 2) in the melt rock samples in comparison with target rocks of such major element composition (granodioritic to dioritic), with average values of 440 ppm for Cr, 50 ppm for Co, 780 ppm for Ni, and 32 ppb for Ir. No variation with depth and no differences between drill cores was found (Koeberl *et al.*, 1997a; Reimold *et al.*, 1999). Koeberl *et al.* (1997a) noted that it was highly unlikely that mafic to ultramafic country rocks, or other mantle-derived sources, were responsible for these high siderophile element and Ir concentrations. The abundances of the PGEs in the various Morokweng impact melt rock samples analyzed were found to have almost chondritic ratios, and, depending on the values used for normalization,

about 2–5 wt.% of a chondritic component is present in the melt rocks (Koeberl *et al.*, 1997a). In contrast to melt rocks from most other impact structures (Koeberl, 1998), the meteoritic component has a high abundance and is uniformly distributed in the Morokweng impact melt rocks.

Koeberl *et al.* (1997a), McDonald *et al.* (2001), and Koeberl and Reimold (2003) found that the chondrite-normalized abundance patterns of the impact melt rock PGE data (cf. Table 1) are flat and indicate the presence of about 2 wt.% of a meteoritic component, whereas the normalized patterns of the various felsic and mafic target rocks are highly fractionated. The latter two studies concluded that an L-chondrite projectile provides the best fit.

To confirm the presence of this component, Koeberl *et al.* (2002b) performed chromium and osmium isotopic studies on a set of melt rocks and basement rocks from Morokweng. The results for Os show high contents of up to about 9 ppb in the impact melt rock samples, with correspondingly low, but remarkably uniform, isotopic ratios of $^{187}Os/^{188}Os$ at 0.1316–0.1341. The breccias show a much wider variation in both isotope ratio and osmium abundance. Figure 17 shows the Os content versus the osmium isotopic composition, with a hyperbolic mixing line that connects the meteorite and target rock fields. The impact melt rock samples plot close to this line at higher meteorite contribution values. The mafic rock samples from the KHK-1 deep drill core, representing the mafic contribution to the Morokweng impact melt rocks, have very low Os abundances (9–42 ppt) and high isotope ratios ($^{187}Os/^{188}Os$ up to 12). This result clearly indicates that the mafic target rocks of the area did not contribute measurably to the high siderophile element abundances observed in the melt rocks. Thus the

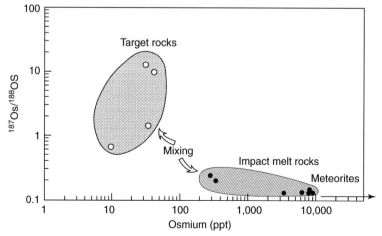

Figure 17 Osmium isotopic ratios versus Os content, showing the mixing relationship between meteoritic compositions and target rock compositions, and melt rocks (from the Morokweng impact structure, South Africa) that plot along a hyperbolic mixing line between these two fields. After Koeberl *et al.* (2002a).

osmium isotope data confirm the presence of a meteoritic component in these melt rocks.

As the meteoritic component in the Morokweng impact melt rocks seems to be fairly abundant, Koeberl *et al.* (2002b) also performed a chromium isotopic study. The results indicated that about half of the chromium in the melt rock is of extraterrestrial origin. It was also possible to show that only an ordinary chondritic source, rather than a carbonaceous chondritic source, can explain the chromium isotope data. Using the constraints of the chromium isotope data together with trace-element data (Shukolyukov *et al.*, 1999), Koeberl *et al.* (2002b) concluded that the Morokweng bolide was most likely of L-chondritic composition. This interpretation agrees well with the conclusions based on PGE abundances (McDonald *et al.*, 2001).

Maier *et al.* (2006) reported finding a large (25 cm), unaltered, fossil meteorite fragment as well as several smaller fragments in a drill core within the Morokweng impact melt rocks. The large fragment (clast) supposedly resembles an LL6 chondrite breccia, but with anomalously iron-rich silicates, iron-nickel sulfides, and no troilite or metal. The rock has chondritic chromium isotope ratios and PGE ratios that are identical with those of the bulk melt rock. This would be the first time that a fragment of the impactor is preserved within a large crater (but note that Kyte, 1998, described what may be a piece of the Chicxulub impactor, albeit outside the crater). The unusual composition of the meteorite is similar to, but not identical with LL chondrites in the known meteorite collections.

5.4.2.3 *Bosumtwi crater, Ghana*

NTIMS was used by Koeberl and Shirey (1993) for the measurement of concentrations and isotopic ratios of osmium and rhenium in four Ivory Coast tektites, two Bosumtwi impact glasses, and five different target rocks from the Bosumtwi crater. The tektites have major- and trace-element compositions as well as large negative ε_{Nd} (-20) and positive ε_{Sr} ($+260$ to $+300$), which are characteristic of old continental crust. From chemical and age data, the Bosumtwi impact crater has been inferred as the source crater for the tektites (e.g., Schnetzler *et al.*, 1966; Gentner *et al.*, 1967; Kolbe *et al.*, 1967, 1997b). The tektites and the target rocks have Os concentrations ranging from 0.09 to 0.30 ppb, and 0.021 to 0.33 ppb, respectively. However, the $^{187}Os/^{188}Os$ ratios in the tektites are close to meteoritic values at about 0.155–0.213 ($^{187}Os/^{186}Os$, 1.29–1.77), whereas the Bosumtwi crater rocks have values of 1.52–5.01. Their $^{187}Os/^{188}Os$ values are typical for old continental crust (12.3–41.4). The low $^{187}Os/^{188}Os$ ratios in the tektites are unambiguous evidence for the existence of up to 0.6% by weight of a meteoritic component (Koeberl and Shirey, 1993) (see Figure 18). More recently, Koeberl *et al.* (2004b) used Cr isotopic data to indicate a possible ordinary chondrite contribution to an Ivory Coast tektite (see also Figure 13).

In these investigations, the geochemistry of target rocks and breccias from the Bosumtwi crater was studied for comparison with Ivory Coast tektites. However, another important line of research, which is necessary for the identification of a meteoritic component in impact breccias and melt rocks, namely the determination of siderophile elements (especially the PGEs) in the target rocks at the crater, has so far been somewhat neglected. Dai *et al.* (2005) tried to calculate the contribution of the meteoritic component from the content of PGEs in the target rock samples,

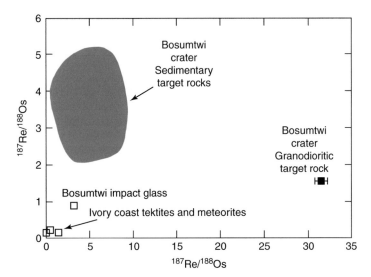

Figure 18 Rhenium–osmium isotopic plot for target rocks and impact glasses from the Bosumtwi impact structure, Ghana, and Ivory Coast tektites that were derived from that crater. The huge difference in the Os isotopic ratios between tektites and impact glasses and the crater basement rocks is obvious. After Koeberl and Shirey (1993).

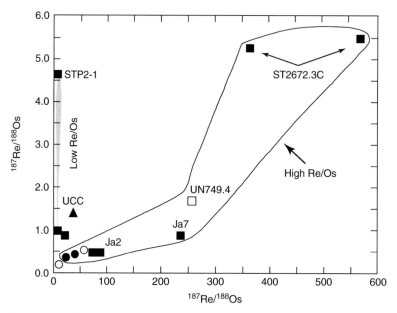

Figure 19 Rhenium–osmium isotopic plot for target rocks and impact breccias from the Chesapeake Bay impact structure, USA, showing for the first time that an extraterrestrial component (probably chondritic) is discernable in suevitic breccia from this crater (Lee *et al.*, 2006).

but found that the concentrations of the PGEs in the target rocks are very high and of do not show a flat abundance pattern when normalized to chondritic (meteorite) abundances (in this context, it is noteworthy that the region around Bosumtwi is a gold exploration area, and elevated siderophile element abundances are found in samples of regional country rock as well; cf. Jones, 1985). Thus, Bosumtwi is a good example for a case where it is difficult, if not impossible, to obtain reliable information on an extraterrestrial component from elemental abundances alone. Another case is the Chesapeake Bay impact structure (Figure 19). Only osmium and chromium isotopic data allowed a positive identification of an extraterrestrial component.

5.4.3 Tektites

Tektites are natural glasses of up to a few centimeters in size (Figure 20) that on Earth occur in just four geographically extended (but well-defined) strewn fields (e.g., Glass *et al.*, 1979, 1991)—the North American strewn field of 35.5 Ma age (associated with the Chesapeake Bay impact structure; cf. Poag *et al.*, 2004), the Central European strewn field of 14.4 Ma age (associated with the Ries crater in southern Germany), the Ivory Coast tektite strewn field, and the 0.8 Ma Australasian strewn field (for which no source crater has been identified so far). For details on these strewn fields and the chemistry and origin of the tektites, see the reviews in, for example, Koeberl (1986) and Montanari and Koeberl (2000). It is well established that the chemical and isotopic

Figure 20 Examples of different tektite forms and shapes (Australasian tektites).

composition of tektites in general is identical to the composition of the upper terrestrial continental crust (e.g., Koeberl, 1994).

In addition to the "classical" tektites on land, microtektites (<1 mm in diameter) from three of the four strewn fields (all but the Central European one) have been found in deep-sea cores (see, e.g., Glass, 1967, 1968, 1969, 1972). Microtektites have been very important for defining the extent of the strewn fields (e.g., Glass *et al.*, 1979; Glass and Zwart, 1979), as well as for constraining the stratigraphic age of tektites, and to provide evidence regarding the location of possible source craters (e.g., Glass and Pizzuto, 1994). Microtektites have been found together with melt fragments, high-pressure phases, and shocked minerals (e.g., Glass, 1989; Glass and Wu, 1993) and, therefore, provide confirming evidence for the association of tektites with an impact event. The variation of the microtektite concentrations in deepsea sediments with location increases towards the assumed or known impact location (Glass and Pizzuto, 1994). Their chemical composition is similar to tektites from the same strewn field found on land, but they show a wider compositional range (e.g., Glass *et al.*, 2004).

There has been some discussion about how to define a tektite, but the following characteristics should probably be included (see Koeberl, 1994; Montanari and Koeberl, 2000): (1) they are glassy (amorphous); (2) they are fairly homogeneous rock (not mineral) melts; (3) they contain abundant lechatelierite; (4) they occur in geographically extended strewn fields (not just at one or two closely related locations); (5) they are distal ejecta and do not occur directly in or around a source crater, or within typical impact lithologies (e.g., suevitic breccias and impact melt breccias); (6) they generally have low water contents and a very small extraterrestrial component; and (7) they seem to have formed from the uppermost layer of the target surface (see below). Thus, it is recommended to use the term "tektite" only for glasses that fulfill (most) of the above points, and, if in doubt, use the (probably much better) general term "impact glass."

An interesting group of tektites are the Muong Nong-type tektites, which, compared to "normal" (or splash-form) tektites, are larger, more heterogeneous in composition, of irregular shape, have a layered structure, and show a much more restricted geographical distribution (for details on these tektites, see Koeberl, 1992a). They are also important because they contain relict mineral grains that indicate the nature of the parent material and contain shock-produced phases that indicate the conditions of formation (e.g., Glass and Barlow, 1979). The occurrence of relict minerals present in some tektites points to sedimentary source rocks. Muong Nong-type tektites contain unmelted relict inclusions, including zircon, chromite, quartz, rutile, and monazite, all showing evidence of various degrees of shock metamorphism. Coesite, stishovite, and shocked minerals were found in the North American and Australasian microtektite layers (Glass, 1989; Glass and Wu, 1993). Some inclusions of metallic spherules were found in some tektites (e.g., El Goresy, 1966; El Goresy *et al.*, 1968); one of the earlier interpretations was that these might represent meteorite-derived material, but Ganapathy and Larimer (1983) used traceelement abundances and ratios to show that these probably formed from terrestrial

material by *in situ* reduction during the impact process. Iridium enrichments are associated with microtektite-layers in the Australasian strewn field (Koeberl, 1993a).

Despite knowing the source craters of three of the four tektite strewn fields, we still do not know exactly when and how during the impact process tektites form. Detailed reviews of the parameters that have to be considered were provided by, for example, Koeberl (1994) and Montanari and Koeberl (2000). These characteristics include in part the following observations: (1) vapor fractionation played no major role in tektite formation; (2) tektites are very poor in water with contents ranging from about 0.002–0.02 wt.%; (3) bubbles in tektites contain residues of the terrestrial atmosphere at low pressures; (4) meteoritic components in tektites are very low or below the detection limit; and (5) the ^{10}Be content of Australasian tektites cannot have originated from direct irradiation with cosmic rays in space or on Earth, but can only have been introduced from sediments that have absorbed ^{10}Be that was produced in the terrestrial atmosphere. Tektites might be produced in the earliest stages of impact, which are poorly understood. It is clear, however, that tektites formed from the uppermost layers of terrestrial target material (otherwise they would not contain any ^{10}Be). However, the question of the exact process that was responsible for tektite production and distribution remains uncertain.

Tektites have been the subject of much geochemical study, which ultimately led to their confirmation as terrestrial distal impact glasses (e.g., Taylor, 1962; Schnetzler *et al.*, 1966, 1967; Kolbe *et al.*, 1967; Taylor and Epstein, 1967). Major and trace-element compositions aided in the identification of the source rocks (e.g., Koeberl, 1992a, 1994). The Rb–Sr and Sm–Nd isotopic systems proved to be very important in the search for tektite source craters (e.g., Shaw and Wasserburg, 1982; Blum *et al.*, 1992; cf. Figure 21). In the case of Ivory Coast tektites, the compositional range of the target rocks at the Bosumtwi crater is significantly wider than that of the Ivory Coast tektites, but overlaps the tektite compositions (Koeberl *et al.*, 1998). A best-fit line for the Bosumtwi crater rocks in a Rb–Sr isotope evolution diagram yielded an "age" of 1.98 Ga, and an initial ^{87}Sr/^{86}Sr ratio of 0.701, which is close to results previously obtained for granitoid intrusions in the Birimian of Ghana. The neodymium isotopic data of Koeberl *et al.* (1998) yielded depleted mantle model ages ranging from 2.16 to 2.64 Ga, and ε_{Nd} values of −17.2‰ to −25.9‰. Similar isotopic and chemical studies for Central European tektites (e.g., Engelhardt *et al.*, 1987, 2005) helped to constrain the exact source rocks from which these tektites formed.

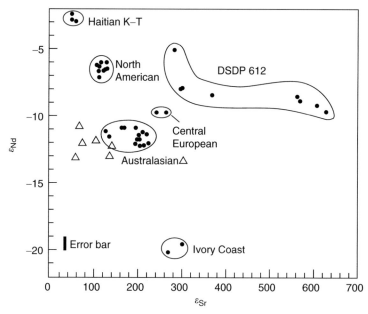

Figure 21 Diagram showing the isotopic compositions of Sr and Nd for the different tektite-strewn fields on Earth; such diagrams are useful in determining the source rocks (and craters) of the various tektites. After Shaw and Wasserburg (1982) and Blum *et al.* (1992).

5.4.4 Libyan Desert Glass

Libyan Desert Glass is an enigmatic type of natural glass, which occurs in an $\sim 2,500\,\mathrm{km}^2$ strewn field located between sand dunes of the southwestern corner of the Great Sand Sea in southwestern Egypt. The glass has an age of about 28 Ma. Glass fragments are found ranging in size from microscopic to several tens of centimeters. In terms of chemical composition, the glass is very silica rich at about 96.5–99 wt.% SiO_2, and shows a limited variation in major- and trace-element abundances. The origin of the glass has been debated for decades, with a variety of alternatives, ranging from low-temperature sol–gel processes to lunar volcanic glass to terrestrial impact glass. Although the origin of the glass is still debated by some workers, an origin by impact agrees best with the available data. There are, however, some differences compared to "classical" impact glasses, which occur in most cases directly at or within an impact crater. Evidence for an impact origin includes the presence of schlieren and partly digested mineral phases, lechatelierite (a high-temperature mineral melt of quartz), and baddeleyite (a high-temperature breakdown product of zircon). High temperature is also indicated by cristobalite inclusion in many of the glass fragments. The rare earth element abundance patterns are indicative of a sedimentary precursor rock, and the trace-element abundances and ratios are in agreement with an upper crustal source. However, there are some good indications for the presence of a meteoritic component in dark streaks or layers

of the desert glass (e.g., Barrat *et al.*, 1996). This was confirmed by Koeberl (2000) in an osmium isotopic study. Crustal Sr and Nd isotopic values exclude a significant mantle component; thus, the osmium abundances and isotopic values confirm the presence of a meteoritic component in LDG. This enigmatic material has recently been the focus of much attention and soon further studies will be presented.

5.4.5 Cretaceous–Tertiary (K–T) Boundary

The impact ejecta at the Cretaceous–Tertiary (K–T) boundary were not the first distal ejecta layer to be described, as microtektite-bearing layers were already discovered in the late 1960s. However, the study of the K–T boundary ejecta provided the greatest impetus in the discussion about the importance of impact events, following the discovery of extraterrestrial PGEs at the K–T boundary (Alvarez *et al.*, 1980; Smit and Hertogen, 1980; Ganapathy, 1980; Kyte *et al.*, 1980) and shocked minerals (e.g., Bohor *et al.*, 1984, 1987, 1993; Krogh *et al.*, 1993; Alvarez *et al.*, 1995). It is easy to distinguish the K–T boundary layer from surrounding strata in the field, because it represents a distinct break in lithology, with a thin (usually up to 1- or 2-cm-thick) layer (cf. Figure 22) commonly composed of clay or claystone at distal marine sections (e.g., Rampino and Reynolds, 1983; Pollastro and Pillmore, 1987; Pollastro and Bohor, 1993). Most of the clay seems to be due to devitrification of impact glass (spherules similar to those still found in Haiti, Mexico, and nearby, see below). Good descriptions of field relations are given, for example, by Kring (1993), Koeberl (1996), and Smit (1999), and references therein. Information on the paleontology of the K–T boundary can be found, for example, in the "Snowbird" series of conference proceedings (Silver and Schultz, 1982; Sharpton and Ward, 1990; Ryder *et al.*, 1996; Koeberl and MacLeod, 2002).

There are a number of different lines of evidence that confirm that there was an impact event at the end of the Cretaceous.

5.4.5.1 *Enrichments and patterns of the PGEs*

The first physical evidence pointing to a contribution of extraterrestrial material that was discovered was the presence of anomalously high PGE abundances in K–T boundary clay in Italy (Alvarez *et al.*, 1980) and other locations around the world (e.g., Smit and Hertogen, 1980). The contents of iridium and other PGEs were found to be enriched in these K–T boundary clay layers by up to four orders of magnitude compared to average terrestrial crustal abundances (e.g., Figure 23). Also, the interelement ratios of the PGEs in K–T boundary clay samples are very similar to the

Figure 22 The famous K–T boundary layer at the Bottacione gorge, Gubbio, Italy, where the original discovery of Alvarez *et al.* (1980) was made. The author is to the left. U. Reimold is to the right.

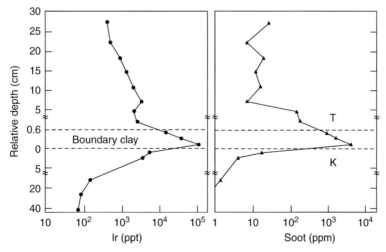

Figure 23 Distribution of the element iridium as well as the concentration of soot at a K–T boundary in New Zealand, showing that both are enriched in the K–T boundary, providing evidence for an extraterrestrial component as well as combustion of large amounts of carbon, probably due to the impact event.

values observed in chondritic meteorites. Other geochemical and isotopic data agree with this observation (e.g., Schmitz, 1985, 1992; Schmitz *et al.*, 1988; Dia *et al.*, 1989; Tredoux *et al.*, 1989; Frei and Frei, 2002).

5.4.5.2 Osmium and chromium isotopic data

Such studies of the K–T boundary sediments provided further evidence of an extraterrestrial component (e.g., Luck and Turekian, 1983; Lichte *et al.*, 1986; Krähenbühl *et al.*, 1988; Esser and Turekian, 1989, 1993; Meisel *et al.*, 1995; Frei and Frei, 2002). Any contribution to the K–T boundary signal from the volcanic eruptions that formed the Deccan Traps can be ruled out on the basis of their Os isotopic signature (Ravizza and Peucker-Ehrenbrink, 2003). A meteoritic component in Chicxulub melt rocks, suggested by Koeberl *et al.* (1994c) and Schuraytz *et al.* (1996), remained somewhat controversial until confirmed recently by Gelinas *et al.* (2004). Shukolyukov and Lugmair (1998) presented chromium isotope data for a meteoritic component that is in better agreement with a carbonaceous chondritic composition of the impactor, rather than an ordinary chondritic composition. More recently, studies by Quitté *et al.* (2003) and Trinquier *et al.* (2006) refined this identification to a CM2 carbonaceous chondrite.

5.4.5.3 A fragment of the K–T bolide?

Kyte (1998) described a 2.5-mm-diameter inclusion recovered from K–T boundary sediments at Deep Sea Drilling Project (DSDP) drill core 576 (western North Pacific), and interpreted it to be a fragment of the K–T boundary impactor. Unfortunately, the piece seems to be completely altered, but may have preserved primary textures. The main minerals present are hematite and clays, with some hematite containing submicrometer-sized nickel–iron metal and nickel–iron sulfide grains. In bulk sample, this altered fragment has iridium, iron, and chromium abundances that are within a factor of about 2 of chondritic values.

5.4.5.4 Presence of evidence for impact-induced wildfires

Many K–T boundary locations around the world show evidence for global wildfires in the form of a charcoal and soot layer (Figure 23) that coincides with the iridium-rich layer (Wolbach *et al.*, 1985, 1990). The insoluble carbon fraction after acid dissolution is dominated by kerogen and elemental carbon, which also show a marked change in their isotopic composition across the K–T boundary (Gilmour *et al.*, 1990). The total amount of soot in the atmosphere due to the global wildfires at the end of the Cretaceous has been estimated at 7×10^{16} g, which must have had a large influence on the environment (e.g., Lyons and Ahrens, 2003). Belcher *et al.* (2005) noted that in North American sites there is a lack of charcoal. This reopens the questions as to the actual source of the carbon that burned and created the soot; there is a chance that there was a contribution from hydrocarbons at the impact site (cf. Grajales-Nishimura *et al.*, 2000).

5.4.5.5 Shock metamorphism

Most important, however, was the discovery of clear evidence of shock metamorphism at the K–T boundary (Bohor *et al.*, 1984, 1987). The shocked quartz grains show multiple intersecting sets of PDFs with shock-characteristic crystallographic orientations. Shocked zircons with planar features were discovered by Bohor *et al.* (1993).

5.4.5.6 Distal impact glasses

Impact glass was found at the K–T boundary (in Haiti) as well (e.g., Sigurdsson *et al.*, 1991a, b; Kring and Boynton, 1991; Izett, 1991; Izett *et al.*, 1991; Blum and Chamberlain, 1992; Koeberl and Sigurdsson, 1992). These impact glasses and spherule deposits are uniquely impact derived because no such spherule layers that would have been derived from volcanic eruptions have ever been found anywhere in the world. Blum *et al.* (1993) showed that the Haitian glasses are mixtures of silicate rocks of upper crustal composition with a high CaO-end member (e.g., limestone). Koeberl (1992b) measured the water content in glasses from Haiti and found a range of 0.013–0.021 wt.% H_2O, which agrees with an origin by impact, as impact glasses are extremely dry (Beran and Koeberl, 1997). From sulfur contents and isotopic ratios, Chaussidon *et al.* (1994) determined that the target rocks contained anhydrite (Figure 24). Using osmium isotope analyses, Koeberl *et al.* (1994c) found a small meteoritic component in the Haitian glasses. Precise age determinations on the Haitian glasses have shown that the materials have an age indistinguishable from that of the K–T boundary, at 65 Ma (e.g., Swisher *et al.*, 1992).

5.4.5.7 Impact-derived diamonds

Small, nanometer-sized diamonds were first reported from K–T boundary sediments in Alberta, Canada (Carlisle and Braman, 1991). Since then, larger diamonds have been found at other K–T boundary locations, including some in Mexico (Hough *et al.*, 1997). These diamonds have a unique carbon- and nitrogen isotopic signature (Gilmour *et al.*, 1992; Gilmour, 1998) and are clearly connected to the impact process (and not

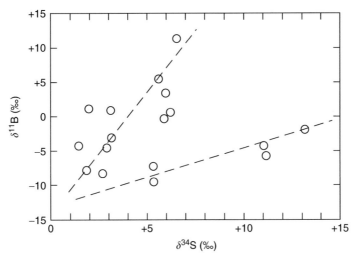

Figure 24 Sulfur versus boron isotopic composition of K–T boundary impact glasses, from Chaussidon *et al.* (1994), showing that there are two distinct compositions of the glasses, which are thought to be derived from marine carbonates and anhydrites at the Chicxulub impact site.

of extraterrestrial origin as had been speculated by Carlisle and Braman, 1991).

5.4.5.8 Spinel

Spinels (magnesioferrite) of various compositions were first reported by Montanari *et al.* (1983) from magnetic spherules at the Petriccio, Italy, K–T boundary section. Spinel at the K–T boundary can be used as an event marker, as it shows an abundance peak similar to that observed for the PGEs. Other important studies of the composition and emplacement of spinels were done by, for example, Kyte and Bostwick (1995) and Ebel and Grossman (2005).

5.4.5.9 Summary

Despite all the evidence supporting a major impact event at the end of the Cretaceous, no large impact structure with a corresponding age was known until the early 1990s, when finally the ~200-km-diameter Chicxulub impact structure was proposed and confirmed as the elusive K–T boundary crater (e.g., Hildebrand *et al.*, 1991; Sharpton *et al.*, 1992, 1993; Schuraytz *et al.*, 1994; Pierazzo and Melosh, 1999). The study of the Chicxulub structure is difficult because it is covered by several hundred meters of postimpact Tertiary sediments, which requires obtaining drill cores. Drill cores from near the center of the structure recovered impact melt breccia and impact melt rock (e.g., Hildebrand *et al.*, 1991; Sharpton *et al.*, 1992; Koeberl, 1993b; Schuraytz *et al.*, 1994). Clasts within the breccia and melt rock samples provided unambiguous evidence of shock metamorphic effects, and a meteoritic component was detected in the melt rocks (Koeberl *et al.*, 1994c; Schuraytz *et al.*, 1996).

Geochemical measurements of the isotopic composition of the melt rocks in comparison with K–T ejecta were of crucial importance to establish the link between the Chicxulub crater and the impact deposits found at the K–T boundary. Blum *et al.* (1993) measured the Rb–Sr, Sm–Nd, and oxygen isotopic composition of melt rocks from Chicxulub and found that these data indicated a common source for the Haitian impact glasses (= K–T boundary impact ejecta) and the Chicxulub crater rocks. A similar result was obtained from the strontium versus the neodymium isotopic compositions of Chicxulub melt rock samples and Haitian impact glasses. The K–T boundary impact glasses and the Chicxulub melt rocks have similar values at about +58 and −3 for ε_{Sr}^{65Ma} and ε_{Nd}^{65Ma}, respectively.

Shocked zircons from distal K–T boundary sites provided another link between the boundary layer samples and the Chicxulub structure. Krogh *et al.* (1993) and Kamo and Krogh (1995) found that the U–Pb ages and isotopic characteristics of single zircons from various K–T boundary sites agree with those of zircons extracted from Chicxulub impact melt breccias. Thus all geochemical and geochronological data (e.g., Sharpton *et al.*, 1992; Swisher *et al.*, 1992; Krogh *et al.*, 1993) confirm the geochemical link between Chicxulub and the impact ejecta marking the K–T boundary. This was one of the largest impact events recorded in geological history (Figure 25) and the environmental effects of this large-scale impact event can be addressed by geochemical approaches (e.g., Sigurdsson *et al.*, 1992; Chen *et al.*, 1994; Yang and Ahrens, 1998; Gupta *et al.*, 2001). The detailed study of a distal impact ejecta layer had led to the discovery of one of the largest impact structures on Earth.

Figure 25   The first seconds of the K–T asteroid impact event at the site of present-day Yucatan peninsula in Mexico. Painting by D. Jalufka (Vienna).

5.4.6   Permian–Triassic (P–Tr) Boundary

The Permian–Triassic (P–Tr) boundary is associated with the largest mass extinction known in Earth history. Following the association of the K–T boundary mass extinction with a large impact event, speculations bloomed that other major mass extinctions might also be related to impact events. However, geochemical data show that no incontrovertible evidence for a P–Tr-age impact event exists. Siderophile element anomalies (e.g., enhanced iridium contents) were found at some P–Tr boundary locations (e.g., Holser *et al.*, 1989), but their source is not clear and confirmation of their extraterrestrial nature is still pending. Recent research, however, succeeded in demonstrating that the P–Tr boundary event was a much shorter event than thought before, and that the severe environmental changes that resulted in the mass extinction were brought on within less than a few hundred thousand years (Bowring *et al.*, 1998). These authors also found that at Meishan, China, a negative excursion in the carbon isotopic composition had a duration of less than about 160,000 years and suggested that it could be the result of the impact of an icy, carbon-rich comet (however, scenarios involving volcanic processes are also conceivable). Retallack *et al.* (1998) reported on the possible discovery of shocked quartz grains from P–Tr boundary locations in Australia and Antarctica, but it was recently shown (Langenhorst *et al.*, 2005) that these grains are definitely not shocked.

Kaiho *et al.* (2001) reported sulfur isotope and chemical data for samples from the Meishan P–Tr boundary section. They interpreted the sulfur isotope data as well as the occurrence of iron- and nickel-rich particles, as evidence for a large-scale impact event that penetrated the Earth's mantle and formed a crater $\sim$1,000 km in diameter. Koeberl *et al.* (2002a) gave a detailed discussion of why the hypothesis of Kaiho *et al.* (2001) is very problematic. Their suggestion that the lack of shocked quartz implies an oceanic impact event is misleading. Shock metamorphic effects are not restricted to quartz, but occur in all rock-forming (and accessory) minerals, which are abundant in ocean floor rocks. Impact-induced volcanism or excavation of mantle material in impact events have been postulated before, but such effects are physically implausible, and that no known impact on Earth has ever had suchconsequences (e.g., Ivanov and Melosh, 2003). In addition, the assumptions of Kaiho *et al.* (2001) require a projectile with a diameter of 750–1,500 km, which is implausible as the largest main belt asteroid has a diameter of 1,000 km and that the largest crater formed on the terrestrial planets in the last 500 Myr is Mead Crater on Venus with a diameter of $\sim$280 km. This would seem to be an upper limit of a crater size we should assume for possible catastrophic impacts during the Phanerozoic on Earth.

Koeberl *et al.* (2002a) noted that none of the points raised by Kaiho *et al.* (2001) provide conclusive evidence—or even vague suggestions—of an impact event at the P–Tr boundary. Attempts to utilize the questionable interpretations by Kaiho *et al.* in an effort to support the equally controversial (cf. Farley and Mukhopadhyay, 2001) claims for the presence of extraterrestrial ^{3}He in fullerenes at the P–Tr boundary represent circular logic. In a detailed geochemical study of two P–Tr boundaries (in Austria and Italy), Koeberl *et al.* (2004a) demonstrated that while there are slight enrichments in iridium and other PGEs, the inter-element ratios are not similar to those of any meteorites. The same layer that contains elevated Ir contents also shows evidence of local anoxia at both boundary sections. Elevated iridium, osmium, platinum, and rhenium concentrations can be produced in anoxic depositional environments (Colodner *et al.*, 1992). Thus, Koeberl *et al.* (2004a) concluded that the widespread development of anoxia may explain the anomalous PGE concentrations in end-Permian marine sedimentary rocks, which is supported by nonchondritic PGE abundance ratios and the high (crustal) initial $^{187}Os/^{188}Os$ ratios. The peak in iridium and drop in $\delta^{13}C$ values also point to a proliferation of fungal remains thought to represent the response of the terrestrial ecosystem to acidifying emissions of the volcanic eruptions of the Siberian Traps, which released large amounts of carbon dioxide to the atmosphere, generating global warming, a reduction of the solubility of oxygen in seawater, and less vigorous oceanic circulation, conditions that would promote oceanic anoxia. Thus, Koeberl *et al.* (2004a) suggested that the biotic crisis, decline in $\delta^{13}C$, and elevated iridium concentrations may all be ultimately linked to massive end-Permian volcanism (cf. also Maruoka *et al.*, 2003; Grard *et al.*, 2005). In contrast to suggestions by Poreda and Becker (2003) and others, Koeberl *et al.* (2004a) did not find any evidence for extraterrestrial helium in their samples.

In an important study, Farley *et al.* (2005) measured helium concentrations and isotopic compositions in a suite of samples across the Permian–Triassic boundary at Opal Creek, Canada, to determine whether high extraterrestrial helium concentrations are associated with a possible extinction-inducing impact event at this time. They found no extraterrestrial ^{3}He, implying that neither fullerene-hosted nor IDP-hosted He is present at or near the boundary. This observation is consistent with similar studies of some Permian–Triassic sections, but contrasts sharply with reports of both fullerene- and IDP-hosted extraterrestrial ^{3}He at other sections (e.g., Poreda and Becker, 2003). Step-heat experiments cited by Farley *et al.* (2005) indicate rapid diffusion of extraterrestrial helium from sediments heated to temperatures above $\sim 70\,°C$, and, as burial and associated heating must have affected P–T age rocks, and makes preservation of IDP-hosted ^{3}He highly unlikely; recent contamination is a much more plausible explanation.

Farley *et al.* (2005) also noted that even though no extraterrestrial ^{3}He was detected in their samples, there is a sharp increase in nucleogenic ^{3}He very close to or at the P–Tr boundary, which may arise from lithologic changes. Increased nucleogenic ^{3}He is associated with increases in both lithium and organic carbon content into the Triassic. Thus, Farley *et al.* (2005) cautioned that care must be taken to eliminate such artifacts before interpreting changes in ^{3}He concentration in terms of fluctuations of extraterrestrial ^{3}He.

Reports of extraterrestrial noble gases in fullerenes (Poreda and Becker, 2003) and unaltered meteorite fragments (Basu *et al.*, 2003; Shukolyukov *et al.*, 2004) in P–Tr boundary sediments as well as an alleged buried impact structure near Australia (Becker *et al.*, 2004) are all very controversial and have largely been refuted. The "shocked quartz" was shown to be not shocked (Langenhorst *et al.*, 2005), the PGE and osmium isotopic signals are terrestrial (Koeberl *et al.*, 2004a), the sulfur isotopic composition is volcanic (Maruoka *et al.*, 2003), the fullerenes and ^{3}He cannot be confirmed (Farley and Mukhopadhyay, 2001; Farley *et al.*, 2005), and the existence of the "impact crater" and its reported P–Tr age have been rejected (e.g., Renne *et al.*, 2004; Müller *et al.*, 2005). A lot of ambiguous data were published, supposedly suggesting an impact event at the P–Tr boundary, but many of those reports could not be confirmed.

5.4.7 Precambrian Spherule Layers

The early Earth was certainly subjected to a lot of impacts—some of it related to the tail end of the accretion of the planet, some related to the postulated (and likely) so-called Late Heavy Bombardment (LHB), a short time of a drastically increased impact flux in the inner solar system (see next section). The first "real" rock record of impact events dates about 400–500 million years after the end of the LHB, in the form of (distal?) ejecta layers (see reviews by Simonson and Glass, 2004; Simonson *et al.*, 2004). Four distinct spherule horizons in the Barberton Greenstone belt, South Africa (designated S1–S4), with ages between about 3.5 and 3.2 Ga, have been proposed as being of impact origin (e.g., Lowe *et al.*, 2003). The spherules are mostly spherical particles, up to a few millimeter across, of quenched melt droplets that supposedly formed by condensation from vapor clouds. The spherule layers generally contain coarser clasts than surrounding strata and show depositional structures indicating they were deposited during unusually high-energy events. The original mineralogical and chemical composition of the spherules has been almost completely changed by alteration. The stratigraphic positions of these layers at different geographic locations are difficult to correlate and the possibility exists that some of the layers represent duplication. Other interpretations (e.g., B. Simonson, personal communication) suggest that there could be even more than four layers. These spherule layers show

Figure 26 Sample from an Early Archean spherule layer in Barberton, South Africa. The spherules show tectonic repetition. Their chemical composition is overprinted by later processes, which made their identification as impact deposits very difficult.

extreme enrichments in the PGEs (in some cases exceeding the PGE abundances found in chondritic meteorites), unlike Phanerozoic impact ejecta deposits (e.g., those at the K–T boundary, or in the late Eocene (see, e.g., Montanari and Koeberl, 2000 for a review), which caused some questions regarding the initial impact interpretation. An example of a sample from a Barberton spherule layer (3.48 Myr old, Byerly *et al.*, 2002) is shown in Figure 26.

The extreme enrichments in the PGEs (e.g., Lowe *et al.*, 1989; Kyte *et al.*, 1992; Koeberl and Reimold, 1995; Reimold *et al.*, 2000), among other problems, caused Koeberl *et al.* (1993) and Koeberl and Reimold (1995) to question the impact interpretation. Figure 27 shows a correlation between the concentrations of Ir and As, a very mobile element, in samples from the Barberton spherule layers, all of which have been subject to pervasive transformation into secondary mineral assemblages (Figure 28). This relationship between Ir and As, thus, indicates remobilization of both elements; this means that the PGE signature in these samples is not primary (e.g., Koeberl and Reimold, 1995; Reimold *et al.*, 2000). More recently, though, Shukolyukov *et al.* (2000, 2002) and Kyte *et al.* (2003) reported chromium isotopic anomalies (using methods and arguments of Shukolyukov and Lugmair, 1998) in samples from several of these layers that seem to support the presence of an extraterrestrial component.

Other occurrences of unusual spherule layers were reported by Simonson (1992) from the Hamersley Basin in Western Australia. On the basis of similarities to microtektites and mikrokrystites, Simonson (1992) interpreted the spherules as having formed in an impact event and having been redeposited in a sediment gravity flow. Later, three additional spherule-bearing layers were found in

the Hamersley Basin sequence, which were also interpreted to be of impact origin (e.g., Simonson *et al.*, 1998). Simonson *et al.* (2000a, b) also reported the discovery of a similar spherule layer (ca. 2.6 Ga) in the Monteville Formation of the Transvaal Supergroup in South Africa, which might be correlated with one of the Australian layers (e.g., Simonson and Hassler, 1997; Simonson *et al.*, 1999, 2004; Rasmussen and Koeberl, 2004). Simonson *et al.* (2006) report a total of three layers in South Africa and three layers in Australia, but not all of them might be correlated, so there are at least three, but possibly four individual spherule layers.

Until recently, no shocked minerals have been reported to be associated with any of these spherule layers. It was suggested that this is because the impacts occurred into oceanic crust, which has little or no quartz, and whatever else there was in terms of shocked minerals had long been destroyed by alteration (Simonson *et al.*, 1998). Rasmussen and Koeberl (2004), however, were able to describe one shocked quartz grain in a sample from the 2.63 Ga Jeerinah impact layer in Australia; this is so far the only evidence of diagnostic shock features. Unfortunately, so far no definitive criteria for the identification of Archean impact deposits are known. For none of the South African (Barberton and Monteville) or Australian spherule layers have source craters been found, and given the scarcity of the early Archean geological record, it is unlikely that they will ever be found. It is not clear why impact events in the Archean would predominantly produce large volumes of spherules, which are mostly absent from post-Archean impact deposits (i.e., those for which source craters are known)—although a recent report about ejecta beds from the 1.85 Ga Sudbury impact structure also included observation

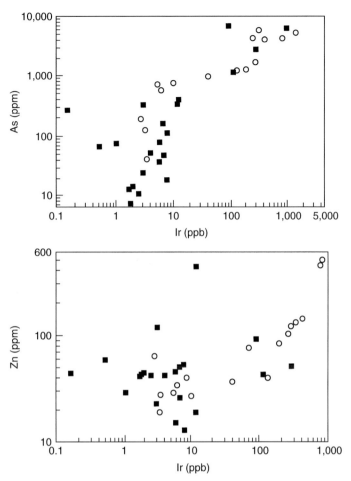

Figure 27 Relationship between Ir and As and Ir and Zn in samples from the Barberton (South Africa) Early Archean impact layers, showing very high contents of Ir (superchondritic), and a correlation between these elements, which indicates remobilization of both elements; this means that the PGE signature in these samples is not primary.

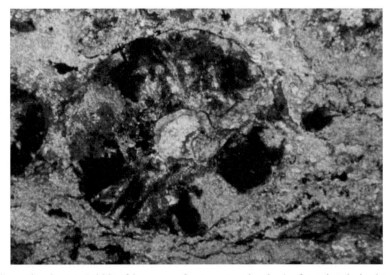

Figure 28 Thin-section image (width of image ca. 2 mm, crossed polars) of a spherule in the Barberton Early Archean spherule layer, showing total replacement of the original mineralogy.

of spherule horizons (Addison *et al.*, 2005). There are spherules in the K–T boundary and Late Eocene impact layers, but they are nowhere near as abundant as in the Archean and Paleoproterozoic layers, and they have different textures as well. These are important issues for future studies. The question regarding how to identify Archean impact deposits remains open and will hopefully be addressed in future studies (but see Simonson and Harnik (2000) and Simonson (2003) for discussions on this subject); the change in atmospheric oxygen content around 2.2–2.4 Ga may have influenced ejecta processes.

5.4.8 The Earth's Earliest Impact History

This section follows a review by Koeberl (2006). Collisions and impact processes have been important throughout the history of the solar system, including that of the Earth. Small bodies in the early solar system, the planetesimals, grew through collisions, ultimately forming the planets. The Earth started growing about 4.56 billion years ago in this way (e.g., Zhang, 2002). Its early history was dominated by violent impacts and collisions, of which we only have circumstantial evidence. The Earth was still growing and had reached about 70–80% of its present mass when, at about 4.5 Ga, a Mars-sized protoplanet collided with Earth, leading to the formation of the moon—at least according to what is currently the most popular hypothesis of lunar origin (cf. Canup and Righter, 2000; Canup and Asphaug, 2001; Canup, 2004). After its formation, the moon was subjected to intense postaccretionary bombardment between about 4.5 and 3.9 billion years ago. In addition, there is convincing evidence that the Moon experienced an interval of intense bombardment with a maximum at $\sim 3.85 \pm 0.05$ Ga (e.g., Tera *et al.*, 1974; Ryder, 1990); subsequent mare plains as old as 3.7 or 3.8 Ga are preserved. It is evident that if such a LHB occurred on the Moon, the Earth must have been subjected to an impact flux at least as intense as that recorded on the Moon.

The accretion of the Earth appears to have been completed about 50–100 Myr after the initial formation of the solar nebula, defining the initiation of the so-called Hadean Eon. Whereas there are almost no terrestrial witnesses to the Hadean Eon, the pre-Nectarian and Nectarian periods cover this time interval on the Moon. Soon after the formation of the Moon, the feldspathic highland crust formed (e.g., Taylor, 1992). The lack of lunar impact melts older than ~ 3.92–3.95 Ga has be taken as evidence that there were no basin-forming impact events prior to that time, or that the LHB erased the record of any prior impact events. Ryder *et al.* (2000) and Stöffler and Ryder (2001) reviewed the evidence that the formation ages of all large-impact basins on the Moon

cluster around 3.8–4 Ga. Cohen *et al.* (2000) and Kring and Cohen (2002) studied the ages of impact melts from the lunar meteorites, which constitute random collections of lunar surface material, and found that they have similar ages as the mare-forming impacts, in agreement with the suggestions of Ryder (1990, 2002) and Ryder *et al.* (2000).

A separate but independent argument in favor of a LHB involves the masses of basin-forming projectiles on the Moon, as discussed in detail by Ryder (2001, 2002). The masses of the Imbrium and Orientale projectiles, for instance, have been estimated at between 8×10^{20}–2×10^{21} g and 4×10^{20}–1.5×10^{21} g, respectively. Thus, the total mass of the about 15 Nectarian and Early Imbrian basin-forming projectiles would be on the order of 10^{21}–10^{22} g. Given that the formation ages of these basins are within about 80 Myr, this leads to a lower limit of the accretion rate on the Moon of about 1.2×10^{13} g yr^{-1}. This is about one to two orders of magnitude above the smoothly declining curve of lunar accretion, and three orders of magnitude above a back-extrapolated current accretion rate. If one assumes that the mass accretion in the first case is the tail-end of the early lunar accretion and not a spike, this means that backextrapolating the masses of basin-forming projectiles accreting onto the Moon between 3.8 and 4 Ga indicates that the current mass of the Moon was exceeded already at about 4.1 Ga instead of 4.45 Ga.

Norman *et al.* (2002) recognized that chemical evidence for multiple impact events, and the clear signatures of specific types of meteoritic impactors in the Apollo 17 melt breccias, show that the lunar crust was not comprehensively reworked by prior impacts from 3.9 to 4.5 Ga, an observation more consistent with a late cataclysm than a smoothly declining accretionary flux. Late accretion of chondritic material during a 3.8–4.0 Ga cataclysm may have contributed to siderophile element heterogeneity on the Earth, but would not have made a significant contribution to the volatile budget of the Earth or oxidation of the terrestrial mantle.

As it seems quite likely that an LHB phase did occur in the inner solar system raises the question regarding the source of the impactors. As summarized by Koeberl (2004), it is conceivable that within a few hundred million years after the formation of Neptune this body had slowly migrated into a planetesimal-rich zone, whereupon it started to accrete more mass and rapidly migrated outwards towards 40 AU, in the process scattering the proto-Kuiper Belt objects all over the solar system, with the LHB in the inner solar system being a consequence of this process. A quantitative treatment of this assumption by Gomes *et al.* (2005), using numerical calculations of the development of the orbits of the planets during the first few hundred million years of the solar system, confirmed that the formation of the outer planets and the delayed migration of Uranus and Neptune outwards can provide a qualitative and quantitative explanation of the LHB. These calculations provide a viable and internally consistent mechanism for the LHB has been found.

During any time period, the Earth is being subjected to a higher impact flux than the Moon, as it has a larger diameter and a much larger gravitational cross-section, thus making it an easier target to hit. If a LHB occurred on the Moon, the Earth was subject to scaling of the same impact flux due to the ratio of the cross-sections (Sleep *et al.*, 1989), which will have resulted in an impact rate ≥20 times greater than the lunar one, comprising both more and larger impact events. The consequences for the hydrosphere, atmosphere, and even the lithosphere of Earth at that time must have been devastating (Zahnle and Sleep, 1997; Frey, 1980; Grieve, 1980).

Considering the evidence for a LHB on the Moon and elsewhere, as discussed above, it is clear that during that time the Earth must have been subjected to an even more intense bombardment. Thus, the question arises if there is any evidence of this bombardment preserved on Earth. The oldest rocks on Earth, from Acasta in Canada and the Isua/Akilia area in Greenland (e.g., papers in Fowler *et al.*, 2003), are the obvious places to conduct such a search. In an early work, Appel (1979) reported finding chromite grains in Isua rocks with sizes of 10–30 μm, of compositions unlike those found in other similar rocks, which he interpreted as being of extraterrestrial origin. It has been noted that chromite grains are the sole, more or less unaltered remnants of the Late Ordovician fossil meteorites

reported by, for example, Schmitz *et al.* (2003). Thus, it seems possible that the chromite grains found by Appel (1979) in Isua rocks could be remnants of the earliest documented flux of extraterrestrial material. However, no detailed follow-up studies of this material have been made. Chromium isotope studies could be helpful, if ever sufficient numbers of these chromites could be extracted.

Another attempt to find a possible earliest Archean extraterrestrial component was reported by Koeberl *et al.* (2000). These authors analyzed samples of some of the oldest rocks on Earth, from Isua, Greenland, for their chemical composition, including the PGE abundances (Figure 29). The goal was to search for a chemical signature similar to that observed in some Archean or Phanerozoic impact ejecta layers (e.g., Koeberl, 1998). Unfortunately, the results were ambiguous, and no clear meteoritic signature was found, as was discussed in detail by Ryder *et al.* (2000). A similar study by Anbar *et al.* (2001), on rocks from the nearby Akilia Island, SW Greenland, also failed to detect any significant amount of iridium that would indicate an extraterrestrial component. In addition, zircon extracted from the rocks at Isua were studied by Koeberl *et al.* (2000) for the possible presence of shock metamorphic effects similar to those found in more recently shocked zircons from impact deposits (e.g., Bohor *et al.*, 1993). Zircon is a very resistant mineral and would be expected to exhibit impact-related shock features if it had been subjected to any impact events in the past. The petrographic studies of Isua zircons failed to show evidence for shock metamorphism (Koeberl *et al.*, 2000).

There are several reasons that might explain the absence of any LHB evidence on Earth in these

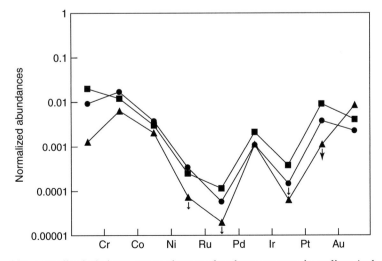

Figure 29 Chondrite-normalized platinum-group element abundance patterns in earliest Archean samples from Greenland, which might have recorded the LHB. However, the nonmeteoritic abundance patterns and very low abundances show that no meteoritic component is included in these rocks. After Koeberl *et al.* (2000).

samples. First, the number of samples was probably too small. Second, calculations that correlate impact melt production with crater dimension (Melosh, 1989; Cintala and Grieve, 1994, 1998) show a breakdown of such a correlation for very large impact structures. As the magnitude of the impact increases, the melt volume relative to the transient crater size increases, with a larger proportion of melt being retained inside the crater, and the depth of melting for large impact structures exceeds the depth of excavation. Thus, the large-scale melting associated with such large impacts will actually reduce the amount of shocked rocks that are formed and preserved. In such impact events, which produce impact structures of several hundred kilometers in diameter, thermal metamorphism could be more important than shock metamorphism. However, craters smaller than a few hundred kilometers in diameter would still produce abundant shocked rocks.

Furthermore, it could be that the Isua rocks are not old enough, in which case the LHB would have long ceased long before and no direct evidence for an extraterrestrial component could be obtained. It is possible that the influence of impact events on early life could have been severe (e.g., Sleep *et al.*, 1989; Zahnle and Sleep, 1997), although Ryder (2002) suggested that this influence did not necessarily lead to sterilization of the planet.

Apart from the intriguing Archean chromite grains discussed by Appel (1979), the only evidence for a LHB on the Earth seems to come from recent tungsten isotopic studies of Schoenberg *et al.* (2002), who found tungsten anomalies in ca. 3.85 Ga metasedimentary rocks from Greenland (Figure 30). Such an anomaly, if real, is difficult to explain by anything but an extraterrestrial component. However, it is difficult to understand why a similar meteoritic signal would not show up in the PGE abundances, as high tungsten abundances are not common in meteorites. Indeed, Frei and Rosing (2005) searched for such a meteoritic signature in similar samples from the Isua Supracrustal Belt in Greenland using the chromium isotopic method and found no trace of an extraterrestrial component; they, thus, comment that they cannot confirm the findings of Schoenberg *et al.* (2002). Thus, little if any evidence of a LHB phase at 3.85 Ga is preserved on Earth, but the reason may well be that the LHB destroyed any previous impact record with the production of enormous amounts of melts on the Earth at that time due to large-scale impact events (Cintala and Grieve, 1994, 1998), erasing any traces even of the LHB.

In terms of evidence for impact on Earth, the first solid evidence exists in the form of various spherule layers found in South Africa and Australia (see previous section). The oldest documented

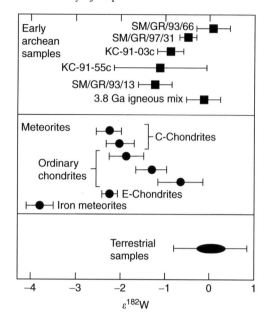

Figure 30 Tungsten isotopic composition of Early Archean samples, compared to meteorites, and terrestrial rocks. The W-isotopic anomalies in the Early Archean rocks were interpreted by these authors to be due to a meteoritic component (as a result of the LHB), but this would require unrealistically large meteoritic contributions, and results by Frei and Rosing (2005) were unable to confirm the earlier findings. After Schoenberg *et al.* (2002).

(and preserved) impact craters on Earth have ages of 2.02 and 1.86 Ga (Vredefort and Sudbury, respectively). Thus, the impact record for more than half of the geological history of the Earth is incomplete and not well preserved (see also Gilmour and Koeberl, 2000), and we mostly have only indirect evidence regarding the impact record and its effects during the first 2.5 Gyr of Earth history.

5.5 SUMMARY

Geochemical and cosmochemical studies are an essential part of the investigation of impact craters, rocks, and processes. At a basic level, such analyses are essential in classifying the various rock types. Geochemical data can be used to determine the source rocks of impactites (and ejecta), to calculate the proportions of various target rocks that go into breccias and melt rocks, to provide evidence regarding extraterrestrial components in melt rocks, or to identify the type of the meteorite that produced a certain crater. Important concepts are also "impact markers," which are all chemical, isotopic, and mineralogical species derived from the encounter of cosmic bodies (such as cometary nuclei or asteroids) with the Earth. Such markers are of prime importance to detect and study accretionary events in the sedimentary record, to identify their origin, and to evaluate their possible role in global change and on the Earth's biotic and climatic evolution throughout geological time.

The chemical composition of distal ejecta layers can be used to study a possible relationship between biotic changes and impact events, because it is possible to study such a relationship in the same outcrops, whereas correlation with radiometric ages of a distant impact structure is always associated with larger errors. The studies of the K–T ejecta layers led to improved detection sensitivities for impact markers, allowing identification of smaller events and the study of their effects. The most commonly used impact markers are elevated contents of siderophile elements (especially iridium and other PGEs), and the presence of shocked minerals (especially quartz). Distal impact ejecta/spherule layers represent high-resolution time planes that could be used to establish an intercontinental stratigraphic framework in the early Precambrian because their geochemistry (including isotopic characteristics) seems to be the only way to establish (or refute) proposed correlations between layers at different locations.

Following the discovery of the evidence for an impact event at the K–T boundary, and its associated environmental consequences, there were numerous suggestions that other extinction events might also be linked to impact events (e.g., Rampino and Haggerty, 1996). Unfortunately, the evidence so far is not encouraging. As discussed above, there is no convincing evidence of an impact event at the P–Tr boundary. The Triassic–Jurassic (Tr–J) boundary is marked by yet another major mass extinction. Bice *et al.* (1992) reported on the discovery of shocked quartz grains from a Tr–J boundary location in northern Italy, although the identification of the PDFs is not certain. The 100-km-diameter Manicouagan impact structure in Quebec has an age that is comparable to that of the Tr–J boundary, but currently available dates indicate that with an age of 214 Ma (Hodych and Dunning, 1992) it slightly predates the Tr–J boundary. Spray *et al.* (1998) discussed the possibility of multiple impacts in the Late Triassic, probably slightly predating that Tr–J boundary. Walkden *et al.* (2002) found evidence for what are probably impact ejecta from Manicouagan in southwestern Great Britain, but there is no connection to the Tr–J boundary. Olsen *et al.* (2002) reported the presence of a minor Ir anomaly at the Tr–J boundary, which was confirmed recently by Tanner and Kyte (2005), but as experience with the P–Tr boundary shows, minor PGE anomalies may or may not be indicative of impacts. Further geochemical studies should provide additional constrains on what happened at the Tr–J (and other) extinction events.

This review summarized some of the important geochemical methods that are used in the investigation of impact craters and processes, and presented a few applications and case studies. As analytical techniques become more sophisticated, more information about physical and chemical processes of impacts will be learned.

ACKNOWLEDGMENTS

I am grateful to many colleagues who, over the years, contributed discussions and data, help with field and lab work, and general encouragement. It is difficult to pick just a few names, but I would like to specifically mention Uwe Reimold (now at the Humboldt-Museum in Berlin, Germany), Joel Blum (Univ. Michigan, Ann Arbor, USA), Steve Shirey (DTM, Carnegie Institution of Washington, DC, USA), Bernhard Peucker-Ehrenbrink (WHOI, USA), and Ross Taylor (ANU, Canberra, Austrialia). I am grateful to Dona Jalufka (Vienna) for drafting the diagrams. Furthermore, I acknowledge steady funding from the Austrian Science Foundation FWF (most recently through grant P17194-N10). Two extremely thorough, constructive, and useful reviews by Jared Morrow (Univ. Northern Colorado) and Bruce Simonson (Oberlin College) are much appreciated. Any remaining errors or omissions are my own.

REFERENCES

Addison A. D., Brumpton G. R., Vallini D. A., McNaughton N. J., Davis D. W., Kissin S. A., Fralick P. W., and Hammond A. L. (2005) Discovery of distal ejecta from the 1850 Ma Sudbury impact event. *Geology* **33**, 193–196.

Allègre C. J. and Luck J. M. (1980) Osmium isotopes as petrogenetic and geological tracers. *Earth Planet. Sci. Lett.* **48**, 148–154.

Alvarez L. W., Alvarez W., Asaro F., and Michel H. V. (1980) Extraterrestrial cause for the Cretaceous-Tertiary extinction. *Science* **208**, 1095–1108.

Alvarez W., Claeys P., and Kieffer S. (1995) Emplacement of Cretaceous-Tertiary boundary shocked quartz from Chicxulub crater. *Science* **269**, 930–935.

Anbar A. D., Zahnle K. J., Arnold G. L., and Mojzsis S. J. (2001) Extraterrestrial iridium, sediment accumulation and the habitability of the early Earth's surface. *J. Geophys. Res.* **106**, 3219–3236.

Anders E. and Grevesse N. (1989) Abundances of the elements: meteoritic and solar. *Geochim. Cosmochim. Acta* **53**, 197–214.

Andreoli M. A. G., Ashwal L. D., Hart R. J., and Huizenga J. M. (1999) A Ni- and PGE-enriched quartz norite impact melt complex in the Late Jurassic Morokweng impact structure, South Africa. In *Large Meteorite Impacts and Planetary Evolution II* (eds. B. O. Dressler and V. L. Sharpton). Geol. Soc. Amer. Spec. Paper 339, pp. 91–108.

Appel P. W. U. (1979) Cosmic grains in an iron-formation from the early Precambrian Isua supercrustal belt, West Greenland. *J. Geol.* **87**, 573–578.

Attrep M., Orth C. J., Quintana L. R., Shoemaker C. S., Shoemaker E. M., and Taylor S. R. (1991) Chemical fractionation of siderophile elements in impactites from Australian meteorite craters. In *Lunar Planet. Sci.* **XXII**, The Lunar and Planetary Institute, Houston, pp. 39–40.

Barrat J. A., Jahn B.-M., Amosse J., Rocchia R., Keller F., Poupeau G., and Diemer E. (1996) Geochemistry of Libyan desert glasses. *Geochim. Cosmochim. Acta* **61**, 1953–1959.

Basu A. R., Petaev M. I., Poreda R. J., Jacobsen S. B., and Becker L. (2003) Chondritic meteorite fragments associated with the Permian-Triassic boundary in Antarctica. *Science* **302**, 1388–1392.

Becker L., Poreda R. J., Basu A. R., Pope K. O., Harrison T. M., Nicholson C., and Iasky R. (2004) Bedout: a possible end-Permian impact crater offshore of Northwestern Australia. *Science* **304**, 1469–1476.

Belcher C. M., Collinson M. E., and Scott A. C. (2005) Constraints on the thermal energy released from the Chicxulub impactor: new evidence from multi-method charcoal analysis. *J. Geol. Soc. Lond.* **162**, 591–602.

Beran A. and Koeberl C. (1997) Water in tektites and impact glasses by FTIR spectrometry. *Meteorit. Planet. Sci.* **32**, 211–216.

Bice D. M., Newton C. R., McCauley S., Reiners P. W., and McRoberts C. A. (1992) Shocked quartz at the Triassic–Jurassic boundary in Italy. *Science* **255**, 443–446.

Birck J.-L., Roy Barman M., and Capmas F. (1997) Re–Os isotopic measurements at the femtomole level in natural samples. *Geostand. Newslett.* **20**, 19–27.

Blum J. D. and Chamberlain C. P. (1992) Oxygen isotope constraints on the origin of impact glasses from; the Cretaceous-Tertiary boundary. *Science* **257**, 1104–1107.

Blum J. D., Chamberlain C. P., Hingston M. P., Koeberl C., Marin L. E., Schuraytz B. C., and Sharpton V. L. (1993) Isotopic comparison of K–T boundary impact glass with melt rock from the Chicxulub and Manson impact structures. *Nature* **364**, 325–327.

Blum J. D., Papanastassiou D. A., Koeberl C., and Wasserburg G. J. (1992) Nd and Sr isotopic study of Australasian tektites: new constraints on the provenance and age of target materials. *Geochim. Cosmochim. Acta* **56**, 483–492.

Boamah D. and Koeberl C. (2002) Geochemistry of soils from the Bosumtwi impact structure, Ghana, and relationship to radiometric airborne geophysical data. In *Meteorite Impacts in Precambrian Shields* (eds. J. Plado and L. Pesonen). Springer, Heidelberg, Berlin, pp. 211–255.

Bodiselitsch B., Montanari A., Koeberl C., and Coccioni R. (2004) Delayed climate cooling in the Late Eocene caused by multiple impacts: high-resolution geochemical studies at Massignano, Italy. *Earth Planet. Sci. Lett.* **223**, 283–302.

Bohor B. F., Betterton W. J., and Krogh T. E. (1993) Impact-shocked zircons: discovery of shock-induced textures reflecting increasing degrees of shock metamorphism. *Earth Planet. Sci. Lett.* **119**, 419–424.

Bohor B. F., Foord E. E., Modreski P. J., and Triplehorn D. M. (1984) Mineralogical evidence for an impact event at the Cretaceous/Tertiary boundary. *Science* **224**, 867–869.

Bohor B. F., Modreski P. J., and Foord E. E. (1987) Shocked quartz in the Cretaceous/Tertiary boundary clays: evidence for global distribution. *Science* **236**, 705–708.

Bottomley R. J., Grieve R. A. F., York D., and Masaitis V. (1997) The age of the Popigai impact event and its relations to events at the Eocene/Oligocene boundary. *Nature* **388**, 365–368.

Bowring S. A., Erwin D. H., Jin Y. G., Martin M. W., Davidek K., and Wang W. (1998) U/Pb zircon geochronology and tempo of the end-Permian mass extinction. *Science* **280**, 1039–1045.

Buchwald V. F. (1975) *Handbook of Iron Meteorites*. University of California Press, Berkeley, 1418pp.

Bunch T. E. and Cassidy W. A. (1972) Petrographic and electron microprobe study of the Monturaqui impactite. *Contrib. Mineral. Petrol.* **36**, 95–112.

Byerly G. R., Lowe D. R., Wooden J. L., and Xie X. (2002) An Archean impact layer from the Pilbara and Kaapvaal Cratons. *Science* **297**, 1325–1327.

Canup R. M. (2004) Dynamics of lunar formation. *Annu. Rev. Astron. Astrophys.* **42**, 441–475.

Canup R. M. and Asphaug E. (2001) Origin of the Moon in a giant impact near the end of the Earth's formation. *Nature* **412**, 708–712.

Canup R. M. and Righter K. (eds.) (2000) *Origin of the Earth and Moon*. University of Arizona Press, Tucson, AZ, 555pp.

Carlisle D. B. and Braman D. R. (1991) Nanometre-size diamonds in the Cretaceous/Tertiary boundary clay of Alberta. *Nature* **352**, 708–709.

Chapman C. R. and Morrison D. (1994) Impacts on the Earth by asteroids and comets: assessing the hazard. *Nature* **367**, 33–40.

Chaussidon M., Sigurdsson H., and Metrich N. (1994) Sulfur isotope study of high-calcium impact glasses from the KT boundary. In *New Developments Regarding the KT Event and Other Catastrophes in Earth History.* LPI Contribution No. 825, Lunar and Planetary Institute, Houston, pp. 21–22.

Chen G., Tyburczy J. A., and Ahrens T. J. (1994) Shock-induced devolatilization of calcium sulfate and implications for K–T extinctions. *Earth Planet. Sci. Lett.* **128**, 615–628.

Cintala M. J. and Grieve R. A. F. (1994) The effects of differential scaling of impact melt and crater dimensions on lunar and terrestrial crater: some brief examples. In *Large Meteorite Impacts and Planetary Evolution* (eds. B. O. Dressler, R. A. F. Grieve, and V. L. Sharpton) Geol. Soc. Am. Special Paper 293, pp. 51–59.

Cintala M. J. and Grieve R. A. F. (1998) Scaling impact melting and crater dimensions: implications for the lunar cratering record. *Meteorit. Planet. Sci.* **33**, 889–912.

Cohen B. A., Swindle T. D., and Kring D. A. (2000) Support for the lunar cataclysm hypothesis from lunar meteorite melt ages. *Science* **290**, 1754–1756.

Colodner D. C., Boyle E. A., Edmond J. M., and Thomson J. (1992) Post-depositional mobility of platinum, iridium and rhenium in marine sediments. *Nature* **358**, 402–404.

Corner B., Reimold W. U., Brandt D., and Koeberl C. (1997) Morokweng impact structure, Northwest Province, South Africa: geophysical imaging and some preliminary shock petrographic studies. *Earth Planet. Sci. Lett.* **146**, 351–364.

Creaser R. A., Papanastassiou D. A., and Wasserburg G. J. (1991) Negative thermal ion mass spectrometry of osmium, rhenium and iridium. *Geochim. Cosmochim. Acta* **55**, 397–401.

Dai X., Boamah D., Koeberl C., Reimold W. U., and McDonald I. (2005) Bosumtwi impact structure, Ghana: geochemistry of impactites and target rocks, and search for a meteoritic component. *Meteorit. Planet. Sci.* **40**, 1493–1511.

Dai X., Koeberl C., and Fröschl H. (2000) Determination of PGEs in impact melt breccias using NAA and USN-ICP-MS after anion exchange preconcentration. *Anal. Chim. Acta* **436**, 79–85.

Dalai T. K., Ravizza G. E., and Peucker-Ehrenbrink B. (2006) The Late Eocene $^{187}Os/^{188}Os$ excursion: chemostratigraphy, cosmic dust flux and the Early Oligocene glaciation. *Earth Planet. Sci. Lett.* **241**, 477–492.

De Paolo D. J., Kyte F. T., Marshall B. D., O'Neil J. R., and Smit J. (1983) Rb–Sr, Sm–Nd, K–Ca, O and H isotopic study of Cretaceous-Tertiary boundary sediments, Caravaca, Spain: evidence for an oceanic impact site. *Earth Planet. Sci. Lett.* **64**, 356–373.

Deutsch A. and Schärer U. (1994) Dating terrestrial impact events. *Meteoritics* **29**, 301–322.

Dia A., Manhès G., Dupre B., and Allègre C. J. (1989) The Cretaceous–Tertiary boundary problem: an assessment from lead isotope systematics. *Chem. Geol.* **75**, 291–304.

Dickin A. P., Richardson J. M., Crocket J. H., McNutt R. H., and Peredery W. V. (1992) Osmium isotope evidence for a crustal origin of platinum group elements in the Sudbury nickel ore, Ontario, Canada. *Geochim. Cosmochim. Acta* **56**, 3531–3537.

Dypvik H. and Attrep M., Jr. (1999) Geochemical signals of the Late Jurassic, marine Mjølnir impact. *Meteorit. Planet. Sci.* **34**, 393–406.

Earth Impact Database (2006) http://www.unb.ca/passc/ImpactDatabase/index.html (accessed on August 2006).

Ebel D. and Grossman L. (2005) Spinel-bearing spherules condensed from the Chicxulub impact-vapor plume. *Geology* **33**, 293–296.

El Goresy A. (1966) Metallic spherules in Bosumtwi crater glasses. *Earth Planet. Sci. Lett.* **1**, 23–24.

El Goresy A., Fechtig H., and Ottemann T. (1968) The opaque minerals in impactite glasses. In *Shock Metamorphism of Natural Materials* (eds. B. M. French and N. M. Short). Mono Book Corp., Baltimore, pp. 531–554.

Engelhardt W.v., Berthold C., Wenzel T., and Dehner D. (2005) Chemistry, small-scale inhomogeneity, and formation of moldavites as condensates from sands vaporized by the Ries impact. *Geochim. Cosmochim. Acta* **69**, 5611–5626.

Engelhardt W.v., Luft E., Arndt J., Schock H., and Weiskirchner W. (1987) Origin of moldavites. *Geochim. Cosmochim. Acta* **51**, 1425–1443.

Esser B. K. and Turekian K. K. (1989) Osmium isotopic composition of the Raton Basin Cretaceous–Tertiary boundary interval. *EOS* **70**, 717.

Esser B. K. and Turekian K. K. (1993) The osmium isotopic composition of the continental crust. *Geochim. Cosmochim. Acta* **57**, 3093–3104.

Evans N. J., Gregoire D. C., Grieve R. A. F., Goodfellow W. D., and Veizer J. (1993) Use of platinum-group elements for impactor identification: terrestrial impact craters and Cretaceous-Tertiary boundary. *Geochim. Cosmochim. Acta* **57**, 3737–3748.

Farley K. A., Montanari A., Shoemaker E. M., and Shoemaker C. S. (1998) Geochemical evidence for a comet shower in the late Eocene. *Science* **280**, 1250–1253.

Farley K. A. and Mukhopadhyay S. (2001) An extraterrestrial impact at the Permian–Triassic Boundary? *Science* **293**, 2343a.

Farley K. A., Vokrouhlicky D., Bottke W. F., and Nesvorny D. (2006) A late Miocene dust shower from the break-up of an asteroid in the main belt. *Nature* **439**, 295–297.

Farley K. A., Ward P., Garrison G., and Mukhopadhyay S. (2005) Absence of extraterrestrial ³He in Permian–Triassic age sedimentary rocks. *Earth Planet. Sci. Lett.* **240**, 265–275.

Faure G. (1986) *Principles of Isotope Geology*, 2nd edn. Wiley, New York, 608pp.

Fehn U., Teng R., Elmore D., and Kubik P. W. (1986) Isotopic composition of osmium in terrestrial samples determined by accelerator mass spectrometry. *Nature* **323**, 707–710.

Fowler C. M., Ebinger C. J., and Hawkesworth C. J. (eds.) (2003) *The Early Earth: Physical, Chemical and Biological Development*. Geol. Soc. Lond. Spec. Pub. No. 199, 352pp.

Frei R. and Frei K. M. (2002) A multi-isotopic and trace element investigation of the Cretaceous–Tertiary boundary layer at Stevens Klint, Denmark-inferences for the origin and nature of siderophile and lithophile geochemical anomalies. *Earth Planet. Sci. Lett.* **203**, 691–708.

Frei R. and Rosing M. T. (2005) Search for traces of the late heavy bombardment on Earth—results from high precision chromium isotopes. *Earth Planet. Sci. Lett.* **236**, 28–40.

French B. M. (1998) *Traces of Catastrophe: A Handbook of Shock-Metamorphic Effects in Terrestrial Meteorite Impact Structures*. LPI Contribution 954, Lunar and Planetary Institute, Houston, 120pp.

French B. M., Koeberl C., Gilmour I., Shirey S. B., Dons J. A., and Naterstad J. (1997) The Gardnos impact structure, Norway: petrology and geochemistry of target rocks and impactites. *Geochim. Cosmochim. Acta* **61**, 873–904.

French B. M. and Nielsen R. L. (1990) Vredefort bronzite granophyre: chemical evidence for an origin as meteorite impact melt. *Tectonophysics* **171**, 119–138.

French B. M., Orth C. J., and Quintana C. R. (1989) Iridium in the Vredefort bronzite granophyre: impact melting and limits on a possible extraterrestrial component. *Proc. 19th Lunar Planet. Sci. Conf.*, pp. 733–744.

Frey H. (1980) Crustal evolution of the early Earth: the role of major impacts. *Precambrian Res.* **10**, 195–216.

Ganapathy R. (1980) A major meteorite impact on the Earth 65 million years ago: evidence from the Cretaceous–Tertiary boundary clay. *Science* **209**, 921–923.

Ganapathy R. and Larimer J. W. (1983) Nickel-iron spherules in tektites: non-meteoritic in origin. *Earth Planet. Sci. Lett.* **65**, 225–228.

Gelinas A., Kring D. A., Zurcher L., Urrutia-Fucugauchi J., Morton O., and Walker R. J. (2004) Osmium isotope constraints on the proportion of bolide component in Chicxulub impact melt rocks. *Meteorit. Planet. Sci.* **39**, 1003–1008.

Gentner W., Kleinmann B., and Wagner G. A. (1967) New K–Ar and fission track ages of impact glasses and tektites. *Earth Planet. Sci. Lett.* **2**, 83–86.

Gilmour I. (1998) Geochemistry of carbon in terrestrial impact processes. In *Meteorites: Flux with Time and Impact Effects* (eds. M. M. Grady, R. Hutchison, G. J. H. McCall, and D. A. Rothery). Geol. Soc. Lond. Spec. Pub. No. 140, pp. 205–216.

Gilmour I., French B. M., Franchi I. A., Abbott J. I., Hough R. M., Newton J., and Koeberl C. (2003) Geochemistry of carbonaceous impactites from the Gardnos impact structure, Norway. *Geochim. Cosmochim. Acta* **67**, 3889–3903.

Gilmour I. and Koeberl C. (eds.) (2000). *Impacts and the Early Earth*. Springer, Heidelberg, 445pp.

Gilmour I., Russell S. S., Arden J. W., Lee M. R., Franchi I. A., and Pillinger C. T. (1992) Terrestrial carbon and nitrogen isotopic ratios from Cretaceous–Tertiary boundary nanodiamonds. *Science* **258**, 1624–1626.

Gilmour I., Wolbach W. S., and Anders E. (1990) Early environmental effects of the terminal Cretaceous impact. In *Global Catastrophes in Earth History* (eds. V. L. Sharpton and P. D. Ward). Geol. Soc. Amer. Spec. Paper 247, pp. 383–390.

Glass B. P. (1967) Microtektites in deep sea sediments. *Nature* **214**, 372–374.

Glass B. P. (1968) Glassy objects (microtektites?) from deep sea sediments near the Ivory Coast. *Science* **161**, 891–893.

Glass B. P. (1969) Chemical composition of Ivory Coast microtektites. *Geochim. Cosmochim. Acta* **33**, 1135–1147.

Glass B. P. (1972) Bottle green microtektites. *J. Geophys. Res.* **77**, 7057–7064.

Glass B. P. (1989) North American tektite debris and impact ejecta from DSDP Site 612. *Meteoritics* **24**, 209–218.

Glass B. P. and Barlow R. A. (1979) Mineral inclusions in Muong Nong-type indochinites: implications concerning parent material and process of formation. *Meteoritics* **14**, 55–67.

Glass B. P., Fredriksson K., and Florensky P. V. (1983) Microirghizites recovered from a sediment sample from the Zhamanshin impact structure. *Proc. 14th Lunar Planet. Sci. Conf., J. Geophys. Res.* **88**, B319–B330.

Glass B. P., Huber H., and Koeberl C. (2004) Geochemistry of Cenozoic microtektites and clinopyroxene-bearing spherules. *Geochim. Cosmochim. Acta* **68**, 3971–4006.

Glass B. P., Kent D. V., Schneider D. A., and Tauxe L. (1991) Ivory Coast microtektite strewn field: description and relation to the Jaramillo geomagnetic event. *Earth Planet. Sci. Lett.* **107**, 182–196.

Glass B. P. and Pizzuto J. E. (1994) Geographic variation in Australasian microtektite concentrations: implications concerning the location and size of the source crater. *J. Geophys. Res.* **99**, 19075–19081.

Glass B. P., Swincki M. B., and Zwart P. A. (1979) Australasian, Ivory Coast and North American tektite strewn field: size, mass and correlation with geomagnetic reversals and other earth events. *Proc. 10th Lunar Planet. Sci. Conf.*, pp. 2535–2545.

Glass B. P. and Wu J. (1993) Coesite and shocked quartz discovered in the Australasian and North American microtektite layers. *Geology* **21**, 435–438.

Glass B. P. and Zwart P. A. (1979) The Ivory Coast microtektite strewn field: new data. *Earth Planet. Sci. Lett.* **43**, 336–342.

Göbel E., Reimold W. U., Baddenhausen H., and Palme H. (1980) The projectile of the Lapajärvi impact crater. *Z. Naturforsch.* **35a**, 197–203.

Gomes R., Levison H. F., Tsiganis K., and Morbidelli A. (2005) Origin of the cataclysmic late heavy bombardment period of the terrestrial planets. *Nature* **435**, 466–469.

Gostin V. A., Haines P. W., Jenkins R. J. E., Compston W., and Williams I. S. (1986) Impact ejecta horizon within late Precambrian shales, Adelaide Geosyncline, South Australia. *Science* **233**, 198–200.

Gostin V. A., Keays R. R., and Wallace M. W. (1989) Iridium anomaly from the Acraman impact ejecta horizon: impacts can produce sedimentary iridium peaks. *Nature* **340**, 542–544.

Grajales-Nishimura J. M., Cadillo-Pardo E., Rosales-Domínguez M., Morán-Zenteno D., Alvarez W., Claeys P., Ruíz-Morales J., García-Hernández J., Padilla-Avila P., and Sánchez-Ríos A. (2000) Chicxulub impact: the origin of reservoir and seal facies in the southeastern Mexico oil fields. *Geology* **28**, 307–310.

Grard A., Francois L. M., Dessert C., Dupré B., and Goddéris Y. (2005) Basaltic volcanism and mass extinction at the Permo-Triassic boundary: environmental impact and modeling of the global carbon cycle. *Earth Planet. Sci. Lett.* **234**, 207–221.

Grieve R. A. F. (1980) Impact bombardment and its role in proto-continental growth on the early Earth. *Precambrian Res.* **10**, 217–247.

Grieve R. A. F., Bottomley R. B., Bouchard M. A., Robertson P. B., Orth C. J., and Attrep M. (1991) Impact melt rocks from New Quebec Crater, Quebec, Canada, *Meteoritics* **26**, 31–39.

Grieve R. A. F., Langenhorst F., and Stöffler D. (1996) Shock metamorphism in nature and experiment. II: Significance in geoscience. *Meteorit. Planet. Sci.* **31**, 6–35.

Grieve R. A. F. and Masaitis V. L. (1994) The economic potential of terrestrial impact craters. *Int. Geol. Rev.* **36**, 105–151.

Grieve R. A. F. and Pilkington M. (1996) The geophysical signature of terrestrial impacts. *J. AGSO Australian Geol. Geophys.* **16**, 399–420.

Grieve R. A. F. and Shoemaker E. M. (1994) The record of past impacts on Earth. In *Hazards Due to Comets and Asteroids* (ed. T. Gehrels). University of Arizona Press, Tucson, AZ, pp. 417–462.

Gros J., Takahashi H., Hertogen J., Morgan J. W., and Anders E. (1976) Composition of the projectiles that bombarded the lunar highlands. *Proc. 7th Lunar Planet. Sci. Conf.*, pp. 2403–2425.

Gupta S. C., Ahrens T. J., and Yang W. (2001) Shock–induced vaporization of anhydrite and global cooling from the K/T impact. *Earth Planet. Sci. Lett.* **188**, 399–412.

Gurov E. P. (1996) The group of Macha craters in Western Yakutia. In *Lunar Planet. Sci.* **XXVII**, The Lunar and Planetary Institute, Houston, pp. 473–474.

Halliday A. N. (2004) Mixing, volatile loss and compositional change during impact-driven accretion of the Earth. *Nature* **427**, 505–509.

Halliday A., Rehkämper M., Lee D.-C., and Yi W. (1996) Early evolution of the Earth and Moon: new constraints on Hf-W geochemistry. *Earth Planet. Sci. Lett.* **142**, 75–89.

Hart R. J., Andreoli M. A. G., Tredoux M., Moser D., Ashwal L. D., Eide E. A., Webb S. J., and Brandt D. (1997) Late Jurassic age for the Morokweng impact structure, southern Africa. *Earth Planet. Sci. Lett.* **147**, 25–35.

Hassler D. R., Peucker-Ehrenbrink B., and Ravizza G. E. (2000) Rapid determination of Os isotopic composition by sparging OsO_4 into a magnetic-sector ICP-MS. *Chem. Geol.* **166**, 1–14.

Henkel H. and Reimold W. U. (2002) Magnetic model of the central uplift of the Vredefort impact structure, South Africa. *J. Appl. Geophys.* **51**, 43–62.

Hildebrand A. R., Penfield G. T., Kring D. A., Pilkington M., Carmargo Z. A., Jacobsen S. B., and Boynton W. V. (1991) Chicxulub crater: a possible Cretaceous–Tertiary boundary impact crater on the Yucatan Peninsula, Mexico. *Geology* **19**, 867–871.

Hodych J. P. and Dunning G. R. (1992) Did the Manicouagan impact trigger end-of-Triassic mass extinction? *Geology* **20**, 51–54.

Holser W. T., Schönlaub H. P., Attrep M., Jr, Boeckelmann K., Klein P., Magaritz M., Orth C. J., Fenninger A., Jenny C., Kralik M., Mauritsch H., Pak E., Schramm J. M., Stattegger K., and Schmöller R. (1989) A unique geochemical record at the Permian–Triassic boundary. *Nature* **337**, 39–44.

Hörz F., Mittlefehldt D. W., See T. H., and Galindo C. (2002) Petrographic studies of the impact melts from Meteor Crater, Arizona, USA. *Meteorit. Planet. Sci.* **37**, 501–531.

Hough R. M., Gilmour I., Pillinger C. T., Langenhorst F., and Montanari A. (1997) Diamonds from the iridium–rich K–T boundary layer at Arroyo el Mimbral, Tamaulipas, Mexico. *Geology* **25**, 1019–1022.

Ivanov B. A. and Melosh H. J. (2003) Impacts do not initiate volcanic eruptions: eruptions close to the crater. *Geology* **31**, 869–872.

Izett G. A. (1991) Tektites in Cretaceous–Tertiary boundary rocks on Haiti and their bearing on the Alvarez impact extinction hypothesis. *J. Geophys. Res.* **96**, 20879–20905.

Izett G. A., Dalrymple G. B., and Snee L. W. (1991) $^{40}Ar/^{39}Ar$ age of Cretaceous–Tertiary boundary tektites from Haiti. *Science* **252**, 1539–1542.

Janssens M.-J., Hertogen J., Takahashi H., Anders E., and Lambert P. (1977) Rochechouart meteorite crater: identification of projectile. *J. Geophys. Res.* **82**, 750–758.

Jones W. B. (1985) Chemical analyses of Bosumtwi crater target rocks compared with Ivory Coast tektites. *Geochim. Cosmochim. Acta* **49**, 2569–2576.

Kaiho K., Kajiwara Y., Nakano T., Miura Y., Kawahata H., Taziki K., Ueshima M., Chen Z., and Shi G. R. (2001) End-Permian catastrophe by bolide impact: evidence of a gigantic release of sulfur from the mantle. *Geology* **29**, 815–818.

Kamo S. L. and Krogh T. E. (1995) Chicxulub crater source for shocked zircons from the Cretaceous–Tertiary boundary layer, Saskatchewan: evidence from new U–Pb data. *Geology* **23**, 281–284.

Kamo S. L., Reimold W. U., Krogh T. E., and Colliston W. P. (1996) A 2.023 Ga age for the Vredefort impact event and a first report of shock metamorphosed zircons in pseudotachylitic breccias and granophyre. *Earth Planet. Sci. Lett.* **144**, 369–388.

Kleine T., Mezger K., Palme H., and Münker C. (2004) The W isotope evolution of the bulk silicate Earth: constraints on the timing and mechanisms of core formation and accretion. *Earth Planet. Sci. Lett.* **228**, 109–123.

Klemm V., Levasseur S., Frank M., Hei J. R., and Halliday A. N. (2005) Osmium isotope stratigraphy of a marine ferromanganese crust. *Earth Planet. Sci. Lett.* **238**, 42–48.

Koeberl C. (1986) Geochemistry of tektites and impact glasses. *Annu. Rev. Earth Planet. Sci.* **14**, 323–350.

Koeberl C. (1992a) Geochemistry and origin of Muong Nong-type tektites. *Geochim. Cosmochim. Acta* **56**, 1033–1064.

Koeberl C. (1992b) Water content of glasses from the K/T boundary, Haiti: indicative of impact origin. *Geochim. Cosmochim. Acta* **56**, 4329–4332.

Koeberl C. (1993a) Extraterrestrial component associated with Australasian microtektites in a core from ODP Site 758B. *Earth Planet. Sci. Lett.* **119**, 453–458.

Koeberl C. (1993b) Chicxulub crater, Yucatan: tektites, impact glasses, and the geochemistry of target rocks and breccias. *Geology* **21**, 211–214.

Koeberl C. (1994) Tektite origin by hypervelocity asteroidal or cometary impact: target rocks, source craters, and mechanisms. In *Large Meteorite Impacts and Planetary Evolution* (eds. B. O. Dressler, R. A. F. Grieve, and V. L. Sharpton). Geol. Soc. Amer. Spec. Paper 293, pp. 133–151.

Koeberl C. (1996) Chicxulub—the K–T boundary impact crater: a review of the evidence, and an introduction to impact crater studies. *Abhandlungen Geologischen Bundesanstalt (Wien)* **53**, 23–50.

Koeberl C. (1998) Identification of meteoritical components in impactites. In *Meteorites: Flux with Time and Impact Effects* (eds. M. M. Grady, R. Hutchison, G. J. H. McCall, and D. A. Rothery). Geol. Soc. Lond. Spec. Pub. No. 140, pp. 133–152.

Koeberl C. (2000) Confirmation of a meteoritic component in Libyan Desert Glass from osmium isotopic data. *Meteorit. Planet. Sci.* **35**, A89–A90.

Koeberl C. (2001) The sedimentary record of impact events. In *Accretion of Extraterrestrial Matter throughout Earth's History* (eds. B. Peucker-Ehrenbrink and B. Schmitz). Kluwer Academic/Plenum, New York, pp. 333–378.

Koeberl C. (2002) Mineralogical and geochemical aspects of impact craters. *Mineral. Mag.* **66**, 745–768.

Koeberl C. (2004) The late heavy bombardment in the inner solar system: is there a connection to Kuiper Belt objects? *Earth Moon Planet.* **92**, 79–87.

Koeberl C. (2006) The Record of impact processes on the early Earth—a review of the first 2.5 billion years. In *Processes of the Early Earth* (eds. W. U. Reimold and R. Gibson). Geol. Soc. Amer. Spec. Paper 405, pp. 1–22.

Koeberl C., Armstrong R. A., and Reimold W. U. (1997a) Morokweng, South Africa: a large impact structure of Jurassic–Cretaceous boundary age. *Geology* **25**, 731–734.

Koeberl C., Bottomley R. J., Glass B. P., and Storzer D. (1997b) Geochemistry and age of Ivory Coast tektites and microtektites. *Geochim. Cosmochim. Acta* **61**, 1745–1772.

Koeberl C., Farley K. A., Peucker-Ehrenbrink B., and Sephton M. A. (2004a) Geochemistry of the end-Permian extinction event in Austria and Italy: no evidence for an extraterrestrial component. *Geology* **32**, 1053–1056.

Koeberl C., Gilmour I., Reimold W. U., Claeys P., and Ivanov B. A. (2002a) Comment on "End-Permian catastrophe by bolide impact: evidence of a gigantic release of sulfur from the mantle" by Kaiho *et al. Geology* **30**, 855–856.

Koeberl C. and Henkel H. (eds.) (2005). *Impact Tectonics*. Springer, Heidelberg, 552pp.

Koeberl C. and Huber H. (2000) Optimization of the multiparameter γ–γ coincidence spectrometry for the determination of iridium in geological materials. *J. Radioanal. Nucl. Chem.* **244**, 655–660.

Koeberl C. and MacLeod K. (eds.) (2002) *Catastrophic Events and Mass Extinctions: Impacts and Beyond.* Geol. Soc. Amer. Spec. Paper 356, 746pp.

Koeberl C. and Martinez-Ruiz F. (2003) The stratigraphic record of impact events: a short overview. In *Impact Markers in the Stratigraphic Record* (eds. C. Koeberl and F. Martinez-Ruiz). Springer, Heidelberg, pp. 1–40.

Koeberl C., Peucker-Ehrenbrink B., Reimold W. U., Shukolyukov A., and Lugmair G. W. (2002b) Comparison of osmium and chromium isotopic methods for the detection of meteoritic components in impactites: examples from the Morokweng and Vredefort impact structures, South Africa. In *Catastrophic Events and Mass Extinctions: Impacts and Beyond* (eds. C. Koeberl and K. G. MacLeod). Geol. Soc. Amer. Spec. Paper 356, pp. 607–617.

Koeberl C., Poag C. W., Reimold W. U., and Brandt D. (1996a) Impact origin of Chesapeake Bay structure and the source of North American tektites. *Science* **271**, 1263–1266.

Koeberl C. and Reimold W. U. (1995) Early Archean spherule beds in the Barberton Mountain Land, South Africa: no evidence for impact origin. *Precambrian Res.* **74**, 1–33.

Koeberl C. and Reimold W. U. (2003) Geochemistry and petrography of impact breccias and target rocks from the 145 Ma Morokweng impact structure, South Africa. *Geochim. Cosmochim. Acta* **67**, 1837–1862.

Koeberl C., Reimold W. U., and Boer R. H. (1993) Geochemistry and mineralogy of Early Archean spherule beds, Barberton Mountain Land, South Africa: evidence for origin by impact doubtful. *Earth Planet. Sci. Lett.* **119**, 441–452.

Koeberl C., Reimold W. U., Blum J. D., and Chamberlain C. P. (1998) Petrology and geochemistry of target rocks from the Bosumtwi impact structure, Ghana, and comparison with Ivory Coast tektites. *Geochim. Cosmochim. Acta* **62**, 2179–2196.

Koeberl C., Reimold W. U., McDonald I., and Rosing M. (2000) Search for petrographical and geochemical evidence for the late heavy bombardment on Earth in Early Archean rocks from Isua, Greenland. In *Impacts and the Early Earth* (eds. I. Gilmour and C. Koeberl). Springer, Heidelberg, pp. 73–97.

Koeberl C., Reimold W. U., and Shirey S. B. (1994a) Saltpan impact crater, South Africa: geochemistry of target rocks, breccias, and impact glasses, and osmium isotope systematics. *Geochim. Cosmochim. Acta* **58**, 2893–2910.

Koeberl C., Reimold W. U., and Shirey S. B. (1996b) A Re–Os isotope and geochemical study of the Vredefort Granophyre: clues to the origin of the Vredefort structure, South Africa. *Geology* **24**, 913–916.

Koeberl C., Reimold W. U., Shirey S. B., and Le Roux F. G. (1994b) Kalkkop crater, Cape Province, South Africa: confirmation of impact origin using osmium isotope systematics. *Geochim. Cosmochim. Acta* **58**, 1229–1234.

Koeberl C., Sharpton V. L., Schuraytz B. C., Shirey S. B., Blum J. D., and Marin L. E. (1994c) Evidence for a meteoritic component in impact melt rock from the Chicxulub structure. *Geochim. Cosmochim. Acta* **58**, 1679–1684.

Koeberl C. and Shirey S. B. (1993) Detection of a meteoritic component in Ivory Coast tektites with rhenium–osmium isotopes. *Science* **261**, 595–598.

Koeberl C. and Shirey S. B. (1996) Re–Os isotope study of rocks from the manson impact structure. In *The Manson Impact Structure, Iowa: Anatomy of an Impact Crater* (eds. C. Koeberl and R. R. Anderson). Geol. Soc. Amer. Spec. Paper 302, pp. 331–339.

Koeberl C. and Shirey S. B. (1997) Re–Os systematics as a diagnostic tool for the study of impact craters and distal ejecta. *Palaeogeog. Palaeoclim. Palaeoecol.* **132**, 25–46.

Koeberl C., Shukolyukov A., and Lugmair G. W. (2004b) An ordinary chondrite impactor composition of the Bosumtwi impact structure, Ghana, West Africa: discussion of siderophile element contents and Os and Cr isotope data. In *Lunar Planet. Sci.* **XXXV**, #1256. The Lunar and Planetary Institute, Houston (CD-ROM).

Koeberl C. and Sigurdsson H. (1992) Geochemistry of impact glasses from the K/T boundary in Haiti: relation to smectites, and a new type of glass. *Geochim. Cosmochim. Acta* **56**, 2113–2129.

Kolbe P., Pinson W. H., Saul J. M., and Miller E. W. (1967) Rb–Sr study on country rocks of the Bosumtwi crater, Ghana. *Geochim. Cosmochim. Acta* **31**, 869–875.

Krähenbühl U., Geissbühler M., Bühler F., and Eberhardt P. (1988) The measurement of osmium isotopes in samples from a Cretaceous/Tertiary (K/T) section of the Raton Basin, USA. *Meteoritics* **23**, 282.

Kramar U., Stüben D., Berner Z., Stinnesbeck W., Philipp H., and Keller G. (2001) Are Ir anomalies sufficient and unique indicators for cosmic events? *Planet. Space Sci.* **49**, 831–837.

Kring D. A. (1993) The Chicxulub impact event and possible causes of K/T boundary extinctions. In *Proceedings of the First Annual Symposium of Fossils of Arizona* (eds. D. Boaz and M. Dornan). Mesa Southwest Museum and Southwest Paleontological Society, Mesa, AZ, pp. 63–79.

Kring D. A. and Boynton W. V. (1991) Altered spherules of impact melt and associated relic glass from the K/T boundary sediments in Haiti. *Geochim. Cosmochim. Acta* **55**, 1737–1742.

Kring D. A. and Cohen B. A. (2002) Cataclysmic bombardment throughout the inner solar system 3.9–4.0 Ga. *J. Geophys. Res.* **107**, 4-1–4-5, doi:10.1029/2001JE001529.

Krogh T. E., Kamo S. L., and Bohor B. F. (1993) Fingerprinting the K/T impact site and determining the time of impact by U-Pb dating of single shocked zircons from distal ejecta. *Earth Planet. Sci. Lett.* **119**, 425–429.

Kyte F. T. (1998) A meteorite from the Cretaceous/Tertiary boundary. *Nature* **396**, 237–239.

Kyte F. T. and Bostwick J. A. (1995) Magnesioferrite spinel in Cretaceous–Tertiary boundary sediments of the Pacific basin: remnants of hot, early ejecta from the Chicxulub impact? *Earth Planet. Sci. Lett.* **132**, 113–127.

Kyte F. T., Shukolyukov A., Lugmair G. W., Lowe D. R., and Byerly G. R. (2003) Early Archean spherule beds: chromium isotopes confirm origin from multiple impacts of projectiles of carbonaceous chondrites type. *Geology* **31**, 283–286.

Kyte F. T., Zhou L., and Lowe D. R. (1992) Noble metal abundances in an Early Archean impact deposit. *Geochim. Cosmochim. Acta* **56**, 1365–1372.

Kyte F. T., Zhou Z., and Wasson J. T. (1980) Siderophile-enriched sediments from the Cretaceous–Tertiary boundary. *Nature* **288**, 651–656.

Lambert P. (1982) Anomalies within the system: Rochechouart target rock meteorite. In *Geological Implications of Impacts of Large Asteroids and Comets on the Earth* (eds. L. T. Silver and P. H. Schultz). Geol. Soc. Amer. Spec. Paper 190, pp. 243–249.

Langenhorst F., Kyte F. T., and Retallack G. J. (2005) Reexamination of quartz grains from the Permian-Triassic boundary section at Graphite Peak, Antarctica. In *Lunar Planet. Sci.* **XXXVI**, # 2358. The Lunar and Planetary Institute, Houston, (CD-ROM).

Lee D.-C., Halliday A. N., Snyder G. A., and Taylor L. A. (1997) Age and origin of the Moon. *Science* **278**, 1098–1103.

Lee S. R., Horton J. W., Jr., and Walker R. J. (2006) Confirmation of a meteoritic component in impact-melt rocks of the Chesapeake Bay impact structure, Virginia, USA—evidence from osmium isotopic and PGE systematics. *Meteorit. Planet. Sci.*, **41**, 819–833.

Leroux H., Reimold W. U., and Doukhan J. C. (1994) A T.E.M. investigation of shock metamorphism in quartz from the Vredefort dome, South Africa. *Tectonophysics* **230**, 223–239.

Lichte F. E., Wilson S. M., Brooks R. R., Reeves R. D., Holzbecher J., and Ryan D. E. (1986) New method for the measurement of osmium isotopes applied to a New Zealand Cretaceous/Tertiary boundary shale. *Nature* **322**, 816–817.

Lodders K. (2003) Solar system abundances and condensation temperatures of the elements. *Astrophys. J.* **591**, 1220–1247.

Lowe D. R., Byerly G. R., Asaro F., and Kyte F. T. (1989) Geological and geochemical record of 3400-million-year-old terrestrial meteorite impacts. *Science* **245**, 959–962.

Lowe D. R., Byerly G. R., Kyte F. T., Shukolyukov A., Asaro F., and Krull A. (2003) Spherule beds 3.47–3.24 billion years old in the Barberton Greenstone Belt, South Africa: a record of large meteorite impacts and their influence on early crustal and biological evolution. *Astrobiology* **3**, 7–48.

Luck J. M. and Turekian K. K. (1983) Osmium-187/Osmium-186 in manganese nodules and the Cretaceous–Tertiary boundary. *Science* **222**, 613–615.

Lugmair G. W. and Shukolyukov A. (1998) Early solar system timescales according to $^{53}Mn–^{53}Cr$ systematics. *Geochim. Cosmochim. Acta* **62**, 2863–2886.

Lyons J. R. and Ahrens T. J. (2003) Terrestrial acidification at the K/T boundary. In *High-Pressure Shock Composition of Solid V: Shock Chemistry with Applications to Meteorite Impacts* (eds. L. Davison, Y. Horie, and T. Sekine). Springer, New York, pp. 181–197.

Ma P., Aggrey K., Tonzola C., Schnabel C., de Nicola P., Herzog G. F., Wasson J. T., Glass B. P., Brown L., Tera F., Middleton R., and Klein J. (2004) Beryllium-10 in Australasian tektites: constraints on the location of the source crater. *Geochim. Cosmochim. Acta* **68**, 3883–3896.

Maier W. D., Andreoli M. A. G., McDonald I., Higgins M. D., Boyce A. J., Shukolyukov A., Lugmair G. W., Ashwal L. D., Graser P., Ripley E. M., and Hart R. J. (2006) Discovery of a 25-cm asteroid clast in the giant Morokweng impact crater, South Africa. *Nature* **441**, 203–206.

Maruoka T., Koeberl C., Hancox P. J., and Reimold W. U. (2003) Sulfur geochemistry across a terrestrial Permian–Triassic boundary section in the Karoo Basin, South Africa. *Earth Planet. Sci. Lett.* **206**, 101–117.

Masaitis V. L. (1994) Impactites from Popigai crater. In *Large Meteorite Impacts and Planetary Evolution* (eds. B. O. Dressler, R. A. F. Grieve, and V. L. Sharpton) Geol. Soc. Amer. Spec. Paper 293, pp. 153–162.

Masaitis V. L. and Raikhlin A. J. (1985) The Popigai crater formed by the impact of an ordinary chondrite (in Russian). *Dokl. Akad. Nauk SSSR* **286**, 1476–1478.

Mason B. (1979) Meteorites. Data of Geochemistry, Chapter B, Part 1. US Geological Survey Professional Paper 440-B-1, 131pp.

McDonald I. (2002) Clearwater East structure: a re-interpretation of the projectile type using new platinum-group element data from meteorites. *Meteorit. Planet. Sci.* **37**, 459–464.

McDonald I., Andreoli M. A. G., Hart R. J., and Tredoux M. (2001) Platinum-group elements in the Morokweng impact structure, South Africa: evidence for the impact of a large ordinary chondrite projectile at the Jurassic–Cretaceous boundary. *Geochim. Cosmochim. Acta* **65**, 299–309.

Meisel T., Krähenbühl U., and Nazarov M. A. (1995) Combined osmium and strontium isotopic study of the Cretaceous–Tertiary boundary at Sumbar, Turkmenistan: a test for impact vs. volcanic hypothesis. *Geology* **23**, 313–316.

Melosh H. J. (1989) *Impact Cratering—A Geologic Process.* Oxford University Press, New York, 245pp.

Mittlefehldt D. W., See T. H., and Hörz F. (1992a) Projectile dissemination in impact melts from Meteor crater, Arizona. In *Lunar Planet. Sci.* **XXIII**, The Lunar and Planetary Institute, Houston, pp. 919–920.

Mittlefehldt D. W., See T. H., and Hörz F. (1992b) Dissemination and fractionation of projectile materials in the impact melts from Wabar crater, Saudi Arabia. *Meteoritics* **27**, 361–370.

Montanari A., Hay R. L., Alvarez W., Asaro F., Michel H. V., and Alvarez L. (1983) Spheroids at the Cretaceous–Tertiary boundary are altered droplets of basaltic composition. *Geology* **11**, 668–671.

Montanari A. and Koeberl C. (2000) *Impact Stratigraphy: The Italian Record.* Springer, Heidelberg, 364pp.

Morgan J. W. (1978) Lonar crater glasses and high-magnesium australites: trace element volatilization and meteoritic contamination. *Proc. 9th Lunar Planet. Sci. Conf.*, pp. 2713–2730.

Morgan J. W., Higuchi H., Ganapathy R., and Anders E. (1975) Meteoritic material in four terrestrial meteorite craters. *Proc. 6th Lunar Planet. Sci. Conf.*, pp. 1609–1623.

Morgan J. W., Janssens M.-J., Hertogen J., Gros J., and Takahshi H. (1979) Ries impact crater: search for meteoritic material. *Geochim. Cosmochim. Acta* **43**, 803–815.

Morgan J. W. and Wandless G. A. (1983) Strangways Crater, Northern Territory, Australia: siderophile element enrichment and lithophile element fractionation. *J. Geophys. Res.* **88**, A819–A829.

Mukhopadhyay S., Farley K., and Montanari A. (2001) A 35 Myr record of helium in pelagic limestones: implications for interplanetary dust accretion from the early Maastrichtian to the Middle Eocene. *Geochim. Cosmochim. Acta* **65**, 653–669.

Müller R. D., Goncharov A., and Kritski A. (2005) Geophysical evaluation of the enigmatic Bedout basement high, offshore northwestern Australia. *Earth Planet. Sci. Lett.* **237**, 264–284.

Nazarov M. A., Barsukova L. D., Badjukov D. D., Kolesov G. M., Nizhegorodova I. V., and Alekseev A. S. (1989) Geology and chemistry of the Kara and Ust-Kara impact craters. In *Lunar Planet. Sci.* **XX**, The Lunar and Planetary Institute, Houston, pp. 764–765.

Nazarov M. A., Barsukova L. D., Badjukov D. D., Kolesov G. M., and Nizhegorodova I. V. (1990) The Kara impact structure: iridium abundances in the crater rocks. In *Lunar Planet. Sci.* **XXI**, The Lunar and Planetary Institute, Houston, pp. 849–850.

Norman M. D., Bennett V. C., and Ryder G. (2002) Targeting the impactors: siderophile element signatures of lunar impact melts from Serenitatis. *Earth Planet. Sci. Lett.* **202**, 217–228.

Olsen P. E., Kent D. V., Sues H.-D., Koeberl C., Huber H., Montanari A., Rainforth E. C., Fowell S. J., Szajna M. J., and Hartline B. W. (2002) Ascent of dinosaurs linked to Ir anomaly at Triassic–Jurassic boundary. *Science* **296**, 1305–1307.

Palme H. (1980) The meteoritic contamination of terrestrial and lunar impact melts and the problem of indigenous siderophiles in the lunar highland. *Proc. 11th Lunar Planet. Sci. Conf.*, pp. 481–506.

Palme H. (1982) Identification of projectiles of large terrestrial impact craters and some implications for the interpretation of Ir-rich Cretaceous/Tertiary boundary layers. In *Geological Implications of Impacts of Large Asteroids and Comets on the Earth* (eds. L. T. Silver and P. H. Schultz). Geol. Soc. Amer. Spec. Paper 190, pp. 223–233.

Palme H., Göbel E., and Grieve R. A. F. (1979) The distribution of volatile and siderophile elements in the impact melt of East Clearwater (Quebec). *Proc. 10th Lunar Planet. Sci. Conf.*, pp. 2465–2492.

Palme H., Grieve R. A. F., and Wolf R. (1981) Identification of the projectile at the Brent crater, and further considerations of projectile types at terrestrial craters. *Geochim. Cosmochim. Acta* **45**, 2417–2424.

Palme H., Janssens M.-J., Takahashi H., Anders E., and Hertogen J. (1978) Meteorite material at five large impact craters. *Geochim. Cosmochim. Acta* **42**, 313–323.

Palme H., Rammensee W., and Reimold W. U. (1980) The meteoritic component of impact melts from European impact craters. In *Lunar Planet. Sci.* **XI**, The Lunar and Planetary Institute, Houston, pp. 848–851.

Parnell J., Osinski G. G., Lee P., Green P. F., and Baron M. J. (2005) Thermal alteration of organic matter in an impact crater and the duration of postimpact heating. *Geology* **33**, 373–376.

Pernicka E., Kaether D., and Koeberl C. (1996) Siderophile element concentrations in drill core samples from the Manson crater. In *The Manson Impact Structure, Iowa: Anatomy of an Impact Crater* (eds. C. Koeberl and R. R. Anderson). Geol. Soc. Amer. Spec. Paper 302, pp. 325–330.

Pesonen L. J., Koeberl C., and Hautaniemi H. (2003) Airborne geophysical survey of the Lake Bosumtwi meteorite impact structure (Southern Ghana)—Geophysical maps with descriptions. *Jahrbuch der Geologischen Bundesanstalt (Wien)* **143**, 581–604.

Peucker-Ehrenbrink B. (1996) Accretion rate of extraterrestrial matter during the last 80 million years and its effect on the marine osmium isotope record. *Geochim. Cosmochim. Acta* **60**, 3187–3196.

Peucker-Ehrenbrink B. and Jahn, B.-M. (2001) Rhenium–osmium isotope systematics and platinum-group element concentrations: loess and the upper continental crust. *Geochem. Geophys. Geosystems* **2**, 22pp., paper number 2001GC000172.

Peucker-Ehrenbrink B., Ravizza G., and Hofmann A. W. (1995) The marine $^{187}Os/^{186}Os$ record of the past 80 million years. *Earth Planet. Sci. Lett.* **130**, 155–167.

Pierazzo E. and Melosh H. J. (1999) Hydrocode modeling of Chicxulub as an oblique impact event. *Earth Planet. Sci. Lett.* **165**, 163–176.

Pilkington M. and Grieve R. A. F. (1992) The geophysical signature of terrestrial impact craters. *Rev. Geophys.* **30**, 161–181.

Poag C. W. (1997) The Chesapeake Bay bolide impact: a convulsive event in Atlantic Coastal Plain evolution. *Sed. Geol.* **108**, 45–90.

Poag C. W., Koeberl C., and Reimold W. U. (2004) *Chesapeake Bay Crater: Geology, and Geophysics of a Late Eocene Submarine Impact Structure.* Springer, Heidelberg, 522pp.

Poag C. W., Powars D., Poppe L. J., and Mixon R. B. (1994) Meteoroid mayhem in Ole Virginny: source of the North American tektite strewn field. *Geology* **22**, 691–694.

Pollastro R. M. and Bohor B. F. (1993) Origin and clay-mineral genesis of the Cretaceous/Tertiary boundary unit, western interior of North America. *Clays Clay Min.* **41**, 7–25.

Pollastro R. M. and Pillmore C. L. (1987) Mineralogy and petrology of the Cretaceous–Tertiary boundary clay bed and adjacent clay-rich rocks, Raton Basin, New Mexico and Colorado. *J. Sed. Petrol.* **57**, 456–466.

Poreda R. J. and Becker L. (2003) Fullerenes and interplanetary dust at the Permian–Triassic boundary. *Astrobiology* **3**, 75–90.

Quitté G., Robin E., Capmas F., Levasseur S., Rocchia R., Birck J.-L., and Allègre C. J. (2003) Carbonaceous or ordinary chondrite as the impactor at the K/T boundary? Clues from Os, W and Cr isotopes. In *Lunar Planet. Sci.* **XXXIV**, #1615. The Lunar and Planetary Institute, Houston (CD-ROM).

Rampino M. R. and Haggerty B. M. (1996) Impact crises and mass extinctions: a working hypothesis. In *The Cretaceous–Tertiary Event and other Catastrophes in Earth History* (eds. G. Ryder, D. Fastovsky, and S. Gartner). Geol. Soc. Amer. Spec. Paper 307, pp. 11–30.

Rampino M. R. and Reynolds R. C. (1983) Clay mineralogy of the Cretaceous-Tertiary Boundary Clay. *Science* **219**, 495–498.

Ramsey M. S. (2002) Ejecta distribution patterns at Meteor Crater, Arizona: on the applicability of lithologic end-member deconvolution for spaceborne thermal infrared data of Earth and Mars. *J. Geophys. Res.* **107**, 3-1–3-14, doi:10.1029/2001JE001827.

Rasmussen B. and Koeberl C. (2004) Iridium anomalies and shocked quartz in a Late Archean spherule layer from the Pilbara craton: new evidence for a major asteroid impact at 2.63 Ga. *Geology* **32**, 1029–1032.

Ravizza G. and Peucker-Ehrenbrink B. (2003) Chemostratigraphic evidence of Deccan Volcanism from the marine osmium isotope record. *Science* **302**, 1392–1395.

Ravizza G. and Pyle D. (1997) PGE and Os isotopic analyses of single sample aliquots with NiS fire assay preconcentration. *Chem. Geol.* **141**, 251–268.

Reimold W. U. (1982) The Lappajärvi meteorite crater, Finland: petrography, Rb–Sr, major and trace element geochemistry of the impact melt and basement rocks. *Geochim. Cosmochim. Acta* **46**, 1203–1225.

Reimold W. U. and Gibson R. L. (1996) Geology and evolution of the Vredefort impact structure, South Africa. *J. Afr. Earth Sci.* **23**, 125–162.

Reimold W. U. and Gibson R. L. (2005) "Pseudotachylites" in large impact structures. In *Impact Tectonics* (eds. C. Koeberl and H. s Henkel). Springer, Heidelberg, pp. 1–53.

Reimold W. U., Horsch H., and Durrheim R. J. (1990) The 'Bronzite'-Granophyre from the Vredefort structure—a detailed analytical study and reflections on the origin of one of Vredefort's enigmas. *Proc. 20th Lunar Planet. Sci. Conf.*, pp. 433–450.

Reimold W. U., Koeberl C., Brandstätter F., Kruger F. J., Armstrong R. A., and Bootsman C. (1999) The Morokweng impact structure, South Africa: geologic, petrographic, and isotopic results, and implications for the size of the structure. In *Large Meteorite Impacts and Planetary Evolution II* (eds. B. O. Dressler and V. L. Sharpton). Geol. Soc. Amer. Spec. Paper 339, pp. 61–90.

Reimold W. U., Koeberl C., Gibson R. L., and Dressler B. O. (2005) Economic mineral deposits in impact structures. In *Impact Tectonics* (eds. C. Koeberl and H. Henkel). Springer, Heidelberg, pp. 479–552.

Reimold W. U., Koeberl C., Johnson J., and McDonald I. (2000) Early Archean spherule beds in the Barberton Mountain Land, South Africa: impact or terrestrial origin? In *Impacts and the Early Earth* (eds. I. Gilmour and C. Koeberl). Springer, Heidelberg, pp. 117–180.

Reimold W. U., Koeberl C., and Reddering J. S. V. (1998) The 1992 drill core from the Kalkkop impact crater, Eastern Cape Province, South Africa: stratigraphy, petrography, geochemistry and age. *J. Afr. Earth Sci.* **26**, 573–592.

Renne P. R., Farley K. A., Reimold W. U., Koeberl C., Rampino M. R., Kelley S. P., and Ivanov B. A. (2004) Is Bedout an impact crater? Take 2. *Science* **306**, 610–612.

Retallack G. J., Seyedolali A., Krull E. S., Holser W. T., Ambers C. P., and Kyte F. T. (1998) Search for evidence of impact at the Permian–Triassic boundary in Antarctica and Australia. *Geology* **26**, 979–982.

Ryder G. (1990) Lunar samples, lunar accretion, and the early bombardment history of the Moon. *EOS* **71**, 313–323.

Ryder G. (2001) Mass flux during the ancient lunar bombardment: the cataclysm. In *Lunar Planet. Sci.* **XXXII**, #1326. The Lunar and Planetary Institute, Houston (CD-ROM).

Ryder G. (2002) Mass flux in the ancient Earth–Moon system and benign implications for the origin of life on Earth. *J. Geophys. Res.* **107**, 6-1–6-14, doi:10.1029/2001JE001583.

Ryder G., Fastovsky D., and Gartner S. (eds.) (1996) *The Cretaceous–Tertiary Event and other Catastrophes in Earth History*. Geol. Soc. Amer. Spec. Paper, vol. 307, 576pp.

Ryder G., Koeberl C., and Mojzsis S. J. (2000) Heavy bombardment on the Earth ~3.85 Ga: the search for petrographic and geochemical evidence. In *Origin of the Earth and Moon* (eds. K. Righter and R. M. Canup). University of Arizona Press, Tucson, AZ, pp. 475–492.

Sawlowicz Z. (1993) Iridium and other platinum-group elements as geochemical markers in sedimentary environments. *Palaeogeogr. Palaeoclimatol. Palaeoecol.* **104**, 253–270.

Schmidt G., Palme H., and Kratz K. L. (1997) Highly siderophile elements (Re, Os, Ir, Ru, Rh, Pd, Au) in impact melts from three European impact craters (Sääksjärvi, Mien, and Dellen): clues to the nature of the impacting bodies. *Geochim. Cosmochim. Acta* **61**, 2977–2987.

Schmidt G. and Pernicka E. (1994) The determination of platinum group elements (PGE) in target rocks and fall-back material of the Nördlinger Ries impact crater (Germany). *Geochim. Cosmochim. Acta* **58**, 5083–5090.

Schmitz B. (1985) Metal precipitation in the Cretaceous/Tertiary boundary clay at Stevns Klint, Denmark. *Geochim. Cosmochim. Acta* **49**, 2361–2370.

Schmitz B. (1992) Chalcophile elements and Ir in continental Cretaceous–Tertiary boundary clays from the western interior of the USA. *Geochim. Cosmochim. Acta* **56**, 1695–1703.

Schmitz B., Andersson P., and Dahl J. (1988) Iridium, sulfur isotopes and rare earth elements in the Cretaceous–Tertiary boundary clay at Stevens Klint, Denmark. *Geochim. Cosmochim. Acta,* **52**, 229–236.

Schmitz B., Haggstrom T., and Tassinari M. (2003) Sediment-dispersed extraterrestrial chromite traces a major asteroid disruption event. *Science* **300**, 961–964.

Schnetzler C. C., Philpotts J. A., and Thomas H. H. (1967) Rare earth and barium abundances in Ivory Coast tektites and rocks from the Bosumtwi crater area, Ghana. *Geochim. Cosmochim. Acta* **31**, 1987–1993.

Schnetzler C. C., Pinson W. H., and Hurley P. M. (1966) Rubidium–strontium age of the Bosumtwi crater area, Ghana, compared with the age of the Ivory Coast tektites. *Science* **151**, 817–819.

Schoenberg R., Kamber B. S., Collerson K. D., and Moorbath S. (2002) Tungsten isotope evidence from ~3.8-Gyr metamorphosed sediments for early meteorite bombardment of the Earth. *Nature* **418**, 403–405.

Schultz P. H., Koeberl C., Bunch T., Grant J., and Collins W. (1994) Ground truth for oblique impact processes: new insight from the Rio Cuarto, Argentina, crater field. *Geology* **22**, 889–892.

Schuraytz B. C., Lindstrom D. J., Marin L. E., Martinez R. R., Mittlefehldt D. W., Sharpton V. L., and Wentworth S. J. (1996) Iridium metal in Chicxulub impact melt: forensic chemistry on the K–T smoking gun. *Science* **271**, 1573–1576.

Schuraytz B. C., Sharpton V. L., and Marin L. E. (1994) Petrology of impact-melt rocks at the Chicxulub multiring basin, Yucatán, Mexico. *Geology* **22**, 868–872.

Serefiddin F., Herzog G. F., and Koeberl C. (2005) Beryllium-10 in Ivory Coast tektites. In *Lunar Planet. Sci.* **XXXVI**, #1466. The Lunar and Planetary Institute, Houston (CD-ROM).

Sharpton V. L., Burke K., Camargo-Zanoguera A., Hall S. A., Lee S., Marién L. E., Suárez-Reynoso G., Quezada-Muñeton J. M., Spudis P. D., and Urrutia-Fucugauchi J. (1993) Chicxulub multiring impact basin: size and other characteristics derived from gravity analysis. *Science* **261**, 1564–1567.

Sharpton V. L., Dalrymple G. B., Marin L. E., Ryder G., Schuraytz B. C., and Urrutia-Fucugauchi J. (1992) New links between the Chicxulub impact structure and the Cretaceous/Tertiary boundary. *Nature* **359**, 819–821.

Sharpton V. L. and Ward, P. D. (eds.) 1990. *Global Catastrophes in Earth History*. Geol. Soc. Am. Spec. Paper 247, 631pp.

Shaw H. F. and Wasserburg G. J. (1982) Age and provenance of the target materials for tektites and possible impactites as inferred from Sm–Nd and Rb–Sr systematics. *Earth Planet. Sci. Lett.* **60**, 155–177.

Shirey S. B. and Walker R. J. (1995) Carius tube digestion for low-blank rhenium–osmium analysis. *Anal. Chem.* **67**, 2136–2141.

Shukolyukov A., Castillo P., Simonson B. M., and Lugmair, G. W. (2002) Chromium in Archean spherule layers from Hamerslay Basin, Western Australia: isotopic evidence for extraterrestrial component. In *Lunar Planet. Sci.* **XXXIII**, #1369. The Lunar and Planetary Institute, Houston (CD-ROM).

Shukolyukov A., Kyte F. T., Lugmair G. W., Lowe D. R., and Byerly G. R. (2000) The oldest impact deposits on Earth— first confirmation of an extraterrestrial component. In *Impacts and the Early Earth* (eds. I. Gilmour and C. Koeberl). Springer, Heidelberg, pp. 99–116.

Shukolyukov A. and Lugmair G. W. (1998) Isotopic evidence for the Cretaceous–Tertiary impactor and its type. *Science* **282**, 927–929.

Shukolyukov A. and Lugmair G. W. (2000) Extraterrestrial matter on Earth: evidence from the Cr isotopes. In *Catastrophic Events & Mass Extinctions: Impacts and Beyond*. LPI Contribution #1053, The Lunar and Planetary Institute, Houston, pp. 197–198.

Shukolyukov A., Lugmair G. W., Becker L., MacIsaac C., Poreda B. (2004) Extraterrestrial chromium in the Permian-Traissic boundary at Graphite Peak, Antarctica. In *Lunar Planet. Sci.* **XXXV**, #1875. The Lunar and Planetary Institute, Houston (CD-ROM).

Shukolyukov A., Lugmair G. W., Koeberl C., and Reimold W. U. (1999) Chromium in the Morokweng impact melt rocks: isotope evidence for extraterrestrial component and type of the impactor. *Meteorit. Planet. Sci.* **34**, A107–A108.

Sigurdsson H., Bonte P., Turpin L., Chaussidon M., Metrich N., Steinberg M., Pradel P., and D'Hondt S. (1991a) Geochemical constraints on source regions of Cretaceous/Tertiary impact glasses. *Nature* **353**, 839–842.

Sigurdsson H., D'Hondt S., Arthur M. A., Bralower T. J., Zachos J. C., van Fossen M., and Channell E. T. (1991b) Glass from the Cretaceous/Tertiary boundary in Haiti. *Nature* **349**, 482–487.

Sigurdsson H., D'Hondt S., and Carey S. (1992) The impact of the Cretaceous/Tertiary bolide on evaporite terrane and generation of major sulfuric acid aerosol. *Earth Planet. Sci. Lett.* **109**, 543–559.

Silver L. T. and Schultz P. H. (eds.) (1982) *Geological Implications of Impacts of Large Asteroids and Comets on the Earth*. Geol. Soc. Amer. Spec. Paper 190, 528pp.

Simonson B. M. (1992) Geological evidence for a strewn field of impact spherules in the early Precambrian Hamersley Basin of Western Australia. *Geol. Soc. Am. Bull.* **104**, 829–839.

Simonson B. M. (2003) Petrographic criteria for recognizing certain types of impact spherules in well-preserved Precambrian successions. *Astrobiology* **3**, 49–65.

Simonson B. M., Byerly G. R., and Lowe D. R. (2004) The early Precambrian stratigraphic record of large extraterrestrial impacts. In *The Precambrian Earth: Tempos and Events* (eds. P. G. Eriksson, W. Altermann, O. Catuneanu, W. U. Mueller, and D. R. Nelson). Elsevier, Amsterdam, pp. 27–45.

Simonson B. M., Davies D., and Hassler S. W. (2000a) Discovery of a layer of probable impact melt spherules in the late Archean Jeerinah Formation, Fortescue Group, Western Australia. *Australian J. Earth Sci.* **47**, 315–325.

Simonson B. M., Davies D., Wallace M., Reeves S., and Hassler S. W. (1998) Iridium anomaly but no shocked quartz from Late Archean microkrystite layer: oceanic impact ejecta? *Geology* **26**, 195–198.

Simonson B. M. and Glass B. P. (2004) Spherule layers—records of ancient impacts. *Annu. Rev. Earth Planet. Sci.* **32**, 329–361.

Simonson B. M. and Harnik P. (2000) Have distal impact ejecta changed through geologic time? *Geology* **28**, 975–978.

Simonson B. M. and Hassler S. W. (1997) Revised correlations in the Early Precambrian Hamersley Basin based on a horizon of resedimented impact spherules. *Australian J. Earth Sci.* **44**, 37–48.

Simonson B. M., Hassler S. W., and Beukes N. J. (1999) Late Archean impact spherule layer in South Africa that may correlate with a Western Australian layer. In *Large Meteorite Impacts and Planetary Evolution II* (eds. B. O. Dressler and V. L. Sharpton). Geol. Soc. Am. Spec. Paper 339, pp. 249–261.

Simonson B. M., Koeberl C., McDonald I., and Reimold W. U. (2000b) Geochemical evidence for an impact origin for a late Archean spherule layer, Transvaal Supergroup, South Africa. *Geology* **28**, 1103–1106.

Simonson B. M., Sumner D. Y., Beukes N. J., Hassler S., Kohl I., Jones-Zimberlin S., Johnson S., Scally A., and Gutzmer J. (2006) Correlating multiple Neoarchean-Paleoproterozoic impact spherule layers between South Africa and Western Australia. In *Lunar Planet. Sci.* **XXXVII**, #1489. The Lunar and Planetary Institute, Houston (CD-ROM).

Sleep N. H., Zahnle K. J., Kasting J. F., and Morowitz H. J. (1989) Annihilation of ecosystems by large asteroid impacts on the early Earth. *Nature* **342**, 139–142.

Smit J. (1999) The global stratigraphy of the Cretaceous–Tertiary boundary impact ejecta. *Annu. Rev. Earth Planet. Sci.* **27**, 75–113.

Smit J. and Hertogen J. (1980) An extraterrestrial event at the Cretaceous-Tertiary boundary. *Nature* **285**, 198–200.

Spray J. G., Kelley S. P., and Rowley D. B. (1998) Evidence for a late Triassic multiple impact event on Earth. *Nature* **392**, 171–173.

Stöckelmann D. and Reimold W. U. (1989) The HMX mixing calculation program. *Mathemat. Geol.* **21**, 853–860.

Stöffler D. (1972) Deformation and transformation of rock-forming minerals by natural and experimental shock processes. 1: Behaviour of minerals under shock compression. *Fortschr. Mineral.* **49**, 50–113.

Stöffler D. (1974) Deformation and transformation of rock-forming minerals by natural and experimental processes. 2: Physical properties of shocked minerals. *Fortschr. Mineral.* **51**, 256–289.

Stöffler D. (1984) Glasses formed by hypervelocity impact. *J. Noncryst. Solids* **67**, 465–502.

Stöffler D. and Grieve R. A. F. (1994) Classification and nomenclature of impact Âmetamorphic rocks: a proposal to the IUGS subcommission on the systematics of metamorphic rocks. In *Post-Âstersund Newsletter, European Science, Foundation (ESF) Scientific Network on Impact Cratering and Evolution of Planet Earth* (eds. A. Montanari and J. Smit). European Science Foundation, Strasbourg, pp. 9–15.

Stöffler D. and Langenhorst F. (1994) Shock metamorphism of quartz in nature and experiment. I: Basic observations and theory. *Meteoritics* **29**, 155–181.

Stöffler D. and Ryder G. (2001) Stratigraphy and isotope ages of lunar geologic units: chronological standard for the inner solar system. *Space Sci. Rev.* **96**, 7–53.

Swisher C. C., Grajales-Nishimura J. M., Montanari A., Margolis S. V., Claeys P., Alvarez W., Renne P., Cedillo-Pardo E., Maurrasse F. J. M. R., Curtis G. H., Smit J., and McWilliams M. O. (1992) Coeval ^{40}Ar/^{39}Ar ages of 65.0 million years ago from Chicxulub crater melt rock and Cretaceous–Tertiary boundary tektites. *Science* **257**, 954–958.

Tagle R. and Claeys P. (2004) Comet or asteroid shower in the late Eocene? *Science* **305**, 492.

Tanner L. H. and Kyte F. T. (2005) Anomalous iridium enrichment at the Triassic–Jurassic boundary, Blomidon Formation, Fundy basin, Canada. *Earth Planet. Sci. Lett.* **240**, 634–641.

Taylor H. P. and Epstein S. (1966) Oxygen isotope studies of Ivory Coast tektites and impactite glass from the Bosumtwi crater, Ghana. *Science* **153**, 173–175.

Taylor S. R. (1962) Fusion of soil during meteorite impact and the chemical composition of tektites. *Nature* **195**, 32–33.

Taylor S. R. (1967) Composition of meteorite impact glass across the Henbury strewn field. *Geochim. Cosmochim. Acta* **31**, 961–968.

Taylor S. R. (1992) *Solar System Evolution*. Cambridge University Press, Cambridge, New York, 307pp.

Taylor S. R. and McLennan S. M. (1985) *The Continental Crust: Its Composition and Evolution*. Blackwell Scientific Publications, Oxford, 312pp.

Taylor S. R. and McLennan S. M. (1995) The geochemical evolution of the continental crust. *Rev. Geophys.* **33**, 241–265.

Tera F., Papanastassiou D. A., and Wasserburg G. J. (1974) Isotopic evidence for a terminal lunar cataclysm. *Earth Planet. Sci. Lett.* **22**, 1–21.

Tredoux M., de Wit M. J., Hart R. J., Lindsay N. M., Verhagen B., and Sellschop J. P. F. (1989) Chemostratigraphy across the Cretaceous–Tertiary boundary and a critical assessment of the iridium anomaly. *J. Geol.* **97**, 585–605.

Trinquier A., Birck J.-L., and Allègre C. J. (2006) The nature of the KT impactor. A 54Cr reappraisal. *Earth Planet. Sci. Lett.* **241**, 780–788.

Turekian K. K. (1982) Potential of ^{187}Os/^{186}Os as a cosmic versus terrestrial indicator in high iridium layers of sedimentary strata. In *Geological Implications of Impacts of Large Asteroids and Comets on the Earth* (eds. L. T. Silver and P. H. Schultz). Geol. Soc. Am. Spec. Paper 190, pp. 243–249.

Vishnevsky S. and Montanari A. (1999) Popigai impact structure (Arctic Siberia, Russia): geology, petrology, geochemistry, and geochronology of glass-bearing impactites. In *Large Meteorite Impacts and Planetary Evolution II* (eds. B. O. Dressler and V. L. Sharpton). Geol. Soc. Am. Spec. Paper 339, pp. 19–60.

Völkening J., Walczyk T., and Heumann K. G. (1991) Osmium isotope ratio determinations by negative thermal ionization mass spectrometry. *Int. J. Mass Spectrom. Ion Proc.* **105**, 147–159.

Walkden G., Parker J., and Kelley S. (2002) A late Triassic impact ejecta layer in southwestern Britain. *Science* **298**, 2185–2188.

Walker R. J., Morgan J. W., Naldrett A. J., Li C., and Fassett J. D. (1991) Re–Os isotope systematics of Ni–Cu sulfide ores, Sudbury Igneous Complex, Ontario: evidence for a major crustal component. *Earth Planet. Sci. Lett.* **105**, 416–429.

Wallace M. W., Gostin V. A., and Keays R. R. (1990) Acraman impact ejecta and host shales: evidence for low-temperature mobilization of iridium and other platinoids. *Geology* **18**, 132–135.

Wasson J. T. and Kallemeyn G. (1988) Compositions of chondrites. *Phil. Trans. Roy. Soc. Lond.* **A325**, 535–544.

Williams G. E. (1986) The Acraman impact structure: source of ejecta in late Precambrian shales, South Australia. *Science* **233**, 200–203.

Williams G. E. (1994) Acraman, South Australia: Australia's largest meteorite impact structure. *Proc. Roy. Soc. Victoria* **106**, 105–127.

Wolbach W. S., Gilmour I., and Anders E. (1990) Major wildfires at the Cretaceous/Tertiary boundary. In *Global Catastrophes in Earth History* (eds. V. L. Sharpton and P. D. Ward). Geol. Soc. Amer. Spec. Paper 247, pp. 391–400.

Wolbach W. S., Lewis R. S., and Anders E. (1985) Cretaceous extinctions: evidence for wildfires and search for meteoritic material. *Science* **230**, 167–170.

Wolf R., Woodrow A., and Grieve R. A. F. (1980) Meteoritic material at four Canadian impact craters. *Geochim. Cosmochim. Acta* **44**, 1015–1022.

Yang W. and Ahrens T. J. (1998) Shock vaporization of anhydrite and global effects of the K/T bolide. *Earth Planet. Sci. Lett.* **156**, 125–140.

Zahnle K. J. and Sleep N. H. (1997) Impacts and the early evolution of life. In *Comets and the Origin and Evolution of Life* (eds. P. J. Thomas, C. F. Chyba, and C. P. Mckay). Springer, New York, pp. 175–208.

Zhang Y. (2002) The age and accretion of the Earth. *Earth Sci. Rev.* **59**, 235–263.

Radioactive Geochronometry
ISBN: 978-0-08-096708-0

6

Geochronology and Thermochronology in Orogenic Systems

K. V. Hodges

Massachusetts Institute of Technology, Cambridge, MA, USA

NOMENCLATURE

a	effective diffusion dimension
C	Euler constant (~ 0.57721547)
C_0	initial concentration of a diffusant
C_ξ	concentration of a diffusant at distance ξ along a gradient
C_{Pb}	concentration of Pb
C_{Th}	concentration of Th
C_U	concentration of U
D_i	diffusivity at infinite temperature
D_t	diffusivity at time t
D_T	diffusivity at temperature T

E	activation energy
erf^{-1}	inverse error function
G_{av}	average value of the closure function for a specific grain geometry
G_x	closure function, evaluated for radial position x in a grain
k	proportionality constant
n_{d}	number of daughter isotopes
$n_{\mathrm{d}0}$	number of non-radiogenic daughter isotopes
$n_{\mathrm{d}*}$	number of radiogenic daughter isotopes
n_{FT}	number of fission tracks
n_{p}	number of parent isotopes
$n_{\mathrm{p}0}$	initial number of parent isotopes
P	pressure
R	universal gas constant ($8.3144\,\mathrm{J\,mol^{-1}\,K^{-1}}$)
t	time
t_{cb}	bulk cooling age
$t_{\mathrm{c}x}$	cooling age for position x in a grain
t_{init}	time of initiation of an unroofing event
t_{p}	present time
t_0	time of crystallization
$t_{1/2}$	half-life of parent isotope
T	temperature
T_{cb}	bulk closure temperature
$T_{\mathrm{c}x}$	closure temperature as a function of position within a grain
T_t	temperature at time t
T_0	crystallization temperature
x	distance, expressed as a fraction of the radius of a crystal
X_{phl}	mole fraction of phlogopite component in biotite
z	depth
γ	geothermal gradient
ε	elevation
λ_{p}	decay constant of parent isotope
ξ	distance along a diffusion gradient
σ	sample standard deviation
τ	duration of a diffusion experiment
Φ	constant of proportionality between synthetic $^{39}\mathrm{Ar}$ and naturally occurring $^{40}\mathrm{K}$, related to the efficiency of the neutron bombardment reaction $^{39}\mathrm{K}\rightarrow{}^{39}\mathrm{Ar}$ and the essentially constant ratio $^{39}\mathrm{K}/^{40}\mathrm{K}$

6.1 INTRODUCTION

An important part of modern tectonic research is determining the rates of deformational, thermal, and erosional processes that define the evolution of orogenic systems. Such endeavors fall into three categories: studies of the crystallization ages of rocks and minerals (geochronology); studies of the thermal history of rocks (thermochronology); and studies of the exposure ages of geomorphic surfaces (cosmogenic nuclide dating). This chapter focuses on the first two types of study; additional material regarding geochronology may

be found in Chapter 7. Here a brief treatment of the basic concepts of radioactive decay and isotopic dating is followed by a discussion of how limited open-system behavior of radiogenic isotopes in minerals, as well as the partial annealing of radiation damage in mineral structures, may be used to explore the thermal histories of orogens. Subsequent sections describe specific applications of these techniques to a variety of problems encountered in orogenic systems, ranging from the frequency of magmatism to the timescales of landscape evolution. The final section suggests potential avenues for future research, both to improve the quality of geochronologic and thermochronologic data and to evaluate their capacity to constrain orogenic processes. Readers interested in learning more about such topics may consult isotope geology textbooks such as those by Faure (1986) or Dickin (1995).

6.2 BASIC CONCEPTS OF GEOCHRONOLOGY

The process of radioactive decay is usually described in terms of the spontaneous transformation of one isotope (the "parent") into another (the "daughter"). It occurs at a rate that is proportional to the number of parent isotopes (n_{p}) present in a rock or mineral at any given time:

$$\frac{-\mathrm{d}n_{\mathrm{p}}}{\mathrm{d}t} = \lambda_{\mathrm{p}} n_{\mathrm{p}} \tag{1}$$

where λ_{p} is the decay constant for the parent and t represents time. This equation may be integrated to yield a second equation that describes the number of parent isotopes at a given time as a function of the number of parent isotopes present at the time of crystallization ($n_{\mathrm{p}0}$):

$$n_{\mathrm{p}} = n_{\mathrm{p}0}\mathrm{e}^{-\lambda_{\mathrm{p}}t} \tag{2}$$

After a period of time indicated by t, the decay process will have produced a number of radiogenic daughter isotopes ($n_{\mathrm{d}*}$) given by

$$n_{\mathrm{d}*} = n_{\mathrm{p}}\left(\mathrm{e}^{\lambda_{\mathrm{p}}t} - 1\right) \tag{3}$$

Most samples contain one (or more) components of the daughter isotope that were not produced by *in situ* decay of the parent. Instead, these "non-radiogenic" isotopes were trapped during crystallization or during postcrystallization alteration. The total number of the daughter isotopes in a sample at time t (or n_{d}) is thus the sum of the radiogenic ($n_{\mathrm{d}*}$) and non-radiogenic atoms ($n_{\mathrm{d}0}$):

$$n_{\mathrm{d}} = n_{\mathrm{d}0} + n_{\mathrm{p}}\left(\mathrm{e}^{\lambda_{\mathrm{p}}t} - 1\right) \tag{4}$$

When solved for t, this relationship yields the *fundamental equation of geochronology*:

$$t = \left(\frac{1}{\lambda_{\mathrm{p}}}\right)\ln\left[\left(\frac{n_{\mathrm{d}} - n_{\mathrm{d}0}}{n_{\mathrm{p}}}\right) + 1\right] \tag{5}$$

which may be used to calculate the elapsed time since crystallization of a rock or mineral when: (i) n_p and n_d are measured, present-day abundances; (ii) n_{d0} can be assumed reasonably; and (iii) the sample has remained a closed system with respect to gain or loss of parent and daughter isotopes since the time of crystallization. Table 1 shows the decay systems that are commonly exploited in geochronology. In general, the age ranges over which these chronometers are used depends on the precision and accuracy with which measurements for a particular system can be made, as well as the decay constant; parent isotopes with larger decay constants produce more daughter isotopes over a given time interval.

6.2.1 Effects of Branched, Sequential, and Multiparent Decay

The preceding equations hold for decay schemes that involve the simple transformation of one isotope to another. Three types of decay schemes useful for geochronology are special cases that yield somewhat more complicated age equations. *Branched decay* leads to the production of two different daughter isotopes from a single radioactive isotope. A familiar example is the spontaneous conversion of ^{40}K to both ^{40}Ar (through electron capture) and ^{40}Ca (through β decay). In such cases, the age equations must be modified by factors that correct for the fact that not all decays result in the production of the daughter isotope of interest. For the ^{40}K$\rightarrow^{40}$Ar decay mechanism, Equation (4) becomes

$$^{40}Ar = {}^{40}Ar_0 + {}^{40}K\left(\frac{\lambda_{40K(ec)}}{\lambda_{40K}}\right)\left(e^{\lambda_{40K}t} - 1\right) \quad (6)$$

where ^{40}Ar is the total number of ^{40}Ar isotopes in the sample, $^{40}Ar_0$ is the number of non-radiogenic ^{40}Ar isotopes, ^{40}K is the number of ^{40}K isotopes, $\lambda_{40K(ec)}$ is the decay constant for the ^{40}K$\rightarrow^{40}$Ar branch (generally assumed to be $5.81\times10^{-11}\text{yr}^{-1}$;

Steiger and Jäger, 1977), and λ_{40K} is the total decay constant for ^{40}K (5.543×10^{-10} yr^{-1}).

Some radioactive isotopes undergo *sequential decay* through a series of intermediate, radioactive daughter isotopes before finally yielding a stable daughter product. (U–Th)/Pb geochronology results from three such decay series: ^{238}U $\rightarrow$ ^{206}Pb, ^{235}U $\rightarrow$ ^{207}Pb, and ^{232}Th $\rightarrow$ ^{208}Pb. Fortunately, the intermediate daughter isotopes are short-lived compared to the typical timescales studied through (U–Th)/Pb geochronology (10^6–10^9 yr), such that a condition of "secular equilibrium" is reached: the rate of decay of the parent and intermediate daughter isotopes is essentially the same. In such cases, the decay series can be treated mathematically like a simple one-step decay mechanism. For the ^{238}U $\rightarrow$ ^{206}Pb mechanism, Equation (4) is thus

$$^{206}Pb = {}^{206}Pb_0 + {}^{238}U(e^{\lambda_{238U}t} - 1) \quad (7)$$

where ^{206}Pb is the total number of ^{206}Pb isotopes in the sample, $^{206}Pb_0$ is the number of non-radiogenic ^{206}Pb isotopes, ^{238}U is the number of ^{238}U isotopes, and λ_{238U} is the decay constant for ^{238}U ($1.55125\times10^{-10}\text{yr}^{-1}$).

Finally, some daughter isotopes of geochronologic interest result from *multiparent decay*. For example, virtually all ^{4}He is produced as particles released from the series decay of ^{232}Th, ^{235}U, and ^{238}U. Taking into consideration the number of α particles released as part of each decay series, Equation (4) for ^{4}He production takes the form

$$\begin{aligned}^4He = {}^4He_0 &+ 8^{238}U(e^{\lambda_{238U}t} - 1) \\ &+ 7^{235}U(e^{\lambda_{235U}t} - 1) \\ &+ 6^{232}Th(e^{\lambda_{232Th}t} - 1)\end{aligned} \quad (8)$$

where ^4He is the total number of ^{4}He isotopes in the sample, $^4He_0$ is the number of non-radiogenic ^{4}He isotopes, ^{238}U is the number of ^{238}U isotopes, ^{235}U is the number of ^{235}U isotopes, and ^{232}Th is the number of ^{232}Th isotopes. The decay constant

Table 1 Decay schemes frequently used for geochronology and thermochronology.

Reaction	Minerals[a]	$\lambda_p(\text{yr}^{-1})$	$t_{1/2}$ (Gyr)[b]	Source
^{147}Sm $\rightarrow$ ^{143}Nd	Grt, Cpx	6.54×10^{-12}	106.0	Lugmair and Marti (1978)
^{87}Rb $\rightarrow$ ^{87}Sr	Ms, Bt, Phl, Kfs	1.42×10^{-11}	48.8	Steiger and Jäger (1977)
^{176}Lu $\rightarrow$ ^{176}Hf	Grt	1.865×10^{-11}	37.2	Scherer *et al.* (2001)
^{232}Th $\rightarrow$ ^{208}Pb[c]	Mz	4.9475×10^{-11}	14.0	Steiger and Jäger (1977)
^{40}K $\rightarrow$ ^{40}Ar	Hbl, Ms, Bt, Phl, Kfs	5.8×10^{-11}	11.9	Steiger and Jäger (1977); Min *et al.* (2000)
^{238}U $\rightarrow$ ^{206}Pb[c]	Zcn, Mz, Xn, Ttn, Rt, Ap	1.55125×10^{-10}	4.5	Steiger and Jäger (1977)
^{238}U fission	Zcn, Ap	$\sim7\times10^{-10}$	~1.0	Naeser *et al.* (1989)
^{235}U $\rightarrow$ ^{207}Pb[c]	Zcn, Mz, Xn, Ttn, Rt, Ap	9.8485×10^{-10}	0.7	Steiger and Jäger (1977)

[a] Abbreviations: Ap—Apatite; Bt—Biotite; Cpx—Clinopyroxene; Grt—garnet; Hbl—Hornblende; Kfs—K-feldspar; Ms—Muscovite; Mz—Monazite; Phl—phlogopite; Rt—rutile; Ttn—titanite; Xn—xenotime; Zcn—zircon. The half-life of a parent isotope, is equal to $0.693/\lambda_p$. [b] $t_{1/2}$, the half-life of a parent isotope, is equal to $0.693/\lambda_p$. [c] These decay schemes also produce ^{4}He, forming the basis for (U–Th)/He geochronology (Equation (7)). Commonly analyzed minerals include Ap, Ttn, and Zcn.

for ^{238}U (λ_{238U}) was given above; for ^{232}Th and ^{235}U (λ_{232Th} and λ_{235U}), they are $4.9475 \times 10^{-11}yr^{-1}$ and $9.8485 \times 10^{-10}yr^{-1}$, respectively.

6.2.2 Fission-track Geochronology

One geochronologic technique is based on the secondary effects of radioactive decay rather than the accumulation of daughter isotopes. Although most ^{238}U decay in a mineral involves α-particle emission, a very small proportion occurs through spontaneous fission. Each fission reaction releases ~200 MeV of kinetic energy that propels the resulting fragments through the mineral, disrupting its crystal structure. The number of these disrupted zones, or *fission tracks*, in a sample (n_{FT}) depends on the amount of ^{238}U present and age:

$$n_{FT} = \left(\frac{\lambda_{238U(FT)}}{\lambda_{238U}} \right)^{238} U \left(e^{\lambda_{238U}t} - 1 \right) \quad (9)$$

where $\lambda_{238U(FT)}$ is the decay constant for ^{238}U fission ($\sim 7 \times 10^{-17}yr^{-1}$). Equation (9) forms the basis for calculating dates for geologic samples by determining uranium content and counting the number of accumulated tracks.

6.2.3 Chemical Pb Dating

Another special technique takes advantage of the fact that high-uranium and high-thorium accessory minerals typically contain very small amounts of non-radiogenic lead. If it is assumed that all lead in a sample is radiogenic, and that the abundances of ^{238}U, ^{235}U, and ^{232}Th are essentially constant in natural uranium and thorium, then the total concentration of lead (C_{Pb}) is related to the concentrations of uranium (C_U) and thorium (C_{Th}):

$$C_{Pb} \approx 0.897 C_{Th} \left(e^{\lambda_{232Th}t} - 1 \right)$$
$$+ 0.006 C_U \left(e^{\lambda_{235U}t} - 1 \right)$$
$$+ 0.859 C_U \left(e^{\lambda_{238U}t} - 1 \right) \quad (10)$$

where concentrations are in parts per million (ppm) (Montel *et al.*, 1996). From Equation (10), an approximate *chemical Pb date* can be calculated knowing only the concentrations of uranium, thorium, and lead.

6.3 ANALYTICAL METHODS

Except for the fission-track and chemical lead methods, all geochronologic techniques involve the use of quadrupole or magnetic-sector mass spectrometers to separate atoms by mass and determine their absolute or relative abundances. Isotopic ratios can be measured more precisely than isotopic abundances on these instruments,

so it is common practice to recast many of the above equations in terms of isotopic ratios by dividing both sides through by the number of some stable, non-radiogenic isotope of the daughter element. For example, Equation (7) can be divided by the number of ^{204}Pb isotopes (^{204}Pb) to yield:

$$\frac{^{206}Pb}{^{204}Pb} = \left(\frac{^{206}Pb}{^{204}Pb} \right)_0 + \frac{^{238}U}{^{204}Pb} \left(e^{\lambda_{238U}t} - 1 \right) \quad (11)$$

where $(^{206}Pb/^{204}Pb)_0$ is the *initial ratio* of non-radiogenic ^{206}Pb to ^{204}Pb. In most cases, however, the absolute abundance of at least one isotope still must be determined for an age calculation, and this is accomplished by spiking the sample with a solution of known, artificial isotopic composition in preparation for isotope dilution analysis (Dickin, 1995; Albarède, 1995).

For the majority of geochronologically important decay schemes, measurements of parent and daughter isotopic ratios and abundances can be made on a single aliquot of sample with the same, solid-source or plasma-source mass spectrometer. The principal complication with such measurements is the need to separate different elements prior to analysis in order to prevent mass interferences, and this is accomplished through standard ion-exchange chemistry (Harland, 1994). For solid-source mass spectrometry, the purified sample is then loaded onto a metallic filament. Under vacuum, this filament is heated by passing an electrical current through it, inducing ionization of the sample in preparation for mass analysis. For plasma-source mass spectrometry, the sample is introduced using an inert carrier gas. Isotope dilution and thermal ionization mass spectrometry (ID-TIMS) is the most widely used method of isotope geochronology when parent and daughter isotopes are in the solid state at Earth surface conditions.

For (U–Th)/He and K/Ar, where the daughter products are noble gases, the analytical process is more involved. Most (U–Th)/He experiments begin by driving off helium (and other gases) from the sample, either by vacuum heating in a furnace or by laser irradiation. After a purification process to remove reactive gases from the mixture, a gas-source mass spectrometer is used to measure the isotopic composition of the helium. Subsequently, the sample is removed from the vacuum system, dissolved, spiked, and analyzed for uranium and thorium with an inductively coupled, plasma-source mass spectrometer (ICP-MS). This approach to (U–Th)/He dating avoids the potential problems with heterogeneity that might arise if the sample was split into separate aliquots for He and U–Th analysis. Such problems are frequently encountered in conventional K–Ar geochronology, where argon isotopes are measured by gas-source, isotope-dilution mass spectrometry and potassium is typically measured on a separate aliquot by flame photometry. As a consequence, most K–Ar studies now employ the $^{40}Ar/^{39}Ar$ method (Merrihue and Turner, 1966; McDougall and Harrison, 1988), which permits the determination of dates by measuring only argon isotopic ratios. If a sample is bombarded with high-energy neutrons in a nuclear reactor prior to analysis, naturally occurring ^{39}K isotopes can be induced to convert to artificial ^{39}Ar with an efficiency that depends on the neutron flux and the duration of irradiation. The ratio $^{39}K/^{40}K$ is effectively constant in geologic materials because the half-life of ^{40}K is so long. Consequently, the amount of ^{39}Ar produced by neutron bombardment is also proportional to the amount of ^{40}K in the sample prior to irradiation. If Φ is the constant of this proportionality, the fundamental equation for $^{40}Ar/^{39}Ar$ geochronology can be written as

$$t = \left(\frac{1}{\lambda_{40K}}\right)$$
$$\ln\left[\frac{(^{40}Ar/^{36}Ar) - (^{40}Ar/^{36}Ar)_0}{\Phi(^{39}Ar/^{36}Ar)(\lambda_{40K(ec)}/\lambda_{40K})} + 1\right] \quad (12)$$

(The stable isotope ^{36}Ar appears in Equation (12) in order to permit the calculation of a date from isotopic ratios rather than isotopic concentrations.) In practice, samples of known $^{40}Ar/^{39}Ar$ age are co-irradiated with the sample to monitor ^{39}Ar production efficiency and allow the determination of Φ using a version of Equation (12).

Fission-track geochronology requires the counting of tracks as well as some method of determining ^{238}U concentration. Because fission tracks are too small to be seen effectively with petrographic microscopes, standard analytical protocols involve the chemical etching of a polished face of a sample, which enlarges tracks intersecting the surface and makes them easier to resolve. Rather than measure ^{238}U with a mass spectrometer, fission-track geochronologists generally employ neutron irradiation. By bombarding the sample with low-energy neutrons, some fraction of ^{235}U can be induced to undergo a fission reaction that produces new tracks, such that the ^{235}U abundance—and thus the ^{238}U abundance, if a constant natural ratio is assumed—can be calculated by counting the new tracks. One of the most popular approaches involves the use of an external detector, in which a uranium-free material is attached to the sample and the two are irradiated together (e.g., Hurford and Carter, 1991; Gallagher et al., 1998). Fission of ^{235}U produces fragments that cross the interface and leave tracks in the external detector. With appropriate corrections for the efficiency of this process, dates can be calculated solely by counting spontaneous ^{238}U fission tracks in the sample prior to irradiation and induced ^{235}U fission tracks in the external detector after irradiation.

6.3.1 Microanalytical Techniques

A major step forward in geochronology since the 1980s has been the development of various microanalytical techniques. Such tools allow us to explore isotopic variations within individual crystals, to distinguish grains of different origins in polygenetic samples, to date mineral inclusions and place absolute age constraints on pressure–temperature histories, and to determine the apparent ages of minerals in petrographic context.

Analytical protocols for ID-TIMS have improved to the point that accurate and precise dates may be obtained for samples as small as a few nanograms, such that even this "conventional" method of geochronology is limited principally by the minimum sample size that can be manipulated for sample dissolution, spiking, and ion-exchange chemistry. Various mechanical microsampling tools, ranging from obsidian knives to microdrills, have been employed to extract datable fragments from single crystals on scales of tens of microns (e.g., Christensen et al., 1989; Hawkins and Bowring, 1997; Müller et al., 2001).

Other microanalytical techniques employ high-energy beams of photons, ions, or electrons to excavate material from the sample or induce the production of characteristic X-rays from the sample surface. *Laser microprobes* of various wavelength are commonly used to melt or ablate samples at spatial resolutions ranging from a few tenths of a micron to (more typically) several hundreds of microns (e.g., York et al., 1981; Maluski and Schaeffer, 1982; Sutter and Hartung, 1984; Feng et al., 1993; Fryer et al., 1993; Halliday et al., 1998; Hodges, 1998a; Horn et al., 2000; Poitrasson et al., 2000; Bruguier et al., 2001; Li et al., 2001; Kosler et al., 2001, 2002; Machado and Simonetti, 2001; Willigers et al., 2002). *Ion microprobes* are based on the principle that ion milling stimulates the emission of secondary ions from a sample's surface. These ions can be filtered and analyzed through a process referred to as secondary-ion mass spectrometry (SIMS). With the advent of ion microprobes having especially high sensitivity and mass resolution, SIMS U/Pb and Th/Pb geochronology of accessory phases has become increasingly popular (e.g., Compston et al., 1986; DeWolf et al., 1993; Harrison et al., 1995b; Williams et al., 1996; Zhu et al., 1997b; Williams, 1998; Stern and Berman, 2001). The spatial resolution afforded by ion microprobes is generally somewhat better than that of laser microprobes, but neither can approach the micron-scale resolution of *electron microprobes*. These instruments enable the nondestructive analysis of uranium, thorium, and lead concentrations for chemical lead dating (Suzuki and Adachi, 1991; Montel et al., 1996; Cocherie et al., 1998; Crowley and Ghent, 1999; Williams et al., 1999; Geisler and Schleicher,

2000; Terry *et al.*, 2000; French *et al.*, 2002; Kempe, 2003).

It is a fundamental principle of analytical geochemistry that higher concentrations of an element or specific isotope can be measured with greater precision than lower concentrations. As a consequence, most microanalytical methods sacrifice high analytical precision for high spatial resolution. Even with the trend toward single-crystal and crystal-fragment studies, ID-TIMS remains the most precise geochronologic method, in part because somewhat larger sample sizes are used and in part because the technique is the oldest and, consequently, the most refined. For example, single crystals of accessory minerals such as monazite often yield ID-TIMS U/Pb dates with 2σ uncertainties of less than 0.5%, even for materials of Late Tertiary age (e.g., Viskupic and Hodges, 2001). SIMS and laser-ablation ICP-MS U/Pb and Th/Pb dates for comparably aged samples typically have 2σ uncertainties that are at least an order of magnitude larger (e.g., Harrison *et al.*, 1996; Kosler *et al.*, 2001). The current generation of electron microprobes has such low abundance sensitivities for uranium, thorium, and lead that the only Tertiary mineral dated by the chemical lead method thus far is the uranium oxide uraninite (Hurtado *et al.*, in preparation), with 2σ uncertainties of ~10%. Most applications of this technique have involved the dating of Paleozoic or Precambrian monazites, with typical uncertainties of a few percent.

6.4 THE INTERPRETATION OF DATES AS CRYSTALLIZATION AGES

The quantity *t* as calculated from the above equations—which will be referred to here as a *date*—may or may not have geologic significance. The interpretation of a date as a *crystallization age* is justified if:

- decay constants do not change with time;
- the non-radiogenic component of the daughter isotope is known or can be determined empirically; and
- the sample has been closed with respect to gain or loss of parent or daughter isotopes since crystallization.

If sequential or multiparent decay is involved in the production of the final daughter isotope, the second criterion must be amended to include all parents and all intermediate daughter isotopes. In some applications, isotopic data for more than one sample are used in an effort to characterize the non-radiogenic daughter component. Examples of this include the analysis of presumably co-genetic, whole-rock samples from a single igneous intrusion, and the analysis of different minerals with different parent/daughter ratios from a single rock sample (e.g., Schreiner, 1958; Fairbairn *et al.*, 1961; Lanphere *et al.*, 1964; Weatherill *et al.*, 1968). Such practices require an additional assumption:

- All samples used for the determination of *t* must have been in isotopic equilibrium with one another at the time of crystallization.

When the date of interest refers to the crystallization of rocks or minerals from a melt and all of the above criteria are met, the interpreted age is best referred to as an *igneous age*. When minerals in a single rock are the products of solid-state recrystallization during metamorphism and all the above criteria are met, then the calculated date may be referred to as a *metamorphic age*.

6.5 OPEN-SYSTEM BEHAVIOR: THE ROLE OF DIFFUSION

Of all the criteria necessary for the interpretation of dates as crystallization ages, the assumption of closed-system behavior is the most problematic. Many daughter isotopes and some important parent isotopes may be mobile during metamorphism, hydrothermal alteration, and deformation, and a surprising number of high-precision isotopic dates for samples from orogenic settings are demonstrably younger than the crystallization ages of the samples. Among the mechanisms potentially responsible for such open-system behavior are:

- metamorphic "net-transfer" reactions that lead to the crystallization of new minerals or the elimination of others as a consequence of changes in temperature and pressure;
- dynamic recrystallization of minerals as a consequence of deformation; and
- metamorphic "exchange" reactions that lead to changes in the compositions of minerals but not their modal proportions, as a consequence of changes in pressure and temperature (Spear, 1993).

The first two mechanisms leave direct textural evidence for open-system behavior. Samples displaying such evidence yield complex geochronologic data that are difficult to interpret without a detailed understanding of the metamorphic and structural context (e.g., Hames and Cheney, 1997; Arnaud and Eide, 2000; Hoskin and Black, 2000; Di Vincenzo *et al.*, 2001; Rubatto *et al.*, 2001; Stern and Berman, 2001; Townsend *et al.*, 2001). However, variable mineral dates also characterize many samples that show no obvious textural evidence of disturbance. In most cases, these dates differ in a predictable way (Hart, 1964; Armstrong *et al.*, 1966; Wagner *et al.*, 1977).

For example, a typical amphibolite-facies pelitic schist will yield a (U–Th)/Pb monazite date older than its $^{40}Ar/^{39}Ar$ muscovite date, which will be, in turn, older than its $^{40}Ar/^{39}Ar$ biotite date. Given the lack of textural evidence for other causes, the most likely mechanism to produce such inconsistencies is isotopic exchange between minerals and their surroundings.

Isotopic exchange can be thought of as a transport process governed by concentration gradients. Consider the simplest case of a mineral that crystallizes with some amount of parent isotopes and that is surrounded by an intergranular medium, such as a fluid, containing none of the parent species. As daughter isotopes are produced in a mineral by radioactive decay, a concentration gradient develops between the mineral and its surroundings and there is a natural tendency for the daughter isotopes in the mineral to move out of the crystal. (This process can be thought of as an exchange of daughter isotope for a crystallographic vacancy in the mineral.) Because the mobility of virtually all isotopes would be greater in an intergranular fluid than in a mineral, any daughter that leaves the mineral is rapidly transported away from the interface and the fluid may be thought of as an infinite sink for the daughter isotope. Under such circumstances, the rate of isotopic exchange is largely dependent on the rate of atomic migration or *diffusion* of the daughter isotope in the mineral. A wealth of experimental evidence (Section 6.5.2) demonstrates that the rate of diffusion (or *diffusivity*) of isotopes in minerals differs in a way that is consistent with the notion that the sequence of older-to-younger dates obtained for many samples reflects a sequence of greater-to-lesser diffusivities for the daughter isotopes used to determine the dates.

Thermochronology is predicated on the notion that minerals that have experienced loss of radiogenic daughter isotopes after crystallization may still provide useful information about the thermal history of a sample because the diffusive-loss mechanism depends strongly on temperature. Establishing a quantitative basis for thermochronology requires developing effective models for the combined effects of three time-dependent processes: the production of daughter isotopes through radioactive decay of parent isotopes (as discussed in Section 6.2), the loss of the daughter isotopes through diffusion, and the cooling of the system. Sections 6.5.1 and 6.5.2 focus on the second process, addressing the nature of diffusion in minerals and presenting diffusion data for mineral-isotopic systems frequently used for thermochronology.

6.5.1 Modes of Diffusion

If mineral crystals were perfect, the only intracrystalline atomic migration we would have to be concerned with would be the three-dimensional or *volume diffusion* of an isotope. Unfortunately, natural samples contain internal imperfections, such as subgrain boundaries or line defects formed by dislocation arrays, that can serve as fast-diffusion pathways. This suggests that a reasonable conceptual model for isotope migration in natural crystals is one that involves interactive volume diffusion at one rate in intact domains and "short-circuit" diffusion at a different rate in the pathways that separate them (e.g., Lee and Aldama, 1992; Lee, 1995). An alternative to this "multipath" model, referred to by Lovera *et al.* (1989) as the "multidiffusion domain" model, regards a crystal as a collection of discrete, non-interactive domains of different size with different volume diffusion characteristics. Recently, Lovera *et al.* (2002) suggested a third, "heterogeneous diffusion" model that features spatially variable diffusivity but no discrete domains. Of the three, the multipath model probably provides the most realistic *physical* model of diffusion in many minerals (Parsons *et al.*, 1999), but the multidiffusion domain model appears to provide an adequate *mathematical* model to explain the diffusive behavior of natural crystals under experimental conditions (Lovera *et al.*, 1997, 2002). This is significant because the multidomain diffusion model has provided a remarkably successful mathematical protocol for extracting cooling histories from $^{40}Ar/^{39}Ar$ K-feldspar data, as will be discussed further in Section 6.7.2.

Despite the fact that open-system behavior in mineral-isotopic systems is governed by a combination of volume and short-circuit diffusional processes, most thermochronologists make the simplifying assumption that volume diffusion alone controls the open- to closed-system transition that is so important to thermochronologic theory. There are a variety of reasons to believe that this assumption is reasonable. First, the structure of short-circuit pathways is such that they should be characterized by much faster diffusion than the intact crystal structure that surrounds them; if so, then the rate of daughter isotope loss should be limited by the rate of diffusion out of intact domains and into short-circuit pathways. Second, the volumetric proportion of short-circuit pathways to intact domains is small in all but the most strongly deformed natural crystals, implying that their contribution to bulk diffusive loss is minor. For example, maps of the distribution of argon isotopes in natural crystals display topologies consistent with volume diffusion at the grain scale with relatively minor modification by fast-diffusion pathways (e.g., Hames and Hodges, 1993; Hodges and Bowring, 1995; Reddy *et al.*, 1996; Pickles *et al.*, 1997; Wartho *et al.*, 1999). Finally, the fact that different thermochronometers yield predictable sequences of apparent ages implies that diffusivity, while temperature-dependent, is largely an intrinsic property of a mineral. We should not expect this to be the case if short-circuit diffusion exerts a controlling influence on mineral ages because the development of fast-diffusion pathways is related to the specific thermal and deformational history of a crystal. For the remainder of the chapter, we will assume that volume diffusion is the principal mode of daughter isotope loss, although practitioners of thermochronology are well-advised to evaluate each data set for evidence of significant loss by other mechanisms, both diffusive and non-diffusive; if important, these processes will have rendered the data set inappropriate for thermochronology.

6.5.2   Experimental Constraints on Daughter-isotope Diffusion for Useful Minerals

The volume diffusivity of isotopes in minerals depends on temperature through the Arrhenius relationship:

$$D_T = D_i e^{-E/RT} \qquad (13)$$

where D_T is diffusivity at the temperature of interest (T), D_i is the diffusivity at infinite temperature, E is the activation energy of the process, and R is the gas constant. Since the mid-1970s, there has been a concerted effort to characterize volume diffusion of daughter elements in minerals commonly used for geochronology and thermochronology by quantifying D_i and E.

The methods employed to do this typically fall into two categories: "bulk" experiments, in which fractional losses of an element or specific isotope are measured; and experiments in which induced diffusion profiles are measured. Both bulk diffusion and diffusion-profile experiments can be done in a vacuum or hydrothermally. Bulk hydrothermal experiments, frequently used to recover the diffusivity of gaseous species such as argon (e.g., Giletti, 1974; Grove and Harrison, 1996), involve the heating of samples at atmospheric or higher pressures for sufficient time to cause the loss of a fraction of the isotope of interest from the bulk sample. This method is required if the sample is hydrous and must be kept stable during the experiment. The fraction of the initial isotopic concentration lost by heating depends on the effective diffusion dimension (a), the diffusion geometry, and D_T (Fechtig and Kalbitzer, 1966; Crank, 1975). A series of experiments is conducted at different temperatures in order to define an array of points in $\ln(D_T/a^2)$ versus $1/T$ space. If loss occurred through volume diffusion over a single diffusion dimension, this array should be linear and diffusion parameters can be recovered by fitting a function derived by linearizing Equation (13) to the array. In many experimental

studies, a is equivalent to the physical grain half-size (e.g., Grove and Harrison, 1996; Reiners and Farley, 1999; Wartho et al., 1999; Farley, 2000). When this is the case, the diffusion parameter D_i can be determined directly from the $\ln(D_T/a^2)$-intercept of the fitted line. A few experiments seem to suggest the existence of an effective diffusion dimension much smaller than the physical grain half-size (Harrison, 1981; Harrison et al., 1985). In those cases, a is generally approximated by conducting experiments on samples crushed to a variety of grain sizes.

For minerals that are thought to be stable at elevated temperatures in a vacuum, an alternative approach to determining diffusivities is to perform bulk *in vacuo* experiments (e.g., Foland, 1994; Wolf et al.,1996). This approach has proved to be especially valuable for studying argon diffusion in K-feldspar, the only mineral commonly used for $^{40}Ar/^{39}Ar$ thermochronology that is stable during vacuum heating. Bulk *in vacuo* diffusion experiments on individual feldspar samples often reveal subparallel linear arrays in $\ln(D_T/a^2)$ versus $1/T$ space, which are consistent with the existence of multiple diffusion domains with different dimensions (Gillespie et al., 1982; Zeitler, 1987; Lovera et al., 1989). The fact that a single stepwise-heating experiment on K-feldspar provides sufficient information to characterize argon diffusivity in these domains, as well as their closure dates, means that $^{40}Ar/^{39}Ar$ data are capable of providing remarkably detailed information about the cooling history of rocks *if* the diffusion characteristics of K-feldspar are the same in the laboratory and in nature during cooling (Lovera et al., 2002).

Many recent diffusion experiments have relied on various microanalytical techniques to quantify diffusion gradients produced by heating under hydrothermal conditions or in a vacuum (e.g., Cherniak, 1993; Cherniak and Watson, 2000). Assuming simple, one-dimensional diffusion, these gradients are related to D_T through the inverse error function (erf^{-1}):

$$\text{erf}^{-1}\left(\frac{C_0 - C_\xi}{C_0}\right) = \left(\sqrt{4D_T\tau}\right)\xi \qquad (14)$$

where C_0 is the initial concentration of the diffusant, C_ξ is the concentration at distance ξ along the gradient, and τ is the duration of the experiment (Crank, 1975). A series of such experiments at different temperatures yields different values of D_T that can be used to recover D_i and E through linear regression of $\ln D_i$ versus $1/T$. Measurements of induced diffusion gradients are particularly useful for two reasons. The first is that heating over timescales and temperatures typically used in diffusion experiments produces little bulk loss of slow-diffusing species such as lead; as a consequence, measurements of bulk fractional losses of such species can have relatively large uncertainties, while careful measurements of diffusion profiles can yield higher-precision data. The second is that diffusion profiles can be measured orthogonal to carefully prepared surfaces or clean crystal faces, minimizing the effects of fast-diffusion pathways and increasing the probability that only volume diffusivity is responsible for loss of the element or isotope of interest.

Table 2 lists diffusion coefficients for many mineral-isotopic systems currently used for thermochronometry. Diffusivities of some important isotopes depend on major-element composition; e.g., ^{40}Ar diffusion in phlogopite is $\sim 88\%$ faster than in iron-rich biotite (Giletti, 1974; Grove and Harrison, 1996). In general, however, such variations are small in comparison with mineral-to-mineral differences in the diffusivity of a specific isotope. These differences for the basis for quantitative thermochronology.

6.6 CLOSURE TEMPERATURE THEORY

As defined in Section 6.4, igneous or metamorphic ages may be obtained for samples that retain all radiogenic daughter isotopes produced subsequent to crystallization. In contrast, the dates measured for samples that experienced some degree of daughter loss substantially underestimate the igneous or metamorphic age. Until the 1960s, these dates were regarded as geologically meaningless, but the eventual realization that daughter isotope loss is a predominantly diffusive process led to the proposition that they may be powerful probes of the cooling history of igneous and metamorphic terrains.

We can explore the concept of *cooling ages* and the temperatures they represent by considering an idealized, spherical mineral grain of radius a, which crystallizes instantaneously from a melt at time t_0 and at temperature T_0. Let us begin by assuming that our hypothetical grain grows with a spatially uniform initial concentration of parent isotopes (n_{p0}), but no non-radiogenic daughter isotopes (n_{d0}). With no atomic migration of parent or daughter within the grain and no loss of either species to the surroundings, the daughter–parent ratio (n_d/n_p) increases with time according to Equation (3), and it will be the same at all positions within the grain at any given time. Figure 1(a) schematically illustrates n_d/n_p as a function of radial position for $t = t_0$ and three subsequent times: t_1, t_2, and t_3. Measurements of n_d/n_p at any of these times would yield the actual crystallization age of the mineral.

Now suppose that T_0, the temperature of crystallization, is sufficiently high to permit diffusive loss of daughter into the intergranular medium surrounding the grain as rapidly as it was produced by radioactive decay. (Since concentration gradients are necessary to drive diffusive transfer, let us surround our hypothetical grain with a fluid that has no initial concentration of daughter isotopes, but a diffusivity of the daughter that is much greater than the corresponding diffusivity in the mineral at any given temperature. This condition ensures that the concentration of daughter in the fluid remains effectively zero through time.) In the absence of cooling from T_0, the grain will never accumulate daughter isotopes, such that n_d/n_p would remain at zero throughout the crystal at t_0, t_1, t_2, and t_3. However, let us allow the grain to cool at some reasonable rate. An assumption commonly made in thermochronologic studies is that reciprocal temperature increases linearly with time during cooling:

$$\frac{1}{T_t} = \frac{1}{T_0} + kt \qquad (15)$$

where T_t is the temperature of the grain at time t, and k is some constant of proportionality. Remembering that volume diffusion depends on temperature through the Arrhenius relationship (Equation (13)), we can see that Equation (15) implies a time dependence for diffusion in a cooling mineral:

$$\ln D_t = \ln D_i - \left(\frac{E}{R}\right)\left(\frac{1}{T_0} + kt\right) \qquad (16)$$

where D_t is diffusivity at time t. Given sufficient time, the diffusivity of the daughter isotope in our hypothetical sample should eventually drop to the point that none is lost to the fluid and the system becomes closed. Just prior to that, the sample will experience a transitional period of partially open-system behavior, during which n_d/n_p will increase, but at a rate less than that dictated by Equation (3).

Although diffusion is a "random-walk" process, the net transfer of matter by diffusion is driven by concentration gradients. In our model sample, the steepest gradients in daughter-isotope concentration will occur near the grain margin, and we should expect that diffusive loss of that isotope over time will be greatest at the grain margin and progressively less toward the center. Thus, the concentration of the daughter in the sample at any given time depends on the competition over time between a spatially independent process (daughter production by radioactive decay) and a spatially dependent process (daughter loss by diffusion). This is illustrated schematically in Figure 1(b) for t_0, t_1, t_2, and t_3; note that, for all times subsequent to t_0 and all positions within the crystal, the value of n_d/n_p is always less than Equation (3) predicts.

Another consequence of the spatial dependence of diffusive loss is that the open- to closed-system transition during cooling will take place at different radial positions in the crystal at different times. Figure 1(c) illustrates the evolution of n_d/n_p with time at three different radial positions (x) within

Table 2 Diffusion data and bulk closure temperatures for important mineral-isotopic systems.

Mineral-isotopic system	D_i (cm²s⁻¹)	E (kcal mol⁻¹)	Geometry	a (μm)[a]	T_{cb}(°C)[b]	References
Monazite (U–Th)/Pb	9.4×10^{3}	141	Spherical(?)	50	987	Cherniak et al. (2002)
Zircon (U–Th)/Pb	7.8×10^{2}	130	Spherical	50	942	Cherniak and Watson (2000)
Garnet Sm–Nd						
Almandine (Alm₇₅Py₂₂)	4.7×10^{-5}	62	Spherical	500	676	Ganguly et al. (1998)
Rutile (U–Th)/Pb	1.6×10^{-6}	58	Spherical	250	671	Cherniak (2000)
Titanite (U–Th)/Pb	1.1×10^{0}	79	Spherical	500	659	Cherniak (1993)
Hornblende ⁴⁰Ar/³⁹Ar						
Magnesio-hornblende	2.4×10^{-2}	64	Spherical	500	557	Harrison (1981)
K-feldspar Rb–Sr	6.0×10^{-3}	68	Spherical	10	487	Cherniak and Watson (1992)
Apatite (U–Th)/Pb	2.0×10^{-4}	55	Spherical	50	446	Cherniak et al. (1991)
Phlogopite ⁴⁰Ar/³⁹Ar	7.5×10^{-1}	58	Cylindrical	500	433	Giletti (1974)
Muscovite ⁴⁰Ar/³⁹Ar	3.9×10^{-4}	43	Cylindrical	500	366	Robbins (1972); Hames and Bowring (1994)
Biotite ⁴⁰Ar/³⁹Ar						
Fe–Mg biotite ($X_{phl}=0.29$)[d]	4.0×10^{-1}	51	Cylindrical	500	359	Grove and Harrison (1996)
Fe–Mg biotite ($X_{phl}=0.46$)[d]	7.5×10^{-2}	47	Cylindrical	500	335	Harrison et al. (1985)
Muscovite Rb–Sr	1.0×10^{-9}	25	Cylindrical	500	316	Chen et al. (1996)
Biotite Rb–Sr	2.0×10^{-9}	25	Cylindrical	500	299	Jenkin et al. (1995)
K-feldspar ⁴⁰Ar/³⁹Ar						
Low sanidine	3.7×10^{-2}	47	Spherical	10	237	Wartho et al. (1999)
Orthoclase	9.8×10^{-3}	44	Spherical	10	218	Foland (1994)
Zircon fission track[c]					227	Brandon et al. (1998)
Titanite (U–Th)/He	5.9×10^{1}	45	Spherical	250	206	Reiners and Farley (1999)
Apatite fission track[c]					110	Laslett et al. (1987); Brandon et al. (1998)
Apatite (U–Th)/He	3.16×10^{1}	33	Spherical	50	63	Farley (2000)

[a] Nominal grain half-sizes or subgrain diffusion dimensions. Other values would raise or lower T_{cb} as described by Equation (19).   [b] T_{cb} calculated assuming a cooling rate of 5°C Myr⁻¹.   [c] Closure temperatures estimated using both experimental and empirical data.   [d] X_{phl}-mole fraction of phlogopite component in Fe–Mg biotite.

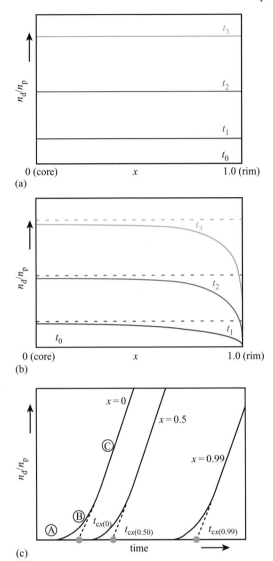

(a)

(b)

(c)

Figure 1 The evolution of daughter/parent (n_d/n_p) ratios with time in a cooling mineral-isotopic system. (a) In a model system that is closed with respect to diffusive loss of daughter isotopes, the ratio n_d/n_p is uniform throughout the crystal at any time. Curves are shown for four successive times (t_0, t_1, t_2, and t_3). In this frame, as in (b), the horizontal axis represents a faction of the radius of an idealized spherical grain (x). (b) With diffusive loss of daughter isotopes during cooling, n_d/n_p at any time is less than the value predicted by Equation (3) by an amount that differs depending on radial position. (c) Changes in the ratio n_d/n_p with time are shown for three different radial positions in the crystal: the core ($x=0$), the rim ($x=0.99$), and an intermediate position ($x=0.5$). The $x=0$ curve is labeled to indicate the three stages of n_d/n_p evolution during cooling: complete open-system behavior (A), partial open-system behavior (B), and complete closed-system behavior (C). Extrapolation of the closed-system behavior segments of three curves to the time axis designates the closure ages for the three positions in the crystal ($t_{cx(0)}$, $t_{cx(0.5)}$, $t_{cx(0.99)}$).

the grain: the core ($x=0$), a near-rim position ($x=0.99$), and half-way between the core and rim ($x=0.5$). (Note the convention of expressing radial distance as a fraction of the radius.) The evolution curve for each of these positions can be thought of in terms of three segments corresponding to: (i) purely open-system behavior, during which $n_d/n_p=0$ (curve section A in Figure 1(c)); (ii) purely closed-system behavior, during which n_d/n_p grows at a rate of $e^{\lambda_p t}-1$ (curve section C in Figure 1(c)); and (iii) transitional behavior characterized by the partial retention of daughter isotopes (curve section B in Figure 1(c)). If we measure the n_d/n_p ratio at any one of these positions at the present time (t_p) and calculate a date, we are, in effect, extrapolating backwards along the closed-system evolution curve in Figure 1(c) to its t-intercept. Note that this date—the *cooling age*—does not correspond to the crystallization age, the initiation age of partial daughter retention, or the initiation age of closed-system behavior. The cooling age (t_{cx}) instead corresponds to the time at which a specific position within the grain cooled through a specific temperature known as its *closure temperature* (also sometimes called the *blocking temperature* after a similar concept in the field of rock magnetism).

6.6.1 Quantitative Estimates of Closure Temperatures

Dodson (1986) presented a mathematical formula for calculating closure temperatures for various positions within a cooling crystal:

$$T_{cx} = \frac{E}{R\left[\ln\left(RD_i T_{cx}^2/Ea^2(dT/dt)\right)+G_x\right]} \quad (17)$$

where T_{cx} is the closure temperature at fractional position x, and dT/dt is the cooling rate. (Note that this equation must be solved iteratively because T_{cx} appears on both sides.) G_x, which Dodson referred to as the "closure function," depends on diffusion geometry; for the case of our hypothetical spherical grain:

$$G_x = C + 4\sum_{n=1}^{\infty}\frac{(-1)^{n+1}\sin(n\pi x)\ln(n\pi)}{n\pi x} \quad (18)$$

where C is the Euler constant (~ 0.57721547) and the trigonometric argument is in radians. Analogous expressions for plane sheet and cylindrical geometries, as well as useful approximations of G_x as a function of x, may be found in Dodson (1986).

Equation (17) provides a theoretical basis for reconstructing a significant proportion of the time–temperature history of a single crystal during cooling. The thermochronologist first measures n_d/n_p at various positions within a crystal using some microanalytical technique, then uses those data to calculate cooling ages, and finally determines the closure temperatures corresponding to those ages. An arbitrary cooling rate is assumed to calculate T_{cx} for two or more positions, these temperatures and their corresponding dates are then used to make a better estimate of the cooling rate (assuming that $dT/dt \approx \Delta T/\Delta t$, where the deltas represent differences), and the process is repeated until convergence.

Although microanalytical dating techniques are developing at a rapid pace, most thermochronologic data still are obtained for entire single crystals or aggregates of crystals; these will be referred to here as *bulk cooling ages* (t_{cb}). Fortunately, we can recover the *bulk closure temperature* (T_{cb}) of a crystal (Dodson, 1973) from the weighted mean of T_{cx}:

$$T_{cb} = \frac{E}{R \ln\left(e^{G_{av}} R D_i T_{cb}^2 / a^2 E(dT/dt)\right)} \quad (19)$$

G_{av}, the means value of G_x, depends only on the assumed geometry of diffusion: 4.00660 for radial diffusion in a sphere, 3.29506 for radial diffusion in a cylinder, and 2.15821 for diffusion across a plane sheet (Dodson, 1986). Equation (19) may be used in conjunction with t_{cb}'s for multiple mineral-isotopic systems, or for multiple grain sizes of the same mineral-isotopic system, in order to determine cooling histories for rocks.

Equations (17) and (19) are strictly valid only if it is assumed that: (i) postcrystallization loss of daughter isotopes occurs exclusively through volume diffusion; (ii) the medium surrounding the mineral serves as an infinite sink for daughter isotopes lost from the grain; (iii) temperature changes linearly with $1/t$, as described by Equation (15), and (iv) that no part of the grain—not even the core—retains all of the daughter isotope produced by radioactive decay of the parent at that position since crystallization. The first of these assumptions is a basic tenet of thermochronology, but the others invite closer scrutiny.

The second assumption is likely to be true in most instances, but exceptions can occur when an intergranular fluid contains high concentrations of the daughter isotope. In some cases, these concentrations are so high that the daughter isotope may diffuse into the mineral as an "excess" component. Excess ^{40}Ar, resulting in anomalously old and geologically meaningless $^{40}Ar/^{39}Ar$ and K/Ar dates, has been documented through many studies, as reviewed by Kelley (2002). Although slight excess ^{40}Ar contamination may be successfully corrected for using various methods (Roddick *et al.*, 1980; Harrison *et al.*, 1994), samples that have been heavily contaminated are unsuitable for thermochronometry. For some mineral-isotopic systems (e.g., Rb/Sr muscovite, biotite, and K-feldspar), the concentrations of non-radiogenic daughter isotopes are sufficiently unpredictable that many geochronologists follow the classical

approach of Nicolaysen (1961) and Fairbairn *et al.* (1961) by combining isotopic data for more than one mineral, or a mineral and the whole rock from which it was separated, to determine both a date and an initial ratio. This *internal isochron* approach requires the assumption that the concentration of daughter isotopes in different minerals are interdependent, a notion that is inconsistent with the second core assumption of Dodson's closure temperature theory. Models more sophisticated than Equation (19) have been proposed to estimate the bulk closure temperatures corresponding to dates obtained with internal isochrons (Giletti, 1991; Jenkin *et al.*, 1995; Chen *et al.*, 1996).

Violations of the third assumption implicit in Equations (17) and (19) are relatively insignificant. Through a series of mathematical experiments, Lovera *et al.* (1989) showed that Dodson's formalism yields correct results for natural samples with virtually any monotonic cooling history. Finally, the fourth assumption may not hold for mineral-isotopic systems characterized by extremely low daughter diffusivities at geologically reasonable temperatures (e.g., Sm/Nd or Lu/Hf in garnet, and (U–Th)/Pb for zircon, monazite, and xenotime). Ganguly and Tirone (1999) have derived alternative forms of Equations (17) and (19) that are more appropriate for such systems, but they require *a priori* knowledge of the crystallization temperature T_0. The Ganguly and Tirone equations yield closure temperature estimates that converge on the results of calculations made with Dodson's equations when T_0 is high ($>750\,°C$) and cooling rates are relatively slow ($<5\,°C\ my^{-1}$). When T_0 is lower than $\sim650\,°C$, low-diffusivity mineral-isotopic systems typically remain closed throughout their cooling history. As a consequence, the Ganguly and Tirone formulations are most useful for thermochronologic studies of igneous and granulite facies metamorphic rocks that cooled quickly from maximum temperatures in the 650–750 °C range.

6.6.2 The Influence of Input Parameters on Closure Temperature Calculations

The specific formulation of G_x used for a mineral-isotopic system depends on crystal morphology, the geometry of subgrain features that might define diffusion domains, and whether or not laboratory experiments on the system yield direct evidence for diffusive anisotropy. For example, experimental studies (Giletti, 1974; Harrison *et al.*, 1985; Grove and Harrison, 1996) and empirical observations (Hames and Bowring, 1994; Hodges *et al.*, 1994; Pickles *et al.*, 1997) suggest that daughter isotope loss from micas is anisotropic, with the most rapid diffusion occurring parallel to the (0 0 1) crystallographic plane. For these minerals, daughter loss is best modeled with a cylindrical diffusion geometry. For most other minerals used in thermochronology, it is common practice to assume a spherical geometry.

The choice of an appropriate diffusion dimension (a) for Equations (17) and (19) naturally depends on diffusion geometry as well. If a spherical geometry is assumed, then half of the crystal's physical grain size is a reasonable estimate of a. For micas with cylindrical geometries, half the crystal dimension measured parallel to the $(0\,0\,1)$ cleavage plane is reasonable. Although the value of a is often assumed to be related to be related to physical grain size, a few careful experiments have documented that diffusive loss in some samples is independent of grain size until the samples are crushed to a dimension much smaller than the natural grain size (e.g., Harrison, 1981; Harrison et al., 1985). A reasonable interpretation of these data is that the experimental starting materials contained intact domains surrounded by fast-diffusion pathways, such that progressive crushing eventually reduced each sample to the dimension of the intact domains. This suggests that using an "effective" diffusion dimension for a may be more appropriate than using the half-grain size *if* that dimension can be known *a priori*. Harrison (1981) and Harrison et al. (1985) have suggested that the effective diffusion dimension for radiogenic ^{40}Ar loss in hornblende and biotite may be intrinsic properties of the minerals themselves, such that a single value of a may be used for all samples of a particular mineral. However, this seems unlikely given that subgrain structures in minerals are demonstrably a function of their thermal and deformational histories. Other researchers have found evidence that a does correspond to the half-grain size for most mineral-isotopic systems (Goodwin and Renne, 1991; Onstott et al., 1991; Hess et al., 1993; Hames and Bowring, 1994; Hodges et al., 1994; Hodges and Bowring, 1995, Farley, 2000; Reiners and Farley, 1999, 2001), but this relationship breaks down in samples subjected to intensive ductile or brittle deformation (e.g., Arnaud and Eide, 2000; Kramar et al., 2001; Mulch et al., 2002). In general, it seems prudent to assume that a is related to the physical grain size when applying Equations (17) and (19) unless samples show textural evidence for the extensive development of subgrain boundaries that may act as fast diffusion pathways, or—in the case of K-feldspar—show direct evidence of the existence of multiple diffusion domains during incremental heating experiments.

Figure 2 illustrates the sensitivity of T_{cb} to variations in parameters a and dT/dt using volume diffusion parameters for radiogenic ^{4}He in apatite (Farley, 2000). Effective diffusion dimension has a strong influence on bulk closure temperature, implying that it is possible to recover a significant proportion of the time–temperature history of a single structural level in an orogen using only

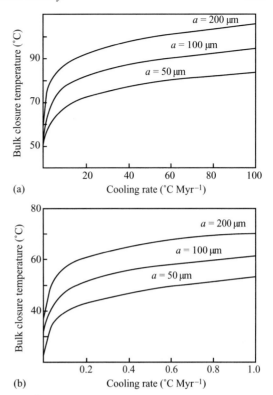

Figure 2 Bulk closure temperature for the (U–Th)/He thermochronometer as a function of cooling rate for effective diffusion dimensions of 50 μm, 100 μm, and 200 μm, based on calculations made with Equation (19) using diffusion coefficients as in Table 2. (a) Variation in T_{cb} for cooling rates from $0\,^{\circ}\mathrm{C}\,\mathrm{Myr}^{-1}$ to $100\,^{\circ}\mathrm{C}$ Myr^{-1}. (b) For cooling rates from $0\,^{\circ}\mathrm{C}$ to $1\,^{\circ}\mathrm{C}$. Note the increased dependence of T_{cb} on dT/dt for very slow cooling rates.

one thermochronometer if a range of grain sizes of the mineral is available for study (e.g., Hess et al., 1993; Reiners and Farley, 2001). Closure temperature is less dependent on dT/dt. Typical postcrystallization cooling rates in orogenic settings range from about $1–100\,^{\circ}\mathrm{C}\,\mathrm{Myr}^{-1}$, and solutions of Equation (19) over this entire range would yield T_{cb} estimates for any specific thermochronometer that vary by less than 50%. However, Equation (19) does imply a strong dependence of T_{c} on dT/dt for metamorphic terrains that cool at $<1\,^{\circ}\mathrm{C}\,\mathrm{my}^{-1}$ (Figure 2(b)). Although documented in only a few settings thus far (e.g., the Proterozoic orogen of the southwestern United States: Hodges et al., 1994; Hodges and Bowring, 1995), such slow cooling rates may have been more common than generally believed in Archean–Proterozoic terrains characterized by long-term tectonic stability.

6.6.3 Qualitative Estimates of Closure Temperatures

Unfortunately, high-quality diffusion data are not available for all useful mineral-isotopic systems.

Approximate closure temperatures for these thermochronometers can be estimated empirically by applying them to natural samples for which cooling ages have been determined using better characterized systems (e.g., Mezger *et al.*, 1992; Scherer *et al.*, 2000; Willigers *et al.*, 2001). Some T_{cb} values approximated in this way are shown in Table 2. An alternative approach is based on the notion that volume diffusion in a mineral is related to its *ionic porosity*. Ionic porosity was defined by Dowty (1980) as the percentage of the unit-cell volume not occupied by ions. Fortier and Giletti (1989) showed that experimental data for the diffusivity of argon and oxygen in silicate minerals defined positive correlations with ionic porosity. Dahl (1996a,b) used such relationships to predict variations in $^{40}Ar/^{39}Ar$ T_{cb} with chemical composition for micas and amphiboles. Ionic porosity also provides a useful tool for predicting the relative closure temperatures for the same daughter isotopes in different minerals. For example, Dahl (1997) found a log–linear relationship between ionic porosity and experimentally determined lead diffusivity in various minerals. Furthermore, he suggested a similar control of ionic porosity on another kinetic process of great importance to thermochronology: fission-track annealing.

6.6.4 Fission-track Closure Temperatures and the Partial Annealing Zone

If a mineral crystallizes at relatively high temperatures, any damage to the crystal structure produced by ^{238}U fission rapidly heals. This annealing occurs by atomic reorganization of the crystal structure and, thus, is a diffusive process. By analogy with the diffusive loss of daughter isotopes from a cooling mineral (Figure 1(c)), we can also define closure temperatures for fission-track thermochronometers based on the concept that the evolution of fission tracks in a mineral with time has three stages: (i) a period of high temperatures when fission tracks anneal as rapidly as they develop; (ii) a period of low temperatures when tracks accumulate and do not anneal perceptibly; and (iii) a period of intermediate temperatures when tracks accumulate but also partially anneal (Haack, 1977; Dodson, 1979; Brandon *et al.*, 1998). Approximate bulk closure temperatures for the commonly used fission-track zircon and apatite thermochronometers, made with the arbitrary assumption that monotonic cooling results in 50% annealing of the tracks in a sample, are provided in Table 1. However, such estimates have proven less valuable in fission-track studies than estimates of the entire temperature range of 0–100% annealing. This *partial annealing zone* has been estimated from experimental data (e.g., Brandon *et al.*, 1998) and from empirical studies of samples from deep bore holes with known

thermal structures (e.g., Gleadow and Duddy, 1981) to be ~60–120 °C for apatite and ~200–300 °C for zircon (Gallagher *et al.*, 1998; Brandon *et al.*, 1998). Examples of how fission-track data are used to identify the temporal evolution of this temperature zone in orogenic settings are discussed in Section 6.7.5.

6.7 APPLICATIONS

The advent of modern methods of geochronology and thermochronology has revolutionized the way earth scientists study the evolution of orogenic systems. This section explores some of the uses of these techniques in tectonics research, and provides specific examples of especially successful applications.

6.7.1 Determining Timescales of Granitic Magmatism

Crustal anatexis and the migration of granitic melts strongly influence the thermal and deformational histories of orogenic systems (DeYoreo *et al.*, 1989; Brown and Solar, 1998: Sandiford *et al.*, 2002). As a consequence, determining the crystallization ages of granitic rocks has become an important part of many research projects in continental tectonics. The principal chronometers used for such studies are (U–Th)/Pb zircon, monazite, and xenotime. All three minerals are common accessory phases in granitic rocks, and all contain large concentrations of uranium and thorium, such that easily measurable concentrations of radiogenic lead are produced after only a few million years of decay. Most importantly, experimental studies of lead diffusivity in zircon and monazite (Cherniak and Watson, 2000; Cherniak *et al.*, 2002) and predictions of lead diffusivity in xenotime from its ionic porosity (Dahl, 1997) imply that nominal closure temperatures for these chronometers greatly exceed the solidus temperatures for granitic rocks commonly encountered in orogenic settings (Thompson, 1982; Vielzeuf and Holloway, 1988; Patiño Douce and Johnston, 1991; Holland and Powell, 2001).

Unfortunately, the interpretation of (U–Th)/Pb geochronologic data for granitic rocks is not always straightforward. Many granitic rocks are derived through the melting of protoliths that contain pre-existing accessory minerals. These refractory phases survive the melting process to be preserved as *inherited* grains in the final melt products. Moreover, the lead diffusivity of minerals like zircon, monazite, and xenotime is such that low-temperature melting may produce little or no resetting of the (U–Th)/Pb chronometer. A typical orogenic granite contains numerous inherited grains in addition to new *magmatic* grains that crystallized from the melt. In many cases, overgrowths of neoblastic magmatic rims on inherited cores are common. Conventional ID-TIMS (U–Th)/Pb analysis of these granites using multigrain separates of zircon or orthophosphates yields highly discordant ^{208}Pb/^{232}Th, ^{207}Pb/^{235}U, and ^{206}Pb/^{238}U dates that, at best, provide low-precision estimates of the granite crystallization age. Single-crystal analyses are more effective, but the results are still sometimes difficult to interpret if inherited grains with neoblastic overgrowths cannot be avoided. The most successful strategies involve: (i) careful backscattered electron, cathodoluminesence, and/or X-ray compositional mapping of grains to understand their internal complexity; (ii) hand picking of crystals with a minimum of structural and chemical complexity; and (iii) the application of microanalytical dating techniques, either the mechanical separation and ID-TIMS dating of "clean" crystal fragments, or ion, electron, or laser microprobe dating of specific regions in individual crystals. The choice of microanalytical method depends on the level of analytical precision desired. As a general rule, ID-TIMS is the favored method unless the accessory phases are so complex at such a small scale that mechanical separation of clean fragments is impossible. Examples of applications of single-crystal and microanalytical (U–Th)/Pb techniques to the dating of granitic rocks in orogenic settings may be found in Parrish and Tirrrul (1989), Parrish (1990), Harrison *et al.* (1995b), Hawkins and Bowring (1997), Coleman (1998), Hodges *et al.* (1998), Anczkiewicz *et al.* (2001), Viskupic and Hodges (2001), and Aleinikoff *et al.* (2002).

6.7.2 Constraining the Cooling Histories of Igneous Rocks

Intrusive rocks of felsic and intermediate composition contain many thermochronologically useful minerals, and their study provides a remarkably detailed accounting of the postmagmatic thermal evolution of continental arc terrains. High-temperature (>400 °C) cooling histories can be constrained with (U–Th)/Pb thermochronometers such as titanite and apatite (Corfu and Stone, 1998; Frost *et al.*, 2000; Chamberlain and Bowring, 2001). Although both minerals are notorious for containing high concentrations of non-radiogenic lead, the analysis of comagmatic feldspars permits adequate corrections for this component in many cases (e.g., Housh and Bowring, 1991). For lower temperatures, the thermal history is typically constrained through a combination of ^{40}Ar/^{39}Ar, (U–Th)/He, and fission-track techniques. Particularly useful in this regard are the ^{40}Ar/^{39}Ar K-feldspar and apatite fission-track thermochronometers. Despite controversies regarding the physical nature of

and developmental mechanism of intragranular diffusion domains in feldspars (Villa, 1994; Parsons *et al.*, 1999), numerous thermochronologic studies of granitic rocks have employed multidomain diffusion modeling of argon data for feldspars to reconstruct detailed temperature–time (*Tt*) paths over tens to hundreds of degrees (e.g., Richter *et al.*, 1991; Krol *et al.*, 1996; Quidelleur *et al.*, 1997; Kirby *et al.*, 2002). The results are, for the most part, both geologically reasonable and consistent with the results of other, less-controversial methods. Unfortunately, feldspars that show evidence for alteration or extensive contamination with excess ^{40}Ar cannot be used to extract reliable thermal histories. In a recent compilation of the results of nearly 200 K-feldspar experiments from a variety of settings, Lovera *et al.* (1989) concluded that only about half were appropriate for thermal history modeling.

Apatite fission-track data also provide substantial segments of *Tt* paths through track length analysis. As fission-tracks anneal, they become progressively shorter, such that the frequency distribution of track lengths in a sample should be a sensitive indicator of cooling history (e.g., Gleadow *et al.*, 1986; Carlson, 1990). Several sophisticated algorithms permit the inversion of track length distributions to recover cooling histories (Corrigan, 1991; Gallagher, 1995; Willett, 1997; Ketcham *et al.*, 2000).

6.7.3 Calibrating Metamorphic Histories

Element partitioning and fluid inclusion thermobarometry, in concert with thermodynamic modeling of porphyroblast zoning, are powerful tools for reconstructing the pressure–temperature (*PT*) paths followed by metamorphic rocks during orogenesis. Placing these paths in temporal context—thus permitting the reconstruction of *PTt* paths—has become a fundamental goal of geochronology and thermochronology in tectonic settings.

The metamorphic samples best suited for thermochronology are those displaying textural and chemical evidence for a single progressive metamorphic event with no appreciable retrograde re-equilibration among the constituent minerals. In such samples, porphyroblasts like garnet typically display major- and trace-element zoning patterns consistent with prograde growth (Hollister, 1966; Tracy *et al.*, 1976; Spear, 1993). Well-conceived thermochronologic studies of these rocks begin with careful petrographic observation, preferably augmented with X-ray compositional mapping of thin sections, to evaluate the paragenetic sequence of minerals suitable for thermobarometry or thermodynamic modeling, as well as those suitable for thermochronology. If peak metamorphic temperatures can be estimated through thermobarometry, this result and the paragenetic sequence can serve as a guide for developing an effective thermochronologic strategy.

For example, suppose mineral rim compositions in a pelitic schist suggest a peak metamorphic temperature of ~500 °C, and textures suggest that the final equilibrium assemblage included the rock-forming minerals garnet + biotite + muscovite and the accessory minerals monazite and apatite. The most reliable way to estimate the age of peak metamorphism is by dating one of these minerals using a decay scheme with a nominal T_{cb} safely above the estimated peak temperature. Both (U–Th)/Pb monazite and Sm/Nd garnet would be appropriate choices, but correct interpretation of the dates obtained using either technique would require care. In the case of (U–Th)/Pb monazite dating, a special concern is that the sample may contain a component of detrital monazite manifested as cores with metamorphic overgrowths; since peak metamorphic temperatures were too low to induce substantial diffusive loss of radiogenic lead in detrital monazite, a conventional (U–Th)/Pb ID-TIMS date of a single polygenetic crystal would be an overestimate of the metamorphic age. Fortunately, the internal structure of accessory minerals can be revealed through backscatter electron, cathodoluminesence, and X-ray compositional mapping. In metamorphic samples, accessory minerals always should be characterized using one or more of these tools prior to dating as a guide to proper interpretation of the results. Under ideal circumstances, lower-precision microanalytical techniques—such as electron microprobe, ion microprobe, or ICP-MS laser microprobe (U–Th)/Pb dating—could be used to establish the approximate ages of different monazite components, and the results might permit us to target specific unzoned metamorphic grains (or grain fragments) for higher-precision ID-TIMS dating. For samples with multicomponent accessory mineral populations, especially those displaying complex, small-scale intergrowths, it may be impossible to extract sufficient material for ID-TIMS analysis and some precision must be sacrificed to ensure having readily interpretable data.

Sm–Nd dating of the garnet in our hypothetical sample also might be problematic. One concern is the likelihood of trace-element growth zoning, which would preclude dating single garnet crystals by Sm/Nd ID-TIMS unless we make the assumption that the growth rate of the garnet was so rapid that the elapsed time permitted no appreciable accumulation of radiogenic Nd. An even greater problem is the probability that the garnet will contain accessory mineral inclusions that would compromise the results; even tiny amounts of monazite would contain very large concentrations of the light rare earths and would dominate

the Sm–Nd isotopic characteristics of the "garnet" (DeWolf *et al.*, 1996; Vance *et al.*, 1998; Prince *et al.*, 2000; Thoni, 2002). As a consequence, the isolation of inclusion-free rims would be required for effective Sm/Nd garnet geochronology. In addition, some method must be employed to estimate the non-radiogenic [143]Nd component in the sample, which in turn has an impact on our confidence in the T_{cb} estimate in Table 2 as well as the precision of the calculated date.

For our hypothetical sample, it is probable that (U–Th)/Pb ID-TIMS dating of the monazite would provide the most reliable and most precise estimate of the age of peak metamorphism *if* the metamorphic component could be physically separated. If the metamorphic monazite could be identified through some imaging technique but could not be physically separated from detrital monazites, (U–Th)/Pb *in situ* microanalytical dating would be the preferred methodology. However, if our petrographic characterization of the rock left doubts regarding the relative age of monazite growth and peak metamorphism, Sm/Nd dating of the garnet rim may be the most reliable way to estimate the age of peak metamorphism. Unfortunately, no *in situ* microanalytical technique with sufficiently high analytical precision has been developed for Sm/Nd dating of garnet, and some method of mechanical separation of the garnet rim would be necessary to enable ID-TIMS dating.

Microanalytical methods may provide age constraints on the *PT* path prior to peak metamorphism. If a modeled garnet contains distributed accessory mineral inclusions useful for microanalytical (U–Th)/Pb dating, the resulting *PT* evolution can be placed in a temporal context (DeWolf *et al.*, 1993; Harrison *et al.*, 1997; Zhu *et al.*, 1997a). However, accessory mineral inclusions are rarely distributed *throughout* their host porphyroblasts, and it would be extremely fortuitous to be able to assign more than one or two absolute ages to *PT* points along a reconstructed path.

An unfortunate reality of *PTt* studies of metamorphic rocks is that the prograde *PT* path can be reconstructed with higher fidelity than the retrograde *PT* path, but the inverse is true for the prograde and retrograde *Tt* paths (Hodges, 1991). Our hypothetical sample provides many opportunities for calibrating its cooling history. In addition to the (U–Th)/Pb dating of monazite discussed above, U/Pb, (U–Th)/He, and fission-track dating of apatite and [40]Ar/[39]Ar and Rb/Sr dating of muscovite and biotite permit *Tt* path reconstruction from peak temperatures to ~60–70 °C. If the sample cooled very slowly, microanalytical studies of isotopic zoning or dating different grain sizes might provide even more detail. In contrast, secondary fluid inclusion thermobarometry is the only tool available for retrograde *PT* path reconstruction for a sample such as this displaying no textual evidence of retrograde resorbtion or re-equilibration of principal rock forming minerals.

Samples that do show evidence for retrogression or polymetamorphism provide special challenges and special opportunities for *PTt* path reconstruction. Most of these involve the topical application of traditional techniques to specific subassemblages of minerals that are thought to have been in equilibrium with one another at a specific time in a sample's history. An analogous approach to geochronology and thermochronology is possible—we can date minerals of different generations. However, as is the case with *PT* path studies, particular attention must be paid to the potential geochemical impact of later prograde or retrograde metamorphic events on minerals crystallized during early events.

As an example, consider the case of a pelitic schist sample (with the same mineral assemblage as our previous hypothetical sample) that experienced early prograde metamorphism (M_1) at $\sim$600 °C and a second episode of metamorphism (M_2) at $\sim$450°C. Suppose further that this sample contains two generations of garnet—expressed as M_2 rims on M_1 cores—and two generations of all other geochronologically important minerals that occur as discrete M_1 and M_2 crystals. In this case, *in situ* microanalysis of inclusions of M_1 and M_2 monazite in the cores and rims of garnets might help elucidate the prograde histories of both M_1 and M_2. Peak M_1 and M_2 temperatures could be dated reliably using the (U–Th)/Pb monazite geochronometer. The age of peak M_2 metamorphism also might be confirmed through (U–Th)/Pb dating of M_2 apatites, but the significance of such dates for M_1 apatites is less clear. Do they date a temperature on the cooling path subsequent to M_1? Were they reset during M_2 metamorphism? Or did partial diffusive lead loss during M_2 render the results geologically meaningless? Thermochronometers with closure temperatures below 450 °C would be useful for reconstructing the post-M_2 cooling history regardless of whether or not the minerals involved grew during M_1 or M_2. Unfortunately, this sample provides no viable options for establishing an unambiguous record of the *Tt* path of the sample between M_1 and M_2.

Such thought experiments emphasize two important points about the thermochronology of metamorphic rocks. The first is that it is virtually impossible to draw robust conclusions about the thermal history of a sample by thermochronologic methods without concomitant petrologic study. The second is that, since multiple generations of datable minerals are likely to be found in metamorphic rocks, the most successful thermochronologic studies employ microanalytical techniques to ensure that the results are placed in proper

petrographic context. Some examples include DeWolf *et al.* (1993), Cocherie *et al.* (1998), Hawkins and Bowring (1999), Schaltegger *et al.* (1999), Vavra and Schaltegger (1999), Simpson *et al.* (2000), Terry *et al.* (2000), Catlos *et al.* (2001), Di Vincenzo *et al.* (2001), Rubatto *et al.* (2001), Rubatto (2002), Catlos *et al.* (2002), and Zeck and Whitehouse (2002).

6.7.4 Calibrating Deformational Histories

In orogenic settings where multiple generations of intrusive or extrusive rocks occur, field relationships frequently provide constraints on their age relative to deformational fabrics and structures. By documenting that an igneous unit is post- or pre-kinematic with respect to a particular deformational feature, and by determining the crystallization age of that unit, it is possible to assign a minimum or maximum age to the feature. Under fortuitous circumstances, it can be possible to "bracket" the age of a structure by dating both pre- and post-kinematic igneous rocks. For landscapes exhumed from deeper structural levels, comparative thermochronometry of the hanging walls and footwalls of major post-metamorphic or post-intrusive thrust and normal faults provides a way to estimate the age of such structures as the time of convergence of their *Tt* paths. If *Tt* paths can be reconstructed only for the hanging wall or footwall but not both, estimates of the developmental temperatures of fault-related rocks might be compared to the hanging wall or footwall *Tt* path to constrain the maximum age of faulting. Such approaches to calibrating deformational histories have been used by Pan and Kidd (1992), Hodges and Applegate (1993), Applegate and Hodges (1995), Harrison *et al.* (1997), Hodges *et al.* (1996, 1998), Coleman and Hodges (1998), Hartz *et al.* (2000), Wells *et al.* (2000), Gans *et al.* (2001), and Murphy *et al.* (2002).

Major normal fault systems that root into the middle and lower crust, sometimes referred to as "detachments," have been identified in tectonic regimes as varied as the Basin and Range Province of western North America to the Himalayan orogen of South Asia (Burchfiel *et al.*, 1992; Wernicke, 1992). Large-magnitude displacement on these structures has the effect of rapidly cooling their footwalls (e.g., Ruppel *et al.*, 1988). Because of this, many researchers have used thermochronologic techniques to date the onset of rapid footwall cooling and have interpreted that date as a close approximation of the inception age of slip on the detachment (e.g., Dokka *et al.*, 1986; Davis and Lister, 1988; Harrison *et al.*, 1995a; Miller *et al.*, 1999; Murphy *et al.*, 2002). However, such interpretations require caution, because both theoretical and empirical studies indicate that they may underestimate the true age of fault initiation by up to several million years (House and Hodges, 1994; Ruppel and Hodges, 1994). Some of the most interesting studies of detachments involve the development of large thermochronologic databases from well-exposed footwalls and the use of cooling age patterns to deduce fault kinematics and slip rates (Foster and John, 1999; Ehlers *et al.*, 2001; Stockli *et al.*, 2001; Brady, 2002).

A more direct approach to dating deformational features involves the application of isotopic geochronology to minerals that grew initially or were dynamically recrystallized during deformation at temperatures below their nominal closure temperatures (Reuter and Dallmeyer, 1989; Gromet, 1991; West and Lux, 1993; Coleman and Hodges, 1995; Resor *et al.*, 1996; Dunlap, 1997; Anderson *et al.*, 2001; Dallmeyer *et al.*, 2001; Müller *et al.*, 2001; Reddy *et al.*, 2001). Many deformed rocks display multiple generations of tectonite fabrics, and techniques such as laser microprobe $^{40}Ar/^{39}Ar$ dating of polished thin sections permit the dating of specific minerals in structural context and thus determination of the absolute ages of different fabric-forming events. Although care must be taken to ensure that the dates for older grains are not compromised by deformation-induced loss of radiogenic ^{40}Ar (Reddy and Potts, 1999; Dunlap and Kronenberg, 2001; Kramar *et al.*, 2001; Mulch *et al.*, 2002), some successful studies have been documented (e.g., Hames and Cheney, 1997; Chan *et al.*, 2000; Di Vincenzo *et al.*, 2001) and this approach ultimately may prove to be particularly valuable for the study of the duration of and interval between fabric-forming events during orogenesis.

A relatively recent development has been the ability to date brittle faulting using the $^{40}Ar/^{39}Ar$ method. Quenched frictional melts (pseudotachylytes) have been dated using both step-heating and laser microprobe techniques (Kelley *et al.*, 1994; Magloughlin *et al.*, 2001; Müller *et al.*, 2001; Sherlock and Hetzel, 2001; White and Hodges, 2002). In addition, van der Pluijm *et al.* (2001) and Parry *et al.* (2001) have shown how careful $^{40}Ar/^{39}Ar$ dating of synkinematic illites can be used to determine the age of fault gouge development during slip on near-surface faults.

6.7.5 Estimating Unroofing Rates

Although thermochronologic studies can yield remarkably detailed reconstructions of the cooling histories of rocks, more useful information—from the perspective of continental tectonics—would be a better understanding of how these rocks changed their positions with respect to Earth's surface over time. Since temperatures are known to increase with depth (z) in the stable continental crust, it seems logical that the Tt path of a sample should reflect the history of its transport toward Earth's surface. When unroofing can be attributed to tectonic denudation, thermochronology can help constrain the kinematic evolution of structurally higher normal fault systems. When it can be attributed to erosion, such data improve our understanding of sedimentary processes during orogenesis.

One approach to estimating unroofing histories has been to relate the cooling rate (dT/dt) determined for a single sample to its unroofing rate (dz/dt) through an assumed geothermal gradient (γ):

$$\frac{dz}{dt} \approx \left(\frac{1}{\gamma}\right)\frac{dT}{dt} \qquad (20)$$

The cooling rate can be determined either by applying at least two thermochronometers and approximating the cooling rate as differences in T_{cb}/t_{cb}, or by determining core–rim variations in t_{cx} and T_{cx} in individual isotopically zoned crystals. A popular alternative is to apply a single thermochronometer to samples collected over a range of modern elevations. For example, suppose samples are collected over a range of elevations ($\varepsilon_1, \varepsilon_2, \varepsilon_3$) and they yield a range of (U–Th)/He closure dates ($t_{cb1}, t_{cb2}, t_{cb3}...$). If T_{cb} is presumed to be the same for all of the apatites, then

$$\frac{dz}{dt} \approx \frac{d\varepsilon}{dt_{cb}} \qquad (21)$$

Age–elevation data also may be helpful in determining when a particular denudation episode began. Consider the case of a stable crustal section in which the upper and lower bounding isotherms of the apatite fission-track partial annealing zone (Section 6.6.4) occur at depths z_a and z_b. If an episode of unroofing begins at time t_{init} and continues to the present day at a uniform rate, modern surface exposures may—if topographic relief is high enough and the unroofing is rapid enough—include samples from below and within the partial annealing zone (Figure 3). Age–elevation profiles developed from samples collected from such a terrain will be characterized by an inflection point that marks the approximate position of paleodepth z_b; rocks below this elevation yield an age–elevation gradient indicative of the denudation rate, whereas the gradient above is a reflection of the pre-unroofing thermal structure. Profiles interpreted as exhumed portions of fission-track partial annealing zones—or conceptually similar (U–Th)/He "partial retention zones" (Wolf *et al.*, 1998)—have been described by Fitzgerald *et al.* (1995), Fitzgerald and Stump (1997), Reiners *et al.* (2000), Stockli *et al.* (2000), Xu and Kamp (2000), Bullen *et al.* (2001), and Crowley *et al.* (2002).

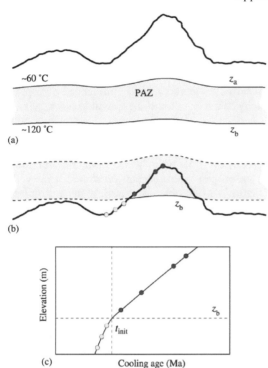

Figure 3 Exhumed apatite fission-track partial annealing zone (PAZ) and its value for unroofing rate studies. (a) Diagrammatic cross-section showing the PAZ (gray) beneath a steady-state topography prior to an unroofing episode. The top and base of the PAZ (z_a and z_b) represent the approximate positions of the 60 °C and 120 °C isotherms, respectively. Only a slight effect of topography on the shape of isotherms is shown for the sake of simplicity. (b) Position of the PAZ after exhumation by erosion. Samples may be collected along a modern age–elevation profile. Those collected above the structural level that once was z_b are shown in red; those below are shown in blue. (c) Elevation versus apatite fission-track cooling ages for samples collected along the transect. An inflection in the slope of the best-fit curve indicates the approximate initiation age of the unroofing event (t_{init}) and the position of the base of the exhumed PAZ.

All methods by which thermochronologic data are used to estimate unroofing rates require a series of assumptions that are unlikely to be valid in all orogenic settings and thus must be made with caution. One is that temperatures in the crust increase monotonically toward deeper structural levels at a constant rate, with isothermal surfaces remaining at least roughly parallel to Earth's surface at all levels. Another is that the thermal structure of the crust is invariant during the unroofing interval. Applications that involve the application of Equation (20) require the assumption of a geothermal gradient. Although the "age–elevation" method avoids this, it requires a different assumption: that the samples collectively moved as a block, without rotation, through the closure isotherm for the applied chronometer.

At the most fundamental level, these assumptions seem inconsistent with our expectations that the thermal structures in active orogens are both complex and transient (e.g., Huerta *et al.*, 1996; Batt and Braun, 1997; Jamieson *et al.*, 2002). Nevertheless, an increasing number of tectonic studies suggest that major structural configurations in orogenic systems may be active over tens of millions of years (e.g., Hodges *et al.*, 2001). This condition naturally leads to the development of a thermodynamic steady state that may persist over relatively long time intervals, a conclusion that should not be surprising given the natural tendency toward self-organization in simpler dynamical systems (Hodges, 1998b). Indeed, steady-state thermal structures are predicted by numerical and analytical models of the thermo-mechanical evolution of orogenic systems (e.g., Huerta *et al.*, 1996; Batt and Braun, 1997; Jamieson *et al.*, 2002).

The topologies of isotherms in orogens at a thermodynamic steady state are more complex than those of isotherms in stable continental crust because they reflect the interplay between both conductive and advective heat transfer, but these differences become less significant toward shallower crustal levels (i.e., towards the temperature boundary condition imposed by Earth's surface). As a consequence, Equations (20) and (21) provide reasonable approximations of the relations between cooling rate and unroofing rate if steady-state conditions have been achieved and if data from thermochronometers with relatively low T_{cb}'s (roughly <500 °C) are used for the calculations. Estimates of denudation rate with higher fidelities may be obtained through more sophisticated inverse methods that account for advective processes explicitly (e.g., Moore and England, 2001).

Unfortunately, it is not possible to know *a posteriori* if, when, and for how long an orogen may have been at steady state. Rather than presume the achievement of steady state for calculating unroofing rates, a better strategy may be to evaluate the consistency of unroofing rates calculated with Equation (20) using samples collected at different structural levels, or with Equation (21) using more than one thermochronometer. Such consistency is a direct indication of the achievement of steady-state conditions and, as noted by Willett and Brandon (2002), is an important monitor of the developmental maturity of an orogenic system.

6.7.6 Monitoring the Evolution of Topography

One manifestation of a thermodynamic steady state during orogeny can be an erosional steady state, during which the rate of rock uplift toward Earth's surface is matched by the erosion rate at the surface. Numerical models of landscape evolution in idealized orogens frequently predict the development of erosional steady states (Kooi and Beaumont, 1996; Braun and Sambridge, 1997; Willett *et al.*, 2001). A true erosional steady state implies that the surface topography is static because landscapes evolve as a consequence of lateral variations in erosion rate with time. While this extreme condition is unrealistic, it offers a valuable point of departure for exploring how topography affects near-surface thermal structures and, thus, how the distributions of cooling ages across a varied topography can be used to understand the antiquity of the landscape.

It has been known for many years that isotherms at shallow crustal levels mimic surface topography (Bullard, 1938; Birch, 1950; Turcotte and Schubert, 1982), but it was not until recently that Stüwe *et al.* (1994) and Mancktelow and Grasemann (1997) demonstrated the effect of this phenomenon on thermochronology. In particular, they showed that high-wavelength topographic variations and high exhumation rates—both of which are characteristic of the orogenic settings where thermochronology is most often practiced—can lead to substantial errors in calculations based on equations such as (20) and (21). This effect is most pronounced in the near-surface, such that exhumation rates estimated from apatite fission-track

and (U–Th)/He apatite data may be especially unreliable unless an appropriate "terrain correction" can be made (e.g., Stüwe *et al.*, 1994).

Like many obstacles in the earth sciences, this one can be turned to our advantage: if apatite closure ages are so strongly influenced by topography, then it should be possible to use spatial patterns in closure ages to reconstruct the evolution of topography. As an example of how this might be done, consider the steady-state thermal effects of periodic topographic variation in an orogen with constant erosion at $1\,mm\,yr^{-1}$ (Figure 4). Sample A, starting at a nominal depth of 9 km beneath a topographic high and uplifted through the illustrated thermal structure at $1\,mm\,yr^{-1}$, would pass through the nominal closure isotherm for (U–Th)/He in apatite after 8 Myr. Sample B, starting at the same structural level but beneath a topographic low, reaches the closure isotherm after 6.6 Myr. The 1.4 Myr difference in (U–Th)/He apatite cooling age between samples A and B—if it could be resolved within the limits of analytical imprecision—would be a direct indication of the magnitude and distribution of topographic relief over the recovered cooling interval. With this technique, House *et al.* (1998) found that the thermal influences of major transverse drainages on the western flank of the Sierra

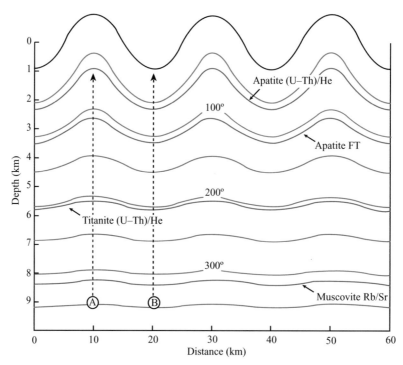

Figure 4 Steady-state temperature structure beneath a periodic topography as calculated using the algorithms of Mancktelow and Grasemann (1997). This simulation assumes a topographic relief of 1.5 km, a topographic wavelength of 20 km, and a uniform denudation rate of $1\,mm\,yr^{-1}$. The topmost sinusoidal line indicates the land surface (defined as having a temperature of 0 °C). Shaded contours are steady-state isotherms from 50 °C to 350 °C at 50 °C intervals. The red contours represent nominal bulk closure isotherms for the muscovite Rb/Sr, apatite fission track, and apatite and titanite (U–Th)/He thermochronometers. Positions A and B and their dashed unroofing paths are discussed in the text.

Nevada of California had been established by Late Cretaceous time. Note that the effects of surface topography on isotherms are damped with depth in Figure 4. As a consequence, the difference between (U–Th)/He titanite closure ages for A and B would be far less than that between (U–Th)/He apatite closure ages, and the difference in muscovite Rb/Sr closure ages would be negligible. At higher exhumation rates and with higher relief, topographic effects extend to deeper levels, such that studies integrating (U–Th)/He, fission-track, $^{40}Ar/^{39}Ar$ and Rb/Sr data for a variety of minerals from the several structural levels can recover not just the magnitude of topography, but also how that topography changed in time and space during the orogenic process.

While simplified, steady-state models with periodic topography are valuable for understanding the fundamental physics of the problem, they do not provide a sufficiently sophisticated mathematical foundation for the themochronologic study of landscape evolution. First steps toward the next generation of models include the work of Stüwe and Hintermüller (2000) and Braun (2002b), who considered the impact of an evolving topography, and Braun (2002a), who introduced a spectral analysis approach to data inversion that provides estimates of mean exhumation rates and relief changes without requiring the assumption of a geothermal gradient.

6.7.7 Reconstructing Regional Patterns of Deformation and Erosion

With the advent of increasingly sophisticated thermal models, many researchers are developing regionally extensive data sets in an effort to invert patterns of bedrock cooling ages for information about how deformational structures and topography co-evolve during orogenesis. Such studies have been done recently along the San Andreas fault zone in the San Bernadino Mountains of southern California (Spotila *et al.*, 2001), the Olympic Mountains of the northwestern United States (Batt *et al.*, 2001), the Southern Alps of New Zealand (Tippett and Kamp, 1993, 1995; Batt *et al.*, 2000), the Coast Mountains of British Columbia, Canada (Farley *et al.*, 2001), the central Alaska Range (Fitzgerald *et al.*, 1995), the Nanga Parbat syntaxis of Pakistan (Zeitler *et al.*, 2001), and the eastern Tibetan Plateau (Xu and Kamp, 2000; Kirby *et al.*, 2002).

Sedimentary deposits provide a valuable record of the erosional history of an orogen. The crystallization or cooling ages of detrital minerals in a basin can be used to determine provenance (e.g., Hurford *et al.*, 1984; Baldwin *et al.*, 1986; Hurford and Carter, 1991; Kelley and Bluck, 1992; von Eynatten *et al.*, 1996; Gray and Zeitler, 1997; Adams and Kelley, 1998; Kalsbeek *et al.*, 2000; Watt and Thrane, 2001; Kosler *et al.*, 2002). By comparing depositional ages with the cooling ages for detrital minerals, it is possible to ascertain the timescales for erosion and transport in orogenic settings (Cerveny *et al.*, 1988; Copeland and Harrison, 1990; Brandon and Vance, 1992; Corrigan and Crowley, 1992; Garver and Brandon, 1994; Bullen *et al.*, 2001). If the distribution of bedrock cooling ages are known for source regions, the thermochronology of detrital minerals provide a valuable tool for the reconstruction of regional patterns of erosion (Renne *et al.*, 1990; DeCelles *et al.*, 1998; Lonergan and Johnson, 1998; Garver *et al.*, 1999; Stuart, 2002). A relatively recent development has been the use of thermochronologic data for modern stream sediment samples to limit the bedrock cooling age distribution and relief (Stock and Montgomery, 1996) or average erosion rate (Brewer, 2001; Bullen *et al.*, 2001) for the stream's catchment. Such studies may yield a sense of how relief developed over time within specific catchments, and catchment-to-catchment variations may, in turn, reveal the nature of topographic change over time.

6.8 DIRECTIONS FOR FUTURE RESEARCH

Modern tectonics research is heavily dependent on the tools of geochronology and thermochronology and, as a consequence, has become an important driver of analytical innovation. Understanding the pace of thermal, deformational, and erosional processes requires the development of chronometers that are both accurate and precise, and it must be possible to apply these tools selectively to features formed during specific tectonic events. These three needs—accuracy, specificity, and precision—help define the frontiers of geochronology and thermochronology research.

The *accuracy* of isotopic chronometry depends, in part, on an exact knowledge of decay constants. While most of the constants shown in Table 1 are reasonably well constrained, the need to resolve small segments of time when studying orogenic processes and the need to compare the results of different mineral-isotopic systems demand that we minimize the uncertainties in decay constants even more. Relatively little attention has been paid over the past two decades to the need to know these values better, and recent comparative geochronology studies suggest that their refinement must be a priority for future research. Especially problematic in this regard are: (i) the $^{40}K \rightarrow {}^{40}Ca$ decay scheme, for which nuclear physicists and geologists typically assume a different decay constant (Min *et al.*, 2000; Renne, 2000); (ii) the $^{147}Sm \rightarrow {}^{143}Nd$ decay scheme, for which there is a large uncertainty in the accepted decay constant (Lugmair and Marti, 1978); and (iii) the $^{176}Lu \rightarrow {}^{176}Hf$ decay scheme, for which the decay constant is actively debated (Scherer *et al.*, 2000, 2001).

Other limits to the accuracy of chronologic data involve uncertainties in the nature of what is being dated, and the corrections necessary for the application of a specific chronometer. Some techniques (e.g., Sm/Nd and Lu/Hf dating of garnet) can be seriously compromised by inclusions of accessory phases, such that accurate geochronology and thermochronology require ultra-pure mineral separates. In addition, many commonly dated minerals display complex compositional zoning and contain defects that influence the intracrystalline transport of daughter isotopes and, thus, closure temperatures. Efforts to develop new microsampling and sample characterization protocols should be encouraged as a way to minimize such problems. A concern specific to $^{40}Ar/^{39}Ar$ thermochronology is the need to have high-purity flux monitors for which ages are known well. Recent studies have cast doubt on the ages of some of the most widely used monitors (Renne *et al.*, 1998; Lanphere and Baadsgaard, 2001; Schmitz and Bowring, 2001). While problems of this kind have no impact on applications that do not require a comparison of $^{40}Ar/^{39}Ar$ results with those made with other isotopic chronometers, they must be resolved through further research before absolute ages based on the $^{40}Ar/^{39}Ar$ technique can be considered truly reliable.

A limitation specific to the (U–Th)/He decay scheme is the need to correct calculated cooling ages for ^{4}He lost due to α-particle emission (Farley *et al.*, 1996). In many cases, these corrections—based on the size and shape of a dated crystal and assumptions about the stopping distances of α-particles in various compounds—amount to several tens of percent of the apparent age, making them the principle contributor to the uncertainty in a (U–Th)/He cooling age. The development of microanalytical methods to eliminate the need for the α-ejection correction—or at least better methods of characterizing crystal morphology and stopping distances in important minerals in order to minimize the uncertainty in the correction—should have high priority for future research.

The accuracy with which we calculate closure temperatures depends on having high-quality diffusion data. For some systems (e.g., Lu/Hf in garnet), reliable diffusion data do not exist. For many others, we have no good understanding of the effects of natural compositional variability or deformational history on diffusivity. The experiments necessary to improve this situation are difficult and time-consuming, but they are of crucial importance.

Specificity refers to our ability to determine crystallization ages for minerals that grew at a specific time during the history of a sample. For a typical metamorphic sample from the internal regions of an orogen, it may be desirable to date minerals that define two or three fabrics, or specific zones within a porphyroblast that has a prolonged growth history, or multiple generations of minerals that grew during different metamorphic episodes. Many of the microanalytical methods reviewed in Section 6.3.1 were developed specifically to address such needs. However, most of these techniques provide high-spatial resolution at the expense of analytical *precision*. This is a significant limitation because, as we learn more and more about the evolution of orogenic systems, it becomes clear that deformational and erosional process transitions typically occur over short timescales that can be resolved only with extremely high-precision chronometers. The problem is particularly acute for Cenozoic orogens. For example, much of the tectonic architecture that has characterized the Himalaya for the past 20 Myr appears to have been established over a period of $\sim$5 Myr, yet many published ages for Himalayan samples are reported with uncertainties in excess of 2 Myr at the $\sim$95% confidence level. Clearly, this level of imprecision is unacceptably large for many avenues of Himalayan research. In the future, we can expect to see more attention paid to the development of analytical techniques that represent a practical compromise between the desire for improved spatial resolution and the equally pressing need for improved analytical precision.

Finally, as the availability of high-quality geochronologic and thermochronologic data increases over the next few years, we can expect to see an increase in the number of inverse modeling techniques developed to extract from those data information about the thermal, deformational, and erosional histories of an orogen. Unfortunately, the success of these initiatives will be limited by the idiosyncratic behavior of complex systems such as orogens. For example, many tectonics researchers currently favor a two-sided developmental model of convergent orogens (Willett *et al.*, 1993) because it predicts many of the first-order characteristics of well-studied active orogens (Beaumont *et al.*, 1996, 2001; Willett, 1999; Jamieson *et al.*, 2002). Without a doubt, such forward models provide valuable insights regarding appropriate interpretations of geochronologic and thermochronologic data, but many time–temperature trajectories predicted for specific structural horizons in these models are dependent on a large number of assumptions that may be impossible to justify for real orogens. As orogenic models become increasingly sophisticated, their sensitivity to specific assumptions becomes more difficult to quantify. Ultimately, they may begin to behave like the complex systems they were designed to simulate. A key characteristic of such systems is their capacity to self-organize into behavioral patterns that are sustained by largely unpredictable process interactions. If such "emergent behavior" occurs, it would be impossible to trace an effect to a specific cause—the principal goal of inverse modeling—in either numerical simulations or real orogens. One of the great challenges before us is to attain the scientific sophistication to understand not only how much we can learn about orogenic processes from geochronology and thermochronology, but also the limitations of what we can learn.

ACKNOWLEDGMENTS

Geochronology and thermochronology are vibrant fields, and no short review can do justice to the work being done today by dozens of research groups worldwide. This chapter is intended as a (relatively!) objective overview of current methods and the many ways in which they are used for tectonic studies. I apologize in advance for any errors of omission.

Much of the review of basic concepts in this chapter derives from the notes for graduate-level subjects that I have taught at MIT, and I appreciate very much how the participating students have helped me to refine the presentation of these concepts. My perceptions about geochronology and thermochronology have benefited greatly from collaborations with my wonderful graduate students and postdoctoral advisees; with my MIT colleagues Sam Bowring, Clark Burchfiel, Leigh Royden, and Kelin Whipple; and with many other colleagues from other universities. In particular, Daniele Cherniak and Ken Farley kindly provided preprints of manuscripts in progress. This work, as well as my research programs in general, have been supported by grants from the US National Science Foundation. Its continuing support is gratefully acknowledged.

REFERENCES

Adams C. J. and Kelley S. (1998) Provenance of Permian–Triassic and Ordovician metagraywacke terranes in New Zealand: evidence from ^{40}Ar/^{39}Ar dating of detrital micas. *Geol. Soc. Am. Bull.* **110**, 422–432.

Albarède F. (1995) *Introduction to Geochemical Modeling*. Cambridge University Press, Cambridge, England.

Aleinikoff J. N., Wintsch R. P., Fanning C. M., and Dorais M. J. (2002) U–Pb geochronology of zircon and polygenetic titanite from the Glastonbury Complex, Connecticut, USA: an integrated SEM, EMPA, TIMS, and SHRIMP study. *Chem. Geol.* **188**, 125–147.

Anczkiewicz R., Oberli F., Burg J. P., Villa I. M., Gunther D., and Meier M. (2001) Timing of normal faulting along the Indus suture in Pakistan Himalaya and a case of major Pa-231/U-235 initial disequilibrium in zircon. *Earth Planet. Sci. Lett.* **191**, 101–114.

Anderson S. D., Jamieson R. A., Reynolds P. H., and Dunning G. R. (2001) Devonian extension in northwestern Newfoundland: ^{40}Ar/^{39}Ar and U–Pb data from the Ming's Bight area, Baie Verte peninsula. *J. Geol.* **109**, 191–211.

Applegate J. D. R. and Hodges K. V. (1995) Mesozoic and Cenozoic extension recorded by metamorphic rocks in the Funeral Mountains, California. *Geol. Soc. Am. Bull.* **107**, 1063–1076.

Armstrong R. L., Jäger E., and Eberhardt P. (1966) A comparison of K–Ar and Rb–Sr ages on Alpine biotites. *Earth Planet. Sci. Lett.* **1**, 13–19.

Arnaud N. O. and Eide E. A. (2000) Brecciation-related argon redistribution in alkali feldspars: an in naturo crushing study. *Geochim. Cosmochim. Acta* **64**(18), 3201–3215.

Baldwin S. L., Harrison T. M., and Burke K. (1986) Fission track evidence for the source of accreted sandstones, Barbados. *Tectonics* **5**, 457–468.

Batt G. E. and Braun J. (1997) On the thermomechanical evolution of compressional orogens. *Geophys. J. Inter.* **128**, 364–382.

Batt G., Braun J., Kohn B., and McDougall I. (2000) Thermochronological analysis of the dynamics of the southern Alps, New Zealand. *Geol. Soc. Am. Bull.* **112**, 250–266.

Batt G. E., Brandon M. T., Farley K. A., and Roden-Tice M. (2001) Tectonic synthesis of the Olympic Mountains segment of the Cascadia wedge, using two-dimensional thermal and kinematic modeling of thermochronological ages. *J. Geophys. Res. Solid Earth* **106**(B11), 26731–26746.

Beaumont C., Kamp P., Hamilton J., and Fullsack P. (1996) The continental collision zone, South Island, New Zealand: comparison of geodynamical models and observations. *J. Geophys. Res.* **101**(B2), 3333–3359.

Beaumont C., Jamieson R. A., Nguyen M. H., and Lee B. (2001) Himalayan tectonics explained by extrusion of a low-viscosity crustal channel coupled to focused surface denudation. *Nature* **414**, 738–742.

Birch F. (1950) Flow of heat in the front range, Colorado. *Geol. Soc. Am. Bull.* **61**, 567–630.

Brady R. J. (2002) Very high slip rates on continental extensional faults: new evidence from (U–Th)/He thermochronometry of the Buckskin Mountains, Arizona. *Earth Planet. Sci. Lett.* **197**, 95–104.

Brandon M. T. and Vance J. A. (1992) Fission-track ages of detrital zircon grains: implications for the tectonic evolution of the Cenozoic Olympic subduction complex. *Am. J. Sci.* **292**, 565–636.

Brandon M. T., Roden-Tice M. K., and Garver J. I. (1998) Late Cenozoic exhumation of the Cascadia accretionary wedge in the Olympic Mountains, northwest Washington State. *Geol. Soc. Am. Bull.* **110**, 985–1009.

Braun J. (2002a) Estimating exhumation rate and relief evolution by spectral analysis of Age–elevation datasets. *Terra Nova* **14**, 210–214.

Braun J. (2002b) Quantifying the effect of recent relief changes on Age-elevation relationships. *Earth Planet. Sci. Lett.* **200**, 331–343.

Braun J. and Sambridge M. (1997) Modelling landscape evolution on geological time scales: a new method based on irregular spatial discretization. *Basin Res.* **9**, 27–52.

Brewer I. D. (2001) Detrital-mineral thermochronology: investigations of Orogenic denudation in the Himalaya of central Nepal. PhD, The Pennsylvania State University.

Brown M. and Solar G. S. (1998) Shear-zone systems and melts: feedback relations and self-organization in orogenic belts. *J. Struct. Geol.* **20**(2–3), 211–227.

Bruguier O., Telouk P., Cocherie A., Fouillac A. M., and Albarede F. (2001) Evaluation of Pb–Pb and U–Pb laser ablation ICP–MS zircon dating using matrix-matched calibration samples with a frequency quadrupled (266 nm) Nd-YAG laser. *Geostand. Newslett.: J. Geostand. Geoanal.* **25**(2–3), 361–373.

Bullard E. C. (1938) The disturbance of the temperature gradient in the Earth's crust by inequalities of height. *Mon. Not. Roy. Astron. Soc.: Geophys. Suppl.* **4**, 360–362.

Bullen M. E., Burbank D. W., Garver J. I., and Abdrakhmatov K. Y. (2001) Late Cenozoic tectonic evolution of the northwestern Tien Shan: new age estimates for the initiation of mountain building. *Geol. Soc. Am. Bull.* **113**(12), 1544–1559.

Burchfiel B. C., Chen Z., Hodges K. V., Liu Y., Royden L. H., Deng C., and Xu J. (1992) *The South Tibetan Detachment*

System, Himalayan Orogen: Extension Contemporaneous with and Parallel to Shortening in a Collisional Mountain Belt. Geological Society of America, Boulder, CO.

Carlson W. D. (1990) Mechanism and kinetics of apatite fission-track annealing. *Am. Mineral.* **75**, 1120–1139.

Catlos E. J., Harrison T. M., Kohn M. J., Grove M., Ryerson F. J., Manning C. E., and Upreti B. N. (2001) Geochronologic and thermobarometric constraints on the evolution of the main central thrust, central Nepalese Himalaya. *J. Geophys. Res.* **106**, 16177–16204.

Catlos E. J., Gilley L. D., and Harrison T. M. (2002) Interpretation of monazite ages obtained via *in situ* analysis. *Chem. Geol.* **188**, 193–215.

Cerveny P. F., Naeser N. D., Zeitler P. K., Naeser C. W., and Johnson N. M. (1988) History of uplift and relief of the Himalaya during the past 18 million years: evidence from sandstones of the Siwalik group. In *New Perspectives in Basin Analysis* (eds. K. L. Kleinspehn and C. Paola). Springer, New York, pp. 43–61.

Chamberlain K. R. and Bowring S. A. (2001) Apatite–feldspar U–Pb thermochronometer: a reliable, mid-range (~450 °C), diffusion-controlled system. *Chem. Geol.* **172**(1–2), 173–200.

Chan Y.-C., Crespi J. M., and Hodges K. V. (2000) Dating cleavage formation in slates and phyllites with the $^{40}Ar/^{39}Ar$ laser microprobe; an example from the western New England appalachians, USA. *Terra Nova* **12**, 264–271.

Chen C.-H., DePaolo D. J., and Lan C.-Y. (1996) Rb–Sr microchrons in the Manaslu granite: implications for Himalayan thermochronology. *Earth Planet. Sci. Lett.* **143**, 125–135.

Cherniak D. J. (1993) Lead diffusion in titanite and preliminary results on the effects of radiation damage on Pb transport. *Chem. Geol.* **110**, 177–194.

Cherniak D. J. (2000) Pb diffusion in rutile. *Contributions to Mineralogy and Petrology* **139**(2), 198–207.

Cherniak D. J. and Watson E. B. (1992) A study of strontium diffusion in K-feldspar, Na–K feldspar and anorthite using Rutherford backscattering spectroscopy. *Earth and Planetary Science Letters* **113**, 411–425.

Cherniak D. J. and Watson E. B. (2000) Pb diffusion in zircon. *Chem. Geol.* **172**, 5–24.

Cherniak D. J., Lanford W. A., and Ryerson F. J. (1991) Lead diffusion in apatite and zircon using ion implantation and Rutherford backscattering techniques. *Geochim. Cosmochim. Acta* **55**, 1663–1673.

Cherniak D. J., Watson E. B., Grove M., and Harrison T. M. (2002) Pb diffusion in monazite. *Geol. Soc. Am. Abstr. Prog.* **34**(6).

Christensen J. N., Rosenfeld J. L., and DePaolo D. J. (1989) Rates of techonometamorphic processes from Rubidium and Strontium isotopes in garnet. *Science* **244**, 1465–1469.

Cocherie A., Legendre O., Peucat J. J., and Kouamelan A. N. (1998) Geochronology of polygenetic monazites constrained by *in situ* electron microprobe Th-U-total lead determination: implications for lead behavior in monazite. *Geochim. Cosmochim. Acta* **62**, 2475–2497.

Coleman M. E. (1998) U–Pb constraints on Oligocene–Miocene deformation and anatexis within the central Himalaya, Marsyandi valley, Nepal. *Am. J. Sci.* **298**(7), 553–571.

Coleman M. and Hodges K. (1995) Evidence for Tibetan Plateau uplift before 14 Myr ago from a new minimum age for east–west extension. *Nature* **374**(6517), 49–52.

Coleman M. E. and Hodges K. V. (1998) Contrasting Oligocene and Miocene thermal histories from the hanging wall and footwall of the South Tibetan detachment in the central Himalaya from $^{40}Ar/^{39}Ar$ thermochronology, Marsyandi Valley, central Nepal. *Tectonics* **17**(5), 726–740.

Compston W., Kinny P. D., Williams I. S., and Foster J. J. (1986) The age and Pb loss behavior of zircons from the Isua supracrustal belt as determined by ion microprobe. *Earth Planet. Sci. Lett.* **80**, 71–81.

Copeland P. and Harrison T. M. (1990) Episodic rapid uplift in the Himalaya revealed by $^{40}Ar/^{39}Ar$ analysis of detrital K-feldspar and muscovite, Bengal Fan. *Geology* **18**, 354–357.

Corfu F. and Stone D. (1998) The significance of titaniite and apatite U–Pb ages: constraints for the post-magmatic thermal-hydrothermal evolution of a batholithic complex, Berens River area, northwestern superior province, Canada. *Geochim. Cosmochim. Acta* **62**, 2979–2995.

Corrigan J. (1991) Inversion of apatite fission track data for thermal history information. *J. Geophys. Res.* **96**, 10347–10360.

Corrigan J. D. and Crowley K. D. (1992) Unroofing of the Himalayas: a view from apatite fission-track analysis of Bengal Fan sediments. *Geophys. Res. Lett.* **19**, 2345–2348.

Crank J. (1975) *The Mathematics of Diffusion.* Oxford University Press, Oxford, England.

Crowley J. L. and Ghent E. D. (1999) An electron microprobe study of the U–Th–Pb systematics of metamorphosed monazite: the role of Pb diffusion versus overgrowth and recrystallization. *Chem. Geol.* **157**, 285–302.

Crowley P. D., Reiners P. W., Reuter J. M., and Kaye G. D. (2002) Laramide exhumation of the Bighorn Mountains, Wyoming: an apatite (U–Th)/He thermochronology study. *Geology* **30**(1), 27–30.

Dahl P. S. (1996a) The crystal-chemical basis for Ar retention in micas: inferences from interlayer partitioning and implications for geochronology. *Contrib. Mineral. Petrol.* **123**, 22–39.

Dahl P. S. (1996b) The effect of composition on retentivity of Ar and O in hornblende and related amphiboles: a field-tested empirical model. *Geochim. Cosmochim. Acta* **60**, 3687–3700.

Dahl P. S. (1997) A crystal-chemical basis for Pb retention and fission-track annealing systematics in U-bearing minerals, with implications for geochronology. *Earth Planet. Sci. Lett.* **150**(3–4), 277–290.

Dallmeyer R. D., Strachan R. A., Rogers G., Watt G. R., and Friend C. R. L. (2001) Dating deformation and cooling in the Caledonian thrust nappes of north Sutherland, Scotland: insights from $^{40}Ar/^{39}Ar$ and Rb–Sr chronology. *J.Geol. Soc. London.* **158**, 501–512.

Davis G. A. and Lister G. S. (1988) Detachment faulting in continental extension: perspectives from the southwestern US Cordillera. In *Processes in Continental Lithospheric Deformation* (eds. S. P. Clark, B. C. Burchfiel, and J. Suppe). Geological Society of America, vol. 218, pp. 133–159.

DeCelles P. G., Gehrels G. E., Quade J., Ojha T. P., Kapp P. A., and Upreti B. N. (1998) Neogene foreland deposits, erosional unroofing, and the kinematic history of the Himalayan fold-thrust belt, western Nepal. *Geol. Soc. Am. Bull.* **110**, 2–21.

DeWolf C. P., Belshaw N., and O'Nions R. K. (1993) A metamorphic history from micron-scale $^{207}Pb/^{206}Pb$ chronometry of Archean monazite. *Earth Planet. Sci. Lett.* **120**, 207–220.

DeWolf C. P., Zeissler C. J., Halliday A. N., Mezger K., and Essene E. J. (1996) The role of inclusions in U–Pb and Sm–Nd garnet geochronology: stepwise dissolution experiments and trace uranium mapping by fission track analysis. *Geochim. Cosmochim. Acta* **60**, 121–134.

DeYoreo J. J., Lux D. R., and Guidotti C. V. (1989) The role of crustal anatexis and magma migration in the in the thermal evolution of regions of thickened crust. In *Evolution of Metamorphic Belts,* Special Publications 43 (eds. J. S. Daly, R. A. Cliff, and B. W. D. Yardley). Geological Society of London, London, England, pp. 187–202.

Di Vincenzo G., Ghiribelli B., Giorgetti G., and Palmeri R. (2001) Evidence of a close link between petrology and isotope records: constraints from SEM, EMP, TEM and *in situ* Ar-40–Ar-39 laser analyses on multiple generations of white micas (Lanterman Range, Antarctica). *Earth Planet. Sci. Lett.* **192**(3), 389–405.

Dickin A. P. (1995) *Radiogenic Isotope Geology.* Cambridge University Press, Cambridge, England.

Dodson M. H. (1973) Closure temperature in cooling geochronological and petrological systems. *Contrib. Mineral. Petrol.* **40**, 259–274.

Dodson M. H. (1979) Theory of cooling ages. In *Lectures in Isotope Geology* (eds. E. Jäger and J. C. Hunziker). Springer, Berlin, pp. 194–202.

Dodson M. H. (1986) Closure profiles in cooling systems. *Mater. Sci. Forum* **7**, 145–154.

Dokka R. K., Mahaffie M. J., and Snoke A. W. (1986) Thermochronologic evidence of major tectonic denudation associated with detachment faulting, northern Ruby Mountains: East Humbolt range, Nevada. *Tectonics* **5**, 995–1006.

Dowty E. (1980) Crystal-chemical factors affecting the mobility of ions in minerals. *Am. Mineral.* **65**, 174–182.

Dunlap W. J. (1997) Neocrystallization or cooling? $^{40}Ar/^{39}Ar$ ages of white micas from low-grade mylonites. *Chem. Geol.* **143**, 181–203.

Dunlap W. J. and Kronenberg A. K. (2001) Argon loss during deformation of micas: constraints from laboratory deformation experiments. *Contrib. Mineral. Petrol.* **141**, 174–185.

Ehlers T. A., Armstrong P. A., and Chapman D. S. (2001) Normal fault thermal regimes and the interpretation of low-temperature thermochronometers. *Phys. Earth Planet. Inter.* **126**, 179–194.

Fairbairn H. W., Hurley P. M., and Pinson W. H. (1961) The relation of discordant Rb–Sr mineral and rock ages in an igneous rock to its time of subsequent Sr87/Sr86 metamorphism. *Geochim. Cosmochim. Acta* **23**, 135–144.

Farley K. A. (2000) Helium diffusion from apatite: general behavior as illustrated by Durango fluorapatite. *J. Geophys. Res.* **105**, 2903–2914.

Farley K. A., Wolf R. A., and Silver L. T. (1996) The effects of long alpha-stopping distances on (U–Th)/He ages. *Geochim. Cosmochim. Acta* **60**(21), 4223–4229.

Farley K. A., Rusmore M. E., and Bogue S. W. (2001) Post-10 Ma uplift and exhumation of the northern Coast Mountains, British Columbia. *Geology* **29**(2), 99–102.

Faure G. (1986) *Principles of Isotope Geology*. Wiley, New York.

Fechtig H. and Kalbitzer S. (1966) The diffusion of argon in potassium-bearing solids. In *Potassium–argon dating* (eds. O. Schaeffer and J. Zähringer). Springer, , pp. 68–107.

Feng R., Machado N., and Ludden J. (1993) Lead geochronology of zircon by laserprobe-inductively coupled plasma mass spectrometry (LP-ICPMS). *Geochim. Cosmochim. Acta* **57**, 3479–3486.

Fitzgerald P. G. and Stump E. (1997) Cretaceous and Cenozoic episodic denudation of the Transantarctic Mountains, Antarctica: new constraints from apatite fission track thermochronology in the Scott Glacier region. *J. Geophys. Res. B: Solid Earth Planets* **102**(4), 7747–7765.

Fitzgerald P. G., Sorkhabi R. B., Redfield T. F., and Stump E. (1995) Uplift and denudation of the central Alaska range: a case study in the use of apatite fission track thermochronology to determine absolute uplift parameters. *J. Geophys. Res.* **100**, 20175–20191.

Foland K. A. (1994) Argon diffusion in feldspars. In *Feldspars and their Reactions* (ed. I. Parsons). Kluwer, Amsterdam, pp. 415–447.

Fortier S. M. and Giletti B. J. (1989) An empirical model for predicting diffusion coefficients in silicate minerals. *Science* **245**, 1481–1484.

Foster D. A. and John B. E. (1999) Quantifying tectonic exhumation in an extensional orogen with thermochronology: examples from the southern basin and range province. In *Exhumation Processes: Normal Faulting, Ductile Flow and Erosion*, Special Publication 154 (eds. U. Ring, M. T. Brandon, G. S. Lister, and S. D. Willett). Geological Society of London, London, pp. 343–364.

French J. E., Heaman L. M., and Chacko T. (2002) Feasibility of chemical U–Th–total Pb baddeleyite dating by electron microprobe. *Chem. Geol.* **188**, 85–104.

Frost B. R., Chamberlain K. R., and Schurnacher J. C. (2000) Sphene (titanite): phase relations and role as a geochronometer. *Chem. Geol.* **172**, 131–148.

Fryer B. J., Jackson S. E., and Longerich H. (1993) The application of laser ablation microprobe-inductively coupled plasma-mass spectrometry (LAM–ICP–MS) *to in situ* (U)–Pb geochronology. *Chem. Geol.* **109**, 1–8.

Gallagher K. (1995) Evolving temperature histories from apatite fission-track data. *Earth Planet. Sci. Lett.* **136**, 421–435.

Gallagher K., Brown R. L., and Johnson C. (1998) Fission track analysis and its application to geological problems. *Ann. Rev. Earth Planet. Sci.* **26**, 519–572.

Ganguly J. and Tirone M. (1999) Diffusion closure temperature and age of a mineral with arbitrary extent of diffusion: theoretical formulation and applications. *Earth Planet. Sci. Lett.* **170**(1–2), 131–140.

Ganguly J., Tirone M., and Hervig R. L. (1998) Diffusional Kinetics of Samarium and neodymium in garnet, and a method for determining cooling rates of rocks. *Science* **281**(5378), 805–807.

Gans P., Seedorff E., Fahey P. L., Hasler R. W., Maher D. J., Jeanne R. A., and Shaver S. A. (2001) Rapid Eocene extension in the Robinson district, White Pine County, Nevada: constraints from $^{40}Ar/^{39}Ar$ dating. *Geology* **29**, 475–478.

Garver J. I. and Brandon M. T. (1994) Erosional exhumation of the British Columbia coast ranges as determined from fission-track ages of detrital zircon from the Tofino basin, Olympic Peninsula, Washington. *Geol. Soc. Am. Bull.* **106**, 1398–1412.

Garver J. I., Brandon M. T., Roden-Tice M., and Kamp P. J. J. (1999) Erosional exhumation determined by fission-track ages of detrital apatite and zircon. In *Exhumation Processes: Normal Faulting, Ductile Flow, and Erosion*, Special Publication 154 (eds. U. Ring, M. T. Brandon, G. S. Lister, and S. D. Willett). Geological Society of London, London, pp. 283–304.

Geisler T. and Schleicher H. (2000) Improved U–Th–total Pb dating of zircons by electron microprobe using a simple new background modelling procedure and Ca as a chemical criterion of fluid-induced U–Th–Pb discordance in zircon. *Chem. Geol.* **163**, 269–285.

Giletti B. J. (1974) Studies in diffusion: I. Argon in phlogopite mica. In *Geochemical Transport and Kinetics,* Carnegie Institute of Washington Publication (eds. A. W. Hofmann, B. J. Giletti, H. S. Yoder, and R. A. Yund), vol. 634, pp. 107–115.

Giletti B. J. (1991) Rb and Sr diffusion in alkali feldspars, with implications for cooling histories of rocks. *Geochim. Cosmochim. Acta* **55**, 1331–1343.

Gillespie A. R., Huencke J. C., and Wasserburg G. J. (1982) An assessment of $^{40}Ar/^{39}Ar$ dating of incompletely degassed xenoliths. *J. Geophys. Res.* **87**, 9247–9257.

Gleadow A. J. W. and Duddy I. R. (1981) A natural long-term track annealing experiment for apatite. *Nuclear Tracks* **5**, 169–174.

Gleadow A. J. W., Duddy I. R., Green P. F., and Lovering J. F. (1986) Confined fission track lengths in apatite: a diagnostic tool for thermal history analysis. *Contrib. Mineral. Petrol.* **94**, 405–415.

Goodwin L. B. and Renne P. R. (1991) Effects of progressive mylonitization on Ar retention in biotites from the Santa Rosa mylonite zone, California, and thermochronologic implications. *Contrib. Mineral. Petrol.* **108**, 283–297.

Gray M. B. and Zeitler P. K. (1997) Comparison of clastic wedge provenance in the Appalachian foreland using U/Pb ages of detrital zircons. *Tectonics* **16**, 151–160.

Gromet L. P. (1991) Direct dating of deformational fabrics. In *Applications of Radiogenic Isotope Systems to Problems in Geology* (eds. L. Heaman and J. N. Ludden). Mineralogical Association of Canada, Ottawa, Canada, pp. 167–189.

Grove M. and Harrison T. M. (1996) $^{40}Ar^*$ diffusion in Fe-rich biotite. *Am. Mineral.* **81**, 940–951.

Haack U. (1977) The closing temperature for fission track retention in minerals. *Am. J. Sci.* **277**, 459–464.

Halliday A. N., Christensen J. N., Lee D.-C., Hall C. M., Ballentine C. J., Rehkämper M., Yi W., Luo X., and Barford D. (1998) ICP multiple-collector mass spectrometry and *in situ* high-precision isotopic analysis. In *Applications of Microanalytical Techniques to Understanding Mineralizing*

Processes, Rev. Econ. Geol. 7 (eds. M. A. McKibben and W. C. Shanks). Society of Economic Geologists, pp. 37–51.

Hames W. E. and Bowring S. A. (1994) An empirical evaluation of the argon diffusion geometry in muscovite. *Earth Planet. Sci. Lett.* **124**, 161–167.

Hames W. E. and Cheney J. T. (1997) On the loss of ^{40}Ar from muscovite during polymetamorphism. *Geochim. Cosmochim. Acta* **61**(18), 3863–3872.

Hames W. E. and Hodges K. V. (1993) Laser (40)Ar/(39)Ar evaluation of slow cooling and episodic loss of (40)Ar from a sample of polymetamorphic muscovite. *Science* **261** (5129), 1721–1723.

Harland C. E. (1994) *Ion Exchange: Theory and Practice*. Royal Society of Chemistry, London.

Harrison T. M. (1981) Diffusion of ^{40}Ar in hornblende. *Contrib. Mineral. Petrol.* **78**, 324–331.

Harrison T. M., Duncan I., and McDougall I. (1985) Diffusion of ^{40}Ar in biotite: temperature, pressure, and compositional effects. *Geochim. Cosmochim. Acta* **49**, 2461–2468.

Harrison T. M., Heizler M. T., Lovera O. M., Chen W., and Grove M. (1994) A chlorine disinfectant for excess argon released from K-feldspar during step heating. *Earth Planet. Sci. Lett.* **123**, 95–104.

Harrison T. M., Copeland P., Kidd W. S. F., and Lovera O. (1995a) Activation of the Nyainquentanghla Shear Zone: implications for uplift of the southern Tibetan Plateau. *Tectonics* **14**, 658–676.

Harrison T. M., McKeegan K. D., and LeFort P. (1995b) Detection of inherited monazite in the Manaslu leucogranite by ^{208}Pb/^{232}Th ion microprobe dating: crystallization age and tectonic implications. *Earth Planet. Sci. Lett.* **133**, 271–282.

Harrison T. M., Ryerson F. J., McKeegan K. D., Le Fort P., and Yin A. (1996) Th–Pb monazite ages of Himalayan metamorphic and leucogranitic rocks: constraints on the timing of inverted metamorphism and slip on the MCT and STD. In *11th Himalaya–Karakoram–Tibet Workshop Abstracts* (eds. A. M. Macfarlane, R. B. Sorkhabi, and J. Quade)., pp. 58–59.

Harrison T. M., Ryerson F. J., Le Fort P., Yin A., Lovera O., and Catlos E. J. (1997) A Late Miocene–Pliocene origin for the central Himalayan inverted metamorphism. *Earth Planet. Sci. Lett.* **146**, E1–E7.

Hart S. R. (1964) The petrology and isotopic-mineral age relations of a contract zone in the front range, Colorado. *J. Geol.* **72**, 493–525.

Hartz E. H., Andresen A., Martin M. W., and Hodges K. V. (2000) U–Pb and ^{40}Ar/^{39}Ar constraints on the Fjord region detachment zone: a long-lived extensional fault in the East Greenland Caledonides. *J. Geol. Soc. London* **157**, 795–809.

Hawkins D. P. and Bowring S. A. (1997) U–Pb systematics of monazite and xenotime: case studies from the Paleoproterozoic of the Grand Canyon, Arizona. *Contrib. Mineral. Petrol.* **127**, 87–103.

Hawkins D. P. and Bowring S. A. (1999) U–Pb monazite, xenotime, and titanite geochronological constraints on the prograde to post-peak metamorphic thermal history of Paleoproterozoic migmatites from the Grand Canyon, Arizona. *Contrib. Mineral. Petrol.* **134**, 150–169.

Hess J. C., Lippolt H. J., Gurbanov A. G., and Michalski I. (1993) The cooling history of the Late Pliocene Eldzhurtinskiy granite (Caucasus, Russia) and the thermochronological potential of grain-size/age relationships. *Earth Planet. Sci. Lett.* **117**, 393–406.

Hodges K., Bowring S., Davidek K., Hawkins D., and Krol M. (1998) Evidence for rapid displacement on Himalayan normal faults and the importance of tectonic denudation in the evolution of mountain ranges. *Geology* **26**, 483–486.

Hodges K. V. (1991) Pressure–Temperature–time paths. *Ann. Rev. Earth Planet. Sci.* **19**, 207–236.

Hodges K. V. (1998a) ^{40}Ar/^{39}Ar geochronology using the laser microprobe. In *Applications of Microanalytical Techniques to Understanding Mineralizing Processes*, Rev. Econ. Geol. (eds. M. A. McKibben and W. C. Shanks). Society of Economic Geologists, Tuscaloosa, AL, pp. 53–72.

Hodges K. V. (1998b) The thermodynamics of Himalayan orogenesis. In *What Drives Metamorphism and Metamorphic Reactions?*. Geological Society of London, Special Publication 138 (eds. P. J. Treloar and P. O'Brien). Geological Society of London, London, pp. 7–22.

Hodges K. V. and Applegate J. D. R. (1993) Age of Tertiary extension, Bitterroot metamorphic core complex, Montana-Idaho. *Geology* **21**, 161–164.

Hodges K. V. and Bowring S. A. (1995) ^{40}Ar/^{39}Ar thermochronology of isotopically zoned micas: insights from the southwestern USA Proterozoic orogen. *Geochim. Cosmochim. Acta* **59**(15), 3205–3220.

Hodges K. V., Hames W. E., and Bowring S. A. (1994) ^{40}Ar/^{39}Ar age gradients in micas from a high-temperature–low-pressure metamorphic terrain: evidence for very slow cooling and implications for the interpretation of age spectra. *Geology* **22**(1), 55–58.

Hodges K. V., Parrish R. R., and Searle M. P. (1996) Tectonic evolution of the central Annapurna range, Nepalese Himalayas. *Tectonics* **15**, 1264–1291.

Hodges K. V., Hurtado J. M., and Whipple K. X. (2001) Southward extrusion of Tibetan crust and its effect on Himalayan tectonics. *Tectonics* **20**, 799–809.

Holland T. and Powell R. (2001) Calculation of phase relations involving haplogranitic melts using an internally consistent thermodynamic dataset. *J. Petrol.* **42**(4), 673–683.

Hollister L. S. (1966) Garnet zoning: an interpretation based on the Rayleigh fractionation model. *Science* **154**, 1647–1651.

Horn I., Rudnick R. L., and McDonough W. F. (2000) Precise elemental and isotope ratio determination by simultaneous solution nebulization and laser ablation-ICP-MS: application to U–Pb geochronology. *Chem. Geol.* **164**(3–4), 281–301.

Hoskin P. W. O. and Black L. P. (2000) Metamorphic zircon formation by solid-state recrystallization of protolith igneous zircon. *J. Metamorph. Geol.* **18**(4), 423–439.

House M. A. and Hodges K. V. (1994) Limits on the tectonic significance of rapid cooling events in extensional settings: insights from the Bitterroot metamorphic core complex, Idaho-Montana. *Geology* **22**, 1007–1010.

House M. A., Wernicke B. P., and Farley K. A. (1998) Dating topography of the Sierra Nevada, California, using apatite (U–Th)/He ages. *Nature* **396**, 66–69.

Housh T. B. and Bowring S. A. (1991) Lead isotopic heterogeneities within alkali feldspars: implications for the determination of lead isotopic compositions. *Geochim. Cosmochim. Acta* **55**, 2309–2316.

Huerta A. D., Royden L. H., and Hodges K. V. (1996) The interdependence of deformational and thermal processes in mountain belts. *Science* **273**, 637–639.

Hurford A. J. and Carter A. (1991) The role of fission track dating in discrimination of provenance. In *Developments in Sedimentary Provenance Studies*, Geological Society of London Special Publication 57 (eds. A. C. Morton, S. P. Todd, and P. D. W. Haughton). Geological Society of London, London, pp. 67–78.

Hurford A. J., Fitch F. J., and Clarke A. (1984) Resolution of the age structure of detrital zircon populations of two lower Cretaceous sandstones from the Weald of England by fission track dating. *Geol. Mag.* **121**, 285–317.

Hurtado J. M., Chatterjee N., Ramezani J., Hodges K. V., and Bowring S. A. (xxxx) Electron microprobe chemical dating of uraninite as a reconaissance tool for leucogranite geochronology. *Earth Planet. Sci. Lett.* (in preparation).

Jamieson R. A., Beaumont C., Nguyen M. H., and Lee B. (2002) Interaction of metamorphism, deformation, and exhumation in large convergent orogens. *J. Metamorph. Geol.* **20**, 9–24.

Jenkin G. R. T., Rogers G., Fallick A. E., and Farrow C. M. (1995) Rb–Sr closure temperatures in bi-mineralic rocks: a mode effect and test for different diffusion models. *Chem. Geol.* **122**, 227–240.

Kalsbeek F., Thrane K., Nutman A. P., and Jepsen H. F. (2000) Late Mesoproterozoic to early Neoproterozic history of the

East Greenland Caledonides: evidence for Grenvillian orogenesis? *J. Geol. Soc.* **157**, 1215–1225.

Kelley S. (2002) Excess argon in K–Ar and Ar–Ar geochronology. *Chem. Geol.* **188**, 1–22.

Kelley S. P. and Bluck B. J. (1992) Laser ^{40}Ar–^{39}Ar ages for individual detrital muscovites in the southern Uplands of Scotland UK. *Chem. Geol.* **101**(1–2), 143–156.

Kelley S. P., Reddy S. M., and Maddock R. (1994) Laser-probe ^{40}Ar/^{39}Ar investigation of a pseudotachylyte and its host rock from the outer Isles thrust, Scotland. *Geology* **22**(5), 443–446.

Kempe U. (2003) *Precise electron microprobe age determination in altered uraninite: consequences on the instrusion age and the metallogenic significance of the Kirchberg granite (Erzgebirge, Germany).* Contributions to Mineralogy and Petrology **145**, 107–118.

Ketcham R. A., Donelick R. A., and Donelick M. B. (2000) AFTSolve: a program for multikinetic modeling of apatite fission-track data. *Geol. Mater. Res.* **2**, 1–32.

Kirby E., Reiners P. W., Krol M. A., Whipple K. X., Hodges K. V., Farley K. A., Tang W. Q., and Chen Z. L. (2002) Late Cenozoic evolution of the eastern margin of the Tibetan Plateau: inferences from ^{40}Ar/^{39}Ar and (U–Th)/He thermochronology. *Tectonics* **21**, 3–22.

Kooi H. and Beaumont C. (1996) Large-scale geomorphology: classical concepts reconciled and integrated with contemporary ideas via a surface processes model. *J. Geophys. Res.* **101**, 3361–3386.

Kosler J., Tubrett M. N., and Sylvester P. J. (2001) Application of laser ablation ICP–MS to U–Th–Pb dating of monazite. *Geostand. Newslett.: J. Geostand. Geoanal.* **25**(2–3), 375–386.

Kosler J., Fonneland H., Sylvester P., Tubrett M., and Pedersen R. B. (2002) U–Pb dating of detrital zircons for sediment provenance studies: a comparison of laser ablation ICPMS and SIMS techniques. *Chem. Geol.* **182**(2–4), 605–618.

Kramar N., Cosca M. A., and Hunziker J. C. (2001) Heterogeneous ^{40}Ar* distributions in naturally deformed muscovite: *in situ* UV-laser ablation evidence for micro structurally controlled intragrain diffusion. *Earth Planet. Sci. Lett.* **192**(3), 377–388.

Krol M. A., Zeitler P. K., Poupeau G., and Pêcher A. (1996) Temporal variations in the cooling and denudation history of the Hunza plutonic complex, Karakoram Batholith, revealed by ^{40}Ar/^{39}Ar thermochronology. *Tectonics* **15**, 403–415.

Lanphere M. A. and Baadsgaard H. (2001) Precise K–Ar, Ar-40/Ar-39, Rb–Sr and U/Pb mineral ages from the 27.5 Ma Fish Canyon Tuff reference standard. *Chem. Geol.* **175**(3–4), 653–671.

Lanphere M. A., Wasserburg G. J., Albee A. L., and Tilton G. R. (1964) Redistribution of strontium and rubidium isotopes during metamorphism, World Beater Complex, Panamint range, California. In *Isotopic and Cosmic Chemistry* (eds. H. Craig, S. L. Miller, and G. J. Wasserburg). North Holland, Amsterdam, pp. 269–320.

Laslett G. M., Green P. F., Duddy I. R., and Gleadow A. J. W. (1987) Thermal annealing of fission tracks in apatite. *Chemical Geology; Isotope Geoscience Section* **65**(1), 1–13.

Lee J. K. W. (1995) Multipath diffusion in geochronology. *Contrib. Mineral. Petrol.* **120**, 60–82.

Lee J. K. W. and Aldama A. A. (1992) Multipath diffusion—a general numerical-model. *Comput. Geosci.* **18**(5), 531–555.

Li X. H., Liang X. R., Sun M., Guan H., and Malpas J. G. (2001) Precise Pb-206/U-238 age determination on zircons by laser ablation microprobe-inductively coupled plasma-mass spectrometry using continuous linear albation. *Chem. Geol.* **175**(3–4), 209–219.

Lonergan L. and Johnson C. (1998) A novel approach for reconstructing the denudation histories of mountain belts: with an example from the Betic Cordillera (S. Spain). *Basin Res.* **10**, 353–364.

Lovera O. M., Richter F. M., and Harrison T. M. (1989) ^{40}Ar/^{39}Ar geochronometry for slowly cooled samples having a distribution of diffusion domain size. *J. Geophys. Res.* **94**, 17917–17936.

Lovera O. M., Grove M., Harrison T. M., and Mahon K. I. (1997) Systematic analysis of K-feldspar ^{40}Ar/^{39}Ar step heating results: I. Significance of activation energy determinations. *Geochim. Cosmochim. Acta* **61**, 3171–3192.

Lovera O. M., Grove M., and Harrison T. M. (2002) Systematic analysis of K-feldspar Ar-40/Ar-39 step heating results: II. Relevance of laboratory argon diffusion properties to nature. *Geochim. Cosmochim. Acta* **66**(7), 1237–1255.

Lugmair G. W. and Marti K. (1978) Lunar initial ^{143}Nd/^{144}Nd: differential evolution of the lunar crust and mantle. *Earth Planet. Sci. Lett.* **39**, 349–357.

Machado N. and Simonetti A. (2001) U–Pb dating and Hf isotopic composition of zircon by laser ablation-MC-ICP-MS. In *Laser-Ablation-ICPMS in the Earth Sciences: Principles and Applications* (ed. P. Sylvester). Mineralogical Association of Canada, Ottawa, pp. 121–146.

Magloughlin J. F., Hall C. M., and van der Pluijm B. A. (2001) ^{40}Ar/^{39}Ar geochronometry of pseudotachylytes by vacuum encapsulation: North Cascade mountains, Washington, USA. *Geology* **29**, 51–54.

Maluski H. and Schaeffer O. A. (1982) ^{39}Ar–^{40}Ar laser probe dating of terrestrial rocks. *Earth Planet. Sci. Lett.* **59**, 21–27.

Mancktelow N. S. and Grasemann B. (1997) Time-dependent effects of heat advection and topography on cooling histories during erosion. *Tectonophysics* **270**, 167–195.

McDougall I. and Harrison T. M. (1988) *Geochronology and Thermochronology by the ^{40}Ar/^{39}Ar Method.* Oxford University Press, New York.

Merrihue C. M. and Turner G. (1966) Potassium–argon dating by activation with fast neutrons. *J. Geophys. Res.* **71**, 2852–2857.

Mezger K., Essene E. J., and Halliday A. N. (1992) Closure temperature of the Sm–Nd system in metamorphic garnets. *Earth Planet. Sci. Lett.* **113**, 397–409.

Miller E. L., Dumitru T. A., Brown R. W., and Gans P. B. (1999) Rapid Miocene slip on the Snake range–Deep Creek range fault system, east-central Nevada. *Geol. Soc. Am. Bull.* **111**, 886–905.

Min K. W., Mundil R., Renne P. R., and Ludwig K. R. (2000) A test for systematic errors in Ar-40/Ar-39 geochronology through comparison with U/Pb analysis of a 1.1-Ga rhyolite. *Geochim. Cosmochim. Acta* **64**(1), 73–98.

Montel J., Foret S., Veschambre M., Nicollet C., and Provost A. (1996) Electron microprobe dating of monazite. *Chem. Geol.* **131**, 37–53.

Moore M. A. and England P. C. (2001) On the inference of denudation rates from cooling ages of minerals. *Earth Planet. Sci. Lett.* **185**, 265–284.

Mulch A., Cosca M. A., and Handy M. R. (2002) *In-situ* UV-laser ^{40}Ar/^{39}Ar geochronology of a micaceous mylonite: an example of defect-enhanced argon loss. *Contrib. Mineral. Petrol.* **142**(6), 738–752.

Müller W., Prosser G., Mancktelow N. S., Villa I. M., Kelley S. P., Viola G., and Oberli F. (2001) Geochronological constraints on the evolution of the Peradriatic fault system (Alps). *Int. J. Earth Sci.: Geol. Rundsch.* **90**, 623–653.

Murphy M. A., Yin A., Kapp P., Harrison T. M., Manning C. E., Ryerson F. J., Ding L., and Guo J. H. (2002) structural evolution of the Gurla Mandhata detachment system, southwest Tibet: implications for the eastward extent of the Karakoram fault system. *Geol. Soc. Am. Bull.* **114**(4), 428–447.

Naeser N. D., Naeser C. W., and McCulloh T. H. (1989) application of fission-track dating to the depositional and thermal history of rocks in the sedimentary basins. In *Thermal history of sedimentary basins: methods and case studies* (eds. N. D. Naeser and T. H. McCulloh). Springer-Verlag, pp. 157–180.

Nicolaysen L. O. (1961) Graphic interpretation of discordant age measurements on metamorphic rocks. *Ann. NY Acad. Sci.* **91**, 198–206.

Onstott T. C., Phillips D., and Pringle-Goodell L. (1991) Laser microprobe measurement of chlorine and argon zonation in biotite. *Chem. Geol.* **90**, 145–168.

Pan Y. and Kidd W. S. F. (1992) Nyainqentanglha shear zone: a late Miocene extensional detachment in the southern Tibetan plateau. *Geology* **20**, 775–778.

Parrish R. (1990) U–Pb dating of monazite and its application to geological problems. *Can. J. Earth Sci.* **27**, 1431–1450.

Parrish R. and Tirrul R. (1989) U–Pb age of the Baltoro granite and implications for zircon inheretance. *Geology* **17**, 1076–1079.

Parry W. T., Bunds M. P., Bruhn R. L., Hall C. M., and Murphy J. M. (2001) Mineralogy, ^{40}Ar/^{39}Ar dating and apatite fission track dating along the Castle Mountain fault, Alaska. *Tectonophysics* **337**, 149–172.

Parsons I., Brown W. L., and Smith J. V. (1999) ^{40}Ar/^{39}Ar thermochronology using alkali feldspars: real thermal history or mathematical mirage of microtexture? *Contrib. Mineral. Petrol.* **136**, 92–110.

Patiño Douce A. E. and Johnston A. D. (1991) Phase equilibria and melt productivity in the pelitic system: implications for the origin of peraluminous granitoids and aluminous granites. *Contrib. Mineral. Petrol.* **107**, 202–218.

Pickles C. S., Kelley S. P., Reddy S. M., and Wheeler J. (1997) Determination of high spatial resolution argon isotope variations in metamorphic biotites. *Geochim. Cosmochim. Acta* **61**(18), 3809–3833.

Poitrasson F., Chenery S., and Shepherd T. J. (2000) Electron microprobe and LA-ICP-MS study of monazite hydrothermal alteration: implications for U–Th–Pb geochronology and nuclear ceramics. *Geochim. Cosmochim. Acta* **64**, 3283–3297.

Prince C. I., Kosler J., Vance D., and Gunther D. (2000) Comparison of laser ablation ICP-MS and isotope dilution REE analyses: implications for Sm–Nd garnet geochronology. *Chem. Geol.* **168**(3–4), 255–274.

Quidelleur X., Grove M., Lovera O. M., Harrison T. M., Yin A., and Ryerson F. J. (1997) Thermal evolution and slip history of the Renbu-Zedong thrust, southeastern Tibet. *J. Geophys. Res.* **102**, 2659–2679.

Reddy S. M. and Potts G. J. (1999) Constraining absolute deformation ages: the relationship between deformation mechanisms and isotope systematics. *J. Struct. Geol.* **21**, 1255–1265.

Reddy S. M., Kelley S. P., and Wheeler J. (1996) A ^{40}Ar/^{39}Ar laser probe study of micas from the Sesia zone, Italian Alps: implications for metamorphic and deformational histories. *J. Metamorph. Geol.* **14**, 493–508.

Reddy S. M., PottsG J., and Kelley S. P. (2001) ^{40}Ar/^{39}Ar ages in deformed potassium feldspar: evidence of microstructural control on Ar isotope systematics. *Contrib. Mineral. Petrol.* **141**, 186–200.

Reiners P. W. and Farley K. A. (1999) Helium diffusion and (U–Th)/He thermochronometry of titanite. *Geochim. Cosmochim. Acta* **63**, 3845–3859.

Reiners P. W. and Farley K. A. (2001) Influence of crystal size on apatite (U–Th)/He thermochronology: an example from the Bighorn Mountains, Wyoming. *Earth Planet. Sci. Lett.* **188**(3–4), 413–420.

Reiners P. W., Brady R., Farley K. A., Fryxell J. E., Wernicke B., and Lux D. (2000) Helium and argon thermochronometry of the Gold Butte block, south Virgin Mountains, Nevada. *Earth Planet. Sci. Lett.* **178**(3–4), 315–326.

Renne P. R. (2000) Ar-40/Ar-39 age of plagioclase from Acapulco meteorite and the problem of systematic errors in cosmochronology. *Earth Planet. Sci. Lett.* **175**(1–2), 13–26.

Renne P. R., Becker T. A., and Swapp S. M. (1990) ^{40}Ar/^{39}Ar laser-probe dating of detrital micas from the Montgomery creek formation, northern California: clues to provenance, tectonics, and weathering processes. *Geology* **18**(6), 563–566.

Renne P. R., Swisher C. C., Deino A. L., Karner D. B., Owens T., and DePaolo D. J. (1998) Intercalibration of

standards, absolute ages and uncertainties in ^{40}Ar/^{39}Ar dating. *Chem. Geol.: Isotope Geosci.* **145**, 117–152.

Resor R. G., Chamberlain K. R., Frost C. D., Snoke A. W., and Frost B. R. (1996) Direct dating of deformation: U–Pb age of syndeformational sphene growth in the Proterozoic Laramie peak shear zone. *Geology,* 623–626.

Reuter A. and Dallmeyer R. D. (1989) K–Ar and ^{40}Ar/^{39}Ar dating of cleavage formed during very low-grade metamorphism: a review. In *Evolution of Metamorphic Belts,* Special Publication 43 (eds. J. S. Daly, R. A. Cliff, and B. W. D. Yardley). Geological Society of London, London, pp. 161–171.

Richter F. M., Lovera O. M., Harrison T. M., and Copeland P. C. (1991) Tibetan tectonics from ^{40}Ar/^{39}Ar analysis of a single K-feldspar sample. *Earth Planet. Sci. Lett.* **105**, 266–278.

Robbins G. A. (1972) Radiogenic argon diffusion in muscorite under hydrothermal conditions. M. S., Brown University.

Roddick J. C., Cliff R. A., and Rex D. C. (1980) The evolution of excess argon in Alpine biotites-A ^{40}Ar–^{39}Ar analysis. *Earth Planet. Sci. Lett.* **48**, 185–208.

Rubatto D. (2002) Zircon trace element geochemistry: partitioning with garnet and the link between U–Pb ages and metamorphism. *Chem. Geol.* **184**, 123–138.

Rubatto D., Williams I. S., and Buick I. S. (2001) Zircon and monazite response to prograde metamorphism in the Reynolds Range, central Australia. *Contrib. Mineral. Petrol.* **140**(4), 458–468.

Ruppel C. and Hodges K. V. (1994) Pressure–Temperature–Time paths from two-dimensional thermal models: prograde, retrograde, and inverted metamorphism. *Tectonics* **13**, 17–44.

Ruppel C., Royden L., and Hodges K. V. (1988) Thermal modeling of extensional tectonics: application to pressure–temperature-time histories of metamorphic rocks. *Tectonics* **7**, 947–957.

Sandiford M., McLaren S., and Neumann N. (2002) Long-term thermal consequences of the redistribution of heat-producing elements associated with large-scale granitic complexes. *J. Metamorph. Geol.* **20**, 87–98.

Schaltegger U., Fanning C. M., Günther D., Maurin J. C., Schulmann K., and Gebauer D. (1999) Growth, annealing and recrystallization of zircon and preservation of monazite in high-grade metamorphism: conventional and *in-situ* U–Pb isotope, cathodoluminesence and microchemical evidence. *Contrib. Mineral. Petrol.* **134**, 186–201.

Scherer E., Munker C., and Mezger K. (2001) Calibration of the lutetium–hafnium clock. *Science* **293**(5530), 683–687.

Scherer E. E., Cameron K. L., and Blichert-Toft J. (2000) Lu-Hf garnet geochronology: closure temperature relative to the Sm–Nd system and the effects of trace mineral inclusions. *Geochim. Cosmochim. Acta* **64**(19), 3413–3432.

Schmitz M. D. and Bowring S. A. (2001) U–Pb zircon and titanite systematics of the Fish Canyon Tuff: an assessment of high-precision U–Pb geochronology and its application to young volcanic rocks. *Geochim. Cosmochim. Acta* **65**(15), 2571–2587.

Schreiner G. D. L. (1958) Comparison of the Rb-87/Sr-87 age of the red granite of the Bushveld complex from measurements on the total rock and separated mineral fractions. *Proc. Roy. Soc. London Ser. A* **245**, 112–117.

Sherlock S. C. and Hetzel R. (2001) A laser-probe ^{40}Ar/^{39}Ar study of pseudotachylite from the Tambach fault zone, Kenya: direct isotopic dating of brittle faults. *J. Struct. Geol.* **23**, 33–44.

Simpson R. L., Parrish R. R., Searle M. P., and Waters D. J. (2000) Two episodes of monazite crystallization during metamorphism and crustal melting in the Everest region of the Nepalese Himalaya. *Geology* **28**, 403–406.

Spear F. S. (1993) *Metamorphic Phase Equilibria and Pressure–Temperature–Time Paths.* Mineralogical Society of America, Washington, DC.

Spotila J. A., Farley K. A., Yule J. D., and Reiners P. W. (2001) Near-field transpressive deformation along the San Andreas fault zone in southern California, based on exhumation constrained by (U–Th)/He dating. *J. Geophys. Res.-Solid Earth* **106**(B12), 30909–30922.

Steiger R. H. and Jäger E. (1977) Subcommission on geochronology: convention on the use of decay constants in geo- and cosmochronology. *Earth Planet. Sci. Lett.* **36**, 359–362.

Stern R. A. and Berman R. G. (2001) Monazite U–Pb and Th–Pb geochronology by ion microprobe, with an application to *in situ* dating of an Archean metasedimentary rock. *Chem. Geol.* **172**(1–2), 113–130.

Stock J. D. and Montgomery D. R. (1996) Estimating paleorelief from detrital mineral age ranges. *Basin Res.* **8**, 317–327.

Stockli D. F., Farley K. A., and Dumitru T. A. (2000) Calibration of the apatite (U–Th)/He thermochronometer on an exhumed fault block, White Mountains, California. *Geology* **28**(11), 983–986.

Stockli D. F., Linn J. K., Walker J. D., and Dumitru T. (2001) Miocene unroofing of the Canyon range during extension along the Sevier desert detachment, west central Utah. *Tectonics* **20**, 289–307.

Stuart F. M. (2002) The exhumation history of orogenic belts from ^{40}Ar/^{39}Ar ages of detrital micas. *Min. Mag.* **66**, 121–135.

Stüwe K. and Hintermüller M. (2000) Topography and isotherms revisited: the influence of laterally migrating drainage divides. *Earth Planet. Sci. Lett.* **184**, 287–303.

Stüwe K., White L., and Brown R. (1994) The influence of eroding topography on steady-state isotherms: application to fission track analysis. *Earth Planet. Sci Lett.* **124**, 63–74.

Sutter J. F. and Hartung J. B. (1984) Laser microprobe ^{40}Ar/^{39}Ar dating of mineral grains *in situ*. *Scan. Electr. Micros.* **4**, 1525–1529.

Suzuki K. and Adachi M. (1991) Precambrian provenance and Silurian metamorphism of the Tsubonosawa paragneiss in the South Kitakami terrane, northeast Japan, revealed by the chemical Th–U–total Pb isochron ages of monazite, zircon and xenotime. *Geochem. J.* **25**, 357–376.

Terry M. P., Robinson P., Hamilton M. A., and Jercinovic M. J. (2000) Monazite geochronology of UHP and HP metamorphism, deformation, and exhumation, Nordoyane, western Gneiss region, Norway. *Am. Mineral.* **85**(11–12), 1651–1664.

Thompson A. B. (1982) Dehydration melting of pelitic rocks and the generation of H_2O-undersaturated granitic liquids. *Am. J. Sci.* **282**, 1567–1595.

Thoni M. (2002) Sm–Nd isotope systematics in garnet from different lithologies (eastern Alps): age results, and an evaluation of potential problems for garnet Sm–Nd chronometry. *Chem. Geol.* **185**(3–4), 255–281.

Tippett J. M. and Kamp P. J. J. (1993) Fission track analysis of late Cenozoic vertical kinematics of continental Pacific crust, South Island, New Zealand. *J. Geophys. Res.* **98**, 16119–16148.

Tippett J. M. and Kamp P. J. J. (1995) Geomorphic evolution of the Southern Alps, New Zealand. *Earth Surf. Process. Landforms* **20**, 177–192.

Townsend K. J., Miller C. F., D'Andrea J. L., Ayers J. C., Harrison T. M., and Coath C. D. (2001) Low temperature replacement of monazite in the Ireteba granite, southern Nevada: geochronological implications. *Chem. Geol.* **172** (1–2), 95–112.

Tracy R. J., Robinson P., and Thompson A. B. (1976) Garnet composition and zoning in the determination of temperature and pressure of metamorphism, central Massachusetts. *Am. Mineral.* **61**, 762–775.

Turcotte D. L. and Schubert G. (1982) *Geodynamics: Applications of Continuum Physics to Geological Problems*. Wiley, New York.

van der Pluijm B. A., Hall C. M., Vrolijk P. J., Pevear D. R., and Covey M. C. (2001) The dating of shallow faults in the Earth's crust. *Nature* **412**, 172–175.

Vance D., Meier M., and Oberli F. (1998) The influence of high U–Th inclusions on the U–Th–Pb systematics of almandine-pyrope garnet: results of a combined bulk dissolution, stepwise-leaching, and SEM study. *Geochim. Cosmochim. Acta* **62**(21–22), 3527–3540.

Vavra G. and Schaltegger U. (1999) Post-granulite facies monazite growth and rejuvenation during Permian to Lower Jurassic thermal and fluid events in the Ivrea zone (southern Alps). *Contrib. Mineral. Petrol.* **134**, 405–414.

Vielzeuf D. and Holloway J. R. (1988) Experimental determination of the fluid-absent melting relations in the pelitic system. Consequences for crustal differentiation. *Contrib. Mineral. Petrol.* **98**, 257–276.

Villa I. M. (1994) Multipath Ar transport in K-feldspar deduced from isothermal heating experiments. *Earth Planet. Sci. Lett.* **122**, 393–401.

Viskupic K. and Hodges K. V. (2001) Monazite–xenotime thermochronometry: methodology and an example from the Nepalese Himalaya. *Contrib. Mineral. Petrol.* **141**, 233–247.

von Eynatten H., Gaup R., and Wijbrans J. R. (1996) ^{40}Ar/^{39}Ar laser-probe dating of detrital white micas from Cretaceous sedimentary rocks of eastern Alps: evidence for Variscan high-pressure metamorphism and implications for Alpine Orogeny. *Geology* **24**, 691–694.

Wagner G. A., Reimer G. M., and Jäger E. (1977) Cooling ages derived by apatite fission track, mica Rb–Sr, and K–Ar dating: the uplift and cooling history of the Central Alps. *Mem. Univ. Padova* **30**, 1–27.

Wartho J.-A., Kelley S. P., Brooker R. A., Carroll M. R., Villa I. M., and Lee M. R. (1999) Direct measurement of Ar diffusion profiles in a gem-quality Madagascar K-feldspar using the ultra-violet laser ablation microprobe (UVLAMP). *Earth Planet. Sci. Lett.* **170**, 141–153.

Watt G. R. and Thrane K. (2001) Early Neoproterozoic events in East Greenland. *Precamb. Res.* **110**(1–4), 165–184.

Weatherill G. W., Davis G. L., and Lee-Hu C. (1968) Rb–Sr measurements on whiole-rocks and separated minerals from the Baltimore Gneiss, Maryland. *Geol. Soc. Am. Bull.* **79**, 757–762.

Wells M. L., Snee L. W., and Blythe A. E. (2000) Dating of major normal fault systems using thermochronology: an example from the Raft river detachment, basin and range, western United States. *J. Geophys. Res.* **105**, 16303–16327.

Wernicke B. (1992) Cenozoic extensional tectonics of the US Cordillera. In *The Cordilleran Orogen: Conterminous United States* (eds. B. C. Burchfiel, P. W. Lipman, and M. L. Zoback). Geological Society of America, vol. G-3 Boulder. Co, pp. 553–581.

West D. P. and Lux D. R. (1993) Dating mylonitic deformation by the ^{40}Ar–^{39}Ar method: an example from the Norumbega fault zone, Maine. *Earth Planet. Sci. Lett.* **120**, 221–237.

White A. P. and Hodges K. V. (2002) Multistage extensional evolution of the central East Greenland Caledonides. *Tectonics* **21**, 101029/2001TC001308.

Willett S., Beaumont C., and Fullsack P. (1993) Mechanical model for the tectonics of doubly vergent compressional orogens. *Geology* **21**, 371–374.

Willett S. D. (1997) Inverse modeling of annealing of fission tracks in apatite 1: a controlled random search method. *Am. J. Sci.* **297**, 939–969.

Willett S. D. (1999) Orogeny and orography: the effects of erosion on the structure of mountain belts. *J. Geophys. Res.* **104**, 28957–28981.

Willett S. D. and Brandon M. T. (2002) On steady states in mountain belts. *Geology* **30**, 175–178.

Willett S. D., Slingerland R., and Hovius N. (2001) Uplift, shortening and steady state topography in active mountain belts. *Am. J. Sci.* **301**, 455–485.

Williams I. S. (1998) U–Th–Pb geochronology by ion microprobe. In *Applications of Microanalytical Techniques to Understanding Mineralizing Processes,* Rev. Econ. Geol. 7 (eds. M. A. McKibben and W. C. Shanks). Society of Economic Geologists, Iuscaloosa, AL, pp. 1–35.

Williams I. S., Buick I. S., and Cartwright I. (1996) An extended episode of early Mesoproterozoic metamorphic fluid flow in the Reynolds range, central Australia. *J. Metamorph. Geol.* **14**, 29–47.

Williams M. L., Jercinovic M. J., and Terry M. P. (1999) Age mapping and dating of monazite on the electron microprobe: deconvoluting multistage tectonic histories. *Geology* **27**, 1023–1026.

Willigers B. J. A., Krogstad E. J., and Wijbrans J. R. (2001) Comparison of thermochronometers in a slowly cooled granulite terrain: Nagssugtoqidian orogen, West Greenland. *J. Petrol.* **42**(9), 1729–1749.

Willigers B. J. A., Baker J. A., Krogstad E. J., and Peate D. W. (2002) Precise and accurate in situ Pb–Pb dating of apatite, monazite, and sphene by laser ablation multiple-collector ICP-MS. *Geochim. Cosmochim. Acta* **66**(6), 1051–1066.

Wolf R. A., Farley K. A., and Silver L. T. (1996) Helium diffusion and low temperature thermochronometry of apatite. *Geochim. Cosmochim. Acta* **60**, 4231–4240.

Wolf R. A., Farley K. A., and Kass D. M. (1998) Modeling of the temperature sensitivity of the apatite (U–Th)/He thermochronometer. *Chem. Geol.* **148**(1–2), 105–114.

Xu G. Q. and Kamp P. J. J. (2000) Tectonics and denudation adjacent to the Xianshuihe fault, eastern Tibetan plateau: constraints from fission track thermochronology. *J. Geophys. Res.-Solid Earth* **105**(B8), 19231–19251.

York D., Hall C. M., Yanase Y., Hanes J. A., and Kenyon W. J. (1981) ^{40}Ar/^{39}Ar dating of terrestrial minerals with a continuous laser. *Geophys. Res. Lett.* **8**, 1136–1138.

Zeck H. P. and Whitehouse M. J. (2002) Repeated age resetting in zircons from Hercynian-Alpine polymetamorphic schists (Betic-Rif tectonic belt S, Spain)—a U–Th–Pb ion microprobe study. *Chem. Geol.* **182**(2–4), 275–292.

Zeitler P. K. (1987) Argon diffusion in partially outgassed alkali feldspars: insights from ^{40}Ar/^{39}Ar analysis. *Isotope Geosci.* **65**, 167–181.

Zeitler P. K., Koons P. O., Bishop M. P., Chamberlain C. P., Craw D., Edwards M. A., Hamidullah S., Jan M. Q., Khan M. A., Khattak M. U. K., Kidd W. S. F., Mackie R. L., Meltzer A. S., Park S. K., Pêcher A., Poage M. A., Sarker G., Schneider D. A., Seeber L., and Shroder J. F. (2001) Crustal reworking at Nanga Parbat, Pakistan: metamorphic consequences of thermal-mechanical coupling facilitated by erosion. *Tectonics* **20**(5), 712–728.

Zhu X. K., O'Nions R. K., Belshaw N. S., and Gibb A. J. (1997a) Lewisian crustal history from *in situ* SIMS mineral chronometry and related metamorphic textures. *Chem. Geol.* **136**, 205–218.

Zhu X. K., O'Nions R. K., Belshaw N. S., and Gibb A. J. (1997b) Significance of *in situ* SIMS chronometry of zoned monazite from the Lewisian granulites, NW Scotland. *Chem. Geol.* **135**, 35–53.

Radioactive Geochronometry
ISBN: 978-0-08-096708-0

7

Ages and Growth of the Continental Crust from Radiogenic Isotopes

P. J. Patchett

University of Arizona, Tucson, AZ, USA

and

S. D. Samson

Syracuse University, NY, USA

7.1 SCOPE OF AVAILABLE METHODS AND DATA

The development and application of radiogenic isotopes to dating of geologic events, and to questions of growth, evolution, and recycling processes in the continental crust are mature areas of scientific inquiry. By this we understand that many of the approaches used to date rocks and constrain the evolution of the continents are well established, even routine, and that the scope of data available on age and evolution of continents is very large. This is not to say that new approaches have not been developed in recent years, or that new approaches and/or insights cannot be developed in the future. However, the science of continental crustal evolution is definitely a domain where many of the problems are well defined, the power of the techniques used to solve them are well known, and the limitations of field and laboratory databases, as well as the preserved geologic record, are understood.

From the very early days of crustal evolution studies, it was innovations and improvements in laboratory techniques that drove the pace of discovery (e.g., Holmes, 1911; Nier, 1939). This remained true through all the increments in capability reviewed in this chapter, up to the present day. Thus, continental crustal evolution is an area of Earth science where a species of very laboratory-oriented investigator, the "radiogenic isotope geologist" or "geochronologist," has made major advances, even breakthroughs, in understanding. This is true in spite of the fact that many of the individuals of the species may have lacked field expertise, or even more than a primitive level of geologic background. Because design and building of instruments like radiation detectors or mass spectrometers requires a knowledge of physics, many of the early practitioners of rock dating were physicists, like Alfred Nier (cited above). Since the 1970s, essentially all mass spectrometers have been constructed by specialized commercial firms, and the level of physics expertise among isotope geologists has been lower. These firms, based mainly (but not exclusively) in and around Manchester, England, and Bremen, Germany, have spearheaded technical innovations in mass spectrometry. Isotope geology researchers are one group of consumers for this technology, along with chemists, nuclear-weapon laboratories and nuclear-power generating facilities. Today, the vast majority of isotope geology researchers are derived from geological backgrounds.

This chapter will briefly review historical aspects of the development of radiogenic isotope geology as applied to continents. Some details, references and cross-references to other chapters in this volume will be provided for most major radiogenic isotopic methods, and for applications of these. However, this chapter will ultimately concentrate on two major approaches that dominate the research field today: (i) crustal tectonic and magmatic ages from U–Pb dating of accessory minerals like zircon; and (ii) crustal differentiation and growth from neodymium isotopic determinations on total rocks.

The sheer amount of data available from continental areas, and the pace of data acquisition today, places any geographically constrained compilation in the impossible category for a chapter like this one. In this chapter, the state of the art in both geochronology and crustal origins will be illustrated by selected examples, not by global compilations or comprehensive discussions of each region that has been studied. Nevertheless, some remarks about the availability of data from different parts of the world need to be made. As with other areas of geology, biology, and botany, the parts of the world that have been longest settled by western civilization have the best data coverage for both ages of continental rocks, and their origins and evolution. Thus, Western Europe, Canada, the USA and Australia have generally somewhat thorough coverage. More limited data have become available from Eastern Europe, Greenland, Central and South America, Africa, Asia, and Antarctica. Some of these data have been produced by groups based in those regions, but much of the data published between 1970 and today have been driven by studies based in, and funded from, western countries. Availability of state-of-the-art results for ages of crust and its evolution are low in Antarctica and Greenland, where climate and ice cover limit work, and in South America, Africa, and parts of Asia, where studies have been sporadic, and are certainly limited in some cases by political instability. This general situation is now changing, however. Two parts of the world where results for ages of orogenic belts and for crustal evolution in general are accumulating more rapidly since about 1990 are China and Russia; this is connected with growth of modern isotope geology facilities in those countries. Understanding of global continent growth and evolution is limited in critical respects by the large regions of the world that are poorly dated. Therefore, gradual improvement of the state of knowledge of less accessible parts of the continents will bring significant benefits, even though the conceptual issues may be understood to a large extent.

7.2 DETERMINATION OF AGES OF IGNEOUS EVENTS

7.2.1 Early Developments in U–Th–Pb Geochronology

Of all the different geochronological decay schemes that have been employed in crustal studies the U–Pb system is unique in that there are two radioactive parent isotopes (^{238}U and ^{235}U) that decay to two different daughter isotopes (^{206}Pb and ^{207}Pb). Because of this unique dual decay scheme the U–Pb system has long been of considerable interest as a geochronological technique. However, because of the extreme mobility of uranium, most U–Pb dates for whole-rock samples were shown to be unreliable and the focus went to analyzing uranium-bearing accessory minerals that were far less susceptible to uranium loss during weathering. Zircon, because of is extremely low initial lead content, its common occurrence in igneous rocks and its extreme chemical and physical resistance were a particular target for geochronological investigation, with the first analyses performed in the 1950s (Vinogradov *et al.*, 1952; Tilton *et al.*, 1955). Because zircon has such an extremely refractory nature, it is one of the most difficult minerals to dissolve; thus early studies used a borax fusion technique to ensure digestion. However, the very high lead content of the borax (200–1,000 ng g^{-1} according to Krogh, 1973) minimized the utility of the technique. In 1971 it was demonstrated that zircon could be completely dissolved in HF in high pressure Teflon$^{®}$-lined vessels (Krogh, 1971a,b) with a reduction in lead blank by three orders of magnitude (Krogh, 1973). That analytical breakthrough, combined with the demonstration that the effects of lead loss in zircons could be minimized by mechanically abrading zircons (Krogh, 1982), resulted in the exponential rise of the technique in a wide array of geological studies. U–Pb dating of accessory minerals, particularly zircon and baddeleyite, is now considered *the* most reliable method for determining the crystallization ages of plutonic rocks, even those that have suffered multiple episodes of metamorphism. U–Th–Pb dating of monazite and U–Pb dating of titanite have become extremely important methods for constraining the timing of metamorphism. *In situ* methods of measuring U–Th–Pb isotopes, combined with imaging techniques such as cathodoluminescence, have made it possible to analyze metamorphic overgrowths on accessory minerals dramatically increasing our ability to precisely date metamorphic events. Overall, the U–Th–Pb technique now dominates the field of geochronology as applied to crustal evolution. Since the 1980s several different methodologies for determining U–Th–Pb dates have been developed and these are briefly reviewed below.

7.2.2 U–Th–Pb Dating by TIMS—The Isotope Dilution Method

The most common method of determining the U–Pb date of an accessory mineral is the determination of uranium and lead isotopic abundances via isotope dilution and thermal ionization mass spectrometry (ID-TIMS). Most measurements in the 1970s and 1980s required that an aliquot of the dissolved mineral solution be made, with one portion being "spiked" with an enriched ^{235}U and ^{208}Pb tracer solution, and the other aliquot directly measured for lead isotopic abundance. Following the first major production of ^{205}Pb from ^{205}Bi, produced by proton bombardment of enriched ^{206}Pb (Krogh and Davis, 1975), a second, larger production and worldwide distribution of high-purity ^{205}Pb occurred (Parrish and Krogh, 1987). Use of ^{205}Pb eliminated the need for aliquoting dissolved samples, and virtually all modern laboratories have since adopted the use of a ^{205}Pb spike. This is of particular use to U–Th–Pb dating of monazite, as it often contains such a large abundance of ^{208}Pb* (the asterisk denotes radiogenic lead) that a spike other than one enriched in ^{208}Pb is needed.

While the general procedures for mineral dissolution, separation of uranium and lead, and mass spectrometry have not changed substantially since 1973, there has been a constant drive to reduce the laboratory contamination level of lead, combined with improvements in detection of very small Pb^{+} ion beams, to allow for an ever-decreasing amount of sample required for high-precision analysis. It is now common for many U–Pb laboratories to report procedural lead blanks of just a few picograms, and some labs have reduced lead blanks to the subpicogram level (e.g., Samson and D'Lemos, 1999; Corfu and Easton, 2000; Ayer *et al.*, 2002; Samson *et al.*, 2003). This reduction in blank level, combined with the development of an emitter solution that produces much stronger and more stable Pb^{+} beams than conventional silica gel (Gerstenberger and Haase, 1997), allows for a relatively precise isotopic measurement of as little as 15–20 pg of radiogenic lead. This amount is equivalent to the typical lead content of very young (<10 Ma) or very small (<50 μm) single zircon crystals, or of carefully extracted portions of single crystals.

7.2.3 Zircon Evaporation Method

A variation of the method for determining ^{207}Pb–^{206}Pb dates of whole single zircons by TIMS analysis that eliminates the need for zircon dissolution and chemical separation of lead was introduced by Kober (1986), with a slight, yet important, modification described a year later (Kober, 1987). In the modified method, usually referred to as the zircon evaporation method, a single zircon crystal is placed into a folded side rhenium filament (the evaporation filament), which is positioned opposite of a blank rhenium ionization filament. The evaporation filament is heated for a short time to evaporate lead onto the target ionization filament. The current to the evaporation filament is then turned off and the ionization filament is heated until Pb^+ ionization begins and lead isotopic ratios are measured in the normal fashion. Current to the ionization filament is then turned off, the evaporation filament heated to a slightly higher temperature than previously, and the process continually repeated until either an adequate number of ratios have been collected or the lead in the zircon is exhausted. The main advantages of the technique are that ultra-clean chemistry is not required, the measured $^{206}Pb/^{204}Pb$ ratio is usually higher than ID-TIMS measurements of single zircons (as minimal lead blank is introduced), and no time is spent waiting for zircons to dissolve or in performing chemical separations. The main disadvantage is that only the age based on $^{207}Pb-^{206}Pb$ can be obtained, with no U–Pb information, and thus the degree of discordance of the analysis cannot be determined. For zircons that experienced only modern-day lead loss, this is not a problem, but for zircons that have suffered non-modern lead loss, the evaporation dates would be inaccurate. Critical to the technique, therefore, is that identical dates are determined from each heating step and that reproducible results are obtained from more than one zircon (Söderlund, 1996). Although some laboratories have embraced the technique, most modern TIMS laboratories involved in U—Pb geochronological studies still use the ID-TIMS method, with the goal of continuing to try to lower lead blanks to allow ever smaller samples to be analyzed with high precision.

7.2.4 U–Th–Pb Dating by Ion Microprobe

Although tremendous advances were made in geochronology with the advent of U–Pb dating using chemical separation procedures and TIMS analyses, there was an obvious need for a method of *in situ* isotopic analysis of accessory minerals to more fully exploit the age information preserved in complex crystals (i.e., ones containing metamorphic overgrowths and/or xenocrystic regions, etc.) The first response to this need was the measurement of $^{207}Pb-^{206}Pb$ dates using a secondary ion microprobe mass analyzer (Anderson and Hinthorne, 1973). Following this work, a very high mass resolution secondary ion mass spectrometer (SIMS) was developed in Australia in the late 1970s (Clement *et al.*, 1977). This instrument, coined the SHRIMP (for sensitive high resolution ion microprobe), is particularly suited for determining the age domains within complex crystals, as very small regions of a mineral can be analyzed (e.g., Stern, 1997 reports that minimal "spot sizes" with surface dimensions of 4×6 µm and depths of <1 µm can be obtained). The first age determinations using the SHRIMP were published in 1982 (Compston *et al.*, 1982), and only a year later the instrument had been used to

determine the age of the first pre-4.0 Ga terrestrial mineral (Froude *et al.*, 1983).

High-resolution ion microprobes (IMP) are used for *in situ* U–Th–Pb dating by bombarding the surface of a highly polished crystal, usually ground to half its original thickness, with a primary beam of O_2^- ions. These bombarding ions, produced by oxygen gas discharge within a hollow nickel cathode, sputter a small portion of the zircon producing a wide variety of secondary ions. The secondary ions are doubly focused, first through an electrostatic analyzer, which filters them by their values of kinetic energy, then these ions are focused a second time by entering a magnetic sector, which discriminates between the ions based solely on their mass. The magnetic sector is similar to that of thermal ionization mass spectrometers, but has a much larger radius and is capable of operating at a much high mass resolution ($>5,000$ compared to ~300) to allow discrimination of the uranium and lead isotopes from ions with nominally similar masses produced during the sputtering process (e.g., $^{96}Zr^{94}Zr^{16}O^+$ ions versus $^{206}Pb^+$ ions for zircon analyses). The secondary ion currents generated are very low and thus are measured using a single secondary electron multiplier. $^{207}Pb-^{206}Pb$ dates can be directly determined for each elliptical region analyzed by directly comparing the $^{207}Pb^+$ and $^{206}Pb^+$ ion beam currents, after correcting for common lead components (usually by measuring the $^{204}Pb^+$ peak). Determining $^{206}Pb^*/^{238}U$ and $^{208}Pb^*/^{232}Th$ dates (Pb^* denotes radiogenic lead) cannot be accomplished by a direct comparison of $^{238}U^+$, $^{232}Th^+$, and $^{206}Pb^+$ ion currents. This is because the measured $^{206}Pb^+/^{238}U^+$ and $^{208}Pb^+/^{232}Th^+$ ratios are not the same as the true ratios but vary between 2–5 times greater than the actual ratios (see Stern (1997) for a discussion of this phenomenon). This variation can be corrected because the ion beams display a systematic relationship between $^{206}Pb^+/^{238}U^+$ and UO^+/U^+, and between $^{208}Pb^+/^{232}Th^+$ and ThO^+/Th^+ (Hinthorne *et al.*, 1979). Thus, by including a mineral of known age and of uniform Pb/U (or Th/Pb) in the target mount with the unknowns, a standardization curve can be established based on the measured Pb/U and UO/U (or Pb/Th and ThO/Th) of the standard, so that a normalization factor can be applied to the measured Pb/U,Th ratios of the minerals of unknown ages. The need for well-calibrated external mineral standards to determine ion microprobe U–Th–Pb dates is thus directly analogous to the determination of $^{40}Ar/^{39}Ar$ dates.

The uncertainty in the Pb/U–UO/U calibration curve must be propagated along with all of the other analytical uncertainties in the estimation of realistic errors of the calculated dates, rather than quoting errors based only on counting statistics, for example. For minerals containing low Pb^*

contents, either because they have low U–Th contents or because they are geologically young, counting statistics may have the dominant influence in error propagation; for minerals containing high Pb[*] content uncertainties in the calibration curve may have a larger influence on the precision of the U–Pb dates. Although a variety of factors may control the ion microprobe precision on any given zircon, typical levels of precision that can be obtained can be estimated. For most zircons the 2σ uncertainty in a *single* $^{238}U/^{206}Pb^*$ date is typically not more precise than 1.5–2%. This level of precision is $\sim$10–20 times larger than that obtainable via ID-TIMS methods (see Figure 1). For zircons younger than 1 Ga $^{207}Pb–^{206}Pb$ dates are typically not more precise than $\sim$5%, and typical values for zircon <500 Ma are more likely to be in the range of 10–20%. However, for zircons $\geq$1.5 Ga the relative precision of a $^{207}Pb–^{206}Pb$ date is considerably higher, with values of $\sim$0.5% obtainable.

7.2.5 U–Th–Pb Dating by ICP-MS

In the early 1980s quadrupole mass spectrometers using argon plasma as the ionization source, i.e., inductively coupled plasma mass spectrometers (ICP-MS), were developed. Although these instruments were designed primarily for measuring the concentrations of trace elements, many studies have employed them for U–Pb dating. $^{206}Pb^*/^{238}U$, $^{207}Pb^*/^{235}U$, and $^{208}Pb^*/^{232}Th$ dates can be determined, in addition to the more simply determined $^{207}Pb–^{206}Pb$ dates, using laser-ablation ICP-MS techniques. The obvious

appeal of U–Th–Pb dating by laser ablation ICP-MS is the elimination of the need for ultra-low blank dissolution and U–Th–Pb separation procedures, the speed of the analysis (<10 min per analysis), and the possibility of *in situ* analysis. However, there are several analytical obstacles to obtaining accurate U–Th–Pb dates via laser ablation that must be overcome. One of the most difficult problems is that there is significant elemental (i.e., U/Pb and Th/Pb) fractionation that occurs during laser ablation. That is, measured Pb/U ratios are lower than actual ratios by tens of percent, and this effect is variable with ablation time (see figure 4 in Horn *et al.*, 2000). A second effect, common to all mass spectrometric measurements, is that there is an instrumental mass bias, or discrimination. This bias is several times higher than the bias that occurs during TIMS measurements and thus would be a significant source of error if not corrected. A third potential difficulty is determining the amount of common lead in an analysis, as the argon gas used in ICP-MS contains enough mercury to cause isobaric interference of ^{204}Pb from ^{204}Hg. However, continued improvements are being made as this technique evolves (see below) and it may begin to approach ID-TIMS analysis in the future, at least for accessory minerals with relatively high radiogenic lead contents.

Early attempts to directly date zircon crystals using ICP-MS techniques involved the use of Nd–YAG lasers, operating at a 1,064 nm wavelength, to ablate the zircon crystals (Feng *et al.*, 1993; Fryer *et al.*, 1993). Because of significant variations of U/Pb isotopic ratios, these early laser-ablation studies concentrated on determining $^{207}Pb–^{206}Pb$ dates, which yielded precision between 0.5–6.0% (e.g., Fryer *et al.*, 1993; Jackson *et al.*, 1996). However, elemental fractionation during laser ablation decreases with decreasing wavelength (Geersten *et al.*, 1994), and thus by quadrupling the frequency of Nd–YAG lasers (266 nm wavelength), or using gas-based lasers operating in the deep UV (such as Ar–F excimer lasers which produce light at 193 nm (e.g., Eggins *et al.*, 1998)), more reproducible U/Pb ratios could be measured compared to the earlier analyses using larger wavelengths. To counterbalance the effects of laser-induced elemental fractionation, as well as minimize temporal variations, many workers analyze externally calibrated standard minerals and unknowns under identical operating conditions, and then apply correction factors to the unknowns (e.g., Fernández-Suárez *et al.*, 1999; Knudsen *et al.*, 2001). In this respect, the technique shares strong similarities with ion microprobe U–Th–Pb age determinations. A major difference between the techniques, however, is that the volume of the pit excavated by a laser is much larger than the spot produced by the ion microprobe (Figure 2), and thus laser ablation must be considered a destructive technique.

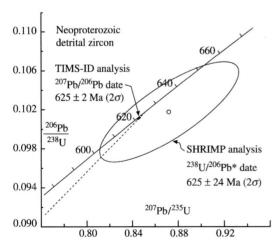

Figure 1 U–Pb Concordia diagram showing the results of an analysis of a detrital zircon crystal by ion microprobe (SHRIMP) followed by analysis of the same crystal using TIMS. Both error ellipses are plotted at 2σ. The best estimate of the age of crystallization of the zircon is identical for both techniques; however, the TIMS analysis is an order of magnitude more precise than that obtained using the ion microprobe (source Samson *et al.*, 2003).

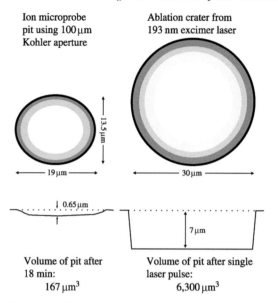

Ion microprobe pit using 100 μm Kohler aperture

Ablation crater from 193 nm excimer laser

Volume of pit after 18 min: 167 μm³

Volume of pit after single laser pulse: 6,300 μm³

Figure 2 Size of a typical pit produced in an accessory mineral using an ion microprobe during an 18 min analytical run (from Stern, 1997) compared to the size of an ablation crater made from a single pulse of an excimer laser (from Horn *et al.*, 2000). Bottom drawings show generalized cross-sections of the spots made from the two techniques. Note the considerably smaller volume of mineral excavated during the ion microprobe analysis.

Precision of $^{206}Pb/^{238}U$ dates using external-standard correction methods with frequency quadrupled lasers is partly dependent on sufficiently high uranium and lead concentrations, but typical values appear to be in the range of several percent. Even higher reproducibility of U/Pb ratios (~2% for $^{206}Pb/^{238}U$ of a Paleozoic zircon standard) from ablated zircons was reported by Horn *et al.* (2000). These workers used an excimer laser (193 nm wavelength), which presumably results in less laser-induced elemental fractionation than studies employing 266 nm wavelength lasers. In addition, Horn *et al.* (2000) combined the laser ablation with solution nebulization of known quantities of Tl and enriched ^{235}U. Comparing the measured $^{205}Tl/^{235}U$ with the known ratio allowed for a correction factor for instrumental mass bias to be used on the measured $^{206}Pb/^{238}U$ ratio. By comparing the measured $^{203}Tl/^{205}Tl$ ratio of the introduced Tl with its natural ratio, a correction factor was applied to the measured lead isotopic ratios to further increase accuracy, an approach first described by Longerich *et al.* (1987). Elemental fractionation produced at the laser-ablation site is still significant, however, and a correction must be applied to obtain accurate data. Horn *et al.* (2000) demonstrated that such fractionation is a function of pit geometry, thus allowing corrections to be made by establishing an empirical correction curve using a standard zircon.

In the 1990s, mass spectrometers with a magnetic sector and a full array of Faraday detectors were coupled with a plasma source. These multi-collector instruments (MC–ICP–MS) produce the same flat-topped peaks produced by TIMS and thus are capable of higher-precision isotope ratio measurements than quadrupole-type instruments. There is considerable interest in using these new-generation instruments for U–Th–Pb dating by directly analyzing zircon and monazite crystals via laser ablation, largely following the techniques originally developed for quadrupole instruments. However, because MC–ICP–MS is such a new technique, few U–Pb geochronological studies have so far been published discussing the results obtainable using these multicollector instruments (Machado and Simonetti, 2001). Based on the limited data currently available, it appears that typical 2σ uncertainties for U–Pb dates on zircon using multicollector instruments (e.g., 1–2%) are better, but not yet substantially so, than those obtainable using quadrupole instruments. However, use of multicollector ICP–MS for U–Pb dating is in its infancy and thus the technique may hold considerable potential as a geochronological tool.

7.2.6 U–Th–Pb Dating of Monazite Using Only Uranium, Thorium, and Lead Concentrations

The mineral monazite, a uranium- and thorium-rich phosphate of a rare-earth element (REE), is a common accessory mineral in a variety of felsic igneous rocks and is a common trace constituent in many metamorphic rocks, particularly metapelites. Because of its high uranium and thorium content, and fairly low common lead content, it has often been dated using ID-TIMS and IMP techniques. In the early 1990s the potential of determining the age of monazite using an electron microprobe (EMP) was investigated (Suzuki and Adachi, 1991; Montel *et al.*, 1994; Suzuki and Adachi, 1994). The so-called "chemical age" of a monazite crystal can be determined solely by measuring its thorium, uranium, and lead contents (no isotopic measurement) if the amount of common lead is negligible and no post-crystallization loss or gain of uranium, thorium and lead occurred (see review in Montel *et al.*, 1996). If these conditions are met, then within ~100–200 million years the amount of accumulated Pb^* in a typical monazite is high enough that it can be measured accurately and an age calculated. Precision of the chemical age is largely governed by lead content (as is true with all *in situ* techniques) and thus is largely age dependent, but 2σ precisions of 10–20 Myr are now obtainable for very lead-rich (>2,000 ppm) crystals, with errors around double those values for monazite with lower lead contents (Williams and Jercinovic, 2002). The main advantage of the

EMP technique is the excellent spatial resolution that can be obtained; *in situ* analysis of monazites as small as 5 μm can be obtained from polished thin sections (Montel *et al.*, 1996). This feature is of limited utility in dating relatively large (radius = 30 μm) monazites with simple histories, but is invaluable for studies where *in situ* analysis is critical, such as determining the age of monazites that occur as inclusions in porphyroblasts (e.g., Williams *et al.*, 1999; Montel *et al.*, 2000). Employing such high spatial resolution must be used with considerable caution, however, as accurate ages can only be obtained if no lead diffusion has occurred within the restricted region of the crystal being analyzed.

The current emphases of the technique are to constrain the timing of multiple metamorphic events, to determine directly the timing of deformational events, and to provide links between metamorphism and deformation (see review by Williams and Jercinovic, 2002). The main limitation of chemical dating of monazite by EMP, aside from the necessity of perfect closed-system behavior of the region of the mineral being analyzed, is the detection limit of lead, typically a few hundred ppm for most instruments. This limitation usually precludes the analysis of monazites younger than $\sim$100–200 Ma. The feasibility of determining chemical ages of young monazites containing only a few tens of ppm lead was demonstrated by Cheburkin *et al.* (1997) by using a newly designed X-ray fluorescence microprobe. Improvements on the original instrument design allow for chemical dating of monazite as young as 15 Ma and as small as 50 μm (Engi *et al.*, 2002); however, the monazites could not be measured *in situ*. In a companion study, Scherrer *et al.* (2002) determined chemical ages of monazites that were first optically examined in thin section, thus still preserving full textural context of the analysis, but were then removed from the thin sections by drilling with a diamond microdrill. Although this is a labor-intensive procedure compared to EMP dating, and still cannot be done on very small monazites, an X-ray microprobe age of 55.3 ± 2 Ma was determined for a 54 ± 1 Ma monazite (^{208}Pb/^{232}Th TIMS date), demonstrating the much higher precision of the X-ray microprobe technique compared to the EMP technique.

7.3 DETERMINATION OF AGES OF METAMORPHISM

From the very beginning of the pioneering days of geochronology, it was noted that different minerals from a single rock had different apparent ages, suggesting that different minerals retained different proportions of radiogenic daughter nuclides (e.g., Wetherill *et al.*, 1955), thus setting the stage for future thermochronologic studies. By 1959, the use of radioactive decay schemes to estimate the timing of metamorphism had been specifically discussed (Compston and Jeffery, 1959; Tilton *et al.*, 1959). Subsequently, a large number of papers were published involving Rb–Sr and K–Ar dating of different minerals within metamorphic rocks, establishing the beginning of a database of the history of continental crustal deformation. With the introduction of the ^{40}Ar/^{39}Ar technique (Merrihue and Turner, 1966) and the seminal discussion of the concept of mineral closure temperature (Dodson, 1973) the field of thermochronology became firmly established and its impact on our understanding of the tectonic evolution of orogenic belts has been profound. Because Chapter 6 is devoted to the discussion of crustal metamorphism, only a brief review of current thermochronologic techniques and their applications to continental crustal evolution is given here. Because the U–Th–Pb system is discussed in detail above, a separate section on its application specifically to constraining the timing of metamorphism is beyond the scope of this chapter. Reviews of U–Pb geochronology applied to metamorphic studies can be found in Heaman and Parrish (1991) and Mezger and Krogstad (1997). Recent discussions of the formation of metamorphic zircon domains and interpretation of geochronologic data can be found in Fraser *et al.* (1997) and Bingen *et al.* (2001), and reference therein.

7.3.1 ^{40}Ar/^{39}Ar Thermochronology

^{40}Ar/^{39}Ar dating is a variation of conventional K–Ar technique in that potassium-bearing samples arc irradiated with neutrons to produce ^{39}Ar from ^{39}K, thereby eliminating the need for separate measurements of potassium and argon on two separate aliquots of a sample (see McDougall and Harrison, 1999 for a detailed review). In the 1970s most studies of metamorphosed crustal regions utilized the technique in a similar fashion to previous K–Ar studies, i.e., determining ^{40}Ar/^{39}Ar dates of different metamorphic minerals and inferring the cooling history based on estimates of the argon closure temperature in the mineral (e.g., Lanphere and Albee, 1974; Dallmeyer, 1975; Dallmeyer *et al.*, 1975, and many others). In these studies milligram-sized mineral separates were step-heated in a furnace and the argon gas released during each temperature step was isotopically analyzed. Lasers were used to heat smaller samples, first on lunar rocks (e.g., Megrue, 1973), and subsequently on terrestrial samples (e.g., York *et al.*, 1981; Maluski and Schaeffer, 1982; Sutter and Hartung, 1984), although most of these early studies produced only total-fusion ^{40}Ar/^{39}Ar dates. Layer *et al.* (1987) demonstrated that detailed age spectra could be determined from single horn-blende and biotite crystals using a defocused continuous laser beam.

The advantage of laser microprobe $^{40}Ar/^{39}Ar$ analyses is the ability of *in situ* analysis. Spatial resolution of 50–100 µm can be achieved using lasers in the visible and near-infrared wavelengths, although these are best for minerals that are strong absorbers of such wavelengths such as biotite, phlogopite, and hornblende (see review by Kelley, 1995). With the employment of UV lasers in $^{40}Ar/^{39}Ar$ thermochronology spatial resolution increased considerably (~10 µm width) as well the ability to analyze most silicate minerals, including white mica and feldspar that are poor absorbers of higher-wavelength energy (Kelley *et al.*, 1994). Important applications of this technique to crustal metamorphic studies include the direct dating of deformation fabrics (e.g., Reddy *et al.*, 1996), dating different portions of single *P–T* paths (e.g., DiVincenzo *et al.*, 2001) and dating of mineral inclusions in porphyroblasts (Kelley *et al.*, 1997). Further discussions of the modern applications of $^{40}Ar/^{39}Ar$ dating to constrain the timing of metamorphic events are given by Hodges in Chapter 6.

7.3.2   Rb–Sr Dating

It was recognized early on that the Rb–Sr system was particularly useful for constraining the timing of metamorphic events because of the significant degree of rubidium and strontium diffusion that occurs between minerals during metamorphism. By constructing Rb–Sr mineral isochrons the timing of diffusion (i.e., metamorphism) can be determined, assuming complete isotopic re-equilibration occurred during a discrete metamorphic event and the system remained closed to any further disturbance (e.g., Fairbairn *et al.*, 1961). Important early work in contact metamorphic zones constrained the behavior of Rb–Sr in mineral systems (Hart, 1964; Hanson and Gast, 1967). The Rb–Sr method continues to play a very important role in studies of the deformational and metamorphic history of crustal regions, but the focus has shifted towards determining the ages of minerals within a well-defined textural context to better interpret the significance of the constructed isochrons. Some recent examples are given below.

Because of its common occurrence in metamorphic rocks, garnet separates have been an obvious choice of one of the components to be incorporated in mineral Rb–Sr isochrons in metamorphic studies (see above). Building on that previous work, Christensen *et al.* (1989) sliced large (3 cm) single garnets into separate pieces for Rb–Sr isotopic analysis along with the rock matrix between the garnets. The outer portions of the garnets were sufficiently higher in $^{87}Sr/^{86}Sr$ compared to central rim portions that growth rates of the garnets (1.0–1.7 mm Myr^{-1}) and duration of total garnet growth (9–13 Myr^{-1}) could be determined. Vance and O'Nions (1990) followed a similar procedure measuring both Rb–Sr and Sm–Nd isotopic parameters on single garnet sections obtained by sawing garnet crystals into inner, middle, and outer portions. The ~448 Ma isochrons that were obtained presumably established the timing of prograde growth of the garnets from that region of Newfoundland.

In a different approach to determining the timing of prograde metamorphism, Burton and O'Nions (1991) analyzed the isotopic composition of small (1 mm) garnets from interlayered metasedimentary rocks that experienced a common *P–T–t* history, but one in which garnets formed at two different *P–T* conditions, as a function of different H_2O activities. Based on the determined Rb–Sr isochrons the lower *P–T* garnets formed at 437.3 ± 11.4 Ma and the higher *P–T* garnets at 423.5 ± 4.7 Ma, in excellent agreement with the determined Sm–Nd isochrons (434.1 ± 1.2 Ma and 424.6 ± 1.2 Ma, respectively). These ages, when combined with the paleothermometric and barometric data, provided a significant amount of information on the rate of metamorphic processes in this region of Norway. These types of studies in similar regions of crustal thickening should provide an advance in our understanding of the lithospheric thermal response during collisional tectonics.

With the demonstration that extremely small amounts of meteoritic material could be dated via Rb–Sr "microchrons" (Papanastassiou and Wasserburg, 1981) the path was opened for microanalysis of terrestrial samples, allowing much more control over the Rb–Sr dating of metamorphic material than previously possible. By analyzing small quantities of white mica from metamorphosed rocks the Rb–Sr system is capable of providing ages of formation, rather than cooling ages, thus establishing the time of at least one specific event in the metamorphic history of an area (e.g., Cliff, 1994; Chen *et al.*, 1996). These types of studies have been further refined and the more recent microchron methods employ a micro-drill and petrographic microscope to allow very specific areas of a geological thick section (~50 µm) to be sampled with complete textural control. Müller *et al.* (2000a) demonstrated the power of this technique by determining Rb–Sr dates of white mica from mylonites, which developed under greenschist facies metamorphic conditions, from shear zones in the eastern Alps.

In an even more novel approach, the timing of the duration of shearing was established by determining Rb–Sr dates of micromilled samples of crystal fibers that developed in the strain fringe around pyrite grains from a fault zone in the northern Pyrenees (Müller *et al.*, 2000b). These types of studies are just at their beginning stages but appear to be poised to revolutionize our ability to determine the timing, and possibly duration, of different deformational events, a critical step to more fully understanding all aspects of crustal evolutionary processes.

7.3.3 Sm–Nd and Lu–Hf Dating

The main target of Sm–Nd and Lu–Hf dating in metamorphic studies continues to be the mineral garnet. Garnet is a major constituent in many metamorphic rocks; it preferentially incorporates heavy rare earth elements, and hence can have very high $^{147}Sm/^{144}Nd$ (e.g., Stosch and Lugmair, 1987) and $^{176}Lu/^{177}Hf$ ratios (e.g., Duchêne *et al.*, 1997), and it has been widely used in thermobarometric studies, thus potentially providing a direct link between time and *P–T* conditions. In addition, garnet has a relatively high closure temperature for both the Sm–Nd (Mezger *et al.*, 1992; Ganguly *et al.*, 1998) and Lu–Hf systems (Scherer *et al.*, 2000), thus increasing its attraction as a useful mineral in determining the timing of metamorphic events. Early studies were geared towards determining the timing of garnet growth during prograde metamorphism and relied on analyzing very large garnets (Vance and O'Nions, 1990; Mezger *et al.*, 1992; Getty *et al.*, 1993). Caution must be used in interpretation of these types of analyses, however, as it may be inclusions of REE-rich accessory minerals, and not the garnet itself, that dominates the Sm–Nd budget (see De Wolf *et al.*, 1996 for a discussion). These types of analyses can provide information about garnet growth only if the inclusions and host garnet grew simultaneously or if the inclusions were isotopically equilibrated with the host matrix. In a more recent study, cores of up to 50 single garnets were mechanically isolated and then combined for Sm–Nd analysis in an effort to minimize the effects of averaging of different zones and thus provide better constraints on the timing of peak metamorphism (Argles *et al.*, 1999). With the increased availability of computer-controlled microdrilling devices, ones capable of isolating very narrow regions of silicate minerals (e.g., Müller *et al.*, 2000a), it is likely that future studies will focus on selecting even more specific regions within single garnets for thermochronology.

The ability to select specific intracrystalline regions for analysis (e.g., Ducea *et al.*, 2003), combined with increasingly sophisticated leaching techniques to minimize the effect of micro-inclusion contamination (e.g., Amato *et al.*, 1999; Anczkiewicz *et al.*, 2002), should increase the accuracy and precision of garnet thermochronology. Similarly, the increasing number of isotope laboratories with MC–ICP–MS instruments capable of analyzing very small quantities of hafnium will likely cause a dramatic increase in the number of metamorphic studies employing Lu–Hf garnet dating. When such studies are combined with the recent advances in experimental studies of REE diffusion and reexamination of concepts of closure ages (Ganguly *et al.*, 1998; Ganguly and Tirone, 1999, 2001; Albarède, 2003), substantial progress should be made in our understanding of the timing and duration of prograde versus retrograde metamorphic reactions. *P–T–t* studies will thus become ever more important to crustal evolution studies as a whole as the link between geochronology and metamorphic textural context becomes increasingly strengthened (e.g., Müller, 2003).

7.4 DETERMINATION OF AGES OF UPLIFT OR EXHUMATION

Determining the magnitude and timing of crustal uplift or exhumation of orogenic belts are critical to our understanding of the crustal evolution of the regions investigated. Under favorable conditions, the magnitude of the exhumation of part of the crust can be estimated by geobarometry. If geobarometric information is combined with measured "cooling ages" of different minerals with very different closure temperatures, then the average rate of crustal exhumation can be estimated. Of most interest to the majority of uplift/exhumation studies are relatively low-temperature (50–300 °C) thermochronologic techniques. Three such techniques, from highest to lowest closure temperature, are discussed.

7.4.1 $^{40}Ar/^{39}Ar$ Dating of Potassium Feldspar

Most common potassium-bearing minerals lose variable amounts of radiogenic argon at geologically modest temperatures, which at first glance would appear to make $^{40}Ar/^{39}Ar$ dating of limited geochronological use. However, argon diffusion appears to be a thermally activated process thus making the $^{40}Ar/^{39}Ar$ technique an excellent and widely used thermochronometer (see Section 7.3.1 and Chapter 6). One of the most important recent applications of $^{40}Ar/^{39}Ar$ dating to apparent uplift/exhumation studies is potassium feldspar thermochronology. An early assumption was that single closure temperatures (T_c) could be determined for potassium feldspars from slowly cooled plutons by using the $^{40}Ar/^{39}Ar$ data from the lower temperature (<600 °C) steps of a step-heating analysis (e.g., Heizler *et al.*, 1988), consistent with single diffusion domains within the feldspars. Lovera *et al.* (1989) demonstrated inconsistencies with cooling rates determined by the specific closure temperatures and those calculated by examining the release spectra and thus suggested that the feldspar diffusion domains were of variable size, a suggestion also made by Zeitler (1988). In subsequent work, Lovera and Richter (1991) further demonstrated the multidomain behavior of feldspars by demonstrating the same phenomenon even when analyzing single crystals. Of particular importance was the demonstration that modeling the thermal history of feldspar was largely independent of domain geometry, size, and volume fraction. Thus, by measuring the ^{39}Ar release spectra from a large number of heating steps of potassium feldspar from a single rock, and generating Arrhenius plots from those data, an apparently very robustly modeled cooling history can be obtained. A wide number of such cooling studies have now been made (see McDougall and Harrison (1999) for full references). Such studies when used in conjunction with other low-temperature techniques such as apatite fission track (FT) analysis and (U–Th)/He dating (see below) can provide a very substantial percentage of the full cooling path encountered by specific regions of crust.

7.4.2 FT Dating of Apatite

In 1962 it was demonstrated that by chemically etching a uranium-bearing mineral the paths, or tracks, traveled by the fragments arising from the spontaneous fission of ^{238}U could easily be viewed with an optical microscope (Price and Walker, 1962). A year later the first FT date of a mineral had been published (Price and Walker, 1963). The track, caused by disruption of the crystal lattice from the oppositely moving fission fragments, is initially ~16–17 μm in length in apatite crystals (after etching), but becomes increasingly shortened, or annealed, with both increasing time and temperature (e.g., Gleadow et al., 1986). The maximum temperatures that can be reached and still retain abundant FTs vary with different minerals, in the same way as does the closure temperature for retention of a specific radiogenic daughter isotope. For apatite, such a closure temperature with regard to FTs is generally quoted at ~100 °C (e.g., 105 ± 10 °C; Parrish, 1983). Assigning an FT closure temperature is much less straightforward than the retention of a daughter isotope, however, as there are no geological temperatures, even 20 °C, at which annealing can be considered negligible (Donelick et al., 1990). Also, chemically different apatites can exhibit significantly different annealing behaviors (e.g., Green et al., 1989).

Because FT formation can be viewed as a continuous, constant process each track has the potential to experience a different segment of the thermal history of the rock containing the apatite crystal. Thus each track could become shortened (annealed) to varying degrees, depending on the thermal history, and if accurate models of FT annealing can be made then the distribution of

track lengths can provide considerable insight into that thermal history (e.g., Green et al., 1986, and references therein). A primary objective of annealing experiments is therefore to establish a thermal model that describes the behavior of FT systematics over geological time scales (e.g., Laslett et al., 1987, and many others). Fitting FT data from natural samples to such models may allow significant constraints to be placed on the past uplift/exhumation history of a crustal region, one of the most significant goals of FT thermochronology. Such studies have been applied to the unroofing history of a large number of orogenic belts (e.g., Fitzgerald et al., 1995; Gallagher et al., 1998).

A potential problem with some of the earlier annealing models, upon which most FT thermal studies of sedimentary basins have been based, is that they characterize only a single type of apatite (i.e., the models are monokinetic), which may not always be applicable given the demonstration of different annealing properties of apatites of different composition (Green et al., 1989; Carlson et al., 1999). Significant advances have being made in multikinetic thermal modeling of apatite FT annealing (Ketcham et al., 1999), which has the potential to significantly advance the level of modeling of sedimentary basin thermal evolution as well as refine further crustal uplift and exhumation studies.

7.4.3 (U–Th)/He Dating of Apatite

Some of the earliest attempts at dating uranium-bearing minerals were made by measuring the accumulation of helium in crystals from the α-decay of uranium and thorium. Two efforts that immediately predated the development of modern geochronology were Hurley (1954) and Damon (1957). Because of the ease of helium diffusion, however, the dates calculated were shown to be too young in most cases and the technique was soon abandoned in favor of U–Th–Pb isotopic techniques. A resurgence of interest in the technique began in the late 1980s with a more thorough quantitative understanding of the diffusive behavior of helium in different minerals (Zeitler et al., 1987). It has now been demonstrated that the mineral apatite has a closure temperature for helium of ~70 °C (Wolf et al., 1996) and thus is well suited as a very low-temperature thermochronometer, that can further extend information relating to a variety of uplift and shallow crustal studies (see reviews by Farley, 2002; Ehlers and Farley, 2003). The typical method of determining a (U–Th)/He date is by extraction of helium gas either with a furnace (e.g., Zeitler et al., 1987; Lippolt et al., 1994; Wolf et al., 1996) or by laser (e.g., House et al., 2000) followed by mass spectrometric analysis, usually with small quadrupole-based instruments. After helium extraction, the apatite grains are recovered and dissolved for measurement of uranium and thorium abundances. There does not appear to be any loss of thorium or uranium during the vacuum extraction of the helium. Before the timing of helium closure can be calculated from the collected data, a correction factor must be applied because of the phenomenon of α ejection (see Farley et al., 1996). The α particles produced from uranium and thorium decay can travel 20 μm through an apatite crystal lattice, thus α particles will be ejected (i.e. helium loss) when the parent nucleus occurs near the edge of the crystal. Corrections for this helium loss must be estimated, and are based on the assumptions of idealized crystal geometry and near homogeneous distribution of uranium and thorium. The correction factors to the age, based on the size of the crystal, are typically between 1.2 and 1.5 (Ehlers and Farley, 2003), with uncertainties in the correction factors increasing with decreasing crystal size. For moderate to large apatite crystals (i.e., those occurring in typical plutonic rocks) the reproducibility of helium cooling ages, combining analytical errors with uncertainties in α ejection correction factors, is $\sim\pm5\%$ (Ehlers and Farley, 2003).

7.5 NEODYMIUM ISOTOPES AND CHEMICAL AGE OF CRUST

7.5.1 Sm–Nd Methodology

The Sm–Nd isotopic method depends upon the decay of ^{147}Sm, comprising $\sim15\%$ of natural samarium, to ^{143}Nd by α-decay. With a reasonably well-known decay constant of $6.54 \times 10^{-12}\,\text{yr}^{-1}$ (Lugmair and Marti, 1978; Begemann et al., 2001), production of ^{143}Nd is slow. The ratio ^{143}Nd/^{144}Nd, which is measured by isotope geologists, changed from ~0.50687 at the birth of the Earth to ~0.51264 today (Jacobsen and Wasserburg, 1984), with the highest values of major rock reservoirs under mid-ocean ridges possessing values ~0.5132. Because the precision of measurement on this ratio is typically $\sim5 \times 10^{-6}$ from modern mass spectrometers, subtle variations of ^{143}Nd/^{144}Nd are actually quite easily discernible.

Samarium and neodymium are both REEs. They therefore belong to a series of elements that have a very important role in geochemistry, due to progressive chemical fractionations that occur between the lighter and heavier REE. Although samarium and neodymium are adjacent elements in the naturally occurring REE series, fractionation between them is a little larger than for two elements with sequential atomic numbers, because there is a missing element between them, promethium, which has no stable isotope. The utility of the Sm–Nd isotopic method in crustal history is driven by the fact that upper continental crust acquires, due to igneous differentiation, a parent/daughter ratio ^{147}Sm/^{144}Nd that is $\sim45\%$ lower than that of undifferentiated Earth, and lower still compared to typical depleted upper mantle sources. This fractionation is due to mantle minerals such as garnet, clinopyroxene, and orthopyroxene having lower distribution coefficients for light REE than for heavy REE, so that neodymium is partitioned into magmas slightly more strongly than samarium when mantle sources are melted. Because the ^{147}Sm/^{144}Nd ratio tends to be reasonably constant in average upper crustal rocks like granite, felsic gneiss and shale, and always $\sim45\%$ lower than upper mantle values, the evolution of ^{143}Nd/^{144}Nd in upper crustal rocks slows down compared to the undifferentiated Earth and to upper mantle reservoirs, always by about the same amount. Thus, the ^{143}Nd/^{144}Nd ratio measured in a crustal rock is usually a good reflection of the average age of mantle separation of the materials in the rock.

The rather constant fractionation of Sm/Nd ratios in upper continental crustal rock reservoirs is the basis for the widely applied neodymium model age that is illustrated in Figure 3. The Sm–Nd systematics of chondritic meteorites serve as a reference for the parent/daughter ratio of the undifferentiated Earth (Jacobsen and Wasserburg, 1984), labeled as CHUR for "chondritic uniform reservoir." The evolution of this undifferentiated Earth is the basis for calculation of CHUR model ages (McCulloch and Wasserburg, 1978), while the neodymium isotopic evolution of the depleted upper part of the mantle is a more valid reference for most crustal materials, resulting in the DM model age (DePaolo, 1981). Neodymium isotopic compositions are usually given by ε_{Nd}, where the deviations of ^{143}Nd/^{144}Nd above or below CHUR are given as parts in 10,000 (see Figure 3 and caption). Because of their Sm/Nd ratios below CHUR, all continental crustal reservoirs evolve towards negative ε_{Nd} with time.

It is important to understand that whether neodymium model ages are an explicit part of the discussion (as in DePaolo, 1981), or are de-emphasized in favor of an interpretation based on ε_{Nd} values (e.g., Patchett et al., 1999), the utility of the Sm–Nd method in crustal evolution is ultimately based on the reproducible Sm/Nd fractionation that occurs between crust and mantle. Note that the Sm/Nd fractionation of more primitive crust of intermediate composition (Figure 3), such as basaltic andesite or andesite, does not evolve to negative ε_{Nd} so rapidly as fully differentiated upper continental felsic crust. In this way, the neodymium isotopic composition of a crustal rock depends on the average mantle differentiation age of the components that went into making it. The origin of some crustal rocks can be very complicated, e.g., a granitoid melted from complex lower crustal sources, or a sedimentary

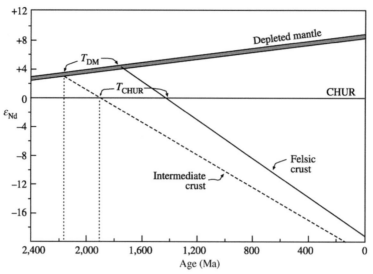

Figure 3 Sm–Nd systematics for crustal model ages and ε_{Nd} values. The parameter $\varepsilon_{Nd} = 10^4 \cdot [^{143}Nd/^{144}Nd(sample) - ^{143}Nd/^{144}Nd(CHUR)]/^{143}Nd/^{144}Nd(CHUR)$, where all $^{143}Nd/^{144}Nd$ values are specified at the age of interest (*t*). CHUR Nd isotopic values can be obtained from the following equation: $^{143}Nd/^{144}Nd(CHUR,t) = 0.512638 - 0.1966 \cdot (e^{\lambda t} - 1)$, where $\lambda = 6.54 \times 10^{-12}$ a^{-1}. All crustal materials evolve towards negative ε_{Nd} with time, because the evolution of $^{143}Nd/^{144}Nd$ is slower than the chondritic or bulk silicate Earth reference (CHUR). Nd model ages of crust can be calculated based on the chondritic reference (T_{CHUR}) or on the intersection with the approximate evolution of depleted upper mantle sources (T_{DM}). T_{DM} is generally more meaningful for juvenile crustal rocks produced during plate-tectonic cycles, because it models the average separation age from the type of mantle that is commonly observed as a source of both oceanic crust and island arcs today. Note that the depleted mantle evolution is here shown as a straight line, but is very slightly concave-upwards in the model of DePaolo (1981).

rock derived from erosion of multiple terrains. Thus, a single model age for a granitoid or a sedimentary rock can be quite misleading in terms of real geological events and processes. Nevertheless, it is the differences in age of separation from mantle sources that bestow usefulness on neodymium isotopic variations, and no matter how complex the origin of a crustal rock, it is those age differences that are the ultimate basis for any interpretation. This is clearly seen, e.g., in the similarity of Sm–Nd systematics in many of the world's major rivers today (Goldstein *et al.*, 1984).

Lu–Hf isotope systematics provide an important complement to Sm–Nd in the study of the crust and mantle (e.g., Patchett *et al.*, 1981; Salters and Hart, 1991; Vervoort and Blichert-Toft, 1999). In the crustal context, Lu–Hf is extremely important because of the ~1% hafnium content of zircon, and the consequent ability to isotopically characterize the hafnium within grains that have been U–Pb dated (Patchett *et al.*, 1981; Corfu and Stott, 1996; Vervoort *et al.*, 1996; Amelin *et al.*, 1999). However, the Lu–Hf isotopic system is currently overshadowed by a controversy over the decay constant. For many years, a value for the ^{176}Lu decay constant of 1.94×10^{-11} yr^{-1}, based on the eucrite meteorite isochron of Patchett and Tatsumoto (1980) and Tatsumoto *et al.* (1981) was used. More recent physical determinations

reviewed by Begemann *et al.* (2001) have high dispersion, but do not seem to corroborate the 1.94×10^{-11} value. At the present time, there is a discrepancy between values based on U–Pb-dated terrestrial Precambrian REE-rich minerals, such as apatite, which suggest a decay constant of 1.865×10^{-11} (Scherer *et al.*, 2001), and meteorite isochrons, that suggest values of $(1.93-1.98) \times 10^{-11}$ (Bizzarro *et al.*, 2003; Blichert-Toft *et al.*, 2002). For this reason, we mostly do not include Lu–Hf isotopic data in discussion of crustal age and origins. Lu–Hf data are of considerable importance in studies of early Archean rocks (see Section 7.6.4), and this uncertainty should be resolved as rapidly as possible.

7.5.2 Juvenile Crust Production versus Intracrustal Recycling

It is fundamental to the neodymium isotopic approach that neodymium isotopes are able to distinguish between material added newly to the Earth's continents and material that is merely recycled older crustal rock. This is important because sedimentary rocks derived from erosion of older continent may appear quite similar to those derived from erosion of young island-arc terrain, and unless they show marked S-type characteristics, granitoids are notoriously similar to each other, regardless of their ultimate origin. Thus, in a world where all regions of continents had been characterized for both orogenic ages and neodymium isotopic characteristics, one could draw two global maps. One would show orogenic-belt ages, representing times of consolidation of regions of the continents, while the other would show generally older neodymium-based average ages, that would represent the true differentiation age of the crust. Some orogenic belts might consist of dominantly juvenile crust, with neodymium ages similar to the orogenic age, but other belts might be dominated by materials recycled from older crustal terrains. Because the sedimentary system is a powerful mover of crustal detritus over large geographic scales, and because melting to produce granitoid batholiths often averages large domains of lower continental crust, orogenic belts very often have mixed origins in terms of the crustal age of their components. Regional maps of crustal age based on neodymium model ages, even without the references to the isotopic work, appear quite often in the literature and in presentations (e.g., Karlstrom *et al.*, 1999). The coverage of U–Pb ages and neodymium isotopic data is not yet sufficient to draw robust global maps, and because the current pace of data accumulation is high, we do not attempt to compile global maps in this chapter. Instead, the approaches will be illustrated with examples.

7.5.3 Juvenile Crust Production at 1.9–1.7 Ga

The abundant crust that was produced in the 1.9–1.7 Ga interval has been the subject of numerous studies, and the evolution of that work illustrates important elements in study of the continental crust. Following the demonstration of a juvenile origin for the ∼1.8 Ga crustal assemblage in Colorado (DePaolo, 1981), there followed a period in which neodymium isotopic data were gathered for numerous terrains in the northern continents (Figure 4). In North America, studies by Nelson and DePaolo (1984, 1985), Bennett and DePaolo (1987), Chauvel *et al.* (1987), Barovich *et al.* (1989), and Bowring and Podosek (1989) added rapidly to the database. In Greenland and Europe, work by Patchett and Bridgwater (1984), Skiöld and Cliff (1984), Kalsbeek and Taylor (1985), Patchett and Kouvo (1986), Huhma (1986), Kalsbeek *et al.* (1987), Claesson (1987) and Patchett *et al.* (1987) performed the same function. All of the studies cited above documented a high proportion of juvenile crust production in the 1.9–1.7 Ga orogenic belts of the northern continents, as highlighted in the reviews by Patchett and Arndt (1986) and Condie (1990). Large volumes of new crust were added, at least to the studied regions, during 1.9–1.7 Ga. It appears that the 1.9–1.7 Ga crust of the northern continents is approximately equal in mass to what could be produced by present-day island-arc generation rates over 200 Myr (Reymer and Schubert, 1984; Patchett and Arndt, 1986). Patchett and Chase (2002) modeled this in terms of accumulation of juvenile terrains in restricted regions of the globe due to transform motions associated with large-scale plate tectonics.

Since the late 1980s, two important changes have occurred in the studies of the growth of

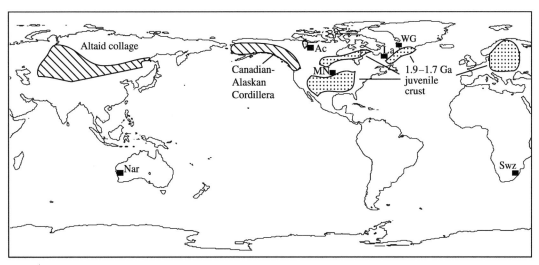

Figure 4 Regions of crust of different ages described in this chapter. Dot pattern represents 1.9–1.7 Ga juvenile crust of North America, Greenland, and Europe. Diagonal lines represent juvenile crust of the Altaid collage of orogenic belts and of the Canadian–Alaskan Cordillera. Black rectangles are locations of small regions of pre-3 Ga crust: WG—west Greenland; La—Labrador; MN—Minnesota; Ac—Acasta Gneiss Complex, northwestern Canada; Swz—Swaziland; Nar—Narryer terrane, western Australia.

1.9–1.7 Ga continental crust. The first is that neo-dymium isotopic data have become a very common component of any tectonic study of orogenic belts, granitoid complexes or sedimentary sequences, par-ticularly in the Precambrian, meaning that neodym-ium isotopic data are now routinely gathered as part of characterization of regions not studied before. The second change is that the pace of data acquisition has slowed a little, because regions that were well-characterized in terms of geology and geochronology were studied in the 1980s, while fresh mapping and geochronology are needed to open up new areas for credible neodymium isotopic study. Thus, more recent studies incorporating neodymium iso-topes are not super-regional reconnaissance studies of regions for which geology and ages are already known, but are targeted to newer topical problems and areas in Proterozoic geology, such as ophio-lites in Quebec and Finland (Hegner and Bevier, 1991; Peltonen *et al.*, 1998), collisional tectonics (Öhlander *et al.*, 1999), or granitoid petrogenesis (Valbracht *et al.*, 1994; Krogstad and Walker, 1996; Rämö *et al.*, 2001).

In regions away from North America, Europe and Australia, the pace at which the volume both of well-dated Precambrian terrain and of neodym-ium isotopic information increases is often much lower. This leaves large parts of the continents poorly known for neodymium crustal age. The very important conclusion, that the 1.9–1.7 Ga period saw a sequence of relatively rapid and high-volume additions to the continental crust, is not likely to change. However, it is not yet really clear if the 1.9–1.7 Ga crust of the northern con-tinents could represent juvenile terrain gathered from the rest of the globe, as suggested by Patchett and Arndt (1986) and Patchett and Chase (2002). That would be a model where the 1.9–1.7 Ga period could have shown normal crustal growth rates, but where plate tectonics had grouped the juvenile crustal products into restricted regions. If the continental regions that are currently poorly known for their crustal age also show significant juvenile crust of 1.9–1.7 Ga age, then this event must have been larger than presently documented in the northern continents, and would represent a distinct, positive spike in crustal growth. Such a situation might correspond to crustal growth initiated by mantle plume activity, as hypothe-sized by authors like Boher *et al.* (1992) and Stein and Hofmann (1994).

7.5.4 Juvenile Crust Production in the Canadian Cordillera

The Canadian–Alaskan Cordillera (Figures 4 and 5) has become the typical example of a Phan-erozoic orogenic belt that contains significant amounts of juvenile crust. It is distinctive because of all the global history of mountain-building

events from Cambrian to Cretaceous time, only three orogenic zones seem to display significant juvenile additions. Two of these are the South Island of New Zealand (e.g., Frost and Coombs, 1989), which is small in size, and the very large Altaid orogenic collage of central Asia (e.g., Sengör *et al.*, 1993), for which isotopic data have only recently become available. The Canadian–Alaskan Cordillera has been studied for radiogenic isotopes by several groups against a backdrop of rather well-constrained geology.

In the 1970s, the Canadian Cordillera became a center of tectonic attention because of the demon-stration of the distinctness of the belts of rocks from which it was constructed (Monger *et al.*, 1972). This region became the typical example for the concept of tectonostratigraphic terranes and their assembly by accretion to form a mountain belt (Coney *et al.*, 1980). Subsequently, discussions were initiated about the apparent transpressional emplacement of many of the terranes, and north-ward transform faulting of them after emplacement (Gabrielse, 1985; Umhoefer, 1987). A recent geo-logical review of Cordilleran geology and the devel-opment of ideas are found in Monger (1993). In parallel with the geological investigations, geochro-nology and strontium isotopic tracer investigations were carried out by Armstrong and co-workers (Armstrong, 1988; Armstrong and Ward, 1993). Because initial $^{87}Sr/^{86}Sr$ ratios of plutonic rocks in the outboard parts of the Canadian Cordillera tended to be low, and because geological associa-tions suggested that a high proportion of intra-oceanic rock assemblages had been accreted to North America, there was a general supposition that juvenile crustal elements from the paleo-Pacific basin would be abundant in the Canadian Cordillera (Monger *et al.*, 1972; Coney *et al.*, 1980; Armstrong, 1988). However, many of the terranes, even outboard ones, contain Paleozoic rock assemblages that were viewed as "continental" (Monger *et al.*, 1972).

It was against this background that the present authors, in collaboration with George Gehrels, initiated neodymium isotopic characterization of the Canadian Cordillera, beginning with the Alexander terrane (Figure 5). In this outboard terrane, tightly folded Neoproterozoic and Paleo-zoic felsic igneous assemblages form the older components, and were some of the rocks referred to by Monger *et al.* (1972) as "continental." At the time when the first research proposal was written, it seemed to the present first author that these assemblages would yield neodymium isotopic data typical of older Precambrian continental crust. The outcome was very different, because all components of the Alexander terrane, includ-ing the oldest ones, turned out to have positive initial ε_{Nd} values (Samson *et al.*, 1989). This result became the keynote for much of subsequent isoto-pic study by our group (Samson *et al.*, 1990,

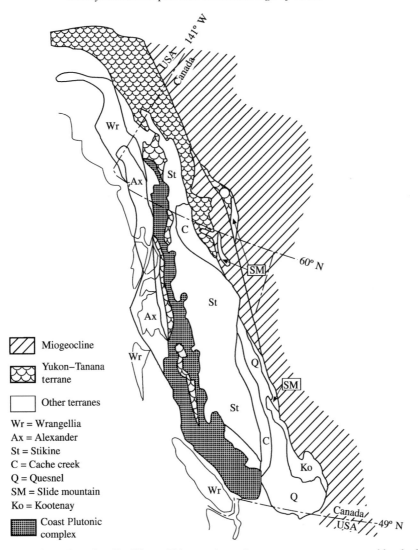

Figure 5 Terranes of the Canadian Cordillera. Older continental components are represented by the Miogeocline, which rests on North American Precambrian basement, and the Yukon–Tanana terrane. Of the other terranes, those near to the Miogeocline, such as Kootenay, have major older crustal components. Juvenile characteristics increase in the direction of the Pacific Ocean (after Patchett *et al.*, 1998; Butler *et al.*, 2001).

1991a,b). The isotopic work was consistent with outboard terranes being generally more juvenile, but in addition it was consistent with ancient sedimentary assemblages having been incorporated into the Coast Plutonic Complex (Samson *et al.*, 1991a,b; Boghossian and Gehrels, 2000; Patchett *et al.*, 1998). By no means did these authors have a monopoly of isotopic study in the Canadian–Alaskan Cordillera, and important data for outboard terranes, the Coast Plutonic Complex (Figure 5) and related areas were published by Arth *et al.* (1988), Farmer *et al.* (1993), Mezger *et al.* (2001), Cui and Russell (1995), Friedman *et al.* (1995) and Mahoney *et al.* (1995). One important finding of the last three cited papers is to show that the southern part of the Coast Plutonic Complex does not contain the older crustal elements seen further north around the Alaska Panhandle.

Further inboard lies a zone of the Canadian Cordillera that incorporates terranes showing well-known oceanic features, like Cache Creek and Slide Mountain, arc assemblages like Quesnel, and also large amounts of clastic sedimentary rocks, along with the Yukon–Tanana terrane of more continental affinity in the northern part of the Cordillera (Figure 5). This zone is characterized by juvenile neodymium signatures in most of its volcanic rocks and mafic intrusives (Smith and Lambert, 1995; Smith *et al.*, 1995; Piercey *et al.*, 2002). Neodymium isotopic signatures showing a mixture of juvenile and Precambrian continental North American materials are shown by both clastic sedimentary assemblages (Creaser *et al.*, 1997; Patchett and Gehrels, 1998; Unterschutz *et al.*, 2002; Erdmer *et al.*, 2002), and granitoid plutons (Ghosh, 1995; Ghosh and Lambert, 1995; Piercey *et al.*, 2003).

Still further inboard towards North America lies the Omineca Crystalline Belt (Monger *et al.*, 1982) and the Miogeocline, where pre-Cordilleran Neoproterozoic through Late Jurassic sedimentary rocks rest on North America basement. The Omineca crystalline belt is not separately distinguished on Figure 5, because it is a metamorphic/plutonic overprint rather than a separate terrane, but it occupies the general area of the Quesnel and Kootenay terranes. Although juvenile neodymium is sometimes seen, the Omineca belt generally shows neodymium isotopic signatures, in both metasedimentary and plutonic rocks, that correspond to older North American continent (Burwash *et al.*, 1988; Ghosh and Lambert, 1989; Stevens *et al.*, 1996; Brandon and Lambert, 1993, 1994; Brandon and Smith, 1994; Driver *et al.*, 2000). The miogeoclinal sequence appears to have been derived from Precambrian basement before ~450 Ma, but after this time to have been supplied by distant mountains of early- to mid-Paleozoic age in the Canadian Arctic (Boghossian *et al.*, 1996; Garzione *et al.*, 1997; Patchett *et al.*, 1999).

The general picture of the Canadian–Alaskan Cordillera is of a mountain belt characterized by accretion to older continental crust of juvenile crustal elements from the ocean floor, or from island-arc environments, whose proportion increases towards the Pacific Ocean. Generally, work in the Canadian–Alaskan Cordillera has evolved in the same way as work on Proterozoic terrains described above. From initial reconnaissance studies and large-scale views of how the crust grew, research has evolved into more detailed studies with more precise geological control. Nevertheless, some important general issues about crustal evolution come to fore by thinking about the Cordillera at the largest scale. Samson and Patchett (1991) reviewed the then-existing neodymium isotopic database, and concluded that ~50% by mass of the Canadian segment of the Cordillera was juvenile crustal material. Clearly, such estimates are dependent on assumptions about whether terranes continue to lower crustal depth. Seismic data often reveal that juvenile belts are underlain by older continental material, both in Proterozoic and Phanerozoic contexts (Lucas *et al.*, 1993; Clowes *et al.*, 1999). The 50%-juvenile estimate of Samson and Patchett (1991) should certainly be revised downwards in light of seismic results summarized by Clowes *et al.* (1999), as well as more recent field and isotopic work such as that of Erdmer *et al.* (2002). Qualitatively however, the status of the Canadian Cordillera as the locus of juvenile crustal growth remains in spite of these uncertainties.

Samson and Patchett (1991) also viewed the Canadian Cordillera as an analogue for Proterozoic crustal growth by accretion of juvenile terranes, a comparison that was explicitly examined in terms of major- and trace-element compositions

by Condie and Chomiak (1996). The present authors, like many others, have always been impressed by the transform faulting that is presently active along the North American margin, which serves to pile juvenile crust into one region that stretches from southern British Columbia to Alaska. Subsequent synthesis of Cordilleran geology (Johnston, 2001) and discussions about crustal growth in general (Patchett and Chase, 2002) have emphasized the northward along-margin transport displayed in the Canadian Cordillera as a critical element in crustal growth models. The rationale for this is that along-margin transport of slices of juvenile crustal material is able to pile them all up in one segment of the continental margin, perhaps even into a restricted region of the globe, creating the impression of a very intense and localized crustal growth period (Patchett and Chase, 2002), for which extraordinary mechanisms might otherwise be required (e.g., Stein and Hofmann, 1994). This process should have been important in most of geologic time, accounting for numerous apparently very intense crustal growth episodes (Patchett and Chase, 2002).

7.5.5 Juvenile Crust Production in the Altaid Collage of Central Asia

Another region of Phanerozoic crustal growth is the Altaid orogenic collage of central Asia, located in the eastern parts of the former Soviet Union, and in western China (Figure 4). The rocks range in age dominantly from Neoproterozoic through Early Carboniferous, with final tectonic movements and post-orogenic igneous activity extending through part of Permian time. Thus, the crustal assemblage is older than the bulk of the material in the Canadian Cordillera. In a simplified view, one might say that the Altaid system represents the bulk of global juvenile crustal accretion of Cambrian through Permian time, while the Canadian Cordillera and New Zealand have that role from Triassic through Cretaceous time.

Early work on the tectonic assembly of Asia, which was published in international journals, took place in the 1980s and early 1990s (e.g., Sengör and Hsü, 1984; Windley *et al.*, 1990; Allen *et al.*, 1992). However, the Altaid system came to the attention of the wider scientific community through the bold syntheses of Sengör and co-workers (Sengör *et al.*, 1993; Sengör and Natalin, 1996a,b). These authors described a collage of belts of arc volcanic rocks, ophiolitic assemblages, greywacke-shale sequences and granitoids, that occupies a very large area in central Asia to the north of the Himalayan system, and lying between the Baltic Shield/Russian Platform to the west, the Siberian Shield to the east, and with the North China Craton providing a partial boundary to the south. Tectonic trends are highly variable, and this seems consistent with the interpretations of Sengör and co-workers, that the collage represents the sweepings of a large ocean, with telescoping of accreted fragments not only by orthogonal collision, but with transpression and transform faulting, essentially along the strike of the developing orogen, over very large geographic scales. This resulted in a collage with variable tectonic trends occupying a large region of roughly triangular shape in the center of the Asian continent.

Isotopic analyses of Altaid-collage rocks are not as abundant as those in the Canadian Cordillera, but have appeared in increasing numbers since about 1996. Kovalenko *et al.* (1999, 2003) presented neodymium isotopic data on juvenile Altaid rocks in Russia and Mongolia, while Wu *et al.* (2000), Jahn *et al.* (2000) and Chen and Jahn (2002) made extensive studies of the abundant granitoid plutons in the Chinese part of the Altaid assemblage. Yarmolyuk *et al.* (1999) and Hu *et al.* (2000) used neodymium isotopic results to study the nature of pre-Altaid basement. Modern U–Pb zircon geochronology, which together with detailed modern documentation of tectonic relations in the field, has also been largely lacking in the Altaid crustal collage, is being actively undertaken (Wilde *et al.*, 2000; Salnikova *et al.*, 2001; Windley *et al.*, 2002). All authors concluded that substantial juvenile material was added to Asia during Altaid events, particularly during Early Ordovician through Devonian time. Post-orogenic granitoids of Late Carboniferous to Early Permian age have sources that may appear juvenile, or to consist of older continental crust, depending on the location with respect to the margins of the Altaid collage (Han *et al.*, 1997; Litvinovsky *et al.*, 2002).

What appears to emerge from the ongoing studies is that the Altaid collage consists of terranes of juvenile material juxtaposed with terranes containing pre-existing Proterozoic crustal rocks over the whole extent of the belt. Sengör *et al.* (1993) painted just such a picture, and also suggested that up to 50% of the entire Altaid collage may represent juvenile crust of Paleozoic age. So far, the neodymium isotopic data appear to support juvenile crust of about this magnitude, but much field data, geochronology and neodymium isotopic data remain to be gathered. In addition, estimates for the volume of juvenile crustal material may need revision when better data are available for whether and how the juvenile terranes project into the lower continental crust, as noted above in the context of the Canadian Cordillera. Estimates of the proportion of juvenile crustal growth represented by the Altaid collage can be expected to be refined in the future.

7.6 ISOTOPES AND PRE-3 Ga CONTINENTAL CRUST

7.6.1 Existence of Ancient Continental Crust

Isotope geology acquires a dominant significance when continental rocks dating from before 3 Ga are studied. This is because fossils are not available, and metamorphism and tectonism are often so severe that original layering of supracrustal sequences may not be visible, and even magmatic contact relationships may be obscured. Gneisses of 3.8 Ga, 2.8 Ga, and 1.8 Ga may look very similar in the field, and isotopic dating is absolutely required to distinguish them. In addition, pre-3 Ga rocks may carry unique information about environments on the early Earth, or about mantle differentiation and layering, and the search for the oldest continental crust has always been a major area of activity. All the areas of pre-3 Ga crust discussed here are indicated on Figure 4.

Early dates for ancient crust were obtained by the Rb–Sr whole-rock isochron method, or by the U–Pb zircon method when it was in its infancy. Approximately 3.5 Ga U–Pb zircon ages were obtained from gneisses in Minnesota by Goldich *et al.* (1970), and gneisses in western Greenland were dated at 3.6–3.8 Ga by Rb–Sr isochrons (Black *et al.*, 1971; Moorbath *et al.*, 1972). These age determinations were the major scientific news of their day in geology, and began a whole field of endeavor in deciphering the history of ancient crustal rocks. Early foci of research on ancient crust were western Greenland (Moorbath, 1978), Labrador (Hurst *et al.*, 1975; Bridgwater and Collerson, 1976) and Swaziland (Davies and Allsopp, 1976). Early studies of these regions using Sm–Nd were by Hamilton *et al.* (1983) and Carlson *et al.* (1983), and these and other isotopic studies documented the juvenile nature of these earliest crustal gneiss complexes.

Presently, most occurrences of pre-3 Ga rocks have been accurately dated using modern approaches of the U–Pb zircon method (e.g., Kröner *et al.*, 1989; Bowring *et al.*, 1989b; Horstwood *et al.*, 1999), and neodymium isotope data have been widely obtained (e.g., Collerson *et al.*, 1991). Hafnium isotopic data, combined with neodymium, have been obtained for critical samples (Vervoort *et al.*, 1996; Vervoort and Blichert-Toft, 1999; Blichert-Toft *et al.*, 1999; Amelin *et al.*, 1999). The focus of neodymium and hafnium isotopic studies has moved away from documentation of juvenile character towards an attempt to constrain the degree of heterogeneity in the early Earth's mantle, with a view to detecting "primordial" rock reservoirs and effects of the primary differentiation of the Earth (e.g., Bowring and Housh, 1995; Albarède *et al.*, 2000). Many arguments have developed concerning the reliability of isotopic values for single samples of polymetamorphic gneisses (e.g., Vervoort *et al.*, 1996; Gruau *et al.*, 1996; Moorbath *et al.*, 1997). The case of the Acasta gneisses, to which the above-cited references apply, is detailed later in this chapter.

7.6.2 Crustal Growth Events and Recycling into the Mantle

Early in the evolution of isotopic studies of ancient crust, Moorbath (1975, 1978) developed the concepts of (i) major Precambrian crustal growth events in which juvenile crust was made, and (ii) the essential indestructibility of continental crust once consolidated by orogenic events. Models based on neodymium and strontium isotopes that grew the continents through time were made possible by the arrival of the meteoritically constrained Sm–Nd system (O'Nions *et al.*, 1979; Jacobsen and Wasserburg, 1979; DePaolo, 1979). A very readable account of the chemical relationships between crust and mantle is by Hofmann (1988), while one of the more mathematically complete treatments is by Allègre *et al.* (1983a,b). Although a neodymium-based cumulative age curve for global continental crust cannot yet be drawn, the estimates for North America and Europe (Patchett and Arndt, 1986; Condie, 1990) may approximate the global picture.

Alternatively, the modeling by Armstrong (1968, 1981) and Zartman and Doe (1981), initially strongly based on lead isotopes, advocated the position that during crustal growth events or during plate movement in general, continent could also be destroyed by the subduction of sedimentary rocks resting on the ocean floor. It was realized by the community that all crust and mantle geochemistry and isotopes were consistent with large amounts of crustal recycling, provided that more or less complete mixing occurred during mantle convection. Much of the discussion concerns the present-day Earth. However, the period before 3 Ga is critical to the discussion of crustal recycling because it is generally agreed that large amounts (perhaps 40–50%) of presently surviving crust came into existence during the immediately following period, between 3.0 Ga and 2.6 Ga (e.g., Condie, 1990). Either the pre-3 Ga Earth had much less felsic crust than existed after 2.6 Ga, or large amounts of pre-3 Ga crust were destroyed before or during the time that the 3.0–2.7 Ga crust was produced. If it could be shown that massive amounts of pre-3 Ga crust were destroyed, then essentially all objections to large-scale continent recycling in the mantle would disappear.

Attempts to test the effects of sediment subduction (Patchett *et al.*, 1984), or whether pre-3 Ga crust was available to be recycled into later crust (Stevenson and Patchett, 1990), founder on two possibilities. One is that in the early Earth, aided by a hotter mantle and continuing bombardment from asteroidal objects, it may have been possible to destroy continental fragments wholesale, so that little trace of them remained at the surface. Another is that large amounts of sediment subduction would result in a very well averaged geochemical signature entering the mantle, so that distinctive features of oceanic sediments, such as enrichment or depletion in zircon and its unradiogenic hafnium, could not be used to argue against the process (Plank and Langmuir, 1998). At the same time, budgets for the mass of rock entering subduction zones appear to suggest that sedimentary material has to disappear into the mantle (von Huene and Scholl, 1993), an argument used forcefully by Armstrong (1981). Consequently, a discussion that was once framed in terms of "growth or no growth" (e.g., Stevenson and Patchett, 1990; Armstrong, 1991; McCulloch and Bennett, 1994) may now focus on how much continent recycling occurs, and the magnitude of recycling compared to the growth seen in places like the Canadian Cordillera. The magnitudes of these fluxes are very important for crust–mantle evolution, but are quite difficult to determine accurately, because of the global basis required, and the deep parts of subduction zones being difficult to image precisely (Reymer and Schubert, 1984; von Huene and Scholl, 1993.

7.6.3 Acasta Gneisses, Northwest Territories, Canada

The Slave craton in northwestern Canada is an Archean granite–greenstone terrain that is bounded on the east and west by Paleoproterozoic orogenic belts (see geological reviews of Henderson (1985) and Padgham (1985)). Crustal regions within the western portion of Slave craton contain granitoids older than 2.8 Ga (see Bowring *et al.*, 1990, and references therein). The antiquity (>3.5 Ga) of the westernmost portion of the Slave craton has been known since the mid-1980s (Bowring and Van Schmus, 1984), but in 1989 it was demonstrated, via ion microprobe U–Pb dating, that zircons within a tonalitic gneiss (sample BGXM) from the Acasta River area were as old as 3.96 Ga (Bowring *et al.*, 1989a), making the gneiss the oldest known terrestrial rock. An identical age was determined for a second gneiss (sample SP-405) with a granitic composition (Bowring *et al.*, 1990). Stern and Bleeker (1998) reported ^{207}Pb–^{206}Pb dates as old as 4.02 Ga from a gneiss collected from within the same outcrop as that sampled by Bowring *et al.* (1989a). More recent U–Pb geochronological work on the gneisses in the Acasta River area (now called the Acasta Gneiss Complex) has extended the known extent of the oldest known surviving continental crustal region, as zircons from two metatonalites and a metagranodiorite have nearly concordant U–Pb dates of 4,002 ± 4 Ma, 4,012 ± 6 Ma, and 4,031 ± 3 Ma, respectively (Bowring and Williams, 1999).

The neodymium isotopic composition of Acasta gneiss BGXM was investigated by Bowring *et al.* (1989b). These authors reported a value of $\varepsilon_{Nd}(3.96\,Ga) = -1.7$ for a whole-rock analysis and $\varepsilon_{Nd}(3.96\,Ga) = +0.7$ for an amphibolitic-rich layer taken from a hand specimen of BGXM. An indistinguishable value, $\varepsilon_{Nd}(3.96\,Ga) = +0.8$, was reported for the granitic gneiss SP-405 (Bowring *et al.*, 1990). The corresponding T_{CHUR} ages of these three samples are 4.1 Ga, 3.92 Ga, and 3.85 Ga, respectively.

The variable ε_{Nd} values of the Acasta samples were interpreted by Bowring *et al.* (1990) as reflecting original protolith heterogeneity, rather than the effects of metamorphic disturbance. The negative ε_{Nd} value and T_{CHUR} age of 4.1 Ga of the tonalitic gneiss was interpreted as evidence that at least this portion of the Acasta Gneiss Complex must have been derived from, or interacted extensively with, a crustal reservoir considerably older than 3.96 Ga. The same argument cannot be easily made for the granite orthogneiss, however, as it has a T_{CHUR} age that is 100 Myr younger than its crystallization age.

Lead isotopic ratios were determined for HF-leaches of alkali feldspars from the granitic gneiss and of plagioclase from the tonalitic gneiss (Bowring *et al.*, 1990). The ^{206}Pb:^{207}Pb:^{208}Pb:^{204}Pb ratios of the least radiogenic leaches, presumed to be closest to the initial lead isotopic composition of the rocks, were 14.12, 15.08, 34.01 and 14.96, 15.29, 33.53, respectively. The $^{206}Pb/^{204}Pb$ and $^{207}Pb/^{204}Pb$ ratios of the three plagioclase leaches lie on a linear array with a slope equivalent to an age of 3.60 ± 0.28 Ga. Because this age overlaps the 3.62 Ga of the unzoned zircon rims from the tonalitic gneiss, the feldspar lead isotopic composition was interpreted as reflecting homogenization of whole-rock lead isotopes during a 3.6 Ga metamorphic event.

The high $^{207}Pb/^{204}Pb$ ratios of the leached feldspars were taken as evidence that the gneisses were derived from a reservoir characterized by a high U/Pb ratio prior to 3.96 Ga, such as significantly older continental crust that had not lost uranium or thorium (Bowring *et al.*, 1990). Implicit in this argument is that the lead isotopic compositions of the HF-leached feldspars accurately represent the lead isotopic composition of the protoliths to the gneisses, and hence accurately reflect the source materials of the protolith.

Hafnium and U–Pb isotopic ratios were determined for single zircons from two gneisses reported to be Acasta gneisses (Amelin *et al.*, 1999). However, no details of the sampling location or composition of the gneisses were given,

except that they were from "granitic" and "amphibolitic" gneisses. Two zircons from the amphibolitic unit have $\varepsilon_{Hf}(t)$ values of -4.7 and -0.6, and the zircon from the granitic unit has a reported $\varepsilon_{Hf}(t)$ value of -1.8. However, the significance of these data to the 4.0 Ga Acasta gneisses are uncertain, as the zircons analyzed by Amelin *et al.* (1999) yielded $^{207}Pb/^{206}Pb$ dates of 3,548 Ma to 3,565 Ma, similar to the 3.6 Ga dates determined for some structureless equant zircons and the unzoned outer portion of some of the 4.0 Ga crystals analyzed by Bowring *et al.* (1989a, 1990), which they interpreted as dating the timing of a second generation of zircon growth. Because of the extreme complexity in the zoning of zircons from the Acasta gneisses, hafnium isotopic compositions of whole zircon crystals might reflect only a weighted average of different isotopic compositions.

All in all, the complexity of the isotopic data and their interpretation in the Acasta Gneiss Complex illustrate the difficulties inherent in study of planetary evolution using polymetamorphic gneisses. This is the ultimate cause for the controversies over reliability of radiogenic isotopic parameters derived from these rocks (e.g., Vervoort *et al.*, 1996; Gruau *et al.*, 1996; Moorbath *et al.*, 1997).

7.6.4 Narryer Terrane, Western Australia

The oldest known portion of the Archean Yilgarn craton is the Narryer terrane, occurring in the northwestern part of the craton. The Narryer terrane contains 3.7 Ga gneisses, 3.7 Ga anorthositic rocks, and 3.6 Ga granites (see Myers, 1995, and references therein). Two important belts of metasedimentary rocks, thought to have been deposited ~3 Ga, occur within the Narryer terrane: the Narryer and Jack Hills belts. The metasedimentary rocks within the Narryer belt include metaconglomerates and quartzites that have preserved cross-bedding despite being metamorphosed at upper amphibolite to granulite facies conditions (Froude *et al.*, 1983). Detrital zircons extracted from one of these quartzites were the first pre–4.0 Ga terrestrial minerals identified (Froude *et al.*, 1983). In the original study, four zircons out of 102 crystals analyzed from one quartzite sample yielded ^{207}Pb–^{206}Pb dates of 4.11–4.18 Ga. The other detrital zircons from that sample and a second quartzite yielded ^{207}Pb–^{206}Pb dates between 3.75–3.3 Ga, with two zircons having nearly concordant 3.1 Ga dates.

The Jack Hills belt, broadly similar to the Narryer belt but metamorphosed to upper greenschist facies, contains metabasalts, chert, and banded iron formation interleaved with clastic metasedimentary rocks (Compston and Pidgeon, 1986). Detrital zircons from a ~3.1 Ga chert pebble conglomerate were analyzed using ion microprobe U–Pb techniques as part of a continuing search for ancient zircons. Compston and Pidgeon (1986) found detrital zircons with [207]Pb–[206]Pb dates ranging between 4.28–4.0 Ga, significantly extending the sampling region of the oldest known minerals. Supporting geochronological evidence for the presence of 4 Ga detrital zircons from these areas was provided by both the zircon evaporation method (Kober *et al.*, 1989) and the ID-TIMS method (Amelin, 1998). In a more recent ion microprobe study of detrital zircons from Jack Hills, Mojzsis *et al.* (2001) confirmed the existence of 4.28 Ga zircons by identifying two grains giving concordant U–Pb dates of 4,279 ± 5 Ma and 4,280 ± 5 Ma. An even older detrital zircon, 4,404 ± 8 Ma, has been discovered from the Jack Hills conglomerate (Wilde *et al.*, 2001), making this crystal only ∼150 Myr younger than the estimated time of Earth formation.

In addition to geochronological studies of the Earth's oldest zircons, geochemical and isotopic studies have also been performed (Kinny *et al.*, 1991; Maas *et al.*, 1992; Amelin *et al.*, 1999; Mojzsis *et al.*, 2001; Peck *et al.*, 2001; Wilde *et al.*, 2001). Maas *et al.* (1992) demonstrated that both the older and younger detrital zircon suites from the Jack Hills conglomerate have trace-element compositions and mineral inclusions consistent with nucleation from a felsic magma. Kinny *et al.* (1991) were the first to attempt to determine the hafnium isotopic composition of the zircons, but the large uncertainties in the ion microprobe measurements (± (5–7) ε_{Hf} units) precluded a detailed discussion of the petrogenesis of the parent magmas. In more recent work, Amelin *et al.* (1999) measured 37 Jack Hills detrital zircons for U–Pb ages and hafnium isotopes. The advantage of measuring U–Pb and hafnium isotopes from the same dissolved zircon crystal is that the degree of discordance can be determined, which in turn is important in assessing the likelihood of the presence of a xenocrystic core and/or metamorphic rims on the zircon. Zircons that are either concordant or not too discordant are the least likely crystals to contain multiple domains. Similarly, zircons that have suffered the least lead loss are the least likely to have experienced open system behavior for lutetium or hafnium (see Samson *et al.*, 2003 for discussion). Of the 10 zircons identified by Amelin *et al.* (1999) that are >3.8 Ga, two have $\varepsilon_{Hf}(t)$ values that are further than ±1 ε_{Hf} units from CHUR. The bulk of these ancient zircons are thus consistent with derivation from a chondritic-like source, which at this age could include very slightly to nondepleted mantle, or juvenile crust that had recently been extracted from such a mantle. Two crystals, with [207]Pb–[206]Pb dates of 3.82 Ga and 3.97 Ga, have $\varepsilon_{Hf}(t)$ values of −2.2 ± 0.6 and −2.7 ± 0.4, respectively. These values are far enough below the CHUR line to suggest that these

zircons crystallized from slightly evolved magmas, assuming that the zircons contain only single-age domains and that the ages and isotopic compositions are accurate. If those assumptions are correct, then we are presented with the intriguing possibility that an isotopically enriched reservoir existed as early as 3.97 Ga, perhaps similar to the one that gave rise to the 4.40 Ga Jack Hills zircon, in contrast to the general conclusions reached by Vervoort and Blichert-Toft (1999). However, all these hafnium isotope arguments are complicated by the present controversy over the decay constant of [176]Lu, described in Section 7.5.1.

The presence of very early evolved crustal sources is reinforced by oxygen isotope data for the ca. 4 Ga Jack Hills detrital zircons. Mojzsis *et al.* (2001) and Wilde *et al.* (2001) presented, independently, the first ion-microprobe oxygen isotope analyses of the >3.9 Ga detrital zircons. Wilde *et al.* (2001) determined $\delta^{18}O$ values of +5‰ and +7.4‰ (mean of two analyses each) for two different regions within the 4.40 Ga crystal and Mojzsis *et al.* (2001) determined $\delta^{18}O$ values of about +6‰ to +10‰ for 4.3–3.9 Ga zircons. Peck *et al.* (2001) confirmed this general range of values, but individual spot measurements did not exceed $\delta^{18}O$ of +8.6‰. The importance of these values is that because there is a fractionation of ∼1.5‰ between zircon and granitic magmas (e.g., Valley *et al.*, 1994), the estimated $\delta^{18}O$ values of the parental magmas to the Jack Hills zircons are in the vicinity of +8.5‰ to +9.5‰ (Peck *et al.*, 2001). Such values are well beyond those of typical mantle (+5.5‰ to +6‰), and are most consistent with granitic magmas having been derived in part from materials that were once exposed to supracrustal conditions (i.e., low-temperature weathering or diagenesis). This suggests the very early presence of some sort of hydrosphere, including possibly a 4.4 Ga ocean (e.g., Valley *et al.*, 2002).

ACKNOWLEDGMENTS

The authors are grateful to Clement Chase and Robert Butler for help with map figures, and to Roberta Rudnick and Mark Schmitz for their thorough reviews of this chapter. Samson acknowledges Syracuse University for granting a sabbatical leave, during which this chapter was written. Patchett was supported by NSF-EAR-0003343, and Samson by NSF-EAR-0106853.

REFERENCES

Albarède F. (2003) The thermal history of leaky chronometers above their closure temperature. *Geophys. Res. Lett.* **30**(1), 1015, doi:10.1029/2002GL016484.
Albarède F., Blichert-Toft J., Vervoort J. D., Gleason J. D., and Rosing M. (2000) Hf–Nd isotope evidence for a transient dynamic regime in the early terrestrial mantle. *Nature* **404**, 488–490.

Allègre C. J., Hart S. R., and Minster J. F. (1983a) Chemical structure and evolution of the mantle and continents determined by inversion of Nd and Sr isotopic data: I. Theoretical methods. *Earth Planet. Sci. Lett.* **66**, 177–190.

Allègre C. J., Hart S. R., and Minster J. F. (1983b) Chemical structure and evolution of the mantle and continents determined by inversion of Nd and Sr isotopic data: II. Numerical experiments and discussion. *Earth Planet. Sci. Lett.* **66**, 191–213.

Allen M. B., Windley B. F., and Chi Z. (1992) Palaeozoic collisional tectonics and magmatism of the Chinese Tien Shan, central Asia. *Tectonophysics* **220**, 89–115.

Amato J. M., Johnson C. M., Baumgartner L. P., and Beard B. L. (1999) Rapid exhumation of the Zermatt-Saas ophiolite deduced from high-precision Sm–Nd and Rb–Sr geochronology. *Earth Planet. Sci. Lett.* **171**, 425–438.

Amelin Y. V. (1998) Geochronology of the Jack Hills detrital zircons by precise U–Pb isotope dilution analysis of crystal fragments. *Chem. Geol.* **146**, 25–38.

Amelin Y. V., Lee D.-C., Halliday A. N., and Pidgeon R. T. (1999) Nature of the Earth's earliest crust from hafnium isotopes in single detrital zircons. *Nature* **399**, 252–255.

Anczkiewicz R., Thirlwall M., and Platt J. P. (2002) Influence of inclusions and leaching techniques on Sm–Nd and Lu–Hf garnet chronology. *Geochim. Cosmochim. Acta (Goldschmidt Conf. Abstr.)* **66**, A19.

Anderson C. A. and Hinthorne J. R. (1973) Thermodynamic approach to the quantitative interpretation of sputtered ion in mass spectra. *Anal. Chem.* **45**, 1421–1438.

Argles T. W., Prince E. I., Foster G. L., and Vauee D. (1999) New garnets for old? Cautionary tales from young mountain belts. *Earth Planet. Sci. lett.* **172**, 301–309.

Armstrong R. L. (1968) A model for the evolution of strontium and lead isotopes in a dynamic Earth. *Rev. Geophys.* **6**, 175–199.

Armstrong R. L. (1981) Radiogenic isotopes: the case for crustal recycling on a near-steady-state no-continental-growth Earth. *Phil. Trans. Roy. Soc. London* **A301**, 443–472.

Armstrong R. L. (1988) Mesozoic and early Cenozoic magmatic evolution of the Canadian Cordillera. *Geol. Soc. Am. Spec. Pap.* **218**, 55–91.

Armstrong R. L. (1991) The persistent myth of crustal growth. *Austral. J. Earth Sci.* **38**, 613–630.

Armstrong R. L. and Ward P. L. (1993) Late Triassic to earliest Eocene magmatism in the North American Cordillera: implications for the western interior basin. In *Evolution of the Western Interior Basin,* Geol. Ass. Can. Spec. Pap. (eds. W. G. E. Caldwell and E. G. Kauffman). Geological Association of Canada, St. Johns, Newfoundland, vol. 39, pp. 49–72.

Arth J. G., Barker F., and Stern T. W. (1988) Coast batholith and Taku plutons near Ketchikan, Alaska: petrography, geochronology, geochemistry and isotopic character. *Am. J. Sci.* **288A**, 461–489.

Ayer J., Amelin Y., Corfu F., Kamo S., Ketchum J., Kwok K., and Trowell N. (2002) Evolution of the southern Abitibi Greenstone belt based on U–Pb geochronology: autochthonous volcanic construction followed by plutonism, regional deformation and sedimentation. *Precamb. Res.* **115**, 63–95.

Barovich K. M., Patchett P. J., Peterman Z. E., and Sims P. K. (1989) Nd isotopes and the origin of 1.9–1.7 Ga Penokean continental crust of the Lake Superior Region. *Geol. Soc. Am. Bull.* **101**, 333–338.

Begemann F., Ludwig K. R., Lugmair G. W., Min K. W., Nyquist L. E., Patchett P. J., Renne P. R., Shih C.-Y., Villa I. M., and Walker R. J. (2001) Call for an improved set of decay constants for geochronological use. *Geochim. Cosmochim. Acta* **65**, 111–121.

Bennett V. C. and DePaolo D. J. (1987) Proterozoic crustal history of the western United States as determined by neodymium isotopic mapping. *Geol. Soc. Am. Bull.* **99**, 674–685.

Bingen B., Austrheim H., and Whitehouse M. (2001) Ilmenite as a source of zirconium during high-grade metamorphism? Textural evidence from the Caledonides of western Norway and implications for zircon geochronology. *J. Petrol.* **42**, 355–375.

Bizzarro M., Baker J. A., Haack H., Ulfbeck D., and Rosing M. (2003) Early history of Earth's crust–mantle system inferred from hafnium isotopes in chondrites. *Nature* **421**, 931–933.

Black L. P., Gale N. H., Moorbath S., Pankhurst R. J., and McGregor V. R. (1971) Isotopic dating of very early Precambrian amphibolite facies gneisses from the Godthaab district, west Greenland. *Earth Planet. Sci. Lett.* **12**, 245–259.

Blichert-Toft J., Albarède F., Rosing M., Frei R., and Bridgwater D. (1999) The Nd and Hf isotopic evolution of the mantle through the Archean. Results from the Isua supracrustals, west Greenland, and from the Birimian terranes of west Africa. *Geochim. Cosmochim. Acta* **63**, 3901–3914.

Blichert-Toft J., Boyet M., Télouk P., and Albarède F. (2002) ^{147}Sm–^{143}Nd and ^{176}Lu–^{176}Hf in eucrites and the differentiation of the HED parent body. *Earth Planet. Sci. Lett.* **204**, 245–259.

Boghossian N. D. and Gehrels G. E. (2000) Nd isotopic signature of metasedimentary pendants in the Coast Plutonic Complex between Prince Rupert and Bella Coola, British Columbia. In *Tectonics of the Coast Mountains,* Southeastern Alaska and Coastal British Columbia (eds. H. H. Stowell and W. C. McClelland). Geol. Soc. Am. Spec. Pap. 343, 77–87.

Boghossian N. D., Patchett P. J., Ross G. M., and Gehrels G. E. (1996) Nd isotopes and the source of sediments in the Miogeocline of the Canadian Cordillera. *J. Geol.* **104**, 259–277.

Boher M., Abouchami W., Michard A., Albarède F., and Arndt N. T. (1992) Crustal growth in West Africa at 2.1 Ga. *J. Geophys. Res.* **97**, 345–369.

Bowring S. A. and Housh T. (1995) The Earth's early evolution. *Science* **269**, 1535–1540.

Bowring S. A. and Podosek F. A. (1989) Nd isotopic evidence from Wopmay Orogen for 2.0–2.4 Ga crust in western North America. *Earth Planet. Sci. Lett.* **94**, 217–230.

Bowring S. A. and Van Schmus W. R. (1984) U–Pb zircon constraints on evolution of Wopmay Orogen N. W. T. *Geol. Assoc. Can.: Mineral. Assoc. Can. Prog. Abstr.* **9**, 47.

Bowring S. A. and Williams I. S. (1999) Priscoan (4.00–4.03 Ga) orthogneisses from northwestern Canada. *Contrib. Mineral. Petrol.* **134**, 3–16.

Bowring S. A., Williams I. S., and Compston W. (1989a) 3.96 Ga gneisses from the Slave province, Northwest Territories, Canada. *Geology* **17**, 971–975.

Bowring S. A., King J. E., Housh T. B., Isachsen C. E., and Podosek F. A. (1989b) Neodymium and lead isotope evidence for enriched early Archaean crust in North America. *Nature* **340**, 222–225.

Bowring S. A., Housh T. B., and Isachsen C. (1990) The Acasta gneisses: remnant of Earth's early crust. In *Origin of the Earth* (eds. H. E. Newsom and J. H. Jones). Oxford University Press, Oxford, pp. 319–343.

Brandon A. D. and Lambert R. StJ. (1993) Geochemical characterization of mid-Cretaceous granitoids of the Kootenay Arc in the southern Canadian Cordillera. *Can. J. Earth Sci.* **30**, 1076–1090.

Brandon A. D. and Lambert R. StJ. (1994) Crustal melting in the Cordilleran interior: the mid-Cretaceous White Creek batholith in the southern Canadian Cordillera. *J. Petrol.* **35**, 239–269.

Brandon A. D. and Smith A. D. (1994) Mesozoic granitoid magmatism in southeast British Columbia: implications for the origin of granitoid belts in the North American Cordillera. *J. Geophys. Res.* **99**, 11879–11896.

Bridgwater D. and Collerson K. D. (1976) The major petrological and geochemical characters of the 3,600 m.y. Uivak gneisses from Labrador. *Contrib. Mineral. Petrol.* **54**, 43–59.

Burton K. W. and O'Nions R. K. (1991) High-resolution garnet chronometry and the rates of metamorphic processes. *Earth Planet. Sci. Lett.* **107**, 649–671.

Burwash R. A., Cavell P. A., and Burwash E. J. (1988) Source terrains for Proterozoic sedimentary rocks in southern British Columbia: Nd isotopic and petrographic evidence. *Can. J. Earth Sci.* **25**, 824–832.

Butler R. F., Gehrels G. E., Crawford M. L., and Crawford W. A. (2001) Paleomagnetism of the Quottoon plutonic complex in the Coast Mountains of British Columbia and southeastern Alaska: evidence for tilting during uplift. *Can. J. Earth Sci.* **38**, 1367–1384.

Carlson R. W., Hunter D. R., and Barker F. (1983) Sm–Nd age and isotopic systematics of the bimodal suite, ancient gneiss complex, Swaziland. *Nature* **305**, 701–704.

Carlson W. D., Donelick R. A., and Ketcham R. A. (1999) Variability of apatite fission-track annealing kinetics: I. Experimental results. *Am. Mineral.* **84**, 1213–1223.

Chauvel C., Arndt N. T., Kielinczuk S., and Thom A. (1987) Formation of Canadian 1.9 Ga old continental crust: I. Nd isotopic data. *Can. J. Earth Sci.* **24**, 396–406.

Cheburkin A. K., Frei R., and Shotyk W. (1997) An energy-dispersive miniprobe multi-element analyzer (EMMA) for direct analysis of trace elements and chemical age dating of single mineral grains. *Chem. Geol.* **135**, 75–87.

Chen B. and Jahn B. M. (2002) Geochemical and isotopic studies of the sedimentary and granitic rocks of the Altai Orogen of northwest China and their tectonic implications. *Geol. Mag.* **139**, 1–13.

Chen C.-H., DePaolo D. J., and Lan C.-Y. (1996) Rb–Sr microchrons in the Manaslu granite: implications for Himalayan thermochronology. *Earth Planet. Sci. Lett.* **143**, 125–135.

Christensen J. N., Rosenfeld J. L., and DePaolo D. J. (1989) Rates of tectonometamorphic processes from rubidium and strontium isotopes in garnet. *Science* **244**, 1465–1469.

Claesson S. (1987) Nd isotope data on 1.9–1.2 Ga old basic rocks and meta-sediments from the Bothnian Basin, central Sweden. *Precamb. Res.* **35**, 115–126.

Clement S. W. J., Compston W., and Newstead G. (1977) Design of a large, high resolution ion microprobe. In *Proceedings of the International Secondary Ion Mass Spectrometry Conference* (ed. A. Benninghoven). Spinger, Berlin, 12pp.

Cliff R. A. (1994) Rb–Sr dating of white mica- a new potential in metamorphic geochronology. *US Geol. Surv. Circular* **1107**, 62.

Clowes R., Cook F., Hajnal Z., Hall J., Lewry J. F., Lucas S., and Wardle R. (1999) Canada's Lithoprobe project (collaborative, mutildisciplinary geoscience research leads to new understanding of continental evolution). *Episodes* **22**, 3–20.

Collerson K. D., Campbell L. M., Weaver B. L., and Palacz Z. A. (1991) Evidence for extreme mantle fractionation in early Archean ultramafic rocks from northern Labrador. *Nature* **349**, 209–214.

Compston W. and Jeffery P. M. (1959) Anomalous common strontium in granite. *Nature* **184**, 1792–1793.

Compston W. and Pidgeon R. T. (1986) Jack Hills, evidence of more very old detrital zircons in Western Australia. *Nature* **321**, 766–769.

Compston W., Williams I. S., and Clement S. W. J. (1982) U–Pb ages within single zircons using a sensitive high mass resolution ion microprobe. *30th Ann. Conf. Am. Soc. Mass Spectrom.* 393–593.

Coney P. J., Jones D. L., and Monger J. W. H. (1980) Cordilleran suspect terranes. *Nature* **288**, 329–333.

Condie K. C. (1990) Growth and accretion of continental crust: inferences based on Laurentia. *Chem. Geol.* **83**, 183–194.

Condie K. C. and Chomiak B. (1996) Continental accretion: contrasting Mesozoic and Early Proterozoic tectonic regimes in North America. *Tectonophysics* **265**, 101–126.

Corfu F. and Easton R. M. (2000) U–Pb evidence for polymetamorphic history of Huronian rocks within the Grenville front tectonic zone east of Sudbury, Ontario, Canada. *Chem. Geol.* **172**, 149–171.

Corfu F. and Stott G. M. (1996) Hf isotopic composition and age constraints on the evolution of the Archean central Uchi Subprovince, Ontario, Canada. *Precamb. Res.* **78**, 53–63.

Creaser R. A., Erdmer P., Stevens R. A., and Grant S. L. (1997) Tectonic affinity of Nisutlin and Anvil assemblage strata from the Teslin tectonic zone, northern Canadian Cordillera: constraints from neodymium isotope and geochemical evidence. *Tectonics* **16**, 107–121.

Cui Y. and Russell J. K. (1995) Nd–Sr–Pb isotopic studies of the southern Coast Plutonic Complex, southwestern British Columbia. *Geol. Soc. Am. Bull.* **107**, 127–138.

Dallmeyer R. D. (1975) Incremental ^{40}Ar/^{39}Ar ages of biotite and hornblende from retrograded basement gneisses of the southern Blue Ridge: their bearing on the age of Paleozoic metamorphism. *Am. J. Sci.* **275**, 444–460.

Dallmeyer R. D., Sutter J. F., and Baker D. J. (1975) Incremental ^{40}Ar/^{39}Ar ages of biotite and hornblende from the northeastern Reading Prong: their bearing on late Proterozoic thermal and tectonic history. *Geol. Soc. Am. Bull.* **86**, 1435–1443.

Damon P. E. (1957) Terrestrial helium. *Geochim. Cosmochim. Acta* **11**, 200–201.

Davies R. D. and Allsopp H. L. (1976) Strontium isotopic evidence relating to the evolution of the lower Precambrian granitic crust in Swaziland. *Geology* **4**, 553–556.

DePaolo D. J. (1979) Implications of correlated Nd and Sr isotopic variations for the chemical evolution of the crust and mantle. *Earth Planet. Sci. Lett.* **43**, 201–211.

DePaolo D. J. (1981) Neodymium isotopes in the Colorado front Range and crust–mantle evolution in the Proterozoic. *Nature* **291**, 193–196.

De Wolf C. P., Zeissler C. J., Halliday A. N., Mezger K., and Essene E. J. (1996) The role of inclusion in U–Pb and Sm–Nd garnet chronology: stepwise dissolution experiments and trace uranium mapping by fission track analysis. *Geochim. Cosmochim. Acta* **60**, 121–134.

DiVincenzo G., Ghiribelli G., and Palmeri R. (2001) Evidence of a close link between petrology and isotope records: constraints from SEM, EMP, TEM and *in situ* Ar-40–Ar-39 laser analyses on multiple generations of white micas (Lanterman range, Antarctica). *Earth Planet. Sci. Lett.* **192**, 389–405.

Dodson M. H. (1973) Closure temperature in cooling geochronological and petrological systems. *Contrib. Mineral. Petrol.* **40**, 259–274.

Donelick R. A., Roden M. K., Mooers J., Carpenter B. S., and Miller D. S. (1990) Etchable length reduction of induced fission tracks in apatite at room temperature (23 °C): crystallographic orientation effects and "initial" mean lengths. *Nucl. Tracks Radiat. Meas.* **17**, 261–266.

Driver L. A., Creaser R. A., Chacko T., and Erdmer P. (2000) Petrogenesis of the Cretaceous Cassiar batholith, Yukon-B. C., Canada: implications for magmatism in the North American Cordilleran Interior. *Geol. Soc. Am. Bull.* **112**, 1119–1133.

Ducea M. N., Ganguly J., Rosenberg E., Patchett P. J., Cheng W., and Isachsen C. (2003) Sm–Nd dating of spatially controlled domains of garnet single crystals: a new method of high temperature thermochronology. *Earth Planet. Sci. Lett.* **213**, 31–42.

Duchêne S., Blichert-Toft J., Luais P., Telouk P., Lardeaux J.-M., and Albarède F. (1997) The Lu–Hf dating of garnets and the ages of the Alpine high-pressure metamorphism. *Nature* **387**, 586–589.

Eggins S. M., Kinsley L. P. J., and Shelley J. M. M. (1998) Deposition and element fractionation processes during atmospheric pressure laser sampling for analysis by ICPMS. *Appl. Surf. Sci.* **127–129**, 278–286.

Ehlers T. A. and Farley K. A. (2003) Apatite (U–Th)/He thermochronometry: methods and applications to problems in tectonic and surface processes. *Earth Planet. Sci. Lett.* **206**, 1–14.

Engi M., Cheburkin A. K., and Köppel V. (2002) Nondestructive chemical dating of young monazite using XRF: 1. Design of a mini-probe, age data for samples from the Central Alps, and comparison to U–Pb (TIMS) data. *Chem. Geol.* **191**, 225–241.

Erdmer P., Moore J. M., Heaman L. M., Thompson R. I., Daughtry K. L., and Creaser R. A. (2002) Extending the ancient margin outboard in the Canadian Cordillera: evidence of Proterozoic crust and Paleocene regional metamorphism in the Nicola horst, southeastern British Columbia. *Can. J. Earth Sci.* **39**, 1605–1623.

Fairbairn H. W., Hurley P. M., and Pinson W. H. (1961) The relation of discordant Rb–Sr mineral and rock ages in an igneous rock to its time of subsequent Sr87/Sr86 metamorphism. *Geochim. Cosmochim. Acta* **23**, 135–144.

Farley K. A. (2002) (U–Th)/He dating: techniques, calibrations, and applications. In *Noble Gases in Geochemistry, Rev. Mineral. Geochem.* (eds. P. D. Porcelli, C. J. Ballentine, and R. Wieler). Mineralogical Society of America, Washington, DC, vol. 47, pp. 819–843.

Farley K. A., Wolf R. A., and Silver L. T. (1996) The effects of long alpha-stopping distances on (U–Th)/He ages. *Geochim. Cosmochim. Acta* **60**, 4223–4229.

Farmer G. L., Ayuso R., and Plafker G. (1993) A Coast Mountains provenance for the Valdez and Orca groups, southern Alaska, based on Nd, Sr, and Pb isotopic evidence. *Earth Planet. Sci. Lett.* **116**, 9–21.

Feng R., Machado N., and Ludden J. (1993) Lead geochronology of zircon by laserprobe-inductively coupled plasma mass spectrometry (LA-ICPMS). *Geochim. Cosmochim. Acta* **57**, 3479–3486.

Fernández-Suárez J., Gutierrez A. G., Jenner G. A., and Tubrett M. N. (1999) Crustal sources in lower Paleozoic rocks from NW Iberia: insights from laser-ablation U–Pb ages of detrital zircons. *J. Geol. Soc. London* **156**, 1065–1068.

Fitzgerald P. G., Sorkhabi R. B., Redfield T. F., and Stump E. (1995) Uplift and denudation of the central Alaska Range: a case study in the use of apatite fission track thermochronology to determine absolute uplift parameters. *J. Geophys. Res.* **100**, 20175–20191.

Fraser G., Ellis D., and Eggins S. (1997) Zirconium abundance in granulite-facies minerals, with implications for zircon geochronology in high-grade rocks. *Geology* **25**, 607–610.

Friedman R. M., Mahoney J. B., and Cui Y. (1995) Magmatic evolution of the southern Coast Belt: constraints from Nd–Sr isotopic systematics and geochronology of the southern Coast Plutonic Complex. *Can. J. Earth Sci.* **32**, 1681–1698.

Frost C. D. and Coombs D. S. (1989) Nd isotope character of New Zealand sediments: implications for terrane concepts and crustal evolution. *Am. J. Sci.* **289**, 744–770.

Froude D. O., Ireland T. R., Kinny P. D., Williams I. S., Compston W., Williams I. R., and Myers J. S. (1983) Ion microprobe identification of 4,100-4,200 Myr-old terrestrial zircons. *Nature* **304**, 616–618.

Fryer B. J., Jackson S. E., and Longerich H. P. (1993) The application of laser ablation microprobe-inductively coupled plasma-mass spectrometry (LAM-ICP-MS) to in situ (U)-Pb geochronology. *Chem. Geol.* **109**, 1–8.

Gleadow A. J. W., Duddy I. R., Green P. F., and Lovering J. F. (1986) Confired track lengths in apatiter—a diagnostic tool for thermal arnlysis. **96**, 1–14.

Gabrielse H. (1985) Major dextral transcurrent displacements along the Northern Rocky Mountain Trench and related lineaments in north-central British Columbia. *Geol. Soc. Am. Bull.* **96**, 1–14.

Gallagher K., Brown R., and Johnson C. (1998) Fission track analysis and its application to geologic problems. *Ann. Rev. Earth Planet. Sci.* **26**, 519–572.

Ganguly J. and Tirone M. (1999) Diffusion closure temperature and age of a mineral with arbitrary extent of diffusion: theoretical formulation and applications. *Earth Planet. Sci. Lett.* **170**, 131–140.

Ganguly J. and Tirone M. (2001) Relationship between cooling rate and cooling age of a mineral: theory and applications to meteorites. *Meteorit. Planet. Sci.* **36**, 167–175.

Ganguly J., Tirone M., and Hervig R. L. (1998) Diffusion kinetics of samarium and neodymium in garnet, and a method of determining cooling rates of rocks. *Science* **281**, 805–807.

Garzione C. N., Patchett P. J., Ross G. M., and Nelson J. (1997) Provenance of sedimentary rocks in the Canadian Cordilleran Miogeocline: a Nd isotopic study. *Can. J. Earth Sci.* **34**, 1603–1618.

Geersten C., Briand A., Chartier F., Lacour J.-L., Mauchient P., and Sjöstrom S. (1994) Comparison between infrared and ultraviolet laser ablation at atmospheric pressure-implications for solid sampling inductively coupled plasma spectrometry. *J. Anal. Atom. Spectrom.* **9**, 17–22.

Gerstenberger H. and Haase G. (1997) A highly effective emitter substance for mass spectrometric Pb isotope ratio determinations. *Chem. Geol.* **136**, 309–312.

Getty S. R., Selverstone J., Wernicke B. P., Jacobsen S. B., Aliberti E., and Lux D. R. (1993) Sm–Nd dating of multiple garnet growth events in an arc-continent collision zone, northwestern US Cordillera. *Contrib. Mineral. Petrol.* **115**, 45–57.

Ghosh D. K. (1995) Nd–Sr isotopic constraints on the interactions of the Intermontane superterrane with the western edge of North America in the southern Canadian Cordillera. *Can. J. Earth Sci.* **32**, 1740–1758.

Ghosh D. K. and Lambert R. St. J. (1989) Nd–Sr isotopic study of Proterozoic to Triassic sediments from southeastern British Columbia. *Earth Planet. Sci. Lett.* **94**, 29–44.

Ghosh D. K. and Lambert R. St. J. (1995) Nd–Sr isotope geochemistry and petrogenesis of Jurassic granitoid intrusives, southeast British Columbia, Canada. In *Jurassic Magmatism and Tectonics of the North American Cordillera*, Geol. Soc. Am. Spec. Pap. (eds. D. M. Miller and C. Busby). The Geological society of America, Boulder, CO, vol. 299, pp. 141–157.

Gleadow A. J. W., Duddy I. R., Green P. F., and Lovering J. F. (1986) Confired track lengths in apatite—a diagnostic tool for thermal analysis. *Contrib. Mineral. Petrol.* **94**, 405–415.

Goldich S. S., Hedge C. E., and Stern T. W. (1970) Age of the Morton and Montevideo gneisses and related rocks, southwestern Minnesota. *Geol. Soc. Am. Bull.* **81**, 3671–3695.

Goldstein S. L., O'Nions R. K., and Hamilton P. J. (1984) A Sm–Nd isotopic study of atmospheric dusts and particulates from major river systems. *Earth Planet. Sci. Lett.* **70**, 221–236.

Green P. F., Duddy I. R., Gleadow A. J. W., Tingate P. R., and Laslett G. M. (1986) Thermal annealing of fission tracks in apatite: 1. A qualitative description. *Chem. Geol. (Isot. Geosci. Sect.)* **59**, 237–253.

Green P. F., Duddy I. R., Laslett G. M., Hegarty K. A., Gleadow A. J. W., and Lovering J. F. (1989) Thermal annealing of fission-tracks in apatite: 4. Quantitative modeling techniques and extension to geological time scales. *Chem. Geol. (Isot. Geosci. Sect.)* **79**, 155–182.

Gruau G., Rosing M., Bridgwater D., and Gill R. C. O. (1996) Resetting of Sm–Nd systematics during metamorphism of >3.7 Ga rocks: implications for isotopic models of early Earth differentiation. *Chem. Geol.* **133**, 225–240.

Hamilton P. J., O'Nions R. K., Bridgwater D., and Nutman A. (1983) Sm–Nd studies of Archaean metasediments and metavolcanics from west Greenland and their implications for the Earth's early history. *Earth Planet. Sci. Lett.* **63**, 263–272.

Han B. F., Wang S. G., Jahn B. M., Hong D. W., Hiroo K., and Sun Y. L. (1997) Depleted mantle source for the Ulungur River A-type granites from North Xinjiang, China: geochemistry and Nd–Sr isotopic evidence, and implications for Phanerozoic crustal growth. *Chem. Geol.* **138**, 135–159.

Hanson G. N. and Gast P. W. (1967) Kinetic studies in contact metamorphic zones. *Geochim. Cosmochim. Acta* **31**, 1119–1153.

Hart S. R. (1964) The petrology and isotopic-mineral age relations of a contact zone in the Front Range, Colorado. *J. Geol.* **72**, 493–525.

Heaman L. M. and Parrish R. (1991) U–Pb geochronology of accessory minerals. In *Applications of Radiogenic Isotope Systems to Problems in Geology*, Short Course Handbook 19 (eds. L. Heaman and J. N. Ludden). Mineralogical Association of Canada, Toronto, pp. 59–102.

Hegner E. and Bevier M. L. (1991) Nd and Pb isotopic constraints on the origin of the Purtuniq ophiolite and Early Proterozoic Cape Smith belt, northern Quebec, Canada. *Chem. Geol.* **91**, 357–371.

Heizler M. T., Lux D. R., and Decker E. R. (1988) The age and cooling history of the Chain of Ponds and Big Island Pond plutons and the Spider Lake Granite, west-central Maine and Quebec. *Am. J. Sci.* **288**, 925–952.

Henderson J. B. (1985) Geology of the Yellowknife-Hearne Lake Area, District of Mackenzie: segment across an Archean basin. *Geol. Surv. Can. Mem.* **414**, 135pp.

Hinthorne J. R., Andersen C. A., Conrad R. L., and Lovering J. F. (1979) Single grain $^{207}Pb/^{206}Pb$ and U/Pb age determinations with a 10-μm spatial resolution using the ion microprobe mass analyzer. *Chem. Geol.* **25**, 271–303.

Hofmann A. W. (1988) Chemical differentiation of the Earth: the relationship between mantle, continental crust, and oceanic crust. *Earth Planet. Sci. Lett.* **90**, 297–314.

Holmes A. (1911) The association of lead with uranium in rock-minerals, and its application to the measurement of geological time. *Proc. Roy. Soc. London* **A85**, 248–256.

Horn I., Rudnick R. L., and McDonough W. F. (2000) Precise elemental and isotope ratio determination by simultaneous solution nebulization and laser-ablation ICP-MS: application to *in situ* U/Pb geochronology. *Chem. Geol.* **164**, 281–301.

Horstwood M. S. A., Nesbitt R. W., Noble S. A., and Wilson J. F. (1999) U–Pb zircon evidence fo an extesive early Archean craton in Zimbabwe: a reassessment of the timing of craton formation, stabilization, and growth. *Geology* **27**, 707–710.

House M., Farley K. A., and Stöckli D. (2000) Helium chronometry of apatite and titanite using Nd-YAG laser heating. *Earth Planet. Sci. Lett.* **183**, 365–368.

Hu A. Q., Jahn B. M., Zhang G. X., and Zhang Q. F. (2000) Crustal evolution and Phanerozoic crustal growth in northern Xinjiang: Nd–Sr isotopic evidence: Part I. Isotopic characterisation of basement rocks. *Tectonophysics* **328**, 15–51.

Huhma H. (1986) Sm–Nd, U–Pb, and Pb–Pb isotopic evidence for the origin of the Early Proterozoic Svecokarelian crust in Finland. *Geol. Surv. Finland Bull.* **337**, 5.

Hurley P. M. (1954) The helium age method and the distribution and migration of helium in rocks. In *Nuclear Geology* (ed. H. Faul). Wiley, New York, pp. 301–329.

Hurst R. W., Bridgwater D., Collerson K. D., and Wetherill G. W. (1975) 3600-m.y. Rb–Sr ages from very early Archaean gneisses from Saglek Bay, Labrador. *Earth Planet. Sci. Lett.* **27**, 393–403.

Jacobsen S. B. and Wasserburg G. J. (1979) The mean age of mantle and crustal reservoirs. *J. Geophys. Res.* **84**, 7411–7427.

Jacobsen S. B. and Wasserburg G. J. (1984) Sm–Nd isotopic evolution of chondrites and achondrites: II. *Earth Planet. Sci. Lett.* **67**, 137–150.

Jackson S. E., Longerich H. P., Horn I., and Dunning G. R. (1996) The application of laser-ablation microprobe (LAM)-ICP-MS to *in situ* U–Pb zircon geochronology. *J. Conf. Abstr. (V. M. Goldschmidt Conference)* **1**, 283.

Jahn B. M., Wu F. Y., and Chen B. (2000) Massive granitoid generation in central Asia: Nd isotope evidence and implication for continental growth in the Phanerozoic. *Episodes* **23**, 82–92.

Johnston S. T. (2001) The great Alaskan train wreck: reconciliation of paleomagnetic and geological data in the northern Cordillera. *Earth Planet. Sci. Lett.* **193**, 259–272.

Kalsbeek F. and Taylor P. N. (1985) Isotopic and chemical variation in granites across a Proterozoic continental margin- the Ketilidian mobile belt of South Greenland. *Earth Planet. Sci. Lett.* **73**, 65–80.

Kalsbeek F., Pidgeon R. T., and Taylor P. N. (1987) Nagssugtoqidian mobile belt of West Greenland: a cryptic 1850 Ma suture between two Archean continents—chemical and isotopic evidence. *Earth Planet. Sci. Lett.* **85**, 365–385.

Karlstrom K. E., Williams M. L., McLelland J., Geissman J. W., and Åhäll K.-I. (1999) Refining Rodinia: geologic evidence for the Australia-western USA connection in the Proterozoic. *GSA Today* **9**(10), 1–7.

Kelley S. P. (1995) Ar–Ar dating by laser microprobe. In *Microprobe Techniques in the Earth Sciences: Mineralogical Society Series 6* (eds. P. J. Potts, J. F. W. Bowles, S. J. B. Reed, and M. R. Cave). Kluwer, Dordrecht, pp. 123–143.

Kelley S. P., Arnaud N. O., and Turner S. P. (1994) High spatial resolution $^{40}Ar/^{39}Ar$ investigations using an ultraviolet laser probe extraction technique. *Geochim. Cosmochim. Acta* **58**, 3519–3525.

Kelley S. P., Bartlett J. M., and Harris N. B. W. (1997) Pre-metamorphic Ar–Ar ages from biotite inclusions in garnet. *Geochim. Cosmochim. Acta* **61**, 3873–3878.

Ketcham R. A., Donelick R. A., and Carlson W. D. (1999) Variability of apatite fission-track annealing kinetics: III. Extrapolation to geological timescales. *Am. Mineral.* **84**, 1235–1255.

Kinny P. D., Compston W., and Williams I. S. (1991) A reconnaissance ion-probe study of hafnium isotopes in zircons. *Geochim. Cosmochim. Acta* **55**, 849–859.

Knudsen T.-L., Griffin W. L., Hartz E. H., Andersen A., and Jackson S. E. (2001) *In-situ* hafnium and lead isotope analyses of detrital zircons from the Devonian sedimentary basin of NE Greenland: a record of repeated crustal reworking. *Contrib. Mineral. Petrol.* **141**, 83–94.

Kober B. (1986) Whole-grain evaporation for $^{207}Pb/^{206}Pb$-age-investigations on single zircons using a double-filament thermal ion source. *Contrib. Mineral. Petrol.* **93**, 482–490.

Kober B. (1987) Single-zircon evaporation combined with Pb^+ emitter-bedding for $^{207}Pb/^{206}Pb$-age investigations using thermal ion mass spectrometry, and implications to zirconology. *Contrib. Mineral. Petrol.* **96**, 63–71.

Kober B., Pidgeon R. T., and Lippolt H. J. (1989) Single-zircon dating by stepwise Pb-evaporation constrains the Archean history of detrital zircons from the Jack Hills, Western Australia. *Earth Planet. Sci. Lett.* **91**, 286–296.

Kovalenko V. I., Yarmolyuk V. V., Kovach V. P., Budnikov S. V., Zhuravlev D. Z., Kozakov I. K., Kotov A. B., Rytsk E. Y., and Salnikova E. B. (1999) Magmatism as factor of crust evolution in the Central Asian Fold Belt: Sm–Nd isotopic data. *Geotectonics* **33**, 191–208.

Kovalenko V. I., Yarmolyuk V. V., Kovach V. P., Kotov A. B., Kozakov I. K., Salnikova E. B., and Larin A. M. (2003) Isotope provinces, mechanisms of generation and sources of the continental crust in the Central Asian Mobile Belt: geological and isotopic evidence. *J. Asian Earth Sci.*, (in press).

Krogh T. E. (1971a) A low contamination method for the decomposition of zircon and the extraction of U and Pb for isotopic age determinations. *Carnegie Inst. Wash. Yearb.* **70**, 258–266.

Krogh T. E. (1971b) A simplified technique for the dissolution of zircon and the isolation of uranium and lead. *Carnegie Inst. Wash. Yearb.* **69**, 342–344.

Krogh T. E. (1973) A low-contamination method for hydrothermal decomposition of zircon and extraction of U and Pb for isotopic age determinations. *Geochim. Cosmochim. Acta* **37**, 485–494.

Krogh T. E. (1982) Improved accuracy of U–Pb zircon ages by the creation of more concordant systems using an air abra sion technique. *Geochim. Cosmochim. Acta* **46**, 637–649.

Krogh T. E. and Davis G. L. (1975) The production and preparation of 205Pb for use as a tracer for isotope dilution analyses. *Carnegie Inst. Wash. Yearb.* **74**, 416–417.

Krogstad E. and Walker R. J. (1996) Evidence of heterogeneous crustal sources: the Harney Peak granite, South Dakota USA. *Trans. Roy. Soc. Edinburgh: Earth Sci.* **87**, 331–337.

Kröner A., Compston W., and Williams I. S. (1989) Growth of early Archaean crust in the Ancient Gneiss Complex of Swaziland as revealed by single zircon dating. *Tectonophysics* **161**, 271–298.

Lanphere M. A. and Albee A. L. (1974) ^{40}Ar/^{39}Ar age measurements in the Worcester Mountians: evidence of Ordovician and Devonian metamorphic events in northern Vermont. *Am. J. Sci.* **274**, 545–555.

Laslett G. M., Green P. F., Duddy I. R., and Gleadow A. J. W. (1987) Thermal annealing of fission tracks in apatite: 2. A quantitative analysis. *Chem. Geol. (Isot. Geosci. Sect.)* **65**, 1–13.

Layer P. W., Hall C. M., and York D. (1987) The derivation of ^{40}Ar/^{39}Ar age spectra of single grains of hornblende and biotite by laser step-heating. *Geophys. Res. Lett.* **14**, 757–760.

Lippolt H. J., Leitz M., Wernicke R. S., and Hagedorn B. (1994) (U + Th)/He dating of apatite: experience with samples from different geological environments. *Chem. Geol.* **112**, 179–191.

Litvinovsky B. A., Jahn B. M., Zanvilevich A. N., Saunders A., Poulain S., Kuzmin D. V., Reichow M. K., and Titov A. V. (2002) Petrogenesis of syenite-granite suites from the Bryansky Complex (Transbaikalia, Russia): implications for the origin of A-type granitoid magmas. *Chem. Geol.* **189**, 105–133.

Longerich H. P., Fryer B. J., and Strong D. F. (1987) Determination of lead isotope ratios by inductively coupled plasma-mass spectrometry (ICP-MS). *Spectrochem. Acta* **42B**, 39–48.

Lovera O. M. and Richter F. M. (1991) Diffusion domains determined by ^{39}Ar released during step heating. *J. Geophys. Res.* **96**, 2057–2069.

Lovera O. M., Richter F. M., and Harrison T. M. (1989) The ^{40}Ar/^{39}Ar thermochronometry for slowly cooled samples having a distribution of diffusion domain sizes. *J. Geophys. Res.* **94**, 17917–17935.

Lucas S. B., Green A., Hajnal Z., White D., Lewry J. F., Ashton K., Weber W., and Clowes R. (1993) Deep seismic profile across a Proterozoic collision zone: surprises at depth. *Nature* **363**, 339–342.

Lugmair G. W. and Marti K. (1978) Lunar initial 143Nd/144Nd: differential evolution of the lunar crust and mantle. *Earth Planet. Sci. Lett.* **39**, 349–357.

Maas R., Kinny P. D., Williams I. S., Froude D. O., and Compston W. (1992) The Earth's oldest known crust: a geochronological and geochemical study of 3,900–4,200 Ma old detrital zircons from Mt. Narryer and Jack Hills, Western Australia. *Geochim. Cosmochim. Acta* **56**, 1281–1300.

Machado N. and Simonetti A. (2001) U–Pb dating and Hf isotopic composition of zircon by laser-ablation MC-ICP-MS. In *Laser-ablation ICP-MS in the Earth Sciences*, Mineral. Ass. Can. Short Course Ser. (ed. P. Sylvester). The mineralogical Association of Canada, Ottawa, ON, vol. 29, pp. 121–146.

Mahoney J. B., Friedman R. M., and McKinley S. D. (1995) Evolution of a Middle Jurassic volcanic arc: stratigraphic, isotopic and geochemical characteristics of the Harrison Lake Formation, southwestern British Columbia. *Can. J. Earth Sci.* **32**, 1759–1776.

Maluski H. and Schaeffer O. A. (1982) ^{39}Ar–^{40}Ar laser probe dating of terrestrial rocks. *Earth Planet. Sci. Lett.* **59**, 21–27.

McCulloch M. T. and Bennett V. C. (1994) Progressive growth of the Earth's continental crust and depleted mantle: geochemical constraints. *Geochim. Cosmochim. Acta* **58**, 4717–4738.

McCulloch M. T. and Wasserburg G. J. (1978) Sm–Nd and Rb–Sr chronology of continental crust formation. *Science* **200**, 1003–1011.

McDougall I. and Harrison T. M. (1999) *Geochronology and Thermochronology by the ^{40}Ar/^{39}Ar Method*, 2nd edn. Oxford University Press, Oxford, 269p.

Megrue G. H. (1973) Spatial distribution of ^{40}Ar/^{39}Ar ages in lunar breccia 14301. *J. Geophys. Res.* **78**, 3216–3221.

Merrihue C. and Turner G. (1966) Potassium–argon dating by activation with fast neutrons. *J. Geophys. Res.* **71**, 2852–2856.

Mezger J. E., Creaser R. A., and Erdmer P. (2001) Cretaceous back-arc basin along the Coast Belt of the northern Canadian Cordillera: evidence from geochemical and Nd isotopic signatures of the Kluane metamorphic assemblage, SW Yukon. *Can. J. Earth Sci.* **38**, 91–103.

Mezger K. and Krogstad E. J. (1997) Interpretation of discordant U–Pb zircon ages: an evaluation. *J. Metamorph. Geol.* **15**, 127–140.

Mezger K., Essene E. J., and Halliday A. N. (1992) Closure temperatures of the Sm–Nd system in metamorphic garnets. *Earth Planet. Sci. Lett.* **113**, 397–409.

Mojzsis S. J., Harrison T. M., and Pidgeon R. T. (2001) Oxygen-isotope evidence from ancient zircons for liquid water at the Earth's surface 4,300 Myr ago. *Nature* **409**, 178–181.

Monger J. W. H. (1993) Canadian Cordilleran tectonics: from geosynclines to crustal collage. *Can. J. Earth Sci.* **30**, 209–231.

Monger J. W. H., Souther J. G., and Gabrielse H. (1972) Evolution of the Canadian Cordillera: a plate-tectonic model. *Am. J. Sci.* **272**, 577–602.

Monger J. W. H., Price R. A., and Tempelman-Kluit D. J. (1982) Tectonic accretion and the origin of the two major metamorphic and plutonic welts in the Canadian Cordillera. *Geology* **10**, 70–75.

Montel J.-M., Veschambre M., and Nicollet C. (1994) Datation de la monazite à la microsonde électronique. *Compt. Rend. Acad. Sci. Paris* **318**, 1489–1495.

Montel J.-M., Foret S., Veschambre M., Nicollet C., and Povost A. (1996) Electron microprobe dating of monazite. *Chem. Geol.* **131**, 37–53.

Montel J.-M., Kornprobst J., and Vielzeuf D. (2000) Preservation of old U–Th–Pb ages in shielded monazite: example from the Beni Bousera Hercynian kinzigites (Morocco). *J. Metamorph. Geol.* **18**, 335–342.

Moorbath S. (1975) Evolution of Precambrian crust from strontium isotopic evidence. *Nature* **254**, 395–398.

Moorbath S. (1978) Age and isotope evidence for the evolution of continental crust. *Phil. Trans. Roy. Soc. London* **A288**, 401–413.

Moorbath S., O'Nions R. K., Pankhurst R. J., Gale N. H., and McGregor V. R. (1972) Further rubidium–strontium age determinations on the very early Precambrian rocks of the Godthaab district, west Greenland. *Nature* **240**, 78–82.

Moorbath S., Whitehouse M. J., and Kamber B. S. (1997) Extreme Nd-isotope heterogeneity in the early Archaean-fact or fiction? Case histories from northern Canada and west Greenland. *Chem. Geol.* **135**, 213–231.

Müller W. (2003) Strengthening the link between geochronology, textures and petrology. *Earth Planet. Sci. Lett.* **206**, 237–251.

Müller W., Aerden D., and Halliday A. N. (2000a) Isotopic dating of strain fringe increments: duration and rates of deformation in shear zones. *Science* **288**, 2195–2198.

Müller W., Mancktelow N. S., and Meier M. (2000b) Rb–Sr microchrons of synkinematic mica in mylonites: an example from the DAV fault of the eastern Alps. *Earth Planet. Sci. Lett.* **180**, 385–397.

Myers J. S. (1995) The generation and assembly of an Archean supercontinent: evidence from the Yilgarn craton, Western Australia. In *Early Precambrian Processes*, Geol. Soc. Spec. Publ. (eds. M. P. Coward and A. C. Ries). Geological Society of London, London, pp. 143–154.

Nelson B. K. and DePaolo D. J. (1984) 1,700-Myr greenstone volcanic successions in southwestern North America and isotopic evolution of Proterozoic mantle. *Nature* **311**, 143–146.

Nelson B. K. and DePaolo D. J. (1985) Rapid production of continental crust 1.7 to 1.9 b.y. ago: Nd isotopic evidence from the basement of the North America mid-continent. *Geol. Soc. Am. Bull.* **96**, 746–754.

Nier A. O. (1939) The isotopic constitution of radiogenic lead and the measurement of geological time. *Phys. Rev. Ser. 2,* **55**, 153–163.

Öhlander B., Mellqvist C., and Skiöld T. (1999) Sm–Nd isotope evidence of a collisional event in the Pre-Cambrian of northern Sweden. *Precamb. Res.* **93**, 105–117.

O'Nions R. K., Evensen N. M., and Hamilton P. J. (1979) Geochemical modeling of mantle differentiation and crustal growth. *J. Geophys. Res.* **84**, 6091–6101.

Padgham W. A. (1985) Observations and speculations on supracrustal successions in the Slave Structural Province. In *Evolution of Archean Supracrustal Sequences,* Geol. Ass. Can. Spec. Pap. 28 (eds. L. D. Ayres, P. C. Thurston, K. D. Card, and W. Weber). Geological Association of Canada, St. Johns, Newfoundland, pp. 156–167.

Papanastassiou D. A. and Wasserburg G. J. (1981) Microchrons: the ^{87}Rb–^{87}Sr dating of microscopic samples. *Proc. 12th Lunar Planet. Sci. Conf. B,* 1027–1038.

Parrish R. R. (1983) Cenozoic thermal evolution and tectonics of the Coast Mountains of British Columbia: 1. Fission track dating, apparent uplift rates, and patterns of uplift. *Tectonics* **2**, 601–631.

Parrish R. R. and Krogh T. E. (1987) Synthesis and purification of 205Pb for U–Pb geochronology. *Chem. Geol. (Isot. Geosci. Sect.)* **66**, 103–110.

Patchett P. J. and Arndt N. T. (1986) Nd isotopes and tectonics of 1.9–1.7 Ga crustal genesis. *Earth Planet. Sci. Lett.* **78**, 329–338.

Patchett P. J. and Bridgwater D. (1984) Origin of continental crust of 1.9–1.7 Ga age defined by Nd isotopes in the Ketilidian terrain of South Greenland. *Contrib. Mineral. Petrol.* **87**, 311–318.

Patchett P. J. and Chase C. G. (2002) Role of transform continental margins in major crustal growth episodes. *Geology* **30**, 39–42.

Patchett P. J. and Gehrels G. E. (1998) Continental influence on Canadian Cordilleran terrains from Nd isotopic study, and significance for crustal growth processes. *J. Geol.* **106**, 269–280.

Patchett P. J. and Kouvo O. (1986) Origin of continental crust of 1.9–1.7 Ga age: Nd isotopes and U–Pb zircon ages in the Svecokarelian terrain of South Finland. *Contrib. Mineral. Petrol.* **92**, 1–12.

Patchett P. J. and Tatsumoto M. (1980) Lu–Hf total-rock isochron for the eucrite meteorites. *Nature* **288**, 571–574.

Patchett P. J., Kouvo O., Hedge C. E., and Tatsumoto M. (1981) Evolution of continental crust and mantle heterogeneity: evidence from Hf isotopes. *Contrib. Mineral. Petrol.* **78**, 279–297.

Patchett P. J., White W. M., Feldmann H., Kielinczuk S., and Hofmann A. W. (1984) Hafnium/rare-earth element fractionation in the sedimentary system and crustal recycling into the Earth's mantle. *Earth Planet. Sci. Lett.* **69**, 365–378.

Patchett P. J., Gorbatschev R., and Todt W. (1987) Origin of continental crust of 1.9–1.7 Ga age: Nd isotopes in the Svecofennian orogenic terrains of Sweden. *Precamb. Res.* **35**, 145–160.

Patchett P. J., Gehrels G. E., and Isachsen C. E. (1998) Nd isotopic characteristics of metamorphic and plutonic rocks of the Coast Mountains near Prince Rupert, British Columbia. *Can. J. Earth Sci.* **35**, 556–561.

Patchett P. J., Ross G. M., and Gleason J. D. (1999) Continental drainage and mountain sources during the Phanerozoic evolution of North America: evidence from Nd isotopes. *Science* **283**, 671–673.

Peck W. H., Valley J. W., Wilde S. A., and Graham C. M. (2001) Oxygen isotope ratios and rare earth elements in 3.3 to 4.4 Ga zircons: ion microprobe evidence for high δ^{18}O

continental crust and the oceans in the Early Archean. *Geochim. Cosmochim. Acta* **65**, 4215–4229.

Peltonen P., Kontinen A., and Huhma H. (1998) Petrogenesis of the mantle sequence of the Jormua ophiolite (Finland): melt migration in the upper mantle during Palaeoproterozoic continental break-up. *J. Petrol.* **39**, 297–329.

Piercey S. J., Mortensen J. K., Murphy D. C., Paradis S., and Creaser R. A. (2002) Geochemistry and tectonic significance of alkalic mafic magmatism in the Yukon-Tanana terrain, Finlayson Lake region, Yukon. *Can. J. Earth Sci.* **39**, 1729–1744.

Piercey S. J., Mortensen J. K., and Creaser R. A. (2003) Neodymium isotope geochemistry of felsic volcanic and intrusive rocks from the Yukon-Tanana terrain in the Finlayson Lake region, Yukon, Canada. *Can. J. Earth Sci.* **40**, 77–97.

Plank T. and Langmuir C. H. (1998) The chemical composition of subducting sediment and its consequences for the crust and mantle. *Chem. Geol.* **145**, 325–394.

Price P. B. and Walker R. M. (1962) Observation of fossil particle tracks in natural micas. *Nature* **196**, 732–734.

Price P. B. and Walker R. M. (1963) Fossil tracks of charged particles in mica and the age of minerals. *J. Geophys. Res.* **68**, 4847–4863.

Rämö T., Vaasjoki M., Mänttäri I., Elliott B. A., and Nironen M. (2001) Petrogenesis of the post-kinematic magmatism of the central Finland granitoid complex: I. Radiogenic isotope constraints and implications for crustal evolution. *J. Petrol.* **42**, 1971–1993.

Reddy S. M., Kelley S. P., and Wheeler J. (1996) Ar-40/Ar-39 laser probe study of micas from the Sesia Zone, Italian Alps: implications for metamorphic and deformation histories. *J. Metamorph. Geol.* **14**, 493–508.

Reymer A. and Schubert G. (1984) Phanerozoic addition rates to the continental crust and crustal growth. *Tectonics* **3**, 63–77.

Salnikova E. B., Kozakov I. K., Kotov A. B., Kröner A., Todt W., Bibikova E. V., Nutman A., Yakovleva S. Z., and Kovach V. P. (2001) Age of Palaeozoic granites and metamorphism in the Tuvino-Mongolian Massif of the Central Asian Mobile Belt: loss of a Precambrian microcontinent. *Precamb. Res.* **110**, 143–164.

Salters V. J. M. and Hart S. R. (1991) The mantle sources of ocean ridges, islands and arcs: the Hf-isotope connection. *Earth Planet. Sci. Lett.* **104**, 364–380.

Samson S. D. and D'Lemos R. S. (1999) A precise late Neoproterozoic U–Pb zircon age for the syntectonic Perelle quartz diorite, Guernsey, Channel Islands, UK. *J. Geol. Soc. London* **156**, 47–54.

Samson S. D. and Patchett P. J. (1991) The Canadian Cordillera as a modern analogue of Proterozoic crustal growth. *Austral. J. Earth Sci.* **38**, 595–611.

Samson S. D., McClelland W. C., Patchett P. J., Gehrels G. E., and Anderson R. G. (1989) Evidence from Nd isotopes for mantle contributions to Phanerozoic crustal genesis in the Canadian Cordillera. *Nature* **337**, 705–709.

Samson S. D., Patchett P. J., Gehrels G. E., and Anderson R. G. (1990) Nd and Sr isotopic characterization of the Wrangellia terrane and implications for crustal growth of the Canadian Cordillera. *J. Geol.* **98**, 749–762.

Samson S. D., Patchett P. J., McClelland W. C., and Gehrels G. E. (1991a) Nd isotopic characterization of metamorphic rocks in the Coast Mountains, Alaskan and Canadian Cordillera: ancient crust bounded by juvenile terranes. *Tectonics* **10**, 770–780.

Samson S. D., McClelland W. C., Patchett P. J., and Gehrels G. E. (1991b) Nd and Sr isotopic constraints on the petrogenesis of the west side of the northern Coast Mountains batholith, Alaskan and Canadian Cordillera. *Can. J. Earth Sci.* **28**, 939–946.

Samson S. D., D'Lemos R. S., Blichert-Toft J., and Vervoort J. D. (2003) U–Pb geochronology and Hf–Nd isotope compositions of the oldest Neoproterozoic crust within the

Cadomian Orogen: new evidence for a unique juvenile terrane. *Earth Planet. Sci. Lett.* **208**, 165–180.

Scherer E. E., Cameron K. L., and Blichert-Toft J. (2000) Lu–Hf garnet geochronology: closure temperature relative to the Sm–Nd system and the effects of trace mineral inclusions. *Geochim. Cosmochim. Acta* **64**, 3413–3432.

Scherer E., Münker C., and Mezger K. (2001) Calibration of the Lutetium–Hafnium clock. *Science* **293**, 683–687.

Scherrer N. C., Engi M., Berger A., Parrish R. R., and Cheburkin A. K. (2002) Nondestructive chemical dating of young monazite using XRF: 2. Context sensitive microanalysis and comparison with Th–Pb laser-ablation mass spectrometric data. *Chem. Geol.* **191**, 243–255.

Şengör A. M. C. and Hsü K. J. (1984) The Cimmerides of eastern Asia: history of the eastern end of paleo-Tethys. *Mem. Soc. Géol. France N. S.* **147**, 139–167.

Şengör A. M. C. and Natalin B. A. (1996a) Turkic-type orogeny and its role in the making of the continental crust. *Ann. Rev. Earth Planet. Sci.* **24**, 263–337.

Şengör A. M. C. and Natalin B. A. (1996b) Paleotectonics of Asia: fragments of a synthesis. In *The Tectonic Evolution of Asia* (eds. A. Yin and M. Harrison). Cambridge University Press, Cambridge, UK, pp. 486–640.

Şengör A. M. C., Natalin B. A., and Burtman V. S. (1993) Evolution of the Altaid tectonic collage and Palaeozoic crustal growth in Eurasia. *Nature* **364**, 299–307.

Skiöld T. and Cliff R. A. (1984) Sm–Nd and U–Pb dating of Early Proterozoic mafic–felsic volcanism in northernmost Sweden. *Precamb. Res.* **26**, 1–13.

Smith A. D. and Lambert R. St. J. (1995) Nd, Sr, and Pb isotopic evidence for cotrasting origins of late Paleozoic volcanic rocks from the Slide Mountain and Cache Creek terranes, south-central British Columbia. *Can. J. Earth Sci.* **32**, 447–459.

Smith A. D., Brandon A. D., and Lambert R. St. J. (1995) Nd–Sr isotope systematics of Nicola Group volcanic rocks, Quesnel terrane. *Can. J. Earth Sci.* **32**, 437–446.

Söderlund U. (1996) Conventional U–Pb dating versus single-grain Pb evaporation dating of complex zircons from a pegmatite in the high-grade gneisses of southwestern Sweden. *Lithos* **38**, 93–105.

Stein M. and Hofmann A. W. (1994) Mantle plumes and episodic crustal growth. *Nature* **372**, 63–68.

Stern R. (1997) The GSC sensitive high resolution ion microprobe (SHRIMP): analytical techniques of zircon U–Th–Pb age determinations and performance evaluation: in radiogenic age and isotopic studies. *Geol. Surv. Can. Current Res.* 1997-F(Report 10), 1–31.

Stern R. and Bleeker W. (1998) Age of the world's oldest rocks refined using Canada's SHRIMP: the Acasta Gneiss Complex, Northwest Territories, Canada. *Geosci. Can.* **25**, 27–31.

Stevens R. A., Erdmer P., Creaser R. A., and Grant S. L. (1996) Mississippian assembly of the Nisutlin assemblage: evidence from primary contact relationships and Mississippian magmatism in the Teslin tectonic zone, part of the Yukon-Tanana terrane of south-central Yukon. *Can. J. Earth Sci.* **33**, 103–116.

Stevenson R. K. and Patchett P. J. (1990) Implications for the evolution of continental crust from Hf isotope systematics of Archean detrital zircons. *Geochim. Cosmochim. Acta* **54**, 1683–1697.

Stosch H.-G. and Lugmair G. W. (1987) Geochronology and geochemistry of eclogites from the Münchberg gneiss massif, FRG. *Terra Cognita* **7**, 163.

Sutter J. F. and Hartung J. B. (1984) Laser microprobe ^{40}Ar/^{39}Ar dating of mineral grains *in situ*. *Soc. Electron. Microsc.* **4**, 1525–1529.

Suzuki K. and Adachi M. (1991) Precambrian provenance and Silurian metamorphism of the Tsunosawa paragneiss in the South Kitakami terrane, northeast Japan, revealed by the chemical Th–U–total Pb isochron ages of monazite, zircon and xenotime. *Geochem. J.* **25**, 357–376.

Suzuki K. and Adachi M. (1994) Middle Precambrian detrital monazite and zircon from the Hida gneiss on Oki-Dogo Island, Japan: their origin and implications for the correlation of basement gneiss of southwest Japan and Korea. *Tectonophysics* **235**, 277–292.

Tatsumoto M., Unruh D. M., and Patchett P. J. (1981) U–Pb and Lu–Hf systematics of Antarctic meteorites. In *Proc. 6th Symp. Antarctic Meteorites*, Natl. Inst. Polar Res., Tokyo, pp. 237–249.

Tilton G., Patterson C., Brown H., Inghram M., Hayden R., Hess D., and Larsen E. S., Jr. (1955) Isotopic composition and distribution of lead, uranium and thorium in a Precambrian granite. *Geol. Soc. Am. Bull.* **66**, 1131–1148.

Tilton G. R., Davis G. L., Wetherill G. W., Aldrich L. T., and Jager E. (1959) Mineral ages in the Maryland piedmont. *Carnegie Inst. Wash. Yearb.* **58**, 171–178.

Umhoefer P. J. (1987) Northward translation of "Baja British Columbia" along the Late Cretaceous to Paleocene margin of western North America. *Tectonics* **6**, 377–394.

Unterschutz J. L. E., Creaser R. A., Erdmer P., Thompson R. I., and Daughtry K. L. (2002) North American margin origin of Quesnel terrane strata in the southern Canadian Cordillera: inferences from geochemical and Nd isotopic characteristics of Triassic metasedimentary rocks. *Geol. Soc. Am. Bull.* **114**, 462–475.

Valbracht P. J., Oen I. S., and Beunk F. F. (1994) Sm–Nd isotope systematics of 1.9–1.8-Ga granites from western Bergslagen, Sweden: inferences on a 2.1–2.0 Ga crustal precursor. *Chem. Geol.* **112**, 21–37.

Valley J. W., Chiarenzelli J. R., and McLelland J. M. (1994) Oxygen isotope geochemsitry of zircon. *Earth Planet. Sci. Lett.* **126**, 187–206.

Valley J. W., Peck W. H., King E. M., and Wilde S. A. (2002) A cool early Earth. *Geology* **30**, 351–354.

Vance D. and O'Nions R. K. (1990) Isotopic chronometry of zoned garnets: growth kinetics and metamorphic histories. *Earth Planet. Sci. Lett.* **97**, 227–240.

Vervoort J. D. and Blichert-Toft J. (1999) Evolution of the depleted mantle: Hf isotope evidence from juvenile rocks through time. *Geochim. Cosmochim. Acta* **63**, 533–556.

Vervoort J. D., Patchett P. J., Gehrels G. E., and Nutman A. P. (1996) Constraints on early Earth differentiation from hafnium and neodymium isotopes. *Nature* **379**, 624–627.

Vinogradov A. P., Zadorozhnyi I. K., and Zykor S. I. (1952) Isotopic composition of lead and the age of the earth. *Dokl. Akad. Nauk. SSSR* **85**, 1107–1110.

von Huene R. and Scholl D. W. (1993) The return of sialic material to the mantle indicated by terrigenous material subducted at convergent margins. *Tectonophysics* **219**, 163–175.

Wetherill G. W., Aldrich L. T., and Davis G. L. (1955) A40/K40 ratios of feldspars and micas from the same rock. *Geochim. Cosmochim. Acta* **8**, 171–172.

Wilde S. A., Zhang X., and Wu F. (2000) Extension of a newly identified 500 Ma metamorphic terrain in North East China: further U–Pb SHRIMP dating of the Mashan Complex, Heilongjiang Province, China. *Tectonophysics* **328**, 115–130.

Wilde S. A., Valley J. W., Peck W. H., and Graham C. M. (2001) Evidence from detrital zircons for the existence of continental crust and oceans on the Earth 4.4 Gyr ago. *Nature* **409**, 175–178.

Williams M. L. and Jercinovic M. J. (2002) Microprobe monazite geochronology: putting absolute time into microstructural analysis. *J. Struct. Geol.* **24**, 1013–1028.

Williams M. L., Jercinovic M. J., and Terry M. P. (1999) Age mapping and dating of monazite on the electron microprobe: deconvoluting multistage tectonic histories. *Geology* **27**, 1023–1026.

Windley B. F., Allen M. B., Zhang C., Zhao Z. Y., and Wang G. R. (1990) Paleozoic accretion and Cenozoic redeformation of the Chinese Tien Shan Range, central Asia. *Geology* **18**, 128–131.

Windley B. F., Kröner A., Guo J. H., Qu G., Li Y., and Zhang C. (2002) Neoproterozoic to Palaeozoic geology of the Altai orogen, Chinese Central Asia: new zircon age data and tectonic evolution. *J. Geol.* **110**, 719–739.

Wolf R. A., Farley K. A., and Silver L. T. (1996) Helium diffusion and low-temperature thermochronometry of apatite. *Geochim. Cosmochim. Acta* **60**, 4231–4240.

Wu F. Y., Jahn B. M., Wilde S., and Sun D. Y. (2000) Phanerozoic crustal growth: U–Pb and Sr–Nd isotopic evidence from the granites in northeastern China. *Tectonophysics* **328**, 89–113.

Yarmolyuk V. V., Kovalenko V. I., Kovach V. P., Budnikov S. V., Kozakov I. K., Kotov A. B., and Salnikova E. B. (1999) Nd-isotopic systematics of western Transbaikalian crustal protoliths: implications for Riphean crust formation in central Asia. *Geotectonics* **33**, 271–286.

York D., Hall C. M., Yanase Y., Hanes J. A., and Kenyon W. J. (1981) $^{40}Ar/^{39}Ar$ dating of terrestrial minerals with a continuous laser. *Geophy. Res. Lett.* **8**, 1136–1138.

Zartman R. E. and Doe B. R. (1981) Plumbotectonics—the model. *Tectonophysics* **75**, 135–162.

Zeitler P. K. (1988) Argon diffusion in partially outgassed alkali feldspars: insights from $^{40}Ar/^{39}Ar$ analysis. *Chem. Geol.* **65**, 167–181.

Zeitler P. K., Herczig A. L., McDougall I., and Honda M. (1987) U–Th–He dating of apatite: a potential thermochronometer. *Geochim. Cosmochim. Acta* **51**, 2865–2868.

Radioactive Geochronometry
ISBN: 978-0-08-096708-0

pp. 223–250

8
Radiocarbon

W. S. Broecker

Lamont-Doherty Earth Observatory, Palisades, NY, USA

8.1 INTRODUCTION

Willard Libby's invention of the radiocarbon dating method revolutionized the fields of archeology and Quaternary geology because it brought into being a means to correlate events that occurred during the past 3.5×10^4 years on a planet-wide scale (Libby *et al.*, 1949). This contribution was recognized with the award of the Nobel prize for chemistry. In addition, radiocarbon measurements have been a boon to the quantification of many processes taking place in the environment, to name a few: the rate of "ventilation" of the deep ocean, the turnover time of humus in soils, the rate of growth of cave deposits, the source of carbon-bearing atmospheric particulates, the rates of gas exchange between the atmosphere and water bodies, the replacement time of carbon atoms in human tissue, and depths of bioturbation in marine sediment. Some of these applications have been greatly aided by the creation of excess ^{14}C atoms as the result of nuclear tests conducted in the atmosphere. Since the 1960s, this so-called bomb radiocarbon

has made its way into all of the Earth's active carbon reservoirs. To date, tens of thousands of radiocarbon measurements have been made in laboratories throughout the world.

8.2 PRODUCTION AND DISTRIBUTION OF ^{14}C

Radiocarbon atoms are produced when protons knocked loose by cosmic ray impacts encounter the nuclei of atmospheric nitrogen atoms. The reaction is as follows:

$$n + {}^{14}N \rightarrow {}^{14}C + p \qquad (1)$$

The half-life of these ^{14}C atoms is 5,730 years. Hence, they have on average 8,270 years to distribute themselves through the Earth's active carbon reservoirs. Radiocarbon atoms decay by emitting an electron, thereby converting a neutron into a proton returning the nucleus to its original ^{14}N form.

Once produced, radiocarbon atoms become oxidized to CO_2 gas and join the Earth's carbon cycle (see Figure 1). CO_2 exchange with the inorganic carbon dissolved in the Earth's surface waters carries these atoms into the sea, lakes, and rivers. Photosynthesis moves the ^{14}C into both terrestrial and aquatic plants and from there to the animals that feed upon them. Radiocarbon also gets incorporated into shell and coral, and into soil and cave carbonates.

8.3 MEASUREMENTS OF RADIOCARBON

Initially, all measurements were made by counting the β-particles emitted during radioactive decay of ^{14}C. The first wave of laboratories followed Libby's lead and using screen wall Geiger counters measured the β-particles emitted by carbon black spun onto the inside of stainless-steel cylinders (Figure 2). But the advent of nuclear testing created a serious problem for this method. Airborne strontium-90, cesium-137, and other fission products became absorbed onto the carbon black adding to the sample's radioactivity. Hans Suess (USGS, Washington, DC), Hessel deVries (The Netherlands), and Gordon Fergusson (New Zealand) pioneered the transition to gas counting.

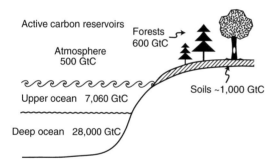

Figure 1 Preanthropogenic distribution of carbon among the Earth's active reservoirs (gigatons of carbon or 10^{15} g of carbon).

Most laboratories filled proportional counters with highly purified CO_2 gas generated by burning or acidifying the sample. A few used acetylene. Other laboratories went a step further converting the sample carbon into liquid benzene and measured its radioactivity in a scintillation counter. In the early 1980s, it was realized that the ^{14}C atoms themselves could be measured using high-energy tandem Van de Graaff accelerators. During the latter part of the 1980s and early 1990s, accelerator mass spectrometry largely replaced the traditional decay-counting techniques. The huge advantage of this new approach was that measurements required only 1 mg of carbon as opposed to 1 g for decay counting. The routine accuracy (± 5 per mil) achievable in atom counting is comparable to that obtained in decay-counting laboratories. Also comparable is the ability to measure the minute amounts of ^{14}C in very old (i.e., $>3.5 \times 10^4$ years) samples. This innovation of atom counting has opened up a large number of applications previously untouchable by decay counting. The convention for expressing ^{14}C results is shown in Table 1.

8.4 TIMESCALE CALIBRATION

It was a biophysicist deVries (1958), who first clearly demonstrated that the radiocarbon timescale was imperfect, that is, that radiocarbon years were not identical to calendar years. Fascinated by the potential of Libby's new method, this brilliant Dutch scientist plunged into all its aspects. deVries devised means to reduce the background of gas counters by greatly increasing the effectiveness of the shielding from cosmic ray mesons. Not only did he strive to extend the method beyond its nominal range (3.5×10^4 years), he also worked to improve the precision of measurements so that he could document small imperfections in the radiocarbon timescale based on measurements on materials of known calendar age.

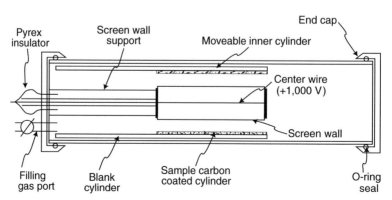

Figure 2 Diagrammatic representation of the Libby screen wall Geiger counter. Half of the moveable inner stainless-steel cylinder was coated with the sample carbon and the other half left bare. By gently tipping the counter, the cylinder could be moved back and forth alternatively exposing first the sample portion and then the blank portion of the cylinder to the active (i.e., screen wall portion) of the counter. In this way, the background count rate could be monitored and subtracted from the total sample plus background count rate.

Table 1 The international convention for expressing radiocarbon results on contemporary materials in delta units. The standard is the age-corrected $^{14}C/C$ ratio in 1850 wood. The ^{13}C correction is designed to remove that part of the radiocarbon variability associated with the isotope separations occurring in nature (e.g., during photosynthesis and air–sea exchange). Also listed are some important characteristics of radiocarbon.

Definitions:

$$\delta^{14}C = \left(\frac{^{14}C/C_{\text{sample}} - {}^{14}C/C_{\text{standard}}}{^{14}C/C_{\text{standard}}} \right) 1,000$$

$$\Delta^{14}C = \delta^{14}C - 2(\delta^{13}C + 25)\left(1 + \frac{\delta^{14}C}{1,000}\right)$$

where

$$\delta^{13}C = \left(\frac{^{13}C/^{12}C_{\text{sample}} - {}^{13}C/^{12}C_{\text{standard}}}{^{13}C/^{12}C_{\text{standard}}} \right)$$

Age correction:

$$^{14}C/C_{\text{age corrected}} = {}^{14}C/C_{\text{measured}}\, e^{\lambda t}$$

where t is the calendar age of the sample (referenced to 1950).

Characteristics:

$$^{14}C/C \text{ (for } \Delta^{14}C = 0\%) = 1.18 \times 10^{-12}$$

$$t_{1/2} = 5,730 \text{ years}$$

$$T_{\text{mean}} = 8,270 \text{ years}$$

$$\lambda = \frac{1}{8,270} \text{ yr}^{-1}$$

$$\text{Decay rate} = \frac{1\%}{82.7 \text{ years}}$$

Note: A computerized calibration program (CALIB) is available.

He was able to document such deviations and he postulated that they reflected changes in the $^{14}C/C$ ratio in atmospheric CO_2 due to perturbations in the flux of cosmic rays entering the Earth's atmosphere. Hessel deVries stands out in my mind as the great genius in this field. Had his life not been snuffed out just as he was reaching his prime, he would have dominated the radiocarbon world for decades. But, fortunately, during his short career, not only did he put his finger on the calibration problem that even today dominates much of the effort in the field, but he also trained a graduate student Minze Stuiver, who would devote much of his career to calibration studies and to measurements of ^{14}C in oceanic carbon.

8.4.1 Calibration Based on Tree Rings

Nature has provided a marvelous set of calibration materials, namely, cellulose in the annual growth rings in trees. Single living trees provide annual ring series extending back 1,000 or more years. More importantly, tree trunks preserved in swamp muck and in riverine alluvium can be cross-dated based on ring-width "fingerprints," thus greatly extending the calibration range. Treasure troves of such material lie in Holocene sediments of the German rivers. Its flood-stage deposits constitute a major source of sand and gravel for the German construction industry. In the course of "mining" these deposits, large trunks of trees are often encountered. Over the years, the late Bernd Becker and members of his Hohenheim dendrochronology team responded to alerts hastening to the site of a new find to cut out a section to add to their ever-growing collection. A single radiocarbon measurement would tell them the approximate growth period of this tree. Measurements of the ring widths would allow them to tie this section precisely into an ever more complete and accurate master chronology. Once this chronology was established, very detailed high-precision ^{14}C measurements were made in the laboratory of Bernd Kromer in Heidelberg. This amazing record extends back to ~8,000 years ago. To extend the growth-ring series further back in time, it was necessary for the Becker team to switch from oaks to pines. This species appeared in northern Europe soon after the abrupt warming which brought to an end the final cold punctuation of the last glacial period (i.e., the Younger Dryas). Working closely with Bernd Kromer, who conducted ultraprecise ^{14}C measurements (to $\pm 2\%$) in his Heidelberg laboratory on both the oak and pine series, Becker and co-workers were satisfactorily able to splice the pine series onto the oak series creating a master chronology covering the last 1.1919×10^4 years (Kromer *et al.*, 1986; Becker and Kromer, 1993; Kromer and Spurk, 1998; Friedrich *et al.*, 1999). Efforts are currently underway using tree trunk series from central Europe and south of the Alps to extend the chronology back into glacial time.

The thousands of measurements which contributed to this calibration effort were largely made in five laboratories: that of Hans Suess in La Jolla, California; that of Paul Damon in Tuscon, Arizona; that of Gordon Pearson in Belfast Northern Ireland; and, of course, those of Minze Stuiver in Seattle, Washington, and Bernd Kromer in Heidelberg, Germany. The resulting calibration curve is a boon to archeologists faced with the task of relating radiocarbon dates to calendar dates for historical events, as well as to geophysicists interested in probing the causes for the $^{14}C/C$ ratio changes. A compilation base on these tree-ring calibration measurements is shown in Figure 3.

Two features of this record stand out, a long-term trend of decreasing ^{14}C culminating ~1,500 years ago and century-duration fluctuations around this trend. Minze Stuiver was the first to demonstrate that the century-duration fluctuations were, at least in part, related to sunspot activity. He did this by showing that during the course of the so-called

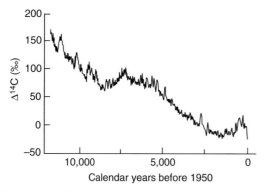

Figure 3 Tree-ring-based reconstruction of the temporal trend of the atmosphere's $^{14}C/C$ ratio over the last 1.1854×10^4 calendar years. $\Delta^{14}C$ is the deviation of the atmospheric ratio from that in pre-Industrial time ($\sim$1,800) (see Table 1). Source: Stuiver *et al.* (1998).

Maunder minimum (a period from 1645 to 1715 when no spots were observed on the Sun), the $^{14}C/C$ ratio in atmospheric CO_2 underwent a significant increase (Stuiver and Quay, 1980, 1981). Stuiver attributed this increase to the great reduction in the ions streaming into space from the Sun's spots (and hence also of the so-called heliomagnetic field). This magnetic field serves to partially shield the solar system from incoming cosmic rays. Similar increases in the ratio $^{14}C/C$ were established by Stuiver to have occurred between 1280 and 1345 and between 1420 and 1540 (Stuiver and Quay, 1980, 1981). By analogy to the Maunder, they have been termed the Spörer and Wolf sunspot minima (see Figure 4).

A reasonably convincing case has been made that the longer-term Holocene trend is a consequence of changes in the Earth's magnetic field which also diverts incoming cosmic rays. This case rests on measurements designed to reconstruct the strength of the Earth field at times in the past. The first such attempt was by Thellier and Thellier (1959), who measured the magnetic field in ancient ceramics and then reheated them beyond the Curie point and allowed them to cool in the same fashion as originally. They then remeasured the magnetic field. These early ceramic measurements have been supplemented by reconstructions based on igneous rocks and deep-sea cores, which extend back many radiocarbon half-lives (see Bard, 1998 for an excellent summary).

However, lurking in the wings is yet another mechanism by which the atmospheric $^{14}C/C$ ratio may have been changed. It involves changes in the rate of ventilation of the deep sea. Around 70% of the Earth's cosmic ray-produced ^{14}C atoms currently reside in the deep sea. Because ^{14}C is lost by radiodecay during residence in the deep sea, the $^{14}C/C$ ratio for the inorganic carbon (i.e., HCO_3^-, CO_3^{2-}, and CO_2) dissolved in these waters

averages $\sim$16% lower than that in the atmospheric CO_2. Hence, were the rate of ocean mixing to have been slower in the past than now, the contrast between the atmospheric and deep-sea $^{14}C/C$ ratios would have increased and, consequently, the ratio in the atmosphere would have been higher than now. Later in the chapter, a dramatic example of such a change will be discussed. For the Holocene (i.e., the last 1.15×10^4 years), there is no concrete evidence that ocean mixing contributed to the $^{14}C/C$ changes. Minze Stuiver has pondered whether the changes in sunspot activity might be linked to the Holocene's small climate changes. If so, then both sunspot activity and ocean mixing may have contributed to the $^{14}C/C$ ratio fluctuations.

An extremely important contribution to this subject was made by an Indian scientist, Devendra Lal, who made calculations aimed at determining the exact dependence on the Earth's magnetic field strength of the influx of cosmic rays to the Earth's atmosphere, and hence also on the rate of production of ^{14}C atoms (Lal, 1988).

While resolving most of the inconsistencies between radiocarbon and historic ages, the calibration curve also reveals a fundamental limitation of the radiocarbon method. During some time periods, the century-duration changes in $^{14}C/C$ ratio created reversals in the age sequence. These reversals give rise to multiple possible calendar ages for a single radiocarbon measurement. Only by measuring several samples in stratigraphic sequence, it is possible to distinguish among these multiple possibilities (see Stuiver, 1982).

In a related manner, radiocarbon dating is powerless to aid in verifying the authenticity of art objects purported to have made in the period 1700–1950. The reason is that during this period, the radiocarbon content of the atmosphere decreased at a rate closely matching the rate of ^{14}C decay (i.e., 1% per 83 years). There are two reasons for this decline. First, the excess ^{14}C atoms produced during the Maunder minimum were being mixed into the ocean and sequestered in the terrestrial biosphere. Second, the burning of fossil fuels added CO_2 free of radiocarbon to the atmosphere, thereby extending this decline into the twentieth century. In the early 1950s, this downward trend was reversed by the production of ^{14}C during nuclear tests (see Section 8.6.2). Hence, based on radiocarbon measurements, a forgery made using wood or parchment dating just prior to 1950 could not be distinguished from the real thing created during the eighteenth century.

8.4.2 Calibration Based on Corals

Working at Columbia University's Lamont–Doherty Earth Observatory with French isotope geochemists Edouard Bard and Bruno Hamelin,

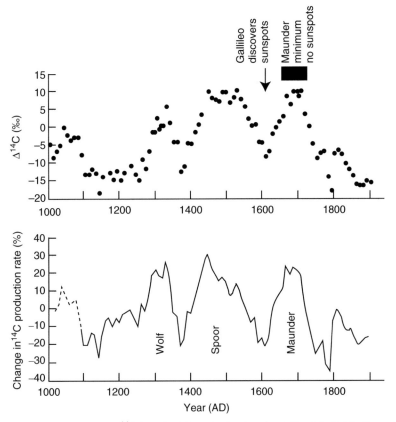

Figure 4 Temporal fluctuations in the ^{14}C/C ratio (given in the Δ units defined in Table 1) as determined from measurements on Douglas fir wood from the Pacific Northwest (upper panel). The record is cut off in the year 1900 because after that time the influence of ^{14}C-free CO_2 released by fossil fuel burning became significant. In the lower panel are shown the changes in the production rate of ^{14}C required to produce the observed temporal changes in atmospheric ^{14}C to C ratio. The production rates are deconvolved from the observed Δ^{14}C changes using a simplified atmosphere–ocean–terrestrial biosphere model. Because of the interchange of carbon atoms with the ocean and the terrestrial biosphere, the changes in atmospheric ^{14}C/C ratio lag the production-rate changes. As can be seen, the last of the peaks in production occurred during a time when no spots were observed on the Sun (i.e., during the Maunder minimum). As telescopic observations of the Sun (by Galileo) commenced in the early 1600s, the existence of the Spörer and Wolf sunspot minima is based on the ^{14}C results. The measurements and modeling were carried out by Stuiver and Quay (1980). Bard *et al.* (1997) showed that the observed ^{14}C variations can be generated from measurements of ^{10}Be in ice cores.

Richard Fairbanks found a means to extend the calibration curve beyond the Holocene into late glacial time. This team adopted a new analytical method developed at Caltech to make very precise uranium and thorium isotope measurements using conventional thermal ionization mass spectrometry (Edwards *et al.*, 1986/1987) instead of the less-precise alpha counting method used previously. Just as the transition from decay to atom counting revolutionized ^{14}C dating, this innovation revolutionized dating based on the uranium decay-series isotope ^{230}Th (half-life 7.5×10^4 years). This new technique was applied to corals obtained from a series of shallow borings Fairbanks conducted off the island of Barbados. These corals formed during the last deglaciation as sea level rose in response to the melting of the glacial age icecaps. The Fairbanks team conducted both ^{230}Th- and ^{14}C-age determinations on these corals, which were ideal for the task as they contained no

original thorium and had never been exposed to CO_2-charged groundwater. The ^{230}Th results produced a big surprise. Rather than revealing a cyclicity in the Earth's magnetic field as a number of authors had predicted, the offset between calendar and radiocarbon ages became larger and larger as one went back in time (Bard *et al.*, 1990a, b) (see Figure 5).

8.4.3 Other Calibration Schemes

The measurements on Fairbanks' Barbados corals spurred efforts to find other means of extending the calibration curves back in time. Several tacks were taken. One obvious strategy was to count annual layers (varves) in lake and marine sediments. Stuiver, the hero of the calibration effort, had adopted this approach way back in the 1960s (Stuiver, 1970, 1971). Another approach was to put to use the annual layering in long ice

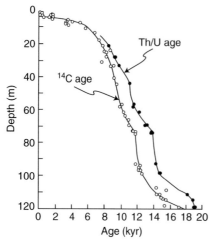

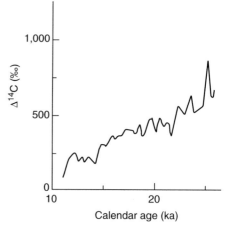

Figure 5 [14]C- (open circles) and [230]Th- (closed circles) based age determinations on corals obtained from shallow drillings off the island of Barbados. As can be seen, the magnitude of the offset between the two ages increases back in time indicating that the ^{14}C/C ratio in atmosphere and upper ocean carbon was declining during the period 18,000–8,000 years ago. As the corals chosen for these measurements were shallow growing, the [230]Th ages portray the rise in sea level during the last deglaciation. These results were reported in papers by Fairbanks (1989) and Bard *et al.* (1990a, b).

Figure 6 Extension of the tree-ring-based and Barbados coral-based calibration curve back to 3.2×10^4 years based on ^{14}C and [230]Th measurements on a stalagmite from a submerged cave in the Bahamas. The growth of the stalagmite came to a halt $\sim 1.1 \times 10^4$ years ago when the rising sea invaded the cave. The upper portion of the record which overlaps in time the Barbados coral record was used to establish the initial [230]Th to [232]Th ratio and also the reservoir correction for the ^{14}C ages. Source: Beck *et al.* (2001).

cores from Greenland. As the ice itself cannot be dated using ^{14}C, this required that the [18]O (i.e., air temperature) record in the ice be correlated with ice-rafting events in radiocarbon-dated sediment cores from the northern Atlantic (Voelker *et al.*, 1998). Finally, coupled [230]Th and ^{14}C measurements on cave formations have been utilized (see e.g., Beck *et al.*, 2001). In my estimation, none of these approaches rests on an entirely firm foundation. Varve counting is often subjective, for example, varves may be missing or doublets may represent a single year. However, Kitagawa and van der Plicht (1998) make an impressive case for a varve sequence in a Japanese lake. Greenland's ice cores provide an excellent example of the difficulty. Counts by a European team and by an American team in two long and virtually identical cores from Greenland's summit locale yield results that differ by as much as several percent (see Southon, 2002). Further, the correlation of the ice record with the marine record is fraught with subjectivity and there are large discrepancies between radiocarbon dates on coexisting planktic and benthic foraminifera. Stalagmites offer perhaps the most promise but, unlike corals, during growth they often incorporate some [230]Th. Also, the initial ^{14}C/C ratio is offset from that for atmospheric CO_2 because the CO_2 in cave waters is derived from the oxidation of organic matter in the overlying soil. But by far the most worrisome aspect of the cave-formation approach is contamination with younger $CaCO_3$ (infilling of pores, recrystallization, etc.). A 3.5×10^4-year-old

sample contains only (1/64)th its original amount of [14]C. This contamination problem is far more severe for $CaCO_3$ than for wood from which chemically inert cellulose can be extracted. Because of these problems, I do not trust any existing extensions of calibration curve beyond 2.5×10^4 years. Much more work will have to be done before this task can be declared a success. And, of course, the chances of ever extending the calibration beyond 4.7×10^4 years (only one part in 256 of the original ^{14}C remains) are indeed slim. However, the ^{14}C and [230]Th measurements of Beck *et al.* (2001) on a Bahamas stalagmite did allow the calibrations curve to be reliably extended back to $\sim 2.5 \times 10^4$ years (see Figure 6). Clearly this effort constituted a major advance.

In a recent paper, Fairbanks *et al.* (2005) provide what is surely the most reliable of these reconstructions. It is based on pristine coral samples recovered from borings made in the shallow waters off Barbados in the Caribbean Sea and Christmas Island in the central equatorial Pacific Ocean. As was the case for his original set of corals, these had never been exposed to groundwater. They bear no hint of diagenetic alteration. However, the reliability of even this record becomes questionable beyond 40×10^4 years.

8.4.4 Cause of the Long-Term [14]C Decline

One might ask what accounts for the decline in atmospheric ^{14}C/C ratio over the last 2.5×10^4 years. The prime suspect is the drop to near-zero values in the Earth's magnetic field $\sim 4 \times 10^4$ years ago (Bonhommet and Zähringer, 1969). This drop, referred to as the Laschamp event, is named after a volcanic field in France where it was first identified. The existence of this event has now been confirmed by measurements of magnetic intensity in deep-sea sediments that reveal a strong minimum close to 4×10^4 years ago. Also, it shows up as a doubling of the abundance of ^{10}Be (another cosmogenic isotope created in the Earth's atmosphere) at this time in the Greenland ice cores (Yiou *et al.*, 1997). Both the drop to near-zero values of magnetic intensity and the doubling of the ^{10}Be concentration are consistent with a temporary shutdown of the Earth's field (presumably associated with a false polarity reversal). The existence of the Laschamp event has provided a strong impetus to extend the calibration curve to times before this temporary shutdown of the Earth's magnetic field. However, as already stated, this challenge will prove to be an extremely difficult one.

It is unlikely that the Laschamp event alone can explain the observed decline in the ^{14}C/C ratio. The duration of this event (several thousand years) was too short and the magnitude of its impact on ^{14}C production was too small to produce a large enough increase in ^{14}C inventory to persist for $\sim 2 \times 10^4$ years. After 2.3×10^4 years had elapsed, the excess would have been diminished by a factor of 16. However, as shown by a number of marine sediment-based magnetic field reconstructions (see Laj *et al.*, 1996), the Laschamp event was followed by a long-term buildup in the strength of the Earth's magnetic field (see Figure 7) (see Bard, 1998, for summary). This slow buildup appears to be adequate to account for the post-4×10^4 years decline in atmospheric ^{14}C.

But there is a fly in this ointment. ^{10}Be measurements on Greenland ice (Muscheler *et al.*, 2005) fail to show the decrease expected were the strength of the magnetic field to have increased in accord with the Laj *et al.* (1996) reconstruction. This anomaly led Chiu (2005) to challenge the accepted half-life for radiocarbon. Were it 6,400 rather than 5,730, the steep downward trend in the ^{14}C to C ratio for atmospheric carbon would be removed. She notes that the measurements (by gas counting) on which the accepted half-life is based were made in the late 1950s. Spurred by Chiu's challenge, an effort is underway to redetermine this important half-life (Southon, personal communication). A long shot, but something that must be done!

8.4.5 Change in Ocean Operation

In addition to attempts to extend the radiocarbon calibration curve back further in time, there have also been attempts to enhance its detail. One

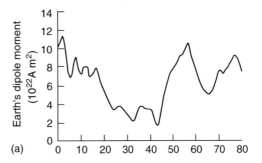

(a)

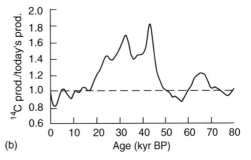

(b)

Figure 7 (a) The reconstruction of the Earth's dipole moment (Laj *et al.*, 1996) based on magnetic measurements made on deep-sea sediments and calibrated against measurements on volcanic rocks of known age. (b) The radiocarbon production rates reconstructed from the magnetic dipole reconstruction (a). Frank *et al.* (1997) have shown that a similar reconstruction is obtained based on ice-core ^{10}Be results.

of these attempts stands out. Konrad Hughen, while a graduate student at the University of Colorado, carried out a detailed set of radiocarbon measurements on planktonic foraminifera shells contained in varved sediment from the Cariaco Basin just off Venezuela. One of the attributes of this core is that the interval correlating with the Younger Dryas cold event is marked by a distinct color change. Further, the boundaries of this interval are very sharp. Hughen used his radiocarbon measurements to splice the Cariaco varve sequence onto the German tree-ring sequence. He then used his varve counts to extend the calibration record several thousand years beyond that established using tree rings (i.e., through the entire Younger Dryas and the underlying Bölling Allerod warm period). His resulting reconstruction revealed a very exciting feature (see Figure 8). During the first 200 years of the Younger Dryas, the ^{14}C/C ratio in upper ocean dissolved inorganic carbon soared by 5%. Then, during the remaining 1,000 years of this cold snap, the radiocarbon content drifted back down. The coincidence of the initiation of the ^{14}C/C ratio rise with the onset of the Younger Dryas led Hughen to propose that this rise was tied to a change in the rate of oceanic mixing rather than to a change in magnetic shielding. Such a ^{14}C rise would be a logical

258   *Radiocarbon*

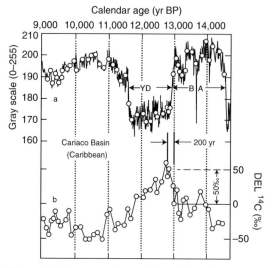

Calendar age (yr BP)

Figure 8 Reconstruction of the radiocarbon content of atmospheric CO_2 during the deglaciation interval (Hughen *et al.*, 1998). This reconstruction was obtained from a varved sediment core from the Cariaco Basin. The position of the individual ^{14}C measurements is shown by the circles. The Younger Dryas (YD) and Bolling-Allerod (BA) intervals are clearly shown in gray-scale record. As can be seen, the ^{14}C content of surface ocean carbon rose by $50 \pm 10‰$ in the first two centuries of the Younger Dryas event. After that, it slowly declined. Assuming that the air–sea difference in the $^{14}C/C$ ratio remained the same as found today, these changes can be transferred to atmospheric CO_2.

consequence of a shutdown in production of new deep water in the northern Atlantic Ocean. In today's ocean, roughly three quarters of the ^{14}C atoms resupplying the deep-sea inventory enter via this route. A number of authors have shown that the sudden release of melt water stored in proglacial Lake Agazzis into the St. Lawrence drainage at the time of the onset of the Younger Dryas would have flooded the northern Atlantic with enough freshwater to shut down the Atlantic's conveyor circulation.

If Hughen's hypothesis proves to be correct then one might ask: why not attribute other aspects of offset of radiocarbon years from calendar years to changes in ocean mixing? In particular, could the higher atmospheric $^{14}C/C$ ratio during peak glacial time ($(2.3-1.5) \times 10^4$ years ago) be attributed to more sluggish ocean circulation? Here the radiocarbon method itself provides an answer. It is based on differences between the radiocarbon age of coexisting planktonic (surface-dwelling) and benthic (bottom-dwelling) foraminifera shells picked from glacial sections of deep-sea sediments. Owing to rapidity of CO_2 exchange with the atmosphere, the $^{14}C/C$ ratio in surface ocean dissolved inorganic carbon is closely linked to that in atmospheric CO_2. Thus, the difference between

the $^{14}C/C$ ratio in surface water and that in deep water recorded by coexisting planktic and benthic foraminifera shells provides a record of how the rate of deep-sea ventilation differed between late glacial time ($(2.3-1.5) \times 10^4$ years ago) and the Holocene (1.1×10^4 years to present). While radiocarbon measurements on coexisting benthic and planktic foraminifera pairs show that during glacial time the deep Atlantic had a ventilation age close to that of today's for the deep Pacific (Robinson *et al.*, 2005), those for samples down to 2,800 m in the Pacific suggest that the ventilation rate in the Pacific was not significantly different from todays (Barker and Broecker, in press; Barker *et al.*, in press). The failure to find evidence for reduced ventilation of the deep Pacific suggests that if isolation of the deep sea's carbon played a significant role in creating the elevated radiocarbon content of the glacial atmosphere, then there must have been an abyssal reservoir highly depleted in radiocarbon. The excessively salty glacial water documented by Adkins *et al.* (2002) in the abyssal Southern Ocean is a prime candidate (see Shackleton *et al.*, 1988; Broecker *et al.*, 1990).

Another radiocarbon-based study showed just how rapidly changes in the patterns of deep ventilation in the ocean can occur. As part of his PhD research with MIT's Ed Boyle, Jess Adkins made a study of a single benthic coralite, which grew at a depth of 1.8 km in the northwestern Atlantic Ocean during the early stages of the last deglaciation. He found an amazing thing. While the base and crest of this several centimeter-high mushroom-shaped coralite yielded nearly identical ^{230}Th ages of 1.54×10^4 years, the ^{14}C age of the base was 1.385×10^4 years and that of the crest 1.452×10^4 years (Adkins *et al.*, 1998). As this coralite probably formed in a period of just a few decades, this difference in radiocarbon age requires that the water in which the coralite grew underwent a sudden and large drop in the $^{14}C/C$ ratio. When Adkins compared his ^{14}C to ^{230}Th age offsets with those obtained by the Fairbanks team, he concluded that the apparent deep Atlantic ventilation age jumped from ~400 to ~1,100 years. Puzzled by how this could have happened, he measured the cadmium content of the calcite along a traverse from the stem to the coralite crest and found that it increased from 0.11 to 0.18 µmol Cd permol Ca (see Figure 9). Only one explanation seems to fit these observations. The production of low-cadmium content and high-radiocarbon water in the northern Atlantic must have come to an abrupt halt allowing the more dense waters with higher cadmium content and lower radiocarbon from the Southern Ocean to flood northward, displacing upward the less-dense waters of the North Atlantic origin.

For all these studies, there is a necessity to establish the reservoir age for the glacial surface

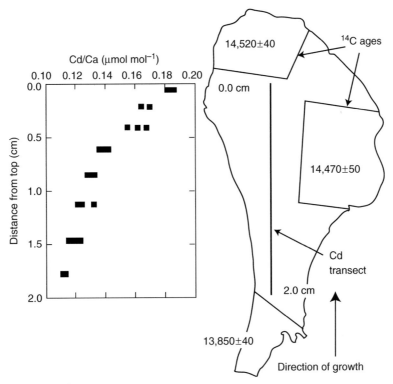

Figure 9 Measurements of ^{14}C/C ratio in a benthic coral from 1.8 km depth in the northern Atlantic reveal that the base has an apparent age 650 radiocarbon years younger than the top. Yet ^{230}Th ages on the base and crest are nearly identical (1.54×10^4 calendar years). Adkins *et al.* (1998) attribute this difference to an abrupt invasion of low ^{14}C and high cadmium content Southern Ocean water. Reproduced from Adkins *et al.* (1998).

ocean. Several authors (Bard *et al.*, 1994; Sikes *et al.*, 2000; Waelbroeck *et al.*, 2001; Siani *et al.*, 2001) have attempted to do this and find puzzling older reservoir ages for glacial times. Clearly, more effort must go into such reconstructions.

8.5 RADIOCARBON AND SOLAR IRRADIANCE

As solar energy output and radiocarbon production are both tied to sunspot activity, it might be possible to reconstruct past irradiance variations from the perturbations in the atmosphere's ^{14}C/C ratio reconstructed from measurements on tree rings of known calendar age. Indeed, based on radiocarbon reconstructions, Stuiver (1980) and Stuiver and Braziunas (1993) suggested that the Little Ice Age might be the result of reduced solar luminosity during periods such as the Maunder minimum when sunspot activity was shut down. However, because the records Stuiver used were quite short and since changes in solar luminosity were, at the time his paper was written, poorly documented, his proposal failed to receive wide acceptance.

The advent of satellite observations of the solar output made it clear that there is indeed a tie between energy output and sunspot activity (see Figure 10). However, as these variations are very

small and the record covers only the last two decades, this finding made only a modest impact with regard to interest in the Sun as a driver of Earth climate fluctuations.

The situation changed with the publication of a paper by Bond *et al.* (2001), in which it was shown that a correlation exists between the cosmic ray flux changes required to generate the tree-ring-based reconstruction of atmospheric ^{14}C/C ratio record for the last 1×10^4 years and an ice-rafting index for the radiocarbon-dated northern Atlantic sediments developed by Bond *et al.* (1997, 1999) (see Figure 11). This index involves the ratio of red-coated (i.e., iron-stained) to total ice-rafted grains. During the Holocene, the index has fluctuated on a timescale on average of 750 years (i.e., 1,500 years for a complete cycle) between highs of ~16% and lows of only a few percent. Bond reasons that times of high index represent cold spells and those of low index, warm spells. His argument is based on the observation that the sources of red-coated grains lie poleward of sources devoid of red-coated grains, and hence, to get them to the site of his deep-sea cores requires colder surface water conditions. As a confirmation that the changes in the ^{14}C/C ratio reflect production rather than changes in ocean circulation, Bond *et al.* (2001) point to the Holocene record of ^{10}Be in Greenland ice cores, which yield a close match in

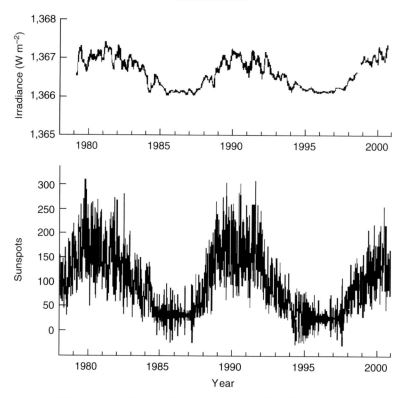

Figure 10 Comparison of the solar irradiance with the sunspot numbers for the last two Schwabe cycles. The irradiance record is a compilation of data from different satellites. During periods of high solar activity there are more sunspots darkening a small part of the solar disk (visible in the negative excursions of the irradiance). However, the brightness of the Sun is increased at the same time, overcompensating the darkening effect of the sunspots.

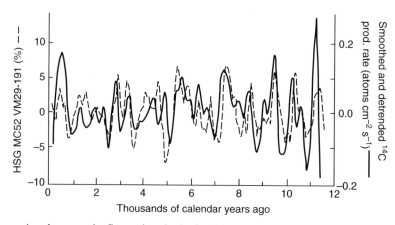

Figure 11 Comparison between the fluctuations in the fraction of red-coated lithic grains (dashed curve) with that of cosmogenic nuclide production (smooth curve) in the Earth's atmosphere. This correspondence provides powerful evidence that the Holocene's small cyclic temperature changes were paced by changes in solar luminosity. Source: Bond *et al.* (2001).

timing and in amplitude of the cosmic ray flux changes reconstructed from ^{14}C measurements.

A confirmation of Bond's climate interpretation has been obtained from the dating of wood and peat being carried by summer melt water from beneath the retreating mountain glaciers in the European Alps (Hormes *et al.*, 1998). The forests and bogs in which these materials were generated must date from warmer times when the glaciers

were even smaller than they are today. Three such times have been documented by radiocarbon dating many tens of wood and peat samples. They match quite well with three of Bond's warm intervals (see Figure 11). Further, evidence in support of Bond's temperature cycles comes from fossil-tree remains found north of the present tree line on Russia's Kola Peninsula (Hiller *et al.*, 2001). They document the existence of a northward expansion of the forest during medieval time (1000–1300). This corresponds to the period during which the Vikings colonized southern Greenland.

8.6 THE "BOMB" ^{14}C TRANSIENT

During the late 1950s and early 1960s, hydrogen bomb tests carried out in the atmosphere by Russia, Britain, and United States led to the production of manmade radiocarbon atoms. In this case, the neutrons that collided with atmospheric nitrogen nuclei were by-products of the fusion reactions. Large-scale anthropogenic production of ^{14}C came to a halt with the implementation of the ban on atmospheric weapons tests on January 1, 1963. At that time, the number of manmade ^{14}C atoms in the atmosphere was roughly equal to the number of natural radiocarbon atoms (i.e., the $^{14}C/C$ ratio nearly doubled). Since then, this atmospheric excess has steadily dwindled until as of the year 2000, the bomb-produced atoms constituted only ~10% of the atmospheric radiocarbon inventory (see Figure 12). This decrease has three causes:

1. exchange with oceanic $\sum CO_2$,
2. exchange with terrestrial biospheric carbon, and
3. continued addition of ^{14}C-free fossil-fuel-derived CO_2 molecules to the atmosphere.

Figure 13 shows an estimate as to how the distribution of bomb radiocarbon atoms among these active carbon reservoirs has evolved.

8.6.1 Radiocarbon as a Tracer for Ocean Uptake of Fossil Fuel CO_2

While $^{14}CO_2$ is in a sense a perfect tracer for fossil fuel CO_2, the quite different time histories for their inputs and the large difference between the times required for the surface ocean to respond to isotopic compared with chemical transients makes the task more complicated than it might seem. Even so, the situation for the ocean turns out to be simpler than that for the land. There are three reasons for this. First, the ocean has a far greater degree of lateral homogeneity. Second, uptake by the sea is governed by basic inorganic

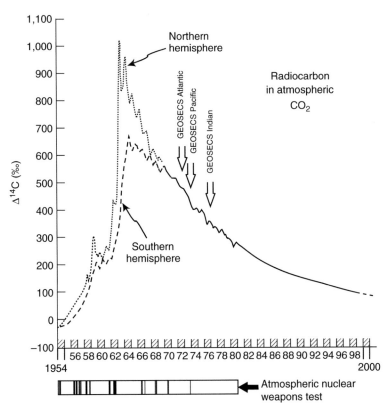

Figure 12 $\Delta^{14}C$ for atmospheric ^{14}C following the onset of H-bomb testing in 1952. The buildup continued until mid-1963 and since then the $\Delta^{14}C$ to C ratio has been declining toward its prenuclear value as the excess bomb testing ^{14}C atoms are transferred to the ocean and terrestrial biosphere. Although China and India, who were not signatories to the test ban treaty explode, conducted atmospheric tests in the late 1960s and early 1970s, the amount of ^{14}C produced was negligible. Within decades, the $\Delta^{14}C$ will drop below zero due to the continued emission of ^{14}C-free fossil fuel CO_2. The times of the GEOSECS ocean surveys are shown. The pre-1980 data are from Nydal and Lövseth (1983).

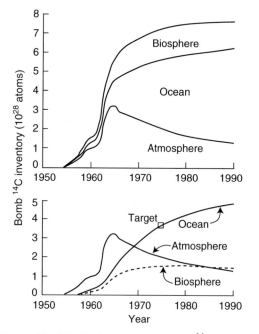

Figure 13 Distribution of bomb testing ^{14}C between the atmosphere, ocean, and terrestrial biosphere as reconstructed by Broecker and Peng (1994). The ocean contribution is obtained by a model constrained by the inventory based on the GEOSECS survey. The contribution of the terrestrial biosphere is based on estimates of the biomass and turnover times for trees and active soil humus.

chemistry while that by the land biota involves complex biotic cycles. Third, the documentation of the radiocarbon transient through direct measurement is far more extensive for the ocean.

8.6.2   Ocean Uptake of $^{14}CO_2$ and CO_2

As of early 2000s, it has not been possible to make an accurate direct assessment of the amount of fossil fuel CO_2 taken up by the sea. The reason is related to the fact that even in the surface mixed layer the increase can have been no more than 2–3% (10% increase in the atmosphere's CO_2 partial pressure produces only a 1% increase in the surface ocean's dissolved inorganic carbon content (i.e., $CO_2 + HCO_3^- + CO_3^{2-}$)). Unfortunately, until the GEOSECS global survey (1972–1978) no measurements of the concentration of dissolved inorganic carbon (i.e., $CO_2 + HCO_3^-$ and CO_3^{2-}) in the ocean of sufficient accuracy were made. Further, as the GEOSECS measurements are accurate to only $\pm 1\%$, even they are not quite up to the task. Only with the TTO surveys in the Atlantic during the early 1980s and the WOCE surveys in the Pacific and Indian oceans during the early 1990s was a base level established with sufficient accuracy (i.e., 0.1%). Hence, the uptake must be estimated using ocean–atmosphere

models rather than direct measurements. Whether these models are of the simple box-diffusion or complex three-dimensional general circulation variety, it is necessary to calibrate (in the case of box models) or constrain (in the case of general circulation models) them with tracer measurements. The most powerful tracer for this purpose is the radiocarbon produced during the atmospheric testing of nuclear weapons.

Two resistances limit the rate at which bomb $^{14}CO_2$ and fossil fuel CO_2 are transferred from the atmosphere to the sea. The first is transfer across the air–sea interface and the second is the rate at which surface waters are mixed into the ocean's interior. Clearly, the relative strengths of these two resistances are of importance. Were the resistance posed by vertical mixing in the ocean to be much smaller than that posed by CO_2 exchange across the air–sea interface, then the air-to-sea difference in CO_2 partial pressure and in $^{14}CO_2$ partial pressure would remain large, and the rate of uptake by the sea would be dominated by the rate of air-to-sea CO_2 exchange. Were the case opposite, then ocean uptake would be limited by the rate of vertical mixing within the sea. It turns out that for CO_2 itself the resistance posed by vertical mixing in the sea dominates. Waters in the tens-of-meters thick wind-mixed layer are able to reach 85% or so of saturation with the atmosphere's fossil fuel CO_2 burden before being mixed into the interior. For $^{14}CO_2$ the situation is quite different. The resistance posed by the interface is comparable with that posed by vertical mixing within the sea. The reason for this difference between $^{14}CO_2$ and CO_2 has to do with the chemical speciation of carbon in surface water. The proportions of CO_2, HCO_3^-, and CO_3^{2-} in the surface ocean are roughly 10:1, 800:200. While isotopic equilibration requires that the ^{14}C in all three species be exchanged, chemical equilibration is accomplished by the reaction

$$CO_2 + CO_3^{2-} + H_2O \rightarrow 2HCO_3^- \qquad (2)$$

Because of this, the time required for chemical equilibration turns out to be roughly an order of magnitude smaller (i.e., 200/1,800) than that for isotopic equilibration.

Taken together, measurements of the vertical distribution of bomb radiocarbon in the sea and the air-to-surface-sea difference in the bomb $^{14}C/C$ ratio permit both the laterally averaged rate of air–sea CO_2 exchange and the laterally averaged rate of vertical mixing to be estimated. As part of the GEOSECS survey, the distribution of radiocarbon throughout the entire world ocean was measured (in the laboratories of Minze Stuiver (University of Washington) and Gote Ostlund (University of Miami)). As this survey was conducted 12 ± 3 years after the implementation of the ban on atmospheric testing, the average bomb

^{14}C atom had a little more than a decade to enter the sea and be mixed into its interior.

To make use of this transient, the contributions of the natural and bomb radiocarbon had to be separated. Three sets of observations went into this separation. First, use was made of $^{14}C/C$ measurements on surface waters collected very early in the nuclear testing era (Broecker *et al.*, 1960) and also with results of measurements on prenuclear growth ring-dated corals and mollusks (Druffel and Linick, 1978; Druffel, 1981, 1989). Second, use was made of the tritium released to the atmosphere during nuclear tests. As this bomb-test tritium swamped the natural tritium present in the ocean, the vertical distribution of tritium in the sea could be used to establish the limit of penetration of bomb radiocarbon. Finally, based on the radiocarbon analyses made on thermocline waters free of bomb tritium, it was shown that there was a close correlation between the natural $^{14}C/C$ ratio and the dissolved silica content of the water (Broecker *et al.*, 1995). More recently, it was shown that there was an even better correlation between natural radiocarbon and salinity-normalized alkalinity. These relationships were essential in separating the two "brands" of ^{14}C in the waters from the upper portion of the main oceanic thermocline.

Taken together, the distributions of natural and bomb radiocarbon in the sea give us an important insight regarding the manner in which the ocean mixes. The steady-state distribution of natural radiocarbon constrains the replacement timescale for waters in the deep sea to be on the order of one millennia. The transient distribution of bomb radiocarbon at the time of the GEOSECS survey tells us that on a timescale of a decade, radiocarbon atoms are able to exchange with $\sim 10\%$ of the ocean's dissolved inorganic carbon (i.e., $CO_2 + HCO_3^- + CO_3^{2-}$). Taken together, these two constraints suggest that the fraction of the ocean volume accessed by vertical mixing increases with the square root of penetration time. This concept gave rise to the early one-dimensional box-diffusion models for fossil fuel CO_2 uptake by the ocean (Oeschger *et al.*, 1975; Seigenthaler *et al.*, 1980). Such models consist of a well-mixed atmosphere and a well-mixed surface ocean layer underlain by a diffusive half-space. The so-called eddy diffusivity assigned to this half-space was set by fitting the horizontally averaged vertical distribution of bomb radiocarbon.

One might conclude that the crude box-model representation of the ocean would have been soon eclipsed by ocean general circulation models capable of simulating, in three dimensions, the full suite of ocean currents and mixing processes. While certainly a critical step in the evolution of our ability to predict the split of fossil fuel CO_2 between the ocean and atmosphere, the task of creating models that match the distributions of not only temperature and salinity, but also those of natural and bomb radiocarbon, is a daunting one. Properly simulating the flow pathways and flow rates in the 600–1,200 m depth range dominated by the ocean's intermediate water masses has proven particularly challenging. Consequently, fossil fuel CO_2 uptake estimates made with these three-dimensional models are, in my estimation, as yet no more reliable than those based on the one-dimensional ^{14}C-calibrated box-diffusion models.

Both types of models suggest that $35 \pm 5\%$ of the CO_2 produced to date by fossil fuel burning has been taken up in the ocean. This represents $\sim 16\%$ of the ocean's capacity for fossil fuel CO_2 uptake (i.e., of the amount of uptake were the entire volume of the ocean to be equilibrated with the atmosphere). One might ask why this percentage is higher than that based on the distribution of bomb radiocarbon. The answer is that while the ^{14}C atoms had at the time of the GEOSECS survey been in existence for only ~ 12 years, the average existence time for fossil fuel CO_2 molecules is more like 30 years. Since the square root of 30/12 is 1.6, hence, to the extent that the diffusive characterization of ocean mixing is correct, fossil fuel CO_2 should have been able to penetrate a 1.6 times larger volume than bomb ^{14}C.

The challenge for ocean modelers is to achieve a match with the observed distributions of bomb ^{14}C not only for the time of the GEOSECS survey (the 1970s) but also for the time of the TTO (1980–1981) and WOCE (early 1990s) surveys. Only then can modelers be confident that they can predict the split in excess CO_2 between atmosphere and ocean for any given future fossil-fuel-use scenario. This is tricky because the uptake capacity for CO_2 by surface ocean waters depends on their carbonate ion concentration. Clearly, as CO_2 builds up in the atmosphere, the carbonate ion concentration in surface waters will decrease, but the rate of decrease will depend on the manner in which the ocean mixes. Advective overturning brings "virgin" water high in CO_3^{2-} concentration to the surface. However, diffusive mixing creates an ever-increasing carbonate ion gradient between "spent" surface water and "virgin" deep water. Different models yield different ratios of diffusive to advective mixing.

8.6.3 Terrestrial Uptake of $^{14}CO_2$ and CO_2

In recent years, the response of the terrestrial biosphere to the increase in the atmosphere's CO_2 content has become a hot topic. Thousands of chamber experiments show that at least on the short term (a year or two) given more CO_2, plants grow faster. Further, global terrestrial carbon inventories based on the rate of the atmosphere's

CO_2 increase and O_2 decline indicate that storage is on the increase. This increase is happening despite large-scale deforestation that, of course, drives down carbon stocks. Three factors appear to be responsible for this increase: (1) the excess CO_2 in the atmosphere; (2) the dispersal of fixed nitrogen evaporated from farmland and given off in automobile exhausts; and (3) regrowth of forests on land previously cleared for farming. Unlike the ocean, the land surface is a highly checkered mosaic of vegetation types and growth histories making the development of models capable of predicting future storage an enormously challenging task.

While radiocarbon is not nearly as valuable to this exercise as it is in the case of the ocean, it does have a role to play. More than half of the terrestrial carbon inventory is stored in soils. The humus in soils consists of a host of complex organic compounds. The evolution of storage in this reservoir will be driven by two competing impacts. Increasing planetary temperature will lead to more rapid oxidation of these humic compounds and hence will tend to drive down the planetary inventory. In contrast, increasing plant growth will lead to increased storage of new humic compounds and hence tend to drive up the inventory.

Radiocarbon's role is to constrain the turnover times for the compounds making up humus. The temporal trend of the $^{14}C/C$ ratio in soil humus (from prenuclear time to the present) makes it possible to separate the contribution of "passive" compounds that have very long soil residence times (measured in hundreds to thousands of years) and hence will be little changed by man's activities, and those "active" compounds which have relatively short chemical lifetimes (measured in decades). It also allows the characterization of the average turnover time of these "active" components. Further, by conducting such measurements on soils from different climate zones, it is possible to get a handle on how the turnover time of active compounds depends on temperature. Clearly, knowledge of this dependence is critical to the prediction of future global humus inventories.

Unfortunately, there was no GEOSECS-type survey of the world's soils during the 1970s. The decay counting method in use at that time required such large soil samples that processing was extremely cumbersome and stored samples were, by and large, far too small. Hence, interest in this subject lagged until the mid-1980s when atom counting became available. And of course, interest was further heightened when the Kyoto Accord permitted excess carbon storage in soils to be deducted from CO_2 emissions. However, unlike the ocean whose bomb ^{14}C inventory continues to rise, that in soils peaked in the 1970s and has subsequently declined. Hence, an enormous opportunity was largely missed—"largely"

because, fortunately, here and there soils have been collected and stored.

As part of his PhD thesis research at Lamont–Doherty, Kevin Harrison summarized his own and published radiocarbon measurements on soils (see Figures 14 and 15). He was able to show three quite important things (Harrison *et al.*, 1993a, b):

1. To reconcile both the lower than modern pre-1950 $^{14}C/C$ ratios in soil humus and the time history of the bomb ^{14}C-induced rise in the $^{14}C/C$ ratio, he concluded that about one quarter of the carbon in soils is in either inactive (i.e., turns over very slowly) or totally inert (immune to destruction) compounds.

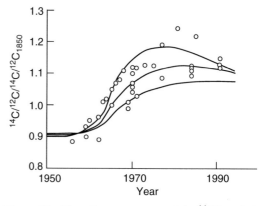

Figure 14 Plot of measurements of the $^{14}C/C$ ratio in untilled soils from a range of locales. The results are expressed as the $^{14}C/C$ ratio in the bulk soil carbon to that in age-corrected 1850 wood. The curves are based on a model which assumes that 25% of the carbon is essentially inert (mean ^{14}C age 3,700 years) and that 75% turns over on timescales of 40 years (lower curve), 25 years (middle curve) and 15 years (upper curve). Source: Harrison *et al.* (1993a).

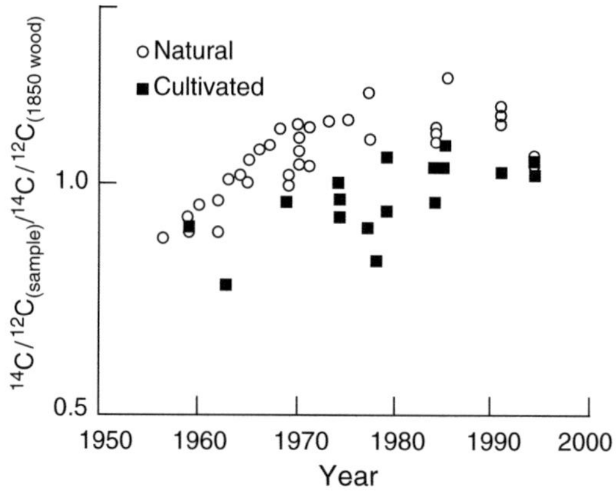

Figure 15 Comparison of ^{14}C to C measurements on natural soils and those on cultivated soils. The lower ratios are presumably the result of the loss of a sizable portion of the active humus component as the result of agricultural practice. Source: Harrison *et al.* (1993a).

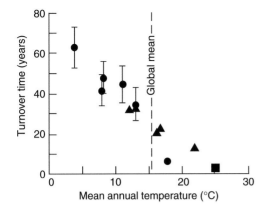

Figure 16 Turnover time for the active humus component of soils from the Brazilian rainforest (square), Hawaii (triangles), and the Sierra Nevada Mountains (circles). The mean global temperature is shown for reference. Source: Trumbore *et al.* (1996).

2. To account for the temporal evolution of bomb ^{14}C in the soil, he concluded that the mean residence time of the active humus component was 25 ± 10 years.

3. To account for the lower ^{14}C/C ratio found for agricultural soil he concluded that a sizable portion of the active component had been lost as the result of farming practices (tilling, harvesting, etc.) thereby enhancing the contribution of the low ^{14}C inactive component.

Sue Trumbore of the University of California, Irvine, demonstrated that the turnover time of humic material is strongly dependent on soil temperature ranging from 100 years at high latitudes to as little as a decade in the tropics (Trumbore *et al.*, 1996) (see Figure 16).

8.7 FUTURE APPLICATIONS

The recent development of a much smaller and hence less-expensive accelerator mass spectrometer capable of state-of-the-art radiocarbon measurements will allow large users to have an in-house unit rather than having to send their samples off to one of the dozen or so existing major centers. These small units operate at 5×10^5 V rather than at several million volts as do the conventional accelerator mass spectrometers. Also they will fit into a standard-size laboratory rather than the one with the size of a small airplane hanger. Further, like conventional mass spectrometers, these so-called tandys (i.e., small tandem Van de Graaff) can be maintained and operated by the users rather than by a team of specialists. Thus, measurements, which now cost from $300 to $600 each, will be carried out for more like $100 each. This will allow scientists to take on problems requiring hundreds of measurements rather than being budgetarily restricted to projects involving only tens of measurements.

As the bomb ^{14}C transient demonstrated the value of radiocarbon as an environmental tracer, availability of "tandy" might lead to small-scale tracer experiments designed to study the allocation of carbon by plants in natural environments. Because of the high accuracy and high sensitivity of ^{14}C measurements, such tracer experiments can be carried out at a few times the ambient atmospheric ^{14}C/C ratio, and hence pose no environmental hazard.

REFERENCES

Adkins J. F., Cheng H., Boyle E. A., Druffel E. R. M., and Edwards R. L. (1998) Deep-sea coral evidence for rapid change in ventilation of the deep North Atlantic 15,400 years ago. *Science* **280**, 725–728.

Adkins J. F., McIntyre K., and Schrag D. P. (2002) The salinity, temperature and δ^{18}O of the glacial deep ocean. *Science* **298**, 1769–1773.

Bard E. (1998) Geochemical and geophysical implications of the radiocarbon calibration. *Geochim. Cosmochim. Acta* **62**, 2025–2038.

Bard E., Arnold M., Mangerud J., Paterne M., Labeyrie L., Duprat J., Mélières M.-A., Sønstegaard E., and Duplessy J.-C. (1994) The North Atlantic atmosphere-sea surface ^{14}C gradient during the Younger Dryas climatic event. *Earth Planet. Sci. Lett.* **126**, 275–287.

Bard E., Hamelin B., Fairbanks R. G., and Zindler A. (1990a) Calibration of ^{14}C timescale over the past 30,000 years using mass spectrometric U–Th ages from Barbados corals. *Nature* **345**, 405–410.

Bard E., Hamelin B., Fairbanks R. G., Zindler A., Arnold M., and Mathieu G. (1990b) U/Th and ^{14}C ages of corals from Barbados and their use for calibrating the ^{14}C time scale beyond 9,000 years BP. In *Proceedings of the 5th International Conference on AMS* (eds. F. Yiou and G. Raisbeck). *Nucl. Inst. Met.* **B52**, 461–468.

Bard E., Raisbeck G. M., Yiou F., and Jouzel J. (1997) Solar modulation of cosmogenic nuclide production over the last millennium: comparison between ^{14}C and ^{10}Be records. *Earth Planet. Sci. Lett.* **150**, 453–462.

Barker S. and Broecker W. A 190 per mil drop in atmosphere's Δ^{14}C during the "Mystery Interval" (17.5 to 14.5 kyrs). *Quat. Sci. Rev.* (in press).

Barker, S., Broecker W., Clark E., Hajdas I., Bonani G., Moreno E., and Stott L. Radiocarbon age of deglacial-age water from 2.8 km depth in the Western Equatorial Pacific. *Geochem. Geophys. Geosys.* (in press).

Beck J. W., Richards D. A., Edwards R. L., Silverman B. W., Smart P. L., Donahue D. J., Hererra-Osterheld S., Burr G. S., Calsoyas L., Jull A. J. T., and Biddulph D. (2001) Extremely large variations of atmospheric ^{14}C concentration during the last glacial period. *Science* **292**, 2453–2458.

Becker B. and Kromer B. (1993) The continental tree-ring record—absolute chronology, ^{14}C calibration, and climatic change at 11 ka. *Palaeogeogr. Palaeoclimatol. Palaeoecol.* **103**, 67–71.

Bond G. C., Kromer B., Beer J., Muscheler R., Evans M. N., Showers W., Hoffmann S., Lotti-Bond R., Hajdas I., and Bonani G. (2001) Persistent solar influence on North Atlantic climate during the Holocene. *Science* **294**, 2130–2152.

Bond G. C., Showers W., Cheseby M., Lotti R., Almasi P., deMenocal P., Priore P., Cullen H., Hajdas I., and Bonani G. (1997) A pervasive millennial-scale cycle in North Atlantic Holocene and glacial climates. *Science* **278**, 1257–1266.

Bond G. C., Showers W., Elliot M., Evans M., Lotti R., Hajdas I., Bonani G., and Johnson S. (1999) The North Atlantic's 1–2 kyr climate rhythm: relation to Heinrich events, Dansgaard/Oeschger cycles, and the Little Ice Age.

In *Mechanisms of Global Climate Change at Millennial Time Scales* (eds. P. Clark, R. Webb, and L. D. Keigwin). Geophysical Monograph Series 112, American Geophysical Union, Washington, DC, pp. 35–58.

Bonhommet N. and Zähringer J. (1969) Paleomagnetism and potassium argon determinations of the Laschamp geomagnetic polarity event. *Earth Planet. Sci. Lett.* **6**, 43–46.

Broecker W. S., Gerard R., Ewing M., and Heezen B. C. (1960) Natural radiocarbon in the Atlantic Ocean. *J. Geophys. Res.* **65**, 2903–2931.

Broecker W. S. and Peng T.-H. (1994) Stratospheric contribution to the global bomb radiocarbon inventory: model versus observation. *Global Biogeochem. Cycles* **8**, 377–384.

Broecker W. S., Peng T.-H., Trumbore S., Bonani G., and Wolfli W. (1990) The distribution of radiocarbon in the glacial ocean. *Global Biogeochem. Cycles* **4**, 103–117.

Broecker W. S., Sutherland S., and Smethie W. (1995) Oceanic radiocarbon: separation of the natural and bomb components. *Global Biogeochem. Cycles* **9**, 263–288.

Chiu, C. (2005) PhD Thesis, Columbia University.

deVries H. L. (1958) Variation in concentration of radiocarbon with time and location on Earth. *Proc. Koninkl. Ned. Akad. Wetenschap.* **61**, 94–102.

Druffel E. R. M. (1981) Radiocarbon in annual coral rings from the eastern tropical Pacific Ocean. *Geophys. Res. Lett.* **8**, 59–62.

Druffel E. R. M. (1989) Decade time scale variability of ventilation in the North Atlantic: high-precision measurements of bomb radiocarbon in banded corals. *J. Geophys. Res.* **94**, 3271–3285.

Druffel E. R. M. and Linick T. W. (1978) Radiocarbon in annual coral rings of Florida. *Geophys. Res. Lett.* **5**, 913–916.

Edwards R. L., Chen J. H., and Wasserburg G. J. (1986/1987) $^{238}U-^{234}U-^{230}Th$–Th systematics and the precise measurement of time over the past 500,000 years. *Earth Planet. Sci. Lett.* **81**, 175–192.

Fairbanks R. G. (1989) A 17,000-year glacio-eustatic sea level record: influence of glacial melting rates on the Younger Dryas event and deep ocean circulation. *Nature* **342**, 637–647.

Fairbanks R. G., Mortlock R. A., Chiu T.-C., Cao L., Kaplan A., Guilderson T. P., Fairbanks T. W., Bloom A. L., Grootes P. M., and Nadeau M.-J. (2005) Radiocarbon calibration curve spanning 0 to 50,000 years BP based on paired $^{230}Th/^{234}U/^{238}U$ and ^{14}C dates on pristine corals. *Quat. Sci. Rev.* **24**, 1781–1796.

Frank M., Schwarz B., Baumann S., Kubik P. W., Sater M., and Mangini A. (1997) A 200 kyr record of cosmogenic radionuclide production rate and geomagnetic field intensity from ^{10}Be in globally stacked deep-sea sediments. *Earth Planet. Sci. Lett.* **149**, 121–129.

Friedrich M., Kromer B., Spurk M., Hofmann J., and Kaiser K. F. (1999) Paleo-environment and radiocarbon calibration as derived from Late Glacial/Early Holocene tree-ring chronologies. *Quat. Int.* **61**, 27–39.

Harrison K. G., Broecker W., and Bonani G. (1993a) A strategy for estimating the impact of CO_2 fertilization on soil carbon storage. *Global Biogeochem. Cycles* **7**, 69–80.

Harrison K. G., Broecker W. S., and Bonani G. (1993b) The effect of changing land use on soil ^{14}C. *Science* **262**, 725–726.

Hiller A., Boettger T., and Kremenetski C. (2001) Medieval climatic warming recorded by radiocarbon dated alpine tree-line shift on the Kola Peninsula, Russia. *Holocene* **11**, 491–497.

Hormes A., Schlüchter C., and Stocker T. F. (1998) Minimal extension phases of unteraar glacier (Swiss Alps) during the Holocene based on ^{14}C analysis of wood. *Radiocarbon* **40**, 809–817.

Hughen K. A., Overpeck J. T., Lehman S. J., Kasgarian M., Southon J., Peterson L. C., Alley R., and Sigman D. M. (1998) Deglacial changes in ocean circulation from an extended radiocarbon calibration. *Nature* **391**, 65–68.

Kitagawa H. and van der Plicht J. (1998) Atmospheric radiocarbon calibration to 45,000 yr BP: late Glacial fluctuations and cosmogenic isotope production. *Science* **279**, 1187–1190.

Kromer B., Rhein M., Bruns M., Schoch-Fisher H., Munnich K. O., Stuiver M., and Becker B. (1986) Radiocarbon calibration data for the eighth millennia BC. In Calibration issue. *Radiocarbon* **28**, 954–960.

Kromer B. and Spurk M. (1998) Revision and tentative extension of the tree-ring based ^{14}C calibration, 9,200–11,855 cal BP. *Radiocarbon* **40**, 1117–1125.

Laj C., Mazaud A., and Duplessy J. C. (1996) Geomagnetic intensity and ^{14}C abundance in the atmosphere and ocean during the past 50 kyr. *Geophys. Res. Lett.* **23**, 2045–2048.

Lal, D. (1988) Theoretically expected variations in the terrestrial cosmic-ray production rates of isotopes. In *Solar–Terrestrial Relationships and the Earth Environment in the Last Millennia* (ed. X. C. V. Corso). Soc. Italiana de Fisica., pp. 216–233.

Libby W. F., Anderson E. C., and Arnold J. R. (1949) Age determination by radiocarbon content: worldwide assay of natural radiocarbon. *Science* **109**, 227–228.

Muscheler R., Beer J., Kubik P. W., and Synal H.-A. (2005) Geomagnetic field intensity during the last 60,000 years based on ^{10}Be and ^{36}Cl from the Summit ice cores and ^{14}C. *Quat. Sci. Rev.* **24**, 1849–1860.

Nydal R. and Lövseth K. (1983) Tracing bomb ^{14}C in the atmosphere 1962–1980. *J. Geophys. Res.* **88**, 3621–3642.

Oeschger H., Siegenthaler U., Gugelmann A., and Schotterer U. (1975) A box-diffusion model to study the carbon dioxide exchange in nature. *Tellus* **27**, 168–192.

Robinson L. F., Adkins J. F., Keigwin L. D., Southon J., Fernandez D. P., Wang S.-L., and Scheirer D. S. (2005) Radiocarbon variability in the Western North Atlantic during the last deglaciation. *Science* **310**, 1469–1473.

Seigenthaler U., Heimann M., and Oeschger H. (1980) ^{14}C variations caused by changes in the global carbon cycle. *Radiocarbon* **22**, 177–191.

Shackleton N. J., Duplessy J.-C., Arnold M., Maurice P., Hall M. A., and Cartlidge J. (1988) Radiocarbon age of the last glacial Pacific deep water. *Nature* **335**, 708–711.

Siani G., Paterne M., Michel E., Sulpizio R., Sbrana A., Arnold M., and Haddad G. (2001) Mediterranean Sea surface radiocarbon reservoir age changes since the Last Glacial maximum. *Science* **294**, 1917–1920.

Sikes E. L., Samson C. R., Guilderson T. P., and Howard W. R. (2000) Old radiocarbon ages in the southwest Pacific Ocean during the Last Glacial period and deglaciation. *Nature* **405**, 555–559.

Southon J. (2002) A first step to reconciling the GRIP and GISP2 Ice-core chronologies, 0–14,500 yr BP. *Quat. Res.* **57**, 32–37.

Stuiver M. (1970) Long-term ^{14}C variations. In *Radiocarbon Variations and Absolute Chronology, 12th Nobel Symposium* (ed. I. U. Olsson). Wiley, New York, pp. 197–213.

Stuiver M. (1971) Evidence for the variation of atmospheric ^{14}C content in the Late Quaternary. In *Late Cenozoic Glacial Ages* (ed. K. K. Turekian). Yale University Press, New Haven, CT, pp. 57–70.

Stuiver M. (1980) Solar variability and climatic change during the current millennium. *Nature* **286**, 868–871.

Stuiver M. (1982) A high-precision calibration of the AD radiocarbon time scale. *Radiocarbon* **24**, 1–26.

Stuiver M. and Braziunas T. F. (1993) Sun, ocean, climate, and atmospheric $^{14}CO_2$: an evaluation of causal and spectral relationships. *Holocene* **34**, 289–305.

Stuiver M. and Quay P. D. (1980) Changes in atmospheric carbon-14 attributed to a variable Sun. *Science* **207**, 11–19.

Stuiver M. and Quay P. D. (1981) Atmospheric ^{14}C changes resulting from fossil fuel CO_2 release and cosmic ray flux variability. *Earth Planet. Sci. Lett.* **53**, 349–362.

Stuiver M., Reimer P. J., Bard E., Beck J. W., Burr G. S., Hughen K. A., Kromer B., McCormac G., van der

Plicht J., and Spurk M. (1997) Intcal98 radiocarbon age calibration, 24,000–0 cal BP. *Radiocarbon* **40**, 1126–1159.

Thellier E. and Thellier O. (1959) Sur l'intensite du champ magnetique terrestre dans le passe historique et geologique. *Annu. Geophys.* **15**, 285–378.

Trumbore S. E., Chadwick O. A., and Amundson R. (1996) Rapid exchange between soil carbon and atmospheric carbon dioxide driven by temperature change. *Science* **272**, 393–396.

Voelker A. H. L., Sarnthein M., Grootes P. M., Erlenkeuser H., Laj C., Mazaud A., Nadeau M.-J., and Schleicher M. (1998) Correlation of marine ^{14}C ages from the Nordic seas with the GISP2 isotope record: implications for radiocarbon calibration beyond 25 ka BP. *Radiocarbon* **40**, 517–534.

Waelbroeck C., Duplessy J.-C., Michel E., Labeyrie L., Paillard D., and Duprat J. (2001) The timing of the last deglaciation in North Atlantic climate records. *Nature* **412**, 724–727.

Yiou F., Raisbeck G. M., Baumgartner S., Beer J., Hammer C., Johnsen S., Jouzel J., Kubik P. W., Lestringuez J., Stievenard M., Suter M., and Yiou P. (1997) Beryllium 10 in the Greenland Ice Core Project ice core at Summit, Greenland. *J. Geophys. Res.* **102**, 26783–26794.

Radioactive Geochronometry
ISBN: 978-0-08-096708-0

pp. 251–268

9

Natural Radionuclides in the Atmosphere

K. K. Turekian and W. C. Graustein

Yale University, New Haven, CT, USA

9.1 INTRODUCTION

Natural radioactive nuclides in the atmosphere have two principal sources—radon and its progeny derived from Earth's surface and cosmic-ray-produced nuclides. Dust from the elevation of soils can also provide secondary sources of these nuclides. Suitable accommodation for these sources must be made if only those species having gaseous precursors are to be considered.

There is a radon isotope in each of the three major natural-decay chains: ^{238}U (^{222}Rn, half-life = 3.8 d), ^{232}Th (^{220}Rn, half-life = 55 s), and ^{235}U (^{219}Rn, half-life = 3.9 s). Almost all radon in the atmosphere is produced in soils and is transported to the atmosphere by diffusion. Because its longer half-life allows for greater diffusive transport, most radon entering the atmosphere is ^{222}Rn; its radioactive decay scheme is presented in Table 1.

An early summary of ^{222}Rn and its progeny in the atmosphere was provided by Turekian *et al.* (1977).

Table 2 lists the radioactive species produced by cosmic rays acting on the gaseous components of the atmosphere, which include nitrogen, oxygen, and all the rare gases as targets. A detailed discussion of the formation of radioactive species by cosmic-ray bombardment has been published by Lal (2001) based on his pioneering work since 1967 in a classic paper by Lal and Peters (1967). Radiocarbon (^{14}C) is not discussed in this chapter. Because of its central role in many Earth's surface processes, a separate chapter is found in this volume (see Chapter 8).

Table 1 The decay chain for ^{222}Rn.

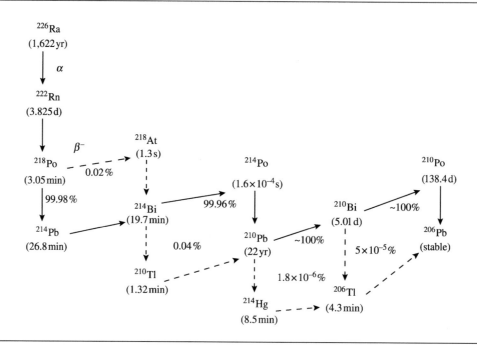

Table 2 Production rates of several isotopes in the Earth's atmosphere; arranged in order of decreasing half-lives.

Isotope	Half-life	Production rate (atoms cm^{-2} s^{-1})		Global inventory
		Troposphere	Total atmosphere	
^{3}He	Stable	6.7×10^{-2}	0.2	3.2×10^3 t^a
^{10}Be	1.5×10^5 yr	1.5×10^{-2}	4.5×10^{-2}	260 t
^{26}Al	7.1×10^5 yr	3.8×10^{-5}	1.4×10^{-4}	1.1 t
^{31}Kr^b	2.3×10^5 yr	5.2×10^{-7}	1.2×10^{-5}	8.5 kg
^{36}Cl	3.0×10^5 yr	4×10^{-4}	1.1×10^{-3}	15 t^d
^{14}C	5,730 yr	1.1	2.5	75 t
^{39}Ar^c	268 yr	4.5×10^{-3}	1.3×10^{-2}	52 kg
^{32}Si	$\sim$150 yr	5.4×10^{-5}	1.6×10^{-4}	0.3 kg
^{3}H	12.3 yr	8.4×10^{-2}	0.25	3.5 kg
^{22}Na	2.6 yr	2.4×10^{-5}	8.6×10^{-5}	1.9 g
^{36}S	87 d	4.9×10^{-4}	1.4×10^{-3}	4.5 g
^{7}Be	53 d	2.7×10^{-2}	8.1×10^{-2}	3.2 g
^{37}Ar	35 d	2.8×10^{-4}	8.3×10^{-4}	1.1 g
^{33}P	25.3 d	2.2×10^{-4}	6.8×10^{-4}	0.6 g
^{32}P	14.3 d	2.7×10^{-4}	8.1×10^{-4}	0.4 g

Source: Based on Lal and Peters (1967).
[a] The inventory of this stable nuclide is based on its atmospheric inventory, which includes an appreciable contribution from crustal degassing of ^{3}He.
[b] Based on atmospheric ^{31}Kr/Kr ratio of $(5.2 \pm 0.4) \times 10^{-13}$.
[c] Based on atmospheric ^{39}Ar/Ar ratio of (0.107 ± 0.004) dpm^{-1} l Ar (STP).
[d] Includes a rough estimate of ^{36}Cl produced by the capture of neutrons at the Earth's surface.

There are more than 30 radionuclides produced in the atmosphere by these two natural processes. This chapter focuses on describing different ways in which the observed distributions of several of these nuclides can be used to infer patterns and rates of transport, mixing, and removal of these nuclides and of other constituents of the atmosphere.

We would like to define here several terms. "Activity" refers to the number of radioactive decays of a nuclide per unit time and is the product of the decay constant, λ, and the number of atoms, N. The SI unit of activity is the Becquerel (Bq), which represents one disintegration per second. The "disintegration per minute" or dpm often appears in the literature. An older representation was the Curie (Ci), which is defined as the activity of 1 g of pure ^{226}Ra or 3.70×10^{10} disintegrations per second. The picocurie (pCi), which is 0.037 Bq or 2.2 dpm, is often used in reporting radon concentrations in air.

In a decay chain shown in Table 1 at "secular equilibrium" the activities of the coupled radionuclides are equal. In a closed system, this occurs after roughly five half-lives of the shorter-lived daughter have elapsed ("parent" and "daughter" refer to the first and subsequent nuclides in the decay series being considered). For example, ^{222}Rn (half-life = 3.8 d) is in secular equilibrium with its parent, ^{226}Ra (half-life = 1,620 yr), in ~20 d.

9.2 RADON AND ITS DAUGHTERS

9.2.1 Flux of Radon from Soils to the Atmosphere

The escape of ^{222}Rn from soils is the source of ~99% of the ^{222}Rn in the atmosphere. Typical radon escape rates are on the order of 1 atom $cm^{-2} s^{-1}$ from the land surface, which result in a radon inventory of the global atmosphere of $\sim 1.5 \times 10^{18}$ Bq. Atmospheric radon itself is a chemically inert and unscavenged, i.e., not removed from the atmosphere by physical or chemical means. Because its half-life is much less than the mixing time of the atmosphere, it is a tracer of atmospheric transport and can be used in a synoptic approach to identify air masses derived from continental boundary layers or in a climatological manner to verify the predictions of numerical models of transport.

^{222}Rn is lost from the atmosphere only by its radioactive decay and it is thus the source of virtually all of the ^{214}Bi, ^{214}Pb, ^{214}Po, ^{210}Pb, ^{210}Bi, and ^{210}Po in the atmosphere. The quantities of radon decay products on soil dust suspended by atmospheric turbulence or volatilized from lava are small by comparison on a global scale, but may be comparable locally and episodically.

The migration of ^{222}Rn through soils into buildings can lead to indoor air concentrations of ^{222}Rn that pose significant radiological health hazards. Knowing the flux and understanding the factors that cause it to vary spatially and temporally improve the utility of ^{222}Rn as an atmospheric tracer and increase the ability to predict potential health hazards National Research Council (1988).

^{238}U is a ubiquitous trace component of rock and soil; typical activities are ~ 30 mBq g^{-1}. In most soils, there is little transport of ^{234}U, ^{230}Th, or ^{226}Ra, so the activities of ^{238}U and ^{226}Ra are nearly equal. When an atom of ^{226}Ra decays in a mineral grain, the energy imparted by the recoil of the alpha particle is sufficient to displace the daughter ^{222}Rn atom by a few tens of nanometers. Although most often ^{222}Rn atoms remain within the grain, a fraction of them emanate from the grain and become free to move through the pore space of the soil. The fraction of ^{226}Ra decays that result in ^{222}Rn atoms in the pore space is termed as the emanating power or fraction. Typical values for emanating fractions are ~25%. Within the top few meters of a soil, a fraction of the radon in the pore space is transported to the surface, where it escapes to the atmosphere.

If one makes the simplifying assumptions that transport of ^{222}Rn is by molecular diffusion alone and that soil has uniform porosity, ^{222}Rn concentration, and emanating power, then the fraction of ^{222}Rn that escapes to the atmosphere as a function of depth, $L_f(z)$, is given by (e.g., Clements and Wilkening, 1974)

$$L_f(z) = ae^{-bz} \qquad (1)$$

where a is the ^{222}Rn emanating power (dimensionless), $b = \sqrt{\lambda \varepsilon / D}$ (cm^{-1}), λ is the decay constant of ^{222}Rn (s^{-1}), ε is the porosity (dimensionless), D is the diffusion coefficient $(cm^2 s^{-1})$ of ^{222}Rn, and z is the depth (cm).

Here $1/b$ has dimensions of length and is called the "mean depth" or "relaxation depth" of radon loss. Values of $1/b$ reported in the literature range from 100 cm to 218 cm (Schery and Gaeddert, 1982; Clements and Wilkening, 1974; Graustein and Turekian, 1990; Dörr and Münnich, 1990). These terms should not be confused with "half-depth" or 0.693 mean depth.

The escape rate of ^{222}Rn, J, can be derived from (1) by multiplying by the ^{226}Ra concentration in soil and integrating over depth, yielding

$$J = \frac{a\rho(^{226}\text{Ra})}{b} \qquad (2)$$

where ρ is the density $(g cm^{-3})$ and $(^{226}$Ra$)$ is the activity (Bq g^{-1}).

Compared to this idealized model, the actual flux of ^{222}Rn may be diminished by the saturation of pore space by water (the mean length of ^{222}Rn diffusion in water is on the order of a millimeter, so saturation diminishes the flux by up to a factor of 1,000) and decreases in porosity with depth. Advection of gas through soil in response to barometric pressure change, soil gas convection, and transpiration of ^{222}Rn saturated soil solution will increase the radon escape rate. All of these processes are difficult to model accurately, so the determination of ^{222}Rn fluxes relies on measurements.

Several methods have been developed to measure the escape rate or flux of radon to the atmosphere; measuring accumulation of ^{222}Rn in chambers placed on the soil surface, modeling the flux from the vertical profile of ^{222}Rn in the atmosphere, modeling the diffusive flux from measurements of ^{222}Rn in the interstitial gas of a

272   *Natural Radionuclides in the Atmosphere*

soil (e.g., Dörr and Münnich, 1990), and measuring the deficiency of the ^{210}Pb daughter with respect to its parent ^{226}Ra (Graustein and Turekian, 1990).

As shown by Equations (1) and (2), the radon loss fraction at shallow depths closely approximates the emanating power, and the escape rate is directly proportional to the emanating power. The data of Graustein and Turekian (1990) put a lower limit on the mean emanating power of US soils at 0.23. The fraction of ^{222}Rn lost from depth between 10 cm and 60 cm for 340 cores from North America (Graustein and Turekian, 1990, in preparation) averages 0.22 and does not vary significantly over that interval. Figure 1 shows the distribution with depth. An exponential curve with a half-depth of 100 cm (mean depth = 144 cm) is plotted for reference. The value of b was assigned; the value for a, 0.28, was obtained from a best fit to the data. With the exception of the sections from above 20 cm and those below 180 cm, the curve fits the data well. The counting errors and small number of samples from below 160 cm diminish our confidence in using this curve to extrapolate to greater depths and we regard extrapolations as representing an upper limit rather

than an estimate. The samples from above 20 cm fall to the right of the reference curve, most probably due to the presence of atmospherically derived ^{210}Pb. The extrapolation of the reference curve to the surface yields an upper limit to the emanating fraction of 0.28.

Using these data the mean ^{222}Rn flux from US soils is at least 1.5 atoms cm^{-2} s^{-1}. Extrapolation using the exponential curve in Figure 1 gives an upper limit of 2.0 atoms cm^{-2} s^{-1}. This range is 50–100% greater than the commonly used estimate of 1.0 atoms cm^{-2} s^{-1} for global radon flux from continents, implying that there are large areas with radon fluxes less than 1 or that the global estimate of 1.0 is low. The survey of ^{222}Rn fluxes from Australia by Schery et al. (1989) averaged 1.05 atoms cm^{-2} s^{-1} with a standard deviation of 0.24 indicating that large areas may have significantly different ^{222}Rn fluxes. Their data were derived from the accumulation of ^{222}Rn in accumulator chambers and are nearly instantaneous compared to the 30 yr mean values determined from the ^{210}Pb deficiency in soil method discussed above. That the two Rn flux measurements yield the same result is justified by the similarity of a ^{222}Rn flux from a soil

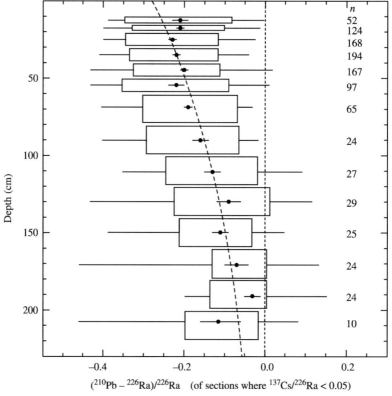

Figure 1 Depth profile of radon loss fraction. Filled circles and error bars represent the mean and standard error of the radon loss fraction from all samples from the indicated depth for which the ^{137}Cs/^{226}Ra ratio is less than 0.05. The boxes represent 1σ deviation and the whiskers the extreme values of each population. The number of samples is given in the column on the right. The curve corresponds to Equation (1) with an emanating power of 0.28 and a mean depth of 144 cm.

sampling site in the same cornfield (Graustein and Turekian, 1990) in which the ^{222}Rn flux was determined by accumulator (Pearson and Jones, 1966).

The spatial variability of the radon fluxes means that many measurements are needed to make precise estimates of the mean value for a region. Conen and Robertson (2002) noted in their compilation of radon fluxes determined by accumulator that there is a correlation with latitude from 30° to 50°, with a decreasing flux with increasing latitude. They ascribed this change as due primarily to the degree of water saturation of the soils, the increase toward higher latitude.

Comparisons of ^{222}Rn flux from the upper 50 cm, based on disequilibrium measurements in soil profiles discussed above, with mean annual temperature and precipitation show only weak relations; neither r^2 exceeded 0.02. The major correlation is with the ^{226}Ra activity of the soil (Figure 2).

9.2.2 Flux of Radon from the Oceans

The radium concentration in the oceans is at least a factor of a thousand less than soils; therefore, the flux is proportionally less. The radon flux from the mixed layer has been determined by Broecker and Peng (1971) and Wilkening and Clements (1975) and is as expected low and varies depending on wind stress. Its contribution to the radon burden of the atmosphere can therefore be taken as negligible. It is less than 1% of the global continental flux. A flux from shoal areas around Hawaii has been measured by Moore et al. (1974). Although the flux is higher than the open ocean, being driven by the higher radium concentration

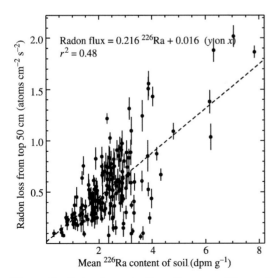

Figure 2 Radon loss rate from the upper 50 cm of soils versus the mean radium concentration. Bars indicate calculated 1σ error of the loss.

of the underlying rocks, on a worldwide basis it too is negligible. The major flux of radon to the atmosphere is clearly soils, mainly on the continents.

9.2.3 Distribution of Radon in the Atmosphere

The half-life of ^{222}Rn (3.8 d) is much less than the mixing time of the atmosphere, so its concentrations are greatest near the land surface and decrease with both altitude and distance from land (e.g., Turekian et al., 1977; Liu et al., 1984).

Although the vertical distribution of radon over the continents is a direct consequence of supply from soils, convection upward (treated as turbulent diffusion), and radioactive decay, the pattern is different over the oceans. This difference is due to the fact that no significant source of radon exists over the oceans and the pattern is set by long-distance transport from continents. Off the northwest coast of the United States, for example, Andreae et al. (1988) show vertical patterns up to $\sim$4 km ranging from constancy with elevation to increases with elevation. The concentration range from $\sim$6 pCi m^{-3} (STP) to 10 pCi m^{-3} (STP) is more typical of upper troposphere air over the continents and not like the $\sim$400 pCi m^{-3} (STP) in the continental boundary layer (Figure 3).

Radon and its progeny ^{210}Pb have been found in the stratosphere in some locations at elevated levels. Although some of ^{210}Pb may be due to explosive volcanic penetration of the tropopause, most of the ^{210}Pb is due to the decay of radon injected convectively mainly in the tropics (Kritz et al., 1993).

Lambert et al. (1970) reported periodic fluctuations in the radon concentrations in boundary layer air sampled on islands and Antarctica between 40° S and 70° S. Being in the geographic zone with no large land masses, all the radon is delivered from lower latitudes by advection. The ambient concentration was found to be $\sim$2 pCi m^{-3} (STP) with occasional "radon storms" raising the concentration briefly to 10 pCi m^{-3} (STP). Lambert et al. (1970) argued that, based on long-term observations, there is a 27–28 d cycle of radon variation in this region. The causes of changes in circulation that would yield this result are not known, but radon measurements obviously put constraints on models of hemispheric circulation.

9.2.4 Short-lived Daughters of ^{222}Rn in the Atmosphere

As can be seen in Table 1, ^{222}Rn decays through a number of short-lived daughters before the long-lived ^{210}Pb is reached. The first of these

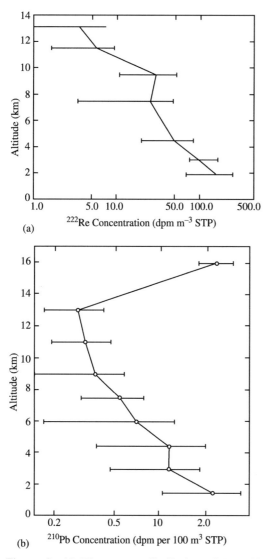

(a)

^{222}Re Concentration (dpm m^{-3} STP)

(b)

^{210}Pb Concentration (dpm per 100 m^3 STP)

Figure 3 (a) The average distribution of Rn with elevation in the North American mid-continental region based on several profiles averaged by Moore *et al.* (1973). (b) Vertical distribution of ^{210}Pb in the atmosphere in the mid-continental region of North America (source Moore *et al.*, 1973).

radionuclides along the decay chain is ^{218}Po with a half-life of 3 min. The 6 MeV energy of the alpha particle emitted by the ^{222}Rn decay causes the ^{218}Po to recoil with a velocity much greater than that of the ambient gas molecules. Collisions with gas molecules or atoms diminish its energy until it is same as that of the ambient air, at which time the ^{218}Po diffuses through the air until it encounters a surface, typically an aerosol particle, to which it attaches.

In an enclosed space within which a large number of aerosol particles exist, such as in a mine or around cigarette smoke, such ^{222}Rn decay derived nuclides attach to particles on the order of seconds to minutes and subsequent radioactive decay

can occur on the aerosol particles. An array of short-lived energetic radioactive species in the chain add to the radioactive burden of the particle. If inhaled, these highly radioactive particles can cause health problems. This was shown in the case of uranium miners and may be a part of the cause of the lung problems of smokers and those affected by secondary particle inhalation around smokers.

Studies of ^{214}Bi (half-life = 19.8 min) activity relative to ^{214}Pb (half-life = 26.8 min) in boundary layer air indicate that these are in equilibrium, indicating that the particles are not removed more rapidly than a couple of hours (Turekian *et al.*, 1999). Attaching of these short-lived daughters on aerosols, and the mean life of the aerosols being much longer than their half-lives means that the nuclide to be tracked on most atmospherically interesting timescales is ^{210}Pb with its half-life of 22 yr.

9.2.5 ^{210}Pb and Its Progeny

After it is produced by the decay of ^{214}Po, ^{210}Pb rapidly associates with aerosols in the 0.1–0.5 μm diameter size range (Knuth *et al.*, 1983). Aerosols in this size range, the so-called "accumulation mode," are large enough that their velocity due to Brownian motion is so small that diffusion is not a significant method of transport. They are also small enough that their gravitational settling velocity is much less than typical rates of vertical motion in the atmosphere. Scavenging by precipitation is the principal mechanism of removal of these aerosols. Many of the chemical species with low vapor pressure that form from gaseous precursors in that atmosphere, such as H_2SO_4 formed from the oxidation of SO_2, are also associated with accumulation-mode aerosols.

As a result, these aerosols follow the motions of the atmosphere and are responsible for the transport of much of the nonvolatile products of photochemical and oxidative reactions. ^{210}Pb is a minor constituent of this aerosol population (one aerosol particle in 10^4 or more will carry an atom of ^{210}Pb) and is a useful tracer of the transport, deposition, and residence time of aerosols.

9.2.5.1   *Distribution of ^{210}Pb in the atmosphere*

Vertical profiles of Rn, ^{210}Pb, ^{210}Bi, and ^{210}Po were measured over the mid-continental United States by Moore *et al.* (1973). These results for ^{222}Rn and ^{210}Pb are shown in Figure 3. A number of surface sites have been locations for long-term studies of ^{210}Pb in conjunction with the measurements of radionuclides from nuclear bomb testing by the Environmental Measurements Laboratory of the Department of Energy and its precursor, the

Atomic Energy Commission (Feely *et al.*, 1981). In addition, there have been measurements made throughout the United States and the islands of the Pacific and Atlantic as parts of the SEAREX and AEROCE programs, respectively. The sampling programs complement each other and, where there are regional overlaps, the results are identical. The values across the Pacific from the SEAREX program (Turekian *et al.*, 1989) are shown in Figure 4. There clearly is a correlation with the size of the land-mass upwind from the island sampling site. This is expected since ^{222}Rn is the precursor to the ^{210}Pb and is derived from soils.

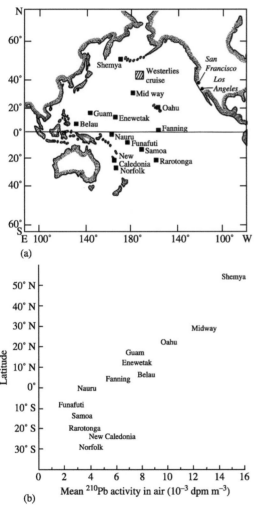

Figure 4 (a) Location map of the sites of the SEAREX network. (b) Mean concentration of ^{210}Pb in air in the SEAREX network. The site name is centered over the point representing the data. The error of the mean is approximated by the length of the name. Data are measured values which should be increased by ~30% to correct for filter capture efficiency (source Turekian *et al.*, 1989).

9.2.5.2 *Flux of* ^{210}*Pb to Earth's surface*

(i) *Methods of determinations.* The most direct method of estimating the flux of ^{210}Pb to the Earth's surface is through bucket collection of precipitation and subsequent analysis of the water for ^{210}Pb obtained over a specific period of time. This method has been used during the assay of precipitation for bomb-produced radionuclides such as ^{90}Sr and ^{137}Cs, notably by the Environmental Measurements Laboratory of the Department of Energy and its precursor, the Atomic Energy Commission. Detailed summaries, mainly along ~80° W longitude in both hemispheres plus some other selected sites in the continental US, are available from the Department of Energy as unpublished reports.

There is great value to the data obtained by such a method especially as it provides seasonal values. The method does not, however, measure the extraction of water and ^{210}Pb from the air by "horizontal" precipitation. Horizontal precipitation is the consequence of impingement of cloud, fog, or dew droplets on leaf surfaces. The water and its constituents are then transferred to the soil by dripping or episodic vertical precipitation. Also the long-term time-averaged flux is limited by the length of time that the sampling program has been in effect.

The accumulation of atmospherically derived ^{210}Pb in soil profiles is another method of assessing the atmospheric flux of ^{210}Pb. This approach has already been referred to in the section on the measurement of ^{222}Rn flux from soils to the atmosphere. Specifically, the atmospherically derived ^{210}Pb in soils is generally found in the topmost organic-rich layer, although there is transport down into the soil profile by illuviation, bioturbation, and possibly chemical transport by chelators.

This method of assaying the long-term total atmospheric ^{210}Pb flux has been studied extensively by Moore and Poet (1976), Nozaki *et al.* (1978), and Graustein and Turekian (1983, 1986, 1989). It requires that the soil profile be undisturbed for ~100 yr. In nonforested areas the directly measured atmospheric fluxes of ^{210}Pb and the fluxes determined by soil profile method are generally in agreement, allowing for the expected short-term variability (Graustein and Turekian, 1986). The relationship of the atmospherically derived standing crop of excess ^{210}Pb in soil profiles and the atmospheric flux is

$$\text{Flux} = \left(^{210}\text{Pb decay constant}\right) \times \text{inventory}$$

(ii) *The flux of* ^{210}*Pb over the oceans.* The flux of ^{210}Pb across the Pacific Ocean based on bucket collections has mainly been studied by Tsunogai *et al.* (1985) for the region around Japan, and by Turekian *et al.* (1989) for the rest of the Pacific as part of the SEAREX program. Figure 5 shows the

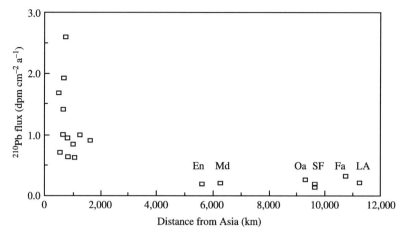

Figure 5 ^{210}Pb deposition flux versus distance from the Asian continent. Unlabeled points are from Tsunogai *et al.* (1985). En, Enewetak; Md, Midway; SF, San Francisco (Fuller and Hammond, 1983; Monaghan *et al.*, 1985/1986); LA, Los Angeles (Fuller and Hammond, 1983); Oa, Oahu; FA, Fanning. For the mid-ocean sites distances from Asia are measured along parallels of latitude (source Turekian *et al.*, 1989).

flux of ^{210}Pb over the North Pacific. The most striking pattern is the very high flux close to Japan and a virtually constant low flux across the rest of the North Pacific including the California coast.

This pattern is different from the model expectation presented by Turekian *et al.* (1977), which predicted an exponential decrease of ^{210}Pb flux across the North Pacific. The observations clearly indicate the efficient removal of the burden of ^{210}Pb from the Asian continent by the time the longitude of Japan is reached. The flux of ^{210}Pb thereafter is determined by the ^{222}Rn concentration in the air and the scavenging efficiency by rain.

(iii) *The flux of ^{210}Pb over the continents.* The flux of ^{210}Pb across the United States has been studied primarily by the Environmental Measurements Laboratory on precipitation (see Feely *et al.*, 1981) and Graustein and Turekian (1986 and references therein) in soil profiles. Figure 6 shows the ^{210}Pb flux pattern across the United States primarily from these sources but from other data as well. The major features are: (i) a general increase in flux from the Pacific coast towards the Great Plains; (ii) an approximately constant flux across the Great Plains to the Appalachians; (iii) marked orographic effects in the Sierras and the Appalachians; and (iv) a decrease along the east coast. The explanation for observation (i) is that air arriving with the prevailing westerly winds is generally depleted in ^{210}Pb and its precursor ^{222}Rn because of scavenging over the oceans and decay during transport, respectively, and the increase towards the center of the continent and general constancy is due to the local nature of both the source of ^{210}Pb and its removal. The marked orographic effects (iii) are due to both the increased rainfall in the mountains and the cloud scavenging alluded to earlier. The decrease

on the east coast is due to precipitation carrying aerosols both from the ^{210}Pb-rich continental interior and ^{210}Pb-poor maritime air.

9.2.5.3 Residence time of ^{210}Pb and associated species in the atmosphere

One approach to measuring the residence time of aerosols is to calculate the time required for presumed initial values in an air mass of two or more nuclides in a decay chain to evolve to the observed values. This approach yields a measure of the effect of removal processes along the path of transport of the air mass being sampled.

If we consider an air parcel as a closed system, then the activity of each nuclide in the ^{222}Rn decay chain is described by

$$dN/dt = P - \lambda N \qquad (3)$$

where N is the number of atoms per unit volume and P is the production rate of the nuclide by the radioactive decay of its parent. Applying (3) to each nuclide in a decay chain and setting initial conditions leads to a set of coupled differential equations. The general form for the solution to such a set of differential equations was obtained by Bateman (1910) and is given in standard references in radiochemistry (e.g., Friedlander *et al.*, 1981).

The dotted line in Figure 7 shows the time evolution of the ratios of radon daughters in closed system in which initially ^{222}Rn is present but its decay products are absent. We refer to this model age as the batch process time.

The atmosphere can also be modeled as a steady-state system, to which ^{222}Rn is added at a constant rate and from which its decay products are removed by a first-order removal process:

$$dN/dt = 0 = P - (\lambda + k)N \qquad (4)$$

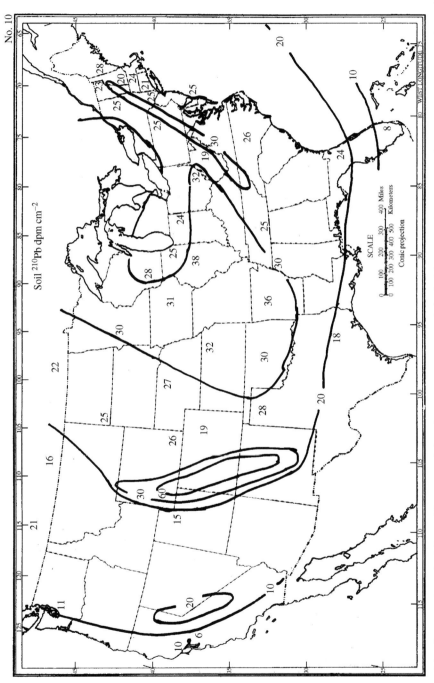

Figure 6 ^{210}Pb deposition flux over the United States. The cartoon is based on measurements made in the laboratory of the authors with accommodation for topography. Standing crop of 32 dpm cm^{-2} equals a deposition flux of 1 dpm ^{210} Pb cm^{-2} yr^{-1}.

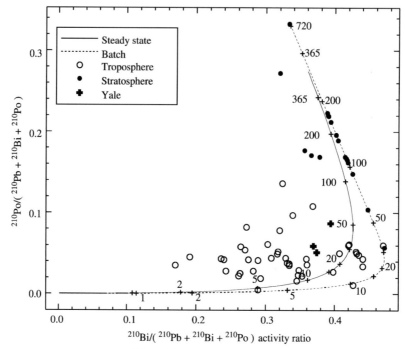

Figure 7 Residence time implications of ^{210}Pb–^{210}Bi–^{210}Po activities in aerosols. The lines represent the evolution of radon daughter product activities for a closed system (dotted line) or a steady-state system (solid line). The numbers indicate the age (for a closed system) or aerosol mean residence time (steady-state system) corresponding to the adjacent + mark. Data from Moore *et al.* (1973) are plotted as filled circles for samples from the stratosphere and open circles for samples from the troposphere. Yale data from the troposphere are marked by crosses.

where k, the removal constant, is the inverse of the mean residence time. It is zero for ^{222}Rn and assumed equal for all of its decay products. The production term, P, for each nuclide is equal to the decay rate of its parent. The solid line in Figure 7 shows the ratio of radon daughters as a function of mean residence time.

The real atmosphere does not conform to either model assumption, but lies somewhere between them. Over the five-day timescale of ^{210}Bi decay, there is too much mixing to allow for the identification and sampling of a parcel that has behaved as a closed system and far too little mixing and transport to allow the homogenization required by the steady-state assumptions. The curves are, however, similar enough to allow meaningful interpretation of data.

Real air samples can be thought of as a mixture of individual small parcels that are small enough to have behaved as closed systems. The plot is therefore constructed with a common denominator for both the *x*- and *y*-axes, so that a mixture of two components will plot along the line connecting the points representing the two components.

Moore *et al.* (1973) reported on a series of samples collected at various altitudes in the troposphere and stratosphere over central North America and discussed their implications for aerosol residence times. We review their data here and add the consideration of parcel mixing. Samples

obtained from the stratosphere are denoted by a filled circle. Most of these samples lie along the parcel evolution line with model ages from 40 d to 2 yr, with the bulk of the samples between 100 d and 180 d. These model ages are consistent with the meteorological understanding of the lower stratosphere, where vertical mixing is suppressed and precipitation scavenging is nil. These ages are also consistent with, though slightly smaller than estimates of stratospheric residence times derived from the behavior of fission products injected into the stratosphere by atmospheric testing of nuclear weapons. A few of the stratospheric samples fall below the evolution curves, suggesting the recent admixture of some "younger" aerosols.

By comparison, few of their tropospheric samples, represented by open circles, lie on either model evolution curve, but all lie within an envelope bounded by the batch model curve on the right. These data are consistent with mixing of aerosols with varying ages, and many of the data strongly indicate that mixing has occurred. The data that fall between the two model curves are most constrained and indicate an age between 20 d and 30 d.

The samples that plot within the envelope bounded by the steady-state evolution curve are less constrained. Although those that do not fall near the model curve show the effects of mixing of aerosols of different ages, the data do not lie along a single trend and therefore do not suggest end-members. Most of the samples are consistent with mixing of aerosols of age 1–2 d with those of up to age 50 d; a few of the samples appear to require an even more aged component.

Martell and Moore (1974) and Moore *et al.* (1973) interpreted these data to yield a mean age of aerosols in the lower troposphere of less than 4 d, increasing by a factor of 3 toward the top of the troposphere. The mixing analysis presented above does not contradict their estimate of the mean and yields information about the range of ages of the parcels that compose their samples. The combination of the approaches suggests that the troposphere is not sufficiently mixed on time-scales of a week to approach homogenization, but is mixed sufficiently that a single aerosol sample almost always contains material from parcels of widely varying history.

Another approach to determining the residence time of ^{210}Pb in the atmosphere is to divide the mean air column inventory of ^{210}Pb by the flux of ^{210}Pb to the surface at a given location. This quotient yields a climatological average for the removal processes at that particular site. Graustein and Turekian (1986) used the atmospheric profiles of ^{210}Pb from Moore *et al.* (1973) and their own measured ^{210}Pb fluxes from soil profiles and bucket collection to obtain a value of ∼6 d over the central and eastern United States. As the source of ^{222}Rn and thus ^{210}Pb is from the ground and the major removal by precipitation is in the lower troposphere, the mean residence time is dominated by the processes of the lower troposphere. Modeling by Balkanski *et al.* (1993) shows that the mean residence time of ^{210}Pb increases with altitude.

9.2.5.4 Use of ^{210}Pb as surrogate for other atmospheric component

(i) *Sulfate*. Of the various atmospheric components for which ^{210}Pb can be considered a surrogate, the most applicable appears to be sulfate. The reason for this is that sulfate, like ^{210}Pb, is derived from a gaseous precursor (SO_2). The conversion to an aerosol-carried species takes place at an approximately similar rate (but, as we shall see later, the oxidation rate of SO_2 actually is variable with season) and ^{210}Pb and SO_4^{2-} attach to similar-size aerosols.

For this reason ^{210}Pb has been used as a tracer of the precipitation fate of SO_4^{2-}. Turekian *et al.* (1989) used the $SO_4^{2-}/^{210}$Pb ratio in aerosols and the flux of ^{210}Pb measured in bucket collections to determine the SO_4^{2-} flux across the Pacific Ocean. Further, they showed that the $SO_4^{2-}/^{210}$Pb in aerosols from regions of high biological productivity was higher than for normal relatively unpolluted air (Table 3) indicating a sulfate source from the oxidation of dimethyl sulfide (DMS). The measured flux of DMS from the oceans at the equator matched the biogenic flux determined from the ^{210}Pb calculation (Table 4). (Actually,

Table 3 Nonmarine $SO_4^{2-}/^{210}$Pb (in μg dpm^{-1}) at continental and oceanic sites as a function of season.

Site	Summer	Winter	Source
Mould Bay(76° N, 119° W)	47	28	Graustein and Barrie (unpublished)
Northeast United States	306	216	Graustein and Turekian (1986)
Shemya	200	51	Turekian *et al.* (1989)
Midway	77	13	Turekian *et al.* (1989)
Oahu	50	39	Turekian *et al.* (1989)
Fanning	123	147	Turekian *et al.* (1989)

Table 4 Comparison of biogenic sulfate deposition flux to Fanning Island (equatorial Pacific) and the DMS flux from the equatorial Pacific Ocean.

	$SO_4^{2-}/^{210}$Pb (μg dpm^{-1})	^{210}Pb flux (dpm cm^{-2} yr^{-1})	Biogenic $SO_4^{2-}/^{210}$flux (mmol m^{-2} yr^{-1})
Sulfate deposition (from Table 3)			
Continental air (Asia)	50		
Fanning Island (equatorial Pacific)			
Total	140		
Biogenic	90[a]	0.33	3.1
Dimethyl sulfide flux			
Equatorial Pacific Ocean			
Cline and Bates (1983)			2.9
Andreae and Raemdonck (1983)			3.3

[a] Biogenic = Total-continental air.

as we shall see below, this concordance is probably due to an underestimate of sulfate flux and an overestimate of the fraction of DMS oxidized to sulfate.)

A similar study for determining sulfate flux from aerosol $SO_4^{2-}/^{210}Pb$ and ^{210}Pb flux across the eastern US yielded a new insight (Graustein and Turekian, 2004a,b, in preparation). Measurements were made at the SURE sites of the Electric Power Research Institute (EPRI). The sulfate flux was measured at each site as part of the SURE sampling program and $SO_4^{2-}/^{210}Pb$ was measured on composite aerosol samples covering a year's collection. The predicted SO_4^{2-} flux from the ^{210}Pb flux (measured in soil profiles, as described above) was generally less than the measured sulfate flux. The results indicate that over a period of a year or longer the increased sulfate flux must have been due to in-cloud conversion of SO_2 to SO_4^{2-}. About 12% of the precipitation flux of sulfate is by in-cloud conversion. A different approach corroborates this estimate. Tanaka *et al.* (1994), using stable isotope analyses of S in SO_2 and SO_4^{2-} associated with rain and air, showed that homogeneous oxidation of SO_2 provided ~90% of the measured sulfate flux and 10% was provided by heterogeneous (in-cloud) oxidation.

(ii) *Lead.* Much of the lead flux to the oceans in recent times has been from pollution sources. This lead was injected into the atmosphere primarily as a volatile compound that subsequently formed an aerosol by photolytic processes. In this sense, much of the lead transported through the atmosphere and ultimately deposited resembles the origin and fate of ^{210}Pb. On this basis, Turekian and Cochran (1981a,b) estimated the flux of lead to Enewetak in the North Pacific. In a more detailed study, Settle *et al.* (1982) were able to calculate the flux of lead from the aerosol $Pb/^{210}Pb$ at a number of diverse sites, including Pigeon Key, Florida, Tahiti, and Bermuda.

(iii) *Mercury.* The primary form of mercury in the atmosphere is as a gaseous element. Only as the mercury is oxidized to ionic species is it removed by precipitation. In this respect it resembles ^{210}Pb, which also arises from a gas (^{222}Rn). Consequently, ^{210}Pb can be used to track mercury precipitation. Lamborg *et al.* (2000) performed such a study (Figure 8) to determine the correlation of ionic mercury and ^{210}Pb in aerosols and precipitation and to determine the flux of mercury from the ^{210}Pb flux determined independently.

(iv) *Other components.* Although other components of aerosols may not have had gaseous precursors, the use of ^{210}Pb as a surrogate may still be used as an approximation assuming the size fraction of the aerosol bearing both ^{210}Pb and the component is about the same. Turekian *et al.* (1989) applied this approach to organic compounds and

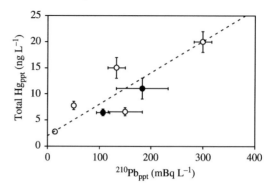

Figure 8 Hg and ^{210}Pb in Wisconsin (circles) and mid-Atlantic (triangles) precipitation (source Lamborg *et al.*, 2000). The filled circles represent side-by-side collections from a single event.

such elements as aluminum. Williams and Turekian (2002) extended this procedure to estimate coastal and open-ocean fluxes of osmium, a metal with isotopic signature impacting the oceans. They were able to show that a high osmium flux was evident close to shore (in New Haven, CT) that could not be characteristic of the whole oceans because of the constancy of the osmium isotope composition in seawater. An analogue to this behavior is the study of ^{210}Pb fluxes in the SEAREX sites discussed above.

9.3 COSMOGENIC NUCLIDES

9.3.1 Atmospheric Production of Cosmogenic Nuclides

The source of cosmogenic nuclides in the atmosphere is the interaction of galactic cosmic rays with the atoms composing the atmosphere. Cosmic rays are primarily composed of protons with energies of billions of electron volts. As these highly energetic charged particles penetrate the magnetic shield of Earth the interactions with the atoms in the atmosphere result in spallation products—fragments of the target nucleus. Some of the secondary particles, especially neutrons, undergo further reactions. This latter process is responsible, for example, for the production of ^{14}C (radiocarbon). The major radioactive nuclides so produced in the atmosphere are given in Table 2.

The production of the radionuclides is controlled by the magnetic pattern of Earth. The production rate of the nuclides is determined by the strength of the magnetic field. The strength of the magnetic field varies as the result of both the intrinsic variations in Earth's dipole magnetic moment and the varying intensity of solar activity. At high magnetic field strengths the production rate is low and at low magnetic field strengths the production rate is high.

The period of magnetic fluctuations due to Earth's intrinsic field is not regular but has been shown to have increased over the past 2×10^4 yr. Fluctuations in the past are also recorded and clearly evident in the changing magnetic polarity of Earth over time.

The solar activity cycles are more complex. The 11 yr sunspot cycle is well known. It is accompanied with charges in the magnetic field strength of Earth. There are fluctuations in the intensity of solar activity on this timescale. As we shall see, periods of fluctuation of ~ 60 yr, 200 yr, and 1,500 yr have been identified through measurements of ^{10}Be in deep-sea sediments and continental ice sheets. These results also indicate possible variations on the 10^5 yr timescale as well.

9.3.2 ^7Be and ^{10}Be

The production rate of ^7Be (half-life = 53 d) as a function of latitude and elevation by Lal and Peters (1967) is shown in Figure 9. Approximately one-third of the nuclide production rate is in the troposphere and two-thirds in the upper atmosphere (stratosphere and higher). This partitioning is valid for all radionuclides except ^{14}C, where most is produced by secondary neutrons in the vicinity of the tropopause.

^{10}Be (half-life = 1.5×10^6 yr), although formed primarily in the atmosphere, is found in surface deposits because of the short residence time of aerosols in the atmosphere and the long half-life. Although the rate of deposition is the same as the rate of production in the atmosphere (4.5×10^{-2} atoms cm^{-2} s^{-1}) its distribution on Earth's surface is determined by the sites of primary stratospheric intrusion into the troposphere ($\sim 40°-50°$ latitude), the focusing in the troposphere due to precipitation controls such as regional climate and orographic effects and, in the oceans, the role of concentration and transport by particles, especially those produced biologically.

Despite these controls, it is possible to use the record of ^{10}Be accumulation at designated sites such as ice cores and certain oceanic areas as recorders of variations in the rate of supply and therefore of production with time.

As the residence time of aerosols in the stratosphere is ~ 2 yr and in the troposphere ~ 1 week, the $^7Be/^{10}Be$ ratio of the two air masses is distinctive. Tropospheric air shows the ratio of ^7Be relative to ^{10}Be of 1.8, whereas stratospheric air has a ratio of 0.13. It is therefore possible to distinguish stratospheric air injected into the troposphere by considering the ratio of $^7Be/^{10}Be$. Of course, the stratospheric air will also be higher in ^{10}Be than the tropospheric air. As stratospheric air will also contain ozone, the interest in this source has been strong to distinguish from pollution-based tropospheric ozone.

9.3.3 ^{35}S and the Kinetics of SO_2 Oxidation and Deposition

^{35}S is produced by the spallation of ^{40}Ar by cosmic rays. After formation it is quickly converted into gaseous $^{35}SO_2$. The subsequent oxidation and removal from the atmosphere can be tracked and the kinetics applied to terrestrial and pollution SO_2. For this reason it has been studied in conjunction with other cosmogenic nuclides, the most useful of which is ^7Be. Both ^{35}S with a half-life of 87 d and ^7Be with a half-life of 53 d, although produced from gaseous precursors by cosmic-ray bombardment, have different initial states. Whereas ^7Be as an ion quickly associates with aerosols, ^{35}S as $^{35}SO_2$ must first be oxidized to be associated with aerosols. Alternatively, $^{35}SO_2$ can react with surfaces in a method called "dry deposition" to distinguish it from ^{35}S removal in precipitation.

There is only one study using these two nuclides for deciphering the kinetics of SO_2 oxidation and removal from the atmosphere (Tanaka and Turekian, 1995, and earlier papers). Their study, performed in New Haven, Connecticut, applied the measurement of these nuclides in air and precipitation to the problem posed. Measurements were made for SO_2, $^{35}SO_2$, SO_4^{2-}, $^{35}SO_4^{2-}$, and ^7Be in air samples and for the same species in precipitation over a year's time period. Figure 10 shows the box model with the kinetic factors listed for each process modifying the composition of the air. Although the coupling of the boxes results in 12 equations with 3 unknowns, the selection of reasonable constraints results in solutions that are comparable with expectations based on experience. Figure 11 shows the variation in one parameter, j_2, the first-order kinetic constant for the homogeneous oxidation of SO_2 in the

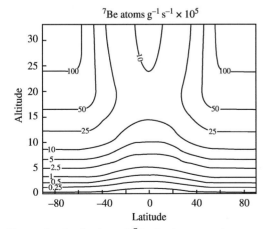

^7Be atoms g^{-1} s^{-1} $\times 10^5$

Figure 9 Production of ^7Be in the atmosphere as a function of latitude and elevation based on data from Lal and Peters (1967) and Lal (personal communication).

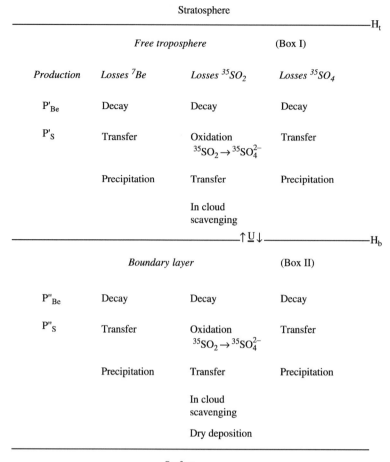

Figure 10 Components of the box model used by Tanaka and Turekian (1995) in determining the parameters for the conversion and removal of SO_2 in the atmosphere based on measurements of the concentrations of cosmogenic ^{35}S and ^7Be. Ps are the cosmic-ray production rates of the radionuclides, U is the mass transport flux between the boundary layer and the free troposphere, H_b and H_t are the heights of the boundary layer and the tropopause, respectively.

boundary layer. The value of j_2 is greatest in the summer months and lowest in the winter months. This prediction is compatible with the observation that SO_2 is the dominant species in winter air and SO_4^{2-} is so in the summer.

The application to the elusive process of dry deposition of SO_2 was evaluated by Tanaka and Turekin (1995) using the S and ^{35}S data on air and precipitation samples and the model used to construct Figure 10. They showed that in New Haven, CT the annual weighted average ratio of dry to wet precipitation was 0.26.

9.3.4 Phosphorus Isotopes

Of the isotopes listed in Table 2 the study of the cosmogenic isotopes of phosphorus ^{32}P (half-life = 14.3 d) and ^{33}P (half-life = 25.3 d) provides insights into the residence time of aerosols in upper troposphere air because of the relatively short half-lives of the nuclides. Waser and Bacon (1995) measured the concentrations of ^{32}P and ^{33}P

in precipitation at Bermuda over three seasons. They concluded that, with the average activity ratio of ^{33}P to ^{32}P of 0.96 and a production activity ratio of 0.7, the average residence time of aerosols in the upper troposphere was ~40 d. This increase in the residence time with height in the troposphere is compatible with the modeling results of Balkanski *et al.* (1993) based on ^{210}Pb. When the upper-troposphere air is transported to the lower troposphere, it is subject to the more efficient removal characteristic of that layer, but with the memory of the original locus of production.

9.4 COUPLED LEAD-210 AND BERYLLIUM-7

9.4.1 Temporal and Spatial Variation

In the absence of atmospheric motion and removal by precipitation, ^7Be and ^{210}Pb would remain where they originated—the upper troposphere and stratosphere and the lowermost meter

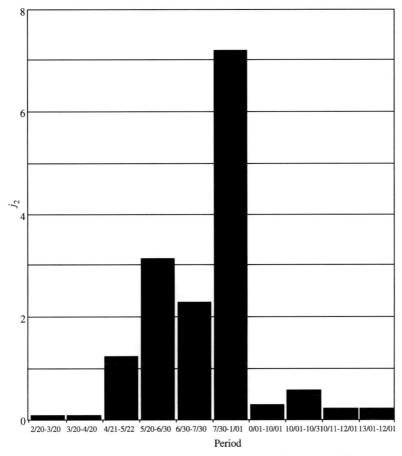

Figure 11 The variation over the year of the homogeneous oxidation coefficient (j_2) of SO_2 in the boundary layer of Figure 10 (source Tanaka and Turekian, 1995).

of the atmosphere over continents and islands, respectively. In the real atmosphere, ^7Be is mixed downward and ^{210}Pb is mixed upwards and both are removed by precipitation. They are distributed through the atmosphere by eddy mixing. The residence time of aerosols is short compared to the time required for homogenization by eddy mixing, but long compared to the life on an individual eddy. Changes in $^7Be/^{210}Pb$ with time and space reflect both vertical and horizontal transport in the atmosphere.

Because of similarity of the behavior of ^7Be and ^{210}Pb in the atmosphere, $^7Be/^{210}Pb$ is little affected by processes other than production and transport. Both ^{210}Pb and ^7Be are formed in the atmosphere as energetic single atoms. Since neither is volatile, each of them attaches to the first particles they encounter. The most abundant aerosols in terms of surface area are typically those with a diameter of 0.1–0.5 µm, the so-called accumulation mode. This size class carries many of the chemical species in the atmosphere that have low volatility and also have gaseous precursors, such as sulfate as discussed above. Accumulation-mode aerosols are most subject to long-distance transport. Scavenging by precipitation is the

principal mechanism of removal of these aerosols from the atmosphere.

The following results are from the work of Graustein and Turekian (1991).

Over continents, seasonal variation in the stability of the atmosphere has a dominant influence on the $^7Be/^{210}Pb$ ratio. Surface heating in the summer increases convective mixing, which reduces ^{210}Pb in surface air by mixing it through a larger volume and simultaneously increases the transport of ^7Be to the surface. Winter stability tends to isolate surface air from the ^7Be source and retain ^{222}Rn and ^{210}Pb near the surface.

The seasonal distributions of ^7Be and ^{210}Pb at Champaign, IL, a mid-continent site, are shown in Figure 12. The highest 24 h mean ^{210}Pb concentrations occur principally in winter when atmospheric stability is greatest; the highest ^7Be concentrations occur in summer when vertical mixing is strongest. The effect is pronounced, and there is little overlap between the summer and winter sets of measurements. The time series of a 45 d moving average of the $^7Be/^{210}Pb$ ratio at Champaign, IL shows a repeating annual cycle (Figure 13) characterized by a late spring and early summer maximum and a December minimum in $^7Be/^{210}Pb$, and is similar to sites at the east and west coasts of North America.

The maritime pattern, determined from islands in the Atlantic, is different from the continental pattern. Island sources of ^{222}Rn are too small to have a significant effect on ^{210}Pb. As a result, virtually all the ^{210}Pb observed at these sites is transported from continents. ^{7}Be/^{210}Pb over the ocean is therefore a measure of the influence of continental sources on the local aerosol. Low values of the ratio reflect a high continental influence; high ratios indicate a relative isolation from continental sources.

At Bermuda (Figure 14) the daily data cluster more tightly than at Champaign, and there is a distinct lower bounding value of ^{7}Be/^{210}Pb. Values that are common in continental winters

are never seen in 24 h samples at Bermuda, indicating that continental boundary layer air is diluted or scavenged in transit to Bermuda in the winter.

The seasonal pattern of ^{7}Be/^{210}Pb at island sites is six months out of phase with continental sites (Figure 15). During the summer, ^{7}Be/^{210}Pb is nearly the same over the continent and the ocean, indicating vigorous mixing of the troposphere. In the winter, however, ^{7}Be/^{210}Pb is higher at oceanic sites than continental sites by a factor of ~4.

Izania is located at an elevation of 2,367 m on Tenerife in the Canary Islands and is in the free troposphere. Figure 16 shows that it also exhibits a strong seasonal cycle in ^{7}Be/^{210}Pb resembling Barbados rather than Champaign. The maximum ^{7}Be/^{210}Pb occurs in December, the same month that the continental sites reach their minima.

Taken together, these observations suggest that there is relatively little long-distance transport of aerosols in the marine boundary layer, and that long-distance transport is relatively efficient in the free troposphere. Wintertime stability over continents inhibits transfer of surface-source aerosols to the free troposphere and limits their transport distance. Spring and summer convective mixing over continents results in an efficient exchange between the boundary layer and free troposphere, which enhances long-range transport of aerosols and greatly increases the uniformity of aerosol composition between the continent and marine boundary layers.

9.4.2 Application of the Coupled ^{7}Be–^{210}Pb System to Sources of Atmospheric Species

As ^{210}Pb has its source in the boundary layer and ^{7}Be has its major source in the troposphere, mainly at higher elevations, the sources of

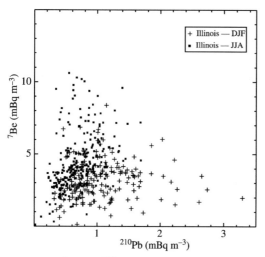

Figure 12 ^{7}Be and ^{210}Pb activities in 24 h air samples from Champaign, Illinois. Plus symbols indicate samples collected in December through February, and filled squares represent samples collected in June–August.

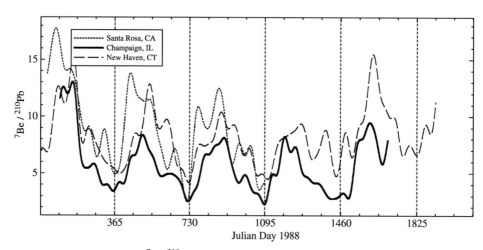

Figure 13 Six-year time series of ^{7}Be/^{210}Pb in near-surface air at three continental sites. The lines represent a Gaussian weighted 45 d moving average of 24 h samples.

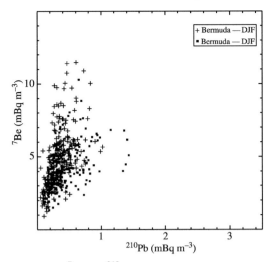

Figure 14 ^{7}Be and ^{210}Pb activities in 24 h air samples from Bermuda. Plus symbols indicate samples collected in December through February and filled squares represent samples collected in June–August.

chemical species in the lower free troposphere can be determined by using the ratio of the two nuclides. Such a study was performed in one of the AEROCE sites at Izania in the Canary Islands. Using a function of the ^{7}Be to ^{210}Pb ratio, Graustein and Turekian (1996) showed that the primary source of ozone in the lower free troposphere of the eastern Atlantic was the upper troposphere (Figure 17). An extension of the study to nitrate using sulfate as a boundary layer index as is ^{210}Pb revealed that the primary source of nitrate in the lower free troposphere was also the upper free troposphere and not the boundary layer (Figure 18). Thus, both chemical species have their origin in the upper troposphere and not anthropogenic pollution source in Europe or biomass burning in Africa. Sampling in the boundary layer downwind of human activity at such sites as Bermuda, Barbados, and Mace Head (Ireland) indicates that both ozone and nitrate are assignable primarily to boundary layer sources.

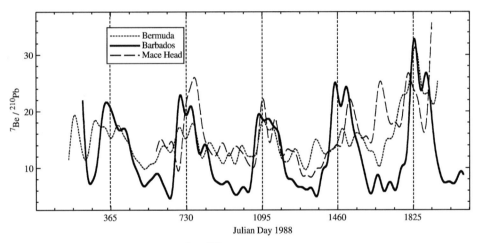

Figure 15 Time series of ^{7}Be/^{210}Pb in near-surface air at three oceanic sites.

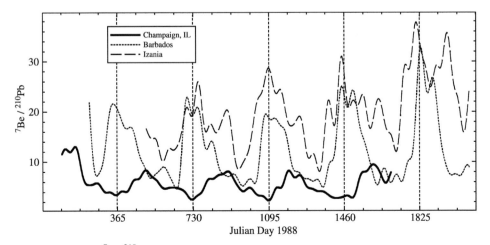

Figure 16 Time series of ^{7}Be/^{210}Pb in surface air at a continental site (Champaign), in surface air at an oceanic site (Bermuda), and in the oceanic free troposphere (Izania). The data plotted are Gaussian weighted 45 d moving averages of 24 h samples.

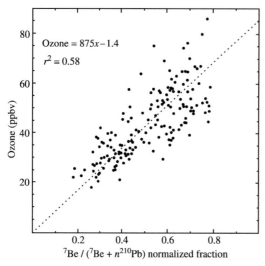

Figure 17 Ozone versus the "normalized fraction" at Izana (Canary Islands) for the summers of 1989–1991. The "normalized fraction" is $(^{7}Be)/[(^{7}Be)+n\ (^{210}Pb)]$, where n is approximated by the ratio of the standard deviation of (^{7}Be) to the standard deviation of (^{210}Pb) in the sample set. The parentheses indicate activities (source Graustein and Turekian, 1996).

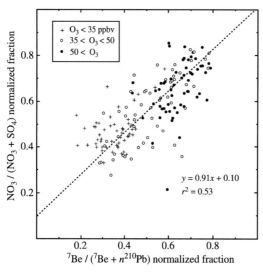

Figure 18 "Normalized fraction" of the nitrate–sulfate system plotted against the "normalized fraction" of the ^{7}Be–^{210}Pb system as defined in the caption for Figure 17 (from unpublished analysis of AEROCE data by the authors).

In another application, Lamborg *et al.* (2000) showed that mercury is distributed homogeneously in the troposphere, as there is no correlation of elemental mercury concentration and $^{7}Be/^{210}Pb$.

REFERENCES

Andreae M. O. and Raemdonck H. (1983) Dimethyl sulfide in the surface ocean and marine atmosphere: a global view. *Science* **221**, 744–747.

Andreae M. O., Berresgeim H., Andreae T. W., Kritz M. A., Bates T. S., and Merrill J. T. (1988) Vertical distribution of dimethylsulfide, sulfur dioxide, aerosol ions, and radon over the northeast Pacific Ocean. *J. Atmos. Chem.* **6**, 149–173.

Balkanski Y. J., Jacob D. J., Gardner G. M., Graustein W. C., and Turekian K. K. (1993) Transport and residence times of continental aerosols inferred from a global three-dimensional simulation of ^{210}Pb. *J. Geophys. Res.* **98**, 20573–20586.

Bateman H. (1910) Solution of a system of differential equations occurring in the theory of radio-active transformations. *Proc. Cambridge Phil. Soc.* **15**, 423–430.

Broecker W. S. and Peng T.-H. (1971) The vertical distribution of radon in the BOMEX area. *Earth Planet. Sci. Lett.* **11**, 99–108.

Clements W. E. and Wilkening M. H. (1974) Atmospheric pressure effects on ^{222}Rn transport across the earth–air interface. *J. Geophys. Res.* **79**, 5025–5029.

Cline J. D. and Bates T. S. (1983) Dimethyl sulfide in the Equatorial Pacific Ocean: a natural source of sulfur to the atmosphere. *Geophys. Res. Lett.* **10**, 949–952.

Conen F. and Robertson L. B. (2002) Latitudinal distribution of radon-222 flux from continents. *Tellus* **54B**, 127–133.

Dörr H. and Münnich K. O. (1990) ^{222}Rn flux and soil air concentration profiles in West-Germany: soil ^{222}Rn as a tracer for gas transport in the unsaturated soil zone. *Tellus* **42B**, 20–28.

Feely H. W., Toonkel L., and Larsen R. (compilers) (1981) *Radionuclides and Trace Elements in Surface Air.* US Dept. Energy, Environ. Q. Rep. EML-395, appendix, New York.

Friedlander G., Kennedy J. W., Macias E. S., and Miller J. M. (1981) *Nuclear and Radiochemistry,* 3rd edn. Wiley, New York, 684 pp.

Fuller C. and Hammond D. E. (1983) The fallout rate of Pb-210 on the western coast of the United States. *Geophys. Res. Lett.* **10**, 1164–1172.

Graustein W. C. and Turekian K. K. (1983) ^{210}Pb as a tracer of the deposition of sub-micrometer aerosols. In *Precipitation Scavenging, Dry Deposition and Resuspension* (eds. H. R. Pruppacher, R. G. Semonin, and W. G. N. Slinn). Elsevier, Amsterdam and Oxford, vol. 2, pp. 1315–1324.

Graustein W. C. and Turekian K. K. (1986) ^{210}Pb and ^{137}Cs in air and soils measure the rate and vertical distribution of aerosol scavenging. *J. Geophys. Res.* **91**, 14355–14366.

Graustein W. C. and Turekian K. K. (1989) The effects of forests and topography on the deposition of sub-micrometer aerosols measured by ^{210}Pb and ^{137}Cs in soils. *Agri. Forest Meteorol.* **47**, 199–220.

Graustein W. C. and Turekian K. K. (1990) Radon fluxes from soils to the atmosphere measured by ^{210}Pb–^{210}Ra disequilibrium in soils. *Geophys. Res. Lett.* **17**, 841–844.

Graustein W. C. and Turekian K. K. (1991) ^{210}Pb and ^{7}Be trace seasonal variations in aerosol transport over North America and the North Atlantic. (Paper presented at CHEMRAWN VII Symposium, Int. Union of Pure and Appl. Chem., Baltimore, MD. Available from the authors).

Graustein W. C. and Turekian K. K. (1996) ^{7}Be and ^{210}Pb indicate an upper troposphere source for elevated ozone in the summertime subtropical free troposphere of the eastern North Atlantic. *Geophys. Res. Lett.* **23**, 539–542.

Graustein W. C. and Turekian K. K. (2004a) Radon flux from soils (in preparation).

Graustein W. C. and Turekian K. K. (2004b) In-cloud scavenging of SO_2 (in preparation).

Knuth R. H., Knutson E. O., Feely H. W., and Volchock H. L. (1983) Size distribution of atmospheric Pb and Pb-210 in rural New Jersey: implications for wet and dry deposition. In *Precipitation Scavenging, Dry Deposition and Resuspension* (eds. H. R. Pruppacher, R. G. Semonin, and W. G. N. Slinn). Elsevier, Amsterdam and Oxford, pp. 1325–1334.

Kritz M. A., Rosner S. W., Kelly K. K., Loewenstein M., and Chan K. R. (1993) Radon measurements in the lower tropical stratosphere: evidence for rapid vertical transport and dehydration of tropospheric air. *J. Geophys. Res.* **98**, 8725–8736.

Lal D. (2001) Cosmogenic isotopes. In *Encyclopedia of Ocean Sciences* (eds. J. H. Steele, S. A. Thorpe, and K. K. Turekian). Academic Press, London, pp. 550–560.

Lal D. and Peters B. (1967) Cosmic rays produced radioactivity on the earth. *Handbuch der Physik* **46**(2), 551–612.

Lambert G., Polian G., and Taupin D. (1970) Existence of periodicity in radon concentrations and in large-scale circulation at lower altitudes between 40° and 70° south. *J. Geophys. Res.* **75**, 2341–2345.

Lamborg C. H., Fitzgerald W. F., Graustein W. C., and Turekian K. K. (2000) An examination of the atmospheric chemistry of mercury using ^{210}Pb and ^{7}Be. *J. Atmos. Chem.* **36**, 325–338.

Liu S., McAfee J. R., and Cicerone R. J. (1984) Radon-222 and tropospheric vertical transport. *J. Geophys. Res.* **89**, 7202–7297.

Martell E. A. and Moore H. E. (1974) Tropospheric aerosol residence times: a critical review. *J. Rech. Atmos.* **8**, 903–910.

Monaghan M. C., Krishnaswami S., and Turekian K. K. (1985/1986) The global-average production rate of ^{10}Be. *Earth Planet. Sci. Lett.* **76**, 279–287.

Moore H. E. and Poet S. E. (1976) ^{210}Pb fluxes determined from ^{210}Pb and ^{226}Ra soil profiles. *J. Geophys. Res.* **81**, 1056–1058.

Moore H. E., Poet S. E., and Martell E. A. (1973) ^{222}Rn, ^{210}Pb, ^{210}Bi, and ^{210}Po profiles and aerosol residence times versus altitude. *J. Geophys. Res.* **78**, 7065–7075.

Moore H. E., Poet S. E., and Martell E. A. (1974) Origin of ^{222}Rn and its long-lived daughters in air over Hawaii. *J. Geophys. Res.* **79**, 5019–5024.

National Research Council (1988) *Health Risks of Radon and other Internally Deposited Alpha-emitters.* National Academy Press, Washington, DC, 602pp.

Nozaki Y., DeMaster D. J., Lewis D. M., and Turekian K. K. (1978) Atmospheric Pb-210 fluxes determined from soil profiles. *J. Geophys. Res.* **83**, 4047–4051.

Pearson J. E. and Jones G. E. (1966) Soil concentrations of "emanating radium-226" and the emanation of radon-222 from soils and plants. *Tellus* **18**, 655–661.

Schery S. D. and Gaeddert D. H. (1982) Measurements of the effect of cyclic atmospheric pressure variation on the flux of ^{222}Rn from the soil. *Geophys. Res. Lett.* **9**, 835–838.

Schery S. D., Whittlestone S., Hart K. P., and Hill S. E. (1989) The flux of radon and thoron from Australian soils. *J. Geophys. Res.* **94**, 8567–8576.

Settle D. M., Patterson C. C., Turekian K. K., and Cochran J. K. (1982) Lead precipitation fluxes at tropical oceanic sites determined from ^{210}Pb measurements. *J. Geophys. Res.* **87**, 1239–1245.

Tanaka N. and Turekian K. K. (1995) The determination of the dry deposition flux of SO_2 using cosmogenic ^{35}S and ^{7}Be measurements. *J. Geophys. Res.* **100**, 2841–2848.

Tanaka N., Rye D. M., Xiao Y., and Lasaga A. C. (1994) Use of stable sulfur isotope systematics for evaluating oxidation pathways and in-cloud-scavenging of sulfur dioxide in the atmosphere. *Geophys. Res. Lett.* **21**, 1519–1522.

Tsunogai S., Shinagawa T., and Kurata T. (1985) Deposition of anthropogenic sulfate and Pb-210 in the western North Pacific area. *Geochem. J.* **19**, 77–90.

Turekian K. K. and Cochran J. K. (1981a) ^{210}Pb in surface air at Enewetak and the Asian dust flux to the Pacific. *Nature* **292**, 522–524.

Turekian K. K. and Cochran J. K. (1981b) ^{210}Pb in surface air at Enewetak and the Asian dust flux to the Pacific: a correction. *Nature* **294**, 670.

Turekian K. K., Nozaki Y., and Benninger L. K. (1977) Geochemistry of atmospheric radon and radon products. *Ann. Rev. Earth Planet. Sci.* **5**, 227–255.

Turekian K. K., Graustein W. C., and Cochran J. K. (1989) Lead-210 in the SEAREX program: an aerosol tracer across the Pacific. In *Chemical Oceanography* (ed. J. P. Riley). Academic Press, London, vol. 10, pp. 51–81.

Turekian V. C., Graustein W. C., and Turekian K. K. (1999) The ^{214}Bi to ^{214}Pb ratio in lower boundary layer aerosols and aerosol residence times at New Haven, Connecticut. *J. Geophys. Res.* **104**, 11593–11598.

Waser N. A. D. and Bacon M. P. (1995) Wet deposition fluxes of cosmogenic ^{32}P and ^{33}P and variations in the ^{33}P/^{32}P ratios at Bermuda. *Earth Planet. Sci. Lett.* **133**, 71–80.

Wilkening M. H. and Clements W. E. (1975) Radon-222 from the ocean surface. *J. Geophys. Res.* **80**, 3828–3830.

Williams G. and Turekian K. K. (2002) The atmospheric supply of osmium to the oceans. *Geochim. Cosmochim. Acta* **66**, 3789–3791.

Radioactive Geochronometry
ISBN: 978-0-08-096708-0

pp. 269–288

10

Groundwater Dating and Residence-time Measurements

F. M. Phillips

New Mexico Tech, Socorro, NM, USA

and

M. C. Castro

University of Michigan, Ann Arbor, USA

10.1 INTRODUCTION

Water is the key to the geochemistry of the upper crust and the surface of the Earth. Water interacts extensively with minerals through hydrolysis, dissolution, precipitation, interface solute layers, etc., and is necessary for life, which directly and indirectly affects the chemistry of Earth systems. The local characterization of aqueous geochemical reactions is a natural and straightforward extension of classical chemistry. However, neither the composition nor the evolution of the upper layers of the Earth can be explained by local-scale geochemistry. It is the open-system nature of geochemistry in this environment that provides most of its explanatory power, as well as its intellectual challenge. The fundamental aspects of near-surface geochemistry such as weathering reactions, precipitation of sedimentary minerals, diagenesis, and metamorphic evolution all depend on transport of mass in and out of these systems. In most cases, moving water is the transport agent.

Geochemical constituents are transported by the circulation of the Earth's three major near-surface dynamic systems: the oceans, the atmosphere, and continental water. This chapter deals with the third of these, specifically with subsurface water. Surface water is an important medium for transporting geochemical constituents on a global scale, and is relatively easily accessed and quantified. In contrast, subsurface water is not easy to access, and fluxes through subsurface systems are very difficult to measure. To some extent, subsurface fluxes can be estimated using methods based on the physics of water flow (e.g., numerical modeling of flow systems), but the very large variability of permeability imparts a high uncertainty to the results. Fortunately, measurement of certain geochemical constituents in water can define residence times and fluxes in subsurface systems. This chapter will explore the geochemical means that are available to help quantify subsurface fluxes, and will evaluate the implications of their application to a variety of specific systems.

10.2 NATURE OF GROUNDWATER FLOW SYSTEMS

10.2.1 Driving Forces

Water moves from areas of higher total fluid energy to areas of lower energy. Kinetic energy, mechanical energy (e.g., compression), gravitational potential, interfacial potential, and chemical potential can drive fluid flow. In general, velocities in the subsurface are low enough, so that kinetic energy can be neglected. Water, therefore, moves from areas of higher to ones of lower fluid potential (Hubbert, 1940). In systems that are only partially saturated with water (e.g., the vadose zone and oil and gas reservoirs), the combination of gravitational potential and interfacial potential (capillary action) serves to drive flow. In relatively shallow (less than a few kilometers) saturated groundwater systems, gravitational potential is usually dominant. In deeper settings, additional forces—such as compression due to compaction or tectonic movement, pressure generated by chemical reactions, or chemical potentials developed across semipermeable shales—may also be important (Neuzil, 1995, 2000).

10.2.2 Topographic Control on Flow

The fundamentals of the major subsurface water cycle are straightforward. Water precipitates out of the atmosphere onto areas of relatively high topography. Some of this water infiltrates into the subsurface. Because it has greater gravitational potential energy than water at lower elevations, it flows downward, eventually re-emerging at the surface either at the lowest elevation within the region, or at higher elevations where low-permeability barriers force it to the surface.

Topographically driven groundwater flow is generally more important for geochemical processes than flow produced by the alternative mechanisms mentioned above. This is because the circulation of meteoric water, driven ultimately by solar energy, is a process that provides a continuous source of subsurface water that can carry large fluxes of solutes. Although other driving forces, such as sediment compaction or tectonic compression, may, for periods of time, provide water velocities comparable to topographically driven flow at similar depths, the total water flux is limited to the pore volume of the rock undergoing compression.

One frequently underappreciated aspect of topographically driven flow is the pervasive extent of water circulation through the shallow crust. Fairly subdued topography can drive appreciable water flow to depths of several kilometers. Under favorable circumstances, more significant relief can produce circulation to depth below 5 km (Person and Baumgartner, 1995; Person and

Garven, 1992). Permeability generally decreases with depth (Manning and Ingebritsen, 1999), and this is one factor that tends to limit groundwater fluxes at considerable depth. Another fairly common limitation is provided by brines at depth; topographically driven freshwater is generally unable to displace brines that are below the discharge point of a groundwater basin.

10.2.3 Hydraulic Conductivity and Its Variability

The flow of groundwater and other subsurface fluids is described by Darcy's law (Darcy, 1856). Hydraulic conductivity, the constant in Darcy's law, has an extremely wide range in natural materials, extending over ~16 orders of magnitude (Freeze and Cherry, 1979; Manning and Ingebritsen, 1999). Although, in any particular subsurface setting, the range will probably be much less, variations of five or more orders of magnitude are common. To compound the difficulties of quantifying the behavior of fluids in such heterogeneous systems, access to the subsurface is generally very limited and actual measurements of hydraulic conductivity are generally sparse. Thus, although the physics of groundwater flow is well understood and amenable to quantitative analysis, the great variability of properties and the lack of constraints on their values render such analyses, subject to great uncertainty. It is in this perspective that the utility of geochemical tracers for water residence time can be appreciated. A sequence of residence-time measurements along a flow path essentially integrates the velocity history of the groundwater and thus provides a very valuable constraint on the permeability structure.

The structure of the heterogeneity in hydraulic conductivity is typically complex. The fundamentals of the geological controls on conductivity have been recognized for many years. For example, Meinzer (1923) codified long-standing ideas of high- and low-conductivity units termed "aquifers" and "aquitards." However, continued research has shown that these are considerable oversimplifications of a typically very complex subsurface permeability structure. An alternative view is to characterize the permeability structure as essentially fractal (Neuman, 1990, 1995); in other words, as the spatial scale of measurement increases, the variability of hydraulic conductivity, and the size of units of differing conductivity, also increases in some proportional fashion, without any apparent upper limit. Further refinement of these ideas would conceptualize hydraulic conductivity as being characterized by nested scales of variability, such that the increase in variability might be large for certain ranges of spatial scale and small for others (Gelhar, 1993). For example, consider an aquifer formed by deposition in the valley of a meandering river, dominated by sand channel deposits, but also containing many fine-grained overbank deposits. At a scale smaller than that of a single channel or overbank unit, the permeability might be relatively uniform, but as the sampling volume increases to incorporate bounding units of contrasting lithology, the variability would increase dramatically. However, further increase of the volume to incorporate many similar units over the thickness of the entire aquifer would result in little additional variability.

This complex and hierarchical variation of hydraulic conductivity results in complex subsurface flow paths. It is not possible to incorporate a complete and reliable description of actual permeability structures in numerical models. Although geochemical tracers cannot provide a map of the permeability variations, they can provide important clues to the nature of the variations, and they can help to discriminate between competing hypotheses.

10.2.4 Scales of Flow Systems

Both permeability distributions and topographic variations can be considered as some form of hierarchical fractal process (Boufadel *et al.*, 2000).

The interaction of hierarchical variability in both the main driving force and the main control on flow results in a nested, hierarchical system of flow regimes (Freeze and Witherspoon, 1966, 1967; Tóth, 1963). These are commonly classified as vadose-zone scale, local scale, aquifer scale, and regional scale, going from smallest to largest (Figure 1).

10.2.4.1 Vadose-zone scale

The vadose zone forms an essential, though frequently overlooked, component of the subsurface hydrologic system (Stephens, 1996). Fluxes are generally vertical (usually downward, but upward in discharge areas and in many desert regions, and also seasonally upward in many recharge areas). The scales of the flow systems range from a few centimeters to as much as several hundred meters in deserts and mountainous karstic regions. Understanding downward fluxes is of considerable importance because of the need to quantify groundwater recharge and to assess the potential for aquifer contamination. However, this quantification is difficult because of the nonlinear nature of the laws governing unsaturated flow and the great degree of seasonal variability in fluxes.

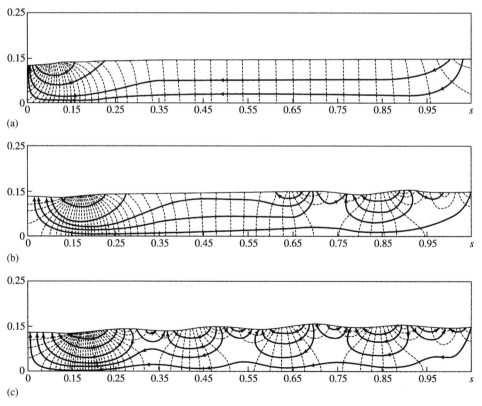

Figure 1 Illustration of the development of increasingly complex flow systems as topography becomes more complex. Contours of hydraulic head are indicated by dashed lines and groundwater flow lines by solid arrows. Scale is arbitrary, but might correspond to 100 km in the horizontal direction. In (a), smooth topography produces a regional-scale flow system. In (b) and (c) increasing local topography creates a mixture of intermediate and local-scale flow systems superimposed on the regional one (Freeze and Witherspoon, 1967) (reproduced by permission of American Geophysical Union from *Water Resour. Res.* **1967**, *3*, 623–634).

10.2.4.2 Local scale

Local flow systems dominate groundwater circulation at shallow depths (down to as much as 100 m) in humid regions (i.e., having significant diffuse areal recharge), unless the topography is very flat (Tóth, 1963). Such systems vary from ~100 m to several kilometers in horizontal scale. It is the discharge from such systems that supports surface-water flow in streams and rivers under all but storm conditions. The primary control on groundwater flow in such systems is generally considered to be topographic, with recharge at higher elevations (ridges or hills) and flow to immediately adjacent streams or rivers (Buttle, 1998). The water table generally represents a subdued replica of the topography. Such systems provide a large fraction of the groundwater supply for drinking and other uses, and are particularly susceptible to contamination.

10.2.4.3 Aquifer scale

Many hydrogeologic investigations are carried out at the scale of the aquifer, a map-scale geological unit capable of yielding significant amounts of water. Aquifer scale may vary from a few kilometers to hundreds of kilometers. Although aquifers may often span the horizontal scale of the relevant flow system, they generally do not encompass the full vertical extent of the system. Although aquifer-scale studies often do not provide a complete representation of the flow system, they are very important because they focus on the units that provide most of the water supply, and for which the most data are available.

10.2.4.4 Regional scale

Many areas exhibit trends of elevation at the continental, or subcontinental scale (hundreds to thousands of kilometers). Unless pervasive permeability barriers are present, the topographic gradients will drive deep circulation systems over these continental scales (Figure 1) (Freeze, and Witherspoon, 1966; Person and Baumgartner, 1995). Although both the depth and the quality of water in such systems limit their human consumption, they are critical for understanding the origins of diagenesis, many ore deposits, and petroleum reservoirs (Ingebritsen and Sanford, 1998).

10.2.5 Sources of Solutes at Various Scales

The chemical (and isotopic) composition of subsurface water shows a fair degree of predictability based on the hydrogeological environment in which it is found (Drever, 1997). This predictability arises from distinctive geochemical reactions and sources of solutes that characterize each of the flow-system scales described above. Although these are treated in more detail elsewhere in the *Treatise on Geochemistry*, they are summarized below because they are an integral part of the origin and interpretation of geochemical indicators of residence time and flow path.

10.2.5.1 Meteoric (recharge)

Meteoric (atmospheric origin) waters that enter the subsurface environment are not devoid of solutes, although they are generally quite dilute (E. K. Berner and R. A. Berner, 1996). They contain ionic solutes, dissolved gases, and isotopic signatures that are derived from atmospheric sources. The most important ionic solutes are generally of marine origin. These especially include Na^+, Cl^-, and SO_4^{2-}. They also contain N_2, O_2, and CO_2 from equilibration with the atmosphere, producing an oxidizing and weakly acidic solution. In areas having minimal soil–water or rock–water interaction after infiltration of precipitation, these characteristics of infiltrating precipitation produce dilute Na^+ or Ca^{2+}–HCO_3^- subsurface waters. These constituents, particularly Cl^-, which tends to act conservatively in groundwater, can under some circumstances be used to infer residence times.

In addition to these "ordinary" constituents, infiltrating meteoric waters may contain more exotic solutes that are particularly useful for tracing of water flow paths and residence times. Prominent among these are radionuclides produced by the action of cosmic rays on the atmosphere. The most commonly employed are ^3H, ^{14}C, and ^{36}Cl, but many others are described in this chapter. Another category of useful tracers is that produced by human activity, whose atmospheric concentration histories are generally known. Those frequently employed for subsurface tracing include ^3H, ^{14}C, and ^{36}Cl produced by atmospheric nuclear-weapons testing, ^{85}Kr and ^{129}I released from nuclear reactors, and chlorofluorocarbons and SF_6 released by industrial and commercial users. Certain other "trace solutes" can be used to estimate residence times by virtue of temporal fluctuations in input concentrations that can be related to known fluctuations in environmental conditions. The trace molecular species of water $^1H_2^{18}O$ and $^2H^1H^{16}O$ are introduced into the atmosphere by oceanic and continental evapotranspiration and are precipitated along with the more common $^1H_2^{16}O$. The proportion of these rare isotopic species is governed by a variety of environmental conditions, primarily by temperature. Similarly, the concentrations of the atmospheric noble gases are temperature dependent. If fluctuations of temperature in the recharge area are known, then subsurface transport times can be inferred if corresponding spatial fluctuations can be observed in the subsurface water. All of these tracers share the characteristic that can then be used to estimate the time since entry into subsurface water systems.

10.2.5.2 Weathering

The chemical composition of most shallow groundwater reflects that of atmospheric precipitation, modified by additional solutes derived from weathering. Weathering reactions generally result from the thermodynamic instability of minerals formed in the deep subsurface (or the oceans) in the low-temperature, low-pressure, dilute, oxidizing, and acidic conditions typical of vadose zones and shallow groundwater. The solutes added to infiltrating precipitation through these reactions vary greatly, depending on the mineralogy of the solid phase and the environmental conditions (Drever, 1997). Owing to the slow kinetics of many weathering reactions (particularly those involving aluminosilicate reactions), solute concentrations can continue to increase and evolve over time (Rademacher et al., 2001). These reactions lead to water at the vadose, local, and often aquifer scales that is more basic, less oxidizing, and more concentrated than precipitation, but still generally containing less than $\sim 1,000$ mg L^{-1}. Solutes introduced from these reactions are generally not used as residence-time tracers, but they can complicate the use of atmospheric tracers that also have weathering sources (e.g., ^{14}C and ^{32}Si).

10.2.5.3 Diagenetic

As the scale of flow systems increases from the local to the regional, groundwater moves to depths where pressures are much higher, temperatures higher, and redox conditions generally more reducing. Under these conditions, waters that have achieved near-equilibrium under vadose or shallow aquifer environments begin to react again with the solid phases present. In some cases these reactions simply take the form of recrystallization to higher $P - T$ phases, but in others solutes are dissolved at one point along a flow path and precipitated at another point, producing cementation or mineralization of the rocks through which they are flowing. Increases of temperature, particularly, tend to produce total solute concentrations that are higher than in groundwaters in shallower flow systems. Anion concentration tends to evolve from HCO_3^- dominance to Cl$^-$ dominance along regional flow paths (Chebotarev, 1955); cation evolution is less predictable. One particularly important aspect of deep rock–water interaction is that highly soluble evaporite minerals (principally halite) are uncommon near the surface of the Earth, because they have generally already been dissolved by circulating meteoric water. Such minerals are more common in the deep subsurface where circulation rates are slow, and thus regional-scale groundwaters frequently acquire very high dissolved solids through interaction with evaporites. Certain distinctive solutes introduced from the matrix along regional-scale flow paths can be used for estimating residence times. These include certain isotopes of He, Ne, and Ar, and ^{129}I.

10.2.5.4 Connate

In addition to evaporites, groundwaters on regional flow paths frequently encounter connate waters, i.e., residual solutions retained in the subsurface from the time that formations in the system were deposited in the ocean. These waters often show Na$^+$– Cl$^-$ dominance. Such waters were previously thought to have been largely swept out of sedimentary basins by circulation of meteoric water (Clayton et al., 1966), but this analysis ignored the high density of these brines, which makes them very resistant to displacement by flowing dilute groundwater (Knauth and Beeunas, 1986). Connate waters (or other saline formation waters) influence the composition of topographically driven groundwater mainly by upward diffusion of solutes. Because of their generally great age and immobility, they are not amenable to most groundwater dating methods, with the possible exception of ^{129}I (Moran et al., 1995).

10.2.5.5 "Basement waters"

Highly saline water of distinctive isotopic composition is often found in environments of such depth and low permeability that flow rates must be extremely low or zero. These waters are often characterized by Ca^{2+}–Cl$^-$ compositions and stable isotope composition "above" the meteoric waterline. They apparently result from water–rock equilibration over very long periods of time. Geochemically, they have little influence on waters in active circulation systems, but mobile isotopes of the noble gases diffusing upward from this environment can be a powerful tool for understanding the flow systems into which they move. The noble-gas isotopes can also provide clues to the histories of these nearly static waters.

10.3 SOLUTE TRANSPORT IN SUBSURFACE WATER

The objective of applying geochemical tracers to subsurface water systems is to understand their dynamics. The tracers themselves are solutes. In order to correctly understand the distribution of the tracers, and to apply them to analyze the system dynamics, a basic understanding of subsurface transport processes is necessary. Transport of subsurface solutes can be ascribed to three processes: advection, diffusion, and dispersion. This section gives a brief introduction to these transport processes. More extensive treatments can be found in Bredehoeft and Pinder (1973), Domenico and Schwartz (1998), and Phillips (1991).

10.3.1 Fundamental Transport Processes

10.3.1.1 Advection

Advection is mechanical transport of solutes along with the bulk flux of the water. It is driven by the gradient in the total mechanical energy of the solution, just as the water flux is driven. For most circumstances, this means the gradient in gravitation potential energy. Advective fluxes are simply the product of the water flux from Darcy's law with the solute concentration:

$$\bar{J}_a = C\bar{\bar{K}}\nabla h = C\bar{q}$$

where $\bar{J}a$ is the advective solute flux, C the volume concentration, $\bar{\bar{K}}$ the hydraulic conductivity tensor, ∇h the hydraulic head gradient, and $\bar{q}$ the specific discharge. Advection transports all solutes at the same rate.

10.3.1.2 Diffusion

Diffusion differs fundamentally from advection in that it is entropy driven, following the principle that solutes within a system will redistribute themselves so as to maximize the entropy of the system. Diffusive transport is, therefore, governed by concentration gradients:

$$\bar{J}_d = D'_d \nabla C$$

where $\bar{J}_d$ is the diffusive flux, D'_d the effective diffusion coefficient (accounting for the effects of porosity and tortuosity), and ∇C the concentration gradient. Since the concentration distributions of various solutes in a single system need not be identical, different solutes generally have different diffusive fluxes.

The relative importance of advection and diffusion is of fundamental significance for understanding solute transport in various settings. This can be assessed using the Peclet number

$$Pe = \frac{q_s d_m}{D'_d}$$

where q_s is the average linear velocity of the water, d_m the mean grain diameter, and D'_d the effective diffusion coefficient. For values of Pe less than ∼1, diffusion dominates, but above this value advection dominates (Perkins and Johnston, 1963). Application of this equation to the various

flow systems described above indicates that for typical local or aquifer-scale systems, advection dominates decisively over diffusion. However, at the very slow flow velocities, characteristic of the deeper portions of regional flow systems, diffusion may become dominant. Diffusion may also dominate in shallower flow systems where permeabilities are very low, such as clay-rich tills or shales, or in unusual circumstances where hydraulic gradients are negligible. In general, diffusion is an effective mechanism for redistributing solutes over short distances under most circumstances, but is a significant transport mechanism at aquifer or regional scales only when the time available for transport is also long.

10.3.1.3 Dispersion

There is an additional mechanism of transport in flowing water termed dispersion. The dispersive flux is given by

$$\bar{J}_{disp} = \bar{\bar{D}}\nabla C$$

where $\bar{\bar{D}}$ is the dispersion coefficient tensor and is given by

$$\bar{\bar{D}} = \bar{\bar{\alpha}}q_s + D'_d$$

where $\bar{\bar{\alpha}}$ is the dispersivity tensor. Although the equation above appears to treat dispersion and diffusion as separate and additive processes, they are, in fact, linked. An increased diffusion coefficient may appear to increase dispersion, but will actually decrease it so long as the linear velocity term in the equation is not very small. Dispersion is a mixing process that is actually driven by differential advection. Small-scale differences in velocity along different flow paths produce increasing variation in the concentration as a function of position. However, diffusion tends to equalize concentrations between high- and low-velocity portions of the flow path and thus limits the amount of differential transport due to mechanical dispersion. The additive diffusion coefficient term on the right-hand side simply ensures that concentration-gradient-driven transport goes to the diffusive flux as the velocity goes to zero.

Dispersivity is treated as a tensor, because the magnitude of velocity variations is greatest in the direction of flow and least transverse to the direction (usually vertically transverse, if the flow is horizontal). This means that, for equal concentration gradients, the magnitude of the dispersive flux will also be greatest in the direction of flow (longitudinal dispersion) and smallest perpendicular to it (transverse dispersion).

10.3.2 Advection–Dispersion Equation

In order to obtain a comprehensive description of the transport of solutes in subsurface water, it is necessary to consider all three transport mechanisms. This can be accomplished by linking the three transport flux equations above through the continuity equation, yielding the advection–dispersion equation for steady-state flow:

$$\overline{\overline{D}} \nabla^2 C - \overline{q_s} \cdot \nabla C + \sum_i R_i = \frac{\partial C}{\partial t}$$

where R_i are chemical or nuclear reactions affecting the constituent modeled. The advection–dispersion equation can be solved analytically for simple boundary conditions, and such solutions have been used frequently in interpreting the distribution of age tracers (Sudicky and Frind, 1981; Zuber, 1986). However, numerical models provide a more flexible and realistic means of implementing the advection–dispersion equation for groundwater systems, which rarely correspond to the simplifications required for analytical solutions.

10.3.3 Interaction between Hydrogeological Heterogeneity and Transport

Section 10.2.3 above emphasized the high degree of heterogeneity of permeability that is characteristic of most geological materials, and the role of hierarchical scales of organization of the permeability in determining average flow paths. These characteristics also play a critical role in the transport of solutes in the subsurface. The role they play depends on the scale of the transport.

10.3.3.1 Small-scale transport and effective dispersion

Dispersion has often been treated as a constant parameter, with values of the dispersivity on the order of centimeters (Perkins and Johnston, 1963). For typical shallow groundwater flow velocities, this places dispersive transport on the same order as diffusion. However, extensive field tracer tests have demonstrated that dispersion is much more effective than these values would indicate and, furthermore, that effective dispersivity tends to increase with the scale of the tracer test (Gelhar, 1986; Schwartz, 1977; Sudicky and Frind, 1981). This behavior derives from the fundamentally quasifractal nature of hydraulic heterogeneity described above. Small-scale variations in pore structure tend to spread solutes over limited distances. This spreading is then amplified as the solute-carrying groundwater encounters progressively larger-scale heterogeneities with increased flow distance. The end result of these processes is that at typical shallow groundwater velocities, longitudinal dispersion is typically one or more orders of magnitude more effective in transporting solutes than is diffusion. Furthermore, the

effectiveness of the process increases as the transport scale increases. The amount of mixing and consequent spreading of concentration fronts is highly dependent on the local hydrogeology, which controls the permeability structure. Methodologies for moving from an understanding of the geological environment to the ability to predict the permeability structure, and from that ability to the prediction of dispersive behavior, have not been achieved (Weissmann *et al.*, 1999).

10.3.3.2 Large-scale transport and mixing

As the scale is increased from the aquifer to the regional, solute transport becomes even more difficult to predict. At smaller scales a somewhat homogeneous spatial distribution of high- and low-permeability facies can often be assumed. This leads to more uniform relative contributions of advection and diffusion to total dispersion. However, as the spatial scale increases, mixing processes extend over units that may have greatly differing distributions of high- and low-permeability facies. For example, transport processes in a sandstone aquifer may be dominated by advection, but in the adjoining shale aquitards, diffusion may dominate. The boundaries between units of differing transport characteristics are not necessarily abrupt or of simple geometry.

Several approaches have been taken to dealing with solute transport in general, and transport of geochemical dating tracers in particular, through these large-scale heterogeneous systems. The earliest approach was to assume such drastic contrasts that low-permeability units could be treated as impermeable and impervious to diffusion (Hanshaw and Back, 1974; Pearson and White, 1967). This is equivalent to treating the permeable units as a sealed tube. This conceptual model is clearly an oversimplification, inasmuch as we know that cross-formational water fluxes are common, and that diffusion coefficients of low-permeability units are not drastically smaller than those of higher permeability units. The second of these shortcomings was acknowledged in models that treated diffusion of tracers into adjacent aquitards as a boundary condition for transport equations (Sanford, 1997; Sudicky and Frind, 1981). Although an improvement, this conceptualization is also limited by the assumption that transport in low-permeability units is diffusion dominated (Phillips *et al.*, 1990). In many cases advection may still dominate, and this carries important implications for the spatial distribution of radionuclide tracers or transient tracer pulses. For example, diffusion will selectively move $H^{14}CO_3^-$ out of aquifers and into aquitards, along the concentration gradient set up by ^{14}C decay within the aquitards, but will not induce any net transport of stable HCO_3^-, thus mimicking the effects of radioactive decay on the $^{14}C/C$ ratio. Advection, in contrast, will transport both species at the same rate.

Any truly comprehensive and accurate mathematical description of large-scale solute transport must incorporate both the effects of encountering progressively larger scales of hydraulic heterogeneity on differential advective transport and the progressive increase in the importance of diffusion as larger volumes of low-permeability rock are contacted. It must account for the three-dimensional interaction between the hydraulic head field and the spatial configuration of heterogeneities, and the effect of this interaction on the fundamental advective and diffusive transport mechanisms. The spatial distributions of permeability in any such model must have a rigorous basis in hydrogeology. This is a formidable undertaking, and we are far from accomplishing it. However, environmental geochemical tracers provide one of the few realistic approaches to obtaining data on solute transport at these scales, and thus represent a critical component to the solution of this fundamental hydrological/geochemical problem.

10.3.4 Groundwater Dating and the Concept of "Groundwater Age"

Suppose one holds an igneous rock specimen in which all of the mineral grains crystallized at very nearly the same time, for which the crystallization time was short compared to the period since solidification, and which has not experienced any reheating or alteration since initial crystallization. One could then measure some time-dependent property of the specimen (e.g., the amount of ^{40}Ar accumulated from the decay of ^{40}K) and infer a reasonably well-defined "crystallization age" for the rock. Unfortunately, an analogous set of conditions is quite unlikely to hold rigorously for any sample bottle full of groundwater. As previously emphasized by Davis and Bentley (1982), the dynamic nature of groundwater systems renders such simplistic scenarios unlikely. This review has emphasized that, due to the complex interaction of irregular topographic forcing with extremely heterogeneous distributions of permeability, groundwater flow paths are complex and difficult to predict, often resulting in mixing on various scales. When large-scale advective mixing is combined with differential transport of different species by diffusion and dispersion, acting over a range of scales, the expectation of simple closed-system behavior for parcels of groundwater is clearly naïve.

Geochemical species that can be used to gain information on groundwater residence times are much better thought of as groundwater "age tracers" than they are as "dating methods." In some cases (e.g., wells sampling local-scale flow systems in relatively homogeneous materials), closed-system behavior may be approximated, and "dates" can be interpreted to reflect residence time in a straightforward fashion. In general, however, concentrations of age tracers should be viewed as data upon which inferences regarding the residence-time structure of the three-dimensional groundwater flow system can be based. Application of several tracers will generally give much better constrained inferences than those using only a single tracer (Castro *et al.*, 2000; Mazor, 1976, 1990). Numerical transport models offer by far the most flexible and comprehensive approach to interpreting the tracer data (Castro and Goblet, 2003; Dinçer and Davis, 1984; Park *et al.*, 2002).

Given the complex flow paths and mixing of waters recharged at different times, even the basic meaning of "groundwater age" can be ambiguous. Recently, several authors (Bethke and Johnson, 2002a,b; Goode, 1996) have addressed this problem by treating groundwater age as a parameter that accumulates in all water in a system by a linear buildup per unit mass per unit time, termed "age mass." The "age" of the groundwater can then be rigorously defined by the accumulated age mass, equivalent to the average of the residence time of every water molecule in the sample. Age mass is a completely conservative quantity whose total does not vary with time in basins that have been at hydraulic steady state for long periods. In such a system, the total age mass is simply "the total water mass" × "the average residence time" (total mass divided by rate of recharge) × "the age mass production rate" (1 kg age mass (kg water)$^{-1}$ yr^{-1}).

More traditional approaches to treating the age of groundwater systems have tended to overemphasize the advective-dominated portions of the system. The age mass approach forces consideration of both advective and diffusive transport through the entire system. This is nicely illustrated by the observation of Bethke and Johnson (2002b) that the accumulation of age mass in the water of an aquifer that is sandwiched between two aquitards is, counter intuitively, independent of the exchange rate with the water in the aquitards. The reason is that if the exchange rate is high, longer residence-time water is rapidly moved into the aquifer, but the average age of the water entering is not great (due to the rapid exchange). Alternatively, if the exchange rate is low, only small amounts of aquitard water enter, but the average age of the aquitard water is great. The average age of the water exiting the aquifer depends only on the total mass of water in the system and the total water flux through the system, not on the proportion of aquifer to aquitard, or the transport characteristics of the two lithologies.

This approach to conceptualizing residence-time distributions in groundwater systems has important practical implications for interpreting age tracer measurements. For example, the pattern of radiogenic helium concentration in the aquifer system described above would strongly resemble that of the theoretical age mass distribution. Specifically, the concentration of helium would increase much more rapidly with distance along the aquifer than a simple calculation based on the linear flow rate through the aquifer would indicate. It would increase more rapidly because helium would diffuse out of the aquitards into the aquifer. If this additional source of helium is not accounted for, the velocity in the aquifer inferred from the helium distribution would be much slower than the actual velocity. Similarly, a radionuclide tracer (such as ^{14}C or ^{36}Cl) would diffuse into the aquitard, along with the water, also producing an incorrectly low apparent linear velocity in the aquifer.

Although a merely conceptual application of the age mass approach can provide important insights into a simple system such as the one described above, its distribution in an actual groundwater system is a complex function of advective and diffusive controls, like any other solute. Thus, for actual applications the most promising approach is to use numerical models to calibrate the age mass distribution using results from a number of age tracers. Regardless of whether the goal of a study is to define groundwater velocity distributions, or to understand subsurface transport processes, the approach must be based on an understanding of the dynamics of the entire groundwater system.

10.4 SUMMARY OF GROUNDWATER AGE TRACERS

10.4.1 Introduction

Age tracers for groundwater and the vadose zone can be divided into three basic classes: (i) radionuclides of atmospheric origin that can be used with the radiometric decay equation to "date" the time since recharge; (ii) stable constituents originating with recharge that have known patterns of age with time and can hence be used to infer residence time; and (iii) constituents that are added to groundwater in the subsurface at an approximately known rate, whose accumulation can thus be used to gain information on the residence time of the water.

10.4.2 Radionuclides for Age Tracing of Subsurface Water

10.4.2.1 Argon-37

Among the unstable isotopes that have been used as natural tracers in groundwater, ^{37}Ar has the shortest half-life (35 d; Lal and Peters, 1967;

Loosli and Oeschger, 1969). Argon-37 is produced by cosmic rays in the atmosphere. Since its half-life is extremely short, most groundwaters have completely lost the atmospheric ^{37}Ar signal that was initially present in the recharge water. Measured ^{37}Ar in groundwater thus reflects the subsurface production rate of ^{37}Ar and the transfer efficiency from rocks to groundwater. The dominant source of subsurface ^{37}Ar is through the following neutron reaction with ^{40}Ca:^{40}Ca $(n,\alpha)^{37}Ar$. The neutron flux originates either directly from spontaneous fission of uranium or from (α,n) reactions within the matrix rock. At present, ^{37}Ar activities can be measured down to 100 μBq per liter (STP) of argon by gas proportional counting together with ^{39}Ar (see below). For comparison, atmospheric ^{37}Ar activity is 50 μBq per liter (STP) of argon (Loosli and Oeschger, 1969); activities as high as 0.2 Bq per liter (STP) of argon have been measured in groundwater in the Stripa granite, in Sweden (Loosli *et al.*, 1989). When used as a natural tracer of groundwater, ^{37}Ar is used in combination with a number of other tracers (e.g., ^{39}Ar, ^{85}Kr, see below) in order to provide more reliable estimates of groundwater residence times (Loosli *et al.*, 1989, 2000).

10.4.2.2 Sulfur-35

Sulfur-35 is formed from the spallation of argon in the atmosphere. It has a half-life of 87 d (Lal and Peters, 1967). After oxidation to sulfate, it is deposited on the land surface. Because of its relatively long residence time in the upper atmosphere and its short half-life, specific activities of ^{35}S are low at the time of deposition. Due to the anionic form of $^{35}SO_4^{2-}$ it is relatively conservative in soil water and groundwater and can be used to estimate residence times in shallow, rapidly circulating systems. It has seen limited application in hydrology because of its low specific activity in groundwater, requiring large sample volumes, and the concomitant requirement for low background sulfate sample waters (Michel, 2000).

Most applications have mainly been to low-sulfate environments with rapid turnover, especially to studying the behavior of sulfate during snowmelt (Michel *et al.*, 2000). It has proved useful in estimating storage time of groundwater contributing to stream flow, showing that even in apparently low-storage alpine systems, a preponderant fraction of the flow can be derived from groundwater with residence times of several years or more (Suecker *et al.*, 1999).

10.4.2.3 Krypton-85

Anthropogenic ^{85}Kr (half-life 10.76 yr) was first released to the atmosphere in considerable amounts in the 1950s and 1960s by atmospheric nuclear-weapons testing. Today, it is continuously produced through fission of uranium and plutonium and released to the atmosphere, mainly during fuel rod reprocessing in nuclear power plants. As shown in Figure 2(d), this artificially produced ^{85}Kr has completely overwhelmed natural cosmogenic production of ^{85}Kr, and has created a steady increase in the activity of this isotope in the atmosphere, particularly in the northern

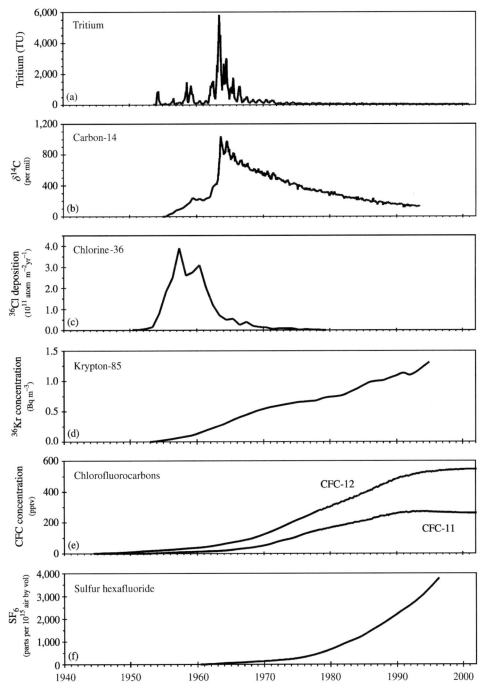

Figure 2 Global environmental histories of hydrologic-cycle tracers that have been strongly affected by human activities: (a) tritium (data from measurements at Ottawa, Canada, by the International Atomic Energy Agency; http://isohis.iaea.org/); (b) carbon-14 (data for 1955–1962 from Kalin (2000), data for 1963–1991 from Nydal *et al.* (1996)) http://cdiac.esd.ornl.gov/epubs/ndp/ndp057/ndp057.htm); (c) chlorine-36 (data from Synal *et al.* (1990); (d) krypton-85 (data from Loosli *et al.*, 2000); (e) chlorofluorocarbons (data for 1944–1978 from Plummer and Busenberg (2000); data for 1978–2003 from Cunnold *et al.* (1997); http://cdiac.esd.ornl.gov/ftp/ale_gage_Agage/); and (f) sulfur hexafluoride (data from Plummer and Busenberg, 2000).

hemisphere (Weiss *et al.*, 1992). The southern hemisphere, without significant sources of this isotope, exhibits an average ^{85}Kr activity 20% lower than the northern hemisphere. Such variations need to be considered when interpreting ^{85}Kr activities in groundwater. Because ^{85}Kr activities in groundwater are very low (1 L of water in equilibrium with the atmosphere has a ^{85}Kr activity of only 8.6 mBq), analysis of this isotope is relatively difficult (Fairbank *et al.*, 1998; Rozanski and Florkowski, 1979; Smethie and Mathieu, 1986; Thonnard *et al.*, 1997). Generally, ^{85}Kr is employed as a natural tracer in combination with other isotopes and, particularly, with ^{3}H, which has a similar half-life. Both tracers are used to estimate the age of rather young ($\leq$40 yr) waters. When dispersion is small in a particular groundwater system, the age of groundwater can be estimated directly from ^{85}Kr activities (e.g., Ekwurzel *et al.*, 1994; Smethie *et al.*, 1992).

10.4.2.4   Tritium

Tritium is produced in the atmosphere by cosmic-ray spallation of nitrogen (Lal and Peters, 1967). Tritium decays to ^{3}He with a half-life of 4,500 $\pm$ 8 d (equivalent to 12.32 $\pm$ 0.02 yr) (Lucas and Unterweger, 2000). Atmospherically produced tritium reacts rapidly to form tritiated water: ^{3}HHO. Tritiated water is precipitated out of the atmosphere together with ordinary water. The tritium concentration in natural water is commonly expressed in Tritium Units (TU), where 1 TU is equal to one molecule of ^{3}HHO per 10^{18} molecules of H_2O. This is equivalent to a specific activity of 0.1181 Bq (kg water)$^{-1}$.

Tritium is commonly measured by two methods: β-counting or ^{3}He ingrowth. Natural levels of tritium are only marginally measurable by direct β-counting methods, but fortunately tritium is highly enriched over ordinary H in the H_2 produced by the electrolysis of water. β-Detection, either by gas proportional counting or liquid scintillation counting, can then easily measure natural tritium (Taylor, 1981). A newer approach, with generally higher sensitivity, is to degass the water, then allow ^{3}He to accumulate for up to a year, and measure the ^{3}He using mass spectrometry (Clarke *et al.*, 1976). The excellent properties of ^{3}HHO as a tracer of water and the relative ease of measurement have encouraged a very wide spectrum of applications of tritium to waters with a residence time of less than 50 yr.

The average global production of tritium is ~2,500 atom m^{-2} s^{-1} (Solomon and Cook, 2000). The deposition rate of the tritium varies with latitude, but it is also mixed with the bulk of precipitation originating from the ocean (which has a very low tritium content), and thus the average tritium content of precipitation tends to vary inversely with annual precipitation. Natural tritium in precipitation varies from ~1 TU in oceanic high-precipitation regions to as high as 10 TU in arid inland areas.

Since the 1950s natural tritium deposition has been swamped by the pulse of tritium released by atmospheric nuclear-weapons testing, followed by other anthropogenic releases (Gat, 1980). Significant tritium was released by the fission and thermonuclear devices tested in the mid-to-late 1950s, but it was the stratospheric thermonuclear-weapons testing of the early 1960s that produced the peak pulse, as large as 5,000 TU in the mid-latitude northern hemisphere (Figure 2(a)). Because most of the testing was in the northern hemisphere, southern hemisphere fallout was much less. This pulse has decayed in a quasi-exponential fashion since that time and is now beginning to approach natural levels. It has provided an invaluable transient tracer for shallow hydrological systems.

Application of tritium to hydrologic problems was first proposed by Libby (Bergmann and Libby, 1957; Libby, 1953). Early applications (Allison and Holmes, 1973; Carlston *et al.*, 1960) focused on tracing the bomb-tritium pulse through aquifers. This was soon extended to estimating recharge by measuring tritium profiles through the vadose zone (Schmalz and Polzer, 1969; Vogel *et al.*, 1974). Measurement of tritium profiles through the English Chalk vadose zone proved an important key to understanding matrix diffusion of solutes in fractured media (Foster, 1975). During the 1980s the bomb-tritium pulse in groundwater began to lose its definition. The combined effects of decay and dispersion tended to reduce bomb-peak levels to values similar to that of precipitation at that time. Fortunately, the limitations imposed by this loss of the tracer pulse were countered by the introduction of the ^{3}H/^{3}He method (Tolstikhin and Kamensky, 1968). By measuring first the tritiogenic ^{3}He and then the ^{3}H content by ^{3}He ingrowth, both the initial ^{3}H content and the time of decay can be calculated. The power of this approach for reconstructing residence times has been elegantly demonstrated by Schlosser *et al.* (1988) and Solomon *et al.* (1992, 1993). The ^{3}H/^{3}He method has since become a standard tool in the investigation of the dynamics of shallow aquifers (Beyerle *et al.*, 1999; Ekwurzel *et al.*, 1994; Puckett *et al.*, 2002). However, ^{3}He is also produced by fission of uranium and nucleogenic processes in the subsurface, and corrections for these sources must sometimes be employed (Andrews and Kay, 1982a,b).

10.4.2.5   Silicon-32

Silicon-32 is produced in the atmosphere by cosmic-ray spallation of ^{40}Ar (Lal *et al.*, 1960).

It decays to ^{32}P with a half-life of 140 ± 6 yr (Morgenstern *et al.*, 1996), although there is some uncertainty regarding the exact value (Nijampurkar *et al.*, 1998). It is oxidized and incorporated into precipitation as silicic acid. The global deposition rate is ~ 2 atom m^{-2} s^{-1} (Kharkar *et al.*, 1966; Morgenstern *et al.*, 1996; Nijampurkar *et al.*, 1998). This produces activities in natural water in the range of 2–20 mBq m^{-3} (Morgenstern, 2000). Silicon-32 does not behave in a conservative fashion after it begins to move through the subsurface. It reacts with silicate minerals, presumably by exchange, and ^{32}Si concentrations are markedly reduced during transport through the vadose zone (Fröhlich *et al.*, 1987; Morgenstern *et al.*, 1995).

Silicon-32 is most sensitively measured using liquid scintillation counting. Silicon must be extracted from ~ 1 m^3 of water. The daughter ^{32}P is allowed to grow into equilibrium over several months, then is milked with stable phosphorus for scintillation counting (Morgenstern, 2000). Very low counting backgrounds are necessary.

Silicon-32 has many attractive aspects for age tracing of groundwater. Its half-life makes it very useful for filling the gap between ^{3}H (12.3 yr) and ^{14}C (5,730 yr). It has a very small subsurface production (Florkowski *et al.*, 1988). However, the disadvantage of unpredictable loss to the solid phase and analytical difficulties have limited the number of its applications (Lal *et al.*, 1970; Nijampurkar *et al.*, 1966). Morgenstern (2000) has proposed that most of the nonconservative behavior is produced in the vadose zone and that ^{32}Si may be useful for residence-time determinations in aquifers if the initial activity in the recharge area can be determined. However, until this can be demonstrated, the applications of the isotope will remain limited.

10.4.2.6 Argon-39

In contrast to ^{37}Ar, ^{39}Ar in groundwater has its origin in the atmosphere where it is produced by cosmic rays, and enters the water cycle as a fraction of the dissolved argon during groundwater recharge. The activity of ^{39}Ar in groundwater is extremely small, a fact that complicates its analysis (Loosli *et al.*, 2000). Modern atmospheric ^{39}Ar activity is 16.7 mBq m^{-3} of air (Loosli, 1983). Two thousand liters of water in equilibrium with the atmosphere at 20 °C have an ^{39}Ar activity of only 1.2 mBq. When subsurface production of ^{39}Ar (resulting from the reaction ^{39}K(n,p)^{39}Ar) can be neglected, this isotope with a half-life of 269 yr has been successfully used in combination with other tracers, in particular with ^{14}C to estimate groundwater ages in sedimentary basins (e.g., Andrews *et al.*, 1984; Purtschert, 1997). However, this procedure can be complicated in crystalline terrains with high uranium and thorium contents and hence a non-negligible subsurface production of ^{39}Ar (Andrews *et al.*, 1989; Lehmann *et al.*, 1993).

10.4.2.7 Carbon-14

Carbon-14 is produced in the atmosphere by a low-energy cosmic-ray neutron reaction with nitrogen. It decays back to ^{14}N with a half-life of 5,730 yr. The production rate is $\sim 2 \times 10^4$ atom m^{-2} s^{-1}, the highest of all cosmogenic radionuclides. The rate is so high, because nitrogen is the most abundant element in the atmosphere and also has a very large thermal neutron absorption cross-section. Radiocarbon activity in the atmosphere is the result of complicated exchanges between terrestrial reservoirs (primarily vegetation), the ocean, and the atmosphere. The combination of varying cosmic-ray fluxes (due to both varying solar and terrestrial magnetic field modulations) and varying exchange between reservoirs has caused the atmospheric activity of radiocarbon to fluctuate by approximately a factor of 2 over the past 3×10^4 yr (Bard, 1998). The current specific activity is 0.23 Bq (g C)$^{-1}$, rendering radiocarbon easily measurable by gas proportional or liquid scintillation counting, or by accelerator mass spectrometry (AMS). Radiocarbon measurements are usually reported as "percent modern carbon," indicating the sample specific activity as a percentage of the 0.23 Bq (g C)$^{-1}$ modern atmospheric specific activity. In addition to the natural cosmogenic production of ^{14}C, the isotope was also released in large amounts by atmospheric nuclear-weapons testing in the 1950s and 1960s (Figure 2(b)).

Natural radiocarbon was first detected by Libby in the mid-1940s (Arnold and Libby, 1949; Libby, 1946), but the first applications to subsurface hydrology were not attempted for another decade (Hanshaw *et al.*, 1965; Münnich, 1957; Pearson, 1966). These early investigators discovered that radiocarbon shows clear and systematic decreases with flow distance that can be attributed to radiodecay, but also exhibits the effects of carbonate mineral dissolution and precipitation reactions. Quantification of residence time is not possible without correction for additions of nonatmospheric carbon. Numerous approaches to this problem were attempted, including simple empirical measurements (Vogel, 1970), simplified chemical equilibrium mass balance (Tamers, 1975), mass balance using δ^{13}C as an analogue for ^{14}C (Ingerson and Pearson, 1964), and combined chemical/δ^{13}C mass balance (Fontes and Garnier, 1979; Mook, 1980).

The correction methods cited above were mainly intended to account for carbonate reactions in the vadose zone during recharge, although in practice they have often been applied to reactions along the groundwater flow path as well. It is now the general consensus that the preferred approach to treating the continuing reactions of dissolved inorganic carbon during flow is geochemical mass transfer/equilibrium models (Kalin, 2000; Zhu and Murphy, 2000). The most commonly employed model is NETPATH (Plummer *et al.*, 1991). This uses a backward-calculated solute mass balance, constrained by simple equilibrium considerations, to calculate mass transfers of solutes from phase to phase between two sample points.

The vast majority of groundwater radiocarbon investigations have extracted the dissolved inorganic carbon from the water for measurement. However, dissolved organic carbon (DOC) has been sampled in a limited number of studies (Murphy *et al.*, 1989; Purdy *et al.*, 1992; Tullborg and Gustafsson, 1999). In principle, sampling DOC could avoid the complex geochemistry of inorganic carbon. In reality, DOC consists of a large number of organic species from various sources (with different original ^{14}C activities), of varying chemical stability, and of varying reactivity with the solid phase. Routine use of DOC for groundwater age tracing may one day show advantages over inorganic carbon, but this will require considerable work separating and identifying the appropriate component of the DOC for this application.

Radiocarbon is an indispensable tool in the age tracing of groundwater. Both its ease of use and its half-life make it the method of choice for many investigations. However, results must always be approached with caution. The reliability of results depends strongly on the complexity of the geochemistry. In some cases groundwater may show little evidence of chemical evolution after recharge and measured radiocarbon activities can be accepted at face value. In other cases, the isotopic composition of carbon may be strongly altered by numerous surface reactions that are difficult to quantify, and interpretations may be speculative, at best. Interpretations must be evaluated on a case-by-case basis.

10.4.2.8 Krypton-81

With a half-life of 2.29×10^5 yr, ^{81}Kr, still in its "infancy" as a natural tracer, has had the potential to be an excellent tool for dating old groundwater ever since it was first detected in the atmosphere (Loosli and Oeschger, 1969). In addition to its constant atmospheric concentration, anthropogenic and subsurface production of ^{81}Kr have been shown to be negligible (e.g., Collon *et al.*, 2000; Lehmann *et al.*, 1993). Nearly all ^{81}Kr in groundwater results from the interaction of cosmic rays with nuclei in the Earth's atmosphere, in particular, through neutron capture by ^{80}Kr and

spallation of heavier krypton isotopes. However, its activity (0.11 nBq in water in equilibrium with the atmosphere) is lower than that of any other tracer that has been used. The analytical challenge to measure this isotope is, therefore, extreme. The first measurements of ^{81}Kr in groundwater were reported in the Milk River aquifer in Canada (Lehmann *et al.*, 1991; Thonnard *et al.*, 1997), but an elaborate multistep enrichment procedure made it difficult to quantify the results, though an estimated groundwater age of 1.4×10^5 yr was obtained. A new measurement technique based on accelerator mass spectrometry (AMS) using positive krypton ions coupled to a cyclotron was reported by Collon *et al.* (2000). In this study, ^{81}Kr was measured in groundwater in the Great Artesian Basin (GAB) in Australia and, for the first time, definite determinations of water residence times were made based on the atmospheric $^{81}Kr/Kr$ ratio. Krypton dissolved in surface water in contact with the atmosphere has an atmospheric $^{81}Kr/Kr$ ratio of $(5.20 \pm 0.4) \times 10^{-13}$. Observed reductions of isotope ratios in groundwater were interpreted as being due to radioactive decay since recharge (Collon *et al.*, 2000; Lehmann *et al.*, 2002b).

10.4.2.9 Chlorine-36

Chlorine-36 is produced in the atmosphere by cosmic-ray spallation of ^{40}Ar (Lal and Peters, 1967). It decays to ^{36}Ar with a half-life of $301,000 \pm 4,000$ yr (Bentley *et al.*, 1986a). The globally averaged production rate is between 20 atom $m^{-2} s^{-1}$ and 30 atom $m^{-2} s^{-1}$ (Phillips, 2000). Meteoric ^{36}Cl then mixes with stable oceanic chlorine in the atmosphere, diluting the ^{36}Cl greatly near coastlines, but less inland. The resultant specific activities of chlorine in atmospheric deposition range from less than 20 μBq $(g\ Cl)^{-1}$ to more than 1,000 μBq $(g\ Cl)^{-1}$. These specific activities are at the lower limits of the sensitivity of any form of decay counting, and require large masses of chlorine that may be difficult to obtain from dilute waters. Routine application of ^{36}Cl to age tracing of groundwater is possible only because of the high sensitivity and much smaller sample size permitted by AMS (Elmore *et al.*, 1979). Because an isotopic ratio is measured by mass spectrometry, the atomic ratio, atoms ^{36}Cl per 10^{15} atoms Cl, is normally used in ^{36}Cl studies instead of specific activity.

Application of ^{36}Cl to earth-science problems was attempted as early as the mid-1950s (Davis and Schaeffer, 1955), but little use was made of the method until the advent of AMS in 1979. Since the analytical barrier was overcome, there have been regular applications to long-residence-time aquifers around the world (mainly large sedimentary basins). These include systems in Australia

(Bentley *et al.*, 1986b; Cresswell *et al.*, 1999; Torgersen *et al.*, 1991), North America (Davis *et al.*, 2003; Nolte *et al.*, 1991; Phillips *et al.*, 1986; Purdy *et al.*, 1996), Europe (Pearson *et al.*, 1991; Zuber *et al.*, 2000), Asia (Balderer and Synal, 1996; Cresswell *et al.*, 2001), and Africa (Kaufman *et al.*, 1990).

The major complications in using ^{36}Cl to estimate the residence time of groundwater are introduction of subsurface chloride and subsurface production of ^{36}Cl. High chloride concentrations arising from connate waters, evaporite dissolution, or possibly rock–water interaction are common in sedimentary basins and deep crystalline rocks. These may either diffuse upward or be advected into aquifers by cross-formational flow. In general, this subsurface-source chloride is not ^{36}Cl free, because ^{36}Cl is produced at low levels (compared to meteoric chloride) in the subsurface by absorption by ^{35}Cl of thermal neutrons produced by uranium and thorium series decay and uranium fission (Lehmann *et al.*, 1993; Phillips, 2000). The effects of radiodecay and of mixing with low ^{36}Cl ratio subsurface chloride must be separated in order to evaluate residence times correctly. Park *et al.* (2002) have recently evaluated in detail the effects of various mixing scenarios. Although simple mixing models may often provide useful approximations of residence time (Bentley *et al.*, 1986a; Phillips, 2000), the best approach, as for all age tracing methods, is comprehensive simulation of flow and transport for the entire groundwater system (Bethke and Johnson, 2002b; Castro and Goblet, 2003).

In addition to the natural cosmogenic production of ^{36}Cl in the atmosphere, a very large pulse of ^{36}Cl was also introduced by the testing of thermonuclear weapons between 1954 and 1958 (Bentley *et al.*, 1982; Phillips, 2000; Zerle *et al.*, 1997). This pulse had much less of a tail than the bomb radiocarbon and tritium pulses, because the atmospheric residence time of chloride is shorter (Figure 2(c)). Chlorine-36 can be applied to groundwater studies in a fashion analogous to tritium, but has proved most useful in tracing water movement through the vadose zone, because, unlike tritium and radiocarbon, it is not volatile and thus does not disperse in the vapor phase (Cook *et al.*, 1994; Phillips *et al.*, 1988; Scanlon, 1992).

10.4.2.10 Iodine-129

Iodine-129 has major sources in both surface and subsurface environments. It is produced in the atmosphere by cosmic-ray spallation of xenon and in the subsurface by the spontaneous fission of uranium. Iodine-129 of subsurface origin is released to the surface environment through volcanic emissions, groundwater discharge, and other fluxes. Due to its very long half-life, 15.7 Ma,

these sources are well mixed in the oceans and surface environment, producing a specific activity of ~5 µBq $(g\ I)^{-1}$ (equivalent to a $^{129}I/I$ ratio of $\sim10^{-12}$). Due to its low activity, the preferred detection method of ^{129}I is AMS (Elmore *et al.*, 1980).

The very long half-life of ^{129}I raises the possibility of dating groundwater on timescales greater than a million years (Fabryka-Martin *et al.*, 1985). For a few subsurface environments that approximate closed systems, including brines incorporated in salt domes (Fabryka-Martin *et al.*, 1985), and connate brine reservoirs (Fehn *et al.*, 1992; Moran *et al.*, 1995), and iodine in oil (Liu *et al.*, 1997), this has proved successful. However, due to the complex subsurface geochemistry of iodine, subsurface fissionogenic production which can be difficult to quantify, and the release of iodine from subsurface organic material, determination of the residence time of water in dynamic groundwater systems has generally not proved practicable (Fabryka-Martin, 2000; Fabryka-Martin *et al.*, 1991).

Iodine-129 has numerous anthropogenic sources. It was globally distributed by the atmospheric nuclear-weapons testing of the 1950s and 1960s, and has continued to be released in large amounts by nuclear technology, especially nuclear fuel reprocessing (Wagner *et al.*, 1996). As a result, environmental levels of the $^{129}I/I$ ratio have increased to values in the range 10^{-10}–10^{-7}. This anthropogenic pulse can be used in a fashion similar to bomb tritium. It has proved to be especially useful in identifying releases from nuclear reprocessing plants and other nuclear facilities (Moran *et al.*, 2002; Oktay *et al.*, 2000; Rao and Fehn, 1999).

10.4.3 Stable, Transient Tracers

10.4.3.1 Chlorofluorocarbons and sulfur hexafluoride

Several industrial chemicals have been created that are highly volatile, resistant to degradation, act conservatively in groundwater, are unusual or absent in the natural environment, and can be detected at very low concentrations. If these compounds show a relatively regular increase in atmospheric concentration, then their concentration in shallow groundwater systems can be correlated with the time of recharge of the water, and residence times can be determined.

One such class of compounds is the chlorofluorocarbons (CFCs), halogenated alkanes that have been used in refrigeration and other industrial applications since the 1930s (Plummer and Busenberg, 2000). The atmospheric lifetimes of the common CFC compounds range from 50 yr to 100 yr. Their concentrations increased in a quasi-exponential fashion from the 1950s through the late 1980s (Figure 2(e)). Unfortunately, CFCs catalyze reactions in the stratosphere that deplete atmospheric ozone. This resulted in international agreements to limit global CFC production drastically. In response the rate of atmospheric CFC increase has leveled off and has started to decrease. This complex concentration history has resulted in a degree of ambiguity in residence-time estimates for water recharged since ~1990.

The advantages of CFC dating are a lack of sensitivity to dispersion and mixing, due to the gradual and monotonic atmospheric concentration history (at least through the 1990s), and the virtual year-to-year dating sensitivity. The major complications of CFC dating are the time lag involved in the diffusion of the atmospheric signal through vadose zones thicker than ~10 m, minor-to-moderate sorption of some CFC species, and microbial degradation of some CFCs under strongly reducing conditions (Plummer and Busenberg, 2000).

Application of CFCs for groundwater studies was first explored in the early 1970s (Thompson, 1976; Thompson *et al.*, 1974), but was not widely employed until important proof-of-concept papers were published in the early 1990s (Böhlke *et al.*, 2002; Busenberg and Plummer, 1991; Dunkle *et al.*, 1993; Ekwurzel *et al.*, 1994). Since that time the method has been widely used in shallow groundwater studies. It has found particular application in studies tracing the source and fate of agricultural contaminants (Böhlke *et al.*, 2002) and surface–groundwater interaction (Beyerle *et al.*, 1999).

Sulfur hexafluoride (SF_6) is a nearly inert gas than has been widely used as a gas-phase electrical insulator. It has a very long atmospheric lifetime, and is detectable to very low levels by gas chromatography using the electron capture detector. These characteristics render it useful for estimating groundwater residence times in a fashion similar to CFCs, for the period from ~1970 to the early 2000s (Busenberg and Plummer, 2000). The major advantage over CFCs is that the atmospheric concentration of SF_6 is continuing to increase monotonically (Figure 2(f)). One potential disadvantage is that relatively high levels of apparently natural background SF_6 have been detected in areas of volcanic and igneous rock (Busenberg and Plummer, 2000).

10.4.3.2 Atmospheric noble gases and stable isotopes

The solubility of ordinary atmospheric noble gases (neon, argon, krypton, and xenon) in water is temperature dependent (Benson, 1973). The stable isotope composition of precipitation (δ^2H, $\delta^{18}O$) also depends on temperature. If variations in these constituents can be related to a known history of temperature variation, then groundwater residence times can be estimated (Stute and Schlosser, 1993).

This approach to residence-time estimation has been most commonly applied on two very different timescales. The first is in utilizing the annual temperature cycle. The second is the global glacial–interglacial transition at ~15 ka. The approach of taking samples along a groundwater flow path and matching it to the seasonal recharge temperature signal generally works best when applied to aquifer recharge from a surface-water body (Beyerle *et al.*, 1999; Sugisaki, 1961), because under conditions of general vertical recharge the vadose zone damps much of the annual signal, and vertical mixing during water-table fluctuations smoothes the annual signal.

The noble-gas concentration method has been applied to investigating the temperature history on the glacial–interglacial timescale around the world (Stute and Schlosser, 1993). Although it is most commonly used in conjunction with independent dating of the groundwater, the magnitude and timing of the noble-gas temperature signal are well enough established, so that, when other dating means are not available or successful, it can be used to define the position of water recharged during the climate transition (Blavoux *et al.*, 1993; Clark *et al.*, 1997; Zuber *et al.*, 2000).

10.4.3.3 Nonatmospheric noble gases

The following sections comprise a brief description of noble-gas isotopes (^3He, ^4He, ^{21}Ne, and ^{40}Ar), where measured concentrations in groundwaters were found to be in excess of solubility equilibrium with the atmosphere (air-saturated water (ASW)). Such excesses allow the use of these isotopes as natural tracers of groundwater flow (as opposed to those such as ^{20}Ne, ^{22}Ne, ^{36}Ar, and ^{38}Ar that exhibit an atmospheric contribution only).

(i)Helium isotopes

Helium-3. The concentration of ^3He in groundwater frequently exceeds that of ASW values (cf. Section 10.4.3.2). These excesses can be up to several orders of magnitude greater than the ASW concentration in old groundwater (e.g., Castro *et al.*, 1998a,b). They can derive from a number of ^3He sources. Typically, with the exception of recharge areas and superficial aquifers with fast-flowing (young) waters, the main source of ^3He over time is nucleogenic. Nucleogenic production of ^3He is the result of a series of reactions (Andrews and Kay, 1982a,b; Lal, 1987; Mamyrin and Tolstikhin, 1984; Morrison and Pine, 1955): (a) spontaneous and neutron-induced fission of uranium isotopes and reaction of α-particles with the nuclei of light elements give rise to a subsurface neutron flux; (b) some of these neutrons reach epithermal energies and react with nuclei of the light isotope of lithium (^{6}Li) to produce α-particles and tritium:

$$^6\text{Li}(3\text{p}, 3\text{n}, 3\text{e}^-) + \text{n} \rightarrow \alpha(2\text{p}, 2\text{n}) + {}^3\text{H}(1\text{p}, 2\text{n}, 1\text{e}^-)$$

(c) ^{3}H decays (half-life of 12.32 yr) by the emission of β^- (decay of a neutron into a proton and electron) to the stable isotope ^{3}He:

$$^3\text{H} \xrightarrow{\beta^-} {}^3\text{He}(2\text{p}, 1\text{n}, 2\text{e}^-)$$

Reactions between the light isotopes of lithium and epithermal neutrons yield almost all terrestrial nucleogenic ^{3}He, including that found in excess of ASW in most groundwater reservoirs. Nucleogenic ^{3}He can also be produced through ^{7}Li reactions (^{7}Li$(\alpha, {}^3\text{H})^8$Be; ^{3}H$(\beta^-)^3$He; ^{7}Li$(\gamma, {}^3\text{H})^4$He, ^{3}H $(\beta^-)^3$He), but the amount produced is negligible compared to that produced through ^{6}Li (Gerling *et al.*, 1971; Kunz and Schintlmeister, 1965; Mamyrin and Tolstikhin, 1984). Excesses of nucleogenic ^{3}He in groundwater can result from *in situ* production (i.e., produced in the reservoir rock of the aquifer itself) and/or can be produced in deeper layers or deeper crust (external source). In the latter case, ^{3}He must be transported to the upper aquifers either through advection, dispersion, and/or diffusion. The concentration of ^{3}He in groundwater will typically increase with distance from the recharge area and/or over time. In tectonically active areas a mantle component of ^{3}He may also be present (Oxburgh *et al.*, 1986), reaching groundwater reservoirs either directly through igneous intrusions or through water transport processes such as advection, dispersion, and diffusion.

In addition to the nucleogenic and mantle component, tritiogenic ^{3}He resulting from β-decay of natural (produced mainly in the upper atmosphere through the bombardment of nitrogen by the flux of neutrons in cosmic radiation) and bomb ^{3}H (as a result of thermonuclear testing in the 1950s and 1960s) (see Section 10.4.2.4) can be present. This bomb-tritiogenic ^{3}He component, when significant, is present in very young (tens of years), fast-flowing waters.

Helium-4. This isotope is also commonly found in excess of ASW concentrations in groundwater; excesses are typically higher in older than in younger groundwaters. Unlike ^{3}He excesses, ^{4}He result directly from radioactive decay of ^{235}U, ^{238}U, and ^{232}Th. The radioactive decay of uranium and thorium series elements yields α-particles (2p, 2n) that rapidly acquire two electrons, thus turning into stable ^{4}He (2p, 2n, 2e$^-$). Because production of ^{4}He is neither dependent on the nuclei of light elements nor on neutron flux, the radiogenic ^{4}He production rate in the crust tends to be greater and more homogeneous than the nucleogenic production of ^{3}He (Martel *et al.*, 1990). As a result, the

observed excesses of radiogenic ^{4}He in groundwater tend to be more uniform than those of nucleogenic ^{3}He. Assuming a homogeneous element distribution in the rocks, the radiogenic production rate of ^{4}He is given by

$$\text{p}(^4\text{He}) = 5.39 \times 10^{-18}[\text{U}] + 1.28$$
$$\times 10^{-18}[\text{Th}]\ \text{mol g}_{\text{rock}}^{-1}\ \text{yr}^{-1}$$

where [U] and [Th] are the concentrations of U and Th in the rock in ppm (Steiger and Jäger, 1977). Radiogenic ^{4}He can be produced *in situ* or have an origin external to the aquifer.

^{3}He/ ^{4}He ratio. The measurement of the ^{3}He/^{4}He ratio (R) in groundwaters is particularly important, because it reflects the relative importance of crustal (radiogenic/nucleogenic) and mantle helium sources. Typical crustal rock ^{3}He/^{4}He production ratios have been estimated to be 1×10^{-8} (Mamyrin and Tolstikhin, 1984). Other authors (e.g., Kennedy *et al.*, 1984) consider a wider range of typical crustal production ratios, with $0.01 < R/R_a < 0.1$ (atmospheric ^{3}He/^{4}He ratio $R_a = 1.384 \times 10^{-6}$; Clarke *et al.*, 1976). Much higher ^{3}He/^{4}He values have been observed in the mantle; the most common is 1.2×10^{-5} (Craig and Lupton, 1981). However, at specific sites such as Hawaii, mantle ^{3}He/^{4}He ratios can be up to a factor of 5 higher than this common mid-ocean ridge basalt (MORB) value (Allègre *et al.*, 1983; Craig and Lupton, 1976; Kyser and Rison, 1982; Rison and Craig, 1983).

In order to use helium isotopes as natural tracers of groundwater flow and, in particular, in order to estimate groundwater ages, it is necessary to separate different nonatmospheric helium components for both isotopes. The major components are ^{3}He and ^{4}He from atmospheric equilibrium and from incorporation of "excess air" (Heaton and Vogel, 1981; Weyhenmeyer *et al.*, 2000), ^{3}He and ^{4}He of crustal origin (i.e., nucleogenic and radiogenic), ^{3}He of mantle origin, and tritiogenic ^{3}He. Methods for separating these components have been described by Stute *et al.* (1992b), Weise (1986), and Weise and Moser (1987).

(ii)Neon-21

Neon-21excesses relative to ASW concentrations have also been observed in old groundwaters. However, the excesses are not comparable to those observed for helium; they are only in the order of a few to a few tens of percent. This excess is attributed to the nucleogenic production of ^{21}Ne in crustal rocks. The reaction ^{18}O$(\alpha, \text{n})^{21}$Ne accounts for 97% of the total nucleogenic ^{21}Ne. Magnesium accounts for the remaining production of nucleogenic ^{21}Ne via the reaction ^{24}Mg$(\text{n}, \alpha)^{21}$Ne (Wetherill, 1954). Although only one study has reported the presence of nucleogenic ^{21}Ne in groundwaters (Castro *et al.*, 1998b), it is possible that, as more measurements of ^{21}Ne in old deep groundwaters become available, the presence of nucleogenic ^{21}Ne will prove to be more common in groundwater reservoirs than is known at present. Numerous studies have reported considerable excesses of nucleogenic ^{21}Ne in oil and gas fields (Ballentine and O'Nions, 1991; Ballentine *et al.*, 1996; Hiyagon and Kennedy, 1992; Kennedy *et al.*, 1985). There is a general consensus that noble gases found in these systems are transported to oil and gas reservoirs by groundwater. If so, similar ^{21}Ne excesses should be expected in neighboring groundwaters in amounts close to those in gas and oil fields.

The presence of nucleogenic ^{21}Ne is generally identified through comparison with the atmospheric ^{21}Ne/^{22}Ne ratio of 0.0290. A typical crustal production ^{21}Ne/^{22}Ne ratio is 0.47 ± 0.02 (Shukolyukov *et al.*, 1973; Kennedy *et al.*, 1990). The total present-day production rate of ^{21}Ne has been calculated for typical crustal rock by Bottomley *et al.* (1984) to be

$$p(^{21}\text{Ne}) = (3.572 \times 10^{-25})[\text{U}] + 1.767 \times 10^{-25}[\text{Th}]$$

where [U] and [Th] are the concentrations of U and Th in the rock in ppm. The values for isotopic composition and decay constants are taken from Steiger and Jäger (1977).

A ^{21}Ne mantle component has not been identified in groundwaters. Its presence was reported for the first time in 1991 in hydrocarbon gas reservoirs within the Vienna Basin (Ballentine and O'Nions, 1991).

(iii) Argon-40

Excess ^{40}Ar, up to 35% of ASW, has also been found in deep, old groundwaters in sedimentary basins (Castro *et al.*, 1998b; Torgersen *et al.*, 1989). Radiogenic production of ^{40}Ar through radioactive decay of ^{40}K by electron capture is responsible for these excesses. Eleven percent of ^{40}K decay produces ^{40}Ar; the remainder produces ^{40}Ca by the emission of β^{-}-particles (Faure, 1986). The presence of a radiogenic ^{40}Ar component is easily identified and quantified by comparison with the ^{40}Ar/^{36}Ar atmospheric ratio (295.5; Ozima and Podosek, 1983). The highest measured value of this ratio due to the presence of radiogenic ^{40}Ar in groundwaters is 471.5. At present, no resolvable mantle component has been identified.

Assuming a homogeneous element distribution in the rocks, the radiogenic production rate of ^{40}Ar is

$$P(^{40}\text{Ar}) = 1.73 \times 10^{-22}[\text{K}] \text{ mol } g_{\text{rock}}^{-1} \text{ yr}^{-1}$$

where [K] is the concentration of K in rocks in ppm, using current values of isotopic composition and decay constants from Steiger and Jäger (1977). In a manner similar to the helium and neon isotopes, ^{40}Ar in groundwater reservoirs can result both from release of ^{40}Ar from minerals forming the aquifer and from production and release deeper in the crust followed by upward transport.

10.5 LESSONS FROM APPLYING GEOCHEMICAL AGE TRACERS TO SUBSURFACE FLOW AND TRANSPORT

10.5.1 Introduction

The most common and traditional tracers used in groundwater studies in the last few decades (e.g., ^{3}H, ^{14}C, and ^{36}Cl), as well as some relatively new and promising tracers for use in quantitative groundwater dating (e.g., ^{32}Si, ^{81}Kr, and ^{4}He), have been discussed above. The practical applications and information to be gained from these environmental tracers are multiple and differ, depending on the tracers to be used and on the hydrogeological problem to be treated. Typically and depending on the scale of the study area (e.g., regional versus local) as well as on the hydraulic properties of the formation(s) (e.g., high versus low to very low hydraulic conductivity), certain tracers will be better suited than others. Radioisotopes with short half-lives (e.g., ^{85}Kr and ^{3}H) are suitable for dating young/fast-flowing waters ($\leq$40 yr). These radioisotopes are, therefore, ideal for the study of aquifers with high hydraulic conductivities or, alternatively, to the study of recharge areas in regional aquifer systems (Schlosser *et al.*, 1989; Solomon *et al.*, 1993). Tracers with intermediate half-lives (e.g., ^{14}C and ^{36}Cl) are typically used to date submodern (>1,000 yr) and old groundwaters up to 10^{6} yr. They are suited to the study of aquifers of intermediate hydraulic conductivity, i.e., study at the aquifer scale and certain regional systems (Jacobson *et al.*, 1998; Pearson and White, 1967; Phillips *et al.*, 1986). In contrast, stable tracers such as ^{4}He have no age limitations and can be used within all age ranges; as a result, they are particularly suited to the study of regional groundwater systems where very old waters may be present, as well as the study of groundwater flow dynamics in aquitards where water can be almost immobile (Andrews *et al.*, 1985; Andrews and Lee, 1979; Bethke *et al.*, 1999; Castro *et al.*, 1998a, 2000; Torgersen and Ivey, 1985). Silicon-32 potentially covers an age from a few tens of years to 1,000 yr, a range that is problematic for most tracers.

Two main approaches to derive information from such tracers in a groundwater system are possible: (i) independent and direct use of one particular tracer by analyzing variations in its concentration in space and/or its evolution over time and (ii) a more indirect use through incorporation of those tracers into analytical or numerical transport models. Although both approaches allow estimation of velocity fields, groundwater ages, recharge rates, and identification of mixing of different water bodies, the second approach will generally provide more complete information for a particular groundwater flow systems, as it usually allows for a better representation of the real system (see below). In all cases, the use of a "multitracer" approach (i.e., the use of several tracers simultaneously) provides more definitive answers compared to those based on only one single tracer.

In the following sections, we will briefly describe both approaches, their limitations and pitfalls, as well as information essential to their proper use. These sections will be followed by a discussion of several case studies that represent groundwater flow at different scales, ranging from regional to local, as well as within the vadose zone.

10.5.2 Approaches

10.5.2.1 Direct groundwater age estimation

This method is commonly used with radioisotope tracers that undergo radioactive decay. The tracer concentration will decrease over time following an exponential decay law:

$$\frac{dN}{dt} = -\lambda N$$

where N is the number of atoms of the radioactive tracer, t the time elapsed since radioactive decay started at some initial time t_0, and λ is the decay constant of the tracer. Integrating this equation yields

$$N(t) = N_0 \, e^{-\lambda t}$$

where N_0 is the initial number of radioactive tracer atoms at time $t = 0$. This equation can be rearranged and expressed in terms of time elapsed since decay started:

$$t = -\frac{1}{\lambda} \ln \frac{N_t}{N_0}$$

If radioactive decay is the dominant process causing reduction/change in tracer concentration, time t represents the "groundwater age" (Cook and Böhlke, 2000). It corresponds to the time that has elapsed since the water became isolated from the Earth's atmosphere, i.e., the time elapsed since recharge took place (Busenberg and Plummer, 1992; Davis and Bentley, 1982). One must be cautious in interpreting such groundwater age estimates. Such ages may represent an "apparent" rather than the "real" groundwater age (Park et al., 2002). A variety of processes may create these discrepancies. For example, ^{14}C-determined groundwater ages may be influenced by the introduction of inorganic "dead" carbon by the dissolution of calcite or dolomite, resulting in apparent groundwater that are too old. Numerous geochemical models have been developed to account for ^{14}C mass transfer and its impact on groundwater ^{14}C concentration (e.g., Fontes and Garnier, 1979; Mook, 1976; Pearson and White, 1967; Plummer et al., 1991; Tamers, 1967). Assumptions inherent in such geochemical models introduce new sources of uncertainty. Some authors, therefore, prefer to the use the term "model age" (Hinkle, 1996). Groundwater mixing, diffusion, or dispersion may also give rise to erroneous age estimates. One way to identify the effects of groundwater mixing when using ^{14}C ages is to use age constraints from the concentration of CFCs or tritium in tandem with the ^{14}C data.

Complications in the estimation of groundwater ages are not unique to ^{14}C. For example, diffusion can create major errors in estimated ages when applying the $^{3}H/^{3}He$ method in areas where the groundwater velocity is less than ~ 0.01 m yr^{-1}. In such cases ^{3}He will diffuse upwards, and little information regarding travel times will be preserved in the ^{3}He concentrations (Schlosser et al., 1989). Similar complications are inherent in the use of other tracers as well (see, e.g., Phillips, 2000; Sudicky and Frind, 1981; Varni and Carrera, 1998).

Some conservative tracers, such as ^{4}He, have also been used for the direct estimation of groundwater ages (Andrews et al., 1982; Bottomley et al., 1984; Marine, 1979; Torgersen, 1980). The general concept of using ^{4}He as a dating tool is simple, and opposite to that of radioactive tracers. As ^{4}He is produced, minerals release it into groundwater. The longer the groundwater in contact with these minerals, the greater the ^{4}He accumulation in groundwater. If ^{4}He release rates can be estimated (either through direct measurement or through comparison of $^{3}He/^{4}He$ ratios in waters and host rocks; e.g., Castro et al., 2000; Solomon et al., 1996); it should be possible to estimate groundwater residence times as the latter should be proportional to ^{4}He concentrations. However, these methods have systematically led to apparent inconsistencies in water ages calculated using other tracers. In particular, ^{4}He methods typically yield water ages that are much older than ^{14}C ages, sometimes by up to several orders of magnitude (e.g., Andrews and Kay, 1982a; Andrews et al., 1982; Bottomley et al., 1984). Numerous authors have observed that the ^{4}He excesses in groundwater are generally larger than can be supported by the steady-state release of ^{4}He from the reservoir rocks to groundwater. Such ^{4}He excesses have been explained by the presence of an upward flux originating mainly in the deep crust (Bethke et al., 1999; Castro et al., 1998a,b, 2000; Marty et al., 1993; Stute et al., 1992b; Torgersen and

Clarke, 1985; Torgersen and Ivey, 1985). Not considering the contribution of ^{4}He from external sources is probably the reason for many of the discrepancies between ages estimated from ^{4}He data and those estimated from other tracers (Castro *et al.*, 1998b). The presence of such an external flux invalidates the use of the ^{4}He method, except in areas where *in situ* production is the only source of the observed excesses, or where its contribution can be determined precisely. Although important information can be gained from such direct age estimation methods, it is important to view these groundwater ages with a critical eye. Accounting for multiple sources of ^{4}He within a complex hydrological system requires the incorporation of data for the concentration of this tracer into a transport model. Helium-4 can be used as a powerful quantitative dating tool for groundwater (Castro *et al.*, 1998a).

10.5.2.2   Modeling techniques: analytical versus numerical

As mentioned above, direct groundwater age estimation methods may not be well suited for all study areas or with all groundwater tracers. Often, groundwater age estimates are more easily and correctly achieved through incorporation data for natural tracers in analytical and/or numerical transport models. This is particularly true when multiple sources contribute to particular tracers, or when an accurate hydrogeological representation is needed for complex groundwater systems.

Traditionally, the use of analytical models in environmental tracer studies has been far more widespread than numerical models. There are a number of reasons for this. Analytical models are easier to use and manipulate than numerical ones; they require less hydrodynamic information and/or field data; and the time required to build a transport model is much less than for a numerical model. Multiple examples of such models can be found in the literature for various tracers: ^{3}H, ^{4}He, ^{14}C, and ^{36}Cl (e.g., Castro *et al.*, 2000; Nolte *et al.*, 1991; Schlosser *et al.*, 1989; Solomon *et al.*, 1996; Stute *et al.*, 1992b; Torgersen and Ivey, 1985). Generally, these models are either applied to a single aquifer in porous or fractured media or to one particular area within the aquifer such as recharge or discharge areas.

The reliability of transport model results depends on how well the model approximates the field situation. They are limited in their ability to incorporate complexities such as heterogeneity in hydraulic properties and three-dimensional flow paths. In order to deal with more realistic situations, or when simulating transport of a tracer at the regional scale (e.g., the scale of an entire sedimentary basin), it is necessary to solve the mathematical model approximately using numerical techniques.

Many different kinds of transport models have been developed, and have been extensively described in the literature (Bethke *et al.*, 1993; Bredehoeft and Pinder, 1973; Goblet, 1999; Konikow and Bredehoeft, 1978; Prickett *et al.*, 1981; Simunek *et al.*, 1999). Although the incorporation of environmental tracers in numerical transport models is not new (see, e.g., Pearson *et al.*, 1983), the effort has been made only recently to simulate the transport of these tracers, in particular stable and conservative tracers such as ^{3}He, ^{4}He, ^{21}Ne, and ^{40}Ar, at a regional scale (e.g., Bethke *et al.*, 1999; Castro *et al.*, 1998a; Zhao *et al.*, 1998). In an attempt to reduce uncertainties present in results obtained independently through groundwater flow models, transport models simulating the distribution of ^{14}C and ^{4}He have been coupled with the latter to obtain recharge rates and groundwater residence times (Castro and Goblet, 2003; Zhu, 2000).

Caution should also be used when estimating hydraulic properties and groundwater ages through transport models. Although analytical and numerical models (in particular) allow for a much better representation of the geometry and the heterogeneity of hydraulic properties of groundwater systems, a number of assumptions and simplifications still have to be made when imposing boundary conditions and hydraulic parameters on the flow and transport domain. These assumptions give rise to sources of uncertainty, resulting from (i) the inability to properly and fully describe the internal properties and boundary conditions of a groundwater system and (ii) the lack of sufficient information for all of the parameters to be included in such models (see, e.g., Konikow and Bredehoeft, 1992; Maloszewski and Zuber, 1993). In this sense, estimated groundwater ages will also carry an error/uncertainty that is difficult to assess. Furthermore, numerical models with a large number of adjustable parameters inherently produce nonunique results. Confidence in model results can be increased through the use/ simulation of multiple tracers within the same groundwater flow system and the use of coupled water flow and transport models.

10.5.2.3   Identification of sources and sinks of a particular tracer

Independent of the method used, proper identification of all sources and sinks of a particular tracer is essential for estimating groundwater residence times. If all important sources (e.g., internal production and external sources) and sinks (e.g., radioactive decay) are not taken into account, erroneous water ages will be obtained. For example, exclusion of an external source of ^{4}He will lead to great discrepancies between ^{4}He and ^{14}C "ages." Commonly, most tracers have multiple sources and sinks (e.g., Loosli *et al.*, 2000; Phillips, 2000; Solomon, 2000; Solomon and Cook, 2000). Some sources have a dominant impact on tracer concentrations; others have only a minor or negligible impact. Moreover, when considering an entire aquifer, one source may be dominant in one part of the system and negligible in others. For example, *in situ* production of ^{3}He and ^{4}He is of major importance near recharge areas, and of lesser importance in basin centers and in discharge areas (Castro *et al.*, 2000). Heterogeneities of hydraulic properties within a formation may also affect the relative impact of a particular source or sink in different parts of the same system; it is, therefore, necessary to account fully for such variations. The use of numerical methods facilitates this task, as variations may be imposed through boundary conditions of the transport model (Bethke *et al.*, 1999; Castro *et al.*, 1998a).

10.5.2.4 Defining boundary conditions (modeling approach)

The imposition of proper boundary conditions is another essential component of groundwater age estimation when using a modeling approach. The task is to account for the effects of areas outside the region of interest on the system being modeled (de Marsily, 1986; Domenico and Schwartz, 1998). Three main types of boundary conditions can be considered in a transport model: (i) imposed concentrations within a particular region—such a boundary condition is independent of the groundwater flow regime; (ii) imposed mass flux entering or exiting the system; and (iii) imposed sink or source terms for a tracer within the modeled domain. When conducting simulations in transient state, the imposition of an initial condition is required. Inappropriate model boundary conditions can lead to important differences between actual and simulated flow regimes and thus can lead to erroneous age estimate in real hydrological systems.

10.6 TRACERS AT THE REGIONAL SCALE

10.6.1 Introduction

This section deals with environmental tracers applied to regional systems, i.e., to the study of groundwater flow regimes in entire sedimentary systems. We will focus on studies employing numerical modeling that have been conducted in two classical sedimentary basin-scale systems: the Paris Basin in France (Castro *et al.*, 1998a,b) and the Great Artesian Basin (GAB) in Australia (Bethke *et al.*, 1999). These multi-layered aquifer systems are ideal to illustrate the behavior of conservative tracers such as noble gases and their

relation to depth and recharge distance. The Paris Basin study illustrates how a conservative tracer such as helium can be used as a quantitative dating tool through the use of coupled transport and groundwater flow models, and how a multitracer approach (e.g., ^{3}He, ^{4}He, ^{21}Ne, and ^{40}Ar) can be used to identify dominant transport processes such as advection, dispersion, and/or diffusion. Both studies illustrate the impact of external ^{4}He fluxes and hydraulic conductivities on ^{4}He concentrations. In addition, Bethke *et al.* (1999) have demonstrated the impact of diffusion coefficients on aquitard concentrations. Applications of the direct dating method using ^{81}Kr, ^{36}Cl, and ^{129}I in the GAB are also presented (e.g., Bentley *et al.*, 1986b; Lehmann *et al.*, 2002). These are an important complement to numerical modeling studies.

10.6.2 Examples of Applications

10.6.2.1 Noble-gas isotopes—Paris Basin

The Paris Basin, situated in northeastern and central France, is an intracratonic depression with a diameter of 600 km and maximum depth of 3,200 m. The system consists of seven major aquifers (from top to bottom: Ypresian, Albian, Neocomian, Portlandian, Lusitanian, Dogger, and Trias) separated by low to very low hydraulic conductivity aquitards (cf. Figures 3(a) and (b)). Aquifers are recharged at outcrops to the east and southeast, and gravitational flow is toward the northwest. Concentrations and isotopic compositions of helium, neon, and argon measured in five of these aquifers show excess ^{3}He (^{3}He)$_{exc}$, ^{4}He (^{4}He)$_{exc}$, and ^{40}Ar, as well as vertical concentration gradients of these isotopes throughout the basin (Castro *et al.*, 1998a,b). Water of the Dogger and the Trias formations also has a ^{21}Ne excess above ASW values (Figure 4). The dominant source of these excesses is radiogenic/nucleogenic production in the deep crust underlying the basin. This external flux is responsible for the vertical concentration gradients. *In situ* production in the basin appears to be a minor source (13% at most) of these isotopes. They are transported vertically throughout the entire basin by advection, dispersion, and diffusion. The spatial and vertical variability throughout the basin of the ^{4}He/^{40}Ar (0.69–70) and the ^{21}Ne/^{40}Ar ($(8–23)\times10^{-7}$) ratios, as compared to the crustal production ratios of 4 ± 3 and 0.96×10^{-7}, respectively, allow for identification of the dominant transport processes of each isotope through aquitards. An interesting feature of the ^{4}He/^{40}Ar ratio was observed in the Dogger, where it exhibits a great spatial variability. Values close to the crustal production ratio occur in water sampled near major faults (where transport by advection is dominant), but much higher values

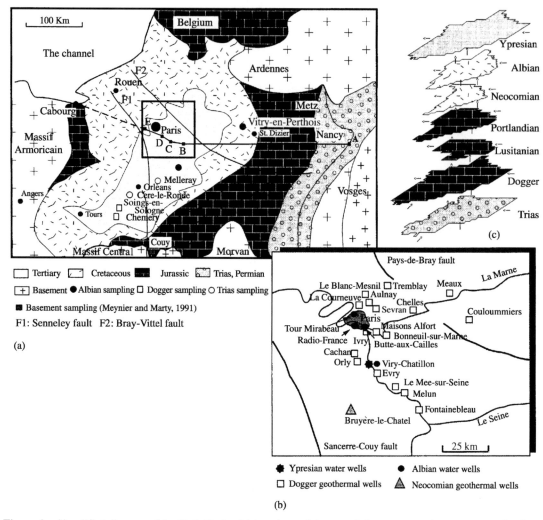

Figure 3 Simplified diagram of the Paris Basin: (a) locations of the sampled areas; central square—samples from the Paris region (Meynier and Marty, 1990); (b) central square enlarged; and (c) simplified diagram of the main aquifers in the Paris Basin—arrows show exchanges of water (entering and leaving) between the aquifers (Castro *et al.*, 1998b) (reproduced by permission of American Geophysical Union from *Water Resour. Res.* **1998**, *34*, 2443–2467).

are observed away from faulted areas (Figure 5). In the Trias, which is the deepest aquifer in direct contact with the bedrock, all ^{4}He/^{40}Ar ratios are similar to the crustal production ratio, suggesting that ^{4}He movement in aquitards is more independent of advection than ^{40}Ar and thus is transported mostly by diffusion, whereas ^{40}Ar is primarily transported by advection (see discussion by Castro *et al.*, 1998a). Neon-21 reflects an intermediate situation. The behavior of this isotope reflects differences in diffusion coefficients in water and differences in vertical concentration gradients. Similar ^{4}He/^{40}Ar deviations have also been observed in the GAB (Torgersen *et al.*, 1989). Numerical simulations of ^{4}He transport in the Paris Basin allowed the calculation of the advective, dispersive, and diffusive flux of ^{4}He throughout the aquitards (cf. Figure 6). They confirmed that diffusion is the main transport mechanism of

^{4}He through these formations, and that these fluxes decrease toward the surface. This decrease is directly related to a decrease in concentration gradient that is related to dilution by recharge water carrying atmospheric ^{4}He. This dilution effect, as well as the variation of the ^{4}He concentration with recharge distance and depth, can be clearly observed in the calibrated two-dimensional transport model along cross-section A–E (Figures 3(a) and 7).

Simulation of ^{4}He transport in the basin coupled with a groundwater flow model allowed the estimation of hydraulic conductivities for the different aquifers and, consequently, the estimation of groundwater residence times. Average turnover times for different aquifers are highly variable, ranging from 8,700 yr for the shallowest aquifer (Ypresian) to 30 Myr for the deepest (Trias). These calculated turnover times were derived

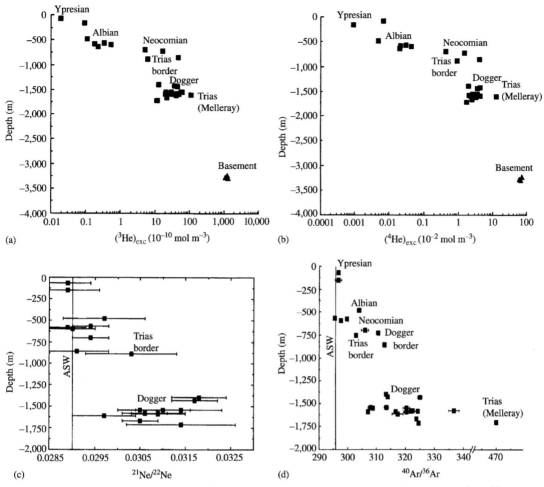

Figure 4 Evolution (versus depth) of: (a) $(^3He)_{exc}$ concentrations in the water; (b) $(^4He)_{exc}$; (c) $(^{21}Ne/^{22}Ne)$ ratio; and (d) $(^{40}Ar/^{36}Ar)$ ratio. The different aquifers are shown. All data are from Castro *et al.* (1998a) (solid squares) except those from the bedrock in figures 6(a) and (b) (Couy) (after Meynier and Marty, 1990), represented by solid triangles. The ASW value is given for $(^{21}Ne/^{22}Ne)$ and $(^{40}Ar/^{36}Ar)$ (Castro *et al.*, 1998b) (reproduced by permission of American Geophysical Union from *Water Resour. Res.* **1998**, *34*, 2443–2467).

from modeling ^4He concentrations in the Dogger aquifer ($\sim10^5$ yr for a corresponding thickness of 20 m). They agree with pure groundwater flow model results (Wei *et al.*, 1990). However, these ^4He groundwater ages are much shorter than those estimated from ^4He data by Marty *et al.* (1993). This discrepancy is due to the neglect of diffusive transport of ^4He by Marty *et al.* (1993). When vertical transport by diffusion is important, the amount of the tracer that moves upward into each aquifer is large compared to transport by advection only. Diffusive transport increases tracer concentrations, and thus leads to "apparent older" residence times if it is assumed that the tracer concentration is only proportional to groundwater age. Diffusion is particularly important in the deeper aquifers within a sedimentary basin. In shallow aquifers the opposite can be true. Significant amounts of ^3He and ^4He can leave the system to the atmosphere by diffusion. This leads

to "apparent younger ages" (e.g., Castro *et al.*, 1998a,b; Schlosser *et al.*, 1989). Simulations of ^3He, ^4He, and ^{40}Ar concentrations in the Paris Basin indicate that their crustal fluxes (in units of mol m^{-2} yr^{-1}) are of 4.33×10^{-13}, 4.0×10^{-6}, and 2.52×10^{-7}, respectively (Castro *et al.*, 1998b). The estimated basin ^4He flux is of the same order of magnitude as the estimated terrestrial crustal flux for this isotope (O'Nions and Oxburgh, 1983). This study also showed how $^4He/^{40}Ar$ ratios can help to determine low to very low hydraulic conductivities. Based on the value of this ratio, an upper limit for the hydraulic conductivity of 10^{-11} m s^{-1} was determined for the Lias aquitard.

10.6.2.2 Great Artesian Basin—noble-gas isotopes, ^{36}Cl, and ^{81}Kr

The GAB, a multi-layered aquifer system with a structure similar to that of the Paris Basin, is

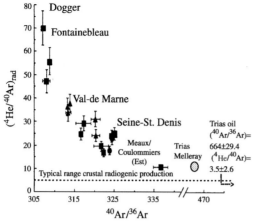

Figure 5 Evolution of $({}^{4}He/{}^{40}Ar)_{rad}$ ratio versus ${}^{40}Ae/{}^{36}Ar$ in Dogger samples; the value of the Trias sample at Melleray is also given, together with the typical range of radiogenic crustal production values; the position of the mean value measured in oil (Pinti and Marty, 1995) is also shown (Castro *et al.*, 1998b) (reproduced by permission of American Geophysical Union from *Water Resour. Res.* **1998**, *34*, 2443–2467).

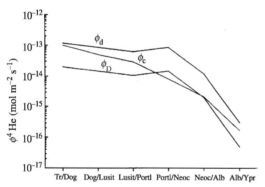

Figure 6 Calculated values of diffusive ϕ_d, advective ϕ_C, and dispersive ϕ_D fluxes of ${}^{4}He$ in each aquitard. Values are estimated on the same vertical line at the center of the basin using ${}^{4}He$ concentrations and vertical velocities obtained by fitting the model. Names on the *x*-axis indicate the aquitards between the named units. For example, Tr/Dog refers to the aquitards between the Trias and Dogger aquifers (Castro *et al.*, 1998a) (reproduced by permission of American Geophysical Union from *Water Resour. Res.* **1998**, *34*, 2467–2484).

located in Australia between the Great Dividing Range to the east and the Australian deserts to the west (Figure 8). The GAB occupies about one-fifth of the area of Australia and is one of the largest contiguous confined aquifer systems in the world. The basin has a long history of investigations using a variety of age tracing methods (Airey *et al.*, 1979; Calf and Habermehl, 1983). The basin was the site of one of the first field tests of ${}^{36}Cl$ behavior in regional-scale basin hydraulics, and ${}^{36}Cl$ data collection and interpretation has continued ever since (Bentley *et al.*, 1986b;

Love *et al.*, 2000; Torgersen *et al.*, 1991). In general, the data show the expected spatial variation: a relatively smooth, quasi-exponential decrease in ${}^{36}Cl$ concentration and ${}^{36}Cl/Cl$ ratio with flow distance (Figure 9). The distribution of ages calculated in these studies appears to be in general agreement with the understanding of flow in the basin developed through physical hydrology. The main source of uncertainty has been discriminating changes in chloride concentration caused by time-variable evapotranspiration in the recharge areas from changes caused by diffusion of chloride out of low-permeability strata, since these affect the age calculations differently (Andrews and Fontes, 1993; Phillips, 2000; Torgersen and Phillips, 1993).

The ${}^{36}Cl$ flow rate data were recently compared with data from the newly advanced ${}^{81}Kr$ method (Lehmann *et al.*, 2002b). Although caution must be used when evaluating a new method such as ${}^{81}Kr$, the correspondence of the residence-time estimates was generally good. Further, application of ${}^{81}Kr$ may help to resolve long-standing uncertainties in the groundwater age distribution of the GAB.

Parallel to the collection of age tracing data using cosmogenic radionuclides, radiogenic subsurface-produced noble gases have also been sampled Torgersen and Clarke, 1985; Torgersen *et al.*, 1992, 1989), allowing an unusually detailed inter-comparison of results. Recently, Bethke *et al.* (1999) have simulated ${}^{4}He$ transport in groundwaters of the GAB, using the ${}^{36}Cl$ results to help constrain the groundwater velocity field. Although (unlike the Paris Basin) ${}^{4}He$ was not directly used to estimate groundwater residence times, the relationship between the groundwater flow regime in the GAB, transport of ${}^{4}He$ within the basin, and the resulting ${}^{4}He$ distribution was examined. In this case, the water flow regime was considered "known," and the groundwater flow model was initially built by deriving all hydraulic parameters from the literature. The groundwater flow model results were used to simulate ${}^{4}He$ transport.

Bethke *et al.* (1999) simplified the hydrogeology of the GAB somewhat by using only four units, the J- and K-aquifers intercalated between two aquitards (Figure 10). However, these simplifications do not affect the response of ${}^{4}He$ to variations in the external fluxes or hydraulic conductivities imposed on aquitards, or on assumed diffusion coefficients in these formations.

The ${}^{4}He$ concentrations in the J-aquifer in simulations by Bethke *et al.* (1999) were obtained from previous investigations (Torgersen and Clarke, 1985; Torgersen *et al.*, 1992, 1989). These show ${}^{4}He$ excesses with respect to the ASW that are larger in the center of the basin than at the margins (Figure 8). For water less than 5×10^{4} yr old, excesses were attributed to *in situ* production; for older waters ($>10^{5}$ yr), the dominant ${}^{4}He$ source

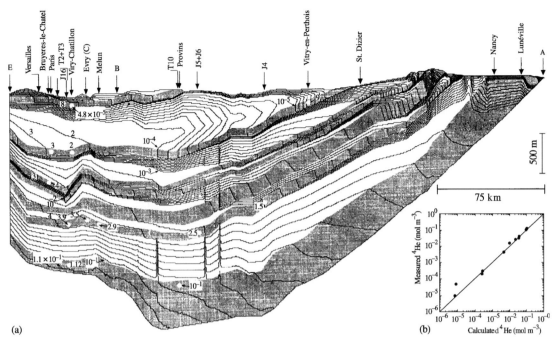

Figure 7 (a) Distribution of ^{4}He concentration contours in mol m^{-3}. Contours for values of 10^{-1}, 10^{-2}, 10^{-3}, 10^{-4}, and 10^{-5} mol m^{-3} are marked. Except for the curve for 1.1×10^{-1} mol m^{-3} in the Trias, all other contours express constant concentration variations of 1 unit inside each order of magnitude. The measured value of each sample is given together with some contours whose values are close to the measured ones. In these cases, only the molar number is given; (b) measured ^{4}He values plotted as a function of calculated values (Castro *et al.*, 1998a) (reproduced by permission of American Geophysical Union from *Water Resour. Res.* **1998**, *34*, 2467–2484).

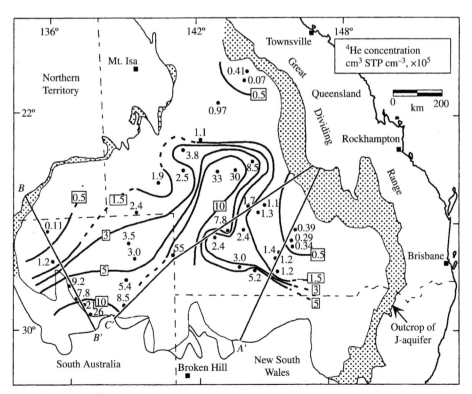

Figure 8 Distribution of ^{4}He in groundwater from J-aquifer of the GAB. Contours and data points show ^{4}He concentrations (10^{-5} cm^3 STP cm^{-3}) measured by Torgersen *et al.* (1992); contour lines are interpretive. Straight lines show flow paths A–A′, B–B′, and C–C′ (Bethke *et al.*, 1999) (reproduced by permission of American Geophysical Union from *J. Geophys. Res.* **1999**, *104*, 12999–13010).

was attributed to crustal degassing. Torgersen and Clarke (1985) estimated the crustal ^{4}He degassing flux in the GAB to be 5.12×10^{-18} mol cm^{-2} s^{-1}. A number of previous studies have reported average ^{4}He continental fluxes, ranging from 0.5×10^{-18} mol cm^{-2} s^{-1} to 13×10^{-18} mol cm^{-2} s^{-1}. The latter is a high estimate of the whole-crustal flux (see Torgersen *et al.*, 1989). Radiogenic ^{40}Ar excesses (compared to the ASW) were also observed, with ^{40}Ar/^{36}Ar ratios up to 325.91 (Torgersen *et al.*, 1989). As in the Paris Basin, high deviations of the ^{4}He/^{40}Ar ratio (up to 55) compared to the crustal production ratio also occur. In this case, differential release of ^{4}He and ^{40}Ar from rock to water, and a spatially and temporally variable release of the two nuclides due to tectonic stresses and igneous intrusions were offered as explanations for these deviations. Transport by diffusion was not considered by Torgersen *et al.* (1989).

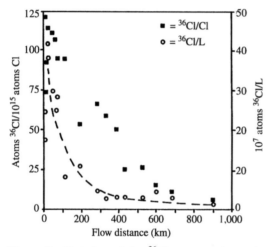

Figure 9 Variation of the ^{36}Cl concentration ratio with distance from the eastern margin of the GAB (Phillips, 1993) (reproduced by permission of Elsevier from *Appl. Geochem.* **1993**, *8*, 643–647).

Hydrologic residence times estimated by Bethke *et al.* (1999) range from 1.5×10^5 yr to 2×10^6 yr, in the central portion of the basin (cross-section C–C′) based on the previous ^{36}Cl studies, and also on physical hydrology. Here, groundwater flows to the west and southwest, and ^{4}He concentrations, calculated assuming a crustal flux to the J-aquifer of 3×10^{-18} mol cm^{-2} s^{-1}, increase to the southeast (Figure 10). The most notable outcome of their simulations was that a set of two-dimensional models, in which hydraulic conductivity was estimated based on ^{36}Cl data, produced a reasonable match to the general trends and irregularities in the ^{4}He distribution. These results support the combined application of multiple tracers and numerical transport modeling as a key to understanding the fluxes of water and geochemically important solutes in the geological environment. Results of sensitivity analyses show the ^{4}He concentrations at the top of the J-aquifer that were predicted for differing fluxes of ^{4}He into the base of the aquifer system (Figure 11). It is clear that the ^{4}He concentration at the top of the J-aquifer increases with higher crustal fluxes. By contrast, an increase in the hydraulic conductivity within aquifers leads to a decrease in ^{4}He concentrations, owing to dilution by recharge water carrying atmospheric ^{4}He. An increase in diffusion coefficients within aquitards will have an opposite effect on upper aquifers, as more ^{4}He will be delivered not only through advection but also through diffusion. Bethke *et al.* (1999) concluded that a crustal flux of 3×10^{-18} mol cm^{-2} s^{-1} best explains the ^{4}He concentrations in the GAB. This value is about half of the production rate for the whole crust.

10.6.2.3 Implications at the regional scale

Simultaneous combination of a number of natural tracer concentrations (e.g., He, Ne, Ar, and ^{36}Cl), isotopic ratios, and modeling of groundwater flow

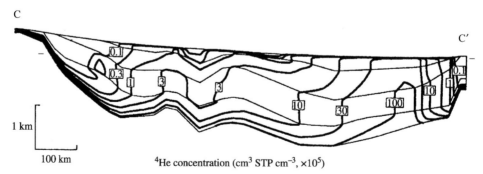

Figure 10 Calculated model of helium transport for cross-section through GAB along flow line C–C′ in the central subbasin, from the Great Dividing Range to near Lake Eyre. Vertical exaggeration is ~100 times. Stratigraphic units, bottom to top, are J-aquifer, confining layers, K-aquifer, and overlying beds. Note logarithmic spacing of ^{4}He contours (Bethke *et al.*, 1999) (reproduced by permission of American Geophysical Union from *J. Geophys. Res.* **1999**, *104*, 12999–13010).

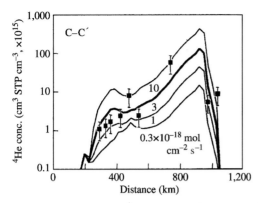

Figure 11 Variation in ^{4}He concentration at top of J-aquifer along flow line C–C′as a function of basal ^{4}He flux. Bold line corresponds to Figure 10. Measured values are from Torgersen and Clarke (1985) and Torgersen *et al.* (1992) (Bethke *et al.*, 1999) (reproduced by permission of American Geophysical Union from *J. Geophys. Res.* **1999**, *104*, 12999–13010).

and mass transport within entire sedimentary systems leads to a tremendous improvement in knowledge of tracer behavior and in the dynamics of these waters as a whole. This understanding would not be possible without such a multitracer modeling approach in a system where extreme contrasts in hydraulic conductivity are present. For example, initial studies of noble gases in the Paris Basin concentrated on analyzing the concentration of gases in one aquifer (Marty *et al.*, 1988). This approach did not indicate the external contribution of the noble gases (helium in particular). Noble gases were thought to be solely the result of *in situ* production. Groundwater ages were estimated based on *in situ* production rates, and the aquifer was thought to be a "closed," nearly static system. Subsequent sampling at multiple levels across the Paris Basin (at different depths and recharge distances) and isotopic analysis revealed that the groundwater system is very dynamic and interactive from a geochemical and hydrodynamic point of view. This dynamism is evident both at the scale of aquifers and aquitards within the system, and at the larger scale of the Earth's continental crust. Indeed, the horizontal and vertical distribution of ^{4}He/^{40}Ar and ^{21}Ne/^{40}Ar within the deepest aquifers emphasize the importance of active, advective "crustal-scale" transport (Castro *et al.*, 1998b). Concentrations and isotopic ratios also reveal combined upward movement via advection, dispersion, and diffusion of gases through successive formations. The horizontal and vertical variations of ^{4}He/^{40}Ar and ^{21}Ne/^{40}Ar highlight a dominant upward diffusive transport of helium isotopes through low and very low permeability formations. This illustrates how such ratios provide insight into the ranges of hydraulic conductivities within such formations, information that is unobtainable from independent

hydrodynamic studies. For example, strongly fractionated ^{4}He/^{40}Ar ratios indicate the presence of underlying aquitards with hydraulic conductivities $\sim 10^{-11}$ m s^{-1} or less. Combined analyses of isotope ratios (actually, absence of fractionated isotope ratios) can identify areas where advective transport dominates over diffusion, as in areas where underlying formations exhibit relatively large vertical hydraulic conductivities or where they are close to important faults. In contrast, strongly fractionated isotope ratios indicate that diffusion is the dominant transport process.

At the origin of these regional studies, there is the question of the "age" of groundwaters and our quest to "accurately determine groundwater ages." Knowledge of the age of a groundwater implies:

- understanding of how a particular groundwater system functions;
- all the hydrodynamic parameters in a particular system, with relative accuracy (e.g., distribution of hydraulic head, hydraulic conductivity, and porosity);
- recharge (infiltration) rates;
- understanding of how our groundwater system relates to the "outside world" through the imposition of boundary conditions represented through numerical modeling; and
- understanding and quantification of water flow and mass interactions taking place between subsystems (aquifers and aquitards) within our major groundwater system of interest.

In one simple phrase, to be able to truly determine the age of a single sample of groundwater implies that we are able to understand precisely and fully the functioning of the system from both physical and geochemical points of view. Such a goal has not been accomplished. The combined regional multitracer/modeling approach represents one new important step in this (iterative) process. Perhaps more than at any other scale, regional studies exemplify the interdependence of geochemical tracers and processes driving the dynamics of groundwater, in that they integrate the long-term response of quasistatic formations (aquitards) with that of much more dynamic formations (aquifers).

10.7 TRACERS AT THE AQUIFER SCALE

10.7.1 Introduction

There are many examples in the literature where a variety of tracers such as ^{37}Ar, ^{39}Ar, ^{36}Cl, ^{14}C, ^{3}He, and ^{4}He have been used to date groundwater at the aquifer scale. Notable examples include the Ojo Alamo aquifer in the San Juan Basin, New Mexico (Phillips *et al.*, 1989; Stute *et al.*, 1995), the Stampriet and Uitenhage aquifers in Southern Africa (Heaton *et al.*, 1983; Vogel *et al.*, 1982), the Carrizo aquifer in Texas (Andrews and Pearson, 1984; Castro and Goblet, 2003; Castro *et al.*, 2000; Pearson and White, 1967), and the Milk River aquifer in Ontario, Canada (Andrews *et al.*, 1991; Fröhlich *et al.*, 1991; Hendry and Schwartz, 1988; Nolte *et al.*, 1991; Phillips *et al.*, 1986; Schwartz and Muehlenbachs, 1979). Here, we illustrate the use of natural tracers in the Carrizo aquifer, where the measured tracers include ^{14}C, ^{3}He, ^{4}He, atmospheric noble gases, and ^{36}Cl. Confined aquifers such as the Carrizo are ideal to observe, analyze, and discuss groundwater mixing with surrounding formations. In addition, while these two aquifers have been the object of extensive study for several decades, our

understanding of tracer behavior in these systems is still an ongoing process (Heaton *et al.*, 1983).

10.7.2 Carrizo Aquifer

The Carrizo aquifer, a major groundwater system, is part of a thick regressive sequence of fluvial, deltaic, and marine terrigenous units in the Rio Grande Embayment area of South Texas along the northwestern margin of the Gulf Coast Basin (Figure 12(a)). This aquifer has been particularly well studied for its content of environmental tracers in Atascosa and McMullen Counties south of San Antonio (Andrews and Pearson, 1984; Castro and Goblet, 2003; Castro *et al.*, 2000; Cowart, 1975; Cowart and Osmond, 1974;

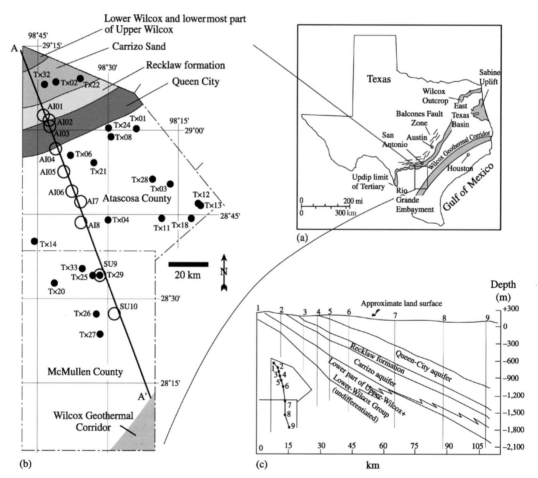

(b)   (c)   km

Figure 12 Simplified diagram of the studied area and formations in southwestern Texas. (a) Geographical and tectonic context of the area of investigation within Texas (after Bebout *et al.*, 1978; Hamlin, 1988). (b) Detailed area of investigation: latitude and longitude are indicated, as well as the location of cross-section A–A' in which simulations of groundwater flow and mass transport were carried out; the locations of all ^{4}He sampled sites in Atascosa and McMullen counties (cf. Castro *et al.*, 2000) are also indicated as well as that of the wells for which hydraulic head measurements are used for calibration of the groundwater flow model, closed and open circles, respectively. (c) General schematic representation of cross-section in the area for the four studied formations: the Lower-Wilcox and lower part of the Upper-Wilcox formations undifferentiated, the Carrizo aquifer, the Recklaw formation, and the Queen-City aquifer (Castro and Goblet, 2003) (reproduced by permission of American Geophysical Union from *Water Resour. Res.* **2003**, *39*, 1172).

Pearson and White, 1967). It is overlain by the Recklaw formation that is composed of shales with thin interbedded sands, and is underlain by the Wilcox Group mainly composed of clay, shale, and lenticular beds of sand (Figure 12(b)). Rainfall recharges the aquifer outcrop at the northern part of Atascosa County, and groundwater flow is to the southeast. Discharge occurs by upward cross-formational leakage driven by high fluid pressure, as well as by fault-related permeability pathways that are mainly present in the south of McMullen County.

The Carrizo was one of the earliest aquifers for which ^{14}C was employed as a tracer (Pearson and White, 1967). The results of the ^{14}C measurements showed a consistent increase of age along flow paths. The early ^{14}C results were checked about 30 years later using measurements of atmospheric noble-gas concentrations (Stute *et al.*, 1992a). These indicated that water recharged prior to a radiocarbon age of $\sim 1.2 \times 10^4$ yr was colder by ~ 5 °C than the modern mean annual temperature. Although this age is $\sim 30\%$ younger than indicated by independent chronology for the last glacial maximum, it generally confirms the radiocarbon results. Further confirmation comes from measurement of ^{36}Cl in the same area (Bentley *et al.*, 1986a). The water is too young to show much ^{36}Cl decay. The $^{36}Cl/Cl$ ratio in the region documenting the noble-gas paleotemperature decrease actually shows an increase in the $^{36}Cl/Cl$. Bentley *et al.* (1986a) explain this by comparison with the curve of distance from the coastline, which varied as sea level decreased during the glacial period. Chloride concentration in precipitation decreases as with increasing distance to the coast. The good correspondence further supports the evidence for distribution of flow velocities in the Carrizo.

This well-supported understanding of flow velocities should aid in the interpretation of the pattern of helium-isotope concentrations.

Helium-3 and helium-4 concentrations are in excess by up to one and two orders of magnitude compared to the ASW, respectively (Castro *et al.*, 2000). Concentrations increase with recharge distance and ^{14}C ages (e.g., Figure 13). Separation of helium components shows that $\sim 99\%$ of the excess helium—$(^3He)_{exc}$ and $(^4He)_{exc}$—is of terrigenic origin due to the addition of nucleogenic ^3He and radiogenic ^4He at a $^3He/^4He$ production ratio of 7×10^{-8}. From this, and through analyses of $^3He/^4He$ ratios in groundwater and in reservoir rocks, it is apparent that most of the helium has an origin external to the aquifer, and that *in situ* production makes a smaller and variable contribution. Simulation and calibration of ^4He transport using a simple analytical model (Castro *et al.*, 2000) illustrates the *in situ* contribution variations with recharge distance and with depth within the aquifer (Figure 14). Modeling results (dashed line, Figure 14) show that *in situ* production can account for ^4He concentrations in the top of the aquifer if the distance from the recharge area is ≤ 40 km. Similar to the Paris Basin (Section 10.6.2.1), this contribution decreases toward the discharge area ($\sim 27\%$). At the bottom of the aquifer the *in situ* contribution represents no more than $\sim 15\%$ of total ^4He.

The analytical model was calibrated with an external flux to the bottom of the aquifer of 3.6×10^{-7} cm^3 STP cm^{-2} yr^{-1}. It is important to note that this flux represents only part of the local integrated crustal production as a consequence of downstream ^4He dilution in the deeper aquifer. Thus, this flux is smaller than the average crustal flux calculated for the GAB by Torgersen and Clarke (1985) (3.62×10^{-6} cm^3 STP cm^{-2} yr^{-1}) or that calculated for the Paris Basin by Castro *et al.* (1998a) (9×10^{-6} cm^3 STP cm^{-2} yr^{-1}). Using an average velocity calculated from previous studies of 1.6 m yr^{-1}, tests of sensitivity to the ^4He external flux were carried out (e.g., Figures 15 (a)–(e)). Some aspects of these tests are worth

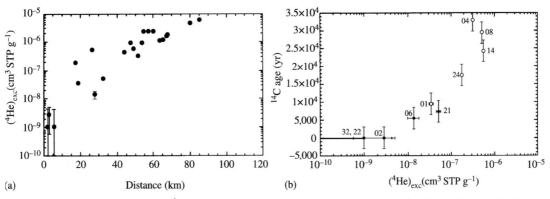

Figure 13 (a and b) Evolution of $(^4He)_{exc}$ concentrations in the water as a function of recharge distance, depth, and calculated ^{14}C ages, respectively, for the Carrizo aquifer (Castro *et al.*, 2000) (reproduced by permission of Elsevier from *Appl. Geochem.* **2000**, *15*, 1137–1167).

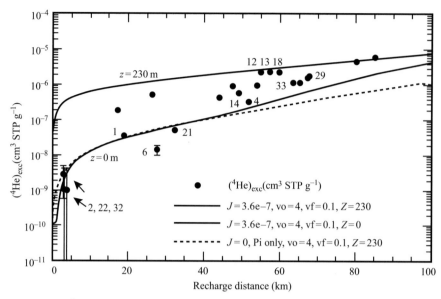

Figure 14 Calculated $(^4He)_{exc}$ concentration curves for the bottom ($z = 230$ m) and for the top of the aquifer ($z = 0$ m) (upper and lower plain line, respectively) as a function of the recharge distance for the Carrizo aquifer. The measured $(^4He)_{exc}$ concentration values are also shown as well as the ^4He concentration curve resulting from *in situ* production only (dashed line) (Castro *et al.*, 2000) (reproduced by permission of Elsevier from *Appl. Geochem.* **2000**, *15*, 1137–1167).

noting: differences between bottom and top concentrations decrease with decreasing flux. As the flux decreases, *in situ* production becomes more important, ^4He accumulation becomes independent of vertical and horizontal position in the aquifer, and varies mainly with time. For the smallest flux of 3.59×10^{-9} cm^3 STP cm^{-2} yr^{-1}, ^4He concentrations are similar in the top and bottom of the aquifer as *in situ* production becomes almost entirely responsible for the ^4He concentrations, and the external flux is negligible. This illustrates the interplay between internal versus external sources of ^4He. In the study by Castro *et al.* (2000), it was only possible to calibrate ^4He concentrations by imposing an exponential decrease in the groundwater velocities between recharge and discharge areas, with initial and final velocities of 4.0 m yr^{-1} and 0.1 m yr^{-1}, respectively.

The analytical modeling of Castro *et al.* (2000) is typical of an aquifer-scale approach. Comparison of a more recent simulation of ^4He transport using a finite element model with a more accurate representation of boundary conditions (i.e., a step toward a comprehensive aquifer-system or even basin-scale approach) has shown representation by a simple analytical model to be inadequate for this system (Castro and Goblet, 2003). Although the finite element model has confirmed an exponential decrease in Carrizo velocities, the external flux value was one order of magnitude higher in the analytical model than that obtained through numerical modeling. Advective groundwater ages estimated from numerical simulation of ^4He

transport (Castro and Goblet, 2003) indicate an increase in Carrizo water ages with distance from the recharge zone (Figure 16). Clearly, even with multiple tracers, issues of model uniqueness continue to affect interpretations.

10.7.3 Implications at the Aquifer Scale

Perhaps more than studies at other scales, modeling of groundwater flow at the aquifer scale has revealed the utility of combining atmospheric-source radioisotope or stable tracers (e.g., ^{14}C, atmospheric noble gases) with subsurface-sourced stable tracers. In comparison to the regional scale, detailed knowledge of the permeability structure is often needed. Independent information can often be obtained from hydraulic tests or from semiquantitative interpretation of aquifer properties based on hydrogeological descriptions. Concentrations of natural tracers reveal not only the history of the tracer itself (e.g., its sources), but also the patterns of water dynamics within a particular aquifer (e.g., variation in water velocity is the horizontal direction). Some generalizations regarding patterns that are very commonly observed from the numerous studies of noble gases in aquifers in sedimentary systems can now be drawn. These are: (i) an increase in concentration with increased recharge distance and depth and (ii) the presence of two main crustal sources (radiogenic or nucleogenic), i.e., *in situ* production and an external flux with variable contributions depending on the depositional context. This is general information that can now be

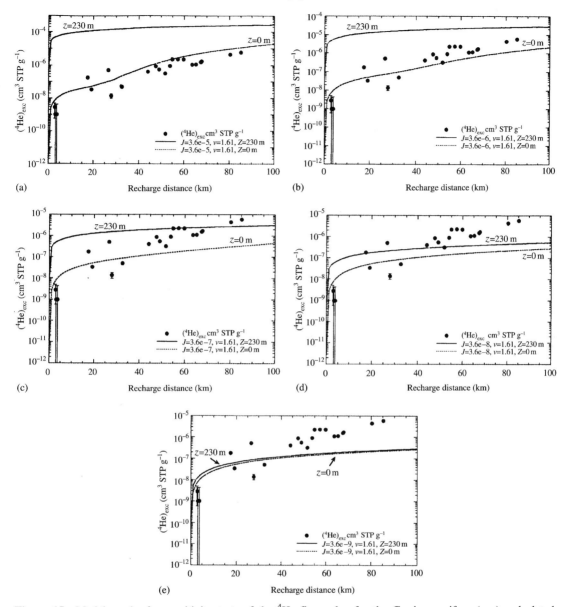

Figure 15 Model results for sensitivity tests of the ^{4}He flux value for the Carrizo aquifer—(a–e): calculated $(^4\text{He})_{exc}$ curves for five different orders of magnitude of the flux value, between 3.6×10^{-5} cm^3 STP cm^2 yr^{-1} and 3.6×10^{-9} cm^3 STP cm^2 yr^{-1} and a constant velocity value of 1.61 m yr^{-1}. The calculated concentration curves for the bottom ($z = 230$ m) and for the top ($z = 0$ m) of the aquifer are represented. The measured $(^4\text{He})_{exc}$ concentration values in the aquifer are also shown (Castro *et al.*, 2000) (reproduced by permission of Elsevier from *Appl. Geochem.* **2000**, *15*, 1137–1167).

considered to be well established, as it has been observed consistently in a variety of aquifer systems around the world in diverse sedimentary settings, as discussed above. If the external flux and *in situ* production of noble gases in general, and ^{4}He in particular, in a specific aquifer system are relatively constant, then the pattern of variation in concentration along one particular aquifer, such as a rapid versus a slow increase in concentrations with increased recharge distance, will yield information on the spatial variation of hydraulic parameters, and those of hydraulic conductivity in particular.

At the aquifer scale, the most important contribution of age tracers is probably reduction in the nonuniqueness of numerical models. Lack of uniqueness stems, among other things, from inadequate knowledge of the distribution of hydraulic properties within groundwater systems and from poor constraints on boundary conditions (Konikow and Bredehoeft, 1992; Maloszewski and Zuber, 1993). Commonly, groundwater flow models are calibrated on measured hydraulic head and/or hydraulic conductivity values. However, a good match does not prove the validity of the model, because the nonuniqueness of model solutions means that a good comparison can be

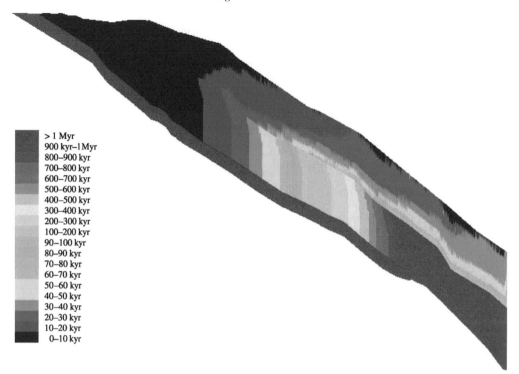

Figure 16 Distribution of calculated advective water ages (kyr) in the system. Water age contours correspond to constant variations of 10^4 yr between 0 Myr and 0.1 Myr and variations of 10^5 yr for time periods varying between 0.1 Myr and 1 Myr, each one of these intervals being represented by a different color, from the youngest (dark blue) to the oldest (red), which corresponds to ages higher than 1Myr (Castro and Goblet, 2003) (reproduced by permission of American Geophysical Union from *Water Resour. Res.* **2003**, *39*, 1172).

achieved with an inadequate or erroneous model. For example, in the case of the Carrizo aquifer, Castro and Goblet (2003) show that variations of hydraulic conductivity up to two orders of magnitude in the Carrizo aquifer and in the overlying confining layer lead to similar calculated hydraulic heads. No clear-cut arguments are available to validate one groundwater flow scenario over a different one. This is due, in part, to lack of information regarding recharge (infiltration) rates in the study area. In contrast, when tested with a ^{4}He transport conceptual model, all groundwater flow calibrated scenarios except one failed to reproduce a coherent ^{4}He transport behavior in the system. This highlights the importance of contributions of a tracer such as ^{4}He for determining which model most closely replicates natural conditions. Without the information provided by this tracer, it would be very difficult to have a precise idea of the groundwater flow conditions of this system, and, as a result, on the velocity distribution of those waters as well as water age distribution.

10.8 TRACERS AT THE LOCAL SCALE

10.8.1 Introduction

There are numerous examples in the literature in which environmental tracers have been applied at a local scale, both in porous media and in fractured crystalline rocks. Among these are studies in Liedern/Bocholt, Germany (Schlosser *et al.*, 1988, 1989), the Stripa Granite, Sweden (Andrews *et al.*, 1989; Loosli *et al.*, 1989), in Sturgeon Falls, Ontario, Canada (Milton *et al.*, 1997; Robertson and Cherry, 1989; Solomon *et al.*, 1993), the Delmarva Peninsula (Dunkle *et al.*, 1993; Ekwurzel *et al.*, 1994; Plummer, 1993), and in the Borden aquifer, Ontario (Smethie *et al.*, 1992; Solomon *et al.*, 1992). Most of these studies were conducted in relatively surficial aquifers, and many focused on the application of the tritium/^{3}He and CFC methods, which are well suited for young, fast-flowing groundwater. Tracers such as ^{37}Ar, ^{39}Ar, ^{85}Kr, ^{36}Cl, and ^{4}He were also investigated. Helium-3 confinement, dispersion, diffusion, sorption, and microbial degradation are some of the processes that have been investigated by means of various tracers. As an example, we focus on studies in the Delmarva Peninsula (eastern seaboard of the US). These studies illustrate in a clear and elegant manner how a multitracer approach using ^{3}H/^{3}He, CFCs (CFC-11 and CFC-12), and ^{85}Kr can give insight into the dynamics of groundwater flow systems and the major transport processes that influence the concentration of different tracers. They also illustrate how the use of a multitracer approach using direct water dating methods at a local scale increases confidence in estimated ages.

10.8.2 $^3H/^3He$, CFC-11, CFC-12, ^{85}Kr—Delmarva Peninsula

The Atlantic Coastal Plain of the Delmarva Peninsula in the East Coast of the US is composed of Jurassic to Holocene sediments deposited during a series of transgressions and regressions (Figure 17). It is covered by a highly permeable surficial Pleistocene aquifer that is directly recharged by local precipitation. Thirty-three wells, some of which tap this formation, were used to sample $^3H/^3He$ (52 samples), CFCs (282 samples), and ^{85}Kr (four samples) (Dunkle *et al.*, 1993; Ekwurzel *et al.*, 1994; Plummer *et al.*, 1993; Reilly *et al.*, 1994).

In wells sampled at several levels, ^3H and $(^3He)_{tri}$ concentrations, as well as apparent $^3H/^3He$, CFC-11, and CFC-12 ages, increase significantly with depth below the water table (Figure 18). Moreover, most of the apparent CFC, $^3H/^3He$, and ^{85}Kr ages agree with each other to within three years over the entire peninsula (e.g., Figures 19(a) and (b)), indicating that these tracers behave fairly conservatively in this sandy unconfined aquifer. Waters located in discharge areas such as those in well KE Be 170, where a mix of a wide range of water ages intersects the well screen, present large differences between CFC and $^3H/^3He$ ages. This illustrates how ages based on different tracers can identify specific recharge versus discharge areas within a groundwater system.

Simulations using different dispersion values and an average vertical velocity of 1 m yr^{-1} in a one-dimensional model (Ekwurzel *et al.*, 1994) showed that $^3H/^3He$ ages are more affected by dispersion than CFC ages (Figure 20). This is because the abrupt concentration front presented by the peak-shaped ^3H source function is subject to greater dispersion than the relatively gradual, monotonic CFC concentration increase (Figure 2). Such dispersion would result in an apparent increase in $^3H/^3He$ ages in postbomb peak waters, and in reduced apparent ages in prebomb peak waters. However, comparison of $(^3H + ^3He)$ concentrations with the ^3H input function shows little smoothing, indicating that dispersion has been negligible (Dunkle *et al.*, 1993; Ekwurzel *et al.*, 1994). This inference is supported by detailed numerical transport modeling (Reilly *et al.*, 1994). The high correlation observed between ages calculated using the three age tracers also supports this inference. Recharge rates calculated from these different tracers are also in good agreement (Ekwurzel *et al.*, 1994).

10.8.3 Implications at the Local Scale

The studies surveyed above, and those at other sites, provide a strong contrast to aquifer and regional scale results. While the results of the larger scale studies typically exhibit many enigmatic features and major issues of nonuniqueness, most of the local-scale flow systems appear to allow straightforward interpretations. Most of the studies conducted at sites investigated previously or, by other approaches (Engesgaard *et al.*, 1996; Reilly *et al.*, 1994; Shapiro *et al.*, 1999; Solomon *et al.*, 1992), hold few surprises.

Several aspects of the results from local flow systems are noteworthy. One is the ease and reliability of estimating recharge rates by measurement of vertical profiles of $^3H/^3He$ or CFCs in recharge areas (Böhlke and Denver, 1995; Ekwurzel *et al.*, 1994; Solomon *et al.*, 1993). Another is the apparently relatively minor level of problems created by degassing of volatile tracers from the water table. Sampling near discharge areas (principally streams), however, is somewhat hazardous because of the mixing of deep, upward flow with shallow flow at the point of discharge.

What is most surprising is the relative homogeneity and the dominance of advective flow in most of the investigated systems. Where multiple tracers have been used, it is possible to estimate dispersivity independently. The longitudinal dispersivity alues obtained for studies at the 10^2-10^3 m flow scale have ranged from 0.1 m to 0 m (Ekwurzel *et al.*, 1994; Engesgaard *et al.*, 1996; Reilly *et al.*, 1994; Solomon *et al.*, 1993; Szabo *et al.*, 1996). These are significantly smaller than many extrapolations would suggest (Gelhar, 1993; Neuman, 1995). Whether these implications can be widely extended, or whether they result from application of the tracing methods to a restricted and hydrologically unusually homogeneous set of study areas is uncertain. Although age tracing techniques for local systems have clearly demonstrated their validity, have increased the general understanding of local flow systems, and have aided in the reconstruction and fate of contaminants introduced into these systems, they have not yet reached their potential for elucidating fundamental transport mechanisms in a variety of geological environments. There is much to be learned by future studies for this purpose.

10.9 TRACERS IN VADOSE ZONES

Vadose zones are of great importance for both hydrological and geochemical reasons. Hydrologically, they represent the portion of the physical system where precipitation partitions the essential elements of the hydrological cycle: evapotranspirative return to the atmosphere, runoff, and deep infiltration (i.e., aquifer recharge). Quantification of the fluxes below the root zone is thus critical for both closing the water balance at the watershed scale and for analyzing groundwater resources. Geochemically, the vadose zone constitutes the primary zone of interaction between

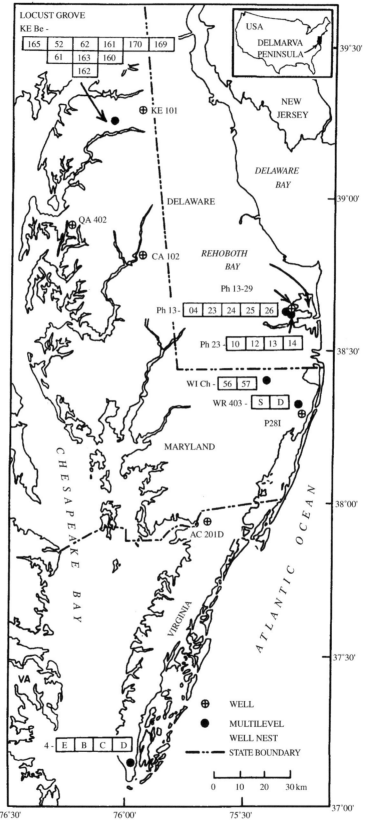

Figure 17   Map of the Delmarva Peninsula, on the east coast of the US. Wells having the same series number (e.g., KE Be) are listed as the series number followed by a dash and well numbers listed in order of increasing depth within the boxes (Ekwurzel *et al.*, 1994) (reproduced by permission of American Geophysical Union from *Water Resour. Res.* **1994**, *30*, 1693–1708).

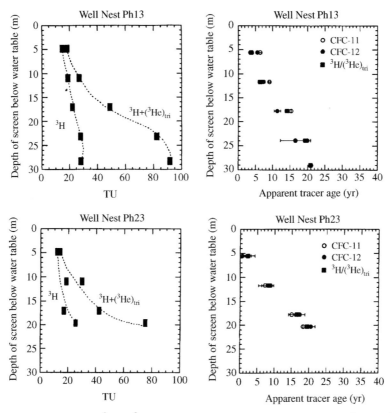

Figure 18 Depth profiles of tritium, $^3H + (^3He)_{tri}$ (tritium units), and apparent tracer ages from well nests Ph13 and Ph23 (Ekwurzel *et al.*, 1994) (reproduced by permission of American Geophysical Union from *Water Resour. Res.* **1994**, *30*, 1693–1708).

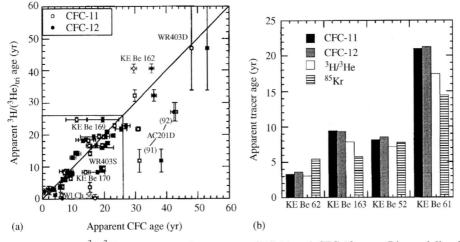

Figure 19 (a) Apparent $^3H/^3He$ age compared to apparent CFC-11 and CFC-12 ages. Diagonal line is a 1 : 1 reference line. The vertical and horizontal reference lines are 26 yr age lines representing the bomb peak and a 2 yr lag for flow through the unsaturated zone. (b) Apparent CFC-11, CFC-12, $^3H/^3He$, and ^{85}Kr ages for Locust Grove wells sampled in November 1991 (Ekwurzel *et al.*, 1994) (reproduced by permission of American Geophysical Union from *Water Resour. Res.* **1994**, *30*, 1693–1708).

earth materials and precipitation. Knowledge of water fluxes through vadose zones is thus a key to understanding the global solute balance, as well as interactions of climate, geochemistry, and tectonics on the continental scale.

Determination of water fluxes under unsaturated conditions by physical monitoring is often very difficult due to the effects of seasonal fluctuations, problems of installing and maintaining instrumentation, and the nonlinear nature of the

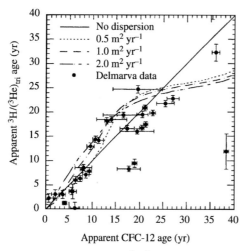

Figure 20 Comparison of ^{3}H/^{3}He and CFC-12 ages from the Delmarva Peninsula, including effects of dispersion. Average linear velocity of 1 m yr^{-1} is plotted with different dispersion coefficients and with a 1:1 reference line (Ekwurzel *et al.*, 1994) (reproduced by permission of American Geophysical Union from *Water Resour. Res.* **1994**, *30*, 1693–1708).

variation of water potential and hydraulic conductivity with water content. More direct means of residence-time estimation by tracing solutes thus provide an important alternative to physical monitoring (Allison *et al.*, 1994; Scanlon *et al.*, 1997). The majority of residence-time tracers that rely on radiodecay for their time constant are volatile; therefore, most vadose-zone dating studies employ some form of signal tracing rather than radioactive decay dating.

Conceptually, vadose zones can be divided into two end-members: humid and arid. In humid climates (or in irrigated areas), vadose zones are generally thin (0–10 m), dynamic, and are characterized by great variability in flow paths. Residence times vary from hours to a few years. In contrast, arid-region vadose zones are usually thick (10–1,000 m), with very small water fluxes, and a high degree of uniformity in flow paths. We will give examples from very humid, subhumid (irrigated), and arid vadose-zone environments.

Bonell *et al.* (1998) described the results of detailed monitoring of soil water and runoff in tropical northeast Queensland, Australia. Figure 21 shows the δ^2H variation of soil water as a function of depth during a three-month period in 1991. Each panel represents a nest of soil water samplers at different positions on a hillslope. Some sites (NC2, NC4, and SC3) show a large range in δ^2H throughout the depth profile. These are in response to heavy precipitation events having differing isotopic compositions. Others (SC1 and SC2) show much variability with time in the upper soil horizons but little at depth, indicating an absence of continuous rapid flow paths to greater than 1 m.

Figure 22 shows δ^2H for well water samples over the same period, compared to stream water samples. The groundwater samples show a high degree of variability in both space and time, but not nearly as much as the soil water. The stream-water depletion on day 47 is in response to a major storm. The groundwater response can be seen on day 52, but the groundwater has returned to near baseline by day 60. These data, therefore, indicate flow times for the vadose zone of ~5 d, but also show that new recharge is mixing with a large reservoir of shallow groundwater. Numerous studies have shown that a short residence time, rapid response and flow rates, and large temporal and spatial variability in fluxes are characteristic of tracer transport in humid vadose zones (Kendall and McDonnell, 1998).

Gvirtzman and Magaritz (1986) and Gvirtzman *et al.* (1986) provide a good example of the power of geochemical tracers for understanding the transport properties of semi-arid vadose zones. The study was conducted beneath an irrigated vineyard in the Negev Desert in Israel. The vadose zone was ~11 m thick and composed of very uniform silt, which was deposited as loess. The vineyard received precipitation containing bomb and natural tritium during the winter months, amounting to 200 mm annual average. During the summer the vineyard was irrigated with ~650 mm yr^{-1} of well water containing less than 1 TU. The authors used the alternating high- and low-tritium inputs to trace 14 annual tritium cycles down to a depth of 8.75 m (Figure 23), for a mean velocity of 0.62 m yr^{-1}. They determined that the spreading of the annual pulses was largely due to exchange by diffusion with immobile regions within the loess that were created by dispersion of clays when the original saline pore water was displaced by dilute irrigation water. By matching the shapes of the annual pulses in transport simulations with those from the data, they estimated an effective dispersion coefficient of 5×10^{-11} m^2 s^{-1}, which, due to the uniformity of the loess, the low velocity, and the high volume of immobile water, is mainly attributable to diffusion rather than dispersion.

Arid or desert vadose zones have been much less studied than those in humid climates. Geochemical dating approaches have played an important role in understanding their dynamics (Allison *et al.*, 1994). Walvoord *et al.* (2002a,b) have employed a combination of liquid–vapor flow modeling and modeling of chloride concentration profiles to demonstrate that pervasive upward total fluid potential gradients below the root zone, caused by desert vegetation, retain all water and solutes that are deposited on the land surface (Figure 24). Measurements of numerous profiles within the southwestern US indicate that most desert vadose zones have been quantitatively

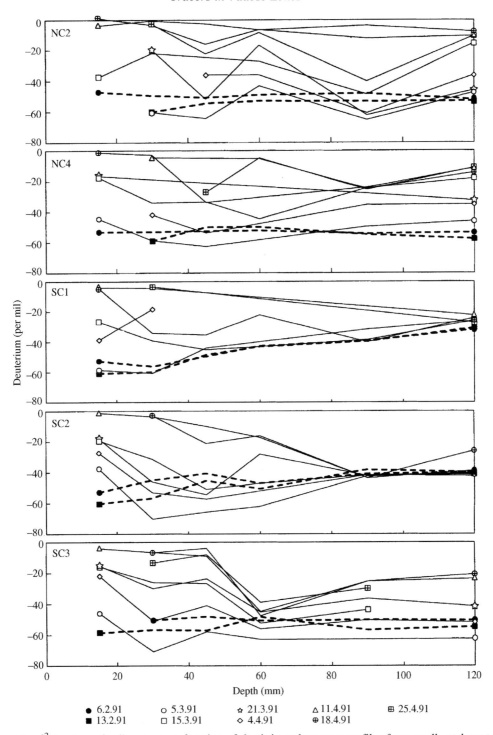

Figure 21 δ^2H values of soil water as a function of depth in vadose-zone profiles from small catchments near Babinda, Queensland, Australia. Numbers in symbol key refer to sampling dates during 1991. The climate is tropical and mean annual precipitation amounts to 4,000 mm (source Bonell *et al.*, 1998).

retaining chloride and other solutes since the end of the last glacial period at ~15 ka (Tyler *et al.*, 1996). This amounts to a "dating" of the duration of a hydrological regime. This history of the vadose-zone hydraulics would be difficult or impossible to ascertain without the use of geochemical tracers.

The examples cited above illustrate the great variability in soil–water movement and resultant geochemical processes in vadose zones. Humid vadose zones are highly dynamic, with large water fluxes and great spatial and temporal variability of those fluxes. Although variability is large at small time and space scales, mixing processes

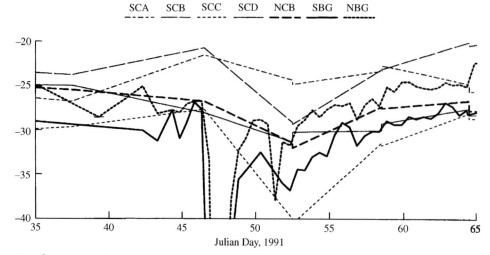

Figure 22 δ^2H of water from wells in the North Creek drainage (NCB) and South Creek drainage (SCA, SCB, SCC, and SCD) and stream water from North Creek (NBG) and South Creek (SBG) as a function of time during 1991. The drainages are near Babinda, Queensland, Australia (Bonell *et al.* (1998) (reproduced by permission of Elsevier from *Isotope Tracers in Catchment Hydrology*, pp. 334–347).

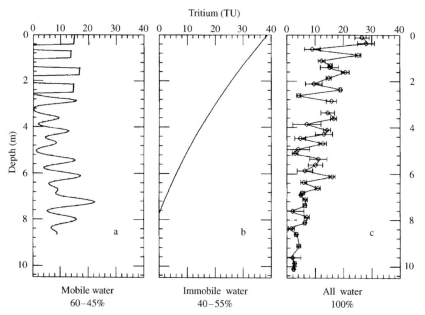

Figure 23 Comparison of reconstructed irrigation water tritium input (left panel), immobile vadose-zone water tritium (center panel), and measured tritium (right panel), all as a function of depth. Study site is an irrigated vineyard in the Negev Desert, Israel (source Gvirtzman and Magaritz, 1986).

tend to rapidly average out these variations and their effects on tracer concentrations. It is clear that these characteristics are very favorable for weathering reactions, and that these are environments of high rates of chemical denudation. Less attention has been devoted to how the small-scale heterogeneity in flow paths and fluxes affects the chemical outcome of weathering averaged over the landscape (Bullen and Kendall, 1998). In contrast, desert landscapes exhibit a remarkable uniformity of hydrologic regime, both in space and time (except in the vicinity of water courses or

areas where runoff is focused). Such vadose zones can be regarded as surfaces of "negative weathering"—there is no export of solutes and rather than being depleted of mobile elements with time; they accumulate them from atmospheric deposition.

10.10 CONCLUSIONS

During most of the second half of the twentieth century, the emphasis in the study of subsurface water age tracers was on the development of new

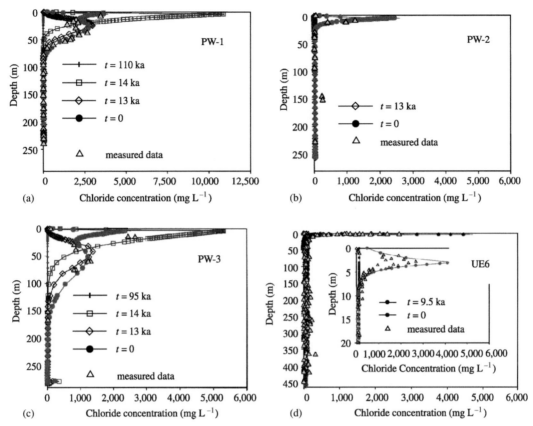

Figure 24 Simulated chloride concentrations as a function of depth, compared to measured data, for deep boreholes at the Nevada Test Site, USA. The simulations indicate that the upper 10–100 m of these vadose zones have been retaining all chloride deposited from the atmosphere onto the land surface for time periods from 9.5 ka to 110 ka (Walvoord *et al.*, 2002a) (Walvoord, 2002) (reproduced by permission of American Geophysical Union from *Water Resour. Res.* **2002**, *38*, 1291).

techniques and methodologies. This phase of the field has slowed greatly. Most tracers not in common use have at least been explored, and most are no longer used, because of major sampling and analytical difficulties. The only tracer that may be poised for a breakthrough in the near future is [81]Kr (Collon *et al.*, 2000; Lehmann *et al.*, 2002b). The inert geochemistry and the half-life of this radionuclide render it sufficiently attractive for regional and aquifer scale investigations in spite of the difficulties in sampling and analysis.

The new frontier in groundwater age tracers lies not in technique development, but rather in interpretation within the hydrogeochemical system. Until the 1990s, interpretation has tended to focus on geochemical issues, while minimizing issues of transport and mixing. With the increasing use of three-dimensional flow and transport computer programs, attention can now be turned to quantifying the dynamics of the systems in which the tracers are measured. This initiative at the boundaries of geochemistry and quantitative hydrogeology will yield major advances for both fields. For hydrogeology, it may prove a key to elucidating the link between hydrogeological structures and transport mechanisms over a wide range of scales. For geochemistry, it will finally enable a quantification of the fluxes through geochemical systems in the shallow crust. The tools have been developed; now let us use them!

REFERENCES

Airey P. L., Calf G. E., Campbell B. L., Hartley P. E., Roman D., and Habermehl M. A. (1979) Aspects of the isotope hydrology of the Great Artesian Basin, Australia. In *Isotope Hydrology 1978*. IAEA, Vienna, vol. 1, pp. 205–219.

Allègre C. J., Staudacher T., Sarda P., and Kurz M. D. (1983) Constraints on evolution of Earth's mantle from rare gas systematics. *Nature* **303**, 762–766.

Allison G. B. and Holmes J. W. (1973) The environmental tritium concentrations of underground water and its hydrological interpretation. *J. Hydrol.* **19**, 131–143.

Allison G. B., Gee G. W., and Tyler S. W. (1994) Vadose-zone techniques for estimating groundwater recharge in arid and semiarid regions. *Soil Sci. Soc. Am. J.* **58**, 6–14.

Andrews J. N. and Fontes J.-C. (1993) Comment on "Chlorine 36 dating of very old groundwater: 3. Further results on the Great Artesian Basin, Australia" by T. Torgerson *et al.* *Water Resour. Res.* **29**, 1871–1874.

Andrews J. N. and Kay R. L. F. (1982a) [234]U/[238]U activity ratios of dissolved uranium in groundwater from a Jurassic limestone aquifer in England. *Earth Planet. Sci. Lett.* **57**, 139–151.

Andrews J. N. and Kay R. L. F. (1982b) Natural production of tritium in permeable rocks. *Nature* **298**, 361–363.

Andrews J. N. and Lee D. J. (1979) Inert gases in groundwater from the Bunter Sandstone of England as indicators of age and paleoclimatic trends. *J. Hydrol.* **41**, 233–252.

Andrews J. N., Giles I. S., Kay R. L. F., Lee D. J., Osmond J. K., Cowart J. B., Fritz P., Barker J. F., and Gale J. (1982) Radioelements, radiogenic helium, and age relationships for groundwaters from the granites at Stripa, Sweden. *Geochim. Cosmochim. Acta* **46**, 1533–1543.

Andrews J. N., Balderer W., Bath A. H., Clausen H. B., Evans G. V., Florkowski T., Goldbrunner J. E., Ivanovich M., Loosli H. H., and Zojer H. (1984) Environmental isotope studies in two aquifer systems: a comparison of ground water dating methods. In *Isotope Hydrology 1983*. IAEA, Vienna, pp. 535–576.

Andrews J. N., Goldbrunner J. E., Darling W. G., Hooker P. J., Wilson G. B., Youngman M. J., Eichinger L., Rauert W., and Stichler W. (1985) A radiochemical, hydrochemical, and dissolved gas study of groundwaters in the Molasse Basin of Upper Austria. *Earth Planet. Sci. Lett.* **73**, 317–332.

Andrews J. N., Davis S. N., Fabryka-Martin J., Fontes J.-C., Lehmann B. E., Loosli H. H., Michelot L., Moser H., Smith B., and Wolf M. (1989) The *in-situ* production of radioisotopes in rock matrices with particular reference to the Stripa granite. *Geochim. Cosmochim. Acta* **53**, 1803–1815.

Andrews J. N., Florkowski T., Lehmann B. E., and Loosli H. H. (1991) Underground production of radionuclides in the Milk River aquifer, Alberta, Canada. *Appl. Geochem.* **6**, 425–434.

Andrews R. W. and Pearson F. J., Jr. (1984) Transport of ^{14}C and uranium in the Carrizo aquifer of South Texas, a natural analog of radionculide migration. *Mater. Res. Soc. Symp. Proc.* **26**, 1085–1091.

Arnold J. R. and Libby W. F. (1949) Age determinations by radiocarbon content: checks with samples of known age. *Science* **110**, 678–680.

Balderer W. and Synal H.-A. (1996) Application of the chlorine-36 method for the characterisation of the groundwater circulation in tectonic active areas: examples from north western Anatolia/Turkey. *Terra Nova* **8**, 324–333.

Ballentine C. J. and O'Nions R. K. (1991) The nature of mantle neon contributions to Vienna Basin hydrocarbon reservoirs. *Earth Planet. Sci. Lett.* **113**, 553–567.

Ballentine C. J., O'Nions R. K., and Coleman M. L. (1996) A Magnus opus: helium, neon, and argon isotopes in a North Sea oilfield. *Geochim. Cosmochim. Acta* **60**, 831–850.

Bard E. (1998) Geochemical and geophysical implications of the radiocarbon calibration. *Geochim. Cosmochim. Acta* **62**, 2025–2039.

Bebout D. G., Gavenda V. J., and Gregory A. R. (1978) Geothermal resources, Wilcox Group, Texas Gulf Coast. Bur. Econ. Geol., University of Texas at Austin, 82pp.

Benson B. B. (1973) Noble gas concentration ratios as paleotemperature indicators. *Geochim. Cosmochim. Acta* **37**, 1391–1396.

Bentley H. W., Phillips F. M., Davis S. N., Gifford S., Elmore E., Tubbs L. E., and Gove H. E. (1982) Thermonuclear ^{36}Cl pulse in natural water. *Nature* **300**, 737–740.

Bentley H. W., Phillips F. M., and Davis S. N. (1986a) Chlorine-36 in the terrestrial environment. In *Handbook of Environmental Isotope Geochemistry* (eds. P. Fritz and J.-C. Fontes). Elsevier, Amsterdam, vol. 2B, pp. 427–480.

Bentley H. W., Phillips F. M., Davis S. N., Airey P. L., Calf G. E., Elmore D., Habermehl M. A., and Torgersen T. (1986b) Chlorine-36 dating of very old ground water: I. The Great Artesian Basin, Australia. *Water Resour. Res.* **22**, 1991–2002.

Bergmann F. and Libby W. F. (1957) Continental water balance, groundwater inventory and storage times, surface ocean mixing rates, and worldwide water circulation patterns from cosmic ray and bomb tritium. *Geochim. Cosmochim. Acta* **12**, 277–296.

Berner E. K. and Berner R. A. (1996) *Global Environment: Water, Air, and Geochemical Cycles*. Prentice Hall, Upper Saddle River, NJ.

Bethke C. M. and Johnson T. C. (2002a) Ground water age. *Ground Water* **40**, 337–339.

Bethke C. M. and Johnson T. C. (2002b) Paradox of groundwater age: correction. *Geology* **30**, 385–388.

Bethke C. M., Lee M. K., Quinodoz H. A. M., and Kreiling W. N. (1993) *Basin Modeling with Basin2: A Guide to Basin2, B2 Plot, B2 Video, B2 View*. University of Illinois at Urbana-Champaign, Urbana.

Bethke C. M., Zhao X., and Torgersen T. (1999) Groundwater flow and the ^4He distribution in the Great Artesian Basin of Australia. *J. Geophys. Res.* **104**, 12999–13010.

Beyerle U., Aeschbach-Hertig W., Hofer M., Imboden D. M., Baur H., and Kipfer R. (1999) Infiltration of river water to a shallow aquifer investigated with $^3H/^3He$, noble gases and CFCs. *J. Hydrol.* **220**, 169–185.

Blavoux B., Dray M., Fehri A., Olive P., Groening M., Sonntag C., Hauquin J. P., Pelissier G., and Pouchan P. (1993) Palaeoclimate and hydrodynamic approach to the Aquitaine Basin deep aquifer (France) by means of environmental isotopes and noble gases. In *Isotope Techniques in the Study of Past and Current Environmental Changes in the Hydrosphere and the Atmosphere*. IAEA, Vienna, pp. 293–305.

Böhlke J. K. and Denver J. M. (1995) Combined use of groundwater dating, chemical, and isotopic analyses to resolve the history and fate of nitrate contamination in two agricultural watersheds, Atlantic coastal plain, Maryland. *Water Resour. Res.* **31**, 2319–2340.

Böhlke J. K., Wanty R., Tuttle M., Delin G., and Landon M. (2002) Denitrification in the recharge area and discharge area of a transient agricultural nitrate plume in a glacial sand outwash aquifer, Minnesota. *Water Resour. Res.* **38**(7), doi: 10.1029/2001WR000663.

Bonell M., Barnes C. J., Grant C. R., Howard A., and Burns J. (1998) High rainfall, response-dominated catchments: a comparative study of experiments in tropical Northeast Queensland with temperate New Zealand. In *Isotope Tracers in Catchment Hydrology* (eds. C. Kendall and J. J. McDonnell). Elsevier, Amsterdam, pp. 334–347.

Bottomley D. J., Ross J. D., and Clarke W. B. (1984) Helium and neon isotope geochemistry of some groundwaters from the Canadian Precambrian Shield. *Geochim. Cosmochim. Acta* **48**, 1973–1985.

Boufadel M. C., Lu S., Molz F. J., and Lavallee D. (2000) Multifractal scaling of the intrinsic permeability. *Water Resour. Res.* **36**, 3211–3222.

Bredehoeft J. D. and Pinder G. F. (1973) Mass transport in flowing groundwater. *Water Resour. Res.* **9**, 194–210.

Bullen T. D. and Kendall C. (1998) Tracing of weathering reactions and water flowpaths: a multi-isotope approach. In *Isotope Tracers in Catchment Hydrology* (eds. C. Kendall and J. J. McDonnell). Elsevier, Amsterdam, pp. 611–646.

Busenberg E. and Plummer L. N. (1991) Chlorofluorocarbons (CCl_3F and CCl_2F_2): use as an age dating tool and hydrologic tracer in shallow ground-water systems. *US Geological Survey, Water Resources Investigations Report 91-4034*.

Busenberg E. and Plummer L. N. (1992) Use of chlorofluorcarbons (CCl_3F and CCl_2F_2) as hydrologic tracers and age-dating tools: the alluvium and terrace system of central Oklahoma. *Water Resour. Res.* **28**, 2257–2284.

Busenberg E. and Plummer L. N. (2000) Dating young groundwater with sulfur hexafluoride: natural and anthropogenic sources of sulfur hexafluoride. *Water Resour. Res.* **36**, 3011–3030.

Buttle J. M. (1998) Fundamentals of small catchment hydrology. In *Isotope Tracers in Catchment Hydrology* (eds. C. Kendall and J. J. McDonnell). Elsevier, Amsterdam, pp. 1–49.

Calf G. E. and Habermehl M. A. (1983) Isotope hydrology and hydrochemistry of the Great Artesian Basin, Australia. In *Isotope Hydrology 1983*. IAEA, Vienna, pp. 397–414.

Carlston C. W., Thatcher L. L., and Rhodehamel E. C. (1960) Tritium as a hydrologic tool, the Wharton tract study. *Int. Assoc. Sci. Hydrol. Publ.* **52**, 503–512.

Castro M. C. and Goblet P. (2003) Calibration of regional groundwater flow models: working toward a better understanding of site-specific systems. *Water Resour. Res.* **39**(6), 1172, doi: 10.1029/2002WR001653.

Castro M. C., Goblet P., Ledoux E., Violette S., and de Marsily G. (1998a) Noble gases as natural tracers of water circulation in the Paris Basin: 2. Calibration of a groundwater flow using noble gas isotope data. *Water Resour. Res.* **34**, 2467–2484.

Castro M. C., Jambon A., de Marsily G., and Schlosser P. (1998b) Noble gases as natural tracers of water circulation in the Paris Basin: 1. Measurements and discussion of their origin and mechanisms of vertical transport in the basin. *Water Resour. Res.* **34**, 2443–2467.

Castro M. C., Stute M., and Schlosser P. (2000) Comparison of ^{4}He ages and ^{14}C ages in simple aquifer systems: implications for groundwater flow and chronologies. *Appl. Geochem.* **15**, 1137–1167.

Chebotarev I. I. (1955) Metamorphism of natural water in the crust of weathering. *Geochim. Cosmochim. Acta* **8**, 22–48, 137–170.

Clark J. F., Stute M., Schlosser P., Drenkard S., and Bonani G. (1997) A tracer study of the Floridan aquifer in southeastern Georgia: implications for groundwater flow and paleoclimate. *Water Resour. Res.* **33**, 281–290.

Clarke W. B., Jenkins W. J., and Top Z. (1976) Determination of tritium by mass spectrometric measurements of ^{3}He. *Int. J. Appl. Radiat. Isotopes* **27**, 515–522.

Clayton R. N., Friedman I., Graf P. L., Mayeda T. K., Meents W. F., and Shimp N. F. (1966) The origin of saline formation waters: I. Isotopic composition. *J. Geophys. Res.* **71**, 3869–3882.

Collon P., Kutschera W., Loosli H. H., Lehmann B. E., Purtschert R., Love A. H., Sampson L., Anthony D., Cole D., Davids B., Morrissey D. J., Sherrill B. M., Steiner M., Pardo R. C., and Paul M. (2000) ^{81}Kr in the Great Artesian Basin, Australia: a new method for dating very old groundwater. *Earth Planet. Sci. Lett.* **182**, 103–113.

Cook P. G. and Böhlke J. H. (2000) Determining time scales for groundwater flow and solute transport. In *Environmental Tracers in Subsurface Hydrology* (eds. P. G. Cook and A. L. Herczeg). Kluwer Academic, Amsterdam, pp. 1–30.

Cook P. G., Jolly I. D., Leaney F. W., Walker G. R., Allan G. L., Fifield L. K., and Allison G. B. (1994) Unsaturated zone tritium and chlorine-36 profiles from southern Australia: their use as tracers of soil water movement. *Water Resour. Res.* **30**, 1709–1719.

Cowart J. B. (1975) Uranium isotope disequilibrium in natural waters, recent studies. *Florida Scientist* **38**, 22.

Cowart J. B. and Osmond J. K. (1974) ^{234}U and ^{238}U in the Carrizo sandstone aquifer of south Texas. In *Isotope Techniques in Groundwater Hydrology.* IAEA, Vienna, vol. II, pp. 149–313.

Craig H. and Lupton J. E. (1976) Primordial neon, helium and hydrogen in oceanic basalts. *Earth Planet. Sci. Lett.* **31**, 369–385.

Craig H. and Lupton J. E. (1981) Helium-3 and mantle volatiles in the ocean and oceanic basalts. In *The Sea* (ed. C. Emiliani). Wiley, New York, pp. 391–428.

Cresswell R. G., Jacobson G., Wischusen J., and Fifield L. K. (1999) Ancient groundwaters in the Amadeus Basin, Central Australia: evidence from the radio-isotope ^{36}Cl. *J. Hydrol.* **223**, 212–220.

Cresswell R. G., Bauld J., Jacobson G., Khadka M. S., Jha M. G., Shrestha M. P., and Regmi S. (2001) A first estimate of ground water ages for the deep aquifer of the Kathmandu Basin, Nepal, using the radioisotope chlorine-36. *Ground Water* **39**, 449–457.

Cunnold D., Weiss R., Prinn R., Hartley D., Simmonds P., Fraser P., Miller B., Alyea F., and Porter L. (1997) GAGE/AGAGE measurements indicating reductions in global emissions of CCl_3F and CCl_2F_2 in 1992–1994. *J. Geophys. Res.* **102**, 1259–1269.

Darcy H. G. (1856) *Les fontaines publiques de la Ville de Dijon.* Victor Dalmont, Paris.

Davis R. J. and Schaeffer O. A. (1955) Chlorine-36 in nature. *Ann. NY Acad. Sci.* **62**, 105–122.

Davis S. N. and Bentley H. W. (1982) Dating groundwater—a short review. In *Nuclear and Chemical Dating Techniques—Interpreting the Environmental Record*, ACS Symposium Series No. 176 (ed. L. A. Currie). American Chemical Society, Washington, DC. pp. 187–222.

Davis S. N., Zreda M., Moysey S., and Cecil L. D. (2003) Chlorine-36 in groundwater of the United States: empirical data. *Hydrogeol. J.* **11**, 217–227.

de Marsily G. (1986) *Quantitative Hydrogeology.* Academic Press, London.

Dinçer T. and Davis G. H. (1984) Application of environmental isotope tracers to modeling in hydrology. *J. Hydrol.* **68**, 95–113.

Domenico P. A. and Schwartz F. W. (1998) *Physical and Chemical Hydrogeology.* Wiley, New York.

Drever J. I. (1997) *The Geochemistry of Natural Waters: Surface and Groundwater Environments.* Prentice Hall, Upper Saddle River, NJ.

Dunkle S. A., Plummer L. N., Busenberg E., Phillips P. J., Denver J. M., Hamilton P. A., Michel R. L., and Coplen T. B. (1993) Chlorofluorocarbons (CCl_3F and CCl_2F_2) as dating tools and hydrologic tracers in shallow groundwater of the Delmarva Peninsula, Atlantic Coastal Plain, United States. *Water Resour. Res.* **29**, 3837–3861.

Ekwurzel B., Schlosser P., Smethie W. M., Jr., Plummer L. N., Busenberg E., Michel R. L., Weppperning R., and Stute M. (1994) Dating of shallow groundwater: comparison of the transient tracers ^{3}H/^{3}He, chlorofluorocarbons, and ^{85}Kr. *Water Resour. Res.* **30**, 1693–1708.

Elmore D., Fulton B. R., Clover M. R., Marsden J. R., Gove H. E., Naylor H., Purser K. H., Kilius L. R., Beukens R. P., and Litherland A. E. (1979) Analysis of ^{36}Cl in environmental water samples using an electrostatic accelerator. *Nature* **277**, 22–25.

Elmore D., Gove H. E., Ferraro R., Kilius L. R., Lee H. W., Chang K. H., Beukens R. P., Litherland A. E., Russo C. J., Purser K. H., Murrell M. T., and Finkel R. C. (1980) Determination of iodine-129 using tandem accelerator mass spectrometry. *Nature* **286**, 138–140.

Engesgaard P., Jensen K. H., Molson J., Frind E. O., and Olsen H. (1996) Large-scale dispersion in a sandy aquifer: simulation of subsurface transport of environmental tritium. *Water Resour. Res.* **32**, 3253–3266.

Fabryka-Martin J. (2000) Iodine-129 as a groundwater tracer. In *Environmental Tracers in Subsurface Hydrology* (eds. P. G. Cook and A. L. Herczeg). Kluwer Academic, Boston, pp. 504–510.

Fabryka-Martin J., Bentley H., Elmore D., and Airey P. L. (1985) Natural iodine-129 as an environmental tracer. *Geochim. Cosmochim. Acta* **49**, 337–348.

Fabryka-Martin J., Whittemore D. O., Davis S. N., Kubik P. W., and Sharma P. (1991) Geochemistry of halogens in the Milk River aquifer, Alberta, Canada. *Appl. Geochem.* **6**, 447–464.

Fairbank W. M., Jr., Hansen C. S., LaBelle R. d., Pan X.-J., Zhang Y., Chamberlin E. P., Nogar N. S., Miller C. M., Fearey B. L., and Oona H. (1998) Photon burst mass spectrometry for the measurement of ^{85}Kr at ambient levels. *Proc. Soc. Photo-Optical Instr. Eng.* **3270**, 174–180.

Faure G. (1986) *Principles of Isotope Geology.* Wiley, New York.

Fehn U., Peters E. K., Tullai-Fitzpatrick S., Kubik P. W., Sharma P., Teng R. T. D., Gove H. E., and Elmore D. (1992) ^{129}I and ^{36}Cl concentrations in waters of the eastern Clear Lake area, California: residence times and source ages of hydrothermal fluids. *Geochim. Cosmochim. Acta* **56**, 2069–2079.

Florkowski T., Morawska L., and Rozanski K. (1988) Natural production of radionuclides in geological formations. *Nucl. Geophys.* **2**, 1–14.

Fontes J.-C. and Garnier J. M. (1979) Determination of the initial ^{14}C activity of the total dissolved carbon—a review of the existing models and a new approach. *Water Resour. Res.* **15**, 399–413.

Foster S. S. D. (1975) The chalk groundwater tritium anomaly—a possible explanation. *J. Hydrol.* **25**, 159–165.

Freeze R. A. and Cherry J. A. (1979) *Groundwater*. Prentice Hall, Englewood Cliffs, NJ.

Freeze R. A. and Witherspoon P. A. (1966) Theoretical analysis of regional groundwater flow: I. Analytical and numerical solutions to the mathematical model. *Water Resour. Res.* **2**, 623–634.

Freeze R. A. and Witherspoon P. A. (1967) Theoretical analysis of regional groundwater flow: II. Effect of water table configuration and subsurface permeability variations. *Water Resour. Res.* **3**, 623–634.

Fröhlich K., Franke T., Gellermann G., Herbert D., and Jordan H. (1987) Silicon-32 in different aquifer types and implications for groundwater dating. In *Proceedings International Symposium on Isotopic Techniques in Water Resources Development*. IAEA, Vienna, pp. 149–163.

Fröhlich K., Ivanovich M., Hendry M. J., Andrews J. N., Davis S. N., Drimmie R. J., Fabryka-Martin J., Florkowski T., Fritz P., Lehmann B., Loosli H. H., and Nolte E. (1991) Application of isotopic methods to dating of very old groundwaters: Milk River aquifer, Alberta, Canada. *Appl. Geochem.* **6**, 465–472.

Gat J. R. (1980) The isotopes of hydrogen and oxygen in precipitation. In *Handbook of Environmental Isotope Geochemistry* (eds. P. Fritz and J. C. Fontes). Elsevier, Amsterdam, pp. 21–48.

Gelhar L. W. (1986) Stochastic subsurface hydrology from theory to applications. *Water Resour. Res.* **22**(suppl.), 135S–145S.

Gelhar L. W. (1993) *Stochastic Subsurface Hydrology*. Prentice Hall, Old Tappan, NJ.

Gerling E. K., Mamyrin B. A., Tolstikhin I. N., and Yaklovleva R. Z. (1971) Isotope composition of helium rocks (in Russian). *Geokhimiya* **5**, 608–617.

Goblet P. (1999) Programme METIS: simulation d'ecoulement et de Transport Miscible in Milieu Poreaux et Fracturé—Notice de conception—Mise à jour $1^{3r}/11/99$. CIG/LHM/RD/99/38.

Goode D. J. (1996) Direct simulation of groundwater age. *Water Resour. Res.* **32**, 289–296.

Gvirtzman H. and Magaritz M. (1986) Investigation of water movement in the unsaturated zone under an irrigated area using environmental tritium. *Water Resour. Res.* **22**, 635–642.

Gvirtzman H., Ronen D., and Magaritz M. (1986) Anion exclusion during transport through the unsaturated zone. *J. Hydrol.* **87**, 267–283.

Hamlin H. S. (1988) Depositional and groundwater flow systems of the Carrizo-Upper Wilcox, South Texas. *Texas Bureau of Economic Geology, Report of Investigations 175*, Austin.

Hanshaw B. B. and Back W. (1974) Determination of regional hydraulic conductivity through use of ^{14}C dating of groundwater. In *Memoirs de l'Association Internationale des Hydrogeologues*. Montpellier, France, vol. 10, pp. 195–196.

Hanshaw B. B., Back W., and Rubin M. (1965) Radiocarbon determinations for estimating groundwater flow velocities in central Florida. *Science* **148**, 494–495.

Heaton T. H. E. and Vogel J. C. (1981) "Excess air" in groundwater. *J. Hydrol.* **50**, 201–216.

Heaton T. H. E., Talma A. S., and Vogel J. C. (1983) Origin and history of nitrate in confined groundwater in the Western Kalahari. *J. Hydrol.* **62**, 243–262.

Hendry M. J. and Schwartz F. W. (1988) An alternative view on the origin of chemical and isotopic patterns in groundwater from the Milk River aquifer, Canada. *Water Resour. Res.* **24**, 1747–1766.

Hinkle S. R. (1996) Age of groundwater in basalt aquifers near Spring Creek National Fish Hatchery, Skkamania County, Washington. *US Geological Survey, Water-Resources Investigations Report 95-4272.*

Hiyagon G. and Kennedy B. M. (1992) Noble gases in CH$_4$-rich gas fields, Alberta, Canada. *Geochim. Cosmochim. Acta* **56**, 1569–1588.

Hubbert M. K. (1940) The theory of groundwater motion. *J. Geol.* **48**, 785–944.

Ingebritsen S. E. and Sanford W. E. (1998) *Groundwater in Geologic Processes*. Cambridge University Press, Cambridge.

Ingerson C. W. and Pearson F. J. (1964) Estimation of age and rate of motion of ground-water by the ^{14}C method. In *Recent Researches in the Field of Hydrosphere, Atmosphere, and Nuclear Geochemistry*, Sugawara Festival Volume. Maruzen Company, Tokyo, Japan, pp. 263–283.

Jacobson G., Cresswell R., Wischusen J., and Fifield K. (1998) Arid-zone groundwater recharge and paleorecharge: insights from the radiosiotope chlorine-36. *AGSO Res. Lett.* **29**, 1–3.

Kalin R. M. (2000) Radiocarbon dating of groundwater systems. In *Environmental Tracers in Subsurface Hydrology* (eds. P. Cook and A. L. Herczeg). Kluwer Academic, Dordrecht, pp. 111–144.

Kaufman A., Magaritz M., Paul M., Hillaire-Marcel C., Hollus G., Boaretto E., and Taieb M. (1990) The ^{36}Cl ages of the brines in the Magadi-Natron basin, East Africa. *Geochim. Cosmochim. Acta* **54**, 2827–2834.

Kendall C. and McDonnell J. J. (1998) *Isotope Tracers in Catchment Hydrology*. Elsevier, Amsterdam.

Kennedy B. M., Reynolds J. H., and Smith S. P. (1984) Helium isotopes: Lower Geyser Basin, Yellowstone National Park. *EOS* **65**, 304.

Kennedy B. M., Lynch M. A., Reynolds J. H., and Smith S. P. (1985) Intensive sampling of noble gases in fluids at Yellowstone: I. Early overview of the data, regional patterns. *Geochim. Cosmochim. Acta* **49**, 1251–1261.

Kennedy B. M., Hiyagon G., and Reynolds J. H. (1990) Crustal neon: a striking uniformity. *Earth Planet. Sci. Lett.* **98**, 277–286.

Kharkar D. P., Nijampurkar V. N., and Lal D. (1966) The global fallout of Si^{32} produced by cosmic rays. *Geochim. Cosmochim. Acta* **30**, 621–631.

Knauth L. P. and Beeunas M. A. (1986) Isotope geochemistry of fluid inclusions in Permian halite with implications for the isotopic history of ocean water and the origin of saline formation waters. *Geochim. Cosmochim. Acta* **50**, 419–433.

Konikow L. F. and Bredehoeft J. D. (1978) Computer model of two-dimensional solute transport and dispersion in ground water. *US Geological Survey, Techniques of Water Resources Investigations, Book 7, Chapter C2.*

Konikow L. F. and Bredehoeft J. D. (1992) Groundwater models cannot be validated. *Adv. Water Resour.* **15**, 75–83.

Kunz W. and Schintlmeister I. (1965) *Tabellen der atomekerne*. Akademie-Verlag, Berlin.

Kyser T. K. and Rison W. (1982) Systematics of rare gases in basic lavas and ultramafic xenoliths. *J. Geophys. Res.* **87**, 5611–5630.

Lal D. (1987) Production of ^{3}He in terrestrial rocks. *Chem. Geol. (Isotope Geosc. Sect.)* **66**, 89–98.

Lal D. and Peters B. (1967) Cosmic ray produced radioactivity on the Earth. In *Handbuch der Physik* (ed. K. Sitte). Springer, Berlin, vol. 46/2, pp. 551–612.

Lal D., Goldberg E. D., and Koide M. (1960) Cosmic-ray-produced ^{32}Si in nature. *Science* **313**, 332–337.

Lal D., Nijampurkar V. N., and Rama S. (1970) Silicon-32 hydrology. In *Isotope Hydrology 1970*. IAEA, Vienna, pp. 847–868.

Lehmann B. E., Loosli H. H., Rauber D., Thonnard N., and Willis R. D. (1991) ^{81}Kr and ^{85}Kr in groundwater, Milk River aquifer, Alberta, Canada. *Appl. Geochem.* **6**, 419–423.

Lehmann B. E., Davis S. N., and Fabryka-Martin J. T. (1993) Atmospheric and subsurface sources of stable and radioactive nuclides used for groundwater dating. *Water Resour. Res.* **29**, 2027–2040.

Lehmann B. E., Love A. H., Purtschert R., Collon P., Loosli H. H., Kutschera W., Beyerle U., Aeschbach-Hertig W., Kipfer R., Frape S., Herczeg A. L., Moran J. E., Tolstikhin I., and Gröning M. (2003) A comparison of groundwater dating with ^{81}Kr, ^{36}Cl, and ^{4}He in 4 wells of the Great Artesian Basin, Australia. *Earth Planet. Sci. Lett.* **211**, 237–250.

Lehmann B. E., Purtschert R., Loosli H. H., Love A. H., Collon P., Kutschera W., Beyerle U., Aeschbach-Hertig W., Kipfer R., and Frape S. (2002b) Kr-81 calibration of Cl-36 and He-4 evolution in the western Great Artesian Basin, Australia. *Geochim. Cosmochim. Acta* **66**(15A(suppl. 1)), A445.

Libby W. F. (1946) Atmospheric helium three and radiocarbon from cosmic radiation. *Phys. Rev.* **69**, 671–673.

Libby W. F. (1953) The potential usefulness of natural tritium. *Proc. Natl. Acad. Sci.* **39**, 245–247.

Liu X., Fehn U., and Teng R. T. D. (1997) Oil formation and fluid convection in Railroad Valley, NV: a study using cosmogenic isotopes to determine the onset of hydrocarbon migration. *Nucl. Instr. Meth. Phys. Res.* **B123**, 356–360.

Loosli H. H. (1983) A dating method with ^{39}Ar. *Earth Planet. Sci. Lett.* **63**, 51–62.

Loosli H. H. and Oeschger H. (1969) ^{37}Ar and ^{81}Kr in the atmosphere. *Earth Planet. Sci. Lett.* **7**, 67–71.

Loosli H. H., Lehmann B. E., and Balderer W. (1989) Argon-39, argon-37 and krypton 85 isotopes in Stripa groundwaters. *Geochim. Cosmochim. Acta* **53**, 1825–1829.

Loosli H. H., Lehmann B. E., and Smethie W. M., Jr. (2000) Noble gas radioisotopes: ^{37}Ar, ^{85}Kr, ^{39}Ar, ^{81}Kr. In *Environmental Tracers in Subsurface Hydrology* (eds. P. Cook and A. L. Herczeg). Kluwer Academic, Dordrecht, pp. 379–396.

Love A. J., Herczeg A. L., Sampson L., Cresswell R. G., and Fifield L. K. (2000) Sources of chloride and implications for ^{36}Cl dating of old groundwater, southwestern Great Artesian Basin, Australia. *Water Resour. Res.* **36**, 1561–1574.

Lucas L. L. and Unterweger M. P. (2000) Comprehensive review and critical evaluation of the half-life of tritium. *J. Res. Natl. Inst. Stand. Technol.* **105**, 541–549.

Maloszewski P. and Zuber A. (1993) Principles and practice of calibration and validation of mathematical models for the interpretation of environmental tracer data in aquifers. *Adv. Water Resour.* **16**, 173–190.

Mamyrin B. A. and Tolstikhin L. N. (1984) *Helium Isotopes in Nature*. Elsevier, Amsterdam.

Manning C. E. and Ingebritsen S. E. (1999) Permeability of the continental crust: implications of geothermal data and metamorphic systems. *Rev. Geophys.* **37**, 127–150.

Marine W. I. (1979) The use of naturally occurring helium to estimate groundwater velocities for studies of geologic storage of radioactive waste. *Water Resour. Res.* **15**, 1130–1136.

Martel D. J., O'Nions R. K., Hilton D. R., and Oxburgh E. R. (1990) The role of element distribution in production and release of radiogenic helium: the Carnmelllis Granite, southwest England. *Chem. Geol.* **88**, 207–221.

Marty B., Criaud A., and Fouillac C. (1988) Low enthalpy geothermal fluids from the Paris sedimentary basin. Characteristics and origin of gases. *Geothermics* **17**, 619–633.

Marty B., Torgersen T., Meynier V., O'Nions R. K., and de Marsily G. (1993) Helium isotope fluxes and groundwater ages in the Dogger aquifer, Paris Basin. *Water Resour. Res.* **29**, 1025–1035.

Mazor E. (1976) Multitracing and multisampling in hydrological studies. In *Interpretation of Environmental Isotope and Hydrochemical Data in Groundwater Hydrology*. IAEA, Vienna, pp. 7–36.

Mazor E. (1990) *Applied Chemical and Isotopic Groundwater Hydrology*. Halsted Press, New York.

Meinzer O. E. (1923) Outline of ground-water hydrology, with definitions. *US Geol. Surv., Water-Supply Pap.* 494.

Meynier V. and Marty B. (1990) Helium isotopes in fluids from the GPF3 scientific borehole at Couy, Cher County, France: implications for deep circulations in the Paris sedimentary

basin (abstract), 4th AGU Chapman Conference, Snowbird Utah.

Michel R. A. (2000) Sulphur-35. In *Environmental Tracers in Subsurface Hydrology* (eds. P. G. Cook and A. L. Herczeg). Kluwer Academic, Dordrecht, pp. 502–504.

Michel R. L., Campbell D., Clow D., and Turk J. T. (2000) Timescales for migration of atmospherically derived sulphate through an alpine/subalpine watershed, Loch Vale, Colorado. *Water Resour. Res.* **36**, 27–36.

Milton G. C. D., Milton G. M., Andrews H. R., Chant L. A., Cornett R. J. J., Davies W. G., Greiner B. F., Imanori Y., Koslowsky V. T., Kotzer T., Kramer S. J., and McKay J. W. (1997) A new interpretation of the distribution of bomb-produced chlorine-36 in the environment, with special reference to the Laurentian Great Lakes. *Nucl. Instr. Meth. Phys. Res.* **B123**, 382–386.

Mook W. G. (1976) The dissolution-exchange model for dating groundwater with carbon-14. In *Interpretation of Environmental Isotope and Hydrochemical Data in Groundwater Hydrology*, IAEA, Vienna, pp. 213–225.

Mook W. G. (1980) Carbon-14 in hydrogeological studies. In *Handbook of Environmental Isotope Geochemistry* (eds. P. Fritz and J. C. Fontes). Elsevier, Amsterdam, vol. 1, pp. 49–74.

Moran J. E., Fehn U., and Hanor J. S. (1995) Determination of source ages and migration patterns of brines from the US Gulf Coast basin using ^{129}I. *Geochim. Cosmochim. Acta* **59**, 5055–5069.

Moran J. E., Oktay S. D., and Santschi P. H. (2002) Sources of iodine and iodine 129 in rivers. *Water Resour. Res.* **38**(8), doi: 10.1029/2001WR000622.

Morgenstern U. (2000) Silicon-32. In *Environmental Tracers in Subsurface Hydrology* (eds. P. G. Cook and A. L. Herczeg). Kluwer Academic, Boston, pp. 499–502.

Morgenstern U., Gellermann R., Hebert D., Borner I., Stolz W., Vaikmae R., Rajamae R., and Putnik H. (1995) ^{32}Si in limestone aquifers. *Chem. Geol.* **120**, 127–134.

Morgenstern U., Taylor C. B., Parrat Y., Gaggeler H. W., and Eichler B. (1996) ^{32}Si in precipitation: evolution of temporal and spatial variation and as a dating tool for glacial ice. *Earth Planet. Sci. Lett.* **144**, 289–296.

Morrison P. and Pine J. (1955) Radiogenic origin of helium isotopes in rocks. *Ann. NY Acad. Sci.* **62**, 69–92.

Münnich K. O. (1957) Messungen des ^{14}C-Gehaltes vom hartem Grundwasser. *Naturwissenschaften* **44**, 32–33.

Murphy E. M., Davis S. N., Long A., Donahue D., and Jull A. J. T. (1989) ^{14}C in fractions of dissolved organic carbon in ground water. *Nature* **337**, 153–155.

Neuman S. P. (1990) Universal scaling of hydraulic conductivities and dispensivities in geologic media. *Water Resour. Res.* **26**, 1749–1758.

Neuman S. P. (1995) On advective transport in fractal permeability and velocity fields. *Water Resour. Res.* **31**, 1455–1460.

Neuzil C. E. (1995) Abnormal pressures as hydrodynamic phenomena. *Am. J. Sci.* **295**, 742–786.

Neuzil C. E. (2000) Osmotic generation of "anomalous" fluid pressures in geological environments. *Nature* **403**, 182–184.

Nijampurkar V. N., Amin B. S., Kharkar D. P., and Lal D. (1966) "Dating" ground waters of ages younger than 1,000–1,500 years using natural silicon-32. *Nature* **210**, 478–480.

Nijampurkar V. N., Rao D. K., Oldfield F., and Renberg I. (1998) The half-life of ^{32}Si: a new estimate based on varved lake sediments. *Earth Planet. Sci. Lett.* **163**, 191–196.

Nolte E., Krauthan P., Korschinek G., Maloszewski P., Fritz P., and Wolf M. (1991) Measurements and interpretations of ^{36}Cl in groundwater, Milk River aquifer, Alberta, Canada. *Appl. Geochem.* **6**, 435–445.

Nydal R., Lövseth K., Zumbrunn V., and Boden T. A. (1996) Carbon-14 measurements in atmospheric CO_2 from northern and southern hemisphere sites, 1962–1991. *Oak Ridge National Laboratory, US Department of Energy, NDP-057*, Oak Ridge, TN.

Oktay S. D., Santschi P. H., Moran J. E., and Sharma P. (2000) The ^{129}I bomb pulse recorded in Mississippi River Delta sediments: results from isotopes of I, Pu, Cs, Pb, and C. *Geochim. Cosmochim. Acta* **64**, 989–996.

O'Nions R. K. and Oxburgh E. R. (1983) Heat and helium in the Earth. *Nature* **306**, 429–431.

Oxburgh E. R., O'Nions R. K., and Hill R. I. (1986) Helium isotopes in sedimentary basins. *Nature* **324**, 632–635.

Ozima M. and Podosek F. A. (1983) *Noble Gas Geochemistry.* Cambridge University Press, Cambridge, 369pp.

Park J., Bethke C. M., Torgersen T., and Johnson T. M. (2002) Transport modeling applied to the interpretation of ground-water ^{36}Cl age. *Water Resour. Res.* **38**(5), doi: 10.1029/2001WR000399.

Pearson F. J. (1966) Ground-water ages and flow rates by the C^{14} method. PhD, University of Texas.

Pearson F. J., Jr., Balderer W., Loosli H. H., Lehmann B. E., Matter A., Peters T., Schmassmann H., and Gautschi A. (1991) *Applied Isotope Hydrogeology: A Case Study in Northern Switzerland.* Elsevier, New York.

Pearson F. J. J. and White D. E. (1967) Carbon-14 ages and flow rates of water in the Carrizo Sand, Atascosa County, Texas. *Water Resour. Res.* **3**, 251–261.

Pearson F. J. J., Noronha C. J., and Andrews R. W. (1983) Mathematical modeling of the distribution of natural ^{14}C, ^{234}U, and ^{238}U in a regional ground-water system. *Radiocarbon* **25**, 291–300.

Perkins T. K. and Johnston O. C. (1963) A review of diffusion and dispersion in porous media. *Soc. Petrol. Eng. J.* **3**, 70–83.

Person M. A. and Baumgartner L. (1995) New evidence for long-distance fluid migration within the Earth's crust. *Rev. Geophys.* (Supplement, July 1995, US National Report to the International Union of Geodesy and Geophysics 1991–1994), 1083–1092.

Person M. A. and Garven G. (1992) Hydrologic constraints on petroleum generation within continental rift basins: theory and application to the Rhine Graben. *Am. Assoc. Petrol. Geologists Bull.* **76**, 468–488.

Phillips F. M. (1993) Comment on "Reinterpretation of ^{36}Cl data: physical processes, hydraulic interconnections and age estimates in groundwater systems" by E Mazor. *Appl. Geochem* **8**, 643–647.

Phillips F. M. (2000) Chlorine-36. In *Environmental Tracers in Subsurface Hydrology* (eds. P. Cook and A. L. Herczeg). Kluwer Academic, Dordrecht, pp. 299–348.

Phillips F. M., Bentley H. W., Davis S. N., Elmore D., and Swannick G. B. (1986) Chlorine-36 dating of very old ground water: II. Milk River aquifer, Alberta. *Water Resour. Res.* **22**, 2003–2016.

Phillips F. M., Mattick J. L., Duval T. A., Elmore D., and Kubik P. W. (1988) Chlorine-36 and tritium from nuclear weapons fallout as tracers for long-term liquid and vapor movement in desert soils. *Water Resour. Res.* **24**, 1877–1891.

Phillips F. M., Tansey M. D., Peeters L. A., Cheng S., and Long A. (1989) An isotopic investigation of groundwater in the central San Juan Basin, New Mexico: carbon-14 dating as a basis for numerical flow modeling. *Water Resour. Res.* **25**, 2259–2273.

Phillips F. M., Knowlton R. G., and Bentley H. W. (1990) Comment on Hendry and Schwartz (1988). *Water Resour. Res.* **26**, 1693–1698.

Phillips O. M. (1991) *Flow and Reactions in Permeable Rocks.* Cambridge University Press, Cambridge.

Pinti D. L. and Marty B. (1995) Noble gases in crude oil from the Paris Basin, France: implications for the orgin of fluids and constraints on oil–water–gas interactions. *Geochem. Cosmochim. Acta* **59**, 3389–3404.

Plummer L. N. (1993) Stable isotope enrichment in paleo-waters of the Southeast Atlantic Coastal Plain, United States. *Science* **262**, 2016–2020.

Plummer L. N. and Busenberg E. (2000) Chlorofluorocarbons. In *Environmental Tracers in Subsurface Hydrology*

(eds. P. Cook and A. L. Herczeg). Kluwer Academic, Dordrecht, pp. 441–478.

Plummer L. N., Prestemon E. C., and Parkhurst D. L. (1991) An interactive code (NETPATH) for modeling *net* geo-chemical reactions along a flow path. *US Geological Survey, Water-Resources Investigations Report 91-4078.*

Plummer L. N., Dunkle S. A., and Busenberg E. (1993) Data on chlorofluorocarbons (CCl_3F and CCl_2F_2) as dating tools and hydrologic tracers in shallow groundwater of the Delmarva Peninsula. *US Geological Survey, Open File Report 93-484.*

Prickett T. A., Naymik T. C., and Lonnquist C. G. (1981) A random-walk solute transport model for selected groundwa-ter quality evaluations. *Illinois State Water Survey, Bulletin 65,* Champaign, Illinois.

Puckett L. J., Cowdery T. J., McMahon P. B., Tornes L. H., and Stoner J. D. (2002) Using chemical, hydrologic, and age dating analysis to delineate redox processes and flow paths in the riparian zone of a glacial outwash aquifer-stream system. *Water Resour. Res.* **38**(8), doi: 10.1029/2001WR000396.

Purdy C. B., Burr G. S., Rubin M., Helz G. R., and Mignerey A. C. (1992) Dissolved organic and inorganic ^{14}C concentrations and ages for coastal plain aquifers in southern Maryland. *Radiocarbon* **34**, 654–663.

Purdy C. B., Helz G. R., Mignerey A. C., Kubik P. W., Elmore D., Sharma P., and Hemmick T. (1996) Aquia aqui-fer dissolved Cl^- and ^{36}Cl/Cl: implications for flow velo-cities. *Water Resour. Res.* **32**, 1163–1172.

Purtschert R. (1997) Multitracer-Studien in der Hydrologie: Anwendugen im Glattal, am Wellenberg und in Vals. PhD Dissertation, University of Bern.

Rademacher L. K., Clark J. F., Hudson G. B., Erman D. C., and Erman N. A. (2001) Chemical evolution of shallow ground-water as recorded by springs, Sagehen basin: Nevada County, California. *Chem. Geol.* **179**, 37–51.

Rao U. and Fehn U. (1999) Sources and reservoirs of anthro-pogenic iodine-129 in western New York. *Geochim. Cos-mochim. Acta* **63**, 1927–1938.

Reilly T. E., Plummer L. N., Phillips P. J., and Busenberg E. (1994) The use of simulation and multiple environmental tracers to quantify groundwater flow in a shallow aquifer. *Water Resour. Res.* **30**, 412–434.

Rison W. and Craig H. (1983) Helium isotopes and mantle volatiles in Loihi Seamount and Hawaiian Island basalts and xenoliths. *Earth Planet. Sci. Lett.* **66**, 407–426.

Robertson W. D. and Cherry J. A. (1989) Tritium as an indica-tor of recharge and dispersion in a groundwater system in central Ontario. *Water Resour. Res.* **25**, 1097–1109.

Rozanski K. and Florkowski T. (1979) Krypton-85 dating of groundwater. In *Isotope Hydrology 1979.* IAEA, Vienna, vol. II, pp. 949–961.

Sanford W. E. (1997) Correcting for diffusion in carbon-14 dating of ground water. *Ground Water* **35**, 357–361.

Scanlon B. R. (1992) Evaluation of liquid and vapor water flow in desert soils based on chlorine-36 and tritium tracers and nonisothermal flow simulations. *Water Resour. Res.* **28**, 285–298.

Scanlon B. R., Tyler S. W., and Wierenga P. J. (1997) Hydro-logic issues in arid, unsaturated systems and implications for contaminant transport. *Rev. Geophys.* **35**, 461–490.

Schlosser P., Stute M., Dörr H., Sonntag C., and Münnich K. O. (1988) Tritium/^{3}He dating of shallow groundwater. *Earth Planet. Sci. Lett.* **89**, 353–362.

Schlosser P., Stute M., Sonntag C., and Münnich K. O. (1989) Tritiogenic ^{3}He in shallow groundwater. *Earth Planet. Sci. Lett.* **94**, 245–256.

Schmalz B. L. and Polzer W. L. (1969) Tritiated water distri-bution in unsaturated soils. *Soil Sci.* **108**, 43–47.

Schwartz F. W. (1977) Macroscopic dispersion in porous media: the controlling factors. *Water Resour. Res.* **13**, 743–752.

Schwartz F. W. and Muehlenbachs K. (1979) Isotope and ion geochemistry of groundwaters in the Milk River aquifer, Alberta. *Water Resour. Res.* **15**, 259–268.

Shapiro S. D., LeBlanc D. R., Schlosser P., and Ludin A. (1999) Characterizing a sewage plume using the ^{3}H–^{3}He technique. *Ground Water* **37**, 861–878.

Shukolyukov Y. A., Sharif-Zade V. B., and Ashkinadze G. S. (1973) Neon isotopes in natural gases. *Geochem. Int.,* **10**(2), 346–354.

Simunek J., Sejna M., and van Genuchten M. T. (1999) The HYDRUS-2D software package for simulating the two-dimensional movement of water, heat and multipe solutes in variably-saturated media—version 2.0. *US Department of Agriculture—US Salinity Laboratory*, Riverside, California, 256pp.

Smethie W. M., Jr. and Mathieu G. (1986) Measurement of krypton-85 in the ocean. *Mar. Chem.* **18**, 17–33.

Smethie W. M., Solomon D. K., Schiff S. L., and Mathieu G. G. (1992) Tracing groundwater flow in the Borden aquifer using krypton-85. *J. Hydrol.* **130**, 279–297.

Solomon D. K. (2000) ^{4}He in groundwater. In *Environmental Tracers in Subsurface Hydrology* (eds. P. Cook and A. L. Herczeg). Kluwer Academic, Dordrecht, pp. 425–440.

Solomon D. K. and Cook P. G. (2000) ^{3}H and ^{3}He. In *Environmental Tracers in Subsurface Hydrology* (eds. P. Cook and A. L. Herczeg). Kluwer Academic, Dordrecht, pp. 397–424.

Solomon D. K., Poreda R. J., Schiff S. L., and Cherry J. A. (1992) Tritium and helium-3 as groundwater age tracers in the Borden aquifer. *Water Resour. Res.* **28**, 741–756.

Solomon D. K., Schiff S. L., Poreda R. J., and Clarke W. B. (1993) A validation of the ^{3}H/^{3}He method for determining groundwater recharge. *Water Resour. Res.* **29**, 2951–2962.

Solomon D. K., Hunt A., and Poreda R. J. (1996) Source of radiogenic helium 4 in shallow aquifers: implications for dating young groundwater. *Water Resour. Res.* **32**, 1805–1813.

Steiger R. H. and Jäger E. (1977) Subcommission on geochronology: convention on the use of decay constant in gas and cosmochronology. *Earth Planet. Sci. Lett.* **36**, 359–362.

Stephens D. B. (1996) *Vadose Zone Hydrology*. Lewis Publishers, Boca Raton, FL.

Stute M. and Schlosser P. (1993) Principles and applications of the noble gas paleothermometer. In *Climate Change in Continental Isotopic Records*, Geophysical Monograph 78 (eds. P. K. Swart; K. C. Lohmann, J. A. McKenzie, and S. Savin). American Geophysical Union, Washington, DC, pp. 89–100.

Stute M., Schlosser P., Clark J. F., and Broecker W. S. (1992a) Paleotemperatures in the southwestern United States derived from noble gases in ground water. *Science* **256**, 1000–1002.

Stute M., Sonntag C., Deak J., and Schlosser P. (1992b) Helium in deep circulating groundwater in the Great Hungarian Plain: flow dynamics and crustal and mantle helium fluxes. *Geochim. Cosmochim. Acta* **56**, 2051–2067.

Stute M., Clark J. F., Schlosser P., Broecker W. S., and Bonani G. (1995) A 30,000 yr continental paleotemperature record derived from noble gases dissolved in groundwater from the San Juan Basin, New Mexico. *Quat. Res.* **43**, 209–220.

Sudicky E. A. and Frind E. O. (1981) Carbon-14 dating of groundwater in confined aquifers: implications of aquitard diffusion. *Water Resour. Res.* **17**, 1060–1064.

Suecker J. K., Turk J. T., and Michel R. L. (1999) Use of cosmogenic sulfur-35 for comparing ages of water from three alpine–subalpine basins in the Colorado Front Range. *Geomorphology* **27**, 61–74.

Sugisaki R. (1961) Measurement of effective flow velocity by means of dissolved gasses. *Am. J. Sci.* **259**, 144–153.

Synal H.-A., Beer J., Bonani G., Suter M., and Wölfli W. (1990) Atmospheric transport of bomb-produced ^{36}Cl. *Nucl. Instr. Meth. Phys. Res.* **B52**, 483–488.

Szabo Z., Rice D. E., Plummer L. N., Busenberg E., Drenkard S., and Schlosser P. (1996) Age dating of shallow groundwater with chlorofluorocarbons, tritium/helium 3, and flow path analysis, southern New Jersey coastal plain. *Water Resour. Res.* **32**, 1023–1038.

Tamers M. A. (1967) Radiocarbon ages of groundwater in an arid zone unconfined aquifer. In *Isotope Techniques in the Hydrologic Cycle*, Monograph No. 11 (ed. G. E. Stout). American Geophysical Union, Washington, DC, pp. 143–152.

Tamers M. A. (1975) Validity of radiocarbon dates on ground water. *Geophys. Surv.* **2**, 217–239.

Taylor C. B. (1981) Present status and trands in electrolytic enrichment of low-level tritium in water. In *Methods of Low-level Counting and Spectrometry*. IAEA, Vienna, pp. 303–323.

Thompson G. M. (1976) Trichloromethane: a new hydrologic tool for tracing and dating ground water. PhD, University of Indiana.

Thompson G. M., Hayes J. M., and Davis S. N. (1974) Fluorocarbon tracers in hydrology. *Geophys. Res. Lett.* **1**, 177–180.

Thonnard N., McKay L. D., Cumbie D. H., and Joyner C. P. (1997) Status of laser-based krypton-85 analysis development for dating of groundwater. *Abstr. Prog., Geol. Soc. Am. Ann. Meet.* **29**(6), 78.

Tolstikhin I. N. and Kamensky I. L. (1968) Determination of groundwater age by the T–^{3}He method. *Geochem. Int.* **6**, 810–811.

Torgersen T. (1980) Controls on pore-fluid concentration of helium-4 and radon-222 and calculation of helium-4/radon-222 ages. *J. Geochem. Explor.* **13**, 57–75.

Torgersen T. and Clarke W. B. (1985) Helium accumulation in groundwater: I. An evaluation of sources and the continental flux of crustal ^{4}He in the Great Artesian Basin, Australia. *Geochim. Cosmochim. Acta* **49**, 1211–1218.

Torgersen T. and Ivey G. N. (1985) Helium accumulation in groundwater: II. A model for the accumulation of the crustal ^{4}He degassing flux. *Geochim. Cosmochim. Acta* **49**, 2445–2452.

Torgersen T. and Phillips F. M. (1993) Reply to "Comment on 'Chlorine 36 dating of very old groundwater: 3. Further results on the Great Artesian Basin, Australia' by T. Torgerson *et al.*" by J. N. Andrews and J.-C. Fontes. *Water Resour. Res.* **29**, 1875–1877.

Torgersen T., Kennedy B. M., Hiyagun H., Chiou K. Y., Reynolds J. H., and Clark W. B. (1989) Argon accumulation and the crustal degassing flux of ^{40}Ar in the Great Artesian Basin, Australia. *Earth Planet. Sci. Lett.* **92**, 43–56.

Torgersen T., Habermehl M. A., Phillips F. M., Elmore D., Kubik P., Jones B. G., Hemmick T., and Gove H. E. (1991) Chlorine-36 dating of very old groundwater: III. Further studies in the Great Artesian Basin, Australia. *Water Resour. Res.* **27**, 3201–3214.

Torgersen T., Habermehl M. A., and Clarke W. B. (1992) Crustal helium fluxes and heat flow in the Great Artesian Basin, Australia. *Chem. Geol.* **102**, 139–152.

Tóth J. (1963) A theoretical analysis of groundwater flow in small drainage basins. *J. Geophys. Res.* **67**, 4375–4387.

Tullborg E.-L. and Gustafsson E. (1999) ^{14}C in bicarbonate and dissolved organics—a useful tracer. *Appl. Geochem.* **14**, 927–938.

Tyler S. W., Chapman J. B., Conrad S. H., Hammermeister D. P., Blout D. O., Miller J. J., Sully M. J., and Ginanni J. M. (1996) Soil–water flux in the southern Great Basin, United States: temporal and spatial variations over the last 120,000 years. *Water Resour. Res.* **32**, 1481–1499.

Varni M. and Carrera J. (1998) Simulation of groundwater age distributions. *Water Resour. Res.* **34**, 3271–3282.

Vogel J. C. (1970) Carbon-14 dating of groundwater. In *Isotope Hydrology 1970*. IAEA, Vienna, pp. 225–239.

Vogel J. C., Thilo L., and Van Dijken M. (1974) Determination of groundwater recharge with tritium. *J. Hydrol.* **23**, 131–140.

Vogel J. C., Talma A. S., and Heaton T. H. E. (1982) The age and isotopic composition of groundwater in the Stampriet Artesian Basin, SWA. *National Physical Research Laboratory*, CSIR, Pretoria, South Africa, 49pp.

Wagner M. J. M., Dittrich-Hannen B., Synal H.-A., Suter M., and Schotterer U. (1996) Increase of ^{129}I in the environment. *Nucl. Instr. Meth. Phys. Res.* **B123**, 367–370.

Walvoord M., Phillips F. M., Tyler S. W., and Hartsough P. C. (2002a) Deep arid system hydrodynamics: Part 2.

Application to paleohydrologic reconstruction using vadose-zone profiles from the northern Mojave Desert. *Water Resour. Res.* **38**(12), 1291, doi: 10.1029/2001WR000925.

Walvoord M., Plummer M. A., Phillips F. M., and Wolfsberg A. V. (2002b) Deep arid system hydrodynamics: Part 1. Equilibrium states and response times in thick desert vadose zones. *Water Resour. Res.* **38**(12), 1308, doi: 10.0129/2001WR000824.

Walvoord M. A. (2002) A unifying conceptual model to describe water, vapor, and solute transport in deep arid vadose zones. PhD, New Mexico Institute of Mining and Technology.

Wei H. F., Ledoux E., and de Marsily G. (1990) Regional modeling of groundwater flow and salt and environmental tracer transport in deep aquifers in the Paris Basin. *J. Hydrol.* **120**, 341–358.

Weise S. M. (1986) Heliumisotopen-Gehalte im Grumdwasser, Messung und Interpretation. PhD Dissertation, University of Munchen.

Weise S. M. and Moser H. (1987) Groundwater dating with helium isotopes. In *Techniques in Water Resources Development*. IAEA, Vienna, pp. 105–126.

Weiss W. H., Sartorius H., and Stockburger H. (1992) The global distribution of atmospheric krypton-85: a data base for the verification of transport and mixing models. In *Isotopes of Noble Gases as Tracers in Environmental Studies*. IAEA, Vienna, pp. 105–126.

Weissmann G. S., Carle S. F., and Fogg G. E. (1999) Three-dimensional hydrofacies modeling based on soil surveys and transition probability geostatistics. *Water Resour. Res.* **35**, 1761–1770.

Wetherill G. W. (1954) Variations in the isotopic abundances of neon and argon extracted from radioactive materials. *Phys. Rev.* **96**, 679–683.

Weyhenmeyer C. E., Burns S. J., Waber H. N., Aesbach-Hertig W., Kipfer R., Loosli H. H., and Matter A. (2000) Cool glacial temperatures and changes in moisture source recorded in Oman groundwaters. *Science* **287**, 842–845.

Zerle L., Faestermann T., Knie K., Korschinek G., and Nolte E. (1997) The ^{41}Ca bomb pulse and atmospheric transport of radionuclides. *J. Geophys. Res.* **102**, 19517–19527.

Zhao X., Fritzel T. L. B., Quinodoz H. A. M., Bethke C. M., and Torgersen T. (1998) Controls on the distribution and isotopic composition of helium in deep ground-water flows. *Geology* **26**, 291–294.

Zhu C. (2000) Estimate of recharge from radiocarbon dating of groundwater and numerical flow and transport modeling. *Water Resour. Res.* **36**, 2607–2620.

Zhu C. and Murphy W. M. (2000) On radiocarbon dating of ground water. *Ground Water* **38**, 802–804.

Zuber A. (1986) On the interpretation of tracer data in variable flow systems. *J. Hydrol.* **86**, 45–57.

Zuber A., Weise S. M., Osenbrück K., Pajnowska H., and Grabczak J. (2000) Age and recharge pattern of water in the Oligocene of the Mazovian basin (Poland) as indicated by environmental tracers. *J. Hydrol.* **233**, 174–188.

Radioactive Geochronometry
ISBN: 978-0-08-096708-0

pp. 289–334

11
Cosmogenic Nuclides in Weathering and Erosion

D. E. Granger

Purdue University, West Lafayette, IN, USA

and

C. S. Riebe

Stillwater Sciences, Berkeley, CA, USA

11.1   INTRODUCTION

Chemical weathering and physical erosion encapsulate a diverse suite of processes that sculpt landscapes, generate soil, and deliver sediment, nutrients, and solutes to streams and the oceans. Quantifying chemical weathering and physical erosion rates is important across a diverse range of disciplines in geology, geomorphology, and biogeochemistry. Yet rates of chemical weathering and physical erosion have until recently been difficult to quantify over the timescales of soil formation and transport. In this chapter, we discuss how cosmogenic nuclide methods have provided a wealth of new opportunities for dating surfaces, measuring denudation rates, and quantifying chemical weathering rates.

Cosmogenic nuclides are produced in mineral grains by secondary cosmic rays that penetrate only the topmost few meters of soil and rock at the ground surface. Because cosmogenic nuclide production rates are rapidly attenuated with depth, the concentration of cosmogenic nuclides in a mineral grain tells us how much time it has spent near the surface, or how rapidly material has been removed from above it (Lal, 1991). From the perspective of cosmogenic nuclide production, weathering and erosion can be considered simply in terms of the translocation of mass as mineral grains are eroded from depth, detached from bedrock, and transported through soils by physical erosion and chemical weathering.

We will discuss four general types of weathering-related problems that can be addressed with cosmogenic nuclides. These include (1) surface exposure dating of rock and soil, (2) determining erosion rates of rock and soil from samples at the surface and at depth, (3) determining spatially averaged erosion rates from sediment, and (4) inferring chemical weathering rates using a geochemical mass-balance approach.

11.1.1   Definitions and Nomenclature

For clarity, and because the terminology is variable across disciplines, it is important to establish the nomenclature that we will use in this chapter. We use the term physical erosion to refer exclusively to mass loss by the physical removal of mineral grains. Examples of physical processes that transport minerals include (but are not limited to) bioturbation by tree throw, rooting, or animal burrowing, rainsplash, freeze–thaw, and landsliding. We use the term chemical weathering to refer to mass loss associated with the alteration and dissolution of minerals by meteoric water. We use the term denudation to refer to the total mass loss due to the sum of physical erosion and chemical weathering, as shown in the following equation:

$$D = E + W \qquad (1)$$

where D is the denudation rate, E the physical erosion rate, and W the chemical weathering rate.

We use the term bedrock to refer exclusively to unweathered rock. The term saprolite refers to weathered rock that retains recognizable structure of the bedrock. Following the usage common in the geochemical literature, we use the term soil instead of regolith to refer to material that has been mobilized, irrespective of its state of weathering or degree of horizonation. When we discuss weathering, the term parent material refers to the source of the weathered soil. Parent material may be either unweathered bedrock, or in the case of a terrace or moraine, partly weathered sediment. Additional definitions and nomenclature are given in Table 1.

11.1.2   Quantifying Weathering and Erosion

Weathering and erosion of soil and bedrock are important across a tremendously broad range of issues. For example, because soils and the nutrients they contain are vital for life, quantifying rates of soil formation, weathering, and erosion is important for assessing the impacts of modern land use and for balancing biogeochemical budgets, including the terrestrial carbon cycle (Yoo *et al.*, 2006). Because weathering and erosion supply sediment and solutes to rivers and the oceans, they are important regulators of aquatic habitat quality, and set the rate of sedimentation in reservoirs and navigational channels. The pace of chemical weathering determines the acid neutralization capacity of soil (e.g., Kirchner and Lydersen, 1995), and over much longer timescales silicate weathering modulates ocean alkalinity, affecting atmospheric CO_2 concentrations and thus regulating global climate via the greenhouse effect (e.g., Raymo *et al.*, 1988).

Despite the importance of weathering and erosion, their rates have been historically difficult

Table 1 Nomenclature.

Symbol	Definition
E	Physical erosion rate ($t\,km^{-2}\,yr^{-1}$)
W	Chemical weathering rate ($t\,km^{-2}\,yr^{-1}$)
D	Total denudation rate ($t\,km^{-2}\,yr^{-1}$)
W_X	Chemical weathering rate of a specific element or mineral (denoted in subscript)
h	Soil thickness
M_W	Total mass lost due to weathering
M_{WX}	Mass lost due to weathering of an individual element X, denoted in subscript
x	Depth beneath surface
ρ	Density
P_n	Cosmogenic nuclide production rate by nucleon spallation
P_1	Constant for cosmogenic nuclide production rate by nucleon spallation
L_1	Penetration length of nucleon spallation reactions
P_μ	Cosmogenic nuclide production rate by negative muon capture
P_2	Constant for cosmogenic nuclide production rate by negative muon capture
P_3	Constant for cosmogenic nuclide production rate by negative muon capture
L_2	Effective penetration length for negative muon capture
L_3	Effective penetration length for negative muon capture
P_f	Production rate due to reactions with fast muons
P_4	Coefficient for production by fast muons
L_4	Effective penetration length for production by fast muons
N	Concentration of cosmogenic nuclides
τ	Radioactive meanlife
t	Time
t_{soil}	Average residence time of material in the soil
N_{inh}	Inherited cosmogenic nuclide concentration
$N_{bedrock}$	Cosmogenic nuclide concentration in bedrock beneath soil cover
f_B	Fractional quartz concentration in bedrock
f_S	Fractional quartz concentration in soil
N_{stream}	Average cosmogenic nuclide concentration in stream sediment
A	Sediment contributing area
$\langle D \rangle$	Spatially averaged denudation rate
z	Altitude
$[X]_{parent}$	Average concentration of element X in parent material
$[X]_{soil}$	Average concentration of element X in soil
$[X]_{saprolite}$	Average concentration of element X in saprolite
$I_{aeolian}$	Aeolian material flux
K	Parameter similar to diffusion coefficient in linear sediment transport model

to measure. Denudation rates are typically far too slow (i.e., at tenths or hundredths of a millimeter per year) for direct observation. Moreover, in places where denudation rates can be determined from yields of sediment and solutes, it is difficult to know whether they are affected by modern land use or rare but important storms and landslides. These problems led Wahrhaftig (1970) to lament "it is questionable that we can find places in temperate climates to measure the rate of erosion uninfluenced by Man [*sic*]."

11.1.2.1 Quantifying physical erosion rates

Estimates of physical erosion rates have historically relied upon extended monitoring of sediment yields over years to decades. Sediment yields have generally been inferred either from sediment-rating curves estimated at stream gauges, or from sediment traps (e.g., Clayton and Megahan, 1986). Although these methods can produce reliable measurements of erosion rates, they can sometimes be dominated by the effects of land use (Milliman *et al.*, 1987) or changes in sediment storage (Trimble, 1977). Moreover, results from sediment yield studies may not be representative of the long-term mean if erosion is driven by episodic processes (such as landsliding) or extreme climatic events. Such episodic erosion may account for a significant fraction of a catchment's long-term average erosion rate, particularly in mountainous terrain (Kirchner *et al.*, 2001).

Over longer timescales, physical erosion rates can sometimes be estimated from the volumes of sedimentary deposits in basins (e.g., Guillaume and Guillaume, 1982), alluvial fans (Beaty, 1970) or colluvial hollows (Dietrich *et al.*, 1982; Reneau *et al.*, 1989). But these methods require both a dateable stratigraphic marker and a sediment reservoir with good trapping efficiency. Alternatively, in special circumstances denudation rates can be estimated from the volume of valleys incised into dateable surfaces such as volcanic flows (Ruxton and McDougall, 1967; Seidl *et al.*, 1994) or marine terraces (Chappell, 1974).

Despite drawbacks and confounding factors, traditional measurements of sediment yield in large rivers and catchments have provided a wealth of information concerning the effects of land use, tectonic setting, mountain relief, and long-term climate change on erosion and denudation. However, it would be helpful to have a more reliable way to measure long-term average denudation rates in a wide range of settings. Cosmogenic nuclides provide just such a method, as will be shown in Section 11.3.

11.1.2.2   *Quantifying chemical weathering rates*

Like physical erosion rates, chemical weathering rates may also be estimated by monitoring fluxes of material—in this case with measurements of solutes in streamwater. To convert solute flux measurements into weathering rates it is necessary to account for chemical additions and/or losses from rainfall, dryfall, litter decomposition, ion exchange, and biomass uptake. Although this approach can provide an accurate picture of weathering, many of the additions and losses in the mass balance are difficult to quantify, with uncertainties that build up rapidly in the overall estimate of weathering rate. Moreover, in many of the large watersheds that have been studied, the effects of secondary weathering in floodplains and colluvial hollows may be difficult to distinguish from primary chemical weathering of rock and soil on hillslopes.

Chemical weathering rates have also been measured from chemical depletion and enrichment measurements in soils of known age (April *et al.*, 1986; Brimhall and Dietrich, 1987; Bain *et al.*, 1993; Taylor and Blum, 1995). This approach, based on mass balance of elements in soils and parent rock, averages weathering rates over the timescales of pedogenesis, thus making it ideal for studying soil development, watershed geochemistry, and the influence of climate on weathering. However, because erosion rates have been difficult to quantify, the mass balance has rarely been fully solved for the weathering rate of an eroding soil. Instead, most studies have considered only soils that are not eroding—a condition that is only rarely met, for example, on terraces or glacial moraines.

Chemical weathering rates should depend on physical erosion rates, which control the supply of fresh mineral surfaces to soils. However, comparisons between rates of physical erosion and chemical weathering have only been made for a few large basins where both types of measurements happen to be available (e.g., Louvat and Allègre, 1997; Gaillardet *et al.*, 1999). The paucity of data, combined with the complications of heterogeneous lithology and climate within large watersheds, has made it difficult to determine the functional relationships between physical erosion and chemical weathering. What is needed is a way to measure rates of physical erosion and chemical weathering together, over comparable timescales, on a hillslope or small catchment scale. As will be shown in Section 11.4, this is now possible with the help of cosmogenic nuclide methods.

11.1.2.3   *The advent of cosmogenic nuclide methods*

Cosmogenic nuclide methods for determining denudation rates are an outgrowth of the development of accelerator mass spectrometry (AMS) in the 1980s. AMS is an ultrasensitive analytical technique capable of measuring isotope ratios as low as 1×10^{-15}, allowing routine measurements of the very low concentrations of *in situ*-produced cosmogenic nuclides in mineral grains. Geoscience applications of AMS rapidly grew to include surface exposure dating (Phillips *et al.*, 1990), the measurement of rock erosion rates (Lal, 1991), and determination of catchment-averaged denudation rates (Brown *et al.*, 1995; Bierman and Steig, 1996; Granger *et al.*, 1996). Over the past 5 years applications have expanded to include the combination of cosmogenic methods with traditional soil mass-balance measurements of chemical depletion and enrichment to determine long-term average chemical weathering rates (Riebe *et al.*, 2003, 2004a, b; Green *et al.*, 2006).

Our goal in this chapter is to explain cosmogenic nuclide methods in a way that introduces the reader to a variety of methods in sufficient detail to understand the fundamental theory, including the various assumptions that must be made to apply them to real landscapes. We do not attempt either a comprehensive review or a detailed discussion of cosmogenic nuclide chemistry and physics. We refer the interested reader instead to more detailed reviews (Nishiizumi *et al.*, 1993; Cerling and Craig, 1994; Bierman, 1994; Bierman *et al.*, 2002; Bierman and Nichols, 2004; Muzikar *et al.*, 2003). A more comprehensive review of cosmogenic nuclide methodology is available in Gosse and Phillips (2001).

To explain cosmogenic nuclide methods, it is first necessary to give an introduction to cosmic rays and cosmogenic nuclide production.

11.2   COSMOGENIC NUCLIDE SYSTEMATICS AT EARTH'S SURFACE

11.2.1   Cosmic Rays

Cosmic rays are energetic particles that continuously bombard the Earth. Cosmogenic nuclides are produced in the atmosphere and in mineral grains near the ground surface as the product of nuclear reactions induced by these cosmic rays (Lal and Peters, 1967). Primary cosmic rays interact with atoms in the upper atmosphere, initiating cascades of secondary particles that shower the ground surface. These secondary particles in turn produce most of the cosmogenic nuclides in the atmosphere and all of the cosmogenic nuclides in mineral grains. These secondary particles can include (1) individual neutrons and protons, (2) larger nuclear fragments, and (3) unstable particles such as pions and muons (Lal and Peters, 1967). Individual protons and neutrons produced in the cascade have energies limited by recoil momentum to $\sim$500 MeV. Recoil nucleons can collide with other nuclei to produce 2–3 additional nucleons. More energetic nuclear fragments produced in the cascade may have enough energy to break up yet another nucleus and thus propagate the nuclear cascade. Unstable pions are also produced in energetic reactions. Free pions have a mean life of $\sim$10 ns. If they are unable to react with a nucleus in that time, they will decay to produce a muon, another unstable particle with a mean life of $\sim$2.2 µs. Pions in the lower atmosphere nearly always collide with a nucleus before they decay, but at altitudes of $\sim$10 km the atmosphere is sufficiently rarefied that many of the pions decay to muons (Lal and Peters, 1967). These muons in turn travel to the ground surface and are able to penetrate many meters into rock.

In the lower atmosphere, near the ground surface, the secondary cosmic ray flux responsible for cosmogenic nuclide production is split almost evenly between neutrons and muons (Rossi, 1948; Hayakawa, 1969). However, because muons are far less likely than neutrons to interact with a nucleus, production at the surface is dominated by neutron reactions. Muons dominate production at depths greater than $\sim$3 m (Heisinger *et al.*, 2002a, b).

11.2.2 Cosmogenic Nuclide Production

The products of cosmic-ray reactions are called cosmogenic nuclides; those in the atmosphere are referred to as meteoric or garden variety, and those within mineral grains are referred to as *in situ*-produced. It has been estimated that there are $\sim 10^2$–10^3 *in situ*-produced cosmogenic nuclides

generated per gram of rock per year at the ground surface (Lal and Peters, 1967). The vast majority of these cosmogenic nuclides, however, are not detectable because they are identical to nuclides already native to the rock. To produce a detectable cosmogenic nuclide requires that a target nuclide be transformed to one that is exceedingly rare. Cosmogenic nuclides in common use include meteoric ^{14}C, ^{10}Be, and ^{36}Cl, and *in situ*-produced ^{14}C, ^{10}Be, ^{26}Al, ^{3}He, ^{21}Ne, and ^{36}Cl (Table 2). They fall into two groups: (1) radionuclides that are not created by decay of other radionuclides (such that any radionuclides incorporated into minerals during rock formation have decayed away) and (2) noble gases that are not incorporated into mineral crystals. Of the nuclides listed in Table 2, *in situ*-produced ^{10}Be has had by far the largest impact on the study of weathering and erosion. We will focus this chapter, therefore, almost entirely on the use of this nuclide. However, most of what we write can be applied directly to the other cosmogenic nuclides, with appropriate modifications to account for differences in production mechanisms and decay constants.

Meteoric cosmogenic nuclides have been used as tracers in studies of weathering and erosion. The systematics of fallout radionuclides as tracers is fundamentally different from that of *in situ*-produced ^{10}Be. We will not discuss meteoric ^{10}Be, but refer the interested reader to an excellent review by Morris *et al.* (2002).

To understand how ^{10}Be and other cosmogenic nuclides can be used to quantify denudation rates, it helps one to first understand how they are produced within a grain. Production of ^{10}Be within the uppermost few meters of rock and soil is dominated by the processes of nucleon spallation and negative muon capture. In a spallation reaction, the kinetic energy of impact breaks up the target nucleus, producing a lighter nuclide. In negative muon capture, a muon is captured into an electronic orbit. Because muons are heavier than electrons, their ground state overlaps with the nucleus. The muon reacts with a proton to form a neutron, and excess energy breaks up the target nucleus. Negative muon capture only accounts for $\sim$2% of ^{10}Be production at the ground surface, but dominates below depths of $\sim$3 m. At still greater depths, fast muon reactions involving bremsstrahlung radiation, pair production,

Table 2 *In situ*-produced cosmogenic nuclides.

Nuclide	Primary targets	Half-life	Primary production mechanisms
^{10}Be	O, C	1.3 Myr	Spallation, μ^- capture
^{14}C	O	5,730 years	Spallation, μ^- capture
^{26}Al	Si	0.7 Myr	Spallation, μ^- capture
^{36}Cl	Ca, K, ^{35}Cl	0.3 Myr	Spallation, μ^- capture, *n* capture
^{3}He	All	Stable	Spallation
^{21}Ne	Si	Stable	Spallation, μ^- capture

ionization, and inelastic scattering become important (Heisinger *et al.*, 2002a).

Both spallation and muon capture reactions produce ^{10}Be from oxygen. An ideal mineral for studying weathering and erosion is therefore one that has abundant oxygen and is also resistant to chemical alteration. Quartz is by far the most common mineral used in cosmogenic studies of weathering and erosion for these reasons (Lal and Arnold, 1985).

11.2.3 Cosmogenic Nuclide Production Profiles

The fundamental reason that cosmogenic nuclides can be used to determine exposure ages and denudation rates is that production rates decline rapidly with depth. Cosmogenic nuclide buildup in a mineral grain occurs almost entirely at shallow depths as it is exhumed by erosion and weathering of overburden. To accurately model cosmogenic nuclide concentrations in an eroding landscape, it is necessary first to describe the depth dependence of production rates.

Secondary cosmic rays have sufficient energy to penetrate matter in a manner that depends solely on mass traversed, irrespective of material composition. The mean free path length of a secondary cosmic ray is therefore determined by density. Penetration lengths can be written as a function of the mass of overburden in units of pressure, typically $g\,cm^{-2}$. (An exception to this rule comes for the low-energy neutrons, which are important for making ^{36}Cl through neutron capture, and are highly dependent on chemical composition. Negative muon capture is also somewhat dependent on composition (Charalambus, 1971).)

Production rates of cosmogenic nuclides by spallation reactions can be written by the following equation:

$$P_n(x) = P_1(0)e^{-\rho x/L_1} \quad (2)$$

where P_n represents the production rate by nucleon spallation, P_1 is a local constant, x the depth below the surface, and ρ the average density of the overburden. The penetration length (L_1) has been determined experimentally to be $\sim 160\,g\,cm^{-2}$, although published values range from 150 to 190 $g\,cm^{-2}$ (Gosse and Phillips, 2001). For production of ^{10}Be at sea level and high latitude, P_1 has a value of ~ 5.1 atoms $g^{-1}\,yr^{-1}$ (Stone, 2000).

Production rates due to negative muon capture cannot be expressed as a pure exponential due to their more complicated energy spectrum. Empirically determined spectra lead to a production rate profile that can be fit over a depth range appropriate for cosmogenic nuclide analysis. To simplify later analyses, we express this curve as the sum of two exponentials:

$$P_\mu(x) = P_2(0)e^{-\rho x/L_2} + P_3(0)e^{-\rho x/L_3} \quad (3)$$

where $P_\mu(x)$ is the production rate due to negative muon capture, P_2 and P_3 are the production rate scaling factors, and L_2 and L_3 the effective penetration lengths fit over a depth range of 100–5,000 $g\,cm^{-2}$ (Granger and Smith, 2000; Granger and Muzikar, 2001). In this parameterization, L_2 and L_3 have values of 738.6 and 2,688 $g\,cm^{-2}$, respectively. For ^{10}Be production at sea level and high latitude, P_2 and P_3 have values of 0.096 and 0.021 atoms $g^{-1}\,yr^{-1}$, respectively (Granger and Smith, 2000; Heisinger *et al.*, 2002a, b).

At depths greater than $\sim 10\,m$, reactions due to fast muons become important. The exact form of the depth profile remains an object of research, but to first order it can be expressed in exponential form as

$$P_f(x) = P_4(0)e^{-\rho x/L_4} \quad (4)$$

where $P_f(x)$ is the production rate due to interactions with fast muons, L_4 is an effective penetration length with a value of $\sim 4,360\,g\,cm^{-2}$, and P_4 a production rate scaling factor. For ^{10}Be production at sea level and high latitude P_4 is ~ 0.026 atoms $g^{-1}\,yr^{-1}$ (Granger and Smith, 2000; Heisinger *et al.*, 2002a, b).

Taken together, Equations (2)–(4) describe to a good approximation the depth dependence of cosmogenic nuclide production in the uppermost 10 m of rock and soil. We emphasize that Equations (3) and (4) are an approximation that is only valid for a depth range of 100–5,000 $g\,cm^{-2}$ (or ~ 0.5–20 m in rock), and that they should not be extrapolated to greater or shallower depths. For a derivation of these curves, see Stone *et al.* (1998) and Heisinger *et al.* (2002a, b). Production rates calculated from Equations (2) to (4) with values appropriate for ^{10}Be production at sea level and high latitude are shown in Figure 1, scaled by depth in terms of mass traversed.

Production rates are governed not only by depth beneath the surface, but also vary with altitude, latitude, and time. The exact scaling parameters remain an object of research; the interested reader is referred to the citations below.

1. Production rates vary strongly with depth in the atmosphere, which is controlled primarily by altitude, and to a lesser degree by temperature and atmospheric circulation (Stone, 2000). Various scaling factors have been proposed, mostly based on measurements of neutron flux in the atmosphere (e.g., Lal, 1991; Dunai, 2000; Desilets and Zreda, 2003). To first order, production rates increase exponentially, with an e-folding length of $\sim 150\,g\,cm^{-2}$ (Desilets and Zreda, 2003), although the attenuation length varies with both latitude and altitude. As a rule of thumb, at air densities typical of the lower atmosphere (i.e., 1.2 $kg\,m^{-3}$) this results in an e-folding length of $\sim 0.8\,km$.

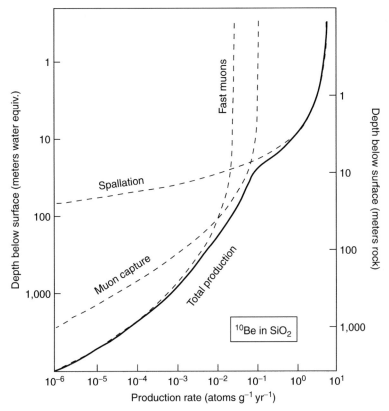

Figure 1 ^{10}Be production rate as a function of depth beneath the surface. Depth is expressed in meters water equivalent; to convert into depth in rock, scale by specific gravity. Production rates calculated from Equations (2) to (4) for sea level and high latitude, based on Heisinger *et al.* (2002a, b).

2. Production rates also vary with latitude, although less strongly than they do with altitude. The latitude dependence is due to effects of the geomagnetic field, which blocks low-energy cosmic rays at equatorial latitudes. Production rates near the equator are ∼30% less than at latitudes greater than 60° (Lal and Peters, 1967).

3. Production rates vary with time, primarily due to changes in Earth's magnetic field strength. Production rates over the past 100 kyr may have differed by up to 30% from the present value due to this effect (Gosse and Phillips, 2001).

Once the production rate profile is known at a particular site, we can use it to model cosmogenic nuclide buildup within mineral grains under a variety of conditions. In each of these models, the concentration is described by solving for buildup and radioactive decay over time, as shown in the following differential equation:

$$dN/dt = \sum P_i(t) - N/\tau \qquad (5)$$

where N is the concentration of the cosmogenic nuclide, τ the radioactive mean life (i.e., the radioactive half-life divided by the natural logarithm of

(2) and $P_i(t)$ represents production rates given in Equations (2)–(4). All terrestrial applications of cosmogenic nuclides are fundamentally based on solutions of Equation (5). The key to using Equation (5) to determine rates of chemical weathering and physical erosion is to meaningfully characterize the production rate within a given mineral grain as a function of time.

Often we think of denudation from an Eulerian perspective, in which the ground surface lowers with respect to some absolute datum. For cosmogenic nuclide analysis, it is often more appropriate to model from a Lagrangian perspective, in which a particular mineral grain is fixed and the surface approaches that grain over time. In a Lagrangian sense, the cosmogenic nuclide buildup within a grain can be modeled as changes in production rate through time.

Perhaps the best way to illustrate the range of applications for cosmogenic nuclides in weathering and erosion is to solve Equation (5) for a number of scenarios, each representing a particular geologic situation. In Section 11.3, we will introduce five different scenarios, each of which illustrates an application of cosmogenic nuclides for determining denudation rates, surface lowering rates, or surface exposure ages.

In Section 11.4, we will then show how to combine cosmogenic nuclide data with chemical weathering indices to quantify chemical weathering rates under these same types of scenarios.

11.3   USING COSMOGENIC NUCLIDES TO DETERMINE RATES OF SURFACE LOWERING AND DENUDATION

11.3.1   No Erosion (Surface Exposure Dating)

The simplest geologic scenario for applying cosmogenic nuclides is the exposure dating of a stable surface. Such a surface is neither eroding nor aggrading, and if a soil is present, it is not mobile. Surface exposure dating is widely used to date glacial moraines (e.g., Owen *et al.*, 2002), river terraces (e.g., Burbank *et al.*, 1996), alluvial fans (e.g., Zehfuss *et al.*, 2001), and other geomorphic surfaces. In studies of weathering and erosion, it is primarily useful for dating soil chronosequences rather than for determining process rates directly. The use of chronosequences in the study of chemical weathering rates will be discussed further in Section 11.4.

11.3.1.1   Theory

The cosmogenic nuclide concentration at depth *x* under a zero-erosion condition can be derived from Equation (5). The result takes the form

$$N(t) = N_{\text{inh}}(x)e^{-t/\tau} + \sum P_i(x)\tau(1 - e^{-t/\tau}) \quad (6)$$

where *t* is the time, and N_{inh} the inherited component of the cosmogenic nuclide, that is, the concentration that was present at time zero. For exposure times that are short with respect to radioactive decay, or for stable nuclides, Equation (6) can be simplified to a linear buildup model, where the concentration is simply proportional to the exposure time, as shown in the following equation:

$$N(t) = N_{\text{inh}}(x) + \sum P_i(x)t \quad (7)$$

Surface exposure dating can be applied to either bedrock or to unconsolidated deposits, provided that the surface has not eroded since the time that it was first created or exposed. In general, bedrock surfaces that can be exposure dated are exhumed from depths greater than a few meters, such that they contain almost no cosmogenic nuclide inheritance. Examples might include glacially scoured bedrock (e.g., Nishiizumi *et al.*, 1989), river-cut strath terraces (e.g., Schaller *et al.*, 2005), volcanic flows (e.g., Fenton *et al.*, 2001), or large landslides (e.g., Kubik *et al.*, 1998). Exceptions occur, for example, if glacial scouring was not deep enough to erase previous

exposure (e.g., Stroeven *et al.*, 2002), or if a volcanic rock contains [10]Be at the time of eruption. In these cases, the inheritance must be determined explicitly.

Unconsolidated deposits may also be amenable to surface exposure dating, as long as they have not eroded significantly since their time of deposition. Examples might include alluvial terraces, colluvial deposits, or glacial moraines. In such cases the cosmogenic nuclide inheritance cannot be ignored, but must be explicitly determined. Because inheritance may vary from clast to clast within a deposit, it is usually necessary to average over many grains. This may be done by collecting 30–50 or more grains and combining them together into a single amalgamated sample (Anderson *et al.*, 1996). If the average inheritance is uniform, then an amalgamated sample from the surface will represent the total concentration in Equation (6) or (7), while a sample from depth will only contain the inherited component. The exposure age is calculated from the difference between the two concentrations (Anderson *et al.*, 1996; Repka *et al.*, 1997; Hancock *et al.*, 1999).

11.3.1.1.1 Complications. There are many caveats and complications with surface exposure dating. The most important limitation is that stable surfaces are difficult to find. If erosion has been significant, cosmogenic nuclides will underestimate the true age of the surface. It is particularly difficult to rule out erosion of surfaces older than 10^5 years in most temperate environments. Care must also be taken to ensure that surfaces were never buried by significant amounts of snow, sand, or water, and that cosmic rays have not been blocked by vegetation such as large trees. Topographic shielding must also be accounted for to properly determine local production rates (Dunne *et al.*, 1999).

11.3.1.2   Examples

To illustrate surface exposure dating, we will discuss two examples: (1) dating bedrock surfaces, and (2) dating boulders on moraines. Both cases illustrate some of the strengths and weaknesses of the exposure dating approach.

11.3.1.2.1 Dating strath terraces. A bedrock or strath terrace is a former streambed that has been abandoned by river incision. Straths are amenable to surface exposure dating if they have been formed by rapid fluvial erosion and if they have been exposed at the surface ever since. Bedrock surfaces on strath terraces often retain smooth, fluted forms indicative of little or no erosion since abandonment by the river. In many cases, strath surfaces are inset many meters below the original ground surface, indicating that the fluvial erosion that exhumed them was rapid enough that inheritance can probably be ignored. If this is the case, then the cosmogenic nuclide concentration of minerals at the surface can be interpreted with either Equation (6) or (7).

As a case in point we consider data from a strath terrace on the Potomac River analyzed by Reusser *et al.* (2004). In most cases of surface exposure dating, only a few samples are collected from any given surface. However, in this case, an especially thorough set of samples was collected from a variety of locations on the same bedrock surface. By allowing sample-to-sample comparison, these data help expose some of the strengths and weaknesses of exposure dating.

Samples were collected from surfaces that appeared fresh, where the strath morphology was well-preserved. Quartz veins within the bedrock did not protrude more than 1 cm from the surface, indicating that there has been little postexposure lowering of the more erodible rock fractions. Figure 2 shows cosmogenic nuclide concentrations for eight samples collected from a prominent strath along the Potomac River, in order of increasing concentration. ^{10}Be concentrations range from (12.6 ± 0.35) to $(18.8 \pm 0.7) \times 10^4$ atoms g^{-1} of quartz. The weighted average ^{10}Be concentration is $(15.0 \pm 0.2) \times 10^4$ atoms g^{-1}, but the standard deviation of the data, at 2.4×10^4 atoms g^{-1}, is significantly larger than the analytical uncertainties. Excess variability in ^{10}Be concentrations suggests that the strath may have a complicated exposure history.

Interpretation of the strath's exposure age is dependent on one's interpretation of the data. For example, inheritance could cause cosmogenic nuclide concentrations in some samples to be too high. In that case, the younger samples would be considered more reliable, and the appropriate ^{10}Be

concentration would be perhaps 13×10^4 atoms g^{-1}. In contrast, partial cover by gravel or sand, or episodic shielding by floodwater could cause cosmogenic nuclide concentrations in some samples to be too low. In that case, the older samples would be more reliable, and the appropriate ^{10}Be concentration would be closer to 18×10^4 atoms g^{-1}. It is difficult to know *a priori* which (if either) interpretation is correct. Nevertheless, the cosmogenic data can be used to bracket the age of the terrace. The most conservative interpretation of the data is that they are distributed normally about their true age, and that the uncertainty is best represented by the standard deviation. At the latitude and altitude of these samples, ^{10}Be production occurs at a rate of 4.8 ± 0.5 atoms g^{-1} yr^{-1}. If we consider the best age of the terrace to be represented by the average plus or minus one standard deviation, then the terrace age becomes 33.3 ± 5.4 kyr (Reusser *et al.*, 2004).

11.3.1.2.2 Dating glacial moraines. Perhaps, the most common application of cosmogenic nuclide surface exposure dating is to boulders on glacial moraines. Boulders can often be assumed to have been plucked from beneath the glacier, and so due to shielding have little cosmogenic nuclide inheritance. However, some boulders may have significant cosmogenic inheritance either because they were derived from supraglacial sources, or because they were eroded from previously exposed rock. Thus, it is commonly the case that 5–10 boulders are collected from a single moraine. Samples that are clearly older than the rest are assumed to have excess inheritance. Samples that are clearly younger than the rest are assumed to have eroded or suffered shielding by snow or till.

An example data set is shown in Figure 3, from the Pomeranian moraine in Poland, an end moraine of the Scandinavian ice sheet. There, Rinterknecht *et al.* (2005) collected 37 samples of moraine boulders for ^{10}Be analysis. Ages are calculated assuming no significant snow cover, and no erosion of the boulder surfaces. Again, as with the example of a strath terrace in the section "Dating glacial moraines," the spread in the data clearly exceeds analytical uncertainty. However, in the case of the moraine boulders, the bulk of the data cluster very closely, while a few samples are clearly older than the rest, and a few samples are clearly younger than the rest. In this case, the samples with excess ^{10}Be likely contain inheritance, and the samples with too little ^{10}Be have been eroded, exhumed from within the moraine, or shielded. A legitimate case can be made for excluding these samples from the age calculation. Doing so yields an average age of 14.8 ± 1.0 kyr for the moraine (including 6% uncertainty in production rate; Rinterknecht *et al.*, 2005).

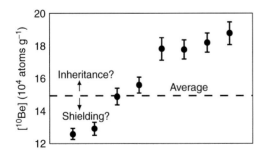

Figure 2 ^{10}Be concentrations from multiple samples of a prominent strath surface alongside the Potomac River, arranged in order of increasing concentration. Spread in the data may be due to either inheritance in samples with high [^{10}Be] or shielding of samples with low [^{10}Be]. Topographic shielding differs by <4% among samples. Using a local production rate of 4.8 atoms g^{-1} yr^{-1} yields a terrace age of 33.3±5.4 kyr. Data from Reusser *et al.* (2004).

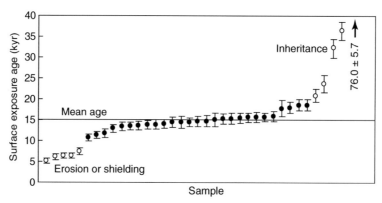

Figure 3 [10]Be surface exposure ages of the Pomeranian moraine, Poland, arranged in order of increasing concentration. Data shown in open symbols were determined to suffer from either significant cosmogenic nuclide inheritance or shielding, and are excluded from the moraine age calculation. Data from Rinterknecht *et al.* (2005).

11.3.2 Rock Erosion

In most landscapes, bedrock is not perfectly preserved at the surface. Instead, chemical weathering and physical erosion remove material from the ground surface, stripping away mineral grains with the highest cosmogenic nuclide concentrations. Hence, cosmogenic nuclide concentrations in an eroding landscape do not reflect the age of the surface, but rather a rate of mass loss (Lal, 1991). Although many particular situations could be modeled, the most common is to assume steady-state denudation. For the steady-state assumption to be valid, denudation rates must have remained constant for long enough to remove several meters of rock and soil.

11.3.2.1 Theory

11.3.2.1.1 Steady erosion. Under steady erosion, the depth of a particular mineral grain can be modeled with the following equation:

$$x(t) = x(0) - (D/\rho)t \quad (8)$$

where D is the denudation rate, and t the time measured forward into the future. (*Note*: Because we use a mass denudation rate here, as is customary in weathering studies, this formulation differs slightly from conventional cosmogenic nuclide notation.) From the Lagrangian perspective of a mineral grain, we can model the production rate as

$$P(t) = \sum P_i e^{-Dt/L_i} \quad (9)$$

assuming that at time $t=0$ the mineral grain had a negligible production rate. Substituting Equation (9) into (5) and solving for steadystate yields

$$N(x) = \sum \{ P_i(0)e^{-\rho x/L_i} / [1/\tau + D/L_i] \} \quad (10)$$

For erosion rates faster than $\sim 10^2 \, \mathrm{mm \, kyr^{-1}}$, radioactive decay is negligible. In this case,

Equation (10) can be simplified for samples collected from the surface to

$$N(0) = \sum P_i(0)L_i/D \quad (11)$$

Equations (10) and (11) show that cosmogenic nuclide concentrations in rock surfaces can be interpreted as denudation rates. But recall that Equations (6) and (7) show that cosmogenic nuclide concentrations can also be interpreted as exposure ages. Although this conundrum has been recognized for almost 50 years (Whipple and Fireman, 1959), it remains very difficult to distinguish exposure and denudation from cosmogenic nuclide data alone. Instead, we often assume that the cosmogenic nuclide concentration represents either a maximum denudation rate or a minimum exposure age. Exceptions can be made when good evidence of surface preservation, such as glacial or fluvial polish, indicates that denudation has been minimal. In other cases, such as soil-mantled hillslopes, it is clear that denudation is active and it is usually more reasonable to assume steady denudation. Nevertheless, one should remain cognizant of the trade-off between minimum exposure ages and maximum denudation rates.

It is important to realize that Equations (8)–(11) apply only to the case where mineral grains approach the surface due to mass loss. We have not yet accounted for vertical mixing within soil or regolith, or weathering within the soil column. Equations (8)–(11) are most appropriate for measuring the denudation rate of a rock surface or of sediment that is not being vertically mixed. Equations (8)–(11) can also be used to measure the denudation rate of rock that is buried beneath mobile regolith, as long as the density and thickness of that regolith are known and accounted for.

11.3.2.1.2 Complications due to unsteady denudation. For a case in which denudation is not steady, but changes from one rate to another, the cosmogenic nuclide concentration will lag the denudation rate. This can be seen in the transient solution to Equations (5) and (9). To illustrate, we consider a case in which denudation rates change from D_1 to D_2 as a step function.

The solution to Equation (4) for a step-change in surface denudation rate at $t = 0$ is given by the equation

$$N(t) = \sum \{N_{1,i} + (N_{2,i} - N_{1,i})$$
$$\times (1 - e^{-t(D/L_i + 1/\tau)})\} \qquad (12)$$

where $N(t)$ is measured at the ground surface and

$$N_{1,i} = \sum P_i(0)/[1/\tau + D_1/L_i] \qquad (13)$$

$$N_{2,i} = \sum P_i(0)/[1/\tau + D_2/L_i] \qquad (14)$$

Equation (12) shows that the response of the cosmogenic nuclide concentration is gradual, and that the concentration asymptotically approaches the new steady-state value. The timescale for response is governed by either radioactive decay or the time taken to erode one penetration length, whichever is faster. Figure 4 shows the response of cosmogenic nuclide concentrations to a periodic change in erosion rates, where the periodicity is comparable to the response time. A similar gradual response is seen when erosion occurs stepwise, rather than continuously (e.g., Small *et al.*, 1997).

11.3.2.2 Examples

To illustrate the use of cosmogenic nuclides for determining denudation rates, we will discuss two examples. In the first, cosmogenic nuclides are used to measure bare rock erosion rates on inselbergs in the Australian desert. In the second, ^{10}Be is used to determine the denudation rate of saprolite beneath a soil cover.

11.3.2.2.1 Denudation of bedrock surfaces. Although most slowly eroding landscapes on Earth are mantled by a soil cover, granitic rock often displays bedrock landforms such as tors or inselbergs that stand above a grussified and soil-mantled plain. These bedrock landforms have long interested geomorphologists, who have traditionally interpreted them as relics from previous cycles of subsurface weathering followed by erosion (e.g., Twidale, 1986). It is thought that weathering of granitic rock is enhanced in areas with high fracture density, such that local aberrant sections of rock with low fracture density remain unweathered compared with the saprolite surrounding them. Stripping of overlying saprolite leaves unweathered rocks as positive landforms. Tors can also form as a consequence of chance exposure of bedrock, if outcrops erode more slowly than rock beneath a soil cover (e.g., Wahrhaftig, 1965; Anderson, 2002). It has been proposed that many inselberg landscapes are tens or hundreds of millions of years old (e.g., Twidale, 1987; Twidale and Vidal Romaní, 2005).

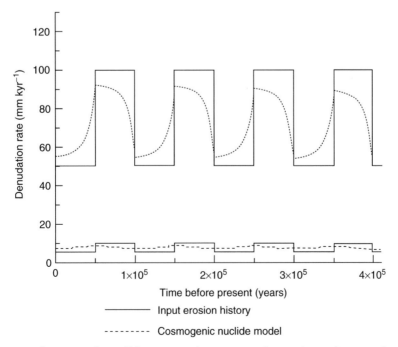

Figure 4 Response of cosmogenic nuclide concentrations to step changes in erosion rate. A periodic forcing function is shown; an individual step would have the exponential form of Equation (12). Reproduced by permission of Elsevier from von Blanckenburg (2005).

To help determine how inselbergs are formed, [10]Be and [26]Al can be used to measure erosion rates and their distribution across the surface (Bierman and Caffee, 2002). Figure 5 shows data from Yarwondutta Rock, a well-studied inselberg of rounded form that stands 15 m in relief. Samples were collected both from the inselberg top and from its flanks. [10]Be concentrations are consistently higher on the top than the flanks, reaching values as high as $(3.60 \pm 0.11) \times 10^6$ atoms g[-1]. With these concentrations, it is possible to interpret either a minimum age or a maximum erosion rate for the landform using Equation (6) or (10). Scaling by local production rates, Equation (6) indicates that the inselberg has a minimum exposure age of 831 ± 106 kyr, much younger than the supposed age of the landform. In this case, it is more reasonable to assume that [10]Be and [26]Al concentrations are dictated by bedrock erosion. Solving Equation (10) indicates that the top of Yarwondutta Rock has an erosion rate as low as 1.56 ± 0.29 t km[-2] yr[-1] (equivalent to 0.60 ± 0.11 m Myr[-1]), among the slowest in the world. It is clear from these data that although the inselberg landscape has changed little over tens of millions of years, the rock surfaces themselves are not older than Pleistocene.

Interpretation of cosmogenic nuclide concentrations on the inselberg flank is potentially more complex, because exhumation of the inselberg from the surrounding saprolite may have occurred in discrete steps, as suggested by correspondence between the heights of benches along inselberg flanks and paleosols preserved nearby. [10]Be and [26]Al concentrations on the flanks may indicate either an exposure age of at least 100–200 kyr, or an erosion rate up to $\sim$13 t km[-2] yr[-1], equivalent to 5 m Myr[-1] (Bierman and Caffee, 2002).

The cosmogenic data are consistent with models of inselberg formation by exhumation, and confirm that erosion rates in the Eyre Peninsula are among the slowest on Earth. By assigning erosion rates to the landforms the cosmogenic nuclide approach removes some of the speculation about the ages of the landforms and thus permits a quantitative discussion of landform generation and decay.

11.3.2.2.2 Rock erosion under soil cover. Just as cosmogenic nuclides measured from a bedrock surface can be used to determine erosion rates, they can also be used to determine erosion rates in samples collected at depth, provided that production rates are corrected for shielding by the overburden. It is irrelevant to cosmogenic nuclide production whether the overburden is rock or soil, as long as its mass is known and invariant through time. This method has been used in a suite of applications to determine erosion rates at the saprolite–soil interface.

The rate of saprolite erosion beneath soil cover is directly tied to the rate of soil production and landscape lowering. It has long been speculated that saprolite erosion depends on soil thickness (e.g., Gilbert, 1877; Ahnert, 1977; Cox, 1980; Dietrich *et al.*, 1995). Processes that disrupt saprolite, such as freeze–thaw cycles, animal burrowing, and plant rooting all depend on soil thickness, and may proceed more quickly under thin soils. The functional form of the relationship between soil production and soil depth, however, remains debated. Some favor a model in which soil production reaches a maximum rate beneath a finite soil depth of perhaps 10–20 cm, because shallower soils are unable to support vigorous vegetation and large populations of burrowing animals. Under such a model, feedbacks would drive landscapes toward a uniform soil depth, except where bare rock is exposed (Cox, 1980). As an alternative, others favor a model in which soil production increases monotonically as soil thickness decreases. Under this model, soil thickness could vary but bare rock would be uncommon except in the most rapidly eroding portions of

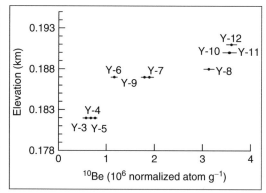

Figure 5 Photo shows Yarwondutta Rock, a smooth inselberg on the Eyre Peninsula, Australia. Labeled points were analyzed for [10]Be and [26]Al by Bierman and Caffee (2001). Results are shown in the graph, and indicate slow erosion of the inselberg top. Adapted by permission of Geological Society of America from Bierman and Caffee (2002).

the landscape (Cox, 1980). Cosmogenic nuclides offer a method for testing these models, by revealing how soil production rates vary with soil depth.

We will focus our discussion on only two particular studies of soil production, but refer the reader to a more extensive literature on the subject (Heimsath *et al.*, 1997, 1999, 2000, 2001a,b; Wilkinson *et al.*, 2005; Small *et al.*, 1999; Anderson, 2002).

To illustrate, we consider two examples from southeastern Australia. The first landscape, at Nunnock River, is developed on the margin of the Great Escarpment (Heimsath *et al.*, 2000, 2001a). Soil depths on that landscape are highly variable, and strongly dependent on hillslope curvature. Soils are thinnest on ridgecrests and systematically thicken downslope. Fifteen samples were collected from saprolite at the soil base, with depths ranging from 0 to 90 cm. Results are shown in Figure 6, and demonstrate a strong dependence of erosion rate on soil depth. The data are well described by an exponential relationship, with

$$D = (138 \pm 8 \,\mathrm{t\, km^{-2}\, yr^{-1}})e^{-(0.020 \pm 0.001)h} \quad (15)$$

where h is the soil depth. The prefactor in Equation (15) is equivalent to $53 \pm 3 \,\mathrm{m\, Myr^{-1}}$ for rock of density $2.6 \,\mathrm{g\, cm^{-3}}$. From the thickest soils down to bare saprolite, the data show no evidence of erosion rates peaking at a discrete soil depth. It is worth noting, however, that unweathered bedrock tors in the same watershed erode much more slowly than exposed saprolite. Two samples from the tops of tors yield fivefold slower erosion rates of $\sim 9 \,\mathrm{m\, Myr^{-1}}$. The discrepancy between erosion rates of bouldery tors and weathered saprolite illustrates an important point. The exponential

relationship of Equation (15) only applies to the physical erosion of preweathered saprolite. A different functional form may be required in cases where bedrock is less weathered prior to soil formation, or where unweathered rock is exposed at the ground surface (e.g., Wahrhaftig, 1965; Granger *et al.*, 2001; Wilkinson *et al.*, 2005).

The data from the Nunnock River landscape imply that it is in a state of decreasing relief. Ridgecrests with shallow soils are eroding more quickly than surrounding hillslopes, suggesting that the topography cannot be maintained in steady-state relief.

A different picture can be seen in a second site from the nearby Blue Mountains Plateau (Wilkinson *et al.*, 2005). There, soil depths are less variable than at the Nunnock River site, and although bare rock is common, there are few places where soil is less than 20 cm thick. In addition, soil depth is not strongly correlated with hillslope curvature. These observations suggest that the functional relationship between erosion and soil depth may be different in this landscape. Figure 6 also shows the apparent rates of soil production at the Blue Mountains site. Erosion rates range from 25 to $70 \,\mathrm{t\, km^{-2}\, yr^{-1}}$ (or $10–27 \,\mathrm{m\, Myr^{-1}}$) at all locations and are nearly invariant with soil depth. Although field data suggest that denudation rates are influenced by vegetation (Wilkinson *et al.*, 2005), the effect is far less than the fivefold variability observed over similar soil depths at Nunnock River (Heimsath *et al.*, 2000). The near-uniformity of erosion rates suggests that this landscape may be near a state of dynamic equilibrium, in which erosion rates are invariant across the topography and landforms maintain their relief (Hack, 1960).

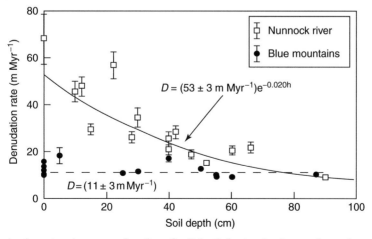

Figure 6 Soil production rates shown as a function of soil depth for two landscapes in southeastern Australia. The Nunnock River site on the Great Escarpment (open squares) shows an exponential relationship that implies decreasing catchment relief. The Blue Mountains site (closed circles) shows no significant variation in denudation rate with soil depth, and is consistent with long-term persistence of relief. Bars represent one standard error, except where error is smaller than symbol. Nunnock River data from Heimsath *et al.* (2000, 2001a); Blue Mountain data from Wilkinson *et al.* (2005).

Similar conditions of uniform erosion have been seen at other locations (e.g., Small *et al.*, 1999; Anderson, 2002).

When comparing the data from the Nunnock River site with that from the Blue Mountains (Figure 6), one naturally asks how such different functional relationships between erosion rate and soil depth could develop. We see two possibilities, both of which highlight natural variability among landscapes.

1. Wilkinson *et al.* (2005) suggest that the Blue Mountains landscape is a shifting mosaic of heath and forest, with temporally varying soil depths. Under these conditions, differences in erosion rate are blurred due to natural variability, and are difficult to detect. It is also possible that soil cover in the Blue Mountains has increased through time, in which case corrections could be made to the erosion rates (Wilkinson *et al.*, 2005).
2. We suggest that the difference in erosional behavior may reflect an underlying difference in the geologic and geomorphic scenario. The Nunnock River site is located on the eastern flanks of the Great Escarpment. In contrast, the Blue Mountains site is atop the Great Escarpment, and has not yet been affected by escarpment retreat. Differences in long-term landscape evolution between the two areas are exhibited by paleosurfaces preserved beneath Miocene basalt flows. Volcanic flows in the Blue Mountains cap a low-relief surface, in contrast to those along the Great Escarpment which cap deeply incised topography (van der Beek *et al.*, 2001). We suggest that the Nunnock River site is in a transient state of declining topography due to its position on the Great Escarpment, while the Blue Mountains site is near a steady-state condition of uniform erosion rates following tens of millions of years of relative stability.

11.3.3 Cosmogenic Nuclides in Vertically Mixed Soils

Thus far, we have discussed strategies for determining rock erosion rates at specific locations and under ideal conditions. The mathematical framework has remained relatively straightforward, because the history of production within any given mineral grain depends only on the removal rate of overlying material. The trajectory of a mineral grain in an eroding rock is always toward the surface. Often, however, we may be interested in determining dates and denudation rates of unconsolidated soils that are vertically mixed, for example by bioturbation. In such cases, the trajectory of any particular mineral grain may be erratic, and its exposure history

complicated by repeated exposure at the surface and burial at depth. It is impossible to accurately model the exposure history of any given clast. We require a new mathematical framework in which to interpret exposure ages and erosion rates from cosmogenic nuclides in vertically mixed sediment. Our approach is to model the average cosmogenic nuclide concentration in the soil, rather than in a particular parcel of rock. One may then analyze bulk samples that are comprised of many grains and representative of the average.

11.3.3.1 Theory

Just as we modeled the two distinct scenarios of rock exposure and erosion, we must also consider two different conditions of soil exposure and erosion. The first condition is that of soil or sediment that is deposited at a particular time, and subsequently the upper part of the profile is mixed. This is conceptually equivalent to the case of rock exposure in that there is no net loss of material. The second condition is that of steady-surface erosion, but we will model cosmogenic nuclides in a vertically mixed soil rather than in undisturbed bedrock or saprolite.

11.3.3.1.1 Exposure dating of vertically mixed sediment. In the case of exposure dating of bedrock, we considered cosmogenic nuclide production to be strictly depth-dependent. As material was advected to the surface, it experienced an exponentially increasing production rate through time. In vertically mixed soils, however, the average production rate in the soil is constant throughout the mixed zone. Although the production rate within any given grain at any given time depends on its depth, each grain spends part of the time at the surface and part of the time at depth. If we assume that the soil is well mixed, so that a large enough sample of grains will accurately reflect the average cosmogenic nuclide concentration, then the average production rate is given by the equation:

$$\langle P \rangle = \sum P_i(0)(L_i/\rho h)(1 - e^{-\rho h/L_i}) \quad (16)$$

where h is the mixing depth and brackets indicate an average value. Following the case of rock exposure given by Equation (6), the simple exposure profile is then given by the following equations:

$$\langle N \rangle = N_{inh}e^{-t/\tau} + \sum P_i(0)\tau L_i/\rho h$$
$$\times \left(1 - e^{-\rho h/Li}\right)\left(1 - e^{-t/\tau}\right) \text{ for } x < h \quad (17)$$

$$N(x) = N_{inh}e^{-t/\tau} + \sum P_i(0)\tau e^{-\rho x/L_i}$$
$$\times (1 - e^{-t/\tau}) \text{ for } x > h \quad (18)$$

Note that Equation (18) is identical to Equation (6) over the appropriate depth interval. The profile determined from Equations (17) and (18) is shown schematically in Figure 8, where it is compared with profiles for eroding soils. Differences between the two results will be discussed in the next section.

Exposure dating of vertically mixed soils can be useful for dating unconsolidated deposits of a specific age, such as alluvial terraces, marine terraces (Perg *et al.*, 2001), or alluvial fan surfaces. Such surfaces are often churned by bioturbation processes such as tree throw and animal burrowing, or by anthropogenic disturbance such as plowing. In such cases, exposure dating of single clasts is virtually impossible, but averaged cosmogenic nuclide profiles may still yield an accurate age. An example follows later in Section 11.3.3.2.

11.3.3.1.2 Denudation rates in vertically mixed soils.

An eroding surface is constantly shedding material from its surface. As we have shown in Section 11.3.2, erosion of a bedrock surface removes material with the highest cosmogenic nuclide concentrations, leading to Equation (10). In contrast, erosion of vertically mixed soil or sediment removes material that samples the entire mixed zone and thus contains the average cosmogenic nuclide concentration (Figure 7).

To model cosmogenic nuclides in an eroding soil, we consider the case shown schematically in Figure 7. The mixed layer has a steady thickness; material is lost from the mixed layer at a mass rate D, and material is introduced into the mixed layer from below at the same mass rate. To simplify the analysis, and to facilitate an analytical solution, we assume that erosion is fast with respect to

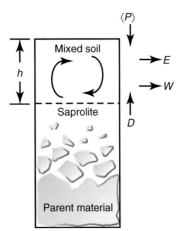

Figure 7 Model of cosmogenic nuclide accumulation and loss in a vertically mixed soil. Production rates are averaged over the mixed layer. Physical erosion adds quartz-containing [10]Be from parent material, and removes quartz from the mixed layer. Chemical weathering enriches insoluble quartz in the mixed layer.

radioactive decay. In that case, the concentration at the soil–bedrock interface is given by

$$N_{\text{bedrock}} \approx \sum P_i(0) e^{-\rho h/L_i} L_i/D$$
$$\text{(ignoring radioactive decay)} \quad (19)$$

The average cosmogenic nuclide concentration within the mixed soil can then be determined by solving differential Equation (20), which describes additions and losses of cosmogenic nuclides from the mixed layer:

$$d\langle N\rangle/dt = \langle P\rangle - \langle N\rangle/\tau$$
$$- \langle N\rangle D/\rho h + N_{\text{bedrock}} D/\rho h \quad (20)$$

Substituting Equations (17) and (19), and solving for steady-state conditions, the concentration within the mixed regolith is then given by Equation (21). Ignoring radioactive decay leads to the approximation shown in Equation (22):

$$\langle N\rangle = \sum P_i(0)(L_i)/\rho h[(1/\tau + D/\rho h)] \quad (21)$$

$$\langle N\rangle \approx \sum P_i(0) L_i/D$$
$$\text{(ignoring radioactive decay)} \quad (22)$$

Losses due to radioactive decay can be safely ignored as long as the residence time of mineral grains within the mixed zone is significantly less than the half-life of the cosmogenic nuclide. For [10]Be ($\tau = 1.93$ Myr) this is normally valid and Equation (22) is appropriate. Remarkably, Equation (22) for mixed soil is exactly equal to Equation (11) at a rock surface. That is, the cosmogenic nuclide concentration in eroding soil is the same as at the surface of eroding bedrock. An important point is that Equation (22) does not depend on soil depth (and Equation (21) is only very weakly dependent on soil depth). This has at least two implications:

1. It is not necessary to determine soil depths when estimating erosion rates from vertically mixed soils. This becomes important when mixing sediment from various sources, as we will show in Section 11.3.4.
2. Cosmogenic nuclides in soils can yield an accurate erosion rate, even if the soil profile has been truncated, for example by land use. It is thus possible to compare long-term erosion rates determined from cosmogenic nuclides with short-term anthropogenic erosion rates in heavily impacted areas (e.g., Brown *et al.*, 1998). A corollary implication is shown in Figure 8, in which we suggest that cosmogenic nuclides can be used to ascertain whether a soil profile has been truncated or not.

Equations (21) and (22) show that the depth-integrated cosmogenic nuclide inventory in an

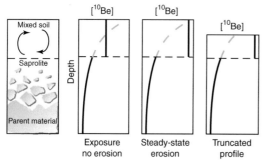

Figure 8 Model predictions of ^{10}Be concentrations in well-mixed soil profiles. Bold line represents actual ^{10}Be concentration, and dashed line represents ^{10}Be concentrations in an unmixed profile for reference. For the case of exposure without erosion, ^{10}Be concentrations correspond to a vertical average, as shown in Equation (16). For the case of steady-state erosion, ^{10}Be concentrations match the value expected for production rates at the surface, as shown in Equation (22). Soils that are truncated by recent erosion have ^{10}Be concentrations higher than would be expected from concentrations in the unmixed parent material.

eroding soil can be substantially higher than in bedrock eroding at the same rate. Although this may at first seem paradoxical, it can be easily explained. Material removed from the surface of eroding bedrock contains the highest cosmogenic nuclide concentration. In contrast, material removed from an eroding soil contains a cosmogenic nuclide concentration averaged over the mixing depth. Although the rate of cosmogenic nuclide production is the same in both bedrock and soil, the rate of cosmogenic nuclide loss through erosion is different in the two cases, leading to different steady-state inventories.

The two distinct cosmogenic nuclide profiles achieved under conditions of no erosion versus steady erosion offer a method for distinguishing whether a surface has eroded or not (Lal and Chen, 2005). Figure 8 shows concentration profiles for the two cases. Although this has not previously been emphasized in the literature, Figure 8 illustrates that there are detectable differences among the cases of (1) no erosion (2) steady-state erosion, and (3) truncation of a soil profile.

11.3.3.1.3 Complications due to selective dissolution and quartz enrichment. In cases where cosmogenic nuclides are used to infer denudation rates, they indicate total mass loss from above the sample, whether by physical erosion or chemical weathering. In the case of rock erosion, chemical weathering is essentially indistinguishable from physical erosion, because both processes lead to mass loss near the surface. However, in vertically mixed soils, chemical weathering may cause mass loss of the bulk soil, while leaving

behind resistant minerals such as quartz. This not only causes enrichment of quartz in the soil, but also causes the residence time of quartz to be longer than the average residence time of other minerals. Cosmogenic nuclide concentrations in the quartz therefore increase. This can lead to an underestimation of true denudation rates.

The effects of quartz enrichment may be accounted for, if the enrichment from bedrock to soil can be quantified (Small *et al.*, 1999). If we consider that quartz is essentially insoluble and only removed by physical erosion, while denudation of the bulk soil occurs by both physical erosion and chemical weathering, then the erosion rate of quartz from the soil is determined by

$$E = D(f_B/f_S) \qquad (23)$$

where f_S is the fraction of quartz in the soil, and f_B the fraction of quartz in the bedrock. We may then rewrite the mass balance of cosmogenic nuclides in quartz from Equation (20) to the form shown in Equation (24), ignoring radioactive decay:

$$d\langle N \rangle/dt \approx f_S\langle P \rangle - f_S\langle N \rangle E/\rho h \\ + f_B N_{bedrock} D/\rho h \qquad (24)$$

Substituting Equations (16), (19), and (23) leads to the form

$$d\langle N \rangle/dt = \sum P_i(0)(L_i/\rho h)[f_S + (f_B - f_S)e^{-\rho h/L_i}] \\ - f_B\langle N \rangle D/\rho \qquad (25)$$

Solving for steady state yields the equation

$$\langle N \rangle = \sum P_i(0)L_i D\{(f_S/f_B) \\ + [1 - (f_S/f_B)]e^{-\rho h/L_i}\} \qquad (26)$$

The cosmogenic nuclide concentration in Equation (26) is lower than that in Equation (19) by a factor shown in brackets that depends on the quartz enrichment factor (f_S/f_B) and soil depth.

The quartz enrichment factor is often difficult to measure, because quartz is a major component of both bedrock and soil. A more easily measured proxy is the enrichment of insoluble elements in the soil, such as zirconium, which can be measured by X-ray fluorescence (XRF). If we assume that both Zr and quartz are enriched together, then erosion rates calculated from cosmogenic nuclides can be corrected using Equation (26). It is important to realize that Zr enrichment is only valid where Zr and quartz are homogeneously distributed. In practice, approximately 10 different soil samples and 3–5 bedrock samples are averaged for [Zr]; sample-to-sample variability is usually 10–15% (Riebe *et al.*, 2001a).

Quartz enrichment can be a major factor in many landscapes, particularly those that are slowly eroding or that are in highly weatherable rocks. Table 3 shows representative Zr-enrichment factors and erosion rate corrections for several granitic watersheds in different environments (Riebe, unpublished results).

11.3.3.2 Examples

11.3.3.2.1 Exposure dating marine terraces. As an example of exposure dating in a vertically mixed soil, we consider a prominent marine terrace at Santa Cruz, California dated by Perg *et al.* (2001). Marine terraces are wave-cut surfaces formed especially during sea-level highstands, which are often mantled with a veneer of sand. The Santa Cruz marine terraces are cut into mudstone and mantled by 1–4 m of coarse sand. Perg *et al.* (2001) collected samples at closely spaced intervals over the entire depth of the marine sands, noting evidence of bioturbation. Figure 9 shows results for one of their profiles. [10]Be concentrations increase smoothly through undisturbed sands at the base, but become more nearly homogenized near the surface. Field observations of bioturbation place the mixed depth in this profile at ~116 cm.

Given the [10]Be profile of Figure 9, is it possible to determine (1) whether the surface is eroding or not, and (2) the age of the terrace? The five samples below the mixed zone show a nearly exponential decrease in [10]Be concentrations with depth. Fitting a curve to these points and extrapolating to the surface, we are able to predict [10]Be concentrations in the mixed zone under the conditions of no erosion (Equation (16)) and steady erosion (Equation (18)). The bold line in Figure 9 represents such conditions. Although there is some structure to the data, indicating incomplete mixing, they conform reasonably well to the case of no erosion, and indicate a terrace age of ~226 kyr. Perg *et al.* (2001) inferred from the cosmogenic nuclide profile that the terrace likely formed during marine isotope stage 7, at ~212 kyr, matching the cosmogenic nuclide exposure age to within 7%.

Table 3 Zirconium concentrations and quartz-enrichment corrections at various watersheds.

Location	Soil depth (cm)	$[Zr]_{soil}/[Zr]_{rock}$	Correction factor[a]
Rio Icacos, Puerto Rico	95	2.5 ± 0.2	1.62 ± 0.08
McNabb Track, New Zealand	100	1.9 ± 0.1	1.61 ± 0.07
Chiapas Highlands, Mexico	150	1.4 ± 0.1	1.33 ± 0.08
Jalisco Highlands, Mexico	25	1.5 ± 0.1	1.12 ± 0.02
Panola Mountain, Georgia, USA	50	1.7 ± 0.1	1.3 ± 0.04
Jalisco Lowlands, Mexico	30	1.3 ± 0.1	1.09 ± 0.03
Sonora Desert, Mexico	15–50	1.2 ± 0.1	1.03–1.09
Fall River, Sierra Nevada, USA	10–50	1.2 ± 0.1	1.02–1.08
Fort Sage Mountains, California, USA	20–30	1.2 ± 0.1	1.04–1.05

[a] Based on estimated soil density of $1.8 \, \text{g cm}^{-3}$, except at Rio Icacos, where data indicate density is $0.9 \, \text{g cm}^{-3}$ (White *et al.*, 1998), and at Fall River and Fort Sage, where soil density was assumed to be $1.6 \, \text{g cm}^3$ (Riebe *et al.*, 2001a). Note that Equation (25) is such that the correction factor is relatively insensitive to plausible errors in soil density, and conversely more sensitive to changes in the enrichment ratio.

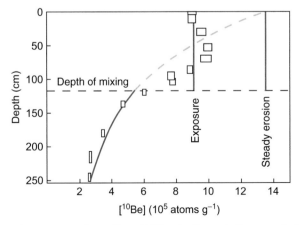

Figure 9 [10]Be profile in marine terrace at Santa Cruz, California. Boxes indicate vertical sampling interval and one standard error in width. Bold line represents model fit to the data for a terrace age of 226 kyr, and dashed line indicates expected profile in an unmixed profile, as in Figure 7. Data are consistent with constant exposure line, and insignificant erosion. Adapted from Perg *et al.* (2001).

11.3.4 Spatially Averaged Denudation Rates

In previous sections, we have shown that cosmogenic nuclides can be used to date surface exposure and to measure erosion rates in both bedrock and soil. The mathematical framework we have developed so far, however, is limited to vertical motion of mineral grains, and so erosion rates can only be determined locally for a particular site. While this is fine for many applications of weathering and erosion, it is not ideal for studying them within the context of an entire watershed. What happens if soil is transported laterally, or if it is mixed from various sources? Variability in soil depth, erosion rate, and degree of weathering within a watershed would make a detailed analysis of many individual sites cumbersome or even misleading. A better way to learn about the average erosion and weathering rate within a watershed is to sample bulk sediment that is derived from the entire watershed. In this section, we will show that cosmogenic nuclide concentrations in stream sediment can be used to determine the spatially averaged erosion rate of the entire sediment contributing area.

11.3.4.1 Theory

We have shown in Equation (22) that the cosmogenic nuclide concentration in vertically mixed regolith is equal to the concentration expected at the surface of eroding bedrock. Under steady-state conditions, sediment collected from a vertically mixed profile can thus be interpreted in terms of an erosion rate, regardless of mixing depth. This has an important implication when determining erosion rates from a watershed: variation in soil depth within a watershed is not important, as long as the soil thickness is in steady state.

The simplest case to consider is a watershed that is eroding everywhere at the same rate. In this case, Equation (22) predicts that the cosmogenic nuclide concentration in vertically mixed soil will be spatially homogeneous. Sediment exported from this watershed will therefore have a cosmogenic nuclide concentration that accurately

reflects the catchment erosion rate. A slightly more complicated situation arises if various portions of the watershed are eroding at different rates. What is the average cosmogenic nuclide concentration in sediment exported from a heterogeneously eroding watershed? We will show here that even in such a situation, the average cosmogenic nuclide concentration in stream sediment accurately reflects the spatially averaged denudation rate of the watershed. We will then discuss various complications that may arise.

Because streams export material from the entire watershed, stream sediment provides a naturally mixed sample derived from a large area. However, it is important to realize that stream sediment, even if it is perfectly mixed, does not represent all portions of the watershed equally. Areas with fast erosion rates supply more sediment to the stream than areas with slow erosion rates (Figure 10). The average cosmogenic nuclide concentration in stream sediment is therefore biased toward that of the most quickly eroding parts of the watershed. Although at first this may seem to bias cosmogenic nuclide concentrations, such mixing actually balances variations in cosmogenic nuclide concentrations toward that of the mean erosion rate. To show this mathematically, consider the mass-averaged cosmogenic nuclide concentration shown in the equation

$$N_{\text{stream}} = \int \{ND\} \, dA \Big/ \int D \, dA \quad (27)$$

where A is sediment contributing area. Substituting Equation (22), and ignoring radioactive decay leads to

$$N_{\text{stream}} = \int \left\{ \sum P_i(0) L_i \right\} dA \Big/ \int D \, dA \quad (28)$$

$$N_{\text{stream}} = \sum P_i(0) L_i / \langle D \rangle \quad (29)$$

where $\langle D \rangle$ is the spatially averaged denudation rate (Granger *et al.*, 1996). We are led to the simple yet remarkable conclusion that the

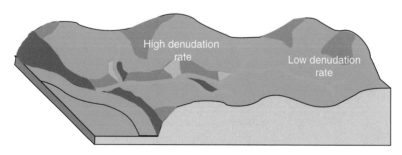

$$N_{\text{stream}} = \sum P_i(0) \, L_i / \langle D \rangle$$

Figure 10 Stream sediment may sample material from a heterogeneously eroding watershed. Despite variation in sediment supply rate, the average cosmogenic nuclide concentration in stream sediment represents the spatially averaged denudation rate of the sediment contributing area, as shown in Equation (29).

cosmogenic nuclide concentration in stream sediment is described by the same equation that governs cosmogenic nuclide concentrations in bedrock surfaces and in vertically mixed soils, despite heterogeneous catchment erosion rates (Brown *et al.*, 1995; Bierman and Steig, 1996; Granger *et al.*, 1996).

11.3.4.1.1 Complications. Although Equation (29) is conceptually simple, it is the culmination of a series of assumptions and approximations, each of which can lead to complications in data interpretation. All of the factors that can complicate interpretation of erosion in rocks and in vertically mixed soil also apply in this case. For example, changes in erosion rate through time will be damped by the residence time of quartz in the soil, but will still affect cosmogenic nuclide concentrations. We have ignored radioactive decay, which will lower cosmogenic nuclide concentrations in cases of very slow erosion rates. Quartz enrichment in the soil will also bias catchment-averaged rates toward lower inferred erosion rates, and should be quantified if possible as shown in Section 11.3.3.1.

Additional complications arise due to the spatial averaging of cosmogenic nuclide concentrations from a broad area. We have assumed in Equation (27) that quartz is evenly distributed throughout the watershed. If quartz is more abundant in one part of the watershed than in another, then erosion rates inferred from cosmogenic nuclides in quartz will naturally be biased toward the quartz-rich regions. Other complications may be less obvious. For example, we know that production rates vary with elevation, with an *e*-folding length of ~0.8 km (see Section 11.2.3). In a watershed with significant relief, the production rates in the upper parts of the catchment will be higher than in the lower parts. The integrand in the numerator of Equation (28) must therefore be weighted exponentially by the hypsometry of the basin:

$$N_{\text{stream}} \approx \int \left\{ \sum P_i(0) e^{z(A)/0.8} L_i \right\} \mathrm{d}A \Big/ \int D \, \mathrm{d}A$$
(30)

where $z(A)$ is the areally dependent elevation, in kilometers. The cosmogenic nuclide concentration can be calculated in terms of the spatially averaged production rate and penetration lengths.

$$N_{\text{stream}} = \left\langle \sum P_i(0) L_i \right\rangle \Big/ \langle D \rangle$$
(31)

Complications may also arise from the assumption that stream sediment is well-mixed and representative of the watershed. Although this is likely a safe assumption in large watersheds, or in landscapes that erode by distributed processes such as surface wash, it can be violated if there are bedrock-seated landslides in the watershed (Niemi *et al.*, 2005). Although the average cosmogenic nuclide concentration in stream sediment will still reflect the average denudation rate over time, a recent bedrock-seated landslide may introduce an inordinate amount of coarse debris into the stream, biasing erosion rates toward values too high. In catchments where bedrock-seated landslides are common, it is often found that the coarse grain fraction is predominantly derived from landslides, and has a lower cosmogenic nuclide concentration than that of the finer fraction (e.g., Brown *et al.*, 1995). Landslides or debris flows that are not bedrock-seated but instead remobilize soil that has accumulated in hollows have less of an effect on cosmogenic nuclide concentrations, because the soil contained within them is representative of the long-term erosion rate of that portion of the catchment.

11.3.4.2 Examples

Although there are many complications that can confound the determination of erosion rates from cosmogenic nuclides in stream sediment, most of these can be avoided through the judicious choice of watersheds, or explicitly acknowledged as a source of uncertainty. In this section, we will show examples of erosion rates in watersheds that illustrate strengths and weaknesses of the method. We will focus on examples that address the variation in erosion rate across landscapes, and the influence of climate on physical erosion rates.

11.3.4.2.1 Test of the method: Comparing cosmogenic rates to absolute erosion rates. Because the cosmogenic nuclide method for determining catchment-wide erosion rates rests on assumptions that are difficult to evaluate, it is important to test the method in situations where the erosion rate is already known. In such a test, erosion rates should be measured over a similar timescale by the two methods, that is, the timescale taken to erode the watershed by approximately one penetration length. However, erosion rates over millennial timescales are often difficult to measure. One such test has been made in two catchments incised into a fault block in northeastern California (Granger *et al.*, 1996).

The Fort Sage Mountains are a fault-bounded block of granodiorite in the northern Basin and Range. During the last glacial period, the basin surrounding this mountain range was submerged by Lake Lahontan. The lake level peaked at ~16,100 ± 500 years ago (Adams and Wesnousky, 1998), and then rapidly subsided, leaving a relatively smooth sheet of well-sorted sand on the former lake floor. Watersheds in the mountains subsequently deposited relatively coarse and poorly sorted alluvial fans atop the former lakebed.

By measuring the accumulation of material in the fan, it is possible to determine the erosion rate of the catchment averaged over the time since lake desiccation.

The two small catchments shown in Figure 11 are incised into a relict upper surface on the Fort Sage Mountains fault block. Alluvial fans in front of the catchments were surveyed and augered to determine fan volume. Zirconium concentrations were measured in soils and bedrock to infer quartz-enrichment factors (Riebe *et al.*, 2001a). [10]Be and [26]Al concentrations were measured in stream sediments to infer erosion rates from cosmogenic nuclides (Granger *et al.*, 1996; Riebe, 2000). Results from these measurements are shown in Table 4.

Cosmogenic nuclide results in Table 4 are highly reproducible. Four different grain size classes were analyzed, and show no apparent difference within each catchment. Replicates collected from different places in the streambed were analyzed for the grain size 0.5–1.0 mm, and were reproducible to within analytical uncertainty. A separate replicate in catchment A was collected 4 years later than the first set of samples, and was reproducible to well within analytical uncertainty. Erosion rates were calculated using Equation (29), and corrected for quartz enrichment following Equation (26). To infer quartz enrichment, [Zr] was measured in 8 samples of fresh bedrock, 9 samples of saprolite, 47 samples of soil, and 15 samples of colluvium, showing an enrichment factor of 1.18 as bedrock is weathered to soil.

Soil depths were measured at 53 locations in the watersheds, and show little to no systematic spatial variation. The total correction for quartz enrichment is 4% for these watersheds.

Erosion rates determined from cosmogenic nuclides can now be compared with erosion rates determined from fan volumes. Table 5 shows the comparison; erosion rates measured from fan volumes are indistinguishable from those determined from cosmogenic nuclides. In these watersheds, where we observed sediment transport by bioturbation, overland flow, and shallow landsliding, intersample variability is ∼25%, even when sample collection spans several years. The cosmogenic nuclide method thus seems reliable, at least to within about 25% for small watersheds.

Now that we have verified this method for determining catchment-wide erosion rates, it can be applied to problems of weathering and landscape evolution.

11.3.4.2.2 Transient versus dynamic equilibrium landscapes. Both conceptual and numerical models of landscape evolution predict hat the distribution of erosion rates within a watershed depends strongly on the history of uplift or river incision. In landscapes where uplift or incision rates have recently changed, erosion rates will be variable as relief grows or shrinks. For example, if river incision has accelerated due to fault motion downstream, then a pulse of rapid incision may migrate upstream. Hillslopes will steepen, and erosion rates will increase first in the downstream

Figure 11 Two study catchments incised into a fault block at the Fort Sage Mountains, California have deposited alluvial fans over lake sediments. Fan thickness is contoured in meters, and catchment boundaries are outlined. Erosion rates determined from [10]Be in stream sediment match accumulation rates of alluvial fans (Table 5). Reproduced by permission of the University of Chicago Press from Granger *et al.* (1996).

Table 4   Denudation rates from cosmogenic nuclides at Fort Sage Mountains, California.

Sample[a]	Nuclide concentration[b] (10^3 atom g^{-1})	Soil depth[c] (cm)	Denudation rate[d] (t km^{-2} yr^{-1})
B5 (0.25–0.5)	290 ± 20 (Be)	26 ± 2 (53)	83 ± 10
	1540 ± 110 (Al)		96 ± 13
B5 (0.5–1)	205 ± 20 (Be)	26 ± 2 (53)	120 ± 16
	1210 ± 80 (Al)		125 ± 16
B5 (0.5–1)	192 ± 20 (Be)	26 ± 2 (53)	138 ± 16
	1240 ± 80 (Al)		151 ± 7
B5 (1–2)	180 ± 14 (Be)	26 ± 2 (53)	127 ± 7
	1010 ± 70 (Al)		122 ± 16
Average B[e]			114 ± 13
A4 (0.25–0.5)	140 ± 14 (Be)	24 ± 2 (38)	177 ± 23
	940 ± 70 (Al)		161 ± 21
A4 (0.5–1)	119 ± 15 (Be)	24 ± 2 (38)	208 ± 34
	670 ± 150 (Al)		229 ± 53
A4 (0.5–1)	N.D.	24 ± 2 (38)	N.D.
	580 ± 50 (Al)		265 ± 36
A4 (1–2)	147 ± 12 (Be)	24 ± 2 (38)	166 ± 21
	780 ± 70 (Al)		195 ± 29
Average A[e]			182 ± 21
A4-1997	128 ± 13 (Be)	24 ± 2 (38)	192 ± 26
	85 ± 6 (Al)		180 ± 23
Average A-1997[e]			185 ± 18

[a] Numbers in parentheses report grain sizes in mm. A4-1997 was prepared from an aggregate of size classes.

[b] Nuclide concentrations reported by Granger *et al.* (1996), except for A4-1997 (Riebe, 2000).

[c] Soil depth for catchment A is an average determined from 38 pits within the catchment. Because only 15 soil depth measurements are available for the larger catchment B, and because they are not spread evenly across the catchment, we use all 53 pits at Fort Sage for the average soil depth of catchment B.

[d] Denudation rates from Equation (26), with $f_{soil}/f_{rock} = 1.18 \pm 0.03$, $\rho_{soil} = 1.6 \pm 0.3$ g cm^{-3} (assumed), and $\rho_{rock} = 2.7$ g cm^{-3}. Production rates scaled to sample latitude (40°N) and altitude (1.45 km for catchment A and 1.44 km for catchment B) are taken as $P_1(0) = 14.3 \pm 1.2$, $P_2(0) = 0.19 \pm 0.02$, $P_3(0) = 0.04 \pm 0$, and $P_4(0) = 0.05 \pm 0.01$ atoms g^{-1} yr^{-1} for ^{10}Be, and $P_1(0) = 86.2 \pm 8.4$, $P_2(0) = 1.41 \pm 0.08$, $P_3(0) = 0.30 \pm 0.04$, and $P_4(0) = 0.38 \pm 0.05$ atoms g^{-1} yr^{-1} for ^{26}Al.

[e] To obtain average denudation rates for each size class and for the aggregate sample A4-1997, denudation rates from each nuclide were first transformed into inverse denudation rates, averaged with inverse-variance weighting, and then reconverted into denudation rates (inverse denudation rates were used in the averaging because the measured quantity, that is, nuclide concentration, is roughly an inverse proportional of denudation rate). Average denudation rates for catchments A and B were then determined from the denudation rates of the different size classes, weighting by size-class mass and inverse variance.

Table 5   Comparison of cosmogenic- and alluvial fan-based denudation rates.

	Catchment A	Catchment B
Fan volume (10^3 m^3)	188 ± 34	310 ± 52
Mass of eroded rock (10^3 t)[a]	355 ± 92	585 ± 148
Contributing area (10^3 m^2)[b]	132 ± 20	408 ± 12
Denudation rate from fan volume (t km^{-2} yr^{-1})[c]	167 ± 50	89 ± 23
Denudation rate from cosmogenic nuclides (t km^{-2} yr^{-1})	185 ± 18	114 ± 13

[a] To obtain mass of eroded rock, fan volumes were multiplied by a bulk density of 1.6 ± 0.3 g cm^{-3}, and a zirconium enrichment factor of (1.18 ± 0.03). Because postdepositional weathering and erosion of the fan itself cannot be ruled out, the weathering losses, and thus fan-based denudation rates, are minimum estimates.

[b] Contributing areas are as reported by Granger *et al.* (1996).

[c] Fan age is 16.1 ± 0.5 kyr.

part of the watershed and later in the upstream part. This transient response will persist until stream incision rates and erosion rates match uplift rates in all parts of the landscape. This latter condition may be referred to as dynamic equilibrium (Hack, 1960), or erosional steady state.

Although equilibrium landscapes are a useful conceptual model, their existence has historically been difficult to confirm or deny. What is needed is a way to measure erosion rates at many points in a watershed to find out if erosion rates are spatially uniform, or if instead they vary systematically throughout a landscape. Cosmogenic nuclides offer just such a method, and can be used to compare erosion rates under various tectonic and climatic conditions.

We begin by contrasting erosion rates in two landscapes in the northern Sierra Nevada, California. Both catchments are in similar granitic rock. The first landscape is in the Diamond Mountains, near Adams Peak. Hillslope gradients there range from 0.2 to 0.7, with varying degrees of bare rock exposure depending on gradient (Granger *et al.*, 2001). Miocene volcanic flows extend from mountaintops to valley bottoms, showing that the Diamond Mountains have maintained similar relief and hillslope gradients over the past 10 Myr (Grose and Mergner, 2000). Eleven samples were collected from different catchments, chosen to represent as wide a range in hillslope gradient as possible.

The second landscape is along the Fall River. Again, bedrock is granodiorite, and hillslope gradients span a range from 0.2 to 0.7, similar to the range at Adams Peak. At this second site, however, relief has locally increased due to incision of the Fall River Canyon. Steep slopes are generally found near the canyon, while more gentle slopes are found away from the canyon. Eight samples were collected from different catchments, again chosen to maximize the range of hillslope gradients (Riebe *et al.*, 2000).

Erosion rates in the two landscapes show very different relationships to hillslope gradient (Figure 12). At the Diamond Mountains, where relief has persisted unchanged for at least 10 Myr, erosion rates are indistinguishable from place to place, regardless of gradient, soil depth, or any other factor (Granger *et al.*, 2001). In contrast, at Fall River, where relief has increased due to more recent canyon incision, erosion rates show a strong relationship with gradient (Riebe *et al.*, 2000).

Similar results have been seen elsewhere (von Blanckenburg, 2005). In areas where recent tectonics or river incision has occurred, erosion rates are correlated with hillslope gradient. For example, erosion rates vary with distance from active faults at the Fort Sage Mountains study site (Granger *et al.*, 1996), and in the Siwalik Range (Wobus *et al.*, 2005). In regions of tectonic stability, erosion rates are uniform across a wide range of hillslope gradient. Examples include the highlands of tropical Sri Lanka, where erosion rates are slow even on slopes up to 30° (Hewawasam *et al.*, 2003), and the Smoky Mountains in the Appalachians, where erosion rates are nearly constant over various lithologies and gradients (Matmon *et al.*, 2003). Cosmogenic nuclides thus provide a method for distinguishing landscapes in a transient state of relief, where erosion rates are spatially variable, from those in dynamic equilibrium where erosion rates are spatially uniform (Riebe *et al.*, 2000; von Blanckenburg, 2005).

11.4 CHEMICAL WEATHERING INFERRED FROM COSMOGENIC NUCLIDES

The methods and examples we have presented so far illustrate how cosmogenic nuclide concentrations in minerals can be used to infer exposure ages and denudation rates. In many cases it would be useful to quantify physical erosion and chemical weathering as separate components of denudation. For example, climate may affect chemical weathering and physical erosion differently. It is also important to understand how chemical weathering and physical erosion interrelate.

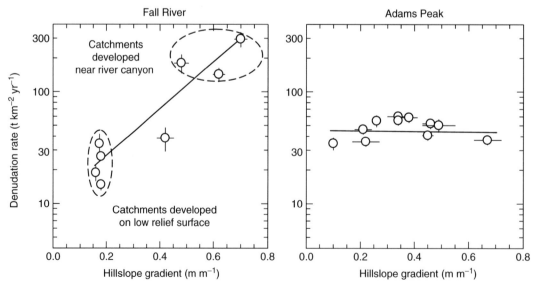

Figure 12 Contrasting relationships between erosion rate and hillslope gradient at two sites in the northern Sierra Nevada, California. At Fall River, rapid erosion on steep slopes is characteristic of transient landscape response to canyon incision. At Adams Peak, uniform erosion rates across a similar range of hillslope gradient reflect long-term stability of mountain relief. Adams Peak data from Granger *et al.* (2001); Fall River data reproduced by permission of the Geological Society of America from Riebe *et al.* (2000).

On the one hand, physical erosion may depend on the chemical breakdown (and thus weakening) of rock. On the other, weathering rates may depend on the availability of highly reactive mineral surfaces supplied by physical breakdown of parent material (Carson and Kirkby, 1972; Stallard and Edmond, 1983).

In this section we present three applications of cosmogenic nuclides to the problem of measuring weathering rates. In the first, weathering rates of a noneroding soil are calculated with the help of cosmogenic exposure ages. In the next, an empirical relationship between erosion rate and soil depth is used to quantify spatial variability in weathering rates along a hillslope. In the last, spatially averaged denudation rates from cosmogenic nuclides are partitioned into rates of chemical weathering and physical erosion. Although the role of cosmogenic nuclides is different in each case, the applications share a foundation in the principles of mass balance.

11.4.1 Chemical Weathering in Noneroding Soils

11.4.1.1 Theory

Mass-balance approaches have been used to study chemical weathering and soil formation for more than a century (e.g., Hilgard, 1860, Merrill, 1906; Larsen, 1948; April *et al.*, 1986; Brimhall and Dietrich, 1987; Chadwick *et al.*, 1990; Bain *et al.*, 1993; Taylor and Blum, 1995; Riebe *et al.*, 2001b; Anderson *et al.*, 2002; Sak *et al.*, 2004; Green *et al.*, 2006). We can write the mass balance of a noneroding soil in terms of mass gains from bedrock and mass loss due to chemical weathering in the equation

$$M_{\text{soil}} = M_{\text{parent}} - M_{\text{W}} \qquad (32)$$

where M_{soil} is the mass of a parcel of soil, M_{parent} the mass of parent material from which the soil was derived, and M_{W} the mass lost in solution. We can also write a similar mass balance for any chemical component of the soil,

$$[X]_{\text{soil}} M_{\text{soil}} = [X]_{\text{parent}} M_{\text{parent}} - M_{\text{WX}} \qquad (33)$$

where $[X]_{\text{soil}}$ is the concentration of element X in the soil, $[X]_{\text{parent}}$ its concentration in parent material, and M_{WX} its mass loss due to chemical weathering. In writing Equation (33) we assume that the soil is derived from a single, chemically homogeneous parent material.

Although Equations (32) and (33) are conceptually simple, solving the mass balance for weathering of a given soil requires an estimate of the original mass of parent material. This quantity cannot be measured directly. The key to solving Equations (32) and (33) lies in recognizing that some elements are immobile during weathering (e.g., April *et al.*, 1986; Brimhall and Dietrich, 1987; Chadwick *et al.*, 1990). These elements become increasingly enriched in the soil as other elements are lost in solution. For example, zirconium is widely recognized to be essentially immobile during chemical weathering (e.g., Colin *et al.*, 1993). If zirconium weathering losses are zero, Equation (33) can be simplified to

$$M_{\text{parent}} = ([Zr]_{\text{soil}}/[Zr]_{\text{parent}}) M_{\text{soil}} \qquad (34)$$

Equation (34) provides a ready way to infer the mass of a given soil's parent material from zirconium enrichment. We can substitute it into Equation (32) or (33) and solve for chemical weathering losses, as shown in the following equations:

$$M_{\text{W}} = M_{\text{soil}} ([Zr]_{\text{soil}}/[Zr]_{\text{parent}} - 1) \qquad (35)$$

$$M_{\text{WX}} = M_{\text{soil}} ([X]_{\text{parent}}[Zr]_{\text{soil}}/[Zr]_{\text{parent}} - [X]_{\text{soil}}) \qquad (36)$$

In practice, element concentrations and density in soil may vary with depth. In that case Equations (35) and (36) can be integrated over the depth of weathering (h) as follows:

$$M_{\text{W}} = \int \rho_{\text{soil}} ([Zr]_{\text{soil}}/[Zr]_{\text{parent}} - 1) \mathrm{d}h \qquad (37)$$

$$M_{\text{WX}} = \int \rho_{\text{soil}} ([X]_{\text{parent}}[Zr]_{\text{soil}}/[Zr]_{\text{parent}} - [X]_{\text{soil}}) \mathrm{d}h \qquad (38)$$

Equations (37) and (38) show that the mass loss due to chemical weathering can be straightforwardly calculated using measurements of the bulk chemical composition of soils and parent material, together with observations of soil density and depth. Bulk chemical compositions of major rock-forming elements and several candidate immobile elements (such as Zr) can be reliably measured using a variety of techniques, including XRF and ICP-MS.

Equations (37) and (38) provide a way to determine chemical weathering losses in a noneroding soil. If there is significant physical erosion, then the mass balance of Equations (32) and (33) must be modified to include terms for erosional losses. We will introduce such a formulation in Section 11.4.2.

To convert weathering losses of Equations (37) and (38) into weathering rates requires an estimate of the amount of time that has elapsed since weathering began. If a soil's age (t) is known, then it can be used to calculate a time-averaged chemical weathering rate (e.g., Bain *et al.*, 1993; Taylor and Blum, 1995) as

$$W_{\text{soil}} = M_{\text{W}}/t \qquad (39)$$

$$W_{X,soil} = M_{WX}/t \qquad (40)$$

where W_{soil} and $W_{X,soil}$ are the weathering rates of the soil and an individual element within it, respectively.

Although soil age has often been difficult to determine using conventional techniques, it can now be much more readily quantified using cosmogenic nuclide surface exposure dating. Even so, because noneroding soils are rare, the approach outlined in Equations (32)–(40) cannot be widely applied, and has therefore yielded only a few measurements of long-term chemical weathering rates.

11.4.1.1.1 Complications. There are many complications and uncertainties that must be considered in any soil mass-balance analysis.

1. It may be difficult to visually recognize the difference between unweathered and slightly weathered parent material, particularly on glacial moraines and terraces which may be formed from preweathered rock. If the presumed parent material has in fact weathered, then Equations (39) and (40) could substantially underestimate the true weathering rate. To minimize the potential for this type of error, sample pits should reach well below the point where bulk chemistry stabilizes at a roughly uniform composition. If element concentrations of the deepest two samples differ substantially, then it is unlikely that true parent material has been reached.

2. Parent material needs to be homogeneous. If a soil is generated from a heterogeneous parent material, then its weathering rate will be difficult to quantify. This is because unweathered samples may not be representative of the average parent material of the soil as a whole. In that case, measurements of immobile element enrichment may reflect variations in parent material composition more than weathering losses. Sedimentary deposits such as moraines and alluvial terraces may have mixed source lithologies and can therefore be particularly susceptible to this problem. Ideally many samples of parent material should be taken, to quantify intersample variability.

3. Even supposedly immobile elements may suffer weathering losses. If they do, the mass balance will underestimate the true weathering rate of the soil. The relative mobility of supposedly immobile tracers has been the subject of some debate. For example, titanium has been used in the past as an immobile tracer (e.g., Taylor and Blum, 1995; April et al., 1986; Merritts et al., 1992). More recently, evidence of Ti mobility in soils has been documented (e.g., White et al., 1998; Riebe et al., 2003; Green et al., 2006). Recent field studies

(Hill et al., 2000; Kurtz et al., 2000) and laboratory experiments (Hodson, 2002) suggest that even highly refractory Zr may not be impervious to chemical weathering, particularly in soils where weathering has been prolonged and intense. However, zirconium dissolution in anything but the most extreme environments is unlikely to be significant enough to affect studies of soil formation. Mass-balance studies of weathering now routinely use Zr as their immobile tracer of choice (Colin et al., 1993; White et al., 1998; Riebe et al., 2001a; Anderson et al., 2002; Green et al., 2006), although titanium immobility is still occasionally assumed (Nezat et al., 2004) and documented (Sak et al., 2004). A potentially important consideration regarding Zr mobility is that zircons may be susceptible to preferential sorting and translocation by physical processes in some deposits (Brimhall et al., 1991). Other candidate immobile elements include yttrium (Hill et al., 2000), niobium (e.g., Hill et al., 2000; Kurtz et al., 2000), cerium (Banfield, 1985 as cited in Green et al., 2006), and tantalum and hafnium (Kurtz et al., 2000).

4. Equations (32)–(38) require that physical erosion has been negligible. This may sometimes be the case on flat surfaces. However, as noted in Section 11.3.1.1, physical erosion is difficult to rule out in temperate environments for surfaces older than about 10^5 years. If significant physical erosion has occurred, then Equations (37) and (38) will underestimate weathering rates. Even a small amount of surface erosion may lead to significant errors in estimated weathering rates for soils with depth-dependent chemical depletion. An adaptation of the mass balance that allows us to account explicitly for physical erosion is presented later.

5. In some circumstances, aeolian additions to the soil may need to be accounted for in Equations (32)–(38). In some chronosequence studies, for example, additions by aeolian fluxes have been judged to be significant compared with weathering losses (Burkins et al., 1999; Chadwick et al., 1999). Implications may be especially severe in the case of extremely old and stable soils. In one such case in Australia, zirconium in the upper soil horizons appears to be derived almost entirely from atmospheric dust (Brimhall et al., 1988, 1991). However, even a small aeolian flux can be an important complication if it is concentrated enough in immobile elements that it alters a weathered soil's immobile element concentration. A detailed discussion of how to quantify the effects of aeolian fluxes is beyond the scope of this chapter. The interested reader is referred to discussions in Brimhall et al. (1988, 1991), Kennedy et al. (1998), and Chadwick et al. (1999).

11.4.1.2 Example

11.4.1.2.1 Weathering rates of glacial moraines.
Glacial moraines can provide excellent conditions for studying soil development. They are made of initially fresh and often homogeneous materials whose ages can be determined by a variety of techniques, including cosmogenic nuclides, radio-carbon, and U-series measurements.

Here we discuss a mass-balance study of a glacial chronosequence in the Wind River Range, Wyoming (Taylor and Blum, 1995). The six moraine surfaces considered appear to have been minimally affected by physical erosion. The ages of the oldest four were determined using *in situ*-produced [10]Be (Gosse *et al.*, 1994, 1995). This example illustrates how mass-balance studies of noneroding soils can be coupled with results from cosmogenic surface exposure dating to infer long-term chemical weathering rates.

To minimize errors introduced by effects of physical erosion, the mass-balance study was limited to the flat tops of the moraines, where evidence of erosion was generally absent (Taylor and Blum, 1997). Pits were excavated until visibly unweathered material was reached, and samples were taken at various depths from the sides and bottoms of the pits. Samples were analyzed for major and trace element concentrations, after removal of coarse (>2 mm) particles. Removal of the coarse fraction of samples is common in soil mass-balance studies (April *et al.*, 1986; Bain *et al.*, 1993; Nezat *et al.*, 2004), but may contribute

to errors if weathering of coarse particles is significant.

Using Ti as an immobile reference, weathering losses were calculated for each moraine in a mass-balance formulation similar to Equation (38). To determine long-term weathering rates of base cations, the elemental mass losses of Ca, Na, Mg, and K were summed and then divided by their respective moraine ages, which range from 0.4 to 297 kyr across the chronosequence (Taylor and Blum, 1995).

Base-cation weathering rates of the moraine surfaces decline sharply over time from 7.9 to $0.1\,t\,km^{-2}\,yr^{-1}$ (Figure 13). Such a decline in weathering rates may reflect either the exhaustion of rapidly weathered phases in the soil or changes in mineral surface area over time. Despite uncertainties and complications due to the potential for physical erosion, aeolian inputs, and Ti mobility (e.g., Dahms *et al.*, 1997; Taylor and Blum, 1997), the results appear to be robust and generally agree with expectations from previous and subsequent research. Both laboratory experiments (e.g., Busenburg and Clemency, 1976; White *et al.*, 1999) and other field-based weathering studies (e.g., Bain *et al.*, 1993) have found similar declines in weathering rates over time.

It is worth noting that many authors—including Taylor and Blum (1995) and, subsequently, others (e.g., Nezat *et al.*, 2004)—express weathering losses in terms of the mass of the parent material and estimate M_{parent} as the product of parent material density and the thickness of the soil:

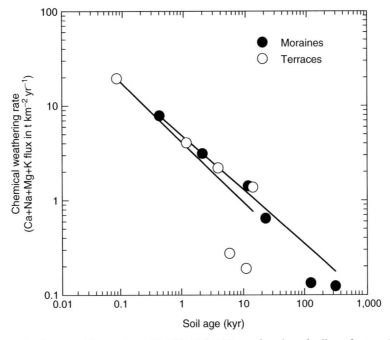

Figure 13 Net weathering rate of base cations (Ca+Na+Mg+K) as a function of soil age for two chronosequences: glacial moraines (closed symbols; after Taylor and Blun, 1995) and fluvial terraces (open symbols; after Bain *et al.*, 1993). In both cases weathering decreases rapidly with soil age.

$$M_W = M_{parent}([X]_{parent} - [X]_{soil}[Zr]_{parent}/[Zr]_{soil})$$ (41)

Although Equation (41) is equivalent to Equation (36), it introduces the unnecessary approximation that the original parent material thickness is equal to the soil thickness. As shown above in Equation (34), the weathering loss can instead be expressed in terms of the mass of the soil, which can be quantified on a per-unit-area basis without approximation, as $\rho_{soil}h$.

11.4.2 Chemical Weathering Rates of Soil along an Eroding Slope

Thus far, the mass-balance formulation requires negligible physical erosion, a condition that is not met in many circumstances. In hilly or mountainous settings in particular, soils nearly always have substantial physical erosion rates. To apply the soil mass-balance approach in settings such as these, it is necessary to explicitly account for losses due to physical transport of material.

11.4.2.1 Theory

In writing the mass balance for a soil column on an actively eroding slope, we need to include all of the inputs and outputs of sediment and solutes (Green *et al.*, 2006). Accounting for mass changes due to physical and aeolian processes, the local change in soil mass per unit time can be written as the following differential equation:

$$\partial(\rho_{soil}h)/\partial t = -\nabla E_{soil} + E_{saprolite} - W_{soil} + I_{aeolian}$$ (42)

where E is the vector form of physical erosion (equivalent to soil flux), $E_{saprolite}$ the rate of incorporation of saprolite into the soil, and $I_{aeolian}$ the aeolian material flux. Use of the vector allows us to account for cross-slope, as well as down-slope transport.

Solving Equation (42) for steady-state soil thickness and ignoring the aeolian fraction leads to

$$W_{soil} = -\nabla E_{soil} + E_{saprolite}$$ (43)

The mass balance developed here can be applied to measure weathering rates of individual elements in the soil ($W_{X,soil}$) as summarized in the following Equation (44) (Green *et al.*, 2006):

$$W_{X,soil} = -[X_{soil}]\nabla E_{soil} + [X_{saprolite}]E_{saprolite}$$ (44)

As discussed in Section 11.3.2, the denudation rate (D) can be determined from cosmogenic nuclide concentrations at the top of the saprolite. If we assume steady state and that most of the mass loss near the saprolite–soil interface is due to physical erosion, then D can be substituted into Equation (43) or (44). All that is then required to quantify the weathering rate of the soil is the divergence of local sediment flux.

Although the divergence of local sediment flux is difficult to measure directly, it can be estimated from topographic data if a relationship between physical erosion and slope is assumed. Many of the dominant soil transport processes of hilly landscapes are thought to be slope-dependent, such that soil transport flux increases linearly with hillslope gradient. In such a case, if soil depths are in steady state, then the divergence of sediment transport can be expressed as

$$\nabla E_{soil} = -\rho_{soil}K\nabla^2 z$$ (45)

where K is a parameter similar to a diffusion coefficient, z the elevation of the landscape surface and $\nabla^2 z$ its local curvature.

Substituting Equation (45) into Equations (43) and (44) yields Equations (46) and (47) for total weathering and elemental weathering:

$$W_{soil} = \rho_{soil}K\nabla^2 z + E_{saprolite}$$ (46)

$$W_{X,soil} = \rho_{soil}K([X_{soil}]\nabla^2 z + \nabla[X_{soil}]\nabla z) + [X_{saprolite}]E_{saprolite}$$ (47)

Equations (46) and (47) show that the chemical weathering flux of a column of soil on a slope can be calculated using measurements of saprolite erosion rates, soil density, element concentrations in soil and saprolite, hillslope gradient and curvature, and the diffusion constant of soil transport. Note that the term $\nabla[X_{soil}]$ requires concentrations of elements in the soil from immediate up- and downslope of the point of interest. Although the list of required values is somewhat longer than the one for weathering of noneroding soils, the analysis is tractable.

The mass-balance approach outlined in Equations (42)–(47), has been developed only recently (Green *et al.*, 2006). It promises to yield a wealth of new information about weathering in landscapes, by permitting quantification of long-term weathering rates from point to point along hillslopes. Below we discuss results from an application of this approach, but first we consider potential complications.

11.4.2.1.1 Complications. At least two complications should be considered.

1. Quantification of the soil transport function is a critical step in this analysis. The formulation of sediment transport in various landscapes remains a subject of active research. Even if the simple transport law leading to Equation (45) is assumed, the mass-balance analysis still requires an empirical estimate of diffusivity (K), which will be a source of uncertainty.

2. The mass-balance approach requires soil mass to remain in steady state, an assumption that is difficult to verify. There is reason to believe that soil thicknesses do not vary substantially through time in many eroding landscapes (Dietrich *et al.*, 1995), but if they do then both the cosmogenic nuclide-based denudation rate and the weathering flux will be affected.

11.4.2.2 Example

11.4.2.2.1 Weathering rates on a hillslope in Australia. The mass-balance approach outlined in Equations (42)–(47) was applied at the Nunnock River site, on the Great Escarpment of Australia (Green *et al.*, 2006), where soil production rates were previously determined by Heimsath *et al.* (2000, 2001a). Soils consist of a 10–12 cm thick A horizon over a 10–100 cm thick B horizon. Drill cores indicate that saprolite is locally developed to >18 m deep.

To quantify the bulk chemistry of soils and saprolite, samples were collected from pits and trenches along four hillslope transects and a prominent ridge near the Nunnock River. A total of 104 soil samples and 33 samples of saprolite were collected from various depths.

The local divergence of physical soil flux along each transect was estimated using a diffusivity constant of $40\,\text{cm}^2\,\text{yr}^{-1}$ previously determined by Heimsath *et al.* (2000, 2001a) together with estimates of topographic curvature and gradient from detailed topographic surveys. At each sampling site, saprolite erosion rates were estimated from observations of soil depth using the empirical soil production function discussed in Section 11.3.2 and given in Equation (15).

Equation (47) was used to estimate long-term weathering rates of Al, Ca, Fe, K, Mg, Na, and Si in the soil. Variability among individual measurements was averaged by binning data along a line of hillslope transport, taken as the steepest rise to the ridge crest. This averaging allowed variation in weathering rates to be viewed as a function of position on the hillslope. Figure 14 shows results for W_{Si}, W_{Na}, and W_{Al}. For each element weathering rates increase with increasing transport distance, to a maximum near the center of the transects, and then decrease to the end of the slopes. Green *et al.* (2006) suggest that the initial increase in weathering rates with distance may be due to a buildup of organic carbon and increases in water flux. They further suggest that at greater distances weathering rates may decrease due to higher chemical saturation in porewater and decreasing availability of readily weathered surfaces, despite continued gains in water flux. Saprolite weathering rates were also determined using the bulk geochemical data in a mass-balance formulation for a column of saprolite. Weathering rates of the saprolite are more uniform than weathering rates of rock, and are 80 times slower. Significantly, Green *et al.* (2006) found that the overall chemical weathering rate at the site is comparable to the physical erosion rate determined by Heimsath *et al.* (2000, 2001a). Any study of denudation, soil development, or landscape evolution at such a site will need to account for both physical erosion and chemical weathering.

11.4.3 Chemical Weathering Rates of Soils and Landscapes

In the previous section, the mass balance expressed the weathering rate of an eroding soil

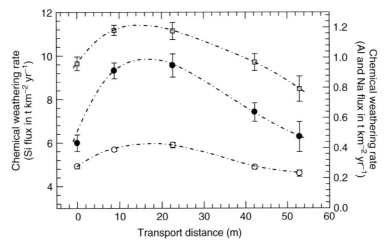

Figure 14 Chemical weathering rates of silicon (closed circles, left axis), aluminum (squares, right axis), and sodium (open circles, right axis) in the soil each as a function of transport distance for slopes at the Nunnock River field site. Dashed lines are not fit to the data, but are meant to guide the eye. Weathering rates are consistently lower near the ridge crest, and peak partway down the hillslope. Adapted from figure 7 of Green *et al.* (2006).

as a function of inputs from physical erosion of saprolite and losses due to the erosional flux of sediment. This requires measurements of bulk chemistry of saprolite and soil, as well as estimates of denudation rates and the local divergence of sediment flux. In the more simple case of a noneroding soil, the mass balance expressed chemical weathering rate in terms of the chemical enrichment of an element that is immobile during weathering. Here, we show how these two approaches can be combined to yield weathering rates of entire catchments, as well as individual points along slopes, with the help of cosmogenic nuclide concentrations in soils or sediment.

11.4.3.1 Theory

As we did in the case of weathering and erosion in a soil column, we will again write the mass balance in terms of inputs and outputs of physical and chemical processes. This time, however, we will consider not just the soil, but the saprolite as well, and thus account for mass gains and losses of the entire weathering column. Under steady state, and ignoring aeolian input, the export of material from a watershed can be summarized by inverting Equation (1), as rewritten here in the following equation:

$$W = D - E_{soil} \qquad (48)$$

On the more local scale of a slope or ridgetop, erosional losses must be written in terms of the divergence of sediment flux, using the vector form of soil erosion as shown in the equation

$$W = D - \nabla E_{soil} \qquad (49)$$

The derivation presented below proceeds from Equation (48) for catchment-averaged fluxes, although we recognize that a similar approach could be applied to Equation (49) for more localized measurements of soil weathering rates.

Provided that the soil is derived from a chemically homogeneous parent material, Equation (48) can also be written for individual elements within the soil as shown in the equation

$$W_X = [X]_{parent}D - [X]_{soil}E_{soil} \qquad (50)$$

For immobile elements, W_X will be zero (as shown in Figure 15) and we can write the equation

$$E_{soil} = ([Zr]_{parent}/[Zr]_{soil})D \qquad (51)$$

recognizing that it is useful to express E_{soil}, an unknown, in terms of D, which can be quantified using cosmogenic nuclides.

We then substitute Equation (51) into Equations (48) and (50) and solve for chemical weathering losses, as shown in the following equations:

$$W = D(1 - [Zr]_{parent}/[Zr]_{soil}) \qquad (52)$$

$$W_X = D([X]_{parent} - [X]_{soil}[Zr]_{parent}/[Zr]_{soil}) \qquad (53)$$

Equations (52) and (53) show that chemical weathering rates can be calculated using measurements of the bulk chemical composition of soils and parent material, together with measurements of denudation rates.

Equations (51)–(53) show how Zr-enrichment measurements in eroding soils permit the partitioning of a given denudation rate into its chemical and physical components.

We can also write the following equation:

$$CDF = (1 - [Zr]_{parent}/[Zr]_{soil}) \qquad (54)$$

where CDF, the chemical depletion fraction, is equal to W/D, and thus is the fraction of denudation that is accounted for by chemical weathering.

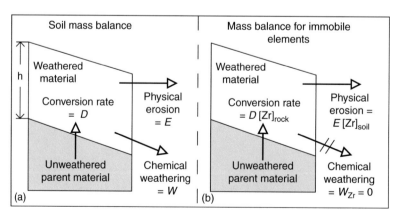

Figure 15 Mass balance of soluble and insoluble material in an eroding soil. For the soluble fraction of the soil (a), conversion of unweathered parent material into weathered by-products is balanced by removal due to physical erosion and chemical weathering such that the mass of the soil is constant. For elements that are immobile during weathering (b), inputs are balanced solely by outputs due to physical erosion. Reproduced by permission of Elsevier from Riebe *et al.* (2003).

We express the CDF for individual elements (CDF_X) in the equation

$$CDF_X = ([X]_{parent} - [X]_{soil}[Zr]_{parent}/[Zr]_{soil})$$

(55)

The approach used to derive Equations (52) and (53) can also be used to derive Equations (56) and (57), which express weathering rates of saprolite $(W_{saprolite})$ and individual elements within it $(W_{X,saprolite})$ in terms of the bulk chemistry of saprolite and parent rock:

$$W_{saprolite} = D(1 - [Zr]_{parent}/[Zr]_{saprolite})$$ (56)

$$W_{X,saprolite} = D([X]_{parent}$$

$$- [X]_{saprolite}[Zr]_{parent}/[Zr]_{saprolite})$$ (57)

Recognizing that $(W - W_{saprolite})$ is equal to W_{soil}, we can obtain weathering rates of soils by subtracting results of Equations (56) and (57) from Equations (52)–(53).

Although the focus in this chapter is on applications of cosmogenic nuclides, they are not required in the mass-balance approach outlined above if conventional estimates of denudation rates are available. In that case, however, any mismatch between the timescales of soil formation and denudation rate will need to be considered in the interpretation of the data. Denudation rates from cosmogenic nuclides are preferred, because they are averaged over the timescales of soil formation. If denudation rates are not available, the approach can still be used provided that the residence time of soil particles (t_{soil}) can be estimated (White *et al.*, 1998; Anderson *et al.*, 2002). To show how, we first recognize that t_{soil} is equal to soil mass divided by the overall rate of removal by chemical weathering and physical erosion of the soil, as expressed in the equation

$$t_{soil} = \rho_{soil}h/(D - W_{saprolite})$$ (58)

Note that $(D - W_{saprolite})$ is equal to $E_{saprolite}$, which appeared in Equation (42). Substituting Equation (56) into (58) and solving for D yields the equation

$$D = \rho_{soil}h[Zr]_{saprolite}/(t_{soil}[Zr]_{parent})$$ (59)

Equation (59) can be then substituted into Equations (52) and (53) to yield

$$W = \rho_{soil}h[Zr]_{saprolite}/(t_{soil}[Zr]_{parent})$$
$$\times (1 - [Zr]_{parent}/[Zr]_{soil})$$ (60)

$$W_X = \rho_{soil}h[Zr]_{saprolite}/(t_{soil}[Zr]_{parent})$$
$$\times ([X]_{parent} - [X]_{soil}[Zr]_{parent}/[Zr]_{soil})$$ (61)

Equations (52)–(55) show that measurements of chemical depletion and enrichment can provide an integrated picture of weathering over catchment scales, with the help of spatially averaged measurements of denudation rates from cosmogenic nuclides. The approach can also be applied locally, to weathering within a soil column, with denudation rates determined from cosmogenic nuclides at the soil–saprolite interface. This mass-balance approach of Equations (48)–(57) is a powerful technique for quantifying long-term chemical weathering rates in diverse settings over a wide range of spatial scales. Below, we will present examples of how this is done, but first we discuss potential complications.

11.4.3.1.1 Complications. In addition to the complications listed under the mass-balance approach for noneroding soils, there are several additional considerations worth noting.

1. Many samples of soil and saprolite are required to characterize weathering at the catchment scale. Samples must be taken from widely distributed surfaces and pits, to characterize spatial variability in bulk chemistry of weathered soil and saprolite (Riebe *et al.*, 2001b, 2003, 2004a).

2. In some catchments, weathering during transport may account for a significant fraction of the total weathering of the soil (Green *et al.*, 2006). In that case, weathering rate can vary with position in the landscape (Figure 14). At the Nunnock River landscape analyzed by Green *et al.* (2006), CDFs vary with depth and transport distance, increasing from about 0.1 at depth along the ridge to nearly 0.5 at the surface near the toe of the slope (Figure 16). Such hillslope-scale variability in weathering rates has the potential to obscure or distort any systematic site-to-site variability in catchment-scale weathering rates if samples are not representative (Green *et al.*, 2006).

3. Like the mass balance for soil weathering rates at points along a slope, the formulation of Equations (50)–(61) stipulates that the mass of the weathering media is in steady state. It is worth noting that even if the steady-state assumption is not valid locally from point-to-point along a slope, it still could be approximately correct (i.e., in an average sense) for the contributing area as a whole.

4. In actively eroding hills and mountains, denudation rates are generally much higher than any plausible aeolian inputs. This will generally make the assumption of negligible aeolian fluxes somewhat more tenable than it is for noneroding soils (Section 11.4.1).

5. The mass-balance approach will be difficult to apply in catchments with variable bedrock

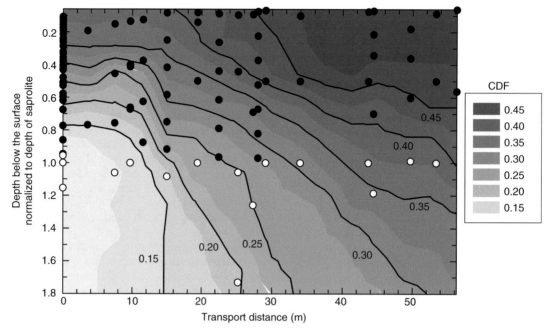

Figure 16 Chemical depletion fraction (CDF) as a function of depth and transport distance for slope transects from the Nunnock River field site in Australia. Depths are normalized to the depth of saprolite, which increases with transport distance along each of the transects. Closed circles show soil sample locations. Open circles show saprolite sample locations. Shading shows contours of equal chemical depletion fraction. Data were smoothed using a running averages algorithm (similar to that used by Green *et al.*, 2006). Chemical depletion fractions increase systematically with increasing proximity to the surface and with increasing transport distance. Calculations of Green *et al.* (2006) assume bedrock is uniform across entire slope. Adapted from figure 5 of Green *et al.* (2006).

composition. Catchments should be selected such that they are underlain by roughly uniform lithologies. To determine whether variability is significant, it is generally necessary to collect samples from widely distributed sources of parent material (Riebe *et al.*, 2001b, 2003). If multiple parent material compositions are identified it may be possible to add terms that account for their individual contribution in Equation (50) and solve for mixing of multiple sources (as well as weathering) using an array of immobile tracer elements.

6. In practice, because much of the unweathered parent material may be at great depth and difficult to access, it will be necessary to rely on outcrop samples for estimates of parent material composition. But outcrops crop out because weathering and erosion have affected them differently than the bedrock that has been converted into soil. This may be due to spatial variability in weathering and erosion processes. It may also be due to variability in some intrinsic property of the rock. In that case, outcrops may be a poor chemical proxy for the parent material that has generated the soil. Ideally a few samples from great depth or a road cut, if available, should be taken to confirm that outcrops are similar to unweathered material at depth.

11.4.3.2 Examples

When carefully applied to minimize errors and other complications, the mass-balance approach can yield a wealth of information about weathering in many eroding landscapes. Because physical erosion and chemical weathering are measured together, it can be used to determine how weathering and erosion interrelate, as we show in one of the examples below. This constitutes an important advance over the conventional mass-balance approach, which required erosion to be negligible (Bain *et al.*, 1993; Taylor and Blum, 1995). Also, because the approach can be applied in diverse settings, it can be used to study how chemical weathering rates vary across landscapes as a function of climate and other factors, as we show in examples below.

11.4.3.2.1 Test of the mass-balance approach. The mass balance for eroding landscapes is a relatively new formulation (Kirchner *et al.*, 1997), and includes several assumptions that are difficult to evaluate. To ascertain whether it yields reliable estimates of long-term weathering rates, it is important to compare results from the cosmogenic nuclide method with independent estimates of weathering flux. We present here an example from Rio Icacos, Puerto Rico, where three independent sets of weathering rate data are available (Riebe *et al.*, 2003). The independent estimates span both short and long timescales and are based on (1) stream solute fluxes estimated from water samples collected at the Rio Icacos gauge (McDowell and Asbury, 1994; White and Blum, 1995), (2) solute concentrations and infiltration rates of soil porewater (Stonestrom *et al.*, 1998; White *et al.*, 1998), and (3) bulk chemical losses from a soil profile (White *et al.*, 1998).

Mean annual temperature in the Rio Icacos watershed is 22 °C and average annual rainfall is 420 cm yr^{-1}. The underlying bedrock is a uniform quartz diorite, which weathers to friable saprolite with depths reaching 8 m and greater (White *et al.*, 1998). Intense weathering at the saprolite-rock boundary converts nearly all of the plagioclase, K-feldspar, and hornblende into secondary minerals, and Ca and Na are essentially absent in the saprolite (White *et al.*, 1998). Soils are 50–150 cm thick, and grade from bioturbated basal B horizons to thin (5–10 cm), organic-rich A horizons. Unweathered bedrock crops out locally on hillslope surfaces as coherent corestones.

Denudation rates were determined from cosmogenic nuclides in amalgamated soils from slopes that appeared to be unaffected by landsliding—a significant geomorphic process at Rio Icacos (e.g., Larsen and Torres-Sanchez, 1992; Brown *et al.*, 1995). To characterize the bulk chemistry of weathered and parent material, soils, saprolite, and outcrops were sampled from widely distributed points in each catchment. A total of 91 samples of soil, 33 samples of saprolite, and 24 samples of outcrops were collected from pits, landslide scars, and auger holes. Bulk chemistry of soils was measured using XRF (Riebe *et al.*, 2003).

CDFs for each subcatchment were estimated from Equations (54) and (55), and reveal that the fraction of denudation that is accounted for by chemical weathering is ~60%. Denudation rates are roughly uniform across the site (range = 79–101 t km^{-2} yr^{-1}; average = 90 ± 21 t km^{-2} yr^{-1}). Combining CDFs with denudation rates using Equation (52) yields chemical weathering rates that are also fairly uniform, agreeing to within a standard error

across the sites (range = 47–59 t km^{-2} yr^{-1}; average = 56 ± 13 t km^{-2} yr^{-1}). Saprolite weathering rates inferred from Equation (56) show that it accounts for ~50% of the overall chemical weathering flux from the catchments.

To compare these results with weathering flux data from previous work, Equation (53) was used to infer long-term chemical weathering rates for each of the major rock-forming elements. The results for Si, Na, Mg, Ca, and K are listed in Table 6 alongside results from the three independent studies. None of the approaches listed in Table 6 provides an unequivocally accurate standard in the comparison. However, the general agreement among the data confirms that the cosmogenic-based approach yields realistic estimates of long-term weathering rates.

11.4.3.2.2 Chemical weathering as a function of altitude. Chemical weathering depends on a wide variety of interrelated factors, including temperature, precipitation, and vegetative cover. All of these factors can vary over short spatial scales on steep mountains. To illustrate how chemical weathering rates vary in a mountain setting, we present an example from a 2-km long ridge spanning 2,090–2,750 m in altitude in the Santa Rosa Mountains, Nevada (Riebe *et al.*, 2004b). The ridge encompasses marked contrasts in vegetative cover, snow depth, and mean annual air temperature (from 3.6 to 0.4 °C), and is underlain by roughly uniform bedrock.

Six sites were selected on the ridgeline itself, and in small steep catchments along it. Each of the sampling areas was small (<1.0 ha). The aim in sampling was to capture local variability and avoid sampling from single, potentially anomalous points, while staying within a narrow elevation range, so that each location would represent a distinct set of climatic conditions. In all, 49 samples of soil, 6 samples of saprolite (from the bases of several soil pits), and 28 samples of parent rock were obtained and analyzed. To quantify denudation rates along the transect, sediments from four sites were analyzed for concentrations of cosmogenic nuclides in sediment and saprolite.

Table 6 Comparison of short- and long-term chemical weathering rates (all in t km^{-2} yr^{-1}).

Element	Long-term: cosmogenic and mass balance (Riebe et al., 2003)	Long-term: regolith profile (White et al., 1998)	Short-term: pore waters (Stonestrom et al., 1998)	Short-term: solute fluxes (McDowell and Asbury, 1994)
Na	1.1 ± 0.3	1.4	3.1	2.5
Mg	1.1 ± 0.3	0.8	1.6	2.8
Si	13.8 ± 3.3	15.4	12.2	22.7
K	0.4 ± 0.1	1.0	1.4	1.3
Ca	4.0 ± 0.9	3.4	4.1	8.5
Na + Mg + Si + K + Ca	20.5 ± 3.4	22.0	22.5	37.7

CDFs decrease rapidly with increasing altitude (Figure 17a), dropping from 0.20 at the lowest sample to essentially 0.0 at the highest. In contrast, denudation rates vary by only a factor of 1.4 and show no clear trends with elevation (Figure 17b). Chemical weathering must therefore be slower at higher altitudes, because weathering is less intense for the same rate of sediment supply. Thus chemical weathering rates decline rapidly with increasing altitude to essentially zero above the upper limits of aspen and woody brush (Figure 17c). The altitudinal decline in weathering rates of individual elements is similarly sharp (Riebe *et al.*, 2004b).

The decrease in elemental chemical weathering rates of Si and Na with increasing altitude is sharper than predicted based on the temperature sensitivity of silicate weathering in granitic watersheds, but is consistent with an earlier report based on solute flux data from an altitude transect in the Swiss Alps (Drever and Zobrist, 1992). Weathering rates at Santa Rosa Mountain appear to be affected by factors other than temperature. For example, the progressive decline in vegetative cover and the increase in snow cover and duration of freezing with altitude are potentially important factors (Riebe *et al.*, 2004b). If that is the case, then weathering rates may be particularly sensitive to differences in elevation at higher sites, where variations in vegetative cover and snow depth may be particularly sharp.

11.4.3.2.3 Quantifying how chemical weathering and physical erosion interrelate. Next, we consider a study which used the mass-balance approach to quantify how weathering and erosion interrelate across granitic sites in the Sierra Nevada, California (Riebe *et al.*, 2001b).

Among the sites considered were the Fort Sage Mountains and Fall River sites introduced earlier in Section 11.3.4.2.

In a series of catchments at each site, samples of soil, saprolite, and parent material were collected from widely distributed surfaces, pits, and outcrops. Bulk and trace element chemistry was used to quantify spatially averaged CDFs for each catchment using Equation (54). Denudation rates for each catchment were available from previous work at the sites.

As noted in Section 11.3.4.2, catchment-wide denudation rates vary substantially across Fort Sage and Fall River, due to differences in the proximity of incising rivers and down-dropping fault throw. This variability in denudation rates makes the sites suitable for studying how weathering and erosion interrelate, because the effects of potentially confounding factors (such as variations in climate and lithology) are minimized by relative homogeneity within each site.

The results of the analysis are shown in Figures 18a and 18b. CDFs are roughly uniform across each site. Thus the fraction of denudation that is accounted for by chemical weathering does not vary systematically with increasing denudation rate. This implies that chemical weathering rates increase almost proportionally with denudation rates (Figure 18b); chemical weathering rates have to be faster to achieve the same degree of depletion when denudation rates are faster. The strong coupling between rates of denudation and chemical weathering was broadly confirmed across the other Sierra Nevada sites (Riebe *et al.*, 2001c) and later over an extended set of 42 catchments and hillslope sites (Riebe *et al.*, 2004a). Results from the later synthesis are shown in Figures 18c and 18d. The coupling of weathering

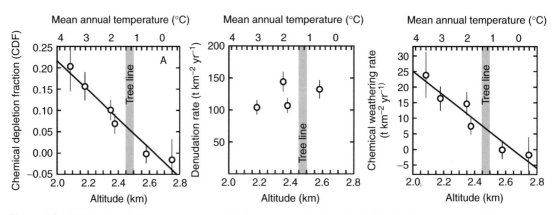

Figure 17 CDFs, denudation rates, and weathering rates plotted against altitude (lower axes) and temperature (upper axes). Vertical bar marks upper limits of trees and brush. CDFs decrease from 0.2 to 0.0 (a), whereas denudation rates vary by a factor of <1.5 across the site (b), consistent with a decrease in chemical weathering rates (c) with increasing altitude. Lines through data are linear regressions. Weathering rates at the highest and lowest locations were calculated using a site-wide average denudation rate. Because the CDF is 0 within error at the highest site, the weathering rate there must be 0 within error as well, irrespective of denudation rate. Reproduced by permission of Elsevier from Riebe *et al.* (2004b).

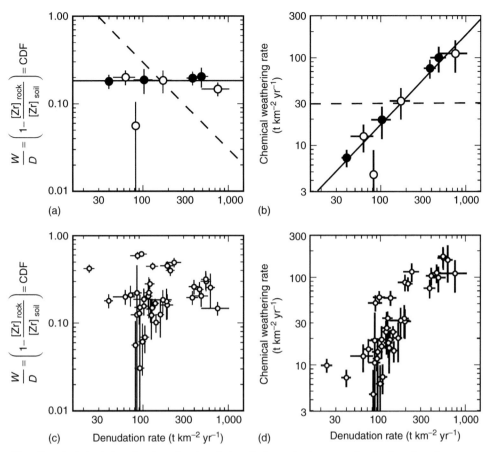

Figure 18 Chemical depletion fractions (a, c) and chemical weathering rates (b, d) plotted against cosmogenic-based denudation rates for catchments at Fort Sage (open circles) and Fall River (closed circles), two Sierra Nevada localities where denudation rates vary substantially and for 42 catchments considered by Riebe *et al.* (2004a) (c, d). CDFs are roughly uniform across Fall River and Fort Sage (a), implying that chemical weathering rates increase with denudation rates (b). Hypothetical dashed lines in (a) and (b) illustrate what would have been observed if chemical weathering rates were uniform at each site (dashed line in b), and thus if soils were chemically fresher at sites with faster denudation rates (dashed line in a). Across the 42 catchments of Riebe *et al.* (2004a), CDFs are not strongly correlated with denudation rates (c), implying that weathering rates are sensitive to the rate of supply of minerals by incorporation of rock into saprolite and soil (d). (a, b) Adapted by permission of Elsevier from Riebe *et al.* (2001b). (c, d) Reproduced by permission of Elsevier from Riebe *et al.* (2004a).

and denudation rates exhibited in Figure 18 could, on the one hand, reflect strong dependence of chemical weathering on the rate of fresh mineral supply from physical breakdown of rock. On the other hand, it could reflect the dependence of rock breakdown on weakening caused by chemical weathering (Riebe *et al.*, 2001b).

The strong coupling of weathering and denudation has important implications. For example, the indication that chemical weathering rates may be higher in areas of rapid uplift (where erosion rates are also rapid) is significant for Earth's long-term climatic evolution. If changes in uplift rates affect chemical weathering rates over long-enough timescales and broad-enough areas, they could modulate the silicate weathering flux to the oceans and thus affect atmospheric CO_2 concentrations. Another implication of the strong coupling of weathering and denudation is that site-to-site

differences in denudation rates are likely to obscure the effects of site-to-site differences in lithology, climate, and other regulators of chemical weathering rates. We will elaborate on this implication and how to deal with it in the next example.

11.4.3.2.4 Climatic effects on chemical weathering. The previous example showed that chemical weathering rates are strongly correlated with physical erosion rates. Any study of climatic effects on chemical weathering must account for site-to-site differences in physical erosion rate. To illustrate how this can be done, we use data from the climatically diverse sites of Figures 18c and 18d. All of the data are from granitic landscapes, including six Sierra Nevada localities (Riebe *et al.*, 2001b), Rio Icacos (Riebe *et al.*, 2003), the altitude transect of northern Nevada

(Riebe *et al.*, 2004b), and catchments and ridge-tops of the South Island, New Zealand, and Panola Mountain, Georgia (Riebe *et al.*, 2004a). Also included are data from several sites in Mexico, including the Sonora Desert, the Chiapas Highlands, and two localities near the Jalisco coast. Overall, the data set includes 42 granitic catchments and hillslope sites clustered in 14 localities that span a diverse range of climates. Mean annual temperatures range from 2 to 25 °C and average precipitation ranges from 22 to 420 cm yr^{-1} across the sites. The sites encompass many combinations of hot/cold and wet/dry climatic regimes, thus enabling distinction between the effects of precipitation and temperature on chemical weathering rates.

Widely distributed samples of soil, saprolite, and rock from each site were analyzed for bulk and trace element chemistry. Samples of stream sediment were analyzed for cosmogenic nuclides to determine spatially averaged denudation rates. At some sites, saprolite was also collected for cosmogenic nuclide analysis of denudation rates based on methods described in Section 11.3.2 for erosion of rock under a soil cover. The bulk chemical measurements and cosmogenic nuclide data were used to estimate chemical depletion fractions and weathering rates using the Equations (52)–(55).

As we have already shown in Figures 18c and 18d, chemical weathering rates calculated from this data set are tightly coupled with denudation rates, because chemical depletion fractions do not vary systematically with denudation rates. This introduces an important complication into the assessment of climatic effects on weathering. It indicates that weathering rates can vary substantially within a given climate regime due to variations in denudation rates. Fort Sage and Fall River provide two cases in point; chemical weathering rates vary by an order of magnitude within each site (Figures 18a and 18b), even though catchment-to-catchment variations in climate are small. This helps explain why chemical weathering rates vary by almost as much at any given precipitation (Figure 19a) and temperature (Figure 19b) as they do across the entire spectrum of sites. As a consequence, chemical weathering rates do not exhibit any clear trends with precipitation or temperature (Figures 19a and 19b).

The confounding effects of site-to-site differences in denudation rate are evident on a case-by-case basis, as well as for the set of 42 weathering rates as a whole. For example, at 11 of the catchments, silicon weathering rates estimated from Equation (53) exceed the range reported in Table 6 for Rio Icacos (Riebe *et al.*, 2004a).

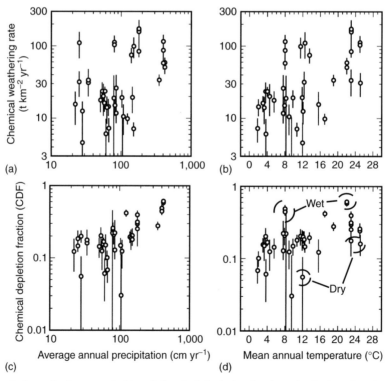

Figure 19 Chemical weathering rates (a, b) and chemical depletion fractions (c, d) plotted against average annual precipitation (a, c) and mean annual temperature (b, d). Significant variations in chemical weathering rates within each site (a and b) make relationships between chemical weathering rates and climate across the sites difficult to see. CDFs increase with average precipitation (c) and temperature (d) when effects of precipitation are taken into account. Reproduced by permission of Elsevier from Riebe *et al.* (2004a).

That makes them the 11 highest silicon weathering rates on record for granitic terrain. This is remarkable because the climates of many of the 11 sites are not very extreme. For example, at two of the sites, average precipitation is more than five times lower than it is at tropical Rio Icacos. Another two of the catchments are in the temperate foothills of the Sierra Nevada, at the Fall River site. Perhaps most remarkable is the catchment at Fort Sage, where the silicon weathering rate is roughly two times higher than it is at Rio Icacos, despite a mean annual temperature that is 10 °C cooler and an average precipitation that is 20 times lower. How could such a relatively mild climate maintain such a fast weathering rate? There is some indication that it may be due at least in part to differences in the rate of supply of fresh minerals for weathering (Riebe *et al.*, 2004a). The overall denudation rate of the Fort Sage catchment is a factor of almost eight times higher than it is at Rio Icacos. The relatively high denudational throughput of material in the soil at Fort Sage keeps the rate of fresh mineral supply relatively high as well. This may contribute to the catchment's relatively high Si flux, by ensuring that abundant readily weathered surfaces are available for chemical attack. In contrast, at Rio Icacos, the throughput of material is slower, and although material is weathered to a greater depth and degree of depletion, the much slower fresh mineral supply rate presumably contributes to the slower overall weathering rate. This same mechanism might help explain the relatively rapid silicon weathering rates of the other 10 sites; the overall rate of denudation in each case is higher than it is at Rio Icacos by a factor of 2–6.

From the analysis above, it is clear that the effects of site-to-site differences in denudation rates need to be taken into account in an analysis of climatic effects on weathering. There are several ways of doing so using measurements from the mass-balance approach.

One way to quantify the climatic effects on weathering is to consider the degree of chemical depletion. CDFs increase with average precipitation (Figure 19c) and temperature (Figure 19d) when effects of precipitation are taken into account. As a case in point, we again consider Rio Icacos, this time in comparison to Fall River. The CDF of Rio Icacos is four times higher, implying that weathering has to account for four times more of the overall denudation rate. Rio Icacos soils are more intensely weathered, and this is at least partly because the ambient climate is hotter and wetter.

Another way to account for variations in denudation rates is to include them explicitly in a regression analysis of climatic effects on weathering rates. For example, weathering rates can be modeled as the product of a power function of denudation rates, an Arrhenius-like function of mean annual temperature, and a power function of average precipitation (Riebe *et al.*, 2004a). Such an analysis of the data in Figure 19 produced regression slopes that are all statistically significant at $p < 0.05$. The model fit explains from 89% to 95% of the variance in chemical weathering rates and yields two key results:

1. For a given precipitation and temperature there is near proportional coupling of rates of chemical weathering and denudation for Na, Ca, Si, and the soil as a whole. This suggests that weathering is supply-limited across the sites, such that fresh material becomes chemically depleted to roughly the same degree, irrespective of its rate of supply from breakdown of rock.
2. The temperature sensitivity of weathering rates, expressed as an apparent activation energy, ranges from 14 to 24 kJ mol^{-1} for long-term weathering of Na, Ca, Si, and the soil as a whole. This is much lower (by a factor of 2–4) than the range of activation energies that has been reported for feldspar weathering rates in laboratory experiments (e.g., Brady and Carroll, 1994; White *et al.*, 1999). This may reflect a difference in the temperature sensitivity of abiotic weathering in the laboratory and biologically mediated weathering in the field (Riebe *et al.*, 2004a). However, this fails to account for why activation energies based on streamwater solute fluxes are consistent with the laboratory results (e.g., White and Blum, 1995; Dalai *et al.*, 2002). Ultimately the results from the solute-based studies and the cosmogenic-based study are difficult to compare. Whereas the cosmogenic nuclide method explicitly accounts for denudation rates, the solute-based studies could not, because denudation rates were not typically available.

More work is needed to clarify the temperature sensitivity of weathering rates in actively eroding landscapes. If a low sensitivity to temperature is confirmed and turns out to apply more generally to other rock types, then it would imply that feedbacks between climate change and silicate weathering may be only a weak buffer against long-term global temperature shifts. The strong coupling between weathering and denudation, however, implies that by regulating denudation rates, tectonic uplift may be an important regulator of chemical weathering rates.

11.5 SUMMARY

This chapter reviews the fundamentals of applying cosmogenic nuclides to problems of physical erosion, chemical weathering, and soil formation. Although cosmogenic nuclides have only been applied to these sorts of problems for the past decade, they are providing many new opportunities for understanding the rates of Earth surface processes.

Cosmogenic nuclides offer important advantages over previous methods for determining physical erosion and chemical weathering rates. Rather than measuring the flux of stream sediment or solutes over years to decades (which may be subject to short-term fluctuations and the confounding affects of land use) or the accumulation of sediment in datable deposits (which are difficult to find), cosmogenic nuclide methods measure near-surface residence times of mineral grains. Cosmogenic nuclides can be measured in rocks, sediment, and soil, if quartz or another suitable mineral is present. Because cosmogenic nuclides measure buildup over thousands of years, they are not very susceptible to short-term fluctuations and thus indicate natural, long-term rates of weathering and erosion, even in landscapes impacted by land use. Because the timescale of cosmogenic nuclide accumulation is comparable to that of pedogenesis, the rates of physical erosion and chemical weathering are relevant for understanding soil-forming processes.

The reason that cosmogenic nuclides can be used to quantify surface process rates is that their production rate declines rapidly with depth. Thus, by modeling the trajectories of mineral grains under conditions of exposure, erosion, and burial, we can predict the cosmogenic nuclide concentrations within them. In this chapter we considered three distinct types of particle trajectories. (1) Mineral grains could be immobile, as in the case of surface exposure dating. (2) They could follow a vertical trajectory toward the surface, as in the case of rock erosion. (3) They could be mixed both vertically within a soil column as well as horizontally across the landscape, as in the cases of soil exposure dating or determination of catchment-wide denudation rates. In each case, cosmogenic nuclide concentrations indicate the residence time of mineral grains near the surface, allowing quantification of not only the ages of soils and landforms, but also rates of erosion and weathering over pedogenic timescales. We illustrated what we consider to be some of the most important contributions to the field using a variety of examples from the literature.

Cosmogenic nuclides can be used to determine denudation rates both locally within a soil column, as well as over entire catchments. These two approaches in many ways complement each other and tell us different things about erosion and weathering. At the local scale, cosmogenic nuclide concentrations at the saprolite–soil boundary indicate the denudation rate,

equivalent to the conversion rate of saprolite to soil. By analyzing the soil production rate across a hillslope and at various soil depths, we can evaluate models of soil formation and landscape evolution. Two examples from southeastern Australia showed very different results. At one site on the Great Escarpment soil production rates decrease exponentially with soil depth, reaching a maximum where bare saprolite is exposed at the surface (Heimsath *et al.*, 2000, 2001a). Soil depths in this landscape are highly correlated with hillslope curvature, being thinnest on ridge crests and systematically thickening downslope. Together, these observations imply that relief in this landscape is decreasing over time as the escarpment retreats. In contrast, at a nearby landscape atop the Great Escarpment that is only beginning to be dissected by escarpment retreat (van der Beek *et al.*, 2001), soil production rates are nearly indistinguishable across the landscape, over a comparable range of soil depths (Wilkinson *et al.*, 2005). This suggests that this landscape has achieved a dynamic equilibrium (sensu; Hack, 1960) in which erosion rates vary but little and relief can persist for long periods of time. In these two examples, cosmogenic nuclides not only provide a method for determining soil production rates, but they may illustrate a fundamental difference in behavior between transient and equilibrium landscapes.

A similar difference in the distribution of erosion rates across the landscape can be seen at a larger scale. In addition to looking at hillslopes, it is possible to use cosmogenic nuclides in stream sediment to ascertain denudation rates over catchment scales. Although such spatial averaging masks variations on the scale of individual hillslopes, it has the advantage of yielding denudation rates that reflect regional, rather than local, variations. This approach lends itself to comparisons of denudation rates under different tectonic and climatic conditions, for example, without requiring details of sediment generation and transport on the hillslope scale. Two examples from the Sierra Nevada, California, illustrate transient versus equilibrium behavior of landscapes. At a site near Adams Peak, erosion rates were indistinguishable from each other, irrespective of differences in gradient, rock cover, or soil thickness (Granger *et al.*, 2001). Erosion rates are therefore consistent with a condition of dynamic equilibrium, in which relief is maintained over long timescales. In contrast, a site near a canyon at Fall River shows a 10-fold increase in erosion rates as gradients increase from 0.2 to 0.7 (Riebe *et al.*, 2000). These erosion rates are indicative of a transient landscape that is responding to recent canyon incision. As with the example from hillslope-scale denudation rates in Australia, catchment-averaged denudation rates at the Sierra

Nevada sites show very different patterns in transient and equilibrium landscapes.

Cosmogenic nuclide determination of spatially averaged denudation rates enables study of the effects of climate on denudation and landscape evolution. Sediments from granitic landscapes over a wide variety of both temperature and precipitation show little apparent dependence of denudation rates on climate (Riebe *et al.*, 2001c, 2004a; von Blanckenburg, 2005). Chemical mass balances show that the export of solutes from a catchment is controlled far more by the physical erosion rate than it is by climate (Riebe *et al.*, 2004a). This is apparently because physical erosion controls the supply of readily weathered minerals, which account for a large fraction of the solute budget. These results are broadly consistent with studies of soil development on none-roding moraines (Taylor and Blum, 1995), which show rapid weathering of fresh parent material but a sharp decline in weathering rate over time.

Although chemical weathering rates are strongly affected by variations in physical erosion rates, climate exerts a measurable effect as well. After accounting for the influence of mineral supply from physical erosion, Riebe *et al.* (2004a) found that the degree of chemical weathering depends on both temperature and precipitation, although to a lesser degree than predicted by laboratory studies.

Just as cosmogenic nuclides can be used to determine denudation rates both at a catchment scale and on a local, hillslope scale, they can be used in geochemical mass balances to ascertain the chemical weathering rate on both spatial scales. An application of a geochemical mass balance to the Nunnock River, Australia landscape shows that chemical weathering not only accounts for a large fraction of total denudation in this landscape, but also depends strongly on position within the landscape (Green *et al.*, 2006). Chemical weathering rates are highest at a point some 10–20 m downslope from the ridge crest, even though sediment supply is most rapid nearer the crest. The spatial variation in weathering rate probably reflects a balance between a requirement for sufficient water and soil depth for weathering processes, both of which increase downslope as the rate of supply of fresh minerals to the soil decreases (Green *et al.*, 2006). Conversely, along an altitude transect where the rate of supply of fresh material is roughly uniform, sharp variations in weathering rate are apparently due to differences in temperature, vegetative cover, and snow depth (Riebe *et al.*, 2004b).

The various applications and examples that we discussed are but a part of what is possible with cosmogenic nuclides. We did not discuss, for example, the determination of paleoerosion rates from cosmogenic nuclide concentrations in sedimentary deposits (Granger and Muzikar, 2001; Schaller *et al.*, 2004; Balco and Stone, 2005). Just as cosmogenic nuclides in modern stream sediments reflect the spatially averaged erosion rate of the sediment contributing area, cosmogenic nuclides in sediments preserved in terraces or basins retain information about erosion rates at the time of sediment deposition. Measurements of paleoerosion rates has great potential for studying the influence of climate change or tectonic uplift on erosion rates within a single watershed. We did not discuss how cosmogenic nuclide profiles can be used to determine solutional mass losses within a soil column (e.g., Braucher *et al.*, 1998). Because production rates in minerals at depth depend on the density of overburden, mass loss or gain within a soil column can affect the shape of the cosmogenic nuclide profile. This provides a way to assess weathering rates in deeply weathered slowly eroding soils such as saprolites (Braucher *et al.*, 1998). We did not discuss the use of cosmogenic nuclides in accumulating soils (e.g., Phillips *et al.*, 1998), where they can indicate accumulation rates or the denudation rate of the source area. Because cosmogenic nuclide production within a mineral grain depends almost entirely on the grain's trajectory and the distribution of mass within the soil column, cosmogenic nuclide methods offer limitless opportunities for quantifying rates of weathering, erosion, deposition, and transport of soil and sediment across landscape.

GLOSSARY

Cosmic ray: An energetic particle from space.

Cosmogenic nuclide: A product of a nuclear reaction induced by a cosmic ray.

Denudation rate: The rate of mass loss from a soil column or catchment.

Inselberg: A rounded hill of bedrock protruding above a plain. Common in granitic landscapes.

Mean life: The time required for a radioactive particle to reduce its abundance to a factor of $1/e$. Equivalent to the half-life divided by the natural logarithm of two.

Muon: An unstable particle with properties similar to that of an electron, but $\sim$200 times heavier.

Negative muon capture: A type of nuclear reaction in which a muon is captured into a nucleus.

Physical erosion rate: The rate of mass loss from a soil column or catchment due to physical removal of particles.

Saprolite: Weathered bedrock that retains original structure.

Spallation: A type of nuclear reaction in which a target nucleus is broken by kinetic energy.

Weathering rate: The rate of mass loss from a soil column or catchment due to removal of solutes.

REFERENCES

Adams K. D. and Wesnousky S. G. (1998) Shoreline processes and the age of the Lake Lahontan highstand in the Jessup Embayment, Nevada. *Geol. Soc. Am. Bull.* **110**(10), 1318–1332.

Ahnert F. (1977) Some comments on the quantitative formulation of geomorphological processes in a theoretical model. *Earth Surf. Proc.* **2**(2–3), 191–201.

Anderson R. S. (2002) Modeling the tor-dotted crests, bedrock edges, and parabolic profiles of high alpine surfaces of the Wind River Range, Wyoming. *Geomorphology* **46**, 35–58.

Anderson R. S., Repka J. L., and Dick G. S. (1996) Explicit treatment of inheritance in dating depositional surfaces using *in-situ* ^{10}Be and ^{26}Al. *Geology* **24**(1), 47–51.

Anderson S. P., Dietrich W. E., and Brimhall G. H. (2002) Weathering profiles, mass-balance analysis, and rates of solute loss; linkages between weathering and erosion in a small, steep catchment. *Geol. Soc. Am. Bull.* **114**, 1143–1158.

April R., Newton R., and Coles L. T. (1986) Chemical weathering in two Adirondack watersheds: past and present-day rates. *Geol. Soc. Am. Bull.* **97**, 1232–1238.

Bain D. C., Mellor A., Robertson-Rintoul M. S. E., and Buckland S. T. (1993) Variations in weathering processes and rates with time in a chronosequence of soils from Glen Feshie, Scotland. *Geoderma* **57**(3), 275–293.

Balco G. and Stone J. O. H. (2005) Measuring middle Pleistocene erosion rates with cosmic-ray-produced nuclides in buried alluvial sediment, Fisher Valley, southeastern Utah. *Earth Surf. Proc. Landforms* **30**, 1051–1067.

Banfield J. F. (1985) The mineralogy and chemistry of granite weathering. MSc Thesis, Australian National University.

Beaty C. B. (1970) Age and estimated rate of accumulation of an alluvial fan, White Mountains, California, USA. *Am. J. Sci.* **268**, 50–77.

Bierman P. R. (1994) Using *in situ*-produced cosmogenic isotopes to estimate rates of landscape evolution: a review from the geomorphic perspective. *J. Geophys. Res.* **99**(B7), 13885–13896.

Bierman P. R. and Caffee M. W. (2002) Cosmogenic exposure and erosion history of Australian rock landforms. *Geol. Soc. Am. Bull.* **114**, 787–803.

Bierman P. R., Caffee M. W., Davis P. T., Marsella K., Pavich M. J., Colgan P., Mickelson D. M., and Larsen J. (2002) Rates and timing of Earth surface processes from *in situ*-produced cosmogenic Be-10. *Rev. Mineral. Geochem.* **50**, 147–205.

Bierman P. R. and Nichols K. K. (2004) Rock to sediment—slope to sea with ^{10}Be—rates of landscape change. *Annu. Rev. Earth Planet. Sci.* **32**, 215–255.

Bierman P. R. and Steig E. J. (1996) Estimating rates of denudation using cosmogenic isotope abundances in sediment. *Earth Surf. Proc. Landforms* **21**, 125–139.

Brady P. V. and Carroll S. A. (1994) Direct effects of CO_2 and temperature on silicate weathering: possible implications for climate control. *Geochim. Cosmochim. Acta.* **58**, 1853–1856.

Braucher R., Colin F., Brown E. T., Bourlès D. L., Bamba O., Raisbeck G. M., Yiou F., and Koud J. M. (1998) African laterite dynamics using *in situ*-produced ^{10}Be. *Geochim. Cosmochim. Acta* **62**, 1501–1507.

Brimhall G. H. and Dietrich W. E. (1987) Constitutive mass balance relations between chemical composition, volume, density, porosity, and strain in metasomatic hydrochemical systems: results on weathering and pedogenesis. *Geochim. Cosmochim. Acta* **51**, 567–587.

Brimhall G. H., Lewis C. J., Ague J. J., Dietrich W. E., Hampel J., Teague T., and Rix P. (1988) Metal enrichment in bauxites by deposition of chemically mature aeolian dust. *Nature* **333**, 819–824.

Brimhall G. H., Lewis C. J., Ford C. R. B., Bratt J., Taylor G., and Warin O. (1991) Quantitative geochemical approach to

pedogenesis: importance of parent material reduction, volumetric expansion, and eolian influx in lateritization. *Geoderma* **51**(1–4), 51–91.

Brown E. T., Stallard R. F., Larsen M. C., Bourlès D. L., Raisbeck G. M., and Yiou F. (1998) Determination of Predevelopment denudation rates of an agricultural watershed (Cayaguás River, Puerto Rico) using *in situ* produced ^{10}Be in river-borne quartz. *Earth Planet. Sci. Lett.* **160**, 723–728.

Brown E. T., Stallard R. F., Larsen M. C., Raisbeck G. M., and Yiou F. (1995) Denudation rates determined from the accumulation of *in situ*-produced ^{10}Be in the Luquillo Experimental Forest, Puerto Rico. *Earth Planet. Sci. Lett.* **129**, 193–202.

Burbank D. W., Leland J., Fielding E., Anderson R. S., Brozovic N., Reid M. R., and Duncan C. (1996) Bedrock incision, rock uplift, and threshold hillslopes in the northwestern Himalayas. *Nature* **379**(6565), 505–510.

Burkins D. L., Blum J. D., Brown K., Reynolds R. C., and Erel Y. (1999) Chemistry and mineralogy of a granitic, glacial soil chronosequence, Sierra Nevada Mountains, California. *Chem. Geol.* **162**(1), 1–14.

Busenburg E. and Clemency C. V. (1976) The dissolution kinetics of feldspars at 25 °C and 1 atm CO_2 partial pressure. *Geochim. Cosmochim. Acta* **40**, 41–49.

Carson M. A. and Kirkby M. J. (1972) *Hillslope Form and Process.* Cambridge University Press, Cambridge, UK.

Cerling T. E. and Craig H. (1994) Geomorphology and *in-situ* cosmogenic isotopes. *Annu. Rev. Earth Planet. Sci.* **22**, 273–317.

Chadwick O. A., Brimhall G. H., and Hendricks D. M. (1990) From black box to a grey box: a mass balance interpretation of pedogenesis. *Geomorphology* **3**, 369–390.

Chadwick O. A., Derry L. A., Vitousek P. M., Huebert B. J., and Hedin L. O. (1999) Changing sources of nutrients during four million years of ecosystem development. *Nature* **397**, 491–497.

Chappell J. (1974) The geomorphology and evolution of small valleys in dated coral reef terraces. *New Guinea J. Geol.* **82**, 795–812.

Charalambus S. (1971) Nuclear transmutation by negative stopped muons and the activity induced by the cosmic-ray muons. *Nucl. Phys. A* **166**, 145–161.

Clayton J. L. and Megahan W. F. (1986) Erosional and chemical denudation rates in the southwestern Idaho Batholith. *Earth Surf. Proc. Landforms* **11**(4), 389–400.

Colin F., Alarcon C., and Vieillard P. (1993) Zircon: an immobile index in soils? *Chem. Geol.* **107**(3–4), 273–276.

Cox N. J. (1980) On the relationship between bedrock lowering and regolith thickness. *Earth Surf. Proc.* **5**, 271–274.

Dahms D. E., Shroba R. R., Gosse J. C., Hall R. D., Sorenson C. J., and Reheis M. C. (1997) Relation between soil age and silicate weathering rates determined from the chemical evolution of a glacial chronosequence: discussion and reply. *Geology* **25**(4), 381–383.

Dalai T. K., Krishnaswami S., and Sarin M. M. (2002) Major ion chemistry in the headwaters of the Yamuna river system: chemical weathering, its temperature dependence and CO_2 consumption in the Himalaya. *Geochim. Cosmochim. Acta* **66**, 3397–3416.

Desilets D. and Zreda M. (2003) Spatial and temporal distribution of secondary cosmic-ray nucleon intensities and applications to *in situ* cosmogenic dating. *Earth Planet. Sci. Lett.* **206**(1–2), 21–42.

Dietrich W. E., Dunne T., Humphrey N. F., and Reid L. M. (1982) Construction of sediment budgets for drainage basins. USDA Forest Service.

Dietrich W. E., Reiss R., Hsu M.-L., and Montgomery D. R. (1995) A process-based model for colluvial soil depth and shallow landsliding using digital elevation data. *Hydrol. Proc.* **9**(3–4), 383–400.

Drever J. I. and Zobrist J. (1992) Chemical weathering of silicate rocks as a function of elevation in the southern Swiss Alps. *Geochim. Cosmochim. Acta* **56**, 3209–3216.

Dunai T. J. (2000) Scaling factors for production rates of *in situ* produced cosmogenic nuclides: a critical reevaluation. *Earth Planet. Sci. Lett.* **176**, 157–169.

Dunne A., Elmore D., and Muzikar P. (1999) Scaling of cosmogenic nuclide production rates for geometric shielding and attenuation at depth on sloped surfaces. *Geomorphology* **27**(1–2), 3–11.

Fenton C. R., Webb R. H., Pearthree P. A., Cerling T. E., and Poreda R. J. (2001) Displacement rates on the Toroweap and Hurricane faults: implications for Quaternary downcutting in the Grand Canyon, Arizona. *Geology* **29**(11), 1035–1038.

Gaillardet J., Dupre B., Louvat P., and Allègre C. J. (1999) Global silicate weathering and CO_2 consumption rates deduced from the chemistry of large rivers. *Chem. Geol.* **159**(1–4), 3–30.

Gilbert G. K. (1877) *Report on the Geology of the Henry Mountains*. US Geographical and Geological Survey.

Gosse J., Klein J., Evenson E., Lawn B., and Middleton R. (1994) *Precision Dating of Glacial Events Based on AMS Measurements of Cosmogenic ^{10}Be Produced in Boulders*. US Geological Survey Circular 1107, p. 114.

Gosse J. C., Klein J., Evenson E. B., Lawn B., and Middleton R. (1995) Beryllium-10 dating of the duration and retreat of the last Pinedale glacial sequence. *Science* **268**, 1329–1333.

Gosse J. C. and Phillips F. M. (2001) Terrestrial *in situ* cosmogenic nuclides: theory and application. *Quat. Sci. Rev.* **20**(14), 1475–1560.

Granger D. E., Kirchner J. W., and Finkel R. (1996) Spatially averaged long-term erosion rates from *in-situ* produced cosmogenic nuclides in alluvial sediment. *J. Geol.* **104**, 249–257.

Granger D. E. and Muzikar P. F. (2001) Dating sediment burial with cosmogenic nuclides: theory, techniques, and limitations. *Earth Planet. Sci. Lett.* **188**(1–2), 269–281.

Granger D. E., Riebe C. S., Kirchner J. W., and Finkel R. C. (2001) Modulation of erosion on steep granitic slopes by boulder armoring, as revealed by cosmogenic ^{26}Al and ^{10}Be. *Earth Planet. Sci. Lett.* **186**, 269–281.

Granger D. E. and Smith A. L. (2000) Dating buried sediments using radioactive decay and muogenic production of ^{26}Al and ^{10}Be. *Nucl. Instr. Meth. Phys. Res. B* **172**, 822–826.

Green E. G., Dietrich W. E., and Banfield J. F. (2006) Quantification of chemical weathering rates across an actively eroding hillslope. *Earth Planet. Sci. Lett* **242**, 155–169.

Grose T. L. T. and Mergner M. (2000) *Geologic Map of the Chilcoot 15' Quadrangle, Lassen and Plumas Counties, California*. California Geological Survey.

Guillaume A. and Guillaume S. (1982) L'erosion dans les Alpes au Plio-Quaternaire et au Miocene. *Eclogae Geol. Helv.* **75**(2), 247–268.

Hack J. T. (1960) Interpretation of erosional topography in humid temperate regions. *Am. J. Sci.* **258-A**, 80–97.

Hancock G. S., Anderson R. S., Chadwick O. A., and Finkel R. C. (1999) Dating fluvial terraces with ^{10}Be and ^{26}Al profiles: application to the Wind River, Wyoming. *Geomorphology* **27**, 41–60.

Hayakawa S. (1969) *Cosmic Ray Physics: Nuclear and Astrophysical Aspects*. Wiley, New York.

Heimsath A. M., Chappell J., Dietrich W. E., Nishiizumi K., and Finkel R. C. (2000) Soil production on a retreating escarpment in southeastern Australia. *Geology* **28**(9), 787–790.

Heimsath A. M., Chappell J., Dietrich W. E., Nishiizumi K., and Finkel R. C. (2001a) Late quaternary erosion in southeastern Australia: a field example using cosmogenic nuclides. *Quat. Int.* **83–85**, 169–185.

Heimsath A. M., Dietrich W. E., Nishiizumi K., and Finkel R. C. (1997) The soil production function and landscape equilibrium. *Nature* **388**, 358–361.

Heimsath A. M., Dietrich W. E., Nishiizumi K., and Finkel R. C. (1999) Cosmogenic nuclides, topography, and the spatial variation of soil depth. *Geomorphology* **27**, 151–172.

Heimsath A. M., Dietrich W. E., Nishiizumi K., and Finkel R. C. (2001b) Stochastic processes of soil production and transport: erosion rates, topographic variation and cosmogenic nuclides in the Oregon Coast Range. *Earth Surf. Proc. Landforms* **26**, 531–552.

Heisinger B., Lal D., Jull A. J. T., Kubik P., Ivy-Ochs S., Knie K., and Nolte E. (2002b) Production of selected cosmogenic radionuclides by muons. 2: Capture of negative muons. *Earth Planet. Sci. Lett.* **200**, 357–369.

Heisinger B., Lal D., Jull A. J. T., Kubik P., Ivy-Ochs S., Neumaier S., Knie K., Lazarev V., and Nolte E. (2002a) Production of selected cosmogenic radionuclides by muons. 1: Fast muons. *Earth Planet. Sci. Lett.* **200**, 345–355.

Hewawasam T., von Blanckenburg F., Schaller M., and Kubik P. (2003) Increase of human over natural erosion rates in tropical highlands constrained by cosmogenic nuclides. *Geology* **31**(7), 597–600.

Hilgard E. W. (1860) *Report on the Geology and Agriculture of the State of Mississippi*. Publisher unknown, 391pp.

Hill I. G., Worden R. H., and Meighan I. G. (2000) Yttrium; the immobility–mobility transition during basaltic weathering. *Geology* **28**, 923–926.

Hodson M. E. (2002) Experimental evidence for mobility of Zr and other trace elements in soils. *Geochim. Cosmochim. Acta* **66**(5), 819–828.

Kennedy M. J., Chadwick O. A., Vitousek P. M., Derry L. A., and Hendricks D. M. (1998) Changing sources of base cations during ecosystem development, Hawaiian Islands. *Geology* **26**(11), 1015–1018.

Kirchner J. W., Finkel R. C., Riebe C. S., Granger D. E., Clayton J. L., King J. G., and Megahan W. F. (2001) Mountain erosion over 10 yr, 10 kyr, and 10 myr time scales. *Geology* **29**(7), 591–594.

Kirchner J. W., Granger D. E., and Riebe C. S. (1997) Cosmogenic isotope methods for measuring catchment erosion and weathering rates. *J. Conf. Abstr.* **2**, 217.

Kirchner J. W. and Lydersen E. (1995) Base cation depletion and potential long-term acidification of Norwegian catchments. *Environ. Sci. Technol.* **29**, 1953–1960.

Kurtz A. C., Derry L. A., Chadwick O. A., and Alfano M. J. (2000) Refractory element mobility in volcanic soils. *Geology* **28**(8), 683–686.

Kubik P. W., Ivy-Ochs S., Masarik J., Frank M., and Schlüchter C. (1998) ^{10}Be and ^{26}Al production rates deduced from an instantaneous event within the dendrocalibration curve, the landslide of Köfels, Ötz valley, Austria. *Earth Planet. Sci. Lett.* **161**, 231–241.

Lal D. (1991) Cosmic ray labeling of erosion surfaces: in situ nuclide production rates and erosion models. *Earth Planet. Sci. Lett.* **104**, 424–439.

Lal D. and Arnold J. R. (1985) Tracing quartz through the environment. *Proc. Ind. Acad. Sci., Earth Planet. Sci.* **94**, 1–5.

Lal D. and Chen J. (2005) Cosmic ray labeling of erosion surfaces. II: Special cases of exposure histories of boulders, soils, and beach terraces. *Earth Planet. Sci. Lett.* **236**, 797–813.

Lal D. and Peters B. (1967) Cosmic ray produced radioactivity on the Earth. In *Handbuch der Physik* (ed. S. Flugge). Springer, Berlin, vol. 46, pp. 551–612.

Larsen E. S., Jr. (1948) *Batholith and Associated Rocks of Corona, Elsinore, and San Luis Rey Quadrangles, Southern California*. Geological Society of America.

Larsen M. C. and Torres-Sanchez A. J. (1992) Landslides triggered by hurricane Hugo in eastern Puerto Rico, September 1989. *Caribb. J. Sci.* **28**, 113–125.

Louvat P. and Allègre C. J. (1997) Present denudation rates on the island of Réunion determined by river geochemistry; basalt weathering and mass budget between chemical and mechanical erosions. *Geochim. Cosmochim. Acta* **61**(17), 3645–3669.

Matmon A., Bierman P. R., Larsen J., Southworth S., Pavich M. J., Finkel R. C., and Caffee M. W. (2003) Erosion of an ancient mountain range, the Great Smoky Mountains, North Carolina and Tennessee. *Am. J. Sci.* **303**, 817–855.

McDowell W. H. and Asbury C. E. (1994) Export of carbon, nitrogen, and major ions from three tropical montane watersheds. *Limnol. Oceanogr.* **39**, 111–125.

Merrill G. P. (1906) *A Treatise on Rocks, Rock Weathering and Soils.* Macmillan, New York.

Merritts D. J., Chadwick O. A., Hendricks D. M., Brimhall G. H., and Lewis C. J. (1992) The mass balance of soil evolution on late Quaternary marine terraces, northern California. *Geol. Soc. Am. Bull.* **104**, 1456–1470.

Milliman J. D., Gin Y.-S., Ren M.-E., and Saito Y. (1987) Man's influence on the erosion and transport of sediment by Asian rivers: the Yellow River (Huanghe) example. *J. Geol.* **95**, 751–762.

Morris J. D., Gosse J., Brachfeld S., and Tera F. (2002) Cosmogenic Be-10 and the solid Earth: studies in geomagnetism, subduction zone processes, and active tectonics. *Rev. Min. Geochem.* **50**, 207–270.

Muzikar P., Elmore D., and Granger D. (2003) Accelerator mass spectrometry in geologic research. *Geol. Soc. Am. Bull.* **15**(5), 643–654.

Nezat C. A., Blum J. D., Klaue A., Johnson C. E., and Siccama T. G. (2004) Influence of landscape position and vegetation on long-term weathering rates at the Hubbard Brook experimental forest, New Hampshire, USA. *Geochim. Cosmochim. Acta* **68**(14), 3065–3078.

Niemi N. A., Oskin M., Burbank D. W., Heimsath A. M., and Gabet E. J. (2005) Effects of bedrock landslides on cosmogenically determined erosion rates. *Earth Planet. Sci. Lett.* **237**, 480–498.

Nishiizumi K., Kohl C. P., Arnold J. R., Dorn R., Klein J., Fink D., Middleton R., and Lal D. (1993) Role of *in situ* cosmogenic nuclides ^{10}Be and ^{26}Al in the study of diverse geomorphic processes. *Earth Surf. Proc. Landforms* **18**, 407–425.

Nishiizumi K., Winterer E. L., Kohl C. P., Klein J., Middleton R., Lal D., and Arnold J. R. (1989) Cosmic ray production rates of ^{10}Be and ^{26}Al in quartz from glacially polished rocks. *J. Geophys. Res.* **94**(B12), 17907–17915.

Owen L. A., Finkel R. C., Caffee M. W., and Gualtieri L. (2002) Timing of multiple late Quaternary glaciations in the Hunza Valley, Karakoram Mountains, northern Pakistan; defined by cosmogenic radionuclide dating of moraines. *Geol. Soc. Am. Bull.* **114**(5), 593–604.

Perg L. A., Anderson R. S., and Finkel R. C. (2001) Use of a new ^{10}Be and ^{26}Al inventory method to date marine terraces, Santa Cruz, California, USA. *Geology* **29**(10), 879–882.

Phillips F. M., Zreda M. G., Smith S. S., Elmore D., Kubik P. W., and Sharma P. (1990) Cosmogenic chlorine-36 chronology for glacial deposits at Bloody Canyon, Eastern Sierra Nevada. *Science* **248**, 1529–1532.

Phillips W. M., McDonald E. V., Reneau S. L., and Poths J. (1998) Dating soils and alluvium with cosmogenic ^{21}Ne depth profiles: case studies from the Pajarito Plateau, New Mexico, USA. *Earth Planet. Sci. Lett.* **160**, 209–223.

Raymo M. E., Ruddiman W. F., and Froelich P. N. (1988) Influence of late Cenozoic mountain building on ocean geochemical cycles. *Geology* **16**, 649–653.

Reneau S. L., Dietrich W. E., Rubin M., Donahue D. J., and Jull J. T. (1989) Analysis of hillslope erosion rates using dated colluvial deposits. *J. Geol.* **97**, 45–63.

Repka J. L., Anderson R. S., and Finkel R. C. (1997) Cosmogenic dating of fluvial terraces, Fremont River, Utah. *Earth Planet. Sci. Lett.* **152**, 59–73.

Reusser L. J., Bierman P. R., Pavich M. J., Zen E.-A., Larsen J., and Finkel R. (2004) Rapid late Pleistocene incision of Atlantic passive-margin river gorges. *Science* **305**, 499–502.

Riebe C. S. (2000) Tectonic and climatic control of physical erosion rates and chemical weathering rates in the Sierra Nevada, California, inferred from cosmogenic nuclides and geochemical mass balance. PhD Thesis, University of California, Berkeley.

Riebe C. S., Kirchner J. W., and Finkel R. C. (2003) Long-term rates of chemical weathering and physical erosion from cosmogenic nuclides and geochemical mass balance. *Geochim. Cosmochim. Acta* **67**(22), 4411–4427.

Riebe C. S., Kirchner J. W., and Finkel R. C. (2004a) Erosional and climatic effects on long-term chemical weathering rates in granitic landscapes spanning diverse climate regimes. *Earth Planet. Sci. Lett.* **224**, 547–562.

Riebe C. S., Kirchner J. W., and Finkel R. C. (2004b) Sharp decrease in long-term chemical weathering rates along an altitudinal transect. *Earth Planet. Sci. Lett.* **218**, 421–434.

Riebe C. S., Kirchner J. W., and Granger D. E. (2001a) Quantifying quartz enrichment and its consequences for cosmogenic measurements of erosion rates from alluvial sediment and regolith. *Geomorphology* **40**(1–2), 15–19.

Riebe C. S., Kirchner J. W., Granger D. E., and Finkel R. C. (2000) Erosional equilibrium and disequilibrium in the Sierra Nevada, inferred from cosmogenic ^{26}Al and ^{10}Be in alluvial sediment. *Geology* **28**, 803–806.

Riebe C. S., Kirchner J. W., Granger D. E., and Finkel R. C. (2001b) Strong tectonic and weak climatic control of long-term chemical weathering rates. *Geology* **29**, 511–514.

Riebe C. S., Kirchner J. W., Granger D. E., and Finkel R. C. (2001c) Minimal climatic control on erosion rates in the Sierra Nevada, California. *Geology* **29**, 447–450.

Rinterknecht V. R., Marks L., Piotrowski J. A., Raisbeck G. M., Yiou F., Brook E. J., and Clark P. U. (2005) Cosmogenic ^{10}Be ages on the Pomeranian Moraine, Poland. *Boreas* **34**(2), 186–191.

Rossi B. (1948) Interpretation of cosmic-ray phenomena. *Rev. Mod. Phys.* **20**(3), 537–583.

Ruxton B. P. and McDougall I. (1967) Denudation rates in northeast Papua from Potassium–Argon dating of lavas. *Am. J. Sci.* **265**, 545–561.

Sak P. B., Fisher D. M., Gardner T. W., Murphy K., and Brantley S. L. (2004) Rates of weathering rind formation on Costa Rican basalt. *Geochim. Cosmochim. Acta* **68**(7), 1453–1472.

Schaller M., Hovius N., Willett S. D., Ivy-Ochs S., Synal H. A., and Chen M. C. (2005) Fluvial bedrock incision in the active mountain belt of Taiwan from in-situ/cosmological produced nuclides. *Earth Surf. Proc. Landforms* **30**(8), 955–971.

Schaller M., von Blanckenburg F., Hovius N., Veldkamp A., van den Berg M. W., and Kubik P. W. (2004) Paleoerosion rates from cosmogenic ^{10}Be in a 1.3 Ma terrace sequence: response of the River Meuse to changes in climate and rock uplift. *J. Geol.* **112**, 127–144.

Seidl M. A., Dietrich W. E., and Kirchner J. W. (1994) Longitudinal profile development into bedrock: an analysis of Hawaiian channels. *J. Geol.* **102**, 457–474.

Small E. E., Anderson R. S., and Hancock G. S. (1999) Estimates of the rate of regolith production using Be-10 and Al-26 from an alpine hillslope. *Geomorphology* **27**, 131–150.

Small E. E., Anderson R. S., Repka J. L., and Finkel R. (1997) Erosion rates of alpine bedrock summit surfaces deduced from *in situ* ^{10}Be and ^{26}Al. *Earth Planet. Sci. Lett.* **150**, 413–425.

Stallard R. F. and Edmond J. M. (1983) Geochemistry of the Amazon 2. The influence of geology and weathering environment on the dissolved load. *J. Geophys. Res.* **88**(C14), 9671–9688.

Stone J. O. (2000) Air pressure and cosmogenic isotope production. *J. Geophys. Res.* **105**(B10), 23753–23759.

Stone J. O. H., Evans J. M., Fifield L. K., Allan G. L., and Cresswell R. G. (1998) Cosmogenic chlorine-36 production in calcite by muons. *Geochim. Cosmochim. Acta* **62**(3), 433–454.

Stonestrom D. A., White A. F., and Akstin K. C. (1998) Determining rates of chemical weathering in soils—solute transport versus profile evolution. *J. Hydrol.* **209**, 331–345.

Stroeven A. P., Fabel D., Hattestrand C., and Harbor J. (2002) A relict landscape in the centre of Fennoscandian glaciation: cosmogenic radionuclide evidence of tors preserved through multiple glacial cycles. *Geomorphology* **44**(1–2), 145–154.

Taylor A. and Blum J. D. (1995) Relation between soil age and silicate weathering rates determined from the chemical evolution of a glacial chronosequence. *Geology* **23**(11), 979–982.

Taylor A. and Blum J. D. (1997) Relation between soil age and silicate weathering rates determined from the chemical evolution of a glacial chronosequence: discussion and reply. *Geology* **25**(4), 381–383.

Trimble S. W. (1977) The fallacy of stream equilibrium in contemporary denudation studies. *Am. J. Sci.* **277**, 876–887.

Twidale C. R. (1986) Granite landform evolution: factors and implications. *Geol. Rundsch.* **75**(3), 769–779.

Twidale C. R. (1987) Etch and intracutaneous landforms and their implications. *Austr. J. Earth Sci.* **34**, 367–386.

Twidale C. R. and Vidal Romaní P. (2005) *Landforms and Geology of Granite Terrains*. A. A. Balkema, Rotterdam.

van der Beek P., Pulford A., and Braun J. (2001) Cenozoic landscape development in the Blue Mountains (SE Australia): lithological and tectonic controls on rifted margin morphology. *J. Geol.* **109**(1), 35–56.

von Blanckenburg F. (2005) The control mechanisms of erosion and weathering at basin scale from cosmogenic nuclids in river sediment. *Earth Planet. Sci. Lett.* **237**, 462–479.

Wahrhaftig C. (1965) Stepped topography of the Southern Sierra Nevada, California. *Geol. Soc. Am. Bull.* **76**, 1165–1190.

Wahrhaftig C. (1970) The trouble with the rate of erosion. *Geol. Soc. Am. Abstr.* **2**(2), 157.

Whipple F. L. and Fireman E. L. (1959) Calculation of erosion in space from the cosmic-ray exposure ages of meteorites. *Nature* **183**, 1315.

White A. F. and Blum A. E. (1995) Effects of climate on chemical weathering in watersheds. *Geochim. Cosmochim. Acta* **59**, 1729–1747.

White A. F., Blum A. E., Bullen T. D., Vivit D. V., Schulz M., and Fitzpatrick J. (1999) The effect of temperature on experimental and natural weathering rates of granitoid rocks. *Geochim. Cosmochim. Acta* **63**, 3277–3291.

White A. F., Blum A. E., Schulz M. S., Vivit D. V., Larsen M., and Murphy S. F. (1998) Chemical weathering in a tropical watershed, Luquillo mountains, Puerto Rico: I. Long-term versus short-term chemical fluxes. *Geochim. Cosmochim. Acta* **62**, 209–226.

Wilkinson M. T., Chappell J., Humphreys G. S., Fifield L. K., Smith B., and Hesse P. (2005) Soil production in heath and forest, Blue Mountains, Australia: influence of lithology and paleoclimate. *Earth Surf. Proc. Landforms* **30**, 923–934.

Wobus C., Heimsath A., Whipple K., and Hodges K. (2005) Active out-of-sequence thrust faulting in the central Nepalese Himalaya. *Nature* **434**(7036), 1008–1011.

Yoo K., Amundson R., Heimsath A. M., and Dietrich W. E. (2006) Spatial patterns of soil organic carbon on hillslopes: integrating geomorphic processes and the biological C cycle. *Geoderma* **130**(1–2), 47–65.

Zehfuss P. H., Bierman P. R., Gillespie A. R., Burke R. M., and Caffee M. W. (2001) Slip rates on the Fish Springs Fault, Owens Valley, California, deduced from cosmogenic ^{10}Be and ^{26}Al and soil development on fan surfaces. *Geol. Soc. Am. Bull.* **113**(2), 241–255.

Radioactive Geochronometry
ISBN: 978-0-08-096708-0

pp. 335–376

12
Geochronometry of Marine Deposits

K. K. Turekian

Yale University, New Haven, CT, USA

and

M. P. Bacon

Woods Hole Oceanographic Institution, MA, USA

12.1   INTRODUCTION

Marine deposits in the form of sediment accumulations, ferromanganese nodules, and crusts and corals are important recorders not only of marine events but also of global environmental changes. Until accurate methods of establishing chronometries of these deposits were developed, however, they were of limited value for coupling global events with the environmental proxies in the deposits. The chronometry of marine deposits ultimately depends on accurate calibration by radioactive methods. The chronometric potential of tracking the change in abundance of a radioactive species or its products lies at the heart of all successful dating methods. After radioactive geochronometry has been established, the system can be further exploited with stratigraphic tools such as periodic changes in properties over time or comparative chronometries with deposits on land.

12.2   PRINCIPLES

12.2.1   Radioactive Geochronometry

All absolute dating methods that have proven dependable are based on radioactive decay. Virtually all of the methods depend on the methodical decrease in the amount of the radioactive nuclide and the growth of the corresponding daughter product. The one exception is in the area of radiation-induced damage in solids, which is the basis of thermoluminescence or ESR dating. This latter scheme will be dealt with later in the chapter. Here the basic principles of canonical radioactive geochronometry as applied to marine deposits is reviewed.

The radioactive decay equation is

$$N = N_0\exp(-\lambda t) \qquad (1)$$

Where N_0 is the number of atoms of the radioactive species at the time of formation of the system under study, N is the amount left after a time t, and λ is the decay constant. Where the ratio of the daughter to parent nuclide is measured after a time t has elapsed, then the equation becomes

$$(N_0 - N)/N = \exp(\lambda t) - 1 \qquad (2)$$

Here N_0-N represents the amount of daughter that has been produced after time t has elapsed.

Where the daughter is also radioactive, the equations become more complex. The full discussion of the complexities can be found in Friedlander *et al.* (1981) or Ivanovich and Harmon (1992).

There are several fundamental requirements for the use of the radioactive systems for geochronometry: (i) there must not be any initial daughter nuclides in the system or, if there are, there must be an adequate method of identifying and accommodating their presence; (ii) there must not be loss or addition of parent or daughter during the life of the system but, if there is, a method for assessing and accommodating the loss or gain of parent or daughter must be devised; and (iii) the half-life of the radioactive species being studied must be compatible with the time range of the system.

12.2.2   Secondary Stratigraphic Procedures

The classical geologic methods for arriving at relative ages based on sequencing and correlation via physical or paleontological markers do not provide chronometry. The stratigraphic arguments can, however, be utilized to extend information on time once suitable radiometric dating has been established in some part of the system. There are several such examples which will be addressed in this chapter:

(i) The $\delta^{18}O$ record (i.e., the $^{18}O/^{16}O$ value normalized to a standard) in foraminifera from deep-sea cores can be used to track changes in climate. The deconvolution of these records into the orbitally driven Milankovitch cycles allows extension of the record beyond the radiometrically calibrated sequence for the past 3×10^5 yr based on the uranium decay chain nuclides.

(ii) The changes, during the Cenozoic, of the global ocean temperatures as recorded in foraminifera from deep-sea sediments can be dated using the K–Ar technique on volcanic-ash layers.

(iii) The calibration of the magnetic-reversal timescale in various geologic settings can be used to date deep-sea deposits that are continuously accumulated.

12.3   RADIOACTIVE SYSTEMS USED IN MARINE GEOCHRONOMETRY

There are four radionuclide systems used in the geochronometry of marine deposits: ^{14}C, the uranium and thorium decay chains, cosmogenic nuclides other than ^{14}C and, for detrital volcanic deposits, ^{40}K–^{40}Ar.

12.3.1 Radiocarbon

Natural ^{14}C (half-life = 5,730 yr) is produced by cosmic-ray interaction with the atmosphere. It enters the oceans by exchange of CO_2 between the ocean and atmosphere. The ratio of ^{14}C/^{12}C in the dissolved inorganic carbon species in seawater is determined by isotopic fractionation, as tracked by following the stable carbon isotopes: ^{12}C and ^{13}C. The ^{14}C/^{12}C ratio is diminished in the deep ocean as the water ages during thermohaline circulation. Upwelling low ^{14}C/^{12}C water equilibrates with the atmospheric CO_2 but this is not instantaneous. This delay causes the ^{14}C/^{12}C value of surface seawater to lag the value in the atmosphere.

Moreover, there are fluctuations in the ^{14}C production rate so that the initial value of ^{14}C/^{12}C ratio in the atmosphere and surface ocean also fluctuates proportionately. This variation has been observed in independently dated tree rings on land and in coral accumulations in the sea (Bard *et al.*, 1990). All these factors plus the intrinsic properties of accumulation, bioturbation, and preservation must be considered to translate a ^{14}C/^{12}C ratio to an age expressed in calendar years.

12.3.2 Uranium and Thorium Decay Chain Nuclides

Table 1 shows the decay schemes for ^{238}U, ^{235}U, and ^{232}Th. Some of the daughters of each of these nuclides have been used in the study of marine deposits, either for rates of sediment accumulation or rates of bioturbation. The fundamental procedure for the use of these nuclides in determining rates depends on the separation of the various members of the uranium and thorium decay series in the aqueous medium as they are produced so that the ratio of daughter to parent nuclide is far from the secular equilibrium expected after the proper lapse of time. For example, the thorium isotopes produced in the uranium and thorium decay chains are quickly removed, whereas radium and uranium remain in solution until included in biological tests.

12.3.3 Cosmogenic Nuclides

Other radionuclides, in addition to ^{14}C, are produced in the atmosphere by the bombardment of the gases in the atmosphere by cosmic rays. The production rate for the Earth as a whole is

Table 1 Uranium and thorium decay chains.

	U-238 SERIES						Th-232 SERIES				U-235 SERIES				
Np															
U	U-238 4.51×10^9 yr	U-234 2.48×10^5 yr									U-235 7.13×10^8 yr				
Pa		Pa-234 1.18 m										Pa-231 3.2×10^4 yr			
Th	Th-234 24.1 d	Th-230 7.52×10^4 yr					Th-232 1.39×10^{10} yr	Th-228 1.90 yr			Th-231 25.6 h		Th-227 18.6 d		
Ac								Ac-228 6.13 h				Ac-227 22.0 yr			
Ra		Ra-226 1622 yr					Ra-228 5.75 yr	Ra-224 3.64 d					Ra-223 11.4 d		
Fr															
Rn		Rn-222 3.825 d						Rn-220 54.5 s					Rn-219 3.92 s		
At															
Po		Po-218 3.05 m		Po-214 1.6×10^{-4} s		Po-210 138.4 d		Po-216 0.158 s	65%	Po-212 30×10^{-7} s			Po-215 1.83×10^{-3} s		
Bi			Bi-214 19.7 m		Bi-210 5.0 d				Bi-212 60.5 m					Bi-211 2.16 m	
Pb		Pb-214 26.8 m		Pb-210 22.3 yr		Pb-206		Pb-212 10.6 h	35%	Pb-208			Pb-211 36.1 m		Pb-207
Tl								Tl-208 3.1 m					Tl-207 4.79 m		

determined by how well the Earth is magnetically shielded. This shield is provided intrinsically by the Earth's magnetic moment and externally by the growth and contraction of the solar magnetic envelope. As in the case of ^{14}C, the production rate of the cosmogenic nuclides will be highest when Earth's magnetic moment is low or when solar activity is small. These fluctuations must be accommodated if the cosmogenic nuclides are to be useful in geochronometry. The short-lived cosmogenic nuclides such as ^{7}Be (half-life = 53 d) are used for the study of bioturbation. The longer-lived radionuclides such as ^{32}Si (half-life = 140 yr) can be used to study both mixing rates and rates of accumulation. The long-lived ^{10}Be (half-life = 1.5 Myr) has been used for dating back to $\sim$10 Myr.

12.3.4 Potassium–Argon

The decay of ^{40}K (half-life = 1.250×10^{9} yr with 10.5% of the decays going to ^{40}Ar) is used in deep-sea deposits in association with volcanic-ash layers. Dating using the ratio of ^{40}Ar to ^{39}Ar (a surrogate for ^{40}K resulting from irradiation of the sample with neutrons) has been widely used in volcanic-ash layers associated with biostratigraphic units and is the basis of the Cenozoic chronology presented by Berggren *et al.* (1995). There have also been efforts to use the scheme for dating low-temperature minerals such as glauconite in continental margin sediments, but these will not be discussed in this chapter.

12.4 COASTAL DEPOSITS

12.4.1 Applicable Methods and Requirements

The rates of accumulation in coastal deposits are commonly greater than deep-sea deposits. This disparity is not true everywhere since coastal areas can be deficient in sediment supply or may be subject to efficient erosive processes. Accumulations can occur in estuaries, coastal depressions, and salt marshes. Attempts at geochronometry with a number of nuclides have been made in all of these areas. Confounding the record for both coastal and deep-sea sediments is the effect of bioturbation. Sediments deposited under anoxic conditions are free of this effect but all other sediments are subject to a variety of scales of bioturbation.

12.4.2 Unbioturbated Deposits

The requirement for obtaining unbioturbated deposits is either the deposition under reducing conditions or deposition in a rigid structure not capable of biological mechanical activity. The former includes anoxic or suboxic basins such as parts of the Gulf of California, Santa Barbara

basin, the Black Sea, or Cariaco Trench. The latter is mainly restricted to "high" salt marshes (cores of middle salt marshes show significant bioturbation according to Saffert and Thomas (1998). Both types of deposits have been dated by radioactive methods.

Anoxic basins in the Gulf of California have been dated by the excess ^{210}Pb decrease with depth (DeMaster and Turekian, 1987). Because of the 22 yr half-life of ^{210}Pb, the dating is restricted to the latest 100 yr of sediment accumulation. DeMaster and Turekian (1987) also used ^{32}Si as a chronometric tool although their main goal initially was to determine the half-life of ^{32}Si by using the ^{210}Pb-dated layers and extension to deeper parts of the sediment pile. Actually the results were flawed as the method for evaluating the half-life of ^{32}Si by independent laboratory determination showed that the half-life was $\sim$140 yr and not as high as the value inferred from the sediment-based measurements. The assumption of constant-sediment accumulation beyond the highly varved ^{210}Pb dated section to greater depths in the sediment was wrong. The use of the correct ^{32}Si half-life shows that the layers accumulated at different rates beyond the 100 yr are datable by ^{210}Pb.

Santa Barbara basin sediments back to 100 yr have been dated by ^{210}Pb (Koide *et al.*, 1972) and back to $\sim$2$\times$10^{4} yr by ^{14}C (Kennett and Ingram, 1995). Santa Barbara basin shows sedimentation changing between a suboxic unbioturbated regime during interglacials and an aerated bioturbated regime during glacials. The pattern tracks other climate-controlled features in ice cores and North Atlantic sediments.

Certain smaller suboxic environments within larger estuarine systems also show unbioturbated sections of sediments. Because of their generally rapid rate of accumulation, the cosmogenic nuclide ^{7}Be (53 d half-life) may be used in addition to ^{210}Pb for sediment accumulation rate assessment for the more recently deposited sediments. Similarly the pattern of bomb-produced ^{137}Cs in sediments can serve as a chronometer. In addition, radiocarbon ages commonly can be obtained from calcareous fractions in the accumulating sediment. All these approaches were used at the FOAM site in Long Island Sound by Krishnaswami *et al.* (1984).

Aside from anoxic or suboxic basins, the other marine environment suitable for radioactive geochronometry is salt-marsh deposits. As sea level has risen over the past 100 years, salt marshes have kept up by vertical growth of a vegetated framework that supports sediment accumulation. In addition, since high salt marshes are inundated by seawater only $\sim$5% of the year, the surface becomes an accumulator of atmospherically derived species including ^{210}Pb. The radioactive decay of ^{210}Pb can then be used to determine the age of levels in the salt marsh and thereby the accumulation rate of the salt marsh and its components. Since the salt-marsh vertical growth depends on the rise in sea level, the ^{210}Pb chronometer becomes a proxy for the rate of rise of sea level recorded along coasts if the tectonic or isostatic upward movement of the land can be corrected for. The procedures have been described by McCaffrey and Thomson (1980) and a plot for the Farm River salt marsh in Connecticut is shown in Figure 1. Other studies of salt-marsh vertical growth and its relation to atmospheric and sediment fluxes have since been published (Varekamp and Thomas, 1998).

12.4.3 Bioturbated Deposits

Under oxic conditions all sediments are mixed as the result of the actions of a variety of types of biota. Depending on the depth of biological activity, the sedimentary record will reflect this mixing process in the distribution of radionuclides. The full equation describing the distribution of a radioactive species in a sediment pile is

$$\frac{\partial A}{\partial t} = D_B \frac{\partial^2 A}{\partial z^2} - S \frac{\partial A}{\partial z} - \lambda A \qquad (3)$$

where A is the radioactivity per unit mass of the nuclide of interest, z is depth in a core, D_B is the particle mixing coefficient treated as a diffusion

phenomenon, and S is the sedimentation rate. The solution to the above equation is

$$A(z) = A_0 \exp\left(\frac{S - \sqrt{S^2 + 4\lambda D_B}}{2D_B}\right) z \qquad (4)$$

Where S is slow compared to the value of λ for the nuclide used the equation is approximated by

$$A(z) = A_0 \exp\left(\frac{-z\sqrt{4\lambda D_B}}{2D_B}\right) \qquad (5)$$

A plot of $\ln A(z)$ against z yields the value of D_B for the system under consideration.

Benninger *et al.* (1979) measured the distribution of ^{234}Th (half-life = 24 d), ^{210}Pb (half-life = 22 yr) and considered the distribution of ^{14}C (half-life = 5,730 yr) in their discussion of the different mixing modes of a core from Long Island Sound. The ^{234}Th measures the mixing D_B of clams, the ^{210}Pb D_B primarily of worms, and the D_B of burrowing crustaceans (*Squilla*). This is shown in Figure 2. The D_B of Squilla bioturbation is small. This fact can explain the decrease in ^{14}C specific activity of the organic component with depth if the sediment accumulation rate was very slow. The ^{14}C was interpreted by Benoit *et al.* (1979) as representing sediment accumulation rate. This interpretation is compromised by the bioturbation effect. Other ^{14}C profiles published for coastal sediments also appear to be

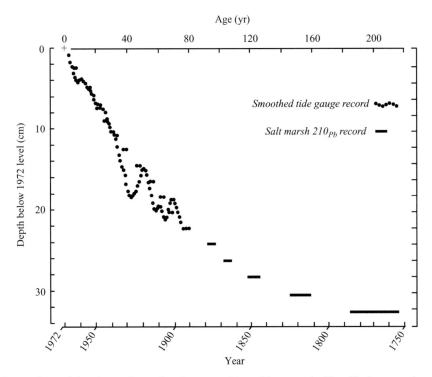

Figure 1 Comparison of the change in sea level measured in a tide gauge in New York versus the excess ^{210}Pb derived from the atmosphere in a core from the Farm River Salt Marsh, Branford, Connecticut (source McCaffrey and Thomson, 1980).

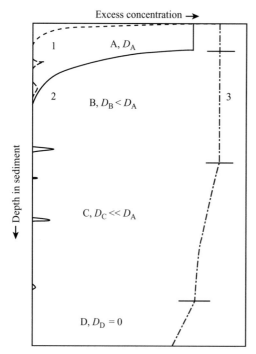

Excess concentration →

1   A, D_A

2   B, $D_B < D_A$ 3

← Depth in sediment

C, $D_C \ll D_A$

D, $D_D = 0$

Figure 2 Schematic representation of depth profiles of excess radionuclides in a sediment column undergoing biological mixing. Curves 1, 2, and 3 represent patterns for different scales of mixing and involving different radionuclides. Decay constants are assumed to decrease with the depths represented and the mixing ("particle diffusion") coefficients (D_A, D_B, D_C, D_D) decrease with depth (zones A, B, C, D). The concentration profiles are continuous over depth intervals where mixing is rapid on the timescale of radioactive decay and discontinuous below except that discontinuities may occur at depths where the mixing regime changes (e.g., curve 2). Mixing and sediment accumulation both influence the shapes of the profiles in the continuous segments, except in zone D which is unmixed.

indicators of bioturbation rather than accumulation (Tanaka *et al.*, 1991).

12.5 DEEP-SEA SEDIMENTS

12.5.1 Radiocarbon

The use of radiocabon as a dating tool depends on the deposition of carbon in the form either of calcareous tests or of organic carbon. The $^{14}C/^{12}C$ ocean ratio is not homogeneous. The deeper waters generally are lower in $^{14}C/^{12}C$ values because the circulation time of the oceans is ~1,000 yr. The results from the GEOSECS (1987) program show the oceanic profiles in all the oceans. The surface oceans are most directly impacted by the $^{14}C/^{12}C$ ratio of the atmosphere, so much so that surface waters have the imprint of the bomb ^{14}C, thus track the atmospheric burden. The GEOSECS profiles clearly show this effect

and the distribution in the oceans has been used to determine the rates of water-mass formation in the North Atlantic.

Prior to the bomb effect, which became important in 1950 and grew to a maximum in 1962, surface seawater had a $^{14}C/^{12}C$ imprint that was generally lower than that expected for equilibrium with the atmosphere. Indeed the average age of surface, inorganic carbon species dissolved in surface seawater appeared to be ~400 yr virtually everywhere in the oceans. The oldest surface ages are found at the sites of upwelling such as the equatorial oceans and the eastern boundaries of the oceans. The "reservoir" age ~400 yr is the consequence of the supply of aged upwelled water to the surface where subsequent exchange with the atmosphere results in the nonzero age surface-carbon value. All dating based on measurements of tests or organic material derived from the surface oceans will have this initial bias in age that must be accommodated independent of other concerns about the radiocarbon dating scheme.

The first uses of radiocarbon in deep-sea core dating were based on few data points and depended on extrapolation assuming the constant rate of titanium deposition (Arrhenius *et al.*, 1951) or interpolation (Suess, 1956) for determination of rates of accumulation and chronology. The first systematic study of radiocarbon incorporating possible changes in accumulation rates with depth in a core was performed by Broecker *et al.* (1958). They showed that accumulation rates of both the carbonate fraction and the detrital fraction varied with time in the equatorial Atlantic and those variations were linked to paleoclimatic indicators inferred from paleontologic data (Figure 3).

It has been observed that the surface deposits in marine cores do not have the expected age of zero. This disparity has been ascribed to loss of core tops at time of core recovery and the reservoir effect discussed above. It has also been discovered that the dominant control of the apparently constant dates in the top 8 cm of a core is bioturbation. Nozaki *et al.* (1977) studied a core obtained by a submersible research vessel and therefore exempt from the artifact of mixing or loss of the top of the sediment pile, which commonly occurs during piston coring and recovery. They measured ^{14}C and ^{210}Pb, using the latter to determine that bioturbation has indeed occurred and establish its rate constant (Figures 4 and 5). Clearly bioturbation has occurred to a depth of ~8 cm, below that depth the absence of bioturbation permits the use of ^{14}C to determine an accumulation rate and establish a chronology. The relationship between depth of bioturbation and sediment accumulation rate is

$$\text{Age} = \frac{1}{\lambda_{14C}} \ln\left[1 + \left(\frac{\lambda_{14C}}{S}\right)X_M\right] \quad (6)$$

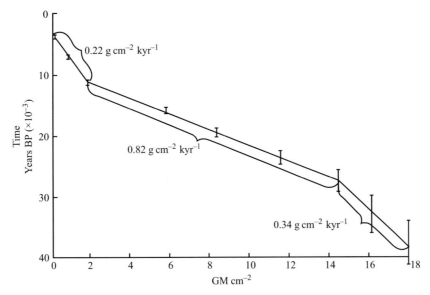

Figure 3 Cumulative curve of weight of acid-insoluble fraction ("clay") as a function of time (BP) in an Atlantic equatorial core from the Mid-Atlantic Ridge (LDEO core A 180-74) (source Broecker *et al.*, 1958).

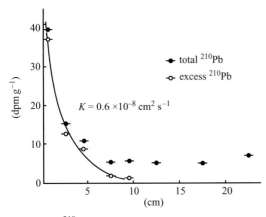

Figure 4 ^{210}Pb distribution in core 527-3 of the FAMOUS expedition at the Mid-Atlantic Ridge west of the Azores. Excess ^{210}Pb was calculated by subtracting ^{226}Ra activity from the total ^{210}Pb activity (source Nozaki *et al.*, 1977).

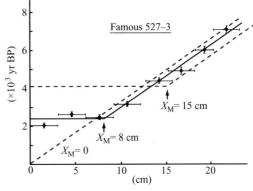

Figure 5 Model fit of ^{14}C distribution in FAMOUS core 527-3. The two dashed lines show the cases of $X_M = 0$ cm and $X_M = 15$ cm and the solid line is the case of $X_M = 8$ cm (source Nozaki *et al.*, 1977).

where the age is the radiocarbon age of the mixed layer, X_M is the depth of the mixed layer, S is the accumulation rate of sediment and λ is the decay constant for ^{14}C. In principle, for the length of the core that is accumulating at the present rate, a radiocarbon age of the surface-mixed layer and the assumption that the mixed layer is 8 cm thick allows the determination of S and therefore a chronometry for the portion of the core below 8 cm.

The organic fraction of deep-sea sediments may also be dated so long as the contribution of older detritus can be excluded. This is a problem in coastal and continental margin sediments primarily (Tanaka *et al.*, 1991). Dating of anoxic or suboxic continental margin sediments by ^{14}C has been discussed above. Generally, radiocarbon dating of

organic carbon is applied to noncalcareous deposits where organic carbon is of sufficient abundance for dating. Turekian and Stuiver (1964) determined the rates of accumulation in the Argentine basin where sediments contained 0.8% C and virtually no calcium carbonate. The accumulation rate of detrital sediment was the highest observed for the deep ocean. The source of sediment, based on the quartz-surface texture (Krinsley *et al.*, 1973), was the circum-Antarctic area by bottom transport rather than stream transport from South America.

12.5.2 ^{230}Th and ^{231}Pa

12.5.2.1 *The basic theory*

The use of ^{230}Th and ^{231}Pa in determining the chronology of deep-sea sediments is based on their production in the oceanic water column from the decay of uranium isotopes (Table 1) dissolved in seawater and their incorporation in the bottom deposit by strong adsorption on particle surfaces followed by the sedimentation of the particles. This combination of the processes leading to removal of particle-reactive radionuclides such as these from the water column is referred to as chemical scavenging. At the time of deposition, sediment contains an initial quantity N_0 of excess (unsupported) ^{230}Th or ^{231}Pa activity along with small quantities supported by decay of the parent uranium isotopes that are present. The decay of the excess nuclides with time and their burial is governed by Equation (1) and leads to an exponential decrease of the unsupported component as a function of the depth in the sediment.

The ^{230}Th and ^{231}Pa methods are not used directly for absolute dating of individual sedimentary horizons, because the assumption that N_0 is constant over time does not hold exactly but can be upset by fluctuations due to changes in sediment-deposition rate and other factors. Instead, the common practice is to use the decreasing activity with depth to derive an average rate of sediment accumulation over the length of a core or some other long interval. If t is the age (time since deposition) of a sediment horizon at depth z, and if S is the average sediment accumulation rate (thickness per unit time), then from Equation (1)

$$\ln N = -\frac{\lambda}{S}z + \ln N_0 \qquad (7)$$

The customary practice is to plot $\ln N$ against z and by regression analysis determine the slope, from which S is calculated. From this result ages can be assigned to events recorded in a core, and uncertainties can be estimated from the regression statistics.

The ^{230}Th and ^{231}Pa methods can each be used independently, and concordance between the two improves confidence in the result. Because of the relative difficulty of analysis, the ^{230}Th method is more often applied alone. Sediment ages of up to 3.5×10^5 yr by the ^{230}Th method and 1.5×10^5 yr by the ^{231}Pa method can be determined. There are several good reviews containing more information on these methods (Goldberg and Bruland, 1974; Ku, 1976; Turekian and Cochran, 1978; Ivanovich *et al.*, 1992; Huh and Kadko, 1992).

12.5.2.2 *The underlying assumptions*

Use of the ^{230}Th and ^{231}Pa methods assumes closed-system behavior, i.e., no diffusional mobility that would change the slope of the $\ln N$ versus z plot. Because of the very strong adsorption of thorium and protactinium on solid phases,

significant mobility is unlikely, and no evidence for it has been reported. The concordance between sediment accumulation rates from ^{230}Th and from ^{231}Pa (Ku, 1976) also argues against significant mobility.

Equation (7) assumes that N_0, the initial amount of ^{230}Th or ^{231}Pa per unit of sediment, remains constant over time. In evaluating this assumption, it is helpful to think of N_0 as the ratio between the flux of the radionuclide and the flux of the particles making up the sediment. For N_0 to remain constant, either the individual fluxes must both remain constant, or they must covary exactly. Because of the long-residence time of uranium in the oceanic water column (4.5×10^5 yr; Cochran, 1992), it is expected that its concentration in seawater should not change very much over the applicable time periods of the ^{230}Th and ^{231}Pa methods, and this is supported by the constancy of the uranium content of fossil corals (Broecker, 1971). Thus, the supply of the two decay products is unlikely to have changed significantly over these periods. Because of the strongly particle-reactive nature of both thorium and protactinium, they are removed from the water column by scavenging on very short timescales ($\sim$10–100 yr) compared to the radioactive half-lives of ^{230}Th (75,200 yr) and ^{231}Pa (32,700 yr), so that there is negligible loss by decay in the water column. Thus, the deposition rates of ^{230}Th and ^{231}Pa are, within a small fraction of 1%, equal to their supply rates and are very unlikely to have changed significantly over time. However, although this simple balance between supply and deposition must hold over the whole ocean, because of horizontal redistributions of the supply, it does not necessarily hold at any particular location, and this is one potential source of this variability in N_0.

A more significant source of variation in N_0 is change in particle flux, which has unquestionably occurred. Indeed, it is the variations in flux that are of greatest interest in paleoclimatic and paleoceanographic studies; the traditional ^{230}Th and ^{231}Pa methods do not resolve them but instead average over them. This point is taken up further in the next section.

12.5.2.3 *Applications*

In spite of their limitations, the ^{230}Th and ^{231}Pa methods have made an important contribution to the establishment of the Late Pleistocene chronology of deep-sea sediments. They provided the timescale upon which the deep-sea δ^{18}O record of global ice volume could be correlated with solar insolation, thus providing strong support for the astronomical theory of climate change (Broecker and Van Donk, 1970).

The more recent work with ^{230}Th and ^{231}Pa has been concerned less with their use to establish absolute chronology and more with the interpretation of their profiles in sediment cores to determine shorter-term variability in particle flux. Particle-flux measurements with sediment traps and other studies of the behavior of ^{230}Th and ^{231}Pa in the oceanic water column have resulted in a better understanding of the extent to which they can be laterally redistributed following their production. It has been shown that, over much of the ocean, the redistribution of ^{230}Th is minimal, so that the flux of particulate ^{230}Th to the seafloor is nearly in balance with the fixed rate of supply from ^{234}U integrated over the water column above (Yu et al., 2001). This is in contrast to the behavior of ^{231}Pa, which shows a stronger tendency toward lateral redistribution and preferential deposition in areas of higher particle flux and higher rates of scavenging around ocean margins and other high-productivity areas (Yang et al., 1986; Bacon, 1988; Walter et al., 1999; Yu et al., 2001).

If it is assumed, because of its constant rate of supply and the minimal potential for lateral redistribution, that the flux of particulate ^{230}Th to any point on the seafloor has remained constant with time, then any variability in N_0 must be due to changes in particle flux, and a simple inverse relationship should hold. This consideration has led to the development of ^{230}Th as a constant-flux reference tracer against which variations in the mass flux or the fluxes of individual sediment components can be measured (Bacon, 1984; Suman and Bacon, 1989; Francois et al., 1990). Downcore profiles of excess ^{230}Th are measured and then converted to profiles of N_0 by removing the primary exponential trend or by decay corrections based on ages determined independently from radiocarbon dating or oxygen-isotope stratigraphy. The normalized flux (rain rate) F_i of any sediment component is then given by

$$F_i = \frac{\beta \cdot Z \cdot f_i}{^{230}\text{Th}_{\text{ex}}^0} \qquad (8)$$

where f_i is the weight fraction of component i, $^{230}\text{Th}_{\text{ex}}^0$ is the activity of excess ^{230}Th decay corrected to the time of deposition, β is the constant rate of production of ^{230}Th from ^{234}U in the water column $(2.63\times10^{-5}$ dpm cm^{-3} kyr$^{-1})$, and Z is the water depth. The ^{230}Th profiling method has been applied to the studies of carbonate, opal, and clay sedimentation (Yang et al., 1990; Francois and Bacon, 1991), pulsed inputs of ice-rafted debris known as Heinrich events (Francois and Bacon, 1994; Thomson et al., 1995, 1999; McManus et al., 1998), the flux of cosmogenic nuclides such as ^{10}Be (Anderson et al., 1990; Frank et al., 1995), and the flux of interplanetary dust particles to Earth as recorded by the ^{3}He content of deep-sea sediments (Marcantonio et al., 1995, 1996, 1998, 1999, 2001).

The approach just described allows records of paleoflux to be inferred, but only to the extent that a component is preserved in the sediment. For example, a carbonate paleoflux determined by this method is the net flux, i.e., carbonate rain, to the seafloor minus dissolution after deposition. Because organic matter is so poorly preserved in deep-sea sediments, there is little possibility of using the method to arrive at an unambiguous record of organic productivity. However, the variations in ^{231}Pa flux and their correlation with variations in particle flux, which is the main factor causing the lateral redistribution of ^{231}Pa, has led to the proposed use of the ^{231}Pa/^{230}Th ratio in sediments as an indicator of past changes in biological productivity or particle export flux of surface waters. The theory is such that the ratio is preserved in the sediment even if the biogenic phases (organic matter, carbonate, opal) are remineralized. However, the ^{231}Pa/^{230}Th ratio in the particle flux depends on other variables such as horizontal transport and particle composition, and these complications may limit the usefulness of the approach. A review of this method and its limitations is given by Walter et al. (1999). It has been suggested that the ^{10}Be/^{26}Al ratio may prove to be a more reliable paleoproductivity indicator (Luo et al., 2001), though many of the same considerations and limitations apply to its use.

12.5.2.4 Problems of erosion and focusing

Deep-sea sediments are not ideal accumulators of the pelagic rain of particles from above. Instead, the action of bottom currents can redistribute the arriving particles so that there is preferential winnowing from topographic highs and accumulation in lows. The accumulation of sediment at a rate that is greater than the local pelagic rain is often called "sediment focusing." Changes in the degree of focusing can cause changes in the rate of sediment accumulation at a given point on the seafloor that are not related to changes in supply by the pelagic rain.

If, as argued above, ^{230}Th can serve as a constant-flux reference, i.e., if its pelagic rain rate remains constant over time and is equal to the integrated production over the water column, then it is possible to quantify the degree of focusing in a core. A focusing factor ψ can be defined as follows:

$$\psi = \frac{\int_{z_2}^{z_1} [^{230}\text{Th}_{\text{ex}}^0] \cdot \rho_b dz}{\beta \cdot Z \cdot (t_2 - t_1)} \qquad (9)$$

where ρ_b is the dry bulk density of the sediment, and t_1 and t_2 are the ages of horizons z_1 and z_2, which can be approximated from the depth profile of $^{230}Th_{ex}$ in the core or can be obtained by independent means such as radiocarbon dating or oxygen-isotope stratigraphy, and the other symbols are as defined before. The numerator on the right-hand side of Equation (9) is then the amount of $^{230}Th_{ex}$ that accumulated between t_2 and t_1, and the denominator is the amount expected from the known rate of supply. Ideally $\psi = 1$. Focusing is indicated if $\psi > 1$ and erosion if $\psi < 1$.

Suman and Bacon (1989) used this method to determine focusing factors over the past 10^4 years ranging from 4 to 13 on the Bermuda rise, a drift deposit where sediment-accumulation rates are unusually high because of lateral input of sediments by bottom currents of the Gulf Stream return flow. Marcantonio et al. (2001) used ^{230}Th to quantify sediment focusing in the equatorial Pacific and to show in this and earlier papers (e.g., Marcantonio et al., 1996) that climate-related variations in the burial rate of extraterrestrial ^3He are due mainly to the variations in sediment focusing and that the supply rate of the ^3He-bearing extraterrestrial dust is relatively constant. Thomas et al. (2000) have argued that the variations in sediment focusing in the equatorial Pacific may be caused not so much by variable bottom currents but more by variations in biological productivity of the upper ocean, which would have caused variation in the scavenging of fine particles and also in the flux of ^{230}Th. More work is needed to resolve the questions that they have raised, including an examination of the ^{231}Pa record in equatorial Pacific cores.

12.5.3 ^{10}Be

The cosmogenic nuclide ^{10}Be (half-life = 1.5Myr) is a logical candidate for dating deep-sea deposits. The production in the atmosphere is primarily in the stratosphere. Its entry into the troposphere from the stratosphere occurs primarily around 40–50° latitude where tropopausal folding occurs. It differs in that regard from the short-lived ^7Be (half-life = 53 d), whose major flux to the ocean surface is from the troposphere and subject to latitudinal production variations due to the Earth's magnetic lines of force (greater production at the poles, least production at the equator). The deposition of ^{10}Be in deep-sea sediments is primarily controlled by efficiency of scavenging by particles as ocean circulation blurs the ^{10}Be concentration variations in ocean water to make it effectively independent of its locus of delivery (Turekian and Cochran, 1978). The mean-residence time of beryllium in the oceans is $\sim$100 yr.

Early measurements (Tanaka et al., 1977; Tanaka and Inoue, 1979) were made by radioactive counting but the use of accelerator mass spectrometry (Raisbeck et al., 1979; Turekian et al., 1979) has improved the quality and quantity of measurements. Tanaka and Inoue (1979) summarized their research on Pacific cores using the radioactive counting technique. Generally the results were compatible, in each core where it could be tested, with magnetic-reversal chronometry. The data up to 1978 have been reviewed by Turekian and Cochran (1978). They indicated that several factors control the deposition of ^{10}Be with time and location in the oceans. Tanaka and Inoue (1979) specifically showed that biological productivity strongly controlled the flux of ^{10}Be to the ocean bottom (Figure 6). Mangini et al. (1984), using accelerator mass spectrometry, studied a deep-sea core from the Pacific and showed that the rate of accumulation of sediments at the site of the core (GPC-3) varied as follows: 2 mm kyr^{-1} from 1.1 Myr to the present, 1.1 mm kyr^{-1} from 3.3 Myr to 1.1 Myr, and 0.5 mm kyr^{-1} from 3.3 Myr to $\sim$14.5 Myr or longer. This range shows that the rate of deposition need not be constant over long time periods even where the viscissitudes of the glacial cycles were not operative. Mangini et al. (1984) also showed that the ^{10}Be flux to sediments was higher in the high-productivity upwelling areas of ocean margins as Tanaka and Inoue (1979) had shown for the equatorial Pacific.

The refinement of ^{10}Be measurements and the fact that its concentrations can also be measured in ice cores for the last several hundred thousand years allows the evaluation of its production variations over time as shown by Frank et al. (1995). They normalized the ^{10}Be concentration with respect to ^{230}Th to accommodate regional focusing of the two nuclides as the result of particle scavenging in much the same way as ^3He accumulation rate was calibrated (see discussion in next section). Sharma (2002) used this approach to show that the ^{10}Be flux changes, once corrected for the Earth's intrinsic magnetic moment changes, were related to solar-magnetic activity. As solar-magnetic activity is coupled to the photon flux, there should be a climate signal linked to the ^{10}Be flux signal, and Sharma (2002) showed that indeed there is a strong correlation in deep-sea sediments between $\delta^{18}O$ record and the ^{10}Be flux.

For long timescales the fluctuations in ^{10}Be flux are damped out by bioturbation in slowly accumulating sediments or averaged out in very slowly growing manganese nodules and crusts as discussed below.

12.5.4 ^3He

The accumulation of cosmic dust on Earth is generally assumed to be constant as it is not a function of any terrestrial process of enrichment or depletion as is known to be the case for cosmic rays (and therefore cosmic-ray produced nuclides such as ^{10}Be discussed above).

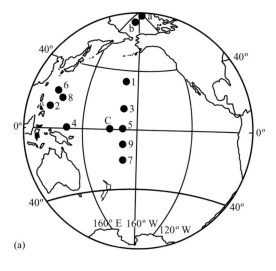

(a)

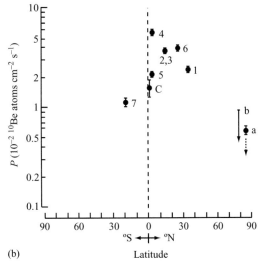

(b)

Figure 6 (a) Sampling locations for [10]Be measurements in Pacific deep-sea cores (source Tanaka and Inoue, 1979). (b) Latitudinal effect on the rates of [10]Be deposition (P) in the Pacific. The sites are those in Figure 6(a). Cores 8 and 9 are excluded because of large scatter of [10]Be concentrations with depth. Note the increased flux of [10]Be at the equator relative to other sites because of high biological productivity and the low values in the Arctic because of low productivity (source Tanaka and Inoue, 1979).

There are characteristic chemical properties of cosmic dust that have been involved in the study of sediment accumulation rates. The platinum group elements, such as iridium and osmium, offer good examples. Attempts to use iridium in this way have had the important result of indicating a giant meteorite impact at the Cretaceous–Tertiary boundary (Alvarez *et al.*, 1980) but it has not been proven important in determining chronometry.

The nuclide that has shown some promise is [3]He as mentioned above. Measurements of [3]He concentration in deep-sea sediments were first made by Ozima *et al.* (1984). Farley and Patterson (1995) measured [3]He concentrations in a deep-sea core and found that the concentration showed an apparent periodicity of 10^5 yr. Marcantonio *et al.* (1996) showed that monitoring the [3]He concentration with the [230]Th concentration produced independently at a constant rate in the ocean-water column (see above) indicated that indeed the [3]He flux was constant at least over the past 4.5×10^5 yr. The concentration changes were the result of both sediment accumulation and sediment focusing. The use of [230]Th helped to separate these two processes. On this basis, in areas not subject to focusing over long periods the [3]He concentration of the sediment would be a direct measure of the flux of cosmic dust and terrigenous sediment. If the cosmic-dust flux has been calibrated with [230]Th and if it has remained constant over the Cenozoic, then it can be used as a chronometer.

One such extended record has been obtained for the Cenozoic for a North Pacific clay core (GPC3) by Farley (1995). There is some indication that the flux of [3]He may have varied, perhaps due to meteorite impacts or possibly focusing, but the uncertainties in calibration of the [3]He flux by independent methods of sediment-accumulation rate determination are far from perfect and more study is called for. Nevertheless, the prospect of using a constant [3]He flux as a chronometric tool remains tantalizing. Farley (2001) provides a summary of results and concepts to the present time.

12.5.5 Volcanic Layers

Volcanic debris from explosive volcanism occurring at convergent plate boundaries can be deposited in deep-sea sediments. As volcanic-ash layers provide the opportunity of dating strata by a number of radiometric methods.

Dymond (1969) obtained four sediment cores from the Pacific that showed the presence of volcanic layers and dated them by the [230]Th method, calibrated paleomagnetic normal/reverse stratigraphy and potassium–argon dating of volcanic fragments. In one core (V19-153, 8° 51′S, 102° 07′E) he obtained ages progressing down the length of the core from 65 cm to 690 cm that spanned K/Ar datable time intervals from 6×10^4 yr to 1.84 Myr.

Macdougall (1971) used fission-track dating of glass shards to determine the ages of volcanic layers in deep-sea sediments. He compared his results to K/Ar dates and showed that both methods gave identical results (Table 2).

The volcanic layer dating has been extended by stratigraphic correlation of diagnostic chemical imprints of volcanic ash dated on land adjacent to the deep-sea sediments of eastern Africa (Brown *et al.*, 1992).

Table 2 Comparison of fission track ages and potassium–argon ages of volcanic material in deep-sea sediments.

Sample	Location	K–Ar age (10^6 yr)	Magnetic reversal age (10^6 yr)	Fission track age (10^6 yr)
V21-145 815 cm	34° 03′ N, 164° 50′ E	1.45 (±0.08)	1.4	1.47 (±0.16)
V21-173 725 cm	44° 22′ N, 163° 33′ W	1.62 (±0.08)	1.6	1.62 (±0.17)
EM 8-13	28° 59′ N, 117° 30′ W	11.4 (±0.6)		10.50 (±0.6)

Source: MacDougall (1971).

12.5.6   Extension of Dating Techniques

12.5.6.1   *The Milankovitch cycles and chronology*

The periodicities of the Milankovitch orbital forcing on environmental parameters has now been established through deep-sea sediment records. The primary proxy has been $\delta^{18}O$ variations in foraminifera, but relative abundance of marine species has also been shown to have the periodicities ascribable to the Milankovitch pattern. Since the periods of the three major components of the Milankovitch cycle, precession, obliquity and eccentricity, have fixed values of $\sim 2 \times 10^4$ yr, 4×10^4 yr, and 10^5 yr, respectively, this pattern can be tracked through a sedimentary record and provide a precise chronology once the periods have been well established and certified in the datable parts of cores. Bassinot *et al.* (1994) showed the value of the procedure by dating the Brunhes/Matuyama magnetic-reversal boundary. At first the Milankovitch reconstruction seemed to give a higher value (7.7×10^5 yr BP) than permissible by the then available K/Ar dates, but subsequent, more precise dating using the $^{40}Ar/^{39}Ar$ method showed that the boundary was actually 7.7×10^5 yr BP as inferred from the Milankovitch reconstruction. Using this technique, dating has been extended through the Oligocene (Shackleton *et al.*, 2000).

12.5.6.2   *Oxygen and carbon isotopes in carbonate tests*

Ocean-drilling campaigns have provided deep-sea sediment cores ranging through the Cenozoic era. These cores often provide continuous records of the changing oxygen and carbon isotope signature of the oceans and the temperature of deposition of the carbonate test. Prior to ~ 33.5 Myr BP there were no large ice caps so that all the oxygen isotopic variations are presumed to be due to the temperature of the ocean in which the foraminiferan grew. Generally, if the bottom waters of the oceans had a high-latitude source the oxygen-isotope signature would reflect water colder than surface waters. At ~ 33.5 Myr BP, the formation of the Antarctic ice cap sequestered water with a light oxygen-isotope signature.

This accumulation of water of low $\delta^{18}O$ drove the $\delta^{18}O$ of the oceans heavier. The climatic cooling associated with this event also cooled the oceans and resulted in a heavier signature in the foraminifera from the temperture effect as well. This situation was further enhanced as the result of the development of the northern hemisphere ice caps since ~ 2.7 MyrBP. These changes in the $\delta^{18}O$ recorded by foraminifera provide a unique and diagnostic pattern that can be used as a chronostratigraphic tool.

In the case of carbon isotopes the $\delta^{13}C$ is the complex result of the relative importance of organic carbon sequestration or release by weathering compared to calcium carbonate deposition since the former has a $\delta^{13}C$ of about -25 and the latter a $\delta^{13}C$ of ~ 0. A further effect can be the episodic release of extremely light carbon ($\delta^{13}C \sim -75$) in the form of methane. Methane is known to be held in clathrate structures in sediments under high-productivity areas at depths and temperatures characteristic of their formation (Dickens *et al.*, 1995). The carbon isotopic record through the Cenozoic combined with the oxygen-isotope record provides a chronostratigraphic tool for dating core sequences (Figure 7).

There have been attempts to use the oxygen-isotope pattern in fish debris to date cores from below the carbonate compensation depth, which are devoid of calcium carbonate tests because of dissolution. Fish teeth are composed of apatite and the phosphate retains its initial oxygen signature since that can be affected seriously only by enzymatic recycling of the phosphate. The dating is established using the pattern of oxygen-isotope variation with time shown in Figure 7. A preliminary attempt to do just that was made on a red-clay core from the north Pacific (GPC3 discussed above in another context) by Blake and Turekian (2000). The analysis was able to discern the characteristic pattern of oxygen isotopes imprinted in a biologically derived test that is in equilibrium with seawater at the ambient temperature of growth. The method has the benefit of providing not only a chronostratigraphic tool but also independent records of surface-ocean temperatures of the times prior to 35 Myr BP (assuming that the fish spent most of their lives in the mixed layer).

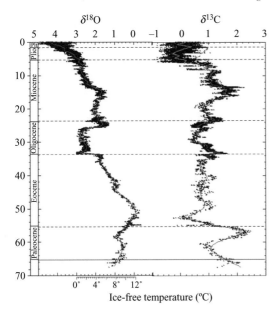

Figure 7 Global deep-sea oxygen and carbon isotope records based on data from more than 40 DSDP and ODP sites (source Zachos *et al.*, 2001).

12.5.6.3 *Magnetic-reversal stratigraphy*

Cande and Kent (1992) have compiled a magnetic stratigraphy for the Earth as recorded from deep-sea sediments and ocean-floor magnetic intensity patterns. As the sequence is accurately dated by independent means, it can provide a chronostratigraphic tool in the absence of other chronometers. Berggren *et al.* (1995) published an integration of bio-magnetostratigraphy with $^{40}Ar/^{39}Ar$ radiometric dates. There has been a joining of the orbital forcing, volcanic dating, and magnetic-reversal stratigraphy to date the Brunhes-Matuyama reversal boundary precisely (Shackleton *et al.*, 1993).

The Fe/Ca ratio in an ODP core from the western North Atlantic has been used as an index of orbitally driven climate change allowing deconvolution of the orbital forcing pattern and extension back to the Late Eocene. In this time period the timing of orbital pattern was used for fine-tuning of the magnetic-reversal record (Palike *et al.*, 2001).

12.5.6.4 *Element accumulation: Titanium and Cobalt*

There have been two attempts to use elemental measurements to determine accumulation rates of marine deposits independent of the cosmic dust ^{3}He approach or the tacitly assumed constant-flux model used in the ^{230}Th and ^{231}Pa approaches. One depends on a trace metal that tracks the fine-grained (clay) fraction of deep-sea sediments, the other on the addition of a hydrogenous element to the accumulating marine deposit.

The so-called titanium method used by Arrhenius *et al.* (1951) assumes a so-called "lutite veil" depositing at a constant rate throughout the deep oceans. Titanium associated with this lutite fraction then is also assumed to accumulate at a constant rate. Variations in the accumulation rates of biogenic components can then be assessed. The initial titanium method was calibrated by a single radiocarbon date as discussed above. The method has not been used since it was discovered that accumulation rates of all detrital and biogenic components of deep-sea sediments are subject to change as a function of climatic history and focusing processes.

Another element used on the assumption of constant accumulation is cobalt. This element is used on the assumption that a component of the sediment contains this element by addition from a hydrogenous source. That is, it is scavenged from seawater as sedimentation occurs but the removal is at a constant rate. The element cobalt was first suggested for ferromanganese oxide crusts (Halbach *et al.*, 1983). It was used for a deep-sea sediment core by Kyte *et al.* (1993). They calibrated the imported constant cobalt addition to the sediments with time by stratigraphically dated horizons such as the 65 Myr old iridium anomaly, due to a large meteorite impact, identified as marking the Cretaceous–Tertiary boundary. They also used epoch boundaries from fish-debris stratigraphy. The assumption of constant cobalt addition to deep-sea sediments assumes that the concentration of cobalt in the oceans has been invariant with time and that the mechanism of sequestration is time independent. These assumptions are difficult to assess. In the absence of any other dating tool, however, this approximation may sometimes be useful.

12.6 FERROMANGANESE DEPOSITS

12.6.1 Applicable Methods

Iron- and manganese-rich deposits on the seafloor are widespread and occur as nodules (the commercially important type), crusts, or thin coatings on rocks. They form authigenically by precipitation of metals from seawater. As they accrete, they incorporate from seawater significant quantities of radionuclides, including ^{230}Th, ^{231}Pa, and ^{10}Be, which allow the determination of their age or rate of growth (accretion). In all cases, the age determination is based on the decay of unsupported radioactivity and is governed by Equation (1). As with sediments, the usual practice is to measure the activity as a function of depth and apply Equation (3) to obtain an average growth rate over depth intervals corresponding to ~1–4 half-lives of the radionuclide. Each of the three radionuclides can be applied individually as an independent method, and variants based on the ratios $^{230}Th/^{232}Th$, $^{231}Pa/^{230}Th$, or $^{10}Be/^9Be$ have often been employed. The long half-life of ^{10}Be, allowing ages of several million years to be determined, makes this nuclide especially attractive for dating slowly growing deep-sea ferromanganese deposits.

Ferromanganese nodules and crusts often grow around a nucleus or upon a substrate consisting of volcanic minerals or glass, which are well suited for dating by the K–Ar method, and provide an alternate approach to determine their age (Barnes and Dymond, 1967). If it is assumed that accretion of the deposit began just after the formation of the nucleus or substrate, then the average rate of accretion over its lifetime can be estimated. This method can give only a maximum age of the deposit, or a minimum rate of accretion, because of the unknown time by which the onset of accretion might have been delayed or dissolution might have occurred (Ku, 1977). Fission-track dating has also been used in a similar way (Aumento, 1969).

A limited amount of work has been done on the dating of shallow-water nodules, which, because of diagenetic remobilization of manganese within the sediment column, generally form more rapidly than deep-ocean deposits. Unlike deep-water nodules, the rapidly growing shallow-water nodules do not contain excess ^{230}Th or ^{231}Pa but instead contain, initially, quantities of ^{230}Th and ^{231}Pa that are less than their equilibrium activities. The ingrowth over time of ^{230}Th and ^{231}Pa toward equilibrium with ^{234}U and ^{235}U provides the basis for dating these deposits (Ku and Glasby, 1972), as it does for corals (see below). An important limitation of the uranium-series methods for dating shallow-water nodules is that not all of the ^{230}Th or ^{231}Pa are necessarily produced by decay *in situ*. Significant amounts may be present initially, so that the derived ages must be interpreted as maximum values. Phosphorites are another type of deposit that can be dated by the ingrowth method (Burnett and Veeh, 1977; Burnett *et al.*, 1982, 2000; Kress and Veeh, 1980; Roe *et al.*, 1983; Kim and Burnett, 1986).

Ku (1977) published an excellent review of the earlier literature on growth rates of ferromanganese deposits.

12.6.2   The Underlying Assumptions

As is the case with deep-sea sediments, all of the dating methods based on decay of unsupported radionuclides assume that N_0—the initial amount of ^{230}Th, ^{231}Pa, or ^{10}Be per unit of deposit—remains constant over time. In general, this requires that both the accretion rate of the deposit and the uptake rate of the radionuclide from seawater have remained constant. The various ratio methods are used primarily so that the latter assumption can

be relaxed, the theory being that the variations in the uptake rate of two isotopes, ^{230}Th and ^{232}Th, e.g., would tend to cancel each other out. Because of variations in seawater chemistry over time, it is not likely that the assumption of constant N_0, or of the initial ratios, is strictly true, but the linearity generally observed in ln N versus z plots indicates that it is a sufficiently good approximation that valid estimates of average accretion rates over long time intervals can be obtained.

The ^{230}Th, ^{231}Pa, and ^{10}Be methods are all based on a concentration gradient with depth below the outer surface of the deposit, which, in principle, would drive a diffusion that would reduce the gradient and lead to apparent growth rates that are too high. This was a controversial point in some of the earlier literature, but recent work has shown quite convincingly that ^{230}Th and ^{10}Be have very low effective diffusivities in ferromanganese crusts and may be regarded as essentially immobile (Mangini *et al.*, 1986; Chabaux *et al.*, 1997; Henderson and Burton, 1999). Because of the inferred large distribution coefficient for uptake of protactinium from seawater by manganese oxides (Anderson *et al.*, 1983), it is likely that a similar immobility (closed-system behavior) applies for ^{231}Pa as well.

12.6.3   Applications

Probably the most important contribution of radiometric dating to the study of deep-sea ferromanganese deposits was the establishment of their very slow growth rates, measured in millimeters per *million* years (Ku, 1977; Burnett and Morgenstein, 1976; Guichard *et al.*, 1978; Krishnaswami and Cochran, 1978; Ku *et al.*, 1979; Moore *et al.*, 1981; Krishnaswami *et al.*, 1982; Huh and Ku, 1984), though manganese crusts from sites near seafloor spreading centers, close to hydrothermal sources of manganese, can grow considerably faster (Moore and Vogt, 1976). The generally slow growth is in contrast to the more rapid accumulation of deep-sea sediments, which is measured in millimeters (or more) per thousand years, and it raises the problem of explaining how manganese nodules avoid burial and remain at the sediment surface. Possibilities include the action of bottom currents or episodic nudging by benthic animals, but the exact mechanism is still not understood.

More detailed sampling has revealed discontinuities in the radionuclide-depth profiles indicating episodic growth (Krishnaswami and Cochran, 1978; Eisenhauer *et al.*, 1992) or relatively sudden changes in rate of growth (Krishnaswami *et al.*, 1982; Mangini *et al.*, 1986). Variations in the rate of growth over time may explain many of the discordant results that have been obtained when ^{230}Th, ^{231}Pa, and

^{10}Be profiles have been compared on the same samples (Krishnaswami *et al.*, 1982), because each radionuclide averages over a different length of time. Comparison of radionuclide distributions, including those of ^{226}Ra, between the top and bottom surfaces of nodules has given evidence they have rolled over at times in the past (Krishnaswami and Cochran, 1978; Huh and Ku, 1984; Moore, 1984). The inferred times between rollover "events" range from 10^3 yr to 10^5 yr.

The more recent work with ferromanganese deposits has focused on further resolving shorter-term variations in their rate of growth (Mangini *et al.*, 1990; Eisenhauer *et al.*, 1992), and there is growing interest in their use as recorders of past changes in seawater chemistry (Huh and Ku, 1990). These studies have focused on crusts more than nodules. The nodules are suspect because of their close contact with the sediments and possible diagenetic supply of metals, and it is believed that crusts, which form, e.g., on the flanks of seamounts away from the sediments, provide a more purely hydrogenous deposit and thus a more accurate record of variations in seawater composition.

Mangini *et al.* (1990) and Eisenhauer *et al.* (1992) have proposed that a constant ^{230}Th flux model, analogous to the one described above for deep-sea sediments, be applied to ferromanganese deposits to determine short-term variations in growth rate. Depth profiles of ^{230}Th are measured and converted to profiles of N_0 by removing the primary exponential trend, and Equation (4) is applied to obtain point-by-point profiles of growth rate. Eisenhauer *et al.* (1992) also showed that very high resolution sampling can be achieved by selecting crusts that have a very flat, laminar structure, and they obtained a resolution of 0.02 mm, corresponding to a time $\sim$5,000 yr, in two crusts from the Pacific Ocean. Both crusts showed systematic variations in ^{230}Th concentrations, which they correlated with the Late Pleistocene glaciation cycle and interpreted as higher growth rates during the post-glacial and the last interglacial and lower growth rates (or growth hiatuses) during glacials. It is clear from this study that such high-resolution records of varying composition contain valuable information, but it remains to be seen whether the constant ^{230}Th flux assumption can be justified. In contrast to the ^{230}Th flux to the sediments, which is limited to a value equal to its rate of supply, the ^{230}Th flux to ferromanganese deposits is typically only 10–20% of the total supply from the water column, so it is far less certain that it would remain constant over time. Chabaux *et al.* (1997) have shown evidence for significant variations in the Th/U and ^{230}Th/^{232}Th ratios recorded in ferromanganese crusts over the past 1.5×10^5 yr.

12.7 CORALS

One of the most important applications of uranium-series methods of age determination has been the dating of fossil corals and other carbonate materials. In contrast to deep-sea sediments, which accumulate excess ^{230}Th and ^{231}Pa that decay over time, carbonates accumulate uranium by co-precipitation from seawater that is essentially free of ^{230}Th and ^{231}Pa. The radioactive ingrowth of ^{230}Th and ^{231}Pa over time toward secular equilibrium with ^{238}U and ^{235}U is the basis of the two methods.

With current measurement techniques, ages as great as 550 ka (^{230}Th) and 200 ka (^{231}Pa) can be determined accurately. The equation for the ingrowth of ^{230}Th is

$$\frac{^{230}\text{Th}}{^{238}\text{U}} = 1 - e^{-\lambda_{230}t} + \left(\frac{^{234}\text{U}}{^{238}\text{U}} - 1\right)\left(\frac{\lambda_{230}}{\lambda_{230} - \lambda_{234}}\right)$$
$$\times (1 - e^{-(\lambda_{230} - \lambda_{234})t}) \qquad (10)$$

where λ_{230} and λ_{234} are the decay constants of ^{230}Th and ^{234}U and the ratios are activity ratios measured in a sample of age t. For ingrowth of ^{231}Pa, the equation is

$$\frac{^{231}\text{Pa}}{^{235}\text{U}} = 1 - e^{-\lambda_{231}t} \qquad (11)$$

where λ_{231} is the decay constant of ^{231}Pa. The terms in Equation (10) involving the ^{238}U/^{234}U ratio are necessary because of the $\sim$15% excess of ^{234}U in seawater caused by the preferential mobility of ^{234}U in natural waters due to alpha recoil during the decay of ^{238}U. The decay of excess ^{234}U provides another independent chronometer if it can be assumed that the initial excess (the seawater value) remains constant over time

$$\frac{^{234}\text{U}}{^{238}\text{U}} = 1 + \left(\left[\frac{^{234}\text{U}}{^{238}\text{U}}\right]_0 - 1\right)e^{-\lambda_{234}t} \qquad (12)$$

(decay of ^{238}U and ^{235}U over the times of interest are negligible). Figure 8, based on Equations (10)–(12), shows how the ^{234}U excess and the ^{230}Th/^{238}U and ^{231}Pa/^{235}Pa activity ratios evolve with time in a closed system.

Until the early 1980s, the development and application of uranium-series methods were based on measurements by alpha counting, which is limited in its precision and sensitivity by the slow disintegration rates (low count rates) of the radionuclides of interest. Beginning in the late 1980s, measurement techniques based on thermal-ionization mass spectrometry (TIMS) were developed (Edwards *et al.*, 1986–1987), markedly improving both precision and sensitivity. Because of the higher precision, dating of older materials

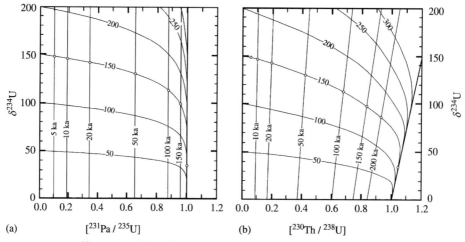

Figure 8 Plots of: (a) δ^{234}U versus ^{231}Pa/^{235}U activity ratio and (b) δ^{234}U versus ^{230}Th/^{238}U activity ratio. The δ expression is defined from the uranium isotope activity ratio as follows: δ^{234}U$=(^{234}$U/^{238}U$-1)\times1{,}000$, modern seawater having a value of $\sim$145 (Henderson, 2002). The nearly vertical lines are contours of constant age, and the curves running from left to right are contours of constant-initial δ^{234}U (source Cheng *et al.*, 1998).

which requires measurement of small departures from equilibrium ratios (Figure 8) is possible, and because of the higher sensitivity smaller and younger samples can be dated. The earlier work based on alpha counting is well reviewed by Ku (1976), and a later review by Burnett and Veeh (1992) includes an extensive discussion of marine phosphorite dating, whose problems are analogous to those of carbonate dating. An account of the development of mass-spectrometric methods is given by Chen *et al.* (1992).

Equations (10) and (11) assume that initial ^{230}Th and ^{231}Pa concentrations are zero or can be corrected for. The ^{232}Th/^{238}U ratio in surface corals is similar to that in seawater, suggesting that thorium does not fractionate from uranium significantly during coral growth (Edwards *et al.*, 1986–1987). Because of the extremely low ^{230}Th/^{238}U ratio in seawater (activity ratio $\sim10^{-5}$), initial ^{230}Th must be negligible. Concordance between ^{230}Th and ^{231}Pa ages indicates that initial ^{231}Pa must also be negligible (Edwards *et al.*, 1997).

As with all radiometric dating methods in applying Equations (10)–(12), a critical requirement, and often the one most difficult to satisfy, is that closed-system behavior has been maintained over the time interval of interest. The ^{234}U/^{238}U ratio is often used as a means of testing the closed-system assumption and assessing the reliability of dates based on the ^{230}Th/^{238}U ratio. If the initial ratio, as inferred from the ^{230}Th/^{238}U age, departs from the seawater value, it can be concluded that the system has been disrupted and that the measured age is unreliable. Use of the ^{234}U/^{238}U ratio for this purpose assumes that the initial ratio (the seawater value) has remained constant, and this has been borne out in recent

studies using high-precision mass-spectrometric measurements, at least for the past 360 ka (Henderson *et al.*, 1993; Gallup *et al.*, 1994; Henderson, 2002), though the possibility of small glacial-to-interglacial variations in the ratio cannot be ruled out completely (Henderson, 2002). Checks for concordancy between ^{230}Th/^{238}U and ^{231}Pa/^{235}U ages are another means of detecting open-system behavior.

Unaltered fossil corals that are free of contaminating materials have proven in general to approximate closed-system behavior closely enough to yield reliable uranium-series ages. This has not, unfortunately, proven to be the case for fossil marine-mollusk shells, which typically contain secondary uranium from the environment that was not present when the animal was alive and also show frequent evidence of extraneous ^{230}Th (Kaufman *et al.*, 1971). The availability of high-precision mass-spectrometric measurements of ^{231}Pa as well as the uranium and thorium isotopes now makes it possible to detect more subtle departures from closed-system behavior in coral samples and other materials, thus providing more stringent tests for validity of ^{230}Th/^{238}U and ^{231}Pa/^{235}U dates and possibly allowing corrections to be derived for certain types of open-system behavior. Cheng *et al.* (1998) have derived in detail the systematics of U–Th–Pa dating for several types of open-system behavior as well as for closed systems.

One of the early triumphs of uranium-series dating was the use of ^{230}Th to establish the chronology of raised coral terraces on Barbados (Broecker *et al.*, 1968; Mesolella *et al.*, 1969) and New Guinea (Veeh and Chappell, 1970). The dates obtained for high sea-level stands (interglacials) provided strong support for the Milankovitch astronomical theory of climate relating ice volume with solar insolation. The more precise measurements and the more stringent concordancy checks obtained by TIMS have given confirmation in several locations and have also provided for more exact correlation with astronomical calculations and for resolution of earlier disputes (Edwards *et al.*, 1987; Gallup *et al.*, 1994; Muhs and Szabo, 1994; Stirling *et al.*, 1995; Szabo *et al.*, 1994; Bard *et al.*, 1996). Precise ^{230}Th dating of submerged corals collected by drilling have provided the best record yet of sea level since the last glacial maximum (Bard *et al.*, 1996; Edwards *et al.*, 1993).

The availability of precise ^{230}Th ages for younger corals allows a comparison with ^{14}C ages. Bard *et al.* (1990) used coral cores raised off Barbados to calibrate the ^{14}C timescale over the past 3×10^4 yr, extending the calibration beyond the range of dendrochronology, which is limited to the past ~9,000 yr. The ^{14}C ages were systematically younger than the ^{230}Th ages by as much as 3,500 yr at 20 kyr BP, indicating a significantly higher ^{14}C/^{12}C ratio in the atmosphere at the time of the last glacial maximum, which they attributed to changes in cosmogenic-nuclide production rate linked to changes in the strength of Earth's magnetic field. Dating of younger corals has also provided a chronology for other paleoclimatic reconstructions such as sea–surface temperature records based on oxygen isotope or Sr/Ca paleothermometry (Beck *et al.*, 1992; Guilderson *et al.*, 1994; McCulloch *et al.*, 1996).

Most of the work with corals has been with reef-building corals, which grow very near the sea surface. An important development, however, is the demonstration that reliable ^{230}Th ages (Adkins *et al.*, 1998; Cheng *et al.*, 2000) and ^{231}Pa ages (Goldstein *et al.*, 2001) can also be obtained on deep-sea corals, and the new field of deep-water coral paleoceanography has begun to grow from this. Apparent ^{14}C ages, coupled with ^{230}Th ages, can be used to infer past variations in the ocean's circulation and rate of convective overturning (Adkins *et al.*, 1998; Goldstein *et al.*, 2001). A difficult problem with deep-sea corals not usually encountered with surface corals is the often serious contamination with extraneous ^{230}Th and ^{231}Pa from manganese oxide coatings or other adsorbing surfaces. It can be overcome by a combination of careful cleaning of samples to remove noncarbonate phases and the use of appropriate corrections based on: (i) ^{232}Th measurements and an assumed ^{230}Th/^{232}Th ratio in the contaminating component (Adkins *et al.*, 1998) or (ii) on a whole-rock isochron method that assumes a two-component mixing of carbonate and contaminant (Lomitschka and Mangini, 1999; Cheng *et al.*, 2000; Goldstein *et al.*, 2001).

12.8 METHODS NOT DEPENDING ON RADIOACTIVE DECAY

12.8.1 Amino Acid Racemization

Chemical production of optically active amino acids yields equal numbers of left-handed and right-handed molecules. The mixture is called racemic. Biological amino acids are all left handed or L-enantiomers. With time the initially L-enantiomers are reconfigured to approach a racemic mixture. The kinetics are primarily determined by temperature but chemical environment can also play a part. In a constant-temperature, constant-chemistry environment of the top 5 m of a deep-sea sediment, the gradual racemization of the amino acid being studied provides a chronometer. Bada *et al.* (1970) analyzed a core from the Atlantic Ocean using the transformation of isoleucine to alloisoleucine, its racemization product, to establish a chronometry (Figure 9). Amino acid racemization dating depends on knowing the temperature history of the sediment pile as sediment accumulates and is subject to errors due to terrestrial heat flow. Further studies using this tool have not been pursued.

12.8.2 Thermoluminescence

When quartz or other suitable mineral detector is deposited into a matrix of minerals containing the radioactive nuclides of the ^{238}U, ^{235}U, and ^{232}Th decay series, and ^{40}K, the detector mineral is subject to radiation damage. If the grain had been cleared of all memory of radiation damage prior to deposition, then the extent of damage is a function of the time the detector mineral has been immersed in the radiation-producing matrix.

The method of assessing is commonly by thermoluminescence, wherein the number of photons released during the annealing process or heating are detected by a phototube. The procedure was developed for pottery where the kilning process

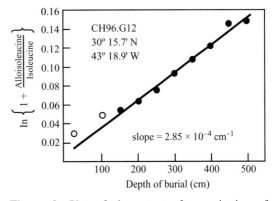

Figure 9 Plot of the extent of racemization of isoleucene against depth in an Atlantic deep-sea sediment core (source Bada *et al.*, 1970).

annealed the detector mineral and therefore provided information on the time since annealing. An alternative method is to use ESR.

In the case of marine deposition, the transport of particles through the air results in the annealing by ultraviolet radiation from the Sun. The reset minerals then act as accumulators of lattice damage once they are buried. The technique has been established for wind-blown deposits such as dunes and loess deposits. Wintle and Huntley (1979) applied it to deep-sea sediments, building on initial studies by Huntley and Johnson (1976). Subsequently, there has not been a great deal of work on deep-sea sediments, and most of the interest has shifted to sand dunes and loess deposits where the method provides a unique dating tool.

ACKNOWLEDGMENTS

Ellen Thomas and Gwyneth Williams provided thoughtful reviews of this chapter. We thank them, with the caveat that they are not responsible for our persistent shortcoming.

MPB is grateful for generous financial support provided by the US Department of Energy (most recently through its Ocean Carbon Sequestration Research Program, Biological and Environmental Reasearch grant #DE-FG02-00ER63020), and the US National Science Foundation (most recently through its Chemical Oceanography Program, grant #OCE-0117922).

REFERENCES

Adkins J. F., Cheng H., Boyle E. A., Druffel E. R. M., and Edwards R. L. (1998) Deep-sea coral evidence for rapid change in ventilation of the deep North Atlantic 15,400 years ago. *Science* **280**, 725–728.

Alvarez L. W., Alvarez W., Asaro F., and Michel H. V. (1980) Extraterrestrial cause for the Cretaceous–Tertiary extinction. *Science* **208**, 1095–1108.

Anderson R. F., Bacon M. P., and Brewer P. G. (1983) Removal of ^{230}Th and ^{231}Pa at ocean margins. *Earth Planet. Sci. Lett.* **66**, 73–90.

Anderson R. F., Lao Y., Broecker W. S., Trumbore S. E., Hofmann H. J., and Wolfli W. (1990) Boundary scavenging in the Pacific Ocean: a comparison of ^{10}Be and ^{231}Pa. *Earth Planet. Sci. Lett.* **96**, 287–304.

Arrhenius G., Kjellberg G., and Libby W. F. (1951) Age determination of Pacific chalk ooze by radiocarbon and titanium content. *Tellus* **3**, 222–229.

Aumento F. (1969) The Mid-Atlantic Ridge near 45° N. V. Fission track and ferro-manganese chronology. *Can. J. Earth Sci.* **6**, 1431–1440.

Bacon M. P. (1984) Glacial to interglacial changes in carbonate and clay sedimentation in the Atlantic Ocean estimated from ^{230}Th measurements. *Isotope Geosci.* **2**, 97–111.

Bacon M. P. (1988) Tracers of chemical scavenging in the ocean: boundary effects and large-scale chemical fractionation. *Phil. Trans. Roy. Soc. London A* **325**, 147–160.

Bada J. L., Luyendyk B. P., and Maynard J. B. (1970) Marine sediments: dating by the racemization of amino acids. *Science* **170**, 730–732.

Bard E., Hamelin B., Fairbanks R. G., and Zindler A. (1990) Calibration of the ^{14}C timescale over the past 30,000 years using mass spectrometric U–Th ages from Barbados corals. *Nature* **345**, 405–410.

Bard E., Jouannic C., Hamelin B., Pirazzoli P., Arnold M., Faure G., Sumosusastro P., and Syaefudin (1996) Pleistocene sea levels and tectonic uplift based on dating of corals from Sumba Island, Indonesia. *Geophys. Res. Lett.* **23**, 1473–1476.

Barnes S. S. and Dymond J. R. (1967) Rates of accumulation of ferromanganese nodules. *Nature* **213**, 1218–1219.

Bassinot F. C., Labeyrie L. D., Vincent E., Quidelleur X., Shackleton N. J., and Lancelot Y. (1994) The astronomical theory of climate and the age of the Brunhes-Matuyama magnetic reversal. *Earth Planet. Sci. Lett.* **126**, 91–108.

Beck J. W., Edwards R. L., Ito E., Taylor F. W., Recy J., Rougerie F., Joannot P., and Henin C. (1992) Sea-surface temperature from coral skeletal strontium/calcium ratios. *Science* **257**, 644–647.

Benninger L. K., Aller R. C., Cochran J. K., and Turekian K. K. (1979) Effects of biological sediment mixing on the ^{210}Pb chronology and trace metal distribution in a Long Island Sound sediment core. *Earth Planet. Sci. Lett.* **43**, 241–259.

Benoit G. J., Turekian K. K., and Benninger L. K. (1979) Radiocarbon dating of a core from Long Island Sound. *Estuar. Coast. Mar. Sci.* **9**, 171–180.

Berggren W. A., Kent D. V., Swisher C. C., III, and Aubry M.-P. (1995) A revised Cenozoic geochronology and chronostratigraphy. *Soc. Sed. Geol. Spec. Publ.* **54**, 129–212.

Blake R. E. and Turekian K. K. (2000) Phosphate oxygen isotope composition of fish debris as a chronostratigraphic tool: results from LL44-GPC3 a Pacific red clay core. *EOS* **81**(suppl. 2), F707.

Broecker W. S. (1971) A kinetic model for the chemical composition of sea water. *Quat. Res.* **1**, 188–207.

Broecker W. S. and Van Donk J. (1970) Insolation changes, ice volumes, and the O^{18} record in deep-sea cores. *Rev. Geophys. Space Phys.* **8**, 169–198.

Broecker W. S., Turekian K. K., and Heezen B. C. (1958) The relation of deep-sea sedimentation rates to variations in climate. *Am. J. Sci.* **256**, 503–517.

Broecker W. S., Thurber D. L., Goddard J., Ku T.-L., Matthews R. K., and Mesolella K. J. (1968) Milankovitch hypothesis supported by precise dating of coral reefs and deep-sea sediments. *Science* **159**, 297–300.

Brown F. H., Sarna-Wojcicki A. M., Meyer C. E., and Haileab B. (1992) Correlation of Pliocene and Pleistocene tephra layers between the Turkana Basin of East Africa and the Gulf of Aden. *Quat. Int.* **13/14**, 55–67.

Burnett W. C. and Morgenstein M. (1976) Growth rates of Pacific manganese nodules as deduced by uranium-series and hydration-rind dating techniques. *Earth Planet. Sci. Lett.* **33**, 208–218.

Burnett W. C. and Veeh H. H. (1977) Uranium-series disequilibrium studies in phosphorite nodules from the west coast of South America. *Geochim. Cosmochim. Acta* **41**, 755–764.

Burnett W. C. and Veeh H. H. (1992) Uranium-series studies of marine phosphates and carbonates. In *Uranium-series Disequilibrium: Applications to Earth, Marine, and Environmental Sciences* (eds. M. Ivanovich and R. S. Harmon). Clarendon Press, Oxford, pp. 487–512.

Burnett W. C., Beers M. J., and Roe K. K. (1982) Growth rates of phosphate nodules from the continental margin off Peru. *Science* **215**, 1616–1618.

Burnett W. C., Glenn C. R., Yeh C. C., Schultz M., Chanton J., and Kashgarian M. (2000) U-series ^{14}C, and stable isotope studies of recent phosphatic "protocrusts" from the Peru margin. In *Marine Authigenesis: from Global to Microbia* (eds. C. R. Glenn, L. Prévôt-Lucas, and J. Lucas). SEPM (Society for Sedimentary Geology), Tulsa, OK, pp. 163–183.

Cande S. C. and Kent D. V. (1992) A new geomagnetic polarity time scale for the Late Cretaceous and Cenozoic. *J. Geophys. Res.* **97**, 13917–13951.

Chabaux F., O'Nions R. K., Cohen A. S., and Hein J. R. (1997) ^{238}U–^{234}U–^{230}Th disequilibrium in hydrogenous oceanic Fe–Mn crusts: palaeoceanographic record or diagenetic alteration? *Geochim. Cosmochim. Acta* **61**, 3619–3632.

Chen J. H., Edwards R. L., and Wasserburg G. J. (1992) Mass spectrometry and applications to uranium-series disequilibrium. In *Uranium-series Disequilibrium: Applications to Earth, Marine, and Environmental Sciences* (eds. M. Ivanovich and R. S. Harmon). Clarendon Press, Oxford, pp. 174–206.

Cheng H., Edwards R. L., Murrell M. T., and Benjamin T. M. (1998) Uranium—thorium–protactinium dating systematics. *Geochim. Cosmochim. Acta* **62**, 3437–3452.

Cheng H., Adkins J., Edwards R. L., and Boyle E. A. (2000) U–Th dating of deep-sea corals. *Geochim. Cosmochim. Acta* **64**, 2401–2416.

Cochran J. K. (1992) The oceanic chemistry of the uranium- and thorium-series nuclides. In *Uranium-series Disequilibrium: Applications to Earth, Marine, and Environmental Sciences* (eds. M. Ivanovich and R. S. Harmon). Clarendon Press, Oxford, pp. 334–395.

DeMaster D. J. and Turekian K. K. (1987) The radiocarbon record in varved sediments of Carmen Basin, Gulf of California: a measure of upwelling intensity variation during the past several hundred years. *Paleoceanography* **2**, 249–254.

Dickens G. R., O'Neil J. R., Rea D. K., and Owen R. M. (1995) Dissociation of oceanic methane hydrate as a cause of the carbon isotope excursion at the end of the Palaeocene. *Paleoceanography* **10**, 965–971.

Dymond J. (1969) Age determinations of deep-sea sediments: a comparison of three methods. *Earth Planet. Sci. Lett.* **6**, 9–14.

Edwards R. L., Chen J. H., and Wasserburg G. J. (1986–1987) ^{238}U–^{234}U–^{230}Th–^{232}Th systematics and the precise measurement of time over the past 500,000 years. *Earth Planet. Sci. Lett.* **81**, 175–192.

Edwards R. L., Chen J. H., Ku T.-L., and Wasserburg G. J. (1987) Precise timing of the last interglacial period from mass spectrometric determination of thorium-230 in corals. *Science* **236**, 1547–1553.

Edwards R. L., Beck J. W., Burr G. S., Donahue D. J., Chappell J. M. A., Bloom A. L., Druffel E. R. M., and Taylor F. W. (1993) A large drop in atmospheric ^{14}C/^{12}C and reduced melting in the Younger Dryas, documented with ^{230}Th ages of corals. *Science* **260**, 962–968.

Edwards R. L., Cheng H., Murrell M. T., and Goldstein S. J. (1997) Protactinium-231 dating of carbonates by thermal ionization mass spectrometry: implications for quaternary climate change. *Science* **276**, 782–786.

Eisenhauer A., Gögen K., Pernicka E., and Mangini A. (1992) Climatic influences on the growth rates of Mn crusts during the Late Quaternary. *Earth Planet. Sci. Lett.* **109**, 25–36.

Farley K. A. (1995) Cenozoic variations in the flux of interplanetary dust recorded by ^{3}He in a deep-sea sediment. *Nature* **376**, 153–156.

Farley K. A. (2001) Extraterrestrial helium in seafloor sediments: identification, characteristics and accretion rates over geologic time. In *Accretion of Extraterrestrial Matter throughout Earth's History* (eds. B. Peucker-Ehrenbrinck and B. Schmitz). Kluwer, New York, pp. 179–204.

Farley K. A. and Patterson D. B. (1995) A 100-kyr periodicity in the flux of extraterrestrial ^{3}He to the seafloor. *Nature* **376**, 600–603.

Francois R. and Bacon M. P. (1991) Variations in terrigenous input into the deep equatorial Atlantic during the past 24,000 years. *Science* **251**, 1473–1476.

Francois R. and Bacon M. P. (1994) Heinrich events in the North Atlantic: radiochemical evidence. *Deep-Sea Res. I* **41**, 315–334.

Francois R., Bacon M. P., and Suman D. O. (1990) Thorium 230 profiling in deep-sea sediments: high-resolution records of flux and dissolution of carbonate in the equatorial Atlantic during the last 24,000 years. *Paleoceanography* **5**, 761–787.

Frank M., Eisenhauer A., Bonn W. J., Walter P., Grobe H., Kubik P. W., Dittrich-Hannen B., and Mangini A. (1995) Sediment redistribution versus paleoproductivity change: Weddell Sea margin sediment stratigraphy and biogenic particle flux of the last 250,000 years deduced from ^{230}Th$_{ex}$, ^{10}Be and biogenic barium profiles. *Earth Planet. Sci. Lett.* **136**, 559–573.

Friedlander G., Kennedy J. W., Macias E. S., and Miller J. M. (1981) *Nuclear and Radiochemistry*, 3rd edn. Wiley-Interscience, New York.

Gallup C. D., Edwards R. L., and Johnson R. G. (1994) The timing of high sea levels over the past 200,000 years. *Science* **263**, 796–800.

GEOSECS (1987) *Atlantic, Pacific, and Indian Ocean Expeditions, 7: Shorebased Data and Graphics*. National Science Foundation, Washington.

Goldberg E. D. and Bruland K. W. (1974) Radioactive geochronologies. In *The Sea, 5* (ed. E. D. Goldberg). Wiley-Interscience, New York, pp. 451–489.

Goldstein S. J., Lea D. W., Chakraborty S., Kashgarian M., and Murell M. T. (2001) Uranium-series and radiocarbon geochronology of deep-sea corals: implications for Southern Ocean ventilation rates and the oceanic carbon cycle. *Earth Planet. Sci. Lett.* **193**, 167–182.

Guichard F., Reyss J.-L., and Yokoyama Y. (1978) Growth rate of manganese nodule measured with ^{10}Be and ^{26}Al. *Nature* **272**, 155–156.

Guilderson T. P., Fairbanks R. G., and Rubenstone J. L. (1994) Tropical temperature variations since 20,000 years ago: modeling interhemispheric climate change. *Science* **263**, 663–665.

Halbach P., Segl M., Puteanus D., and Mangini A. (1983) Co-fluxes and growth rates in ferromanganese deposits from Central Pacific seamount areas. *Nature* **304**, 716–719.

Henderson G. M. (2002) Seawater (^{234}U/^{238}U) during the last 800 thousand years. *Earth Planet. Sci. Lett.* **199**, 97–110.

Henderson G. M. and Burton K. W. (1999) Using (^{234}U/^{238}U) to assess diffusion rates of isotope tracers in ferromanganese crusts. *Earth Planet. Sci. Lett.* **170**, 169–179.

Henderson G. M., Cohen A. S., and O'Nions R. K. (1993) ^{234}U/^{238}U ratios and ^{230}Th ages for Hateruma Atoll corals: implications for coral diagenesis and seawater ^{234}U/^{238}U ratios. *Earth Planet. Sci. Lett.* **115**, 65–73.

Huh C.-A. and Kadko D. C. (1992) Marine sediments and sedimentation processes. In *Uranium-series Disequilibrium: Applications to Earth, Marine, and Environmental Sciences* (eds. M. Ivanovich and R. S. Harmon). Clarendon Press, Oxford, pp. 460–486.

Huh C.-A. and Ku T.-L. (1984) Radiochemical observations on manganese nodules from three sedimentary environments in the north Pacific. *Geochim. Cosmochim. Acta* **48**, 951–963.

Huh C.-A. and Ku T.-L. (1990) Distribution of thorium 232 in manganese nodules and crusts: paleoceanographic implications. *Paleoceanography* **5**, 187–195.

Huntley D. J. and Johnson H. P. (1976) Thermoluminescence as a potential means of dating siliceous ocean sediments. *Can. J. Earth Sci.* **13**, 593–596.

Ivanovich M. and Harmon R. S. (ed.) (1992). *Uranium Series Disequilibrium: Applications to Earth, Marine, and Environmental Problems*. Clarendon Press, Oxford.

Ivanovich M., Latham A. G., and Ku T.-L. (1992) Uranium-series disequilibrium applications in geochronology. In *Uranium-series Disequilibrium: Applications to Earth, Marine, and Environmental Sciences* (eds. M. Ivanovich and R. S. Harmon). Clarendon Press, Oxford, pp. 62–94.

Kaufman A., Broecker W. S., Ku T.-L., and Thurber D. L. (1971) The status of U-series methods of mollusk dating. *Geochim. Cosmochim. Acta* **35**, 1155–1183.

Kennett J. P. and Ingram B. L. (1995) A 20,000-year record of ocean circulation and climate change from Santa Barbara basin. *Nature* **377**, 510–514.

Kim K. H. and Burnett W. C. (1986) Uranium-series growth history of a quaternary phosphatic crust from the Peruvian continental margin. *Chem. Geol.* **58**, 227–244.

Koide M., Soutar A., and Goldberg E. D. (1972) Marine geochronology with ^{210}Pb. *Earth Planet. Sci. Lett.* **14**, 442–446.

Kress A. G. and Veeh H. H. (1980) Geochemistry and radiometric ages of phosphatic nodules from the continental margin of northern New South Wales. *Australia. Mar. Geol.* **36**, 143–157.

Krinsley D., Biscaye P. E., and Turekian K. K. (1973) Argentine Basin sediment sources as indicated by quartz surface textures. *J. Sedim. Petrol.* **43**, 251–257.

Krishnaswami S. and Cochran J. K. (1978) Uranium and thorium series nuclides in oriented ferromanganese nodules: growth rates, turnover times and nuclide behavior. *Earth Planet. Sci. Lett.* **40**, 45–62.

Krishnaswami S., Mangini A., Thomas J. H., Sharma P., Cochran J. K., Turekian K. K., and Parker P. D. (1982) ^{10}Be and Th isotopes in manganese nodules and adjacent sediments: nodule growth histories and nuclide behavior. *Earth Planet. Sci. Lett.* **59**, 217–234.

Krishnaswami S., Monaghan M. C., Westrich J. T., Bennett J. T., and Turekian K. K. (1984) Chronologies of sedimentary processes in sediments of the FOAM site, Long Island Sound, Connecticut. *Am. J. Sci.* **234**, 706–733.

Ku T.-L. (1976) The uranium-series methods of age determination. *Ann. Rev. Earth Planet. Sci.* **4**, 347–379.

Ku T. L. (1977) Rates of accretion. In *Marine Manganese Deposits* (ed. G. P. Glasby). Elsevier, New York, pp. 249–267.

Ku T. L. and Glasby G. P. (1972) Radiometric evidence for the rapid growth rate of shallow-water, continental margin manganese nodules. *Geochim. Cosmochim. Acta* **36**, 699–703.

Ku T. L., Omura A., and Chen P. S. (1979) ^{10}Be and U-series isotopes in manganese nodules from the central North Pacific. In *Marine Geology and Oceanography of the Pacific Manganese Nodule Province* (eds. J. L. Bischoff and D. Z. Piper). Plenum, , pp. 791–814.

Kyte F. T., Leinen M., Heath G. R., and Zhou L. (1993) Cenozoic sedimentation history of the central North Pacific: inferences from the elemental geochemistry of core LL44-GPC3. *Geochim. Cosmochim. Acta* **57**, 1719–1740.

Lomitschka M. and Mangini A. (1999) Precise Th/U-dating of small and heavily coated samples of deep sea corals. *Earth Planet. Sci. Lett.* **170**, 391–401.

Luo S., Ku T.-L., Wang L., Southon J. R., Lund S. P., and Schwartz M. (2001) ^{26}Al, ^{10}Be and U–Th isotopes in Blake Outer Ridge sediments: implications for past changes in boundary scavenging. *Earth Planet. Sci. Lett.* **185**, 135–147.

Macdougall D. (1971) Fission track dating of volcanic glass shards in marine sediments. *Earth Planet. Sci. Lett.* **10**, 403–406.

Mangini A., Segl M., Bonani G., Hofmann H. J., Morenzoni E., Nessi M., Suter M., Wölfli W., and Turekian K. K. (1984) Mass-spectrometric ^{10}Be dating of deep-sea sediments applying the Zurich tandem accelerator. *Nucl. Instr. Meth. Phys. Res. B* **5**, 353–358.

Mangini A., Segl M., Kudrass H., Wiedicke M., Bonani G., Hofmann H. J., Morenzoni E., Nessi M., Suter M., and Wölfli W. (1986) Diffusion and supply rates of ^{10}Be and ^{230}Th radioisotopes in two manganese encrustations from the South China Sea. *Geochim. Cosmochim. Acta* **50**, 149–156.

Mangini A., Eisenhauer A., and Walter P. (1990) Response of manganese in the ocean to the climatic cycles in the Quaternary. *Paleoceanography* **5**, 811–821.

Marcantonio F., Kumar N., Stute M., Anderson R. F., Seidl M. A., Schlosser P., and Mix A. (1995) A comparative study of accumulation rates derived by He and Th isotope analysis of marine sediments. *Earth Planet. Sci. Lett.* **133**, 549–555.

Marcantonio F., Anderson R. F., Stute M., Kumar N., Schlosser P., and Mix A. (1996) Extraterrestrial ^{3}He as a

tracer of marine sediment transport and accumulation. *Nature* **383**, 705–707.

Marcantonio F., Higgins S., Anderson R. F., Stute M., Schlosser P., and Rasbury E. T. (1998) Terrigenous helium in deep-sea sediments. *Geochim. Cosmochim. Acta* **62**, 1535–1543.

Marcantonio F., Turekian K. K., Higgins S., Anderson R. F., Stute M., and Schlosser P. (1999) The accretion rate of extraterrestrial ^{3}He based on oceanic ^{230}Th flux and the relation to Os isotope variation over the past 200,000 years in an Indian Ocean core. *Earth Planet. Sci. Lett.* **170**, 157–168.

Marcantonio F., Anderson R. F., Higgins S., Stute M., and Schlosser P. (2001) Sediment focusing in the central equatorial Pacific Ocean. *Paleoceanography* **16**, 260–267.

McCaffrey R. J. and Thomson J. (1980) A record of the accumulation of sediment and trace elements in a Connecticut salt marsh. *Adv. Geophys.* **22**, 165–236.

McCulloch M., Mortimer G., Esat T., Xianhua L., Pillans B., and Chappell J. (1996) High resolution windows into early Holocene climate: Sr/Ca records from the Huon Peninsula. *Earth Planet. Sci. Lett.* **138**, 169–178.

McManus J. F., Anderson R. F., Broecker W. S., Fleisher M. Q., and Higgins S. M. (1998) Radiometrically determined sedimentary fluxes in the sub-polar North Atlantic during the last 140,000 years. *Earth Planet. Sci. Lett.* **155**, 29–43.

Mesolella K. J., Matthews R. K., Broecker W. S., and Thurber D. L. (1969) The astronomical theory of climate change: Barbados data. *J. Geol.* **77**, 250–274.

Moore W. S. (1984) Thorium and radium isotopic relationships in manganese nodules and sediments at MANOP Site S. *Geochim. Cosmochim. Acta* **48**, 987–992.

Moore W. S. and Vogt P. R. (1976) Hydrothermal manganese crusts from two sites near the Galapagos spreading axis. *Earth Planet. Sci. Lett.* **29**, 349–356.

Moore W. S., Ku T.-L., MacDougall J. D., Burns V. M., Burns R., Dymond J., Lyle M. W., and Piper D. Z. (1981) Fluxes of metals to a manganese nodule: radiochemical, chemical, structural, and mineralogical studies. *Earth Planet. Sci. Lett.* **52**, 151–171.

Muhs D. R. and Szabo B. J. (1994) New uranium-series ages of the Waimanalo Limestone, Oahu, Hawaii: implications for sea level during the last interglacial period. *Mar. Geol.* **118**, 315–326.

Nozaki Y., Cochran J. K., Turekian K. K., and Keller G. (1977) Radiocarbon and ^{210}Pb distribution in submersible-taken deep-sea cores from Project FAMOUS. *Earth Planet. Sci. Lett.* **34**, 167–173.

Ozima M., Takayanagi M., Zashu S., and Amari S. (1984) High ^{3}He/^{4}He ratio in ocean sediments. *Nature* **311**, 448–450.

Palike H., Shackleton N. J., and Rohl U. (2001) Astronomical forcing in Late Eocene marine sediments. *Earth Planet. Sci. Lett.* **193**, 589–602.

Raisbeck G. M., Yiou F., Fruneau M., Loiseau J. M., and Lieuvin M. (1979) ^{10}Be concentration and residence time in the oceans. *Earth Planet. Sci. Lett.* **43**, 237–240.

Roe K. K., Burnett W. C., and Lee A. I. N. (1983) Uranium disequilibrium dating of phosphate deposits from the Lau Group, Fiji. *Nature* **302**, 603–606.

Saffert H. L. and Thomas E. (1998) Living foraminifera in salt marsh peat cores: Kelsey marsh (Clinton, CT) and the Great Marshes (Barnstable, MA). *Mar. Micropaleontol.* **33**, 175–202.

Shackleton N. J., Hagelberg T. K., and Crowhurst S. J. (1993) Evaluating the success of astronomical tuning: Pitfalls of using coherence as a criterion for assessing Pre-Pleistocene timescales. *Paleoceanography* **10**, 693–697.

Shackleton N. J., Hall M. A., Raffi I., Tauxe L., and Zachos J. (2000) Astronomical calibration age for the Oligocene–Miocene boundary. *Geology* **28**, 447–450.

Sharma M. (2002) Variations in solar magnetic activity during the lst 200,000 years: is there a Sun-climate connection? *Earth Planet. Sci. Lett.* **199**, 459–472.

Stirling C. H., Esat T. M., McCulloch M. T., and Lambeck K. (1995) High-precision dating of corals from Western Australia and implications for the timing and duration of the Last Interglacial. *Earth Planet. Sci. Lett.* **135**, 115–130.

Suess H. E. (1956) Absolute chronology of the last glaciation. *Science* **123**, 355–357.

Suman D. O. and Bacon M. P. (1989) Variations in Holocene sedimentation in the North American Basin determined from Th-230 measurements. *Deep-Sea Res.* **36**, 869–878.

Szabo B. J., Ludwig K. R., Muhs D. R., and Simmons K. R. (1994) Th-230 ages of corals and duration of the Last Interglacial sea level high stand on Oahu Hawaii. *Science* **266**, 93–96.

Tanaka N., Turekian K. K., and Rye D. M. (1991) The radiocarbon, $\delta^{13}C$, ^{210}Pb, and ^{137}Cs record in box cores from the continental margin of the Middle Atlantic Bight. *Am. J. Sci.* **291**, 90–105.

Tanaka S. and Inoue T. (1979) ^{10}Be dating of north Pacific sediment cores up to 2.5 million years B.P. *Earth Planet. Sci. Lett.* **45**, 181–187.

Tanaka S., Inoue T., and Imamura M. (1977) The ^{10}Be method of dating marine sediments—comparison with the paleomagnetic method. *Earth Planet. Sci. Lett.* **37**, 55–60.

Thomas E., Turekian K. K., and Wei K.-Y. (2000) Productivity control of fine particle transport to equatorial Pacific sediment. *Global Biogeochem. Cycles* **14**, 945–955.

Thomson J., Higgs N. C., and Clayton T. (1995) A geochemical criterion for the recognition of Heinrich events and estimation of their depositional fluxes by the $(^{230}Th_{excess})_0$ profiling method. *Earth Planet. Sci. Lett.* **135**, 41–56.

Thomson J., Nixon S., Summerhayes C. P., Schönfeld J., Zahn R., and Grootes P. (1999) Implications for sedimentation changes on the Iberian margin over the last two glacial/interglacial transitions from $(^{230}Th_{excess})_0$ systematics. *Earth Planet. Sci. Lett.* **165**, 255–270.

Turekian K. K. and Cochran J. K. (1978) Determination of marine chronologies with natural radionuclides. In *Chemical Oceanography, 7* (eds. J. P. Riley and R. Chester). Academic Press, London, pp. 313–360.

Turekian K. K. and Stuiver M. (1964) Clay and carbonate accumulation rates in three South Atlantic deep-sea cores. *Science* **146**, 55–56.

Turekian K. K., Cochran J. K., Krishnaswami S., Lanford W. A., Parker P. D., and Bauer K. A. (1979) The measurement of ^{10}Be in manganese nodules using a tandem Van de Graaff accelerator. *Geophys. Res. Lett.* **6**, 417–420.

Varekamp J. C. and Thomas E. (1998) Sea level rise and climate change over the last 1000 years. *EOS* **79**, 69–75.

Veeh H. H. and Chappell J. (1970) Astronomical theory of climatic change: support from New Guinea. *Science* **167**, 862–865.

Walter H.-J., Rutgers van der Loeff M. M., and François R. (1999) Reliability of the $^{231}Pa/^{230}Th$ activity ratio as a tracer for bioproductivity of the ocean. In *Use of Proxies in Paleoceanography: Examples from the South Atlantic* (eds. G. Fischer and G. Wefer). Springer, pp. 393–408.

Wintle A. G. and Huntley D. J. (1979) Themoluminescence dating of a deep-sea sediment core. *Nature* **279**, 710–712.

Yang H.-S., Nozaki Y., Sakai H., and Masuda A. (1986) The distribution of ^{230}Th and ^{231}Pa in the deep-sea surface sediments of the Pacific Ocean. *Geochim. Cosmochim. Acta* **50**, 81–99.

Yang Y.-L., Elderfield H., and Ivanovich M. (1990) Glacial to Holocene changes in carbonate and clay sedimentation in the equatorial Pacific Ocean estimated from thorium 230 profiles. *Paleoceanography* **5**, 789–809.

Yu E.-F., Francois R., Bacon M. P., and Fleer A. P. (2001) Fluxes of ^{230}Th and ^{231}Pa to the deep sea: implications for the interpretation of excess ^{230}Th and $^{231}Pa/^{230}Th$ profiles in sediments. *Earth Planet. Sci. Lett.* **191**, 219–230.

Zachos J., Pagani M., Sloan L., Thomas E., and Billups K. (2001) Trends, rhythms, and aberrations in global climate 65 Ma to present. *Science* **292**, 686–692.

Radioactive Geochronometry
ISBN: 978-0-08-096708-0

pp. 377–398

13

Chronometry of Sediments and Sedimentary Rocks

W. B. N. Berry

University of California, Berkeley, CA, USA

13.1 INTRODUCTION

Sedimentary rocks are commonly organized into discrete strata. The strata are composed of materials, diverse particles of inorganic and/or organic origin, that reflect aspects of the environmental conditions under which they got accumulated. Sequences of sedimentary-rock layers were seen and studied initially in cliffs, man-made exposures, and sites where the vegetation was not thick enough to obscure the rock layers. It was in mines, however, that sequences of strata came to be examined closely. Miner's observations of the succession of sedimentary rock layers they saw and quarried below the Earth's surface gave birth to a domain, viz. stratigraphy, and an understanding of sedimentary rocks. Berry (1968, 1987) discusses—from a historical perspective, the geological timescale—how an economic imperative became a significant force in the development of chronometry of sedimentary rocks.

13.2 CHRONOMETRY BASED ON THE FOSSIL RECORD—FIRST STEPS

Coal miners came to regard many layers of sedimentary rock as "friends," commonly giving them interesting names (Stinking Vein, the Dungy Drift, Kingswood Toad, etc.), or in some cases, simply numbers. Miners working in the British coal mines of the late 1700s understood the orderly succession of coal-bearing layers as well as the layers above and below them so well, that even in the dim candlelight of a mine shaft, they could discern where they were in a sequence of layers. Bits of clam or snail shell, wood, or fossils commonly helped them distinguish one layer from another. William Smith—in his professional work as a surveyor—has successfully applied his knowledge of the orderly succession of strata in many mines, and understanding that fossils were useful in distinguishing one stratum from another.

Winchester (2001, p. 71), in his wonderfully incisive discussion on William Smith's life and contribution to geology, pointed out: "The stratigraphical order in which the different types of rock were arranged in the coalfield, as the local miners knew and as William Smith learned from them all too rapidly, had an utterly predictable regularity to it." Winchester (2001, pp. 71–72) deftly portrayed development of Smith's understanding of the consistent orderliness of strata as follows:

> Smith would see and come to know the strata intimately as he saw them one by one, again and again, as the great winding chain (of the bucket in he stood to be lowered down into a mine) lowered him still further down through the measures. Smith learned both from records of the arrangement of strata in coal mining areas, such as that documentated by John Strachey in 1719 (see Fig. 1) and from talks with miners that each coal bed was considered unique. Because miners could distinguish one coal bed from others, they had a basic understanding of the succession of strata in a coal mine, and they could recite the sequence of coal beds from mine entry downwards. Smith became interested in and made observations of rocks intercalated between the coals. For example, Smith saw that certain sandstones bore features similar to those in modern dunes or beaches. He could see that these sands gave way to rocks made of muds similar to those found in the banks or beds of rivers. He noted, as Strachey (1719) had before him, (see Fig. 1), that some rock layers were crowded with fossils. Certain of these fossils were similar to those found in rivers, others were similar to creatures living in modern nearshore ocean environments (cockles in Fig. 1), and others resembled modern ferns. Smith realized that, in most mines, a sequence of sandstones–mudstones–coals was repeated over and over, as Strachey depicted (see Fig. 1) and described (1725).

Smith visited a number of coal mines, finding in each the same stratal sequence. The numerous mine tours led him to realize that the miners could recognize individual coal seams by certain unique characteristics, which included the fossil content.

An opportunity to explore further on his observations came when he was employed to conduct surveys needed to plan a route for a canal that would carry coal from the mines to markets. Smith realized that cutting a new lengthy canal would mean slicing open sequences of rock layers he had not studied previously. By close scrutinizing the rock layers in the pathway of the canal, Smith observed rocks, called Lias, lying above red rocks that he had seen above the coal in many mines.

Fossil collecting was an attractive pastime in Europe in the 1700s. Generally, collectors displayed their treasures extracted from diverse rock layers in a glass-fronted cabinet so that visitors could admire these treasures. Seldom were the specific rock layers from which the fossils were obtained noted. The desired goal was to have a collection of visually interesting objects. In some households, the most treasured specimens were given the same esthetic value as fine china. Indeed, china and fossils were displayed side by side. Smith viewed many such collections in the course of duties as a land surveyor. Among the collections he examined in the town of Bath were those of the Reverend Joseph Townsend and the Reverend Benjamin Richardson. Townsend was educated as a doctor and in the training had acquired a considerable interest in science. Those interests led him to purchase an interest in certain Irish mines which, in turn, led to mingling religious activities with observations of natural processes. He acquired a considerable amount of knowledge on rocks and their associated minerals and fossils. Richardson was an avid fossil collector, who met Smith at a meeting of the Agricultural Society. Richardson had assembled an unusually large fossil collection of which he was quite proud. He invited Smith to his home in Bath to see this collection. Smith saw that Richardson's fossil collection was organized on the basis of the types of the organisms. For example, ammonites were in one drawer and corals were in another. Richardson told Smith that he had no knowledge of the rock layers from which his treasured fossils had come. Smith responded by saying that he could arrange the fossils according to the sedimentary-rock layers from which each had been obtained. That organization, Smith suggested, would enable Richardson to see the progression of life forms from primitive and less advanced to the modern and relatively more sophisticated and advanced. Richardson agreed to Smith's suggested reorganization of his collection. Within about a day's time, Smith had completed it. Richardson was amazed at how rapidly the task had been accomplished. He questioned Smith on how he did it. Smith said that he had several years' experience observing the orderly succession of sedimentary-rock layers and their contained fossils. These observations have been made both beneath the ground in mines and above the ground in various exposures, including those made in cutting through rock layers for canals. Smith pointed out that as he went to different parts of Britain, the same rock layers occurred in the same superpositional order and each stratum bore essentially the same fossils in each area studied. Because the fossil content of each layer was consistent from area to area, Smith recognized that fossils could be used to identify the presence of any given rock layer. Smith was well aware that miners had been doing essentially that for many years in their quest for economically valued resources.

Richardson realized that Smith's observations were revolutionary. He also understood that they needed confirmation. Richardson contacted his long-time friend Townsend. Together they devised a test of Smith's idea. They pointed to a nearby hill on which a church had been built. They asked Smith to predict the rock layers and the fossil content of each rock layer that lay beneath the church. Smith told them what fossils should be found on the hill slopes and the superpositional order of each of the fossil-bearing rock layers they should encounter as they climbed up the hill. Excitedly, the three men went to the hill, climbed it, and examined the rock layers and collected fossils from many layers. At every step, they confirmed the predictions Smith had made.

The three men dined at Townsend's home in Bath on the evening of June 11, 1799. After dinner, Richardson and Townsend spread a large piece of paper on the dining table and asked Smith to tell them what he knew of the order of strata in the vicinity of Bath and the fossils found in each stratum. Smith recited the orderly superposition of rock layers that he knew so well, from chalk at the top as the youngest to coal at the base as the oldest. He cited 23 rock layers and he indicated the fossil content, the basic rock characteristics and the thickness of each layer. The tabulation that Smith dictated was designated "The Order of Strata and their Embedded Organic Remains, in the vicinity of Bath; examined and proved prior to 1799." One copy of that table of strata is preserved in the files of the Geological Society of London. This table is a milestone in development of an understanding of Earth history and in man's ability to reckon the passage of time. Winchester (2001, p.134) commented on the significance of the table: "For the first time the earth had a provable history, a written record that paid no heed or obeisance to religious teaching and dogma, that declared its independence from the kind of faith that is no more than blind acceptance of absurdity. A science—an elemental basic science that would in due course allow mankind to exploit the almost limitless treasures of the underworld—had at last broken free from the age-old constraints of doctrine and canonical instruction."

The basic principle embodied in the table of strata that Smith dictated to his ecclesiastic associates in June 1799 has come to be known as the principle of faunal (and floral) succession (see discussions in Berry, 1968 and 1987 and in Kleinpell, 1979). Smith used that principle to predict what rock layers and the fossils they contained lay unseen underground in his everyday work. Smith realized that as searches were carried out for more coal and other natural resources—e.g, ores, building stones, sand and gravel, etc.—the principle would be invaluable in finding what lies underneath the surface of the earth. Predictions on the underneath strata at any site could be based on analyses of rocks and faunas.

The principle of faunal succession proved economically valuable in the quest and recovery of many natural resources used by mankind. The principle also led to the understanding that certain fossil aggregates are unique in the time of their occurrence and, therefore, could be used as the basis for a time unit. Two elements were required for such a use: (i) uniqueness of the fossil aggregate and (ii) its position in the overall succession of rock layers. The time units founded upon such fossil aggregates in the years after recognition of the principle of faunal succession were defined relatively broadly and even somewhat vaguely. However, they were fit together to form the basic elements of a geological time scale still in use. Darwin used that timescale in his discussion of organic evolution through natural selection.

Although Smith was encouraged by his friends to publish his table of strata and the information on which it was based, he was not able to do so. He had to continue his active professional work as a surveyor to provide a living for his family. As he pursued his surveying duties, however, he added to his knowledge the rock and fossil successions in previously unexplored areas. His ever-expanding knowledge of sedimentary rock sequences and the fossils found in many rock layers led him to set the goal of using his knowledge to make a geological map of England and Wales. That map would be based upon faunal succession and the active tracing of such rock layers. Smith envisioned a colored map that would depict the rock units as they are spread throughout England and Wales. Inasmuch as most of the rocks, he wished to depict, lay unseen under forests, plowed fields, ponds, and streams, he had to base his map upon his principle of faunal succession. The basic observations were time-consuming and, commonly, relatively tedious. Nevertheless, Smith pressed onward with his project on producing a map (see Simon Winchester's book *The Map that Changed the World* for a comprehensive and insightful account of Smith's career leading to production of the map). Smith hand-colored initial copies of the map and delivered a copy to the Board of Agriculture on May 23, 1815. Copies became available to the public by August of that year. The map clearly demonstrated the applicability and validity of the principle of faunal (and floral) succession. Not long after the map was published, the fundamental units of a timescale based upon faunal aggregates began to be developed. Berry (1968, 1987) wrote a brief history of that timescale. This history indicates that many of the units were initially relatively broadly defined, being founded upon accumulations of large groups of fossils obtained from relatively large aggregates of strata. Each unit was characterized by its superpositional relationship to subjacent and superjacent units and its unique fossil aggregate.

Initially, Smith's map demonstrated that the fossil content of each unit was essential to its recognition over a wide geographic area. Subsequently, the identity of the contained fossils came on focus for study. Miners knew most of the fossil plants or animal remains that typified the various layers, and they could predict where to dig to find more coal or other resources. Many private citizens interested in collecting some of "nature's wonders" also had a basic understanding of the fossils that were characteristic of certain rock layers. That knowledge was gradually integrated by a number of fossil collectors into a general understanding of the spectrum of fossils that characterized each major rock aggregate, the superpositional relationships of which were known. Fundamentally, most of the units of the timescale came to be recognized by bulk faunal-and/or floral-fossil aggregates obtained from large-scale clusters of sedimentary rocks.

Some of the units, now considered Periods in what is known as the geological timescale, were recognized because their rock or mineral contents were economically valuable. The need for building materials and coal led to numerous detailed studies of sedimentary rocks in many areas. Fossils were obtained from a number of layers of these rocks and were used to identify the positions of the most economically valuable layers, as they were traced from productive areas to prospective sites for more raw materials. The Carboniferous Period as a unit in the geologist's timescale is an example. Coal-bearing strata had been examined in many western European countries for almost a century before Smith clambered down into the mines in Somerset County, England to see the rock layer successions that the local miners knew so well (Figure 1). Miners in other western European countries, notably Belgium, France, and Germany knew the rock-layer superpositional order in mines in which they worked. The Belgian J. J. D'Omalius d'Halloy, described coal seams and subjacent and superjacent strata to the coal seen in Belgian mines. He designated the coal-bearing sequence as the Terrain Bituminifere. A few years later Conybeare and Phillips (1822) wrote a remarkable summary of British geology, *Outlines of the Geology of England and Wales.* Smith's map, available less than seven years earlier than publication of this remarkable summary, provided the essential chronometric understanding for the work. In the Conybeare–Phillips analysis, coal-bearing strata were grouped with superjacent lime rocks and subjacent red sandstones within a Carboniferous Order. Fossils from Carboniferous Order rocks in Britain and Ireland were described, respectively, by John Phillips and Frederick McCoy. Laurent G. de Koninck devoted much of his professional life in recording the fossils from Belgian Carboniferous rocks. Miners were very familiar with the remains of plants they found in the coals. Some of them collected fossil

plants they found interesting or decorative. Some coal-measure plants were obtained by collectors for display in their own cabinets of nature's interesting objects. The French geologist Brogniart published a precise account of certain of these fossil floras in 1821. The many studies of floras and faunas from Carboniferous order rocks led to a fundamental understanding of the bulk aggregate of fossils by which the Carboniferous Period could be recognized as a unique interval in the history of the earth. The superpositional relationships of the group of rocks bearing those fossils had been documented in numerous mines throughout western Europe. Therefore, the fossil content could be acknowledged as indicative of a certain, unique, interval of deep or geologic time.

Giovanni Arduino described rocks that formed certain mountains in Italy as Tertiary because not only did they overlie older rocks but also they contained shell and rock particles that came from older, subjacent rocks. Because the subjacent rocks were termed "Secondary," the rocks above them were called "Tertiary." The Tertiary rocks seen in Italy commonly contained numbers of relatively modern-looking shells of marine clams, snails and sea urchins. Fossils bearing close resemblance have had been reported to be found at many sites in western Europe. Tertiary rocks were recognized widely throughout western Europe. The area around Paris includes one of the most extensively examined successions of Tertiary rocks. Cuvier and Brongniart provided a detailed account of these strata and their contained fossils early in the nineteenth century. Similar faunas were found and recorded in areas near London. The presence of Tertiary fossils and the allure of important scientific "work" in and around Paris as well as in sedimentary rock sequences in French resorts drew attention from Charles Lyell. Although he studied law, earthly processes fascinated Lyell. He became friendly with Roderick Murchison at Geological Society of London meetings, an association that led to an invitation from Murchison to join him and his wife on a geological tour of the continent for several months in 1828. Lyell accepted the invitation. He and the Murchisons traveled across France studying rocks and collecting fossils. Lyell parted from the Murchisons in the Fall of 1828 to tour Italy before returning to Paris, and, eventually, London by the end of February 1829. Lyell's observations of his collections of Tertiary fossils from many sites in France and Italy led him to realize that they could be divided. The divisions could be established upon the degree of modernity of fossil clam and snail shells. Lyell's idea for division was founded upon analyses of a large number of marine-mollusc shells obtained from many sedimentary-rock successions from a number of areas. Lyell did not consider possible boundaries between the divisions he proposed.

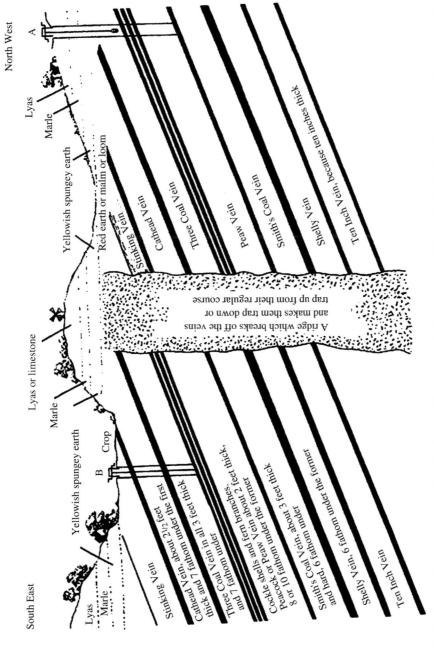

Figure 1 John Strachey's sketch of the occurrence of coal beds and adjacent strata in Somerset near Bristol, England. Note that each coal bed is named and that the occurrence of fossils ("cockle shells and fern branches") is indicated (source Strachey, 1719).

Lyell (1833, pp. xii–xiii) stated in his introduction to the third volume of his *Principles of Geology* that by January 1829 he:

> had fully decided on attempting to establish four subdivisions of the great Tertiary epoch, the same which are fully illustrated in the present work. I considered the basin of Paris and London to be the type of the first division; the beds of the Superga, of the second; the Subapennine strata of northern Italy, of the third; and Ischia and Val di Noto, of the fourth.

Lyell found that most Tertiary sedimentary rocks bore abundant fossils, enabling him to collect masses of fossils from virtually every site visited. When he arrived in Paris, Lyell told Jules Desnoyers of his thoughts on Tertiary divisions. Desnoyers commented to Lyell that Gerard Deshayes had reached similar conclusions based upon his studies of fossil clam and snail shells housed in museum collections. Lyell contacted Deshayes and persuaded him to examine more than 4×10^4 specimens of Tertiary and modern clam and snail shells that represented more than 8,000 species. Deshayes compiled tables of the species using the stratigraphic position of each as the guide. Initially, Lyell and Deshayes agreed to three Tertiary divisions, which were, from oldest to youngest, Eocene, Miocene, and Pliocene. Subsequently, Lyell divided the Pliocene to an older and a newer Pliocene. The divisions were based upon the percentage of still-living marine-mollusc species. The percentages had relatively broad ranges and the faunas were so numerous that the percentage of still-living marine-mollusc species method for dividing the Tertiary came to be widely used across Europe. Darwin found that he could use it as well in his studies of Tertiary sedimentary rocks and their fossil content in South America.

Lyell's newer Pliocene contained ∼90% still-living species. Lyell (1833, p. 54) noted that in his Older Pliocene "the proportion of recent species varies from upwards of a third to somewhat more than half the entire number." The Miocene was typified by "rather less than eighteen to one hundred" still extant species (Lyell, 1833, p. 54), and the Eocene had ∼3% and 0.5% extant mollusc species. In 1846, Edward Forbes studied floral and faunal changes that took place during the last glaciation and used the designation Pleistocene for that glacial interval. Pleistocene quickly thereafter became the designation used by most students of the last major glaciation. Lyell concluded that the newer Pliocene was synonymous with Pleistocene and made that change in print in 1873. Lyell (1833) commented that his units might be divided through more precise floral and faunal studies. Divisions of certain of Lyell's Tertiary divisions were recognized, as predicted. The lower part of Lyell's Eocene was split off to form

the Paleocene by Schimper in a study of fossil floras in 1874. The sedimentary rocks and fauna of Lyell's upper part of the Eocene were designated as Oligocene by von Beyrich in 1854.

13.3 REFINEMENTS IN CHRONOMETRY USING FOSSILS

Lyell (1833, p. 59) analyzed the stratigraphic occurrences of species from the lower to upper part of the Tertiary stratigraphic succession and commented:

> This increase of existing species, and gradual disappearance of the extinct, as we trace the series of formations from the older to the newer, is strictly analogous, as we before observed, to the fluctuations of a population such as might be recorded at successive periods, from the time when the oldest of the individuals now living was born to the present moment. The disappearance of persons who never were contemporaries of the greater part of the present generation, would be seen to have kept pace with the birth of those who now rank amongst the oldest men living, just as the Eocene and Miocene species are observed to have given place to those Pliocene testacea which are now contemporary with man.

Lyell's remarks foreshadowed the next steps in developing a timescale using fossils.

The French paleontologist Alcide d'Orbigny proposed stages as chronometric divisions that were more precise than Lyell's Tertiary divisions. D'Orbigny was engaged in preparing a comprehensive analysis of fossils and rocks in which they were found in France during the 1830s and 1840s. In the course of that work, he became dismayed at the plethora of local rock-unit names and successions of fossiliferous strata that had been described in exquisite detail. What concerned d'Orbigny was that few of the successions as well as local rock units had been related or correlated with each other. In his monumental compendium on Jurassic strata, *Terrains Jurassiques*, d'Orbigny (1842, p. 9) commented:

> Geologists, in their classifications permit themselves to be influenced by mineralogic composition of beds, whereas I take for my point of study ... the annihilation of an assemblage of organisms and replacement by another. I proceed solely on the identity of faunal composition ...

D'Orbigny recognized that the similarity of faunal assemblages was the key to correlating the many sedimentary-rock units and their contained faunas. Furthermore, he was well aware that certain groups of strata were characterized by aggregates of fossils that were unique to them. A fossil aggregate might occur in a single layer or in several layers of rock. The essential ingredient to d'Orbigny was that each aggregate could be recognized over a broad area. Furthermore, d'Orbigny asserted that in his studies of Jurassic rocks throughout France, the same aggregates of fossils always occurred in the same superpositional relationships. D'Orbigny gave the designation *stage* to each group of strata that bore the same fossil aggregate. He named most of the stages after geographic sites at which rocks bearing many of the fossils that typified each stage could be found.

D'Orbigny consulted geologists/paleontologists who had collected fossil aggregates closely similar to those he found in France in other parts of Europe as well as in Russia, India, and central America. Their comments suggested to him that the fossil aggregates characteristic of most stages could be widely found outside France. That idea led d'Orbigny to the conclusion that the faunas that characterized each stage were divisions that "nature has delineated with bold strokes across the whole earth." Comprehensive studies of Jurassic faunas in countries outside France demonstrated that d'Orbigny was overenthusiastic in his assertion that his stage faunas were global in their distribution. Nonetheless, most elements in his faunal aggregates could be recognized relatively widely across some of western Europe and, therefore, many of d'Orbigny's stage names were used by practicing geologists in many countries. Ultimately, because at least a few fossils in aggregates that characterize each stage were found in many different areas across the globe, stages came to be accepted as the basic unit for international correlations and chronometric discussions.

Hancock (1977, p. 11) analyzed d'Orbigny's concept of stage, and concluded that each of d'Orbigny's stages "is the major body of strata less than a system (anywhere in the world) that contains at least some of a long list of fossils which are peculiar to that piece of the total stratigraphic column." Hancock's (1977) analyses of d'Orbigny's stages makes clear that each stage is based on its fossil content and that the faunas that characterize each stage occur in a stratigraphic superpositional context.

The basic ingredient for making as precise, short-duration chronometric time units as is possible using fossils in superpositional order was elucidated by the German geologist F. A. Quenstedt. He advocated careful, detailed measurements of sequences of rock layers coupled with precise collecting of fossils layer by layer (see discussion in Berry, 1968). When Quenstedt examined occurrences of fauna in Jurassic strata in German sites, he found that he had difficulty using d'Orbigny's stages because certain fossil taxa indicated by d'Orbigny to be found only in one stage could be found occurring with fauna indicative of the subjacent and/or superjacent stage fauna. Quenstedt criticized d'Orbigny's work because it did not involve precise measurements of strata and careful positioning of each fossil collected from each stratal layer. Quenstedt found that many of the associations of fossils said by d'Orbigny to characterize a stage were actually faunal aggregates from a cluster of strata. Quenstedt maintained that geologists interested in precise chronometry should make a very detailed study of all the layers through which any fossil species could be found. Indeed, Quenstedt suggested, rock layer and faunal studies should be conducted at the scale of centimeters. Such detailed observations should be made in as many rock layer sequences as possible in any given area. After such observations had been made, Quenstedt believed that such analyses would lead, through comparison of the ranges of many species in many stratigraphic sections at several localities over a somewhat broad area, to a relatively precise understanding of the succession of individual species. That understanding, Quenstedt indicated, would result in relatively precise chronometry.

Precise measurements of rock-layer successions and collection of faunas, as Quensedt advocated, were being carried out by many geologists working in the 1830s and 1840s in a number of areas in Europe. The majority of these studies involved collecting ammonites from Jurassic strata (see Arkell, 1933).

Although Quenstedt asserted that detailed measurements of sequences of rock layers and their contained fossils could lead to relatively precise chronometries, it was one of his students, Albert Oppel, who demonstrated the validity and applicability of his mentor's assertion. Oppel initially studied the Jurassic rocks and faunas in the Schwabian area of the Jura Mountains in Germany, following the precise rock-layer measuring and fossil collecting proclaimed necessary by his mentor. Then, he traveled to many exposures of Jurassic strata in Europe, obtaining and studying fossils, and carefully measuring successions of rock layers. After extensive studies of a number of fossil species and the layers in which each occurred throughout a broad area in western Europe, Oppel formulated the principle which permits establishment of the most-precise, short-duration chronometric units possible using fossil occurrences in layered rocks. Oppel called these units zones. He grouped the zones to broader units, which were essentially stages.

In his studies, Oppel noted as carefully as possible, the ranges of every species he found through all strata in which it occurred in every sequence of rock layers he examined. Oppel (1856–1858, p. 3) pointedly stated that he took care to "investigate the vertical distribution of each individual species at many different places ignoring the mineralogic character of the beds." From analyses of the plots of the vertical ranges of every species in all succession of rock layers examined, he realized that there were groups of strata characterized by closely similar aggregates of fossil species. He believed such bodies were time synchronous over the area in which they occurred. Oppel saw from his plots of vertical occurrences or stratigraphic ranges that some species occurred only in a short vertical distance in a few rock layers. Other taxa occurred through many layers. Still other species were found to have intermediate stratigraphic ranges. The differences among species in their occurrences in rock-layer sequences led to recognition of a pattern of overlapping ranges. Oppel discovered that, using certain appearances of new species, he could quite clearly point out boundaries between rock layers bearing distinctive and unique fossil aggregates. His studies resulted in recognition of a succession of 33 aggregates. He gave the designation *zone* to such bodies of rock bearing a unique aggregate of species. In Oppel's study, the base of each zone was chosen at the initial occurrence of one or a few species and the top of each zone was the base of the superjacent zone.

Hancock (1977, p. 12) pointed out that although the word "zone" had been used by geologists for many years with a number of different meanings, it was Oppel who gave the term a chronometric identity. Hancock (1977, p. 12) went on to state:

Even today a brief perusal of Oppel's book impresses with its spread of detail. In eight separate districts of western Europe, the Jurassic rocks are subdivided into 33 zones correlated on the basis of their fossil content. Oppel's contemporaries outside Germany were completely bowled over; even the French admitted that it was pertinent to France and published a Tabular Summary (Laugel, 1858).

Arkell (1933) wrote a masterful review of the British Jurassic rocks and faunas. In that work, he considered Oppel's zones so vital in understanding British Jurassic history that he provided the following translation of the passage he considered as Oppel's most concise statement of his method of recognizing zones.

Comparison has often been made between whole groups of beds, but it has not been shown that each horizon, identifiable in any place by a number of peculiar and constant species, is to be recognised with the same degree of certainty in distant regions. This task is admittedly a hard one, but it is only by carrying it out that an accurate correlation of a whole system can be assured. It necessarily involves exploring the vertical range of each separate species of the beds; by this means will be brought into prominence those zones which, through the constant and exclusive occurrence of certain species, mark themselves off from their neighbors as distinct horizons. In this way is obtained an ideal profile, of which the component parts of the same age in the various districts are characterised always by the same species (Oppel, 1856, p. 3; translated in Arkell, 1933, p. 16).

Hancock's (1977, p. 12) discussion of chronometry drew attention to Oppel's careful, precise rock-layer measurements and fossil collecting in establishing a method by which the "record of irreversible evolution of life on earth" could be documented. Hancock (1977, p. 12) pointed out that Oppel's complete work was published in the same year that "Alfred Russell Wallace and Charles Robert Darwin read their joint paper to the Linnean Society of London, 'On the Tendency of Species to Form Varieties, and on the Perpetuation of Varieties and Species by Natural Means of Selection.'" Hancock (1977, p. 12) commented that "Oppel himself remarked that the more accurately the fossils are examined and species defined, the greater the number of zonal divisions that could be recognized."

Oppel grouped his zones into stages. He pointed out (Oppel, 1856–1858, pp. 814–815) that "the stages which have been introduced by d'Orbigny were first supposed to represent stages or zones, but only later did it become clear that the majority of his stages could be further subdivided into more zones." Oppel clustered his zones to form groups of zones or stages, finding that, in general, he could recognize and use d'Orbigny's Jurassic Stages. Oppel (1856–1858) did not simply divide d'Orbigny's stages, rather he grouped his zones to form stages. When Oppel determined that certain zones did not fit into one of d'Orbigny's stages the stage definition was modified.

Hancock (1977) pointed out that the methods d'Orbigny and Oppel used were accepted slowly and reluctantly by many geologists. In the United States, e.g., the entities stages and zones were ignored until the 1930s. Many of the fundamental European studies of Jurassic rocks and fossils involved use of zones, founded primarily upon studies of ammonites. Charles Lapworth introduced Oppel's zone methodology to the Paleozoic through his work with graptolites. Lapworth and his students, E. M. R. Wood and Gertrude Elles, carried out the precise stratigraphic measuring advocated by Quenstedt and Oppel in collecting graptolites and examining their stratigraphic ranges in rock-layer sequences throughout the British Isles. Using Oppel's methodology, Elles and Wood delineated nineteen graptolite zones within the Ordovician and Silurian in Britain. Lapworth (1879, p. 3), in a masterful analyses of Lower Paleozoic faunas, commented that "... we have no reliable chronological scale in geology but such as is afforded by the relative magnitude of zoological change." Graptolite zones, documented using Oppel's methodology, proved to be extraordinarily valuable tools in working out geologic structures in the Victoria (Australia) goldfields. Recognition of graptolite zones guided miners to noses of anticlines on which gold was concentrated. Similarly, graptolite zones have been used successfully in the search for gold in Nevada, USA. Both the Australian and Nevada gold occurrences proved to be among the world's most valuable at the time of their exploitation.

Use of zones based upon Oppel's methodology has proven to be of significant economic value in the 1920s and 1930s. On both the North American Gulf Coast and West Coast, oil finding was enhanced through the study of fossil benthic foraminiferans obtained through precise determinations of rock-layer positions of each species. Whole cores of rock layers were obtained in drilling for oil. These cores through numbers of rock layers gave precise rock–layer position of fossils encountered underground. Complete cores from any well proved difficult and expensive to obtain. Development of well side-wall cores reduced the need to obtain complete cores yet they could result in sampling most levels of the strata through which the driller's bit penetrated.

13.4 OIL RECOVERY IN CALIFORNIA USING FOSSIL-BASED CHRONOMETRY

R. M. Kleinpell (1938) employed Oppel's techniques in a study of the occurrences of fossil benthic foraminiferans in Miocene oil-bearing succession in California. Kleinpell entered the domain of oil exploration in California at a time when geologists were somewhat perplexed at how most effectively to ascertain geologic ages of the complexly folded and faulted Tertiary strata that contained California's oil resources (see Berry, 2000). Geologists concerned with finding oil soon realized that fossils were the only tools the oil-well driller had to determine rock unit and rock-layer position information when drilling beneath the ground surface. Nearly all of the rocks encountered in the initial exploration of California's oil-producing areas were fine-grained and didn't have obvious fossils. The problem was, simply, if oil was found in one well at some depth beneath the ground within a relatively thick, homogenous-appearing succession of rocks, how could one identify this oil-bearing position in other wells, given the folded and faulted aspects of California's oil-bearing sedimentary rocks? Fossils simply had to be obtained from the rock layers lying unseen underground to answer this question. In contrast to initial observations of strata encountered in drilling for oil in California, R. M. Kleinpell did see that the tiny shells of benthic foraminiferans are relatively plentiful in these rocks. He examined many rocks in cores from numerous wells drilled for oil. A well-read California geologist, Ralph Reed, was concerned greatly with expanding California's petroleum industry. When he learned that Kleinpell had found multitudes of fossil foraminiferans in

rocks in wells already drilled for oil, he suggested to Kleinpell that the German Jurassic ammonite specialist, Albert Oppel, had published a methodology for analyzing fossil occurrences in precise stratigraphic positions that might "work" in California's oil exploration. Kleinpell studied Oppel's methods as a graduate student at Stanford and, working under the guidance of Hubert Schenck, Kleinpell used Oppel's methodology in his doctoral dissertation study of California's Miocene strata. Those strata were the most highly oil-bearing known at the time.

Kleinpell (1938) followed Oppel's procedures by tabulating the precise stratigraphic level at which every foraminiferan species he identified was found in every well core and every surface section he examined. That information permitted him to ascertain the stratigraphic ranges of benthic foraminiferan species found in mid-Tertiary rocks studied throughout much of central California. In all, Kleinpell analyzed the stratigraphic ranges of ~200 benthic foraminiferan species in ~200 stratigraphic sequences. His study resulted in delineation of six stages in strata considered to be approximately of Miocene age. Two or three zones were recognized within most of the stages. Each zone is recognized by a unique association of species. Every zone bears the name of one species in the unique association.

Documentation of the California Miocene zones and stages comprised Kleinpell's doctoral dissertation at Stanford which he completed in 1933. Upon fulfilling the doctoral degree requirements, Kleinpell went to work as a consulting oil geologist using his zones and stages to find oil successfully (see discussion in Berry (2000)). The book, *Miocene Stratigraphy of California*, which is based on Kleinpell's doctoral dissertation, was published by the American Association of Petroleum Geologists in 1938.

Kleinpell's zones and stages have been used with great success in California oil exploration. As wells were drilled and surface samples taken, foraminiferans were extracted from rock matrix. The associations were studied and species in each identified. Then the species associations found were compared with those unique associations of taxa that characterized each zone. When a match was recognized, the presence of the zone was documented. Certain zones were divided into subzones using the same (Oppel's methodology) procedures used to recognize the zones. Many geologists were employed to sit beside the well being drilled to obtain and prepare samples for foraminiferan study. Careful matching of species associations found at ever-deeper stratigraphic positions led to precise pinpointing of oil-bearing strata. Kleinpell and others involved in oil exploration simply matched associations of species found with those unique associations of taxa that

Kleinpell indicated were characteristic of his zones. Use of Kleinpell's zones and stages resulted in recovery of many millions of barrels of oil from California's oil-bearing strata. A great deal of that oil was used to fuel American and Allies' efforts in World War II. At one time, California's oil, most of which was found using Kleinpell's zones and stages, amounted to about one-fourth of the world's production.

Because it seemed to Kleinpell that, among Americans and certainly among Californians, perhaps only Ralph Reed was aware of Oppel's work, Kleinpell included a number of lengthy quotes in German from Oppel in his 1938 book on the California Miocene stratigraphy. Over the years, based on comments from friends and associates, Kleinpell realized that few, if any, actually read these German passages. Therefore, he (Kleinpell, 1979) translated certain passages he thought were especially pertinent from Oppel's work (1856–1858). Certain portions of the passages Kleinpell (1979) translated are included herein. They include the same passage that Arkell (1933) cited in his work on the British Jurassic. Both Arkell and Kleinpell considered page 3 of Oppel's compendium as Oppel's most significant statement on zones. Kleinpell's (1979) translations are:

Page 3 of Oppel (1856–1858): "... it becomes necessary to explore, without regard to the mineralogic nature of the beds, the vertical distribution of every single species in the various localities, and then to erect those zones which may be distinguished from the adjacent one by the constant and unique occurrences of certain species. Thereby one obtains an ideal profile of the contemporaneous subdivisions which are always again characterized, in different areas, by the same species."

Page 4 "Since individual horizons may often be distinguished one from another more clearly than can one entire stage be distinguished from another stage, I have still retained the groupings of Jurassic strata into stages, since thereby the piecing together of the occurrences of the less well known fossils is facilitated. At the conclusion of every stage, I list those species upon which the classification and correlation of the beds is preeminently based."

Page 13 "... when I erect the zone of Am. raricostatus as the uppermost division of the lower Liassic. Above it begins the first Paxillose (Bel. Elongatus), as well as other species, which characterize the lowest zone of the middle Liassic. Although the paleontologic distinctions between adjacent (juxtaposed) bounding beds of the two stages is seldom pronounced, much as is the case of the distinctions between two neighboring zones within the same stage, so even here these distinctions in most cases can be readily accomplished."

Pages 814–815 "So, first of all we had to distinguish between 3 Jurassic zones. Many of them showed a remarkable uniformity through the entire terrain under consideration here, others are, on the contrary, only recognized with difficulty, partly because of the change in prevailing facies in some localities, ... the stages which have been introduced by D'Orbigny were first supposed to represent stages or zones, but only later did it become clear that the majority of his stages could be further subdivided into more zones. D'Orbigny has almost exclusively selected only locality names for the designation of his Jurassic stages."

In post-World War II America, need for oil grew tremendously. To help meet that need, Kleinpell not only used but also, perhaps more significantly, taught a number of students and professional associates how to use his California version of Oppel's methodology. In time, certain California oil geologists realized that Kleinpell's application of Oppel's methods, which had proven so significant in California petroleum exploration, were being overlooked by many involved in oil exploration. Accordingly, they persuaded Kleinpell to review his ideas and thoughts on Oppel's methods in a work entitled *Criteria in Correlation: Relevant Principles of Science* which was published in 1979 by the Pacific Division of the American Association of Petroleum Geologists. In that work, Kleinpell (1979, p. 12) pointed out:

> Oppel was the first to fully recognize and use the overlapping ranges of a few taxa that remain diagnostic regardless of the particular facies in which they occur. This provides the finest level of refinement in biochronology and is made possible by recognizing the fractions of taxonomic ranges where consistent joint occurrences of such faithful taxa are found. Such units of fractional ranges are called Zones and constitute the bases of the smallest chronologic units that can be designated using sound scientific principle.

Kleinpell (1979, p. 12) stated that zones characterized according to Oppel's method may be recognized throughout a faunal province, and that zones established are not applicable throughout the world "except fortuitously or during the occasional flourishing of cosmopolitan faunas." Kleinpell (1979, p. 12) went on to point out that Davies (1934, p. 56) had described Oppel's zone method in a textbook for oil-field geologists as follows:

> Molluscan species are, as a rule, of small value as zonefossils; but associations of species may be of value. A long-lived species has as contemporaries in its youth other species that preceded it in extinction, and new species arose to be its contemporaries in old age. Consequently, if the time-range of a great number of species were tabulated, it might be found that even small divisions in time were characterized by a particular association or overlap of species, some of which lived no later, others no earlier. Even then, it is not to be supposed that such an overlap was contemporaneous over very wide areas.

Kleinpell (1979, pp. 15–16) concluded his discussion of Oppel's zonation methodology as follows:

> The chronologic dimension of an Oppelian Zone, then, is diagnosed by the unique congregation of fossil species, the vertical stratigraphic ranges of which consistently overlap. The boundaries of such a Zone are two horizons; in paleontological correlations two critical horizons, not simply one, are involved: a *lower* horizon than which the age-diagnostic congregation cannot have occurred any earlier, and an *upper* horizon than which the age-diagnostic congregation cannot have occurred any later. Thus, Oppelian Zonation provides the greatest possible refinement of the general prehistoric time scale that has been made available to us on the basis of paleontological evidence.
>
> The strengths of Opplian Zonation lie in: (1) emphasizing careful induction from many stratigraphic sequences throughout the province so that a disciplined generalization is achieved; (2) use of the refined time-stratigraphic phenomenon of overlapping ranges of selected taxons (congregations); (3) acknowledging the limitations placed on any faunal chronology by the biogeographic facies problem; and (4) minimizing dependence of a Zone's units on ecological factors by considering all biofacies within a province.

Zones and stages based upon Oppel's procedures have been used by students for studying Mesozoic rocks and faunas for many years. Many of these studies involve very detailed analyses of the occurrences and evolution of ammonites. Minutely detailed layer-by-layer ammonite collecting at many European localities at which Jurassic strata are exposed has led to recognition of subzones within certain zones that are based upon ammonite faunas. Oppel predicted that such subzones might be recognized as divisions of his zones using the same basic methodology that he had used to distinguish zones. Callomon (1995) pointed out that certain Jurassic subzones based on ammonite faunas may have had durations of less than a million years (Figure 2).

Harland *et al.* (1990) comprehensively reviewed the development and use of many of the stages recognized in the Paleozoic, Mesozoic, and Cenozoic. They cited certain zonal successions based upon different organisms for each of the systems (Figure 3).

13.5 PRINCIPLES OF CHOROLOGY: THE SCIENCE OF THE DISTRIBUTION OF ORGANISMS

As Kleinpell (1979) pointed out, Oppel made an attempt to consider fossil taxa from a number of different facies which represented different environmental conditions in ascertaining the association or congregation of taxa that are used to denote a zone. Kleinpell (1979, p. 18) stated specifically: "one of the most important elements in Oppelian stages and zones is that ecologic facies are selected before a reliable chronology is attained." Kleinpell (1979) indicated that species from rocks that accumulated under different environmental conditions should be included within the faunas used to characterize a zone. In his discussion of the science of chorology, Kleinpell (1979) attempted to draw attention to the fact that ecologic and biogeographic factors constrain the distribution of organisms and that these constraints must be factored into any attempt to use zones characterized using Oppel's methodology.

Oppel's faunas, which were primarily ammonites, came from a large portion of western Europe. Oppel realized, however, that certain zones could not be recognized throughout the entire area and that, in such instances, only a stage or group of zones, could be recognized. This realization led, ultimately, to an understanding that zones delineated using Oppel's methodology are confined to a biogeographic province.

When d'Orbigny discussed his concept of stage, he thought the faunas that typified each were "global" in distribution. That idea was not consistent with the discoveries made in studies of distribution patterns of modern organisms by Alfred Russell Wallace and many others since the mid 1850s which demonstrated that most organisms are limited in their distribution to certain geographic areas. Modern land and marine organisms are members of faunal or floral aggregates that collectively comprise the fauna or flora of a biogeographic province. Analyses of the distribution of fossil floras and faunas have demonstrated that biogeographic provinces essentially similar to those in the modern world have existed throughout the Phanerozoic. Biogeographic provinces are separated by climatic and/or physical barriers. Obviously, e.g., land masses are barriers to distributions of marine organisms. Among land-dwelling organisms, physical barriers such as mountains and deserts are barriers to distribution. Water temperatures commonly are barriers to distribution of whole marine provincial faunas. Availability of certain food resources or nutrient supplies may serve as barriers to distribution of many organisms.

Biogeographic provinces limit the distributions of those species that characterize zones delineated following Oppel's methodology. Stages, which are characterized by faunas that are more inclusive than the faunas that characterize zones, may not be limited to a single biogeographic province. Stage faunas may occur more widely. Some of them may be recognized throughout a biogeographic region or realm.

Many oil company-sponsored studies conducted in the quest for oil resources demonstrated the great value inherent in assessing environmental and ecologic relationships within the context of stages and zones delineated using Oppel's methodology. One such study is cited herein to

Period	Epoch	Stage			AMMONITE ZONES (from Cope. *et al* 1980 a and b)	DINOFLAGELLATE ZONES (from Woollam and Riding 1983)	NANNOFOSSIL ZONES (from van Hinte, 1978b)
Jurassic	Malm	Tithonian	Volgian (Vol)	Portlandian (Por)	Subcraspedites lamplughi	Gochteodinia villosa	Nannoconus colomi
					Subcraspedites preplicomphalus		
					Subcraspedites primitivus		
					?Titanites (Paracraspehtes) oppressus		
					Titanites anguilormis	Ctenidodinium cumulum/Ctenidodinium panneum	
					Galbanites (Kerberites) kerbeus		
					Galbanites okusensis		
					Glaucolithites glaucolithus		
					Progalbanifes alban!		
			Kimmeridgian	Late (Kim)	Virgatopavlovia fittoni	Glossodinium dimorphum/Dingodinium tuberosum	Parhabdolithus embergeri
					Pavlovia rotunda		
					Pavlovia pallasioides		
					Pectinatites (Pectinatites) pectinatus		
					Pectinafites Arkellites hudlestoni		
					Pect (Virggatosphinctoides) wheatleyensis		
					Pectinatites (Virgato.) scitulus		Watznaueria communis
					Pectinatites (Virgato.) elegans		
		Kimmeridgian (Tth)		Early	Aulacostephanus autissiodorensis		
					Aulacostephanus eudoxus		
					Aulacostephaodies mutabilis	Scnniodinium luridum	
				(Kim)	Rasenia cymodoce		
					Pictonia baylei		
	M/m	Oxfordian		Late	Amoebocews rosenkrantzi	Gonyaulacysta jurassica Scriniodinium crystallinium	Vekshinella stradneri
					Amoeboceras regulare		
					Amoeboceras serratum		
					Amoeboceras glosense		
				Mid	Cardioceras tenuiserratum		
					Cardioceras densiplicatum	Acanthaulax senta	
				Early	Cardioceras cordatum		Actinozygus geometricus / Diadozgus dorsetense
			(Oxf)		Ouenstedtoceras mariae	Wanaea fimbriata	
	Dogger	Callovian			Ouenstedtoceras (Lamberticeras) lamberti	Wanaea thysanota	Discorhabdus jungi
					Peltoceras athleta		Podorhabdus rahla
					Erymnoceras coronatum	Ctenidodinium ornatum/Ctenidodinium continum	Podorhabdus escaigi
					Kosmoceras (Gulielmites) jason		Stephanolithion bigoti
					Sigaloceras calloviense		Stephanolithion hexum
			(Clv)		Macrocephalites (M.) macrocephalus		
		Bathonian		Late	Clydoniceras (Clydoniceras) discus	Ctenidodinium combaziril Ctenidodium sellwoodii	Stephanolithion speciosum var. cotum
					Oppelia (Oxycerites) aspidoides		
					Procerites hodsoni		
				Mid	Morrisiceras (morrisiceras)morrisi		Diazomatolithus lehmani
					Tulites (Tulites) subcontractus		
					Procerites progracilis		
			(Bth)	Early	Asphinctites tenuiplicatus		
					Zigiagiceras (Zigiagiceras) zigzag		
		Bajocian		Late	Parkinsonia parkinsoni	Acanthaulax crispa	
					Strenoceras (Garantiana)garantiana		
					Strenoceras sublurcatum		
				Early	Stephanoceras humphriesianum	Nannoceratopsis gracilis	Stephanolithion speciosum s.s.
					Emileia (Otoites) souzei		
			(Baj)		Witcheiia laeviuscula		
					Hyperlioceras discites		
	Dog	Aalenian		(Aal)	Graphoceras concavum		
					Ludwigia murchisonae		
					Leioceras opalinum		Discorhabdus tubus
	Lias	Toarcian			Dumortieria levesquei	Mancodinium semitabulatum	
					Grammoceras thouaisense		
					Haugia variabilis		
					Hildoceras bifrons		Podorhabdus cylindratus
					Harpoceras falciferum		
			(Toa)		Dactylioceras tenuicostaum		
		Pliensbachian			Pleuroceras spinatum	Luehndea spinosa	
					Amaltheus margaritatus		
					Prodactylioceras davoei		
			(Plb)		Tragophylloceras ibex	Liasidium variabile	Crepidolithus crassus
					Uptonia jamesoni		
		Sinemurian		Late	Echioceras raricostatum		
					Oxynoticeras oxynotum		Palaeopontosphaera dubia
					Asteroceras obtusum		
				Early	Caenisiles turnei		Parhabdolithus liasicus
			(Sin)		Arnioceras semicostamum	Dapcodinium priscum	Parhabdolithus marthae
					Arietites bucklandi		Crucirhabdus primulus
J	Lia	Hettangian		(Het)	Schlotheimia angula		Annulithus arkelli
					Alsatites liasicus		
					Psiloceras planorbis		

(Vertical label spanning Bajocian–Pliensbachian dinoflagellate column: *Nannoceratopsis gracilis*)

Figure 2 Jurassic stages, zones, and subzones recognized using Oppel's methods using stratigraphic ranges of ammonite species. Synchroneity between ammonite-bearing rocks that accumulated in shelf environments and sediments that accumulated in deep-ocean settings is also shown. Zones are based on dinoflagellates and coccolithophores (source Calloman, 1995).

document the economic value of using chronometric units delineated following Oppel's methods. Bandy and Arnal had access to a large volume of detailed stratigraphic, sedimentologic, and fossil-occurrence data from rocks encountered in numbers of wells drilled into oil-bearing Middle Tertiary strata in the San Joaquin Basin in California. They (Bandy and Arnal, 1969) used Kleinpell's zones and stages as chronometric units. Their study was part of an extensive program sponsored by the Gulf Oil Corporation to reconstruct the environmental history of the San Joaquin Basin with the goal of enhancing oil exploration and, ultimately, recovery. The basin

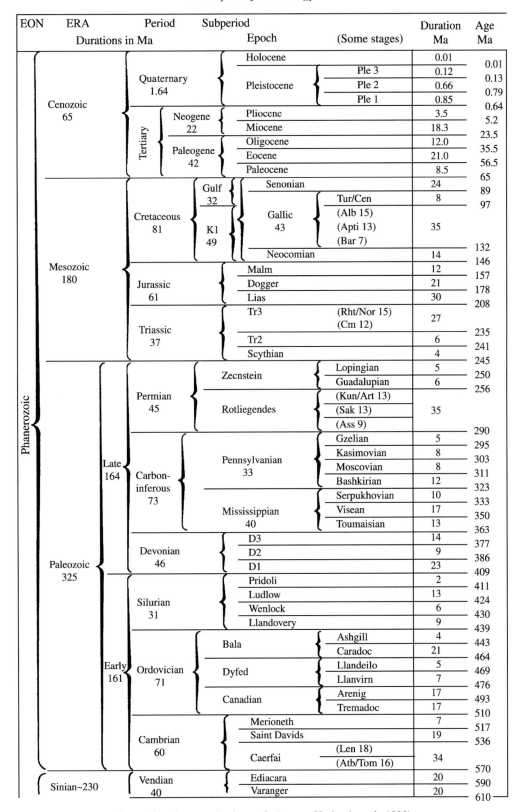

Figure 3 Phanerozoic timescale (source Harland *et al.*, 1990).

lies in the southern part of California's Central Valley. The basin is bounded on the south by the Tehachapi Mountains, on the west by the California Coast Ranges, on the east by the Sierras, and it passes laterally northward into the Sacramento valley.

Bandy and Arnal (1969) examined more 5,000 mid-Tertiary foraminifer-bearing samples taken from 109 wells drilled for petroleum. The foraminifers were identified to species and a chronometric determination was made for every association recovered using Kleinpell's (1938) zones and stages as the chronometric scale (see Figure 4). Each sample was assigned to a zone and stage in the Kleinpell scheme. Most of the zone and stage assignments in the Bandy and Arnal study (see Figures 5–7) made by a consulting firm were headed by Stanley Beck. Then, using studies of the sedimentary constituents of each sample as well as ecologic analyses of the benthic foraminiferan species, Bandy and Arnal (1969) assigned each sample to a biofacies which they suggested was indicative of certain environmental conditions. They (Bandy and Arnal, 1969, p. 787)

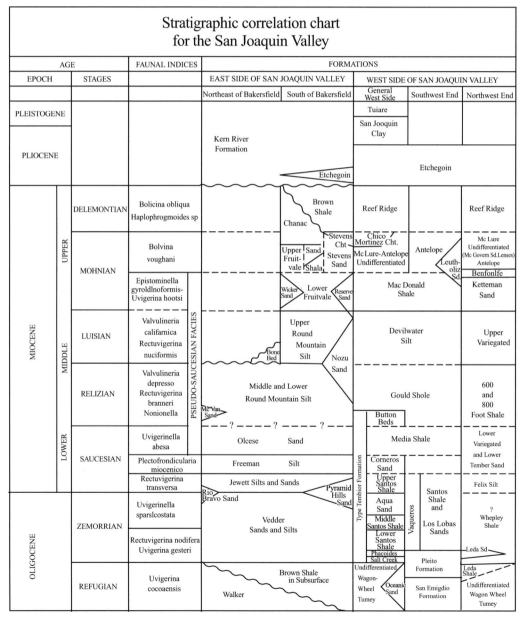

Figure 4 Kleinpell's mid-Tertiary stages and zones and rock unit correlations using these chronometric units for the southern San Joaquin Valley, California. This table was developed by Stanley Beck for the Gulf Oil Company for use by Bandy and Arnal (1969) in their discussion of the mid-Tertiary sedimentary history and petroleum potential of the area.

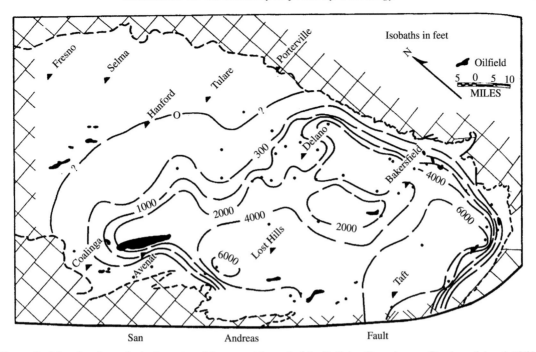

Figure 5 Map showing paleobathymetry of the area at the end of the Relizian Stage (source Bandy and Arnal, 1969, figure 10).

reported that "recent studies have indicated that many important bathyal species have rather similar upper depth limits in contrasting oceanic areas, regardless of differences in temperature, oxygen, and salinity." They compared the fossil foraminiferan associations indicative of a biofacies with those from modern environments and concluded that the biofacies they recognized could be deduced to be indicative of the depth of the seafloor. They (Bandy and Arnal, 1969, p. 787) noted that "examination of the various biofacies in the Tertiary strata of the San Joaquin Basin illustrates the entire gamut from the marsh habitat with a totally arenaceous fauna of *Ammobaculites* and *Ammobaculites* to the other extreme, an abyssal fauna." Because they (Bandy and Arnal, l969) used biofacies as indicative of seafloor depth, they could plot a depth for each association within each stage. Those plots were used to construct paleobathymetric maps for each stage (see Figure 5). They evaluated fossil aggregates obtained from debris flows to indicate sites at which such flows had carried faunas down from shallower environments. Changes in bathymetry during each stage, as well as the areal distribution pattern for each depth interval were discussed. For example, Bandy and Arnal (l969, pp. 797–798) commented in their discussion on the Saucesian stage that "changes in bottom depth during the Saucesian include one prominent area of shoaling along the entire northern side of the basin and a few areas of subsidence generally within the deeper areas. There were 1000 to 2000 feet of shoaling in the

northern area and as much as 2000 feet of deepening in a few places." Inasmuch as the study was of oil-bearing sites, Bandy and Arnal (1969, p. 798) pointed out that "location of oil fields with respect to basin topography suggests that major producing areas are in deeper waters, especially near the steeper slopes of the sea floor and in areas of rapid tectonism." Bandy and Arnal (1969) discussed basin subsidence for each stage, which they termed "paleotectonism." For the Saucesian Stage (see Figures 6 and 7), they inferred that the greatest thicknesses of Saucesian sediments were the sites of greatest subsidence, and that they were in the deepest waters at the south end of the basin. They (Bandy and Arnal, 1969) assessed changes in bathymetry in their analyses of sediment accumulations for each stage. For the Saucesian, e.g., they (Bandy and Arnal, 1969, p. 799) noted that the relevant data showed that "uplift was prominent in at least three areas within the basin." They (Bandy and Arnal, 1969, p. 816) concluded this study by pointing out that the oil-producing strata were "laid down on or near slopes of the basin where there are rather rapid changes in facies."

13.6 CONSTRAINTS ON CHRONOMETRY IMPOSED BY CHOROLOGY

As analyses of the stratigraphic distributions of organisms developed, the need to understand the constraints imposed by chorology on the extent to which zones defined using Oppel's methodology

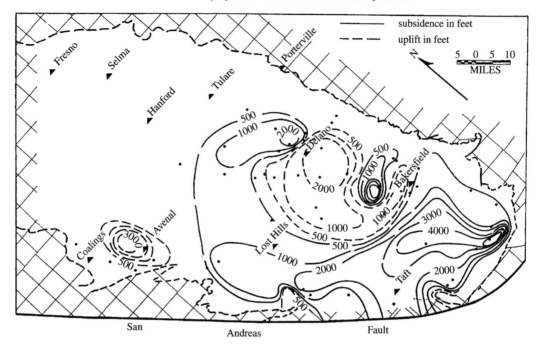

Figure 6   Paleotectonic map of the Saucesian stage indicating areas of uplift and of subsidence (source Bandy and Arnal, 1969, figure 9).

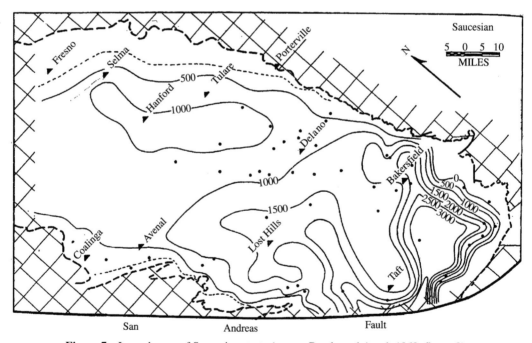

Figure 7   Isopach map of Saucesian strata (source Bandy and Arnal, 1969, figure 8).

could be recognized and used became more and more apparent. In the case of graptolites, e.g., not only were graptolite zones found to be limited to a single faunal province, but also, within a province, they were found to be most recognizable in strata that formed on the margins of shelves or platforms. Rarely are graptolites found in inner shelf settings. Conodonts, an extinct fossil group, are abundant in carbonate rocks that formed in a spectrum of shelf environments, primarily those that lay within the tropics. Conodont zones are limited to a biogeographic province. Within a province, relatively inner- or shallow-shelf fauna differ from deeper-shelf faunas.

Studies of benthic foramineferans revealed certain clusters of taxa which were primarily limited to certain depths on the seafloor (e.g., the faunas that typify the pseudo-Saucesian of Bandy and Arnal (1969) which reflects a deep bathyal seafloor environment). Remains of planktic foraminiferans were primarily recovered from strata that accumulated on the slopes of platforms as well as on the floors of the open ocean. Bandy and Arnal's (1969) study indicated that planktic foraminiferans are relatively rare in most basinal strata. When cores of sediments beneath the ocean floors were taken from many areas, the sediments were found to be composed primarily of shells of organisms that were planktic in life. Shells of planktic foraminiferans were found to be plentiful in many deep ocean-floor sites. Plates of the planktic algae, coccolithoporids, are recovered in large numbers at many deep ocean-floor sites, especially those under tropical oceans. Locally, especially in sediments near the equatorial region of the Pacific, radiolarian shells occur in large numbers. Shells of these three types of planktic organisms occur in stratal succession in cores of wells drilled into deep ocean-floor sediments that are similar to stratal sequences in cores recovered in drilling for oil. The sedimentary record in many deep ocean-floor cores may extend back in time from the present day into the early Tertiary and, in some sites, into the Cretaceous and Jurassic. Shells are so plentiful in most of the deep ocean-floor cores and recovery may be made at such closely spaced intervals that evolutionary development of certain lineages may be traced closely. In initial studies of the deep-ocean sediment record, zones were delineated using essentially Oppel's methodology. As more cores were taken, subzones of these initial zones were recognized, certain of them also following Oppel's procedures. As more and more cores were taken from deep-ocean sediments, specific evolutionary developments in certain taxonomic lineages, e.g., among globorotalid (a type of planktic foraminiferan) species, have been recognized and used to divide the original zones. Additionally, studies of the occurrences of planktic taxa in numerous cores have led to the recognition that the first and last appearances of many species occur at approximately the same stratigraphic level within a province. That realization has led to use of first and last appearance datums as a means of dividing zones. These datums are termed first appearance datums (FADs) and last appearance datums (LADs). The FADs and LADs are considered to be essentially time synchronous within a province. Time of migration of an organism from place to place within a province is thought to have been relatively slight when viewed in the context of geologic time. Datums are used for precise chronometry in studies of Cenozoic sediments (Berggren *et al.*, 1995; Aubry, 1995).

Complete recovery of sediments in cores taken from deep-ocean drilling sites proved difficult to achieve in certain instances, thus care must be taken in examining cores from the deep ocean to ascertain if core recovery has been complete. If not, those levels at which materials are missing are noted and gaps or missing portions of the sediment and fossil record are considered when chronometric analyses are made.

The cores do provide such vast numbers of fossil shells the stratigraphic occurrences of which may be so precisely noted that many divisions of planktic foraminiferan and coccolithophorid zones seem to be only a few tens of thousands of years in duration (see Figure 8). The core record is somewhat similar to stratigraphic sections studied on land and to similar sequences seen in mines, but the age duration of most cores is significantly longer and the fossils in them far more numerous. The deep-ocean sediments bearing these fossiliferous successions lie on top of volcanic rocks the radiometric dates of which may be obtained and the magnetic polarity of which can be determined. The combined record of fossil remains, magnetic polarities, and links to radiometric chronologies provided by drilling into the seafloor opened the way for developing more precise chronometries than seems possible to achieve in most land-based successions.

Studies of fossils recovered from ocean cores indicate that zones based on planktic organisms are limited to biogeographic provinces. These provinces appear to be reflective of latitudinal distributions of water temperatures, and, to some extent, water masses.

13.7 RADIOCHRONOMETRY

Chronometry based on the fossil record gave geologists, especially those involved in the search for natural resources, a valuable tool with which they could direct recovery of natural resources. William Smith demonstrated that the use of chronometric units were used to guide routes for shipping those resources to markets. Being as valuable as the chronometry, it cannot indicate absolute ages of rocks. That development had to wait until about the turn of the 1800s to the 1900s. The French physicist Antoine Henri Becquerel discovered in 1896 that uranium spontaneously emits rays that can cloud photographic plates in the dark. He called this process radioactivity. The American chemist, B. B. Boltwood noted that radioactive decay of uranium leads ultimately to lead. Armed with that observation, Boltwood analyzed uranium minerals obtained from many rocks found at numerous sites around the globe. He found that uranium minerals from older rocks contained more lead than uranium minerals in younger rocks. Using an estimated rate of decay of uranium to lead, Boltwood generated some preliminary ages for certain Paleozoic and older rocks in 1907.

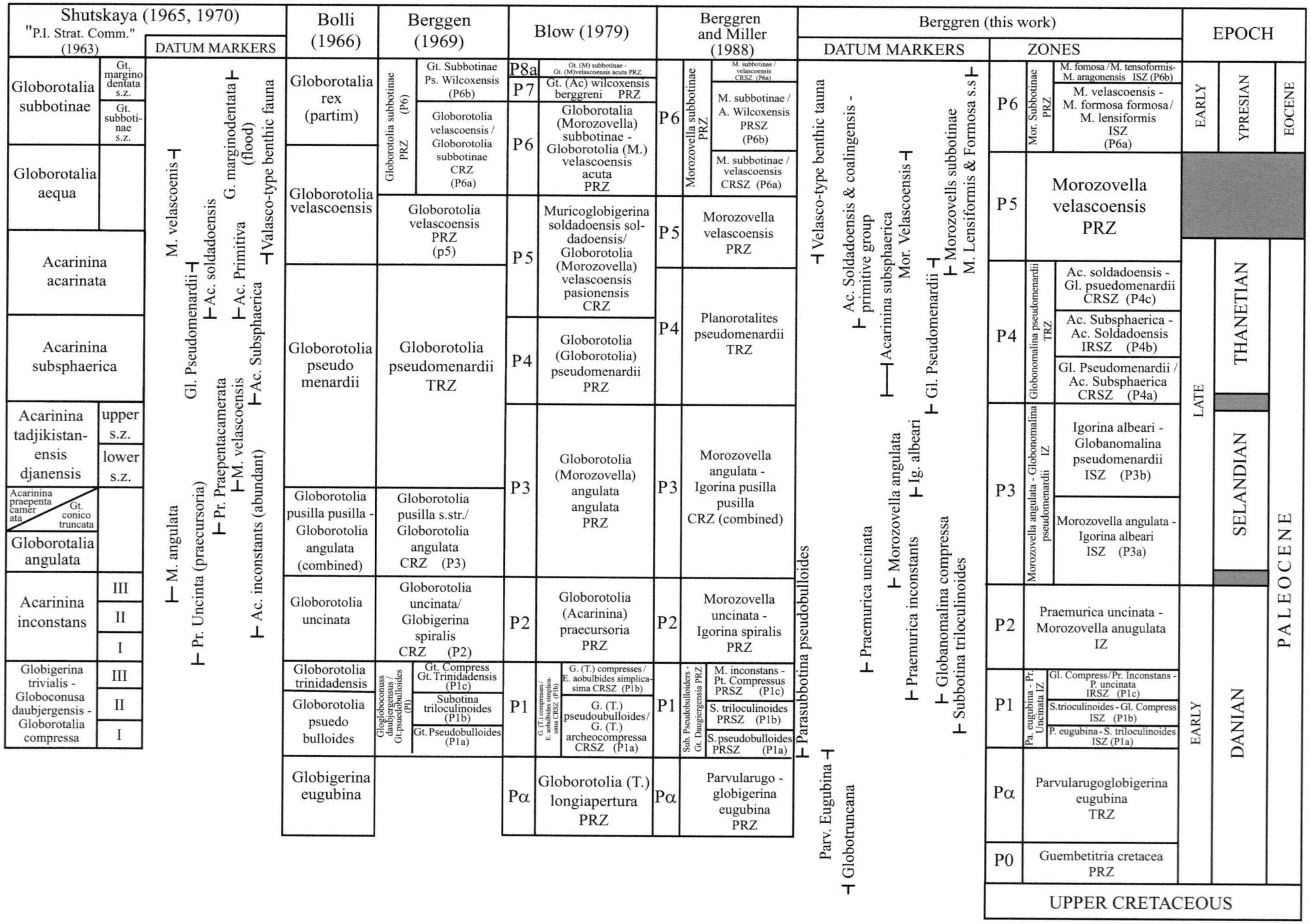

Figure 8 Comparative Paleocene planktic foraminiferan zonations illustrating zones from the work of several authors based on fossil aggregates and datums (source Berggren et al., 1995, figure 7).

Boltwood's assessments were followed by those of Arthur Homes who was able to refine a uranium to lead decay rate and to propose absolute ages for many parts of the timescale based not only upon the fossil record but also older rocks, those that comprise the pre-Phanerozoic or Precambrian. Improvements in technology and in ascertaining naturally occurring daughter elements produced by radiometric decay from a parent element occurring in rocks followed the initial studies. These methods are discussed in significant depth by Harland *et al.* (1990, pp. 73–103). Radiochronometry is discussed in many textbooks. Of them, Boggs (2001, pp. 592–601) is as inclusive as any.

Potassium, which occurs in many volcanic rocks, has an isotope, potassium-40, which decays to argon-39. This parent–daughter ratio proved widely useful in studies of a number of volcanic rocks and became perhaps the most widely used radiochronometric method. It has been used to date many parts of the sedimentary-rock record through analyses of the relationships of volcanic rocks with sedimentary rocks. In an effort to achieve greater precision in radiochronometry, specialists in the study of decay of potassium-40 to argon-39 found that if they converted potassium-39 to argon-39 by irradiation with neutrons in a nuclear reactor they could let argon-39 proxy for potassium-40. To accomplish this, they had to know the ratio of potassium-39 to potassium-40. The process involving irradiation permitted analysis just for argon isotopes, viz., argon-40 and argon-39. Age of the containing mineral and, presumably, rock may then be determined from the argon-40 to argon-39 ratio. The assumption in this procedure is that that the potassium-39 to potassium-40 ratio in nature has been essentially constant and that the amount produced by irradiation depends on the duration of the irradiation and the total amount of potassium present. Inasmuch as the amount of potassium-39 that is converted to argon-39 during irradiation is not known, samples are irradiated with what is called a "fluence monitor mineral" or, in essence, a standard mineral the age of which is known. This procedure allows calculation of an irradiation coefficient which may then be used in calculations of the age of the unknown sample. Clearly, calibration of the "fluence monitor mineral" or standard influences the accuracy of argon-40 to argon-39 ages. In discussing the argon–argon method, Swisher (in Berggren *et al.* (1995, p. 133)) commented that the precision of argon-40/argon-39 ages "far exceeds the accuracy of the age of any currently available monitor mineral." Swisher (in Berggren *et al.* (1995, p. 133)) discussed details of the argon/argon methodology and noted that, for purposes of intercalibration among laboratories, one of two monitor minerals are commonly used. The argon/argon procedure has proven to be valuable because small samplescan be analyzed and corrections may be made for argon leaks from the sample.

Another widely used radiochronometric method involves study of the isotopes of carbon. Studies on these isotopes demonstrated that carbon-14 is taken in consistently by living organisms but that when the organism dies, carbon-14 starts to decay. Thus, a ratio of carbon-14 to the stable carbon isotopes, one of which is carbon-12, can provide an age in years since death of the organism. The ages of organic materials from the present as far back in time to $\sim 5 \times 10^4$ yr of age can be determined, with appropriate corrections for materials somewhat greater than 10^4 yr old.

Sediments have been dated by examination of the potassium/argon ratio in certain glauconites. Glauconite systems take time to close off the loss of argon, thus dates derived from glauconites in sediments are significantly younger than the age of the enclosing sediments.

Another radiochronometric method used with sedimentary rocks is the decay of uranium-238, which decays through several daughters one of which is thorium-230. Uranium-238 is relatively soluble in seawater and so is detectable in seawater analyses. Thorium-230 becomes included in certain minerals that are incorporated into sea-bottom sediment. Thorium-230 decays, with a half-life of 7.5×10^4 yr, to radium-226. Cores of seafloor sediment exhibit a decrease in throrium-230. By comparing the amount of thorium at any given depth in a sediment core with the amount in the surface sediment, an age for different sediment levels in the core can be determined.

Radiochronologic methods are especially valuable for many volcanic and plutonic rocks and organic substances, but, in general, they have limited applicability in determining the absolute ages of sedimentary rocks. Organic materials that occur as components of relatively young sediments may be studied to indicate a minimum age for the rock or sediment. For most sediments and sedimentary rocks, especially those older than a few tens of thousands of years, the absolute age must be ascertained by examination of the relationships with volcanic rocks. For much of the fossil-based timescale, especially the Paleozoic and early Mesozoic, absolute ages have to be worked out by looking at nearby volcanic rocks or, perhaps, obtaining a minimum age from a plutonic rock that cuts through the sediments. Because of experimental errors involved in radiochronologic methodology, zones and some stages, notably those in the Paleozoic, provide more precise estimates of age than radiochronology in many studies of sedimentary rocks.

Dating of the sedimentary and, indeed, the record of other types of rocks for the pre-Phanerozoic or Precambrian essentially requires use of radiochronologic methods. In portions of the Precambrian, notably the latter part, fossil remains of algae and bacteria have been useful in certain cases. For most of the Precambrian, however, radiochronology is essentially the only useful method. Harland *et al.* (1990) reviewed Precambrian radiochronology and ages of Phanerozoic rocks.

13.8 MAGNETIC FIELD POLARITY AND CHRONOMETRY

Remanent magnetism studies developed during the 1960s. The study was led primarily by two groups of geophysicists, one in California and the other in Australia. These scientists studied the remanent magnetism of a number of volcanic rocks on land and discovered that the magnetic field in certain ones of them was the reverse of the modern magnetic field. As they studied volcanic rocks of different ages that had formed over the last few thousand years, they realized that the Earth's magnetic field did reverse itself at intervals throughout the past. From that initial recognition, a sequence of normal and reversed magnetic field polarities was documented. The next step taken was to use potassium–argon radiometric dating methodology to determine the ages of the volcanic rocks the remanent magnetism of which had been determined. Following that step, the absolute or age in years of each reversal could be established. Initially, the polarity reversal scale was developed for only about the last 5 Myr. At the time the dating was carried out, the error percent in the potassium–argon dates was greater than a number of the shorter polarity reversals. Enhanced techniques in the potassium–argon dating method, notably development of the argon–argon method, allowed the polarity reversal scale to be extended back in time into the Cretaceous and even the latter part of the Jurassic.

Although the polarity reversal time scale was based initially on studies of volcanic rocks on land, geophysicists realized that the ocean floors seemingly provided a marvelously continuous record of volcanic activity that extended back in time through the Middle Jurassic. Fluctuations in magnetic field intensity they saw in records of seafloor volcanic rocks came to be interpreted as being indicative of the polarity reversals seen in the volcanic rocks on land. The records obtained in the initial studies started with newly formed volcanic rocks near the crests of submarine oceanic ridges and went from the ridges into older and older rock. The ages of the seafloor volcanic rocks could, to some degree, be determined from the paleontologic ages of the oceanic sediments superjacent to the volcanic-rock basement.

The seafloor magnetic field intensity records began with towing a magnetometer from a ship back and forth over a large segment of the northeastern part of the Pacific Ocean. Analysis of the results from that study revealed that the earth's magnetic field intensities were aligned in relatively long, linear "stripes" that lay parallel with a deep ocean ridge crest from which new crustal material was emanating. This discovery provided the means by which the polarity reversal scale documented on land could be extended back in time and, where possible, calibrated with both radiochronology and fossil-based chronometry.

Originally, the ages of oceanic magnetic field fluctuations were determined by extrapolating their ages based on an assumption of the rate of movement of the volcanic material away from the place it formed, the oceanic ridge crest. Initially, the rate of spreading since the Late Cretaceous was thought to be constant at ~ 1.9 cm yr^{-1}. Subsequent studies indicated that the rate has not been constant but rather that it has decreased from 70 mm yr^{-1} late in the Cretaceous to ~ 32 mm yr^{-1} at present. Furthermore, even that rate of decrease has not been constant.

With the use of sensitive detectors of magnetism, the oceanic-sediment record has come into focus in polarity reversal analyses. Although the magnetism signal is relatively weak in sediment, it can be detected, notably in iron-bearing oceanic sediments. Polarity determinations from oceanic sediment have an advantage because the sediments commonly are fossiliferous. In instances where they are, a paleontologic age may be determined for the same sediment yielding a magnetic field determination. Cores taken from deep-ocean sediments have provided age and magnetic polarity information back in time into about the mid-Jurassic.

Harland *et al.*, 1990, p. 142) commented that the magnetic field reversals seen in the ocean-floor volcanic rock record is the "richest single source of information about magnetic reversals." They went on to point out that "gaps and duplications are commonly present on the profiles" of seafloor volcanic rock magnetic field changes "because of the jumping of ridges to new positions." They also pointed out (Harland *et al.*, 1990, p. 246) that the fossil record in oceanic sediments and/or radiochronologic methods have to be used to calibrate the seafloor magnetic field fluctuations.

Geophysicists number the magnetic polarity reversal scale starting at 1 for the present-day ridge-center volcanism. The polarity time intervals are called chrons. The Cenozoic chrons are designated as C and a number. The Mesozoic chrons are designated M followed by a number. Where extensive studies of remanent magnetism of oceanic materials reveal reversals within a time of normal or reversed polarity, the number of a chron is followed by a letter, e.g., in C5a, C5b, C5c.

Because many of the chrons may be dated paleontologically, study of such chrons has resulted in recognition of an Integrated Magnetobiostratigraphic Scale (IMBS). Berggren *et al.* (1995, pp. 130–132) discussed terminology of chrons and chron divisions (see Figure 9) as components of "The Integrated Magnetobiostratigraphic Scale."

Aubry (1995) drew attention to certain problems that could arise in using only magnetic chronology units. Two normal polarity units could be separated by an unconformity but the unconformity might not be detected. For example, an unconformity could eliminate detection of a normal magnetic unit if two reversed units lay on each side of that unconformity. Aubry (1995, p. 215) pointed out that only by use of the fossil evidence for dating in conjunction with the magnetic reversal data could such uncertainties be eliminated. Aubry stated cogently (1995, p. 215) that "biostratigraphy remains an essential component in the science of stratigraphy. In most instances, magnetozones and stable isotope events can only be confidently identified with the support of biostratigraphy." Both the comments made by Aubry (1995) and by Harland *et al.* (1990, pp. 144–147) concerning uncertainties in calibrating the seafloor magnetic field fluctuations indicate the value of and, in fact, necessity of using the fossil record in sediments to have confidence in magnetic chronometry.

13.9 ORBITAL CHRONOMETRY

Herbert *et al.* (1995) pointed out from a review of Cretaceous to Pleistocene hemipelagites and pelagites that these deposits display cyclical variations that "ripple in an almost unbroken wave train from the Pleistocene Ice Age world into the warm Cretaceous Period." The cyclical deposits commonly display alternations in color, from light to dark, or show up as subtle changes in grain size and sorting. The forces that lead to this cyclical variation appear to be "variations in insolation as a function of latitude and season, cause by quasicyclical changes in the Earth's orbit" (Herbert *et al.*, 1995, p. 81). These changes in the Earth's orbit, commonly called the "Milankovitch cycles," "can be partitioned into the precessional index (modern mean period circa 21 ky), obliquity, (modern mean period circa 41 ky), and eccentricity (modern mean periods of 95, 123, and 413 ky)" (Herbert *et al.*, 1995, p. 81). Herbert *et al.* (1995, p. 81) describe it, using orbital chronometry as a means of providing enhanced refinement of other chronometric methods, as follows:

Just as tick marks of a yardstick measure distance, sedimentary cycles of orbital origin count time at a high precision as compared with most chronological methods. Furthermore, because cyclic marine sediments generally contain stratigraphically useful microfossils and often retain polarity reversal information, the "Milankovitch" Clock could be widely applicable to time scale problems. When anchored accurately in a magnetochronologic and biochronologic framework (which must ultimately be calibrated by radiometric methods), orbital cycles in sediments provide a chronometer that functions at one to two orders of magnitude better resolution than standard techniques."

Having made that general statement of the potential inherent studies of cylicity in hemielagites and pelagites, Herbert *et al.* (1995, p. 82) caution: "The orbital rhythms imprinted on sediments do not constitute perfect chronometers. Repeat times are only quasi-periodic, due to the complicated modulation patterns of the basic cycles." They (Herbert *et al.*, 1995, p. 82) suggested that, at present, orbital cycles may be most useful for dating sediments of late Miocene–Pleistocene age. Herbert *et al.* (1995, p. 82) commented calculations used in celestial mechanics suggest that eccentricity periods "should stay constant over time" and that the "mean periods of obliquity and precessional obliquity should increase gradually over Earth's history due to tidal friction."

Precise times of duration of certain Cretaceous stages (defined by their diagnostic faunas) seem to be possible, based on integration of orbital chronometry with radiometric techniques. Orbital chronometry is a significant component in relatively precise Pliocene–Pleistocene age analyses (Berggren *et al.*, 1995, pp. 131–132).

13.9.1 Aurichorology—The Golden Spikes and Global Statotype Section and Points

One development aimed at creating greater precision in defining boundaries of chronometric units, especially those based on near-shore marine and land dwelling organisms, is the process called "aurichorology" by Walsh (2000). In that process a so-called "golden spike" is driven into the stratigraphic level of the lowest occurrence of a selected taxon of a fauna that characterizes a zone or stage. The stratigraphic section at which this particular occurrence is seen commonly has been selected after considerable careful study. This procedure was initiated in the process of selecting the boundary between the Silurian and Devonian. Intensive study of graptolite zone faunas within that boundary interval focused ultimately on the evolutionary development of a single taxon, *Monograptus uniformis uniformis*, on which to found the base of a graptolite zone as well as the boundary between two systems. That taxon is one of the significant components of the graptolite fauna that characterizes the basal zone

Middle–late miocene timescale

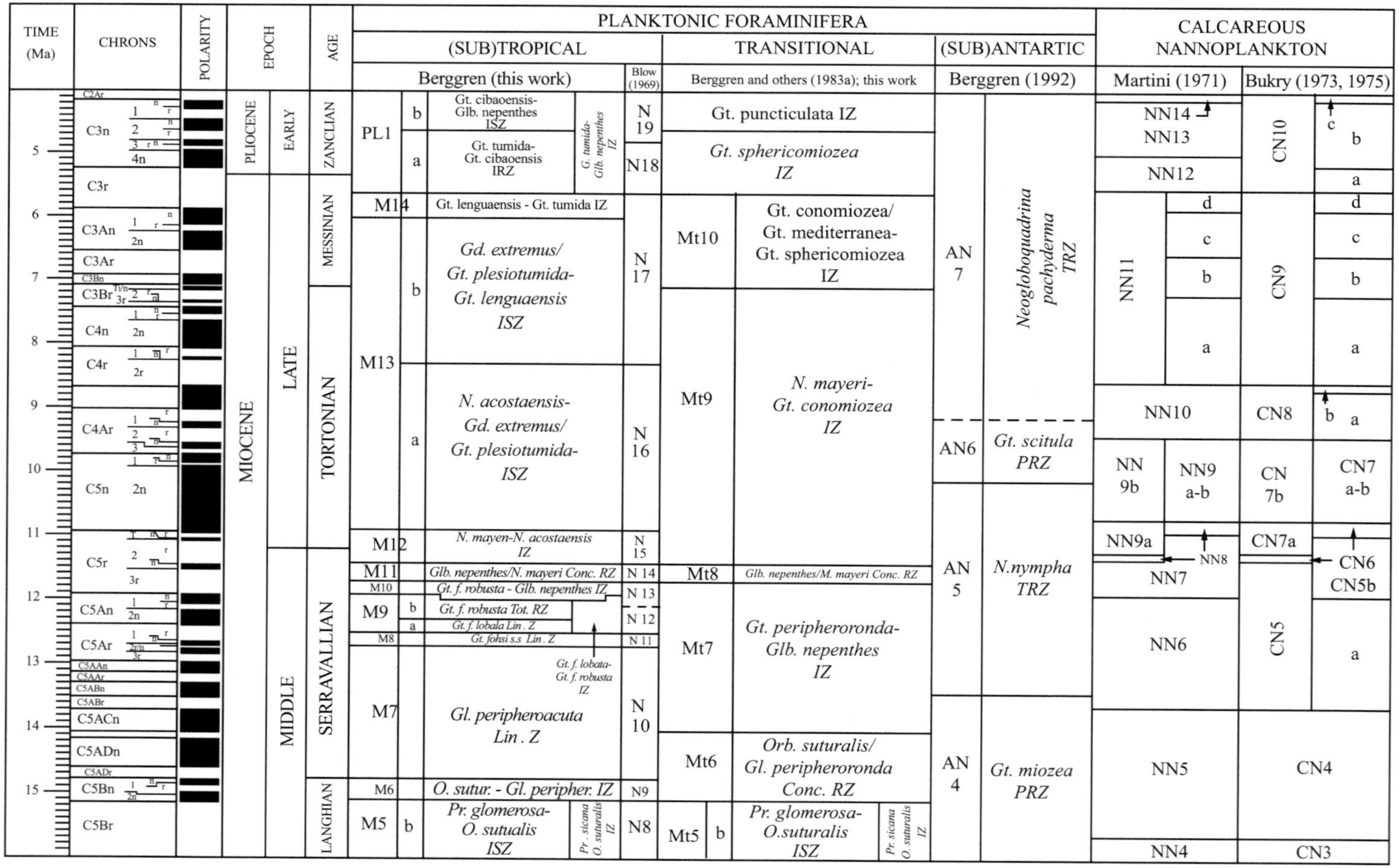

Figure 9 Middle-Late Miocene timescale indicating relationships among fossil-based, radiometric and magnetic field reversal chronometries. Note that three zone successions based on developments among planktic foraminiferans are indicated. These three different zonal sequences reflect biogeographic provincialism in these organisms. That provincialism resulted from ocean surface-water temperature differences. The calcareous nannoplankton or coccolithophore zonal scales of two different specialists are shown. These reflect different perceptions of the fossil aggregates that characterize each zone (source Berggren *et al.*, 1995, figure 5).

of the Devonian System. The next step, following that decision, was to identify a stratigraphic section at which that faunal occurrence could be found relatively easily. One other consideration in that selection was co-occurrence of other taxa in close stratigraphic proximity that could be used to enhance correlation. The procedure approved by the International Stratigraphic Commission involved driving a marker (called, perhaps euphemistically, the "golden spike") into the specific rock layer which bears the stratigraphically lowest *M. uniformis uniformis* in one thoughtfully selected section. That stratigraphic section, chosen after considerable debate among international specialists in the study of graptolites and the rocks and other faunas of both the Silurian and the Devonian, is at Klonk near Prague, Czech Republic. The base of the Devonian defined by this process became the top of the Silurian by definition. Geologists have adopted the procedures used in selecting the Silurian–Devonian boundary as the fundamental steps to be followed by which boundaries between other systems and divisions of systems, series, stages, and zones (characterized in the Oppel methodology) are identified.

The many discussions involving the procedures to define the base of the Devonian and, therefore, the Silurian–Devonian boundary ultimately led to widespread acceptance of the procedures in defining boundaries between units in the chronometric scale. Accordingly, studies of sedimentary rocks and their contained fauna and flora were focused on those taxa whose lowest stratigraphic occurrences can be documented as precisely as possible to be used as the base of zones, stages, and systems as well as the stratigraphic sections. That stratigraphic position, once chosen by vote of a group of stratigraphic paleontologists who are specialists in the study of the fossil organisms and time interval under analysis, is considered to be unmovable for a number of years. This procedure is thought, by many, to be the most appropriate for defining chronometric units because it specifies a particular evolutionary event seen in a specific stratigraphic and geographic place irreversible in time. The stratigraphic section and point in it chosen for the "golden spike" is referred to as a global statotype section and point (GSSP). Although the process has been formally codified in international stratigraphic practice (Remaine, 2000), its usefulness and even validity have been challenged (Fortey, 1993; Aubry *et al.*, 2000). Remaine (2000, p. 213) stated that chronometric units are "*characterized*" by their fossil content and are "*defined* by their lower boundary which is in the same time the upper boundary of the underlying" unit. Fortey (1993) commented on the procedure involved in selecting GSSPs, both from the viewpoint of understanding the realities of the stratigraphic and fossil record and the fragility of human nature. One development arising from the quest for the "best" GSSPs, and the evolutionary development upon which a definition of a chronometric unit was laid is, as Fortey (1993) pointed out:

> ... the spread of an idea of the ideal biostratigraphic section, a kind of platonic rock section equipped with perfect properties for international correlation: continuous, confacial and conformable, fossiliferous throughout, yet with cryptic breaks minimized, replete with fossils of several groups, which are arranged in evolutionary series. Furthermore, such sections have an horizon suitable for hammering in a 'golden spike' for the base of a chronostratigraphic interval to immediate international satisfaction. Such sections rarely, if ever, exist in nature. Yet their pursuit has been one of the motivating forces behind various Working Groups of the International Geological Correlation Programme. It is curious to find such an idealistic concept holding sway in the geological sciences, which are so generally pragmatic. The importance of sound criteria for international correlation is not to be gainsaid, but this is, perhaps, a different matter from the relentless pursuit of a perfect section which is likely to prove a chimera.

The search for GSSPs has also led many to overlook the realities of how organisms are constrained in their distribution by faunal/floral provinces, and by the environments in which they live. Chorology is a significant component of chronometry using the evolution of organisms regardless of the—essentially imagined—precision inherent in GSSPs.

13.10 TERMINOLOGIES

An almost overwhelming plethora of terms—accompanied by a multitude of misunderstandings—accompanies the subject of the chronometry of sedimentary rocks. Many terms have been codified (see Hedberg, 1965, 1976, 1978; Murphy and Salvador, 1999; Remaine, 2000; Remaine *et al.*, 1996; Berggren *et al.*, 1995; Walsh, 2000, 2001). The terms and their meanings have been so hotly contested and debated that understanding of the real value inherent in chronometry of sedimentary rocks has essentially been lost. As the recovery of petroleum in California based on studies such as that of Bandy and Arnal, discussed herein, using the chronometric units elucidated by Kleinpell, sedimentary-rock chronometry has been and continues to be of great value to the global economy. Those concerned with recovery of natural resources kept their attention directed toward that goal. Those not so directly concerned with recovery of natural resources entered into debates on terminology.

13.11 SUMMARY

It was the need to tap increasingly larger volumes of nature's resources and get them to treatment facilities and, ultimately, to consumers that motivated William Smith to lead the way in documenting the need for a potential chronometry. As the practical geologic work carried out by William Smith, Albert Oppel, Charles Lapworth, Robert Kleinpell, and others concerned with recovery of natural resources has revealed that the procedures in chronometry using fossils are basically simple. Many fossiliferous rock sequences in a relatively broad area must be studied and measured carefully and precisely. Numerous fossils must be extracted, from these stratigraphic sections, from as many layers as possible. These fossils must be identified and analyzed critically, seeking unique associations that characterize certain intervals in the stratigraphic record. Such associations, if selected after careful analyses of range of species, are unique in time and so may be used in chronometry. This is the basic procedure used by Oppel, who followed in the footsteps of his mentor, August Quenstedt, in documenting the need for detailed measurements of strata and precision in locating stratigraphic positions of species. As Kleinpell (1979) pointed out, and as work with Jurassic ammonite zones and subzones, especially, demonstrated, Oppel's methodology provides the students of the Phanerozoic sedimentary rock record with the most effective means of chronometry. Units in that chronometry are limited in the areas in which they may be used by distribution patterns of organisms that are in the domain of chorology.

Fossil-based chronometry has been enhanced through studies of radiochronometry and magnetobiochronometry. Today, explorations of deep ocean-floor sediments provides data that enables precise chronometry of the Mesozoic to modern geologic history of the oceans. Orbital chronometry seems to promise enhanced precision in chronometry of the Late Mesozoic and younger sedimentary record.

The need for sedimentary-rock chronometry and its application remains as valuable a tool in finding and recovering nature's resources today as it was in William Smith's day. Deep ocean-floor explorations have added an interesting, exciting, and still-expanding dimension to sedimentary-rock chronometry.

REFERENCES

Arkell W. J. (1933) *The Jurassic System in Great Britain*. Clarendon Press.

Aubry M.-P. (1995) From chronology to stratigraphy: interpreting the Lower and Middle Eocene stratigraphic record in the Atlantic Ocean. In *Geochronology, Time Scales and Global Stratigraphic Correlation*, SEPM Special Publication 54 (eds. W. A. Berggren, D. V. Kent, M.-P. Aubry, and J. Hardenbol), pp. 213–274.

Aubry M.-P., Berggren W. A., Van Couvering J. A., and Steininger F. (2000) Should the golden spike glitter? *Episodes* **23**, 203–210.

Bandy O. L. and Arnal R. E. (1969) Middle Tertiary basin development in California. *Geol. Soc. Am. Bull.* **80**, 783–820.

Berggren W. A., Kent D. V., Swisher C. C., III, and Aubry M.-P. (1995) A revised Cenozoic geochronology and chronostratigraphy. In *Geochronology, Time Scales and Global Stratigraphic Correlation*, SEPM Special Publication 54 (eds. W. A. Berggren, D. V. Kent, M.-P. Aubry, and J. Hardenbol), pp. 129–212.

Berry W. B. N. (1968) *Growth of a Prehistoric Time Scale*. W. H. Freeman.

Berry W. B. N. (1987) *Growth of a Prehistoric Time Scale*, revised edn. Blackwell.

Berry W. B. N. (2000) R. M. Kleinpell's zones and stages: an oppelian biostratigraphic solution to a challenge in the search for petroleum in California. *Earth Sci. Hist.* **19**, 161–174.

Boggs S., Jr. (2001) *Principles of Sedimentology and Stratigraphy*, 3rd edn. Prentice Hall.

Callomon J. H. (1995) Time from Fossils: S.S. Buckman and Jurassic high-resolution geochronology. In *Milestones in Geology*, Geol. Soc. London Mem. (ed. M. J. LeBas), vol. 16, pp. 127–150.

Conybeare W. D. and Phillips W. (1822) *Outlines of the Geology of England and Wales*. William Phillips, George Yard.

Davies A. M. (1934) *Tertiary Faunas, a Text-book for Oilfield Paleontologists and Students of Geology*. Thomas Murby and Co.

d'Orbigny A. D. (1842) *Paleontologie Francaise, Terraines Jurassiques: Part 1*. Cephalopodes. Masson.

Fortey R. A. (1993) Charles Lapworth and the biostratigraphic paradigm. *Geol. Soc. London J.* **150**, 209–218.

Hancock J. M. (1977) The historic development of concepts of biostratigraphic correlation. In *Concepts and Methods of Biostratigraphy* (eds. E. G. Kauffman and J. E. Hazel). Hutchinson and Ross, pp. 3–22.

Harland W. B., Armstrong R. L., Cox A. V., Craig L. E., Smith A. G., and Smith D. G. (1990) *A Geologic Time Scale 1989*, 2nd edn. Cambridge University Press.

Hedberg H. D. (1965) Chronostratigraphy and biostratigraphy. *Geol. Mag.* **102**, 451–461.

Hedberg H. D. (ed.) (1976) *International Stratigraphic Guide* Wiley.

Hedberg H. D. (1978) Stratotypes and a geochronologic scale. In *Contributions to the Geologic Time Scale*, Am. Assoc. Petrol. Geol. Studies in Geology (ed. G. V. Cohee), vol. 6, pp. 33–38.

Herbert T. D., Premoli-Silva I., Erba E., and Fischer A. G. (1995) Orbital chronology of Cretaceous–Paleocene marine sediments. In *Geochronology, Time Scales and Stratigraphic Correlation*, SEPM Spec. Publ. (eds. W. A. Berggren, D. V. Kent, M.-P. Aubry, and J. Hardenbol), vol. 54, pp. 81–93.

Kleinpell R. M. (1938) *Miocene Stratigraphy of California*. Am. Assoc. Petrol. Geol.

Kleinpell R. M. (1979) *Criteria in Correlation: Relevant Principles of Science*. Pacific Section Am. Assoc. Petrol. Geol.

Lapworth C. (1879) On the tripartite classification of the Lower Paleozoic rocks. *Geol. Mag.* **6**, 1–15.

Laugel F. (1859) Un tableau resume de la classification du terrain jurassique etablier par le docteur Albert Oppel. *Soc. Geol. France 2* **15**, 657–664.

Lyell C. (1833) *Principles of Geology*. vol. 3, J. Murray.

Murphy M. A. and Salvador A. (1999l) International stratigraphic guide—an abridged version. *Episodes* **22**, 255–271.

Oppel A. (1856–1858) *Die Juraformation Englands, Frankreichs und des sudwestlichen Deutschlands*. Wurttemb. Naturwiss. Verein.

Remane J. (2000) Comments on the paper of Aubry *et al.* (2000). *Episodes* **23**, 211–213.

Remane J., Bassett M. G., Cowie J. W., Gohrbandt K. H., Lane R., Michelsen O., and Naiwen W. (1996) Revised guidelines for the establishment of global chronostratigraphic standards by the international commission on stratigraphy. *Episodes* **19**, 77–81.

Strachey J. (1719) A curious description of the strata observed in the coal mines of mendip in Somersetshire, being a letter of John Strachey Esq. To Dr. Robert Welsted, M.D. and R.S. Soc. and by him communicated to the society. *Roy. Soc. (London) Phil. Trans.* **30**(360), 968–973.

Strachey J. (1725) An account of the strata in coal mines, etc. *Royal Society (London) Philos. Trans.* **33**(391), 395–398.

Walsh S. L. (2000) Eubiostratigraphic units, quasibiostratigraphic units, and "assemblage zones." *J. Vert. Paleo.* **20**, 761–775.

Walsh S. L. (2001) Notes on geochronologic and chronostratigraphic units. *Geol. Soc. Am. Bull.* **113**, 704–713.

Winchester S. (2001) *The Map that Changed the World.* Harper Collins.

Radioactive Geochronometry
ISBN: 978-0-08-096708-0

pp. 399–424

14

The Early History of Life

E. G. Nisbet and C. M. R. Fowler

Royal Holloway, University of London, Egham, UK

14.1   INTRODUCTION

14.1.1   Strangeness and Familiarity— The Youth of the Earth

The youth of the Earth is strange to us. Many of the most fundamental constraints on life may have been different, especially the oxidation state of the surface. Should we suddenly land on its Hadean or early Archean surface by some sci-fi accident, we would not recognize our home. Above, the sky may have been green or some other unworldly color, and above that the weak young Sun might have been unrecognizable to someone trying to identify it from its spectrum. Below, seismology would show a hot, comparatively low-viscosity interior, possibly with a magma ocean in the deeper part of the upper mantle (Drake and Righter, 2002; Nisbet and Walker, 1982), and a core that, though present, was perhaps rather smaller than today. The continents may have been small islands in an icy sea, mostly frozen with some leads of open water, (Sleep *et al.*, 2001). Into these icy oceans, huge protruding Hawaii-like volcanoes would have poured out vast far-spreading floods of komatiite lavas in immense eruptions that may have created sudden local hypercane storms to disrupt the nearby icebergs. And meteorites would rain down.

Or perhaps it was not so strange, nor so violent. The child is father to the man; young Earth was mother to Old Earth. Earth had hydrogen, silicate rock below and on the surface abundant carbon, which her ancient self retains today. Moreover, Earth was oxygen-rich, as today. Today, a tiny part of the oxygen is free, as air; then the oxygen would have been in the mantle while the surface oxygen was used to handcuff the hydrogen as dihydrogen monoxide. Oxygen dihydride is dense, unlikely to fly off to space, and at the poles, rock-forming. Of all the geochemical features that make Earth unique, the initial degassing (Genesis 2:b) and then the sustained presence of liquid water is the defining oddity of this planet. Early Earth probably also kept much of its carbon, nitrogen, and sulfur as oxide or hydride. And, after the most

cataclysmic events had passed, ~4.5 Ga ago, for the most part the planet was peaceful. Even the most active volcanoes are mostly quiet; meteorites large enough to extinguish all dinosaurs may have hit as often as every few thousand years, but this is not enough to be a nuisance to a bacterium (except when the impact boiled the ocean); while to the photosynthesizer long-term shifts in the solar spectrum may be less of a problem than cloudy hazy days. Though, admittedly, green is junk light to biology, the excretion from the photosynthetic antennae, nevertheless even a green sky would have had other wavelengths also in its spectrum.

Most important of all, like all good houses, this planet had location: Earth was just in the right spot. Not too far from the faint young Sun (Sagan and Chyba, 1997), it was also far enough away still to be in the comfort zone (Kasting *et al.*, 1993) when the mature Sun brightened. As many have pointed out, when Goldilocks arrived, she found everything just right. But what is less obvious is that as she grew and changed, and the room changed too, she commenced to rearrange the furniture to make it ever righter for her. Thus far, the bears have not arrived, though they may have reclaimed Mars from Goldilocks's sister see (Figure 1).

14.1.2   Evidence in Rocks, Moon, Planets, and Meteorites—The Sources of Information

The information about the early history of life comes from several sources: ancient relics, modern descendants, and models. The ancient material is in the rocks, in meteorites, in what we can learn from other planets, and in solar system and stellar science. The Lucretian view of a planet, ramparts crumbling with age, may apply to Mars, but as Hutton realized, virtually all of the surface of the Earth is renewed every few hundred million years, and if it were not so, life would die from lack of resources. But in the tiny fragment that is not renewed, relics of early life remain. Some of these relics are direct—specks of carbon, or structures of

Habitable zones, 4.5 Ga ago and today

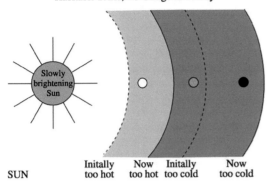

SUN

| Initally too hot | Now too hot | Initally too cold | Now too cold |

Figure 1 The habitable zone (Kasting *et al.*, 1993). Too close to the Sun, a planet's surface is too hot to be habitable; too far, it is too cold. Early in the history of the solar system, the Sun was faint and the habitable zone was relatively close; 4.5 Ga later, with a brighter Sun, planets formerly habitable are now too hot, and the habitable zone has shifted out. Note that boundaries can shift. By changing its albedo and by altering the greenhouse gas content of the air, the planet can significantly widen the bounds of the habitable zone (Lovelock, 1979, 1988).

biogenic origin. Other relics are indirect: changes in the isotopic ratio of inorganic material or oxidation states of material that is of inorganic origin. Yet other information is simply scene-setting: evidence, for example, that water was present, or that volcanism was active.

Extraterrestrial sources of evidence are also important. From Venus there is evidence that a planet can have water oceans and then lose all its hydrogen (Donahue *et al.*, 1982; Watson *et al.*, 1984). From Mars there is evidence that planets can die geologically, and become unable to renew their surface by tectonics and volcanism. Perhaps they can also die biologically. From moons of outer planets comes evidence that a wide variety of early conditions was possible. Meteorites (Ahrens, 1990; Taylor, 2001) provide clear signs that in the early part of the history of the solar system there could have been significant exchange of surface material between the inner planets. Study of the Sun and of sun-like stars demonstrates that even stable stars do change, and over the past 4.6 billion years the Sun has significantly increased in power (Sagan and Chyba, 1997), and altered in spectrum.

14.1.3 Reading the Palimpsests—Using Evidence from the Modern Earth and Biology to Reconstruct the Ancestors and their Home

"Ontogeny," the old saying went, "recapitulates phylogeny." We each start as a couple of lengths of DNA, one loose with a few attendants, the other comfortable in a pleasant container full

of goodies, itself held in a warm and safe maternal universe. The DNA-, the RNA-based processing, and information-transfer systems, and the protein machinery of the cell all carry historical information. Every human cell lives in its own seawater, the blood—we came from a warm kindly ocean. Every oxygen-handling blood cell carries iron: we learned this trick somewhere that our ancestors could acquire iron, surely without the sophisticated metal-gathering equipment that is provided by modern biochemistry. At the very heart of the information transfer in the cell is the ribosome: a massive (compared to other enzymes) RNA-based super-enzyme that in a strange way is both chicken and egg and, though much modified by evolution, is surely of the very greatest antiquity.

Modern life comes in very many forms: animals, plants, and single-celled eukaryotes in the *Eucarya* domain; prokaryotes in two great domains, *Archea* and *Bacteria* (Woese, 1987; Woese *et al.*, 1990), and *not-life* viruses. Some not-life is even anthropogenic: the wild-type polio virus that used to be found in water bodies is now replaced in the pools and rivers of America and Europe by the altered vaccine-type virus. From all this information, deductions can be made. Clearly, multicelled life came from single-celled life; less obviously but most probably each of our cells carries mitochondria that are descended from symbiotic purple bacteria. Plants, in addition, carry chloroplasts that are descended from partner cyanobacteria.

There is an enormous wealth of this type of information that is only just beginning to be deciphered. Indeed, deciphering the molecular record (Zuckerkandl and Pauling, 1965) may be the best route to understanding of Archean palaeontology. Geological study interacts with this, both by calibrating the timing of the evolutionary steps (e.g., by dating the arrival of the multicelled organisms), and secondly by identifying the impact of each step (e.g., the onset of oxygenic photosynthesis). Both molecular and rock-based studies are needed: without the rock information, molecular evidence can lead to (and has done so) very erroneous deduction; equally, the geological evidence cannot of itself give much detail about major steps. But there is also a danger of circularity of reasoning: just because something looks plausible biochemically, it is possible to reinterpret the geological evidence to fit, but wrongly; conversely, weakly supported geological models can on occasion unduly sway interpretation of complex and nonunique molecular evidence.

14.1.4 Modeling—The Problem of Taking Fragments of Evidence and Rebuilding the Childhood of the Planet

Model building is part of all science: lovely falsifiable hypotheses are built and then broken on the cold facts. Certain key components are common to all models of life's origins—water (though not necessarily in an ocean: aerosols are possible hosts of proto-life); inorganic supplies of thermodynamic drive (i.e., interface settings where two or more different conditions are accessible); ambient temperatures in the 0–110 °C range.

Models of the early history of life come in two broad categories—models of the origin and first development of life itself and models of the environmental settings of that life. The geologist can contribute much more to the second class of model than to the first: from the geological evidence it is possible to make reasonable models of the early planets, including their surface condition and the supply of chemicals and nutrients from the interior and from space.

All information is fragmental, and the further back in time, the less the information. But enough is left that reasonable guesses may be made about the surface conditions of the four inner planets and the Moon as they evolved in the first two billion years of the solar system's history. These models set the scene for the biochemists: without them, the biochemical deductions are unconstrained and can be wrong (e.g., the primaeval "soup" is geologically unlikely). Thus, the debate over the various models of origin is avoided here: for that, seek out google.com. The focus instead is on what geologists and geochemists can usefully contribute.

14.1.5 What Does a Planet Need to be Habitable?

Venus may have been in the right place once, and is the right size, but in the long term it was too close to the Sun. Mars may be in a tolerable place, and on occasion with liquid water (Baker, 2001; Carr, 1996), but has a small heart, almost dead of cold. Only Earth had long-term location and is large enough to keep an active interior.

What are the requirements? First, liquid water. It is difficult to imagine biology that does not include water. It may be possible (indeed, someday computers may produce some sort of water-free inorganic sentience that achieves genetic take-over from organic life) but not in Nature as we know it. Externally, the planet needs to be far enough from the Sun not to overheat, close enough not to freeze entirely. The allowable bounds of the habitable zone (Kasting *et al.*, 1993) are wide at any one moment, given the range of temperature control provided by changes in atmospheric greenhouse heat trapping, but these bounds progressively shift outwards as the Sun evolves and brightens. Thus, while Venus was probably within the bounds of the habitability zone early in the history of the solar system with a faint young Sun, and covered by oceans, it may today be too close to the bright old Sun to allow life. Even if water

were added it would be difficult to sustain liquid water on the planet with any plausible life-supporting atmosphere. However, that is not to say it is uninhabitable: some day humanity may well add water supplies from an outer solar system source and hang aluminum foil mirrors around the planet to reduce sunlight input. Mars, alternatively, is too cold to sustain liquid water, but could in future be warmed by chlorofluorocarbon and methane greenhouse gases, such that it sustained puddle oceans. These thought-experiments with Venus and Mars demand teleological action, not possible in Darwinian evolution, but there are persuasive arguments that feedbacks from nonteleological life have carried out very similar processes on Earth over 4 Ga (Lovelock and Margulis, 1974; Lovelock, 1979, 1988).

Planets also need to be geologically active to sustain life over long periods. Nature needs to renew her face continuously, or the chemical and thermodynamic resources behind life, especially early life, are rapidly exhausted. For example, DNA-based life is built on phosphates. If the available surface phosphorus supply is exhausted, and not renewed continually by volcanism and tectonics, then life must become hungry for phosphorus and eventually die out. Life depends on a small number of essential house-keeping proteins, and many of these proteins use metals: if the geological metal supply ended, the proteins would not be formed and life would be unsustainable. This need for "supply" places constraints on the physical evolution of a planet, if it is to be capable of sustaining life over many aeons. Planets vary (Taylor, 2001). The Moon is too small. It was once active, but now has died. Mars is just about dead. Mercury may once have been larger but now seems to be a barren metal-rich relict of the innards of a planet. Jupiter and the outer gas giants are too large. Some of the tidally heated great moons (some with radii comparable to Mars) of the gas giants do remain very active geologically and offer possible homes. But of the internally warm bodies that have a Sun-warm outer surface, only Earth and Venus are just the right size.

14.1.6 The Power of Biology: The Infinite Improbability Drive

What is just right in one moment becomes wrong in the next. The porridge that Goldilocks tasted would have been perfect in the first mouthful, but a little later, especially if she ate slowly as her mother would have taught her, it would have cooled.

Biology has the power to sustain, to draw out, its environmental conditions (Lovelock, 1979, 1988), and indeed to remake them in an improbable path. Swiss travelers do not descend peaks by jumping over the cliffs. Instead they use cable cars, and as they descend they help others to ascend: only a small input of energy is needed to overcome frictional losses. Indeed, consider a hypothetical cable car that had an attached snow-tank, filled from a snowfield at the top. At the bottom the snow would be dropped off, so that the rising car was always somewhat lighter than the descending car. This system could work without extra input perpetually carrying tourists up to the peak and down again, as the potential energy transfer would make up for the frictional losses. Indirectly, this is solar power: the Sun lifts the water, evaporating it from the bottom and replacing it back on the top as snow.

Most microbial processes are like that—they move enormous numbers of traveling chemical species on cogways up and down the thermodynamic peaks and valleys with only small extra inputs of externally sourced energy. Moreover, at the intermediate stations part-way up (or down) the peaks, the microbial processes link with innumerable smaller cable car systems that scatter metabolic tourists around the ecological mountain sides in a complex web of ascents, descents, and lateral movements. Thus, biology creates local order, primarily by using the high quality of sun-given energy, to exploit and create redox contrast between the surface of the Earth and its interior.

14.2 THE HADEAN (~4.56–4.0 Ga AGO)

14.2.1 Definition of Hadean

The Hadean is the first of the four aeons of Earth history (Nisbet, 1991). Aeons are the largest divisions of geological time: Hadean, Archean, Proterozoic, Phanerozoic. The first and last aeons are "short" (relatively, if 560 Ma can be called short); the middle two are billions of years long. The Hadean was the period of the formation of the Earth, from the first accretion of planetesimals at the start of the Hadean, to the end of the aeon, when the Earth was an ordered, settled planet, with a cool surface under oceans and atmosphere, and with a hot active interior mantle and core.

The bounds of the Hadean have never been properly defined. The birth of the Earth is the start of the Hadean, but is this the moment of the beginning of the solar system, or the moment when the first significant planetesimals collected to begin the accretion? Fortunately, this is not much more than an academic discussion—the time of formation of the oldest material in meteorites is usually taken to be representative of the start of accretion. Very roughly, 4.56 Ga is taken as the start, where 1 Ga is one thousand million (10^9) years.

The end of the Hadean is more difficult to define, though the definition is more useful to the geologist, as this is within the terrestrial geological record. For the interim, as a rough guide, 4 Ga is used as a working definition of the Hadean/Archean boundary. But this is unsatisfactory—a random number. The choice of the start of life as the defining moment for the boundary between the Hadean and Archean has great appeal and should surely be preferred. Fortunately, at present the best guess for the origin of life is also "somewhen around 4 Ga ago."

14.2.2 Building a Habitable Planet

The solar system accreted from a dust cloud, formed after a supernova explosion. From this primitive solar nebula condensed the Sun and the planets. Some of the oldest objects in the solar system yet found are Ca–Al-rich inclusions in meteorites, ~4.566 Ga old (Allegre *et al.*, 1995). It is possible that these grains predate the solar nebula and may have been formed in the expanding envelope of the supernova explosion (Cameron, 2002).

The formation of planetesimals may have been very rapid after the initial formation of the solar nebula. Objects as large as Mars would have grown within 10^5 yr (Weatherill, 1990). The core of the asteroid Vesta may have formed within only 3–4 Myr, and lavas flows on its surface may have occurred at this time also (Yin *et al.*, 2002). Bodies like Vesta would have collided rapidly, aggregating their cores to form larger planetoids and then planets. The date of core formation in the Earth remains controversial but may have been as little as 30 Myr or less after the birth of the solar system (Kleine *et al.*, 2002). Yin *et al.* (2002) suggest that the aggregation of the Earth's core took place within 29 Myr. The core of Mars may have formed as early as within 13 Myr.

The special events of this planet's accretion (Newsom and Jones, 1990; Weatherill, 1990; Ahrens, 1990; Taylor, 2001) were crucial in making Earth habitable over billions of years. Segregation of the core physically separated reduction power in the iron-rich center of the Earth, from a more oxidized mantle. Simultaneously, the early events controlling the surface environment made possible the development of a habitable ocean/atmosphere system.

The most important single physical event took place roughly 4.5 Ga ago, 25–30 Myr after the birth of the solar system. At this stage, Earth was probably a substantial fraction of its present mass, with a segregated core. Sunwards of Earth, Venus and Mercury had formed; outwards, were Mars-like planets. Then, the Earth suffered its largest collision: a defining moment in habitability.

One model is that a planet at least double the mass of Mars hit a half-formed Earth in a double collision (Cameron, 2002); an alternative model is that a Mars-sized body hit the 90%-formed Earth (Canup and Asphaug, 2001). When the impact took place, the Earth was transformed. Internally, it would have been melted, even if primordial radiogenic and infall heat had not already melted it. The already-formed iron core of the impactor would have crashed to the center to join the core of the Earth. This large core, with its solid center and molten outer region, gave Earth its distinctive magnetic field, and life-protecting van Allen belts. Arguably, planets without a strong protective field (e.g., Venus), are initially uninhabitable as the surface environment may be too severe for unstable early genomes.

The surface of Earth was completely changed. Any deep primordial atmosphere/ocean, possibly rich in noble gases, would have been removed by the impact. Presumably, the event was followed by further cometary infall and further degassing from the interior to produce our present thin water-dominated inventory of volatiles. By this stage, the inner solar system was probably swept clear of volatiles and was a relatively gas-poor environment. Volatile influx would have come perhaps in larger planetesimals infalling from the outer solar system.

The mechanical effect of the impact was that the Earth was tilted, creating winter and summer. This is very important in distributing heat evenly across the surface, as the intensity of solar radiation falling on any particular place varies in the annual cycle. Even more important for the habitability of the Earth is the spin: much faster immediately after the impact but now slowed by aeons of tidal friction to give the 24 h day. Thus there is no hot day-side and cold night-side, but an even illumnation. Moreover, the night-day cycle allows a variety of photosynthetic/respiratory cycles in cells, and contributes greatly to the diversity of biota.

The Moon was created from the mantle-derived ejecta. Physically, over the aeons this may have played a useful sheltering role in protecting the planet from some meteorite impacts. Arguably more important, the presence of the Moon leads to the tides. These create the intertidal and near-subtidal habitat with rapidly varying geochemical settings, from wet submarine to dry subaerial, in which sediment is repeatedly flushed with fluid. Such cyclically varying habitats may have been vital in the early evolution of microbial biofilms and eventually microbial mats.

Other planets had varying histories (Taylor, 2001). Mercury also had a major collision, possibly being hit by an object ~0.2 Mercury masses, removing much of its silicate mantle and leaving a planet of high intrinsic density, with a major core and a thin rocky mantle, an uninhabitable planet. Mars and Venus had kinder gentler histories. On neither did a great impact eject splat; neither planet gained a significant Moon. Though subject to geological or impact catastrophe, both planets

evolved sustainable systems within the constraints of kinetics and the thermodynamics of equilibrium; only Earth produced an intrinsically unsustainable disequilibrium system.

On Mars (a tenth of Earth mass and 38% of its radius), the present water inventory is much less, enough to cover the planet to a few tens of meters: puddle oceans (Carr, 1996). On Venus, which must have been very nearly Earth's twin prior to the giant impact on Earth (0.815 modern Earth mass, 95% of its radius), the atmosphere evolved to its present runaway CO_2 greenhouse. There has been much speculation about early Venusian oceans, perhaps some kilometers deep, but possibly only a few meters if Venus formed too close to the Sun to inherit a large water inventory (see Taylor, 2001 for a brief summary of this dispute).

The main part of the accretion of the Earth can be considered complete by ~4.45 Ga. By this stage most of what now makes up the Earth was in place. The then much nearer Moon orbited close by. The Earth would have been molten except for a thin rocky outer carapace, possibly of broadly basaltic composition (komatiitic basalt; or even komatiite?). A large magma ocean may have persisted in the mantle for some time in the Hadean or even longer. Within the Earth, ongoing late precipitation of the core may have continued, with reaction between water in the mantle and infalling iron, adding oxygen to the iron, and giving a mantle source of hydrogen that may have made its way eventually to the surface via mantle plumes and thence volcanoes.

The composition of the Earth is unique, subtly different from the other rocky planets, and this suggests that different parts of the material of the inner solar system went to make each planet (Drake and Righter, 2002). The origin of Earth's water is particularly interesting (Yung et al., 1989). A significant fraction of Earth's early hydrogen endowment may have been lost to space in short-lived steam greenhouse events. Seawater has a D/H ratio of 150×10^{-6} in contrast to Mars water which has D/H of 300×10^{-6}. Perhaps Mars lost more hydrogen to space, enriching D, but it also may be possible that the Martian interior has water of very different D/H, since cooler Mars has outgassed less than Earth. One possibility is that temperatures were high in the inner part of the accretion disk: thus the Earth may have accreted as a dry planet, with water and carbon compounds delivered after the main accretion by comets and meteorites. Alternately, Earth did indeed accrete with a significant water content, and some geochemical evidence suggests the early magma ocean was hydrous (Drake and Righter, 2002).

Late Hadean Earth (say 4.2 Ga to 4 Ga ago) was thus very unusual among the inner rocky planets, in its Moon, spin, tilt, likely magnetic field, and especially in its inventory of water and its location in the "habitable band." Such a planet is not improbable, given the allowable common accidents of accreting planets by collision, but perhaps may be found to be rare as knowledge of distant extra-solar planetary systems increases.

14.2.3 The Hadean Record

Jack Hills and Mt. Narryer, Western Australia. Some of the oldest material found on Earth consists of a few crystals of detrital zircon that are now preserved in quartzites in the Mt. Narryer and nearby Jack Hills area, Western Australia (Compston and Pidgeon, 1986, Wilde *et al.*, 2001, Halliday, 2001). The host sediment is ~3.3–3.5 Ga old, but the some of the zircons themselves are up to 4.2–4.4 Ga old (Figure 2). There are several implications of the discovery. First (also shown in many other successions) that by 3.3 Ga ago, in the mid-Archean, there was already old continental crust being eroded and redeposited by water. Second, abundant zircons are typical of rocks broadly characteristic of continental crust. This line of reasoning thus suggests granitoid rocks and continental crust were present in the Hadean. Intuitive reasoning would suggest komatiitic and basaltic rocks would be expected to be typical of

Figure 2 Zircon grain, in part ~4.3 Ga old (Compston and Pidgeon, 1986). Jack Hills, Western Australia. Scale bar is 100 μm long.

4.4 Ga ago crust, rather than granitoids, but the existence of Hadean zircon implies otherwise, at least locally in what is now Western Australia. Moreover, to form granitoid nowadays, subduction of old hydrated oceanic plate is needed: water is needed to make granites, and subduction is needed to supply the water (Campbell and Taylor, 1983). Did subduction occur as early as 4.4 Ga ago, and did oceans of water exist to hydrate the crust? Oxygen isotope evidence (Wilde *et al.*, 2001; Mojzsis *et al.*, 2001) supports the deduction that oceans of liquid water were indeed present. The zircons contain isotopically heavy oxygen: suggesting derivation from liquid surface water. This is speculation, and just as one swallow does not make a summer, one zircon does not make either a continent or an ocean of water (Moorbath, 1983). Yet the question remains open: did Hadean continents exist, and oceans, and were hydrothermal systems present on continental land surfaces around andesitic volcanoes, fed by water-mediated subduction?

Acasta Gneiss, Canada. Next oldest is the Acasta Gneiss, close to 4 Ga old (Bowring *et al.*, 1989). This is a rock, of sorts, though highly deformed and metamorphically recrystallized. The oldest rocks form a small part of a 20 km^2 terrain of old rocks. There are various such terrains worldwide: examples include the Nain province in Labrador (~3.9 Ga); the Napier complex in Antarctica (up to ~3.7 Ga); and the Narryer complex, Australia (host rock up to ~3.7 Ga, hosting the older zircons). Some of these terrains are up to several thousand square kilometers, though the datable older rocks may only be a small proportion of the whole. The implication is that massifs of continental crust at least up to the size of, say, Luxembourg or Rhode Island existed in the latest Hadean and earliest Archean.

There is no evidence for the existence of life before 4 Ga ago. Even if a living organism had appeared, life would probably have been obliterated within a few million years, killed in the intense Hadean bombardment. This was a time when from time to time (say every few million to tens of millions of years, large meteorite impact events would have occurred that so heated the oceans and the atmosphere as to make the Earth briefly uninhabitable, sterilized at several hundred °C (Sleep *et al.*, 2001).

14.2.4 When and Where Did Life Start?

Enough has been said of the origin of life to show that the problem is as far from solution as it was in Charles Darwin's time. The debate continues (Line, 2002). The geologist can make little contribution to this debate, except to point out possible habitats where the first life could have been born.

There are many possibilities: in the air, in the sea, on the shallow seafloor, on the deep seafloor, near on-land hydrothermal systems around andesite volcanoes (variable, intermediate to low pH), near on-land hydrothermal systems around komatiite volcanoes and hot ultramafic rocks (alkaline), near deep-water hydrothermal systems (acid), near carbonatite-driven hydrothermal systems (which could be phosphorus-rich), in hydrothermal systems under ice caps, in shallow-water tidal muds, anywhere else that is fancied.

There are also five planets on which life could have begun (Nisbet and Sleep, 2001, 2003). Earth is the most likely, as it is the only place where Cartesian logic suggests life exists today. Next most likely on the list is Mars, which could at one stage have had an early wet environment under a strong greenhouse. Mars would have been hit by many impacts capable of ejecting relatively unshocked rocks that could have carried a living cell to Earth, surviving the transit frozen in space. There would have been a numerically vast early flux of such rocks in the Hadean, and it is thus very reasonable to infer that *if* life had begun on Mars, it would have been transferred to Earth. The logic that applies to Mars also applies to Venus, except that it is a very much deeper gravity well, and thus the outward flux of ejecta would have been much less, and those ejecta would be more shocked. The Moon is a possible though unlikely candidate, early on. Finally, a candidate is the impactor planet that hit the Earth. This Mars-sized object could have hosted life. On the great impact, ejected cells could have gone into space, either seeding Mars or much later falling back to Earth or new Moon. The most likely first homes are Earth or Mars; the other candidates are varying shades of improbable, only entertained because life is itself so improbable.

14.3 THE ARCHEAN (∼4–2.5 Ga AGO)

14.3.1 Definition of Archean

The inter-aeon boundary between the Hadean and Archean is presently not defined (Nisbet, 1991). There are various options: (i) the date of the first life on Earth; (ii) the date of the last common ancestor; (iii) a "round" number, such as exactly 4 Ga—4,000,000,000 years ago; (iv) the oldest record of a terrestrial rock (∼4 Ga ago); (v) the oldest record of a terrestrial mineral crystal (∼4.3–4.4 Ga ago).

Each option has attractions and problems. The choice of a "round number" goes sharply against long-held stratigraphic logic, which firmly maintains any definition should be "in the rock." Dating calibrations shift when decay constants are remeasured and can be made more precise: such changes would reclassify material across the boundary. But a definition rooted in rock does not shift. The choice of a particular "oldest" rock

or "oldest" mineral has more logic, but inevitably the candidate would be supplanted as a new "oldest" is discovered.

Life-based definitions are more satisfying. After all, the word "Archean" comes from the Greek for beginning: St. John's gospel starts with the words "In the Archae...." One option is the start of life: it is not clear when this was, yet, but given life's impact on carbon isotopes, it is perhaps not over-optimistic to hope that the geological record may eventually provide some insight into when life began. A second option—perhaps better—is suggested by phylogenetic studies that infer a *last common ancestor* of life—the cell or group of cells from which all modern cells are descended (Woese, 1987, 1999). Any such successful cell would spread rapidly across the globe to inhabit all accessible habitats within a geological moment—and thus there is a hope that a global signature of its metabolism could be found. Moreover, there are clocks in the genetic divergence, and the rRNA record has already been used for this. The clocks may not be very accurate at present, but there is the hope that they can be calibrated better. The date of the last common ancestor is thus perhaps the most attractive candidate for the definition of the Hadean/Archean boundary.

Once life had begun, the early Archean bombardment during later phases of accretion would have imposed a major constraint on its survival (Sleep *et al.*, 2001; Gogarten-Boekels *et al.*, 1995).

14.3.2 The Archean Record

14.3.2.1 Greenland

The most informative old sequence is the Itsaq gneiss complex of southern West Greenland (Nutman *et al.*, 1996). This complex includes a wide variety of rocks older than 3.6 Ga and ranging up to 3.9 Ga (early Archean): components are the Isua Belt, the Amitsoq gneisses, and the Akilia association. The Isua belt is especially interesting because it is supracrustal: it was laid down on the surface of the planet. The rocks include mafic pillow lavas, felsic volcanics, and volcaniclastic rocks, some of which were deposited from turbidity currents. The ensemble is reminiscent of material deposited today in volcanic island arcs, for example, in the western Pacific volcanic island chains. The implications are profound. There was clearly an ocean present, and land masses (at least volcanoes, possibly other older crust). Erosion occurred, sediments were deposited; volcanic eruptions must have been normal features of the geological setting. Moreover, this was a time early enough that the Earth was still under heavy bombardment by meteorites (the face of the Moon, like a ravaged battlefield, dates from this time). There is good evidence from Isua of a meteoritic component in sediment (Schoenberg *et al.*, 2002).

With volcanoes come hydrothermal systems, and there is good evidence for these in Isua. Localized low-strain zones in ~3.75 Ga rocks show many primary features (Appel *et al.*, 2001), including mafic lavas with fine-grained cooling rims, and in pillow breccias, quartz globules occur. These globules are interpreted as former gas vesicles, infilled with quartz from hydrothermal veins that formed during and immediately after volcanism. These quartz infills contain rare fluid inclusions. Appel *et al.* (2001) describe inclusions containing remnants of two independent fluid/mineral systems, comprising pure methane and highly saline (25% NaCl) aqueous fluids, and co-precipitating calcite. These fluids strongly resemble modern sea-floor hydrothermal fluids. The conclusion reached by Appel *et al.* (2001) is thus that methane-brine hydrothermal systems operated 3.75 Ga ago, in the early Archean. If correct, the implications are twofold: that, as common sense already tells us, hydrothermal systems existed, and that they emitted methane, useful for metabolism.

There have been various claims of evidence for life in the rocks of west Greenland. These have been reviewed by Myers and Crowley (2000), and also studied by van Zuilen *et al.* (2002) and Fedo and Whitehouse (2002). Significantly, they contest claims (Mojzsis *et al.*, 1996) for evidence of very early life at Akilia island. Fedo and Whitehouse (2002) showed that the rock studied by Mojzsis *et al.* was not sedimentary but an ultramafic igneous rock. They further considered that the isotopic ratios of the carbon particles at Akilia recorded high temperature metamorphic processes, not life, and yielding abiotic hydrocarbons. Thus the Akilia rock, though interesting, is not a guide to early life.

Rosing (1999) reported carbon microparticles from >3,700 Ma rocks in Isua that are strongly depleted in ^{13}C relative to bulk Earth. δ^{13}C in these particles is in the range of −10‰ to −20‰, strongly indicative of organic fractionation though inorganic processes can also fractionate carbon isotopes (Pavlov *et al.*, 2001). This work is not contested by Fedo and Whitehouse (2002). The carbon is present as 2–5 μm graphite globules, that appear to be biogenic detritus. They are hosted in turbiditic sediments and in pelagic muds. The simplest interpretation is that these carbon particles were originally (before deformation and metamorphism) organic remains, and represent the bodies of settled planktonic organisms. The implication is that plankton, and hence mesothermophilic organisms, were present globally before 3.7 Ga ago. Currently, this is the oldest claimed evidence for life on Earth that has as yet withstood critical scepticism.

14.3.2.2 Barberton

Evidence for early life comes from the Barberton Mountain land of South Africa (Byerly *et al.*, 1986), in material from the 3.3 Ga to 3.5 Ga Swaziland Supergroup.

Byerly *et al.* (1986) described probable stromatolites in the Fig Tree Group, preserved in grey-black finely laminated chert. The structures are made primarily of microcrystalline chert, forming low-relief laterally linked domes and in places pseudo-columnar structures. Byerly *et al.* did not find evidence of microfossils but inferred an organic origin from the morphology of the structures. However, Lowe (1994) disputed this evidence and concluded that the structures were not demonstrably of biotic origin.

Elsewhere in the Barberton Mountain Land is a wide array of mid-Archean volcanic and sedimentary rocks, ranging up to >3.5 Ga old. Some material is clearly biogenic (Westall *et al.*, 2001), with highly fractionated carbon isotopes (δ^{13}C −27‰), but may be of non-Archean age. Thus the case for mid-Archean biotic material in Barberton remains open.

14.3.2.3 Western Australia

Rocks of similar age to Barberton occur in the 3.4–3.5 Ga Warrawoona Group, Pilbara, Western Australia. A wide range of rock types is present, both lavas and sediments. There is strong controversy as to whether or not microfossils are present in the Apex cherts of the Warrawoona Group (Buick *et al.*, 1981): this controversy is summarized by the debate between Schopf *et al.* (2002) and Brasier *et al.* (2002) (see also Gee (2002) and Kerr (2002) for excellent reporting on the debate, and Buick, 1990). Lowe (1994) also dismisses claims that structures described as stromatolites in the Warrawoona Group are actually of organic origin.

Schopf *et al.* (2002) and earlier work cited therein, found evidence for microbial fossils in Pilbara and Barberton material. The laser-Raman imagery reported by Schopf *et al.* (2002) demonstrated that the material was made of kerogen and they interpreted this as evidence for remains microbial life. Brasier *et al.* disputed the earlier work by Schopf and Packer (1987) and Schopf (1993) on Warrawoona material, constructing a detailed case in which they reinterpreted the supposed microfossils of the earlier study as secondary artifacts of graphite in hydrothermal veins. However, Brasier *et al.* (2002) did report C isotopic results that are most easily (though not conclusively) interpreted as microbial. Thus although the "microfossils" earlier reported by Schopf may not be organic, there is isotopic evidence suggesting biological activity, though of uncertain age (possibly later than the host country rock).

Several notable pieces of evidence for early life come from Western Australia. Shen *et al.* (2001) found isotopic evidence for microbial sulfate reduction in 3.47 Ga barites from North Pole in the Pilbara. Intuitively, sulfate reduction would be expected to be very old: this confirmatory evidence is strong. Also notable is the discovery by Rasmussen (2000) of filamentous microfossils in a 3.235 Ga old volcanogenic massive sulfide deposit, a type of deposit that only forms under deep water. The implication is that hyperthermophile microbial life was certainly present on Earth by this date, and in deep water. One diversion is of interest here. The abundant microbial life around mid-ocean ridge vents would have meant that considerable amounts of reduced carbon were preserved under the lava flows. This would have affected the net balance of the atmosphere, leaving an excess of oxygen. It would also have introduced reduced carbon down subduction zones. Interestingly, some diamonds have light carbon isotopes that may have "organic" ratios prior to metamorphism, and also contain "ophiolite like" inclusions, palimpsests of a mid-ocean ridge origin. Just possibly, some diamonds may be carbon from ancient microbial colonies (Nisbet *et al.*, 1994).

There is also evidence for the presence of methanotrophs in the ~2.8 Ga old Mount Roe palaeosol. This contains highly fractionated organic carbon, probably recording the activity of methanotrophs living near ephemeral ponds: this implies that significant biological methane sources existed in the late Archean (Rye and Holland, 2000). Oil is also present in some Archean sandstone (Dutkeiwicz *et al.*, 1998, Rasmussen and Buick, 2000).

In the late Archean of Western Australia, there is much evidence of life, both macroscopic and microscopic. Of particular interest are stromatolites from the Tumbiana Formation, in the 2.7 Ga Fortescue Group (Buick, 1992). These have diverse morphology and occur in lacustrine sediments. Texturally, they closely resemble younger microbialites, and they are most probably the product of phototrophic microbial life, living by oxygenic photosynthesis in shallow water with negligible sulfate concentrations. Slightly younger, the 2.5 Ga Mt. McRae shale yields bitumens that contain biomolecules characteristic of cyanobacteria (Summons *et al.*, 1999). This evidence strongly supports the notion that cyanobacterial oxygenic photosynthesis was fully established.

The late Archean of Australia contains many carbonate rocks with $\delta^{13}C \sim 0‰$. This is strong circumstantial evidence for global oxygenic photosynthesis. The logic depends on the strong fractionation imposed by rubisco as it selects carbon from the ocean/atmosphere system to incorporate it into living organisms (Schidlowski and Aharon, 1992; Schidlowski, 2002). Though some rubisco-using cells are not photosynthetic, most are, and

the energy that allows rubisco to incorporate carbon into life is photosynthesis. Carbon emitted from the mantle has $\delta^{13}C \sim -5‰$ to $-7‰$. This is emitted into the air and ocean mainly as carbon dioxide. From this mantle-derived carbon, carbon is acquired into organic matter by rubisco, using the harvest of thermodynamic reduction power from the apparatus of oxygenic photosynthesis in the presence of abundant ambient atmospheric CO_2. This carbon chosen by life is strongly selected for ^{12}C and thus has $\delta^{13}C \sim -28‰$ to $-30‰$. Thus, the residue left in the air/sea system is enriched in ^{13}C. In modern-day carbonates, $\delta^{13}C \sim 0‰$, implying by balance ($-7‰$ source, partitioning into $-28‰$ organic life and $0‰$ inorganic sinks) that about a quarter to a fifth of primitive carbon is captured by organic matter, and three-quarters to four-fifths is left as carbonate with $\delta^{13}C \sim 0‰$. Because carbon dioxide is globally mixed, the presence of carbonate with $\delta^{13}C \sim 0‰$ implies a global fractionation of carbon by oxygenic photosynthesis. This indeed is what is recorded in the late Archean.

14.3.2.4 Steep Rock, Ontario, and Pongola, South Africa

The evidence from the 3.0 Ga sequence at Steep Rock, Northwest Ontario, Canada, is very different (Wilks and Nisbet, 1985; Nisbet, 1987) (Figure 3). Here is a large limestone reef, some kilometers long, displaying a wide variety of structures interpreted as formed by life, and also with a range of isotopic evidence that is not greatly dissimilar to modern sequences. The structures vary from large stromatolites (several meters long) to smaller (1–20 cm) stromatolitic structures (sadly some of the loveliest of these have been fractured recently by unknown collectors), deposited close to a major unconformity. These are among the oldest unchallenged examples of stromatolites: claims of older examples have been strongly criticized (Lowe, 1994). Isotopic evidence from Steep Rouch (Abell, Grassineau, and Nisbet, unpublished) indicates that rubisco-mediated carbon capture (i.e., oxygenic photosynthesis) controlled the global carbon partitioning between carbon dioxide and carbonate: this is some of the oldest evidence for global oxygenic photosynthesis.

The Pongola sequence in South Africa (Matthews and Scharrer, 1968; von Brunn and Hobday, 1976) also includes stromatolites above a major unconformity, and is uncannily like Steep Rock both in age and sequence: it is tantalizing to wonder if they were once contiguous before the vagaries of continental breakup and re-assembly.

14.3.2.5 Belingwe

The evidence for life in the sediments of the Belingwe belt, Zimbabwe, has been described by

Figure 3 (a) The surface of the 3 Ga Earth, Steep Rock, NW Ontario, Canada. The hill-face is very close to a 3 Ga unconformity surface, and the rocks (granitoids and mafic dikes) exposed on the hill-face are immediately below the unconformity. Above them are assorted sediments, including thick stromatolitic limestones. (b) Stromatolitic limestone, Steep Rock, Ontario, Canada (ca. 3 Ga old). The palaeohorizontal surface dips ~70°. Stromatolitic domes are up to 4–5 m long and 2 m high.

Martin *et al.* (1980), Nisbet (1987), Grassineau *et al.* (2001, 2002), and Nisbet (2002). The Belingwe Greenstone Belt has a wide and diverse array of evidence for late Archean life. In this it is not unique—many Australian and South African sequences also have abundant evidence of life. What makes the Belingwe belt fascinating is the range of features outcropping in a small area, coupled with some extremely well-preserved igneous rocks (Bickle *et al.*, 1975; Nisbet, 1987).

The rocks of the Belingwe belt span a range of ages, but the sequence that carries the most detailed evidence for life (Figures 4 and 5), the Ngesi Group, is 2.7 Ga old. The base of the Group includes shallow-water sediment locally rich in carbon and sulfur that is highly fractionated isotopically, suggesting the original presence of methanogens, as well as the operation of complex sulfur fractionating processes (Grassineau *et al.*, 2001, 2002). Oil is present in some rocks (Grassineau and Nisbet, own observations). Locally associated with this stratigraphic unit are stromatolites made of calcite with $\delta^{13}C \sim 0\%$, with kerogen that contains carbon which is strongly fractionated

isotopically, implying the selection of carbon by rubisco (see Section 14.7.2). Immediately above the basal sediments are komatiite pillow lavas and flows (Figure 5). Close to the contact with the lavas, in the uppermost sediments, are sediments that in places are very rich in kerogen and sulfides, with highly variable fractionated carbon and sulfur isotopes, different in very small physical distances. The simplest interpretation of this (Grassineau *et al.*, 2002; Nisbet, 2002) is that the complex isotopic fractionation is a record of consortia of prokaryotes, some reducing sulfate and some perhaps oxidizing sulfur, others generating methane, some photosynthesizing and capturing carbon by using rubisco, and perhaps carrying out other microbial biochemistry (using metal enzymes). Both in shallow-water photo-settings and in deeper water below the photic zone, microbial mats may have cycled sulfur in sulfureta, as in modern parallels (Fenchel and Bernard, 1995).

Above the komatiites are thick basalt pillows and flows. At the top of the sequence is a further sequence of shallow-water sediments, including limestones that locally have extensive and very well preserved stromatolites (Figure 4). These too have evidence for rubisco fractionation (see Section 14.7.2), both in kerogen carbon, and in carbonate with $\delta^{13}C \sim 0\%$.

14.4 THE FUNCTIONING OF THE EARTH SYSTEM IN THE ARCHEAN

14.4.1 The Physical State of the Archean Planet

The map of the surface of the Archean planet remains largely blank, populated by imagined beasts and perhaps some features seen dimly but truly (Macgregor, 1949). The main input from the mantle to the surface is via volcanism. Late Hadean and early Archean volcanism would have provided thermodynamic contrast, placing material that had equilibrated with the mantle in contact with the ocean-atmosphere system that was open at the top to space and light. In the latest Hadean and earliest Archean this contrast would have been most likely thermodynamic basis of life.

Early Archean volcanism was probably largely basaltic or komatiitic, but perhaps with some andesitic and alkaline centers. The mantle may have been somewhat hotter than today (Nisbet *et al.*, 1993), and thus the primary melt at mid-ocean ridges would likely have been more magnesian than today. Moreover, a hotter mantle would likely have sourced more plume volcanoes than today. These volcanoes would have been comparable to modern Hawaii but may have ranged up to much larger sizes. The plumes would have emitted komatiite lava flows. These are less viscous than

Figure 4 Stromatolitic limestone, Cheshire Formation, Belingwe belt, Zimbabwe (2.6–2.7 Ga old): (a) outcrop surface—structures occur on a variety of scales, from microscopic to metre relief; (b) detail of one outcrop (from Nisbet, 1987); and (c) shallow-water shale associated with Cheshire stromatolites.

basalt, and would have flowed long distances on relatively flat surfaces, creating huge flat shields, perhaps as large wide islands emerging as the upper fraction of enormous volcanic platforms resting on oceanic plate.

Komatiite lava flows are very rich in MgO. They contain significant iron oxide, and are typically associated with nickel sulfides and chromite. Hydrothermal systems in highly magnesian rocks can be very alkaline, with very high pH. Thus, it would be expected that rain falling onto komatiite flows, or flows into shallow (low-pressure) seawater, would generate very alkaline outflows of hot or warm water.

The zircon evidence and the existence of 4 Ga gneiss provide evidence for the existence of continental crust, but this may have been of limited areal extent. Significantly, by the early Archean there had probably been inadequate time for deep continental lithosphere to develop, yet by 2.7 Ga, late Archean diamonds are known in the Witwatersrand record (Nisbet, 1987). Diamonds imply lithosphere at least 150 km or so thick, and suggest kimberlite and probably a spectrum of alkali volcanism on land. Alkaline volcanism is indeed known to have occurred, a source of high pH and perhaps phosphatic environments. There is a small but significant record of Archean alkali volcanics (Nisbet, 1987), for example, in the Timiskaming Group in Northern Ontario (Cooke and Moorhouse, 1969), which includes leucitic flows and pyroclasts. Just possibly phosphatic volcanics did occur—arguably the most likely

setting for constructing sugar–phosphate chains in an inorganic process.

Early Archean continents were subject to erosion. Rocks from Isua include sediments, implying the action of rain and the existence of subaerial exposure, as well as the presence of wide oceans capable of evaporating the rainwater. The nature of the sediment was different from today, however. Nowadays, most surface rock is actually recycled previous sediment, and aluminous clay-containing muds (mature sediments) are common. Most of what little there is of the early Archean sedimentary record is not mature: primary volcanic terrains were being eroded. Clays would have been widely present, but were probably mainly magnesium-rich clays derived from weathering of volcanic rock, not aluminum-rich material. This scarcity of mud may be important in considering likely biological host environments.

14.4.2 The Surface Environment

The sedimentary evidence implies the existence of oceans. Although the initial deep volatile inventory of the planet would have been removed by the late great impact that formed the Moon, much of the water presently in Earth's oceans would have degassed from the hot mantle or infallen as comets soon after that great impact, and the ongoing volcanism would have added more. However, at ridges water is rapidly returned to ocean crust by serpentinization and metamorphic hydration of basalt. As soon as old oceanic plate

Figure 5 (a) Thin section of 2.7 Ga komatiite lava, Reliance Fm, Belingwe belt, Zimbabwe. About 6.5 mm across photo. Olivine crystals set in fine grained to once-glassy groundmass. For details see Nisbet *et al.* (1987). Photo W. E. Cameron. (b) Alternating iron-rich and carbon-rich shales. White bands are chert: this lithology is transitional to banded ironstone. Approximately, 20 cm across picture. Belingwe belt, Zimbabwe.

developed, cold plate and hence crustal water would begin to fall back in to the interior down subduction zones, returning more water than mid-ocean ridge volcanism emitted. Given the high mantle temperature, subduction zone volcanism probably rapidly restored the subducted water to the surface. Nisbet and Sleep (2003) suggest that in effect the Earth's mantle is self-fluxing. The net annual contribution of primary new water to from the deep interior the surface (ocean) would thus be set by the inputs of volcanism at ridges and plumes, plus infall of cometary material, minus net loss back into the interior from the small net amount of water that was carried down into the deep interior, and net loss by loss of hydrogen to space.

The controls on carbon dioxide would have been somewhat different. Today, carbon dioxide is stored in carbonate minerals in the ocean floor and on the continental shelf. Subduction, followed by volcanism, cycles the carbon dioxide to the mantle and then restores the CO_2 to the air. Metamorphic decarbonation of the lower crust also returns carbon dioxide. The carbon dioxide is then cycled back to the water, some via rain, some dissolved via wave bubbles. Erosion provides calcium and magnesium, eventually to precipitate the carbonate. In the earliest Archean, parts of this cycle may have been inefficient. The continental supply of calcium may have been limited; however, subseafloor hydrothermal systems would have been vigorous and abundant, exchanging sodium for calcium in spilitization reactions, and hence providing calcium for *in situ* precipitation in oceanic crust.

Before significant thicknesses of lithosphere had cooled over large areas, the subduction may have been limited, and hence the return of carbonate-held carbon dioxide to air, via subduction volcanoes, would have been hindered in the earliest Hadean: by late Hadean subduction should have become the general fate of old oceanic plate. Cooling of plate depends on having a cool surface. The temperature of the late Hadean Earth's surface is unknown, but Sleep and Zahnle (2001) and Sleep *et al.* (2001) have made an excellent circumstantial case that the ambient surface environment was glacial, ice over cold ocean. The crustal Urey cycle buffers carbon dioxide in the air. In the Urey cycle, if global warming occurs, silicate weathering is speeded up, more calcium, magnesium, strontium cations are released and hence carbonate is formed: thus the carbon dioxide greenhouse is reduced, ending the warming. Carbon dioxide is also cycled via the mantle: outgassing at the mid-ocean ridges adds carbon dioxide to the air, while alteration of ocean floor basalt precipitates carbonate, and the subsequent subduction of carbonated oceanic crust returns carbon dioxide to the Earth's interior.

To return now to the carbon dioxide question, early in Earth's history, degassing would have been vigorous but so would have been the return of carbon to the interior, and it is likely that the mantle cycle would have dominated (Sleep and Zahnle, 2001). Moreover, frequent meteorite impacts would have created vast quantities of basalt ejecta that would also have reacted with carbon dioxide to precipitate carbonate. Sleep and Zahnle (2001) concluded that so much carbon dioxide would have been held in the mantle that the greenhouse warming would have been small: the Earth was probably heavily glaciated—the Hadean was probably a Norse ice-hell. Possibly early Hadean Earth risked loss of atmospheric carbon dioxide to the interior more than dehydration by hydrogen loss to space, though this would depend on how much methane was in the air.

If so, the likely ambient conditions (Sleep *et al.*, 2001) would have included a dry troposphere with little water vapor (and hence little OH) in the low temperatures, and wide ice cover only locally broken by water leads on the sea surface (Figure 9(a)). The air would have had very high dust content from the volcanic eruptions and meteorite impacts, possibly being so dusty as to inhibit rainfall (especially given the dryness). Continents would have been covered by dirty ice or perhaps dry permafrost (given the very low humidity). If conditions were cold enough and CO_2 concentration high enough, possibly carbon dioxide was present in polar ice. From time to time, perhaps millions of years apart, massive meteorite impacts would have ejected huge amounts of water and dust, melting ice and changing albedo (Sleep *et al.*, 2001). Brief warm episodes would result, with water-aided greenhouse conditions, and then slowly the ice cover would return. In this oscillating climate there would be many local oases of warmth around volcanic hydrothermal systems. Some of these would operate under ice cover (as in Iceland today) offering an interesting and very diverse variety of chemical settings very closely juxtaposed in space and perhaps repeatedly replacing each other in time as the hydrothermal systems fluctuated. These settings would include all possible phases: warm rock surfaces in warm or hot water/brine; fumaroles with various vapor phases; and locations in ice; and in warm water (Nisbet and Sleep, 2001a,b).

In addition to plate boundary and plume-related hydrothermal systems, the chemistry of the prebiotic world would have had strong redox contrasts in the restricted areas that had tidal coasts, and perhaps within the oceans where differing water masses interacted, or under ice. These redox contrasts were ultimately driven by photolysis in the atmosphere/ocean (presumably made of water and carbon gases) and escape from the top of the system, and also by magmatic interaction at the bottom (around komatiitic vents), where reduced species such as H_2 would have been generated. Likely terrestrial sources of redox contrast included: hydrogen emitted from serpentinization reactions when water reached hot ultramafic rock; sulfates in air and water versus sulfides in hydrothermal deposits; carbon dioxide in air versus methane or CO in hydrothermal systems; nitrogen oxides in air and water versus ammonia in hydrothermal emissions; as well as the contrast

between water and magmatic hydrogen. Meteorites would have provided reduced iron and carbon particles. Hot iron, falling in to water, could generate hydrogen.

A world is only interesting to biology if it offers a way of making a living. The first life must have been unskilled, not equipped to search out the necessities of life: Thus it must have existed where a strong redox contrast was accessible, either spatially (over a few microns) or temporally (in a fluctuating setting, where regular variation took place between one redox regime and another, within hours or even minutes (e.g., as in a geyser). Obviously, the late Hadean Earth offered these on a plate.

14.5   LIFE: EARLY SETTING AND IMPACT ON THE ENVIRONMENT

14.5.1   Origin of Life

Over the origin of life, Nature has chosen to draw a veil. A basic criterion in science is that the result should be reproducible, falsifiable. Not one of the notions of the origin of life has led to reproduction, yet not one can be falsified. No doubt success will soon come in the effort to understand the detailed step-by-step molecular controls of reproduction. There are many notions about the origin of life. Where there is little fact, imagination is allowable and profitable, but where there is no fact, then even imagination is best left unimagined here. Similarly, the question "what is life?" is perhaps best left to the consideration of Hades by trouser-role opera singers, of uncertain reproductive ability, seeking Eurydice. Life is more than reproduction, which clay minerals also achieve. Defining the boundary between life and nonlife is, to quote N. H. Sleep, like searching for the world's smallest giant.

Nevertheless, despite these warnings, the questions are of supreme interest. Given that life bends the rules, a slight digression is warranted. A definition of life is perhaps best approached via thermodynamics (Nisbet, 1987). Life is growth—it is always in disequilibrium with its surroundings, and its actions are such as to increase that disequilibrium. Sustainable, equilibrium molecules are dead molecules. In practice the boundary is set between the cell and the virus: the cell can in principle reproduce and thus increase the scale of the disequilibrium, while in contrast the virus can crystallize and thus set itself in a fixed point on the entropy scale.

There are several favorite notions of the site of the origin of life (Nisbet, 1987). The best known is the Marxist hypothesis of the "primaeval soup"—that the early ocean was a soup of organic molecules that had fallen in from meteorites (which frequently contain complex carbon-chain compounds: organic chemicals, but made by prebiotic inorganic processes). In this soup, lipid blobs somehow evolved into living cells. The discovery of hydrothermal systems led to the realization that early oceans would have pervasively reacted with basalt, both in hydrothermal systems and also with basalt ejecta after impacts. Thus, the late Hadean ocean was most unlikely to be a festering broth, but more likely a cool clean ocean not greatly dissimilar to the modern ocean: exit the primaeval soup.

Other hypotheses note the properties of minerals, especially clay minerals (Bernal, 1951, 1967), iron oxides and zeolites. Hooker, in a letter to Darwin that provoked the "warm little pond" hypothesis (Darwin, 1959), noted the characteristics of modern hydrothermal systems: abiotic formation of hydrocarbons may occur today in mid-ocean ridge systems (Holm and Charlou, 2001). An interesting variant is the idea of "genetic take-over" (Cairns-Smith, 1982). This is based on the notion that some minerals are not greatly different from viruses—as Schrodinger (1944) pointed out, life is based on molecules that can be crystallized as aperiodic crystals. Mineral crystals reproduce, in a sense, when they grow—each crystal surface seeds new copies of itself. In one version of the genetic takeover hypothesis, the earliest replicating structures were simply minerals, that replicated just as clays minerals grow. These structures bound proteins, which helped in the reproduction. Then nucleic acid took over the role of the mineral template, and occupied the central direction of the reproducing body (Figure 6).

The "panspermia" hypothesis is simple (Crick, 1981)—Earth was seeded by little green men from outer space, who spread life cells by sending rockets throughout the galaxy. This hypothesis has the attraction of avoiding the impossible task of elucidating how life began on Earth by transferring the problem to another planet far away and long ago; it also achieves a happy congruence with Star Trek's DNA-based universe. However, it is not discussed why the men were green, or why they were men: pan-oo would be perhaps more likely than pan-sperm.

14.5.2 RNA World

Of the many origin of life ideas, the "RNA-first" idea (Gilbert, 1986) is worth noting in more detail: the idea that prior to DNA, the genetic code was held in RNA. This does not necessarily mean that life began as RNA (a takeover is possible), but at some stage it seems likely that life was RNA-based. All cells today use ribosomes—a giant RNA enzyme—to read the DNA tape, and RNA retains the key role of carrying messages in the cell. It may be that at one stage life was a few self-replicating RNA molecules.

If so, how did these RNA molecules exist? Possibly they were sophisticated enough already

to have outer bags and thus containers for the protein they made. But it is also possible to imagine an early RNA world (Gilbert, 1986; Nisbet, 1986) in vesicles in a rock, where the container was provided either by the rock itself, or by minerals with large tubular shapes, such as faujasitic zeolites or some of the iron oxide minerals. Chemicals and redox drive would be provided by fluids flushing through the setting. Any RNA molecules that accidentally managed to self-replicate would be protected and would propagate; one might next accidentally develop the ability to synthesize proteins that could be assembled to act as enzymes aiding replication, increasing the population. Volcanic accident could spread the molecules from the first container into other parts of the system. Finally, any molecule that accidentally acquired the ability to enclose itself with a lipid bag would be pre-adapted to life in the open environment, away from the rock vesicle. But this is a notion—many other notions have equal or greater validity.

Geologically, some inferences can be made. The setting of the first life to use nucleic acids would presumably have had abundant local phosphate sources and accessible phosphorus, as well as sugars and nitrogen bases. Here the evidence of the existence of komatiite plumes and the antiquity of continents is just possibly relevant. Alkali volcanism is a feature of plume volcanoes (e.g., Mauna Kea in Hawaii). Carbonatite volcanism and associated very unusual rocks (such as phosphatites) occur today mainly on ancient continental crust. Whether phosphate-rich volcanism could have been possible as early as the Hadean is a moot point. Then the lithosphere may have been thin and limited to a segregated cooled–melt earliest crust, plus giant plume volcanic centers, fractionated in their upper stages. Assuming phosphate-rich igneous rocks did exist, then phosphorus-rich hydrothermal systems may have occurred.

More generally, alkaline hydrothermal systems would have occurred around the widespread cooling ultramafic rocks, such as the enormous komatiite flows that would have issued from komatiitic plume volcanoes, and also at distal sites near early mid-ocean ridges (themselves possibly fed by komatiitic basalt liquid). These hydrothermal systems would emit high-pH hot fluids. Here ammoniacal hydrothermal systems (Hall, 1989, Hall and Alderton, 1994) would probably have occurred. Under such high-pH conditions metal atoms (e.g., iron, copper) can form compounds within cages of four nitrogen atoms. Possibly the cytochrome family of proteins, which is clearly very ancient, may have had its origins in such a setting. These proteins have at their heart a metal surrounded by four nitrogen atoms: haem with iron and four nitrogens; chlorophyll with magnesium surrounded by four nitrogens.

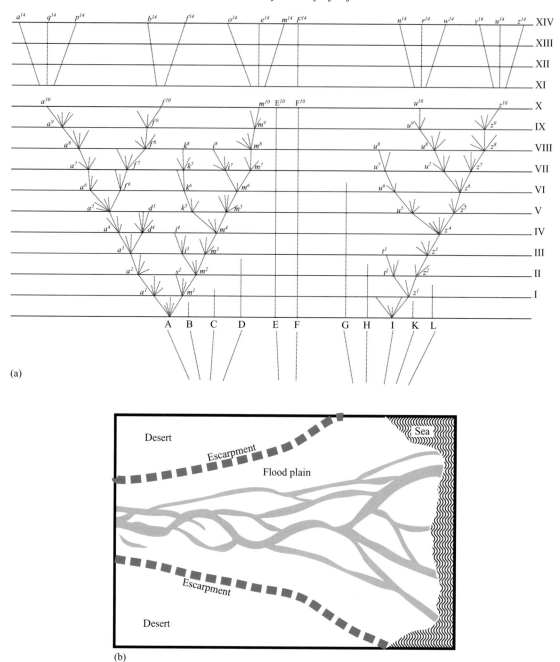

Figure 6 Models of the descent of life: (a) after Darwin's single illustration in Origin of Species (chapter IV) (Darwin, 1859, 1872) and (b) braided delta model, assuming large-scale lateral gene transfers and boundaries of nonviability.

14.5.3 The Last Common Ancestor

The *last* common ancestor is more accessible to geology and molecular biology than the first ancestor. Though not less controversial than the first ancestor, it is at least the subject of testable hypotheses.

The *last* common ancestor is the notional cell, or population of cells, from which all modern living cells are descended (Woese, 1999). One definition of the Hadean/Archean boundary is the date of the last common ancestor. This last ancestor would have been a DNA-based organism, already complex, with many of the so-called housekeeping proteins that are broadly common to nearly all modern types of cell. Note however, that viruses, especially RNA viruses, may (or may not) be separately descended from an earlier ancestor.

There is much debate about the habitat—and hence metabolic processes—of the last common ancestor.

The majority view is that the root was a prokaryote, more bacterial than anything else, from which diverged the sister domains of Archea and Eucarya (Woese, 1987) (Figure 7). In this view, complex eukaryotes evolved from simple prokaryotes. This interpretation also leads to the inference that the last common ancestor was a hyperthermophile, living in hot conditions (>85 °C) probably in close proximity to a hydrothermal system (Stetter, 1996; Nisbet and Fowler, 1996a,b; Miyazaki *et al.*, 2001). In standard microbial phylogenies (e.g., Woese, 1987; Barnes *et al.*, 1996; Pace, 1997), the most deeply rooted organisms all appear to live in high-temperature settings. This view makes abundant geological sense, as the diversity and fluctuation of chemical settings in hydrothermal systems offers readily accessible thermodynamic drive for prephotosynthetic life; while the deep

involvement of metals in the ubiquitous (and thus presumably ancient) enzymes responsible for the ousekeeping biochemistry of cells strongly suggests hydrothermal supply. Moreover, heat shock proteins are integral to protein shaping, suggesting the speculation that heat shock was a general problem for early life (Figure 8).

However, the argument in favor of a mesophile last common ancestor is equally strong (Forterre, 1995; Forterre and Philippe, 1999; Galtier *et al.*, 1999). Heat is a threat to life: it cooks it, and cells have heat-shock proteins to restore them if slightly cooked. It seems counterintuitive to imagine that life started in a place so risky, before it could evolve protective mechanisms. Forterre suggested that life began in milder mesophile settings, with an initially poorly organized and complex structure. Then, when bacteria and archaea spread to the more dangerous but thermodynamically

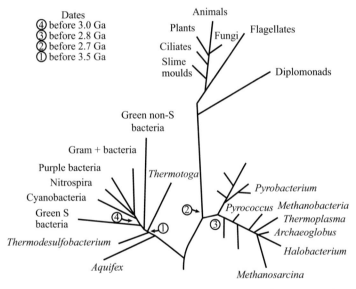

Figure 7 Model of the descent of life following the "standard" model of Woese (1987), as calibrated by the geological evidence (source Shen *et al.*, 2001, and other evidence). See Figure 11 for alternative model.

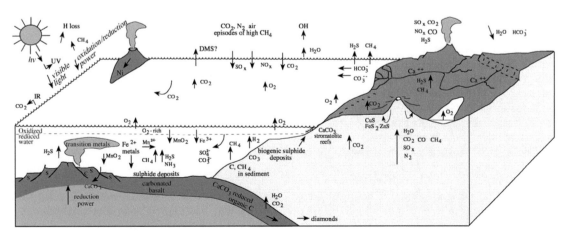

Figure 8 Sources of disequilibrium—possible geochemical (redox) resources for life in the early to mid-Archean.

advantageous hyperthermophile settings, those that prospered were cells descended from lines that had evolved more efficient, streamlined genomes ("thermoreduction"). Forterre (1995) considered the RNA-world idea incompatible with the notion (e.g., Stetter, 1996; Nisbet and Fowler, 1996a) that early life was hyperthermophile. RNA is unstable at very high temperatures. Moreover, modern hyperthermophiles have very sophisticated mechanisms to sustain them in hot environments: unlikely in very primitive cells. Forterre (1995) suggested that early cells were complex mesophiles, and those that strayed into hotter settings slowly adapted to the conditions by selection for reduced and streamlined genotypes, to produce the hyperthermophiles.

An analogy would be the comparison between geometrically complex early multiple-winged aircraft such as Sopwith biplanes of the 1914–1918 era, the streamlined Hawker Hurricane monoplanes of the Battle of Britain with protruding propellers and tail wheel, and the Hawker Harrier still today in service: all built under the direction of Thomas Sopwith. These airplanes simplified in shape as they became more powerful and internally complex. Yet they are all part of a single line. Or, continuing the analogy, the officer commanding those Battle of Britain fighters, Marshal of the RAF Sholto Douglas, was brother-in-law of author J. D. Salinger: close relations, utterly different careers.

Derivation of molecular phylogeny from rRNA suffers from various mathematical pitfalls, especially the difficulty of dealing with branches of the tree of life that evolve especially rapidly. Any model that assumes uniform rates of evolution will make these branches appear inaccurately ancient. Moreover, there is massive evidence for multiple gene transfer between distinct lines within domains, and across domain boundaries. For example, up to 18% of *E. coli's* genes may be relatively recent foreign acquisitions (Martin, 1999). This complicates interpretation enormously (Doolittle, 1999, 2000), and leads to models not so much of "trees" of descent but of "mangrove roots" (Martin, 1999), or analogies with braided deltas (Nisbet and Sleep, 2001; Figure 6). Woese (1999) concluded that the communal ancestor was not so much a single discrete organism but a diverse community of cells that evolved together as a biological unit. In this view, the universal phylogenetic tree is not an organismal tree at its base, but becomes one as the peripheral branchings emerge.

The choice between explanations suggesting (i) shared ancestry between the lines, rather than (ii) lateral transfer of information between contemporary but unrelated lines, is not easy. Thus the molecular record is very "noisy" and the interpretation of descent is ambiguous. It is very difficult to be sure of the limited number of positions

in an amino acid or nucleotide sequence that actually record true antiquity.

Initially, it was thought that derivation of phylogeny from molecular information is intrinsically superior to phenotypic information. Forterre and Philippe (1999), Penny and Poole (1999), and Glansdorff (2000) point out that this is not necessarily true. The microsporidae, for example, were originally misclassified as very ancient. More recently, these have been shown to be closely related to fungi, a much younger line (Hirt *et al.*, 1999). The discovery of the error in placing the microsporidae increases awareness that massive lateral gene transfer has occurred between the three domains of life. Each domain is distinct and monophyletic, but members of each domain have obtained genetic information from other domains.

In this view, the eukaryotes may well preserve some very primitive characteristics that are not seen in prokaryotes. Glansdorff (2000) reappraised claims for lateral gene transfer and concluded that the extent of transfer was overemphasized; moreover, Glansdorff inferred that the last common ancestor was probably nonthermophilic and perhaps a protoeukaryote, from which the thermophilic archaea may have been the first divergent branch.

Conceivably, if the last common ancestor were mesophile, the majority of bacteria (except perhaps planctomycetes (Brochier and Philippe, 2002)) may descend from an early mesophile prokaryote, perhaps via a genetically streamlined descendant that occupied a hyperthermophile setting. Archea too may descend from the last common ancester via a streamlined cell that had evolved to inhabit hyperthermophile settings. In contrast, the Eucarya may be directly descended from a mesophile, as may the planctomycetes.

A possible geological scenario for this process may be that the last common ancestor lived on the warm ($\sim$40 °C) but not hot periphery of a hydrothermal system, on a glaciated planet. Descendants of the last common ancestor may have evolved to occupy hyperthermophile habitats, with sophisticated biochemical processes to ensure their survival. Other descendants may have spread to occupy planktonic mesophile habitats. Large meteorite impacts, capable of heating the oceans to near 100 °C, would have occurred occasionally prior to 3.8 Ga ago. Such impacts would have destroyed all life except two types of organism: those forms capable of living in high-temperature conditions, and perhaps also those organisms that had been accidentally preserved in especially thick ice caps. Modern organisms can survive up to half a million years or more in ice (Reeve, 2002), and there are cells preserved in ice that has crystallized from Lake Vostok, the great ancient lake under the Antarctic ice cap. Just possibly, early relatives of the planctomycetes, a bacterial branch which may be of the greatest antiquity, may have been distributed in the glacial oceans, and would have been subject to freezing in thick ice cap, and thus preferentially likely to survive a global heating event after a meteorite impact.

14.5.4 A Hyperthermophile Heritage?

Whatever the setting of the last common ancestor, there are many aspects of modern cells that have a possible or likely hyperthermophile origin. To possess such a heritage, it is not necessary that a cell's primary ancestral line once occupied a hyperthermophile habitat. There has been much genetic exchange between organisms both within lines and even massively between domains (Figure 6(b)).

Candidates for biochemical processes or molecules with hyperthermophile origins include the heat shock proteins, and the metal enzymes (Nisbet and Fowler, 1996b). Heat-shock proteins are ubiquitous in all domains of life. They help repair damage after heat shock, but more generally they help to shape new protein molecules so they can carry out their proper functions. The heat-shock proteins are clearly of the greatest antiquity, given their involvement in very basic housekeeping processes. Their role as heat-shock repairers may of course simply be a relatively late adaptation to life in hot settings. Alternately, however, heat-shock proteins may indeed descend from an original function evolved to enable life to enter hyperthermophile settings around hot-water vents.

Like the heat-shock proteins, the metal enzymes are central to many very basic cell functions. The Metal-4N and Ni proteins have already been mentioned. Many other metal proteins involve metals such as iron, copper, or zinc, often associated with four sulfur atoms. Such metals are characteristic of hydrothermal systems hosted by basaltic and andesitic volcanism. More generally, easily available metals in hydrothermal systems play a key role in many vital housekeeping proteins, often but not always associated with four sulfurs. Examples include zinc in carbonic anhydrase, alcohol dehydrogenase, and RNA and DNA polymerases; copper in proteins used in respiration, such as cytochrome c oxidase; cobalt in transcarboxylase; Mo in many enzymes participating in the nitrogen cycle, in sulfite oxidase, in some dehydrogenases, and in Dimethylsulfoxide-trimethylamine oxide reductase (which may have had an important role in early methane-linked atmospheric chemistry); selenium in hydrogenases; and iron in a wide range of catalases, peroxidases, ferredoxins, oxidases, and all nitrogenases.

Nickel, in particular, is interesting to the geologist. For example, carbon monoxide dehydrogenase, which is at the center of the acetyl-coA pathway of reducing carbon dioxide, characteristically contains nickel, zinc, iron, and molybdenum. Both coenzyme F_{430} of methanogens and hydrogenase contain nickel. Consequently, nickel is essential to methanogens. Moreover, urease, a key part of the nitrogen cycle, converting urea to carbon dioxide and ammonia, is based on nickel. The most obvious supply of nickel in nature is komatiite: highly magnesian high-temperature lavas that would have been widespread in the late Hadean and early Archean. Around komatiites nickel sulfide would have been freely available. It could be that it was in this setting that nickel metal proteins evolved: perhaps it was around komatiite flows that hydrogenases, carbon monoxide dehydrogenase, and urease began. It may be that it was in such settings that methanogens first appeared, exploiting the hydrogen made from serpentinization reactions (see Section 14.6.5) (Figure 9). It is interesting to wonder if the cytochromes, methanogens, and the nitrogen cycle all first evolved on the flanks of komatiite volcanoes.

Today, metals are scavenged from water by extremely sophisticated biochemical processes (Morel and Price, 2003). Thus, seawater can have very low ambient levels of metal ions. Early Archean seawater would likely have been much richer in trace metals. But given that early organisms presumably had very unsophisticated processes for capturing metals, even in seawater rich in metal it would have been difficult to access the metal. Perhaps the earliest distribution of organisms was very restricted, with few cells living away from locations such as volcanoes that had readily accessible metals. Only the evolution of effective metal-gaining siderophores would have allowed the spread of life. There is thus reason to believe that, even if the last common ancestor was not hyperthermophile but lived in somewhat cooler conditions, from it came volcano-hosted hyperthermophile ancestral lines, living in and around hydrothermal systems, that led to the Archeal domain and perhaps also to most bacteria (excluding perhaps the planctomycetes). There has been much gene exchange since then, and consequently enzymes of hyperthermophilic origin are ubiquitous in the housekeeping chemistry of all cell lines. The volcanic signature is written deeply into all life.

14.5.5 Metabolic Strategies

It is likely that the oldest organisms were not photosynthetic (see discussion in Nisbet *et al.*, 1995; Nisbet and Sleep, 2001). Prephotosynthetic organisms would have depended on natural redox contrasts, and would thus have lived in habitats where such contrasts were accessible, either spatially or temporally (in fluctuating conditions). Air is never in chemical equilibrium, and always

Hadean

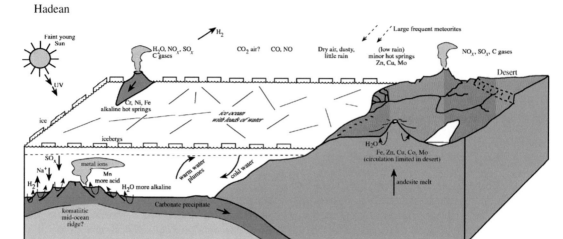

(a)

Early Archean: prephotosynthesis

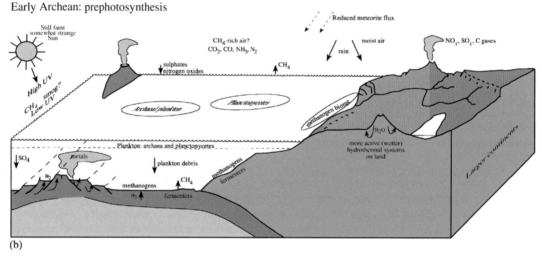

(b)

Late Archean: with oxgenic photosynthesis

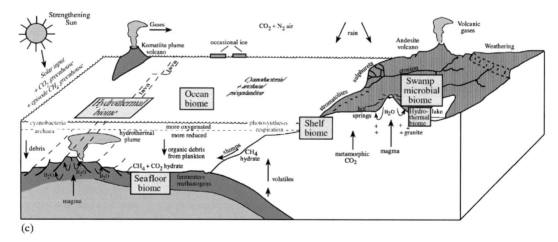

(c)

Figure 9   Model of the evolution of the planetary surface: (a) Hadean surface, possibly glacial (apart from rare very hot events after major meteorite impacts); (b) early Archean surface, before the onset of photosynthetic processing of the air; and (c) late Archean surface, assuming that the major biochemical pathways had evolved, and that the main groups of prokaryotes had evolved.

contains both reduced and oxidized species. At the top of the atmosphere and in the higher levels there would have been radiation-induced sources of oxidation power: the oxygen left after loss of hydrogen knocked out by UV, cosmic rays or solar wind; and also OH formed from water vapor in the lower air. The flux of UV in particular has major biological impact (Cockell, 2000). In addition, sulfate and nitrate from volcanic eruptions would have been present (Kasting *et al.*, 1989; see also Alt and Shanks, 1998). Such transient species would have contributed vital oxidation power to the oceans; simultaneously reduced species such as CO, H_2, and perhaps NH_3 would have been present also. The chief source of reduction power would be hydrothermal exchange with magma, providing reduced sulfur species, H_2, methane, CO, and ammonia.

Major reactions supporting prephotosynthetic life (Reysenbach and Shock, 2002) may have included a series of processes that depended on molecular hydrogen that was formed inorganically. Water/rock reaction at high temperature (Stevens and McKinley, 1995; Kaiser, 1995) produces molecular hydrogen when circulating groundwater reacts with ferromagnesian minerals (FeO silicate), producing iron oxide (e.g., Fe_2O_3) and quartz. Deeper in the earliest Earth hydrogen may also have been formed by water reaction with iron, as the iron precipitated to the core, producing an oxide component to the core and releasing molecular hydrogen to the mantle.

Such inorganically released hydrogen would have been available to be exploited by microbial life. Some archaea and bacteria use the "knallgas" reaction:

$$\frac{1}{2}O_2(\text{aqueous}) + H_2(\text{aqueous}) \rightarrow H_2O$$

Others reduce sulfur:

$$S + H_2(\text{aqueous}) \rightarrow H_2S(\text{aqueous})$$

Methanogenesis (Thauer, 1998) is another process that involves H_2; in this,

$$CO_2(\text{aqueous}) + 4H_2(\text{aqueous})$$
$$\rightarrow CH_4(\text{aqueous}) + 2H_2O$$

These processes then allow sulfate chemistry to give many microbial possibilities: for example, an extreme option is

$$CH_4 + SO_4^- \rightarrow HCO_3^- + HS^- + H_2O$$

Similarly, some planctomycetes can exploit ammonium and nitrogen oxides, likely to be found around volcanoes, to make dinitrogen:

$$NO_2^- + NH_4^+ \rightarrow N_2 + 2H_2O$$

Nature evolves by processing waste dumps. It is possible to imagine, for example, an early community of cells, as a single biofilm located on the site of a redox contrast, making its living from one of the hydrogen-using reactions. This would produce a waste of dead cells—reduced carbon—and also the by-products of metabolism such as sulfur or bicarbonate. Various specialist cells would evolve to tap into the new opportunities afforded either by oxidizing the dead carbon, or using the by-product. These new cells would form a substrate, thickening the biofilm. Then, in turn, the waste of the new cells would be utilized, until the whole network resembled a complex clock with innumerable wheels cycling and recycling the thermodynamic possibilities provided by the basic metabolic redox-driven winding of the spring.

14.6 THE EARLY BIOMES

14.6.1 Location of Early Biomes

Replication of the last common ancestor would lead to mutation: in turn, mutation would create accidental pre-adaptation to life in diverse new habitats. Whatever the habitat of the last common ancestor, the spread of life across the more accessible other locations on Earth was probably rapid, when compared to a geological timescale.

The habitats available were disparate. Examples include: hydrothermal high-temperature ($>85\,^\circ$C) settings; moderate thermophile settings ($\sim40\,^\circ$C) on the fringes of hot areas, or in cooler, probably more alkaline springs; very cool ($\sim0\,^\circ$C) water distal to hydrothermal vents but in the flux of metals and geochemical contrasts from hydrothermal plumes; in tidal waters where currents create a flux of nutrient; around terrestrial volcanoes; under ice; or even in air in dust clouds in frequent eruptions and meteorite impacts.

The first organisms to replicate in each habitat would immediately create new habitat by their very existence. Nature spreads on its own ordure. Dead cells would provide reduced organic matter that could be exploited by reoxidation by other cells, and specialist cells would rapidly evolve. Within a number of generations, mutation would lead to a diversified biofilm, relatively reduced at its base and relatively oxidized above (assuming that the redox gradient is between more reduced rock substrate and more oxidized air/ocean system).

This diversification would lead to distinct types of biofilms in specific habitats—the earliest biomes. They may have been only a few microns thick, but these would have been the first complex communities: the ancestors of interdependent ecologies.

14.6.2 Methanogenesis: Impact on the Environment

Life operates on a global scale. On a geological timescale, once the first cell had replicated, all habitats on the planet would immediately be filled. This would rapidly have consequences for the atmosphere. In particular, methanogens are likely very ancient, and may long predate methanotrophic bacteria. Methanogens most probably predate photosynthesizers if the evolutionary lengths in the standard models of molecular palaeontology (Woese, 1987; Barnes *et al.*, 1996; Pace, 1997) have value. Possibly they also predate the methane oxidizing archaea. These operate by anaerobic oxidation of methane against sulfate, to produce bicarbonate, HS^- and water: their impact would have been limited by the supply of sulfate oxidant. Once methanogens had evolved, they would have occupied proximal and distal hydrothermal habitats, and then perhaps wider habitats such as open ocean (Sansone *et al.*, 2001) and tidal habitats. Possibly methanotrophs evolved quickly following the arrival of methanogens, to exploit the new opportunity: but, in the likely absence of abundant free molecular oxygen, they would have been severely limited by the supply of oxidant.

Methanogenesis on a scale large enough to affect the atmosphere would have been possible if the hydrogen supply from inorganic and organic sources (and hence methanogenesis) had been adequate: given the likely abundance of ultramafic rock near the surface, interacting with hydrothermal water, it is not unreasonable to suppose a major flux of inorganic hydrogen. If so, and there was a surplus of methane, then much of the methane formed by the first methanogens would have been emitted directly to atmosphere. In the dry air on a cold glacial planet, this methane might rapidly overwhelm the OH. Over a few tens of millennia, the atmospheric methane burden would build up and have a major greenhouse impact (see Pavlov *et al.*, 2000), until enough ice melted to permit OH in air and thus control the methane.

Methane may have played a crucial role in allowing the early Earth to be habitable (Pavlov *et al.*, 2000, 2001). Methane emitted by organisms would have had a substantial greenhouse effect, and if the methane/carbon dioxide ratio in the air were high, methane could have fostered an organic smog that protected shallow-level life against ultraviolet radiation in sunlight (Lovelock, 1988). Thus, there is a possible progression here, from the first methanogens, few in total number and confined to the immediate vicinity of hydrothermal systems on a very cold planet, then a warming trend, then development of planktonic life and much more widely spread methanogens, increasing the warming.

Catling *et al.* (2001) pointed out that in the early Archean, biogenic methane may have saved the Earth from permanent glaciation. On the modern Earth, on a 20 yr timescale, emission of methane has an incremental greenhouse impact nearly 60 times, weight-for-weight, or 21 times,

molecule-for-molecule, that of carbon dioxide. On the Archean planet, this ratio would have been very different, and the difference is nonlinear with burden. But whatever the greenhouse impact was, methane is a very powerful greenhouse gas. Indeed, unless abundant methane existed in the air it is difficult to imagine how intense global glaciation was avoided. Thus geologically likely models of the early Archean atmosphere, that are consistent with the Isua evidence for water-eroded and water-transported sediment, would be expected to invoke high methane concentrations (10^2–10^3 p pmv—compared to modern air with less than 2 ppmv CH_4 and $\sim$375 ppmv of CO_2). Such high levels of methane would lead to hydrogen escape by photolysis and loss from the top of the atmosphere, and hence irreversible net oxidation of the planetary surface environment (Catling *et al.*, 2001), though not necessarily to significant ambient O_2 at any particular time.

Methanogenesis may have had the interesting consequence of triggering the evolution of nitrogen fixation (Navarro-Gonzalez *et al.*, 2001; Kasting and Siefert, 2001). On an early planet with CO_2 present in the air, nitrogen fixation would have occurred in lightning strikes, which would have used oxygen atoms from the carbon dioxide (or from water) to form NO. However, if CO_2 levels declined and CH_4 rose, the oxygen supply would be reduced, limiting the synthesis of NO. This would have created a crisis for the biosphere as usable nitrogen is essential. Out of this crisis, Navarro-Gonzalez *et al.* suggested, may have come what now appears to be the essentially "altruistic" process of nitrogen fixation, which is very expensive in energy.

Another, not necessarily incompatible hypothesis is that nitrogenase first evolved as a manager of excess ammonia in the lower, anaerobic part of microbial mats, where hydrogen is present. The product, dinitrogen, could be safely bubbled away. Had a crisis occurred, in which there was a shortage of fixed nitrogen, any cell or consortium of cells able to reverse the process would have been advantaged. It is perhaps notable that in nitrogenase the N_2 is bound to a cluster of Mo–3Fe–3S. Molybdenum, iron and sulfur are likely to be abundant together at hydrothermal systems, especially around andesite volcanoes, and this may be a protein with a hydrothermal heritage. Falkowski (1997) points out that the requirement for iron, and the need for anoxia, would have put severe limits on nitrogen fixation, such that fixed nitrogen supply (and hence the availability of iron), not phosphorus, may be the chief limitation on the productivity of the biosphere. Indeed, the vast scale of human fixation of nitrogen, and perhaps the pH change of the ocean, may some day be seen as the greatest peril of global climate change: not the greenhouse.

14.6.3 Prephotosynthetic Ecology

Early life most likely depended on exploiting the transient redox contrasts available from two sources: within the inorganic geological system—especially at hydrothermal vents (Reysenbach and Shock, 2002); and secondly from inorganic light-driven reactions, such as the formation of transient oxidizing and reducing species in the atmosphere by incident radiation.

These sources of redox contrast would have been limited. The hydrothermal contrasts depend on local thermally driven juxtaposition (e.g., in vent fluids) of chemical species from differing environments. From the vents would come H_2, H_2S, and probably CH_4. The size and activity of the hydrothermal biosphere, and hence its impact, would have been considerable, as early Archean volcanism was probably much more common than today, with a higher heat flow out of the Earth. Nevertheless, the total potential productivity of an early hydrothermal biosphere would have been small on a global scale compared to the modern photosynthetically driven biosphere. Moreover, modern biota at hydrothermal vents depend on the supply of sulfate, oxidized in the photosynthetic biosphere: before photosynthesis, the sulfate supply may have been limited. Thus, as a first guess, with a planetary heat flow higher than today but not massively so, and with a limited supply of oxidation power, it is unlikely that the early Archean chemolithotrophic biosphere would have been vastly greater than the sum of today's hydrothermal communities.

In addition, there would have been redox input from transient chemical species formed in the air. The solar radiation, acting on an atmosphere containing water vapor, would likely have produced OH, and probably some O_2. Volcanic gases, taking part in atmospheric chemistry, would produce a small but important supply of sulfur oxides—and hence sulfate and sulfide in the sea, as well as nitrates and nitrites. Moreover, H_2 and CO would have been present. Together, the inorganic sources of redox contrast probably would have been capable of sustaining a small global biological community.

Life must be continuous—it must always have habitat. Volcanoes, however, become extinct. Thus, life must either have been able to live in the open ocean or must have hopped from dying volcano to new volcano. Volcanic vents were probably abundant enough, close enough and accessible enough (especially to cells capable of floating in cool water, or blowing in wind) that they could host gypsy-like cells that were perpetually seeking a new home as the old one was exhausted.

Nonphotosynthetic plankton are abundant today (Karl, 2002). Many of these are eukaryote zooplankton, but there is also a massive population of planktonic archaea, that live near the base of the photic zone. Indeed, in the Pacific, the archaea dominate the deeper waters below $\sim$1,000 m depth, where pelagic crenarchaeota are abundant (Karner *et al.*, 2001). In the early Archean, there may have been a significant boundary between deeper, more reduced water, and shallower water in sunlight. This boundary, as it shifted diurnally, would provide a fluctuating redox contrast for organisms that could exploit it. For example, the planctomycetes, form macroscopic aggregates ($>$0.5 mm) of detritus, in which they create tiny microaerobic or microanaerobic habitats in otherwise aerobic environments (Fuerst, 1995). They can thus exploit local redox contrast. Among the diverse and interesting properties of the planctomycetes is their ability to react nitrate with ammonia, evolving dinitrogen (the anammox process: Jetten *et al.*, 2001; Fuerst, 1995). This too may be of the greatest antiquity.

Most intriguing of all, they are bacteria that have babies.

14.6.4 Geological Settings of the Early Biomes

Geological evidence for the early distribution of life is fragmentary. In the early Archean of the Isua belt, Rosing (1999) reported isotopic and textural evidence of planktonic life, presumably occupying mesophile or cool, even near-freezing habitats, from prior to 3.7 Ga. A possible (though not robust) inference is that from the last common ancestor, fairly early in Earth's history, came the occupation of a diversity of habitats. If Rosing's evidence is correctly interpreted, by $\sim$3.7 Ga, mesophile plankton existed. On the modern Earth, archaeal plankton are abundant in the deeper parts of the upper ocean, in the deep photic zone and below. Though ill-studied, the planctomycetes have marine examples. Thus, a marine biome, occupied by free-living cells, was probably well established and diversified by the mid-Archean.

The geological evidence for the presence of sulfur-processing microbial life and for methanogens goes back at least as far as the late Archean, and probably earlier. Rocks containing highly fractionated sulfur isotopes, closely spatially associated with highly fractionated carbon, are known from many localities (Goodwin *et al.*, 1976). For example, in the late Archean 2.7 Ga sediments of the Belingwe belt, Grassineau *et al.* (2001, 2002) describe what is interpreted as evidence for a complex biological sulfur cycle. Fractionated pyrite, implying sulfur-processing bacteria, is also known from 3.4 Ga Barberton rocks in South Africa (Ohmoto *et al.*, 1993).

Strong evidence for Archean methanogens comes from highly fractionated carbon isotopes. As mentioned above, these have been found in

2.7 Ga material from Belingwe (Grassineau *et al.*, in press), and also from similarly aged rocks in Australia (Rye and Holland, 2000).

Standard rRNA molecular phylogeny (Woese, 1987; Barnes *et al.*, 1996; Stetter, 1996; Pace, 1997) implies the antiquity of hyperthermophile organisms. Though there has been much dispute about the rRNA interpretation, there is some consensus that, whether or not it is the very most ancient, life around hot-water vents is certainly of great antiquity. The implication is that by mid-Archean, hyperthermophile habitats around hot vents were populated by microbial mats, and the waters around hot vents were occupied by free-swimming cells. Mesophile prephotosynthetic plankton probably existed in the open seas, and, distal to the thermophile life in the surroundings of vents, the mesophile habitats further from the hot springs were also occupied.

The reactions that involve sulfur oxidation states leave isotopically fractionated sulfur and hence sulfide, a target for investigation by the geologist. Though there is controversy about sulfur isotope fractionation (Farquhar *et al.*, 2000), the strong fractionation of $\delta^{34}S$ seen in the best-preserved Archean organo-sedimentary rocks can only be biological. Sulphate reducers are probably very old, present 3.5 Ga ago in the early Archean (Shen *et al.*, 2001), and may have provided sulfur deposits, which in turn supported an increase in the supply of HS and H_2S at the bottom of the biofilm: the biofilm would have thickened, diversified, and turned to a microbial mat, created by structured consortia of prokaryotes (Fenchel and Bernard, 1995; Nisbet and Fowler, 1999). Such mats could have had a large impact on the production of reduced gases added to the air (Hoehler *et al.*, 2001), and could have had a global significance in keeping the planet warm (Kasting and Siefert, 2002). Methane generated at the bottom may have been recycled nearer the top of the mat, in processes such as those described in the modern ocean by Boetius *et al.* (2000) in which archaea and sulfate-reducing bacteria consort.

The evolution of photosynthetic oxidation of sulfur compounds permitted the development of the full microbial sulfur cycle in sulfureta. In this cycle, some bacteria and archaea reduce oxidized sulfur compounds, pumping them downward in the microbial mat, while other bacteria reoxidize them photosynthetically. The development of this cycle, coupled with the use of stored sulfur as a redox bank balance that could be exploited either way the redox budget swung during tidal and diurnal cycles, would have greatly expanded the thermodynamic power of the biosphere.

The thermodynamic drive for this life would have come from various sources. In hot-spring settings, reduced species such as CH_4, H_2S, and H_2 would have emanated from inorganic reactions

around hot magma. These could have provided the basis of methanogenic life; quickly the supply of dead biomass would provide opportunity for other organisms to generate H_2 organically, thus multiplying the opportunities of the methanogens. At the top of the biofilms, sulfate was probably available in water. In the open seas, prephotosynthetic archaeal and planctomycete planktonic life probably spread ubiquitously even before the advent of photosynthesis—it is a small evolutionary hop from a cell loosely bound to a microbial biofilm and a cell that lives in the sea, floating up and down between redox setting. Possible sources of life support, though limited in total flux, would have been widespread. They would have come from volcanic sources, especially in plumes of hot water, creating the contrast between, above, SO_x and NO_y chemical species dissolved in seawater from the atmosphere, and below, reduced chemical species emanating from hydrothermal vents on the seafloor. Structured consortia of archaea and sulfate-reducing bacteria (Boetius *et al.*, 2000) may have had global distribution.

14.7   THE EVOLUTION OF PHOTOSYNTHESIS

14.7.1   The Chain of Photosynthesis

Photosynthesis is the source of the redox power that allowed life to escape from the very restricted early settings where inorganic redox contrast existed, and occupy the planet. Without access to light energy, life would have been permanently restricted to a few narrow settings, probably as thin biofilms, and as plankton near upwellings.

Photosynthesis involves a complex chain of events, each of which must have its roots in the remote Archean (Blankenship, 2001). The chain is of great interest, as each unit presents a separate puzzle in explaining its evolutionary history. Light is captured by pigments, such as chlorophylls (in oxygenic photosynthesis by eukaryotes and cyanobacteria) or bacteriochlorophylls (in other bacteria), as well as accessory pigments such as phycobiliproteins. The light is harvested by an array of chlorophyll molecules (say 300) that form an antenna, around a light-harvesting complex. This array passes the energy of the absorbed photon from molecule to molecule until it reaches a photosynthetic reaction center. In purple bacteria, the photosynthetic reaction center consists of special bacteriochlorophyll molecules, linked to other molecules and a central Fe(II) atom. In the overall process in purple bacteria, the net result of two photons hitting the reaction center is the transfer of four H^+ from the interior cytoplasm to the external medium.

In oxygenic photosynthesis, in cyanobacteria and chloroplasts in plants, there are two linked reaction centers. One (photosystem II; PSII) is similar to that in purple bacteria. At PSII, an oxygen evolving complex based on manganese oxide splits two water molecules into $4H^+$ and dioxygen, O_2, which is evolved as waste. The other, PSI, is electrically connected to the PSII production of H^+, and, with two further electrons, generates NADPH; in addition, ADP synthesis occurs on the membrane, driven by proton flow turning the ADP synthase motor. Thus, the products of light capture are NADPH and ATP.

Then in the biosynthesis reactions, the NADPH and ATP are used to capture carbon from the environment, for use in biology. Three ATP and two NADPH, with two H^+ combine with a water and a CO_2 molecule to form carbohydrate. In sum, a dozen quanta of light energy are needed to incorporate one molecule of CO_2. This process is accomplished by the enzyme ribulose-1,5-bisphosphate carboxylase oxygenase, or rubisco, which can in effect work both ways, either capturing carbon dioxide from the air, or oppositely to return it, depending on the $O_2 : CO_2$ ratio it is exposed to (Lorimer and Andrews, 1973, Lorimer, 1981). On rubisco hangs the balance of the atmosphere (Tolbert, 1994).

14.7.2 The Rubisco Fingerprint

Geologically, photosynthesis presents several quarries to be hunted down in the geological record. The distinctive isotopic fingerprint of rubisco, which presumably must predate oxygenic photosynthesis, is the most obvious target—it is very selective in the carbon atoms it accepts and hence the organic molecules it creates are highly depleted in ^{13}C. There are two main types of rubisco. Rubisco I is used in oxygenic photosynthesis, it operates in aerobic or microaerobic conditions, not anaerobic settings. Rubisco II is characteristic of organisms that fix CO_2 anaerobically. It may be more ancient, and is today found in deep-sea rent organisms (Elsaied and Nagunama, 2001). The oxygen-evolving complex is also a target for the geologist, as it is based on manganese oxide, as are the transition metal isotopes that are likely to have been fractionated by capture in key enzymes.

More subtly, the isotopic signatures of photosynthesis in inorganic sediment are also valuable. Rubisco depletes the environment of ^{12}C. Hence, inorganic carbonate is enriched in ^{13}C if rubisco operates on a planetary scale. Carbon dioxide emitted from the mantle is about $\delta^{13}C - 5‰$ to $-7‰$, on the arbitrary PDB scale. About a quarter to a fifth of carbon in the environment is captured by rubisco to make organic matter: kerogen (rubisco-fractionated organic matter) has about $\delta^{13}C \sim -28‰$ to $-30‰$ when fractionated by rubisco I, but around $-11‰$ when fractionated

by rubisco II (e.g., Guy *et al.*, 1993; Robinson *et al.*, 1998). Three-quarters to four-fifths is residue precipitated as carbonate at $\delta^{13}C \sim 0‰$. The presence of $\delta^{13}C \sim 0‰$ carbonate is thus testimony that rubisco I was capturing carbon on a global scale, in aerobic conditions: this is known as the rubisco fingerprint.

14.7.3 The Evolutionary Chain

Respiration most probably evolved before photosynthesis (Xiong and Bauer, 2002). Each step in this chain must have a long and complex evolutionary history (Pierson, 1994): the puzzle is similar to Darwin's puzzle—what use is half of an eye? And half of photosynthesis? The debate is vigorous and is addressed by Blankenship (2001), and references cited therein. How did the full chain evolve, given that half a chain is useless? The challenge to the geologist is to identify the small steps of pre-adaptive advantage on which evolutionary change worked, to date those steps, and to explain the way the individual links in the chain were incorporated.

There is much debate about the origin of photosynthesis, and little agreement. Among the many hypotheses, Nisbet *et al.* (1995) suggested that photosynthesis began in organisms that were pre-adapted by their ability to use IR thermotaxis to detect hot sources (Figure 10). This hypothesis offers a set of small incremental steps, each immediately advantageous, each depending on accidental pre-adaptation that led to the very sophisticated electron management that occurs in photosynthesis. The steps begin with accidental IR sensitivity in cells that had pigments in their outer surfaces. Deep-water hot vents emit IR radiation at around $350–400\,°C$, slightly below the temperature of a hot plate on a kitchen cooker before it becomes a visible cherry-red. Detection of this radiation would have been very advantageous to motile organisms, and such organisms that possessed IR detection and the ability to move towards the source (or away if it became too powerful) would have gained survival advantage. Evolutionary survival would then have favored those cells that were increasingly finer-tuned to the IR. Then, in the next step, organisms that had spread to a shallow-water vent would be pre-adapted to use solar IR as a supplementary energy source; finally, full dependence on abundant and energetic visible light energy would follow. The hypothesis suggested by Nisbet *et al.* (1995) invokes IR phototaxis: the cells that use IR depend on bacteriochlorophyll and are anoxygenic and usually act in anaerobic settings.

However, other hypotheses suggest that oxygenic photosynthesis came first, depending on chlorophyll, and that from chlorophyll evolved bacteriochlorophyll. Bacteriochlorophyll absorbs

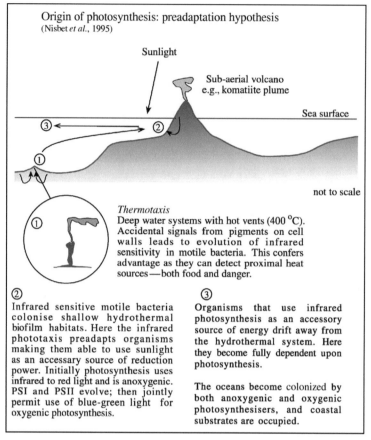

Origin of photosynthesis: preadaptation hypothesis
(Nisbet *et al.*, 1995)

Sunlight

Sub-aerial volcano
e.g., komatiite plume

Sea surface

not to scale

Thermotaxis
Deep water systems with hot vents (400 °C).
Accidental signals from pigments on cell
walls leads to evolution of infrared
sensitivity in motile bacteria. This confers
advantage as they can detect proximal heat
sources—both food and danger.

② Infrared sensitive motile bacteria
colonise shallow hydrothermal
biofilm habitats. Here the infrared
phototaxis preadapts organisms
making them able to use sunlight
as an accessary source of reduction
power. Initially photosynthesis uses
infrared to red light and is anoxygenic.
PSI and PSII evolve; then jointly
permit use of blue-green light for
oxygenic photosynthesis.

③ Organisms that use infrared
photosynthesis as an accessory
source of energy drift away from
the hydrothermal system. Here
they become fully dependent upon
photosynthesis.

The oceans become colonized by
both anoxygenic and oxygenic
photosynthesisers, and coastal
substrates are occupied.

Figure 10 Possible evolutionary chain leading to photosynthesis: hypothesis of pre-adaptation for infrared thermotaxis (Nisbet *et al.*, 1995).

into the IR. Chlorophyll *a* is green in color as it absorbs red and blue light and reflects green: thus green, though much loved, is waste light: the biosphere's chief excretion. If chlorophyll came first, the hypothesis of "evolution via IR thermotaxis" would be invalid.

Which came first—bacteriochlorophyll or chlorophyll? Much of the debate centers on the long-held Granick hypothesis (Granick, 1965). The steps to the synthesis of chlorophyll being simpler, it would intuitively be expected to have come first. Recent work supports the notion that bacteriochlorophyll predates chlorophyll (Xiong *et al.*, 2000), refuting the Granick hypothesis, so the "IR thermotaxis" hypothesis remains tenable.

Chlorophyll and bacteriochlorophyll are closely related, and both center around a porphyrin ring that contains an magnesium atom surrounded by four nitrogen atoms (see Section 14.5.2 for the argument that these originated in alkaline fluids from hydrothermal systems in ultramafic lavas such as komatiites). Similar porphyrin rings lie at the heart of haem (where the central metal is iron) and the enzyme catalase, that helps split hydrogen peroxide to water and dioxygen (thereby allowing the excretion of the poison, either to the external

environment or to attack neighboring cells), as well as in the cytochromes. Many of these must be of the very greatest antiquity and probably predate the last common ancestor. They were clearly exploited in the ancestry of photosynthesis, which may have been via evolutionary tweaking of respiratory processes. Xiong and Bauer (2002) concluded that cytochrome *b* may have been the ancestor of type II photosynthetic reaction centers. Inorganically, linking metal with nitrogen occurs at very high pH.

14.7.4 Anoxygenic Photosynthesis

Anoxygenic photosynthesis is carried out by a wide range of bacteria. The chief groups are green sulfur bacteria, such as *Chlorobium* (which do not use rubisco), green nonsulfur bacteria, such as *Chloroflexus*, purple sulfur bacteria, such as *Thiospirillum*, and purple nonsulfur bacteria (e.g., *Rhodobacter*). The purple bacteria (proteobacteria) are classified by 16S rRNA study into several major evolutionary groups (Woese, 1987). Green sulfur bacteria are strict anaerobes and obligate phototrophs, using hydrogen sulfide, hydrogen, or elemental sulfur, and are unable to respire in the dark. Some have gas vesicles that allow them to float up and down in lakes, adjusting their level with the movement of the redox boundary. Green nonsulfur bacteria are thermophiles, and *Chloroflexus* is typically found as gliding bacteria in mats in hot springs. Purple sulfur bacteria are strict anaerobes, oxidizing hydrogen sulfide to sulfur and often eventually to sulfate. They typically inhabit deeper, anaerobic, parts of the photic layer of lakes, where IR light penetrates. Purple nonsulfur bacteria (many of which are nonphototrophic) are very flexible in life. Normally anaerobic photosynthesizers that use organic molecules as electron acceptors and carbon sources, some species can also oxidize low (nontoxic) levels of sulfide to sulfate. In dark, most purple nonsulfur bacteria can grow in aerobic or microaerobic conditions.

The linking characteristic between these groups is the use of various types of bacteriochlorophyll in a single stage process, involving either photosynthetic reaction center II (e.g., purple bacteria) or photosynthetic reaction center I (e.g., green sulfur bacteria). This photosynthetic process uses electron donors such as H_2, H_2S, S, or organic matter, and does not, as a consequence, evolve waste oxygen. Many green and purple bacteria can grow phototrophically using H_2 as the sole electron donor and CO_2 as the carbon source, using hydrogenase (a nickel enzyme) for CO_2 reduction.

The two photosystems are structurally related, and Xiong *et al.* (2000) concluded from a study of sequence information in photosynthesis genes that green sulfur and green nonsulfur bacteria are each other's most closely related groups. Phototrophic purple bacteria use the Calvin cycle, and utilize rubisco, with its characteristic (and geologically identifiable) strong fractionation of carbon. Green bacteria, however, do it differently and do not produce the same isotopic signature: *Chlorobium* uses the reverse citric acid cycle, and *Chloroflexus* the hydroxyproprionate pathway. Geologically, these should be distinguishable in the kerogen record from rubisco-captured carbon.

When the first photosynthetic sulfur-compound oxidizers first appeared, the development of full sulfureta would have been possible. Sulfate reducers would take sulfate from the external environment and eventually produce H_2S. Then the photosynthetic oxidizers would reverse the steps, e.g.,

$$6CO_2 + 12H_2S \rightarrow C_6H_{12}O_6 + 6H_2O + 12S^0$$

depositing the sulfur either outside the cells (as in phototrophic green bacteria) or inside them (in most of the purple bacteria). This is a trail the geologist can hunt.

The isotopic evidence of Shen *et al.* (2001) is not inconsistent with a full sulfur cycle but does not prove it. The wider fractionation observed in 2.7 Ga material by Grassineau *et al.* (2002) (which also includes carbon isotope support) is very suggestive of a full S cycle. Grassineau *et al.* (2002) reported abundant evidence for highly fractionated carbon in kerogen, the signature of rubisco. Such evidence suggests either the presence of cyanobacteria, or anoxygenic photosynthesizers: by extrapolation, this implies the presence of bacteria capable of oxidizing sulfur compounds.

Once anoxygenic photosynthesis had evolved, the planet would have become widely habitable and the biosphere much more productive. Tidal and shallow-water environments around the globe would have been immediately occupied by life. Sulfureta would have cycled sulfur, derived from oceanic volcanogenic sulfate, between the upper more oxidized layers of mats and the lower H_2S-rich layers, creating a complex microbial mat habitat.

Plankton today are very diverse, including archaea, bacteria, and eukaryotes (Beja *et al.*, 2002). It is very likely that anoxygenic photosynthesizers would rapidly have spread as plankton, limited only by the availability of reducing chemicals. It is possible to imagine a microbial biosphere dependent on anoxygenic photosynthesis, with widespread abundance of oxidized sulfur and nitrogen chemical species in the uppermost few tens of meters of the sea, above an anoxic deeper mass of the seas and oceans. At this stage, the oceanic biomes may have been stratified with anoxygenic purple and green photosynthesizers, as well as planctomycetes (Fuerst, 1995), which may be a very ancient prephotosynthetic branch of the bacteria (Brochier and Philippe, 2002), as discussed earlier.

Complex global-scale nitrogen cycles would also have become possible at this stage. The planctomycetes, if presumed ancient, are capable of emitting N_2 by reacting NH_3 with NO_2. In the reverse direction, supplying nitrogen from the air, the inorganic sources are mainly volcanoes and lightning. But this source could have been restricted (e.g., see Navarro-Gonzalez *et al.*, 2001). The modern nitrogen cycle is dominated by bacteria. Some bacteria such as *Pseudomonas* release N_2. Many purple and green bacteria can fix nitrogen, using the Fe–Mo enzyme nitrogenase. Anoxic nitrification can occur, coupled with manganese reduction (Hulth *et al.*, 1999). A nitrogen cycle may have become possible very early on, some species emitting gaseous nitrogen, others capturing nitrogen from air/ocean. Nitrogen fixation may be closely connected with hydrogen emission: in reducing N_2 to NH_3, eight electrons are consumed, six for producing $2NH_3$ and two to make H_2: hydrogen production and nitrogen fixation appear closely linked. Possibly nitrogenase originally evolved to manage ammonia in close association with methanogenesis using H_2.

With the evolution of the anoxygenic bacteria, the global-scale biosphere would have been greatly enriched. It would have been capable of cycling sulfur, carbon, and nitrogen on a global, scale, and presumably with fluxes that were on a much greater scale than the inorganic volcanogenic fluxes—over a geologically brief time, bacterial emissions would thus have used photosynthetic energy to reconstruct the atmosphere. From this date also N_2 has been a biological product in the main, produced and consumed by organisms.

14.7.5 Oxygenic Photosynthesis

The development of oxygenic photosynthesis created the modern biosphere. The use of ubiquitous ingredients, water, carbon dioxide, and light, to capture carbon into life, was the final metabolic step that made the entire planet habitable by life. The waste product was simply dumped—indeed, it may originally have been a deliberate toxic by-product in toxin warfare between cyanobacteria and their neighbors. Cyanobacteria achieve this by a multicomponent system. Most likely (though not if the Granick hypothesis is correct), oxygenic photosynthesis came *after* the development of photosynthetic reaction system II in purple bacteria and reaction system I in green sulfur bacteria (Nisbet and Fowler, 1999).

There are many notions about how oxygenic photosynthesis evolved (e.g., Blankenship, 2001; Nisbet and Fowler, 1999). All photosystems are basically alike and must have had a common origin (Jordan *et al.*, 2001; Kuhlbrandt, 2001). Heliobacteria, which are anoxygenic phototrophs living in tropical soils, utilize a modified form, bacteriochlorophyll *g*, that is related to cyanobacterial chlorophyll *a*. They may be the microbial branch with the photosynthetic genes, which are most closely related to the ancestral cyanobacteria (Xiong *et al.*, 2000). Perhaps the photosynthetic reaction system in green gliding bacteria, such as *Chloroflexus*, is ancestral to both. What of the host cell, apart from the photosynthetic process? One possibility is that the cyanobacterial cells themselves are chimaera, created by genetic transfer (perhaps lunch) between close-living or symbiotic purple and heliobacteria. This would imply that purple bacteria evolved before cyanobacteria. Perhaps a primitive reaction system evolved first, in the mutation that produced the common ancestor of the purple bacteria, then a further mutation led to the ancestor of the green sulfur bacteria and of the heliobacteria. It is possible that the first O_2-evolving photoreaction center originated in green nonsulfur bacteria, and that this was later incorporated into cyanobacteria (Dismukes *et al.*, 2001). Then, to speculate further, possibly the two lines formed a symbiotic partnership across a redox boundary and eventually became so close

that the genes for PSI and PSII were incorporated into the cell. Another possibility is that, following the development of photosynthesis in the purple bacteria, transfer of Mg-tetrapyrrole genes occurred to the line leading to the cyanobacteria occurred, plus gene duplication, to produce the cyanobacterial reaction center II in the ancestral cyanobacterium (Xiong and Bauer, 2002). The puzzle remains open.

The evolution of the cyanobacteria massively changed the ability of the biosphere by harvesting sunlight, and using it to sequester reduced chemical species from the waste oxidation power dumped into the air. These cells would be able, in a single cell, to photosynthesize with the most available of ingredients, water, light and air, and to fix nitrogen (e.g., Zehr *et al.*, 2001), and even to grow anaerobically if need be. Nitrogen supply is a key limitation on productivity (Falkowski, 1997). Cyanobacteria fix nitrogen. This is a process that needs low oxygen tension in heterocysts, yet the cyanobacteria can also use the oxygen-evolving complex to excrete waste oxygen. The formation of nitrate from ammonium needs molecular oxygen (Falkowski, 1997): it is reasonable to suppose that this could not have evolved until oxygenic photosynthesis appeared. But, conversely, productive oxygenic photosynthesis could not have become global unless there was a good supply of biologically accessible nitrogen. Cyanobacterial plankton still today occupy the tropical oceans in vast numbers (Capone *et al.*, 1997), and the chloroplast in a modern plant is in effect a cyanobacterium in a space suit. Given that respiration today is still carried out by mitochondria, which are in effect proteobacteria also in space suits, the modern cycle of life had begun.

When did this occur? The key signature (see Section 14.7.2) is in the $\delta^{13}C \sim 0‰$ isotopic signature of rubisco I in carbonate rocks (Schidlowski and Aharon, 1992; Schidlowski, 1988, 2002; Nisbert, 2002). This is the modern fingerprint imposed by the chloroplast, still a member of the cyanobacterial line. Carbon dioxide in the atmosphere and ocean is well mixed. For the $\delta^{13}C \sim 0‰$ fingerprint to occur, carbon dioxide must have been managed by rubisco I on a global scale. The only process that could perform this is photosynthesis: Although purple bacteria use rubisco, arguably only oxygenic photosythesis can drive the Calvin cycle to capture carbon dioxide on a scale large enough to create the isotopic signature.

The evidence for the $\delta^{13}C \sim 0‰$ signature is strong around 2.7 Ga (e.g., Grassineau *et al.*, 2002). Buick (1992) in the 2.7 Ga Tumbiana formation in Western Australia presents strong textural evidence for oxygenic photosynthesis in stromatolites growing in shallow lakes. However, older evidence for oxygenic photosynthesis is problematic. The ca. 3 Ga Steep Rock carbonates have $\delta^{13}C$ not far from 0‰ (Abell, Grassineau and Nisbet, unpublished), but in older material there is strong controversy (e.g., Brasier *et al.*, 2002; see also Schopf *et al.*, 2002).

14.7.6 Archean Oxygen

By 2.7 Ga ago, the modern carbon cycle was in operation: the oxygen production must have been considerable. Did it build up in the air? For contrasting views on this vexed problem, see Holland (1999) and Ohmoto (1997). Catling *et al.* (2001) argue persuasively for a high-methane atmosphere, or Earth would have frozen over. Towe (2002), commenting on Catling *et al.* (2001) presented strong arguments that it would be very difficult for the Earth system to scavenge back the free dioxygen released by the cyanobacteria, and argued equally persuasively for a low-O_2 but oxic atmosphere in the late Archean. Catling *et al.* in response (see Towe, 2002), with somewhat different assumptions, defended the methane-rich model of the air, though agreeing that local high-O_2 "oases" (presumably water masses rich in dissolved oxygen) and high-O_2 events could occur just as today methane accumulates in swamps despite the O_2-rich air. Phillips *et al.* (2001), in a careful review of the actual rock evidence, based on much field knowledge, consider that some of the mineralogical and field evidence can be interpreted as supporting an oxidized Archean atmosphere but conclude that the geological evidence for a reducing atmosphere remains ambiguous. In particular, postdepositional processes may need far more examination. Similar conclusions can be drawn from the rocks of Steep Rock and Belingwe.

Kasting (2001) argues in support of the view of Farquhar *et al.* (2000) (but see also Ohmoto *et al.*, 2001) that sulfur isotope fractionation changed around 2.3 Ga. This opinion is based on the claim, from comparison of sulfur isotopes, that so-called "mass independent" fractionation occurred as a result of gas-phase photochemical reactions, particularly photolysis of SO_2. Such fractionation would be much more likely to occur in a low-O_2 atmosphere in which sulfur was present in a variety of oxidation states. Thus, the claim that fractionation changed around 2.3 Ga ago can be seen as supporting the notion that there was a substantial rise in O_2 around this time. This, however, raises the question: if cyanobacterial oxygen production had been sufficient to create the rubisco fingerprint in carbonates as early as 2.7–3.0 Ga ago, why did the rise of free O_2 only occur 400–700 Myr later?

The implications of the Catling *et al.* (2001) suggestion that the air had high methane concentrations (>0.1%) in the late Archean are worth further thought. If so, then consequently, as methane mixed into the stratosphere and upwards through the mesosphere, the Earth would have lost much hydrogen through the thermosphere at the top of the atmosphere. Loss of hydrogen from biologically produced methane equates to surplus of oxygen. This would have produced a substantial net accumulation of oxygen, consumed by oxidation of crust and perhaps by the creation of an upside-down biosphere (Walker, 1987), in which the sediment was more oxidized than the water or air above. The debate continues.

14.8 MUD-STIRRERS: ORIGIN AND IMPACT OF THE EUCARYA

14.8.1 The Ancestry of the Eucarya

The origin of the *Eucarya* remains deep mystery. Some (e.g., Forterre, 1995, 1996) would place it very early indeed; yet it has also been ascribed to a time as recently as 850 Ma ago, in the later Proterozoic (Cavalier- Smith, 2002).

The geological evidence for Archean eukaryotes can be dealt with swiftly. Brocks *et al.* (1999) found organic molecules (sterols) in Archean sediment, that they ascribed to the presence of eukaryotes. This is permissive but not necessarily persuasive evidence, as some bacteria (e.g., methanotrophs, planctomycetes) may leave similar molecular records; thus the interpretation by Brocks *et al.* (1999) is contested (Cavalier-Smith, 2002). Nevertheless, the simplest interpretation is that this is a just-plausible record of Archean eukaryotes.

The molecular evidence for the descent of the eukaryotes (Hartman and Fedorov, 2002) is deeply controversial (see Section 14.6.3). Standard models (Woese, 1987) suggest an ancestral line among the Archea, with massive transfers and symbioses from the bacteria. The standard model (e.g., see summaries in Pace (1997); Nisbet and Fowler, 1996a,b) is that early archaea and bacteria diverged from a hyperthermophile last common ancestor. Then, a sequence of symbiotic events took place between a stem–cell line, among the archaea, that developed partnerships with symbiotic purple and cyanobacteria, either in separate events, or in a single moment of fusion. This produced the eukaryote cell, with the mitochondria derived (Bui *et al.*, 1996 from within the α-proteobacteria such as *Rickettsia*. The other great acquisition of the eukaryotes, the chloroplast, is clearly related to the cyanobacteria (Figure 11).

Much discussion followed on the timing of the event or events, especially as some eukaryotes lack mitochondria. Could they be more primitive? At first it was thought so, but recently it has become clear that even eukaryotes, such as the microsporidae, that are no longer capable of aerobic respiration still have relict mitochondrial proteins (Williams *et al.*, 2002; Roger and Silberman, 2002). The ancestral eukaryote did probably

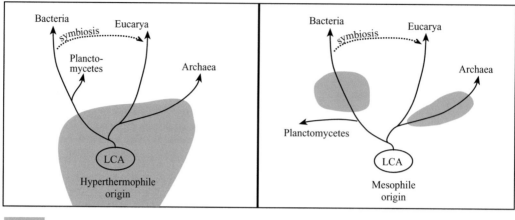

Figure 11 Standard and alternative models of eukaryote evolution (see Figure 7) (with thanks to J. Fuerst). Alternative model assumes a last common ancestor that was mesophile, and that the divergence of the planctomycetes was very ancient (see Brochier and Philippe, 2002).

possess mitochondria, and the amitochondrial eukaryotes lost them. Even simple eukaryotes that today do not have mitochondria (e.g., some parasites) appear to have once had them and then lost them; moreover, mitochondria and hydrogenosomes (distinctive hydrogen-producing organelles in some amitochondrial organisms) appear to have had a common ancestor (Bui *et al.*, 1996). Thus it appears that the ancestors of all modern eukaryotes diverged after the mitochondrion symbiosis. Likewise, animals may have descended from an ancestral photosynthesizer by loss of the chloroplast.

Whatever the explanation of the stem eukaryote, the eukaryote organelles, both mitochondria and chloroplasts, are best explained as symbiont bacteria. Explanations of the mitochondrial symbiosis mostly invoke an early Archean stem that incorporated a bacterial symbiont. One explanation of the mitochondrion is that the origin of the mitochondrion was simultaneous with the origin of the eukaryote nucleus (Grey *et al.*, 1999). In the "hydrogen hypothesis" (Martin and Muller, 1998), the symbiosis is seen as the end product of a tight physical association between anaerobic archaea and heterotrophic proteobacteria capable of producing molecular H_2 through anaerobic fermentation. In another version of the close-association idea, anaerobic archaea may have evolved the ability to survive in oxidizing settings by incorporating respiring proteobacteria (Vellai and Vida, 1999).

In contrast, Penny and Poole (1999) suggested that the last universal common ancestor may have been a mesophile with many features of the eukaryote genome, and the first distinct eukaryote may thus also have been mesophile—a distal inhabitant of a hydrothermal system, or a planktonic form.

However, if the "eukaryote-like" view of the last common ancestor (Forterre and Philippe, 1999;

Glansdorff, 2000) is correct, then the sequence of events may have been greatly different. In this view, an ancient common ancestor may have been a fairly complex organism, living in mesophile conditions, possibly some hundreds of meters distal to a shallow hydrothermal system, in water between, say, 35–45 °C (blood-temperature, optimal for DNA-based life), perhaps in water with pH around 7 or more alkaline (to account for the cytochromes).

From this, in Forterre's thermoreduction hypothesis (1995), came several lines of descendants. Some became colonists of much hotter environments. Of these lines, some necessarily streamlined both their genomes and their physiology, in order to survive, while others, also with reduced physiology, developed heat shock proteins to correct damage. This led to the distinct domains, the *Bacteria* and *Archea*. In contrast, other descendants retained the complex more primitive physiology—biplanes or triplanes, as opposed to prokaryote monoplanes. This third line became the *Eucarya*. This third line may share some characteristics with the planctomycetes (Fuerst, 1995; Lindsay *et al.*, 2001).

At some stage in this hypothesis came the key acquisition by the ancestor of the modern *Eucarya* of the mitochondrion. Possibly: (i) this was very early—perhaps not so much an acquisition as a primitive characteristic; alternately (ii) it may have been a later product of a symbiosis between a mesophile eukaryote stem-organism directly descended from a mesophile eukaryote-like last common ancestor and a proteobacterium that had evolved from a line that had passed through a hyperthermophile bottleneck; or, (iii) it could have been a later product of a symbiosis between two organisms that had both been through a hyperthermophile stage, an achaea-like host and a proteobacterial symbiont.

Chloroplasts may have been acquired at the same time as the mitochondrion, or much later. Some lines of dinoflagellates appear to have had multiple gains and losses of plastids (Saldarriga *et al.*, 2001), although perhaps from a single ancient endosymbiotic origin (Fast *et al.*, 2001). The same hypotheses apply. It is not clear whether all modern eukaryotes are descended from organisms that possessed chloroplasts, but the hypothesis is attractive. Some have multiply acquired and lost chloroplasts.

At many stages in their evolution each of the three domains gave and received genetic material with the other two lines, so that major innovations were acquired by sharing between all three domains.

As for the antiquity of the *Eucarya* there is no consensus. Those who support a eukaryote-like last common ancestor, of course, propose that the eukaryotes date to the very start of the Archean and end of the Hadean. It is not improbable to those who consider that the *Eucarya* were the last domain to appear, that Eukaryotes first evolved in the Archean aeon. There is, however, little support in the rock record for the hypothesis of a "very late" origin of both the archaea and eukaryotes, proposed by Cavalier-Smith (2002), especially as the evidence for early methanogens is strong (Grassineau *et al.*, 2002; Rye and Holland, 2000). However, a proterozoic origin of the eukaryotes is not yet excluded, as the sterols found by Brocks *et al.* (1994) could be of prokaryote origin.

14.8.2 Possible Settings for the Eukaryote Endosymbiotic Event

What was the purpose of symbiosis? And were the organelles incorporated simultaneously or sequentially? The answers are not known, but an argument can be made that simultaneous acquisition of both organelles took place, as they in effect balance.

The most likely setting of symbiosis is a microbial mat community, in which a complex community of cells is clustered across a redox boundary, cycling and recycling redox power (Nisbet and Fowler, 1999; Nisbet, 2002). The aerobic top of the mat would include photosynthetic cyanobacteria, above photosynthesizing purple bacteria. There would be a very sharply focused redox boundary. Below would be the green photosynthetic bacteria, and at the base the methanogens and the hydrogen producers.

In the Archean, such prokaryotic mats would be limited to some extent by diffusion gradients, in the absence of multicelled organisms like worms capable of physical movement of fluid on a large scale. However, microbes are motile and, moreover, they can move fluid, so the thickness of the mat would be substantial compared to the dimensions of a single cell, despite lack of physical power.

In such a setting there is great benefit from being very close to the redox boundary between aerobic and anaerobic conditions, where the greatest thermodynamic power is to be had. Any cluster of cells that straddled this boundary, or incorporated it within itself, would possess great advantage. To some extent, some cyanobacteria already do so within their cell, as they include heterocysts, which protect nitrogenase, which enzyme needs to function within the cell in anaerobic conditions despite the emission of molecular oxygen from the cell during photosynthesis. Any cluster of cells that carried out oxygenic photosynthesis and yet managed to control redox levels by respiration also, would be greatly advantaged. Oxygen is dangerous, and mitochondria may have evolved to manage it (Abele, 2002). There would be much advantage to a symbiotic association, located just above the redox boundary, of a host cell that linked cyanobacteria and purple bacteria, alternately providing useful redox waste to each.

Photosynthesis by a symbiont cyanobacterium would produce reactive oxygen species in the fluid. The buildup of oxidation power in the near environment would be a nuisance to the cyanobacterium as its rubisco would begin to work in reverse. Thus, it would have to wait until the oxygen diffused away before continuing photosynthesis. Moreover, the oxygen would be damaging to the nearby symbiont cell. However, if nearby, on the redox boundary, there were an α-proteobacterium, this would mop up the excess oxidation power immediately, allowing the cyanobacterium to keep photosynthesizing. Thus, the host cell would be protected, the rubisco in the cyanobacterium would operate, and the respiring α-proteobacterium would flourish. Such an arrangement is beneficial to all: thus it is tempting to imagine that both incorporations took place by single lucky improbable accident. Arguably, however, it is more likely that initially only one partner was incorporated, then the second.

14.8.3 Water and Mud Stirring— Consequences

Multi-celled eukaryotes have some unique advantages and some disadvantages. The disadvantage is that they evolve by Darwinian evolution. Genetic change can only occur when parent organisms have a number of different offspring, some of which are better suited than others to the environment in which they find themselves: these are more likely to survive and in turn have offspring, so natural selection chooses the genes most suited. Only females reproduce, so the ability to "bloom" is slightly restricted by one generation time. For single-celled eukaryotes, rapidly passing through the generations, the evolutionary process of adapting to a changed environment (e.g., the arrival of a virus) can be quick and population recovery fast, but for an elephant that lives for decades, or a tree that lasts for centuries, adaptation can be slow and the population can be brought dangerously near extinction before it can respond to the new challenge.

Human cultures evolve as bacteria do, by swapping genetic information among living individuals. This is a rapid and highly advantageous method of adaptation. Most readers of this (except Scots) are likely to wear some variety of undergarment, but the habit is only a few generations old: prior to that it was thought unclean. The change was non-Darwinian. Those human families (probably most) who did not wear such attire a few lifetimes ago did not become extinct—they mysteriously acquired the habit from contemporaries by a hidden process of cultural infection. Eukaryotes do, to some extent, adopt such quasi-Lamarkian evolution, in the immune system. This has an extraordinarily bacterial-like ability to learn in life—perhaps it alone has ensured the domain's survival in the face of viral challenge and microbial attack.

The advantages of the eukaryotes are that they can mix genomes over long distances (males wander), and also create striking cellular architecture, and thereby link together colonies of cells so that they form a single unit with distributed tasks. This may have major consequences in the Proterozoic when multicelled eukaryotes became capable of moving water and stirring mud. Bacteria do this to some extent, but only slightly. By stirring mud and water, the eukaryotes expand the range of the biosphere. The bacterial biosphere is at most only a few millimeters thick—the growing biofilm in a microbial mat. The physical structure of a prokaryote mat may include a debris layer a meter or more thick, but most of the action is close to the redox boundary. In contrast, eukaryotes can move redox power up and down, and widen the environment—they become capable of more reducing power, more photosynthesis, limited only by nutrients such as iron. Eventually, they even become able to send roots down into the soil and rock to extract the nutrient, or, in humans, to dig for potash to put on fields, or to fix nitrogen directly.

The expansion of the productivity of the biosphere by the eukaryotes must have begun slowly, but it probably started in late Archean. Around 2,300 Ma, much evidence suggests (but does not conclusively prove: see Section 14.8.6) that oxygen levels rose sharply. Possibly the eukaryotes were beginning to muscle the world.

14.9 THE BREATH OF LIFE: THE IMPACT OF LIFE ON THE OCEAN/ATMOSPHERE SYSTEM

14.9.1 The Breath of Life

The modern atmosphere is the breath of life—a biological construction. Excepting the argon, the balance of the exosphere is entirely fluxed by life: the gases are emitted and taken up by living organisms. This does not necessarily mean that life manages the air, as life may simply be fast-tracking inorganic processes that would happen anyway, but there is also the possibility that life is maintaining thermodynamic disequilibrium (e.g., the sharp contrast between reducing sediment and oxidizing air; or on an even finer scale, the kinetically improbable presence of ammonia and methane in oxygen-rich lower tropospheric air).

The controls on the atmosphere's operation (Walker, 1977) are complex and poorly understood, yet have been robust enough to keep the planet habitable over 4 Ga. The greenhouse effect adds $\sim$33 °C to the temperature (Lewis and Prinn, 1984). Without the atmosphere, the temperature of the planet would be $\sim$−18 °C; with the atmosphere it is a pleasant +15 °C. But is the control pure inorganic chance, or is it somehow implemented because the Earth, uniquely, is inhabited? And when did the control begin?

14.9.2 Oxygen and Carbon Dioxide

Since the Archean the oxygen that has been emitted into the atmospheric reserve by the oxygen-evolving complex as the waste product of oxygenic photosynthesis, has been taken up again by respiration. The carbon is the other side of this coin: it is stored for the most part in the biosphere and crustal reserves, forming a well of reduction power that matches the surficial oxidation power, with just enough carbon dioxide is sustained in the air to allow rubisco to operate in balance (Tolbert, 1994). There are complex inorganic controls and buffers in the carbon dioxide content of the air and the partitioning of carbon between air/water and surface/crust (Walker, 1994), but the extent to which biological processes exert the fundamental control remains very controversial. One argument that may be made that the most basic control may lie within the cell itself, in within-cell controls (Joshi and Tabita, 1996). The control in photosynthetic cells would lie in the balance between chloroplasts and mitochondria in eukaryote cells. The debate remains open.

14.9.3 Nitrogen and Fixed Nitrogen

Dinitrogen, like dioxygen and carbon, is almost entirely a biological product: there is roughly

3.8×10^{21} g presently in air, and each year $\sim 3 \times 10^{14}$ g are added and subtracted from this reservoir by denitrifying and nitrifying bacteria (lifetime in air ~ 10–100 Ma). Nitrogen is emitted to the air by guilds of denitrifying bacteria (e.g., *Pseudomonas, Bacillus*), which reduce nitrate to N_2 as an alternative respiration process in anaerobic settings when oxygen is absent. This raises the inference that this process may be very old, and could have evolved before oxygenic photosynthesis, predating the time of abundant oxygen supply. As argued above (Section 14.6.3), the anammox process may also be of great antiquity, perhaps long predating oxygenic photosynthesis.

The nitrogen cycle has a controlling role on the carbon dioxide/oxygen cycle (Falkowski, 1997). Were all fixed nitrogen to be evolved as dinitrogen, the biosphere would rapidly be reduced to the nitrogen oxides and ammonia emitted by volcanoes and hydrothermal systems. Nitrogen fixation reverses this, and is carried out by cyanobactria, by free-living bacteria such as *Azotobacter* and *Clostridium*, by archaea such as *Methanococcus*, and (perhaps since late Palaeozoic) by plant symbionts. Fixation is very expensive energetically, requiring a large ATP price. At least six electrons and 12 ATP are required to fix one dinitrogen molecule. The use of nitrogen fixation by cyanobacteria is especially interesting geologically, as cyanobacterial picoplankton are very important today and presumably were from the geological moment the cyanobacteria evolved. Given the clear evidence for cyanobacteria (albeit bound in mats) in the Archean (Summons *et al.*, 1999), it is reasonable to assume that cyanobacterial plankton were presumably ubiquitous in the Archean oceanic photic zone. Thus, there is a reasonable presumption that in the Archean the dinitrogen atmospheric burden was organically fluxed.

14.9.4 Methane

Of the lesser gases, methane is the most interesting. Today, the natural methane sources are primarily archaea, but operating in eukaryote hosts (e.g., archaea symbiotically cooperating with plants in wetland, termite stomachs, cows, and sheep). In the Archean, methane was probably managed by complex microbial communities, comparable to those in modern oceans (Boetius *et al.*, 2000). Walker's (1987) surprising hypothesis that the Archean biosphere at times may have been inverted, with a relatively reducing atmosphere and a relatively oxidizing sediment is not as absurd as it seems. Today the oxygen-rich air is sustained by photosynthesis. Prior to oxygenic photosynthesis, the air would have contained relatively oxidized species (carbon dioxide and water, as well as dinitrogen), but also substantial methane and probably ammonia emissions occurred, that

would have had multiyear atmospheric lifetimes. In the continental slope sediments, huge methane hydrate reserves would have built up, as they do today (Kvenvolden, 1988), In these circumstances, episodes of major atmospheric methane burden could occur, as perhaps happened at the Archean–Proterozoic transition (Hayes, 1994). For example, this could occur after massive release of geological methane stores (e.g., see Kvenvolden, 1988; Harvey and Huang, 1995).

Today a large part of the biosphere is reducing—much of the soft sediment. It is possible, especially if methanotrophy were absent or ineffective in the absence of abundant oxidant, that Archean "Walkerworld" events may have occurred, when the biosphere was inverted: relatively reducing air and oxidizing sediment. Once such an event was established, it might be stable for long periods, until reversed by the combined impact of volcanic degassing of carbon dioxide and nitrogen oxides, and of methanotrophy. With methane, other reduced gases such as ammonia would build up in the air, reversing the nitrogen cycle also.

14.9.5 Sulfur

The oxidation states of sulfur may have been the core tool by which life bootstrapped its way to a global biosphere (Kasting *et al.*, 1989; Kasting, 1993, 2001). Sulfur offers a wide range, from H_2S through HS and sulfur to the oxidized species up to SO_3 (H_2SO_4). Moreover, dimethyl-sulfide is two methyls linked with one sulfur.

For bacteria living close to a redox boundary, sulfur is a marvellous reservoir. Should conditions become reducing, they can tap it and make H_2S. Conversely, if conditions become strongly oxidizing, they can make SO_x species. Thus, the bacteria can sequester sulfur rather as in a piggy bank, saved for a needful day: it becomes a redox currency. Even better, sulfur-bearing chemical species are common components of hydrothermal fluids—readily available!

In the inorganic world, sulfur would have been available in a variety of oxidation states. Even in a reduced atmosphere, transient SO_x would have been present from volcanic sources, supplemented by interaction between sulfur-bearing aerosols and oxidants produced by photolytic chemistry in the early UV flux, or from escape of hydrogen to space. Reduced sulfur species would have been widely available in lavas and volcanic vents. Thus, for the early organisms, shuffling sulfur between various oxidation states would have been the best way of exploiting redox ratchets. When anoxygenic photosynthesis started, a full sulfuretum cycle would have been possible in sediments, fluxing sulfur endlessly up and down to capture a living from oxidizing decaying organic matter, or reducing available oxidant.

Once abundant oxygenic photosynthesis began, the sulfureta would have become much more productive, and sulfur would have become the chief currency of redox transactions on the bottom. Finally, as in most piggy banks, the contents are lost, buried in the mass of reduced sediment as pyrite or sulfide mineral, or even as sulfur, eventually to return via the plate cycle to the volcanoes or groundwater as oxidized sulfur.

14.10 FEEDBACK FROM THE BIOSPHERE TO THE PHYSICAL STATE OF THE PLANET

The planet shapes life, but life also shapes the planet (Nisbet, 2002). The maintenance of surface temperature is managed by the air: hence, as life controls the composition of the air and the atmospheric greenhouse, then life sets the surface temperature (Lovelock and Margulis, 1974; Lovelock, 1979, 1988).

What would be the nature of the air if life did not exist? If for the past 4Ga, life had not captured carbon and sequestered it, and cycled nitrogen back from soluble compounds, returning it as atmospheric dinitrogen, and evolved oxygen and hence permitted ozone to form in the stratosphere, what would the atmosphere be? It is almost impossible to say. Reasonable guesses include a nitrogen–carbon dioxide atmosphere; or perhaps a nitrogen atmosphere over ice, with the carbon dioxide removed as carbonate after volcanic paroxysm. One possibility is that over time the air would have evolved as Venus's air may have evolved, first as a steam greenhouse, then after hydrogen loss, to a dry hot carbon dioxide greenhouse over a dehydrated planet. Alternately, the surface could have become very cold and icy. This would have interesting consequences, as it may have changed the operation of the erosional cycle and the plate system, perhaps leading to periods of long quiescence, followed by volcanic resurfacing. The persistence of oceans of liquid water is closely interwoven with the long-term history of the continents and oceans (Hess, 1962), and the controls on water depth may be linked to the physical properties of water (Kasting and Holm, 1992) as well as to the nature of the atmosphere and its greenhouse impact. In an inanimate planet, whether or not liquid water would have persisted for as long as 4Ga is a moot point. If it had disappeared, would Earth have had plate tectonics? Perhaps not: perhaps the life has shaped the face of the Earth.

ACKNOWLEDGMENTS

Thanks to the community of Archean field geologists in Zimbabwe and Canada for many years of discussions, especially Mike Bickle and Tony Martin, Jim Wilson, John Orpen, and Tom Blenkinsop; to Freeman Dyson, Jim Lovelock, Norm Sleep, Crispin Tickell, and Kevin Zahnle for much thought; and to the late Preston Cloud, Teddy Bullard, Harold Jeffreys, and Drum Matthews for tuition and insight. Finally, thanks to Mabel of Yale for e-mail encouragement during the task of writing this, other matters being more pressing.

REFERENCES

Abele D. (2002) The radical life-giver. *Nature* **420**, 27.

Ahrens T. J. (1990) Earth accretion. In *Origin of the Earth* (eds. H. E. Newsom and J. H. Jones). Oxford University Press, Oxford, pp. 211–227.

Allegre C., Manhes G., and Gopel C. (1995) The age of the Earth. *Geochim. Cosmochim. Acta* **59**, 1445–1456.

Alt J. C. and Shanks W. C. (1998) Sulfur in serpentinized oceanic peridotites: serpentinization processes and microbial sulfate reduction. *J. Geophys. Res.* **103**, 9917–9929.

Appel P. W. U., Rollinson H. R., and Touret J. L. R. (2001) Remnants of an early Archean (>3.75 Ga) sea-floor hydrothermal system in the Isua greenstone belt. *Precamb. Res.* **112**, 27–49.

Baker V. R. (2001) Water and the Martian landscape. *Nature* **412**, 228–236.

Barnes S. M., Delwiche C. F., Palmer J. D., and Pace N. R. (1996) Perspectives on archaeal diversity, thermophyly and monophyly from environmental rRNA sequences. *Proc. Natl. Acad. Sci. USA* **93**, 9188–9193.

Beja O., Suzuki M. T., Heidelberg J. F., Nelson W. C., Preston C. M., Hamada T., Eisen J. A., Fraser C. M., and deLong E. F. (2002) Unsuspected diversity among marine aerobic anoxygenic phototrophs. *Nature* **415**, 630–632.

Bernal J. D. (1951) *The Physical Basis of Life*. Routledge and Kegan Paul, London.

Bernal J. D. (1967) *The Origin of Life*. Weidenfeld and Nicholson, London.

Bickle M. J., Martin A., and Nisbet E. G. (1975) Basaltic and peridotitic komatiites, stromatolites, and a basal unconformity in the Belingwe Greenstone belt, Rhodesia. *Earth Planet. Sci. Lett.* **27**, 155–162.

Blankenship R. E. (2001) Molecular evidence for the evolution of photosynthesis. *Trends Plant Sci.* **6**, 4–6.

Boetius A., Ravenschlag K., Schubert C., Rickert D., Widdel F., Gieseke A., Amann R., Jorgensen B. B., Witte U., and Pfannkuche O. (2000) A marine microbial consortium apparently mediating anaerobic oxidation of methane. *Nature* **407**, 623–626.

Bowring S. A., Williams I. S., and Compston W. (1989) 3.96 Ga gneisses from the Slave province, Canada. *Geology* **17**, 760–764.

Brasier M. D., Green O. R., Jephcoat A. P., Kleppe A., van Kranedonk M. J., Lindsay J. F., Steele A., and Grassineau N. V. (2002) Questioning the evidence for Earth's oldest fossils. *Nature* **416**, 76–81.

Brochier C. and Philippe H. (2002) A non-hyperthermophilic ancestor for Bacteria. *Nature* **417**, 244.

Brocks J. J., Logan G. A., Buick R., and Summons R. E. (1999) Archean molecular fossils and the early rise of eukaryotes. *Science* **285**, 1033–1036.

Bui E. T. N., Bradley P. J., and Johnson P. J. (1996) A common evolutionary origin for mitochondria and hydrogenosomes. *Proc. Natl. Acad. Sci. USA* **93**, 9651–9656.

Buick R. (1990) Microfossil recognition in Archean rocks: an appraisal of spheroids and filaments form a 3,500 MY old chert-barite unit at North Pole, Western Australia. *Palaios* **5**, 441–459.

Buick R. (1992) The antiquity of oxygenic photosynthesis: evidence from stromatolites in sulphate-deficient Archean lakes. *Science* **255**, 74–77.

Buick R., Dunlop J. S. R., and Groves D. I. (1981) Stromatolite recognition in ancient rocks: an appraisal of irregularly laminated structures in an Early Archean chert-barite unit from North Pole, Western Australia. *Alcheringa* **5**, 161–181.

Byerly G. R., Lowe D. R., and Walsh M. M. (1986) Stromatolites from the 3,300–3,500 Myr Swaziland supergroup, Barberton Mountain Land, South Africa. *Nature* **319**, 489–491.

Cairns-Smith A. G. (1982) *Genetic Take-over and the Mineral Origins of Life*. Cambridge University Press, Cambridge.

Cameron A. G. W. (2002) Birth of a solar system. *Nature* **418**, 924–925.

Campbell I. H. and Taylor S. R. (1983) No water, no granites, no oceans, no continents. *Geophys. Res. Lett.* **10**, 1061–1064.

Canup R. M. and Asphaug E. (2001) The lunar-forming giant impact. *Nature* **412**, 708–712.

Capone D. G., Zehr J. P., Paerl H. W., Bergman B., and Carpenter E. J. (1997) *Trichodesmium*, a globally significant marine cyanobacterium. *Science* **276**, 1221–1229.

Carr M. (1996) *Water on Mars*. Cambridge University Press, Cambridge.

Catling D. C., Zahnle K. J., and McKay C. P. (2001) Biogenic methane, hydrogen escape, and the irreversible oxidation of the Earth. *Science* **293**, 839–843. (see also Towe below).

Cavalier-Smith T. (2002) The phagotrophic origin of eukaryotes and phylogenetic classification of Protozoa. *Int. J. Sys. Evol. Microbiol.* **52**, 297–354.

Cockell C. S. (2000) The ultraviolet history of the terrestrial planets—implications for biological evolution. *Planet. Space Sci.* **48**, 203–214.

Compston W. and Pidgeon R. T. (1986) Jack Hills, evidence of more very old detrital zircons in Western Australia. *Nature* **321**, 766–769.

Cooke D. L. and Moorhouse W. W. (1969) Timiskaming volcanism in the Kirkland Lake area, Ontario, Canada. *Can. J. Earth Sci.* **6**, 117–132.

Crick F. H. C. (1981) *Life Itself: Its Origin and Nature*. Simon and Schuster, Touchstone, New York.

Darwin C. (1859, 1872) *On the Origin of Species by Means of Natural Selection or the Preservation of Favoured Races in the Struggle for Life,* diagram from 1872 5th edn. D. Appleton and Co., New York.

Darwin C. (1959) *Some Unpublished Letters* (1871) (ed. Sir Gavin de Beer). Notes and Records of the Royal Society, London. **14**, p. 1.

Dismukes G. C., Klimov V. V., Baranov S., Kozlov Yu. N., DasGupta J., and Tryshkin A. (2001) The origin of atmospheric oxygen on Earth: the innovation of oxygenic photosynthesis. *Proc. Natl. Acad. Sci. USA* **98**, 2170–2175. (www.pnas.org/cgi/doi/10.1073/pnas.061514798).

Donahue T. M., Hoffman J. H., Hodges R. R., Jr., and Watson A. J. (1982) Venus was wet: a measurement of the ratio of deuterium to hydrogen. *Science* **216**, 630–633.

Doolittle W. F. (1999) Phylogenetic classification and the universal tree. *Science* **284**, 2124–2128.

Doolittle W. F. (2000) Uprooting the tree of life. *Sci. Am.* (February), 72–77.

Drake M. J. and Righter K. (2002) Determining the composition of the Earth. *Nature* **416**, 39–44.

Dutkeiwicz A., Rasmussen B., and Buick R. (1998) Oil preserved in fluid inclusions in Archean sandstone. *Nature* **395**, 885–888.

Elsaied H. and Nayunama T. (2001) Phylogenetic diversity of Ribulose-1,5-Birphosphate Carboxylase/Oxygenase large-subunit genes from deep sea microorganisms. *Appl. Environ. Microbiol.* **67**, 1751–1765.

Falkowski P. G. (1997) Evolution of the nitrogen cycle and its influence on the biological sequestration of CO_2 in the ocean. *Nature* **387**, 272–275.

Farquhar J., Bao H., and Thiemans M. (2000) Atmospheric influence of Earth's earliest sulfur cycle. *Science* **289**, 756–758, see also comment by Ohmoto *et al.* (2001), and response by Farquhar *et al.* (1959a), *Science*, **292**.

Fast N. M., Kissinger J. C., Roos D. S., and Keeling P. J. (2001) Nuclear-encoded, plastid targeted genes suggest a single common origin for apicomplexan and dinoflagellate plastids. *Mol. Biol. Evol.* **18**, 418–426.

Fedo C. M. and Whitehouse M. J. (2002) Metasomatic origin of quartz-pyroxene rock, Akilia, Greenland, and implications for Earth's earliest life. *Science* **296**, 1448–1452.

Fenchel T. and Bernard C. (1995) Mats of colourless sulphur bacteria: I. Major microbial processes. *Mar. Ecol. Prog. Ser.* **128**, 161–170.

Forterre P. (1995) Thermoreduction, a hypothesis for the origin of prokaryotes. *C. R. Acad. Sci. Paris: Sci. de la vie* **318**, 415–422.

Forterre P. (1996) A hot topic: the origin of hyperthermophiles. *Cell* **85**, 789–792.

Forterre P. and Philippe H. (1999) Where is the root of the universal tree of life? *BioEssays* **21**, 871–879.

Fuerst J. A. (1995) The planctomycetes: emerging models for microbial ecology, evolution, and cell biology. *Microbiology* **141**, 1493–1506.

Galtier N., Tourasse N., and Gouy M. (1999) A non-hyperthermophile common ancestor to extant life forms. *Science* **283**, 220–221.

Gee H. (2002) That's life? *Nature* **416**, 28.

Gilbert W. (1986) The RNA world. *Nature* **319**, 618.

Glansdorff N. (2000) About the last common ancestor, the universal life-tree and lateral gene-transfer: a reappraisal. *Mol. Microbiol.* **38**, 177–185.

Goodwin A. M., Monster J., and Thode H. G. (1976) Carbon and sulphur isotope abundances in iron-formations and early Precambrian life. *Econ. Geol.* **71**, 870–891.

Gogarten-Boekels M., Hilario E., and Gogarten J. P. (1995) The effects of heavy meteorite bombardment on the early evolution the emergence of the three domains of Life. *Origins Life Evol. Biosphere* **25**, 251–264.

Granick S. (1965) *Evolving Genes and Proteins* (eds. V. Bryson and H. J. Vogel). Academic Press, New York.

Grassineau N. V., Nisbet E. G., Bickle M. J., Fowler C. M. R., Lowry D., Mattey D. P., Abell P., and Martin A. (2001) Antiquity of the biological sulphur cycle: evidence from sulphur and carbon isotopes in 2700 million year old rocks of the Belingwe belt, Zimbabwe. *Proc. Roy. Soc. London* **B268**, 113–119.

Grassineau N. V., Nisbet E. G., Fowler C. M. R., Bickle M. J., Lowry D., Chapman H. J., Mattey D. P., Abell P., Yong J., and Martin A. (2002) Stable isotopes in the Archean Belingwe belt, Zimbabwe: evidence for a diverse prokaryotic mate ecology. In *The Early Earth: Physical, Chemical and Biological Development*, Geological Society London, Special Publication 199, (eds. C. M. R. Flower, *et al.*), pp. 309–328.

Grey M. W., Burger G., and Lang B. F. (1999) Mitochondrial evolution. *Science* **283**, 1476–1481.

Guy R. D., Fogel M., and Berry J. A. (1993) Photosynthetic fractionation of the stable isotopes of oxygen and carbon. *Plant Physiol.* **101**, 37–47.

Hall A. (1989) Ammonium in spilitized basalts of southwest England and its implications for the recycling of nitrogen. *Geochem. J.* **23**, 19–23.

Hall A. and Alderton D. H. M. (1994) Ammonium enrichment associated with hydrothermal activity in the granites of south-west England. *Proc. Ussher Soc.* **8**, 242–247.

Halliday A. N. (2001) In the beginning. *Nature* **409**, 144–145.

Hartman H. and Fedorov A. (2002) The origin of the eukaryotic cell: a genomic investigation. *Proc. Natl. Acad. Sci.* **99**, 1420–1425. www.pnas.org/cgi/doi/10.1073/pnas.032658599.

Hayes J. M. (1994) Global methanotrophy at the Archean-Proterozoic transition. In *Early Life on Earth* (ed. S. Bengtson). Columbia University Press, New York, pp. 220–236.

Hirt R. P., Logsdon J. M., Healy B., Dorey M. W., Doolittle W. F., and Embley T. M. (1999) Microsporidae are related to fungi: evidence from the largest subunit of

RNA polymerase II and other proteins. *Proc. Natl. Acad. Sci. USA* **96**, 580–585.

Harvey L. D. D. and Huang Z. (1995) Evaluation of the potential impact of methane clathrate destabilisation on future global warming. *J. Geophys. Res.* **100**, 2905–2926.

Hess H. H. (1962) History of ocean basins. In *Petrological Studies: A Volume in Honour of A. F. Buddington* (eds. A. E. J. Engel, H. L. James, and B. F. Leonard). Geological Society of America, pp. 599–620.

Hoehler T. M., Bebout B. M., and Des Marais D. J. (2001) The role of microbial mats in the production of reduced gases on the early earth. *Nature* **412**, 324–327.

Holland H. D. (1999) When did the Earth's atmosphere become oxic? A reply. *Geochem. News* **100**, 20–22.

Holm N. G. and Charlou J. L. (2001) Initial indications of abiotic formation of hydrocarbons in the Rainbow ultramafic hydrothermal system, mid-Atlantic ridge. *Earth Planet. Sci. Lett.* **191**, 1–8.

Hulth S., Aller R. C., and Gilbert F. (1999) Coupled anoxic nitrification/manganese reduction in marine sediments. *Geochim. Cosmochim. Acta* **63**, 49–66.

Jetten M. S. M., Wagner M., Fuerst J., van Loosdrecht M., Kuenen G., and Strous M. (2001) Microbiology and application of the anaerobic ammonium oxidation (anammox) process. *Curr. Opinion Biotechnol.* **12**, 283–288.

Jordan P., Fromme P., Tobias Witt H., Klukas O., Saenger W., and Krauss N. (2001) Three dimensional structure of cyanobacterial photosystem: I at 2.5. A resolution. *Nature* **411**, 909–917.

Joshi H. M. and Tabita F. R. (1996) A global two-way component signal transduction system that integrates the control of photosynthesis, carbon dioxide assimilation and nitrogen fixation. *Proc. Natl. Acad. Sci. USA* **93**, 14515–14520.

Kaiser J. (1995) Can deep bacteria live on nothing but rocks and water? *Science* **270**, 377.

Karl D. M. (2002) Hidden in a sea of microbes. *Nature* **415**, 590–591.

Karner M. B., DeLong E. F., and Karl D. M. (2001) Archeal dominance in the mesopleagic zone of the Pacific Ocean. *Nature* **409**, 507–510.

Kasting (1993) Earth's early atmosphere. *Science* **259**, 920–926.

Kasting J. F. (2001) The rise of atmospheric oxygen. *Science* **293**, 819–820.

Kasting J. F. and Holm N. G. (1992) What determines the volume of the oceans? *Earth Planet. Sci. Lett.* **109**, 507–515.

Kasting J. F. and Siefert J. L. (2001) The nitrogen fix. *Nature* **412**, 26–27.

Kasting J. F. and Siefert J. L. (2002) Life and the evolution of the Earth's atmosphere. *Science* **296**, 1066–1067.

Kasting J. F., Zahnle K. J., Pinto J. P., and Young A. T. (1989) Sulfur, ultraviolet radiation, and the early evolution of life. *Origins Life Evol. Biosphere* **19**, 95–108.

Kasting J. F., Whitmire D. P., and Reynolds R. T. (1993) Habitable zones around main sequences stars. *Icarus* **101**, 108–128.

Kerr R. A. (2002) Reversals reveal pitfalls in spotting ancient and E. T. life. *Science* **296**, 1384–1385.

Kleine T., Munker C., Mezger K., and Palme H. (2002) Rapid accretion and early core formation on asteroids and the terrestrial planets from Hf–W chronometry. *Nature* **418**, 952–955.

Kuhlbrandt W. (2001) Chlorophylls galore. *Nature* **411**, 896–898.

Kvenvolden K. (1988) Methane hydrate—a major reservoir of carbon in the shallow geosphere? *Chem. Geol.* **29**, 159–162.

Lewis J. S. and Prinn R. G. (1984) *Planets and their Atmospheres.* Academic Press, Orlando.

Lindsay M. R., Webb R. I., Strous M., Jetten M. S. M., Butler M. K., Forde R. J., and Fuerst J. A. (2001) Cell compartmentalisation in plactomycetes: novel types of structural organization for the bacterial cell. *Arch. Microbiol.* **175**, 413–429.

Line M. A. (2002) The enigma of the origin of life and its timing. *Microbiology* **148**, 21–27.

Lorimer G. H. (1981) The carboxylation and oxygenation of ribulose 1,5-Bisphosphate: the primary events in photosynthesis and photorespiration. *Ann. Rev. Plant Physiol.* **32**, 349–383.

Lorimer G. H. and Andrews T. J. (1973) Plant photorespiration-an inevitable consequence of the existence of atmospheric oxygen. *Nature* **243**, 359.

Lovelock J. E. (1979) *Gaia: A New Look at Life on Earth.* Oxford University Press, Oxford.

Lovelock J. E. (1988) *Ages of Gaia.* Norton, London.

Lovelock J. E. and Margulis L. (1974) Homeostatic tendencies of the Earth's atmosphere. *Origins Life* **5**, 93–103.

Lowe D. R. (1994) Abiological origin of described stromatolites older than 3.2 Ga. *Geology* **22**, 387–390.

Macgregor A. M. (1949) The influence of life on the face of the Earth. In *Presidential Address. Rhodesia Scientific Association.* Proceedings and Transactions, XLII, 5–11 (now Zimbabwe Scientific Association).

Martin A., Nisbet E. G., and Bickle M. J. (1980) Archean stromatolites of the Belingwe greenstone belt, Zimbabwe (Rhodesia). *Precamb. Res.* **13**, 337–362.

Martin W. (1999) Mosaic bacterial chromosomes: a challenge en route to a tree of genomes. *BioEssays* **21**, 99–104.

Martin W. and Muller M. (1998) The hydrogen hypothesis for the first eukaryote. *Nature* **392**, 37–41.

Matthews P. E. and Scharrer R. H. (1968) A graded unconformity at the base of the early Precambrian Pongola system. *Trans. Geol. Soc. S. Afr.* **71**, 257–272.

Miyazaki J., Nakaya S., Suzuki T., Tamakoshi M., Oshima T., and Yamagishi A. (2001) Ancestral residues stabilising 3-isopropylmalate dehydrogenase of an extreme thermophile; experimental evidence supporting the thermophilic common ancestor hypothesis. *J. Biochem.* **129**, 777–782.

Mojzsis S. J., Arrhenius G., McKeegan K. D., Harrison T. M., Nutman A. P., and Friend C. R. L. (1996) Evidence for life on Earth before 3,800 million years ago. *Nature* **384**, 55–59.

Mojzsis S. J., Harrison T. M., and Pidgeon R. T. (2001) Oxygen-isotope evidence from ancient zircons for liquid water at the Earth's surface 4,300 Myr ago. *Nature* **409**, 178–181.

Morel F. M. M. and Price N. M. (2003) The biogeochemical cycles of trace metals in the oceans. *Science* **300**, 944–947.

Moorbath S. (1983) The most ancient rocks. *Nature* **304**, 585–586.

Myers J. S. and Crowley J. L. (2000) Vestiges of life in the oldest Greenland rocks? a review of early Archean geology in the Godthabsfjord region, and reappraisal of field evidence for the >3,850 Ma life on Akilia. *Precamb. Res.* **103**, 101–124.

Navarro-Gonzalez R., McKay C. P., and Mvondo D. N. (2001) A possible nitrogen crisis for Archean life due to reduced nitrogen fixation by lightning. *Nature* **412**, 61–64.

Newsom H. E. and Jones J. H. (1990) *Origin of the Earth.* Oxford University Press, New York.

Nisbet E. G. (1986) RNA and hydrothermal systems. *Nature* **322**, 206.

Nisbet E. G. (1987) *The Young Earth.* George Allen and Unwin, London, 402pp.

Nisbet E. G. (1991) Of clocks and rocks the four aeons of Earth. *Episodes* **14**, 327–331.

Nisbet E. G. (1995) Archean ecology: a review of evidence for the early development of bacterial biomes, and speculations on the development of a global scale biosphere. In *Early Precambrian Processes,* Geological Society of London Spec. Publ. 95 (eds. M. P. Coward and A. C. Ries). Geological Society of London, London, pp. 27–51.

Nisbet E. G. (2002) The influence of life on the face of the Earth: garnets and moving continents, Geol. Soc. London Fermor lecture. In *The Early Earth: Physical, Chemical and Biological Development,* Geological Society of London Spec. Publ. 199 (ed. C. M. R. Fowler, *et al.*). Geological Society of London, London, pp. 275–307.

Nisbet E. G. and Fowler C. M. R. (1996a) Some liked it hot. *Nature* **382**, 404–405.

Nisbet E. G. and Fowler C. M. R. (1996b) The hydrothermal imprint on life: did heat shock proteins, metalloproteins and photosynthesis begin around hydrothermal vents? In *Tectonic, Magmatic, Hydrothermal, and Biological Segmentation of Mid-Ocean Ridges,* Geological Society of London Spec. Publ. 118 (eds. C. J. MacLeod, P. A. Tyler, and C. L. Walker). Geological Society of London, London, pp. 239–251.

Nisbet E. G. and Fowler C. M. R. (1999) Archean metabolic evolution of microbial mats. *Proc. Roy. Soc. London* **B266**, 2375–2382.

Nisbet E. G. and Sleep N. H. (2001) The habitat and nature of early life. *Nature* **409**, 1083–1091.

Nisbet E. G. and Sleep N. H. (2003) The early earth. In *The Physical Setting for Early Life* (eds. A. Lister and L. Rothschild). Academic Press, San Diego, pp. 3–24.

Nisbet E. G. and Walker D. (1982) Komatiites and the structure of the Archean mantle. *Earth Planet. Sci. Lett.* **60**, 105–113.

Nisbet E. G., Arndt N. T., Bickle M. J., Cameron W. E., Chauvel C., Cheadle M., Hegner E., Kyser T. K., Martin A., Renner R., and Roedder E. (1987) Uniquely fresh 2.7 Ga komatiites from the Belingwe greenstone belt, Zimbabwe. *Geology* **15**, 1147–1150.

Nisbet E. G., Cheadle M. J., Arndt N. T., and Bickle M. J. (1993) Constraining the potential temperature of the Archean mantle: a review of the evidence from komatiites. *Lithos* **30**, 291–307.

Nisbet E. G., Mattey D. P., and Lowry D. (1994) Can diamonds be dead bacteria? *Nature* **367**, 694.

Nisbet E. G., Cann J. R., and van Dover C. L. (1995) Origins of photosynthesis. *Nature* **373**, 479–480.

Nutman A. P., Macgregor V. R., Friend C. R. L., Bennett V. C., and Kinny P. D. (1996) The Itsaq geniss complex of southern West Greenland: the world's most extensive record of early crustal evolution (3,900–3,600 Ma). *Precamb. Res.* **78**, 1–39.

Ohmoto H. (1997) When did the Earth's atmosphere become oxic? *Geochem. News* **93**, 12–13.

Ohmoto H., Kakagawa T., and Lowe D. R. (1993) 3.4 billion year old pyrites from Barberton, South Africa: sulfur isotope evidence. *Science* **262**, 555–557.

Ohmoto H., Yamaguchi K. E., and Ono S. (2001) Questions regarding Precambrian sulfur fractionation, and Response by Farquhar *et al. Science* **292**, 1959a.

Pace N. R. (1997) A molecular view of biodiversity and the biosphere. *Science* **276**, 734–740.

Pavlov A. A., Kasting J. F., Brown L. L., Rages K. A., and Freedman R. (2000) Greenhouse warming by CH_4 in the atmosphere of early Earth. *J. Geophys. Res.* **105**, 11981–11990.

Pavlov A. A., Kasting J. F., Eigenbrode J. L., and Freeman K. H. (2001) Organic haze in Earth's early atmosphere: source of low ^{13}C kerogens? *Geology* **29**, 1003–1006.

Phillips G. N., Law J. D. M., and Myers R. E. (2001) Is the redox state of the Archean atmosphere constrained? *SEG Newslett.: Soc. Econ. Geol.,* Oct. 2001, No. 47, 1–19.

Pierson B. K. (1994) The emergence, diversification, and role of photosynthetic eubacteria. In *Early Life on Earth,* Nobel Symposium 84 (ed. S. Bengtson). Columbia University Press, New York, pp. 161–180.

Rasmussen R. (2000) Filamentous microfossils in a 3,235 million year old volcanogenic massive sulphide deposit. *Nature* **405**, 676–679.

Rasmussen R. and Buick R. (2000) Oily old ores, evidence for hydrothermal petroleum generation in an Archean volcanogenic massive sulpide deposit. *Geology* **27**, 115–118.

Reeve J., Christner B. C., Kvitko B. H., Mosley-Thompson E., and Thompson L. G. (2002) Life in glacial ice. In *Extremophiles 2002: 4th International Congress on Extremophiles, Naples,* L5, 27.

Reysenbach A.-L. and Shock E. (2002) Merging genomes with geochemistry in hydrothermal ecosystems. *Science* **296**, 1077–1082.

Robinson J. J., Stein J. L., and Cavanaugh C. M. (1998) Cloning and sequencing of a Form II Ribulose-1,5-Bisphosphate

Carboxylase/Oxygenase from the bacterial symbiont of the hydrothermal vent tubeworm *Riftia pachyptila. J. Bacteriol.* **180**, 1596–1599.

Roger A. J. and Silberman J. D. (2002) Mitochondria in hiding. *Nature* **418**, 827–829.

Rosing M. T. (1999) ^{13}C-depleted carbon in >3700 Ma seafloor sedimentary rocks from West Greenland. *Science* **283**, 674–676.

Rye R. and Holland H. D. (2000) Life associated with a 2.76 Ga ephemeral pond? *Geology* **28**, 483–486.

Sagan C. and Chyba C. (1997) The early Sun paradox: organic shielding of ultraviolet-labile greenhouse gases. *Science* **276**, 1217–1221.

Saldarriaya J. F., Taylor F. J. R., Keeling P. J., and Cavalier-Smith T. (2001) Dinoflagellate nuclear SSU rRNA phylogeny suggests multiple plastid losses and replacements. *J. Mol. Evol.* **53**, 204–213.

Sansone F. J., Popp B. N., Gasc A., Graham A. W., and Rust T. M. (2001) Highly elevated methane in the eastern tropical North Pacific and associated isotopically enriched fluxes to the atmosphere. *Geophys. Res. Lett.* **28**, 4567–4570.

Schidlowski M. (1988) A 3,800 million year record of life from carbon in sedimentary rocks. *Nature* **333**, 313–318.

Schidlowski M. (2002) Sedimentary carbon isotope archives as recorders of early life: implications for extraterrestrial scenarios, chap. 11. In *Fundamentals of Life,* Editions Scientifiques et Medicales (eds. G. Palyi, C. Zucchi, and L. Caglioti). Elsevier, Amsterdam, pp. 307–329.

Schidlowski M. and Aharon P. (1992) Carbon cycle and carbon isotopic record: geochemical impact of life over 3.8 Ga of earth history. In *Early Organic Evolution: Implications for Mineral and Energy Resources* (ed. M. Schidlowski, *et al.*). Springer, Berlin, pp. 147–175.

Schoenberg R., Kambeer B. S., Collerson K. D., and Moorbath S. (2002) Tungsten isotope evidence from ~3.8-Gyr metamorphosed sediments for early meteorite bombardment of the Earth. *Nature* **418**, 403–405.

Schopf J. W. (1993) Microfossils of the early Archean Apex chert: new evidence of the antiquity of life. *Science* **260**, 640–646.

Schopf J. W. and Packer B. M. (1987) Early Archean (3.3 billion to 3.5 billion-year-old) microfossils from Warrawoona group, Australia. *Science* **237**, 70–73.

Schopf J. W., Kudryavtsev A. B., Agresti D., Wdowiak T., and Czaja A. D. (2002) Laser-Raman imagery of Earth's earliest fossils. *Nature* **416**, 73–76.

Schrodinger E. (1944) *What is Life?* Cambridge University Press, Cambridge.

Shen Y., Buick R., and Canfield D. E. (2001) Isotopic evidence for microbial sulphate reduction in the early Archean era. *Nature* **410**, 77–81.

Sleep N. H. and Zahnle K. (2001) Carbon dioxide cycling and implications for climate on ancient Earth. *J. Geophys. Res.* **106**, 1373–1399.

Sleep N. H., Zahnle K., and Neuhoff P. S. (2001) Initiation of clement surface conditions on the earliest Earth. *Proc. Natl. Acad. Sci. USA* **98**, 3666–3672.

Stetter K. O. (1996) Hyperthermophiles in the history of life. In *Evolution of Hydrothermal Systems on Earth and Mars?* Ciba Foundation Symposium 202 (eds. G. R. Bock and J. A. Goode). Wiley, Chichester, pp. 1–10.

Stevens T. O. and McKinley J. P. (1995) Lithoautotrophic microbial ecosystems in deep basalt aquifers. *Science* **270**, 450–454.

Summons R. E., Jahnke L. L., Hope J. M., and Logan G. A. (1999) 2-Methylhopanoids as biomarkers for cyanobacterial oxygenic photosynthesis. *Nature* **400**, 554–557.

Taylor S. R. (2001) Solar System Evolution. In *A New Perspective,* 2nd edn. Cambridge University Press, Cambridge.

Thauer R. K. (1998) Biochemistry of methanogenesis: a tribute to Marjory Stephenson. *Microbiology* **144**, 2377–2406.

Tolbert N. E. (1994) Role of photosynthesis and photorespiration in regulating atmospheric CO_2 and O_2. In *Regulation of*

Atmospheric CO₂ and O₂ by Photosynthetic Carbon Metabolism (eds. N. E. Tolbert and J. Preiss). Oxford University Press, Oxford, pp. 8–33.

Towe K. M. (2002) The problematic rise of Archean oxygen. *Science* **295**, 1419a (see also reply by Catling et al., p1419b).

van Zuilen A., Lepland A., and Arrhenius G. (2002) Reassessing the evidence for the earliest traces of life. *Nature* **418**, 627–630.

Vellai T. and Vida G. (1999) The origin of eukaryotes: the difference between eukaryotic and prokaryotic cells. *Proc. Roy. Soc. London* **B266**, 1571–1577.

von Brunn V. and Hobday D. K. (1976) Early Precambrian tidal sedimentation in the pongola supergroup of South Africa. *J. Sedim. Petrogr.* **46**, 670–679.

Walker J. C. G. (1977) *Evolution of the Atmosphere*. Macmillan, New York.

Walker J. C. G. (1987) Was the Archean biosphere upside down? *Nature* **329**, 710–712.

Walker J. C. G. (1994) Global geochemical cycles of carbon. In *Regulation of Atmospheric CO₂ and O₂ by Photosynthetic Carbon Metabolism* (eds. N. E. Tolbert and J. Preiss). Oxford University Press, Oxford, pp. 75–89.

Watson A. J., Donahue T. M., and Kuhn W. R. (1984) Temperatures in a runaway greenhouse on the evolving Venus. *Earth Planet. Sci. Lett.* **68**, 1–6.

Weatherill G. W. (1990) Formation of the Earth. *Ann. Rev. Earth Planet. Sci.* **18**, 205–256.

Westall F., de Wit M. J., Dann J., van der Gaast S., de Ronde C. E. J., and Gerneke D. (2001) Early Archean fossil bacteria and biofilms in hydrothermally-influenced sediments from the Barberton greenstone belt, South Africa. *Precamb. Res.* **106**, 93–116.

Wilde S. A., Valley J. W., Peck W. H., and Graham C. M. (2001) Evidence from detrital zircons for the existence of continental crust and oceans on the Earth 4.4 Gyr ago. *Nature* **409**, 175–178.

Wilks M. E. and Nisbet E. G. (1985) Archean stromatolites from the Steep Rock group NW Ontario. *Can. J. Earth Sci.* **22**, 792–799.

Williams B. A. P., Hirt R. P., Lucocq J. M., and Embley T. M. (2002) A mitochondrial remnant in the microsporidian *Trachipleistophora hominis*. *Nature* **418**, 865–869.

Woese C. R. (1987) Bacterial evolution. *Microbiol. Rev.* **51**, 221–271.

Woese C. R. (1999) The universal ancestor. *Proc. Natl. Acad. Sci.* **95**, 6854–6859.

Woese C. R., Kandler O., and Wheelis M. I. (1990) Towards a natural system of organisms: proposals for the domains Archea, Bacteria, and Eucarya. *Proc. Natl. Acad. Sci. USA* **87**, 4576–4579.

Xiong J. and Bauer C. E. (2002) A cytochrome *b* origin of photosynthetic reaction centers: an evolutionary link between respiration and photosynthesis. *J. Molecul. Biol.* **322**, 1025–1037.

Xiong J., Fischer W. M., Inoue K., Nakahara M., and Bauer C. E. (2000) Molecular evidence for the early evolution of photosynthesis. *Science* **289**, 1724–1730.

Yin Q., Jacobsen S. B., Yamashita K., Blichert-Toft J., Telouk P., and Albarede F. (2002) A short timescale for terrestrial planet formation from Hf–W chronometry of meteorites. *Nature* **418**, 949–952.

Yung Y., Wen J.-S., Moses J. I., Landry B. M., and Allen M. (1989) Hydrogen and deuterium loss from the terrestrial atmosphere: a quantitative assessment of non-thermal escape fluxes. *J. Geophys. Res.* **94**, 14971–14989.

Zehr J. P., Waterbury J. B., Turner P. J., Montoya J. P., Omoregie E., Steward G. F., Hansen A., and Karl D. M. (2001) Unicellular cyanobacteria fix N₂ in the subtropical North Pacific Ocean. *Nature* **412**, 635–638.

Zuckerkandl E. and Pauling L. (1965) Molecules as documents of evolutionary history. *J. Theor. Biol.* **8**, 357–366.

Radioactive Geochronometry
ISBN: 978-0-08-096708-0

pp. 425–462

15

Heavy Metals in the Environment—Historical Trends

E. Callender

US Geological Survey, Westerly, RI, USA

15.1 INTRODUCTION

15.1.1 Metals: Pb, Zn, Cd, Cr, Cu, Ni

These six metals, commonly classified as heavy metals, are a subset of a larger group of trace elements that occur in low concentration in the Earth's crust. These heavy metals were mined extensively for use in the twentieth century Industrial Society. Nriagu (1988a) estimated that between 0.5 (Cd) and 310 (Cu) million metric tons of these metals were mined and ultimately deposited in the biosphere. In many instances, the inputs of these metals from anthropogenic sources exceed the contributions from natural sources (weathering, volcanic eruptions, forest fires) by several times (Adriano, 1986). In this chapter, heavy metals (elements having densities greater than 5) and trace elements (elements present in the lithosphere in concentrations less than 0.1%) are considered synonymous.

It has been observed in the past that the rate of emission of these trace metals into the atmosphere is low due to their low volatility. However, with the advent of large-scale metal mining and smelting as well as fossil-fuel combustion in the twentieth century, the emission rate of these metals has increased dramatically. As most of these emissions are released into the atmosphere where the mammals live and breathe, we see a great increase in the occurrence of health problems such as lead (Pb) poisoning, cadmium (Cd) Itai-itai disease, chromium (Cr), and nickel (Ni) carcinogenesis.

In this chapter, the author has attempted to present a synopsis of the importance of these metals in the hydrocycle, their natural and anthropogenic emissions into the environment, their prevalent geochemical form incorporated into lacustrine sediments, and their time-trend distributions in watersheds that have been impacted by urbanization, mining and smelting, and other anthropogenic activities. These time trends are reconstructed from major–minor–trace–element distributions in age-dated sediment cores, mainly from reservoirs where the mass sedimentation rates (MSRs) are orders of magnitude greater than those in natural lakes, the consequences of which tend to preserve the heavy-metal signatures and minimize the metal diagenesis (Callender, 2000). This chapter focuses mainly on the heavy metals in the terrestrial and freshwater environments whilst the environmental chemistry of trace metals in the marine environment is discussed in Volume 6, Chapter 3 of the Treatise on Geochemistry.

The data presented in Tables 2–5 are updated as much as possible, with many of the references postdate the late 1980s. Notable exceptions are riverine particulate matter chemistry (Table 2), some references in Table 3, and references concerning the geochemical properties of the six heavy metals discussed in this chapter. There appears to be no recent publication that updates the worldwide average for riverine particulate matter trace metal chemistry (Martin and Whitfield, 1981; Martin and Windom, 1991). This is supported by the fact that two recent references (Li, 2000; Chester, 2000) concerning marine chemistry still refer to this 1981 publication. As for references in Table 3, there is a very limited data available concerning the pathways of heavy-metal transport to lakes. Some of the important works have been considered and reviewed in this chapter. In addition, the analytical chemistry of the sedimentary materials has changed little over the past 30 years until the advent and use of inductively coupled plasma/mass spectrometry (ICP/MS) in the late 1990s. Extensive works concerning the geochemical properties of heavy metals have been published during the past 40 years and to the author's knowledge these have survived the test of time.

15.1.2 Sources of Metals

There are a variety of natural and anthropogenic sources of these heavy metals (Pb, Zn, Cd, Cr, Cu, Ni) in the environment.

15.1.2.1 Natural

The principal natural source of heavy metals in the environment is from crustal material that is either weathered on (dissolved) and eroded from (particulate) the Earth's surface or injected into the Earth's atmosphere by volcanic activity. These two sources account for 80% of all the natural sources; forest fires and biogenic sources, account for 10% each (Nriagu, 1990b). Particles released by erosion appear in the atmosphere as wind-blown dust. In addition, some particles are released by vegetation. The natural emissions of the six heavy metals are 12,000 (Pb); 45,000 (Zn); 1,400 (Cd); 43,000 (Cr); 28,000 (Cu); and 29,000 (Ni) metric tons per year, respectively (Nriagu 1990b). Thus, we can conclude that an abundant quantity of metals are emitted into the atmosphere from natural sources. The quantity of anthropogenic emissions of these metals is given in the next section.

15.1.2.2 Anthropogenic

There are a multitude of anthropogenic emissions in the environment. The major source of these metals is from mining and smelting. Mining releases metals to the fluvial environment as tailings and to the atmosphere as metal-enriched dust whereas smelting releases metals to the atmosphere as a result of high-temperature refining processes. In the lead industry, Pb–Cu–Zn–Cd

Table 1 Global primary production and emissions of six heavy metals during the 1970s and the 1980s.

Metal	Metal production		Emissions to air		Emissions to soil	Emissions to water
	1970s	1980s	1970s	1980s	1980s	1980s
Pb	3,400	3,100	449	332	796	138
Zn	5,500	5,200	314	132	1,372	226
Cd	17	15	7.3	7.6	22	9.4
Cr	6,000	11,250	24	30	896	142
Cu	6,000	7,700	56	35	954	112
Ni	630	760	47	56	325	113

Source: Nriagu (1980a), Pacyna (1986), and Nriagu and Pacyna (1988).
All values are thousand metric tons.

are released in substantial quantities; during Cu and Ni smelting, Co–Zn–Pb–Mn as well as Cu–Ni are released; and in the Zn industry, sizeable releases of Zn–Cd–Cu–Pb occur (Adriano, 1986). Table 1 shows that the world metal production during the 1970s and the 1980s has remained relatively constant except for Cr production that substantially increased during the 1980s due to the technological advances and increased importance (Faust and Aly, 1981). Much of the demand for Cr was due to steel and iron manufacturing and the use of Cr in pressure-treated lumber (Alloway, 1995). Table 1 also shows that anthropogenic emissions to the atmosphere, to which mining and smelting are major contributors, are in the interval of two times (Cu, Ni), five times (Zn, Cd), and 33 times (Pb) greater than the natural emissions of metals to the atmosphere. Anthropogenic atmospheric emissions decreased substantially from the 1970s to the 1980s for Pb, Zn, and Cu (Table 1). On the other hand, Cd and Cr have remained the same and Ni emissions have increased in the 1980s. In addition, anthropogenic emissions of Cr are only about one-half of those from the natural sources. The major contributor of Cr to natural atmospheric emissions is windblown dust (Nriagu and Pacyna, 1988).

Other important sources of metals to the atmosphere include fossil-fuel combustion (primarily coal), municipal waste incineration, cement production, and phosphate mining (Nriagu and Pacyna, 1988). Important sources of metals to the terrestrial and aquatic environment include discharge of sewage sludges, use of commercial fertilizers and pesticides, animal waste and wastewater discharge (Nriagu and Pacyna, 1988). Table 1 shows that metal emissions to soil are several times those to air, suggesting that land disposal of mining wastes, chemical wastes, combustion slags, municipal wastes, and sewage sludges are the major contributors of these emissions. Emissions to water are only about twice those relative to air (except for Pb and Cd) suggesting that direct chemical and wastewater releases to the aquatic environment are the only additional inputs besides the atmospheric emissions (Table 1).

Table 2 gives a comparison of the six heavy-metal contents of a variety of natural earth materials that annually impact atmospheric, terrestrial, and aquatic environments. The primary data of metals are also normalized with respect to titanium (Ti). Titanium is a very conservative element that is associated with crustal rock sources. Normalization with respect to Ti compensates for the relative percentage of various diluents (non-crustal rock sources) and allows one to see more clearly metal enrichment due to anthropogenic inputs. For instance, in Table 2, recent lacustrine sediment is clearly enriched in metal content relative to pre-Industrial lacustrine sediment.

It is obvious that there is a progressive enrichment in the metal content of the earth materials as one migrates from the Earth's upper crust to the soils to river mud to lacustrine sediments, and finally to the river particulate matter. This is especially true for Zn and Cd. If we consider the recent lacustrine sediments, then Pb, Zn, Cd, and Cu are all highly enriched compared to the upper crust and soils. Chromium and Ni, on the other hand, are not especially enriched when compared to the crust and soils (Table 2). The metal content of the river particulate matter is also highly enriched in relation to the crust and soils. It is obvious that anthropogenic activities have a pronounced effect on the particulate matter chemistry of lakes and rivers. It is also obvious that much of the enriched portion of the riverine particulates are deposited near river mouths and in the coastal zone (continental shelf) as the Ti-normalized metals for estuarine sediments and hemipelagic mud are less enriched than riverine particulates but still enriched relative to the crust and soils. Table 2 also shows the effect of diagenetic remobilization and reprecipitation of ferromanganese oxides in surficial pelagic clays as both Cu and Ni (major accessory elements in ferromanganese nodules) are significantly enriched in these marine deposits relative to the precursor earth materials. Finally, Table 2 shows the effects of high-temperature combustion on the enrichment of metals in coals as they are concentrated in fly ash. This is especially true for Pb, Zn, Cr, Cu, and Ni.

Table 2 Average concentration of six heavy metals in natural earth materials.

Material	Pb (ppm)	Zn (ppm)	Cd (ppm)	Cr (ppm)	Cu (ppm)	Ni (ppm)	Ti (wt.%)	References
Upper crust	17(52)	67(203)	0.1(0.30)	69(209)	39(118)	55(167)	0.33	Li (2000)
Average soils	26(68)	74(195)	0.1(0.26)	61(160)	23(60)	27(71)	0.38	Li (2000)
River mud	23(42)	78(142)	0.6(2.0)	85(155)	32(58)	32(58)	0.55	Govindaraju (1989)
Pre-industrial, baseline lacustrine sediment	22(69)	97(303)	0.3(0.55)	48(150)	34(106)	40(125)	0.32	Shafer and Armstrong (1991); Forstner (1981); Heit et al. (1984); Mudroch et al. (1988); Eisenreich (1980); Kemp et al. (1976, 1978); Wren et al. (1983); Wahlen and Thompson (1980)
Recent lacustrine sediment	102(316)	207(640)	2.2(6.8)	63(195)	60(186)	39(121)	0.32	Above references plus: Dominik et al. (1984); Rowell (1996); Mecray et al. (2001)
River particulate matter	68(120)	250(446)	1.2(2.1)	100(178)	100(178)	90(161)	0.56	Martin and Windom (1991); Martin and Whitfield (1981)
Estuarine sediment	54(108)	136(272)	1.2(2.4)	94(188)	52(104)	35(70)	0.50	Alexander et al. (1993); Coakley and Poulton (1993); Anikiyev et al. (1994); Hanson (1997)
Hemipelagic mud	23(49)	111(236)	0.2(0.44)	79(168)	43(91)	44(94)	0.47	Li (2000); Chester (2000)
Pelagic clay	80(174)	170(370)	0.4(0.9)	90(196)	250(543)	230(500)	0.46	Li (2000)
Coal	15(24)	53(84)	0.4(0.6)	27(43)	16(25)	17(27)	0.63	Tillman (1994); Adriano (1986)
Fly ash	43(70)	149(245)	0.5(0.8)	115(189)	56(92)	84(137)	0.61	Hower et al. (1999); Adriano (1986)

Values in parentheses are Ti-normalized.

15.1.3 Source and Pathways

The two main pathways for heavy metals to become incorporated into air–soil–sediment–water are transport by air (atmospheric) and water (fluvial). In the previous section it was shown that heavy-metal emissions to air and water (Table 1) are a significant percentage of the amounts of metals that are extracted from the Earth's crust by mining. Ores are refined by smelting thus releasing large amounts of metal waste to the environment (primary source). Relatively pure metals are incorporated into a multitude of technological products which, when discarded, produce a secondary, but important, source of metals to the environment. Metals are also incorporated naturally and technologically into foodstuffs which, when consumed and discarded by man, result in an important metal source to the aquatic environment (sewage wastewater), soils, and sediments (sewage sludge).

We can see from Table 3 that except for Pb in the terrestrial environment and Cd in the marine environment, metal transport to the lakes and to the oceans via water (fluvial) is many times greater (2–10) than that by air (atmospheric). This undoubtedly reflects the prevalence of wastewater discharges from sewage–municipal–industrial inputs that are so common in our industrialized society. The prevalence of Pb atmospheric emissions is probably due to the burning of leaded gasoline which was phased out in North America and Western Europe by the early 1990s but is still occurring in the Third World countries. Natural atmospheric emissions of Cd (volcanoes) are most likely the cause of substantial atmospheric Cd fluxes to the marine environment (Nriagu, 1990b).

15.2 OCCURRENCE, SPECIATION, AND PHASE ASSOCIATIONS

15.2.1 Geochemical Properties and Major Solute Species

15.2.1.1 Lead

Lead (atomic no. 82) is a bluish-white metal of bright luster, is soft, very malleable, ductile, and a poor conductor of electricity. Because of these properties and its low melting point (327 °C), and resistance to corrosion, Pb has been used in the manufacture of metal products for thousands of years. In fact, the ancient world technology for smelting Pb–Ag alloys from PbS ores was developed 5,000 years ago (Settle and Patterson, 1980). Lead has a density of $11.342 \, \text{g cm}^{-3}$, hence finds extensive use as a shield for radiation; its atomic weight is 207.2. Lead has two oxidation states, +2 and +4. The tetravalent state is a powerful oxidizing agent but is not common in the Earth's surficial environment; the divalent state, on the other hand, is the most stable oxidation level and most Pb^{2+} salts with naturally-occurring common

Table 3 Relative percentage of atmospheric (%A) and fluvial (%F) inputs of six heavy metals to lakes, a coastal zone, and the ocean.

Lake/Ocean	Metal												References
	Pb		Zn		Cd		Cr		Cu		Ni		
	%A	%F	%A	%F	%A	%F	%A	%F	%A	%F	%A	%F	
Lake IJsselmeer	NA	NA	7	93	2	98	0.1	99.9	6	94	1	99	Salomons (1983)
Southern Lake Michigan	47	53	22	78	NA	NA	NA	NA	13	87	NA	NA	Dolske and Sievering (1979)
Lake Michigan	60	40	35	65	10	90	41	59	15	85	NA	NA	Eisenreich (1980)
Lake Erie	40	60	12	88	NA	NA	NA	NA	9	91	NA	NA	Nriagu et al. (1979)
South Atlantic Bight	2	98	1	99	41	59	NA	NA	7	93	9	91	Chester (2000)
Ocean	15	85	5	95	17	83	3	97	2	98	5	95	Chester (2000)

NA = not available.

anions are only slightly soluble. It is composed of four stable isotopes ($^{208}Pb = 52\%$) and several radioisotopes whose longest half-life is 15 Myr (Reimann and de Caritat, 1998). Lead belongs to group IVa of the periodic table which classifies it as a heavy metal whose geochemical affinity is chalcophilic (associated with sulfur).

In a simple freshwater system, exposed to atmospheric CO_2 and containing 10^{-3} M Cl^-, 10^{-4} M SO_4^{-2}, and 10^{-6} M HPO_4^{-2}, it is predicted that Pb will be complexed by the carbonate species $Pb(CO_3)_2^{-2}$ in the pH range of 6–8 (Hem and Durum, 1973). The complex $PbSO_4^0$ is stable below pH 6 (or in low sulfate waters Pb^{2+}) and the complex $Pb(OH)_2$ is stable above pH 8 (Hem, 1976). In oxygenated stream and lake environments the concentration of dissolved Pb is less than 1 $\mu g\,L^{-1}$ over the pH range of 6–8 (Reimann and de Caritat, 1998) while its average concentration in world river water is 0.08 $\mu g\,L^{-1}$ (Gaillardet et al., 2003). The dissolved Pb concentration in ocean water (0.002 $\mu g\,L^{-1}$) is an order of magnitude lower than that in river water (Chester, 2000).

Adsorption and aggregation-complexation with organic matter appear to be the most important processes that transform dissolved Pb to particulate forms in freshwater systems. Krauskopf (1956) originally suggested that the concentration of Pb, as well as certain other trace metals, could be controlled by adsorption onto the ferric and manganese oxyhydroxides–clay mineral–organic matter. The extent of Pb adsorption onto hydrous Fe and Mn oxides is influenced by the physical characteristics of the adsorbent (specific surface, crystallinity, etc.) and the composition of the aqueous phase (pH, Eh, complexation, competing cations). In a recent study of Fe and Pb speciation, reactivity, and cycling in a lacustrine environment, Taillefert et al. (2000) determined that Pb is entrained during the formation of Fe-exocellular polymeric substances (EPS) that aggregate in a water column near the chemocline. It is not yet clear whether the metal is complexed to the EPS or adsorbed directly to the Fe oxide. However, extraction data from lake sediments suggest that the Pb–FeO_x phase is available to chemical attack (see below).

The average concentration of Pb in the lithosphere is about 14 $\mu g\,g^{-1}$ and the most abundant sources of the metal are the minerals galena (PbS), anglesite ($PbSO_4$), and cerussite ($PbCO_3$). The most important environmental sources for Pb are gasoline combustion (presently a minor source, but in the past 40 years a major contributor to Pb pollution), Cu–Zn–Pb smelting, battery factories, sewage sludge, coal combustion, and waste incineration.

15.2.1.2   Zinc

Zinc (atomic no. 30) is a bluish-white, relatively soft metal with a density of 7.133 g cm^{-3}.

It has an atomic weight of 65.39, a melting point of 419.6 °C, and a boiling point of 907 °C. Zinc is divalent in all its compounds and is composed of five stable isotopes ($^{64}Zn = 49\%$) and a common radioisotope, ^{65}Zn, with a half-life of 245 days. It belongs to group IIb of the periodic table which classifies it as a heavy metal whose geochemical affinity is chalcophilic.

In freshwater, the uncomplexed Zn^{2+} ion dominates at an environmental pH below 8 whereas the uncharged $ZnCO_3^0$ ion is the main species at higher pH (Hem, 1972). Complexing of Zn with SO_4^{-2} becomes important at high sulfate concentrations or in acidic waters. Hydrolysis becomes significant at pH values greater than 7.5; hydroxy complexes of $ZnOH^-$ and $Zn(OH)_2^0$ do not exceed carbonate species at typical environmental concentrations of 15 $\mu g\,L^{-1}$ for world stream water (Reimann and de Caritat, 1998). More recent data of Gaillardet et al. (2003) places the concentration of dissolved Zn in average world river water at 0.60 $\mu g\,L^{-1}$. Significant complexing with organic ligands may occur in stream and lake waters with highly soluble organic carbon concentrations. The concentration of Zn in ocean water is 0.39 $\mu g\,L^{-1}$ (Chester, 2000), which is close to its value in world river water.

There are several factors that determine the relative abundance of dissolved and particulate Zn in natural aquatic systems. These include media pH, biogeochemical degradation processes that produce dominant complexing ligands, cation exchange and adsorption processes that control the chemical potential of solid substrates, and the presence of occluded oxyhydroxide compounds (Adriano, 1986). At pH values above 7, aqueous complexed Zn begins to partition to particulate Zn as a result of sorption onto iron oxyhydroxide. The clay mineral montmorillonite is particularly efficient in removing Zn from solution by adsorption (Krauskopf, 1956; Farrah and Pickering, 1977).

The average Zn content of the lithosphere is ~80 $\mu g\,g^{-1}$ and the most abundant sources of Zn are the ZnS minerals sphalerite and wurtzite and to a lesser extent smithsonite ($ZnCO_3$), willemite (Zn_2SiO_4), and zincite (ZnO) (Reimann and de Caritat, 1998). The smelting of nonferrous metals and the burning of fossil fuels and municipal wastes are the major Zn sources contributing to air pollution.

15.2.1.3   Cadmium

Cadmium has an atomic number of 48, an atomic weight of 112.40 consisting of eight stable isotopes ($^{112,114}Cd$ are most abundant), and a density of 8.65 g cm^{-3} (Nriagu, 1980a). In several aspects Cd is similar to Zn (it is a neighbor of Zn in the periodic table); in fact it is almost always associated with Zn in mineral deposits and other earth materials. Cadmium is a soft, silvery white, ductile metal with a faint bluish tinge. It has a melting point of 321 °C and a boiling point of 765 °C. It belongs to group IIb of elements in the periodic table and in aqueous solution has the stable 2+ oxidation state. Cadmium is a rare element (67th element in order of abundance) with a concentration of $\sim 0.1 \mu g\, g^{-1}$ in the lithosphere and is strongly chalcophilic, like Zn.

In a natural, aerobic freshwater aquatic system with typical Cd–S–CO_2 concentrations (Hem, 1972), Cd^{2+} is the predominant species below pH 8, $CdCO_3^0$ is predominant from pH 8 to 10, and $Cd(OH)_2^0$ is dominant above pH 10. The solubility of Cd is minimum at pH 9.5 (Hem, 1972). The speciation of Cd is generally considered to be dominated by dissolved forms except in cases where the concentration of suspended particulate matter is high such as "muddy" rivers and reservoirs and near-bottom benthic boundary layers, and underlying bottom sediments in rivers and lakes (Li *et al.*, 1984). The distribution coefficient between the particulate and the dissolved Cd is remarkably consistent for a wide range of riverine and lacustrine situations (Lum, 1987). The sorption of Cd on particulate matter and bottom sediments is considered to be a major factor affecting its concentration in natural waters (Gardiner, 1974). Pickering (1980) has quantitatively evaluated the role clay minerals, humic substances, and hydrous metal oxides in Cd adsorption and concludes that some fraction of the particle-bound Cd is irreversibly held by the solid substrate. The concentration of dissolved Cd in average world river water is $0.08 \mu g\, L^{-1}$ (Gaillardet *et al.*, 2003). This concentration is identical to that of Cd in ocean water ($0.079 \mu g\, L^{-1}$; Chester, 2000).

15.2.1.4   Chromium

Chromium has an atomic number of 24, an atomic weight of 51.996 consisting of four stable isotopes ($^{52}Cr = 84\%$), and a density of 7.14 g cm^{-3} (Adriano, 1986). Crystalline Cr is steel-gray in color, lustrous, hard metal that has a melting point of 1,900 °C and a boiling point of 2,642 °C. It belongs to group VIb of the transition metals and in aqueous solution Cr exists primarily in the trivalent (+3) and hexavalent (+6) oxidation states. Chromium, as well as Zn, are the most abundant of the "heavy metals" with a concentration of about $69 \mu g\, g^{-1}$ in the lithosphere (Li, 2000).

In most natural waters at near neutral pH, Cr^{III} is the dominant form due to the very high redox potential for the couple Cr^{VI}/Cr^{III} (Rai *et al.*, 1989). Chromium(III) forms strong complexes with hydroxides. Rai *et al.* (1987) report that the dominant hydroxo species are $CrOH^{2+}$ at pH values 4–6, $Cr(OH)_3^0$ at pH values from 6 to 11.5, and $Cr(OH)_4^-$ at pH values above 11.5.

The OH^- ligand was the only significant complexer of Cr^{III} in natural aqueous solutions that contain environmental concentrations of carbonate, sulfate, nitrate, and phosphate ions. The only oxidant in natural aquatic systems that has the potential to oxidize Cr^{III} to Cr^{VI} is manganese dioxide. This compound is common on Earth's surface and thus one can expect to find some Cr^{VI} ions in natural waters. The predominant Cr^{VI} species at environmental pH is CrO_4^{2-} (Hem, 1985). The principal Cr^{III} solid compound that is known to control the solubility of Cr^{III} in nature is $Cr(OH)_3^0$. However, Sass and Rai (1987) have shown that $Cr/Fe(OH)_3$ has an even lower solubility. This compound is a solid solution and thus its solubility is dependent on the mole fraction of Cr; the lower the mole fraction, the lower the solubility (Sass and Rai, 1987). Most Cr^{VI} solids are expected to be relatively soluble under environmental conditions. In the absence of solubility-controlling solids, Cr^{VI} aqueous concentrations under neutral pH conditions will primarily be controlled by adsorption/desorption reactions (Rai *et al.*, 1989). Under environmental conditions, iron oxides are the predominant adsorbents of chromate (Cr^{VI}) in acidic to neutral pH range and oxidizing environments. The Cr concentration in average world river water is $0.7 \mu g\, L^{-1}$ (Gaillardet *et al.*, 2003) and that in ocean water is $0.21 \mu g\, L^{-1}$ (Chester, 2000).

Chromium occurs in nature mainly in the mineral chromite; Cr also occurs in small quantities in many minerals in which it replaces Fe^{3+} and Al^{3+} (Faust and Aly, 1981). The metallurgy industry uses the highest quality chromite ore whilst the lower-grade ore is used for refractory bricks in melting furnaces. Major atmospheric emissions are from the chromium alloy and metal producing industries. Smaller emissions come from coal combustion and municipal incineration. In the aquatic environment, the major sources of Cr are electroplating and metal finishing industries. Hexavalent Cr^{VI} is a potent carcinogen and trivalent Cr^{III} is an essential trace element (Krishnamurthy and Wilkens, 1994).

15.2.1.5   Copper

Copper has an atomic number of 29, an atomic weight of 63.546, consists of two stable isotopes ($^{63}Cu = 69.2\%$; $^{65}Cu = 30.8\%$), and has a density of 8.94 g cm^{-3} (Webelements, 2002). Metallic Cu compounds (sulfides) are typically brassy yellow in color while the carbonates are a variety of green- and yellow-colored. The metal is somewhat malleable with a melting point of 1,356 °C and a boiling point of 2,868 °C. It belongs to group Ib of the transition metals and in aqueous solution Cu exists primarily in the divalent oxidation state although some univalent complexes and compounds of Cu do occur in nature (Leckie and

Davis, 1979). Copper is a moderately abundant heavy metal with a concentration in the lithosphere of about $39 \mu g\,g^{-1}$ (Li, 2000).

Chemical models for the speciation of Cu in freshwater (Millero, 1975) predict that free $Cu^{2+}(aq)$ is less than 1% of the total dissolved Cu and that $Cu(CO_3)_2^{2-}$ and $CuCO_3^0$ are equally important for the average river water. Leckie and Davis (1979) showed that the $CuCO_3^0$ complex is the most important one near the neutral pH. At pH values above 8, the dihydroxo–Copper(II) complex predominates. The chemical form of Cu is critical to the behavior of the element in geochemical and biological processes (Leckie and Davis, 1979). Cupric Cu forms strong complexes with many organic compounds.

In the sedimentary cycle, Cu is associated with clay mineral fractions, especially those rich in coatings containing organic carbon and manganese oxides. In oxidizing environments (Cu–H_2O–O_2–S–CO_2 system), Cu is likely to be more soluble under acidic than under alkaline conditions (Garrels and Christ, 1965). The mineral malachite is favored at pH values above 7. Under reducing conditions, Cu solubility is greatly reduced and the predominant stable phase is cuprous sulfide (Cu_2S) (Leckie and Nelson, 1975). In natural aquatic systems, some of the Cu is dissolved in freshwater streams and lakes as carbonate and organic complexes; a larger fraction is associated with the solid phases. Much of the particulate Cu is fixed in the crystalline matrix of the particles (Gibbs, 1973). Some of the riverine reactive particulate Cu may be desorbed as the freshwater mixes with seawater. The biological cycle of Cu is superimposed on the geochemical cycle. Copper is an essential element for the growth of most of the aquatic organisms but is toxic at levels as low as $10 \mu g\,L^{-1}$ (Leckie and Davis, 1979). Copper has a greater affinity, than most of the other metals, for organic matter, organisms, and solid phases (Leckie and Davis, 1979) and the competition for Cu between the aqueous and the solid phases is very strong. Krauskopf (1956) noted that the concentration of copper in natural waters, 0.8–$3.5 \mu g\,L^{-1}$ (Boyle, 1979), is far below the solubility of known solid phases. Davis *et al.* (1978) found that the adsorption behavior of Cu in natural systems is strongly dependent on the type and concentration of inorganic and organic ligands. Recent data of Gaillardet *et al.* (2003) places the concentration of dissolved Cu in average world river water at $1.5 \mu g\,L^{-1}$ and that in ocean water at $0.25 \mu g\,L^{-1}$ (Chester, 2000).

The most common Cu minerals, from which the element is refined into the metal, are Chalcocite (Cu_2S), Covellite (CuS), Chalcopyrite ($CuFeS_2$), Malachite and Azurite (carbonate compounds). It is not surprising that Cu is considered to have a chalcophillic geochemical affinity. In the past, the major source of Cu pollution was smelters that contributed vast quantities of Cu–S particulates to the atmosphere. Presently, the burning of fossil fuels and waste incineration are the major sources of Cu to the atmosphere and the application of sewage sludge, municipal composts, pig and poultry wastes are the primary sources of anthropogenic Cu contributed to the land surface (Alloway, 1995).

15.2.1.6 Nickel

Nickel has an atomic number of 28, an atomic weight of 58.71 consisting of five stable isotopes of which ^{58}Ni (67.9%) and ^{60}Ni (26.2%) are the most abundant, and a density of $8.9\,g\,cm^{-3}$ (National Science Foundation, 1975). Nickel is a silvery white, malleable metal with a melting point of $1,455\,^{\circ}C$ and a boiling point of $2,732\,^{\circ}C$. It has high ductility, good thermal conductivity, moderate strength and hardness, and can be fabricated easily by the procedures which are common to steel (Nriagu, 1980b). Nickel belongs to group VIIIa and is classified as a transition metal (the end of the first transition series) whose prevalent valence states are 0 and 2+. However, the majority of nickel compounds are of the Ni^{II} species.

Morel *et al.* (1973) showed that the free aquo species (Ni^{2+}) dominates at neutral pH (up to pH 9) in most aerobic natural waters; however, complexes of naturally occurring ligands are formed to a minor degree ($OH^- > SO_4^{2-} > Cl^- > NH_3$). Under anaerobic conditions that often occur in the bottom sediments of lakes and estuaries, sulfide controls the solubility of Ni. Under aerobic conditions, the solubility of Ni is controlled by either the co-precipitate $NiFe_2O_4$ (Hem, 1977) or $Ni(OH)_{2(s)}$ (Richter and Theis, 1980). The latter authors performed laboratory adsorption experiments for Ni in the presence of silica, goethite, and amorphous manganese oxide and found that manganese oxide removed 100% of the Ni over the pH range 3–10. The iron oxide began to adsorb Ni at pH 5.5, the oxide's zero point of charge. Hsu (1978) found that Ni was associated with both amorphous iron and manganese oxides that coated silica sand grains.

In 1977, Turekian noted that the calculated theoretical concentrations of Ni and other trace metals in seawater were in orders of magnitude higher than the measured values. Turekian (1977) hypothesized that the role of particulate matter was most important in sequestering reactive elements and transporting them from the continents to the ocean floor. For lakes, Allan (1975) demonstrated that atmospheric inputs were responsible for Ni concentrations in sediments from 65 lakes surrounding a nickel smelter. As Jenne (1968) and Turekian (1977) note, hydrous iron and manganese oxides have a large capacity for sorption or co-precipitation with trace metals such as Ni. These hydrous oxides exist as coatings on the particles, particularly clays, and can transport sequestered metals to great distances (Snodgrass, 1980). In the major rivers of the world, Ni transport is divided into the following phases (Snodgrass, 1980): 0.5% solution, 3.1% adsorbed, 47% as precipitated coating, 14.9% complexed by organic matter, and 34.4% crystalline material.

The concentration of Ni in the lithosphere is $55 \mu g \ g^{-1}$ (Li, 2000) and the concentration of dissolved Ni in stream water is $2 \mu g \ L^{-1}$ (Turekian, 1971). More recent data on the concentration of dissolved Ni in average world river water indicates the value to be $0.8 \mu g \ L^{-1}$ (Gaillardet *et al.*, 2003) and the Ni concentration in ocean water to be $0.47 \mu g \ L^{-1}$ (Chester, 2000). Natural emissions of Ni to the atmosphere are dominated by windblown dusts while anthropogenic sources that represent 65% of all emission sources are dominated by fossil-fuel combustion, waste incineration and nonferrous metal production (Nriagu, 1980b). Major uses of Ni include its metallurgical use as an alloy (stainless steel and corrosion-resistant alloys), plating and electroplating, as a major component of Ni–Cd batteries, and as a catalyst for hydrogenating vegetable oils (National Science Foundation, 1975).

15.2.2 Occurrence in Rocks, Soils, Sediments, Anthropogenic Materials

Table 4 presents the average concentration of six heavy metals (Pb, Zn, Cd, Cr, Cu, Ni) in a variety of earth materials, soils, sediments, and natural waters. For Pb it can be seen that the solid-phase concentration increases little along the transport gradient from the Earth's crust to world soils to lake sediments ($14 < 22 < 23 \mu g \ g^{-1}$; Table 4). However, stream sediment and particularly riverine particulate matter is substantially enriched ($50–68 \mu g \ g^{-1}$) suggesting that anthropogenic inputs from the past use of leaded gasoline, the prevalent burning of fossil fuels and municipal waste, and land disposal of sewage sludge are mobilized from soils and become concentrated in transported particulate matter. The Pb content of soils in England and Wales (UK) is much higher ($74 \mu g \ g^{-1}$) than that ($12 \mu g \ g^{-1}$) found in remote soils of the USA (Alloway, 1995). This is due in part to the more densely populated regions of the UK that were sampled and the inclusion of metalliferous mining areas. Shallow marine sediments appear not to be enriched in Pb related to source materials (crustal rocks and world soils; Table 4) and deep-sea sediments appear to be the final repository of Pb that becomes concentrated in a variety of authigenic phases.

Zinc and Cd show a similar pattern with riverine particulate Zn ($250 \mu g \ g^{-1}$) greatly exceeding average Zn in terrestrial earth materials ($68 \pm 32 \mu g \ g^{-1}$) and world soils ($66 \pm 17 \mu g \ g^{-1}$), and particulate Cd ($1.2 \mu g \ g^{-1}$) greatly exceeding the terrestrial earth materials and soil concentrations (0.14 ± 0.08 and $0.23 \pm 0.15 \mu g \ g^{-1}$). As for Pb, UK soils are significantly greater in Zn and Cd concentrations relative to USA soils, a fact that reflects the urban and metalliferous character of the

Table 4 Heavy metals in the Earth's crustal materials, soils, freshwater sediments, and marine sediments.

Material	Pb	Zn	Cd	Cr	Cu	Ni	References
Crust	14.8	65	0.10	126	25	56	Wedepohl (1995)
Granite	18, 17	40, 50	0.15, 0.13	20, 10	15, 20	8, 10	Adriano (1986); Drever (1988)
Basalt	8, 6	100, 105	0.2, 0.2	220, 170	90, 87	140, 130	″
Shale	23, 20	100, 95	1.4, 0.3	120, 90	50, 45	68, 68	″
Sandstone	10, 7	16, 16	<0.03,	35, 35	2, 2	2, 2	″
Limestone	9, 9	29, 20	0.05, 0.03	10, 11	4, 4	20, 20	″
Soils (general)	19	60	0.35	54	25	19	Adriano (1986)
Soils (World)	30	66	0.06	68	22	22	Kabata-Pendias (2000)
Soils, UK	74	97	0.8	41	23	25	Alloway (1995)
Soils, USA	12	57	0.27		30	24	″
Stream sediments	51 ± 28	132 ± 67	1.57 ± 1.27	67 ± 24	39 ± 13	44 ± 19	Various sources[a]
Lake sediment	22	97	0.6	48	34	40	Table 2
River particulates	68	250	1.2	100	100	90	″
Shallow marine sediment	23	111	0.2	79	43	44	Li (2000); Chester (2000)
Deep-sea clay	80	170	0.4	90	250	230	Li (2000)
Streams	1	30	0.01	1	7	2	Drever (1988)
Ocean	0.03	2	0.05	0.2	0.5	0.5	Drever (1988)

Units are $\mu g \ g^{-1}$ dry weight. Dissolved metal data for streams and ocean water are expressed in units $\mu g \ L^{-1}$.
[a] Various Sources: Dunnette (1992), Aston *et al.* (1974), Presley *et al.* (1980), Olade (1987), Mantei and Foster (1991), Zhang *et al.* (1994), Osintsev (1995), Chiffoleau *et al.* (1994), Borovec *et al.* (1993), Gocht *et al.* (2001).

UK soils (Alloway, 1995). Stream–lake–shallow marine sediments are all more concentrated in Zn ($113 \pm 18\,\mu g\,g^{-1}$) than crustal rocks and soils ($64 \pm 3\,\mu g\,g^{-1}$). As in the case of Pb, deep-sea clays are the ultimate repository for Zn also.

Chromium has the highest concentration of all the six heavy metals in the Earth's crust (Table 4), mainly due to a very high concentration in basalt and shale. Average crustal rocks ($72 \pm 75\,\mu g\,g^{-1}$) are similar in Cr concentration to world soils ($73 \pm 19\,\mu g\,g^{-1}$) and the average Cr concentration in stream sediment–riverine particulates–lake sediment–shallow marine sediment ($74 \pm 22\,\mu g\,g^{-1}$). Only deep-sea clay is slightly enriched relative to all the other earth materials (Table 4). From these data it is apparent that natural Cr concentrations of various earth materials that constitute the weathering-transport continuum from continent to oceans have not been seriously altered by man's activities. As has been seen before, this is not the case for Pb, Zn, and Cd. These metals, along with Cu and Ni, are the backbone of the world's metallurgical industry and thus man's mining and smelting activities that have gone on for centuries have greatly altered the natural cycles.

The Cu concentration of crustal rocks (32 ± 34 $\mu g\,g^{-1}$) is approximately equivalent to that for average soils ($25 \pm 4\,\mu g\,g^{-1}$). However, as the earth material is weathered and transported to streams–lakes–shallow marine sediments there is a minimal enrichment in Cu concentration ($39 \approx 34 \approx 43\,\mu g\,g^{-1}$) (Table 4). And, as for Pb–Zn–Cd, riverine particulate matter is greatly enriched ($100\,\mu g\,g^{-1}$) relative to the other sedimentary materials. While the Pb–Zn–Cd concentrations of deep-sea clay are enriched 1.5 times that of the continental sedimentary materials, Cu is enriched approximately five times. The substantial enrichment of Cu in oceanic pelagic clay relative to terrestrial earth materials is due to the presence of ubiquitous quantities of ferromanganese oxides in surficial ocean sediments (Drever, 1988).

The Ni concentration of crustal rocks (58 ± 53 $\mu g\,g^{-1}$) is substantially greater than the average world soils ($23 \pm 3\,\mu g\,g^{-1}$), but essentially equal to continental sedimentary materials ($49 \pm 13\,\mu g\,g^{-1}$). Riverine particulate matter ($90\,\mu g\,g^{-1}$) is nearly twice the Ni concentration of these continental sedimentary materials and deep-sea clay is nearly three times ($230\,\mu g\,g^{-1}$) that concentration. As noted for Cu, the substantial Ni enrichment of deep-sea clays is due to the presence of ferromanganese micronodules in the oxidized surficial sediment column (Drever, 1988).

Table 5 gives the average concentration of six heavy metals in anthropogenic by-products; that is, materials refined from natural materials such as fly ash from coal and smelting of metal ores or by-products from man's use such as sewage sludge and animal waste. It is evident that smelting of the metal ores is a major contributor to the environmental pollution caused by atmospheric transport of heavy metals (Table 5). However, fly ash emissions from coal-fired power plants is probably a more important source of atmospheric heavy-metal pollution due to the fact that these power plants are the main sources of electricity for much of the world's population. In addition, sewage sludge is a major contributor of heavy-metal pollution in soils as land disposal of human waste becomes the only practical solution. It is not surprising that riverine particulates are so enriched in Pb, Zn, Cd, Cu, and Ni as soils polluted with atmospheric emissions from mining and smelting activities, and those altered by the addition of sewage sludge are swept into streams and rivers that eventually empty into the ocean.

15.2.3 Geochemical Phase Associations in Soils and Sediments

Not all metals are equally reactive, toxic, or available to biota. The free ion form of the metal is thought to be the most available and toxic (Luoma, 1983). With regards to reactivity, it is

Table 5 Average concentration of six heavy metals in anthropogenic by-products.

By-Product	Pb	Zn	Cd	Cr	Cu	Ni	References
Coal	15	53	0.4	27	16	17	Tillman (1994), Adriano (1986)
Fly ash	43	144	0.5	115	56	84	Hower et al.(1999), Adriano (1986)
Soils down-wind of smelters	28, 2200	61, 3000	25, 91		184	306	Adriano (1986), Alloway (1995)
Fertilizers	235	288, 371	32, 35	151, 60	18, 84	36, 20	,,
Sewage sludge	1049, 820	3025, 2490	72, 18	1221, –	1085, –	319, –	,,
Animal waste	45, 11	93, 130	0.36, 0.55	16, 30	20, 31	29, 19	Adriano (1986), Kabata-Pendias and Pendias (2001)

Units are $\mu g\,g^{-1}$ dry weight.

generally thought that different metal ions display differing affinities for surface binding sites across the substrates (Warren and Haack, 2001). The speciation or dissolved forms of a metal in solution is of primary importance in determining the partitioning of the metal between the solid and solution phases. Mineral surfaces, especially those of Fe oxyhydroxides, have been studied well by aquatic chemists. This is due to their ubiquitous and abundant nature and their proven geochemical affinity (Honeyman and Santschi, 1988). Metals can be incorporated into solid minerals by a number of processes; nonspecific and specific adsorption, co-precipitation, and precipitation of discrete oxides and hydroxides (Warren and Haack, 2001). Furthermore, Fe and Mn oxyhydroxides form surface coatings on other types of mineral surfaces such as clays, carbonates, and grains of feldspar and quartz. The three most common environmental solid substrates are Fe-oxides, Mn-oxides, and natural organic matter (NOM) (Warren and Haack, 2001).

Sediments are an important storage compartment for metals that are released to the water column in rivers, lakes, and oceans. Because of their ability to sequester metals, sediments can reflect water quality and record the effects of anthropogenic emissions (Forstner, 1990). Particles as substrates of pollutants originate from two sources; (a) particulate materials transported from the watershed that are mostly related to soils and (b) endogenic particulate materials formed within the water column. Since adsorption of metal pollutants onto air- and waterborne particles is the primary factor in determining the transport, deposition, reactivity, and potential toxicity of these metals, analytical techniques should be related to either the chemistry of the particle surface or to the metal species that is highly enriched on the particle surface (Forstner, 1990). In the absence of highly-sophisticated solid-state techniques, chemical methods have been devised to characterize the reactivity of metal-rich phases adsorbed to solid particle surfaces. Single leaching and combined sequential extraction schemes have been developed to estimate the relative phase associations of sedimentary metals in various aquatic environments (Pickering, 1981). The most widely applied extraction scheme was developed by Tessier *et al.* (1979) in which the extracted components were defined as exchangeable, carbonates, easily-reducible Mn oxides, moderately-reducible amorphous Fe oxides, sulfides and organic matter, and lithogenic material.

Partition studies on river sediments were first reported by Gibbs (1973) for suspended loads of the Amazon and Yukon rivers. Nickel was the main heavy metal bound to hydroxide coatings while a lithogenic crystalline phase concentrated the Cr and Cu. Salomons and Forstner (1980), in an extraction study of river sediments from different regions of the world, found that less polluted or unpolluted river systems exhibit an increase in the relative amount of the metals' lithogenic fraction and that the excess of metal contaminants released to the aquatic environment by man's activities exist in relatively unstable chemical associations such as exchangeable and reducible. With the exception of Cd and Mn, the amount of heavy metals in exchangeable positions is generally low (Salomons and Forstner, 1984). In addition to this, Zn is often concentrated in the easily reducible phase (amorphous Fe/Mn oxyhydroxides), and Fe–Pb–Cu–Cr are concentrated in the moderately reducible phase (crystalline Fe/Mn oxyhydroxides) (Salomons and Forstner, 1984). As can be seen later, for reservoir and lake sediments, Pb is almost completely extracted by the mildly acidic hydroxylamine hydrochloride but Zn is only partially extracted by this chemical that defines the easily-reducible phase.

In a series of landmark papers by Tessier and coworkers, the role of hydrous Fe/Mn oxides in controlling the heavy-metal concentrations in natural aquatic systems has been defined by careful field and laboratory studies by comparing with theory (Tessier *et al.*, 1985). They concluded that the adsorption of Cd, Cu, Ni, Pb, and Zn onto Fe-oxyhydroxides is an important mechanism in the lowering of heavy-metal concentrations in oxic pore waters of Canadian-Shield lakes. These heavy-metal concentrations were below the concentrations prescribed by equilibrium solubility models. In a more recent study, Tessier *et al.* (1989) concluded that Zn is sorbed onto Fe oxyhydroxides and that their field data fit reasonably well into a simple model of surface complexation. They also concluded that other substrates (Mn oxyhydroxides, organic matter, clays) can sorb Zn. Also, removal of Zn by phytoplankton has been shown to be an important mechanism for controlling the dissolved Zn concentrations in the eutrophic Lake Zurich (Sigg, 1987). Finally, Tessier *et al.* (1996) expanded their studies to include adsorbed organic matter. Their results strongly suggest that pH plays an important role in determining which types of particle surface binding sites predominate in the sorption of heavy metals in lakes. In circumneutral lakes metals are bound directly to hydroxyl groups of the Fe/Mn oxyhydroxides, and in acidic lakes metals are bound indirectly to these oxyhydroxides via adsorption of metals complexed by NOM.

Some words of caution should be included concerning these "solid speciation" sediment extraction techniques. Kersten and Forstner (1987) noted that "useful information on solid speciation influencing the mobility of contaminants in biogeochemically reactive sediments by the chemical leaching approach requires proper and careful handling of the anoxic sediment samples." Martin *et al.* (1987)

showed that the specificity and reproducibility of the extraction method greatly depends on the chemical properties of the element and the chemical composition of the samples. They state that "these methods provide, at best, a gradient for the physicochemical association strength between trace elements and solid particles rather than their actual speciation." The problem of post-extraction readsorption of As, Cd, Ni, Pb, and Zn has been addressed by Belzile *et al.* (1989) who found that by using the "Tessier method" (Tessier *et al.*, 1979) on trace-element spiked natural sediments it is possible to recover the added trace elements within the limits of experimental error.

In a recent study of extraction of anthropogenic trace metals from sediments of US urban reservoirs, Conko and Callender (1999) showed that Pb had the highest anthropogenic content accounting for 80–90% of the total metal concentration. Three extractions were used: (i) easily-reducible 0.25 M Hydroxylamine HCL in 0.25 N HCl (Chao, 1984); (ii) weak-acid digest (Hornberger *et al.*, 1999); and (iii) Pb-isotope digest of 1 N HNO_3 + 1.75 N HCl (Graney *et al.*, 1995). Chao (1984) extraction, originally thought to extract only amorphous Mn oxyhydroxides, is now considered to be an acid-reducible extraction that solubilizes amorphous hydrous Fe and Mn oxides (Sutherland and Tack, 2000). Hornberger *et al.*'s (1999) weak-acid digest (0.6 N HCl) is thought to represent the bioavailable fraction of the metal (Hornberger *et al.*, 2000). Graney *et al.*'s (1995) HNO_3 + HCl acid digest is the most aggressive of the three extracts and has been shown to represent, using Pb isotopes, the anthropogenic fraction of Pb in lacustrine sediments. A plot of extractable Pb

and total Pb for a 1997 sediment core from the suburban Lake Anne watershed in Reston, Virginia (Callender and Van Metre, 1997) is presented in Figure 1(a). It can be seen that between 85 and 95% of the total Pb is extracted by these chemicals. In general, the Chao extraction recovered 95% of the total Pb and since this technique is thought to specifically extract amorphous Fe and Mn oxyhydroxides, Conko and Callender (1999) postulated that most of the Pb is bound by these amorphous oxides. Figure 1(b) is a plot of various extractions and total Zn in the same Lake Anne core. Only 70% of the total Zn was extracted by any of the three techniques mentioned before; thus the remaining Zn must be associated with other sedimentary phases. Conko and Callender (1999) suggest Zn fixation by 2:1 clay minerals (i.e., montmorillonite), whereby sorbed Zn is fixed in the alumina octahedral layer, is an important phase (Pickering, 1981). An additional phase could be biotic structures that are postulated by Webb *et al.* (2000) as substrates where Zn occurs in intimate combination with Fe and P. While Lake Anne sediment is a typical siliclastic material, Lake Harding (located south of Atlanta, Georgia) sediment is reddish in color and consists of appreciably more iron and aluminum oxides. Much of the iron oxides are undoubtedly crystalline in character and may not be attacked by the mild extraction techniques listed above (especially the Chao extraction). Figure 2 shows a plot of the extractable and total Pb in a sediment core from Lake Harding. Contrary to the Lake Anne Pb data where 95% is extractable, only 75% of the total Pb in Lake Harding is extractable. The Chao easily-reducible extraction yielded the lowest

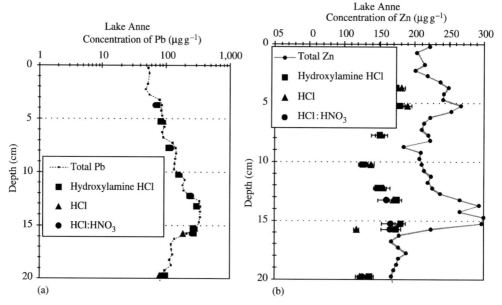

Figure 1 Temporal distribution of total and extractable lead (a) and zinc (b) in a sediment core from Lake Anne, Reston, Virginia, USA.

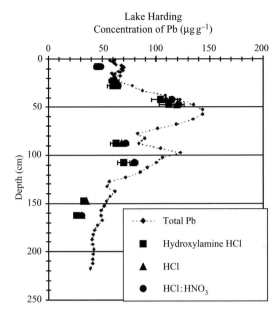

Lake Harding
Concentration of Pb (µg g⁻¹)

Figure 2 Temporal distribution of total and extractable lead in a sediment core from Lake Harding, Atlanta, Georgia, USA.

extraction efficiencies. It is clear from these data that the type and nature of phase components that comprise natural aquatic sediments are most important for understanding the efficiency of any extraction scheme. Very little is known about the relationship between easily-extracted phases removed by sequential extraction (Tessier *et al.*, 1979) and those liberated by single leaches. Sutherland (2002) compared the two approaches using soil and road deposited sediment in Honolulu, Hawaii. The results indicated that the dilute HCl leach was slightly more aggressive than the sequential procedure but that there was no significant difference between the Pb and Zn concentrations liberated by the two approaches. Further, the data also indicated that a dilute HCl leach was a valuable, rapid, cost-effective analytical tool for contamination assessment. The Hawaii data also indicated that between 75% and 80% of the total Pb is very labile and anthropogenically enhanced (Sutherland, 2002; Sutherland and Tack, 2000). On the other hand, while labile Zn comprises 75% of the total, it is equally distributed between acid extractable and reducible (Sutherland *et al.*, 2000). The extractable Pb data agrees well with the lacustrine Pb data presented in Figure 1(a). The single HCl leach method on Hawaii sediments (Sutherland, 2002) extracts about 50% of the total Zn; a figure even lower than the 70% for lake sediments. Unfortunately, no information was available concerning the phase distribution of Zn in these sediments.

An important reason for testing "selective" leach procedures on sediments that are subjected to anthropogenic influence is to determine whether

such leaches can be used to measure the anthropogenic metal content of sedimentary materials. For Lake Anne sediment, Conko and Callender (1999) calculated the anthropogenic Pb and Zn content by subtracting the background metal concentrations from the total metal content. These were then compared to the "anthropogenic" leach concentrations. For both Pb and Zn, there was essentially no difference between the two procedures. These techniques were applied to several other lake sediments with similar successes suggesting that a mild acid leach might be used to estimate the labile, anthropogenic metal content of a variety of sedimentary materials.

Terrestrial materials (river sediments, lake sediments, and urban particulate matter) appear to have between 50% and 70% exchangeable Pb and Zn while marine sediments contain very little exchangeable metal but appreciably more reducible and much more residual Pb and Zn (Kersten and Forstner, 1995). This may not be too surprising as exchangeable metals are released once freshwater mixes with salt water and redistribution in the marine environment results in some precipitated phases (carbonates, Fe/Mn oxyhydroxides) and the relative increase in the lithogenic fraction. In future, the solid-phase identification techniques should be used to classify the sediments that are to be subjected to "selective" extraction techniques for the purpose of understanding the heavy metal phase associations.

15.3 ATMOSPHERIC EMISSIONS OF METALS AND GEOCHEMICAL CYCLES

Both natural and perturbed geochemical cycles include several subcycle elements, not the least of which is the emission of metals into the atmosphere. Atmospheric metals deposited on the land and ocean surface are a part of the runoff from land into the ocean and become incorporated in marine sediments. Thus, the two major pathways whereby heavy metals are injected into the natural geochemical cycles are atmospheric and fluvial. Considering the land surface, atmospheric emissions from stationary and mobile facilities and aqueous emissions from manufacturing and sewage disposal facilities are the primary sources of heavy metal contamination. As for the ocean, atmospheric deposition and continental runoff are the primary inputs. Duce *et al.* (1991) summarized the global inputs of metals to the ocean for the 1980s and these data are presented in Table 6. Riverine inputs are substantially greater than the atmospheric inputs, especially particulate riverine inputs that account for 95% of the total (Chester, 2000). For Pb and Zn, riverine inputs are 20 and 30 times greater than the corresponding atmospheric inputs. For Cd the factor is only 5, while

Table 6 Global deposition of metals to the ocean for the 1980s.

Pb	Zn	Cd	Cu	Ni
Atmospheric				
90	137	3.1	34	25
Riverine				
1,602	3,906	15.3	1,510	1,411
World atmosphere				
342	177	8.9	63	86

Source: Duce *et al.* (1991).
All deposition values are thousand metric tons per year.

Table 7 Heavy metal deposition fluxes at Summit, Central Greenland.

Age	Pb	Zn	Cd	Cu
BP 7760	1.3	53	0.6	3.9
1773	18	37	0.6	5.0
1850	35	70	0.6	5.3
1960s–1970s	250	200	4.1	22
1992	39	120	1.8	17

Source: Candelone *et al.* (1995).
All values are in picograms per cm^2 per year.

for Cu and Ni the factors are 45 and 30, respectively. Global atmospheric inputs to land and ocean for the same time period are substantially greater (2–3 times, Table 6) than atmospheric deposition to the ocean. This is due to the presence of major pollution sources (mining, smelting, fossil-fuel combustion, waste incineration, manufacturing facilities) on the land masses. In fact, for Pb during the 1970s and 1980s, the use of leaded gasoline in vehicles resulted in the emission of four times the metal to the land surface compared to the ocean (Table 6).

From the above data it is obvious that atmospheric emissions on land are a major source of heavy-metal contamination to our natural environment. In the following sections the focus will be on these emissions due to the fact that there are numerous data available to construct emission estimates (Nriagu and Pacyna, 1988) and that historical atmospheric emissions have been archived in continental ice accumulations (Greenland and Antarctica). The metal emission estimates of Nriagu and Pacyna (1988) are the most complete, and recent data are available for worldwide metal emissions.

15.3.1 Historical Heavy Metal Fluxes to the Atmosphere

Claire Patterson and his co-workers have pioneered the study of natural earth materials to uncover the "secrets of the ages". As early as 1963, Tatsumoto and Patterson (1963) related the high concentrations of Pb in surface seawater off Southern California to automotive aerosol fallout. In the United States it was found that Pb in gasoline was the largest single source of air pollution. Aerosols account for about one-third of the industrial Pb added to the oceans (Patterson *et al.*, 1976). Murozumi *et al.* (1969) provided a very convincing argument that airborne Pb particulates can be transported over vast distances in their classic study of the Greenland ice sheet. Their data indicated that before 1750 the concentration of Pb in the atmosphere began to increase above "natural" levels and that this was mainly due to the lead smelters, and that the sharp increase in the

atmospheric Pb occurred around 1950 due to the burning of Pb alkyls in gasoline after 1940.

More recently, Claude Boutron and his co-workers in France have published high-quality data for Pb–Zn–Cd–Cu in Greenland snows (Boutron *et al.*, 1991; Hong *et al.*, 1994; Candelone *et al.*, 1995). Table 7 gives heavy metal deposition fluxes for the Summit Central Greenland Icesheet sampling locality. Lead increased dramatically between BP 7760 and AD 1773 (Industrial Revolution), and subsequently through 1850–1960 (Pb alkyl additives to automobile gasoline) (Nriagu, 1990a). Candelone *et al.* (1995) have successfully extended the uncontaminated metal record in ice from Central Greenland. Besides the above Pb record, for Zn, Cd, and Cu there is a clear increasing trend from 1773 to the 1970s (Table 7). However, between BP 7760 and AD 1773, there is essentially no change in metal flux. In fact, Zn decreased slightly; there is no change in Cd; and Cu increased slightly (Table 7). Over the past 200 years, Zn fluxes started to increase but it was not until 1900s that the increase became more rapid. On the other hand, for Cd and Cu, it was not until after the 1850s that their atmospheric concentrations and fluxes increased substantially (Candelone *et al.*, 1995). The maximum remote atmospheric concentrations of Zn occurred around 1960 while those for Cd and Cu occurred around 1970 (Candelone *et al.*, 1995). Finally, 1992 icesheet data indicate that the remote atmospheric Pb fluxes (Table 7) decreased by 6.5-fold in response to the banning of leaded gasoline throughout most of the world. Zinc and Cu decreased only 1.5 times while Cd decreased about 2.5 times (Table 7). These large increases in historical metal fluxes to the remote atmosphere are undoubtedly related to the major changes in the large scale anthropogenic emissions to the atmosphere in the northern hemisphere.

There is a wealth of data available on the world production of heavy metals during the past century or so (Nriagu, 1990b). Candelone *et al.* (1995) present historical Zn–Cd–Cu concentrations in snow/ice deposited at Summit, Central Greenland from 1773 to 1992. If one assumes that the 1773 concentrations are the result of natural atmospheric emissions, then the ratio of 1980s concentrations to 1773 concentrations are a measure of anthropogenic contamination of the remote atmosphere to that date. These ratios are 4, 7, and 3 for Zn, Cd, and Cu, respectively (figure 3 in Candelone *et al.*, 1995). Compare this to the 1983 total emissions divided by the natural emissions (Nriagu, 1990b). These values are 3.9, 6.4, and 2.3 for Zn, Cd, and Cu, respectively. It appears that historical changes in Zn–Cd–Cu deposition in the Greenland icesheet are consistent with the estimates of metal emissions to the global atmosphere (Candelone *et al.*, 1995). These emissions are primarily a result of smelting/ refining, manufacturing processes, fossil-fuel combustion, and waste incineration (Nriagu, 1990b).

A similar analysis for Pb yields the following results: icesheet concentration ratio is about 15 and atmospheric emission ratio is about 20. While this is not too bad a comparison, it is not as good as for the other three metals (Zn–Cd–Cu). It is clear that most of the Pb increase in snow/ice samples from Greenland is due to the use of leaded gasoline after the 1950s (Murozumi *et al.*, 1969).

Going back to the Holocene era (BP 7760 years) where dated ice cores give metal concentrations that reflect a time when man's impact on the global environment was minimal, Candelone *et al.* (1995) measured Zn–Cd–Cu concentrations that were comparable to values for ice dated at AD 1773. Even by AD 1900 the concentrations of Zn and Cu were only 1.3 and 1.5 times those recorded for the AD 1773 date (Candelone *et al.*, 1995). Cadmium concentrations had increased more than four times during this period and Pb concentrations had increased nearly 10 times. In fact for Pb, the concentrations recorded in the icesheet have increased at least 30 times between BP 7760 and AD 1900. (Candelone *et al.*, 1995). It is obvious that much Pb was emitted to the atmosphere long before the Industrial Revolution and that some Cd was emitted during the early stages of the Industrial Revolution. It is possible that Cd was a by-product of the Pb mining and smelting during the Greco-Roman civilization (2500–BP 1700 years).

15.3.2   Perturbed Heavy Metal Cycles

In this discussion of heavy metals, geochemical cycles are treated in a simple manner; emissions from land and oceans to the global atmosphere and subsequent deposition on the land and ocean surface, and runoff from the land to the ocean and eventual deposition in marine sediments. Only two components of this simple cycle will be discussed due to the availability of relatively accurate and complete data; deposition of metals from the atmosphere to the land and ocean surface, and continental runoff to the ocean.

Figure 3 presents these data for two simple scenarios: minimal human disturbances and maximum human disturbances. For the minimal human disturbances senario, it was assumed that deposition from the atmosphere was due to natural sources (Nriagu, 1990b) and that there was minimal anthropogenic impact on the Earth's surface. The continental runoff (riverine) data was taken from Bertine and Goldberg (1971) who calculated the amounts of metals entering into the world's oceans as a result of the weathering cycle. They accounted for both the dissolved and particulate phases by using the marine rates of sedimentation. It can be seen from Figure 3 that for Pb–Zn–Cr–Cu–Ni, continental runoff was 5–10 times greater than the natural atmospheric inputs. For Cd, the atmospheric fluxes are greater than the continental runoff suggesting that continental rocks are depleted in Cd or that there are poor quality Cd data for these two sources. The latter explanation seems to be the most likely.

For the maximum human disturbances scenario, riverine inputs were calculated with the data of Martin and Whitfield (1981). As can be seen from Table 2 in this chapter, Pb–Zn–Cd–Cu–Ni are strongly enriched (by man's activities) when compared to the average Earth's upper crust and soils and the Cr enrichment is found to be only somewhat enriched (Table 2). Atmospheric input data was computed as the average of global emissions data for the 1970s and 1980s (Garrels *et al.*, 1973; Lantzy and Mackenzie, 1979; Nriagu and Pacyna, 1988; and Duce *et al.*, 1991), and was assumed to be the time of maximum anthropogenic emissions to the atmosphere. For the maximum human disturbances scenario, riverine inputs are still larger than the atmospheric inputs except that Pb–Cr are only three times greater, Zn is six times greater, Cu is 10 times greater, Ni is 13 times greater, and Cd is about equal. These differences are undoubtedly due to the magnitude of different source functions.

A calculation of maximum/minimum ratio from the atmospheric input data in Figure 3 yields the following results: Pb = 33, Zn = 9, Cd = 17, Cr = 1.5, Cu = 5, Ni = 4. We know that the burning of leaded gasoline is responsible for the large increase of Pb. Enormous metal production of Zn and Cd ores as well as refuse incineration are responsible for the increases of these metals. In addition, marine aerosols are an important source of Cd (Li, 1981). Obviously, Cu–Ni production from ores increased during this period but not nearly as much as for Zn–Cd. Also, combustion of fossil fuels contributed somewhat to the increase of Cu and Ni. The main source of Cr is steel and iron manufacturing which appears to not be as important an impact on the atmospheric environment as sources for the other metals. The pollution sources of Cr are minimal as reflected in the balance between riverine input and marine sediment output (Li, 1981).

A similar calculation for the riverine inputs (Figure 3) yields the following results: Pb = 14, Zn = 5, Cd = 40, Cr = 7.5, Cu = 6, Ni = 8. With

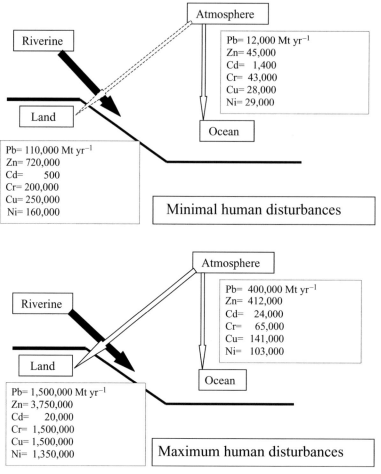

Figure 3 Schematic diagrams of perturbed heavy metal cycles representing prehistoric times of minimal human disturbances and modern times of maximum human disturbances. Data sources for minimum human disturbances: Nriagu (1990b), Bertine and Goldberg (1971). Data sources for maximum human disturbances: Martin and Whitfield (1981), Garrels *et al.* (1973), Lantzy and Mackenzie (1979), Nriagu and Pacyna (1988), Duce *et al.* (1991).

the exception of Pb and Cd, the increases for Zn–Cr–Cu–Ni are similar. Smelting wastes and coal fly ash releases are the common sources of these four metals. Gasoline residues are an obvious source of the Pb increases and urban refuse incineration is a major source of the Cd increase (Nriagu and Pacyna, 1988).

15.3.3 Global Emissions of Heavy Metals

Table 8 presents the data on the global emissions of heavy metals to the atmospheric and terrestrial environments for the 1970s and 1980s. The atmospheric and riverine input (weathering mobilization) data are the same as that used for the minimum and maximum human disturbances to the geochemical cycling of the heavy metals presented in the previous section. Total industrial discharges of heavy metals are the calculated discharges into soils and water minus the emissions to the atmosphere (Nriagu, 1990b). Only a fraction of the heavy-metal production from mines is

released into the atmosphere in the same year (Nriagu, 1990b). For instance for Pb, in the year 1983, about 30% of the metal produced from mining is used for metal production, other sources, and is wasted as industrial discharges (Table 8): Zn 27%, Cd 190%, Cr 16%, Cu 14%, and Ni 57%. It is not surprising that the price of base metals fluctuate so widely in that there appears to be a substantial excess of supply over demand (Table 8). This is not the case for Cd; the data presented in Table 8 suggest that there may be a deficit in the supply of Cd. It appears unlikely but it may be that there is a sufficient demand for Cd that can just about balance the mine production. Another explanation is that the estimate of Cd from industrial discharges might be in error. Other discrepancies in Cd estimates have also been noted and it is reasonable to think that since the concentrations of Cd are so low in natural earth materials, the analytical data may not be good.

In order to assess the internal consistency of the emissions, as shown in Table 8, a calculation was

Table 8 Global emissions of heavy metals to the atmosphere and terrestrial environment during 1970s and 1980s.

Element	Atmospheric input[a]	Weathering mobilization[b]	Total industrial discharges[c]	Production from mines[d]	World Metal production (Atmos.)[e]	Other sources (Atmos.)[f]	Emissions H_2O, soil (Atmos.)[g]	Global natural emissions[h]
Pb	400,000	295,000	565,000	3,077,000	83,800	292,000	875,000	12,000
Zn	412,000	1,390,000	1,427,000	6,040,000	125,800	67,000	2,083,000	45,000
Cd	24,000	15,000	24,000	19,000	8,500	3,500	43,000	1,400
Cr	65,000	1,180,000	1,010,000	6,800,000	28,500	25,000	1,397,000	43,000
Cu	141,000	635,000	1,048,000	8,114,000	35,400	15,500	1,428,000	28,000
Ni	103,000	540,000	356,000	778,000	15,900	71,000	614,000	29,000

Units are metric tons per year. Atmospheric Input (a) = World Metal Production (e) + Other Sources to the Atmosphere (f) + Natural Emissions to the Atmosphere (h). Pb: 400,000 = 388,000; Zn: 412,000? 238,000; Cd: 24,000 ≅ 13,400; Cr: 65,000 ≅ 96,500; Cu: 141,000 ≅ 79,000; Ni: 103,000 ≅ 116,000.
[a] Source: Lantzy and Mackenzie (1979), Garrels *et al.* (1973), Nriagu and Pacyna (1988), Duce *et al.* (1991).
[b] Bertine and Goldberg (1971).
[c,d,h] Nriagu (1990b).
[e–g] Nriagu and Pacyna (1988).

made whereby the mean atmospheric input was equated to the world metal production emitted to the atmosphere plus natural emissions and other sources to the atmosphere. With the exceptions of Cu and Zn, the quantities of emissions balance rather well. There is no obvious reason why Cu is out of balance by nearly a factor of 2 (atmospheric input>sources). For Zn, with an imbalance of 1.7 for atmospheric input>sources, there is an obvious problem with other sources in that the impact of rubber tire wear. This source term will be addressed in the next section. However, even with this term, the right side of the equation would increase to a maximum emissions figure of $300,000\,t\,yr^{-1}$ (Table 8). It is possible that maximum Cu and Zn emissions to the atmosphere have been overestimated but there is no way to check this with the available data.

15.3.4 US Emissions of Heavy Metals

While there is a reasonable amount of data pertaining to global emissions of heavy metals during the last half of the twentieth century, there is a wealth of data available for emissions of heavy metals to the US atmosphere. Most of this has been calculated from USEPA and US Bureau of Mines materials production data combined with emission factors for a variety of source functions (Pacyna, 1986). In this section data plots will be presented to show the calculated emissions of several heavy metals to the US atmosphere over a decade of time. Some of the data, such as that for Pb, are from the published literature. On the other hand, much of the data for Zn has been calculated by the author and his colleagues and is presented for the first time.

15.3.4.1 Lead

With the scientific realization that Pb had contaminated the global atmosphere (Murozumi *et al.*, 1969), scientists set out to identify the major sources of this contamination. The late Claire Patterson, formerly of the California Institute of Technology, was the leader in this field. In an earlier paper concerning Pb contamination and its effect on human beings, Patterson (1965) wrote "the industrial use of lead is so massive today that the amount of lead mined and introduced into our relatively small urban environments each year is more than 100 times greater than the amount of natural lead leached each year from soils by streams and added to the oceans over the entire earth". This conclusion was reached by Chow and Patterson (1962) in their landmark study of Pb isotopes in pelagic sediments. This information, coupled with the well-known health impacts of Pb (USEPA, 2000a,b), arose the interest of toxicologists worldwide and prompted detailed studies of the cycling of this element in the environment. Ingested Pb (food, water, soil, and dust) damages organs, affects the brain and nerves, the heart and blood, and particularly affects young children and adults (USEPA, 2000a,b). With the use of leaded gasoline that began in the 1930s (Nriagu, 1990a), the public outcry about the outbreak of severe lead poisoning, and the drastic increase in the US in automobile miles traveled, the US Congress passed an amendment to the Clean Air Act (Callender and Van Metre, 1997) banning the use of leaded gasoline. The USEPA (2000a,b), in their most recent air pollutant emission trends report, showed that since 1973 the quantity of Pb emitted to the environment (Table 9) has decreased drastically from about 200,000 t to about 500 t in 1998. As a comparison, European Pb emissions for 1979/1980 were released at a rate of 80,800 t yr^{-1} (Pacyna and Lindgren, 1997) while those for the US were 66,600 tons per year (USEPA, 2000a,b).

Figure 4 presents the important EPA emissions data on a five-year time scale from 1970 to 1995. In the 1970s and 1980s, it is clear that leaded gasoline consumption was the overwhelming

emitter of Pb to the environment with metal processing a far second. Presently, the total amount of Pb emitted to the environment is a paltry 2,500 t (USEPA, 2000a,b), with metal processing and waste disposal being the main emitters. While the US consumption of leaded gasoline has all but stopped, it is not the case for the rest of the world. As of 1993 when leaded gasoline consumption in North America (mostly Mexico) emitted 1,400 t of Pb to the atmosphere, the rest of the world emitted 69,000 tons of Pb to the atmosphere (Thomas, 1995).

15.3.4.2 Zinc

The US atmospheric emissions data for Zn are somewhat sparse. Nriagu (1979) published data on the worldwide anthropogenic emission of Zn to the atmosphere during 1975 (Table 10). The author has taken this report as a model for the type of Zn emissions that appear to be important and has added several categories such as cement and fertilizer production and automobile rubber tire wear.

The reason why the emission data for Zn are sparse is that until recently it was thought that Zn

Table 9 Total US emissions of lead (Pb) to the atmosphere.

Source category	1970	1975	1980	1985	1990	1995
Waste incineration	1,995	1,447	1,097	790	729	552
Fossil fuel combustion	9,269	9,385	3,899	469	454	446
Metals processing	21,971	9,000	2,745	1,902	1,968	1,864
Gasoline consumption	164,800	123,657	58,688	17,208	1,086	475

Source: USEPA (1998, 2000a,b).
Units are metric tons.

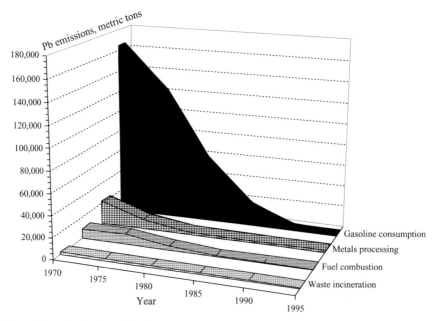

Figure 4 Three-dimensional plot of lead emissions to the US atmosphere for the period 1970–1995. Data from USEPA (2000a).

was not harmful to the environment and that health risks were minimal compared to other heavy metals. Zinc is an essential micronutrient and plays a role in DNA polymerization (Sunda, 1991) and nervous system functions (Yasui *et al.*, 1996). Zinc is generally less toxic than other heavy metals (Nriagu, 1980a); however, it is known to cause a variety of acute and toxic effects in aquatic biota. Several studies have established links between human activities and environmental Zn enrichment (Pacyna, 1996).

Figure 5 is a plot of second tier Zn emissions to the atmosphere for the period 1960–1995 in five-year time intervals. The only important Zn emission category not included in Figure 5 is Zn mining. This is and has been the largest Zn emission category with 102,000 t in 1960 to 112,000 in 1970, declining to 48,000 in 1980, and stabilizing at about 32,000 t in the 1990s (Nriagu and Pacyna, 1988; http://www.minerals.usgs.gov). Obviously, emissions from Zn mining and smelting are the overwhelming sources. Total US Zn emissions for

Table 10 Total US Emissions of zinc (Zn) to the atmosphere.

Source category	1960	1965	1970	1975	1980	1985	1990	1995
Cement production[a]	617	716	729	667	735	754	752	846
Fertilizer production[b]	1,000	1,000	1,054	1,329	1,632	1,525	1,390	1,365
Copper mining[c]	1,035	1,163	1,200	983	915	795	1,185	1,448
Iron and steel[d]	1,628	2,160	2,236	1,952	1,682	1,223	1,342	1,374
Fossil Fuel combustion[e]	1,532	1,719	2,141	1,878	1,916	1,984	1,658	1,298
Rubber tire wear[f]	3,747	4,901	5,503	5,044	5,983	7,258	8,329	8,847
Waste incineration[g]	6,367	7,280	5,920	5,006	3,232	5,659	7,298	7,941
Zinc mining[h]	101,500	126,280	111,440	55,580	47,600	36,540	36,820	32,480

Units are metric tons.
[a] Source: Nriagu and Pacyna (1988), http://minerals.usgs.gov/minerals/pubs/commodity/cement/stat/tbl1.txt.
[b] Source: Nriagu and Pacyna (1988), http://minerals.usgs.gov/minerals/pubs/commodity/phosphate_rock/stat/tbl1.txt.
[c] Source: Nriagu and Pacyna (1988), http://minerals.usgs.gov/minerals/pubs/commodity/of01-006/copper.
[d] Source: Nriagu and Pacyna (1988), http://minerals.usgs.gov/minerals/pubs/commodity/of01-006/ironandsteel
[e] Source: Pacyna (1986), Statistical Abstracts of the United States (1998) (Coal and Oil production data).
[f] Source: Councell *et al.* (2003).
[g] Source: Pacyna (1986), USEPA (1998).
[h] Source: Nriagu and Pacyna (1988)http://minerals.usgs.gov/minerals/pubs/commodity/of01-006/zinc.

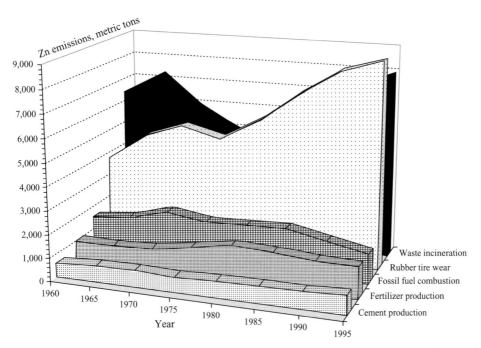

Figure 5 Three-dimensional plot of second tier zinc emissions to the US atmosphere for the period 1960–1995. Data from Councell *et al.* (2003), Nriagu and Pacyna (1988), http://minerals.usgs.gov/minerals/pubs/commodity/cement/stat/tbl1.txt, http://minerals.usgs.gov/minerals/pubs/commodity/phosphate_rock/stat/tbl1.txt, Pacyna (1986), USEPA (1998), Statistical Abstracts of the United States (1998), author's calculations.

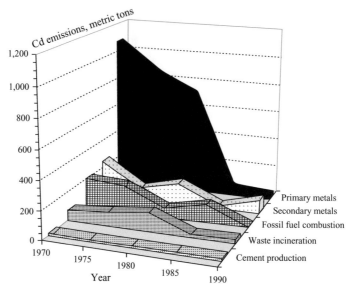

Figure 6   Three-dimensional plot of cadmium emissions to the US atmosphere for the period 1970–1990. (sources Davis and Associates, 1970; USEPA, 1975a,b, 1976, 1978; Wilber *et al.*, 1992).

the 1980s amount to approximately 60,000 t yr^{-1} while European Zn emissions total 43,000 t yr^{-1} (Pacyna and Lindgren, 1997). Mining–smelting emissions overwhelm others that are important but it is difficult to plot these clearly on Figure 5 if Zn mining is also included.

Of the five Zn emission categories plotted in Figure 5, waste incineration and rubber tire wear are the most important. Note that in general these emissions have increased during the last 40 years such that the second-tier emissions total approximately one-half of the Zn mining–smelting emissions.

15.3.4.3   *Cadmium*

Cadmium has received a wide variety of uses in American industries with the largest being electroplating and battery manufacture. Its emission from natural sources (erosion and volcanic activity) are negligible. The dominant sources of Cd emissions to the atmosphere are primary metals smelting (Cu and Pb), secondary metals production, fossil-fuel combustion, waste incineration, iron and steel production, and rubber tire wear. Figure 6 is a plot of Cd emissions to the US atmosphere for five-year time periods from 1970 to 1990. Between 1970 and 1980, primary metals smelting was the primary source of Cd emissions to the atmosphere. Then fossil-fuel combustion became the primary emitter (60%) with Cu–Pb smelting accounting for much of the remainder (30%) (Wilber *et al.*, 1992). These emissions were concentrated in the central part of the US (Wilber *et al.*, 1992). By 1990, fossil fuel emissions decreased significantly, a fact that is probably related to the increased efficiency of stack emission controls; and secondary metal production became

Table 11   Total U.S. emissions of cadmium (Cd) to the atmosphere.

Source category	1970[a]	1975[b]	1980[c]	1985[d]	1990[e]
Rubber tire wear	6	6	5	5	5
Cement production	14	13	14	14	15
Waste incineration	72	97	131	21	28
Fossil fuel combustion	200	186	60	121	7
Secondary metals	245	75	143	15	96
Primary metals	1,075	860	728	54	32

Units are metric tons.
[a] Source: Davis and Associates (1970).
[b] Source: USEPA (1975a,b, 1976).
[c] Source: USEPA (1978).
[d] Source: Wilber *et al.* (1992).
[e] Source: USEPA (2000b).

the major source of Cd to the US atmosphere (Figure 6). In the 1980s, Cd emissions to the US atmosphere amounted to 650 t yr^{-1} (Table 11). European emissions were nearly double this amount, i.e., 1,150 t yr^{-1} (Pacyna and Lindgren, 1997).

15.4   HISTORICAL METAL TRENDS RECONSTRUCTED FROM SEDIMENT CORES

15.4.1   Paleolimnological Approach

Many governmental agencies collect the data routinely for the assessment of the quality of rivers, streams, lakes, and coastal oceans. Water-quality monitoring involves temporal sampling of water resources that are affected by natural random events, seasonal phenomena, and anthropogenic forces. Testing water quality, monitoring data at regular intervals has become a common feature and an important exercise for water managers who are interested in checking whether the investment of large sums of money has improved the water quality of various water resources during the past 30 years. In addition, trends available by such monitoring can provide a warning of degradation of water quality.

Historical water quality databases suffer from many limitations such as lack of sufficient data, changing sampling and analytical methods, changing detection limits, missing values and values below the analytical reporting level. With regard to trace metal data, the problem is even more difficult in that many metals occur at environmental concentrations so low (parts per billion or less) that current (up until the early 1990s) routine analytical methodology was unable to detect ambient concentrations with adequate sensitivity and precision. Even the best statistical techniques, when applied to questionable data, can produce misleading results. For example, Pb in the Trinity River, south of Dallas, TX, USA. Abundant dissolved Pb data for the period 1977–1992 (Van Metre and Callender, 1996) indicate that there were no trends in Pb; in fact, the concentrations were scattered from 0 to 5 ppb. On the other hand, Pb in sediment cores from Lake Livingston, downstream from the Trinity River sampling station, showed a decline in Pb concentration from 1970 to 1993 (Van Metre and Callender, 1996). Thus, from the core data, one can conclude that there is a declining trend in Pb in the Trinity River.

For these very reasons, the US Congress supported the US Geological Survey's National Water Quality Assessment Program with its goals to describe the status and trends in water quality of our Nation's surface and groundwater, and to provide an understanding of the natural and human factors that affect the observed conditions and trends. The US public is eager to know whether the water quality of US rivers and lakes has benefited from the expenditure of billions of dollars since the passage of the Clean Air and Clean Water Acts in the 1970s.

An alternate approach to statistical analyses of historical water-quality data is to use metal distributions in dated sediment cores to assess the past trends in anthropogenic hydrophobic constituents that impact watersheds (paleolimnological approach). It is well known that marine and lacustrine sediments often record natural and anthropogenic events that occur in drainage basins, local and regional air masses, or are forced upon the aquatic system (Valette-Silver, 1993). A good example of the former is the increase in erosional inputs to lakes in response to anthropogenic

activities in the drainage basin (Brush, 1984). Atmospheric pollution resulting from the cultural and industrial activities (Chow *et al.*, 1973) is a compelling example of the latter. Thus, aquatic sediments are archives of natural and anthropogenic change. This is especially true for hydrophobic constituents such as heavy metals.

Because of their large adsorption capacity, fine-grained sediments are a major repository for the contaminants and a record of the temporal changes in contamination. Thus, sediments can be used for historical reconstruction. To guarantee a reliable age dating, and, therefore, to be useful in the historical reconstruction, the core sediment must be undisturbed, fine-grained, and collected in an area with a relatively fast sedimentation rate. These conditions are often found in lakes where studies in the 1970s by Kemp *et al.* (1974) and Forstner (1976) used lake sediments to understand the pollution history of several Laurentian Great Lakes and some European lakes. However, there is a serious limitation in using sediment cores from many natural lakes in that the sedimentation rate is generally too slow in providing the proper time resolution to discern modern pollution trends. Lacustrine sediments usually accumulate at rates less than $1 cm\,yr^{-1}$ (Krishnaswami and Lal, 1978) and often at rates less than $0.3\,cm\,yr^{-1}$ (Johnson, 1984). Thus, there may be sufficient time for early diagenesis, such as microbiologically-mediated reactions, to occur. On the other hand, in lacustrine environments where sediments accumulate at rates exceeding 1 and may exceed $5–10\,cm\,yr^{-1}$ (Ritchie *et al.*, 1973), such as in surface-water reservoirs, rapid sedimentation exerts a pronounced influence on sedimentary diagenesis.

A brief discussion of sedimentary diagenesis is warranted as post-depositional chemical, and physical stability is probably the most important factor in preserving heavy-metal signatures that may be recorded in aquatic sediments. For sediments to provide a historical record of pollution, the pollutant must have an affinity for the sedimentary particles. It is well known that most of the metals, and certainly the heavy metals discussed in this chapter, are hydrophobic in nature and allow partition to the solid phase. Once deposited in the sediment, the pollutants should not undergo chemical mobilization within the sediment column nor should the sediment column be disturbed by physical and biological processes.

Natural lacustrine and estuarine sediments whose accumulation rates are low, generally below $0.25\,cm\,yr^{-1}$, often do not satisfy the above requirements. The biophysical term bioturbation refers to surficial sediments mixed by the actions of deposit feeders, irrigation tube dwellers, and head-down feeders (Boudreau, 1999). In general, these bioturbation processes do not occur in reservoirs where sediment accumulation rates exceed 1 and often 5 cm yr^{-1} (Callender, 2000). At these rates, the sediment influx at the water–sediment interface is too great for benthic organisms to establish themselves.

On the other hand, geochemical mobility affects every sedimentary environment; varying in degree, from slowly accumulating natural lacustrine and estuarine sediments to rapidly accumulating reservoir sediments. The major authigenic solid substrates for adsorption and co-precipitation of heavy metals in aquatic sediments are the hydrous oxides of iron (Fe) and manganese (Mn) (Santschi et al., 1990). These primary metal oxides sorb/co-precipitate Pb–Cr–Cu (Fe oxyhydroxides) and Zn–Cd–Pb (Mn oxyhydroxides) (Santschi et al., 1990). Manganese oxides begin to dissolve in mildly oxidizing sediments while Fe oxides are reduced in anoxic sediments (Salomons and Forstner, 1984). In the mildly oxidizing zone, Mn^{2+} diffuses upward and precipitates as Mn oxide in the stronger oxidizing part of the sediment column. At greater sediment depths, Fe oxide reduction to Fe^{2+} begins and ferrous iron diffuses upward and precipitates as Fe oxide in the mildly oxidizing part of the sediment sequence (Salomons and Forstner, 1984).

An example from a slowly-accumulating (0.01–0.1 cm yr^{-1}) sediment profile in a freshwater lake in Scotland (Williams, 1992) should suffice to illustrate the formation of diagenetic metal profiles. Early diagenetic processes, such as those described before, have promoted extensive metal enrichment immediately beneath the water–sediment interface. The oxic conditions, near the water–sediment interface, that promote metal precipitation and enrichment (Mn, Fe, Pb, Zn, Cu, Ni) are entirely confined to strata of post-industrial age (Williams, 1992).

Callender (2000) extensively studied the geochemical effects of rapid sedimentation in aquatic systems and postulated that rapid sedimentation exerts a pronounced influence on early sedimentary diagenesis. The following are two case studies that illustrate this point. The Cheyenne River Embayment of Lake Oahe, one of the several impoundments on the upper Missouri River, accumulates sediment at an average rate of 9 cm yr^{-1} (Callender and Robbins, 1993). Three interstitial-water Fe profiles from the same site taken over a three-year period (August 1985, August 1986, June 1987), when superimposed on the same depth axis, show the effects of inter-annual variations in sediment inputs such that in 1986 a rapid input of oxidized material suppressed the dissolved Fe concentration to less than 0.1 mg L^{-1} to a depth of 8 cm. In 1985 when there was a drought and sediment inputs were reduced substantially, near-surface sediment became nearly anoxic and the interstitial Fe concentration rose to a very high 26 mg L^{-1} (Callender, 2000).

In Pueblo Reservoir on the upper Arkansas River in central Colorado, cores of bottom sediments showed distinct reddish-brown layers that indicate rapid transport and sedimentation of Fe-rich colloids formed by the discharge of acid-mine waters from abandoned mines upstream (Callender, 2000). The amorphous sedimentary Fe profile from a sediment core near the river mouth shows two peaks at depths that correspond to the dates of heavy metal releases from the mines. Although the amorphous Fe oxyhydroxide concentrations are only 10% of the total Fe concentrations, they are adequate to adsorb Pb (Fergusson, 1990) and produce the anthropogenic Pb concentrations found in the core (Callender, 2000). Copper and Zn show similar distributions in this core whose sedimentation rate is 5 cm yr^{-1}.

In these examples as well as for most aquatic sediments, the principal diagenetic reactions that occur in these sediments are aerobic respiration and the reduction of Mn and Fe oxides. Under the slower sedimentation conditions in natural lakes and estuaries, there is sufficient time (years) for particulate organic matter to decompose and create a diagenetic environment where metal oxides may not be stable. When faster sedimentation prevails, such as in reservoirs, there is less time (months) for bacteria to perform their metabolic functions due to the fact that the organisms do not occupy a sediment layer for any length of time before a new sediment is added (Callender, 2000). Also, sedimentary organic matter in reservoir sediments is considerably more recalcitrant than that in natural lacustrine and estuarine sediments as reservoirs receive more terrestrial organic matter (Callender, 2000).

The author hopes that this discussion of sedimentary diagenesis, as it applies to heavy-metal signatures in natural lacustrine and reservoir sediments, will help the reader interpret the results presented in the following sections on reconstructed metal trends from age-dated reservoir sediment cores.

The approach that Callender and Van Metre (1997) have taken is to select primarily reservoir lakes that integrate a generally sizeable drainage basin that is impacted by a unique landuse such as agriculture, mining, stack emissions, suburban "sprawl", or urban development with some commercial and light industrial activity. Sediment cores are taken to sample the post-impoundment section as much as possible and to penetrate the pre-impoundment material. Core sampling is accomplished with a variety of coring tools (box cores, push cores, piston cores) in order to recover a relatively undisturbed sediment section. The recovered sediment is sampled on approximately an annual sediment thickness and samples are preserved (chilled, then frozen) for future analytical determinations. In the laboratory, sediment samples are weighed, frozen, freeze-dried, weighed again, and ground to a fine powder. Elemental concentrations are determined on concentrated acid digests (nitric and hydrofluoric in microwave pressure vessels) by inductively coupled plasma-atomic emission spectrometry (ICP/AES) or by graphite furnace atomic adsorption spectrometry (GF/AAS).

For reservoirs to be a good medium for detecting the trends in heavy metals, several conditions need to be satisfied. First, the site sampled should be continuously depositional over the life of the reservoir. This condition is most easily satisfied by sampling in the deeper, lacustrine region of the reservoir where sedimentation is slower but more uniform and the sediments predominantly consist of silty clay material. The second condition is that the sediments sampled should not be subject to significant physical and chemical diagenesis; that is, mobilization of chemical constituents after deposition. Callender (2000) has written an extensive paper indicating that rapid sedimentation promotes minimal diagenesis and preserves historical metal signatures. The third condition is that the chemical quality of reservoir bottom sediments should be related to the water quality of the influent river and that the influent water quality be representative of the drainage basin.

15.4.2 Age Dating

In general, reservoir sediments can be dated by several techniques. In one technique, the sediment surface is dated by the time of coring while in the other the date is derived from a visual inspection of the cored sediment column which often penetrates the pre-impoundment surface. The primary age dating tool for reservoir sediments is by counting the radioactive isotope ^{137}Cs which has a half-life period of 30 years (Robbins and Edgington, 1975; McCall *et al.*, 1984). The ^{137}Cs activity of freeze-dried sediment samples is measured by counting the gamma activity in fixed geometry with a high-resolution, intrinsic germanium detector gamma-spectrometer (Callender and Robbins, 1993). Depending on the penetration depth of the core and the age of the reservoir, ^{137}Cs can provide one or two date markers and can be used to evaluate the relative amount of postdepositional mixing or sediment disturbance (Van Metre *et al.*, 1997). The peak ^{137}Cs activity in the sediment core is assigned a date of 1964, consistent with the peak in atmospheric fallout levels of ^{137}Cs for 1963–1964. In reservoirs constructed prior to or around 1950, the first occurrence of ^{137}Cs, if it did not appear to have been effected by postdepositional sediment mixing, was assigned a date of 1953 which is consistent with the generally accepted date of 1952 for the first large-scale

atmospheric testing of nuclear weapons by the US in Nevada (Beck *et al.*, 1990). This is also the date of the first globally-detectable levels of ^{137}Cs in the atmosphere. In some cases dates for samples between the known date-depth markers were assigned using constant mass accumulation rates (MARs), and in other cases the MARs were varied.

In natural lacustrine and slowly-accumulating reservoir sediments, core dating with the isotope ^{210}Pb has been used extensively (Schell and Barner, 1986). Appleby and Oldfield (1983) found that the constant rate of ^{210}Pb supply model (CRS) provides a reasonably accurate sedimentation chronology. The basic assumption of the CRS model is that the rate of supply of excess ^{210}Pb to the lake is constant. This model, thus, assumes that the erosive processes in the catchment are steady and give rise to a constant rate of sediment accumulation (MAR) (Appleby and Oldfield, 1983). In practice, for reservoirs, this assumption is rarely met because, for example, an increase in the MAR caused by land disturbances, such as those associated with the urban development, transports additional surficial soils and sediments to the lake. This additional erosion increases the MAR and also increases the rate of supply of ^{210}Pb to the lake. In general, because excess ^{210}Pb is an atmospheric fallout radionuclide, the model works better in low sedimentation rate, atmospherically dominated lakes with undisturbed watersheds, than in high sedimentation rate, fluvially dominated urban lakes and reservoirs.

Another problem with age dating of reservoir sediment is the concept of sediment focusing. This concept was developed to correct for postdepositional resuspension and redistribution of sediment in parts of the lake (Hermanson, 1991). A common focus correction factor is derived from the inventory of ^{137}Cs in the sediment column compared to the estimated total ^{137}Cs fallout at the sampling site (Hermanson, 1991). The same concept was found to not work well for lakes and reservoirs where the catchment area far exceeds the lake area. Such is the case for most reservoirs (Van Metre *et al.*, 2000). In these cases where the catchment area is 10–100 times the lake area, sediment focusing in the lake basin is overwhelmed by the concentration effect of atmospheric fallout over the catchment area being funneled into the lake or reservoir. The catchment area focus corrections are calculated the same way as lake basin focus corrections except that there may be some variation in the ^{137}Cs flux to large catchment areas and that there will almost always be a correction factor greater than 1. These focus corrections must be calculated in order to compare the contaminant fluxes between the sites within a lake basin and between lake basins.

15.4.3 Selected Reconstructed Metal Trends

15.4.3.1 Lead and leaded gasoline: consequence of the clean air act

Of the six heavy metals discussed in this chapter, Pb has been studied extensively with respect to the environmental effects. Clair Patterson, the father of environmental Pb studies, in one of his many major publications concerning the global Pb cycle (Patterson and Settle, 1987), noted that during pre-industrial times Pb in the troposphere originated from soil dusts and volcanic gases. In modern times (1950–1980) the proportion of natural Pb in the atmosphere is overwhelmed by the industrial sources of smelter emissions and automobile exhausts. Lead air pollution levels measured near our Nation's roadways decreased 97% between 1976 and 1995 due to the consequence of the Clean Air Act that eliminated leaded gasoline which interfered with the performance of catalytic converters.

For remote locations on a more global scale, Boyle *et al.* (1994) showed that the stable Pb concentration in North Atlantic waters decreased at least three-fold from 1979 to 1988. Wu and Boyle (1997) confirmed and extended this time series to 1996 whereby the concentration of stable Pb apparently stabilized at $50 \, \text{pmol} \, \text{kg}^{-1}$ in surface waters near Bermuda. Shen and Boyle (1987) presented a 100-year record of Pb concentration in corals from Bermuda and the Florida Straits showing that Pb peaked in the 1970s and declined thereafter. Veron *et al.* (1987) found high Pb concentrations in northeast Atlantic surficial sediments and noted that the quantity of Pb stored in these sediments is of the same order of magnitude as the amount of pollutant Pb present in the water column.

On a more local level, man's activities in the urban/suburban environment have produced a strong imprint of Pb on the land surface. In the US, automobile and truck travel are the primary means of moving people and goods around the continent. With the introduction of leaded gasoline in the 1950s, the mean annual atmospheric concentration of Pb nearly tripled in value, especially near population centers (Eisenreich *et al.*, 1986). A substantial proportion of these atmospheric emissions of Pb have been deposited relatively close to the source. Figures 7(a)–(c) presents the age-dated sedimentary Pb profiles for reservoirs and lakes from urban–suburban–rural localities. One can see that the peak concentrations decrease from 700 to 300 to $100 \, \mu\text{g} \, \text{g}^{-1}$ as the distance from urban centers increase. All but two of the Pb peak concentrations date between 1970 and 1980, and are consistent with the decline in US atmospheric Pb concentrations following the ban of unleaded

gasoline in 1972 (Callender and Van Metre, 1997). Sedimentary Pb data from many urban centers around the US (Boston, New York, New Jersey, Atlanta, Orlando, Minneapolis, Dallas, Austin, San Antonio, Denver, Salt Lake City, Las Vegas, Los Angeles, Seattle, and Anchorage) have been subjected to statistical trend analysis (B.J. Mahler, personal communication, 2002) and the results plotted in Figure 8. It is obvious that essentially all urban reservoirs and lake records show a very significant decline in Pb since 1975 and that this trend is most probably a result of the ban on leaded gasoline that was instituted in 1972.

15.4.3.2 Zinc from rubber tire wear

Contrary to the distribution of Pb in sediment cores whereby peak concentrations occurred during the 1970s, the concentration of sedimentary Zn often increases to the 1990s. It was observed in the atmospheric emissions of metals that waste incineration was one of the major contributors to the second tier of Zn emissions to the US atmosphere. Figure 9(a) presents age-dated sedimentary Zn data from a spectrum of urban/suburban sites around the US It is obvious that the general trend of sedimentary Zn is one of increasing concentrations from 1950s to 1990s. However, the general increasing trend for Zn is not as prevalent as that for Pb and at a few of the urban/suburban/reference sites noted for the Pb trend map there is no significant trend in sedimentary Zn concentration (B.J. Mahler, personal communication, 2002). Figure 5 shows that rubber tire wear is the most important and increasing contributor to the second tier of Zn emissions to the US atmosphere. Tire tread material has a Zn content of about 1% by weight. A significant quantity of tread material is lost to road surfaces by abrasion prior to tire replacement on a vehicle. In Figure 9(b) the anthropogenic Zn data for urban/suburban core sites is regressed against the mass sedimentation rate (MSR) for each core site. When MSR-normalized anthropogenic Zn is plotted against average annual daily traffic (AADT) data for the various metropolitan areas shown in Figure 9(a), a significant regression results (Figure 9(c)) suggesting that there is a causal relationship between anthropogenic Zn and vehicle traffic. Councell *et al.* (2003) produced data that estimates the magnitude of the Zn releases to the environment from rubber tire abrasion. Two approaches, wear rate ($g \, \text{km}^{-1}$) and tread geometry (abrasion to wear bars), were used to assess the magnitude of this nonpoint source of Zn in the US for the period 1936–1999. For 1999, the quantity of Zn released by tire wear in the US is estimated to be between 10,000 and 11,000 t.

Two specific case studies focused on the impact of vehicle tire wear to the Zn budget of

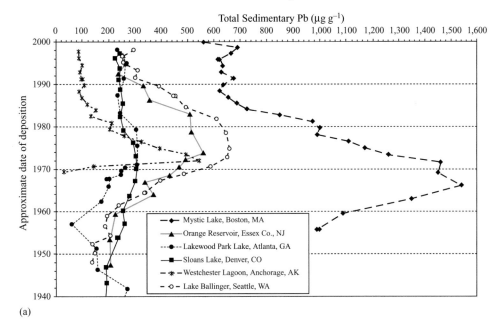

(a)

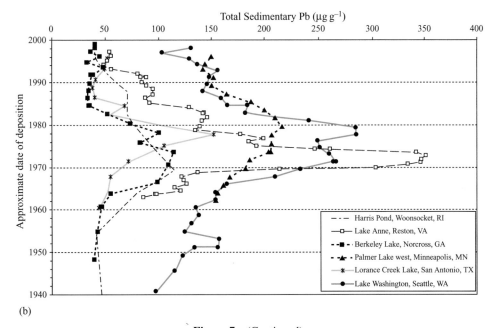

(b)

Figure 7 (Continued).

watersheds in the Washington, DC metropolitan area. For Lake Anne, a suburban watershed located 40 km southwest of Washington, DC, the wet deposition atmospheric flux of Zn was $8 \, \mu g \, cm^{-2} \, yr^{-1}$ (Davis and Galloway, 1981) and the flux of Zn estimated from tire wear was $31 \, \mu g \, cm^{-2} \, yr^{-1}$ (Landa *et al.* 2002). The measured accumulation rate of Zn in age-dated sediment cores from Lake Anne is $41 \, \mu g \, cm^{-2} \, yr^{-1}$ (Landa *et al*, 2002) suggesting that tire-wear Zn inputs to suburban watersheds can be significantly greater than atmospheric inputs. In a rural/atmospheric reference site watershed, located ~90 km northwest of Washington, DC, the atmospheric Zn flux is $12 \, \mu g \, cm^{-2} \, yr^{-1}$ (Davis and Galloway, 1981) and that from tire wear is only $1 \, \mu g \, cm^{-2} \, yr^{-1}$ (Landa *et al.*, 2002). There are only dirt roads leading to cabins in this protected watershed and it is obvious that vehicle tire wear is only a minor component of the Zn flux in this remote watershed. One conclusion drawn from these case studies and the substantial set of age-dated sediment core Zn profiles is that those watersheds that are impacted by vehicular traffic receive significant amounts of Zn via tire abrasion and that this Zn-enriched particulate matter is fluvially transported to lakes and reservoirs.

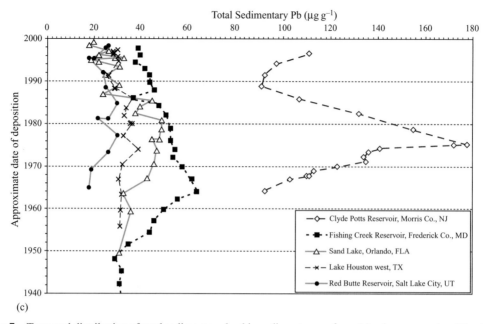

Figure 7   Temporal distribution of total sedimentary lead in sediment cores from (a) urban reservoirs, (b) suburban reservoirs and lakes, and (c) atmospheric reference site reservoirs.

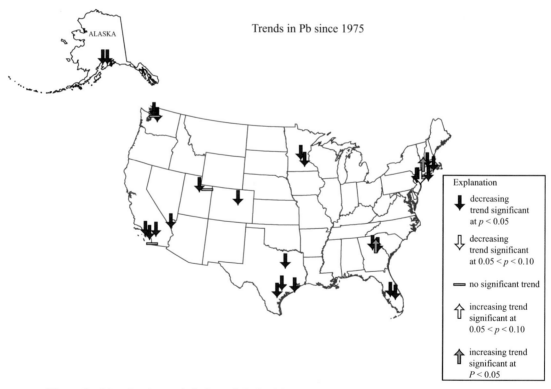

Figure 8   Map showing statistical trends in lead from sediment cores located throughout the US.

15.4.3.3   *Metal processing and metal trends in sediment cores*

While the relationship between Pb and Zn distributions in reservoir and lake sediment cores and environmental forcing functions (leaded gasoline use and vehicular traffic) are clear, the same is not true in the case of other metals such as Cd, Cr, Cu, and Ni. These metals do have one common, major source: nonferrous and ferrous metal production. Approximately 70% of the Cd and Cr anthropogenic emissions, 50% of the Cu, and 21% of the Ni anthropogenic emissions come from

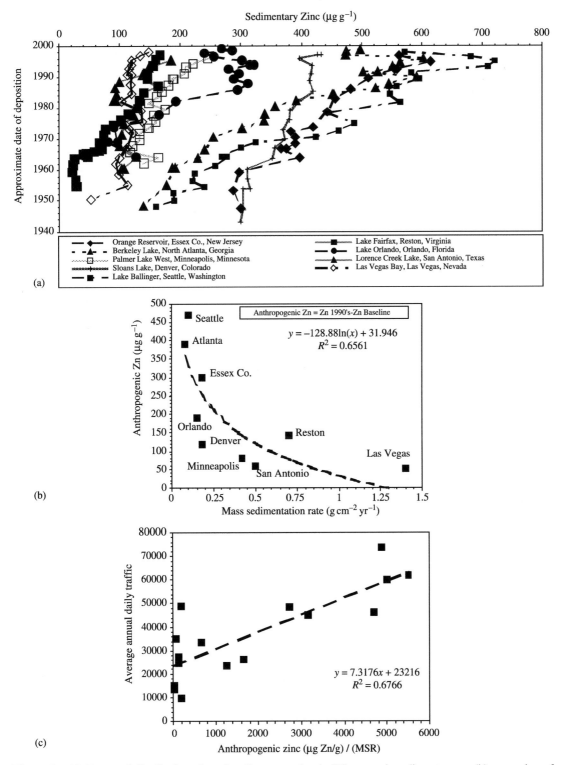

Figure 9 (a) Temporal distribution of total sedimentary zinc in US reservoir sediment cores, (b) regression of anthropogenic zinc versus MSR for US reservoir sediment cores, (c) MSR-normalized anthropogenic zinc versus average annual daily traffic for urban and suburban watersheds throughout the US

mining–smelting–metal processing. In an attempt to interpret metal trend maps for these four metals, historical US metal mining and production statistics have been shown in Figure 10. Copper is the

only metal whose production increased during the past 25 years. Starting in 1985, the primary production of Cu has increased from about one million metric tons to a maximum of about two

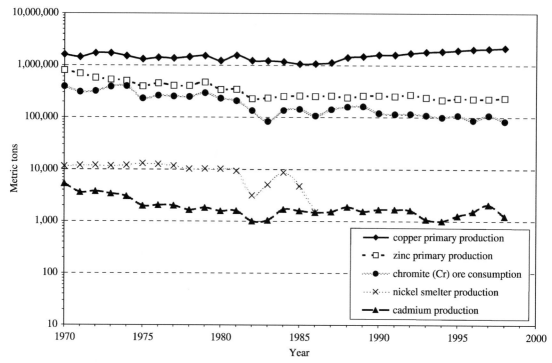

Figure 10 Historical production of nonferrous metals in the US Data from US Geological Survey, Mineral Resources Program.

million metric tons in 1998 (Figure 10). Arizona has the largest Cu mining production followed by Utah, New Mexico, and Montana. Thus, all the major point sources of Cu exist in the western states. Figure 11 shows the statistical trends (since 1975) for Cu in sediment cores from 30 reservoir and lake sites around the US (B.J. Mahler, personal communication, 2002). Increasing or decreasing trends in the upper part of sediment cores were tested statistically for eight trace metals. Significant trends in sediments deposited since 1975 were tested using a Spearman's rank correlation (Helsel and Hirsch, 1992). Trends were determined to be significantly increasing or decreasing based on a p value of less than 0.05 (B.J. Mahler, personal communication, 2002). The Cu trend indicators in Figure 11 show that there are increasing trends significant at $p < 0.05$ at the core sites in Washington–California–Nevada and that there are no significant trends in Cu at core sites in the Midwest and Southwest. Such a pattern suggests that an increase in Cu mining in the Rocky Mountain States may cause airborne emissions that impact the western US but that these emissions are not transported east to the mid-continent. Along the Atlantic coast, from Georgia to Massachusetts, there are some sites that show an increasing trend in sedimentary Cu. It should be noted that many of the east coast sites are small ponds used for water supply and recreation and that some have been treated with $CuSO_4$ to control the algae. In addition, many of these sites are located in the east coast

urban/suburban corridor and obviously receive a multitude of anthropogenic contaminants.

Limited space does not allow for the presentation of all metal trend maps such as those presented for Pb and Cu. Of the three remaining metals (Cd, Cr, Ni), the trends in Ni since 1975 is representative of the other metals as well. For the western half of the US, there are nine sites where there is a significant decreasing trend in Ni (Figure 12). This corresponds to the decrease in Ni smelter production for the 1970s and 1980s (Figure 10). There are a few increasing Ni trends along the east coast of the US, a pattern that may reflect the location of many nickel consumption facilities in Pennsylvania, West Virginia, and New Jersey. The trends in Cr since 1975 are very similar to those for Ni; for the western half of the US there are eight sites where there is a significant decreasing trend. Along the east coast of the US there is only one site in the Boston area that shows an increasing trend. Thus, the overall decreasing Cr trend for the US reflects the four-fold decrease in chromite (Cr ore) consumption by metallurgical and chemical firms in the US since 1970 (Figure 10). The trends in Cd since 1975 are not as strongly skewed toward decreasing trends as those for Ni and Cr. This is probably due to the fact that much of the Cd in the US is recovered by the processing of Zn ore and as one can see from Figure 10, Zn production has leveled out in the 1980s and 1990s. In fact, the preponderance of coring sites in the US (20 out of 30) show no significant trend in Cd.

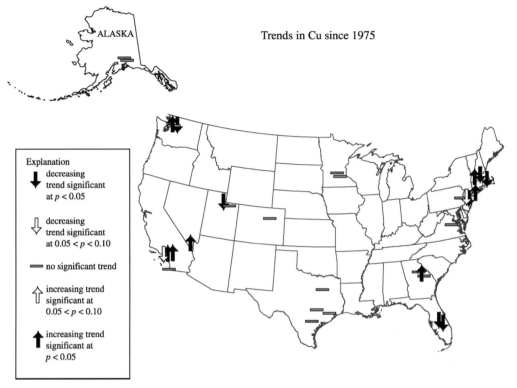

Figure 11   Map showing statistical trends in copper from sediment cores located throughout the US.

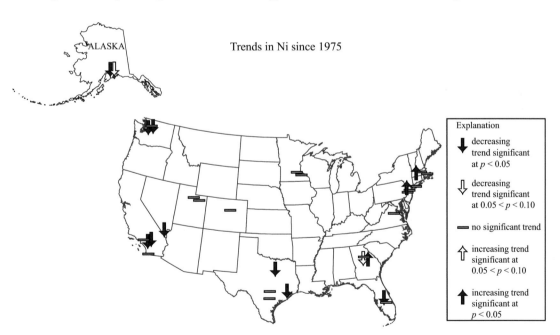

Figure 12   Map showing the statistical trends in nickel from sediment cores located throughout the US.

15.4.3.4   *Reduction in power plant emissions of heavy metals: clean air act amendments and the use of low sulfur coal*

Only recently have electric utility power plant emissions been included on the US Environmental Protection Agency's (USEPA) Toxic Release Inventory which reported that electric utilities ranked highest for industrial toxic air emissions in 1998. These emissions were likely to be an important component of toxic air releases in the past, particularly prior to the passage of the Clean Air Act of 1970.

A sediment core from one reservoir (Mile Tree Run Reservoir) located in southwestern West

Virginia was selected for the purpose of identifying particulate signatures of coal-fired power plant emissions from the nearby Ohio River Valley region. Arsenic (As) and some Pb and Zn releases to the environment may reflect coal-related geochemical processes. Stream sediments from the Appalachian Basin Region are particularly enriched in As compared to the sediments outside the Basin. While some As enrichment may come from the weathering of As-rich coal, this area of West Virginia is underlain primarily by sandstone which is low in trace-element composition. Thus, the elevated As contents are not likely to have an origin in the regional country rock but might have originated from numerous large coal-fired power plants situated along the Ohio River.

Figure 13(a) shows the temporal distribution of Ti-normalized As, Pb, and Zn in the sediment core from Mile Tree Run Reservoir. These Ti-normalized metal peaks date back to 1987, 1966, and 1946, respectively (Figure 13(a)). Figure 13(b) shows the temporal distribution of Ti-normalized sulfur (S), isothermal remnant magnetization (IRM) (a magnetite proxy), and Fe in the same core. Peaks in these constituents correspond to the aforementioned dates. These dates match the maximum values in combined coal production for the states of West Virginia, Pennsylvania, and Ohio (M.B. Goldhaber, personal communication, 2002). The temporal profile for the magnetic property IRM that is indicative of the mineral magnetite is shown in Figure 13(b) (M.B. Goldhaber, personal communication, 2002).

Figures 14 (a) and (b) are scatter plots between Zn and coal production, and Zn and IRM (magnetite). Note the excellent correlations suggesting that Zn relates to atmospheric input (fly ash) from power plants. A glance at Table 2 shows that Zn in fly ash is substantially enriched compared to average soils. One can also see this relationship for Mile Tree Run watershed soils in Figure 13(a). Arsenic also has a very strong positive correlation with IRM in the Mile Tree Run Reservoir core. This element is known to be strongly associated with magnetite, an important fly ash mineral that is formed in the high temperature combustion of coal (Locke and Bertine, 1986). Such a geochemical association is not surprising in that it is thought that magnetite is formed from the pyrite in coal that is subjected to high temperature combustion. It is also well known that As is a minor element associated with pyrite.

The simultaneous decline after 1985 of the correlated Fe–As–Pb–Zn and magnetite peaks (Figures 13(a) and (b) suggests that the amount of power plant particulate emissions decreased since the 1977 amendment of the Clean Air Act that mandated reduced amounts of sulfur in the feed coal (Hower *et al.*, 1999). It appears that this action resulted in lower metal and magnetite

quantities in the fly ash combusted residue. Despite this decrease, soil samples from the Mile Tree Run Reservoir watershed are strongly enriched in As, Pb, and Zn when compared to average soils (Table 2) and local bedrock (Callender *et al.*, 2001), indicating a regional power plant emissions impact on the geochemical landscape.

15.4.3.5 *European lacustrine records of heavy metal pollution*

Much of the recent literature pertaining to European studies of heavy metal pollution using sediment cores to track time trends focused on Pb. Petit *et al.* (1984) used the stable isotope geochemistry to identify Pb pollution sources and to evaluate the relative importance of anthropogenic sources to total Pb fluxes in a semi-rural region of western Europe. Thomas *et al.* (1984) showed that metal enrichment factors for Pb–Zn–Cd were significantly above unity, indicating a diffuse contribution through atmospheric transport from industrialized areas. A comparison with atmospheric fluxes showed good agreement for diffuse atmospheric supply of Pb, Zn, and Cd in the lake sediments (Thomas *et al.*, 1984).

Much work has been done on Lake Constance situated on the borders of Germany, Austria, and Switzerland. The lake has undergone extensive cultural eutrophication due to the surrounding population, industrial activity, and the input of River Rhine. Muller *et al.* (1977) found that sedimentary Pb and Zn concentrations peaked around 1965. They also noted that there was a strong positive correlation between heavy-metal content and PAHs in the sediment core. Muller *et al.* (1977) suggested that coal burning was the source of this relationship. In a more recent study of Lake Constance sediments, Wessels *et al.* (1995) also noted high concentrations of Pb and Zn that began around 1960. However, they postulated that the origin of the Pb increase was emissions by regional industry and the origin of the Zn increase was a combination of urban runoff and coal burning.

Two groups, one in Switzerland and the other in Sweden, have used age-dated sediment cores from peat bogs and natural lake sediments to record the history of atmospheric Pb pollution dating back to several thousand years.

In Switzerland, Shotyk and co-workers (Shotyk *et al.*, 1998) rebuilt the history of atmospheric Pb deposition over the last 12,000 years. They cored ombrotrophic peat bogs that are hydrologically isolated from the influence of local groundwaters and surfacewaters, and receive their inorganic solids exclusively by atmospheric deposition. Whereas slowly-accumulating natural lake sediments appear to be affected by chemical diagenesis, studies have shown that peat bogs provide a reliable record of changes in atmospheric metal deposition

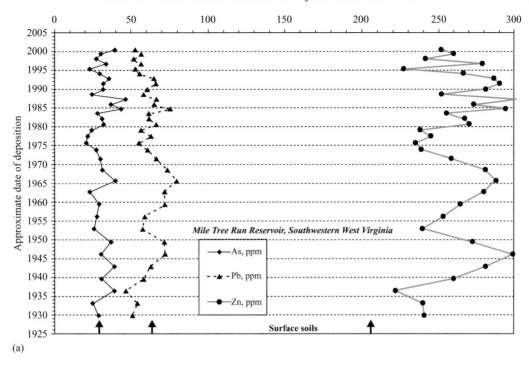

(a)

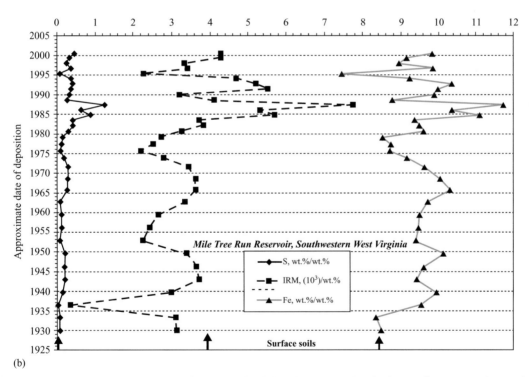

(b)

Figure 13 (a) Temporal distribution of Ti-normalized As–Pb–Zn (ppm/wt.%) in a sediment core from Mile Tree Run Reservoir, Southwestern West Virginia, USA; (b) Temporal distribution of Ti-normalized S–IRM–Fe (wt.%/wt.%) in the Mile Tree Run Reservoir sediment core.

(Roos-Barraclough and Shotyk, 2003). Their radiocarbon dated core profiles of stable Pb and $^{206}Pb/^{207}Pb$ isotopic ratios indicate that enhanced fluxes of Pb were caused by climatic changes between 10,500 and 8,250 years before present

(BP). Soil erosion, caused by forest clearing and agricultural tillage, increased Pb deposition subsequent to this time. Beginning 3,000 yr BP, Pb pollution from mining and smelting was recorded by a significant increase in normalized

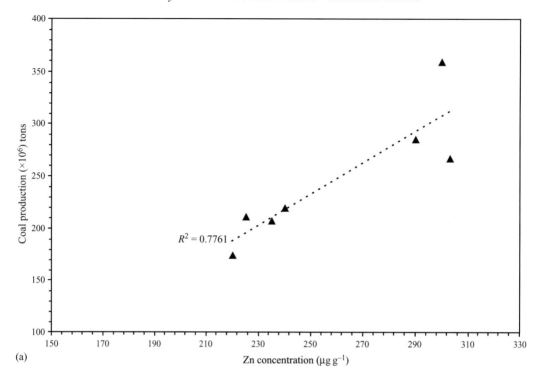

(a)

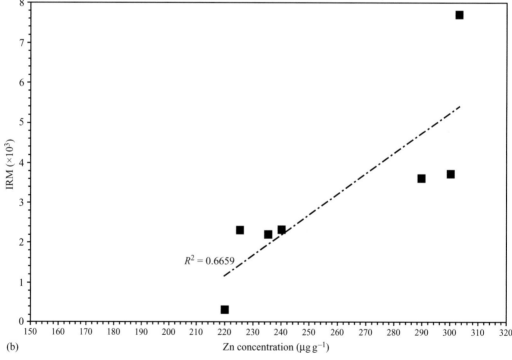

(b)

Figure 14 (a) Scatter plot of Appalachian Basin temporal coal production versus temporal concentration of Zn in a Mile Tree Run Reservoir sediment core, (b) Scatter plot of the temporal concentration of IRM (magnetite) versus the temporal concentration of Zn in the Mile Tree Run Reservoir sediment core.

Pb concentrations (Pb/Sc) and a decreasing $^{206}Pb/^{207}Pb$ ratio. Around BP 2100, Roman Pb mining became the most important source of atmospheric Pb pollution; and in AD 1830 the effects of the Industrial Revolution were recorded by a very large peak in Pb enrichment. The $^{206}Pb/^{207}Pb$ ratio declined significantly around AD 1940 indicating the use of leaded gasoline which was subsequently discontinued in AD 1979 (Shotyk *et al.*, 1998). In an earlier paper, Shotyk *et al.* (1996) noted that there were significant enrichments in As and Sb as well as for Pb dating back to Roman times (BP 2100). These enrichments in As and Sb were thought to be related to Pb mining and smelting.

In Sweden, Brannvall and his colleagues published several papers culminating in a summary paper (Brannvall *et al.*, 2001) describing four thousand years of atmospheric Pb pollution in northern Europe. They cored 31 lakes throughout Sweden; some were age-dated with radiocarbon while others had varved sediments. Their stable Pb and ^{206}Pb/^{207}Pb isotope data indicate that the first influx of noncatchment atmospheric Pb occurred between 3,500 and 3,000 years ago. The large world production of Pb ($80,000\,t\,yr^{-1}$) during Greek and Roman times 2,000 years ago caused widespread atmospheric Pb pollution. There was a decline in the atmospheric Pb flux between AD 400 and 900. Brannvall *et al.* (2001) note that the Medieval period, rather than the Industrial Revolution, was the real beginning of the contemporary Pb pollution era. This era extended from AD 1000 to 1800. Lead peaked in the mid-twentieth century in Sweden (1950s–1970s) due to the use of leaded gasoline and fossil-fuel combustion. The recent decline in atmospheric Pb deposition since 1980 is very steep and significant. Johansson (1989) analyzed the heavy-metal content of some 54 lakes in central and northern Sweden, and noted that the Pb content of surface sediment was 50 times greater than the background concentration and that this Pb enrichment decreased substantially from south to north. Ek *et al.* (2001) studied the environmental effects of one thousand years of Cu production in Central Sweden. Metal analyses of the lake sediments showed that Cu pollution was restricted to a smaller area near the emission sources and that Pb, Zn, and Cd pollution was widespread. Sedimentary metal enrichments began about AD 1000 and peaked in the seventeenth century when Central Sweden produced two-thirds of the world's Cu supply.

Sedimentary lacustrine records of heavy metal pollution in Lough Neagh (northern Ireland) and Lake Windermere (England) suggest that there have been two periods of metal disturbance since the AD 1600. Both lakes are situated within rural catchments. In Lough Neagh (Rippey *et al.*, 1982), a change in the catchment erosion regime during the seventeenth century produced an increase in the sedimentary Cd–Cu–Pb concentrations. This change in erosion was a result of widespread and comprehensive forest clearance. A second and larger change occurred about AD 1880 when the concentrations of Cr, Cu, Zn, and Pb increased toward the sediment surface. Sediments from Lake Windermere also show a pronounced increase in Cu–Pb–Zn concentrations within the upper part of the sediment column (Hamilton-Taylor, 1979). Since the catchments are not proximal to the local anthropogenic sources such as mining, smelting, or wastewater inputs, it is possible that a substantial part of these metal enrichments are due to a more regional or even global atmospheric input that was generated by many anthropogenic processes related to the Industrial Revolution.

REFERENCES

Adriano D. C. (1986) *Trace Elements in the Terrestrial Environment*. Springer, New York.

Alexander C. R., Smith R. G., Calder F. D., Schropp S. J., and Windom H. L. (1993) The historical record of metal enrichment in two Florida estuaries. *Estuaries* **16**, 627–637.

Allan R. J. (1975) Natural versus unnatural heavy metal concentrations in lake sediments in Canada. *Proc. Int. Conf. Heavy Metals in the Environ.* **2**, 785–808.

Alloway B. J. (1995) *Heavy Metals in Soils* 2nd edn. Blackie Academic and Professional, London, UK.

Anikiyev V. V., Perepelitsa S. A., and Shumilin Ye N. (1993) Effects of man-made and natural sources on the heavy-metal patterns in bottom sediments in the Gulf of Peter the Great, Sea of Japan. *Geokhimiya* **9**, 1328–1340.

Appleby P. G. and Oldfield F. (1983) The assessment of ^{210}Pb data from sites with varying sediment accumulation rates. *Hydrobiologia* **103**, 29–35.

Aston S. R., Thornton I., Webb J. S., Purves J. B., and Milford B. L. (1974) Stream sediment composition: an aid to water quality assessment. *Water Air Soil Pollut.* **3**, 321–325.

Beck H. L., Helfer I. K., Bouville A., and Deicer M. (1990) Estimates of fallout in the continental US from Nevada weapons testing based on gummed-film monitoring data. *Health Phys.* **59**, 565–576.

Belzile N., Lecomte P., and Tessier A. (1989) Testing readsorption of trace elements during partial chemical extractions of bottom sediments. *Environ. Sci. Technol.* **23**, 1015–1020.

Bertine K. K. and Goldberg E. D. (1971) Fossil fuel combustion and the major sedimentary cycle. *Science* **173**, 233–235.

Borovec Z., Tolar V., and Mraz L. (1993) Distribution of some metals in sediments of the central part of the Labe (Elbe) River, Czech Republic. *Ambio* **22**, 200–205.

Boudreau B. P. (1999) Metals and models: diagenetic modeling in freshwater lacustrine sediments. *J. Paleolimnol.* **22**, 227–251.

Boutron C. F., Gorlach U., Candelone J.-P., Bolshov M. A., and Deimas R. J. (1991) Decrease in anthropogenic lead, cadmium and zinc in Greenland snows since the late 1960s. *Nature* **353**, 153–156.

Boyle E. A. (1979) Copper in natural waters. In *Copper in the Environment, Part I: Ecological Cycling* (ed. J. O. Nriagu). Wiley, New York, pp. 77–88.

Boyle E. A., Sherrell R. M., and Bacon M. P. (1994) Lead variability in the western North Atlantic Ocean and central Greenland ice: implications for the search for decadal trends in anthropogenic emissions. *Geochim. Cosmochim. Acta* **58**, 3227–3238.

Brannvall M. L., Bindler R., Emteryd O., and Renberg I. (2001) Four thousand years of atmospheric lead pollution in northern Europe: a summary from Swedish lake sediments. *J. Paleolimnol.* **25**, 421–435.

Brush G. S. (1984) Patterns of recent accumulation in Chesapeake Bay (Virginia-Maryland, USA) tributaries. *Chem. Geol.* **44**, 227–242.

Callender E. (2000) Geochemical effects of rapid sedimentation in aquatic systems: minimal diagenesis and the preservation of historical metal signatures. *J. Paleolimnol.* **23**, 243–260.

Callender E., Goldhaber M. B., Reynolds R. L., and Grosz A. (2001) Geochemical signatures of power plant emissions as revealed in sediment cores from West Virginia reservoirs. *Ann. Meeting Geol. Soc. Am.* (Abstracts).

Callender E. and Robbins J. A. (1993) Transport and accumulation of radionuclides and stable elements in a Missouri River reservoir. *Water Resour. Res.* **29**, 1787–1804.

Callender E. and Van Metre P. C. (1997) Reservoir sediment cores show US lead declines. *Environ. Sci. Technol.* **31**, 424A–428A.

Candelone J.-P., Hong S., Pellone C., and Boutron C. F. (1995) Post-industrial revolution changes in large-scale atmospheric pollution of the northern hemisphere by heavy metals as documented in central Greenland snow and ice. *J. Geophys. Res.* **100**, 16,605–16,616.

Chao T. T. (1984) Use of partial dissolution techniques in geochemical exploration. *J. Geochem. Explor.* **20**, 101–135.

Chester R. (2000) *Marine Geochemistry,* 2nd edn. Oxford, Malden.

Chiffoleau J.-F., Cossa D., Auger D., and Truquet I. (1994) Trace metal distribution, partition and fluxes in the Seine estuary (France) in low discharge regime. *Mar. Chem.* **47**, 145–158.

Chow T. and Patterson C. C. (1962) The occurrence and significance of lead isotopes in pelagic sediments. *Geochim. Cosmochim. Acta* **26**, 263–308.

Chow T. J., Bruland K. W., Bertine K. K., Soutar A., Koide M., and Goldberg E. D. (1973) Lead pollution: records in southern California coastal sediments. *Science* **181**, 551–552.

Coakley J. P. and Poulton D. J. (1993) Source-related classification of St. Lawrence Estuary sediments based on spatial distribution of adsorbed contaminants. *Estuaries* **16**, 873–886.

Conko K. M. and Callender E. (1999) Extraction of anthropogenic trace metals from sediments of two US urban reservoirs. *Abstracts, 1999 Fall Meeting Am. Geophys. Union.*

Councell T. B., Duckenfield K. U., Landa E. R., and Callender E. (2003) Tire-wear particles as a source of zinc to the environment. *Environ. Sci. Technol.* (submitted).

Davis W. E. and Associates (1970) *National Inventory of Sources and Emissions of Cadmium.* Nat'l Tech. Information Service Rept. PB 192250, Springfield, VA.

Davis A. O. and Galloway J. N. (1981) Atmospheric lead and zinc deposition in lakes in the eastern United States. In *Atmospheric Pollutants in Natural Waters* (ed. S. J. Eisenreich). Ann Arbor Science, Ann Arbor, pp. 401–408.

Davis J. A., James R. O., and Leckie J. O. (1978) Surface ionization and complexation at the oxide/water interface: 1. Computation of electrical double layer properties in simple electrolytes. *J. Colloid Interface Sci.* **63**, 480.

Dolske D. A. and Sievering H. (1979) Trace Element loading of southern Lake Michigan by dry deposition of atmospheric aerosol. *Water Air Soil Pollut.* **12**, 485–502.

Dominik J., Mangini A., and Prosi F. (1984) Sedimentation rate variations and anthropogenic metal fluxes into Lake Constance sediments. *Environ. Geol.* **5**, 151–157.

Drever J. I. (1988) *The Geochemistry of Natural Waters,* 2nd edn. Prentice-Hall, New York.

Duce R. A., Liss P. S., Merrill J. T., Atlas E. L., Buat-Menard P., Hicks B. B., Miller J. M., Prospero J. M., Arimoto R., Church T. M., Ellis W., Galloway J. N., Hansen L., Jickells T. D., Knap A. H., Reinhardt K. H., Schneider B., Soudine A., Tokos J. J., Tsunogai S., Wollast R., and Zhou M. (1991) The atmospheric input of trace species to the world ocean. *Global Biogeochem. Cycles* **5**, 193–259.

Dunnette D. A. (1992) Assessing global river quality, overview and data collection. In *The Science of Global Change: the Impact of Human Activities on the Environment,* Am. Chem. Soc. Symposium 483 (eds. D. A. Dunnette and R. J. O'Brien). ACS, pp. 240–286.

Eisenreich S. J. (1980) Atmospheric input of trace elements to Lake Michigan. *Water Air Soil Pollut.* **13**, 287–301.

Eisenreich S. J., Metzer N. A., Urban N. R., and Robbins J. A. (1986) Response of atmospheric lead to decreased use of lead in gasoline. *Environ. Sci. Technol.* **20**, 171–174.

Ek A. S., Lofgren S., Bergholm J., and Qvarfort U. (2001) Environmental effects of one thousand years of copper production at Falun Central Sweden. *Ambio* **30**, 96–103.

Farrah H. and Pickering W. F. (1977) Influence of clay-solute interactions an aqueous heavy metal ion levels. *Water Air Soil Pollut.* **8**, 189–197.

Faust S. D. and Aly O. M. (1981) *Chemistry of Natural Waters.* Ann Arbor Science, Ann Arbor.

Fergusson J. E. (1990) *The Heavy Elements: Chemistry, Environmental Impact, and Health Effects.* Pergamon, Oxford, 614pp.

Forstner U. (1976) Lake sediments as indicators of heavy-metal pollution. *Naturwissenschaften* **63**, 465–470.

Forstner U. (1981) Recent heavy metal accumulation in limnic sediments. In *Handbook of Strata-Bound and Stratiform Ore Deposits* (ed. K. H. Wolf). Elsevier, pp. 179–269.

Forstner U. (1990) Inorganic sediment chemistry and elemental speciation. In *Sediments: Chemistry and Toxicity on In-Place Pollutants* (eds. R. Baudo, J. P. Giesy, and H. Mantau). Lewis, pp. 61–105.

Gaillardet J., Viers J., and Dupre B. (2003) Trace Elements in River Waters. In *Treatise on Geochemistry.* Elsevier, Amsterdam, vol. 5, Chapter 6.

Gardiner J. (1974) The chemistry of cadmium in natural water: II. The adsorption of cadmium on river muds and naturally occurring solids. *Water Res.* **8**, 157–164.

Garrels R. J. and Christ C. L. (1965) *Solutions, Minerals and Equilibria.* Harper and Row, New York.

Garrels R. M., Mackenzie F. T., and Hunt C. (1973) *Chemical Cycles and the Global Environment: Assessing Human Influences.* William Kaufmann, Los Altos.

Gibbs R. J. (1973) Mechanisms of trace metal transport in rivers. *Science* **180**, 71–73.

Gocht T., Moldenhauer K.-M., and Puttmann W. (2001) Historical record of polycyclic aromatic hydrocarbons (PAH) and heavy metals in floodplain sediments from the Rhine River (Hessisches Ried, Germany). *Appl. Geochem.* **16**, 1707–1721.

Govindaraju K. (1989) 1989 compilation of working values and sample description for 272 geostandards. *Geostand. Newslett.* **13**, 1–113.

Graney J. R., Halliday A. N., Keeler G. J., Nriagu J. O., Robbins J. A., and Norton S. A. (1995) Isotopic record of lead pollution in lake sediments from the northeastern United States. *Geochim. Cosmochim. Acta* **59**, 1715–1728.

Hamilton-Taylor J. (1979) Enrichments of Zinc Lead, and Copper in recent sediments of Windermere, England. *Environ. Sci. Technol.* **13**, 693–697.

Hanson P. J. (1997) Response of hepatic trace element concentrations in fish exposed to elemental and organic contaminants. *Estuaries* **20**, 659–676.

Heit M., Klusek C., and Baron J. (1984) Evidence of deposition of anthropogenic pollutants in remote Rocky Mountain lakes. *Water Air Soil Pollut.* **22**, 403–416.

Helsel D. R. and Hirsch R. M. (1992) *Statistical Methods in Water Resources, Studies in Environmental Science.* **49**, Elsevier, Amsterdam.

Hem J. D. (1972) Chemistry and occurrence of cadmium and zinc in surface water and ground water. *Water Resour. Res.* **8**, 661–679.

Hem J. D. (1976) Geochemical controls on lead concentrations in stream water and sediments. *Geochim. Cosmochim. Acta* **40**, 599–609.

Hem J. D. (1977) Reactions of metal ions at surfaces of hydrous iron oxide. *Geochim. Cosmochim. Acta* **41**, 527–538.

Hem J. D. (1985) *Study and Interpretation of the Chemical Characteristics of Natural Water.* US Geol. Survey Water-Supply Paper 2254, US Govt. Printing Office.

Hem J. D. and Durum W. H. (1973) Solubility and occurrence of lead in surface water. *Am. Water Works Assoc. J.* **65**, 562–568.

Hermanson M. H. (1991) Chronology and sources of anthropogenic trace metals in sediments from small, shallow arctic lakes. *Environ. Sci. Technol.* **25**, 2059–2064.

Honeyman B. D. and Santschi P. H. (1988) Metals in aquatic systems. *Environ. Sci. Technol.* **22**, 862–871.

Hong S., Candelone J.-P., Patterson C. C., and Boutron C. F. (1994) Greenland ice evidence of hemispheric lead pollution two millennia ago by Greek and Roman civilizations. *Science* **265**, 1841–1843.

Hornberger M. I., Luoma S. N., Van Geen A., Fuller C. C., and Anima R. J. (1999) Historical trends of metals in the sediments of San Francisco Bay, California. *Mar. Chem.* **64**, 39–55.

Hornberger M. I., Luoma S. N., Cain D. J., Parchaso F., Brown C. L., Bouse R. M., Wellise C., and Thompson J. K. (2000) Linkage of bioaccumulation and biological effects to changes in pollutant loads in south San Francisco Bay. *Environ. Sci. Technol.* **34**, 2401–2409.

Hower J. C., Robl T. L., and Thomas G. A. (1999) Changes in the quality of coal combustion by-products produced by Kentucky power plants, 1978 to 1997: consequences of Clean Air Act directives. *Fuel* **78**, 701–712.

Hsu C. L. (1978) Heavy metal uptake by soils surrounding a fly ash pond. MS Thesis, University of Notre Dame.

Jenne E. A. (1968) Controls on Mn, Fe, Co, Ni, Cu and Zn concentrations in soils and water: the significant role of hydrous Mn and Fe oxides. In *Trace Inorganics in Water*, Am. Chem. Soc. Adv. Chem. Ser. 73, ACS, pp. 337–387.

Johansson K. (1989) Metals in sediment of lakes in Northern Sweden. *Water Air Soil Pollut.* **47**, 441–455.

Johnson T. C. (1984) Sedimentation in large lakes. *Ann. Rev. Earth Planet. Sci.* **12**, 179–204.

Kabata-Pendias A. (2000) *Trace Elements in Soils and Plants*, 3rd edn. CRC Press, Boca Raton.

Kabata-Pendias A. and Pendias H. (1992) *Trace Elements in Soils and Plants*, 2nd edn. CRC Press, Boca Raton.

Kabata-Pendias A. and Pendias H. (2001) *Trace Elements in Soils and Plants*, 3rd edn. CRC Press, Boca Raton.

Kemp A. L. W., Anderson T. W., Thomas R. L., and Mudrochova A. (1974) Sedimentation rates and recent sediment history of Lake Ontario, Erie, and Huron. *J. Sed. Petrol.* **44**, 207–218.

Kemp A. L. W. and Thomas R. L. (1976) Impact of man's activities on the chemical composition in the sediments of Lakes Ontario, Erie, and Huron. *Water Air Soil Pollut.* **5**, 469–490.

Kemp A. L. W., Williams J. D. H., Thomas R. L., and Gregory M. L. (1978) Impact of man's activities on the chemical composition of the sediments of Lakes Superior and Huron. *Water Air Soil Pollut.* **10**, 381–402.

Kersten M. and Forstner U. (1987) Effects of sample pretreatment on the reliability of solid speciation data of heavy metals-implications for the study of early diagenetic processes. *Mar. Chem.* **22**, 299–312.

Kersten M. and Forstner U. (1995) Speciation of trace metals in sediments and combustion waste. In *Chemica Speciation in the Environment* (eds. A. M. Ure and C. M. Davidson). Blackie Academic and Professional, London, pp. 234–275.

Krauskopf K. B. (1956) Factors controlling the concentration of thirteen rare metals in sea-water. *Geochim. Cosmochim. Acta* **9**, 1–32.

Krishnamurthy S. and Wilkens M. M. (1994) Environmental chemistry of chromium. *Northeastern Geol.* **16**, 14–17.

Krishnaswami S. and Lal D. (1978) Radionuclide limnochronol. In *Lakes-Chemistry, Geology, Physics* (ed. A. Lerman). Springer-Verlag, New York, pp. 153–177.

Landa E. R., Callender E., Councell T. B., and Duckenfield K. V. (2002) Where the rubber meets the soil: tire wear particles as a source of zinc to the environment. *Abstracts, Ann. Meeting Soil Sci. Soc. America.*

Lantzy R. J. and Mackenzie F. T. (1979) Atmospheric trace metals: global cycles and assessment of man's impact. *Geochim. Cosmochim. Acta* **43**, 511–525.

Leckie J. O. and Davis J. A. (1979) Aqueous environmental chemistry of copper. In *Copper in the Environment* (ed. J. O. Nriagu). Wiley, New York, pp. 90–121.

Leckie J. O. and Nelson M. B. (1975) Role of natural hetrerogeneous sulfide systems in controlling the concentration and distribution of heavy metals. Paper presented at the Second International Symposium on Environmental Biogeochemistry, Ontario, Canada.

Li Y.-H. (1981) Geochemical cycles of elements and human perturbation. *Geochim. Cosmochim. Acta* **45**, 2073–2084.

Li Y.-H. (2000) *A Compendium of Geochemistry*. Princeton University Press, Princeton.

Li Y.-H., Burkhardt L., and Teraoka H. (1984) Desorption and coagulation of trace elements during estuarine mixing. *Geochim. Cosmochim. Acta* **48**, 1879–1884.

Locke G. and Bertine K. K. (1986) Magnetite in sediments as an indicator of coal combustion. *Appl. Geochem.* **1**, 345–356.

Lum R. R. (1987) Cadmium in freshwaters: the Great Lakes and St. Lawrence River. In *Cadmium in the Aquatic Environment* (eds. J. O. Nriagu and J. B. Sprague). Wiley, New York, pp. 35–50.

Luoma S. N. (1983) Bioavailability of trace metals to aquatic organisms: a review. *Sci. Total Environ.* **28**, 1–23.

Mantei E. J. and Foster M. V. (1991) Heavy metals in stream sediments: effects of human activities. *Environ. Geo. Water Sci.* **18**, 95–104.

Martin J.-M. and Meybeck M. (1979) Elemental mass balance of material carried by major world rivers. *Mar. Chem.* **7**, 173–206.

Martin J.-M., Nirel P., and Thomas A. J. (1987) Sequential extraction techniques: promises and problems. *Mar. Chem.* **22**, 313–341.

Martin J.-M. and Whitfield M. (1981) The significance of the river input of chemical elements to the ocean. In *Trace Elements in Seawater* (eds. C. S. Wong, E. Boyle, K. W. Bruland, J. D. Burton, and E. D. Goldberg). Plenum Press, New York, pp. 265–296.

Martin J.-M. and Windom H. L. (1991) Present and future roles of ocean margins in regulating marine biogeochemical cycles of trace elements. In *Ocean Margin Process in Global Change* (eds. R. F. C. Mantoura, J.-M. Martin, and R. Wollast). Wiley-Interscience, New York, pp. 45–67.

McCall P. L., Robbins J. A., and Matisoff G. (1984) ^{137}Cs and ^{210}Pb transport and geochronologies in urbanized reservoirs with raoidly increasing sedimentation rates. *Chem. Geol.* **44**, 36–65.

Mecray E. L., King J. W., Appleby P. G., and Hunt A. S. (2001) Historical trace metal accumulation in the sediments of an urbanized region of the Lake Champlain watershed Burlington, Vermont. *Water Air Soil Pollut.* **125**, 201–230.

Millero F. J. (1975) The physical chemistry of estuaries. In *Marine Chemistry in the Coastal Environment*, ACS Symposium Series 18 (ed. T. Church). American Chemical Society, pp. 25–55.

Morel F. M. M., McDuff R. E., and Morgan J. J. (1973) Interactions and chemostasis in aquatic chemical systems: role of pH, pE, solubility, and complexation. In *Trace Metals and Metal-Organic Interactions in Natural Waters* (ed. P. C. Singer). Ann Arbor Science, Ann Arbor, pp. 157–200.

Mudroch A., Sarazin L., and Lomas T. (1988) Summary of surface and background concentrations of selected elements in the Great Lakes sediments. *J. Great Lakes Res.* **14**, 241–251.

Muller G., Grimmer G., and Bohnke H. (1977) Sedimentary record of heavy metals and polycyclic aromatic hydrocarbons in Lake Constance. *Naturwissenschaften* **64**, 427–431.

Murozumi M., Chow T. J., and Patterson C. C. (1969) Chemical concentrations of pollutant lead aerosols, terrestrial dusts and sea salts in Greenland and Antarctic snow strata. *Geochim. Cosmochim. Acta* **33**, 1247–1294.

National Science Foundation (1975) *Nickel.* National Academy of Sciences, Washington, DC.

Nriagu J. O. (1979) Global inventory of natural and anthropogenic emissions of trace metals to the atmosphere. *Nature* **279**, 409–411.

Nriagu J. O. (1980a) Global cadmium cycle. In *Cadmium in the Environment, Part I: Ecological Cycling* (ed. J. O. Nriagu). Wiley, New York, pp. 1–12.

Nriagu J. O. (1980b) Global cycle and properties of nickel. In *Nickel in the Environment* (ed. J. O. Nriagu). Wiley, New York, pp. 1–26.

Nriagu J. O. (1988a) A silent epidemic of environmental metal poisoning? *Environ. Pollut.* **50**, 139–161.

Nriagu J. O. (1990a) The rise and fall of leaded gasoline. *Sci. Total Environ.* **921**, 13–18.

Nriagu J. O. (1990b) Global metal pollution. *Environment* **32**, 7–33.

Nriagu J. O. and Pacyna J. M. (1988) Quantitative assessment of worldwide contamination of air, water, and soils by trace metals. *Nature* **33**, 134–139.

Nriagu J. O., Kemp A. L. W., Wong H. K. T., and Harper N. (1979) Sedimentary record of heavy metal pollution in Lake Erie. *Geochim. Cosmochim. Acta* **43**, 247–258.

Olade M. A. (1987) Dispersion of Cadmium, Lead, and Zinc in soils and sediments of a humid tropical ecosystem in Nigeria. In *Lead, Mercury, Cadmium and Arsenic in the Environment* (eds. T. C. Hutchinson and K. M. Meema). Wiley, New York, SCOPE 31, pp. 303–312.

Osintsev S. P. (1995) Heavy metals in the bottom sediments of the Katun' River and the Ob' upper reaches. *Water Resour.* **22**, 42–49.

Pacyna J. M. (1986) Atmospheric trace elements from natural and anthropogenic sources. In *Toxic Metals in the Atmosphere* (eds. J. O. Nriagu and C. I. Davidson). Wiley, New York, pp. 33–50.

Pacyna J. M. (1996) Monitoring and assessment of metal contaminants in the air. In *Toxicology of Metals* (ed. L. W. Chang). CRC Press, Boca Raton, pp. 9–28.

Pacyna J. M. and Lindgren E. S. (1997) Atmospheric transport and deposition of toxic compounds. In *The Global Environment* (eds. D. Brune, D. V. Chapman, M. D. Gwynne, and J. M. Pacyna). Wiley, New York, pp. 386–407.

Patterson C. C. (1965) Contaminated and natural lead environments of man. *Arch. Environ. Health* **11**, 344–360.

Patterson C. C. and Settle D. M. (1987) Review of data on eolian fluxes of industrial and natural lead to the lands and seas in remote regions on a global scale. *Mar. Chem.* **22**, 137–162.

Patterson C. C., Settle D., Schaule B., and Burnett M. (1976) Transport of pollutant lead to the ocean and within ocean ecosystems. In *Marine Pollution Transfer* (eds. H.l. Windon and R. A. Duce). Heath, pp. 23–38.

Petit D., Mennessier J. P., and Lamberts L. (1984) Stable lead isotopes in pond sediments as a tracer of past and present atmospheric lead pollution in Belgium. *Atmos. Environ.* **18**, 1189–1193.

Pickering W. F. (1980) Cadmium retention by clays and other soil or sediment components. In *Cadium in the Environment, Part I: Ecological Cycling* (ed. J. O. Nriagu). Wiley, New York, pp. 365–397.

Pickering W. F. (1981) Selective chemical extraction of soil components and bound metal species. *CRC Critical Rev. Anal. Chem. Nov.*, 233–266.

Presley B. J., Trefry J. H., and Shokes R. F. (1980) Heavy metal inputs to Mississippi Delya sediments. *Water Air Soil Pollut.* **13**, 481–494.

Rai D., Eary L. E., and Zachara J. M. (1989) Environmental chemistry of chromium. *Sci. Tot. Environ.* **86**, 15–23.

Rai D., Sass B. M., and Moore D. A. (1987) Chromium (III) hydrolysis constants and solubility of chromium (III) hydroxide. *Inorg. Chem.* **26**, 345–349.

Reimann C. and de Caritat P. (1998) *Chemical Elements in the Environment.* Springer, Berlin.

Richter R. O. and Theis T. L. (1980) Nickel speciation in a soil/water system. In *Nickel in the Environment* (ed. J. O. Nriagu). Wiley, New York, pp. 189–202.

Rippey B., Murphy R. J., and Kyle S. W. (1982) Anthropogenically derived changes in the sedimentary flux of Mg Ni, Cu, Zn, Hg, Pb, and P in Lough Neagh, Northern Ireland. *Environ. Sci. Technol.* **16**, 23–30.

Ritchie J. C., McHenry J. R., and Gill A. C. (1973) Dating recent reservoir sediments. *Limnol. Oceanogr.* **18**, 254–263.

Robbins J. A. and Edgington D. N. (1975) Determination of recent sedimentation rates in Lake Michigan using Pb-210 and Cs-137. *Goechim. Cosmochim. Acta* **39**, 285–304.

Roos-Barraclough F. and Shotyk W. (2003) Millennial scale records of atmospheric mercury deposition obtained from ombrotrophic and minerotrophic peatlands in the Swiss Jura Mountains. *Environ. Sci. Technol.* **37**, 235–244.

Rowell H. C. (1996) Paleolimnology of Onondaga Lake: the history of anthropogenic impacts on water quality. *Lake Reserv. Mgmt.* **12**, 35–45.

Salomons W. (1983) Trace metal cycling in a polluted lake (Ijsselmeer, the Netherlands). *Delft Hydraulics Laboratory Rept.* S 357, 50pp.

Salomons W. and Forstner U. (1980) Trace metal analysis on polluted sediments: II. Evaluation of environmental impact. *Environ. Technol. Lett.* **1**, 506–517.

Salomons W. and Forstner U. (1984) *Metals in the Hydrocycle.* Springer-Verlag, New York.

Santschi P., Hohener P., Benout G., and Buchholtz-ten Brink M. (1990) Chemical processes at the sediment-water interface. *Mar. Chem.* **30**, 269–315.

Sass B. M. and Rai D. (1987) Solubility of amorphous chromium (II)-iron (III) hydroxide solid solutions. *Inorg. Chem.* **26**, 2228–2232.

Schell W. R. and Barner R. S. (1986) Environmental isotope and anthropogenic tracers of recent lake sedimentation. In *Handbook of Environmental Isotope Geochemistry, The Terrestrial Environment* (eds. J. C. Fontes and P. Fritz). Elsevier, Amsterdam, Netherlands, pp. 169–206.

Scudlark J. R. and Church T. M. (1997) Atmospheric deposition of trace elements to the mid-Atlantic bight. In *Atmospheric Deposition of Contaminants to the Great Lakes and Coastal Waters* (ed. J. E. Baker). SETAC Press, pp. 195–208.

Settle D. M. and Patterson C. C. (1980) Lead in albacore: guide to lead pollution in Americans. *Science* **207**, 1167–1176.

Shafer M. M. and Armstrong D. E. (1991) Trace element cycling in southern Lake Michigan: role of water column particle components. In *Organic Substances and Sediments in Water* (ed. R. A. Baker). Lewis Publishers, Chelsea, vol. 2, pp. 15–47.

Shen G. T. and Boyle E. A. (1987) Lead in corals: reconstruction of historical industrial fluxes to the surface ocean. *Earth Planet. Sci. Lett.* **82**, 289–304.

Shotyk W., Cheburkin A. K., Appleby P. G., Frankhauser A., and Kramers J. D. (1996) Two thousand years of atmospheric arsenic, antimony, and lead deposition recorded in an ombrotrophic peat bog profile, Jura Mountains, Switzerland. *Earth Planet. Sci. Lett.* **145**, E1–E7.

Shotyk W., Weiss D., Appleby P. G., Cheburkin A. K., Frei R., Gloor M., Kramers J. D., Reese S., and Van Der Knaap W. O. (1998) History of atmospheric lead deposition since 12,370 ^{14}C yr BP from a peat bog, Jura Mountains, Switzerland. *Science* **281**, 1635–1640.

Sigg L. (1987) Surface chemical aspects of the distribution and fate of metal ions in lakes. In *Aquatic surface Chemistry* (ed. W. Stumm). Wiley, New York, pp. 319–349.

Snodgrass W. J. (1980) Distribution and behavior of nickel in the aquatic environment. In *Nickel in the Environment* (ed. J. O. Nriagu). Wiley, New York, pp. 203–274.

Sunda W. G. (1991) Trace metal interactions with marine phytoplankton. *Biol. Oceanogr.* **6**, 411–442.

Sutherland R. A. (2002) Comparison between non-residual Al Co, Cu, Fe, Mn, Ni, Pb and Zn released by a three-step sequential extraction procedure and a dilute hydrochloric acid leach for soil and road deposited sediment. *Appl. Geochem.* **17**, 353–365.

Sutherland R. A. and Tack F. M. G. (2000) Metal phase associations in soils from an urban watershed, Honolulu, Hawaii. *Sci. Total Environ.* **256**, 103–113.

Sutherland R. A., Tack F. M. G., Tolosa C. A., and Verloo M. G. (2000) Operationally defined metal fractions in road deposited sediment Honolulu, Hawaii. *J. Environ. Qual.* **29**, 1431–1439.

Taillefert M., Lienemann C.-P., Gaillard J. F., and Perret D. (2000) Speciation, reactivity, and cycling of Fe and Pb in a meromictic lake. *Geochim. Cosmochim. Acta* **64**, 169–183.

Tatsumoto M. T. and Patterson C. C. (1963) The concentration of common lead in some Atlantic and Mediterranean waters and in snow. *Nature* **199**, 350–352.

Tessier A., Cambell P. G. C., and Bisson M. (1979) Sequential extraction procedure for the speciation of particulate trace metals. *Anal. Chem.* **51**, 844–851.

Tessier A., Carignan R., Dubreuil B., and Rapin F. (1989) Partitioning of zinc between water column and the oxic sediments in lakes. *Geochim. Cosmochim. Acta* **53**, 1511–1522.

Tessier A., Fortin D., Belzile N., DeVitre R. R., and Leppard G. G. (1996) Metal sorption to diagenetic iron and manganese oxyhydroxides and associated organic matter: narrowing the gap between field and laboratory measurements. *Geochim. Cosmochim. Acta* **60**, 387–404.

Tessier A., Rapin F., and Carignan R. (1985) Trace metals in oxic lake sediments: possible adsorption onto iron oxyhydroxides. *Geochim. Cosmochim. Acta* **49**, 183–194.

Thomas V. M. (1995) The elimination of lead in gasoline. *Ann. Rev. Energy Environ.* **20**, 301–324.

Thomas M., Petit D., and Lamberts L. (1984) Pond sediments as historical record of heavy metals fallout. *Water Air Soil Pollut.* **23**, 51–59.

Tillman D. A. (1994) *Trace Metals in Combustion Systems.* Academic Press, New York.

Turekian K. K. (1971) Rivers, tributaries, and estuaries. In *Impingement of Man on the Oceans* (ed. D. W. Hood). Wiley, New York, pp. 9–74.

Turekian K. K. (1977) The fate of metals in the oceans. *Geochim. Cosmochim. Acta* **41**, 1139–1144.

Turekian K. K. and Wedepohl K. H. (1961) Distribution of the elements in some major units of the Earth's crust. *Bull. Geol. Soc. Am.* **72**, 175–192.

United States Environmental Protection Agency (1975a) *Scientific and Technical Assessment Report on Cadmium.* EPA-600/6-6-75-003, US Govt. Printing Office.

United States Environmental Protection Agency (1975b) *Technical and Microanalysis of Cadmium and its Compounds.* EPA-560/3-75-005, US Govt. Printing Office.

United States Environmental Protection Agency (1976) *Cadmium: Control Strategy Analysis.* EPA-GCA-TR-75-36-G, US Govt. Printing Office.

United States Environmental Protection Agency (1978) *Sources of Atmospheric Cadmium.* EPA-68-02-2836, US Govt. Printing Office.

United States Environmental Protection Agency. (1998) *Characterization of Municipal Solid Waste in the United States: 1997 Update.* EPA530-R-98-007.

United States Environmental Protection Agency (2000a) *National Air Pollutant Emission Trends, 1900–1998.* EPA-454/R-00-002, US Govt. Printing Office.

United States Environmental Protection Agency (2000b) *Deposition of air Pollutants to the Great Waters.* EPA-453/R-00-005, US Govt. Printing Office.

United States Department of Commerce (1998) *Statistical Abstracts of the United States 1998.* US Govt. Printing Office.

Valette-Silver N. J. (1993) The use of sediment cores to reconstruct historical trends in contamination of estuarine and coastal sediments. *Estuaries* **16**, 577–588.

Van Metre P. C. and Callender E. (1996) Identifying water-quality trends in the Trinity River Texas, USA, 1969–1992, using sediment cores from Lake Livingston. *Environ. Geol.* **28**, 190–200.

Van Metre P. C., Callender E., and Fuller C. C. (1997) Historical trends in organochlorine compounds in river basins identified using sediment cores from reservoirs. *Environ. Sci. Technol.* **31**, 2339–2344.

Van Metre P. C., Mahler B. J., and Furlong E. T. (2000) Urban sprawl leaves its PAH signature. *Environ. Sci. Technol.* **34**, 4064–4070.

Veron A., Lambert C. E., Isley A., Linet P., and Grousset F. (1987) Evidence of recent lead pollution in deep north-east Atlantic sediments. *Nature* **326**, 278–281.

Wahlen M. and Thompson R. C. (1980) Pollution records from sediments of three lakes in New York State. *Geochim. Cosmochim. Acta* **44**, 333–339.

Warren L. A. and Haack E. A. (2001) Biogeochemical controls on metal behavior in freshwater environments. *Earth-Sci. Rev.* **54**, 261–320.

Webb S. M., Leppard G. G., and Gaillard J.-F. (2000) Zinc speciation in a contaminated aquatic environment: characterization of environmental particles by analytical electron microscopy. *Environ. Sci. Technol.* **34**, 1926–1933.

WebElements. http://www.webelements.com/webelements/elements/text/Cu.html (accessed June 16, 2002).

Wedepohl K. H. (1968) Chemical fractionation in the sedimentary environment. In *Origin and Distribution of the Elements* (ed. L. H. Ahrens). Pergamon Press, New York, pp. 999–1015.

Wedepohl K. H. (1995) The composition of the continental crust. *Geochim. Cosmochim. Acta* **59**, 1217–1232.

Wessels M., Lenhard A., Giovanoli F., and Bollhofer A. (1995) High resolution time series of lead and zinc in sediments of Lake Constance. *Aquatic Sci.* **57**, 291–304.

Wilber G. G., Smith L., and Malanchuk J. L. (1992) Emissions inventory of heavy metals and hydrophobic organics in the Great Lakes Basin. In *Fate of Pesticides and Chemicals in the Environment* (ed. J. L. Schnoor). Wiley, New York, pp. 27–50.

Williams T. M. (1992) Diagenetic metal profiles in recent sediments of a Scottish freshwater loch. *Environ. Geol. Water Sci.* **20**, 117–123.

Wren C. D., Maccrimmon H. R., and Loescher B. R. (1983) Examination of bioaccumulation and biomagnification of metals in a Precambrian shield lake. *Water, Air, Soil Pollut.* **19**, 277–291.

Wu J. and Boyle E. A. (1997) Lead in the western North Atlantic Ocean: completed response to leaded gasoline phaseout. *Geochim. Cosmochim. Acta* **61**, 3279–3283.

Yasui M., Strong M. J., Ota K., and Verity M. A. (1996) *Mineral and Metal Neurotoxicology.* CRC Press, Boca Raton.

Zhang J., Huang W. W., and Wang J. H. (1994) Trace-metal chemistry of the Huanghe (Yellow River), China— Examination of the data from *in situ* measurements and laboratory approach. *Chem. Geol.* **114**, 83–94.

Published by Elsevier Ltd.

Radioactive Geochronometry
ISBN: 978-0-08-096708-0

pp. 463–500

APPENDIX 1. Periodic Table of the Elements.

Legend:
Atomic number
Element symbol
Atomic mass

1	2	3	4	5	6	7	8	9	10	11	12	13	14	15	16	17	18
1 H 1.00794																	2 He 4.00260
3 Li 6.941	4 Be 9.01218											5 B 10.811	6 C 12.011	7 N 14.0067	8 O 15.9994	9 F 18.9984	10 Ne 20.1797
11 Na 22.9898	12 Mg 24.3050											13 Al 26.9815	14 Si 28.0855	15 P 30.9738	16 S 32.066	17 Cl 35.4527	18 Ar 39.948
19 K 39.0983	20 Ca 40.078	21 Sc 44.9559	22 Ti 47.88	23 V 50.9415	24 Cr 51.9961	25 Mn 54.9380	26 Fe 55.847	27 Co 58.9332	28 Ni 58.69	29 Cu 63.546	30 Zn 65.39	31 Ga 69.723	32 Ge 72.61	33 As 74.9216	34 Se 78.96	35 Br 79.904	36 Kr 83.80
37 Rb 85.4678	38 Sr 87.62	39 Y 88.9059	40 Zr 91.224	41 Nb 92.9064	42 Mo 95.94	43 Tc (98)	44 Ru 101.07	45 Rh 102.906	46 Pd 106.42	47 Ag 107.868	48 Cd 112.411	49 In 114.82	50 Sn 118.710	51 Sb 121.75	52 Te 127.60	53 I 126.905	54 Xe 131.29
55 Cs 132.905	56 Ba 137.327	57 La 138.906 (★)	72 Hf 178.49	73 Ta 180.948	74 W 183.85	75 Re 186.207	76 Os 190.2	77 Ir 192.22	78 Pt 195.08	79 Au 196.967	80 Hg 200.59	81 Tl 204.383	82 Pb 207.2	83 Bi 208.980	84 Po (209)	85 At (210)	86 Rn (222)
87 Fr (223)	88 Ra 226.025	89 Ac 227.028 (▲)	104 (261)	105 (262)	106 (263)	107 (262)	108 (265)	109 (267)									

★ Lanthanides

58 Ce 140.115	59 Pr 140.908	60 Nd 144.24	61 Pm (145)	62 Sm 150.36	63 Eu 151.965	64 Gd 157.25	65 Tb 158.925	66 Dy 162.50	67 Ho 164.930	68 Er 167.26	69 Tm 168.934	70 Yb 173.04	71 Lu 174.967

▲ Actinides

90 Th 232.038	91 Pa 231.036	92 U 238.029	93 Np 237.048	94 Pu (244)	95 Am (243)	96 Cm (247)	97 Bk (247)	98 Cf (251)	99 Es (252)	100 Fm (257)	101 Md (258)	102 No (259)	103 Lr (260)

APPENDIX 2. Table of Isotopes[a].

A	Element	Abundance/half-life	Source
1	H	99.985%	
2	H	0.015%	
3	H	12.33 yr	C, B
3	He	0.000137%	
4	He	99.999863%	
6	Li	7.5%	
7	Li	92.5%	
7	Be	53.12 d	C
9	Be	100%	
10	Be	1.51e+6 yr	C
10	B	19.9%	
11	B	80.1%	
12	C	98.90%	
13	C	1.10%	
14	C	5,730 yr	C, B
14	N	99.634%	
15	N	0.366%	
16	O	99.762%	
17	O	0.038%	
18	O	0.200%	
19	F	100%	
20	Ne	90.48%	
21	Ne	0.27%	
22	Ne	9.25%	
22	Na	2.6019 yr	C
23	Na	100%	
24	Mg	78.99%	
25	Mg	10.00%	
26	Mg	11.01%	
26	Al	7.17e+5 yr	C
27	Al	100%	
28	Si	92.23%	
29	Si	4.67%	
30	Si	3.10%	
32	Si	150 yr	C
31	P	100%	
32	P	14.262 d	C
33	P	25.34 d	C
32	S	95.02%	
33	S	0.75%	
34	S	4.21%	
35	S	87.32 d	C
36	S	0.02%	
35	Cl	75.77%	
36	Cl	3.01e+5 yr	C, B
37	Cl	24.23%	
36	Ar	0.337%	
37	Ar	35.04 d	
38	Ar	0.063%	
39	Ar	269 yr	C
40	Ar	99.600%	

APPENDIX 2. (Continued).

A	Element	Abundance/half-life	Source
39	K	93.2581%	
40	K	1.277e+9 yr	
		0.0117%	
41	K	6.7302%	
40	Ca	96.941%	
41	Ca	1.03e+5 yr	C
42	Ca	0.647%	
43	Ca	0.135%	
44	Ca	2.086%	
46	Ca	0.004%	
48	Ca	6e+18 yr	
		0.187%	
45	Sc	100%	
46	Ti	8.0%	
47	Ti	7.3%	
48	Ti	73.8%	
49	Ti	5.5%	
50	Ti	5.4%	
50	V	1.4e+17 yr	
		0.250%	
51	V	99.750%	
50	Cr	1.8e+17 yr	
		4.345%	
51	Cr	27.7025 d	B
52	Cr	83.789%	
53	Cr	9.501%	
54	Cr	2.365%	
53	Mn	3.74e+6 yr	E
54	Mn	312.3 d	C
55	Mn	100%	
54	Fe	5.8%	
56	Fe	91.72%	
57	Fe	2.2%	
58	Fe	0.28%	
60	Fe	1.5e+6 yr	E
59	Co	100	
60	Co	5.2714 yr	B
58	Ni	68.077%	
59	Ni	7.6e+4 yr	C
60	Ni	26.223%	
61	Ni	1.140%	
62	Ni	3.634%	
63	Cu	69.17%	
65	Cu	30.83%	
64	Zn	48.6%	
65	Zn	244.26 d	B
66	Zn	27.9%	
67	Zn	4.1%	
68	Zn	18.8%	
69	Ga	60.108%	
71	Ga	39.892%	
70	Ge	21.23%	
72	Ge	27.66%	
73	Ge	7.73%	

(Continued)

(Continued)

APPENDIX 2. (Continued).

A	Element	Abundance/half-life	Source
74	Ge	35.94%	
76	Ge	7.44%	
75	As	100%	
74	Se	0.89%	
76	Se	9.36%	
77	Se	7.63%	
78	Se	23.78%	
79	Br	50.69%	
81	Br	49.31%	
78	Kr	0.35%	
80	Kr	2.25%	
81	Kr	2.29e+5 yr	C
82	Kr	11.6%	
83	Kr	11.5%	
84	Kr	57.0%	
85	Kr	10.756 yr	B
85	Rb	72.165%	
87	Rb	4.75e+10 yr	
		27.835%	
84	Sr	0.56%	
86	Sr	9.86%	
87	Sr	7.00%	
88	Sr	82.58%	
90	Sr	28.79 yr	B
89	Y	100%	
90	Zr	51.45%	
91	Zr	11.22%	
92	Zr	17.15%	
93	Zr	1.53e+6 yr	
94	Zr	17.38%	
96	Zr	3.8e+19 yr	
		2.80 2%	
93	Nb	100%	
92	Mo	14.84%	
94	Mo	9.25%	
95	Mo	15.92%	
96	Mo	16.68%	
97	Mo	9.55%	
98	Mo	24.13%	
100	Mo	1.00e+19 yr	
		9.63%	
99	Tc	2.111e+5 yr	E
98	Ru	1.88%	
99	Ru	12.7%	
100	Ru	12.6%	
101	Ru	17.0%	
102	Ru	31.6%	
104	Ru	18.7%	
103	Rh	100%	
102	Pd	1.02%	
104	Pd	11.14%	
105	Pd	22.33%	
106	Pd	27.33%	

APPENDIX 2. (Continued).

A	Element	Abundance/half-life	Source
107	Pd	6.5e+6 yr	E
108	Pd	26.46%	
110	Pd	11.72%	
107	Ag	51.839%	
109	Ag	48.161%	
106	Cd	1.25%	
108	Cd	0.89%	
110	Cd	12.49%	
111	Cd	12.80%	
112	Cd	24.13%	
113	Cd	7.7e+15 yr	
		12.22%	
114	Cd	28.73%	
113	In	4.3%	
115	In	4.41e+14 yr	
		95.7%	
112	Sn	0.97%	
114	Sn	0.65%	
115	Sn	0.34%	
116	Sn	14.53%	
117	Sn	7.68%	
118	Sn	24.23%	
119	Sn	8.59%	
120	Sn	32.59%	
122	Sn	4.63%	
124	Sn	5.79%	
121	Sb	57.36%	
123	Sb	42.64%	
120	Te	0.096%	
122	Te	2.603%	
123	Te	1e+13 yr	
		0.908%	
124	Te	4.816%	
125	Te	7.139%	
126	Te	18.95%	
128	Te	2.2e+24 yr	
		31.69%	
130	Te	7.9e+20 yr	
		33.80%	
127	I	100%	
129	I	1.57e+7 yr	E,C,B
124	Xe	1.6e+14 yr	
		0.10%	
126	Xe	0.09%	
128	Xe	1.91%	
129	Xe	26.4%	
130	Xe	4.1%	
131	Xe	21.2%	
132	Xe	26.9%	
134	Xe	10.4%	
136	Xe	2.36e+21 yr	
		8.9%	
133	Cs	100%	
134	Cs	2.0648 yr	B

(Continued)   (Continued)

APPENDIX 2. (Continued).

A	Element	Abundance/half-life	Source
137	Cs	30.07 yr	B
130	Ba	0.106%	
132	Ba	0.101%	
134	Ba	2.417%	
135	Ba	6.592%	
136	Ba	7.854%	
137	Ba	11.23%	
138	Ba	71.70%	
138	La	1.05e+11 yr	
		0.0902%	
139	La	99.9098%	
138	Ce	0.25%	
140	Ce	88.48%	
142	Ce	5e+16 yr	
		11.08%	
141	Pr	100%	
142	Nd	27.13%	
143	Nd	12.18%	
144	Nd	2.29e+15 yr	
		23.80%	
145	Nd	8.30%	
146	Nd	17.19%	
148	Nd	5.76%	
150	Nd	1.1e+19 yr	
		5.64%	
–	Pm	no stable or long-lived isotope	
144	Sm	3.1%	
146	Sm	1.03e+8 yr	E
147	Sm	1.06e+11 yr	
		15.0%	
148	Sm	7e+15 yr	
		11.3%	
149	Sm	2e+15 yr	
		13.8%	
150	Sm	7.4%	
152	Sm	26.7%	
154	Sm	22.7%	
151	Eu	47.8%	
153	Eu	52.2%	
152	Gd	1.08e+14 yr	
		0.20%	
154	Gd	2.18%	
155	Gd	14.80%	
156	Gd	20.47%	
157	Gd	15.65%	
158	Gd	24.84%	
160	Gd	21.86%	
159	Tb	100%	
156	Dy	0.06%	
158	Dy	0.10%	
160	Dy	2.34%	
161	Dy	18.9%	
162	Dy	25.5%	
163	Dy	24.9%	

APPENDIX 2. (Continued).

A	Element	Abundance/half-life	Source
164	Dy	28.2%	
165	Ho	100%	
162	Er	0.14%	
164	Er	1.61%	
166	Er	33.6%	
167	Er	22.95%	
168	Er	26.8%	
170	Er	14.9%	
169	Tm	100%	
168	Yb	3.05%	
171	Yb	14.3%	
172	Yb	21.9%	
173	Yb	16.12%	
174	Yb	31.8%	
176	Yb	12.7%	
175	Lu	97.41%	
176	Lu	3.78e+10 yr	
		2.59%	
174	Hf	2.0e+15 yr	
		0.162%	
176	Hf	5.206%	
177	Hf	18.606%	
178	Hf	27.297%	
179	Hf	13.629%	
180	Hf	35.100%	
180	Ta	1.2e+15 yr	
		0.012%	
181	Ta	99.988%	
180	W	0.13%	
182	W	26.3%	
183	W	1.1e+17 yr	
		14.3%	
184	W	3e+17 yr	
		30.67%	
186	W	28.6%	
185	Re	37.40%	
187	Re	4.35e+10 yr	
		62.60%	
184	Os	5.6e+13 yr	
		0.02%	
186	Os	2.0e+15 yr	
		1.58%	
187	Os	1.6%	
188	Os	13.3%	
189	Os	16.1%	
190	Os	26.4%	
192	Os	41.0%	
191	Ir	37.3%	
193	Ir	62.7%	
190	Pt	6.5e+11 yr	
		0.01%	
192	Pt	0.79%	
194	Pt	32.9%	
195	Pt	33.8%	
196	Pt	25.3%	

(Continued)

(Continued)

APPENDIX 2. (Continued).

A	Element	Abundance/half-life	Source
198	Pt	7.2%	
197	Au	100%	
196	Hg	0.15%	
198	Hg	9.97%	
199	Hg	16.87%	
200	Hg	23.10%	
201	Hg	13.18%	
202	Hg	29.86%	
204	Hg	6.87%	
203	Tl	29.524%	
205	Tl	70.476%	
206	Tl	4.199 min	U238
207	Tl	4.77 min	U235
208	Tl	3.053 min	Th232
210	Tl	1.30 min	U238
204	Pb	$1.4e+17$ yr 1.4%	
207	Pb	22.1%	
208	Pb	52.4%	
210	Pb	22.3 yr	U238
211	Pb	36.1 min	U235
212	Pb	10.64 h	Th232
214	Pb	26.8 min	U238
209	Bi	100%	
210	Bi	5.013 d	U238
211	Bi	2.14 min	U235
212	Bi	60.55 min	Th232
214	Bi	19.9 min	U238
215	Bi	7.6 min	U235
210	Po	138.376 d	U238
211	Po	0.516 s	U235
212	Po	0.299 μs	Th232
214	Po	164.3 μs	U238
215	Po	1.781 ms	U235

(Continued)

APPENDIX 2. (Continued).

A	Element	Abundance/half-life	Source
216	Po	0.145 s	Th232
218	Po	3.10 min	U238
215	At	0.10 ms	U235
218	At	1.5 s	U238
219	Rn	3.96 s	U235
220	Rn	55.6 s	Th232
222	Rn	3.8235 d	U238
223	Fr	21.8 min	U235
223	Ra	11.435 d	U235
224	Ra	3.66 d	Th232
226	Ra	1600 yr	U238
228	Ra	5.75 yr	Th232
227	Ac	21.773 yr	U235
228	Ac	6.15 h	Th232
228	Th	1.9116 yr	Th232
230	Th	$7.538e+4$ yr	U238
232	Th	$1.405e+10$ yr 100%	
234	Th	24.10 d	U238
231	Pa	32760 yr	U235
234	Pa	6.70 h	U238
234	U	$2.455e+5$ yr 0.0055%	
235	U	$7.038e+8$ yr 0.7200%	
238	U	$4.468e+9$ yr 99.2745%	

Sources of short-lived radionuclides: B, bomb or reactor sources; C, cosmogenic; E, extinct radioactivities; U235, U238, Th232—nuclides in respective decay chains.
Note: the symbol e indicates that the number following is that raised to the power of 10.
[a] Modified from: Lawrence Berkeley Laboratory web site: http://ie.lbl.gov/education/isotopes.htm

APPENDIX 3. The Geologic Timescale.

Eon	Era	Period	Epoch	Millions of years ago
Phanerozoic				
	Cenozoic	(Quaternary)	Holocene	
				0.011
			Pleistocene	
				1.82
		(Tertiary)	Pliocene	
				5.32
			Miocene	
				23
			Oligocene	
				33.7
			Eocene	
				55
			Paleocene	
				65
	Mesozoic	Cretaceous		
				144
		Jurassic		
				200
		Triassic		
				250
	Paleozoic	Permian		
				295
		Carboniferous Pennsylvanian		
				320
		Mississippian		
				355
		Devonian		
				410
		Silurian		
				440
		Ordovician		
				500
		Cambrian		
				543
Proterozoic				
				2,500
Archean		Oldest rock		4,400
		Age of the solar system		4,550

APPENDIX 4. Useful Values.

Molecular mass of dry air, $m_a = 28.966$
Molecular mass of water, $m_w = 18.016$
Universal gas constant, $R = 8.31436$ J mol^{-1} K^{-1}
Gas constant for dry air, $R_a = R/m_a = 287.04$ J kg^{-1} K^{-1}
Gas constant for water vapor, $R_v = R/m_w = 461.50$ J kg^{-1} K^{-1}
Molecular weight ratio $\varepsilon \equiv m_w/m_a = R_a/R_v = 0.62197$
Stefan's constant $\sigma = 5.67 \times 10^{-8}$ W m^{-2} K^{-4}
Acceleration due to gravity, g (m s^{-2}) as a function of latitude φ and height z (m)

$$g = (9.78032 + 0.005172 \sin^2\varphi - 0.00006 \sin^2 2\varphi)(1 + z/a)^{-2}$$

Mean surface value, $\bar{g} = \int_0^{\pi/2} g \cos\varphi \, d\varphi = 9.7976$
Radius of sphere having the same volume as the Earth, $a = 6{,}371$ km (equatorial radius $= 6{,}378$ km, polar radius $= 6{,}357$ km)
Rotation rate of Earth, $\Omega = 7.292 \times 10^{-5}$ s^{-1}
Mass of Earth $= 5.977 \times 10^{24}$ kg
Mass of atmosphere $= 5.3 \times 10^{18}$ kg
Mass of ocean $= 1400 \times 10^{18}$ kg
Mass of groundwater $= 15.3 \times 10^{18}$ kg
Mass of ice caps and glaciers $= 43.4 \times 10^{18}$ kg
Mass of water in lakes and rivers $= 0.1267 \times 10^{18}$ kg
Mass of water vapor in atmosphere $= 0.0155 \times 10^{18}$ kg
Area of Earth $= 5.10 \times 10^{14}$ m^2
Area of ocean $= 3.61 \times 10^{14}$ m^2
Area of land $= 1.49 \times 10^{14}$ m^2
Area of ice sheets and glaciers $= 1.62 \times 10^{13}$ m^2
Area of sea ice $= 1.9 \times 10^{13}$ m^2 in March and 2.9×10^{13} m^2 in September (averaged between 1979 and 1987)

INDEX

NOTES:

Page numbers suffixed by *t* and *f* refer to Tables and Figures respectively. vs. indicates a comparison.

M

Printed in the United States
By Bookmasters